AF341976

# NETWORK ANALYSIS AND SYNTHESIS

## (Including Linear System Analysis)

# NETWORK ANALYSIS AND SYNTHESIS

## (Third Edition)

C. L. Wadhwa

**ANSHAN LTD**
6 Newlands Road
Tunbridge Wells, Kent.
TN4 9AT. UK

Co-published in the U. K. by

ANSHAN LTD, 6 Newlands Road, Tunbridge Wells, Kent TN4 9AT
In 2008

Tel/Fax: +44(0)1892557767
e-mail: info@anshan.co.uk
Web Site: www.anshan.co.uk



ISBN: 978-1848-290-020



British Library Cataloguing in Publication Data

A Catalogue record for this book is available from the British Library

# Preface to the Third Edition

In Chapter 1 articles on single phase and three phase circuits have been eliminated and some basic continuous time signals e.g. Impulse, unit step, unit ramp etc. and periodic signals with their mathematical representation have been added. Concepts of various types of systems e.g. linear, lumped, passive, bilateral etc. have been discussed.

Chapter 15 on Frequency Response and Bode Plot has been added.

The Appendix on z-transform has been converted into chapter 16 on z-transforms by adding more information on z-transform e.g. Discrete time convolution, pulse transfer function, initial and final values of transfer functions, stability of linear discrete time systems etc.

The author feels that these changes incorporated in this edition shall enhance the utility of the book.

Last but not the least, I wish to express my gratitude to my wife Usha, daughter Meenu and son Sandeep for their patience and encouragement during the preparation of third edition of the book.

C.L. Wadhwa

# Preface to the First Edition

Network Analysis and Synthesis has been written for the undergraduate students in Electrical, Electronics and Communication and Computer Engineering. The book covers conventional topics like basic circuit elements, Kirchhoff's laws, formulation of network equations and the more advanced topics like state variable analysis, modern filters, Active RC filters, sensitivity considerations etc.

Circuit theory and Field theory are the two basic courses for Electrical, Electronics and Communication and Computer Engineering students. A good understanding of these subjects helps the students to grasp other subjects comfortably. In fact, circuit theory has revolutionized the process of analysis in all branches of engineering. It has been possible to represent an electrical analog of a real physical system in terms of electric circuit elements and analysis made accurately.

It has been the endeavour of the author to explain various analytical techniques with simple description and illustrations through typical numerical problem solving.

Chapter 1 describes various basic circuit elements, basic continuous time signals, periodic signals and various types of systems.

Chapter 2 deals with formulation and solution of network equations using Kirchhoff's laws. Source transformation and source shifting have been clearly explained.

In case of a large network, if we are interested in the analysis of a part of the network, network theorems are more useful which are the subject of study in chapter 3. While discussing these theorems, the applicability of these theorems to certain of networks only has been underlined. Network equations have been developed using Graph theory in chapter 4.

Driving point and Transfer functions for various types of two-port networks have been discussed in chapter 5. The interconnections of two-port network and the performance of these networks in terms of their poles and zeros have been explained in detail.

Chapter 6 deals with Fourier Analysis of Continuous time signals. It is found advantageous to begin with Fourier series and proceed to the Fourier transform of non-periodic signals by considering the latter as the limiting case when the period becomes infinitely large.

Chapter 7 develops the theory of Laplace transform relating to Fourier Transform. Laplace transform solves differential equations in a much simpler way as compared to the conventional method. Concept and physical significance of complex frequency has been underlined. Laplace transform has been applied to a large number of electric networks to demonstrate its utility in the solution of complex networks.

Resonance is a very important phenomenon in electric and other systems and has been described nicely in chapter 8.

Chapter 9 covers Passive Network Synthesis. Here elements of realisability theory for such networks have been presented and both Foster and Cauer forms of realisation have been explained.

The state variable method of analysis has been covered in chapter 10. With the advent of computers this technique has assumed great importance in analysis of systems.

Chapter 11 deals with Attenuators where various types of attenuators including L-type minimum attenuation attenuators have been described. The concept of image parameters and matched systems has been nicely developed.

Chapter 12 covers classical filters theory based on the concept of image parameters matching and cascading in filter design. Though these filters are of little practical importance now, yet they provide good academic values and serve as a link in the development of modern filters which are the subject of chapter 13.

This chapter is devoted to modern filter theory and RC filters using operational amplifiers. Butterworth and Chebychev filters have been studied and their performance compared for low-pass filters. Sensitivity considerations for active filters have been developed.

Chapter 14 on Analogous Systems helps electrical engineers to solve interdisciplinary problems using analogous electric networks.

Chapter 15 deals into Frequency Response and Bode Plot.

Appendix A deals with the basics of matrices. Appendix B gives a brief introduction to SPICE (Simulation Program with Integrated Circuit Emphasis).

At the end of the book, a large number of typical multiple choice questions have been included covering almost all aspects of Network Theory. They would enable the reader to test himself. An extensive bibliography will help the reader to locate detailed information on various topics of his interest.

While writing this book I had to refer to some books and journals on Network Theory. I hereby express my gratitude to all the authors of the books and the articles in the journals.

Last but not the least, I wish to express my gratitude to my wife Usha, daughter Meenu and son Sandeep for their patience and encouragement during the preparation of the book.

Constructive suggestions for the further improvement of the book will be gratefully received and acknowledged.

**C.L. Wadhwa**

# Contents


*Preface to the Third Edition*     *v*

*Preface to the First Edition*     *vii*

**1. Basic Circuit Elements**     1

    1.1    Introduction     1

    1.2    Circuit Elements     2

    1.3    Network Classification     9

         1.3.1    Linearity     9

         1.3.2    Passivity     9

         1.3.3    Lumped     10

         1.3.4    Bilateral     10

         1.3.5    Time Invariant and Time Varying Systems     10

         1.3.6    Reciprocity     10

         1.3.7    Non-Linear Systems     11

         1.3.8    Static and Dynamic Systems     11

         1.3.9    Continuous Time and Discrete-time Systems     12

         1.3.10   Deterministic and Stochastic Systems     12

    1.4    Representation of Time Signals     13

         1.4.1    Unit Step Function     14

         1.4.2    Unit Ramp Function     14

         1.4.3    Unit Impulse     15

         1.4.4    Exponential Function     16

    1.5    Periodic Signals     17

         1.5.1    Sinusoidal Waveform     17

         1.5.2    Square Wave     18

         1.5.3    Saw Tooth Waveform     19

         1.5.4    Impulse Train     20

    1.6    The Ideal Transformer     24

         1.6.1    The Perfect Transformer     26

| | | |
|---|---|---|
| 1.7 | The Gyrator | 28 |
| 1.8 | The Negative Converter | 30 |
| 1.9 | Independent Sources | 31 |
| | 1.9.1   Ideal Voltage Source | 31 |
| | 1.9.2   Ideal Current Source | 31 |
| 1.10 | The Dependent or Controlled Sources | 32 |
| 1.11 | Source Transformation | 32 |
| 1.12 | Dot Convention for Coupled Circuits | 36 |
| | Problems | 42 |

**2. Network Equations** — **45**

| | | |
|---|---|---|
| 2.1 | Introduction | 45 |
| 2.2 | Kirchhoff's Laws | 45 |
| 2.3 | Star Delta Transformation | 48 |
| 2.4 | Loop Analysis | 51 |
| 2.5 | Nodal Analysis | 55 |
| 2.6 | Duality | 67 |
| | 2.6.1   Source Orientation | 69 |
| 2.7 | The Concept of Super Mesh (Loop) and Super Nodes | 70 |
| | Problems | 72 |

**3. Network Theorems** — **78**

| | | |
|---|---|---|
| 3.1 | Introduction | 78 |
| 3.2 | Superposition Theorem | 78 |
| 3.3 | Reciprocity Theorem | 84 |
| 3.4 | Substitution Theorem | 85 |
| 3.5 | Thevenin's Theorem | 87 |
| 3.6 | Norton's Theorem | 92 |
| 3.7 | Millman's Theorem | 95 |
| 3.8 | Tellegen's Theorem | 98 |
| 3.9 | Maximum Power Transfer Theorem | 101 |
| 3.10 | Compensation Theorem | 105 |
| | Problems | 112 |

**4. Graph Theory and Network Equations** — **118**

| | | |
|---|---|---|
| 4.1 | Introduction | 118 |
| 4.2 | Definitions | 118 |
| 4.3 | Incidence Matrix | 121 |
| 4.4 | Loop Matrix | 124 |
| 4.5 | Cut Sets and the Cut Set Matrix | 126 |
| | Planar Graph | 130 |
| 4.6 | Interrelation Among Various Matrices | 131 |
| | 4.6.1   Relation between Incidence Matrix and f-cut-set Matrix | 133 |
| | 4.6.2   Relation between f-loop and f-cut-set Matrices | 133 |

|  |  |  |
|---|---|---:|
| 4.7 | Generalised Element | 135 |
| | 4.7.1 Mesh Equations | 136 |
| | 4.7.2 Node Equations | 137 |
| | 4.7.3 Cut-set Equations | 141 |
| 4.8 | Network with Mutual Inductance | 143 |
| | Problems | 157 |

**5. Differential Equations**    **160**

| | | |
|---|---|---:|
| 5.1 | Introduction | 160 |
| 5.2 | Homogeneous Linear Differential Equations | 161 |
| 5.3 | Non-homogeneous Equations | 164 |
| 5.4 | Initial Conditions in Circuits | 165 |
| | 5.4.1 Initial Conditions for the Resistor | 166 |
| | 5.4.2 Initial Conditions for an Inductor | 166 |
| | 5.4.3 Initial Condition for a Capacitor | 166 |
| 5.5 | First Order Circuits | 169 |
| | 5.5.1 First Order Circuit with d.c. Excitation | 172 |
| | 5.5.2 AC Source Switched in an RL Series Circuit | 175 |
| 5.6 | Second Order Circuits | 177 |
| 5.7 | Analysis of Transformer | 184 |
| | Problems | 190 |

**6. Fourier Series and Fourier Transforms**    **192**

| | | |
|---|---|---:|
| 6.1 | Introduction | 192 |
| 6.2 | Fourier Theorem | 193 |
| 6.3 | Even and Odd Functions | 200 |
| 6.4 | Half Wave Symmetry | 202 |
| 6.5 | Exponential Fourier Series | 208 |
| 6.6 | Change of Interval | 209 |
| 6.7 | RMS or Effective Value of a Function | 211 |
| | 6.7.1 Power | 212 |
| 6.8 | Fourier Transform | 220 |
| 6.9 | Addition Theorem | 222 |
| | 6.9.1 Shift Theorem | 222 |
| 6.10 | Transform of an Impulse Function $\delta(T)$ | 225 |
| 6.11 | Fourier Transform of a Train of Impulses | 226 |
| 6.12 | Differentiation and Integration in Time Domain | 228 |
| | 6.12.1 Time and Frequency Scaling | 228 |
| 6.13 | Duality | 232 |
| 6.14 | Fourier Transform of a Sinusoid | 234 |
| 6.15 | Causal Functions | 240 |
| | Problems | 241 |

**7. Laplace Transform**                                                              **245**

   7.1   Introduction                                                          245

   7.2   Properties of Laplace Transform                                        246

   7.3   Laplace Transform of Some Typical Functions                            249

   7.4   Gate Function                                                          255

   7.5   The Concept of Complex Frequency                                       263

   7.6   Partial Fraction Expansion                                             264

   7.7   Operational Representation (Laplace Transform) of Circuit Elements      268

       7.7.1   Initial and Final Value Theorem                               273

   7.8   Application of Laplace Transform to Electric Networks                   274

   7.9   Convolution Integral                                                   282

       Problems                                                               301

**8. Network Functions**                                                             **308**

   8.1   Introduction                                                           308

   8.2   Driving Point and Transfer Functions                                   308

       8.2.1   Driving Point Function                                        310

       8.2.2   Transfer Functions                                            311

   8.3   The Calculation of Network Functions                                   316

   8.4   Two Port Networks                                                      322

       8.4.1   Short Circuit Admittances                                     322

       8.4.2   Open Circuit Impedances                                       324

       8.4.3   Hybrid Parameters                                             326

       8.4.4   Chain Parameters                                              329

       8.4.5   Inverse Transmission Parameters                               331

   8.5   Interrelation between Parameters                                       346

       8.5.1   $z$-Parameters in Terms of other Parameters                   346

   8.6   Interconnection of Two Port Networks                                   349

       8.6.1   Cascade Connection                                            349

       8.6.2   Series Connection                                             352

       8.6.3   Parallel Connections                                          353

       8.6.4   Series-Parallel Connection                                    353

       8.6.5   Parallel-Series Connection                                    354

       8.6.6   Permissibility of Interconnection                             356

   8.7   Two Port Symmetry                                                      372

   8.8   Poles and Zeros of Network Function                                    373

       8.8.1   Physical Significance of Poles and Zeros                      374

       8.8.2   Necessary Conditions for Driving Point Impedance              377

       8.8.3   Necessary Conditions for Transfer Function                    377

       8.8.4   Response of the System vis-a-vis Location of Poles and Zeros   377

       Problems                                                               388

**References**     **757**

**Multiple Choice Questions**     **758**

**Answers to Multiple Choice Questions**     **858**

**Answer to Problems**     **861**

Chapter I     861
Chapter II     861
Chapter III     862
Chapter IV     863
Chapter V     863
Chapter VI     864
Chapter VII     866
Chapter VIII     869
Chapter IX     870
Chapter X     871
Chapter XI     872
Chapter XII     875
Chapter XIII     876
Chapter XIV     877

*Bibliography*     *878*

*Index*     *879*

# 1

# Basic Circuit Elements

## 1.1 INTRODUCTION

The rapid progress made in Electrical Engineering during the past few decades has been possible due to the fact that an Electrical Engineer is able to assess, analyse and predict with greater accuracy the performance of various systems with the help of a simple but powerful tool known as circuit theory. In fact circuit theory has revolutionised the process of analysis in all branches of engineering be it Civil, Mechanical, Electronices or Computer engineering. It has been possible to represent an electrical analog of a real physical system in terms of electric circuit elements and analysis made accurately. Circuit theory is a simplified approximation of a more exact field theory.

Electromagnetic field theory deals directly with electric and magnetic field vectors E and H which are due to the electric charge and current whereas circuit theory deals with voltages and currents that are integrated effects of electric and magnetic fields. Of course an electrical engineer is more interested in voltage and current from which charge, power energy fields etc. can be obtained if required.

The following approximations are made while writing circuit equations:

(*i*) The distributed capacitance effects or displacement currents from one conductor to another have been ignored.

(*ii*) The inductance is assumed to be independent of frequency.

(*iii*) The radiation effects are ignored.

These approximations are quite excellent at power frequencies.

The two basic quantities to define the functions of circuits are charge and energy.

### Charge and Energy

The concept of charge is based on atomic theory. An atom has positive charges (protons) in its nucleus and an equal number of electrons (negative charges) surround the nucleus making the atom neutral. Removal of an electron leaves the atom positive charged and addition of an e lectron makes the atom negatively charged. The basic unit of charge is the charge on an electron. The mks unit of charge is coulomb. An electron has a charge of $1.062 \times 10^{-19}$ C.

When a charge is transferred from one point in the circuit to another point it constitutes what is known as electric current. An electric current is defined as the time rate of flow of charge through a certain section. Its unit is ampere. A current is said to be of one ampere when a charge of 1 coulomb flows through a section per second.

Mathematically,

$$i = \frac{dq}{dt} \qquad \qquad ...(1.1)$$

If charge $q$ is expressed in coulomb and time in second, 1 amp flow of current through a section is equivalent to approx. flow of $6.24 \times 10^{18}$ electrons per second through the section.

Law of conservation of energy states that energy can neither be created nor destroyed. However, its form can be converted i.e. energy can be converted from one form of energy to another e.g. Electro-mechanical energy conversion, Electro-chemical energy conversion, MHD energy conversion, photoelectric energy conversion etc. In all these cases the function of each of these sources of electric energy is the same in terms of energy and charge i.e. the energy is spent as work for transporting charge from one point to another in a circuit. The movement of charges contribute to current and the amount of work done per unit charge is the potential difference between the two points. The electronic charges flow from a lower potential to a higher potential and these contribute to electronic current, whereas the conventional current is considered to flow from higher potential to lower potential.

If a differential charge $dq$ is given a differential energy $dw$, the rise in potential of the charge

$$v = \frac{dw}{dq} \qquad \qquad ...(1.2)$$

If potential is multiplied by the current $\dfrac{dq}{dt}$

$$v \times i = \frac{dw}{dq} \times \frac{dq}{dt} = p \qquad \qquad ...(1.3)$$

which gives rate of change of energy with time and is equal to power. Thus,

$$\text{power } p = v \times i$$

$$\text{Since } p = \frac{dw}{dt}$$

In general

$$dw = p\,dt$$

$$W = \int v i\,dt \qquad \qquad ...(1.4)$$

## 1.2  CIRCUIT ELEMENTS

Basically there are three circuit elements in electric circuits. These are resistance, inductance and capacitance. We shall study these elements in terms of the physical phenomenon, occurring in these elements, the field phenomenon associated with them and finally its relevance in circuit theory.

### Resistance

When electrons flow through a material, they collide with other electrons and atomic particles. These collisions being inelastic, these electrons lose energy. This loss of energy per unit charge is considered as a drop in potential across the material. The time rate of flow of electrons through the section of the material constitutes the current.

From field theory point of view, the change of energy per unit distance per unit charge is the force per unit charge which is the electric field intensity $E$. It has been observed experimentally that if $J$ is the current density in A/m$^2$ $\sigma$ is the conductivity of the material then

$$J = \sigma\, E \qquad\qquad ...(1.5)$$

Consider a wire of length $l$ and cross-section $A$, the total current through the wire

$$I = JA \qquad\qquad ...(1.6)$$

and voltage drop across the wire

$$v = E.l \qquad\qquad ...(1.7)$$

Therefore,

$$\frac{v}{I} = \frac{E.l}{J.A} = \frac{l}{\sigma\, A} \qquad\qquad ...(1.8)$$

The quantity on the right hand side is constant for a fixed geometry of the conductor and is termed as resistance and is denoted by $R$. The unit is ohm after the name of the discoverer who derived this relation experimentally.

## Inductance

A current carrying conductor produces a field surrounding this conductor. The field was found to be magnetic in nature by Oersted when he brought compass needle in its vicinity which experienced force and got deflected. In terms of Field Theory this could be considered as force per unit magnetic pole which is termed as magnetic field density denoted by the letter B. The direction of force is found to be experimentally perpendicular to the current carrying conductor. Field lines are found to be as shown in Fig. 1.1 in the form of closed circular loops.

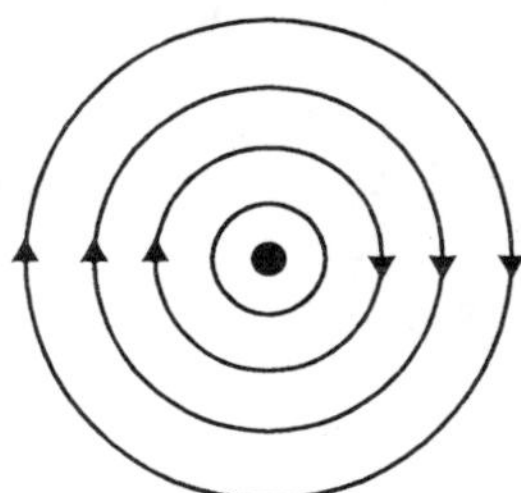

**Fig. 1.1**  Magnetic field lines surrounding a current carrying conductor.

The direction of the magnetic field density with respect to the direction of current in the conductor is given by right hand rule. With the thumb pointing in the direction of the current, the fingers of the right hand encircling the wire, point in the direction of the magnetic field. It is to be noted that these field lines are conceptual and these should not be considered as actually being present. A magnetic compass could be used to measure approximate direction of these lines.

Magnetic flux surrounding a general geometry conductor is given by

$$\phi = \int_s B \cdot ds \qquad\qquad ...(1.9)$$

If all the $N$ conductors are surrounded by these flux lines, the total flux linkages would be

$$\Psi = N\phi \qquad \qquad \text{...(1.10)}$$

Faraday observed that when two electrically conducting circuits are placed close to each other, change in flux in one circuit induced voltage in the other circuit. The flux could be changed by either changing (*i*) the relative position between the two circuits continuously or by changing the (*ii*) current in the circuit. According to Faraday's law

$$v = k\,\frac{d\Psi}{dt} \qquad \qquad \text{...(1.11)}$$

If $\Psi$ is in wb-turns, time in second and $v$ in volts then $k$ is unity.

Therefore, $\qquad \qquad \Psi = \int v\,dt \qquad \qquad \text{...(1.12)}$

Also $\qquad \qquad q = \int i\,dt \qquad$ in electrostatic field. $\qquad \text{...(1.13)}$

Therefore, we see that $\Psi$ is to voltage as charge is to current. Since flux linkages are related to the magnetic field $B$ which in turn depends on current $i$, we can write

$$\Psi = N\Phi = \left[ N\int\left(\int dB\right)ds \right] i \qquad \qquad \text{...(1.14)}$$

where the quantity in the bracket is defined as inductance of the circuit. If $\Psi$ and $i$ refer to the same circuit i.e. $i$ produces $\Psi$ flux linkages in the same circuit, it is defined as self inductance and is denoted by $L$ and expressed as $\Psi = L\,i$. However, if $i$ produces flux linkages $\Psi_2$ in another circuit, the parameter is termed as mutual inductance, denoted by $M$ and is expressed as

$$\Psi_2 = M_2\,i \qquad \qquad \text{...(1.15)}$$

We know that $\qquad \qquad v = d\,\Psi/dt \qquad \qquad \text{...(1.16)}$

and $\qquad \qquad \Psi = L\,i \qquad \qquad \text{...(1.17)}$

Therefore, $\qquad \qquad d\,\Psi/dt = L\,\dfrac{di}{dt} = v \qquad \qquad \text{...(1.18)}$

if $L$ is constant.

Sometimes when the coil is wound on a core whose $B$–$H$ curve is not linear and gets saturated after certain current ($H$) the inductance decreases and is not constant. In fact, the inductance variation is shown as in Fig. 1.2.

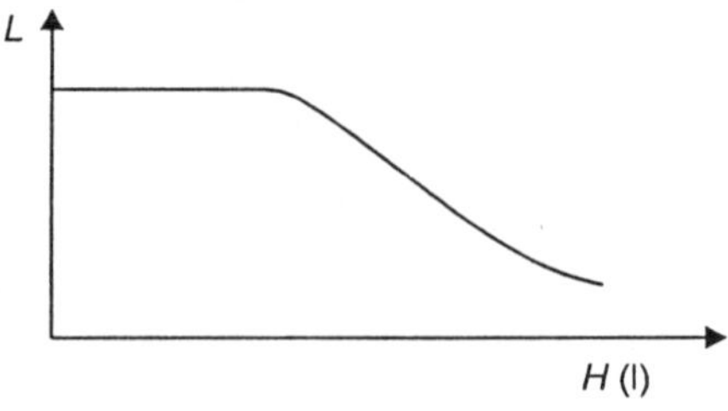


**Fig. 1.2**  Variation of inductance as a function of current.

Now taking equation (1.16)

$$\Psi = \Psi_0 + \int_0^t v\,dt \qquad \qquad \text{...(1.17)}$$

In a system consisting of an inductance connected to a source of voltage $v(t)$ when the source is switched in, at $t = 0$ when $v$ is finite the flux linkage is $\Psi_0$ as the integral does not contribute any flux linkages at $t = 0$, which means that the flux linkages can't be changed instantaneously in a given system which again means that in a fixed inductance circuit the current can't be changed instantaneously.

The differential equations describing the circuit in Fig. 1.3 when the switch is closed is

$$L\frac{di}{dt} = v$$

or

$$di = \frac{v}{L}dt$$

$$i(t) = \frac{v}{L}t + K$$

At $\quad t = 0 \quad i(t) = 0$ and hence $K = 0$

and

$$i(t) = \frac{v}{L}t \qquad \qquad ...(1.18)$$

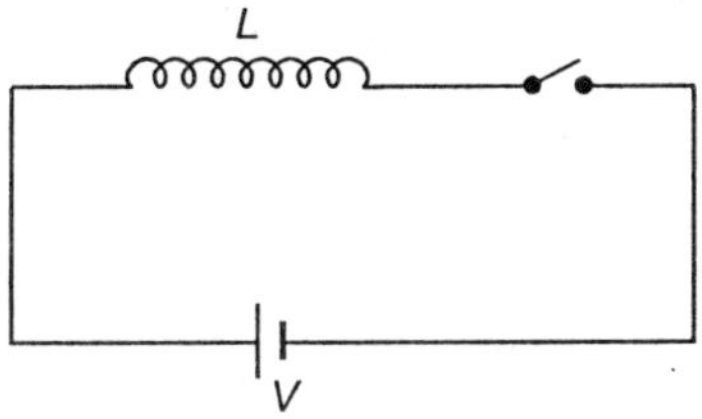

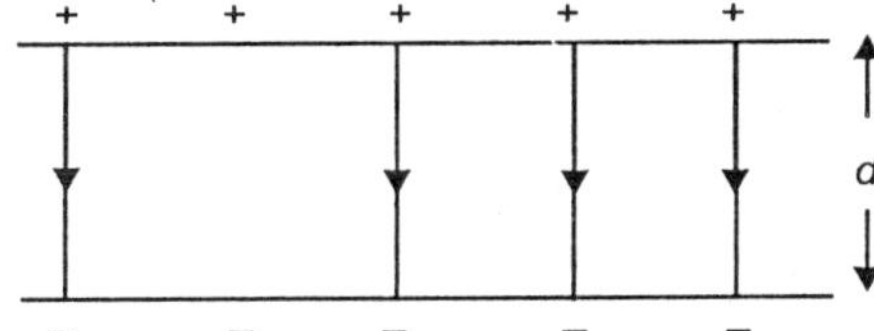

**Fig. 1.3** An inductor switched into a d.c. source.

Even though under steady state condition the current through the inductor is theoretically infinite but at $t = 0^+$ the current is zero and the increase in current is gradual with time.

## Capacitance

If a charge $q$ is placed in space, this gives rise to an electric field. Again this field is not visible but its presence can be felt by bringing another charge in the vicinity of this charge. It is found that the charge experiences a force of attraction if it is of opposite polarity to that of $q$ and vice versa. Also, if there are two metallic parallel plates separated by some distance $d$ as shown in Fig. 1.4 and if one of the plates is charged with $q^+$ charges, the other plate is charged with $q^-$ charges by induction and an electric field is set up between the plates and the two plates experience a force between each other. The force per unit charge is defined as the electric field strength. Now $D = \frac{q}{A}$ where $D$ is the field flux density and $A$ is the area of plates. Also $D = \epsilon E$ where $E$ is the electric field intensity.

**Fig. 1.4** Electric field across plates of parallel capacitor.

Therefore $E = \frac{q}{\epsilon A} = \frac{V}{d}$ where $V$ is the voltage across the plates when the charge on the plate is $q$.

Therefore

$$\frac{q}{V} = C = \frac{\epsilon A}{d} \qquad \qquad ...(1.19)$$

Here $C$ is known as capacitance between the parallel plates separated by a distance $d$. So, capacitance depends upon the geometry of the conductors and $\in$. If $q$ is measured in coulombs and $V$ in volts, $C$ is measured in Farads. If $q_0$ is the initial charge on the plates of the capacitor and while charging if the charge increases linearly with time, the charge at any instant is given as

$$q = q_0 + Kt$$

Hence,

$$i = \frac{dq}{dt} = K$$

or

$$q = q_0 + \int_0^t i \, dt \qquad \qquad ...(1.20)$$

Also we know that $q = C\,V$

Assuming $C$ to be constant, it is seen that, an instantaneous change in $q$ would mean an instantaneous change in voltage $V$. But from the integral equation for charge $q$, it is found that for finite value of $i$, the charge can't change at $t = 0^+$ and it remains $q_0$. Therefore, the voltage across a capacitor can't change instantaneously.

Let us take up a few examples to illustrate the application of some of the relations obtained in the previous discussions.

**Example 1.1:** The current wave form in a series circuit of 500 $\Omega$ resistance and a 500 mH inductor is given in Fig. E1.1 (a). Determine the voltage wave form in the resistor, in the inductor and the total voltage.

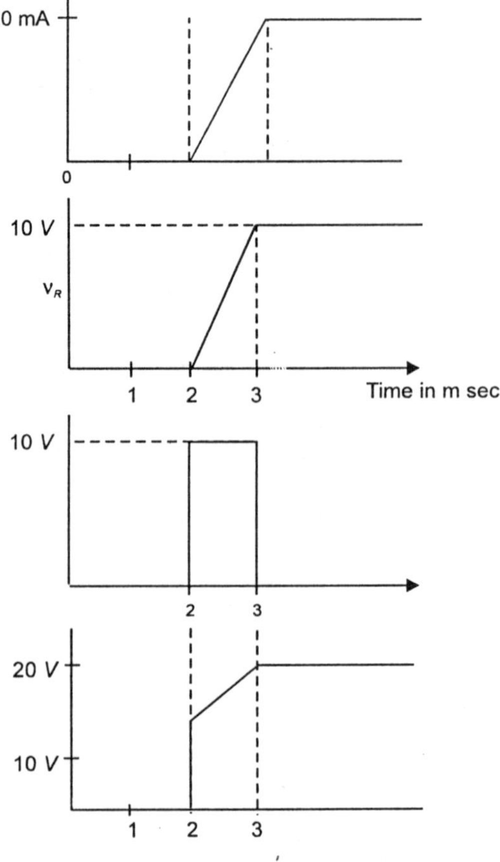


**Fig. E.1.1**

**Solution:** Since the voltage drop across resistance $= iR$, the shape of the voltage is exactly same as that of the current and the magnitude
$$= 20 \times 10^{-3} \times 500 = 10 \text{ volt.}$$

*The voltage across the inductor*

From 0 to 2 m sec, the change in current $di$ is zero hence $\dfrac{di}{dt} = 0$. Hence the voltage across inductor from 0–2 m sec is zero volt.

From 2 to 3 m sec over 1 m sec, the change in current is 20 mA

i.e.
$$= \frac{20 \times 10^{-3}}{1 \times 10^{-3}} = 20 \text{ A/sec}$$

Hence voltage
$$L\frac{di}{dt} = 500 \times 10^{-3} \times 20$$
$$= 10 \text{ volt.}$$

From 3 m sec onwards since the change in current is zero. Hence voltage across inductor is zero. Therefore, the total voltage is the sum of two voltages $v_R + v_L$.

**Example 1.2:** Suppose if the voltage across an inductor is given, it is required to find out the current wave form through the inductor. Fig. E.1.2 (a) shows voltage wave form across an inductor of 0.1 H.

Determine the current wave form:

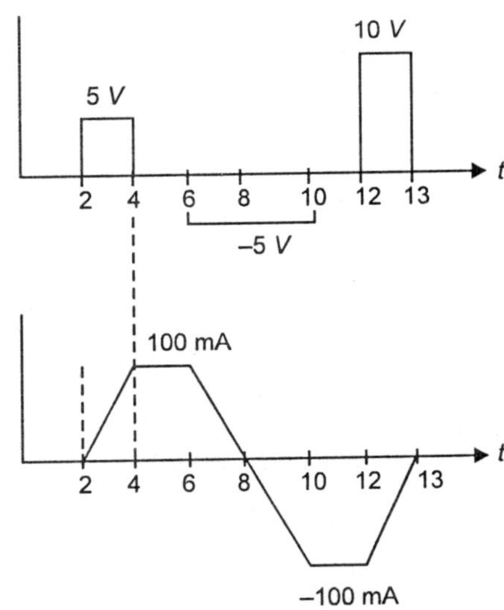


**Fig. E1.2**   (a) Voltage wave form, (b) Current wave form.

**Solution:** In an inductor current $i$ is given
$$i = \frac{1}{L}\int v\, dt$$

From 0 to 2 m sec $v = 0$. Therefore $i = 0$

From 2 to 4 m sec
$$i = \frac{1}{0.1}\int_{2}^{4} 5\, dt = 10\,[5t]_{2}^{4} = 100 \text{ mA}$$

From 4 to 6 m sec, $v = 0$ since there is no change in the current it continues to be same i.e. 100 mA

From 6 to 10 m sec.

$$i = \frac{1}{0.1} \int\limits_{6}^{10} (-5)\, dt = -200 \text{ mA}$$

It decreases linearly between 6 to 10 m sec. Hence the current at 10 m sec. is

$$-200 + 100 = -100 \text{ mA}$$

From 10 to 12 m sec. since voltage is zero there is no change in current and it is −100 mA, from 12 to 13 m sec, the voltage is 10 volts. Hence current increases linearly by an amount

$$\frac{1}{0.1} \int\limits_{12}^{13} 10\, dt = 100 \text{ mA}$$

Beyond 13 m sec the current is zero.

Let us next consider application of non-sinuoidal voltages and currents in a capacitive circuit.

**Example 1.3:**  Consider the application of a voltage wave form shown in Fig. E1.3 (*a*) to a 0.2 µF capacitor.

Determine the current wave form.

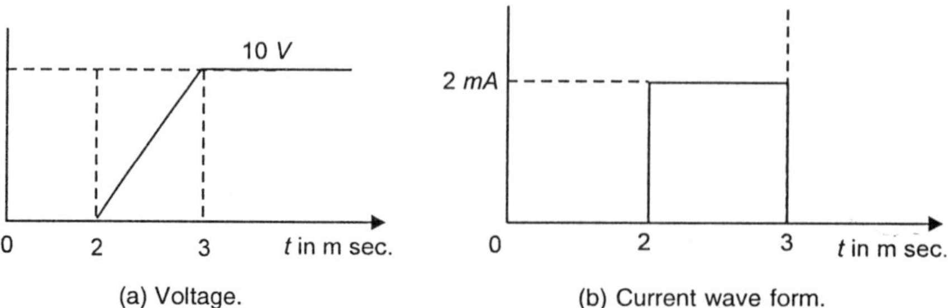



(a) Voltage.  (b) Current wave form.

**Fig. E.1.3**

**Solution:** Since current in a capacitor is given by

$$i = C \frac{dv}{dt}$$

From 0 to 2 m sec $dv = 0$ hence $\dfrac{0}{2} = 0$ Hence the current in this period is zero.

From 2 to 3 m sec the voltage increases linearly from 0 to 10 volts in 1 m sec. Hence

$$v = kt,\ k = \frac{10}{1 \times 10^{-3}} = 10 \times 10^3 \text{ v/sec} \quad \text{or} \quad 10 \text{ V/m sec}$$

$$i = C \frac{dv}{dt} = 0.2 \times 10^{-6} \times k$$

$$= 0.2 \times 10^{-6} \times 10 \times 10^3$$

$$= 2 \text{ mA}$$

Beyond 3 m sec since $v$ is constant $\dfrac{dv}{dt} = 0$

Hence $\qquad\qquad\qquad\qquad\qquad\qquad i = 0$

The variation of current is shown in Fig. E1.3($b$)

## 1.3  NETWORK CLASSIFICATION

Electric networks can be classified mainly in two ways. The first one is based on the kind of elements in the network e.g. time-invariant or nonlinear or time variant etc. The second one is based on excitation and response i.e. for a given excitation the response decides the type of network.

Here we shall study classification based on the second approach.

### 1.3.1  Linearity

Suppose the network to be classified is under relaxed condition i.e. the initial condition is zero and an excitation $v_1(t)$ is applied for which response is $c_1(t)$ and for excitation $v_2(t)$ the response is $c_2(t)$. Then the network is classified as linear if for excitation $v_1(t) + v_2(t)$ the response is $c_1(t) + c_2(t)$. This shows that a linear network follows superposition principle.

### 1.3.2  Passivity

Some networks have the property of absorbing/dissipating or storing energy. These are able to return the energy previously stored to the external network. However, this energy returned is never more than the energy stored. Such networks are known as passive networks.

Let network $N$ be connected to a source $S$. Suppose the network is initially relaxed. If $v(t)$ is the voltage of the source and $i(t)$ is the resulting current into the network $N$, the average power delivered into the network is $p(t) = v(t)\, i(t)$ and the energy delivered $w(t)$ over time $t$ is given by

$$w(t) = \int_{-\infty}^{t} v(t)\, i(t)\, dt \qquad\qquad ...(1.21)$$

This energy will always be non-negative for a passive network i.e. over time $t$ there will always be some energy delivered to the network or at least there will not be any energy when the network would have returned all the energy back to the source or it would have stored some energy. It will never be a situation for a passive network when it would supply more energy what it had drawn previously. Mathematically for a passive network.

$$w(t) = \int_{-\infty}^{t} v(t)\, i(t)\, dt \geq 0 \qquad\qquad ...(1.22)$$

However, if the network has some initial energy stored equal to $w(t_0)$ before it is switched on to a supply at $t = t_0$, then for a passive network energy delivered to the network from the source should satisfy the inequality at any time $t$

$$w(t) = \int_{-\infty}^{t} v(t)\, i(t)\, dt + w(t_0) \geq 0 \qquad \qquad ...(1.23)$$

A circuit which does not satisfy this condition is known as active circuit. That is for an active circuit

$$\int_{-\infty}^{t} v(t)\, i(t)\, dt < 0 \qquad \qquad ...(1.24)$$

### 1.3.3  Lumped

Many devices in electric system are distributed in space e.g. transmission lines, the windings of transformers or that of the generators, are distributed in a way. Whenever these devices are energized, the effect is not experienced by the line length or winding lengths instantly because of finite velocity of electric signals. However, if we are interested in steady state and terminal quantities, it is sufficient to assume the parameters to be lumped rather than distributed. Sometimes when we are interested in the intermediate values and point to point variation of electric signals, we talk true nature of these devices i.e. the distributed nature. Otherwise, we assume the parameters to be lumped. In this book we treat these parameters as lumped unless it is specified otherwise.

### 1.3.4  Bilateral

The elements to be considered for electric network are assumed to be bilateral which means the voltage and current relations are same irrespective of direction of flow of current e.g. resistance, inductance, etc. However, for unilateral elements the voltage and current are different for two possible direction of flow of currents e.g. diodes.

### 1.3.5  Time Invariant and Time Varying Systems

A network is said to be time-invariant when there is some response to a certain excitation irrespective of time of application of excitation. Suppose for certain excitation $v_1\,(t)$ the response is $c_1\,(t)$, if now the excitation is $v\,(t + t_0)$ the response would be $c\,(t + t_0)$. Here the values of the parameters are assumed to be constant at all times and do not change with time. Therefore, a system in which one or more parameters are function of time, the system is known as time varying system. The terms stationary system and non-stationary systems are also used for such systems

### 1.3.6  Reciprocity

A network is said to possess the property of reciprocity when excitation and response terminals could be interchanged. Consider network $N$ with terminals 1 and 2 as shown in Fig. 1.12.

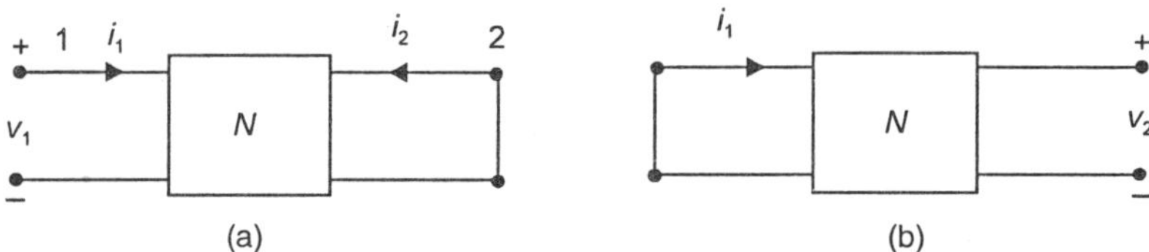


**Fig. 1.5**  Networks to describe reciprocity.

Say if $v_1$ is excitation and $i_2$ response as in Fig. 1.5 (*a*) and if $v_2$ is excitation then $i_1$ is the response as in Fig. 1.5 (*b*) The network is reciprocal if for $v_1 = v_2$, $i_1 = i_2$. A transmission line is a reciprocal network. The network which does not satisfy the condition is a non-reciprocal network. $R$, $L$, $C$, are in general linear, passive, bilateral, time invariant and lumped elements. Since $L$ and $C$ can store energy in the form of electro-magnetic and electro-static fields respectively their representation needs special mention. In general voltage across capacitor is given by

$$v(t) = \frac{1}{C} \int i(t)\, dt + v(o) \qquad \ldots(1.25)$$

where $v(o)$ indicates the voltage due to initial charge. The capacitor which has initial voltage $v(o)$ can be represented as shown in Fig. 1.6(*a*).

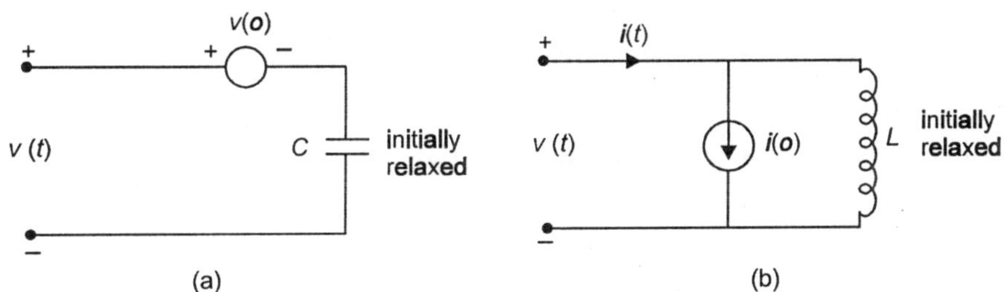

(a)                    (b)

**Fig. 1.6**  Initial condition for a capacitor and inductor.

Similarly if an inductor initially has a current $i(o)$ the current at anytime $t$ is given by

$$i(t) = \frac{1}{L} \int v(t)\, dt + i(o) \qquad \ldots(1.26)$$

Therefore, $i(o)$ could be considered as a d.c. current source in parallel with an initially relaxed (initial condition zero) inductor as shown in Fig 1.6(*b*).

### 1.3.7  Non-Linear Systems

Some of the non-linearities commonly encountered in engineering systems are (i) saturation (ii) Friction (iii) Hysteresis (iv) Dead zone.

Presence of any of the non-linearities would make the system non-linear. The resulting mathematical model representing such a system would obviously be non-linear and analysis of such non-linear model is quite complex.

### 1.3.8  Static and Dynamic Systems

When only steady state operating condition is of interest many systems can be modelled by algebraic equations and such systems are known as static systems. Here the output waveform is a replica of the input waveform, its magnitude at any time being equal to the input magnitude at that time multiplied by a constant. K. Fig 1.7 shows a static system.

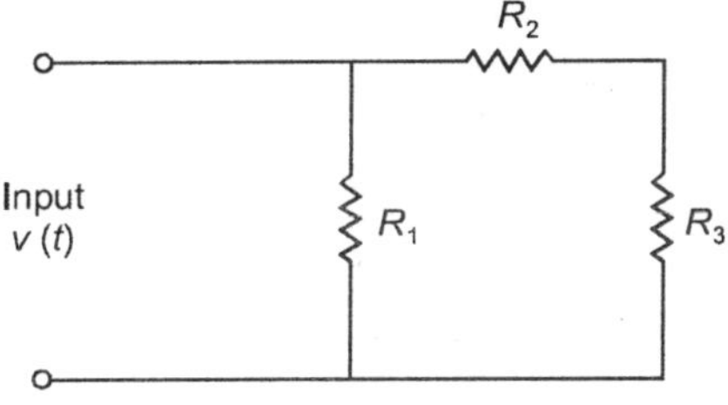

**Fig 1.7**  A resistive network (static system).

Thus the output at any instant depends upon the value of the input at that instant only. Such systems are known as static systems.

However, dynamic systems are those which are represented by differential equations or difference equations. In such differential equations time is the independent variable. The output is always a function of time even when the input is a constant. Thus there is 'change' or 'motion' and hence the name dynamic system. In such system the output not only depends upon the input but upon the initial conditions also. We know that solution of nth order differential equations requires the knowledge of the initial value of the dependent variables and first (n-1) derivatives. These initial values of the system variables must be present due to the past inputs to the system.

Any electrical circuit with resistive elements will be static however complex the circuit may be. However, a circuit with a single inductor or a capacitor or both will always be dynamic as an inductor or a capacitor can store energy over a period of time whereas a resistor dissipates energy. Thus we conclude that the presence of an energy storage element-makes the system dynamic.

## 1.3.9  Continuous Time and Discrete-time Systems

Continuous time systems are those for which the system variables like current and voltages are defined for every instant of time. However, their values can change with time. In contrast to this, in the mathematical modelling of many processes we come across system variable which are either defined at only specified instants of time or can change their values only at these specified instants as in a digital system. Such signals are called discrete-time function (e.g. impulse train) and the system to which they belong are called discrete-time system. A plot of both types of function is shown in Fig. 1.8.

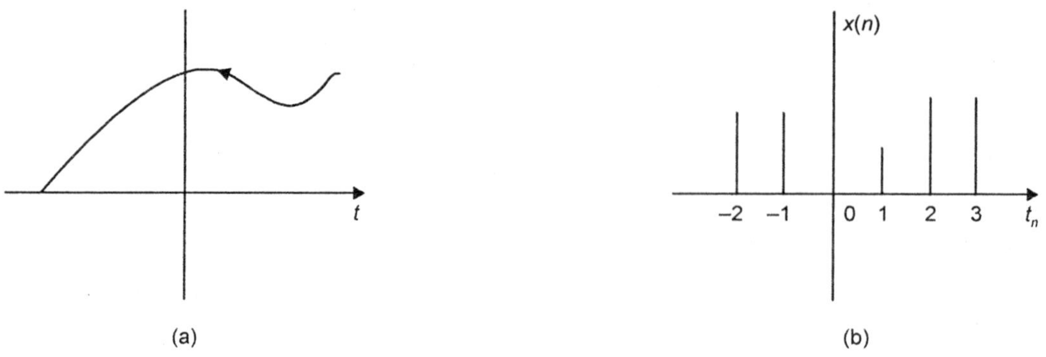


**Fig. 1.8**  (a) Continuous time  (b) Discrete time function

For a continuous function x, its functional dependence on variable time is indicated as $x(t)$ whereas for discrete since the function is defined only at the sequence of instant $t_n$ it is expressed as $x(t_n)$ or simply $x(n)$. Discrete time functions find applications in the modelling of instrumentation and control systems where multiplexing digital signal processing or computer controls are used.

## 1.3.10 Deterministic and Stochastic Systems

A mathematical model which takes into account the uncertainty in the output and relates the probability of the output with the input is called stochastic model. In such a model the actual output for a given input is uncertain, only its' probability of occurrence can be predicted. In contrast the deterministic systems are those for which output can be exactly determined for a given input. The uncertainties in stochastic system could be grouped into two categories: (i) The random noise signals picked up at the input and other points of the system. The effect of noise signals is quite harmful is communication systems (ii) Uncertainties in the exact values of the parameters of the system. These uncertainties are taken care of by the stochastic model of the system and use these models to design filters to reduce the effect of noise.

## 1.4 REPRESENTATION OF TIME SIGNALS

We have so far studied circuit elements and classification of networks. We now consider specific forms and shapes of voltage and current signals which drive these circuit elements and networks. We discuss several basic types of input signals which are quite useful in network analysis.

Consider a general independent voltage source as shown in Fig 1.9. The source has a switch in series with it which is closed at-time $t = O$. The terminals marked $a$, $b$ are to be

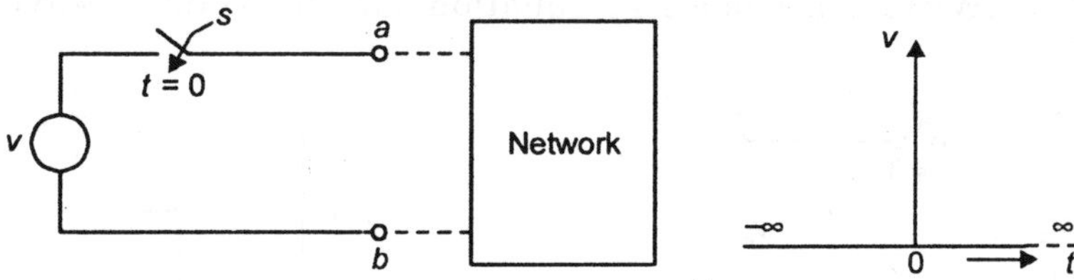

**Fig. 1.9**   Independent voltage source

connected to network so that the voltage that is applied to the network is $V_{ab}(t)$. Our objective is to describe the voltage source in mathematical form. We consider dividing the time axis $-\infty < t < \infty$ into two parts $-\infty < t < 0$ and $0 \leq t < \infty$. Since for the time interval $-\infty < t < 0$ the switch is open and the voltage $V_{ab}(t)$ is zero i.e.

$$v_{ab}(t) = 0 \quad -\infty < t < 0 \qquad \qquad ...(1.27)$$

Whereas for the interval $0 \leq t < \infty$, the switch is closed and the voltage $V_{ab}(t)$ will have its value $V(t)$. Suppose the voltage source is a battery of terminal voltage $V$, then

$$v(t) = V \quad 0 \leq t < \infty \qquad \qquad ...(1.28)$$

Combining the two equations (1.27) and (1.28) in a single relationship describing $V_{ab}$ as a function of time

$$v_{ab}(t) = \begin{cases} 0 & -\infty < t < 0 \\ V & 0 \leq t < \infty \end{cases} \qquad \qquad ...(1.29)$$

A plot of equation (1.29) is shown in Fig 1.10

It can be seen that the voltage $V_{ab}(t)$ is a step of size $V$ at $t = 0$. We now describe this mathematically by defining a function which exhibits this property.

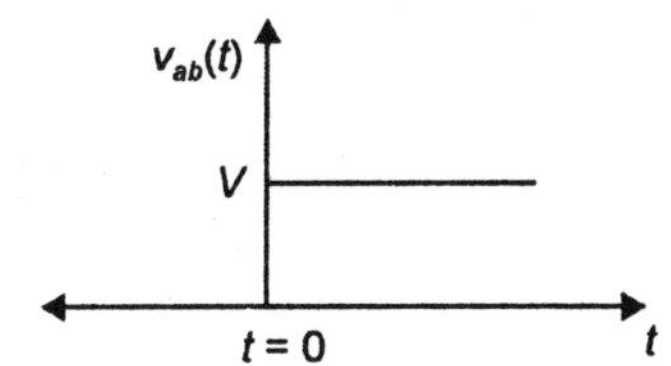

**Fig 1.10**   Time signal generated by battery and switch in series

### 1.4.1  Unit Step Function

The unit step function $u_0(t)$ is defined by the relationship

$$u_0(t) = \begin{cases} 1 & t \geq 0 \\ 0 & t < 0 \end{cases}$$

This function is plotted in Fig 1.11. It is to be noted that $u_0(t)$ at $t = 0$ is unity. Since the function is discontinuous at the point $t = 0$, the value we assign it at this point is arbitrary. We have defined the value of the function at the discontinuity as the value the function assumes immediately after the discontinuity i.e.

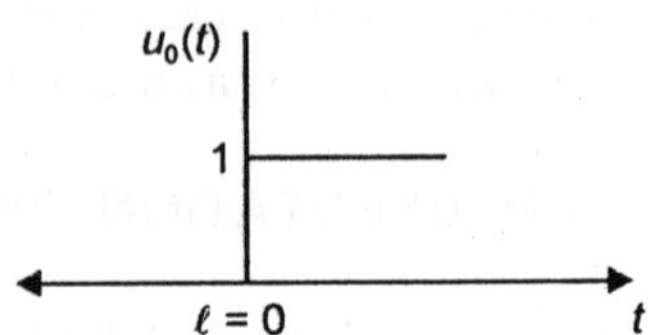

**Fig 1.11**   Unit step function

$$vu_0(0_-) = 0 \text{ and } vu_0(0_+) = 1$$

However, if the strength of the step is $V$, $V_{ab} = Vu_0(t)$ ...(1.30)

**Example 1.4:**  Write mathematical equation for the voltage waveform shown in Fig 1.4

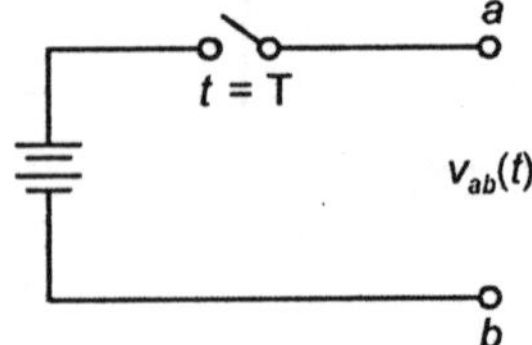


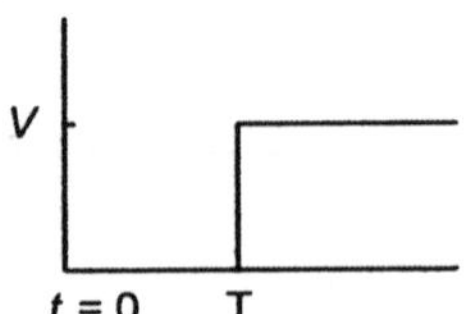


**Fig. E 1.4**   Unit step applied at $t = T$

*Solution:*  The voltage $V_{ab}(t)$ is zero for $t < T$ and is V volts for $t \geq T$
Thus we write $\qquad\qquad\qquad V_{ab}(t) = Vu_0(x)$
For $x < T$, the voltage signal is zero and $x \geq T$ it attains a value $V$, therefore

$$V_{ab}(t) = Vu_0(t - T)$$

Which is shown in Fig. E1.4 (b)

It is to be noted that if the signal appears at $t = 0$, we have $u_0(t)$, if it appears to the right of $t = 0$ axis we have $u_0(t - T)$ and if it appears to the left of $t = 0$ axis then we have $u_0(t + T)$.

### 1.4.2  Unit Ramp Function

The unit ramp function usually denoted by $u_1(t)$ or $r(t)$ is defined as the integral of the unit step function

$$u_1(t) \underline{\Delta} \int_{-\infty}^{t} u_0(x)dx \qquad\qquad\qquad ...(1.31)$$

As is seen in Fig 1.12, for $t < 0$ the area between $u_0(x)$ and time axis is zero whereas for $t \geq 0$, it is equal to the area contained in a rectangle of height unity and base t

Hence, we have

$$r(t) = u_1(t) = \begin{cases} 0 & t < 0 \\ t & t \geq 0 \end{cases} \qquad \ldots(1.32)$$

and it is shown in Fig. 1.13

The unit ramp function $r(t)$ or $u_1(t)$ is, thus, the indefinite integral of the unit step $u_0(t)$ and has a slope of unity for $t \geq 0$. Since $u_1(t)$ is the integral of $u_0(t)$, we must have

$$\frac{du_1(t)}{dt} = u_0(t)$$

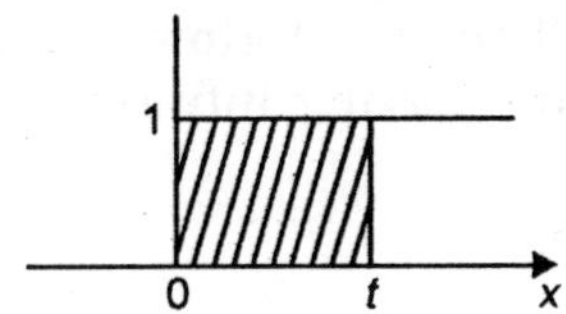

Fig.1.12   Graphical evaluation of integral of equation (1.31)

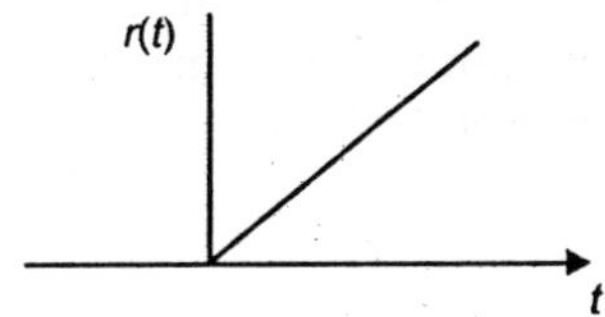

Fig 1.13   unit ramp (slope unity) function

i.e. the unit step function is the derivative of the unit ramp function. If we continue to integrate the unit step function to generate higher order functions of $t$, for example, integrating the unit step function twice and multiplying by a constant 2 we can define $u_2(t)$ as

$$u_2(t) = \begin{cases} 0 & t < 0 \\ t^2 & t \geq 0 \end{cases} \qquad \ldots(1.33)$$

or equivalently

$$u_2(t) = 2\int_{-\infty}^{t} u_1(x)dx \qquad \ldots(1.34)$$

Equation 1.33 represents a parabola starting from the origin. In general we define the function

$$u_n(t) \triangleq n\int_{-\infty}^{t} u_{n-1}(x)dx \qquad \ldots(1.35)$$

so that $U_n(t)$ has the property

$$u_n(t) = \begin{cases} 0 & t < 0 \\ t^n & t \geq 0 \end{cases} \qquad \ldots(1.36)$$

Also

$$u_{n-1}(t) = \frac{1}{n}\frac{d}{dt}u_n(t) \qquad \ldots(1.37)$$

### 1.4.3   Unit Impulse

We have considered the integral of the unit step. Next we study the derivative of the unit step function. The function $u_0(t)$ is zero for $t < 0$ and is unity for $t > 0$ and hence $\dfrac{du_0(t)}{dt} = 0$

for $t < 0$ and $t > 0$. However at $t = 0$ the function $u_0(t)$ is discontinuous, with a step of height unity and having infinite slope. The derivative of $u_0(t)$ is denoted as $\delta(t)$

i.e.
$$\frac{du_0(t)}{dt} = \delta(t) \qquad \qquad \qquad \dots(1.38)$$

or
$$u_0(t) = \int_{-\infty}^{t} \delta(t)dt \ \text{ and } \ \delta(t) = 0 \text{ for all } t \neq 0.$$

Here $\delta(t)$ is known as impulse function or unit impulse or unit delta function.

The signal $\delta(t)$ can also be obtained as the limit of the derivative of the waveform shown in Fig 1.14 (a).

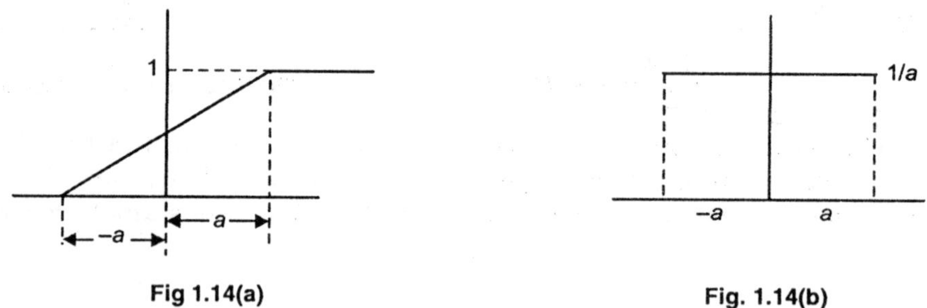

Fig 1.14(a)        Fig. 1.14(b)

As $a \to 0$ the waveform in Fig. 1.14(a) becomes the unit step. The derivative of this waveform is shown in Fig. 1.14(b). As $a \to 0$ the waveform in Fig 1.14(b) approaches zero everywhere except the origin where the amplitude becomes infinite. However, we note that the area contained under the curve remains unity as $a \to 0$. Thus in the limit we can visualize $\delta(t)$ as approaching a very large spike or impulse at the origin with infinite amplitude and zero width yet area equal to unity

Unit impulse can be mathematically expressed as

$$\delta(t) = 0 \ t \neq 0 \text{ and } \int_{-\infty}^{\infty} \delta(t)dt = 1 \qquad \qquad \dots(1.39)$$

Impulse function is also known as delta function. A very important property of the unit impulse is its sampling property. If we consider multiplying the unit impulse by a function $f(t)$ we have from equation (1.39)

$$f(t)\delta(t) = 0 \ t \neq 0$$
$$= f(0) \text{ at } t = 0 \qquad \qquad \dots(1.40)$$

Which means the multiplication of a function by a unit impulse corresponds to sampling the function at the time instant where the impulse is non zero.

### 1.4.4  Exponential function

Another function which we will use frequently is the exponential function $e^{at}$. This function has the property

$$\frac{de^{at}}{dt} = ae^{at} \qquad \qquad \dots(1.41)$$

and
$$\int_{-\infty}^{t} e^{ax}dx = \frac{1}{a}e^{at}$$
...(1.42)

In general
$$\frac{d^n e^{at}}{dt^n} = a^n e^{at}$$
...(1.43)

And
$$\underbrace{\int \cdots \int}_{n \text{ times}} e^{at}dt = \frac{1}{a^n}e^{at}$$
...(1.44)

**Example 1.5:**  Write down the mathematical equation for the function shown in Fig. E 1.5.

*Solution:*  From the Fig. E 1.5 we have

$$f(t) = \begin{cases} 0 & t < 0 \\ 2t^2 & 0 \leq t < 3 \\ 18 & 3 \leq t < 6 \\ 4.5(t-8) & 6 \leq t \leq 8 \\ 0 & t \geq 8 \end{cases}$$

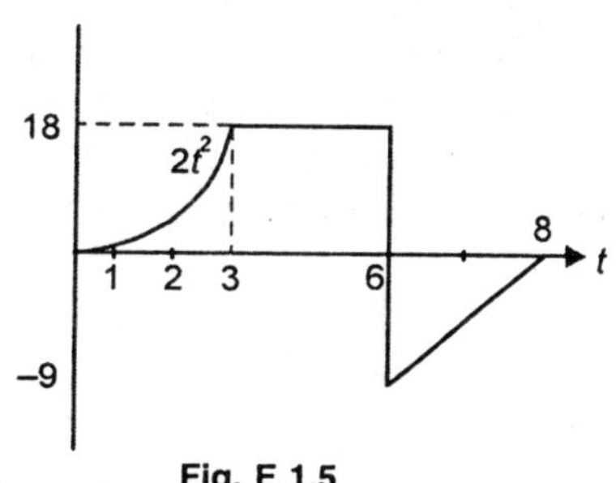


**Fig. E 1.5**

The function can be regarded as sum of three functions as shown in Fig. E 1.5.1.

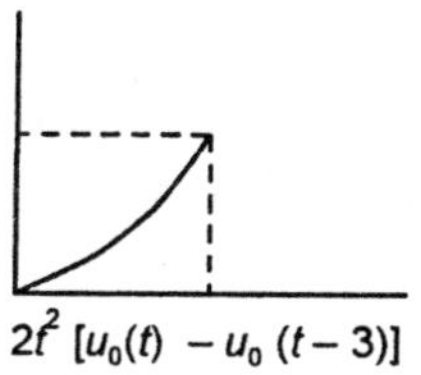
$$2t^2\,[u_0(t) - u_0(t-3)]$$

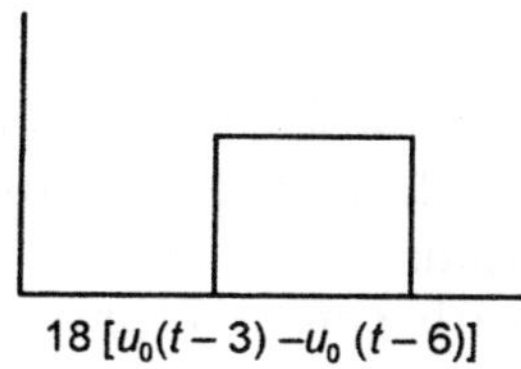
$$18\,[u_0(t-3) - u_0(t-6)]$$

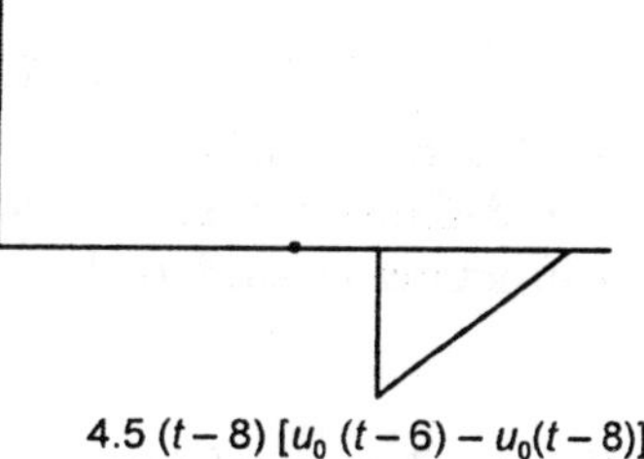
$$4.5\,(t-8)\,[u_0(t-6) - u_0(t-8)]$$

**Fig. E 1.5.1**

## 1.5  PERIODIC SIGNALS

The signals so far considered are not repetitive in time. We next consider signals which are repetitive or periodic.

### 1.5.1  Sinusoidal Waveform

The sinusoidal signal is given as $\sin 2\pi ft$ where $f$ is the frequency of the signal which is measured in Hertz or cycles per second i.e. the no. of times the signal is repeated per second or the time period of the signal is $T = \dfrac{1}{f}$. Which means that whatever be the frequency of the signal, the resulting waveform remains periodic representing the same value over each interval of time $T$. Fig 1.15 represents the shape of the sinusoidal waveform for $f = 1$ Hz

and $f = 5$ Hz

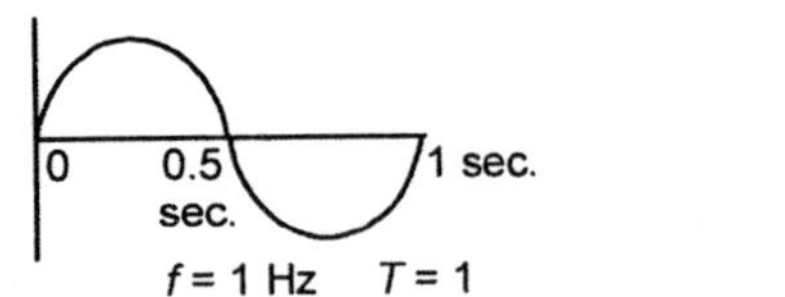


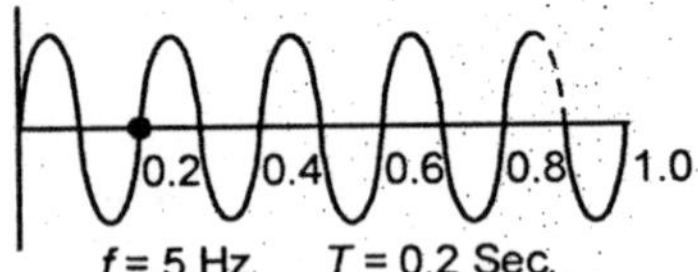


**Fig. 1.15**   Plot-of Sine wave for $f = 1$ *Hz*, $f = 5$ Hz

The sinusoidal waveform need not necessarily start at $t = 0$. It may start from the left of $t = 0$ axis or from the right of the $t = 0$ axis. In terms of angles one complete cycle corresponds to $2\pi$ radians or 360°. Fig. 1.16 shows three sinusoidal waveform with three different phase parameter $\phi$. From the Fig. 1.16 it can be seen that periodic waveforms can be generated by shifting the single cycle of the sinusoidal shown with solid line in the figure through all integer multiples of the period.

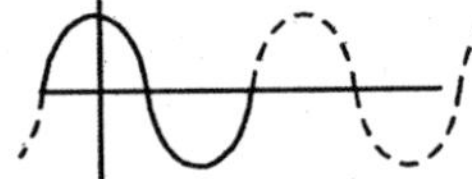
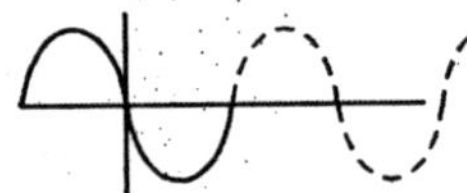

**Fig. 1.16**   Plot of sinusoidal waveform (a) $\phi = 0$  (b) $\phi = \dfrac{T}{4}$  (c) $\phi = \dfrac{T}{2}$

The phase parameter determines the portion of the period of that fundamental cycle which lies to the left of the axis $t = 0$

The periodic sinusoidal has the property

$$\sin(2\pi\ ft + \phi) = \sin(2\pi\ ft + T + \phi) \qquad\qquad \text{...(1.45)}$$

as $T$ correspond to time for one cycle of the waveform. This property suggests the following definition for a periodic function:

A function is said to be periodic with period $T$ provided

$$g(t) = g(t + T) \quad -\infty < t < \infty \qquad\qquad \text{...(1.46)}$$

The fundamental frequency $f_0$ of $g(t)$ defined as $f_0 = \dfrac{1}{T}$ and represents the number of repetitive cycles of the waveform which occur in a period of 1 sec.

### 1.5.2  Square Wave

A square wave signal is shown in Fig. 1.17. This wave form satisfies the definition of a periodic function as given in equation (1.46). The period is $T$ sec

and hence the fundamental frequency $f_0 = \dfrac{1}{T}$ Hz.

Now consider another square wave shown in Fig 1.18

This wave can be decomposed as follows and hence square wave can be mathematically represented,

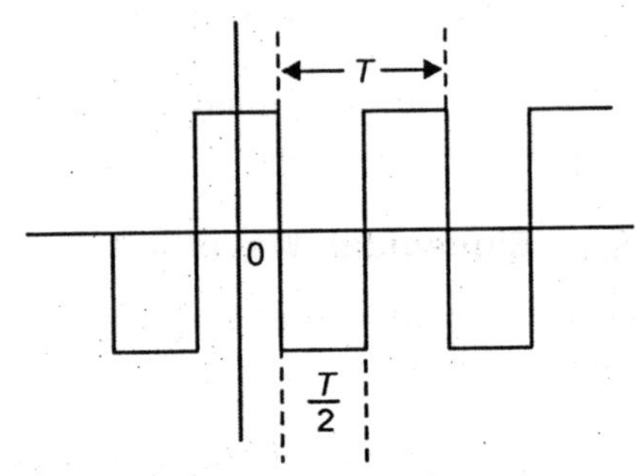


**Fig. 1.17**   Square wave signal

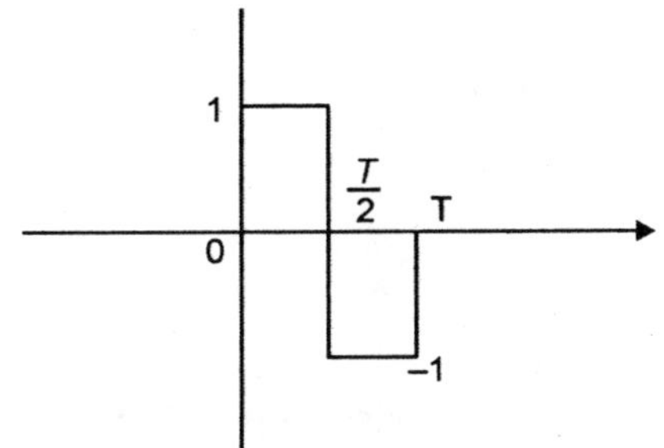

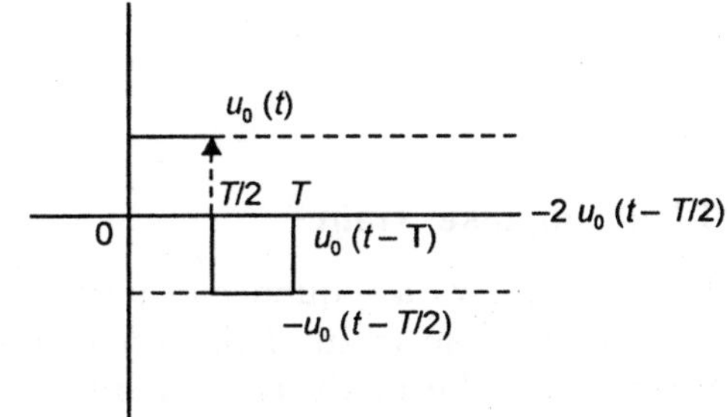

**Fig. 1.18** One cycle of square wave
signal

**Fig. 1.19** Decomposition of wave in
Fig. 1.18

as
$$s(t) = u_0(t) - 2u_0(t - T/2) + u_0(t - T) \qquad \ldots(1.47)$$

Now for Fig. 1.17 using the shifting property of the unit step we have

$$g(t) = \sum_{n=-\alpha}^{\infty} s(t + nT + \phi) \qquad \ldots(1.48)$$

Where $\phi$ is the phase shift from the axis $t = 0$. Here it is $\dfrac{T}{4}$ on the left-side of the $t = 0$ axis and hence for fig. 1.17.

$$g(t) = \sum_{n=-\infty}^{\infty} s(t + nT + T/4) \qquad \ldots(1.49)$$

### 1.5.3 Saw Tooth Waveform

This waveform is usually used in television and Oscilloscope circuits and one such waveform is shown in Fig 1.20

An analytical expression for this waveform can be obtaining by recalling that, this is similar to a ramp function and ramp function is the first integral of unit step function and hence the expression would be

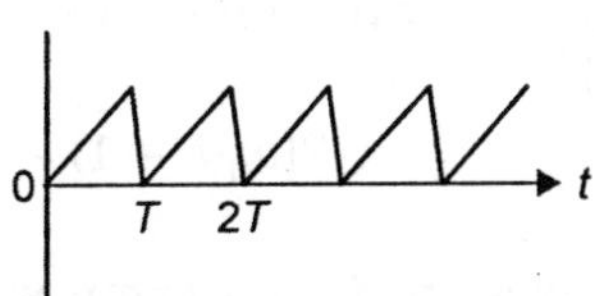

**Fig. 1.20** Sawtooth wave form

$$r(t) = u_1(t) - u_1(t)\, u_0(t - T) \qquad \ldots(1.50)$$

However, if the saw tooth wave is shifted on to left of $t = 0$ axis by a phase shift $\phi = \dfrac{T}{4}$ then the general expression will be

$$g(t) = \sum_{n=-\infty}^{\infty} s(t + nT + \phi) \qquad \ldots(1.51)$$

and hence for $\phi = \dfrac{T}{4}$ we have

$$g(t) = \sum_{n=-\alpha}^{\infty} s(t+nT+T/4) \qquad\qquad .....(1.52)$$

## 1.5.4  Impulse Train

Fig. 1.21 shows a sequence of unit impulses known as unit impulse train. This signal plays an important role in the analysis of steady state circuit response to a periodic signal input. It is obvious that $g(t)$ in general for impulse train is given

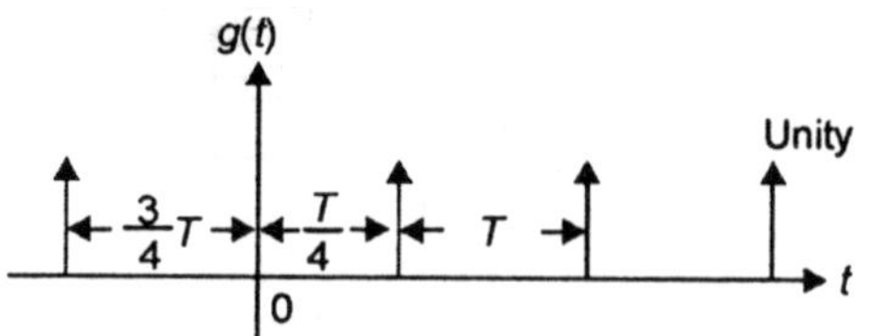

Fig. 1.21   Unit impulse train

as
$$g(t) = \sum_{n=-\infty}^{\infty} \delta(t+nT+\phi) \qquad\qquad ....(1.53)$$

and $g(t)$ for the specific case shown in Fig 1.21 is given as

$$g(t) = \sum_{n=-\infty}^{\infty} \delta\left(t+nT+\frac{3}{4}T\right) \qquad\qquad ....(1.54)$$

**Example 1.6:**  Write mathematic expression for the waveform shown in Fig E 1.6

*Solution:*  The waveform $A$ is a ramp function with slope $\dfrac{2}{1} = 2$ and starting from $t = 0$ to $t = 1$ hence   $2t[u_0(t) - u_0(t - 1)]$

Waveform B is below the time axis and is a square wave and starts from $t = 1$ to $t = 3$. Hence
$$-2[u_0(t - 1) - u_0(t - 3)]$$

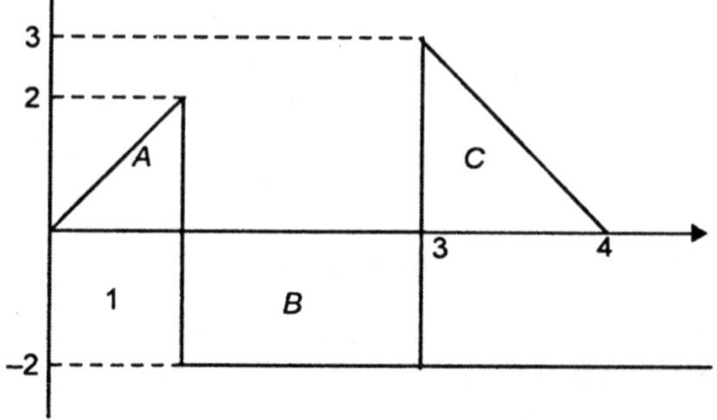

Fig. E1.6

Waveform $C$ is a ramp with negative slope and slope is $-3$ and hence the waveform is
$$-3(t - 4)[u_0(t - 3) - u_0(t - 4)]$$
and hence the complete waveform is given as
$$g(t) = t[u_0(t) - u_0(t - 1)] - 2[u_0(t - 1) - u_0(t - 3)] - 3(t - 4)[u_0(t - 3) - u_0(t - 4)]$$

**Example 1.7:**  Determine mathematical expression for the waveform shown in Fig E 1.7

*Solution:*  These are three square waves. Wave $A$ is
Given as $\qquad\qquad\qquad 3[u_0(t) - u_0(t - 2)]$
Wave B $\qquad\qquad [u_0(t - 2) - u_0(t - 4)]$ and
Wave C $\qquad\qquad 3[u_0(t - 4) - u_0(t - 6)]$ Hence
The waveform $g(t)$ is given as

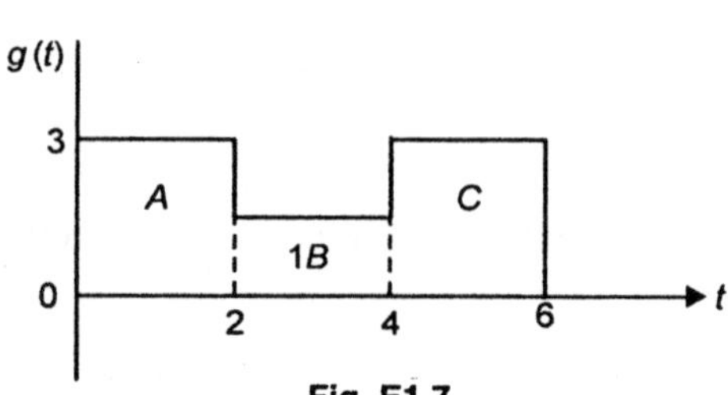

Fig. E1.7

$$g(t) = 3[u_0(t) - u_0(t - 2)] + [u_0(t - 2) - u_0(t - 4)] + 3[u_0(t - 4) - u_0(t - 6)] \textbf{ Ans.}$$

**Example 1.8:** Write the mathematical expression for the waveform shown in Fig. E 1.8

*Solution:* Using the shifting property of the step function the mathematical expression for the above wave is given as

$$g(t) = \sin \frac{\pi t}{T} \, [u(t - 2T) - u(t - 3T)] \textbf{ Ans.}$$

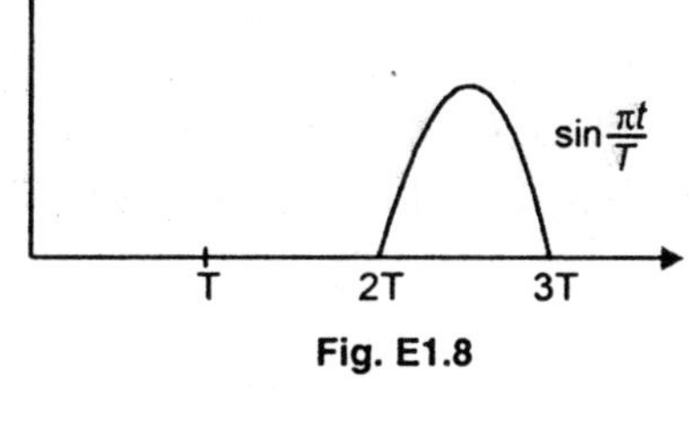


**Fig. E1.8**

**Gate Function:** A rectangular pulse of duration $\tau$ and height unity commencing at $t = t_0$ is shown in Fig. 1.21 and is represented mathematically as $u(t - t_0) - u(t - t_0 - \tau)$ and is known as a Gate function. It is known as Gate function because any function if multiplied by Gate function will have its natural value during the duration of the Gate function and zero elsewhere.

The expression for the gate function starting from the origin is $G_0(t) = u(t) - u(t - \tau)$

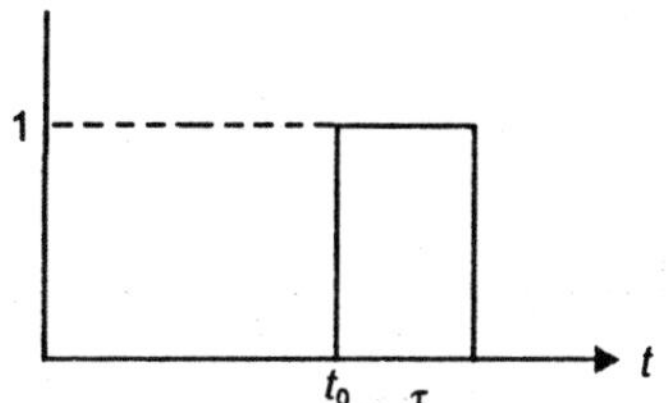


**Fig. 1.21  Gate function**

**Example1.9:** Represent the given wavefrom as shown in Fig. E 1.9 in mathematical form

*Solution:* The first part is a ramp with slope

$\dfrac{2}{1} = 2$ and using the gate function at origin $[u(t) - u(t - 1)]$, it is represented as $2t[u(t) - u(t - 1)]$
Next is a gate function. Where $\tau = 2$ and $t_0 = 1$ and Hence $2u(t - 1) - 2u(t - 1 - 2) = 2u(t - 1) - 2u(t - 3)$
Combining the two we have

$$F(t) = 2tu(t) - 2tu(t - 1) + 2u(t - 1) - 2u(t - 3)$$

$$2tu(t) - (2t - 2)u(t - 1) - 2u(t - 3)$$

$$= 2tu(t) - 2(t - 1)u(t - 1) - 2u(t - 3) \textbf{ Ans.}$$

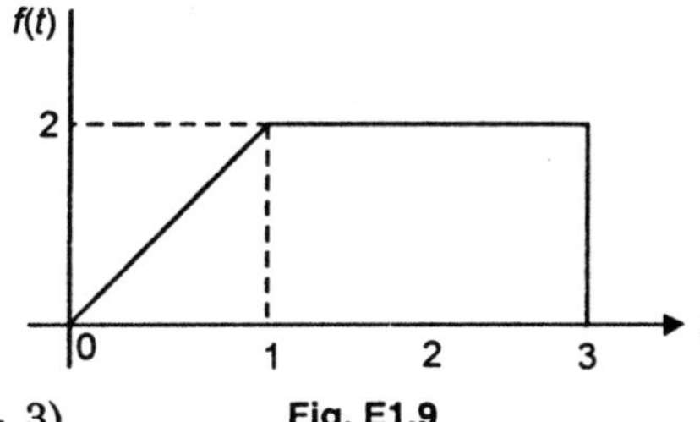


**Fig. E1.9**

**Example 1.10:** Represent the signal given in Fig.E1.10 using gate function or in mathematical form

*Solution:* The given signal is sinusoidal and is represented as $V_m \sin wt$ and the gate functions is $u(t) - u(t - T/2)$. Hence the mathematical representation is

$$V_m \sin wt[u(t) - u(t - T/2)]$$

$$= V_m \sin wt \, u(t) - V_m \sin wt \, u(t - T/2)$$

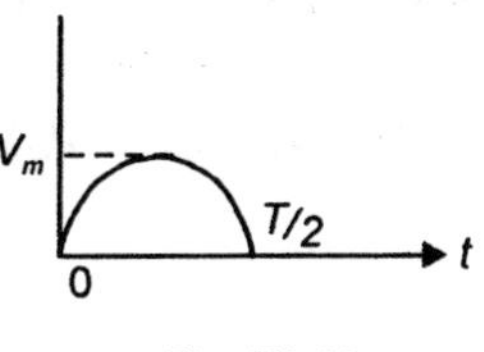


**Fig. E1.10**

Alternatively the waveform can be decomposed into the waveform as shown in Fig.E1.10.1

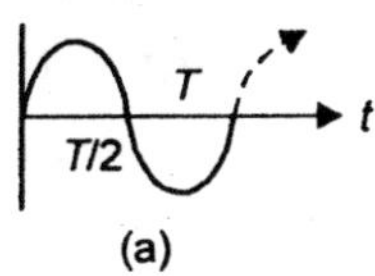


(a)

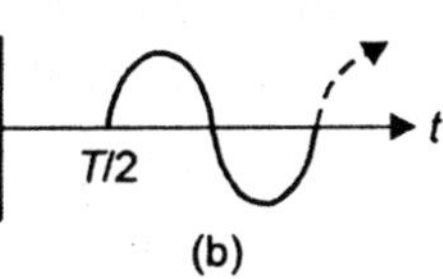


(b)

**Fig.  E1.10.1**

The first part is $V_m \sin wt\, u(t)$ as the sine wave is starting from origin. The second part is shifted to the right of the $t = 0$ axis by a time $T/2$ hence it is represented as

$$V_m \sin w(t - T/2)\, u(t - T/2)$$

Hence the representation for the half sinusoid of the example is
$V_m[\sin wt\, u(t) + \sin w(t - T/2)\, u(t - T/2)]$ **Ans.**

**Example 1.11:**   Represent the waveforms given in Fig. E1.11 as mathematical expressions.

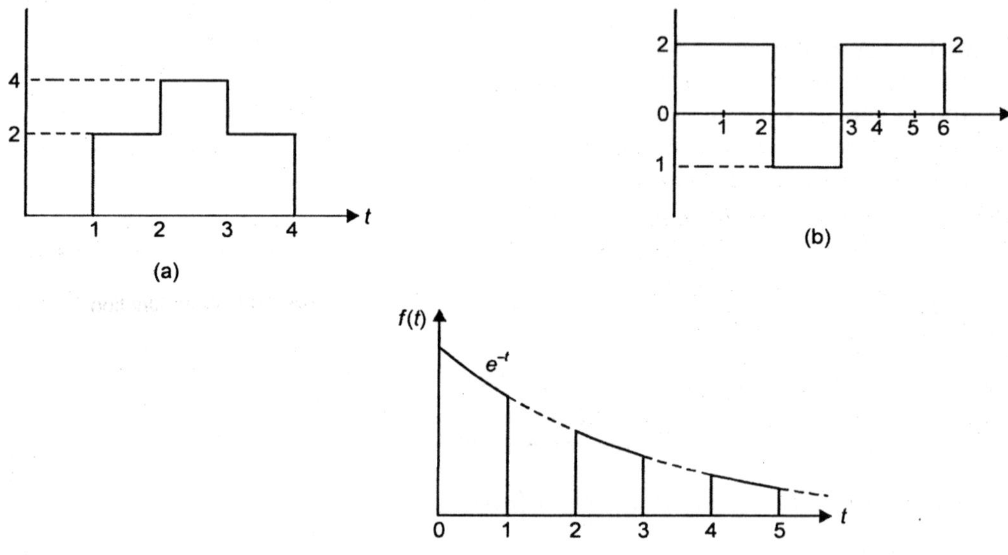

(a)

(b)

(c)

**Fig. E1.11**

*Solution:* (a) The first step is shifted to right of $t = 0$ axis by $t = 1$ and is upto $t = 2$ and magnitude 2.

Hence    $2u(t - 1) - 2u(t - 2)$

The second step of magnitude 4 and begins at $t = 2$ and gets upto t = 3.
Hence    $4u(t - 3) - 4u(t - 4)$

Hence the function $f(t) = 2u(t - 1) - 2u(t - 2) + 4u(t - 2)$

$$-4u(t - 3) + 2u(t - 3) - 2u(t - 4)$$

$$= 2u(t - 1) + 2u(t - 2) - 2u(t - 3) - 2u(t - 4)$$

(b) Using the logic as given in (a) part of this problem

$$f(t) = 2u(t) - 2u(t - 2) - u(t - 2) + u(t - 3) + 2u(t - 3) - 2u(t - 6)$$

$$= 2u(t) - 6u(t - 2) - 6u(t - 3) - 2u(t - 6)$$

(c) It is an exponentially decaying waveform with discontinuities. Using gate function.

$$f(t) = e^{-t}[u(t) - u(t - 1)] + e^{-t}[u(t - 2) - u(t - 3)] + e^{-t}[u(t - 4) - u(t - 5) + \ldots\ldots]$$

$$= e^{-t}[u(t) - u(t - 1) + u(t - 2) - u(t - 3) + u(t - 4) - u(t - 5) \ldots\ldots] \textbf{ Ans.}$$

**Example E1.12:** Draw the waveforms for the following functions.

(i) $\sin w(t - t_0)$

(ii) $\sin w(t - t_0) \, u(t)$

(iii) $\sin w(t - t_0) \, u(t - t_0)$

(iv) $\sin wt \, u(t - t_0)$

*Solution:* (i) This is a simple sinsoidal curve which is shifted to right of $t = 0$ axis by $t_0$ and at $t = 0$, it has its value $-\sin w \, t_0$ and hence the wave will be as shown below

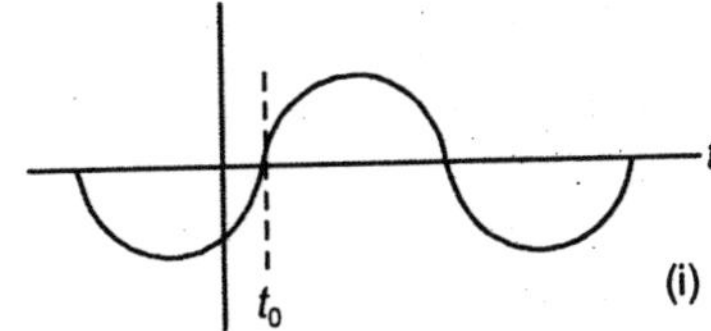

**Fig. E1.12**  $\sin w(t - t_0)$

(ii) Sin $w(t - t_0) \, u(t)$. This is similar to the previous problem except that the wave has zero value for $t < 0$ as it is associated with $u(t)$ and, therefore, the waveform will be as shown here

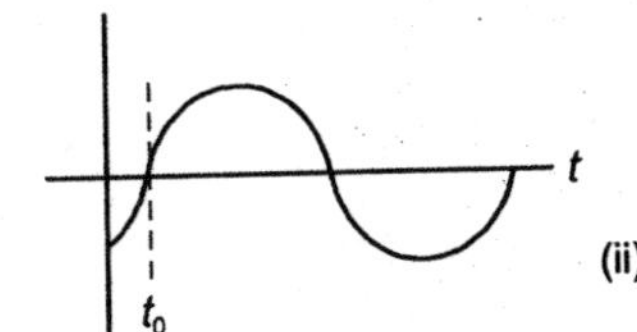

**Fig. E1.12(ii)**  $\sin w(t - t_0) \, u(t)$

(iii) Sin $w(t - t_0) \, u(t - t_0)$
This is again sinusoidal waveform limited to first quadrant and has value zero for $t < t_0$ because of $u(t - t_0)$ and also it has zero value at $t = t_0$ as then sin $w \, 0 = 0$. Therefore, the waveform will be as shown here.

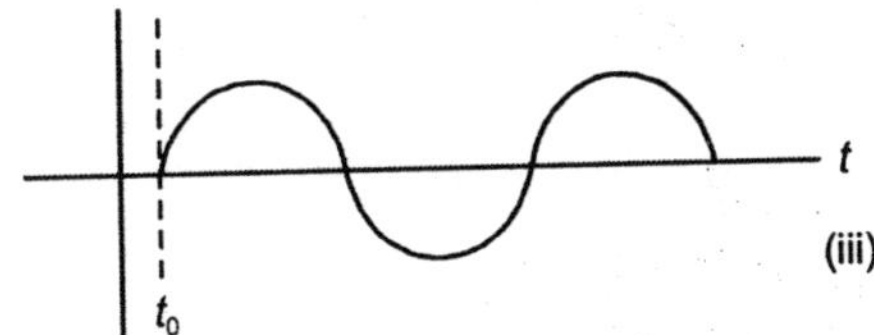

**Fig. E1.12(iii)**  $\sin w (t - t_0) \, u(t - t_0)$

(iv) Sin $wt \, u(t - t_0)$
This is again a sinusoidal curve which has zero value for $t < t_0$ and at $t = t_0$ it is sin $w \, t_0$ and then it follows its usual shape

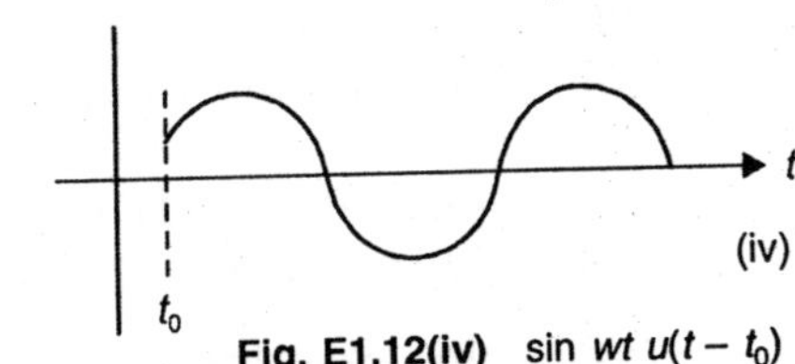

**Fig. E1.12(iv)**  $\sin wt \, u(t - t_0)$

**Example 1.13:** The first derivative of a functions is shown in Fig. E1.13 by the impulse train. Draw the function $f(t)$

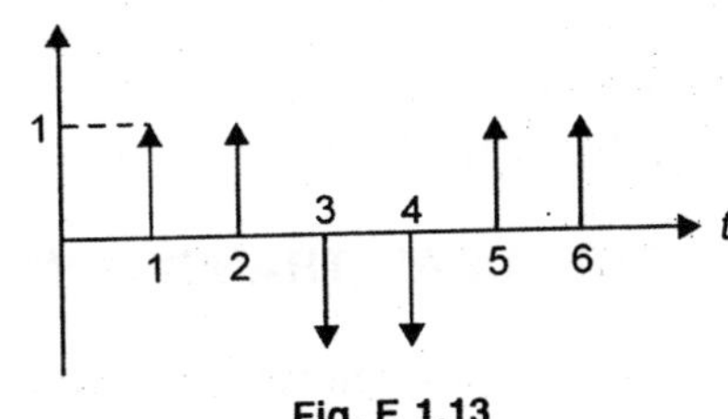

**Fig. E 1.13**

*Solution:* Since the first derivative of a function is an impulse i.e. $\dfrac{df(t)}{dt} = \delta(t)$.

Hence $f(t) = u(t)$.

The step function is discontinuous between $t = 1$ and 2
$t = 3$ and 4 and $t = 5$ and 6 and taking into account the sign of the magnitude
$f(t) = u(t - 1) + u(t - 2) - u(t - 3) - u(t - 4) + u(t - 5) + u(t - 6)$

The waveform is shown here

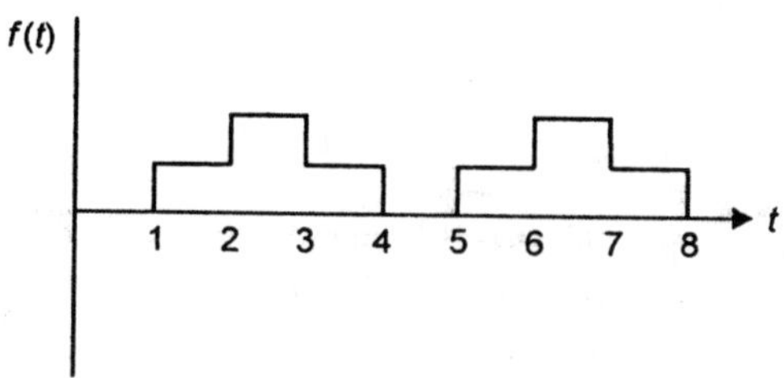

**Example 1.14:** Write mathematical expressions for the following waveforms

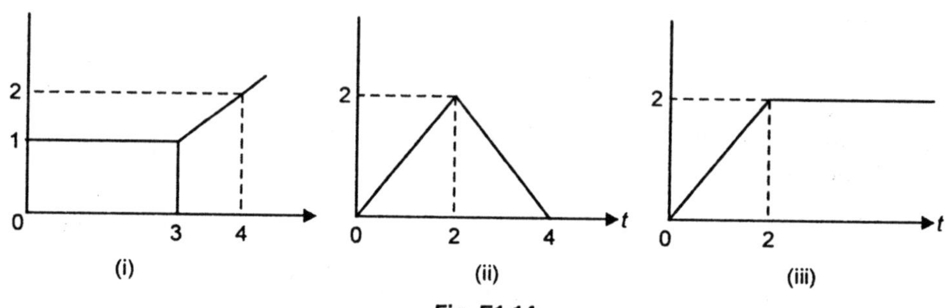

Fig. E1.14

*Solution:* (*i*) The first part is a unit step over which a ramp of $\dfrac{2-1}{4-3}=1$ slope at $t = 3$. Hence $f(t) = u(t) + t\,u(t-3)$

(*ii*) In this the first ramp has slope of plus unity

Whereas the other should have slope of –2 so as to annual the plus one slope of the first ramp and the final ramp should have slope of minus one at the instant of subtraction i.e.

Net ramp = Ramp (1) ± Ramp (2)
Hence $f(t) = t\,u(t) - 2t\,u(t-2)$          **Ans.**

(*iii*) In this case the first ramp is starting from origin hence $f_1(t) = t\,u(t)$

The second part is a step starting from $t = 2$  i.e. the slope of the first should be made zero by superposing a ramp with negative slope of the first ramp at the instant of change and hence $f_2(t) = -t\,u(t-2)$ and hence

$\quad f(t) = f_1(t) + f_2(t) = t\,u(t) - tu(t-2)$     **Ans.**

## 1.6  THE IDEAL TRANSFORMER

It is a two port network. An ideal transformer shown in Fig. 1.22 is described by the following equations:

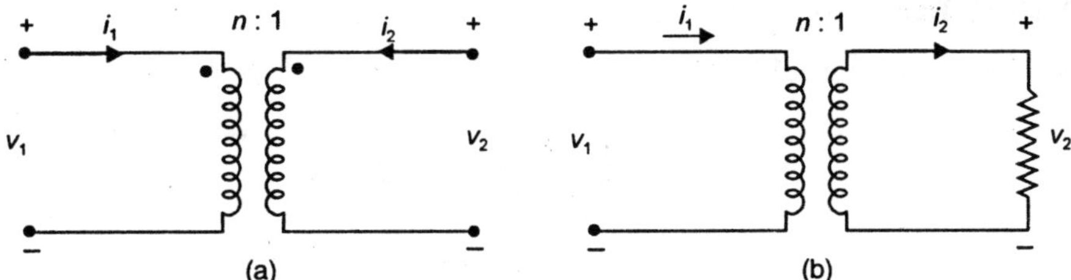

**Fig. 1.22** The ideal transformer.

For Fig. 1.22 (*a*)

$$v_1 = n\, v_2 \quad \text{and} \quad i_2 = -n\, i_1 \qquad \text{...(1.55)}$$

If the direction of $i_2$ is reversed as what is given in Fig. 1.22 (*b*), $i_2 = ni_1$
In matrix notation

$$\begin{bmatrix} v_1 \\ i_2 \end{bmatrix} = \begin{bmatrix} 0 & n \\ -n & 0 \end{bmatrix} \begin{bmatrix} i_1 \\ v_2 \end{bmatrix} \qquad \text{...(1.56)}$$

It can be seen that an ideal transformer is characterized by a single parameter $n$ known as turns ratio. The $v$-$i$ relationships indicate idealized relations expressing Faraday's law ($v_1 = n\, v_2$) and Ampere's law ($i_2 = -n\, i_1$). Another characteristic of an ideal transformer is that a resistance $R$ connected on to secondary side is reflected as $n^2 R$ when referred to the primary side.

Also it can be seen easily that the total energy input to the ideal transformer is zero.

$$\text{Total energy input} = \int_{-\infty}^{t} (v_i i_i + v_2 i_2)\, dt$$

$$= \int_{-\infty}^{t} \left\{ n\, v_2 \left( -\frac{i_2}{n} \right) + v_2\, i_2 \right\} dt$$

$$= \int_{-\infty}^{t} (-v_2 i_2 + v_2 i_2)\, dt$$

$$= 0 \qquad \text{...(1.57)}$$

This shows that an ideal transformer is a passive device, it transmits energy, it neither stores nor dissipates the energy. It can be summarized that an ideal transformer is

- (*i*) an abstract concept;
- (*ii*) characterized by a single parameter $n$, the turns ratio;
- (*iii*) the voltage and current are idealized relations derived out of Faraday's law and Ampere's law respectively;
- (*iv*) In case of an ideal transformer the primary and secondary inductive reactances are extremely large ($L_1$ and $L_2 \to \infty$) as compared to the terminating impedance and hence if the terminating impedance is $Z_L$, it is reflected on the primary side as $n^2 Z_L$.
- (*v*) One where a resistance $R$ connected to secondary side is reflected as $n^2 R$ when referred to primary side; and
- (*vi*) a passive device.

## 1.6.1  The Perfect Transformer

A less abstract model of a real life transformer is shown in Fig. 1.23.

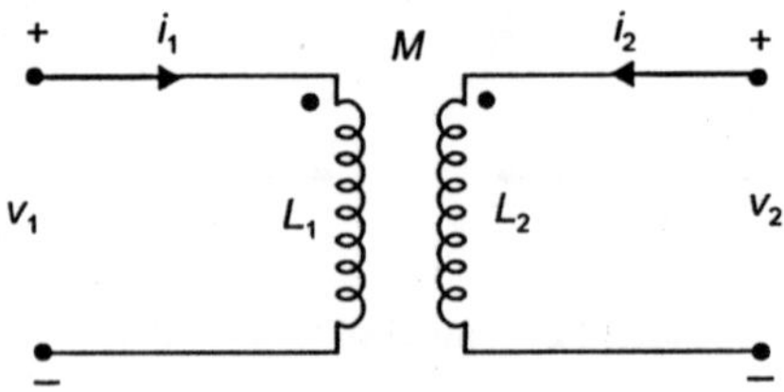

**Fig. 1.23**  The perfect transformer.

The *v-i* relations for the perfect transformer are given as

$$v_1 = L_1 \frac{di_1}{dt} + M \frac{di_2}{dt} \qquad \qquad \text{...(1.58)}$$

$$v_2 = L_2 \frac{di_2}{dt} + M \frac{di_1}{dt} \qquad \qquad \text{...(1.59)}$$

Thus this is characterized by three parameters $L_1$, $L_2$ and $M$ the two self inductances and one mutual inductance in contrast to ideal transformer which is characterized by a single parameter $n$. The total energy delivered to the transformer is

$$= \int_{-\infty}^{t} (v_1 i_1 + v_2 i_2)\, dt$$

$$= \int_{-\infty}^{t} \left( L_1 \frac{di_1}{dt} + M \frac{di_2}{dt} \right) i_1\, dt + \int_{-\infty}^{t} \left( L_2 \frac{di_2}{dt} + M \frac{di_1}{dt} \right) i_2\, dt$$

$$= \int_{0}^{i_1} L_1 i_1\, di_1 + \int_{0}^{i_1, i_2} M (i_2\, di_1 + i_1\, di_2) + \int_{0}^{i_2} L_2 i_2\, di_2$$

$$= \frac{1}{2} L_1 i_1^2 + M i_1 i_2 + \frac{1}{2} L_2 i_2^2 \qquad \qquad \text{...(1.60)}$$

For perfect transformer to be passive, the energy stored should be non-negative i.e.

$$L_1 i_1^2 + 2 M i_1 i_2 + L_2 i_2^2 \geq 0$$

To prove that it is non-negative, we prove that the minimum of this is non-negative and hence the expression will be non-negative for all other values.

Let $i_2 = -x i_1$ where $x$ is a real positive number.

The above expression reduces to

$$f(x) = L_1 i_1^2 - 2 Mx i_1^2 + L_2 x^2 i_1^2$$

$$\frac{df(x)}{dx} = -2M i_1^2 + 2 x L_2 i_1^2 = 0$$

or
$$x = \frac{M}{L_2}$$

Substituting value of $x$ in $f(x)$ we have

$$f(x) = \left( L_1 - 2\,\frac{M^2}{L_2} + \frac{M^2}{L_2} \right) \geq 0$$

or $\qquad\qquad L_1 L_2 - M^2 \geq 0$

or $\qquad\qquad\quad M^2 \leq L_1 L_2$

or $\qquad\qquad\quad \dfrac{M^2}{L_1 L_2} \leq 1$

Let $\qquad\qquad K^2 = \dfrac{M^2}{L_1 L_2} \quad\text{or}\quad K = \sqrt{\dfrac{M}{L_1 L_2}}$

Since this inequality is true, the inequality (1.55) is also true.

Here $K$ is known as co-efficient of coupling and its maximum value is unity when this transformer is known as a perfect or perfectly coupled transformer. It can be seen that a perfect transformer is not same as ideal transformer. To establish the difference we know that in case of ideal transformer $\dfrac{v_1}{v_2} = n$

In case of perfect transformer

$$\frac{v_1}{v_2} = \frac{L_1\, di_1/dt + M\, di_2/dt}{L_2\, di_2/dt + M\, di_1/dt}$$

$$= \frac{L_1\, di_1/dt + \sqrt{L_1 L_2}\, di_2/dt}{L_2\, di_2/dt + \sqrt{L_1 L_2}\, di_1/dt}$$

Multiplying and dividing the RHS by $\sqrt{\dfrac{L_2}{L_1}}$ we have

$$\frac{v_1}{v_2} = \frac{\sqrt{L_2/L_1}\cdot L_1\, di_1/dt + \sqrt{\dfrac{L_2}{L_1}}\,\sqrt{L_1 L_2}\, di_2/dt}{\sqrt{\dfrac{L_2}{L_1}}\cdot L_2\,\dfrac{di_2}{dt} + \sqrt{\dfrac{L_2}{L_1}}\,\sqrt{L_1 L_2}\, di_1/dt}$$

$$= \sqrt{\frac{L_1}{L_2}} \qquad\qquad\qquad\qquad \text{...(1.61)}$$

This shows for perfect transformer to be the same as ideal transformer $n = \sqrt{L_1/L_2}$. Again, using current relations it can be shown that if in a perfect transformer $i_1$ and $i_2 \to \infty$ but such that their ratio remains constant, the result will be an ideal transformer. Fig. 1.24 shows relation between a perfect and an ideal transformer.

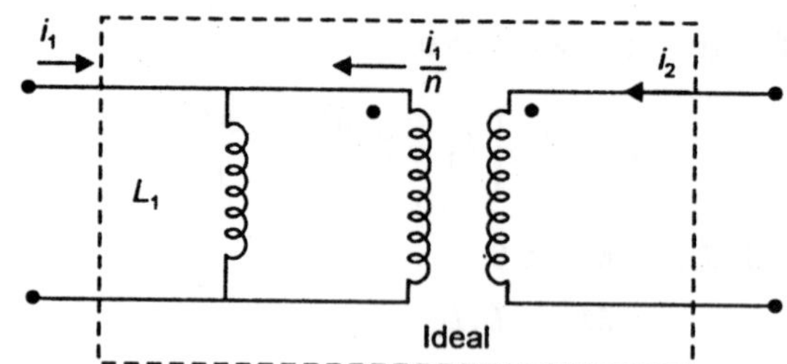

**Fig 1.24**   A perfect transformer and an ideal transformer relations.

## 1.7  THE GYRATOR

A gyrator is another two-port terminal device shown in Fig. 1.25

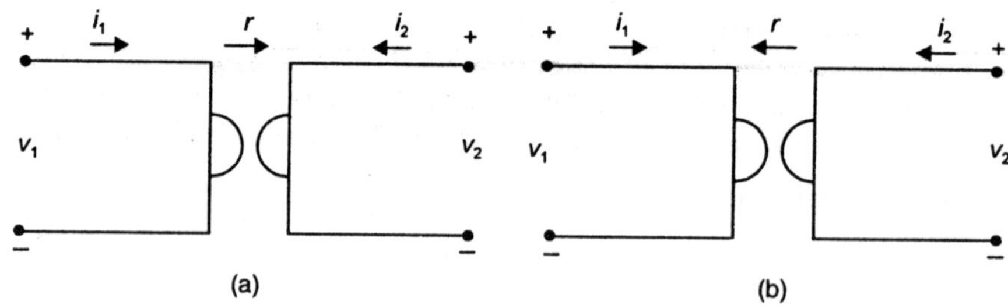

(a)                                    (b)

**Fig. 1.25**   A Gyrator.

The *v-i* relations for the gyrator are given as
For Fig. 1.25 (a)

$$v_1 = -r\, i_2 \quad \text{or} \quad \begin{bmatrix} v_1 \\ v_2 \end{bmatrix} = \begin{bmatrix} 0 & -r \\ r & 0 \end{bmatrix} \begin{bmatrix} i_1 \\ i_2 \end{bmatrix} \qquad \text{...(1.62)}$$
$$v_2 = r\, i_1$$

For Fig. 1.25 (b)

$$v_1 = r\, i_2 \quad \text{or} \quad \begin{bmatrix} v_1 \\ v_2 \end{bmatrix} = \begin{bmatrix} 0 & r \\ -r & 0 \end{bmatrix} \begin{bmatrix} i_1 \\ i_2 \end{bmatrix} \qquad \text{...(1.63)}$$
$$v_2 = -r\, i_1$$

Here $r$ is known as gyration resistance. The direction of arrow associated with $r$ gives the direction of gyration.

The gyrator is a hypothetical device which is introduced to account for physical situations where the reciprocity condition does not hold. It can be shown that gyrator is not a reciprocal device, it is anti-reciprocal.

The total energy input to the gyrator

$$E(t) = \int_{-\infty}^{t} (v_1 i_1 + v_2 i_2)\, dt$$

$$= \int_{-\infty}^{t} \{(-r i_2)\, i_1 + r i_1\, i_2\}\, dt$$

$$= 0 \qquad \text{...(1.64)}$$

Hence it is a passive device. Suppose the gyrator is terminated through a resistance $R$, it's equivalent on the input side is given by

$$v_1 = -ri_2 = -r\left(\frac{-v_2}{R}\right) = r\,\frac{r\,i_1}{R} = (r^2\,G)i_1 \qquad \text{...(1.65)}$$

Thus, the equivalent resistance at the input terminals equals $r^2$ times the conductance terminating at the output terminals. The gyrator, therefore, has the property of inverting.

A very interesting case is found when the gyrator is terminated through an inductor or a capacitor. Say it is terminated through a capacitor. The equivalent of this when referred to input side is obtained as follows.

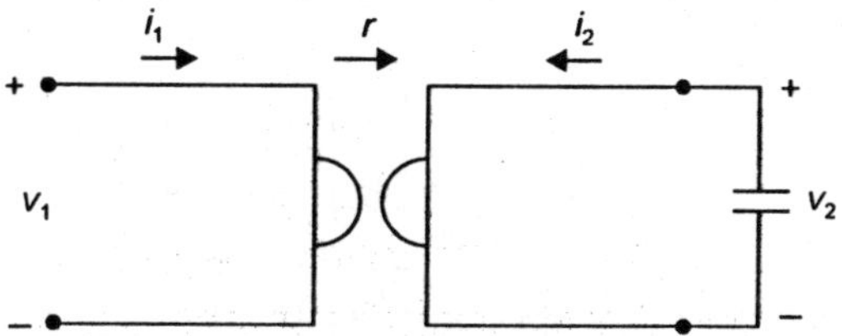

**Fig. 1.26**  Gyrator terminated through a capacitor.

$$v_1 = -ri_2 = (-r)\left(-C\,\frac{dv_2}{dt}\right) = rC\,\frac{d}{dt}\,(r\,i_1)$$

$$= r^2C\,\frac{di_1}{dt} = \text{Leq}\,\frac{di_1}{dt} \qquad \text{...(1.66)}$$

This shows that through inversion a $r^2$ times capacitor is equivalent to an inductor and it can be shown that an inductor is inverted as a capacitor.

It is found that in electronic circuits it is difficult to have inductors of suitable values. These can be simulated through the use of a gyrator terminated through a suitable capacitor.

Fig. 1.27  shows a general representation of a gyrator and its $v$-$i$ relations are given as

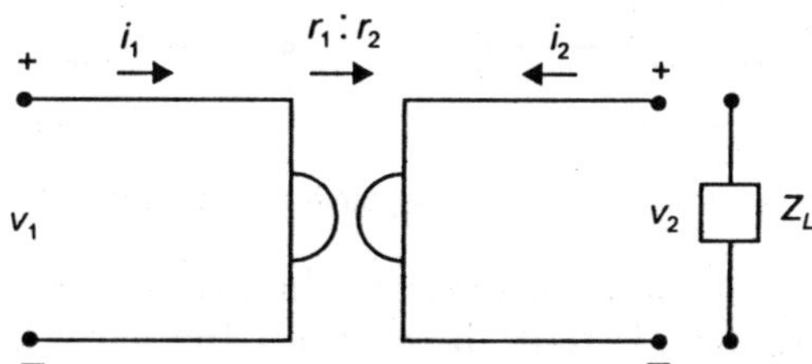

**Fig. 1.27**  A general gyrator.

$$V_1 = -I_2\,r_1 \quad \text{and} \quad V_2 = I_1\,r_2$$

The arrow here indicates that the device has a forward transmission path and this has no backward transmission.

If $r_1 = r_2 = r$ it is known as an ideal gyrator and it is a passive device. However if $r_1 \neq r_2$ it is an active device as power can be made negative as shown here. Power delivered to the gyrator is

$$V_1\,I_1 + V_2\,I_2 = (r_2 - r_1)\,I_1\,I_2 \qquad \text{...(1.67)}$$

Therefore, with this it is an active device. The input impedance of the gyrator is

$$\text{Zin} = \frac{V_1}{I_1} = \frac{-I_2\, r_1}{V_2/r_2} = \left(\frac{-I_2}{V_2}\right) r_1\, r_2$$

$$= \frac{r_1\, r_2}{Z_L} \qquad\qquad\qquad \text{...(1.68)}$$

## 1.8  THE NEGATIVE CONVERTER

The next two pair terminal device is the negative convertor which is characterized by the following *v-i* relations

$$\begin{aligned}v_1 &= kv_2 \\ i_2 &= ki_1\end{aligned} \quad \text{or} \quad \begin{bmatrix} v_1 \\ i_2 \end{bmatrix} = \begin{bmatrix} o & k \\ k & o \end{bmatrix}\begin{bmatrix} i_1 \\ v_2 \end{bmatrix}$$

or

$$\begin{aligned}v_1 &= -kv_2 \\ i_2 &= -ki_1\end{aligned} \quad \text{or} \quad \begin{bmatrix} v_1 \\ i_2 \end{bmatrix} = \begin{bmatrix} o & -k \\ -k & o \end{bmatrix}\begin{bmatrix} i_1 \\ v_2 \end{bmatrix} \qquad \text{...(1.69)}$$

Fig. 1.28 shows a general representation of negative converter, It is characterised by a single parameter $k$ known as conversion ratio.

It can be seen from the first set of equations that when $i_1$ is in the reference direction $i_2$ is also in the reference direction and hence the current is said to be inverted in going through the negative converter. However, the voltage is not inverted. Therefore, the first set of equations characterise a current negative converter. However the second set of equations show that the voltage is inverted but the current is not and therefore, these equations characterise a voltage negative converter.

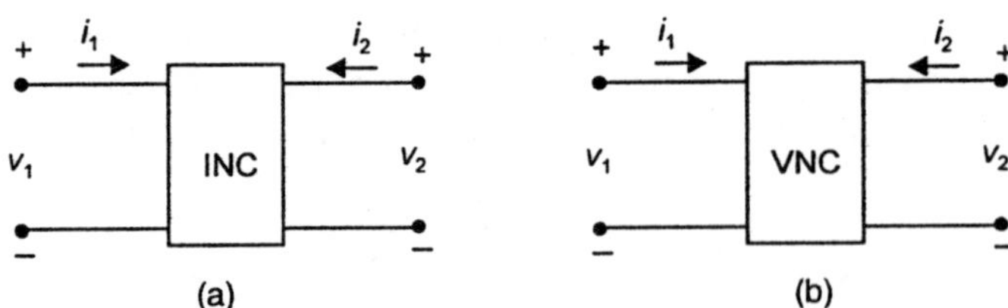


**Fig. 1.28**  Negative converters (a) Current negative converter (b) Voltage negative converter.

Let us study the behaviour of these devices when these are terminated through a passive element. Suppose these are terminated through an inductor $L$. Then the voltage $v_2 = -L\, di_2/dt$ Inserting the value of $v_2$ in the v-i relations of the two negative converters we have

$$v_1 = \pm k\, v_2 = \pm k\left(-L\,\frac{di_2}{dt}\right) = \pm k\left[-Lk\,\frac{d\,(\pm i_1)}{dt}\right]$$

$$= -L\, k^2\, \frac{di_1}{dt} \qquad\qquad \text{...(1.70)}$$

Here $\pm$ signs indicate the application to INC and VNC respectively.

Thus at the input terminals, the equivalent inductance is $-k^2 L$ i.e. negative of the $k^2 L$.

Hence the device is termed as negative converter.

Similar conclusions can be drawn by terminating the device with $R$ or $C$.

## 1.9  INDEPENDENT SOURCES

So far we have considered passive elements when interconnected, form a network. A network without an active source is meaningless. Mainly there are two types of sources (*i*) Voltage source (*ii*) Current source. Again these are classified as ideal or independent and dependent sources.

### 1.9.1  Ideal Voltage Source

An ideal voltage source is a two terminal device whose terminal voltage is independent of the current drawn by the network connected to its terminals. Both the magnitude and wave form of voltage remain unaffected. This means an ideal voltage source should have zero internal resistance. It is to be noted that short circuit at the terminals of an ideal voltage source makes no sense as this would impose two conflicting requirement. Fig. 1.29 shows the symbols and reference conventions for such a source and the voltage current characteristic.

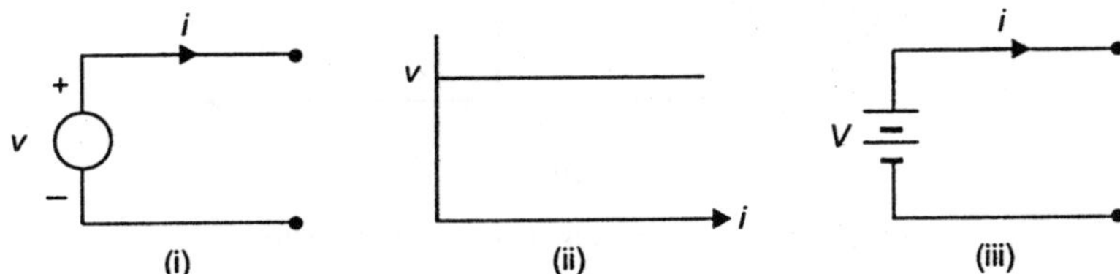

**Fig. 1.29**  (i) Ideal voltage source (ii) *v-i* characteristic (iii) d.c. source.

When the voltage generated does not change with time, it is represented by a battery as shown in Fig. (*iii*) otherwise for all the cases by Fig. (*i*)

However, in actual practice, there is no voltage source which does not have internal resistance, however small it could be and, therefore, an actual voltage source is always associated with an internal resistance as shown in Fig. 1.30.

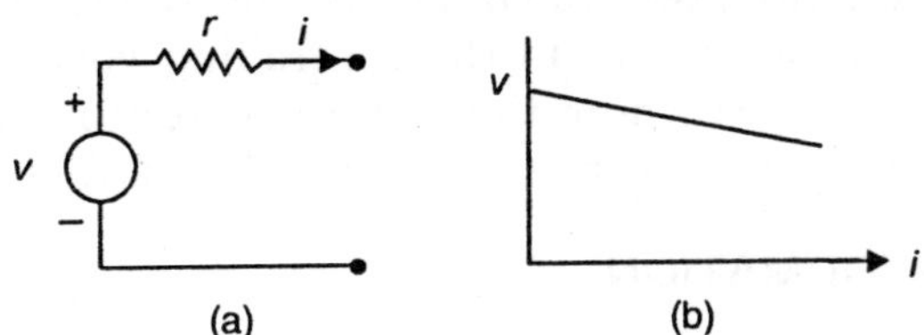

**Fig. 1.30**  (a) Voltage source (b) *v-i* characteristic.

Therefore, the terminal voltage decreases as the current drawn by the external network increases and the voltage current relation is as shown in Fig. 1.30(*b*).

### 1.9.2  Ideal Current Source

It is a two terminal device which delivers a constant current to the network connected across its terminals irrespective of the elements of the network i.e. the current is independent of the voltage across its terminals. Here again it makes no sense to consider open circuiting of the terminals of a current source as this again imposes two conflicting requirements at the terminals. The voltage at the terminals will be determined by the network connected across its terminals. Fig. 1.31 shows the symbol and reference convention for the current source and also the v-i characteristic of such a source.

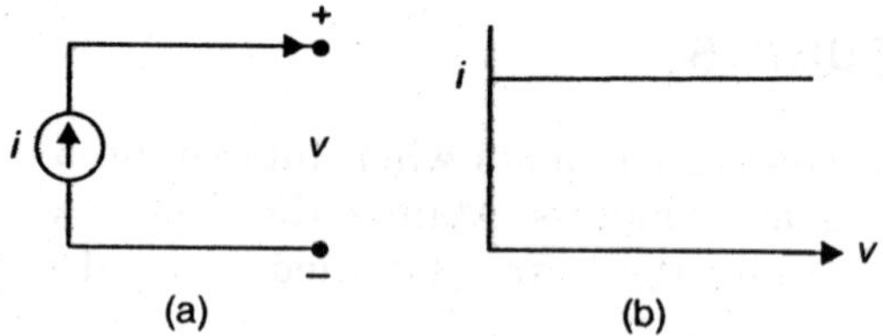

(a)           (b)

**Fig. 1.31** (a) Ideal current source (b) *i-v* characteristic.

Certain devices may be represented by the model shown in Fig. 1.32 wherein a resistor is connected across the current source. The corresponding variation of terminal current with voltage at the terminals is shown in Fig. 1.31($b$) Such devices as transistors and photo electric cells are examples of current sources.

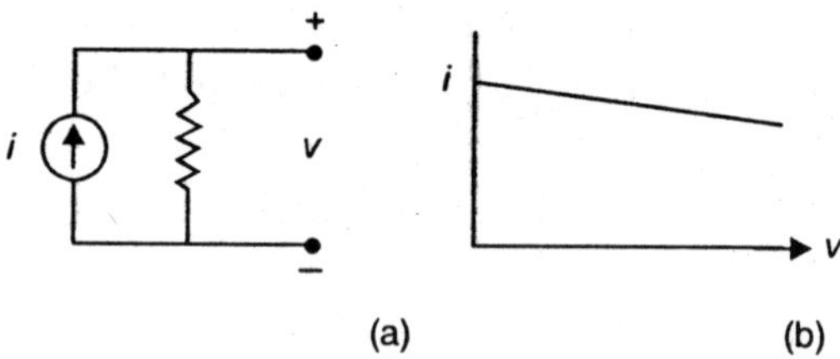

(a)           (b)

**Fig. 1.32** (i) Current source (ii) *v-i* characteristic.

It can be seen that the voltage and current sources are dual in the sense that the roles for current and voltage are interchanged in the two sources.

## 1.10  THE DEPENDENT OR CONTROLLED SOURCES

A controlled voltage/current source is one whose terminal voltage/current is a function of some other voltage or current. These devices have two pairs of terminals, one pair corresponds to the controlling quantity and the other the controlled quantity. Here the controlled voltage/current is directly proportional to the controlling voltage or current.

There are controlled sources where the voltage/current is proportional to derivative of some other voltage or current. Here controlled source will not be considered for the present.

## 1.11  SOURCE TRANSFORMATION

In order to make the network manipulations direct and simple, following aspect would be considered:

    ($i$) Transformation of voltage source into a current source and vice versa.
    ($ii$) Shifting the positions of the sources in the network.

In doing this, it is to be remembered that the terminal conditions i.e. $v$ and $i$ of these sources remain unchanged with respect to the external network to be connected to these sources.

Suppose there are two voltage sources in series as shown in Fig. 1.33, the equivalent will be one source with voltage $(v_1 + v_2)$.

However, it is to be noted that two voltage sources can't be connected in parallel unless these are identical in magnitude and wave shape and similarly two current sources can't be connected in series unless these are identical. We know that two generators can't be

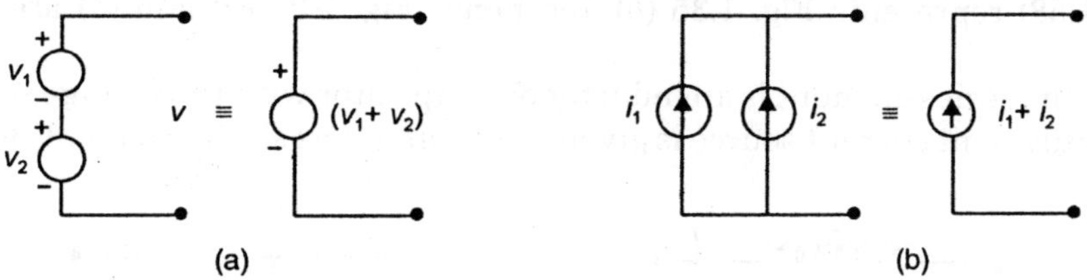

Fig. 1.33   (a) Two voltage sources in series, (b) Two current sources in parallel.

connected in parallel unless their voltages are exactly identical otherwise, this results in a short circuit and the generators get damaged.

Fig. 1.34(*a*) shows a resistance $R$ connected across a voltage source $v$ and Fig. 1.34(*b*) shows a resistance $R$ in series with a current source when an external network is connected across the  terminals of these source circuits. From computation point of view, the resistor across the voltage source and/or a resistor in series with the current source can be ignored as shown in Fig. 1.34 (*a*) and (*b*) respectively.

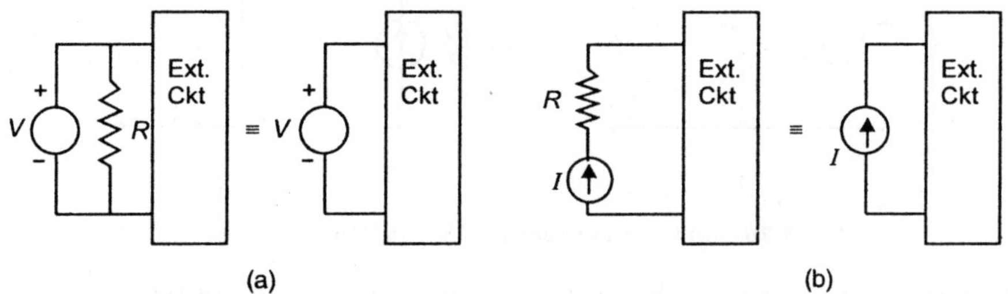

Fig. 1.34   (a) *R* connected across a voltage source,        (b) *R* connected in series with a current source.

In network analysis, sometimes it may be more convenient to transform a voltage source into a current source and vice versa from computation point of view. To this, the basic criterion is that the terminal conditions ($v$ and $i$) of the two networks, the source and the external network must remain same before and after the transformation.

Let us transform a voltage source into a current source. Refer to Fig 1.35.

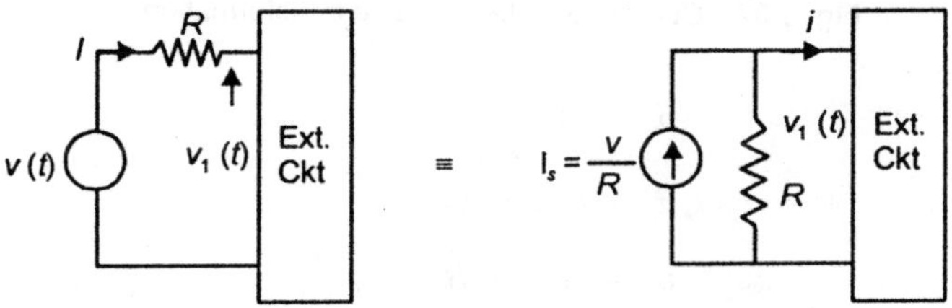

Fig. 1.35   Voltage to current source transformation.

$$v = v_1 + I\,R$$

...(1.71)

or
$$I = \frac{v}{R} - \frac{v_1}{R} \qquad\qquad ...(1.72)$$

Equation (1.68) represents Fig. 1.35 (*a*) whereas equation (1.69) which has been derived from equation (1.68) represents Fig. 1.35 (*b*) and hence Fig. 1.35 (*a*) and (*b*) are equivalent of each other.

Similarly if the series element is an inductor or a capacitor with the voltage source Fig. 1.36 (*a*) and (*b*), the equivalent current source is given by the circuit as shown in Fig. 1.36.

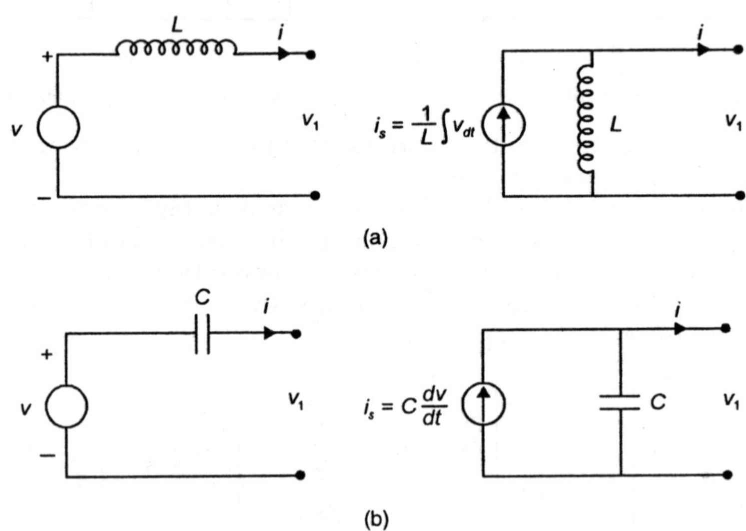

**Fig. 1.36**   (a) Inductor in series,   (b) Capacitor in series.

Similarly current sources can be transformed into voltage sources

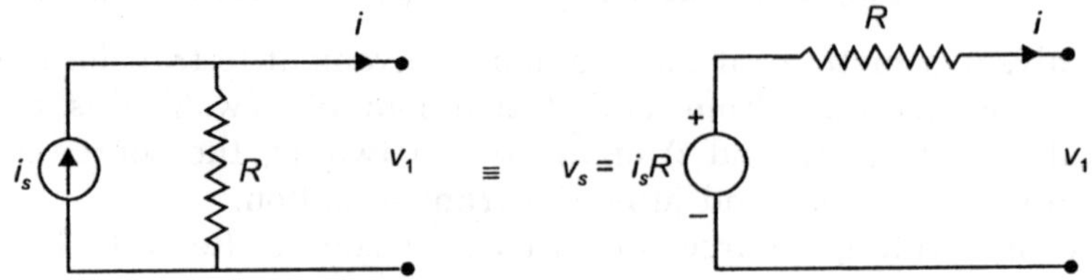

**Fig. 1.37**   Current to voltage source transformation.

$$i_s = i + \frac{v_1}{R} \qquad\qquad ...(1.73)$$

or
$$i_s R = i R + v_1$$

or
$$v_1 = v_s - i R \qquad\qquad ...(1.74)$$

If an inductor is connected across a current source (Fig. 1.38)

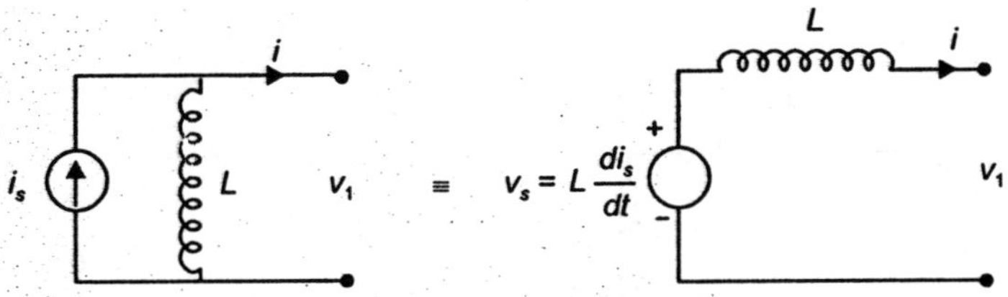

**Fig. 1.38**  Inductor across current source.

$$i_s = i + \frac{1}{L} \int v_1 \, dt \qquad\qquad \ldots(1.75)$$

or
$$L\,i_s = Li + \int v_1 \, dt$$

Differentiating the equation gives

$$L \frac{di_s}{dt} = L \frac{di}{dt} + v_1$$

or
$$v_1 = L \frac{di_s}{dt} - L \frac{di}{dt} \qquad\qquad \ldots(1.76)$$

Similarly an equivalent voltage source can be obtained when a current source is connected across a capacitor.

It is to be noted that a single passive element when connected in series with a voltage source or across a current source its equivalent current or voltage source can be obtained by the method explained above. However for more than one passive elements when considered, above method can't be used, as the determination of the source equivalent involves the solutions of the differential equation.

In many situations a voltage source may not be associated with a series passive element, and a current source with a parallel passive element. Here, before transformation could be carried out, it is necessary to shift the source within the network. A voltage source is shifted from one branch to the network by the process known as "push the voltage source through the node" with a new identical source appearing in every branch connected to the node such that it does not affect the current distribution in the network as shown in Fig 1.39.

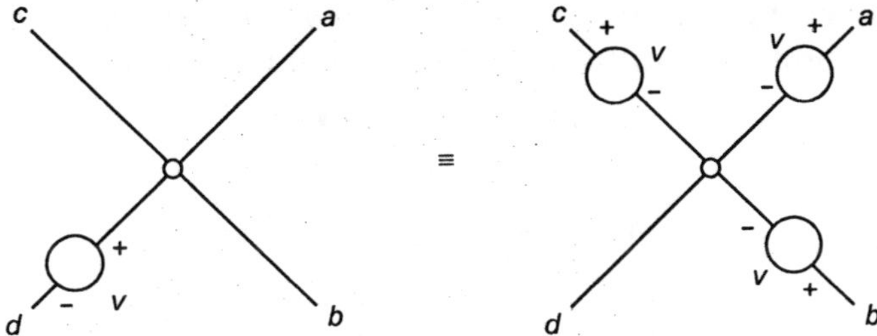

**Fig. 1.39**  Voltage source shift in a network.

It is to be noted that whereas current distribution remains unaffected, the voltage distribution in the network is changed since node 'o' is at the same potential as 'd' after the shift.

In order to shift current sources, it is required to maintain the same currents at all nodes of the network by adding and then subtracting the same current quantity as shown in Fig. 1.40.

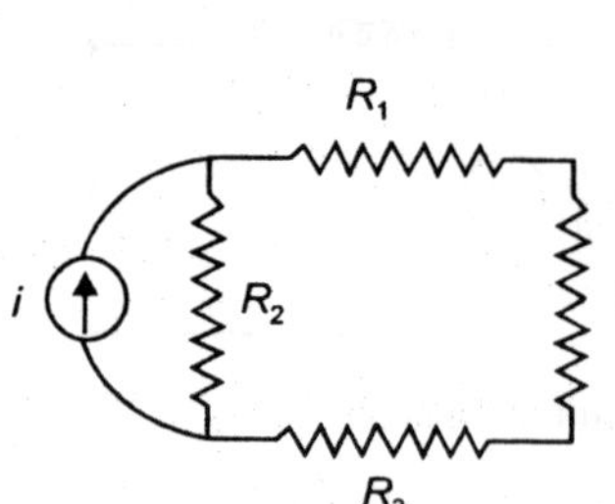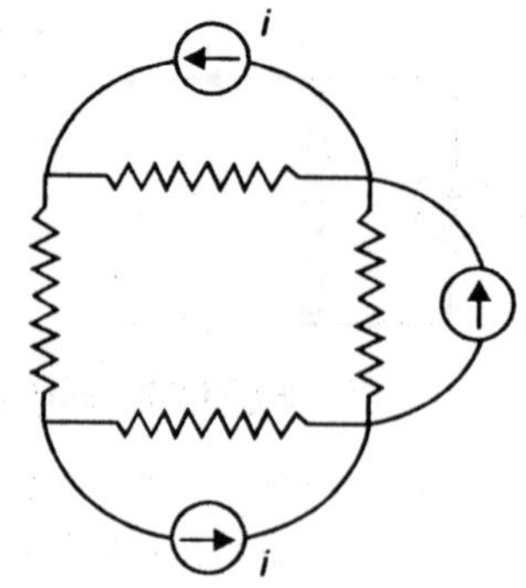

**Fig. 1.40**  Shifting of current sources.

In the current source shifting process, the voltages are not changed by the transformation even though the currents in the active source branches are changed.

While drawing graph from a network, the source transformation makes following suggestions:

(*i*) Elements in parallel with voltage source or in series with a current source can be eliminated from the graph.

(*ii*) When voltage source is shifted and hence pushed through the node, the voltage source may be short-circuited and the current source may be open circuited and so eliminated from the graph which represents the network.

While preparing a network before analysis following thumb rules are to be followed for convenience. If the network is to be analysed on nodal basis, the network should contain current sources and if it is to be analysed on loop basis, the network should contain voltage sources. However, if it is to be analysed based on state variable approach, the network may contain both kinds of sources.

## 1.12  DOT CONVENTION FOR COUPLED CIRCUITS

The significance of dot convention is explained with the help of Fig. 1.41.

Suppose a time varying voltage source is connected across the primary of the transformer and at any given instant the voltage source has the polarity shown and current $i$ (*t*) is in the direction shown by the arrow and is increasing with time entering the dotted terminal, this current induces a voltage in the secondary which is positive at the dotted

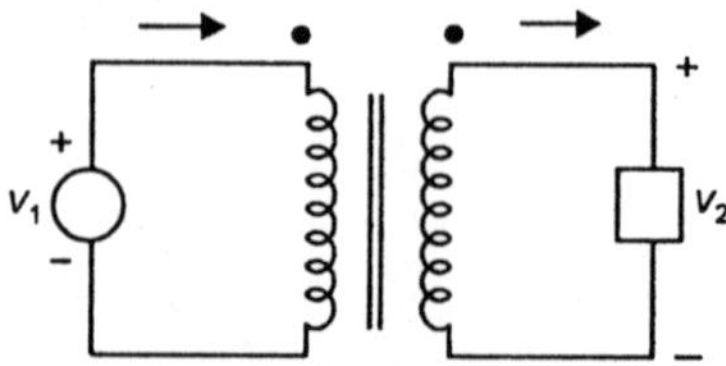

**Fig. 1.41**  A circuit for dot convention.

terminal. Conversely if the current in the primary is increasing and is leaving the dotted terminal this will induce the voltage polarity in the secondary such that the undotted end will become positive and hence dotted terminals of both the windings will be negative simultaneously. The following experimental procedure can be used to establish dotted ends of the transformer windings. Mark a dot arbitrarily on one end of the primary winding and connect a d.c. source with positive terminal connected to the dotted end and the negative terminal to the undotted end of the winding through a switch. Connect a MC voltmeter across the secondary winding. The end of the secondary winding which goes positive momentarily on closing the switch on the primary side as measured by the voltmeter, is the terminal to be dotted.

If there are more than two windings, similar procedure can be followed for identifying relationship between each pair of windings. For each pair of winding different forms of dots (● ❖ ■ ❀) should be used to avoid any confusion.

Suppose we are required to write loop equations for a mutually coupled circuit shown in Fig 1.42.

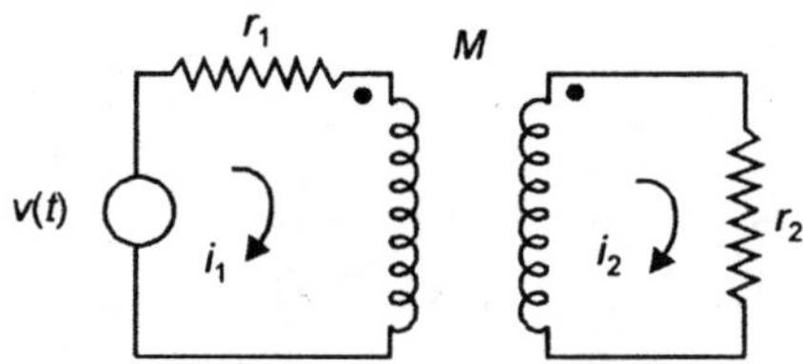

**Fig. 1.42** Coupled circuit.

For primary loop the drops are $i_1 r_1 + L_1 \dfrac{di_1}{dt}$ + drop in primary loop due to current in secondary loop. The voltage induced in primary will be determined by the direction of $i_2$ with respect to its dot in secondary loop. Since current is going away from the dot, the polarity of voltage on the primary side will be positive at the undotted terminal. Therefore, for current $i_1$ it will be rise in voltage and the equation will be

$$i_1 r_1 + L_1 \frac{di_1}{dt} - M \frac{di_2}{dt} = v(t)$$

and for the secondary loop, similarly

$$i_2 r_2 + L_2 \frac{di_2}{dt} - M \frac{di_1}{dt} = 0$$

The equivalent two loops are as shown in Fig. 1.43.

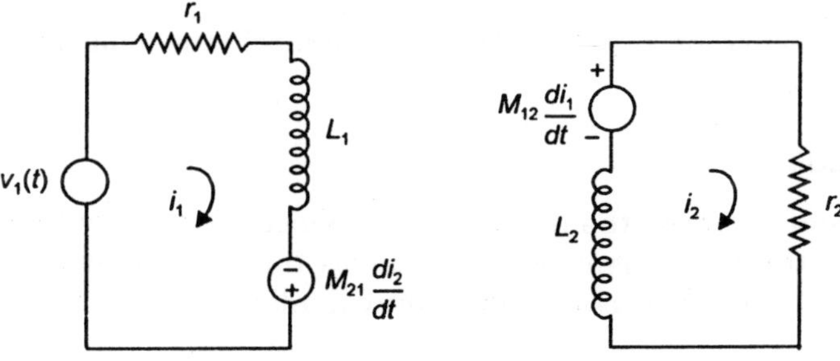

**Fig. 1.43** Equivalent of Fig. 1.34.

Consider Fig. 1.44 which is a magnetically coupled network

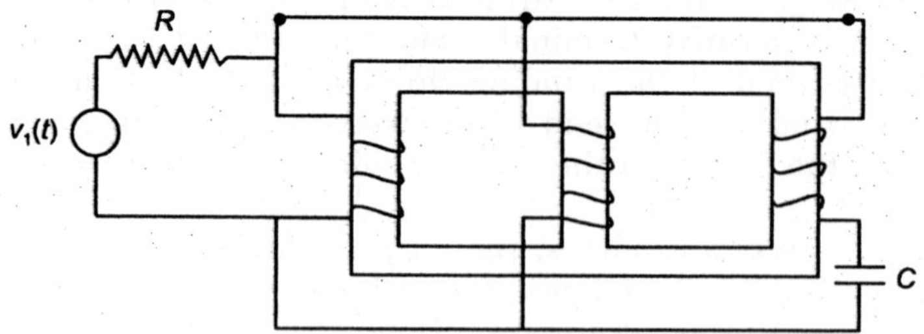

**Fig. 1.44**  Magnetically coupled network.

The equivalent with dots is given in Fig. 1.45.

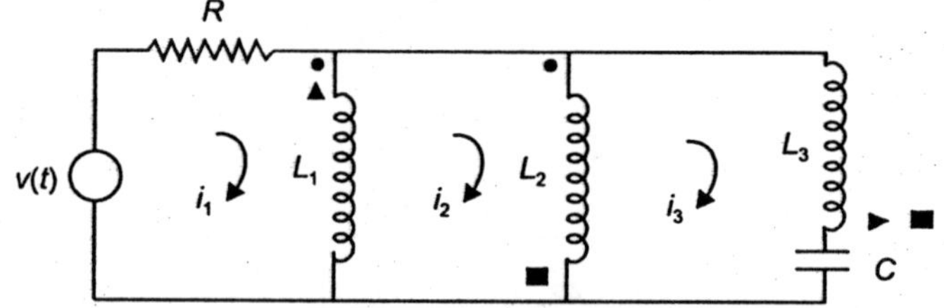

**Fig. 1.45**  Equivalent of Fig. 1.36.

Since all the three coils are mutually coupled, each coil will induce voltage in the other two coils, the magnitude will depend upon the current flowing through the coil and the polarity its relative direction of flow with respect to dot of the coil.

### Voltage induced in coil 1

Due to coil 2, the current through coil 2 is $(i_2 - i_3)$ entering the circular dot, therefore the voltage induced in coil 1 will be with positive polarity at the dot. Due to coil 3, since the current $i_3$ is leaving the triangular dot, the polarity will be −ve at the corresponding dot in coil 1.

### Voltage induced in coil 2

Due to coil 1, the current $(i_1 - i_2)$ enters the dot, therefore positive polarity at the dot of coil 2.

Due to coil 3, the current $i_3$ leaves the dot (square), the polarity +ve will be at the undotted terminal of coil 2.

### Voltage induced in coil 3

Due to coil 1 the current entering the dot is $(i_1 - i_2)$ hence the positive polarity at the dotted terminal of coil 3. Due to coil 2 current $(i_2 - i_3)$ leaves the dot (square), the positive polarity will be at the undotted terminal. Hence the equivalent circuit given as follows:

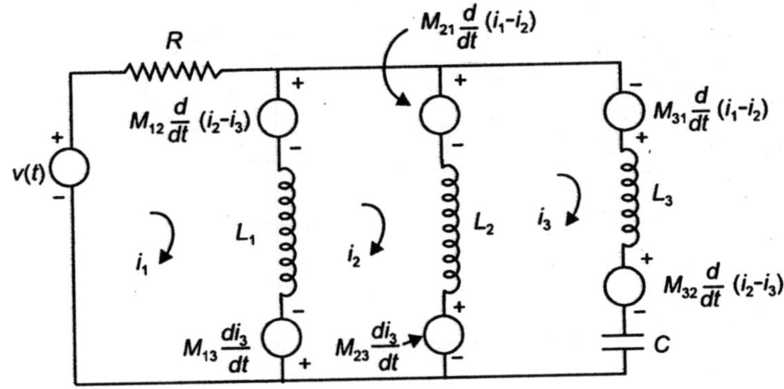

**Fig. 1.46** Equivalent of Fig. 1.36 with voltage sources.

The loop equations can be written as follows:

*Loop 1*

$$Ri_1 + M_{12}\frac{d\,(i_2 - i_3)}{dt} + L_1\frac{d\,(i_1 - i_2)}{dt} - M_{13}\frac{di_3}{dt} = v(t)$$

*Loop 2*

$$M_{13}\frac{di_3}{dt} + L_1\frac{d\,(i_2 - i_1)}{dt} - M_{12}\frac{d\,(i_2 - i_3)}{dt} + M_{21}\frac{d\,(i_1 - i_2)}{dt}$$

$$+ L_2\frac{d\,(i_2 - i_3)}{dt} + M_{23}\frac{di_3}{dt} = 0$$

*Loop 3*

$$L_2\frac{d\,(i_3 - i_2)}{dt} - M_{23}\frac{di_3}{dt} - M_{21}\frac{d\,(i_1 - i_2)}{dt} - M_{31}\frac{d\,(i_1 - i_2)}{dt}$$

$$+ L_3\frac{di_3}{dt} + M_{32}\frac{d\,(i_2 - i_3)}{dt} + \frac{1}{C}\int i_3\,dt = 0$$

The above example explains clearly how voltage sources (corresponding to induced voltages) are inserted in the coils which are magnetically coupled and the dots have been assigned.

When the two coils are to be interconnected it is important to know whether the mutual inductance $M$ is aiding or opposing. However, it is known that the effect on each of the two coils must be the same. Refer to Fig. 1.47(a) where the mutual inductance is aiding their self inductances whereas in (b) it is opposing. If $L_1$ and $L_2$ are the self inductances of the two coils and $M$ the mutual inductance, the total inductance say if is $L_A$ in case of Fig. 1.47(a) and $L_B$ that in Fig 1.39(b) then,

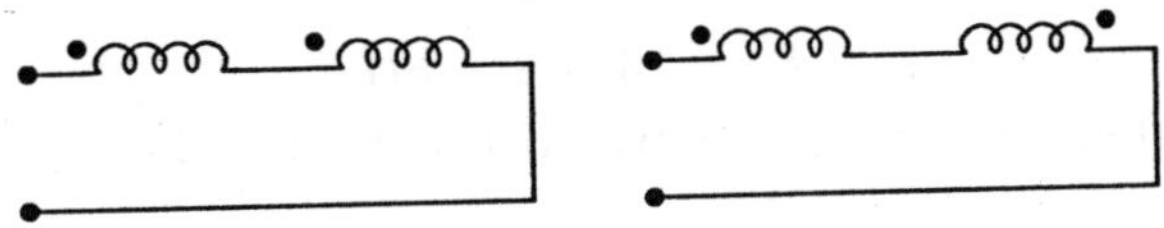

**Fig. 1.47** (a) Two coils aid and (b) Oppose.

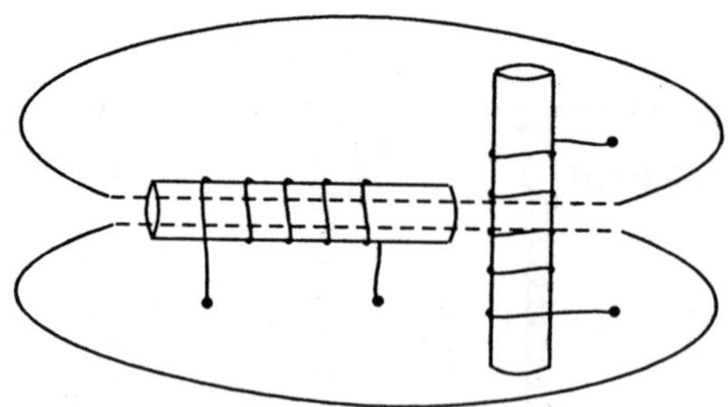

**Fig. 1.48** The coils at right angles.

$$L_A = L_1 + L_2 + 2M$$
$$L_B = L_1 + L_2 - 2M$$

and

Therefore,

$$M = \frac{L_A - L_B}{4}$$

This method provides a very convenient way of determining the value of mutual inductance between the coils.

The co-efficient of coupling and the mutual inductance between the two coils shown in Fig 1.48 are both zero. Although the flux of one coil appears to pass through the other coil, it does not actually link the turns of the other coil.

Therefore, the linkages are zero and the value of $M$ is zero. It is only when one coil is rotated with respect to other that a linking occurs.

**Example 1.15:** Determine the inductance of the individual winding and the equivalent inductance when mutual inductance is 8H.

Since the flux through the two coils opposes each other, the mutual inductance is substractive from each coil and hence

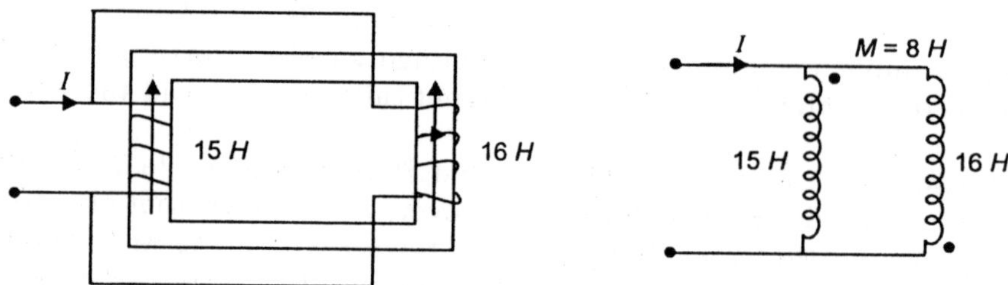


**Fig. E.1.15**

$$L_1 = 15 - 8 = 7\text{H}$$

and

$$L_2 = 16 - 8 = 8\text{H}$$

Therefore equivalent inductance between the terminals

$$L = \frac{L_1 L_2}{L_1 + L_2} = \frac{7 \times 8}{15} = 3.73\text{H}$$

**Example 1.16:** Determine the inductance between the terminals for a three coil system shown below:

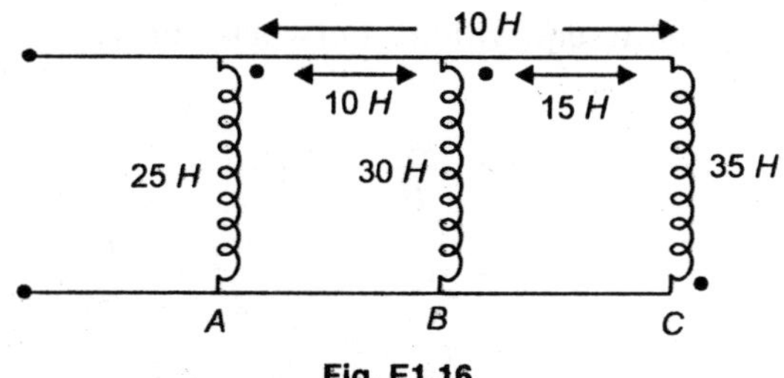

**Fig. E1.16**

From the Fig.

$$L_A = 25 + 10 - 10 = 25\text{H}$$
$$L_B = 30 + 10 - 15 = 25\text{H}$$
$$L_C = 35 - 15 - 10 = 10\text{H}$$

Therefore,

$$\frac{1}{L_{eq}} = \frac{1}{25} + \frac{1}{25} + \frac{1}{10} = \frac{2 + 2 + 5}{50}$$

or

$$L_{eq} = \frac{50}{9} = 5.55\,\text{H}$$

A given network will always have a unique solution in terms of its nodal voltages and loop currents. It is, therefore, required to write down a set of linearly independent simultaneous equations equal to the number of unknowns, and smaller the number of unknowns fewer the linear simultaneous equations and less time consuming will be the solution of these equations. It would be seen that sometimes it is more convenient to use nodal voltages as independent variables, sometimes loop currents as independent variables and yet sometimes a suitable combination of both from the point of view of minimum independent variables required to describe the netowrk and hence the number of equations to seek solution of the network.

If 'n' is the number of nodes and 'b' the number of branches in a given network, the minimum number of independent nodal voltages (nodal analysis) minimum number of independent loop currents (loop analysis) to describe the network uniquely is given by $(n - 1)$ or $(b - n + 1)$ respectively. Therefore, from solution point of view, whichever is smaller of the two should be used for analysis of a given network. However, it may be mentioned here that there could be situations when it could be mandatory to use only a particular kind of variables (viz. nodal vollages) as we may not be interested in the other kind of variables (viz. loop currents).

**Example 1.17:** Fig. E1.17(a) shows a gyrator which is described by the equation

$$v_1 = i_2 r$$
$$v_2 = -i_1 r$$

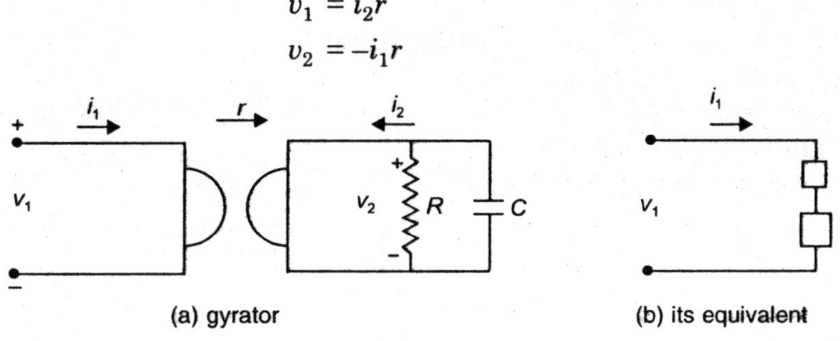

(a) gyrator

(b) its equivalent

**Fig. E1.17**

Determine the two element equivalent network shown in Fig. E1.17(*b*)
**Solution:**

$$i_2 = -\frac{v_2}{R} - C\frac{dv_2}{dt}$$

Now

$$v_2 = -i_1 r \quad \text{or} \quad \frac{dv_2}{dt} = -\frac{di_1}{dt}r$$

Hence

$$i_2 = \frac{v_1}{r} = +\frac{i_1 r}{R} + rC\frac{di_1}{dt}$$

or

$$v_1 = i_1\frac{r^2}{R} + r^2 C\frac{di_1}{dt}$$

Hence the two elements are $Req = \dfrac{r^2}{R}$

and

$$L_{eq} = r^2 C$$

**Example 1.18:** Two gyrators are connected in cascade.
Derive the terminal relations for the network.

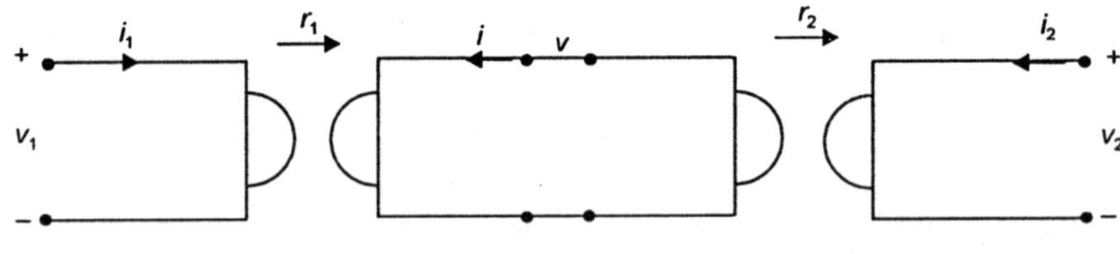

**Fig. E1.18**

$$v_1 = -ir_1$$
$$v = -i_1 r_1 = i_2\, r_2$$
$$v_2 = -ir_2$$

To eliminate $v$ and $i$

$$v_2 = \frac{v_1}{r_1}\, r_2 = \frac{r_2}{r_1}v_1$$

and

$$i_2 = \frac{-r_1}{r_2}i_1$$

The relations show that two gyrators connected in cascade behave as an ideal transformer with ratio as $\dfrac{r_2}{r_1}$.

## PROBLEMS

**1.1.** Show that the rate at which energy is storted by the inductor equals its reactive power requirements.

**1.2.** A circuit has voltage and current given as

$$v = 100 \sin\left(\omega t - \frac{\pi}{6}\right) \quad \text{and} \quad i = 10 \sin\left(\omega t + \frac{\pi}{6}\right) \quad \text{Determine the power loss.}$$

**1.3.** A circuit has voltage and current given as
$v = (50 + j\,60)$ volts and $i = (25 + j\,40)$ A
Determine the active and reactive power. Is the circuit inductive or capacitive.

**1.4.** An impedance $(5 + j\,12)$ is connected across a 26 volt rms voltage. Determine and compare the power in the circuit using the formulae (*i*) $P = \dfrac{V^2}{R}$ (*ii*) $P = I^2 R$.

**1.5.** Two capacitors of capacities 5 μF and 10 μF are charged with 100 volts and 200 volts respectively. Determine the energy stored by the capacitors. If now the two capacitors are joined together, what is the energy stored by the two capacitors? Is there any change in energy before joining and after joining of the capacitors?

**1.6.** A variable capacitor has its capacitance given as $C = C_0\,(1- \cos \omega t)$ and it is connected across a voltage source $V = V_m \sin \omega t$. Determine the equation for the current in the circuit.

**1.7.** For the gyrator shown in Fig. P1.7, obtain the nodal admittance matrix.

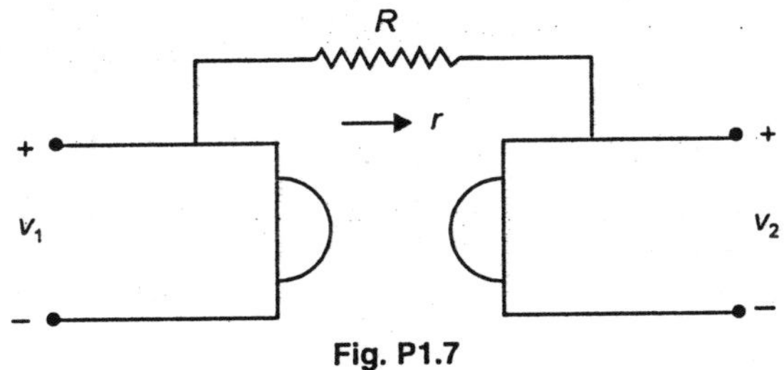

**Fig. P1.7**

**1.8.** A gyrator is terminated through a series $R$, $L$ circuit as shown in Fig. (*a*) here. Determine an equivalent two element network shown in Fig. P1.8(*b*)

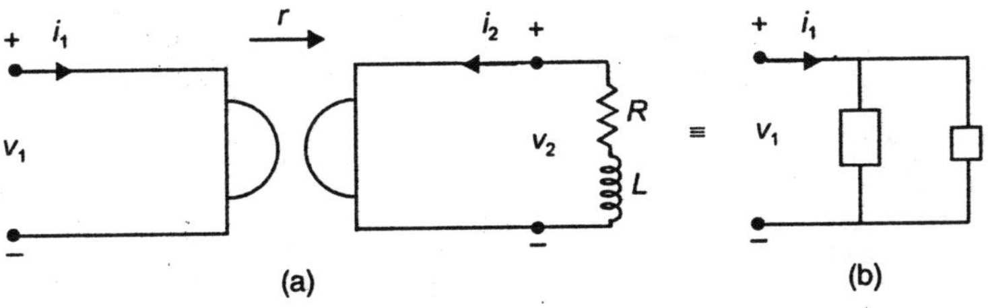

(a)

(b)

**Fig. P1.8**

**1.9.** Three coils are interconnected with winding sense shown by dots. The self inductance of the coils are 2 H each and the mutual inductance 1 H each. Determine $L_{eq}$ between 1 1′

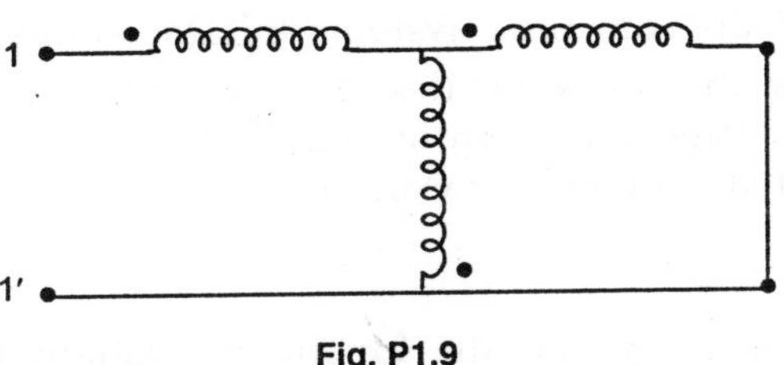

**Fig. P1.9**

**1.10.** Two coils are wound on the same core and have negligible resistance as shown in the Fig. P1.10. Determine the current in the 1000 ohm resistor and its phase angle with respect to the applied voltage of 100 volt at a frequency of $\dfrac{5000}{2\pi}$ Hz when the coils act (*i*) in the same sense (*ii*) in apposite sense.

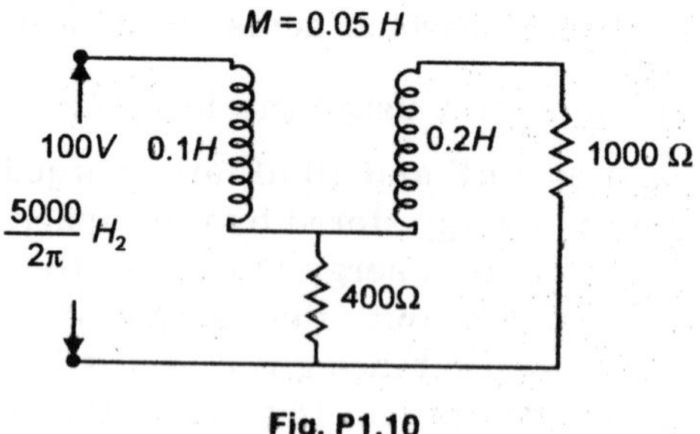


**Fig. P1.10**

**1.11** Write in mathematical form the following waveforms:

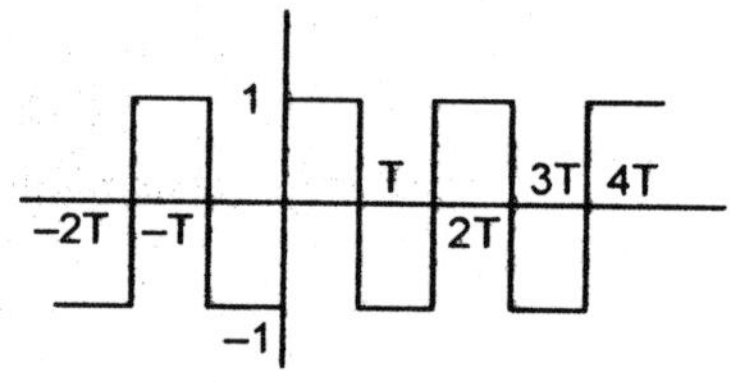


(*a*) Square wave

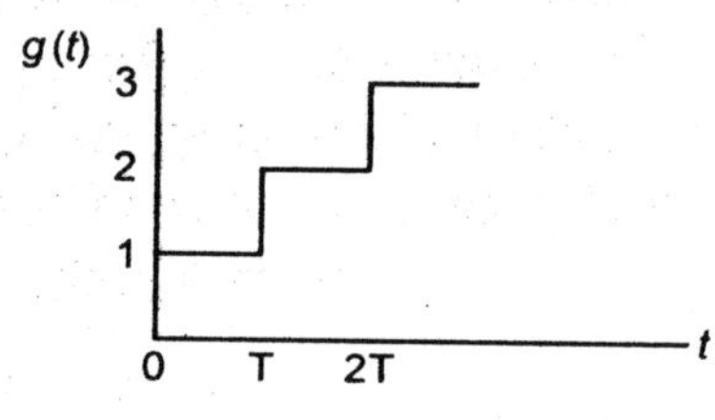


(*b*) Stair case

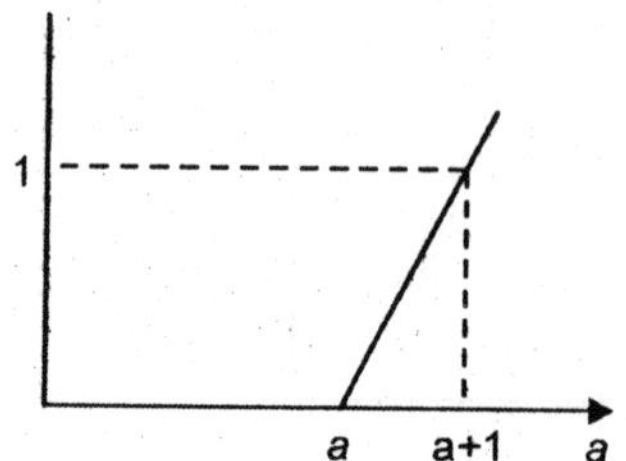


(*c*) Ramp function with time shift a

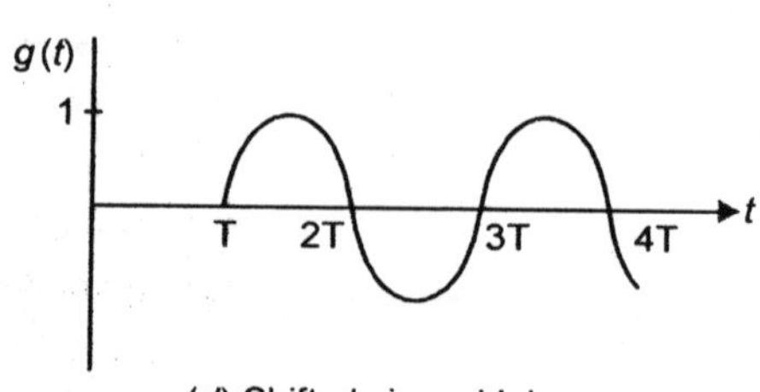


(*d*) Shifted sinusoidal wave

**1.12** Differential between
   (*i*)  Linear and nonlinear systems (*ii*) Lumped and distributed systems
   (*iii*) Time invariant and time variant systems (*iv*) Static and dynamic systems
   (*v*)  Continuous and discrete time systems (*vi*) Deterministic and stochastic systems

**1.13** Explain clearly how the following time signals are represented mathematically.
   1.   Unit step and shifted unit step by period T
   2.   Unit ramp shifted unit ramp by period T
   3.   Unit Impulse
   4.   Impulse train

**1.14** Define what you mean by periodic functions. Explain how the following periodic functions are represented mathematically:
   (*i*) Square wave (*ii*)  Sinusoidal wave   (*iii*) Saw-tooth wave

**1.15** If the impulse response of a linear time invariant system is $h(t) = 2e^{-t}$. Determine the step response.

# 2
# Network Equations

## 2.1 INTRODUCTION

Kirchhoff's postulated two basic laws way back in 1845 which are used for writing network equations. These laws concern the algebraic sum of voltages around a loop and currents entering or leaving a node. The word algebraic is used to indicate that summation is carried out taking into account the polarities of voltages and direction of currents. While traversing a loop we will take voltage drops as positive and voltage rises as negative. Also while considering currents at a node, the currents entering the node will be taken as positive and those leaving would be taken as negative.

## 2.2 KIRCHHOFF'S LAWS

Kirchhoff's voltage law usually abbreviated as KVL is stated as follows:

The algebraic sum of all branch voltages around any closed loop of a network is zero at all instants of time. Alternatively, Kirchhoff's voltage law can be stated in terms of voltage drops and rises as follows: The sum of voltage rises and drops in a closed loop at any instant of time are equal. KVL is a consequence of law of conservation of energy as voltage is energy or work per unit charge. If we start from one node in a loop and move along the closed loop and comeback to the same node, obviously the total potential difference or sum of potential rises must equal the total sum of potential falls. Just as, if we start from one point on the surface of the earth and after travelling through valleys and hills come back to the same point, the total displacement is zero. We talk elevation's and depression on the earth with respect to the sea level. Similarly, in case of voltages we take ground as the reference which is shown in Fig. 2.1(a). Here potential of node $A$ is above the ground and that of $B$ is below the ground potential.

We know that electronic current flows from negative potential to positive potential, the conventional current flows from positive potential to negative potential. Therefore, when current $i$ flows in the circuit of Fig. 2.1(b) it produces voltages polarites in various elements as shown in the Fig. Applying Kirchhoff's voltage law

$$v_1 + v_2 + v_3 - v = 0$$

Or in terms of voltage drops and voltage rises

$$v_1 + v_2 + v_3 = v \qquad \ldots(2.1)$$

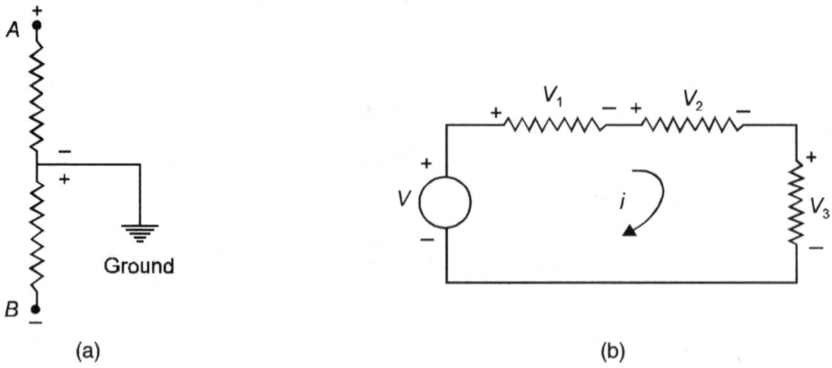

**Fig. 2.1** (a) Potential reference, (b) KVL application.

Kirchhoff's current law states that the algebraic sum of all currents terminating at a node equals zero at any instant of time. Alternatively, this states that sum of all currents entering a node equals the sum of currents leaving the same node at any instant of time.

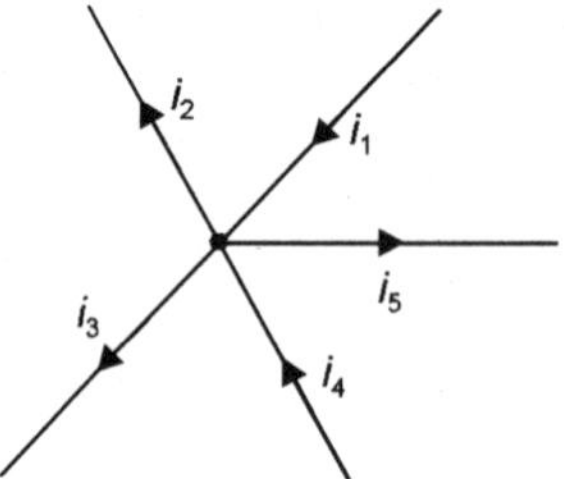

**Fig. 2.2** One node with terminating branches.

Fig 2.2 shows one node along with various branches terminating in it, in a network and the currents with directions in the various branches are shown by the arrows. Applying Kirchhoff's current law abbreviated by KCL we get

$$i_1 + i_4 - i_2 - i_3 - i_5 = 0$$

or
$$i_1 + i_4 = i_2 + i_3 + i_5$$

KCL is a consequence of law of conservation of charge. The charge that enters a node, must leave that node as it can't be stored there. Since the algebraic sum of charges at a node must be zero, it's time derivative must also be zero at any instant of time.

The two basic laws by Kirchhoff's can be applied to solve any network irrespective of its complexity. One typical application could be to find out equivalence between two networks or given a network in some configurations, how to find its equivalent so that it could be used more conveniently.

By definition, two networks are said to be equivalent at a pair of terminals if the voltage current relationships for the two networks are identical at these terminals. Consider a network having two resistances in series as shown. The objective is to replace it by a single equivalent resistance.

From Fig. 2.3(a), we have

$$v = v_1 + v_2 = ir_1 + ir_2 \qquad \qquad ...(2.3)$$

and from Fig. 2.3(b) we have

$$v = ir$$

Therefore, $\qquad \qquad r = r_1 + r_2$

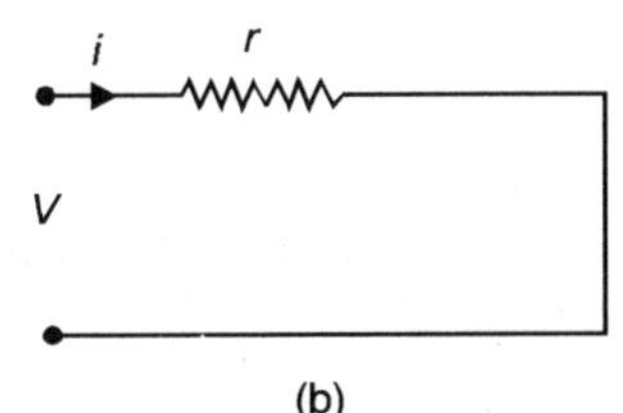

(a)             (b)

**Fig. 2.3**   (a) Series connection of resistances, (b) Equivalent of (a).

Similarly if there are $n$ number of resistances in series, the equivalent single resistance would be

$$r = r_1 + r_2 + ... + r_n \qquad \qquad ...(2.4)$$

Similarly if the elements are inductances and $v_1\ v_2$ are the drops across $L_1, L_2$ respectively, and $i$ is the current then applying Kirchhoff's voltage law we have

$$v = v_1 + v_2 = L_1 \frac{di}{dt} + L_2 \frac{di}{dt}$$

$$= (L_1 + L_2) \frac{di}{dt} = L \frac{di}{dt}$$

Therefore, $\qquad \qquad L = L_1 + L_2$

If there are $n$ inductors in series it can be shown that single equivalent inductor

$$L = L_1 + L_2 + ...... + L_n \qquad \qquad ...(2.5)$$

If two capacitors are connected in series then

$$v = v_1 + v_2 = \frac{1}{C_1} \int i\ dt + \frac{1}{C_2} \int i\ dt$$

$$= \left(\frac{1}{C_1} + \frac{1}{C_2}\right) \int i\ dt = \frac{1}{C} \int i\ dt$$

Therefore, $\qquad \qquad \dfrac{1}{C} = \dfrac{1}{C_1} + \dfrac{1}{C_2} \qquad \qquad ...(2.6)$

Similarly by using KCL, law of parallel combination of these elements can be obtained. We will try here for inductors. Say there are two inductors connected in parallel. From Fig. 2.4(a)

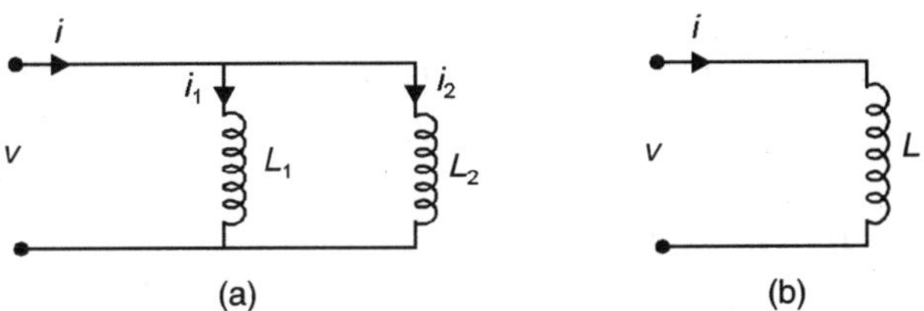

(a)             (b)

**Fig. 2.4**   (a) Inductor in parallel (b) Equivalent of (a).

$$i = i_1 + i_2 = \frac{1}{L_1} \int v \, dt + \frac{1}{L_2} \int v \, dt = \left( \frac{1}{L_1} + \frac{1}{L_2} \right) \int v \, dt$$

$$= \frac{1}{L} \int v \, dt$$

Therefore, $\qquad \dfrac{1}{L} = \dfrac{1}{L_1} + \dfrac{1}{L_2}$ $\qquad\qquad$ ...(2.7)

Similarly it can be proved that if there are $n$ resistors connected in parallel its single equivalent resistance is given by

$$\frac{1}{r} = \frac{1}{r_1} + \frac{1}{r_2} + \cdots\cdots + \frac{1}{r_n} \qquad\qquad ...(2.8)$$

and if $n$ number of capacitors are connected in parallel, then single equivalent capacitance is given as

$$C = C_1 + C_2 + \cdots + C_n \qquad\qquad ...(2.9)$$

## 2.3 STAR DELTA TRANSFORMATION

By using the same definition of equivalence of two networks i.e. the two networks are equivalent between two terminals if the $v\text{-}i$ relations are same i.e. the impedances as seen between the two terminals are same.

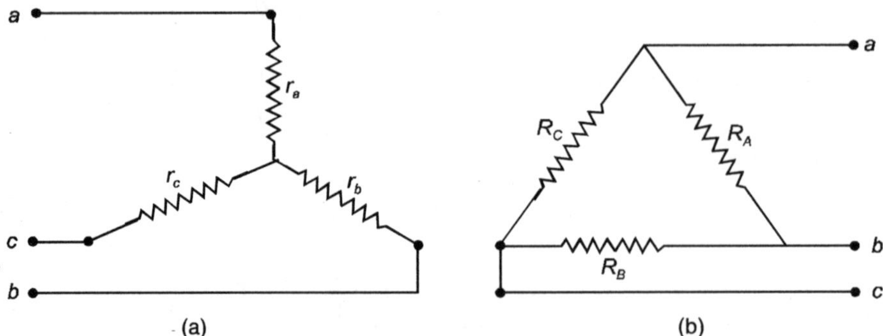


**Fig. 2.5** (a) Star, (b) Delta.

Suppose we are given star connected elements as shown in Fig. 2.5(a) and we have to find it's equivalent delta connected elements then we should equal the equivalent resistance between the same two terminals.

Between $a$ and $b$

$$\frac{R_A \left( R_B + R_C \right)}{R_A + R_B + R_C} = r_a + r_b \qquad\qquad ...(2.10)$$

Between $b$ and $c$

$$\frac{R_B \left( R_A + R_C \right)}{R_A + R_B + R_C} = r_b + r_c \qquad\qquad ...(2.11)$$

Between $c$ and $a$

$$\frac{R_C\left(R_A + R_B\right)}{R_A + R_B + R_C} = r_c + r_a \qquad \qquad ...(2.12)$$

Similarly shorting $bc$ and measuring resistance between $a$ and $bc$, we have

$$\frac{R_A\,R_C}{R_A + R_C} = r_a + \frac{r_b\,r_c}{r_b + r_c} \qquad \qquad ...(2.13)$$

Shorting $ca$ and resistance between $b$ and $ca$

$$\frac{R_A\,R_B}{R_A + R_B} = r_b + \frac{r_c\,r_a}{r_c + r_a} \qquad \qquad ...(2.14)$$

Shorting $ab$ and resistance between $c$ and $ab$

$$\frac{R_B\,R_C}{R_B + R_C} = r_c + \frac{r_a\,r_b}{r_a + r_b} \qquad \qquad ...(2.15)$$

Substracting equation (2.13) from (2.10) we have

$$r_b - \frac{r_b\,r_c}{r_b + r_c} = \frac{R_A\left(R_B + R_C\right)}{R_A + R_B + R_C} - \frac{R_A\,R_C}{R_A + R_C}$$

$$\frac{r_b^2}{r_b + r_c} = \frac{R_A^2\,R_B}{\left(R_A + R_C\right)\left(R_A + R_B + R_C\right)}$$

Subtstituting for $(r_b + r_c)$ from equation (2.11)

$$\frac{r_b^2\left(R_A + R_B + R_C\right)}{R_B\left(R_A + R_C\right)} = \frac{R_A^2\,R_B}{\left(R_A + R_C\right)\left(R_A + R_B + R_C\right)}$$

or

$$r_b^2 = \frac{R_A^2\,R_B^2}{\left(R_A + R_B + R_C\right)^2}$$

or

$$r_b = \frac{R_A\,R_B}{R_A + R_B + R_C} \qquad \qquad ...(2.16)$$

$r_a$ and $r_c$ can be obtained using equation (2.10) and (2.11) along with the expression for $r_b$ and it can be seen that

$$r_a = \frac{R_A\,R_C}{R_A + R_B + R_C} \qquad \qquad ...(2.17)$$

and

$$r_c = \frac{R_B\,R_C}{R_A + R_B + R_C} \qquad \qquad ...(2.18)$$

To transform a given star into delta following steps are used.
From equation (2.13) we have

$$\frac{R_A\,R_C}{R_A + R_C} = \frac{r_a\,r_b + r_b\,r_c + r_c\,r_a}{r_b + r_c}$$

using equations (2.11) and (2.16) we have

$$r_a\, r_b + r_b\, r_c + r_c\, r_a = R_c\, r_b$$

Therefore,
$$R_C = \frac{r_a\, r_b + r_b\, r_c + r_c\, r_a}{r_b} \qquad \text{...(2.19)}$$

Similarly,
$$R_B = \frac{r_a\, r_b + r_b\, r_c + r_c\, r_a}{r_a} \qquad \text{...(2.20)}$$

$$R_A = \frac{r_a\, r_b + r_b\, r_c + r_c\, r_a}{r_c} \qquad \text{...(2.21)}$$

An easier way to remember these formulae is as follows. Suppose we want to transform star into delta. The numerator is same for all the three branches and is equal to the product of impedance taken two at a time i.e. $Z_a\, Z_b + Z_b\, Z_c + Z_c\, Z_a$ and in the denominator it will be the star branch opposite to the equivalent delta branch i.e. if $Z_A$ is required the denominator will have impedance $Z_C$. Similarly for $Z_B$ it will be $Z_a$ and for $Z_C$ it will be $Z_b$, Fig. 2.6.

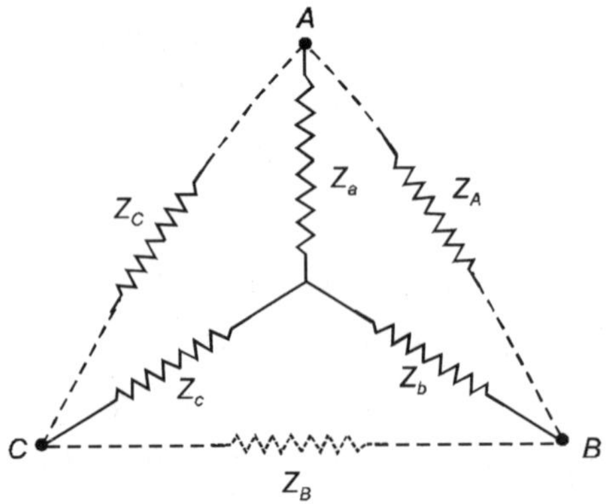


**Fig. 2.6**

Again for transforming delta into star consider the Fig. 2.7.

To find $Z_a$ open the delta vertex at $A$. The two impedances associated at vertex $A$ are $Z_A$ and $Z_C$. The equivalent star.

$$Z_a = \frac{Z_A\, Z_C}{Z_A + Z_B + Z_C}$$

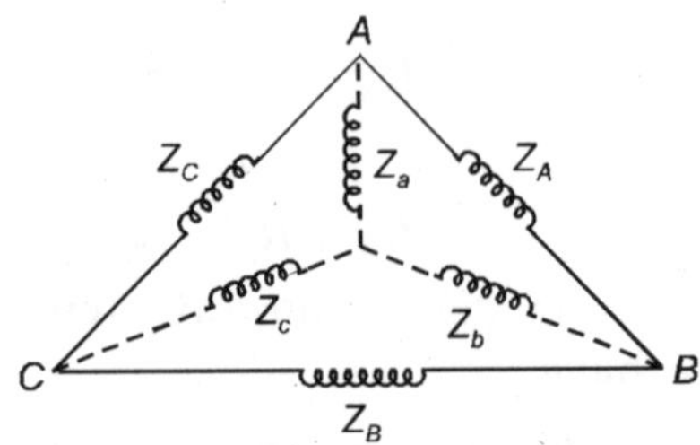


**Fig. 2.7**  Delta to star transformation.

Here denominator is same for all the three legs of the equivalent star.

## 2.4  LOOP ANALYSIS

Let us consider a few application of KVL in terms of loop or mesh analysis.

**Example 2.1:**  Determine the currents $i_1$, $i_2$, and $i_3$ and the Voltage $V_1$ and $V_2$ in the given network.

Using KVL we have

$$0.2\, i_1 + (i_1 - i_2)\, 3.75 = 120 \qquad\qquad \textit{Loop I}$$
$$3.75\,(i_2 - i_1) + 0.3\, i_2 + 5.45\,(i_2 - i_3) = 0 \qquad\qquad \textit{Loop II}$$
$$5.45\,(i_3 - i_2) + 0.1\, i_3 = 110 \qquad\qquad \textit{Loop III}$$

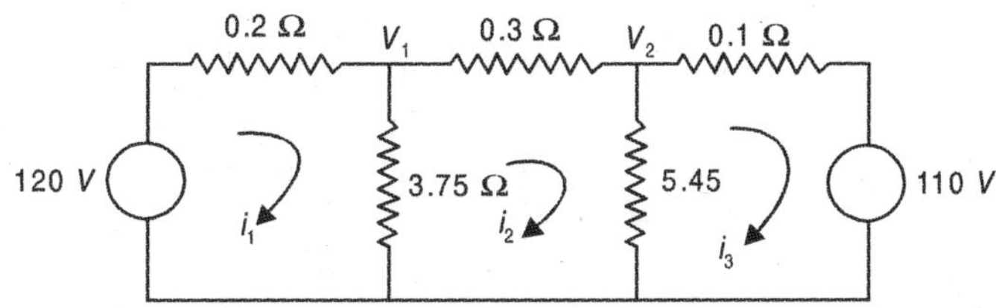


**Fig E2.1**

Rearranging the equations, we have

$$(0.2 + 3.75)\, i_1 - 3.75\, i_2 = 120$$
$$-3.75\, i_1 + (3.75 + 0.3 + 5.45)\, i_2 - 5.45\, i_3 = 0$$
$$-5.45\, i_2 + (5.45 + 0.1)\, i_3 = 110$$

These equations can be written in matrix form as follows

$$\begin{bmatrix} (0.2 + 3.75) & -3.75 & 0.0 \\ -3.75 & (3.75 + 0.3 + 5.45) & -5.45 \\ 0.0 & -5.45 & 5.45 + 0.1 \end{bmatrix} \begin{bmatrix} i_1 \\ i_2 \\ i_3 \end{bmatrix} = \begin{bmatrix} 120 \\ 0.0 \\ 110 \end{bmatrix}$$

or
$$Z\,I = V$$

Here $Z$ matrix is the loop impedance matrix and is found to be a symmetric matrix along its leading diagonal.

It can be seen that the loop impedance matrix can be written by inspection of the loops.

(*i*) The diagonal elements are the sum of all the branch impedances in the corresponding loop.

(*ii*) The off-diagonal elements say $Z_{12}$ is the impedance of the branches common to both the loop 1 and 2. The sign is positive if current $i_1$ and $i_2$ in the common branch flow in the same direction and is taken as negative if the two currents oppose each other.

The above set of these equations can again be written in matrix form as follows:

$$\begin{bmatrix} 3.95 & -3.75 & 0.0 \\ -3.75 & 9.50 & -5.45 \\ 0.0 & -5.45 & 5.55 \end{bmatrix} \begin{bmatrix} i_1 \\ i_2 \\ i_3 \end{bmatrix} = \begin{bmatrix} 120 \\ 0.0 \\ 110 \end{bmatrix}$$

Using Cramer's rule

$$i_1 = \cfrac{\begin{vmatrix} 120 & -3.75 & 0.0 \\ 0.0 & 9.50 & -5.45 \\ 110 & -5.45 & 5.55 \end{vmatrix}}{\begin{vmatrix} 3.95 & -3.75 & 0.0 \\ -3.75 & 9.50 & -5.45 \\ 0.0 & -5.45 & 5.55 \end{vmatrix}} = \Delta$$

Similarly

$$i_2 = \cfrac{\begin{vmatrix} 3.95 & 120 & 0.0 \\ -3.75 & 0.0 & -5.45 \\ 0.0 & 110 & 5.55 \end{vmatrix}}{\Delta}$$

Similarly expression for $i_3$ can be written. On solution the approximate values of $i_1$, $i_2$ and $i_3$ are found to be 40A, 10A and −10A. Here −10A in the third loop means the direction of current in the third loop should be opposite to what has been assumed at the beginning of the solution.

The voltage
$$v_1 = (i_1 - i_2)\,3.75$$
$$= 30 * 3.75$$
$$= 112.5 \text{ Volts}$$
and
$$V_2 = 20 \times 5.45 = 109 \text{ volts}$$

**Example 2.2:** Determine the current in the battery, the current in each branch and the p.d. across *AB* in the network shown.

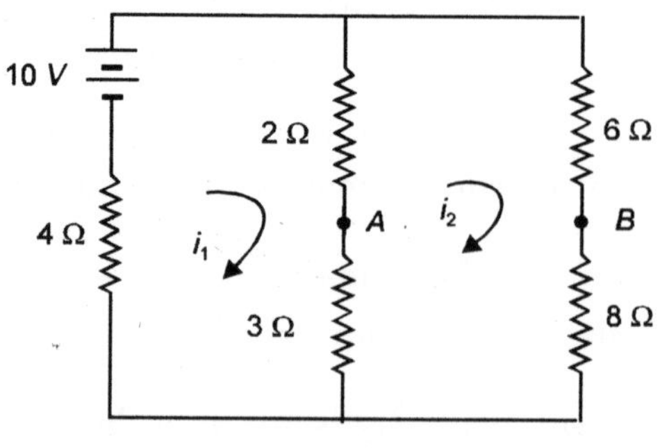


**Fig. E2.2**

Using KVL we have the equations in the matrix form.

$$\begin{bmatrix} 9 & -5 \\ -5 & 19 \end{bmatrix} \begin{bmatrix} i_1 \\ i_2 \end{bmatrix} = \begin{bmatrix} 10 \\ 0 \end{bmatrix}$$

From second equation
$$-5\,i_1 + 19\,i_2 = 0$$

or
$$i_1 = \frac{19}{5}\,i_2$$

From first equation

$$9\,i_1 - 5\,i_2 = 10$$

$$\frac{171}{5}\,i_2 - 5\,i_2 = 10$$

or $\qquad i_2 = \dfrac{10 \times 5}{146}\qquad$ or $\quad i_1 = \dfrac{190}{146} = 1.3A$

**Alternative solution:**

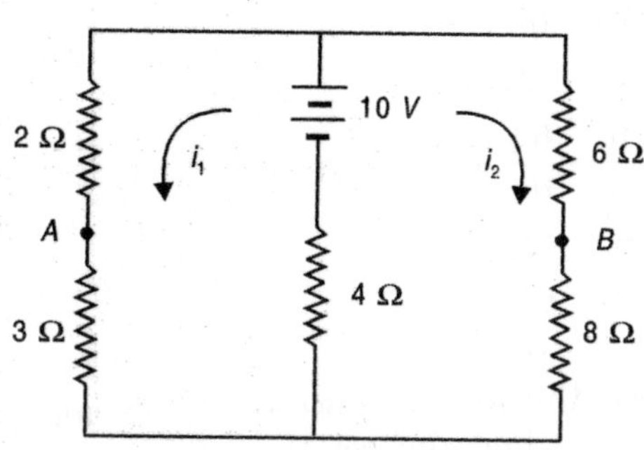


**Fig. E2.2.1**

Using KVL we have

$$9\,i_1 + 4\,i_2 = 10 \qquad\qquad\qquad\qquad \ldots(1)$$

$$4\,i_1 + 18\,i_2 = 10$$

or $\qquad 9\,i_1 + 4\,i_2 = 4\,i_1 + 18\,i_2$

or $\qquad i_1 = \dfrac{14}{5}\,i_2$

Substituting in (1) we have

$$9 \times \frac{14}{5} + 4i_2 = 10$$

or $\qquad\qquad i_2 = 0.343$

and $\qquad\qquad i_1 = \dfrac{14}{5} \times \dfrac{50}{146} = 0.96\,\text{A}$

Therefore current through the battery 1.303 A.
Potential of $A$ w.r. to $C$ $0.96 \times 3 = 2.88\ v$
and that of $B$ w.r. to $C$ $0.343 \times 8 = 2.744$
$V_{AB} = V_A - V_B = 2.88 - 2.74 = 0.14\ v$ **Ans.**

**Example 2.3:** Determine the current through the branch $AB$ of the network shown

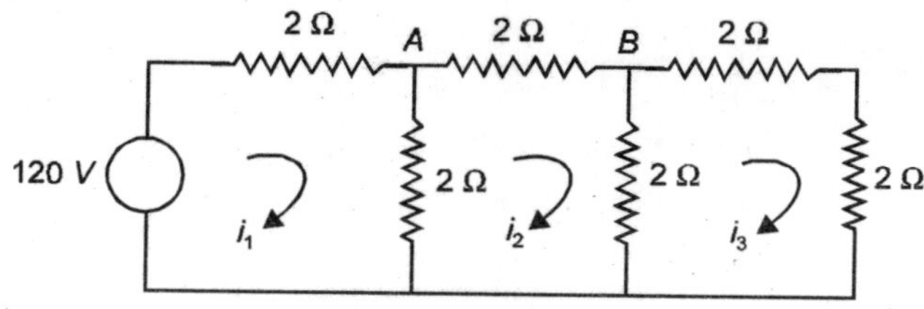


**Fig. E2.3**

Using **KVL** we have

$$\begin{bmatrix} 4 & -2 & 0 \\ -2 & 6 & -2 \\ 0 & -2 & 6 \end{bmatrix} \begin{bmatrix} i_1 \\ i_2 \\ i_3 \end{bmatrix} = \begin{bmatrix} 10 \\ 0 \\ 0 \end{bmatrix}$$

Using Cramer's rule

$$i_2 = \frac{\begin{vmatrix} 4 & 10 & 0 \\ -2 & 0 & -2 \\ 0 & 0 & 6 \end{vmatrix}}{\begin{vmatrix} 4 & -2 & 0 \\ -2 & 6 & -2 \\ 0 & -2 & 6 \end{vmatrix}} = \frac{15}{13} \ A \ \textbf{Ans.}$$

**Example 2.4:**  Write down the loop equations of the network shown and determine the voltage across the branch *AB* when the source voltage is 10 volt rms

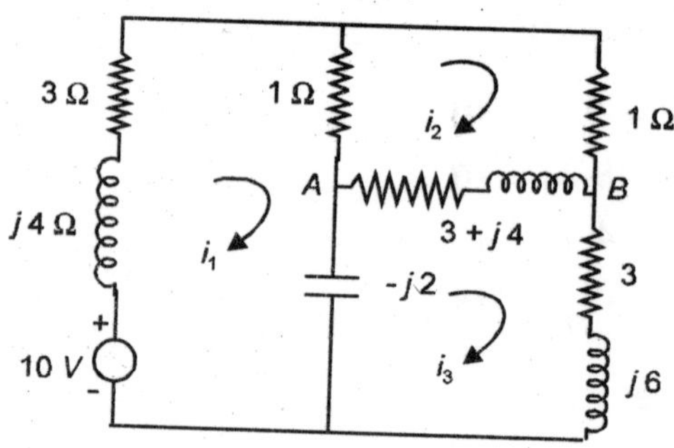


**Fig. E2.4**

Using KVL equation we have

$$\begin{bmatrix} 4 + j2 & -1 & + j2 \\ -1 & 5 + j4 & -3 - j4 \\ j2 & -3 - j4 & 6 + j8 \end{bmatrix} \begin{bmatrix} i_1 \\ i_2 \\ i_3 \end{bmatrix} = \begin{bmatrix} 10 \\ 0 \\ 0 \end{bmatrix}$$

To calcultate voltage across *AB* we should find out $i_2$ and $i_3$

$$i_2 = \frac{\begin{vmatrix} 4 + j2 & 10 & j2 \\ -1 & 0 & -3 - j4 \\ j2 & 0 & 6 + j8 \end{vmatrix}}{\begin{vmatrix} 4 + j2 & -1 & j2 \\ -1 & 5 + j4 & -3 - j4 \\ j2 & -3 - j4 & 6 + j8 \end{vmatrix} = \Delta}$$

$$i_3 = \frac{\begin{vmatrix} 4 + j2 & -1 & 10 \\ -1 & 5 + j4 & 0 \\ j2 & -3 - j4 & 0 \end{vmatrix}}{\Delta} \qquad \text{Hence } V_{AB} = (i_3 - i_2)(3 + j4)$$

## 2.5  NODAL ANALYSIS

We now illustrate the application of KCL for solving the networks. Here whenever a voltage source is given it should be converted into a current source before nodal equations using KCL are written.

**Example 2.5:**  Consider example 2.3. The figure is reproduced here

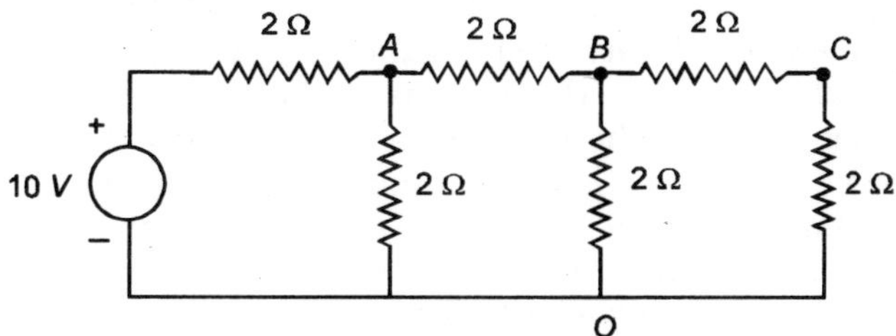


**Fig. E2.5**

First we transform the voltage source into a current source and the diagram becomes

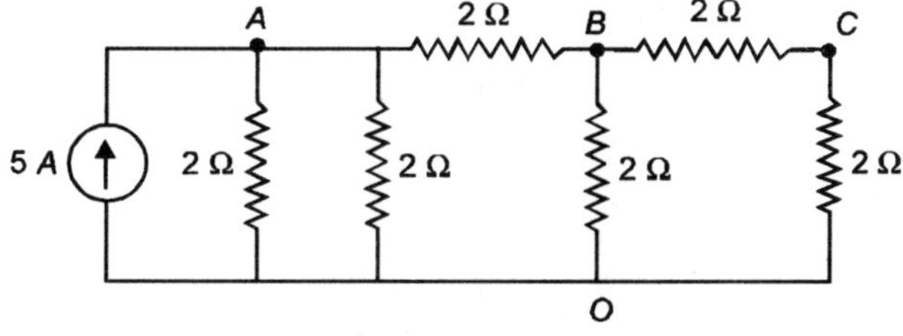


**Fig. E2.5.1**

Taking node $O$ as reference, we write down KCL equations
At node $A$

$$\frac{V_A}{2} + \frac{V_A}{2} + \frac{V_A - V_B}{2} = 5$$

$$V_A\left[\frac{1}{2} + \frac{1}{2} + \frac{1}{2}\right] - V_B/2 + 0.V_c = 5$$

At node $B$ $\qquad -V_A/2 + \left(\frac{1}{2} + \frac{1}{2} + \frac{1}{2}\right)V_B - V_C/2 = 0$

At node C $\qquad 0.V_A - \frac{V_B}{2} + V_C\left(\frac{1}{2} + \frac{1}{2}\right) = 0$

These equations can be written in matrix form as follows:

$$\begin{bmatrix} \dfrac{1}{2}+\dfrac{1}{2}+\dfrac{1}{2} & -\dfrac{1}{2} & 0 \\[2mm] -\dfrac{1}{2} & \dfrac{1}{2}+\dfrac{1}{2}+\dfrac{1}{2} & -\dfrac{1}{2} \\[2mm] 0 & -\dfrac{1}{2} & \dfrac{1}{2}+\dfrac{1}{2} \end{bmatrix} \begin{bmatrix} V_A \\ V_B \\ V_C \end{bmatrix} = \begin{bmatrix} 5 \\ 0 \\ 0 \end{bmatrix}$$

Here the first matrix is known as nodal admittance matrix and is denoted by $Y$ and the two column matrices are known as nodal voltage and nodal current matrices denoted by $V$ and $I$ respectively.

Hence, the above matrices can be written in compact form as

$$Y\,V = I \qquad\qquad ...(2.22)$$

It can be seen that the nodal admittance matrix can be written simply by looking at the interconnections of the network using the following procedure:

(*i*) The diagonal element of each node is the sum of the admittances connected to it.

(*ii*) The off diagonal element is the negative admittance between the nodes. If there is no connection between the two nodes, the entry in the off-diagonal for both the nodes is zero i.e. if node $i$ is not connected to the node $j$, then $Y_{ij} = Y_{ji} = 0$. In general every node will not be connected to all other nodes in the network and hence normally nodal admittance matrix is a highly sparse matrix i.e. it will have very small number of non-zero elements. In fact, in a large network consisting of say 100 nodes, these non zero elements would be as small as 2% of the total elements.

We come back to original problem of finding out $V_{AB}$
Using Cramer's rule

$$V_A = \frac{\begin{vmatrix} 5 & -\dfrac{1}{2} & 0 \\[2mm] 0 & \dfrac{3}{2} & -\dfrac{1}{2} \\[2mm] 0 & -\dfrac{1}{2} & 1 \end{vmatrix}}{\begin{vmatrix} \dfrac{3}{2} & -\dfrac{1}{2} & 0 \\[2mm] -\dfrac{1}{2} & \dfrac{3}{2} & -\dfrac{1}{2} \\[2mm] 0 & -\dfrac{1}{2} & 1 \end{vmatrix}} = \frac{5 \times \dfrac{5}{4}}{\dfrac{13}{8}}$$

Therefore, $\qquad V_A = \dfrac{50}{13}$ volts.

Similarly,

$$V_B = \frac{\begin{vmatrix} \dfrac{3}{2} & 5 & 0 \\[2mm] -\dfrac{1}{2} & 0 & -\dfrac{1}{2} \\[2mm] 0 & 0 & 1 \end{vmatrix}}{\Delta}$$

Therefore, $V_B = \dfrac{5}{2} \times \dfrac{8}{13} = \dfrac{20}{13}$ and $V_{AB} = \dfrac{50}{13} - \dfrac{20}{13} = \dfrac{30}{13}$ volts

**Example 2.6:** Determine the current through the branch $AB$ of the given network using (*i*) nodal (*ii*) Loop analysis.

(*i*) Nodal analysis

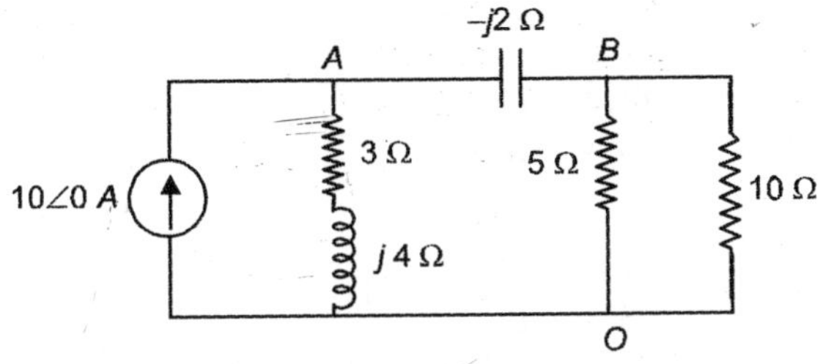


**Fig. E2.6**

Using KCL and writing in matrix form taking $O$ as the reference node

$$\begin{bmatrix} \dfrac{1}{3+j4} + j0.5 & -j0.5 \\[2ex] -j0.5 & \dfrac{1}{5} + \dfrac{1}{10} + \dfrac{j}{2} \end{bmatrix} \begin{bmatrix} V_A \\ V_B \end{bmatrix} = \begin{bmatrix} 10 \\ 0 \end{bmatrix}$$

$$V_A = \frac{\begin{vmatrix} 10 & -j0.5 \\ 0 & (3+j5)/10 \end{vmatrix}}{\begin{vmatrix} -1+j1.5 & -j0.5 \\ -j0.5 & (3+j5)/10 \end{vmatrix}}$$

Similarly,

$$V_B = \frac{\begin{vmatrix} -1+j1.5 & 10 \\ -j0.5 & 0 \end{vmatrix}}{\begin{vmatrix} -1+j1.5 & -j0.5 \\ -j0.5 & (3+j5)/10 \end{vmatrix}}$$

$$V_{AB} = V_A - V_B = \frac{60(3+j4)}{-6+j19}$$

Using loop analysis, we first convert current source into a voltage source.

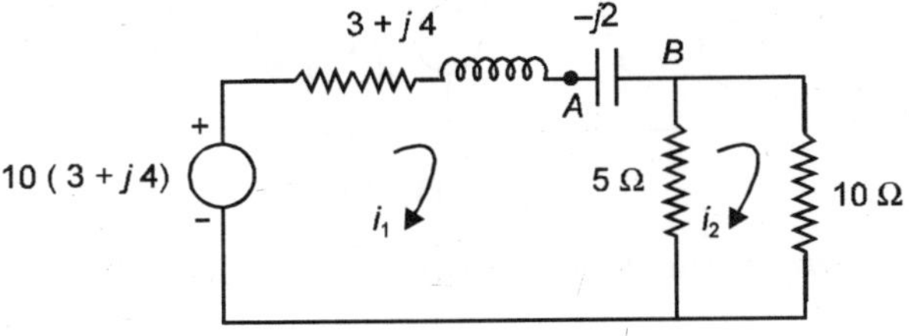


**Fig. E2.6.1**

$$\begin{bmatrix} 8+j2 & -5 \\ -5 & 15 \end{bmatrix} \begin{bmatrix} i_1 \\ i_2 \end{bmatrix} = \begin{bmatrix} 10(3+j4) \\ 0 \end{bmatrix}$$

To find out $V_{AB}$, it is enough to find out $i_1$

$$i_1 = \frac{\begin{vmatrix} 10(3 + j4) & -5 \\ 0 & 15 \end{vmatrix}}{\begin{vmatrix} 8 + j2 & -5 \\ -5 & 15 \end{vmatrix}} = \frac{150(3 + j4)}{15(8 + j2) - 25}$$

$$= \frac{150(3 + j4)}{95 + j30}$$

Hence,

$$V_{AB} = \frac{150(3 + j4)}{95 + j30} \cdot (-j2) = \frac{-j\,300(3 + j4)}{95 + j30}$$

Dividing numerator and denominator by $-j5$ we have

$$V_{AB} = \frac{60(3 + j4)}{-6 + j19}$$

It is found that $V_{AB}$ is same by both the methods i.e. nodal and loop analysis.

**Example 2.7:** For the network shown determine the current through the branch *AB*

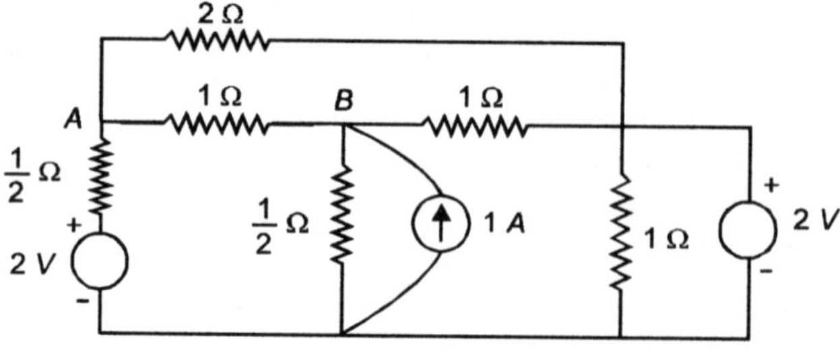

**Fig. E2.7**

(*i*) Converting the current source into a voltage source we have

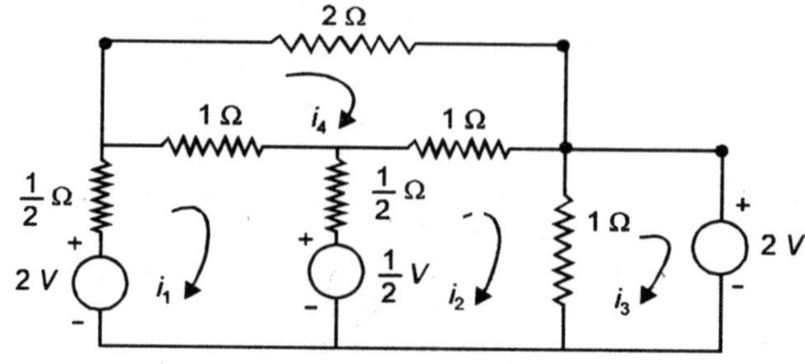

**Fig. E2.7.1**

Using KVL, we write the loop equations in matrix form as follows:

$$\begin{bmatrix} 2 & -1/2 & 0 & -1 \\ -1/2 & 5/2 & -1 & -1 \\ 0 & -1 & 1 & 0 \\ -1 & -1 & 0 & 4 \end{bmatrix} \begin{bmatrix} i_1 \\ i_2 \\ i_3 \\ i_4 \end{bmatrix} = \begin{bmatrix} 3/2 \\ 1/2 \\ 2 \\ 0 \end{bmatrix}$$

Using Cramer's rule we have

$$i_1 = \frac{\begin{vmatrix} 1.5 & -1/2 & 0 & -1 \\ 0.5 & 5/2 & -1 & -1 \\ 2.0 & -1 & 1 & 0 \\ 0.0 & -1 & 0 & 4 \end{vmatrix}}{\begin{vmatrix} 2 & -1/2 & 0 & -1 \\ -1/2 & 5/2 & -1 & -1 \\ 0 & -1 & 1 & 0 \\ -1 & -1 & 0 & 4 \end{vmatrix}} = 0.46$$

Similarly other currents can be obtained

$$i_2 = -0.923 \quad i_3 = -2.923 \quad i_4 = -0.115$$

Current through $AB$ branch is $i_1 - i_4 = 0.46 + 0.115 = 0.575$

(*ii*) In order to use KCL equations, voltage sources are to be converted to current sources. Since voltage source on the right side does not have a series resistance we can't convert it straightaway. However, we assume a resistance $R$ to be connected in series with the source and while finding the final solutions we would let this resistance tend to zero. The circuit becomes.

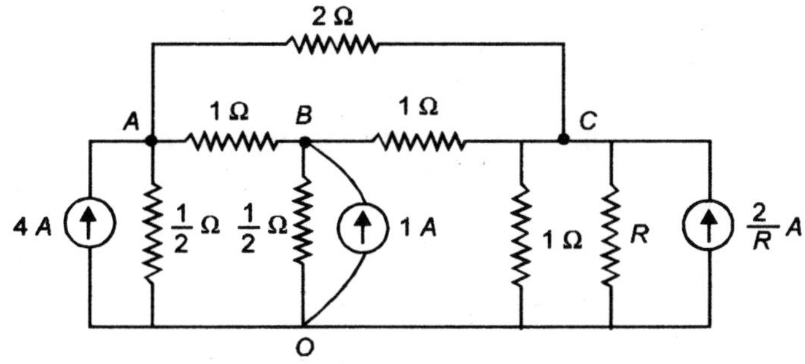

**Fig. E2.7.2**

Taking node $O$ as reference we have

$$\begin{bmatrix} 3.5 & -1 & -1/2 \\ -1 & 4 & -1 \\ -1/2 & -1 & 1+1/R+1+1/2 \end{bmatrix} \begin{bmatrix} V_A \\ V_B \\ V_C \end{bmatrix} = \begin{bmatrix} 4 \\ 1 \\ 2/R \end{bmatrix}$$

$$V_A = \frac{\begin{vmatrix} 4 & -1 & -1/2 \\ 1 & 4 & -1 \\ 2/R & -1 & 5/2+1/R \end{vmatrix}}{\begin{vmatrix} 3.5 & -1 & -1/2 \\ -1 & 4 & -1 \\ -1/2 & -1 & 5/2+1/R \end{vmatrix}}$$

$$V_A = \frac{39+23/R}{30.5+13/R}$$

Similarly,
$$V_B = \frac{20.5 + 15.5/R}{30.5 + 13/R}$$

Hence,
$$V_{AB} = V_A - V_B = \frac{18.5R + 7.5}{30.5R + 13} = \frac{7.5}{13} = 0.577 \text{ volt.}$$

Lt $\qquad\qquad\qquad R \to 0$

The current through branch $AB = 0.577\ A$

Alternatively, we simplify the circuit and we find that it becomes a very simple problem to solve. We remove 1 ohm resistance across the voltage source $2v$ as it is redundant for the external circuit, external to the voltage source.

We push the voltage source of $2v$ into 1 ohm and 2 ohm branches and we convert another $2V$ source into an equivalent current source and we get the following circuit:

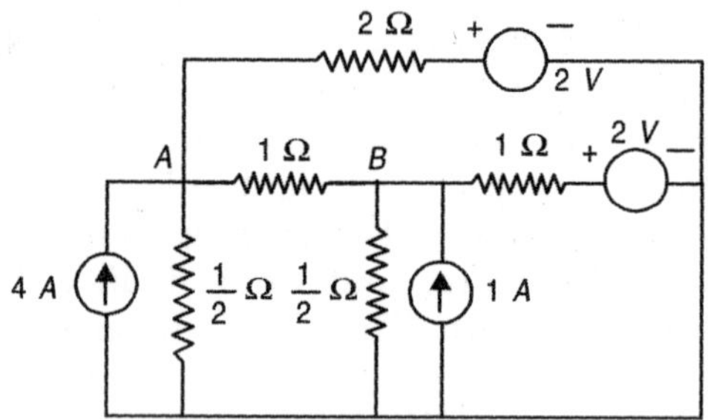


**Fig. E2.7.3**

Next we transform the voltage sources into current sources, we have the network as shown.

Applying KCL at node $A$

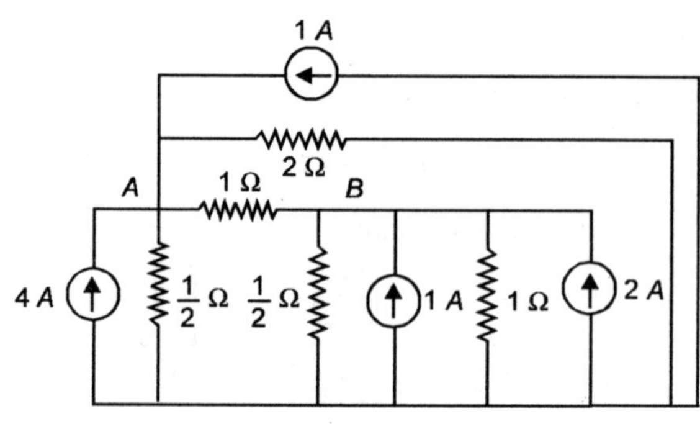


**Fig. E2.7.4**

$$\left(\frac{1}{1/2} + \frac{1}{1} + \frac{1}{2}\right)V_a - V_b = 5$$

$$3.5\ V_a - V_b = 5$$

At node $B$

$$-V_a + 4V_b = 3$$

The solution of these equations gives
$$V_a = 1.77 \text{ volt.} \quad V_b = 1.19$$

Hence,
$$V_{ab} = V_a - V_b = 1.77 - 1.19 = 0.58 \text{ volt.}$$

**Example 2.8:** Determine the current through the branch $AB$ of the network using
(*i*) Simple Logic (*ii*) Loop Method of analysis (*iii*) Nodal Method of analysis

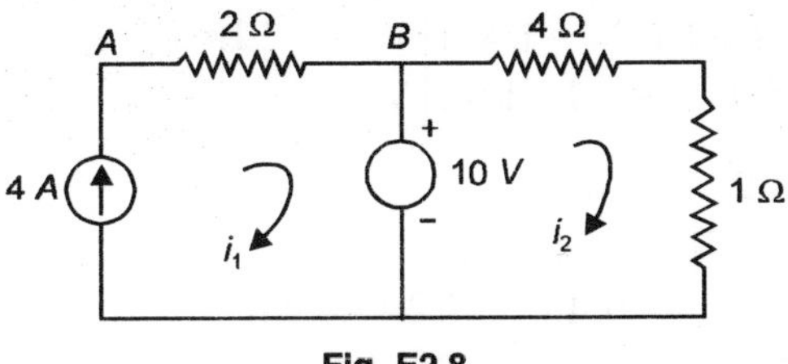

**Fig. E2.8**

(*i*) Since current source of 4$A$ is connected in series with the branch $AB$, the current through branch $AB$ will be 4$A$.

Similarly since a resistance of $(4 + 1) = 5$ is connected across a constant voltage source of 10$V$, current $i_2 = \dfrac{10}{5} = 2A$.

(*ii*) *Loop analysis:* Here the current source should be converted into a voltage source. Since no resistance is connected across the current source, the transformation is not possible. Let us connect resistance $R$ across the current source, and after the final result is obtained we will tend $R$ to infinity. The circuit therefore, with voltage source will be as follows.

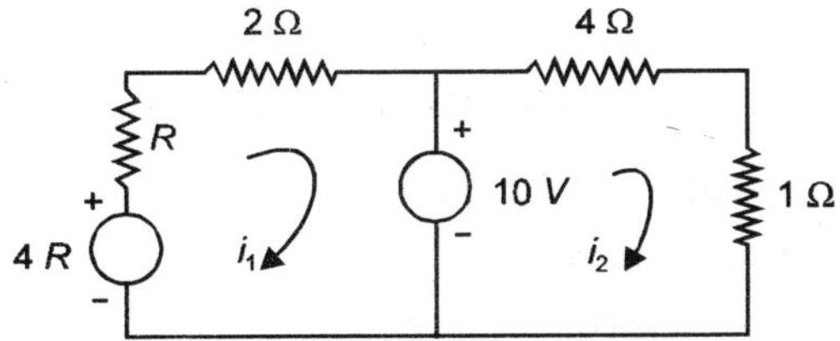

**Fig. E2.8.1**

$$\begin{bmatrix} R+2 & 0 \\ 0 & 5 \end{bmatrix}\begin{bmatrix} i_1 \\ i_2 \end{bmatrix} = \begin{bmatrix} 4R - & 10 \\ 10 \end{bmatrix}.$$

$$i_1 = \frac{4R-10}{R+2} = \frac{4-10/R}{1+2/R} = 4A$$

$$Lt\ R \rightarrow \infty$$

(*iii*) Nodal Analysis: Here voltage source should be converted into current source.

Consider a resistance $R$ connected in series with the voltage and towards the end we will have the result when $R \rightarrow O$.

The circuit is

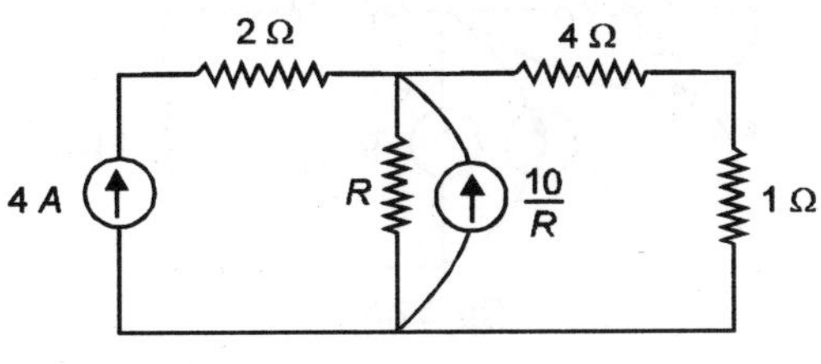

**Fig. E2.8.2**

Writing nodal equations

$$\begin{bmatrix} 1/2 & -1/2 & 0 \\ -1/2 & 1/2+1/4+1/R & -1/4 \\ 0 & -1/4 & 1+1/4 \end{bmatrix} \begin{bmatrix} V_1 \\ V_2 \\ V_3 \end{bmatrix} = \begin{bmatrix} 4 \\ 10/R \\ 0 \end{bmatrix}$$

$$V_1 = \dfrac{\begin{vmatrix} 4 & -1/2 & b \\ 10/R & 3/4+1/R & -1/4 \\ 0 & -1/4 & 5/4 \end{vmatrix}}{\begin{vmatrix} 1/2 & -1/2 & 0 \\ -1/2 & 3/4+1/R & -1/4 \\ 0 & -1/4 & 5/4 \end{vmatrix}} = \dfrac{7/2+45/4R}{1/8+5/8R}$$

$$V_2 = \dfrac{\begin{vmatrix} 1/2 & 4 & 0 \\ -1/2 & 10/R & -1/4 \\ 0 & 0 & 5/4 \end{vmatrix}}{\Delta} = \dfrac{25/4R+5/2}{1/8+5/8R}$$

$$V_1 - V_2 = \dfrac{5/R+1}{1/8+5/8R} = \dfrac{8(R+5)}{5+8R} = 8$$

$$Lt\ R \to 0$$

The voltage difference is 8 volt. Therefore current is $\dfrac{8}{2}$ = 4A

Similarly current $i_2$ can be obtained.

(*iv*) By shifting the voltage source

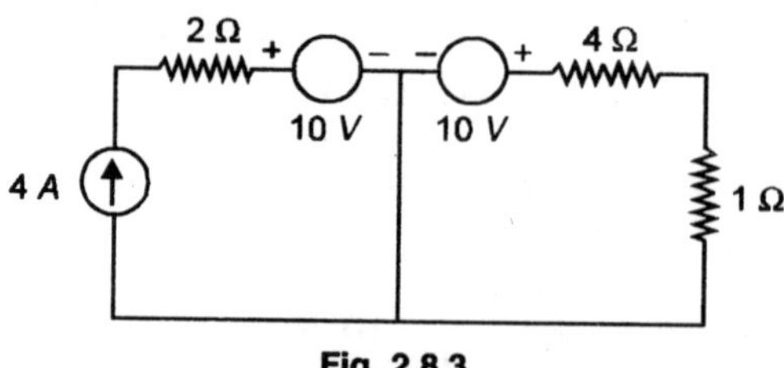


**Fig. 2.8.3**

Again through 2 ohm resistor the current is 4A and that through 4 ohm resistor it is $\dfrac{10}{5}$ = 2A.

By shifting the current source and assuming a resistance $R$ in series with the voltage

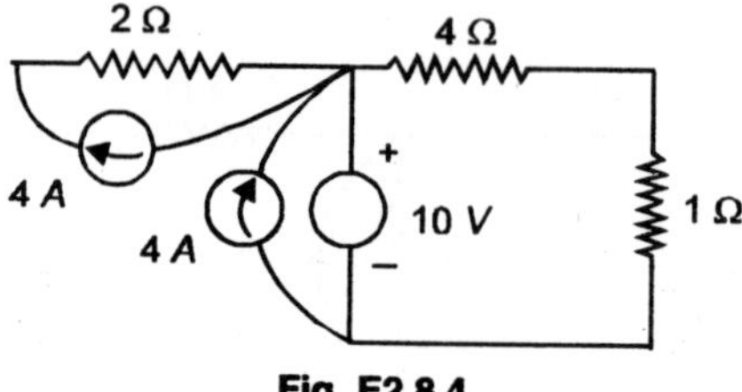


**Fig. E2.8.4**

source and then finally letting $R \to 0$ we can have the final solution.

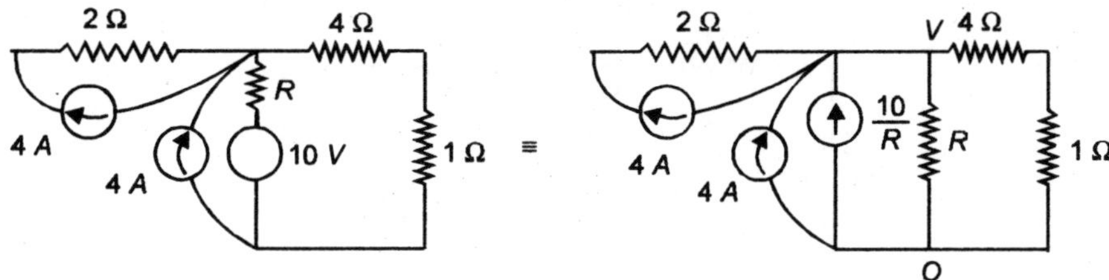

**Fig. E.2.8.5**

Using nodal analysis we have

$$4 + \frac{10}{R} = \frac{V}{R} + \frac{V}{5} = V\left(\frac{5+R}{5R}\right)$$

$20R + 50 = 5V + VR$ Let $R \to 0$, $5V = 50$  or  $V = 10$ volts.

Therefore, current through 2 ohm resistor is $4A$ and that through 4 ohm is $2A$.

**Example 2.9:** Determine the current through the 4 ohm resistor branch of the given network.

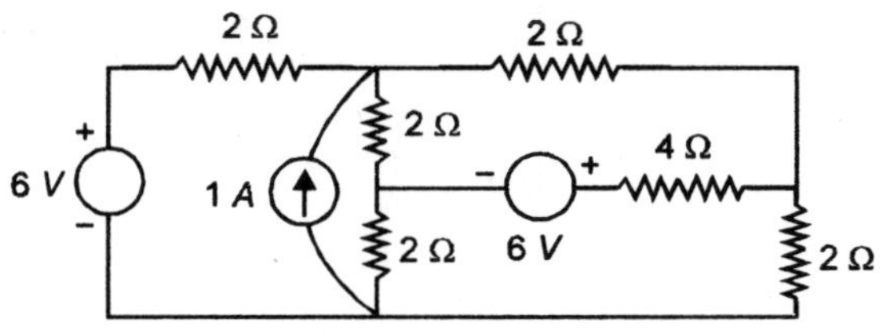

**Fig. E.2.9**

Shifting the current source, the network becomes

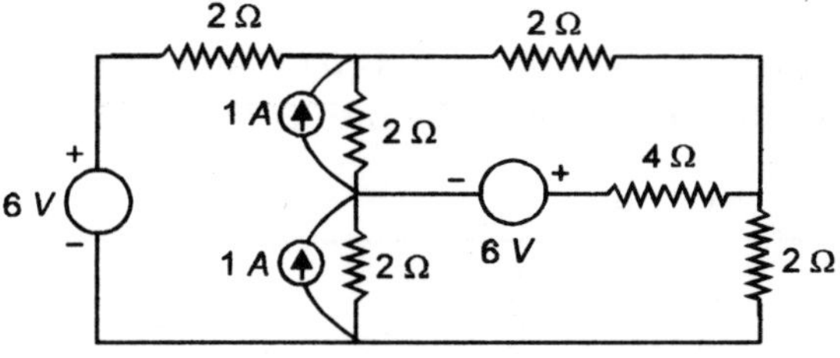

**Fig. E2.9.1**

Again current sources are converted into voltage sources, we have

$$\begin{bmatrix} 6 & -2 & -2 \\ -2 & 8 & -4 \\ -2 & -4 & 8 \end{bmatrix}\begin{bmatrix} i_1 \\ i_2 \\ i_3 \end{bmatrix} = \begin{bmatrix} 2 \\ -4 \\ 8 \end{bmatrix}$$

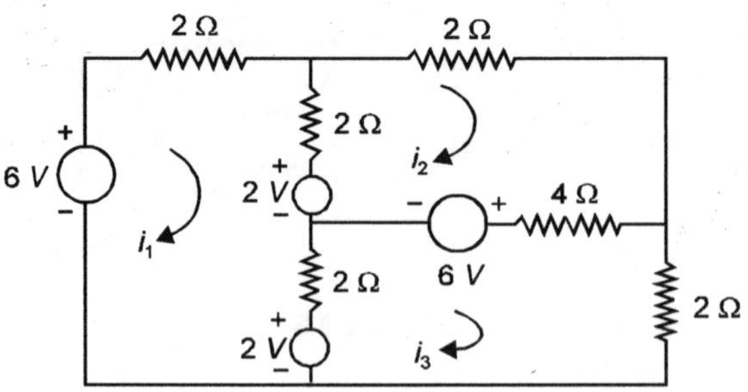

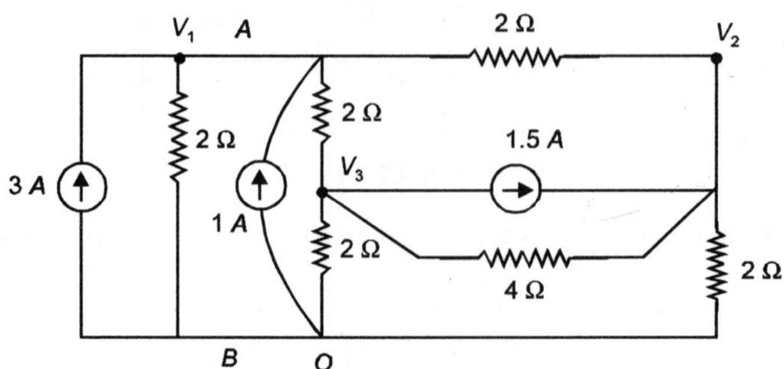

**Fig. E2.9.2**

To find out current through the branch of $4\,\Omega$ resistor we need find $i_2$ and $i_3$ only.
Using Cramer's rule

$$i_2 = \frac{\begin{vmatrix} 6 & 2 & -2 \\ -2 & -4 & -4 \\ -2 & 8 & 8 \end{vmatrix}}{\begin{vmatrix} 6 & -2 & -2 \\ -2 & 8 & -4 \\ -2 & -4 & 8 \end{vmatrix}} = \frac{96}{48 \times 4} = 0.5A$$

Similarly,

$$i_3 = \frac{\begin{vmatrix} 6 & -2 & 2 \\ -2 & 8 & -4 \\ -2 & -4 & 8 \end{vmatrix}}{\Delta} = \frac{48 \times 6}{48 \times 4} = 1.5A$$

Current through $4\Omega$ resistor is $i_3 - i_2 = 1.0\ A$ (*ii*) Using nodal analysis the original circuit becomes after transforming the voltage sources into current sources.

**Fig. E2.9.3**

Taking node $O$ as reference

$$\begin{bmatrix} 3/2 & -1/2 & -1/2 \\ -1/2 & 5/4 & -1/4 \\ -1/2 & -1/4 & 5/4 \end{bmatrix} \begin{bmatrix} V_1 \\ V_2 \\ V_3 \end{bmatrix} = \begin{bmatrix} 4 \\ 1.5 \\ -1.5 \end{bmatrix}$$

To calculate current through $4\Omega$ resistor branch we need calculate $V_2$ and $V_3$.

$$V_2 = \frac{\begin{vmatrix} 1.5 & 4 & -0.5 \\ -0.5 & 1.5 & -0.25 \\ -0.5 & -1.5 & 1.25 \\ 1.5 & -0.5 & -1/2 \\ -1/2 & 5/4 & -1/4 \\ -1/2 & -1/4 & 5/4 \end{vmatrix}}{} = \frac{36/8}{24/16} = 3 \text{ volts}$$

Similarly,

$$V_3 = \frac{\begin{vmatrix} 3/2 & -1/2 & 4 \\ -1/2 & 5/4 & 3/2 \\ -1/2 & -1/4 & -3/2 \end{vmatrix}}{\Delta} = \frac{12/8}{24/16} = \frac{12}{8} \times \frac{16}{24} = 1 \text{ volts}$$

Therefore $V_2 - V_3 = 3{-}1 = 2$ volt and current through $4\Omega$ resistor $= 1.5 - 0.5 = 1A$.

Yet another way of transforming the current sources into the voltage sources we take the same problem.

The voltage source $6V$ with series resistance $2\ \Omega$ could be considered between the nodes $A$ and $B$ as a current source of $3A$ across a resistance of $2\ \Omega$ and which again is an equivalent to current source of $3A + 1A = 4A$ across resistance of $2\ \Omega$ and is equivalent to a voltage source of $4 \times 2 = 8$ volts in series with a resistance of $2\ \Omega$. The equivalent circuit, therefore, becomes as shown Writing loop equations.

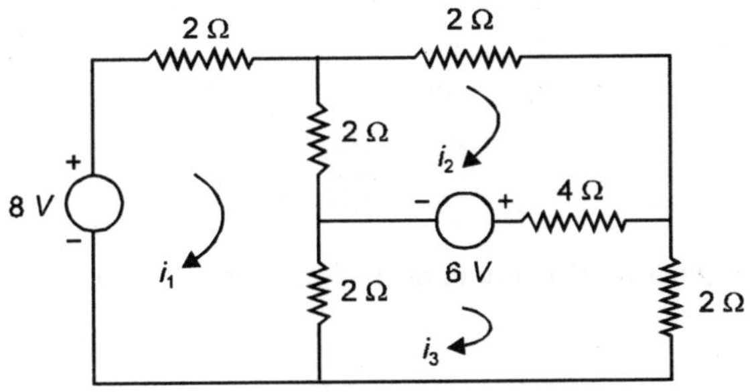


**Fig. E2.9.4**

$$\begin{bmatrix} 6 & -2 & -2 \\ -2 & 8 & -4 \\ -2 & -4 & 8 \end{bmatrix} \begin{bmatrix} i_1 \\ i_2 \\ i_3 \end{bmatrix} = \begin{bmatrix} 8 \\ -6 \\ +6 \end{bmatrix}$$

$$i_2 = \frac{\begin{vmatrix} 6 & +8 & -2 \\ -2 & -6 & -4 \\ -2 & 6 & 8 \end{vmatrix}}{48 \times 4} = \frac{24 \times 4}{48 \times 4} = 0.5A$$

Similarly,

$$i_3 = \frac{\begin{vmatrix} 6 & -2 & 8 \\ -2 & 8 & -6 \\ -2 & -4 & 6 \end{vmatrix}}{48 \times 4} = \frac{24 \times 12}{48 \times 4} = 1.5A$$

Hence current through $4\Omega$ resistor is $i_3 - i_2 = 1.5 - 0.5 = 1A$.
If we choose different loops as shown in Fig. E 2.8.6, the loop equations are

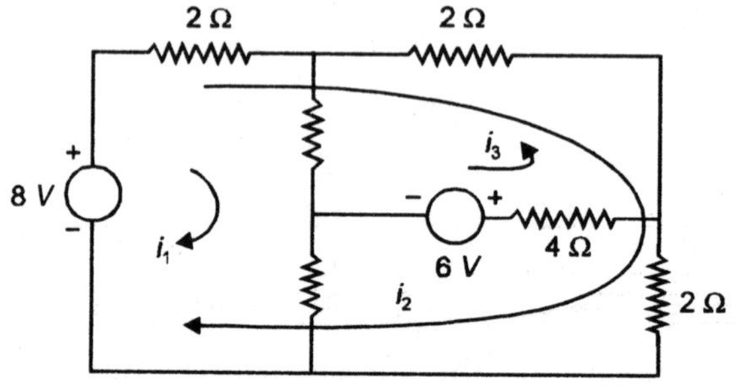

**Fig. E2.9.5**

$$\begin{bmatrix} 6 & 2 & 2 \\ 2 & 6 & -2 \\ 2 & -2 & 8 \end{bmatrix} \begin{bmatrix} i_1 \\ i_2 \\ i_3 \end{bmatrix} = \begin{bmatrix} 8 \\ 8 \\ 6 \end{bmatrix}$$

$$i_3 = \frac{\begin{vmatrix} 6 & 2 & 8 \\ 2 & 6 & +8 \\ 2 & -2 & 6 \end{vmatrix}}{\Delta} = \frac{192}{192} = 1A$$

**Example 2.10:**   Determine the current through the 2 $\Omega$ resistor branch of the given network.

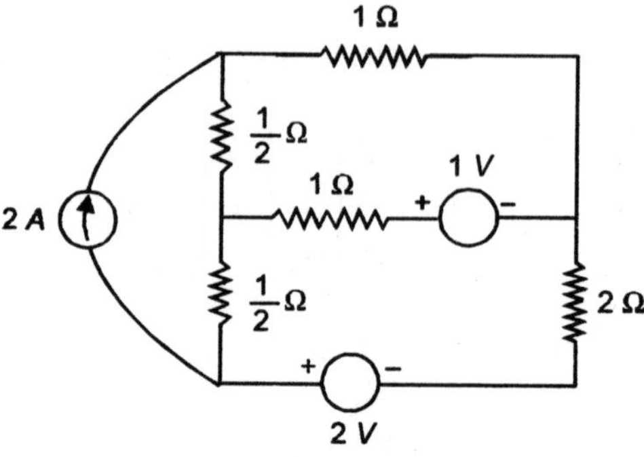

**Fig. E2.10**

**Solution:** Transforming the current source into a voltage source we have

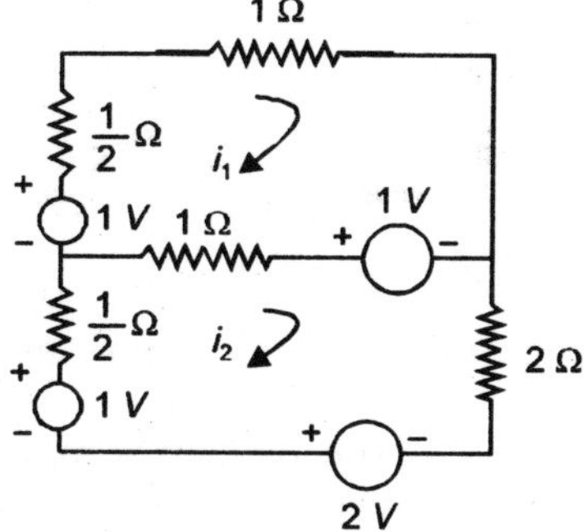



**Fig. E2.10.1**

$$\begin{bmatrix} 5/2 & -1 \\ -1 & 7/2 \end{bmatrix} \begin{bmatrix} i_1 \\ i_2 \end{bmatrix} = \begin{bmatrix} 2 \\ 2 \end{bmatrix}$$

$$\frac{\begin{vmatrix} 5/2 & 2 \\ -1 & 2 \end{vmatrix}}{\begin{vmatrix} 5/2 & -1 \\ -1 & 7/2 \end{vmatrix}} = \frac{7}{31/4} = \frac{28}{31} A$$

## 2.6  DUALITY

If we look at the two Kirchhoff's laws carefully, we find that the two laws are almost similar word for word when voltage is substituted for current, independent Loop for independent node pair. Similarly, the intergo-differential equations that result from the application of Kirchhoff's laws have similar appearance. This similarity in all its implication is known as duality. Consider two figures 2.8.

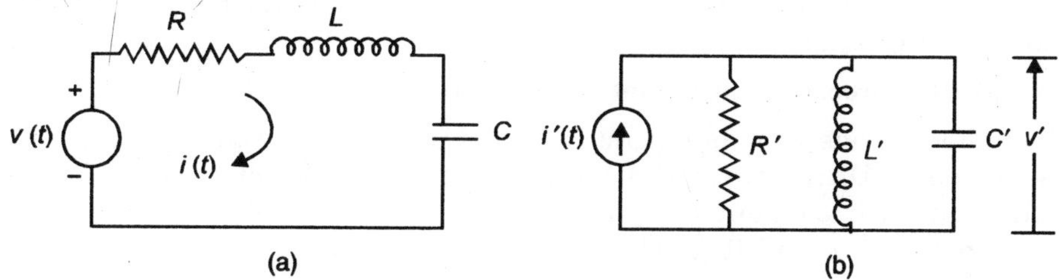



**Fig. 2.8**  (a) RLC series, (b) RLC parallel.

The integro-differential equations for the two networks are:

$$V(t) = Ri(t) + L\frac{di}{dt} + \frac{1}{C}\int i\, dt \qquad \qquad ...(2.22)$$

$$i'(t) = Gv'(t) + C'\frac{dv'}{dt} + \frac{1}{L'}\int v'\, dt \qquad \qquad ...(2.23)$$

The two equations specify identical mathematical operations, the only difference being in letter symbols. The solution of one equation is the solution of the other. The two networks are said to be dual of each other. It is to be noted that the two networks are not equivalent to each other. From the observations of the two equations analogies between various quantities are found.

$$R\,i \quad \text{and} \quad G'v'$$

$$L\frac{di}{dt} \quad \text{and} \quad C'\frac{dv'}{dt}$$

$$\frac{l}{C}\int i\,dt \quad \text{and} \quad \frac{1}{L'}\int v'\,dt$$

i.e. $R$ and $G$, $L$ and $C$, $i$ and $V$, loop and node pair, short circuit and open circuit and charge and flux etc.

In terms of matrix notations if $A$ is the incidence matrix of network $N_1$ and $B$ is loop matrix of network $N_2$, the conditions for two networks $N_1$ and $N_2$ to be dual of each other is that

$$r\,[A] = r\,[B]$$

i.e. $\qquad$ rank of $A$ = rank of $B$

with this $\qquad b_1 = b_2$ and $\quad n_1 = b_2 - n_2$

where $b_1$ and $b_2$ are the branches of the networks $N_1$ and $N_2$ and $n_1 + 1$ and $n_2 + 1$ are the number of nodes of the two networks respectively.

The following procedure is followed in constructing dual of a given network:

1. Inside each loop place a node and number them for all loops. Place a datum node outside the network say 'O' node. Place the same number of nodes on a separate space on the paper for construction of dual.
2. Draw a line between two nodes of the original network traversing only one element at a time. For such element traversed in the original network connect the dual element from the listing just given (nodes on the separate space on the paper.)
3. Continue this process until the number of possible paths through single elements is exhausted. If we go through a connecting lead assumed to be a short, the dual element is an open circuit. The network so constructed is a dual network. The construction can be verified by writing loop equations for one network and the node for the other.

To illustrate the procedure consider Fig. 2.9 (a)

1. Since there is one loop, one node is placed in this loop and one datum node outside the network. Draw two nodes for the dual network separately. Join node 1 and 0 through each element and therefore in the dual network draw dual elements between nodes 1 and 0 as shown in Fig. 2.9. (b)

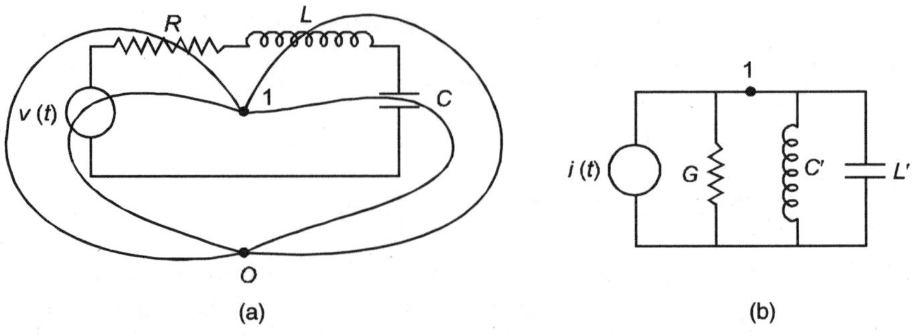



**Fig. 2.9** (a) Original Network, (b) Dual of (a).

Consider another network as shown in Fig. 2.10 and it is desired to draw its dual.

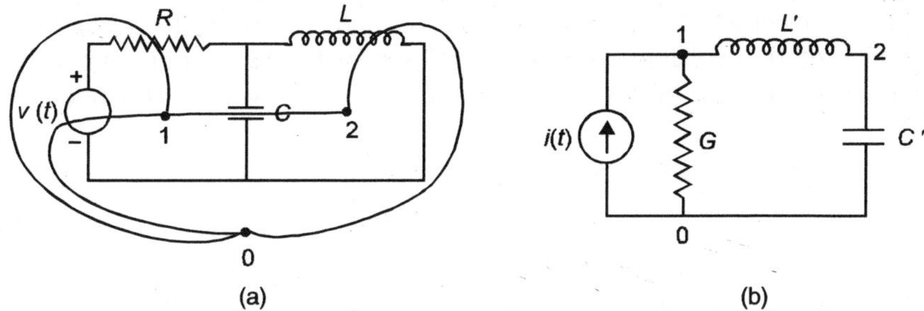

**Fig. 2.10**  (a) Original, (b) dual of (a).

Following above procedure the dual will be as shown in Fig. 2.10(b).

It is to be noted that the above procedure will always determine dual of a given network if the graph of the given network is planar. For non-planar networks the method does not provide dual of the given network. By planar graph is meant the one which may be drawn on a piece of paper without crossing lines. Fig. 2.11 shows a typical non-planar graph.

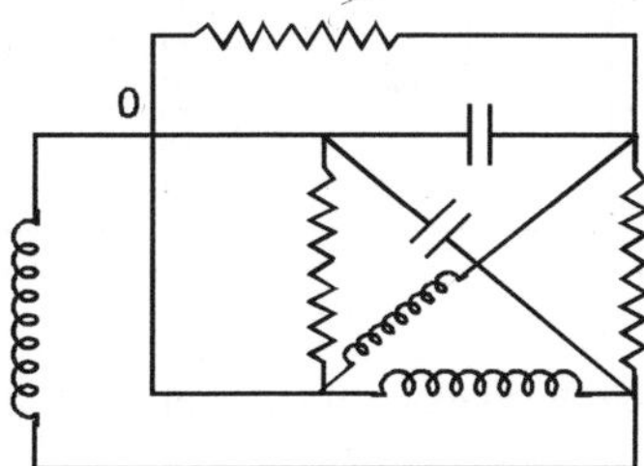

**Fig. 2.11**  Non-planar network.

Since at $O$ the two branches are crossing each other, it is a circuit having non-planar graph.

## 2.6.1  Source Orientation

If a network has two or more than two sources voltage or current or both voltage and curent sources the nodal voltages and loop currents are affected by the orientation of the sources. Therefore, it is very important to know conventions for providing orientation to voltage and /or current sources when dual network of a given network is obtained. The following procedure is adopted for finding the orientation of the sources.

We assueme a clockwise direction for all loop currents in the original network say $N_1$. Each loop current encloses a dot corresponding to the node of the dual network say $N_2$.

1. The orientation of the voltage source is identified by the direction of the loop current. If the direction of the loop current and the voltage is same then the orientation of the current source in the dual network will be towards the node (dot). However, if the direction of voltage source (polarity) and the clockwise loop current are opposing each other, the current source in the dual network will be directed away from the node. This is made clear with the following example.

Consider Fig. 2.12(a) and (b), the original network and dual network respectiely.

There are two loops hence two dots (nodes) $A$ and $B$ are placed in the loops and current $i_A$ and $i_B$ are shown clockwise in the two loops $A$ and $B$. Since loop current $i_A$ is in the direction of polarity of the voltage source, therefore, in the dual network voltage source of $5V$ appears as current source of $5A$ and current directed towards the node $A$.

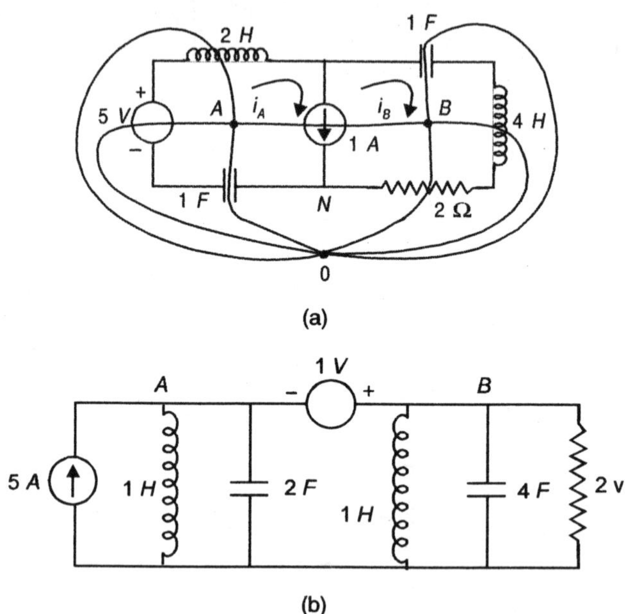

(a)

(b)

**Fig. 2.12**  Source orientation (a) Original graph (b) Dual of (a).

Now current source of $1A$ is a common branch to loops $A$ and $B$ and hence it is going to be connected between nodes $A$ and $B$ of the dual network. Current source in the original network is directed towards node $N$ which could be considered a dot in the dual network and when we traverse the original network along the nodes $A$, $B$, $O$ clockwise, the node $N$ is enclosed and since the current source is directed towards $N$, the polarity of the voltage source in the dual network will be clockwise from node $A$ to $B$ and shall be the one as shown in the dual network. The passive elements, of course, have been taken care of according to the procedure outlined earlier.

## 2.7  THE CONCEPT OF SUPER MESH (LOOP) AND SUPER NODES

If a given network is to be analysed based on loop (mesh) method, there is no problem as long as the network has only voltage sources. However, if the network has one or more current sources in the network, KVL equations as such cannot be applied to the loops containing current sources as voltage across the current source is not known.

Similarly, if the network is to be analysed based on nodal method and if between two nodes, there is a voltage source, again nodal analysis as such is not possible as the current through the voltage source is not known.

In order to take into account these eventualities super mesh (loop) and super node methods which are explained below are used respectively.

*Super Mesh Method*
Consider the network shown in Fig. 2.13.

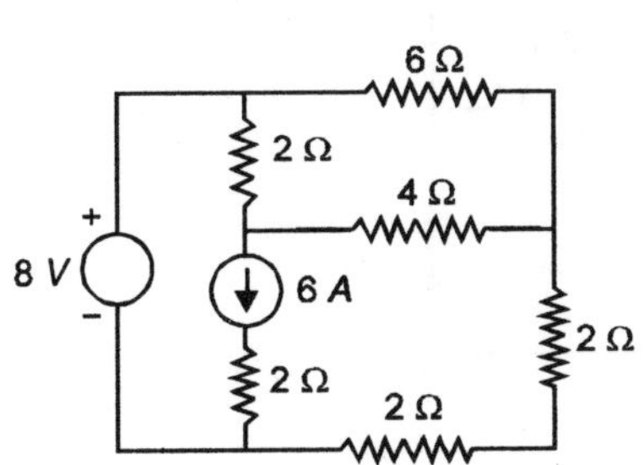 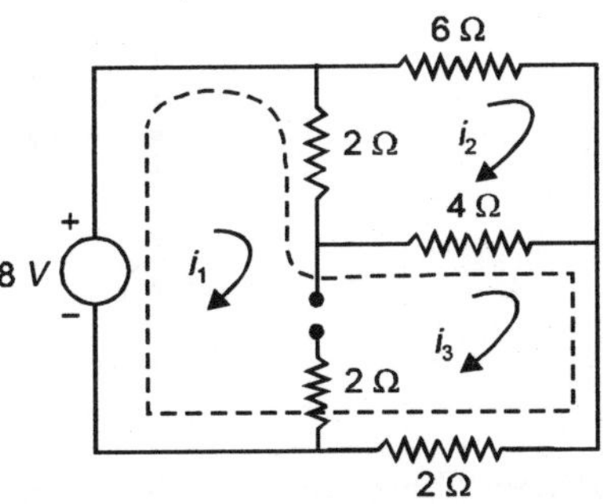

**Fig. 2.13**  Super mesh analysis.　　　　　　　　**Fig. 2.14**

Here we open circuit the current source and create a new mesh called the super mesh as shown in Fig. 2.14.

Here the super mesh consists of the interior of meshes 1 and 3. So there are virtually two meshes now. Writing KVL equation for the two loops (meshes) we have

Super mesh (loop)

$$2\,(i_1 - i_2) + 4\,(i_3 - i_2) + 2\,i_3 = 8$$
$$2i_1 - 6i_2 + 6i_3 = 8$$
$$i_1 - 3i_2 + 3i_3 = 4$$

Mesh 2

$$4\,(i_2 - i_3) + 2\,(i_2 - i_1) + 6\,i_2 = 0$$

or

$$12\,i_2 - 4\,i_3 - 2i_1 = 0$$

or

$$-\,i_1 - 2\,i_3 + 6\,i_2 = 0$$

The current source branch is common to loop 1 and 3 and hence

$$i_1 - i_3 = 6$$

The solution of these equations give

$$i_1 = 4A\ \ i_2 = 1 \text{ and } i_3 = 1A$$

Next we consider the solution of a given network by nodal analysis when there is a voltage source between two nodes. As mentioned earlier such a node is termed as super node. We generate a super node by short circuiting the two nodes. This node is known as super node. In general if $n$ is the number of nodes in a network we need $(n - 1)$ number of equations to solve the network as we take one of the nodes as reference. Under the super node condition if there is only one super node in the network, we will be able to write $(n - 2)$ equations for $(n - 1)$ number of variables using KCL equation. One more equation is required to solve the network. This equation is obtained from the fact that the difference of the nodal voltage across which the voltage source is connected, is the magnitude of the voltage source. Hence we have $(n - 1)$ equations and we can solve for the network. This is illustrated with the following examples:

**Examples 2.11:**  Solve the given network

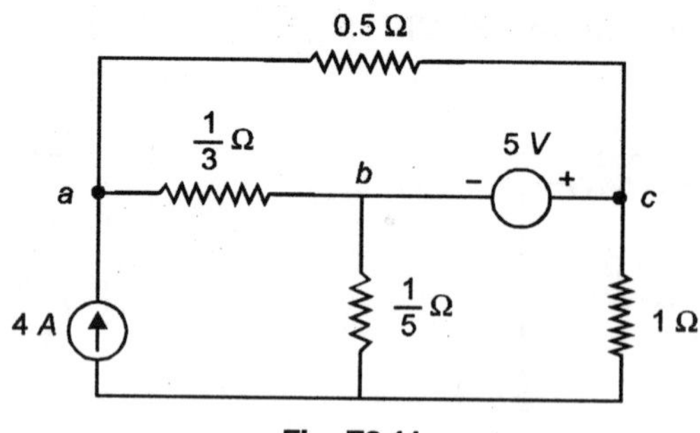

**Fig. E2.11**

**Solution:** There are 4 nodes in all. By shorting $b$ and $c$, we reduce the nodes to 3 and we take node $O$ as the reference. Writing *KCL* at Super node $bc$. There are four branches terminated at the super node

$$3 (V_b - V_a) + 2 (V_c - V_a) + V_c + 5 V_b = 0$$

or

$$8V_b - 5V_a + 3 V_c = 0$$

At node $a$

$$2 (V_a - V_c) + 3 (V_a - V_b) = 4$$
$$5V_a - 3V_b - 2V_c = 4$$

or

$$5V_a = 4 + 3 V_b + 2 V_c$$

or

$$8V_b - 4 - 3V_b - 2V_c + 3V_c = 0$$

or

$$5V_b + V_c = 4 \quad \text{or} \quad V_c = 4 - 5 V_b$$

and the third equation from the node $b$ and $c$

$$V_c = V_b + 5 = 4 - 5V_b$$

or

$$6V_b = -1 \quad \text{or} \quad V_b = -1/6$$

Hence,

$$V_c = -1/6 + 5 = 29/6$$

Now

$$5V_a = 4 + 3 V_b + 2V_c$$
$$= 4 - 3/6 + 29/3$$
$$= 24 - 3 + 58/6 = 79/6$$
$$V_a = 2.633$$

Hence,

$$V_a = 2.633, \quad V_b = -0.1667, \quad V_c = 4.833 \text{ volt}$$

## PROBLEMS

**2.1.** Determine the resistance between the terminals *AB* of the network shown

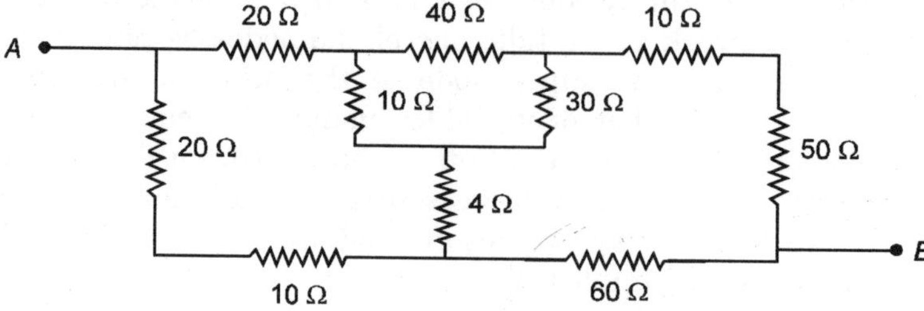

**Fig. P2.1**

**2.2.** Determine the resistance between the terminals $A$ and $B$ if the resistance of each branch is 6 $\Omega$ in the given circuit.

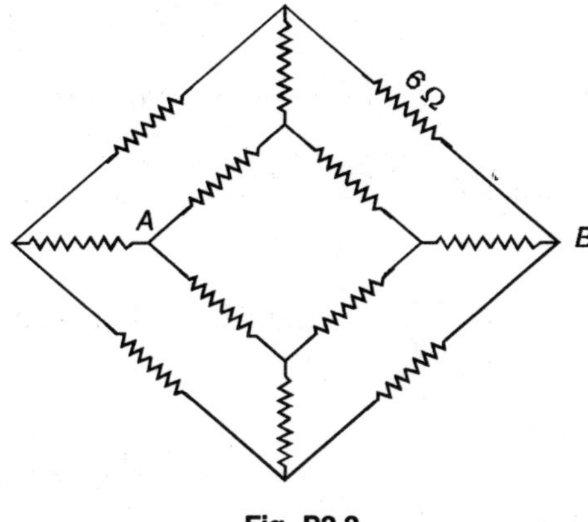


**Fig. P2.2**

**2.3.** Determine the nodal voltages of the circuit shown here if each resistance branch has $R = \dfrac{1}{2}\Omega$

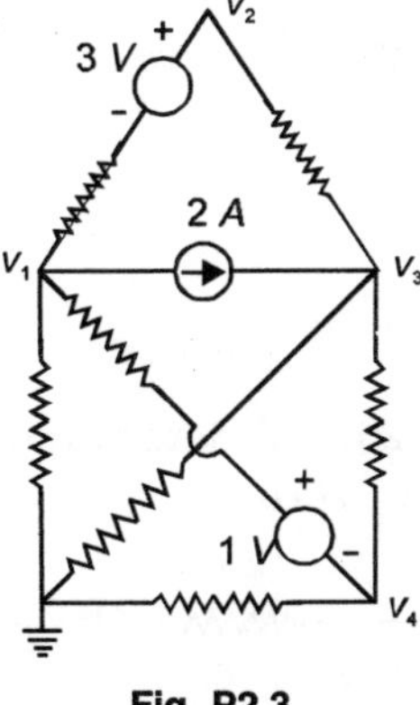


**Fig. P2.3**

**2.4.** Determine the current through the inductor in the Fig. P2.4 using nodal analysis.

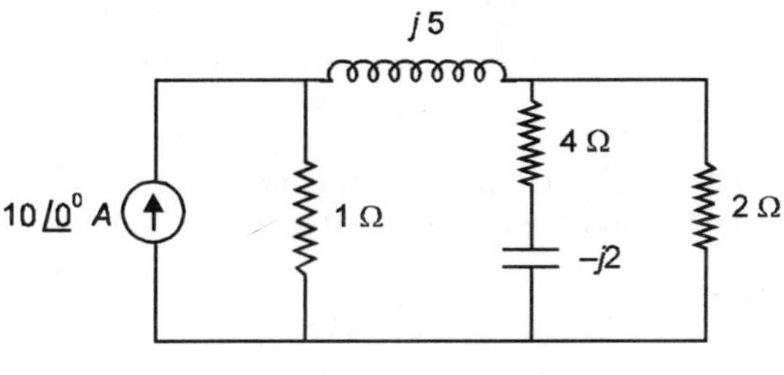


**Fig. P2.4**

**2.5.** Determine the current in each branch and the current drawn from the supply for the network shown when it is connected across 200 V, 50 Hz supply.

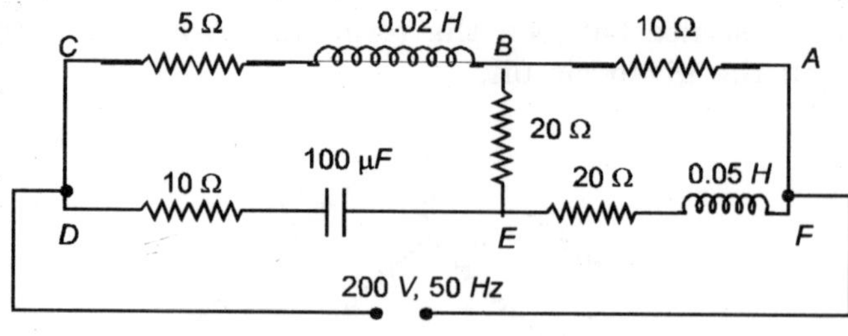

**Fig. P2.5**

**2.6.** In the networks shown: show that the impedance between terminals $I, I'$ is independent of frequencies and equals $R$ if $\dfrac{L}{C} = R^2$

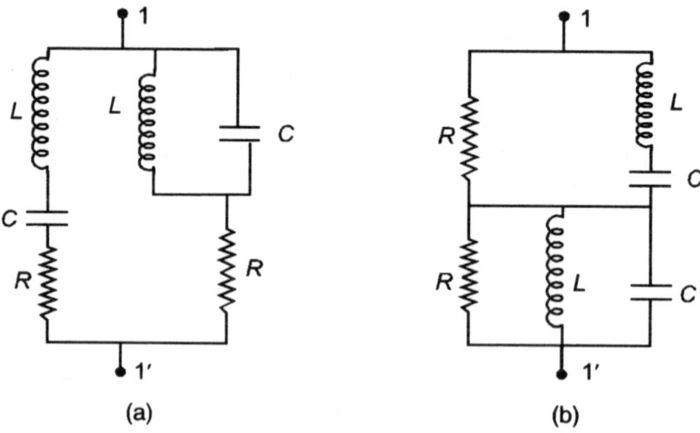

(a)  (b)

**Fig. P2.6**

**2.7.** In the network shown if $V_2 = 20 \angle 0^\circ$ determine $V_1$ for which the current through the branch containing source $V_2$ is zero.

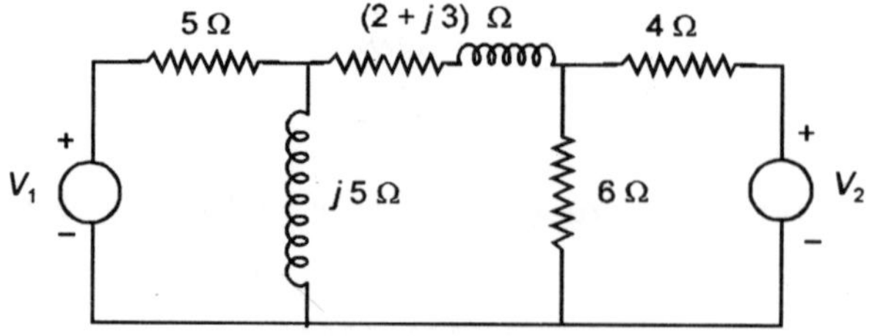

**Fig. P2.7**

**2.8.** Determine the voltage $V_{ab}$ by nodal voltage method for the circuit shown in Fig. P2.8 and verify the result by loop current method.

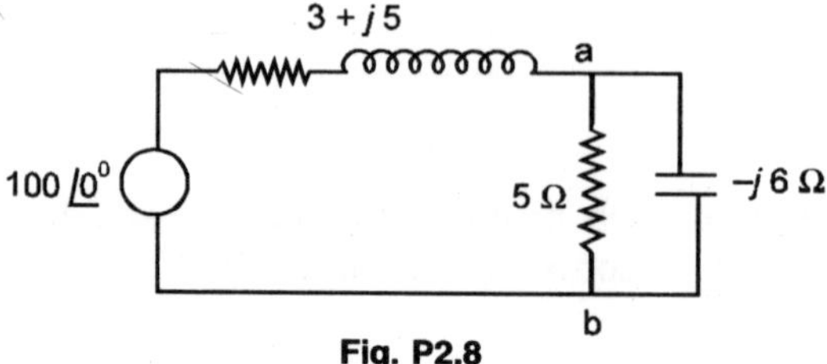

**Fig. P2.8**

**2.9.** Determine the current in 10 Ω resistor in the circuit shown.

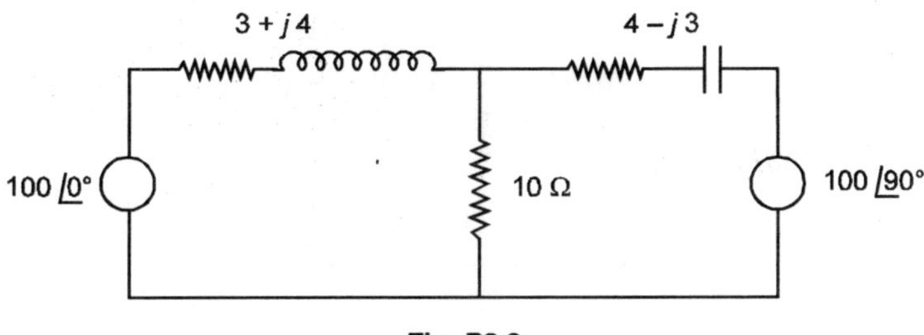



**Fig. P2.9**

**2.10.** Determine the currents in all the branches of the network shown in Fig. P2.10.

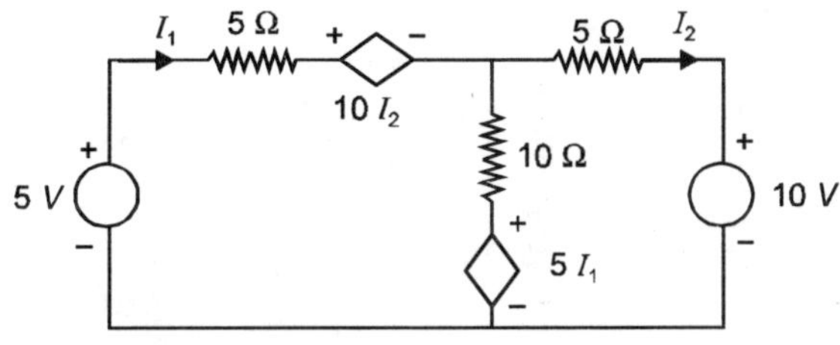



**Fig. P2.10**

**2.11.** Write nodal equation for the network shown in Fig. P2.11 by inspection.

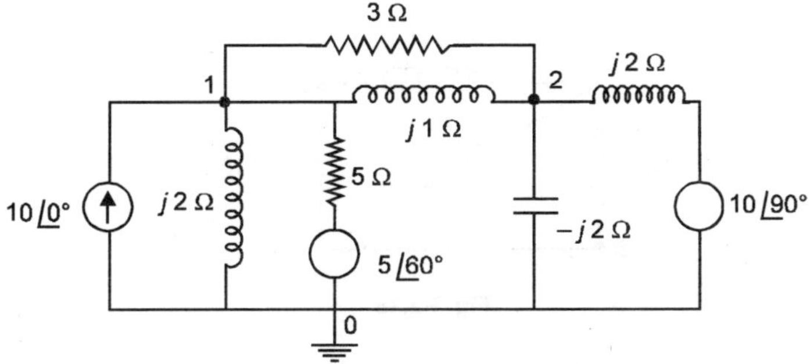



**Fig. P2.11**

**2.12.** Find the dual of the given networks.

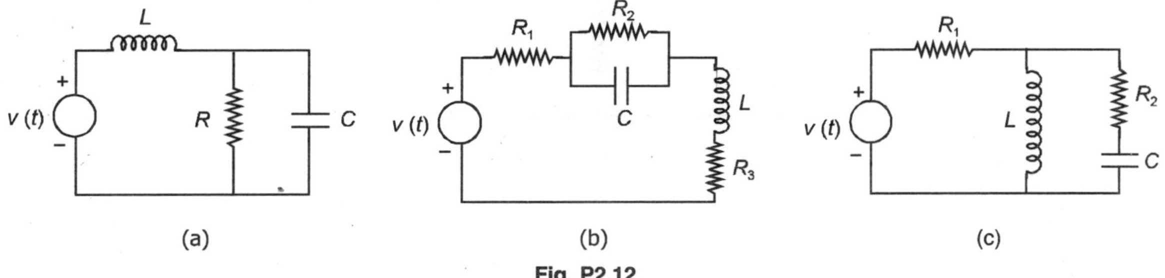



**Fig. P2.12**

**2.13.** In the network shown determine the voltage Vab.

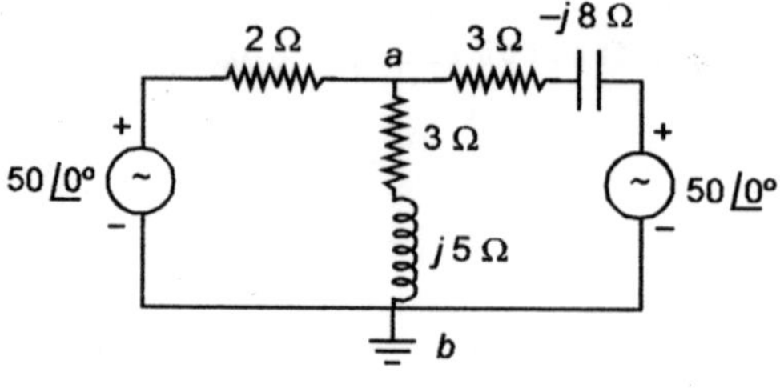

**Fig. P2.13**

**2.14.** Determine the current in each circuit and the current from the source.

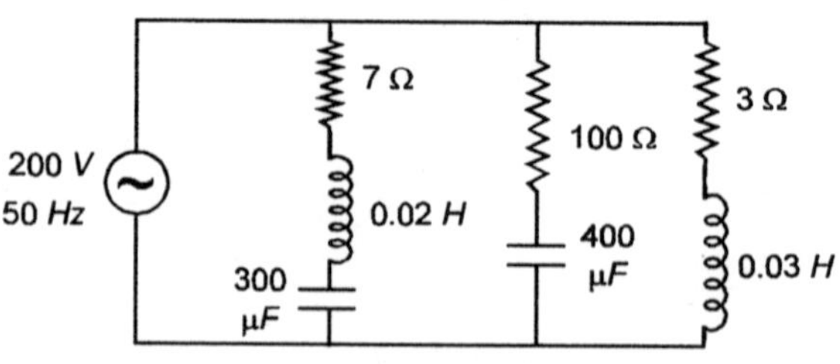

**Fig. P2.14**

**2.15.** Determine the impedance between the terminals *AB* of the network shown.

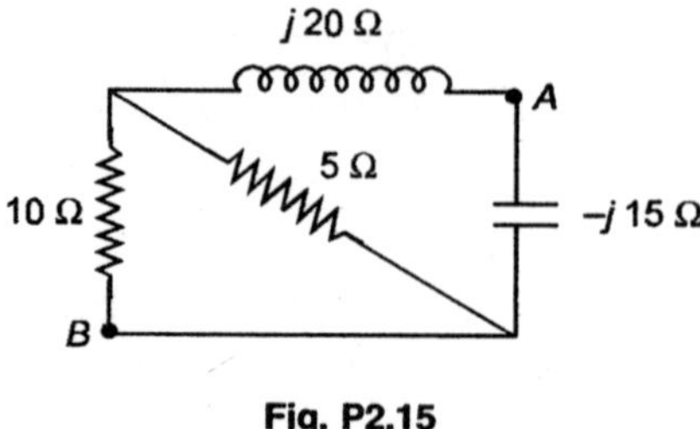

**Fig. P2.15**

**2.16.** Transform the circuit in the Fig. 2.16 into $\pi$ form

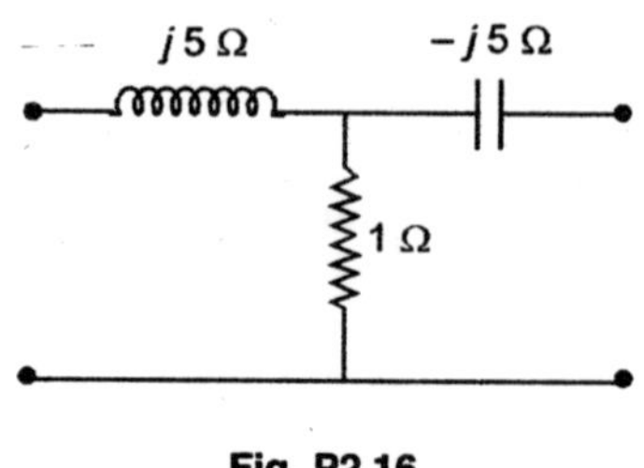

**Fig. P2.16**

**2.17.** For the circuit shown determine the equivalent impedance and the current *I*. Also calculate *Ir*.

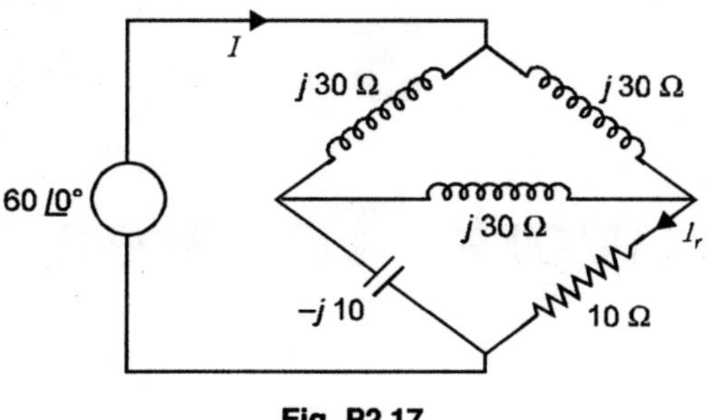


**Fig. P2.17**

# 3

# Network Theorems

## 3.1  INTRODUCTION

We have studied various methods of analysing a given network. Depending upon the configuration of a network, loop analysis or nodal analysis may be used. In case it is a large network and if we were interested in the analysis of a part of the network (e.g. current through a particular branch or voltage across a branch) these methods require lot of computational effort. Also these methods do not give better insight into the functioning of the network. To handle such situations more effectively, certain techniques known as "Network Theorems" are used which are the subject of this chapter. Various theorems studied here are Superposition, Maximum power transfer, Substitutions theorems, Thevenins theorem, Norton's theorem, Tellegan's theorem, Millman theorem and Reciprocity theorem. While discussing these theorems the applicability of these theorems to certain kinds of networks only, must be clearly understood. As discussed in chapter 1 the network element could be classified as linear, Lumped, finite, passive and bilateral usually abbreviated as LLFPB elements. These qualities will be mentioned under different theorems.

## 3.2  SUPERPOSITION  THEOREM

This theorem is applicable to linear networks. A network is said to be linear where the response is proportional to the excitation. If $X_1$ is the excitation and $Y_1$ the response of a given network related by $Y_1 = CX_1$ and for an excitation $X_2$ the response $Y_2$ is related by $Y_2 = CX_2$ then for excitation $(X_1 + X_2)$ the response will be $Y_1 + Y_2$ for a linear network. It is to be noted that if the excitation and response are related by the linear equation $Y_1 = CX_1 + d_1$ the network is not linear as for double the excitation; response will not be doubled. Thus a circuit with $R, L, C$, elements with constant values and initially relaxed (for $L\ i(o) = O$ and for $C\ V(o) = O$) is a linear circuit. However, the initial conditions could be considered as sources or excitation and the overall response of the circuit can be obtained.

Superposition theorem can be stated as follows: In a linear network, if more than one number of sources (excitation) are acting, the response in a certain branch can be obtained by superimposing algebraically the responses due to various sources taking one excitation at a time and considering the remaining voltage sources (ideal) short circuited or replaced by their equivalent internal impedances and current sources (ideal) open circuited or replaced by their equivalent admittances.

The theorem is further illustrated with a few examples.

**Example 3.1:** Determine the current through the voltage source and the voltage across the current source in the network shown here.

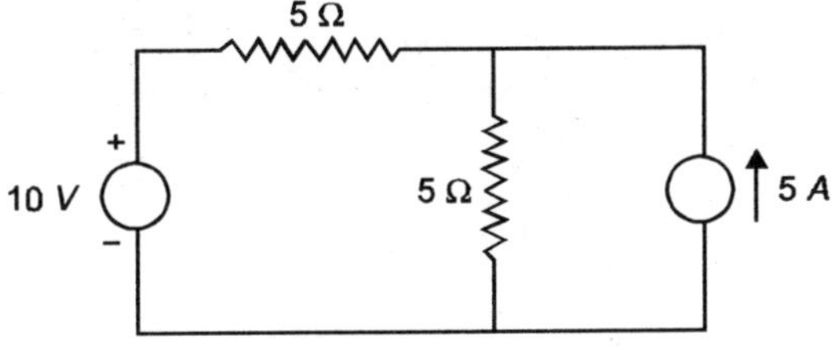

**Fig. E3.1**

Since we have two sources (excitations) one voltage and another current we have to look for two circuits where in one circuit, voltage source will be short circuited and in another, current source will be open circuited.

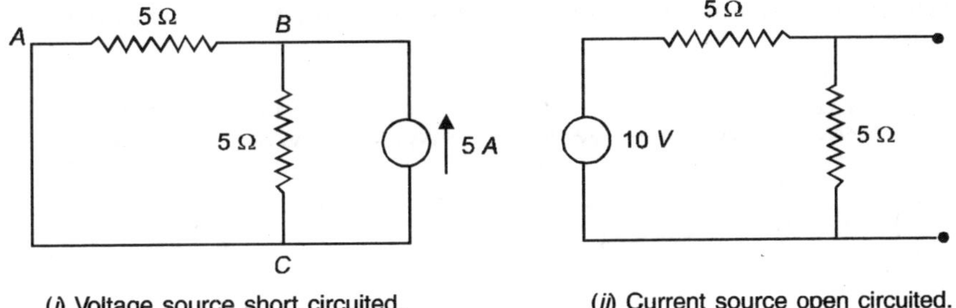

(*i*) Voltage source short circuited.　　　(*ii*) Current source open circuited.

**Fig. E3.1.1**

To find out current through the voltage source i.e. through branch *AB* we proceed as follows:

(*a*) (*i*) When current source alone is acting
Current through *AB* is

$$5 \times \frac{5}{10} = 2.5 \text{ A}$$

in the direction from *B* to *A*

(*ii*) When voltage source alone is acting in the circuit. Current through *AB* is
$$\frac{10}{10} = 1 \text{ Amp. from } A \text{ to } B.$$
Therefore the current through *AB* branch when both sources are simultaneously acting, is given by

$$
\begin{array}{ccc}
BA\rightarrow & AB\rightarrow & \\
2.5 & + \quad 1.0 & = \quad 2.5 - 1.0 \\
& & = \quad 1.5 \text{ Amp.}
\end{array}
$$

The direction of the current is from *B* to *A*.

(*b*) To find out voltage across the current source we find current through the branch *BC*.

Current through branch *BC* due to current source alone is 2.5 *A* from *B* to *C*. Due to voltage source it is 1 Amp. again from *B* to *C*. Therefore total current is 2.5 + 1 = 3.5 Amp. Therefore voltage across the current source is 3.5 × 5 = 17.5 Volts.

**Example 3.2:** Using superposition principle determine the current through 2 Ω resistor connected between *A* and *B* shown in the circuit here.

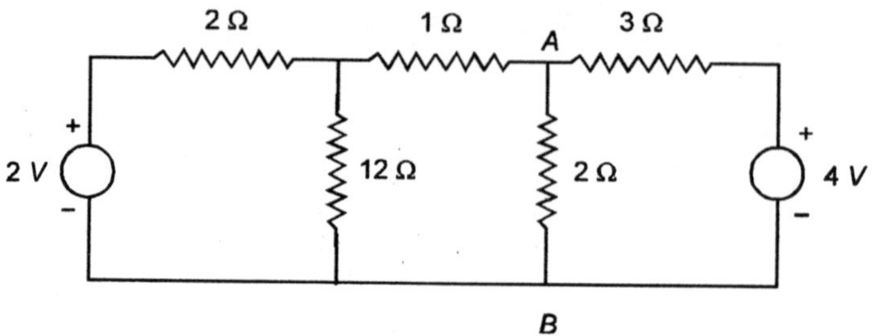

**Fig. E3.2**

**Solution:** Since there are two sources in the circuit, there will be two circuits for analysis purposes where in one source will be present and the other short circuited being the voltage source.

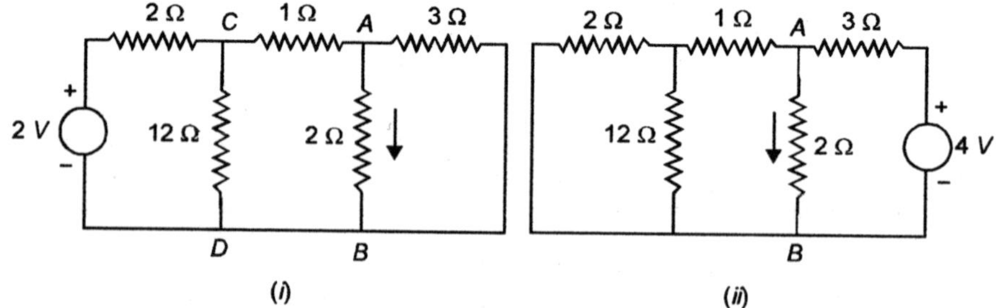

**Fig. E3.2.1**

The resistance between *CD* excluding 12 Ω resistance

$$1 + \frac{6}{5} = 1 + 1.2 = 2.2 \ \Omega, \text{ including } 12 \ \Omega \text{ resistance}$$

$$\frac{1}{R} = \frac{1}{12} + \frac{10}{22} = \frac{(11 + 60)}{132} = \frac{71}{132}$$

Total resistance $2 + \dfrac{132}{71} = \dfrac{(142 + 132)}{71} = \dfrac{274}{71}$

Therefore, current through the source is $\dfrac{2 \times 71}{274}$

The current through 1 Ω resistor is

$$\frac{2 \times 71}{274} \times \frac{12 \times 10}{142} = \frac{120}{274}$$

Therefore current through 2 Ω resistor

$$\frac{120}{274} \times \frac{3}{5} = \frac{72}{274} \ \text{A}$$

(*ii*)
$$1 + \frac{24}{14} = \frac{38}{14}\,\Omega$$

$$\frac{1}{R} = \frac{1}{2} + \frac{14}{38} = \frac{(19 + 14)}{38} = \frac{33}{38}$$

or
$$R = \frac{38}{33}\,\Omega$$

Total resistance $3 + \dfrac{38}{33} = \dfrac{137}{33}\,\Omega$

Current through the source $\dfrac{4 \times 33}{137}\,$A

Current through $2\,\Omega$ resistor

$$\frac{4 \times 33}{137} \times \frac{38}{14} \times \frac{14}{66} = \frac{76}{137}$$

Therefore, total current is

$$\frac{72}{274} + \frac{76}{137} = \frac{224}{274} = 0.8\,\text{A}$$

**Example 3.3:** Determine the current through the branch *AB* of the network shown using principle of superposition

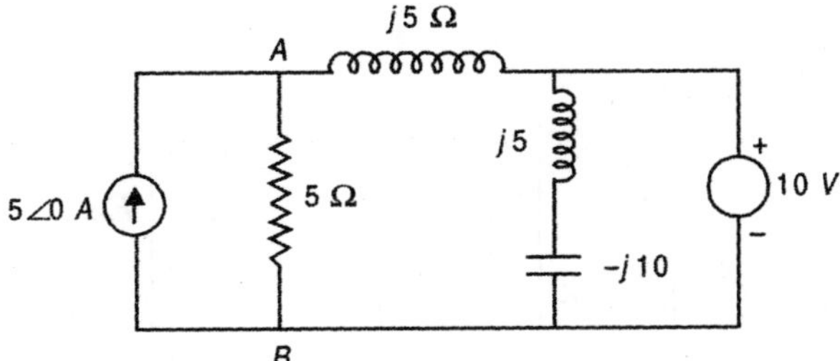

**Fig. E3.3**

The two required networks are:

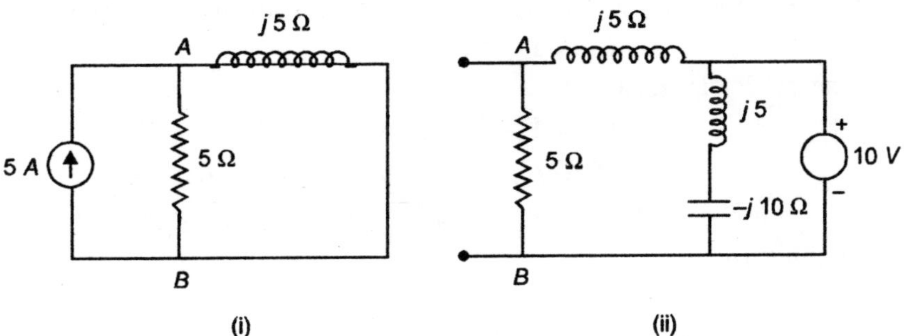

(i)                              (ii)

**Fig. E3.3.1**

($i$) Current through $AB$ is

$$\frac{5 \times j5}{5 + j5} = \frac{j5(1 - j1)}{(1 + j1)(1 - j1)}$$

$$= \frac{5 + j5}{1 + 1} = 2.5 + j2.5 \text{ A}$$

($ii$) Current through $AB$

$$\frac{10}{5 + j5} = \frac{2}{1 + j1} = 1 - j1$$

Considering both the sources acting simultaneously, the total current

$$2.5 + j2.5 + 1 - j1.0 = 3.5 + j1.5 \text{ Amp.}$$

The applicability of superposition principle for large networks can be demonstrated using loop and nodal analysis.

Consider a linear network with $l$ no. of loops having voltage sources in each loop (In practice if source in a particular loop is absent $E_i = O$).

If $Z$ is the loop impedance matrix, $E$ the column excitation matrix and $I$ the loop current column matrix. These are related as

$$[E] = [Z]\,[I] \qquad \qquad ...(3.1)$$

or
$$I = [Z]^{-1}\,[E] = [Y]\,[E] \qquad \qquad ...(3.2)$$

Where $Y$ is loop admittance matrix. Writing the above equations under the summation sign.

$$I_i = \sum_{j=1}^{l} Y_{ij}\, E_j \quad i = 1, 2, \,..... \, l \qquad \qquad ...(3.3)$$

In expanded matrix from

$$\begin{bmatrix} I_1 \\ I_2 \\ . \\ . \\ . \\ I_l \end{bmatrix} = \begin{bmatrix} Y_{11} & Y_{12} & . & . & Y_{1l} \\ Y_{21} & Y_{22} & . & . & Y_{2l} \\ . \\ . \\ . \\ Y_{l1} & Y_{l2} & . & . & Y_{ll} \end{bmatrix} \begin{bmatrix} E_1 \\ E_2 \\ . \\ . \\ . \\ E_l \end{bmatrix} \qquad \qquad ...(3.4)$$

To find out current $I_1$, let us find out response to various sources $E_1$ through $E_l$ taken one at a time. Say $E_j = O$ for all $j \neq 1$ i.e $E_1$ is present and all other sources short circuited.

$$I_{11} = Y_{11}\, E_1$$

When all $E_s$ except $E_2$ is zero

Current
$$I_{12} = Y_{12}\, E_2 \text{ so on}$$

Therefore,
$$I_1 = I_{11} + I_{12} + I_{13} + ...... + I_{1l}$$

$$= Y_{11}\, E_1 + Y_{12}\, E_2 + ..... + Y_{1l}\, E_l$$

which is equivalent to

$$I_1 = \sum_{j=1}^{l} Y_{1j}\, E_j$$

Which is same as equation (3.3) for $I_i$ above. Similarly using nodal analysis superposition principle can be demonstrated. The only difference is that instead of loop admittance matrix we will have nodal admittance matrix and nodal voltage rather than loop voltages i.e. the basic equation is

$$I = YV$$

$$I_i = \sum_{j=1}^{n} Y_{ij} V_j \quad i = 1, 2, \dots n \qquad \dots(3.5)$$

where $n$ is the number of nodes in the network.

The properties of additivity and homogeneity are fundamental to superposition principle. Mathematically these properties are stated as follows:

Additivity if $\qquad\qquad Y_1 = cx_1 \quad$ and $\quad Y_2 = cx_2$

Then $\qquad\qquad Y_1 + Y_2 = c(x_1 + x_2)$

Homogeneity if $\qquad\qquad Y_1 = cx_1, \quad$ for $\quad mcx_1$

$$Y = mY_1 = mcx_1$$

A few more examples will make things clearer.

Consider the system of Fig. 3.2 where the capacitor is charged to a voltage $V_c(0)$ before the switch is closed and say it is equal to the source voltage.

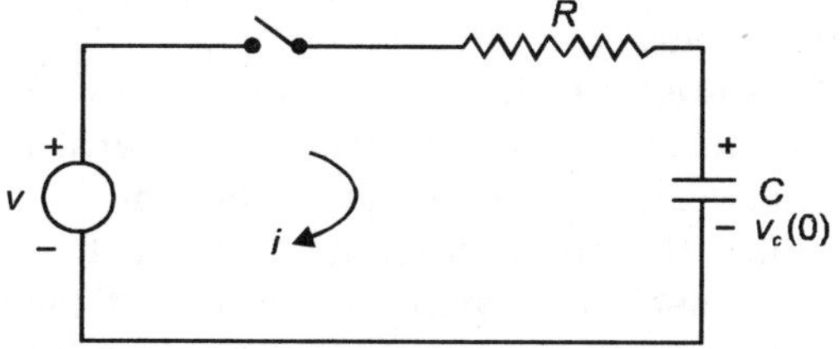

**Fig. 3.2**

The loop equation is given as

$$Ri + \frac{1}{C}\int idt = V$$

Differentiating equation w.r.t. $t$ we have

$$R\frac{di}{dt} + \frac{i}{C} = 0$$

or $\qquad\qquad R\dfrac{di}{dt} = -\dfrac{i}{C} \quad$ or $\quad \dfrac{di}{i} = -\dfrac{dt}{RC}$

$$\log i = -\frac{t}{RC} + \log k$$

At $\qquad\qquad t = 0, \quad i(0) = \dfrac{v - v_0}{R}$

Hence $\qquad\qquad K = \dfrac{v - v_0}{R}$

and $\qquad\qquad \dfrac{i \cdot R}{v - v_0} = e^{-t/RC}$

or
$$i(t) = \frac{v - v_0}{R} e^{-t/RC} \qquad \qquad ...(3.6)$$

However, if
$$v = v_0 \text{ then}$$
$$i(t) = 0$$

If
$$v = 2v_0, \text{ then from equation (3.6) we have}$$

$$i(t) = \frac{2v_0 - v_0}{R} e^{-t/RC} = \frac{v_0}{R} e^{-t/RC} \qquad \qquad ...(3.7)$$

Suppose both the sources are multiplied by $K$ then the current is found to be

$$i(t) = \frac{2Kv_0 - Kv_0}{R} e^{-t/RC}$$

$$= K \frac{v_0}{R} e^{-t/RC} \qquad \qquad ...(3.8)$$

i.e. the current becomes $K$ times its previous value which shows the homogeneity aspect of superposition principle.

## 3.3  RECIPROCITY THEOREM

Reciprocity is an important property of linear, passive, time invariant bilateral networks which implies that the input and output can be interchanged without altering the response of the network to a certain input.

Reciprocity theorem is stated as "If any source of emf $E$, located at one point in a network consisting of linear, passive, time invariant, bilateral elements produces a current $I$ at a second point in the network, the same source of emf $E$ acting at the second point of the same network will produce the same current $I$ at the first point".

This means suppose in a network having $l$ number of loops, we consider loop $m$ and $n$. First we connect a voltage source in loop $m$, $E = E_m$ and find the current in loop $n$, Let this be $I_n$, then
$$I_n = Y_{nm} E_m \qquad \qquad ...(3.9)$$

Now we remove the voltage source from loop $m$ and insert it in loop $n$ and measure the current in loop $m$. Let it be $I_m$

Then
$$I_m = Y_{mn} E_n \qquad \qquad ...(3.10)$$

Since $Y_{nm}$ and $Y_{mn}$ are the loop admittance matrices which are symmetrical hence
$$Y_{mn} = Y_{nm} \qquad \qquad ...(3.11)$$

and since in our assumption
$$E_m = E_n = E$$

Therefore,
$$I_m = I_n$$

i.e. the excitation (source) and response can be interchanged in a linear passive bilateral network.

The reciprocity theorem can also be proved by considering current sources and nodal voltage responses. In this case also since the source and response are connected through nodal admittance matrix which is a symmetric matrix hence the reciprocity theorem can be proved.

Reciprocity theorem is applicable to resistors, capacitors and inductors with and without coupling and transformers. However, both independent and dependent sources are not included. Also we assume an initially relaxed system.

We illustrate now the application of reciprocity theorem with the help of an example.

**Example 3.4:** Verify reciprocity theorem for the network shown

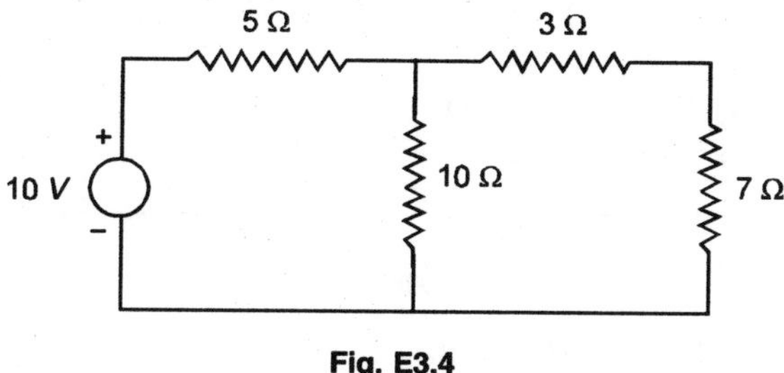

**Fig. E3.4**

**Solution:** Let the voltage source be connected in loop 1, current in loop 2 is given as $I_2$

Now
$$I_1 = \frac{10}{10} = 1 \text{ A}$$

Hence
$$I_2 = 1 \times \frac{10}{20} = 0.5 \text{ A}$$

If we now connect the voltage source in loop 2, the current in loop 1 must come out to be 0.5 A if the network obeys reciprocity property. The circuit is as follows:

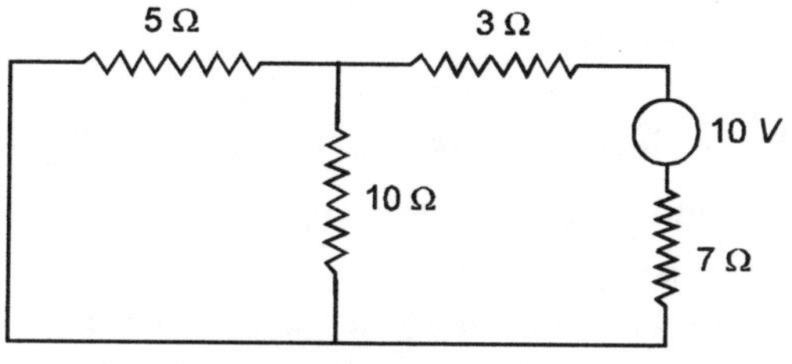

**Fig. E3.4.1**

Current
$$I_1 = \frac{10}{\left(10 + \dfrac{50}{15}\right)} \times \frac{10}{15} = \frac{10 \times 16}{200} \times \frac{15}{15} = 0.5 \text{ A}$$

Hence reciprocity property is proved.

## 3.4 SUBSTITUTION THEOREM

This theorem helps in replacing an impedance branch in a network by another branch with different circuit component without disturbing the voltage current relations in the whole network. This theorem states that:

"Any branch in a network can be substituted by a different branch without disturbing the voltages and currents in the entire network provided the new branch has the same set of terminal voltage and current as the original network".

Consider Fig. 3.3 for illustration. Consider branch $a\,b$ which has voltage $V_{ab} = 1$ volt and current of 1 A. If we replace branch $ab$ by a voltage source of 1 V with proper polarity we find that the current comes out to be 1 A.

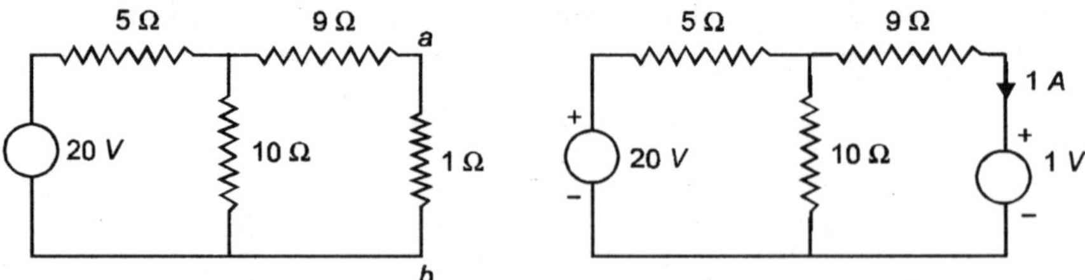

**Fig. 3.3**

$$15i_1 - 10i_2 = 20$$

or $$3i_1 - 2i_2 = 4 \qquad ...(1)$$

and $$19i_2 - 10i_1 = -1$$

or $$10i_1 - 19i_2 = 1 \qquad ... (2)$$

Solution of (1) and (2) gives $i_1 = 2$ A

and $$i_2 = 1 \text{ A}$$

This means the branch can be replaced by a voltage source of $I_2R$ with suitable polarity based on the current flow in the original network.

In general we can represent the voltage across the branch *ab* as $V_{ab}$ and is given by

$$V_{ab} = Z_{ab}\, I_{ab} + E \qquad ...(3.12)$$

This branch could be replaced by any other branch so that the branch voltage is given by

$$V_{ab} = Z'_{ab}\, I_{ab} + E' \qquad ...(3.13)$$

when $Z'_{ab}$ and $E'$ are so chosen that $V_{ab}$ and $I_{ab}$ are not disturbed. In the above illustration we have chosen $E' = 1$ volt and $Z'_{ab} = 0$. We could choose $E' = \dfrac{1}{2}$ volt in series with $Z'_{ab} = 0.5\ \Omega$. This substitution is also possible.

We could also replace the branch *ab* by a current source of 1 A in this case directed in the direction of original network. Some of the alternatives for substitution for this particular case are given here.

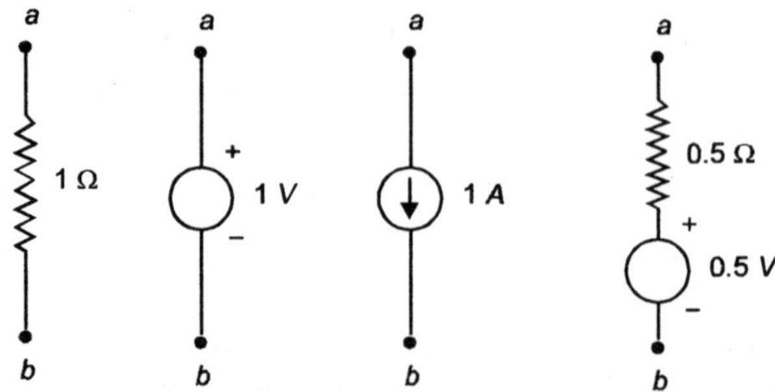

**Fig. 3.4**   Various substitutions for branch *ab*.

The following points are important in regard to substitution theorem:
1. It is a general theorem applicable for any arbitrary network.
2. The modified network must have a unique solution. For linear network, of course, it is always true. However, for non-linear network due consideration must be given.
3. This theorem is quite useful for analysis in networks having one non-linear element.

**Proof:** Say, a network has $b$ number of branches. The complete solution requires (voltages across and current through each branch) $2b$ number of equations. If one branch is substituted, $(b-1)$ branches have $(2b-1)$ equations which remain unaltered. The equation of the substituted branch is altered from (3.12) to (3.13). However, the voltage and current of the substituted branch remain unaltered, it implies that the set of $2b$ simultaneous equations will still be satisfied with the same voltages and currents as before. Hence the substitution theorem is proved.

**Example 3.5:** In the illustration if we want to replace the resistance across branch $ab$ ($1\,\Omega$) by a voltage source in series with a reactance of $j0.5\,\Omega$. The calculations are as follows:

$$V_{ab} = Z'_{ab}\, I_{ab} + E'$$
$$1 = j0.5 \times 1 + E'$$
$$E' = 1 - j0.5 = 1.1181\ \angle{-26.6^\circ}$$

Therefore, the branch $ab$ is shown here

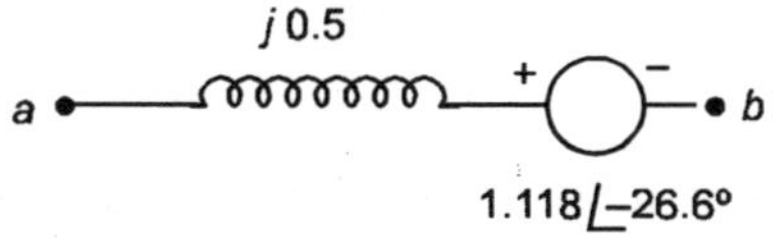

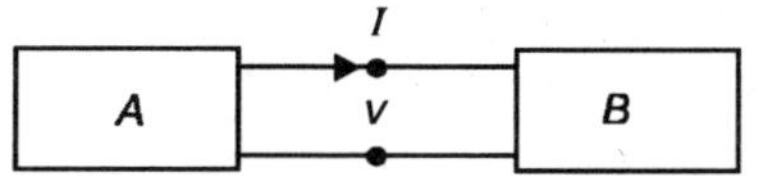


**Fig. E3.5**

## 3.5 THEVENIN'S THEOREM

This theorem is most useful as many a times we may be interested in the current/voltage in a part of the otherwise a much more complex network. We may not be interested in the current/voltage through all the branches of the network e.g. determining fault current in a particular branch of a complex power system or finding out the value of load resistance for an electronic circuit which will result in the maximum power transferred to this load.

Thevenin's theorem can be explained with the help of two networks $A$ and $B$ connected as shown in Fig. 3.5.



**Fig. 3.5**  Two networks $A$ and $B$ connected.

The original network can be considered as a combination and connection of networks $A$ and $B$. Network $A$ contains linear elements which may have initial conditions. It may also have an independent and dependent sources. However, there is no mutual magnetic coupling between the networks $A$ and $B$ nor does a controlled source in $A$ couple to $B$.

For network $B$, the elements need not necessarily be linear. These could be non-linear or time varying or both. However, in our analysis we will assume that the network elements are linear and many a times this will be a single branch (load) of the network.

Network $A$ is to be replaced by an equivalent network under the condition that the current $I$ and voltage $V$ identified in the Fig. remain invariant when the replacement is made. This amounts to replacing network $A$ by an equivalent Thevenin's voltage source and an equivalent Thevenin's series impedance (resistance) as shown in Fig. 3.5($a$).

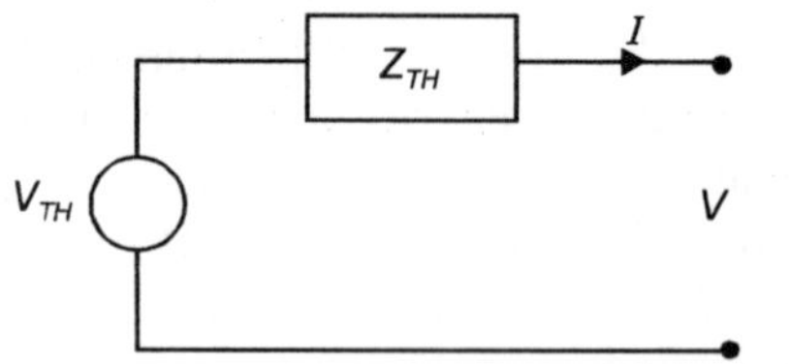

**Fig. 3.5(a)** Thevenin equivalent of Fig. 3.5.

The following procedure is followed:

To determine $V_{TH}$ we remove the network $B$ (could be a load resistor or the element through which current is required) and solve the network $A$ to find out voltage across the open circuited terminal. This voltage is the Thevenin's equivalent voltage source $V_{TH}$. Also $Z_{eq}$ is the impedance as seen in the network between the two open circuited terminals replacing the voltage sources by their equivalent series impedance and the current sources by their equivalent shunt admittances. The impedance as seen is known as Thevenin's equivalent series impedance. While finding Thevenin's equivalent, dependent sources continue to operate in the network.

If $Z_L$ is the load impedance and $Z_{eq}$ the Thevenin's equivalent series impedance (Fig. 3.6) the current through the load impedance is given by

$$I = \frac{V_{TH}}{Z_{eq} + Z_L} \qquad \qquad ...(3.14)$$

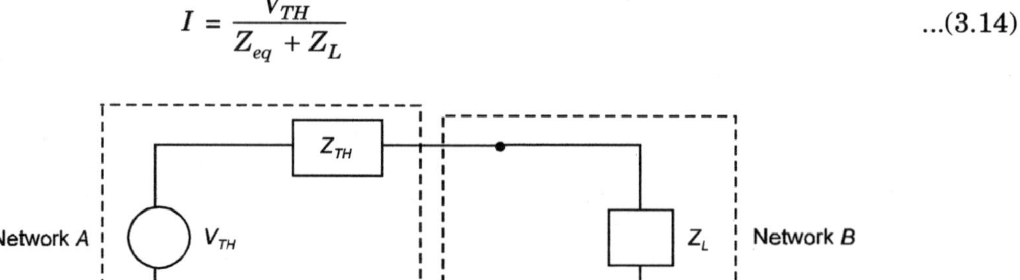

**Fig. 3.6** Thevenin's equivalent with load.

A few examples will further illustrate the application of Thevenin's theorem.

**Example 3.6:** Determine the current through $2\,\Omega$ resistor connected between $A$ and $B$ in the circuit shown using Thevenin theorem.

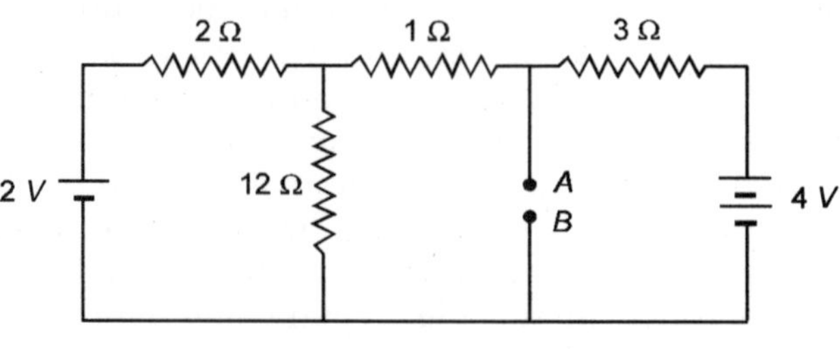

**Fig. E3.6**

**Solution:** To obtain $V_{TH}$ refer Fig. below:

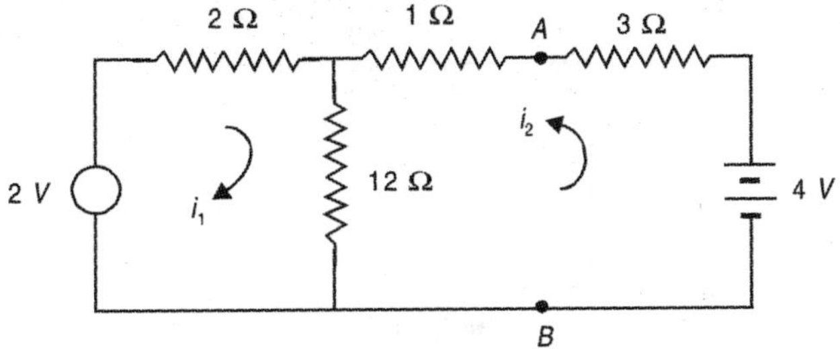

**Fig. E3.6.1**

$$2i_1 + 12\,(i_1 + i_2) = 2, \quad 4i_2 + 12i_1 + 12\,i_2 = 4$$
$$14i_1 + 12\,i_2 = 2 \quad \text{or} \quad 7i_1 + 6i_2 = 1$$
$$12i_1 + 16i_2 = 4 \quad \text{or} \quad 3i_1 + 4i_2 = 1 \qquad \text{...(1)}$$

or
$$7i_1 + 6i_2 = 3i_1 + 4i_2$$

or
$$4i_1 = -2i_2$$

or
$$i_1 = \frac{-i_2}{2}$$

Substituting in (1) we have

$$\frac{-3i_2}{2} + 4i_2 = 1$$

or
$$\frac{5i_2}{2} = 1$$

or
$$i_2 = \frac{2}{5}\ \text{Amp.}$$

Therefore,
$$V_{TH} = V_{AB} = 4 - \frac{2}{5} \times 3 = 2.8\ \text{volt}$$

The equivalent series resistance

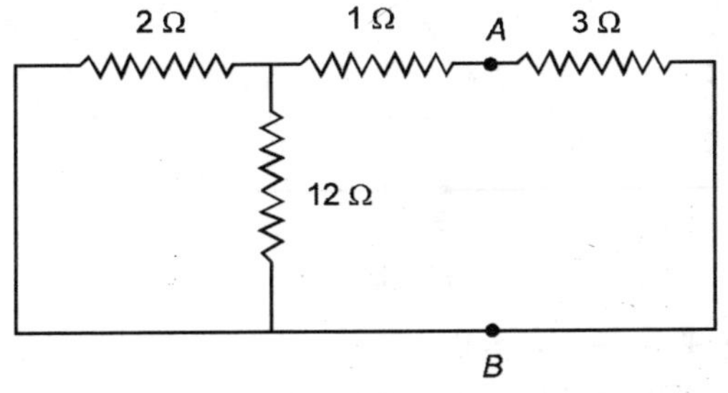

**Fig. E3.6.2**

$$\left(1 + \frac{24}{14}\right) \| 3 \quad \text{or} \quad \frac{\dfrac{38}{14} \times 3}{3 + \dfrac{38}{14}} = \frac{57}{40}\ \Omega$$

Thevenin's equivalent circuit is, therefore, as follows:

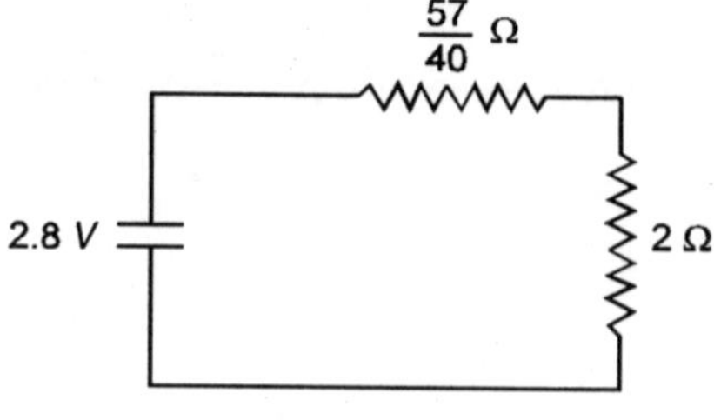

**Fig. E3.6.3**

Therefore, current through $2\,\Omega$ resistor

$$\frac{2.8}{\dfrac{57}{40} + 2} = 0.82 \text{ A}$$

**Example 3.7:** Determine the current through the branch $AB$ of the network shown here using Thevenins equivalent.

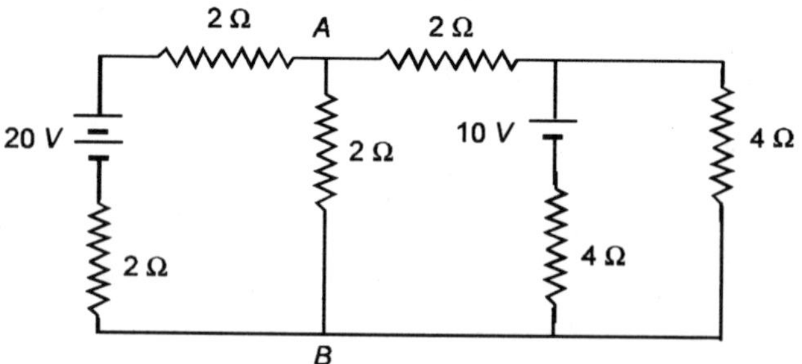

**Fig. E3.7**

**Solution:** To obtain $V_{TH}$ the equivalent circuit is given as

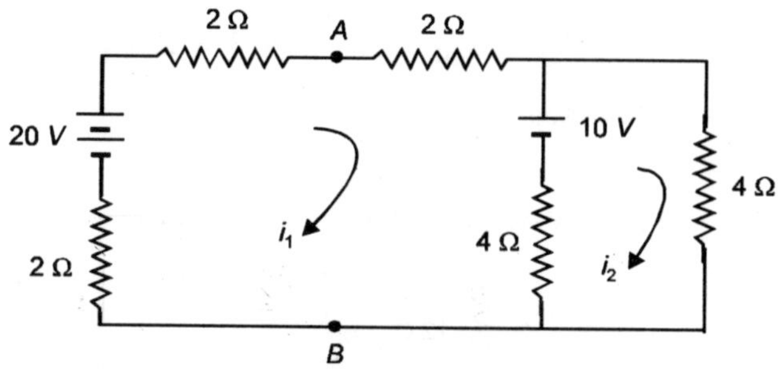

**Fig. E3.7.1**

$$2i_1 - 20 + 4i_1 + 10 + 4\,(i_1 - i_2) = 0$$

$$10i_1 - 4i_2 = 10 \quad \text{or} \quad 5i_1 - 2i_2 = 5$$

$$4\,(i_2 - i_1) - 10 + 4i_2 = 0$$

$$-4i_1 + 8i_2 = 10$$

$$-2i_1 + 4i_2 = 5$$

or
$$-2i_1 + 4i_2 = 5i_1 - 2i_2$$

or
$$6i_2 = 7i_1, \quad -2i_1 + \frac{14}{3}i_1 = 5$$

or
$$i_2 = \frac{7}{6}i_1, \quad \frac{8}{3}i_1 = 5$$

$$V_{AB} + \frac{15}{2} = 20, \quad i_1 = \frac{15}{8}$$

$$V_{AB} = 12.5 \text{ V}$$

Therefore
$$V_{AB} = V_{TH} = 12.5 \text{ volts}$$

To calculate $R_{TH}$ or $R_{eq}$

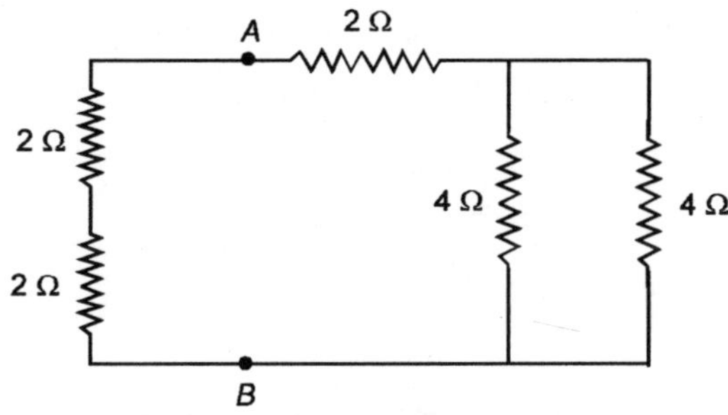

**Fig. E3.7.2**

Therefore current through the $AB$ branch

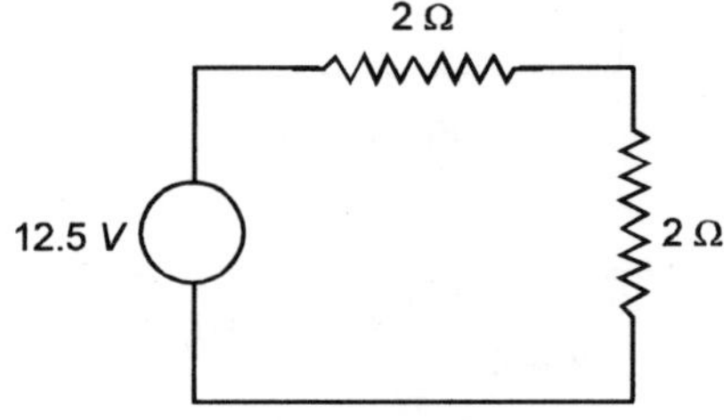

**Fig. E3.7.3**

$$I_{AB} = \frac{12.5}{4} = 3.125 \text{ A.}$$

**Example 3.8:** Determine the current $i_2(t)$ through $10\,\Omega$ resistor of the given network using Thevenins theorem if the network is initially relaxed.

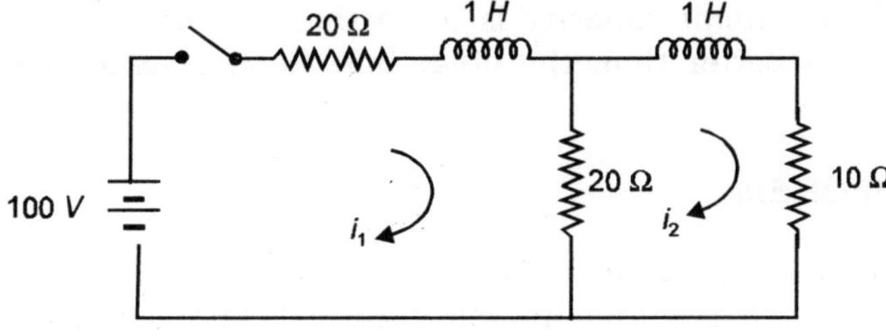

**Fig. E3.8**

**Solution:** Figure below shows the transform impedance with the voltage source also transformed.

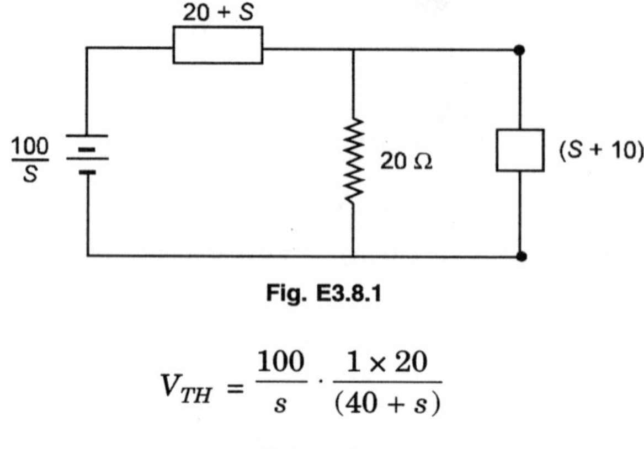

**Fig. E3.8.1**

$$V_{TH} = \frac{100}{s} \cdot \frac{1 \times 20}{(40 + s)}$$

and

$$R_{TH} = \frac{(20 + s)20}{40 + s}$$

Thevenin's equivalent network

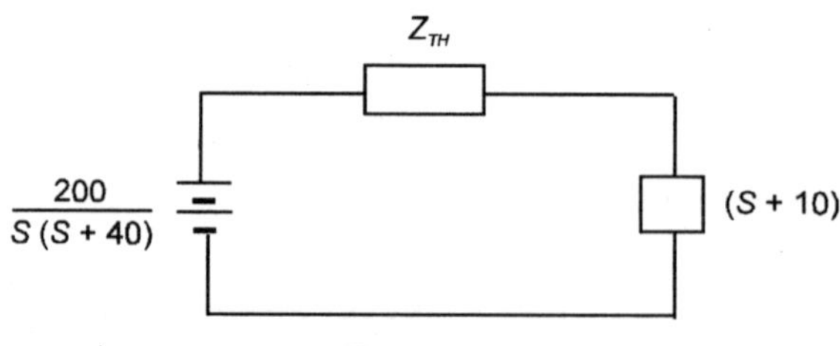

**Fig. E3.8.2**

$$I_2(s) = \frac{2000}{s(s + 40)} \cdot \frac{1}{\dfrac{20\,(s + 20)}{s + 40} + (s + 10)} = \frac{2000}{s\,(s^2 + 70s + 800)}$$

The expression can be further simplified using partial fraction and inverse transform obtained to find out $i_2(t)$.

It is to be noted here that even though we want current through $10\,\Omega$ resistor, we have included the inductor also to simplify our calculations.

It is to be noted that Thevenin's theorem is useful not only for simplification of calculation but more so it is for conceptual visualization of the operation of networks e.g. if the Thevenins equivalent of a particular bus in a large power system is less as compared to some other bus, its short circuit capacity is more than the other bus and also the voltage regulations of this bus is better than the other bus. So it gives better visualization of the system.

## 3.6  NORTON'S THEOREM

Norton's theorem is dual of Thevenins theorem and can be stated as follows:

It states that any two terminal linear network can be replaced by an equivalent network consisting of a current source $I_0$ in parallel with a network of impedance $Z_0$. The impedance is same as that found for Thevenins equivalent and $I_0$ is the current between the two terminals when these terminals are short circuited. The following example will further clarify the concepts associated with the theorem.

**Example 3.9:** Solve the network of example E3.6 using Norton's theorem.

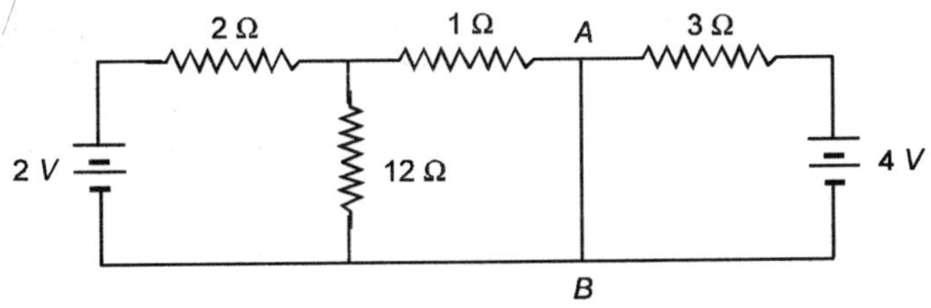

Fig. E3.9

Norton's equivalent current source will be the current through the short circuited branch $AB$.

Using superposition principle to find current through branch $AB$ we first short circuit $2V$ battery, the current through $AB$ due to 4 volt will be $\dfrac{4}{3}$ A. When 4 volt battery is shorted, the current supplied by 2 V battery is

$$\frac{2}{2 + \dfrac{12}{13}} = \frac{26}{38}$$

Therefore current through $AB$ will be $\dfrac{26}{38} \times \dfrac{12}{13} = \dfrac{12}{19} A$

so total current is $\dfrac{4}{3} + \dfrac{12}{19} = \dfrac{112}{57} A$

So that Norton's equivalent current source is 112/57 A. The Norton's equivalent impedance is same as in case of Thevenins equivalent impedance except that it is to be connected across the current source. Along with this the resistance through which current is required should be connected across the current source. This forms the complete equivalent Norton's circuit and is drawn here.

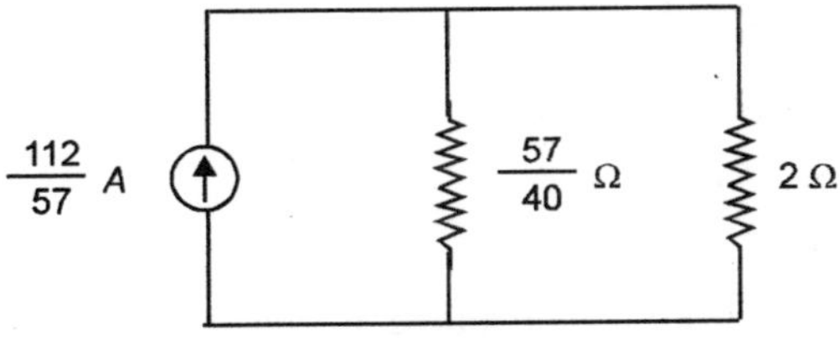

Fig. E3.9.1

Current through 2 $\Omega$ resistor is

$$\frac{112}{57} \times \frac{57}{40} \times \frac{1}{\dfrac{57}{40} + 2}$$

$$= \frac{112}{137} = 0.82 \ A$$

**Example 3.10:** Solve example 3.7 using Norton's theorem

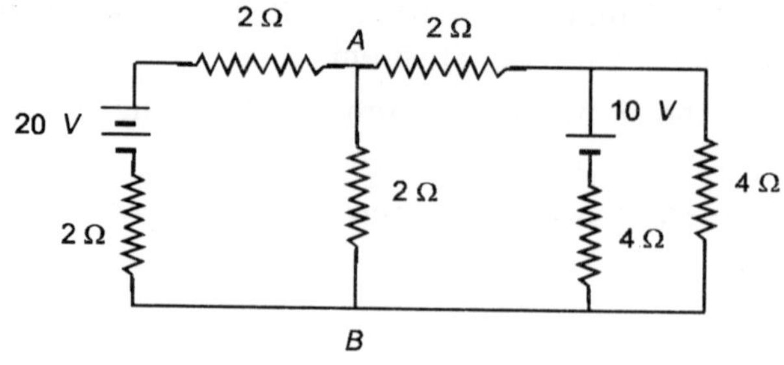

**Fig. E3.10**

**Solution:** Shorting branch $AB$ and using superposition principle to find current through $AB$.

20 Volts in the circuit and shorting 10 volt source

The current is $\dfrac{20}{4}A = 5\ A$

10 volts in the circuit and shorting 20 V

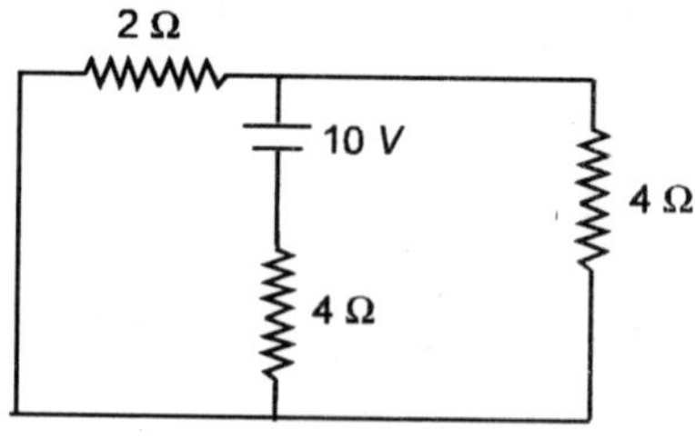

**Fig. E3.10.1**

Current supplied by the battery

$$\frac{10}{4+4/3} = \frac{3 \times 10}{16} = \frac{15}{8}A$$

Current through $2\ \Omega$ resistor

$$\frac{15}{8} \times \frac{4}{6} = \frac{5}{4}A$$

Total current through $AB$ is

$$5 + \frac{5}{4} = \frac{25}{4}A$$

The Norton's equivalent circuit, therefore is

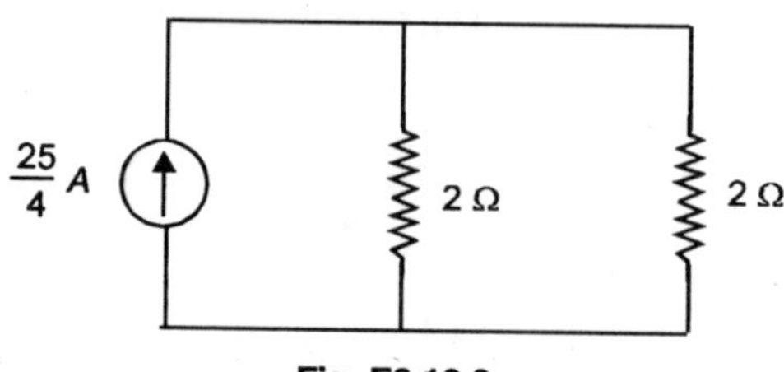

**Fig. E3.10.2**

Therefore, current through $AB$ branch is

$$\frac{25}{4} \times \frac{1}{2} = \frac{25}{8} = 3.125 \ A$$

**Example 3.11:** Determine the current through $AB$ branch of the example 3.8 using Norton's theorem.

**Solution:** We include the element $1H$ inductor also as current through the two elements is same. Shorting the two elements, the transformed current through the shorted branch is

$$I_0 \ (s) = \frac{100/s}{20+s}$$

which is the Norton's equivalent current transformed source and the equivalent impedance is $\dfrac{20\left(20+s\right)}{40+s}$ and Norton's equivalent network is

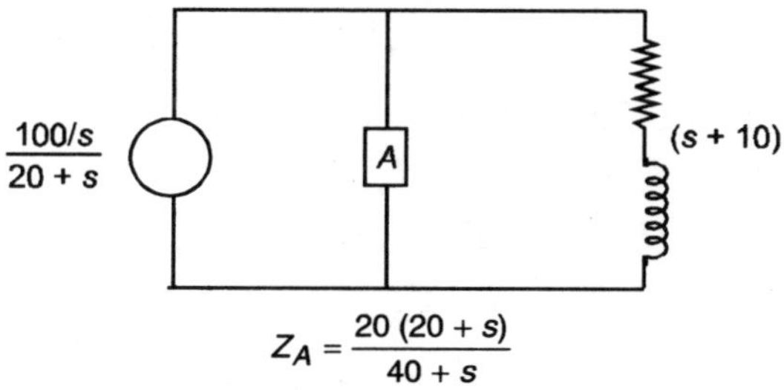


**Fig. E3.11**

Transformed current through $(s + 10)$ impedance is

$$I_{AB}(s) = \frac{100}{s(s+20)} \cdot \frac{20(s+20)}{s+40} \cdot \frac{1}{\left(s+10\right)+\dfrac{20(s+20)}{s+40}}$$

$$= \frac{2000}{s} \cdot \frac{1}{s^2+70s+800}$$

The results obtained are same as through the Thevenins theorem.

## 3.7 MILLMAN'S THEOREM

If there are large number of sources with emfs $E_1...E_n$ and internal impedance $Z_1...Z_n$ and these are connected in parallel as shown, the potential of the common point is given by Millman's theorem as

$$V = \frac{E_1 Y_1 + E_2 Y_2 + ... + E_n Y_n}{Y_1 + Y_2 + ... + Y_n} \tag{3.16}$$

where

$$Y_n = \frac{1}{Z_n}$$

Consider Fig. 3.7

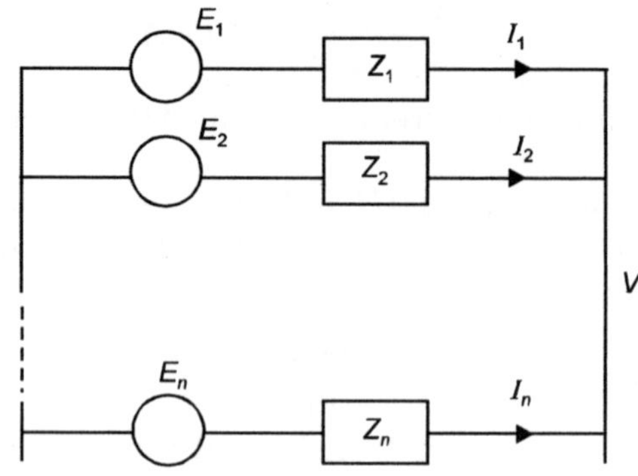


**Fig. 3.7** Sources connected in parallel.

From Fig. 3.7  $I_1 = \dfrac{E_1 - V}{Z_1}$, $I_2 = \dfrac{E_2 - V}{Z_2}$ and $I_n = \dfrac{E_n - V}{Z_n}$

According to Kirchhoff's current law

$$I_1 + I_2 + \dots + I_n = 0$$

$$(E_1 - V)\,Y_1 + (E_2 - V)\,Y_2 + \dots + (E_n - V)\,Y_n = 0$$

or

$$E_1\,Y_1 + E_2\,Y_2 + \dots + E_n\,Y_n = V\,(Y_1 + Y_2 + \dots + Y_n)$$

or

$$V = \frac{E_1 Y_1 + E_2 Y_2 + \dots + E_n Y_n}{Y_1 + Y_2 + \dots + Y_n}$$

Millman's theorem can also be derived using source transformation technique. Fig. 3.7 shows voltage sources with series impedances therefore, their equivalent current sources in general are $E_n Y_n$ across the admittance $Y_n$ and the equivalent circuit becomes as shown in Fig. 3.8.

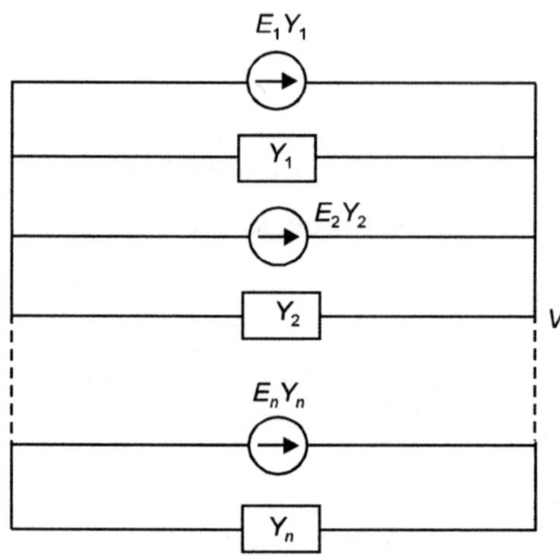


**Fig. 3.8** Equivalent of Fig. 3.7.

which is equivalent to a current source with current $I_s = Is_1 + Is_2 \dots Is_n$ across an admittance $Y = Y_1 + Y_2 + \dots + Y_n$

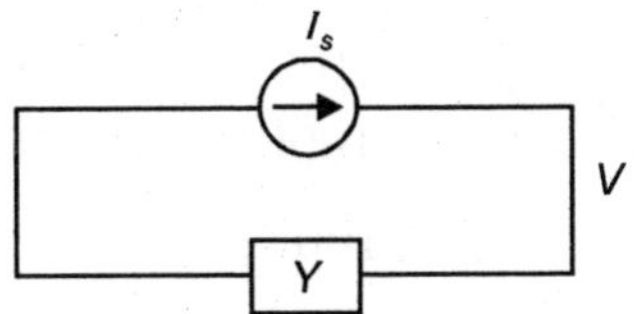

**Fig. 3.9**  Equivalent of Fig. 3.8.

or
$$I_s = YV \qquad\qquad ...(3.17)$$

where
$$I_s = E_1\,Y_1 + E_2\,Y_2 + ... + E_n\,Y_n$$

**Example 3.12:** Determine the current through the branch $AB$ using Millman's theorem.

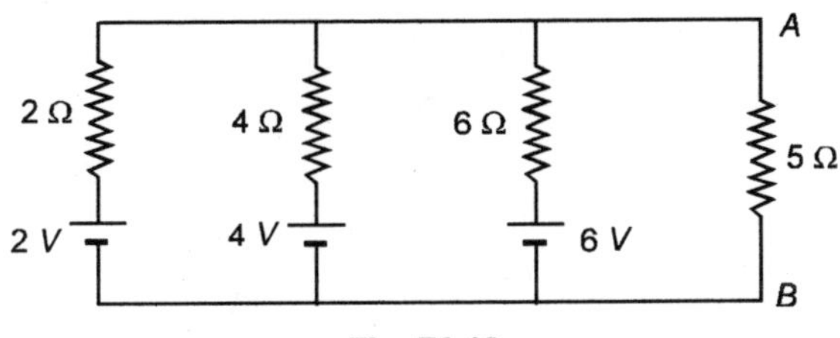



**Fig. E3.12**

**Solution:** Using Millman theorem

$$V_{AB} = \frac{2/2 + 4/4 + 6/6}{1/2 + 1/4 + 1/6} = \frac{36}{11} \text{ volt}$$

and the equivalent impedance between $AB$

$$Z = \frac{12}{11}\ \Omega$$

Therefore current
$$I_{AB} = \frac{36/11}{5 + 12/11} = \frac{36}{67} \text{ A \textbf{Ans.}}$$

**Example 3.13:** A 3 phase 3-wire delta connected load consisting of the following impedances is connected to a 3 phase 3–wire 50 Hz balanced supply of 400 volts $Z_{12} = 100\,(2 + j)$

$$Z_{23} = Z_{31} = 100\,(1 - j2)$$

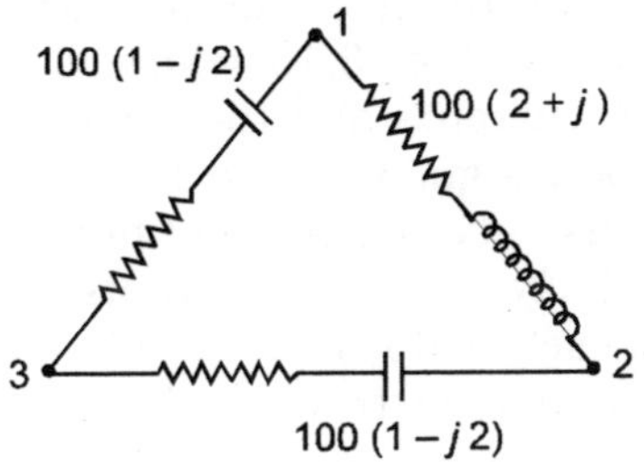

**Fig. E3.13**

Determine the impedances of the three legs of the equivalent $Y$ connected load and find the potential of the neutral of the load.

**Solution:**

$$Z_{10} = \frac{100(1-j2).100(2+j)}{100(2-j4)+100(2+j)} = 100 \ \Omega$$

$$Z_{20} = 100 \ \Omega$$

$$Z_{30} = \frac{10^4(1-j2)^2}{10(4-j3)} = -j100 \ \Omega$$

So the equivalent star is

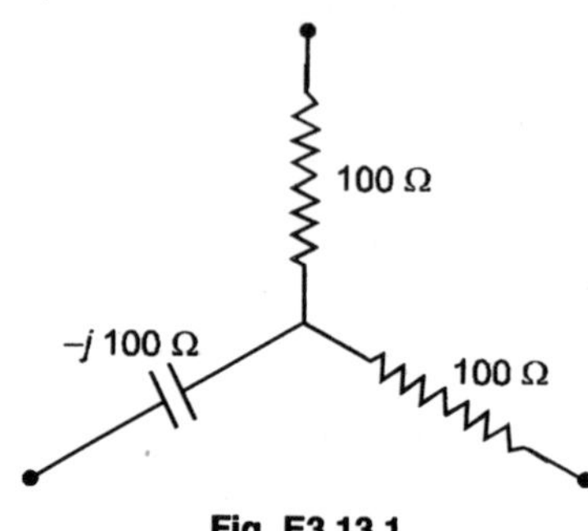


**Fig. E3.13.1**

Using Millman theorem to obtain potential of star point

$$V = \frac{E_1 Y_1 + E_2 Y_2 + E_3 Y_3}{Y_1 + Y_2 + Y_3}$$

$$= \frac{231\angle 0.\dfrac{1}{100} + 231\angle -120.\dfrac{1}{100} + \dfrac{231}{100}\angle 120^\circ +90^\circ}{0.02 + j0.01} = 146 \ \text{volt}$$

## 3.8  TELLEGEN'S THEOREM

Since Tellegen's theorem depends only on the two Kirchhoffs laws it is applicable to a very general class of lumped network consisting of elements that are linear or non-linear, passive or active, time invariant or time varying. This generality is the main reason for Tellegen's theorem to be one of the most powerful tools for network solutions.

Consider the network shown in Fig. 3.10 where arbitrary reference directions have been selected for all of the branch currents and the corresponding branch voltages such that the positive reference direction is assigned at the tail of the current arrow.

We shall now select out of this network, a set of branch voltages such that these satisfy Kirchhoffs voltage law and similarly a set of currents for the branches such that these satisfy Kirchhoffs current law at each node without any consideration of the previous choice of branch voltages. We shall now demonstrate that these arbitrarily chosen voltages and currents satisfy the equation

$$\sum_{k=1}^{b} V_k i_k = 0 \qquad \qquad ...(3.18)$$

subject to the condition that KCL and KVL. have been satisfied while selecting these branches.

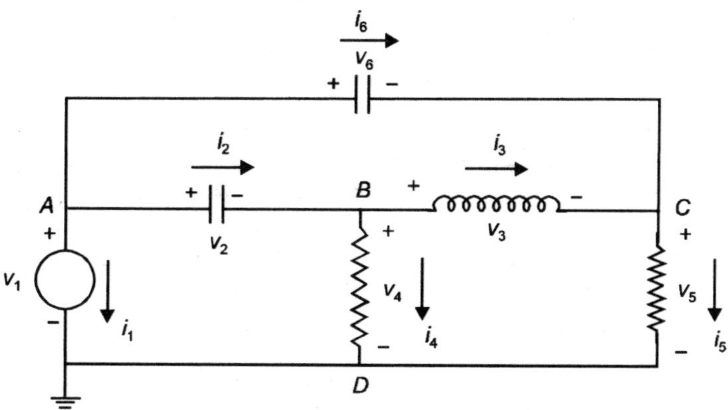

**Fig. 3.10.**

In loop *ABDA* let $V_1$ = 5 volt $V_2$ = 3 volts. Using KVL, therefore $V_4$ = 2 volts. Loop *BCDB* let $V_3$ = 3 volts, $V_5$ = $-1V$, loop *ABCA* $V_2$ = 3 V, $V_3$ = 3 volts $V_6$ = 6 volts. Using KCL, Node *A* Let $i_2$ = 4 A, $i_6$ = 2 A. Hence $i_1$ = $-6$ A.

Node *B* Let $i_4$ = 1A $i_3$ = 3 A as at B $i_2$ = $i_3$ + $i_4$

Node *C* $i_3$ + $i_6$ = $i_5$ = 3 + 2 = 5 A

$$\text{Hence} \quad \sum_{k=1}^{6} V_k i_k = 5 \times (-6) + 3 \times 4 + 3 \times 3 + 2 \times 1 + (-1) \times 5 + 6 \times 2$$

$$= -30 + 12 + 9 + 2 - 5 + 12$$

$$= 0$$

Hence Tellegen's theorem is verified.

**Proof:** Equation (3.18) which is a mathematical statement of Tellegen's theorem observes that the rate of supply of energy by the active elements of a network equals the rate of energy dissipated or stored by the passive element of the same network.

If the network is divided into two parts as illustrated in Fig. 3.11 one part with all of the energy sources and the other with all of the passive elements then one can say that the power delivered by the sources must equal the power dissipated or stored by passive elements.

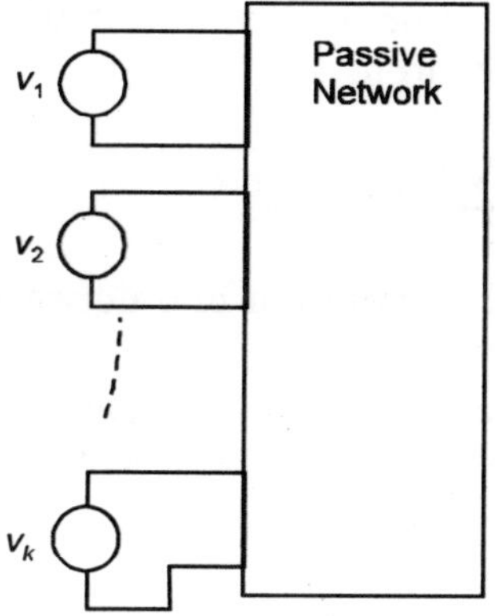

**Fig. 3.11**   General network divided.

Let us consider a general branch $K$ between nodes $p$ and $q$. Now

$$p \underline{\quad\quad^{K_{th}}\quad\quad} q$$

current can be assumed from $p$ to $q$ or $q$ to $p$ and hence it could be written as $i_{pq}$ or $i_{qp}$. With $i_{pq}$ we write voltage of $K_{th}$ branch as $(e_p - e_q)$ as the positive reference direction of voltage is to be assigned at the tail of the current arrow. Hence for $i_{qp}$ the voltage of $K_{th}$ branch is $(e_q - e_p)$.

The power through the branch $K$ can be taken as the average of the power flow from $p$ to $q$ and $q$ to $p$ i.e.

$$V_k i_k = \frac{1}{2} \left[ (e_p - e_q)\, i_{pq} + (e_q - e_p)\, i_{qp} \right] \qquad \text{...(3.19)}$$

Therefore for all the $b$ branches of the network

$$\sum_{k=1}^{b} V_k i_k = \frac{1}{2} \sum_{p=1}^{b} \sum_{q=1}^{b} (e_p - e_q)\, i_{pq} \qquad \text{...(3.20)}$$

This covers all the nodes.

The equation (3.20) can be rewritten as

$$\sum_{k=1}^{b} V_k i_k = \frac{1}{2} \sum_{p=1}^{b} e_p \sum_{q=1}^{b} i_{pq} - \frac{1}{2} \sum_{q=1}^{b} e_q \sum_{p=1}^{b} i_{pq} \qquad \text{...(3.21)}$$

Now at each node as per KCL the algebraic sum of current is zero. Hence,

$$\sum_{q=1}^{b} i_{pq} = \sum_{p=1}^{b} i_{pq} = 0 \qquad \text{...(3.22)}$$

Therefore, 
$$\sum_{K=1}^{b} V_k i_k = 0$$

Which proves Tellegen's theorem.

Tellegen's theorem is not only applicable to one individual network, it is applicable to two networks having same graph. The elements in the two networks could be different.

If the two networks have the same graphs and if the voltages and currents satisfy the Kirchhoffs laws, then Tellegen's theorem states that

$$\sum_{K=1}^{b} V_{1k}\, i_{2k} = \sum_{K=1}^{b} V_{2k}\, i_{1k} = 0$$

where $V_{1k}$, $i_{1k}$ are the voltages and currents of one network and $V_{2k}$ and $i_{2k}$ are the respective quantities for the second network.

Tellegen's theorem is so general that even though the voltages and currents in the network may correspond to different times, still the theorem holds good i.e.

$$\sum V_k(t_1)\, i_k(t_2) = 0 \qquad \text{...(3.24)}$$

**Example 3.14:** Verify Tellegen's theorem in respect of the two networks given here which have identical graphs.

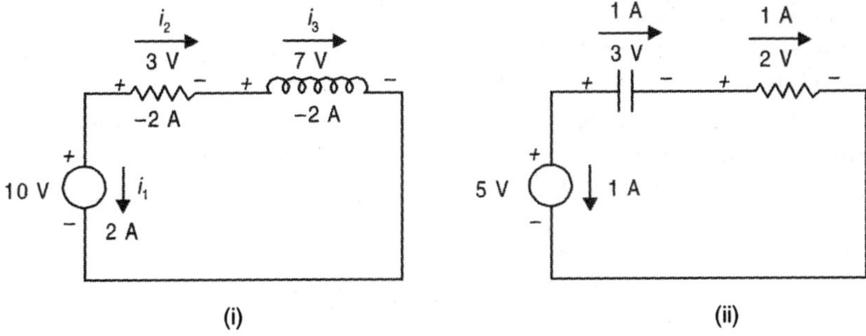

(i)               (ii)

**Fig. E3.14**

For the individual networks.

$(i)\ \sum_{i=1}^{3} V_k\, i_k = 10 \times 2 - 3 \times 2 - 7 \times 2 = 0$

$(ii)\ \sum_{k=1}^{3} V_k\, i_k = 5 \times 1 - 3\,(1) - 2\,(1) = 0$

Voltage of network 1 and current of network 2

$$\sum_{k=1}^{3} V_{1k}\, i_{2k} = 10 \times 1 - 3 \times 1 - 7 \times 1 = 0$$

$$\sum V_{2k}\, i_{1k} = 5 \times 2 - 3 \times 2 - 2 \times 2 = 0$$

Hence varified.

## 3.9 MAXIMUM POWER TRANSFER THEOREM

Sometimes in electronics circuits it is desirable to transfer maximum power through a source to a load e.g. impedance matching is carried out in a loud speaker so that maximum sound could be heard. However in power system maximum power transfer is never tried as huge amount of power in terms of MW are transferred from source to the load and the efficiency of transfer is 50% as will be seen shortly if power transferred is based on maximum power transfer criterion. Also under this, the terminal voltage becomes half of the source voltage and hence this results in poor voltage regulation, therefore, maximum power transfer criterion is never considered in power system.

We will consider the theorem both for d.c. and a.c. circuits.

*d.c circuits:*

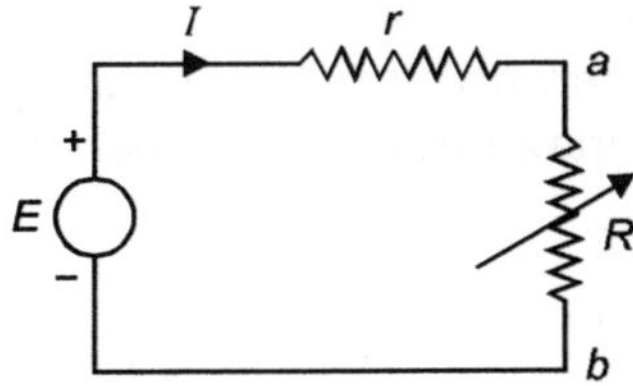

**Fig. 3.12**  Maximum power transfer-d.c. circuit.

Here $r$ is the internal resistance of the source or the equivalent resistance as seen between $a\,b$ or the load resistance $R$. Usually $r$ in a circuit or in a system is fixed, the load resistance $R$ is variable.

The problem is: what should be $R$ expressed in terms of $r$ so that power delivered to the load resistance is maximum.

The current in the circuit $I = E/(r + R)$ and power transferred is

$$P = \frac{E^2}{(r+R)^2}.R$$

For this power to be maximum $\dfrac{dP}{dR} = 0$

$$= \frac{E^2\left[(r+R)^2 - R.2(R+r)\right]}{(r+R)^4} = 0$$

or
$$r + R - 2R = 0$$

or
$$R = r \qquad\qquad\qquad\text{...(3.25)}$$

This shows that for maximum power transfer the load resistance $R$ should equal the internal resistance $r$ or the equivalent resistance $r$. Under this condition

Power
$$P_{max} = \frac{E^2}{(2r)^2}.r = \frac{E^2}{4r}$$

Same amount of power is dissipated in the internal resistance $r$ and hence the efficiency is 50%. Also it can be seen that voltage across the load is $\dfrac{E}{2}$ and voltage regulation is 50% which is very poor in power system operation.

*a.c.Circuit:*

(*i*) Load is $R + jX$ and internal impedance $(r + jx)$

Therefore,
$$I = \frac{E}{(r+R) + j(x+X)}$$

Power transfer $P = I^2R = \dfrac{E^2\,R}{(r+R)^2 + (x+X)^2} \qquad\qquad\text{...(3.26)}$

For power to be maximum $\dfrac{\partial P}{\partial X} = 0$

$$= \frac{E^2R.\{-2(x+X)\}}{D^2} = 0$$

or
$$x + X = 0 \quad\text{or}\quad X = -x \qquad\qquad\qquad\text{...(3.27)}$$

which means the load reactance must be equal to the negated value of the source internal reactance. The maximum power becomes

$$P_{max} = \frac{E^2R}{(r+R)^2} \qquad\qquad\qquad\text{...(3.28)}$$

Again taking $\dfrac{dP}{dR} = 0$ for maximum power

We have just now seen the condition is $R = r$ and, therefore, in a.c. circuits the load impedance should be complex conjugate of the internal impedance that is $Z_L = Z_I^*$ where $Z_L$ is the load impedance and $Z_I$ is source impedance. Usually source impedance is inductive $r + jx$ and, therefore, the load impedance should be capacitive for maximum power transfer condition. Here again the efficiency is 50%.

(*ii*) Load resistance $R$ and internal impedance $(r + jx)$

$$\text{Here current} \qquad I = \frac{E}{(r+R)+jx}$$

$$\text{And power} \qquad P = \frac{E^2\,R}{(r+R)^2+x^2} \qquad\qquad \text{...(3.29)}$$

$$\text{Therefore,} \qquad \frac{dP}{dR} = 0 \text{ for maximum power transfer}$$

$$E^2\,[\{(r+R)^2 + x^2\} - R.2\,(R+r)] = 0$$

or $\qquad\qquad r^2 + x^2 = R^2$

or $\qquad\qquad R = \sqrt{r^2+x^2} = Z_I \qquad\qquad \text{...(3.30)}$

This means when load resistance $R$ equals the impedance of the source the power transfer is maximum. Let us find out the magnitude of power under this condition

$$P = \frac{E^2\,Z_I}{(r+Z_I)^2+x^2} = \frac{EI_{sc}\,Z_I^2}{r^2+Z_I^2+2r\,Z_I+x^2}$$

$$= \frac{EI_{sc}\,Z_I^2}{2Z_I^2+2r\,Z_I} = \frac{EI_{sc}\,Z_I}{2(Z_I+r)} \qquad\qquad \text{...(3.31)}$$

This power is certainly less than the power under the previous two conditions.

**Example 3.15:** Determine the value of $R_L$ in the network shown for maximum power transfer and calculate the value of power.

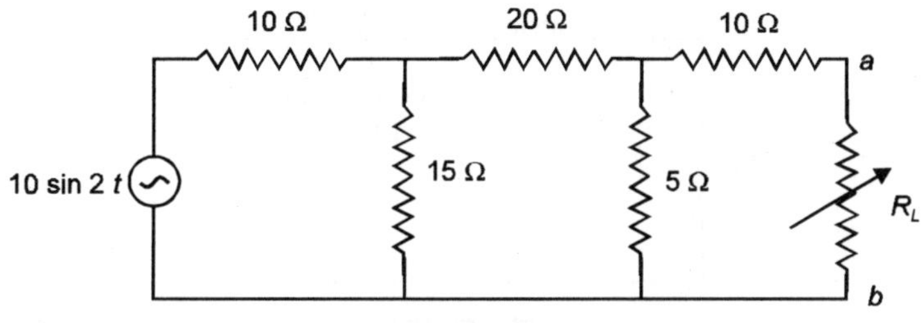


**Fig. E3.15**

**Solution:** For maximum power transfer, the resistance should equal the resistance between $a\,b$ when looked into the network replacing the voltage source by a short circuit.

$$\frac{10\times15}{25} + 20 = 26\ \Omega$$

$$r_{ab} = \frac{26\times5}{31} + 10 = 14.2\ \Omega$$

$$V_{TH} = \frac{10}{\sqrt{2}} \times \frac{8}{155} \times \frac{15}{40} \times 5$$

$$= 0.684\text{ volt}$$

$$\text{Power transfer} = \frac{V^2}{4r_{ab}} = \frac{(0.684)^2}{4 \times 14.2}$$

$$= 8.24 \text{ mW } \textbf{Ans.}$$

**Example 3.16:** Determine the value of $Z_L$ for maximum power transfer, in $Z_L$, in the given network.

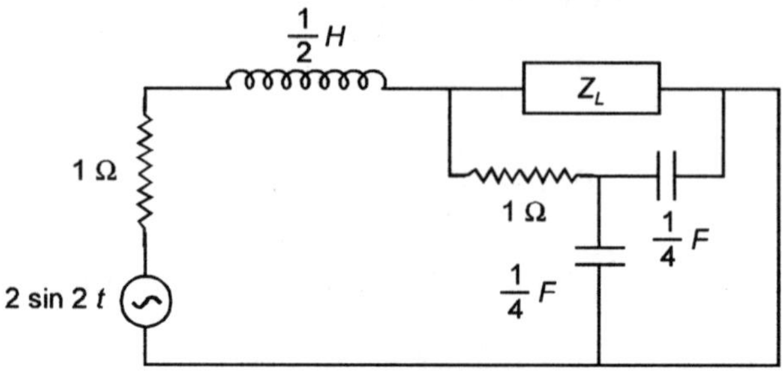

**Fig. E3.16**

Since $(\omega) = 2$ rad/sec, the network is drawn as

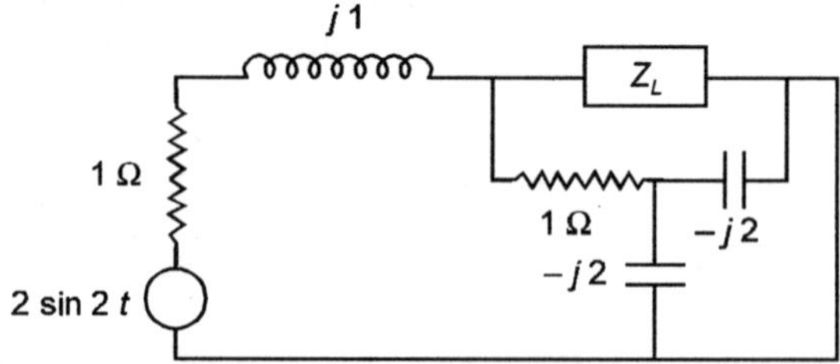

**Fig. E3.16.1**

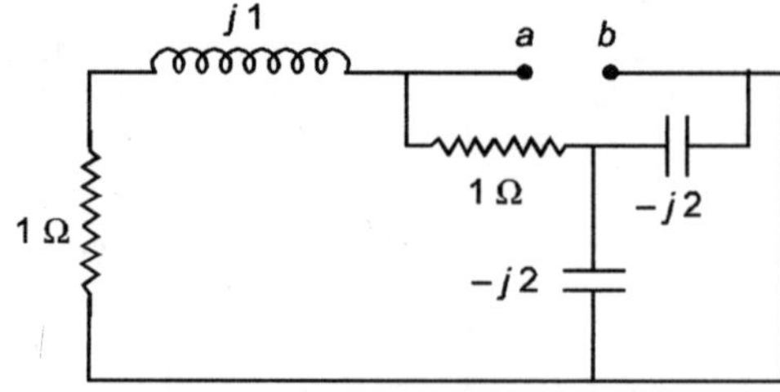

**Fig. E3.16.2**

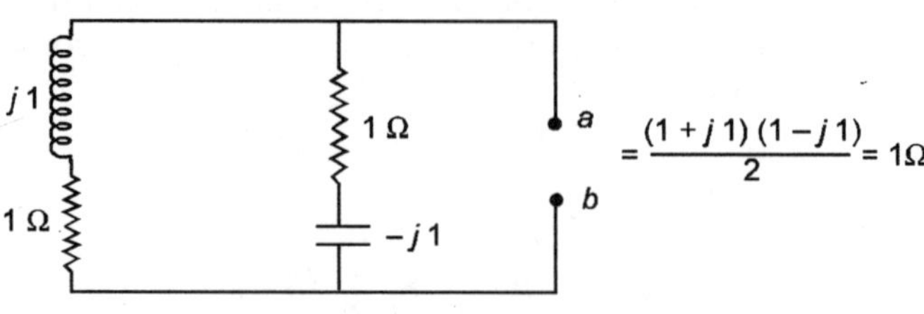


$$= \frac{(1 + j\,1)\,(1 - j\,1)}{2} = 1\Omega$$

**Fig. 3.16.3**

$$Z_L = \frac{(1+j1)(1-j1)}{2}$$

Hence for maximum power through $Z_L$, it should be $Z_L = 1\,\Omega$.

**Example 3.17:** In the network shown determine the value of impedance $Z_L$ for maximum power and calculate the maximum power.

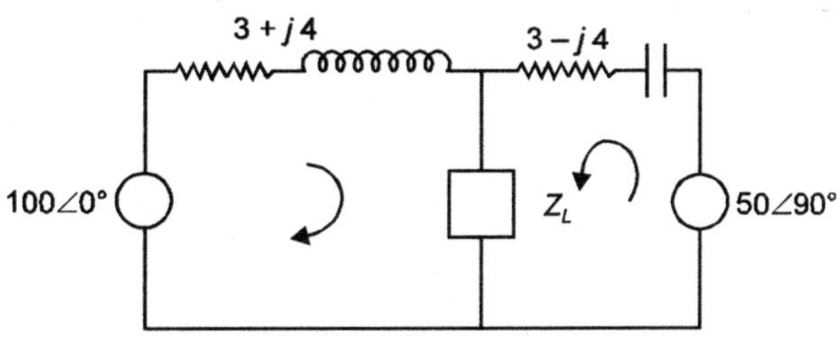


**Fig. E3.17**

**Solution:** The equivalent impedance

$$\frac{(3+j4)(3-j4)}{6} = \frac{25}{6} = 4.16\,\Omega$$

To obtain maximum power through this resistance we obtain current through this branch. To find out that we find $V_{TH}$. The current in the circuit when $Z_L$ is removed is

$$I = \frac{100\angle 0^\circ - 50\angle 90^\circ}{6} = \frac{100}{6} - \frac{50}{6} = \frac{50}{6}\ \text{Amp}$$

Hence,

$$V_{TH} = 100\ \angle 0^\circ - \frac{50}{6}(3+j4) = 100 - 25 - j33.3$$

The current through $Z_L$

$$I = \frac{V_{TH}}{2 \times 4.16} = \frac{75 - j33.3}{8.32} = \frac{82.06\angle -23.94^\circ}{8.32}$$

$$= 9.86\ \angle -23.94^\circ$$

The power (maximum) through $Z_L$ = is $4.16I^2$

$$= 404.6\ \text{watts}\ \textbf{Ans.}$$

## 3.10  COMPENSATION THEOREM

If a network has certain distribution of current and voltages through and across various branches and due to some reason if the impedance of some of the branch changes, redistribution of current and voltages in the branches will take place. This situation can be taken care of by a theorem known as compensation theorem. Since the calculation of redistribution is based on superposition theorem, the compensation theorem is applicable to linear networks.

Suppose we have a network as shown in Fig. 3.13($a$)

The change to be considered is in the branch with $Z$ impedance. $Z_{th}$ is the Thevenin's equivalent and $V$ is the Thevenin's equivalent voltage source.

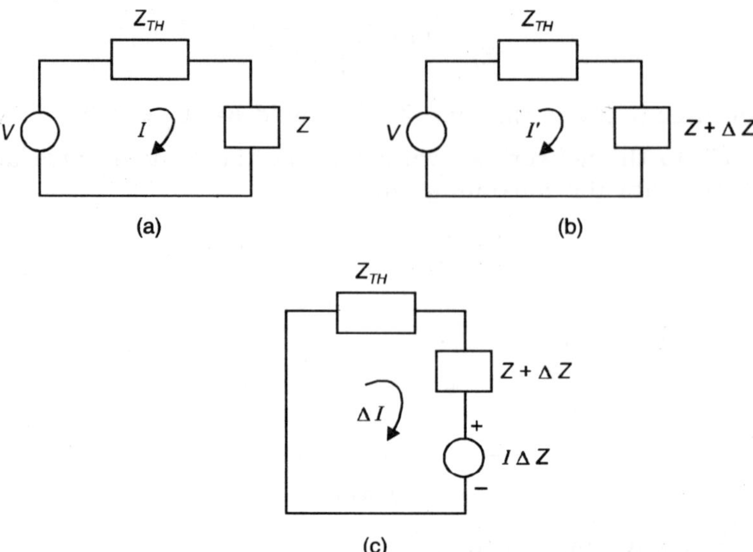



**Fig. 3.13**

The current through the branch where change in impedance is envisaged is given by

$$I = \frac{V}{Z + Z_{TH}} \qquad \qquad \text{...(3.32)}$$

Now if $Z$ changes to $Z + \Delta Z$, the current in the same branch is

$$I' = \frac{V}{Z + Z_{TH} + \Delta Z} \qquad \qquad \text{...(3.33)}$$

The change in current

$$\Delta I = I' - I = \frac{V}{Z + Z_{TH} + \Delta Z} - \frac{V}{Z + Z_{TH}}$$

$$= \frac{-V\,(\Delta Z)}{(Z + Z_{TH} + \Delta Z)\,(Z + Z_{TH})}$$

$$= \frac{-I\,(\Delta Z)}{Z + Z_{TH} + \Delta Z} = \frac{-V_C}{Z + Z_{TH} + \Delta Z} \qquad \qquad \text{...(3.34)}$$

Where $\qquad\qquad V_C = I\Delta Z$

The direction of $I$ is same as that of $I'$. This Compensation source is to be inserted in series with $Z + \Delta Z$ and the current through various branches of the network can be calculated with the compensation voltage source in place and replacing the voltage and current source of the original network by their equivalent impedance. The algebraic sum of the current calculated by the compensation source and the current of the original network will give the net value of current in various branches of the network if a change in $Z$ by $\Delta Z$ takes place.

This is illustrated with the following example.

**Example 3.18:** Consider the circuit shown. If the resistance of 5 Ω branch increases to 6 Ω. Determine the compensation source and verify the results.

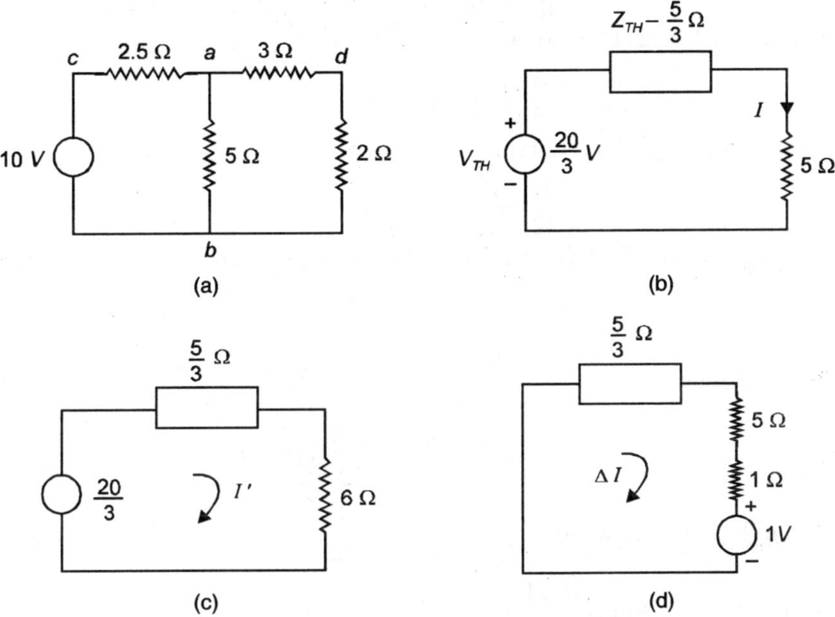

**Fig. E3.18**

The current in 5 Ω resistor before the change takes place is

$$I = \frac{20/3}{5/3 + 5} = 1\,\text{A}$$

Now

$$I' = \frac{20/3}{5/3 + 6} = \frac{20/3}{23/3} = \frac{20}{23}\,\text{A}$$

$$\partial I = I' - I = -\frac{3}{23}$$

Therefore compensation source voltage is $I\partial Z$

$$= 1 \times 1 = 1 \text{ volt.}$$

The polarity of this source is as shown, as $\partial I$ is negative. If $\partial I$ were positive the polarity of the source will be opposite to what is given in the Fig.

Now let us superimpose the current due to compensation source on various branches of the network and find out the branch current and verify the result by solving the original network with 6 Ω resistance rather than 5 Ω Initially the current through 5 Ω branch is from $a$ to $b$ and with the current $\partial I$ will be from $b$ to $a$, the new current in branch $ab$ is

$$1 - \frac{3}{23} = \frac{20}{23}\,\text{A.}$$

Now $\partial I$ current to be distributed among the parallel combination of 2.5 Ω and 5 Ω.

Current through 2.5 Ω branch $\dfrac{3}{23} \times \dfrac{5}{7.5} = \dfrac{2}{23}\,\text{A}$

from $a$ to $c$ hence net current

$$2 - \frac{2}{23} = \frac{44}{23} \text{ A}$$

Current through $(3 + 2)\,\Omega$ branch

$$\frac{3}{23} \times \frac{2.5}{7.5} = \frac{1}{23} \text{ A}$$

with direction $a$ to $d$ i.e. the net current

$$1 + \frac{1}{23} = \frac{24}{23} \text{ A}$$

The results can be verified by taking $6\,\Omega$ resistance in the original network across branch $ab$.

Net input resistance

$$= \frac{5}{2} + \frac{5 \times 6}{11}$$

$$= \frac{5}{2} + \frac{30}{11} = \frac{115}{22} \ \Omega$$

Current through $2.5\,\Omega$ resistance $\approx \dfrac{10 \times 22}{115} = \dfrac{44}{23}$

Current through $6\ \Omega$ branch $\dfrac{44}{23} \times \dfrac{5}{11} = \dfrac{20}{23}$ A

($a$) Current through $(3 + 2)$ branch $\dfrac{44}{23} \times \dfrac{6}{11} = \dfrac{24}{23}$ A

Hence the results are varified.

**Example 3.19:** For the network shown determine the Thevenin's and Nortons equivalent circuit.

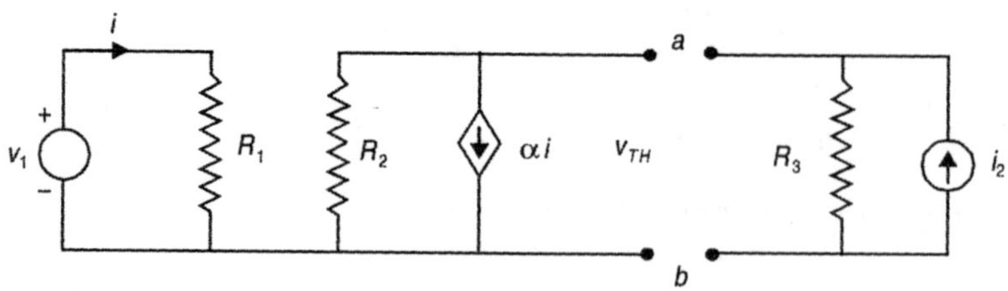

**Fig. E3.19**

**Solution:** On open circuiting $a\,b$ terminal, left side of $a\,b$ termials.

$$V_{TH} = -\,\alpha\,iR_2 = -\,\alpha\,R_2\,\frac{v_1}{R_1}$$

and to find $R_{th}$, we short circuit the independent voltage source. Hence $i$ is zero, therefore, the current source $\alpha\,i = 0$ which means it is open circuited and hence $R_{TH} = R_2$

Taking R.H.S. of the terminal $a\,b$ we have $V_{TH} = i_2\,R_3$ and $R_{TH} = R_3$ when we open circuit the independent current source. Hence the equivalent Thevenin's circuit has been obtained by what is known as repeated application of Thevenin's theorem whenever the theorem can be repeated within a complex network.

The circuit is shown here.

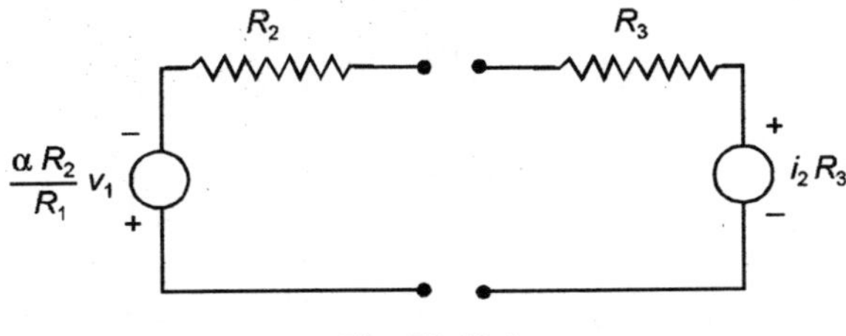

**Fig. E3.19.1**

To find out Norton's equivalent we use repeated Norton's theorem. We short circuit the left and right side of the circuit as shown here.

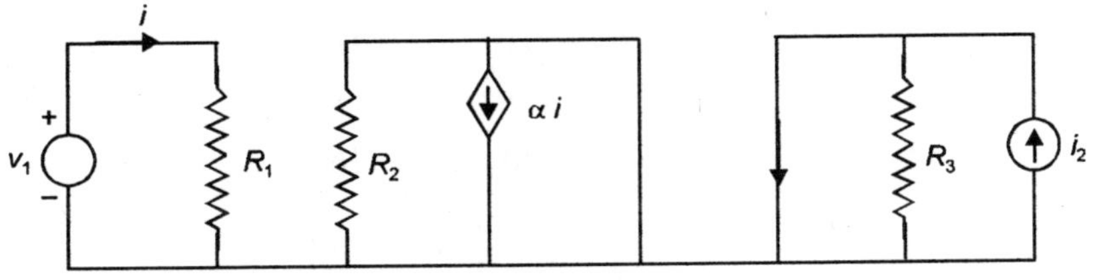

**Fig. E3.19.2**

*Left hand side current source*

$$i_s = \alpha\, i = a\frac{v_1}{R_1} \text{ and } R_{TH} \text{ is of course}$$

same as in Thevenin's equivalent i.e. $R_2$

Taking right hand side the Norton's equivalent current source is $i_s = i_2$ and $R_{TH} = R_3$. Hence Norton's equivalent circuit is

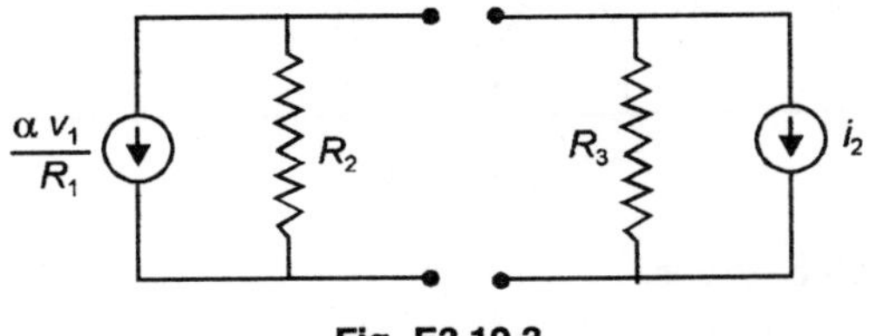

**Fig. E3.19.3**

**Example 3.20:** Determine current $I$ in the network using Thevenin's theorem

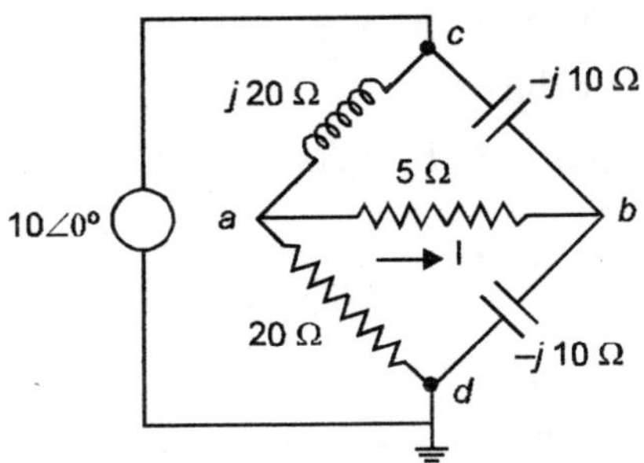



**Fig. E3.20**

To obtain $V_{TH}$, we remove branch between $a\ b$ and obtain $V_a$ and $V_b$ and the difference of the two voltages will give $V_{TH}$

$$V_a = \frac{10}{20 + j20} \cdot 20 = 5 - j5$$

$$V_b = \frac{10}{-j20} \cdot -j10 = 5 + j0$$

$$V_{TH} = V_a - V_b = -j5 = 5\ \angle -90º$$

To obtain $Z_{TH}$ we short circuit the voltage source and see the impedance between $a$ and $b$

$$Z_{TH} = Z_{ab} = Z_{ac} \parallel Z_{ad} + Z_{cb} \parallel Z_{bd}$$

$$= \frac{20 \times j20}{20 + j20} + \frac{(-j10)(-j10)}{-j20}$$

$$= 10 + j10 - j5 = 10 + j5$$

Therefore,
$$I = \frac{V_{TH}}{Z_{TH} + 5} = \frac{5\ \angle -90º}{10 + j5 + 5} = \frac{5\ \angle -90º}{15 + j5}$$

$$= 0.316\ \angle -108º\ \text{A}$$

**Example 3.21:** For the network shown in Fig. E3.21 determine Thevenin's equivalent voltage source and the equivalent series resistance. Also determine Norton's equivalence

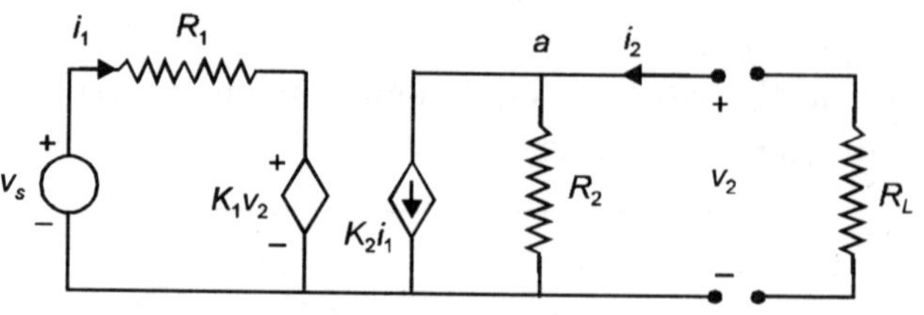


**Fig. E3.21**

**Solution:** In loop 1 KVL gives
$$v_s = i_1 R_1 + K_1 v_2 \qquad\qquad ...(1)$$
If we convert dependent current source $K_2 i_1$ and shunt resistance into a voltage source in series with $R_2$ we have
$$v_2 = -K_2 i_1 R_2 = V_{TH}$$
Substituting for $v_2$ in (1) we have
$$v_s = i_1 R_1 - K_1 K_2 i_1 R_2 = i_1 (R_1 - K_1 K_2 R_2)$$

Now
$$V_{TH} = -K_2 i_1 R_2 = \frac{-K_2 R_2 v_s}{R_1 - K_1 K_2 R_2}$$

or
$$V_{TH} = \frac{K_2 R_2}{K_1 K_2 R_2 - R_1} v_s$$

To obtain $R_{TH}$ we short circuit the independent voltage source and we apply voltage $v_2$ at the output terminal and find out $i_2$, the ratio $\dfrac{v_2}{i_2}$ will give us $Z_{TH}$.

In the independent source which is short circuited.
The KVL gives

$$i_1 R_1 = -K_1 \, v_2$$

Now KCL at node '$a$'

$$i_2 = \frac{v_2}{R_2} + K_2 i_1$$

$$= \frac{v_2}{R_2} - \frac{K_2 K_1}{R_1} v_2$$

$$= v_2 \left[ \frac{1}{R_2} - \frac{K_1 K_2}{R_1} \right]$$

or
$$Z_{TH} = \frac{v_2}{i_2} = \frac{1}{\dfrac{1}{R_2} - \dfrac{K_1 \, K_2}{R_1}} = \frac{R_1 \, R_2}{R_1 - K_1 \, K_2 \, R_2}$$

Hence the Thevenin's circuit is

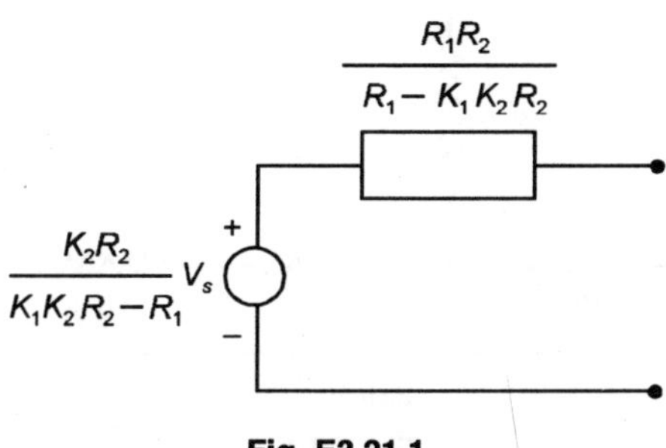

**Fig. E3.21.1**

To obtain Norton's equivalent $Z_{TH}$ will remain same as in Thevenin's theorem, only it is to be connected across Norton's equivalent current source. To obtain $i_s$ we short circuit the output terminals. Let the current be $i_s$ and $v_2 = 0$ we replace $v_s$ in series with $R_1$ by a current source $\dfrac{v_s}{R_1}$ in parallel with $R_1$.

Since $v_2 = 0$, the depedent source in the first loop $K_1 \, v_2 \rightarrow 0$ hence the circuit is reduced to the one shown in Fig. E3.21.2.

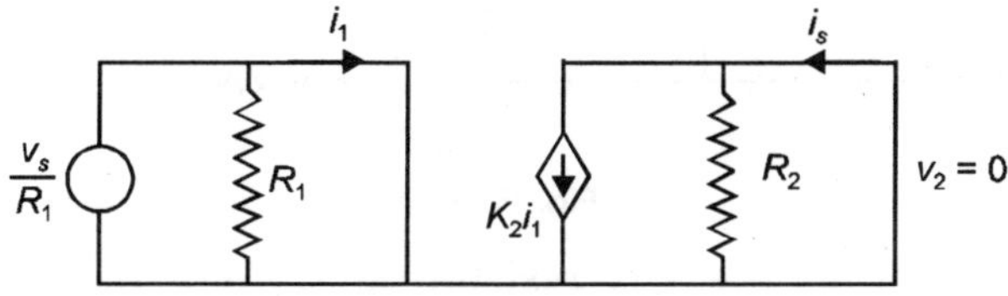

**Fig. E3.21.2**

Here
$$i_1 = \frac{v_s}{R_1} \quad \text{and} \quad i_s = K_2\, i_1$$

$$= K_2\, \frac{v_s}{R_1}$$

Therefore, Norton's equivalent circuit is given as

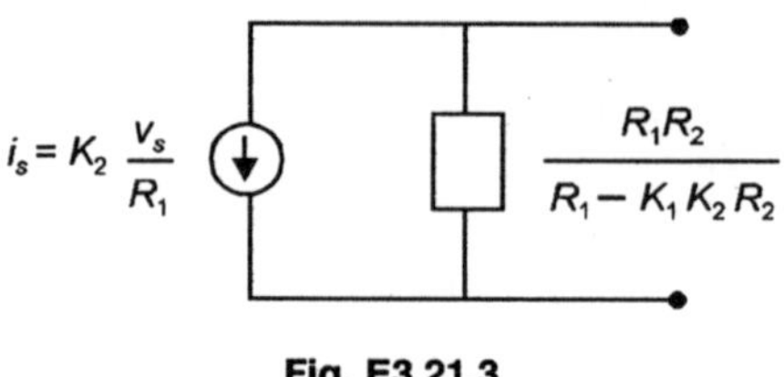

**Fig. E3.21.3**

## PROBLEMS

**3.1.** Determine the current through $10\,\Omega$ resistor in the network by (*i*) star-delta transformation (*ii*) Thevenins theorem.

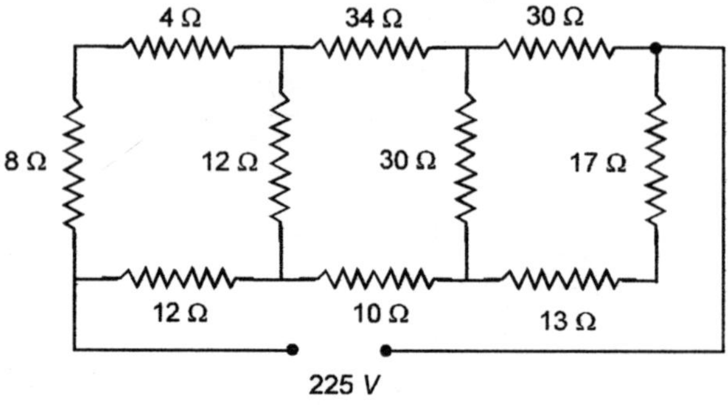

**Fig. P3.1**

**3.2.** Determine the current through $2.5\,\Omega$ resistor using Thevenin's and Norton's theorem in the network shown.

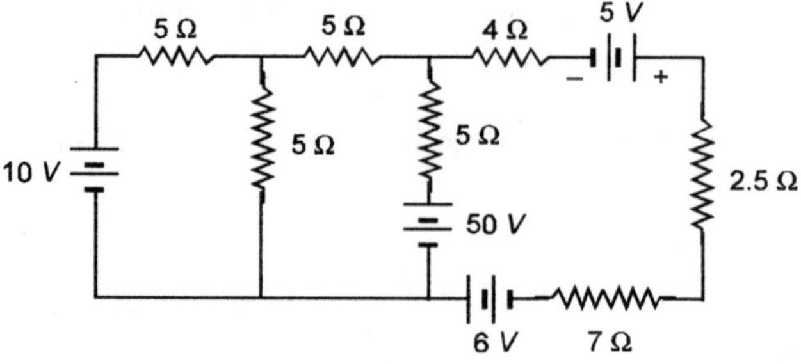

**Fig. P3.2**

**3.3.** Determine the current in the $4\,\Omega$ resistor in the circuit shown using Thevenin's theorem and superposition theorem.

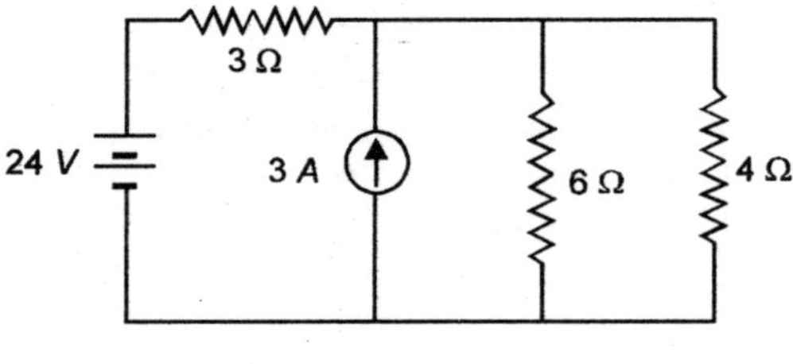

**Fig. P3.3**

**3.4.** Determine $X_1$ and $X_2$ in terms of $R_1$ and $R_2$ to give maximum power dissipation in $R_2$.

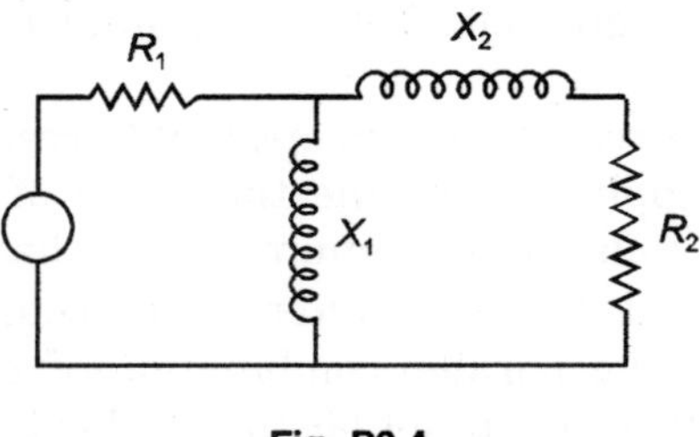

**Fig. P3.4**

**3.5.** Determine the current through the branch $a$ $b$ using Thevenin's theorem and superposition theorem.

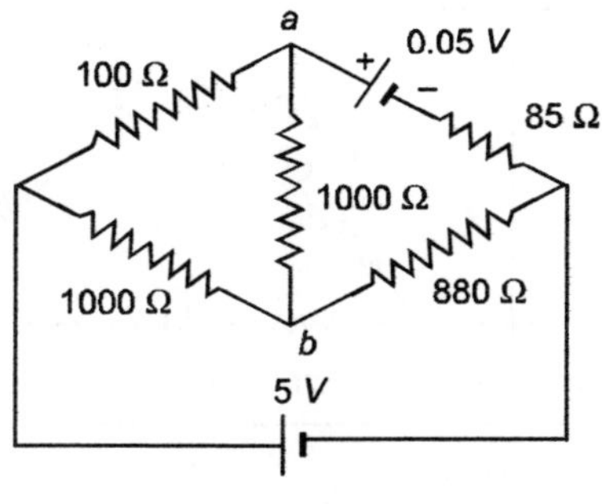

**Fig. P3.5**

**3.6.** For the networks shown determine Thevenin's equivalent source and the series impedance.

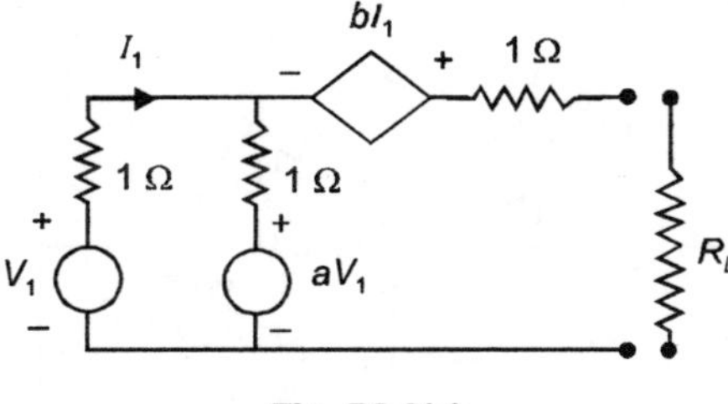

**Fig. P3.6(*a*)**

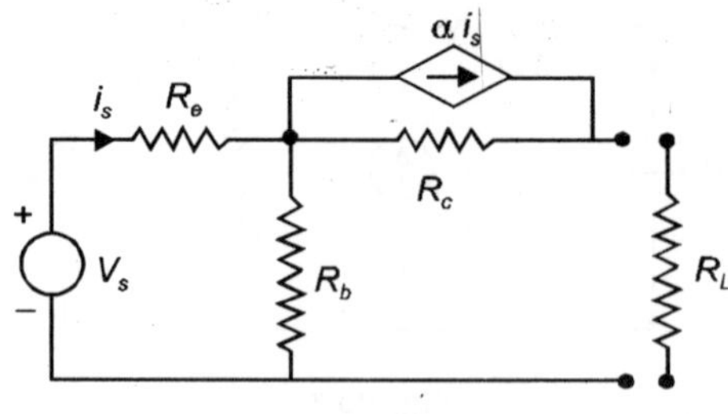

**Fig. P3.6(b)**

**3.7.** A load of $(20 - jX_c)$ is supplied from a source of 10 V rms and internal impedance $(10 + j20)$. Determine the value of $X_C$ for maximum power supplied to the load and determine this power.

**3.8.** Two reactors each of 50 Ω reactance and negligible resistance are connected in series across a 200 V 50 Hz supply. Determine the value of resistance to be connected in parallel with one of the coils for maximum power dissipation.

**3.9.** A source of voltage $2\sqrt{2}\ \sin 2t$ has an internal impedance $R = 1\,\Omega$ and $L = 1/3$ H. It is connected to a load having parallel combination of $c$ and resistance of $1\,\Omega$. Determine $c$ for maximum power consumption in $1\,\Omega$ load resistance.

**3.10.** A 3-phase unbalanced mesh connected load is connected across a balanced 250 V 3-phase supply as shown in Fig. P3.10.

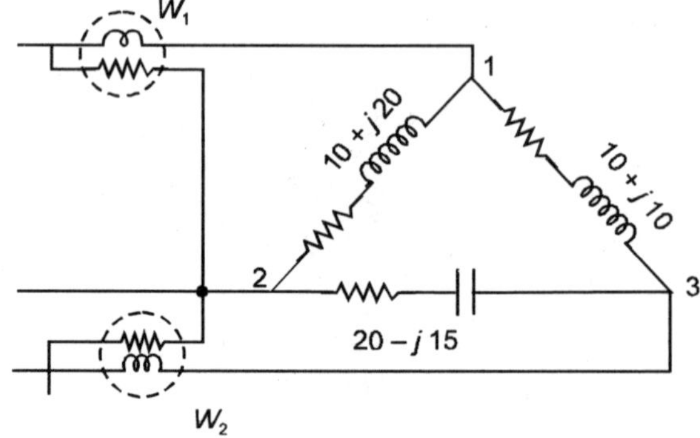



**Fig. P3.10**

Determine the current in each branch, the current in each line, the reading of each wattmeter, phase sequence RYB.

**3.11.** Using Thevenin's theorem determine the current through branch $a\ b$.

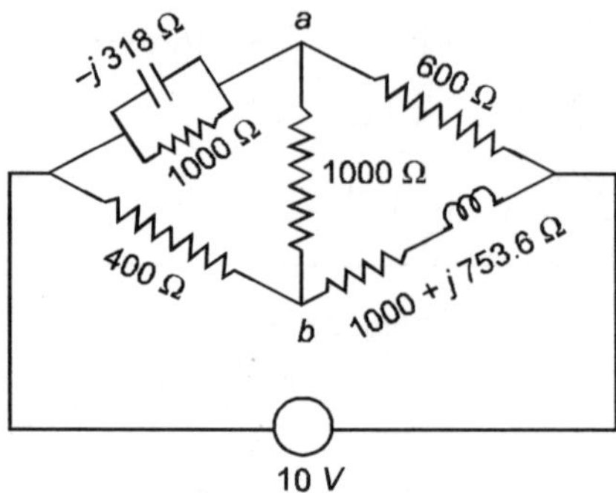



**Fig. P3.11**

**3.12.** A 3-phase 4-wire system has a star connected load $Z_R = (3 - j1)\,\Omega$ $Z_Y = (1 - j0.5)$ and $Z_B = (3 + j4)\,\Omega$ is connected across a 3-phase 4-wire 400 volt supply. The leads connecting the supply and the load have impedance of $(2 + j1)$. The wire connecting the star point of load and system neutral also has impedance $(2 + j1)$. Using Millman Theorem, determine the potential of the neutral.

**3.13.** Verify Tellegen theorem for the pair of networks shown. Select suitable values in the two circuits.

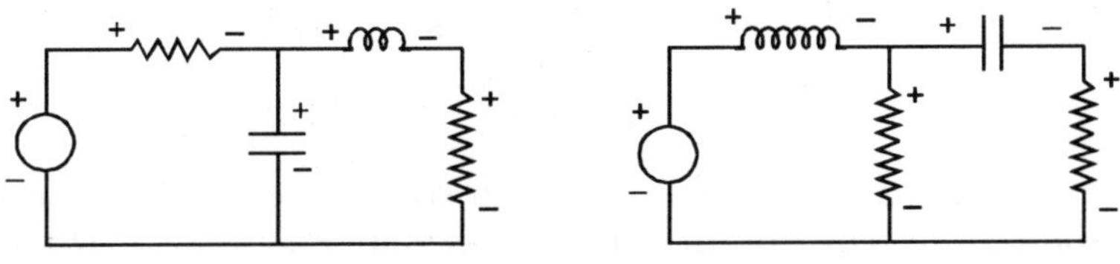

**Fig. P3.13**

**3.14.** Three non-reactive resistances are connected in star across a 440 V 3 – phase supply. $Z_R = 10$, $Z_Y = 15$ and $Z_B = 20\,\Omega$. Determine the potential of the star point of the load using Millman theorem and calculate the line currents.

**3.15.** If in the network the resistance is changed from $10\,\Omega$ to $5\,\Omega$, determine $V_C$ using compensation theorem. Also find the change in current through $R$.

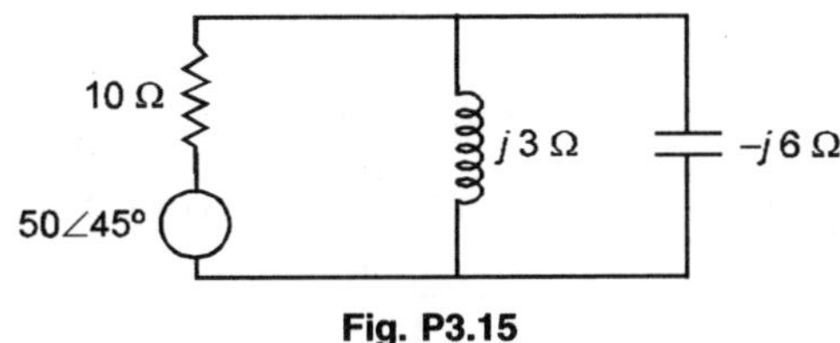

**Fig. P3.15**

**3.16.** For a certain network Thevenin's equivalent resistance equals its load resistance. Now if we want the load to be twice its previous value what should be the load resistance?

**3.17.** Using compensation theorem determine the change in current in the circuit when the reactance is changed to $j35\,\Omega$.

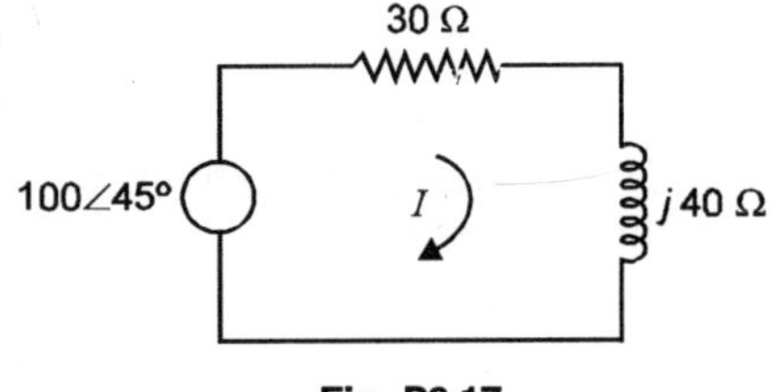

**Fig. P3.17**

**3.18.** Determine the current flowing through the $3\,\Omega$ resistor and verify substitution theorem for the network by replacing the $3\,\Omega$ resistor by (*a*) a current source (*b*) a voltage source.

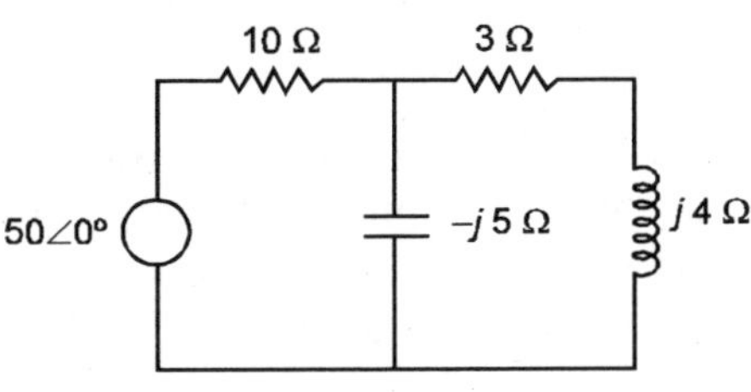

**Fig. P3.18**

**3.19.** In the Fig. P3.19 shown use substitution theorem to replace $L$ by a current source $i_L$ and $C$ by a voltage source $V_C$. Use superposition to obtain $V_L$ and $I_C$.

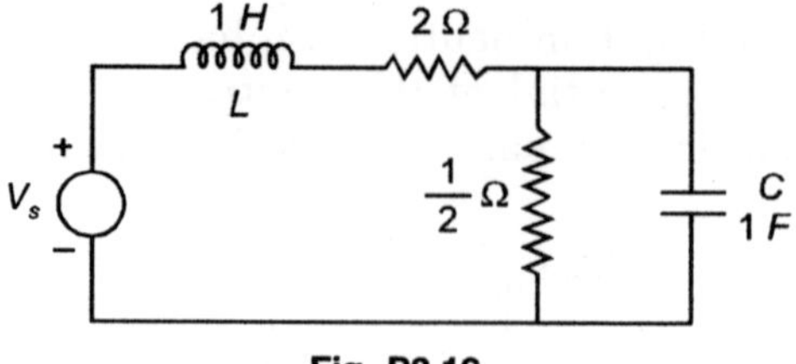

**Fig. P3.19**

**3.20.** If in the Fig. P. 3.20 $(a)$, the current through the ammeter when the voltage is $20\angle30º$ is 4 $\angle{-}45º$ determine current in the ammeter in Fig. P 3.20 $(b)$ when the voltage applied in the circuit is $15\angle90º$.

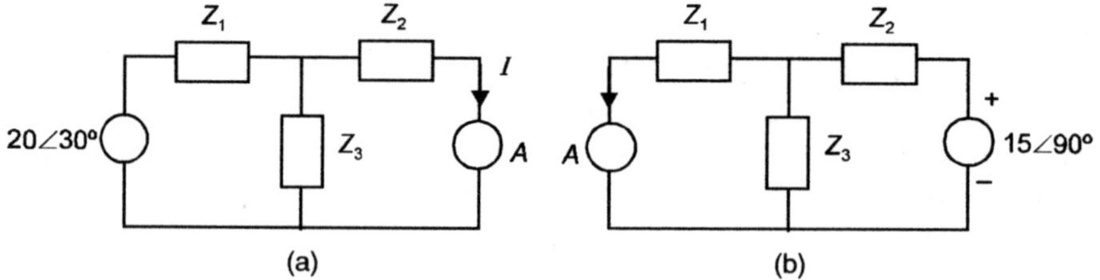

(a)                              (b)

**Fig. P3.20**

**3.21.** Using superposition principle determine the current through 10 $\Omega$ branch.

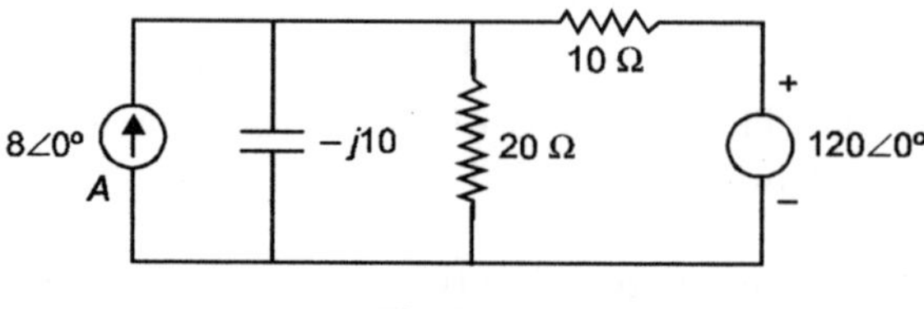

**Fig. P3.21**

**3.22.** Using Thevenin's theorem determine the current through 10 $\Omega$ inductive branch of the network.

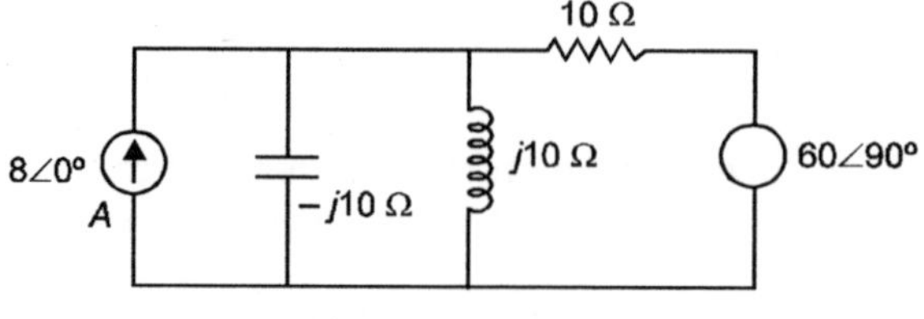

**Fig. P3.22**

**3.23.** A current $I = 2\angle0ºA$ flows between terminals 1 and 2 in a complex linear network when a voltage $v = 10\angle90º$ volt is applied between terminals 3 and 4. If a voltage $v = 25\angle45º$ is applied between 1 and 2 what current be observed to flow between terminals 3 and 4. State the theorem used.

**3.24.** In the circuit shown $R_g$ is variable between 2 Ω and 55 Ω. What value of $R_g$ results in maximum power transfer to the load $R_L$?

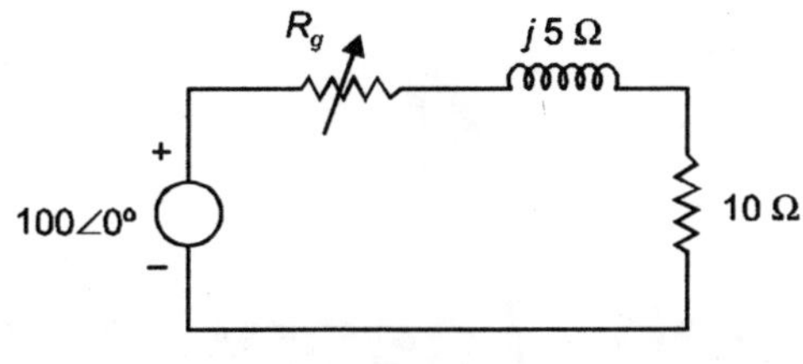

**Fig. P3.24**

**3.25.** A 10 volt generator has fixed loss of 2 watt and a Thevenin resistance of 5 Ω. What value of $R_L$ should be used for (*a*) maximum efficiency $P_L/P_T$ (*b*) maximum power transfer?

Find the load power in each case.

**3.26.** In the network shown determine the impedance $z$ to absorb maximum power. Also determine the power absorbed by the impedance.

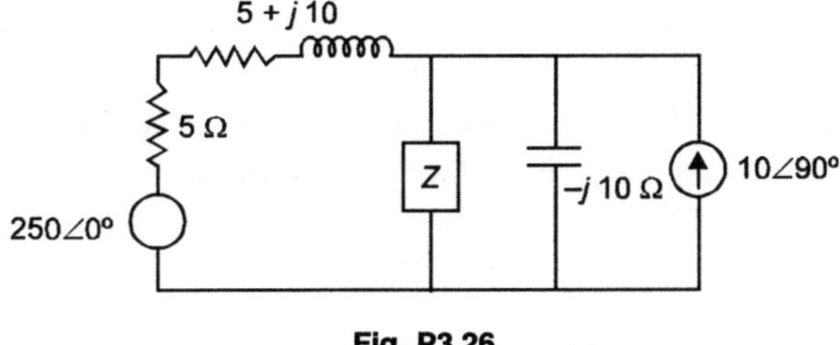

**Fig. P3.26**

**3.27.** For the networks of problem 3.6 determine Norton's equivalent networks.

# 4

# Graph Theory and Network Equations

## 4.1  INTRODUCTION

We have studied in chapter 1 that when two or more than two components or elements are interconnected, the result is an electric network. A component or an element between two terminals to which connections can be made is called a branch. The position where two or more than two branches are connected together is called a node. A single closed path in a network is called a loop.

Kirchhoffs two laws express constraint imposed in the currents and voltages of the network elements by the very arrangements of these elements into a structure. Network topology is a generic name that refers to all properties arising from the structure or geometry of a network.

Since the topological properties of an electric network are independent of the types of elements or component ($R$, $L$ or $C$) which constitute the branches, it is convenient to replace each network element by a simple line segment without caring for v-i relation of the component. The resulting structure, therefore, consists of line segments and nodes. The theory of linear graphs handles such structures very efficiently.

## 4.2  DEFINITIONS

A linear graph is a collection of points known as nodes and line segments known as branches, the nodes being joined together by branches. It is to be noted here that the voltage and current sources have to be replaced by their internal impedances. An ideal voltage source is to be replaced by a short circuit and an ideal current source by an open circuit i.e. there will not be any branch corresponding to an ideal current source. Fig. 4.1 shows (a) an electric network and (b) linear graph.

The branches and nodes of the graph have been numbered. The current source branch does not appear. There are 4 nodes and 6 branches. The arrows on the branches indicate the orientation of the branches and the graph is known as oriented graph. The orientation of the branches will be assumed to coincide with the current reference.

**Incidence of a branch:** Branches whose end fall on a node are said to be incident at the node. In Fig. 4.1(b) branches 1, 2, 3 are said to be incident at node 1.

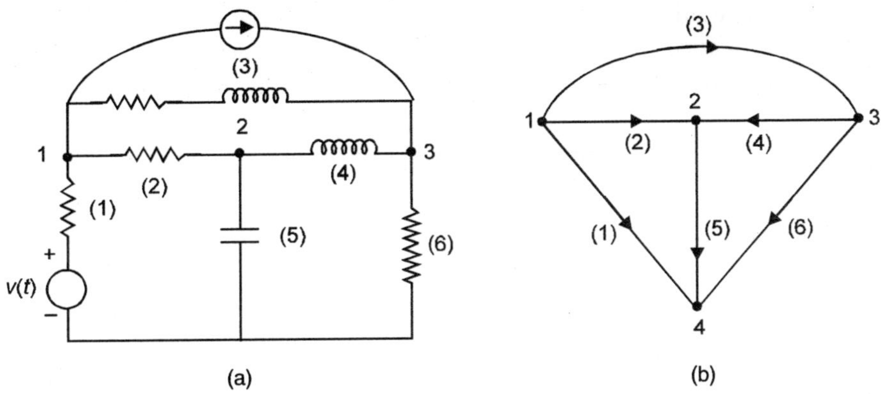

Fig. 4.1 (a) Electric network, (b) Linear graph of (a).

**Subgraph:** A subgraph is a subset of nodes and branches of a graph. The subgraph is said to be proper if it has number of nodes and branches stricitly less than that of the original graph.

**A path:** It is a particular subgraph consisting of an ordered sequence of branches having the following properties:

1. At all the internal nodes, there are incident exactly two branches of the subgraph. However, at the two terminal nodes only one branch is incident.
2. No proper subgraph of this subgraph having the same two terminal nodes has the property at 1.

Figure 4.2 is a subgraph of the original graph which has node 2 and 4 as the internal nodes and 1 and 3 as the terminal nodes.

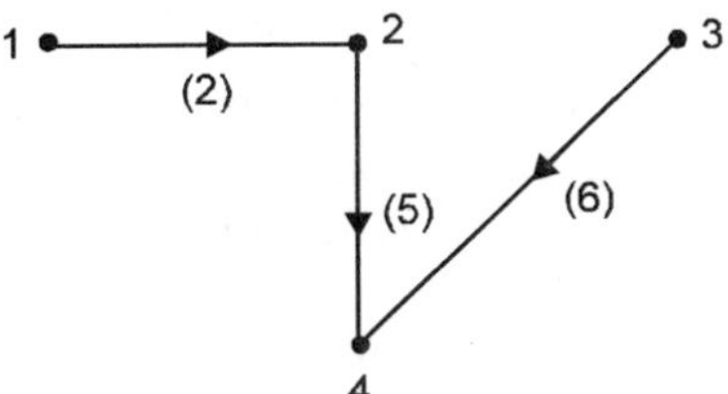

Fig. 4.2 Subgraph of graph in Fig. 4.1(b).

**Connected graph:** A graph is said to be connected when there exists at least one path between any two nodes. The graph having a transformer as one of the elements is an unconnected graph.

**Loop:** It is a subgraph of a graph wherein at each node exactly two branches are incident. While specifying a loop, a set of nodes or a set of branches consisting the loop are to be listed e.g. the loop containing nodes 2, 3, 4 can be specified with branches 4, 6, 5 and so on. A loop in a graph has the following properties:

1. There are at least two branches in a loop.
2. The number of branches equal the number of nodes.
3. There are exactly two paths between any pair of nodes in the loop.

**Tree:** A tree is a connected subgraph of a connected graph having all the nodes of the graph but without any loop. Since a tree has all the nodes of the original graph it is specified by listing the branches. Some of the trees of the graph of Fig. 4.1(b) are given here:

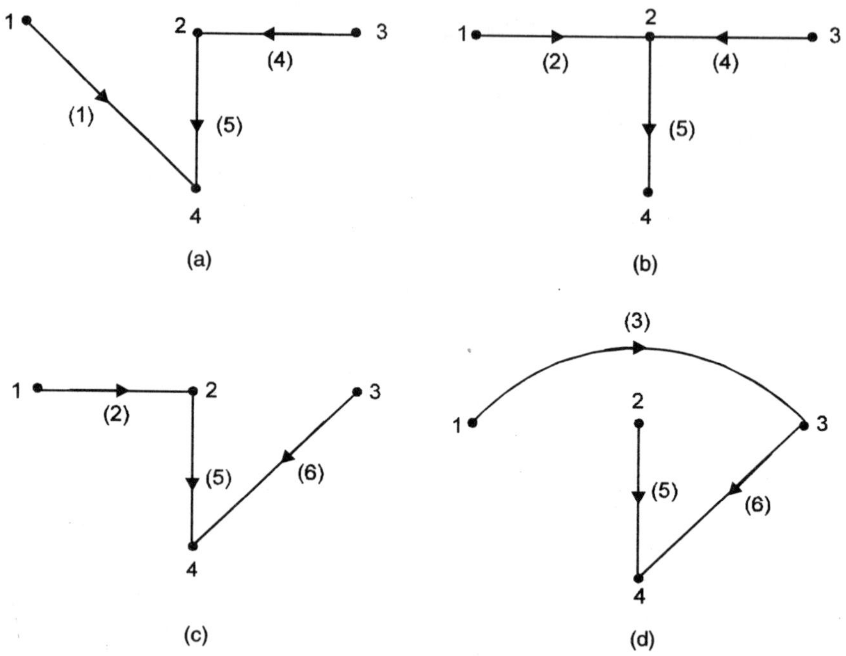

Fig. 4.3   Some of the trees of graph in Fig. 4.1(b).

The concept of TREE as we will see soon is very important in graph theory.

**TWIGS:** Branches of the tree are called TWIGS.

**LINKS:** Those branches of the graph which are removed from the tree are called LINKS. Since these lines are the complement of the twigs, they constitute what is known as Co-tree. It is clear that Twigs and Links can be specified only when a tree is selected. Twigs in one tree may form links in another tree.

In Fig. 4.3(b) all the twigs are incident at a common node (node 2), such a tree is known as a star like tree or a star-tree. Here only one node is internal node, all others are terminal nodes.

In Fig. 4.3(a) (c) and (d) the nodes or branches can be so ordered that the tree consists of a single path extending form one node to the last node. Such a tree is known as a linear tree. Here there are only two terminal nodes, all others are internal nodes.

Since a tree has all the nodes of the graph the number of branches in the tree is one less than the number of nodes i.e. if $n$ is the number of nodes, the number of branches in a tree is $(n - 1)$ and hence number of branches in the co tree is $b - (n - 1)$. If a graph is unconnected, the concept of tree is extended to each separate parts and then the graph consisting of one tree for each separate part is known as a forest. If $p$ is the number of separate parts and $n$ the number of nodes of a graph, the forest will have $(n - p)$ twigs. The complement of a forest is a co-forest.

While drawing a graph out of a given network all passive elements between nodes are represented by line segments whereas for independent current and voltage sources if these have accompanied passive elements viz a shunt admittance branch across a current source and a series impedance branch in a voltage source, the current and voltage sources are to be replaced by the segments corresponding to the passive accompanied element of source, and the current source is to be open circuited and voltage source short circuited. However, if the sources do not have the accompanied passive elements, these sources should be shifted into the network as explained in Chapter 1.

## 4.3  INCIDENCE MATRIX

The information regarding the incidence of branches at various nodes and their orientation with respect to the nodes is nicely described in the form of a matrix known as Incidence matrix and is obtained as follows for a given graph.

For a graph having $n$ number of nodes and $b$ number of branches the complete incident matrix $Aa$ is an $n \times b$ rectangular matrix, the elements of which are given as:

$a_{ij} = 1$ if branch $j$ is incident at node $i$ and is oriented away form the node $i$.

$a_{ij} = -1$ if branch $j$ is incident at node $i$ and is oriented towards the node $i$.

$a_{ij} = 0$ if branch $j$ is not incident at node $i$. The subscript '$a$' to $A$ stands for all nodes of the graph. The incidence matrix for the graph for Fig. 4.1 (b) is given below:

$$Aa = \begin{array}{c} \\ 1 \\ 2 \\ 3 \\ 4 \end{array}\begin{array}{c} \begin{array}{cccccc} 1 & 2 & 3 & 4 & 5 & 6 \end{array} \\ \left[\begin{array}{cccccc} -1 & 1 & 1 & 0 & 0 & 0 \\ 0 & -1 & 0 & -1 & 1 & 0 \\ 0 & 0 & -1 & 1 & 0 & 1 \\ 1 & 0 & 0 & 0 & -1 & -1 \end{array}\right] \end{array} \qquad ...(4.1)$$

From the matrix $Aa$ it can be seen that each column corresponding to branches has a single +1 and −1 entries which is obvious as each branch will be connected between two nodes and the orientation of the branch will be towards one node and hence away from the other node. Thus if all the other rows are added to the last row the net result will be a row of zeros which shows that all rows are not independent. At least one of these rows can be represented as a linear combination (negative sum of other rows) of the remaining rows. Therefore, one of these rows can be eliminated. Thus the rank of $Aa$ can be no more than $(n - 1)$. The matrix so obtained after eliminating one of the rows is known as reduced incidence matrix or incidence matrix and is denoted by $A$. This concept has physical significance also in that one of the nodes should always be taken as a reference and, therefore, the row corresponding to reference node should be eliminated from the incidence matrix. The incidence matrix is given as

$$A = \begin{array}{c} \\ 1 \\ 2 \\ 3 \end{array}\begin{array}{c} \begin{array}{cccccc} 1 & 2 & 3 & 4 & 5 & 6 \end{array} \\ \left[\begin{array}{cccccc} -1 & 1 & 1 & 0 & 0 & 0 \\ 0 & -1 & 0 & -1 & 1 & 0 \\ 0 & 0 & -1 & 1 & 0 & 1 \end{array}\right] \end{array} \qquad ...(4.2)$$

Rearranging the incidence matrix by putting elements corresponding to twigs of tree in Fig. 4.3(a) first, followed by the links, we have

$$
A = \begin{array}{c} \\ 1 \\ 2 \\ 3 \end{array}
\begin{array}{cccccc}
\overbrace{\phantom{1\ \ 5\ \ 4}}^{\text{Twigs}} & & & \overbrace{\phantom{2\ \ 3\ \ 6}}^{\text{Links}} & & \\
1 & 5 & 4 & 2 & 3 & 6 \\
\end{array}
$$

$$
A = \begin{array}{c} 1 \\ 2 \\ 3 \end{array}
\left[\begin{array}{cccccc}
-1 & 0 & 0 & 1 & 1 & 0 \\
0 & 1 & -1 & -1 & 0 & 0 \\
0 & 0 & 1 & 0 & -1 & 1
\end{array}\right] \qquad \text{...(4.3)}
$$

or
$$
A = [A_t, \ A_l] \qquad\qquad \text{...(4.4)}
$$

The determinant of the sub matrix $A_t$ is $-1\,(1) = -1$ which is non- zero and hence the sub matrix $A_t$ is non singular and, therefore, the rank of $A$ is 3 or in general $(n-1)$. In fact, in general the rank of the incidence matrix is $(n-1)$ where $n$ is the total number of nodes in a graph. Conversely it can be said that if a non-singular submatrix of $(n-1) \times (n-1)$ is given, out of a reduced incidence matrix, the columns of the matrix will correspond to twigs of a particular tree of the original graph. By taking various trees it can be shown that the determinant of $A_t$ is either $+1$ or $-1$. It can be seen that the number of trees that can be obtained out of a graph is given by det $[AA^T]$.

In the present example

det $[A\ A^T]$

$$
= \det \left[\begin{array}{cccccc}
-1 & 0 & 0 & 1 & 1 & 0 \\
0 & 1 & -1 & -1 & 0 & 0 \\
0 & 0 & 1 & 0 & -1 & 1
\end{array}\right]
\left[\begin{array}{ccc}
-1 & 0 & 0 \\
0 & 1 & 0 \\
0 & -1 & 1 \\
1 & -1 & 0 \\
1 & 0 & -1 \\
0 & 0 & 1
\end{array}\right]
$$

$$
= \det \left[\begin{array}{ccc}
3 & -1 & -1 \\
-1 & 3 & -1 \\
-1 & -1 & 3
\end{array}\right]
$$

$$
= 3[9-1] + 1[-3-1] - 1[1+3]
$$
$$
= 24 - 4 - 4 = 16
$$

This shows that not all combinations of any three branches will constitute a tree. Some of them will turn to form loops and, therefore, even though the total combinations possible in this case are $^6C_3 = \dfrac{6 \times 5 \times 4}{1 \times 2 \times 3} = 20$, the possible combinations which constitute a tree are 16 rather than 20. We have been able to write an incidence matrix given a graph. However, at times, it may be the converse situation. Given an incidence marix how to draw a graph. If a reduced incidence matrix is given, add to it one more node and corresponding to this row, add the element 0, 1 or $-1$ to make the sum of entries in various columns as zero.

For example let the given reduced incidence matrix be

$$
A = \left[\begin{array}{cccccccc}
1 & 1 & 0 & 0 & 0 & 0 & 0 & -1 \\
0 & -1 & 1 & 1 & 0 & 0 & 0 & 0 \\
0 & 0 & 0 & -1 & 1 & 1 & 0 & 0 \\
0 & 0 & 0 & 0 & 0 & -1 & 1 & 1
\end{array}\right] \qquad \text{...(4.5)}
$$

The augmented or complete matrix will be

$$A = \begin{bmatrix} 1 & 1 & 0 & 0 & 0 & 0 & 0 & -1 \\ 0 & -1 & 1 & 1 & 0 & 0 & 0 & 0 \\ 0 & 0 & 0 & -1 & 1 & 1 & 0 & 0 \\ 0 & 0 & 0 & 0 & 0 & -1 & 1 & 1 \\ -1 & 0 & -1 & 0 & -1 & 0 & -1 & 0 \end{bmatrix} \qquad \text{...(4.6)}$$

Place 5 nodes (corresponding to $Aa$) on the paper and draw segments in between various nodes as given by the column. For example branch 1 to be connected between node 1 and 5, oriented away from node 1 and towards node 5. Following this procedure following graph is obtained:

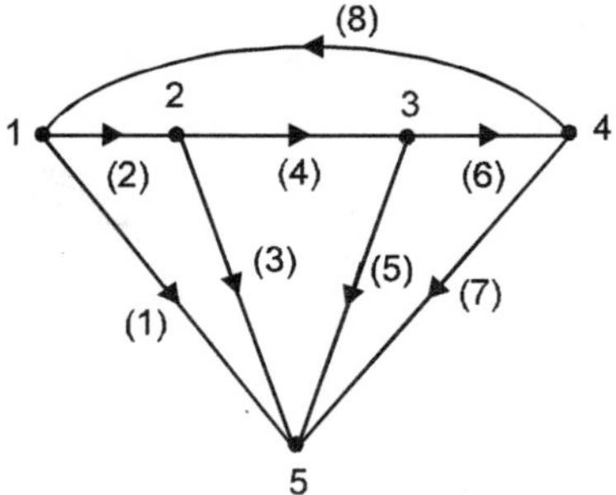

**Fig. 4.4**   One possible graph of the incidence matrix.

The shape of the graph may turn out to be different depending upon initial placement of the different nodes e.g. the other shape with the same $Aa$ would be

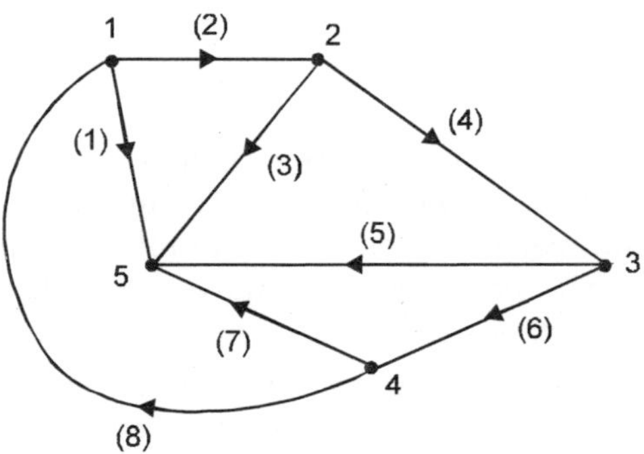

**Fig. 4.5**   Another graph of incidence matrix.

Two graphs are said to be isomorphic if they have the same incidence matrix which means that they have same number of nodes, branches and the same orientation of branches towards various nodes.

Following observations are made about the incidence matrix.

(*i*) The unit entries (1 or –1) made in a row identify the branches incident at a node. The degree of the node equals the no. of branches incident at a node.

(*ii*) Columns of $A$ with unit entries in two identical rows correspond to two branches with same end nodes and hence they are in parallel.

(*iii*) If the degree of a node is two that is if there are two unit entries in a row which means there are exactly two branches incident at the node and hence the branches are in series.

(*iv*) If the degree of a node is one i.e. if there is one unity entry in a row, there is only one branch incident at that node which is possible in a tree only.

(*v*) The incidence matrix is a highly sparse matrix i.e. it has few no. of non-zero elements. For a 100 node system the non-zero elements are about 2 to 3% and the remaining are zeros. For this reasons this requires a small memory for storage in a computer.

## 4.4 LOOP MATRIX

The incidence matrix gives information regarding the number of nodes, number of branches and their connection and orientation of branches to the corresponding nodes. It does not give any idea about the inter connection of branches which constitutes loops. It is, however, possible to obtain this piece of information in a matrix form. For this we first assign each loop of the graph with an orientation with the help of a curved arrow as shown in Fig. 4.6.

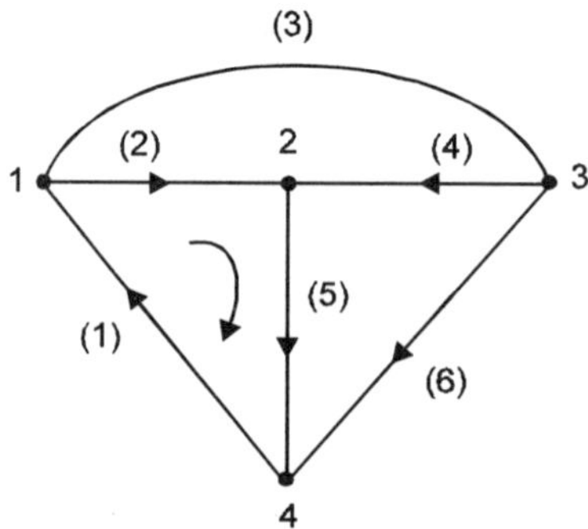

**Fig. 4.6**  Loop orientation.

In order to avoid clustering of arrows it is desirable to list all the loops with ordered nodes e.g. the oriented loop can be listed as [1, 2, 4] and [1, 3, 2].

For a graph having $n$ number of nodes and $b$ number of branches, the complete loop matrix or complete circuit matrix $Ba = [b_{ij}]$ will have $b$ number of columns and as many number of rows as the number of possible loops and its elements $b_{ij}$ have the values.

$b_{ij}$ = 1 if branch $j$ is in loop $i$ and their orientations coincide
  = –1 if branch $j$ is in loop $i$ and their orientations do not coincide
 = 0 if branch $j$ is not in loop $i$.

In case of complete incidence matrix it is possible to know the number of rows (equal to number of nodes) and number of columns (equal number of branches) in a graph. It is not true in case of complete loop matrix. Even though the number of columns are known equal to number of branches the number of rows can be known only if all loops of the graph have been identified. For the example chosen there are seven possible loops listed as here.

Loop 1 [1, 2, 4]     Loop 2 [1, 3, 2]     Loop 3 [2, 3, 4]
Loop 4 [1, 3, 4]     Loop 5 [1, 2, 3, 4]     Loop 6 [1, 2, 4, 3] and
Loop 7 [3, 2, 4, 1]

The loop matrix, therefore, is

$$Ba = \begin{array}{c c} & \begin{array}{c} \text{Branches} \\ \begin{array}{c c c c c c} \text{Loops } 1 & 2 & 3 & 4 & 5 & 6 \end{array} \end{array} \\ \begin{array}{c} 1 \\ 2 \\ 3 \\ 4 \\ 5 \\ 6 \\ 7 \end{array} & \left[ \begin{array}{rrrrrr} 1 & 1 & 0 & 0 & 1 & 0 \\ 0 & -1 & 1 & 1 & 0 & 0 \\ 0 & 0 & 0 & -1 & -1 & 1 \\ 1 & 0 & 1 & 0 & 0 & 1 \\ 1 & 1 & 0 & -1 & 0 & 1 \\ 0 & 1 & -1 & 0 & 1 & -1 \\ 1 & 0 & 1 & 1 & 1 & 0 \end{array} \right] \end{array} \qquad \ldots(4.7)$$

The complete loop matrix appears to be quite unwieldy for a network having only 4 nodes and, therefore, to solve a network of this characteristic (4 nodes and 6 branches) does not require such a large matrix. In fact there exists a submatrix of the complete loop matrix which has interesting characteristics and more useful from the viewpoint of solution of networks and is studied next.

As is said earlier the concept of tree is very important in graph theory or solution of network. We use this concept here. Suppose we take a tree and remove all the links. Next we connect one link at a time which will constitute one loop and orientation of the loop will be determined by the orientation of the link for that loop. Thus there will be as many loops as the number of links. Such loops are known as tie-set fundamental loops or f-loops for short. Therefore, in a graph having $n$ nodes and $b$ branches, the number of f-loop will be $b - (n - 1)$ and the order of the f-loop matrix will be

$$[b - (n - 1)] \times b.$$

Let us consider the tree as given in Fig. 4.3(b) reproduced here as Fig. 4.7.

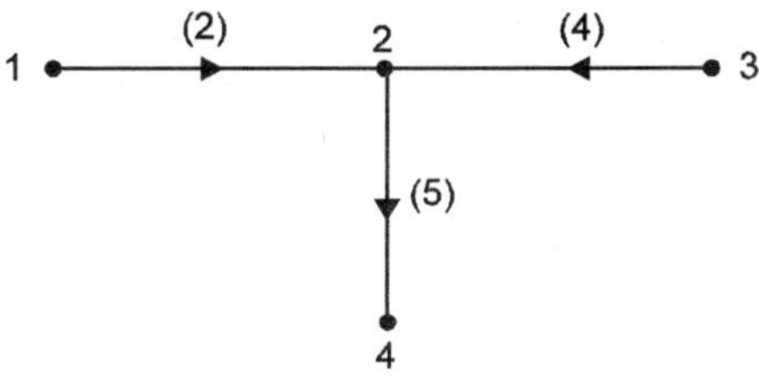


**Fig. 4.7**

Therefore f-loop matrix is going to be $3 \times 6$ matrix

$$B_f = \begin{array}{c c} & \begin{array}{c} \begin{array}{c c c c c c} & \text{Twigs} & & & \text{Links} & \\ \text{Loops } 2 & 4 & 5 & 1 & 3 & 6 \end{array} \end{array} \\ \begin{array}{c} 1 \\ 2 \\ 3 \end{array} & \left[ \begin{array}{rrrrrr} 1 & 0 & 1 & 1 & 0 & 0 \\ -1 & 1 & 0 & 0 & 1 & 0 \\ 0 & -1 & -1 & 0 & 0 & 1 \end{array} \right] \end{array} \qquad \ldots(4.8)$$

It is seen that the submatrix corresponding to links (of order $[b - (n - 1)] \times \{b - (n - 1)\}$ here $3 \times 3$ is an identity matrix and therefore the rank of $B_f$ is $3 \times 3$ or in general $\{b - (n - 1\} \times \{b - (n - 1)\}$.

In f-loop, matrix can be partitioned as

$$B_f = [B_t, B_l] = [B_t \; U]$$

It can be seen that the sub matrix $B_l$ will always be an identity matrix.

Now since $B_f$ is a submatrix of $B_a$, the rank of $B_a$ is not less than that of $B_f$. In fact it is found that rank of $B_a$ is same as that of $B_f$ namely $b - (n - 1)$. To prove this we follow the procedure as given here.

Suppose for a graph the columns in the two matrices $A_a$ and $B_a$ be in the same order then it can be shown that $A_a B_a^T = 0$ and $B_a A_a^T = 0$. These relations are known as orthogonality relations and are proved as follows. The two matrices are written as

$$
\begin{array}{cccccc}
1 & 2 & 3 & 4 & 5 & 6
\end{array}
\qquad\qquad\qquad \text{Loop}
$$

$$
\begin{array}{c}
1 \\ 2 \\ 3 \\ 4
\end{array}
\begin{bmatrix}
-1 & 1 & 1 & 0 & 0 & 0 \\
0 & -1 & 0 & -1 & 1 & 0 \\
0 & 0 & -1 & 1 & 0 & 1 \\
1 & 0 & 0 & 0 & -1 & -1
\end{bmatrix}
\begin{bmatrix}
1 & 0 & 0 & 1 & 1 & 0 & 1 \\
1 & -1 & 0 & 0 & 1 & 1 & 0 \\
0 & 1 & 0 & 1 & 0 & -1 & 1 \\
0 & 1 & -1 & 0 & -1 & 0 & 1 \\
1 & 0 & -1 & 0 & 0 & 1 & 1 \\
0 & 0 & 1 & 1 & 1 & -1 & 0
\end{bmatrix} = 0 \qquad \text{...(4.10)}
$$

After multiplication it is found that $AB^T = 0$. This can in general be reasoned out as follows. Consider any one column of $B_a$ (say 3) and any one of the rows of $A_a$ (say 2) i.e. consider a loop and a node. There are two possibilities that a node may be a part of the loop or it may not be on the loop. Say we take node and loops 2 and 4 separately. Node 2 lies on loop 3 and loop and node 2 have branches 4 and 5 and branch 4 is away from 2 and 5 is towards node 2 and both the branches are oppositely oriented to the loop 3 Therefore, the sum of product of elements of row 2 and column three are zero. If we take loop 4 and node 2, since node 2 is not in the loop 4, for every non-zero element in the row 2 of $A_a$ matrix there is a zero in the 4th loop of $B_a$ matrix giving product again to be zero. Thus it can be said that two matrices are orthogonal to each other. To obtain the rank of matrix $B_a$ we state here Sylvestre's law of nullity. It states that if the product of two matrices is zero, the sum of ranks of the two matrices can't be greater than the columns of the first matrix in the product.

Since $AB^T$ is zero, therefore the rank of $A+$ rank of $B \leq$ column of $A \leq b$

Since rank of $A = (n - 1)$

Therefore, rank of $B \leq b - (n - 1)$

Since it was pointed out earlier that the rank of $B_a$ can't be less than $b - (n - 1)$, therefore rank of $B_a = b - (n - 1)$

The $B$ matrix or loop matrix is also known as Tie set matrix.

## 4.5  CUT SETS AND THE CUT SET MATRIX

Consider again the graph of Fig. 4.1 which is reproduced here.

If we remove branches [1, 2, 3] Fig. 4.1 (b) or branches [3, 4, 6] Fig. 4.8 (c), the graph is cut into two parts. The branches which are removed are said to form a cut set. Therefore, a cut set is defined as a minimum set of branches of a connected graph whose removal causes the graph to become unconnected into exactly two connected subgraphs with further stipulation that the removal of any proper subset of this set leaves the graph connected. For example [1, 2, 3] is a cut set. However, if only branches 2 and 3 are removed, the graph gets connected through the branch 1. It is seen that set of branches incident at a node form a cut set.

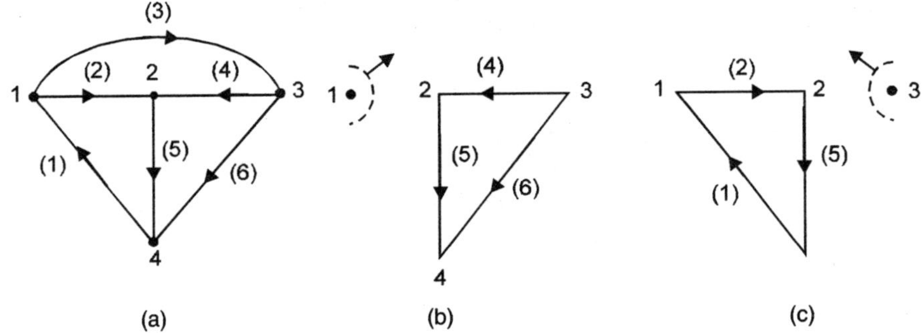

**Fig. 4.8** (a) Oriented graph (b) and (c) cut-sets.

Let us see removal of branches [1, 2, 3, 6], the graph is as shown in Fig. 4.9.

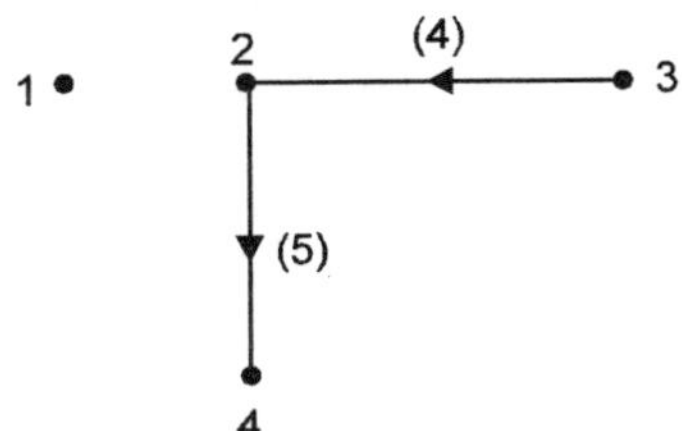

**Fig. 4.9** Example of a non-cut-set.

This also cuts the graph into two parts but [1, 2, 3, 6] does not form a cut set, as subset of this is a cut set [1, 2, 3]. Since it is the minimum number of branches which when removed should cut the graph into two unconnected subgraphs.

The cut set divides the nodes of a graph into two groups or parts, every node of the original graph being in one of these two parts. Also each branch of the cut set has one of its terminals at a node in one group and its other end incident at a node in the other group. A cut set is oriented by selecting an orientation from one of the two groups of connected nodes to the other group of connected nodes. The orientation can be shown as in Fig. 4.8 (b) and (c). The orientation of cut set branches may coincide with the orientation of the cut set or it may not.

Just as one incidence matrix gives the incident or orientation of the branches at various nodes the cut set matrix should provide all the branches in a cut set and their orientation with respect to the cut set. We define here the cut-set matrix $Q_a = [q_{ij}]$, which has its rows corresponding to cut set branches and the columns corresponding to all the branches of the graph. The elements have the following values:

$q_{ij} = 1$ if branch $j$ is in cut set $i$ and the orientation coincide.

$\quad = -1$ if branch $j$ is in cut set $i$ and the orientation do not coincide.

$\quad = 0$ if branch $j$ is not in cut set $i$.

The subscript '$a$' stands for all the cut-sets of the graph.

The various cut sets of the given graph are.

The four cut sets corresponding to 4 nodes are

$$[1, 2, 3], [2, 4, 5], [3, 4, 6], [1, 5, 6]$$

The other cut sets are

$$[1, 3, 4, 5], [2, 3, 5, 6] \text{ and } [1, 2, 4, 6]$$

The graphs for the last three cut sets are shown in Fig. 4.10

$$[1, 3, 4, 5] \quad [2, 3, 5, 6] \quad [1, 2, 4, 6]$$

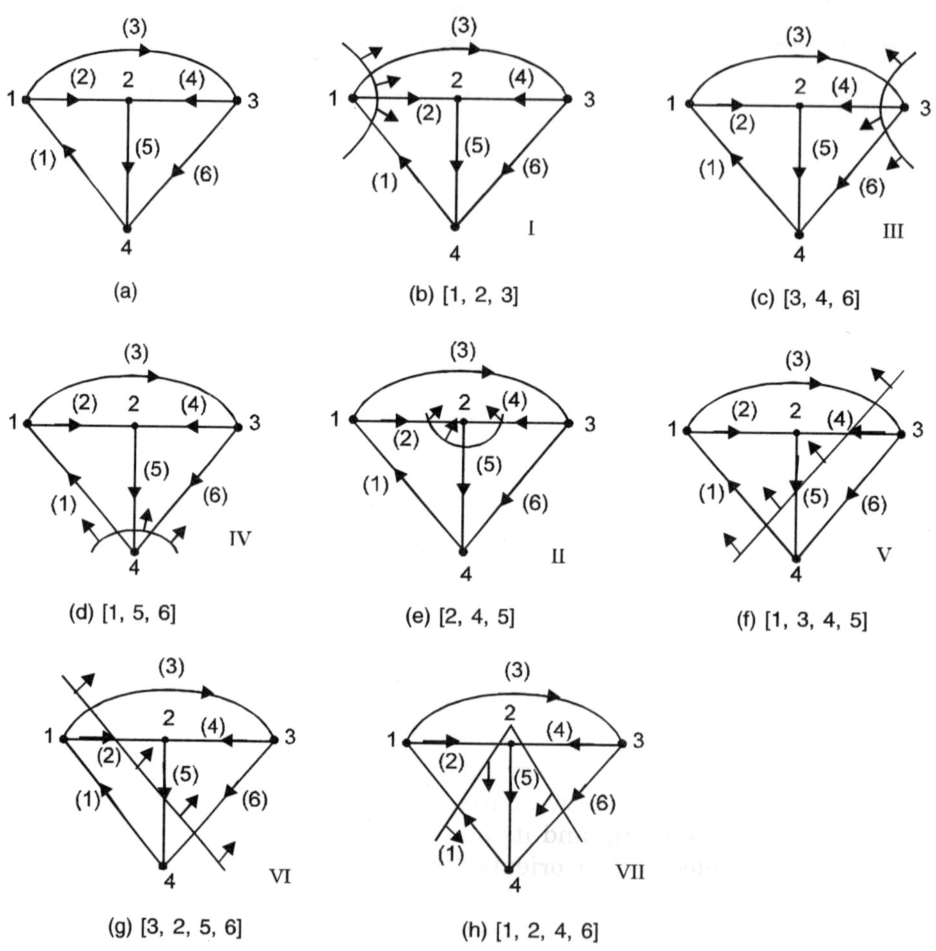

**Fig 4.10   (a) Oriented graph (b)-(h)** Various cut-sets of Fig 4.8.

$$
\begin{array}{c}
\text{Cut-set} \hspace{4em} \text{Branches} \\[4pt]
\begin{array}{c c}
 & \begin{array}{cccccc} 1 & 2 & 3 & 4 & 5 & 6 \end{array} \\
\begin{array}{c} 1 \\ 2 \\ 3 \\ 4 \\ 5 \\ 6 \\ 7 \end{array} &
\left[ \begin{array}{rrrrrr}
-1 & 1 & 1 & 0 & 0 & 0 \\
0 & -1 & 0 & -1 & 1 & 0 \\
0 & 0 & -1 & 1 & 0 & 1 \\
1 & 0 & 0 & 0 & -1 & -1 \\
1 & 0 & -1 & 1 & -1 & 0 \\
0 & 1 & 1 & 0 & -1 & -1 \\
-1 & 1 & 0 & 1 & 0 & 1
\end{array} \right]
\end{array}
\end{array}
$$

$$...(4.11)$$

For cut-set 1 corresponding to branches at node 1, the orientation is shown in Fig. 4.8(b), branches 2 and 3 are along the direction of the cut set whereas 1 is away from the cut set, therefore entries are 1, 1 and –1 respectively. Similarly entries for 2, 3 and 4 nodes can be carried out. For 5th cut-set as shown above, the direction of the cut set is shown and with reference to this, entries for 1 and 4 are positive 1 and for 3 and 5 it is –1. It is seen that the number of rows in the $Q_a$ matrix is much more than those in the incidence matrix. Again complete cut set matrix is normally not required for solution of networks. The cut set matrix corresponding to smaller cut sets is normally more useful and the formulation of this matrix is explained here.

Select a tree and remove one of the branches (twig) out of this, which will divide the tree into two connected subgraphs. All the links including the twig removed which go from one part of this graph to unconnected tree to the other part will constitute a cut set. We call this a fundamental cut set or cut-set for short. Therefore, for each twig there will be a fundamental cut set and for a graph having $n$ number of nodes, the number of twigs is $(n - 1)$, therefore, there will be $(n - 1)$ f-cut sets. The orientation of an f-cut set is chosen as the orientation of the corresponding twig. Consider the tree of Fig. 4.8(a).

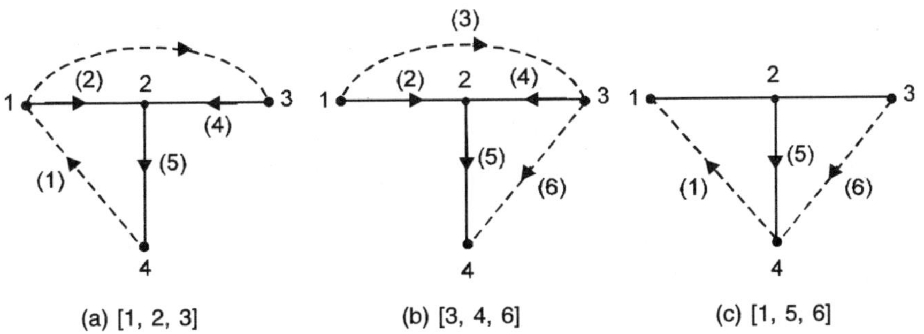



**Fig. 4.11**   (a) Branch 2 as twig of first cut-set,
(b) Branch 4 as twig of second cut-set,
(c) Branch 5 as twig of third cut-set.

$$Q_f = \begin{array}{c} \\ 2 \\ 4 \\ 5 \end{array} \begin{array}{c} \begin{array}{cccccc} 2 & 4 & 5 & 1 & 3 & 6 \end{array} \\ \left[ \begin{array}{cccccc} 1 & 0 & 0 & -1 & 1 & 0 \\ 0 & 1 & 0 & 0 & -1 & 1 \\ 0 & 0 & 1 & -1 & 0 & 1 \end{array} \right] \end{array} \qquad \text{...(4.12)}$$

The submatrix corresponding to first three columns of the matrix is an identity matrix and hence it is non-singular and the rank of the matrix equals the number of twigs in a tree i.e. the rank is $(n - 1)$.

It can be proved after algebrabric manipulation that $A = A_t\, Q_f$ where $A$ is the reduced incidence matrix $A_t$ is the submatrix of $A$ corresponding to twigs of a tree and $Q_f$ is the fundamental cut set matrix corresponding to the same tree. We know that $A_t$ is a non-singular matrix, therefore the incidence matrix of a graph is row equivalent to the fundamental cut-set for same tree. Thus the rows of $A$ are linear combination of the row of $Q_f$ and vice versa.

Let us obtain $A_t$ out of $A$ from the previous description for the twig [2, 4, 5].

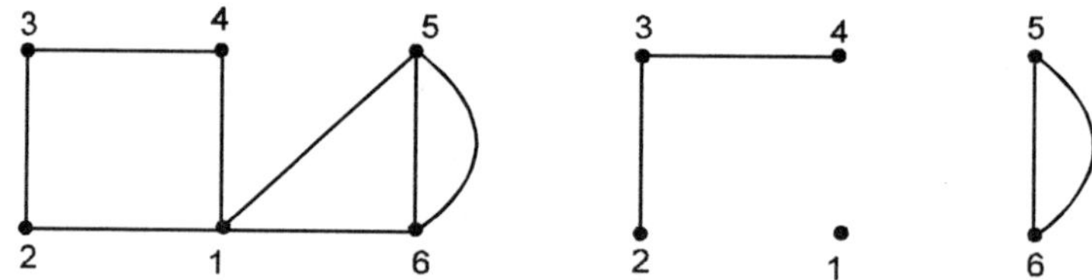

Consider the Fig. 4.12. Here if branches at node 1 are removed, this divides the graph into three parts

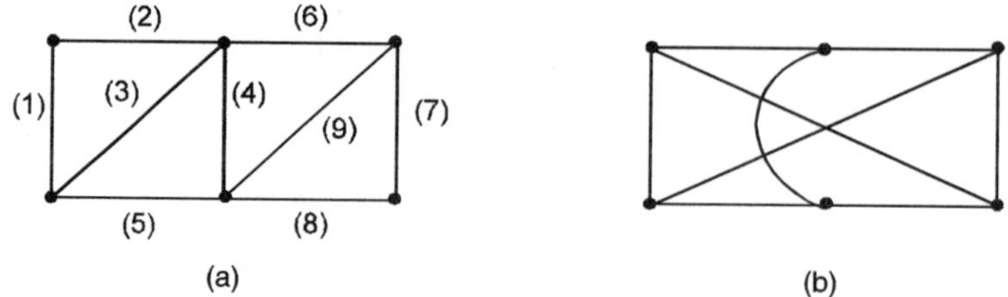

**Fig. 4.12** Graph not having a cut-set.

Such a graph is known as a hinged graph and is defined a graph in which there is at least one subgraph which has only one node in common with its complement subgraph in the graph. A node having this property is known as a hinged node. Therefore, the branches removed from node 1 will not form a cut-set.

## Planar Graph

A planar graph is one which can be mapped on a plane in such a way that no two branches cross each other Fig. 4.13 shows two graphs having the same number of nodes and branches but one of them is a planar graph the other one is not

**Fig. 4.13** (a) Planar, (b) Non planar graph.

The branches of a planar graph separate a plane into small regions; each of these is called a mesh. A mesh is defined as a sequence of branches of a planar graph that encloses no other branch of the graph with the boundary formed by these branches. In Fig. 4.13 (a) [1, 2, 3] form a mesh whereas [1, 2, 4, 5] do not form. Therefore, every mesh is a loop but every loop is not necessarily a mesh.

*The relation between Branch Voltages and Twig Voltages*

The two voltages are connected through the f-cut set matrix and is given as

$$V_b = Q^T_f V_t \qquad \qquad ...(4.14)$$

where $V_b$ is the branch voltage $Q_f$ the f-cut- set matrix and the $V_t$ the twig voltage. Consider one of the trees of the graph given in Fig. 4.14. Considering twigs as branches 2, 4, 5.

$$V_1 = -V_{t_2} - V_{t_5}, \quad V_2 = V_{t_2}, \quad V_3 = V_{t_2} - V_{t_4}$$

$$V_4 = V_{t_4}, \quad V_5 = V_{t_5} \text{ and } V_6 = V_{t_4} + V_{t_5}$$

This can be written in the matrix form as

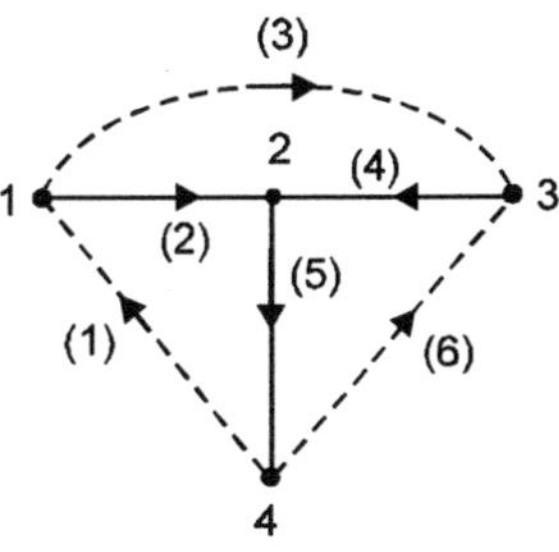

Fig. 4.14

$$\begin{bmatrix} V_1 \\ V_2 \\ V_3 \\ V_4 \\ V_5 \\ V_6 \end{bmatrix} = \begin{bmatrix} -1 & 0 & -1 \\ 1 & 0 & 0 \\ 1 & -1 & 0 \\ 0 & 1 & 0 \\ 0 & 0 & 1 \\ 0 & 1 & 1 \end{bmatrix} \begin{bmatrix} V_{t_2} \\ V_{t_4} \\ V_{t_5} \end{bmatrix} \qquad ...(4.15)$$

If we compare this matrix with the matrix earlier obtained for this tree, column by column, according to the branch number it is found that the two matrices are indentical. Hence,

$$V_b = Q^T_f V_t \qquad \qquad ...(4.14)$$

## 4.6 INTERRELATION AMONG VARIOUS MATRICES

We have studied three different types of matrices from the graph of a network. These are: (*i*) incidence matrix $A$ or $A_r$ (*ii*) The loop matrix $B$ or fundamental loop matrix $B_f$ and (*iii*) the cut-set matrix $Q$ or f-cut-set matrix $Q_f$.

These are not independent matrices i.e. these are interrelated through some relation such that if one of these is known, the other two can be obtained. We will not give proof of these relations. We mention the relations and use them as and when required.

$$A_r B^T_f = 0 \qquad \qquad ...(4.16)$$

Here column of the two matrices are to be arranged in the same order i.e. first the twigs and then the links, the two matrices are said to be orthogonal to each other.

If we write these matrices in terms of their submatrices corresponding to twigs and links we have

$$[A_t, A_l] \begin{bmatrix} B^T_t \\ I \end{bmatrix} = 0 \qquad \qquad ...(4.17)$$

or $\qquad A_t B^T_t + A_l = 0$

$$B^T_t = -A_t^{-1} A_l \qquad \qquad ...(4.18)$$

This means that the column of $B_t$ are a linear combination of the rows of $A_l$.

Taking 2, 4, 5 as twigs and 1, 3, 6 as links

$$A_t = \begin{pmatrix} 1 & 0 & 0 \\ -1 & -1 & 1 \\ 0 & 1 & 0 \end{pmatrix}$$

Det $[A_t] = -1$

$$Ad; \, A_t = \begin{pmatrix} -1 & 0 & -1 \\ 0 & 0 & -1 \\ 0 & -1 & -1 \end{pmatrix}^T = \begin{pmatrix} -1 & 0 & 0 \\ 0 & 0 & -1 \\ -1 & -1 & -1 \end{pmatrix}$$

$$A_t^{-1} = \begin{pmatrix} 1 & 0 & 0 \\ 0 & 0 & 1 \\ 1 & 1 & 1 \end{pmatrix}$$

$$A_l = \begin{pmatrix} -1 & 1 & 0 \\ 0 & 0 & 0 \\ 0 & -1 & 1 \end{pmatrix}$$

Hence

$$B_t^T = -A_t^{-1} A_l = -\begin{pmatrix} 1 & 0 & 0 \\ 0 & 0 & 1 \\ 1 & 1 & 1 \end{pmatrix}\begin{pmatrix} -1 & 1 & 0 \\ 0 & 0 & 0 \\ 0 & -1 & 1 \end{pmatrix}$$

$$= -\begin{pmatrix} -1 & 1 & 0 \\ 0 & -1 & 1 \\ -1 & 0 & 1 \end{pmatrix} = \begin{pmatrix} 1 & -1 & 0 \\ 0 & 1 & -1 \\ 1 & 0 & -1 \end{pmatrix}$$

$$B_t = \begin{pmatrix} 1 & 0 & 1 \\ -1 & 1 & 0 \\ 0 & -1 & -1 \end{pmatrix}$$

Hence we have been able to obtain $B_t$ if matrix $A_r$ is known through the relation given in equation 4.18.

By taking $A_r$ and $B_f$ for the same network we can prove that the two matrices are orthogonal.

$$A \, B^T_f = \begin{pmatrix} 1 & 0 & 0 & -1 & 1 & 0 \\ -1 & -1 & 1 & 0 & 0 & 0 \\ 0 & 1 & 0 & 0 & -1 & 1 \end{pmatrix}\begin{pmatrix} 1 & -1 & 0 \\ 0 & 1 & -1 \\ 1 & 0 & -1 \\ 1 & 0 & 0 \\ 0 & 1 & 0 \\ 0 & 0 & 1 \end{pmatrix}$$

$$A \, B^T_f = [0]$$

### 4.6.1  Relation between Incidence Matrix and f-cut-set Matrix

We know that KCL equations can be obtained from a graph by using incidence as well as f-cut-set matrices and are given as

$$A\,I_b = 0 \qquad \qquad \text{...(4.19)}$$

and

$$Q_f\,I_b = 0 \qquad \qquad \text{...(4.20)}$$

Let

$$I_b = \begin{vmatrix} I_t \\ I_l \end{vmatrix} \qquad \qquad \text{...(4.21)}$$

Therefore, the above equations are written as

$$[A_t\,,A_l]\begin{bmatrix} I_t \\ I_l \end{bmatrix} = 0$$

and

$$[Q_t\,,Q_l]\begin{bmatrix} I_t \\ I_l \end{bmatrix} = 0$$

or

$$[I\,,Q_l]\begin{bmatrix} I_t \\ I_l \end{bmatrix} = 0$$

$$A_t\,I_t + A_l\,I_l = 0$$

or

$$I_t = -A_t^{-1}\,A_l\,I_l \qquad \qquad \text{...(4.22)}$$

whereas from the f-cut-set matrix we have

$$I_t = -\,Q_l\,I_l \qquad \qquad \text{...(4.23)}$$

Hence,

$$Q_l = A_t^{-1}\,A_l \qquad \qquad \text{...(4.24)}$$

This means that the rows of f-cut-set matrix are a linear combination of incidence matrix. Therefore, if incidence matrix for a tree is given, the f-cut-set matrix can be obtained.

### 4.6.2  Relation between f-loop and f-cut-set Matrices

The relation between f-loop matrix and f-cut-set matrix hence can be obtained from the two relations derived i.e. equation (4.18) and (4.24).

$$Q_l = -B_t^T \qquad \qquad \text{...(4.25)}$$

Now since

$$B_f = [B_t\,\,I] = [-Q^T{}_l,\,I] \qquad \qquad \text{...(4.26)}$$

and

$$Q_f = [Q_t\,,\,Q_l] = [I,\,Q_l] \qquad \qquad \text{...(4.27)}$$

Hence,

$$B_f\,Q_f^T = [-Q_l^T,\,I]\begin{bmatrix} I \\ Q_l^T \end{bmatrix} = -Q_l^T + Q_l^T = 0 \qquad \qquad \text{...(4.28)}$$

Similarly,

$$Q_f\,B_f^T = [I\,,Q_l]\begin{bmatrix} -Q_l \\ I \end{bmatrix} = -Q_l + Q_l = 0 \qquad \qquad \text{...(4.29)}$$

This shows that the f-cut-set and f-loop matrices are also orthogonal to each other.

It can be seen from previous manipulations that when 2,4,5 are taken as twigs $Q_l = -B_t^T = A_t^{-1} A_l$. Similarly for any graph for a particular tree these relations hold good which means if any one of the three matrices is known, the other two can be evaluated using these relations. If the system to be analysed is large the formulation of these matrices is very simple and if one of the matrices is known other large size matrices can be handled with ease on digital computers as standard program on matrix manipulation are readily available.

As is evident from equations, (4.22) and (4.23) if we are interested in all the branch currents the minimum number of independent currents correspond to the link currents. Once link currents are known twig currents can be obtained through the use of either equation (4.22) (i.e. incidence matrix) or equation (4.23) i.e. f-cut set matrix. Hence the currents through all the branches can be obtained.

We have also seen that in order to obtain branch voltages in a network, the minimum number of independent variables is the twig voltages. Once twig voltages are known, then through the use of the following relation all branch voltages can be obtained:

$$V_b = Q_f^T V_t \qquad \qquad ...(4.30)$$

In general for a network with $n$ nodes and $b$ branches KVL equations are written as

$$B_f V_b = 0 \qquad \qquad ...(4.31)$$

where $B_f$ is the fundamental loop matrix of order $(b - n + 1) \times b$ and $V_b$ is the column matrix of $b \times 1$. Now partitioning $B_f$ into $B_l$ and $B_t$, we have

$$[B_t , I]\begin{bmatrix} V_t \\ V_l \end{bmatrix} = 0 \qquad \qquad ...(4.32)$$

or
$$B_t V_t + V_l = 0$$

or
$$V_l = -B_t V_t \qquad \qquad ...(4.33)$$

Therefore, if $V_t$ is known i.e. if the twig voltages are known, the remaining branches of the network which are links, their voltages can be obtained using the above relation. Therefore, to obtain branch voltages of the network, the minimum number of independent variables equals the twig voltages.

We know that from equation (4.25)

$$Q_l = -B_t^T$$

or
$$B_t = -Q_l^T$$

$$V_l = +Q_l^T V_t$$

Now
$$V_b = \begin{bmatrix} V_t \\ V_l \end{bmatrix} = \begin{bmatrix} U\, V_t \\ Q_l^T\, V_t \end{bmatrix} = \begin{bmatrix} U \\ Q_l^T \end{bmatrix} V_t$$

$$= Q_f^T V_t \qquad \qquad ...(4.30)$$

This means the branch voltages $V_b$ are a linear combination of twig voltages, Thus there are independent variables corresponding to twig voltages.

We can also obtain the branch voltages using nodal voltages as independent variables through the use of incidence matrix and is given as

$$V_b = A^T V_n \qquad \qquad ...(4.34)$$

where $V_b$ the branch voltage and $A^T$ is the transpose of the incidence matrix and $V_n$ is the voltages of the various nodes.

We know that the twig currents are a linear combination of link currents through the use of f-cut-set matrix or f-loop matrix equation (4.23) and is given as

$$I_t = -Q_l \, I_l = B^T_{\,t} \, I_l$$

Also
$$I_l = UI_l$$

Therefore,
$$I_b = \begin{bmatrix} I_t \\ I_l \end{bmatrix} = \begin{bmatrix} B^T_t \, I_l \\ U \, I_l \end{bmatrix} = \begin{bmatrix} B^T_t \\ U \end{bmatrix} I_l$$

$$= B^T_{\,f} \, I_l \qquad\qquad\qquad ...(4.35)$$

This shows that the branch currents are expressed as a linear combination of link currents.

Similarly branch currents can be expressed as a linear combination of mesh currents and is given as

$$I_b = B^T_m \, I_m \qquad\qquad\qquad ...(4.36)$$

where $I_m$ is the mesh current column vector.

The observations made above are summarized for ready reference.

If $b$ is the number of branches and $n$ the number of nodes then the dimensions of various matrices are:

$$A \rightarrow (n-1) \times b$$
$$B_f \rightarrow (b-n+1) \times b$$
$$Q_f \rightarrow (n-1) \times b$$

The number of trees in a graph are given by
$$\det [A \; A^T]$$

When these matrices are partitioned corresponding to twigs and links.
$A_t$ and $A_l$ are non-unit matrices.
$B_t$ is non-unit. $B_l$ is unit matrix.
$Q_t$ is unit matrix $Q_l$ is non-unit matrix.
Twig currents and link currents are related as
$$I_t = -A_t^{-1} \, A_l \, I_l = -Q_l \, I_l = B^T_t \, I_l$$

The branch currents in terms of link currents are given as
$$I_b = B^T_f \, I_l$$

The link voltages are specified as a function of twig voltages as
$$V_l = -B_t V_t = Q^T_l \, V_t$$

and the branch voltages are given as
$$V_b = Q^T_f \, V_t$$

Branch voltages in terms of node voltages are given as
$$V_b = A^T \, V_n$$

## 4.7 GENERALISED ELEMENT

While drawing graph out of a given network all passive elements between nodes are represented by line segments whereas for independent current and voltage sources if these have accompanied passive element viz. a shunt admittance branch in a current source or a series impedance branch in a voltage source the current and voltage sources are to be represented by line segments corresponding to the source impedance and the current source is to be open circuited, and voltage source short circuited. However, if the sources do not have the accompanied passive elements, these sources must be shifted into the network as explained in Chapter 1.

A branch in general can be indicated as shown in Fig. 4.15. The general branch consists of an accompanied voltage source and a current source in parallel with the accompanied voltage source. Thus, the current of this $K$th branch that is to be used in KCL will be $i_k - i_{sk}$ and the voltage to be used in KVL will be $v_k - v_{sk}$

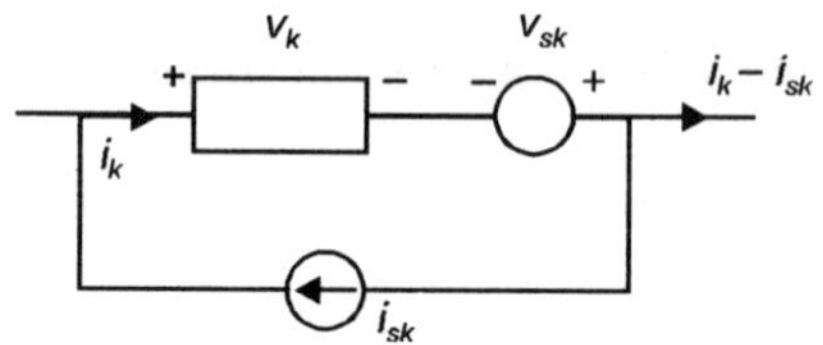

**Fig. 4.15**   Representation of a general branch.

If the graph didn't have sources such as shown in the general branch, KCL equations are:

$$A \, I_b = 0 \qquad \qquad \text{...(4.19)}$$

and
$$Q_f \, I_b = 0 \qquad \qquad \text{...(4.20)}$$

However, with the sources shifted into the network, will modify these equations and are given as

$$AI_b - AI_s = 0 \qquad \qquad \text{...(4.37)}$$

$$Q_f \, I_b - Q_f \, I_s = 0 \qquad \qquad \text{...(4.38)}$$

and the modified KVL equation would be

$$B_f \, V_b = B_f \, V_s \qquad \qquad \text{...(4.39)}$$

If we take loop currents or link currents, the twig voltages or the nodal voltages as independent variables depending upon the method of analysis, the modified equations when sources are inserted in the branches are given as

$$I_b - I_s = B_f^T \, I_l \qquad \qquad \text{...(4.40)}$$

$$V_b - V_s = Q_f^T \, V_t \qquad \qquad \text{...(4.41)}$$

and
$$V_b - V_s = A^T \, V_n \qquad \qquad \text{...(4.42)}$$

Since we are able to take care of the sources as mentioned above, independent of the passive part of the branch, we shall now concentrate on the passive component.

Hence, sources will be replaced by their internal impedances.

The branch relationships for the passive network in matrix form can be written as

$$V_b = Z_b \, I_b \quad \text{(Loop or mesh analysis)} \qquad \qquad \text{...(4.43)}$$

and
$$I_b = Y_b \, V_b \quad \text{(Node analysis)} \qquad \qquad \text{...(4.44)}$$

### 4.7.1   Mesh Equations

Substituting the value of $V_b$ from equation 4.43 in equation (4.39) we have

$$B_f \, [Z_b \, I_b] = B_f \, V_s \qquad \qquad \text{...(4.45)}$$

Substituting for $I_b$ from equation (4.40)

$$B_f Z_b [B^T_f I_l + I_s] = B_f V_s$$

or
$$B_f Z_b B^T_f I_l = B_f [V_s - Z_b I_s]$$

or
$$ZI_l = Q_f E \qquad \qquad ...(4.46)$$

where $Z$ is the loop impedance matrix and is of order $(b - n + 1)$ and is a symmetric matrix. $Z_b$ is the branch impedance matrix of order $b \times b$ and is a diagonal matrix, $V_s$ is a source voltage column vector of order $b \times 1$, $I_s$ is the source current column vector of order $b \times 1$. The elements of $Z_b$ in general could be

$$Z_b = R_b + sL_b + \frac{1}{C_b s} \qquad \qquad ...(4.47)$$

and for steady state solutions

$$Z_b = R_b + jwL_b + \frac{1}{jwc_b} \qquad \qquad ...(4.48)$$

From equation (4.46) loop currents (or link currents) can be obtained as follows:

$$I_l = Z^{-1} . E \qquad \qquad ...(4.49)$$

and hence all branch currents can be obtained.

## 4.7.2  Node Equations

Rewriting equation (4.37) we have

$$A I_b = A I_s$$

Now
$$I_b = Y_b V_b, \text{ and } V_b = A^T V_n + V_s$$

Hence,
$$A I_b = A Y_b V_b = AI_s$$

$$A Y_b [A^T V_n + V_s] = AI_s$$

$$AY_b A^T V_n = AI_s - AY_b V_s$$

$$= A [I_s - Y_b V_s] \qquad \qquad ...(4.50)$$

$$Y V_n = AJ \qquad \qquad ...(4.51)$$

where $Y = AY_b A^T$ is an $(n - 1) \times (n - 1)$ nodal admittance matrix, $Y_b$ is the branch admittance matrix of order $b \times b$ and is a diagonal matrix. $V_n$ is a nodal voltage column vector of $(n - 1) \times 1$ order, $J$ is the equivalent node-current source vector whose entries are the algebraic sum of current source including the Norton's equivalents of voltage incident at the corresponding node. If transients behaviour of the network is to be studied then each branch $Y_b$ in general may be given as

$$Y_b = \frac{1}{R_b + \dfrac{1}{C_b s} + sL_b}$$

However, if steady state behaviour of the network is to be studied then each branch admittance is given as

$$Y_b = \frac{1}{R_b + \dfrac{1}{jwc_b} + jwL_b} \qquad \qquad ...(4.53)$$

However, if each branch has individual elements either $R$, or $L$ or $C$ then

$$Y_b = G_b = \frac{1}{R_b}$$

$$Y_b = \frac{1}{sL_b} \quad \text{or} \quad Y_b = sC_b \qquad \qquad ...(4.54)$$

and for steady state operation

$$Y_b = G_b = \frac{1}{R_b}, \quad Y_b = \frac{1}{jwL_b} \quad \text{or} \quad Y_b = jwc_b$$

From equation (4.51) we can obtain $V_n$ as

$$V_n = Y^{-1}J \qquad \qquad ...(4.55)$$

**Example 4.1:** Consider the network shown here, draw its graph, obtain $B_f$ taking branches 2, 4, 5 as tree branches. Hence determine the loop impedance matrix and find the loop equations.

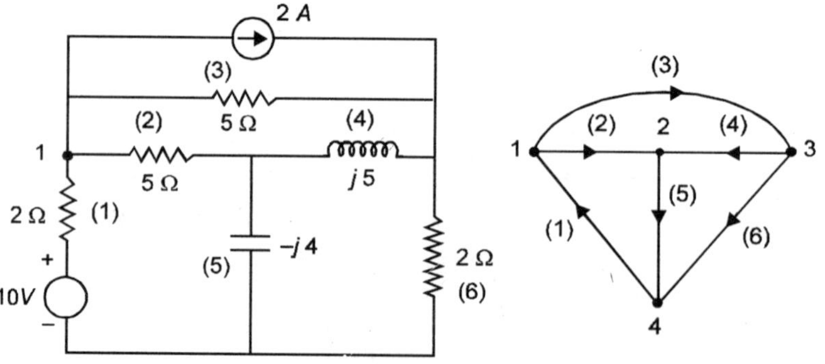


**Fig. E4.1**

**Solution:** Short-circuiting the voltage source and open circuiting the current source we have the graph as shown. It is possible as both the voltage source and current sources are accompanied sources. Selecting branch 2, 4, 5 we have the tree as shown.

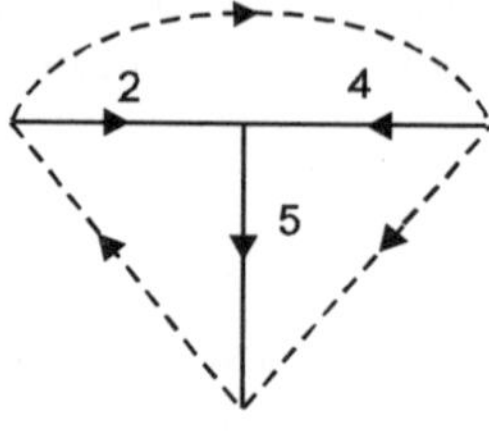


**Fig E4.1.1**

Taking one link at a time and thus forming a loop and taking the orientation of the link as the orientation of the loop we have $B_f$ as follows:

$$B_f = \begin{array}{c} \\ 1 \\ 2 \\ 3 \end{array} \begin{pmatrix} 1 & 2 & 3 & 4 & 5 & 6 \\ 1 & 1 & 0 & 0 & 1 & 0 \\ 0 & -1 & 1 & 1 & 0 & 0 \\ 0 & 0 & 0 & -1 & -1 & 1 \end{pmatrix}$$

with column heading **Loops** (rows) and **Branches** (columns).

Now

$$Z_b = \begin{array}{c} 1 \\ 2 \\ 3 \\ 4 \\ 5 \\ 6 \end{array} \begin{pmatrix} 1 & 2 & 3 & 4 & 5 & 6 \\ 2 & 0 & 0 & 0 & 0 & 0 \\ 0 & 5 & 0 & 0 & 0 & 0 \\ 0 & 0 & 5 & 0 & 0 & 0 \\ 0 & 0 & 0 & j5 & 0 & 0 \\ 0 & 0 & 0 & 0 & -j4 & 0 \\ 0 & 0 & 0 & 0 & 0 & 2 \end{pmatrix}$$

Hence loop impedance matrix

$$B_f Z_b B_f^T$$

$$\begin{pmatrix} 1 & 1 & 0 & 0 & 1 & 0 \\ 0 & -1 & 1 & 1 & 0 & 0 \\ 0 & 0 & 0 & -1 & -1 & 1 \end{pmatrix} \begin{pmatrix} 2 & 0 & 0 & 0 & 0 & 0 \\ 0 & 5 & 0 & 0 & 0 & 0 \\ 0 & 0 & 5 & 0 & 0 & 0 \\ 0 & 0 & 0 & j5 & 0 & 0 \\ 0 & 0 & 0 & 0 & -j4 & 0 \\ 0 & 0 & 0 & 0 & 0 & 2 \end{pmatrix}$$

$$= \begin{pmatrix} 2 & 5 & 0 & 0 & -j4 & 0 \\ 0 & -5 & 5 & j5 & 0 & 0 \\ 0 & 0 & 0 & -j5 & j4 & 2 \end{pmatrix} \begin{pmatrix} 1 & 0 & 0 \\ 1 & -1 & 0 \\ 0 & 1 & 0 \\ 0 & 1 & -1 \\ 1 & 0 & -1 \\ 0 & 0 & 1 \end{pmatrix}$$

$$Z = \begin{pmatrix} 7-j4 & -5 & j4 \\ -5 & 10+j5 & -j5 \\ j4 & -j5 & j1+2 \end{pmatrix} = B_f Z_b B_f^T$$

The result can be verified by inspection method and it is found to be alright. Here of course the current source should be first converted into a voltage source.

It is to be noted that whenever there is mutual coupling between the branches of the network, the $Z_b$ is not a diagonal matrix. Otherwise it is a diagonal matrix.

The overall network equations will be

$$Z I_l = E \quad \text{where} \quad E = B_f [V_s - Z_b I_s]$$

$$\begin{pmatrix} 1 & 1 & 0 & 0 & 1 & 0 \\ 0 & -1 & 1 & 1 & 0 & 0 \\ 0 & 0 & 0 & -1 & -1 & 0 \end{pmatrix} \begin{pmatrix} 10 \\ 0 \\ 10 \\ 0 \\ 0 \\ 0 \end{pmatrix} = \begin{pmatrix} 10 \\ 10 \\ 0 \end{pmatrix}$$

Hence the final equations are:

$$\begin{pmatrix} 7-j4 & -5 & j4 \\ -5 & 10+j5 & -j5 \\ j4 & -j5 & 2+j1 \end{pmatrix}\begin{pmatrix} I_1 \\ I_3 \\ I_6 \end{pmatrix} = \begin{pmatrix} 10 \\ 10 \\ 0 \end{pmatrix}$$

In this problem the column vector

$$V_s = [10 \quad 0 \quad 0 \quad 0 \quad 0 \quad 0]^T$$

and since the current source is in parallel with branch 3

$$-Z_b\, I_s = [0 \quad 0 \quad 10 \quad 0 \quad 0 \quad 0]^T$$

**Example 4.2:** For the network shown, obtain the incidence matrix, the nodal admittance matrix and the node equation.

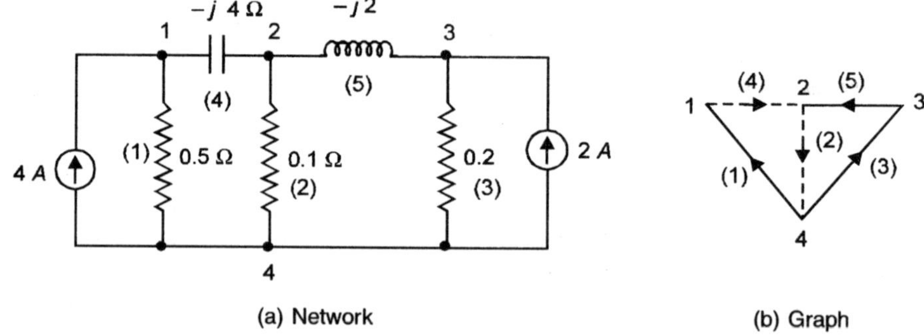


(a) Network    (b) Graph

**Fig. E4.2**

**Solution:** For the network, the tree is shown alongside. The current source branches are open-circuited. Orientation of branches 1 and 3 is same as that of the current sources and for branches 2, 4 and 5 the orientation is arbitrary. Node 4 is taken as the reference. The incidence matrix is given as

$$\begin{array}{cc} & \text{Branches} \\ \text{Nodes} & \begin{array}{ccccc} 1 & 2 & 3 & 4 & 5 \end{array} \end{array}$$

$$A = \begin{array}{c} 1 \\ 2 \\ 3 \end{array}\begin{pmatrix} -1 & 0 & 0 & 1 & 0 \\ 0 & 1 & 0 & -1 & -1 \\ 0 & 0 & -1 & 0 & 1 \end{pmatrix}$$

The branch admittance matrix is given as

$$Y_b = \begin{array}{c} 1 \\ 2 \\ 3 \\ 4 \\ 5 \end{array}\begin{pmatrix} 2 & 0 & 0 & 0 & 0 \\ 0 & 10 & 0 & 0 & 0 \\ 0 & 0 & 5 & 0 & 0 \\ 0 & 0 & 0 & j0.25 & 0 \\ 0 & 0 & 0 & 0 & -j0.5 \end{pmatrix}$$

Hence      $Y = A\, Y_b\, A^T$

Now
$$AY_b = \begin{pmatrix} -1 & 0 & 0 & 1 & 0 \\ 0 & 1 & 0 & -1 & -1 \\ 0 & 0 & -1 & 0 & 1 \end{pmatrix} \begin{pmatrix} 2 & 0 & 0 & 0 & 0 \\ 0 & 10 & 0 & 0 & 0 \\ 0 & 0 & 5 & 0 & 0 \\ 0 & 0 & 0 & j0.25 & 0 \\ 0 & 0 & 0 & 0 & -j0.5 \end{pmatrix}$$

$$A\,Y_b\,A^T = \begin{pmatrix} -2 & 0 & 0 & j0.25 & 0 \\ 0 & 10 & 0 & -j0.25 & j0.5 \\ 0 & 0 & -5 & 0 & -j0.5 \end{pmatrix} \begin{pmatrix} -1 & 0 & 0 \\ 0 & 1 & 0 \\ 0 & 0 & -1 \\ 1 & -1 & 0 \\ 0 & -1 & 1 \end{pmatrix}$$

$$Y = \begin{pmatrix} 2 + j0.25 & -j0.25 & 0 \\ -j0.25 & 10 + j0.25 - j0.5 & j0.5 \\ 0 & j0.5 & 5 - j0.5 \end{pmatrix}$$

Now $\qquad V_n = [V_1 \quad V_2 \quad V_3]^T$

and $\qquad J = A\,[I_s - Y_b\,V_s]$

Since $\qquad V_s = [0]$

Taking currents entering the node as $-$ve

$$J = \begin{pmatrix} -1 & 0 & 0 & 1 & 0 \\ 0 & 1 & 0 & -1 & -1 \\ 0 & 0 & -1 & 0 & 1 \end{pmatrix} \begin{pmatrix} -4 \\ 0 \\ -2 \\ 0 \\ 0 \end{pmatrix} = \begin{pmatrix} 4 \\ 2 \\ 0 \end{pmatrix}$$

Hence the equations are:

$$\begin{pmatrix} 2 + j0.25 & -j0.25 & 0 \\ -j0.25 & 10 - j0.25 & j0.5 \\ 0 & j0.5 & 5 - j0.5 \end{pmatrix} \begin{pmatrix} V_1 \\ V_2 \\ V_3 \end{pmatrix} = \begin{pmatrix} 4 \\ 0 \\ 2 \\ 0 \\ 0 \end{pmatrix}$$

### 4.7.3  Cut-set Equations

Considering every element of the network to be a generalized element we can write one such equation for each element. Then the overall equation for the network can be written as

$$I_b = Y_b\,V_b + J \qquad\qquad\qquad \text{...(4.56)}$$

Now $\qquad Q_f\,I_b = 0$

or $\qquad Q_f\,[Y_b\,V_b + J] = 0 \qquad\qquad\qquad \text{...(4.57)}$

Also from equation 4.30

$$V_b = Q_f^T V_t$$

Substituting for $V_b$ in equation (4.57) we have

$$Q_f [Y_b Q_f^T V_t + J] = 0$$

or
$$Q_f Y_b Q_f^T V_t = -Q_f J$$

or
$$Y_c V_t = -Q_f J$$

...(4.58)

where $Y_c = Q_f Y_b Q_f^T$ is the cut-set admittance matrix.

**Example 4.3:** For the network in example 4.2 determine the cut-set matrix, the cut-set admittance matrix and the cut set equations.

**Solution:** The graph is drawn here and the tree is selected same as in example 4.2

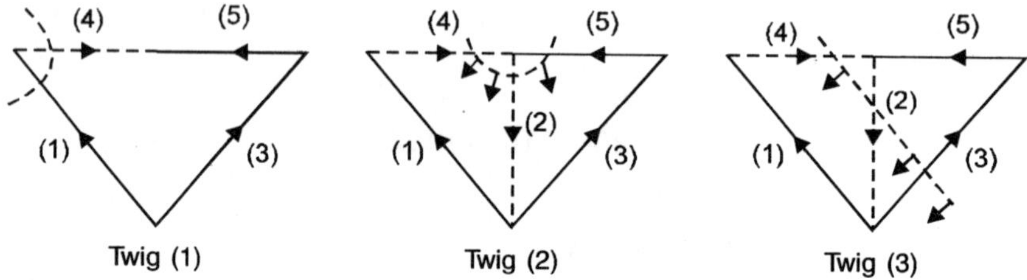


**Fig. E4.3**

The cut-set matrix is obtained as follows:

$$Q_f Y_b = \begin{array}{c} \\ \text{Twig} \\ 1 \\ 2 \\ 3 \end{array} \begin{array}{ccccc} \text{Branches} & & & & \\ 1 & 2 & 3 & 4 & 5 \\ \end{array} \begin{pmatrix} -1 & 0 & 0 & 1 & 0 \\ 0 & 1 & -1 & -1 & 0 \\ 0 & 1 & 0 & -1 & -1 \end{pmatrix} \begin{pmatrix} 2 & 0 & 0 & 0 & 0 \\ 0 & 10 & 0 & 0 & 0 \\ 0 & 0 & 5 & 0 & 0 \\ 0 & 0 & 0 & j0.25 & 0 \\ 0 & 0 & 0 & 0 & -j0.5 \end{pmatrix}$$

$$Q_f Y_b Q_f^T = \begin{pmatrix} -2 & 0 & 0 & j0.25 & 0 \\ 0 & 10 & -5 & -j0.25 & 0 \\ 0 & 10 & 0 & -j0.25 & +j0.5 \end{pmatrix} \begin{pmatrix} -1 & 0 & 0 \\ 0 & 1 & 1 \\ 0 & -1 & 0 \\ 1 & -1 & -1 \\ 0 & 0 & -1 \end{pmatrix}$$

$$Q_f Y_b Q_f^T = \begin{pmatrix} 2 + j0.25 & -j0.25 & -j0.25 \\ -j0.25 & 15 + j0.25 & 10 + j0.25 \\ -j0.25 & 10 + j0.25 & 10 - j0.25 \end{pmatrix}$$

Assuming current entering the node as $-$ve

Now
$$-Q_f J = \begin{pmatrix} 1 & 0 & 0 & -1 & 0 \\ 0 & -1 & 1 & 1 & 0 \\ 0 & -1 & 0 & +1 & 1 \end{pmatrix} \begin{pmatrix} -4 \\ 0 \\ -2 \\ 0 \\ 0 \end{pmatrix}$$

$$= \begin{pmatrix} -4 \\ -2 \\ 0 \end{pmatrix}$$

Hence the cut set equations are:

$$\begin{pmatrix} 2 + j0.25 & -j0.25 & -j0.25 \\ -j0.25 & 15 + j0.25 & 10 + j0.25 \\ -j0.25 & 10 + j0.25 & 10 - j0.25 \end{pmatrix} \begin{pmatrix} V_1 \\ V_3 \\ V_5 \end{pmatrix} = \begin{pmatrix} -4 \\ -2 \\ 0 \end{pmatrix}$$

It is to be noted that out of the three matrices, the loop impedance matrix, the nodal admittance matrix and cut-set admittance matrix, the nodal admittance matrix is a sparse matrix whereas the other two matrices are full matrices i.e. in the two matrices each entry is a non-zero entry whereas for nodal admittance matrix with three nodes there are two entries with zeros. In fact for a large system nodal admittance matrix is highly sparse. It is estimated that for a 100 node nodal admittance matrix, the non-zero elements are only 2%, rest all are zeros and hence from computer memory point of view and other manipulations associated, nodal admittance approach is preferred.

## 4.8  NETWORK WITH MUTUAL INDUCTANCE

As mentioned earlier, if there is no mutual coupling between the networks the branch impedance or admittance matrix is a diagonal matrix otherwise it may have non-zero elements in the off-diagonal locations also. Consider Fig. 4.16 (a) and (b) with different polarity magnetic couplings.

Here $M = K\sqrt{L_1 L_2}$ where $K$ is the coefficient of coupling and lies between 0 and 1. With $K = 1$ the coupling is perfect i.e. there is no leakage. In Fig. 4.16 (a) the branch inductance gets modified to

$$L_b = \begin{pmatrix} L_1 & M \\ M & L_2 \end{pmatrix}$$

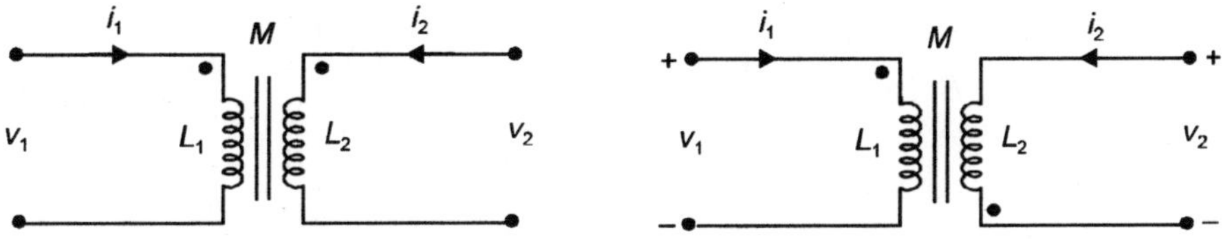

(a) With positive coupling



(b) Opposing coupling

**Fig. 4.16**

whereas in Fig. 4.16 (b) it becomes

$$L_b = \begin{pmatrix} L_1 & -M \\ -M & L_2 \end{pmatrix}$$

**Example 4.4:** For the network shown determine the f-loop matrix, the loop impedance matrix and f-loop equations.

$$j\,\omega\,L_1 = j\,4\;\Omega \qquad\qquad j\,\omega\,L_2 = j\,5\;\Omega \qquad\qquad M = j\,2\;\Omega$$

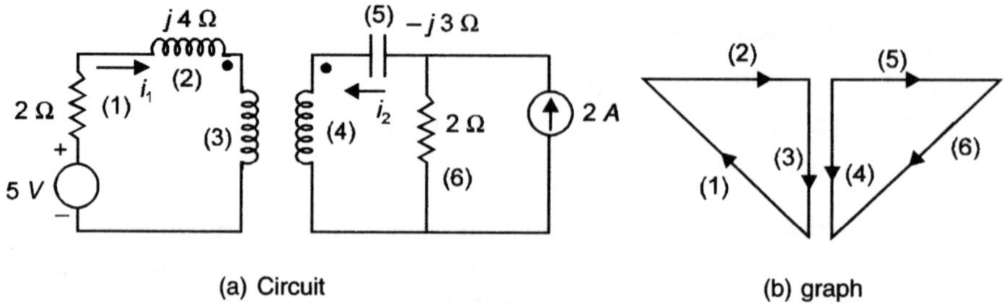

(a) Circuit                                   (b) graph

**Fig. E 4.4**

The current source is converted to a voltage source of 4 V in series with $2\,\Omega$ resistor. The branch impedance matrix is

$$Z_b = \begin{pmatrix} 2 & 0 & 0 & 0 & 0 & 0 \\ 0 & j4 & 0 & 0 & 0 & 0 \\ 0 & 0 & j4 & j2 & 0 & 0 \\ 0 & 0 & j2 & j5 & 0 & 0 \\ 0 & 0 & 0 & 0 & -j3 & 0 \\ 0 & 0 & 0 & 0 & 0 & 2 \end{pmatrix}$$

The f-loop matrix is given as

$$B_f = \begin{array}{c} 1 \\ 2 \end{array}\begin{pmatrix} 1 & 2 & 3 & 4 & 5 & 6 \\ 1 & 1 & 1 & 0 & 0 & 0 \\ 0 & 0 & 0 & +1 & -1 & -1 \end{pmatrix}$$

$$B_f\,Z_b\,B^T{}_f = \begin{pmatrix} 2 & j4 & j4 & j2 & 0 & 0 \\ 0 & 0 & +j2 & +j5 & j3 & -2 \end{pmatrix}\begin{pmatrix} 1 & 0 \\ 1 & 0 \\ 1 & 0 \\ 0 & 1 \\ 0 & -1 \\ 0 & -1 \end{pmatrix}$$

$$B_f\,Z_b\,B^T{}_f = Z = \begin{pmatrix} 2 + j8 & j2 \\ +j2 & 2 + j2 \end{pmatrix}$$

Hence the loop equations are:

$$\begin{pmatrix} 2 + j8 & j2 \\ j2 & 2 + j2 \end{pmatrix}\begin{pmatrix} I_{l_1} \\ I_{l_2} \end{pmatrix} = \begin{pmatrix} 5 \\ 4 \end{pmatrix}$$

Let us now take a simple but typical problem and try to solve by different methods.

**Example 4.5:** For the network shown determine
(*a*) f-loop matrix, loop impedance matrix and loop currents.
(*b*) Node admittance matrix and nodal voltages.

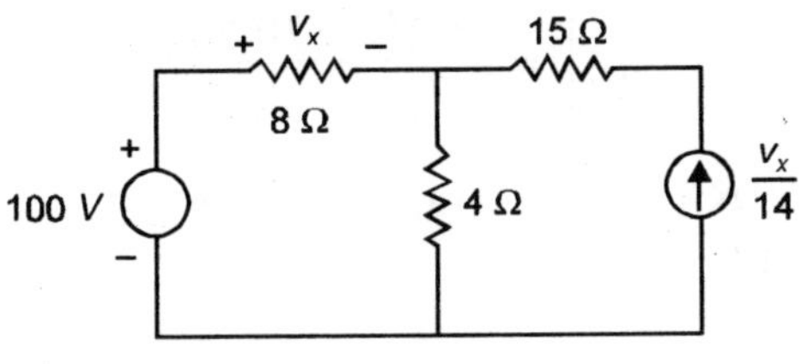

**Fig. E4.5**

**Solution:** Here we have one voltage controlled current source.

*f-loop matix*: Since the current source is not accompanied we assume a resistance $R$ across the current source and convert this into a voltage source and finally we shall let $R \to \infty$. The circuit becomes

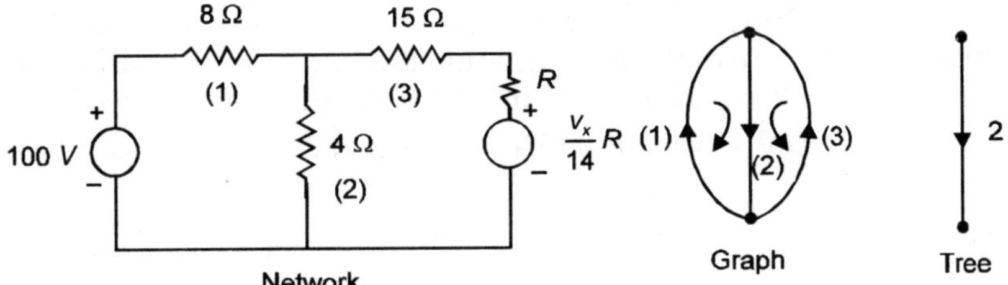

**Fig. E4.5.1**

$$B_f = \begin{array}{c} \\ 1 \\ 2 \end{array}\begin{array}{c} 1 \;\; 2 \;\; 3 \end{array}\begin{pmatrix} 1 & 1 & 0 \\ 0 & 1 & 1 \end{pmatrix} \qquad Z_b = \begin{pmatrix} 8 & 0 & 0 \\ 0 & 4 & 0 \\ 0 & 0 & R+15 \end{pmatrix}$$

$$B_f\, Z_b = \begin{pmatrix} 8 & 4 & 0 \\ 0 & 4 & R+15 \end{pmatrix}$$

$$B_f\, Z_b\, B_f^T = \begin{pmatrix} 8 & 4 & 0 \\ 0 & 4 & R+15 \end{pmatrix}\begin{pmatrix} 1 & 0 \\ 1 & 1 \\ 0 & 1 \end{pmatrix} = \begin{pmatrix} 12 & 4 \\ 4 & R+19 \end{pmatrix}$$

The loop equations are:

$$\begin{pmatrix} 12 & 4 \\ 4 & R+19 \end{pmatrix}\begin{pmatrix} I_1 \\ I_2 \end{pmatrix} = \begin{pmatrix} 100 \\ \dfrac{V_x R}{14} \end{pmatrix}$$

$$12I_1 + 4I_2 = 100 \quad \text{or} \quad 3I_1 + I_2 = 25$$

$$4I_1 + (R+19)\,I_2 = \frac{V_x R}{14} = \frac{8I_1 R}{14}$$

$$\frac{4\,(25-I_2)}{3} + (R+19)\,I_2 = \frac{8R}{14}\,\frac{(25-I_2)}{3}$$

$$I_2\left[R+19-\frac{4}{3}+\frac{8R}{42}\right] = \frac{100R}{21} - \frac{100}{3}$$

$$I_2\left[\frac{50R}{42}+\frac{53}{3}\right] = \frac{100R-700}{21}$$

$$I_2 = \underset{R\to\infty}{Lt}\ \frac{\left(100-700/R\right)}{\dfrac{50}{42}+\dfrac{53}{3R}} = 4\,A$$

$$I_1 = \frac{25-I_2}{3} = 7\,A$$

(*b*) *Nodal analysis:* Since 8 $\Omega$ in series with 100 V source can't be considered as accompanied as the current source is a function of the voltage drop across $V_x$, therefore we connect a series resistance $R_1$ in series with voltage source and $R_2$ across the current source. Finally we will let $R_1 \to 0$ and $R_2 \to \infty$ to obtain final result.

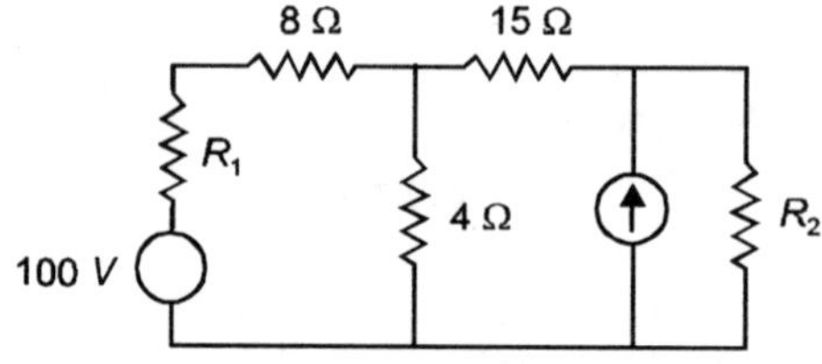


**Fig. E4.5.2**

Converting the voltage source into current source we have

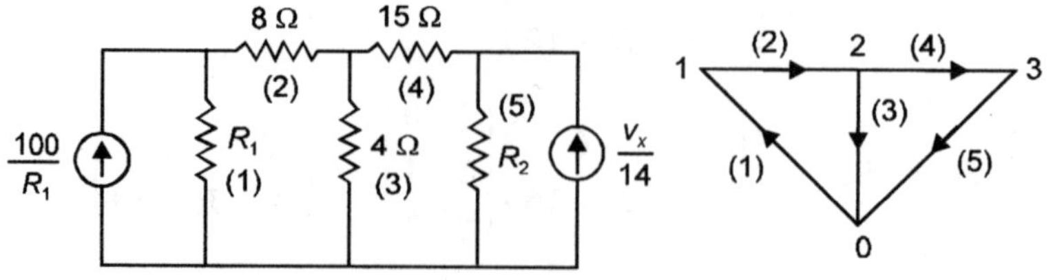


**Fig. E4.5.3**

$$\begin{array}{c|ccccc}
\text{Nodes} & 1 & 2 & 3 & 4 & 5 \\
\hline
1 & -1 & 1 & 0 & 0 & 0 \\
A = 2 & 0 & -1 & 1 & -1 & 0 \\
3 & 0 & 0 & 0 & 1 & 1
\end{array}$$

$$Y_b = \begin{pmatrix} G_1 & 0 & 0 & 0 & 0 \\ 0 & 1/8 & 0 & 0 & 0 \\ 0 & 0 & 1/4 & 0 & 0 \\ 0 & 0 & 0 & 1/15 & 0 \\ 0 & 0 & 0 & 0 & G_2 \end{pmatrix}$$

$$A\,Y_b\,A^T = \begin{pmatrix} -G_1 & 1/8 & 0 & 0 & 0 \\ 0 & -1/8 & 1/4 & -1/15 & 0 \\ 0 & 0 & 0 & 1/15 & G_2 \end{pmatrix} \begin{pmatrix} -1 & 0 & 0 \\ 1 & -1 & 0 \\ 0 & 1 & 0 \\ 0 & -1 & 1 \\ 0 & 0 & 1 \end{pmatrix}$$

$$A\,Y_b\,A^T = \begin{pmatrix} G_1 + 1/8 & -1/8 & 0 \\ -1/8 & \left(\dfrac{1}{8}+\dfrac{1}{4}+\dfrac{1}{15}\right) & -1/15 \\ 0 & -1/15 & G_2 + \dfrac{1}{15} \end{pmatrix} = Y$$

The nodal equations are:

$$\begin{pmatrix} G_1 + 1/8 & -1/8 & 0 \\ -1/8 & \left(\dfrac{1}{8}+\dfrac{1}{4}+\dfrac{1}{15}\right) & -1/15 \\ 0 & -1/15 & G_2 + \dfrac{1}{15} \end{pmatrix} \begin{pmatrix} V_1 \\ V_2 \\ V_3 \end{pmatrix} = \begin{pmatrix} 100\,G_1 \\ 0 \\ V_x/14 \end{pmatrix}$$

$$\left(G_1 + \frac{1}{8}\right)V_1 - \frac{V_2}{18} = 100\;G_1 \qquad\qquad \text{...(1)}$$

$$-\frac{V_1}{8} + \frac{53}{120}V_2 - \frac{1}{15}V_3 = 0 \qquad\qquad \text{...(2)}$$

$$-\frac{1}{15}V_2 + \left(G_2 + \frac{1}{15}\right)V_3 = \frac{V_x}{14} \qquad\qquad \text{...(3)}$$

Dividing equation 1 by $G_1$ and letting $G_1 \to \infty$, we have $V_1 = 100$ Equation 2 becomes

$$0.44167\;V_2 - \frac{V_3}{15} = 12.5 \quad \text{and let}\quad G_2 \to 0 \quad \text{equation 3 becomes}$$

$$-\frac{V_2}{15} + \frac{V_3}{15} = \frac{V_x}{14}$$

$$0.44167\;V_2 - \frac{V_3}{15} = 12.5$$

Addition of the two equations

$$0.375 \ V_2 = 12.5 + \frac{V_x}{14} = 12.5 + \frac{100 - V_2}{14}$$

$$= 19.6428 - \frac{V_2}{14}$$

$$0.44642 \ V_2 = 19.6428$$

$$V_2 \simeq 44 \ \text{volt}$$

Hence, $\qquad\qquad V_3 = 104 \ \text{volt}$

The solution of these equations gives $V_1 = 100$ v, $V_2 = 44$ volt and $V_3 = 104$ volts.

**Example 4.6:** The given matrix is a reduced incidence matrix of a graph. Draw the graph and obtain the f-cut-set matrix. Also verify the result by obtaining f-cut matrix mathematically.

$$
A = \begin{array}{c c} & \begin{array}{c} \text{Nodes} \quad \text{Branches} \\ \begin{array}{cccccc} 1 & 2 & 3 & 4 & 5 & 6 \end{array} \end{array} \\ \begin{array}{c} 1 \\ 2 \\ 3 \end{array} & \left( \begin{array}{cccccc} 1 & 0 & 0 & -1 & 0 & 1 \\ -1 & 0 & 1 & 0 & 1 & 0 \\ 0 & 1 & -1 & 0 & 0 & 0 \end{array} \right) \end{array}
$$

The augmented incident matrix is given as

$$
A_a = \left( \begin{array}{cccccc} 1 & 0 & 0 & -1 & 0 & 1 \\ -1 & 0 & 1 & 0 & 1 & 0 \\ 0 & 1 & -1 & 0 & 0 & 0 \\ 0 & -1 & 0 & 1 & -1 & -1 \end{array} \right)
$$

The fourth row is written so that the sum of the entries in each column is zero.

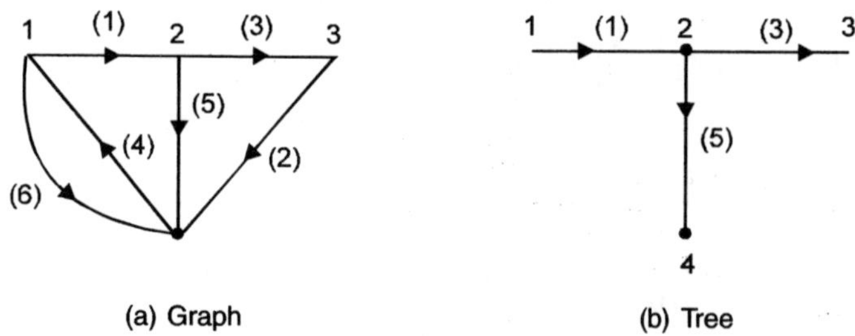


(a) Graph          (b) Tree

**Fig. E4.6**

Taking $1, 3, 5$ as twigs, the tree is shown above, the reduced incidence matrix can be rewritten as

$$
A = \begin{array}{c c} & \begin{array}{cccccc} 1 & 3 & 5 & 2 & 4 & 6 \end{array} \\ \begin{array}{c} 1 \\ 2 \\ 3 \end{array} & \left( \begin{array}{cccccc} 1 & 0 & 0 & 0 & -1 & 1 \\ -1 & 1 & 1 & 0 & 0 & 0 \\ 0 & -1 & 0 & 1 & 0 & 0 \end{array} \right) \end{array}
$$

Now $\qquad Q_f = [I,\ Q_l]$

Where $\qquad Q_l = A_t^{-1}\, A_l$

Now $\qquad A_t = \begin{pmatrix} 1 & 0 & 0 \\ -1 & 1 & 1 \\ 0 & -1 & 0 \end{pmatrix}$ $\quad$ det $[A_t] = 1$

$$C_{ij} = \begin{pmatrix} 1 & 0 & 1 \\ 0 & 0 & 1 \\ 0 & -1 & 1 \end{pmatrix} \quad C_{ij}^{T} = \begin{pmatrix} 1 & 0 & 0 \\ 0 & 0 & -1 \\ 1 & 1 & 1 \end{pmatrix} = A_t^{-1}$$

$$Q_l = A_t^{-1}\, A_l = \begin{pmatrix} 1 & 0 & 0 \\ 0 & 0 & -1 \\ 1 & 1 & 1 \end{pmatrix} \begin{pmatrix} 0 & -1 & 1 \\ 0 & 0 & 0 \\ 1 & 0 & 0 \end{pmatrix}$$

$$= \begin{pmatrix} 0 & -1 & 1 \\ -1 & 0 & 0 \\ 1 & -1 & 1 \end{pmatrix}$$

Hence $\qquad Q_f = \begin{array}{c} \begin{array}{cccccc} 1 & 3 & 5 & 2 & 4 & 6 \end{array} \\ \begin{pmatrix} 1 & 0 & 0 & 0 & -1 & 1 \\ 0 & 1 & 0 & -1 & 0 & 0 \\ 0 & 0 & 1 & 1 & -1 & 1 \end{pmatrix} \\ \begin{array}{cccccc} & & Q_t & & Q_l & \end{array} \end{array}$

Now from the tree, $Q_f$ is obtained as

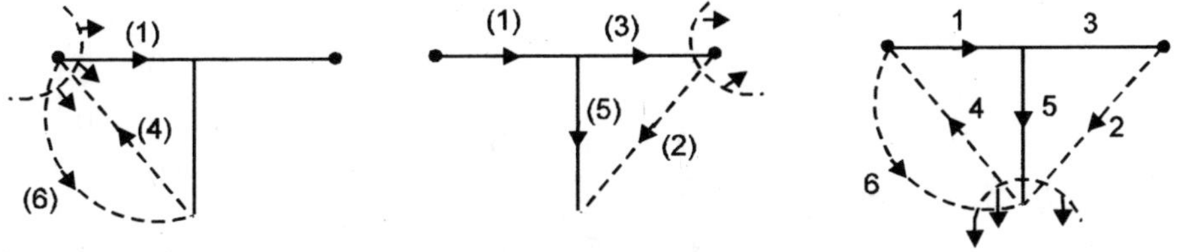

Fig. E4.6.1

$$Q_f = \begin{array}{c} \begin{array}{cccccc} 1 & 2 & 3 & 4 & 5 & 6 \end{array} \\ \begin{array}{c} 1 \\ 3 \\ 5 \end{array} \begin{pmatrix} 1 & 0 & 0 & -1 & 0 & 1 \\ 0 & -1 & +1 & 0 & 0 & 0 \\ 0 & +1 & 0 & -1 & +1 & +1 \end{pmatrix} \end{array}$$

**Example 4.7:** For the network shown write the tie-set matrix and determine the loop current and the branch currents.

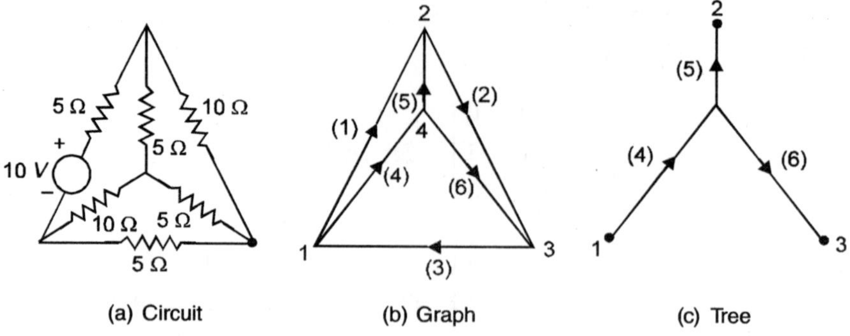

(a) Circuit  (b) Graph  (c) Tree

**Fig. E 4.7**

$$
B_f = \begin{array}{c} \\ 1 \\ 2 \\ 3 \end{array}
\begin{pmatrix}
1 & 0 & 0 & 1 & -1 & 0 \\
0 & 1 & 0 & 0 & 1 & -1 \\
0 & 0 & 1 & -1 & 0 & +1
\end{pmatrix}
$$

*Link   Branches*

1  2  3  4  5  6

The branch impedance matrix

$$
Z_b = \begin{pmatrix}
5 & 0 & 0 & 0 & 0 & 0 \\
0 & 10 & 0 & 0 & 0 & 0 \\
0 & 0 & 5 & 0 & 0 & 0 \\
0 & 0 & 0 & 10 & 0 & 0 \\
0 & 0 & 0 & 0 & 5 & 0 \\
0 & 0 & 0 & 0 & 0 & 5
\end{pmatrix}
$$

$$
B_f \, Z_b \, B_f^T =
\begin{pmatrix}
5 & 0 & 0 & 10 & -5 & 0 \\
0 & 10 & 0 & 0 & 5 & -5 \\
0 & 0 & 5 & -10 & 0 & 5
\end{pmatrix}
\begin{pmatrix}
1 & 0 & 0 \\
0 & 1 & 0 \\
0 & 0 & 1 \\
1 & 0 & -1 \\
-1 & 1 & 0 \\
0 & -1 & 1
\end{pmatrix}
$$

$$
= \begin{pmatrix}
20 & -5 & -10 \\
-5 & 20 & -5 \\
-10 & -5 & 15
\end{pmatrix}
$$

The loop equations are:

$$
\begin{pmatrix}
20 & -5 & -10 \\
-5 & 20 & -5 \\
-10 & -5 & 15
\end{pmatrix}
\begin{pmatrix}
I_1 \\ I_2 \\ I_3
\end{pmatrix}
=
\begin{pmatrix}
10 \\ 0 \\ 0
\end{pmatrix}
$$

$$I_1 = \frac{\begin{vmatrix} 10 & -5 & -10 \\ 0 & 20 & -5 \\ 0 & -5 & 15 \end{vmatrix}}{2625} = \frac{2750}{2625} = 1.047$$

$$I_2 = \frac{\begin{vmatrix} 20 & 10 & -10 \\ -5 & 0 & -5 \\ -10 & 0 & 15 \end{vmatrix}}{2625} = \frac{+10 \times 125}{2625} = 0.47$$

$$I_3 = \frac{10 \times 225}{2625} = 0.857 \text{ **Ans.**}$$

$$I_b = B_f^T I_l = \begin{pmatrix} 1 & 0 & 0 \\ 0 & 1 & 0 \\ 0 & 0 & 1 \\ 1 & 0 & -1 \\ -1 & 1 & 0 \\ 0 & -1 & 1 \end{pmatrix} \begin{pmatrix} 1.047 \\ 0.476 \\ 0.857 \end{pmatrix}$$

$$I_4 = 1.047 - 0.857 = 0.19 \text{ A}$$
$$I_5 = -1.047 + 0.476 = -0.571 \text{ A}$$
$$I_6 = 0.857 - 0.476 = 0.381 \text{ A}$$

**Example 4.8:** For the network shown write f-cut matrix and determine the nodal voltages.

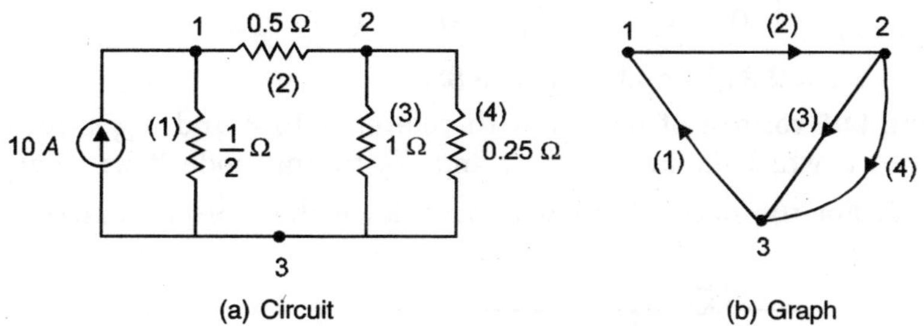


Fig. E4.8

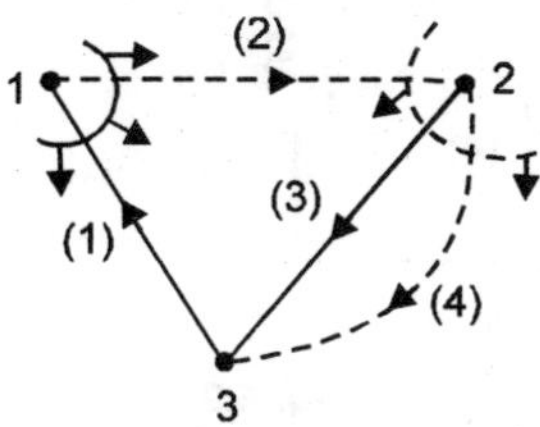


Fig. 4.8.1

$$Q_f = \begin{array}{c} \\ 1 \\ 3 \end{array}\begin{array}{cccc} 1 & 2 & 3 & 4 \\ \begin{pmatrix} -1 & 1 & 0 & 0 \\ 0 & -1 & 1 & 1 \end{pmatrix} \end{array}$$

$$Q_f Y_b Q_f^T = \begin{pmatrix} -1 & 1 & 0 & 0 \\ 0 & -1 & 1 & 1 \end{pmatrix} \begin{pmatrix} 2 & 0 & 0 & 0 \\ 0 & 2 & 0 & 0 \\ 0 & 0 & 1 & 0 \\ 0 & 0 & 0 & 4 \end{pmatrix} \begin{pmatrix} -1 & 0 \\ 1 & -1 \\ 0 & 1 \\ 0 & 1 \end{pmatrix}$$

$$= \begin{pmatrix} -2 & 2 & 0 & 0 \\ 0 & -2 & 1 & 4 \end{pmatrix} \begin{pmatrix} -1 & 0 \\ 1 & -1 \\ 0 & 1 \\ 0 & 1 \end{pmatrix} = \begin{pmatrix} 4 & -2 \\ -2 & 7 \end{pmatrix}$$

Here equations are:

$$\begin{pmatrix} 4 & -2 \\ -2 & 7 \end{pmatrix}\begin{pmatrix} V_1 \\ V_2 \end{pmatrix} = \begin{pmatrix} 10 \\ 0 \end{pmatrix}$$

$$V_1 = \frac{\begin{vmatrix} 10 & -2 \\ 0 & 7 \end{vmatrix}}{24} = \frac{70}{24} = 2.917$$

$$V_2 = \frac{\begin{vmatrix} 4 & 10 \\ -2 & 0 \end{vmatrix}}{24} = \frac{20}{24} = 0.833 \text{ volt } \textbf{Ans.}$$

or using the relation given in equation 4.58

$$Q_f\, Y_b\, Q_f^T\, V_t = -Q_f\, J$$

$$\begin{pmatrix} 4 & -2 \\ -2 & 7 \end{pmatrix}\begin{pmatrix} V_{t_1} \\ V_{t_3} \end{pmatrix} = \begin{pmatrix} -1 & 1 & 0 & 0 \\ 0 & -1 & 1 & 1 \end{pmatrix}\begin{pmatrix} -10 \\ 0 \end{pmatrix} = \begin{pmatrix} 10 \\ 0 \end{pmatrix}$$

$$V_{t1} = 2.917 \quad \text{and} \quad V_{t2} = 0.833$$

$V_{t1}$ is nothing but voltage of node 1 with reference to 3 and $V_{t3}$ is also voltage of node 2 with reference to 3 which is same as $V_1$ and $V_2$ taking node 3 as reference.

**Example 4.9:** For the network shown write down the f-cut-set matrix and solve for the nodal voltage.

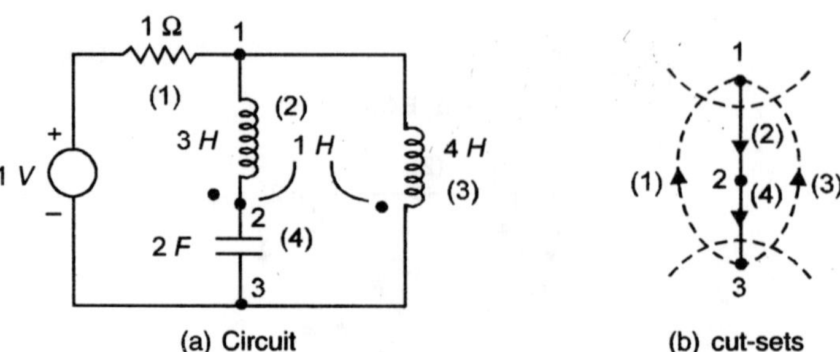



(a) Circuit          (b) cut-sets

**Fig. 4.9**

Since the dots are on the same side and assuming current direction out of the dotted terminal, the mutual inductance $M$ would be positive.

Taking branch 2, 4 as twigs we write the f-cut-set matrix

$$Q_f = \begin{array}{c} 2 \\ 4 \end{array}\begin{pmatrix} \overset{1}{-1} & \overset{2}{1} & \overset{3}{1} & \overset{4}{0} \\ 1 & 0 & -1 & -1 \end{pmatrix}$$

$$Y_b = Z_b^{-1} = \begin{pmatrix} [1]^{-1} & 0 & 0 & 0 \\ 0 & \begin{pmatrix} 3s & s \\ s & 4s \end{pmatrix}^{-1} & 0 \\ 0 & 0 & 0 & \left(\dfrac{1}{2s}\right)^{-1} \end{pmatrix}$$

$$Y_b = \begin{pmatrix} 1 & 0 & 0 & 0 \\ 0 & 4/11s & -1/11s & 0 \\ 0 & -1/11s & 3/11s & 0 \\ 0 & 0 & 0 & 2s \end{pmatrix}$$

$$Q_f Y_b Q_f^T = \begin{pmatrix} -1 & \dfrac{3}{11s} & \dfrac{2}{11s} & 0 \\ 1 & \dfrac{1}{11s} & \dfrac{-3}{11s} & -2s \end{pmatrix}\begin{pmatrix} -1 & 1 \\ 1 & 0 \\ 1 & -1 \\ 0 & -1 \end{pmatrix}$$

$$= \begin{pmatrix} 1+\dfrac{5}{11s} & -1-\dfrac{2}{11s} \\ -1-\dfrac{2}{11s} & 1+\dfrac{3}{11s}+2s \end{pmatrix}$$

Now $Q_f Y_b Q_t^T V_t = Q_f [I_s - Y_b V_b] = -Q_f Y_b V_b$

as $\qquad\qquad\qquad I_s = 0$

$$-Q_f Y_b V_b = \begin{pmatrix} -1 & \dfrac{3}{11s} & \dfrac{2}{11s} & 0 \\ 1 & \dfrac{1}{11s} & \dfrac{-3}{11s} & -2s \end{pmatrix}\begin{pmatrix} -1 \\ 0 \\ 0 \\ 0 \end{pmatrix} = \begin{pmatrix} 1 \\ -1 \end{pmatrix}$$

Hence the nodal equations are:

$$\begin{pmatrix} 1+\dfrac{5}{11s} & -\left(1+\dfrac{2}{11s}\right) \\ -\left(1+\dfrac{2}{11s}\right) & 1+2s+\dfrac{3}{11s} \end{pmatrix}\begin{pmatrix} V_{t_2} \\ V_{t_4} \end{pmatrix} = \begin{pmatrix} 1 \\ -1 \end{pmatrix}$$

**Example 4.10:** The reduced incidence matrix of a linear graph is given below.

$$A = \begin{pmatrix} 1 & 2 & 3 & 4 & 5 & 6 & 7 \\ 0 & 0 & 1 & 1 & 1 & 0 & -1 \\ 0 & 1 & 0 & 0 & -1 & 1 & 1 \\ -1 & 0 & -1 & 0 & 0 & -1 & 0 \end{pmatrix}$$

Assume branches 2,3,4 to constitute twigs of a tree, determine $Q_f$ and $B_f$ and verify the results by drawing the graph selecting branches 2,3,4 as twigs and obtain these matrices.

**Solution:** We know that $Q_l = A_t^{-1} A_l$ and $Q_f = [U \ Q_l]$ Hence we need find $A_t^{-1}$ first and then pre-multiply to $A_l$ to obtain $Q_l$ and hence $Q_f$ can be obtained.

$$A_t = \begin{pmatrix} 2 & 3 & 4 \\ 0 & 1 & 1 \\ 1 & 0 & 0 \\ 0 & -1 & 0 \end{pmatrix}$$

$$\det [A_t] = 0 - 0 - 1 = -1$$

$$C_{ij} = \begin{pmatrix} 0 & 0 & -1 \\ -1 & 0 & 0 \\ 0 & +1 & -1 \end{pmatrix} \quad \text{Hence} \quad C_{ij}^T = \begin{pmatrix} 0 & -1 & 0 \\ 0 & 0 & +1 \\ -1 & 0 & -1 \end{pmatrix}$$

Here
$$A_t^{-1} = \frac{C_{ij}^T}{\Delta} = \begin{pmatrix} 0 & +1 & 0 \\ 0 & 0 & -1 \\ 1 & 0 & 1 \end{pmatrix}$$

$$Q_l = A_t^{-1} A_l = \begin{pmatrix} 0 & 1 & 0 \\ 0 & 0 & -1 \\ 1 & 0 & 1 \end{pmatrix} \begin{pmatrix} 0 & 1 & 0 & -1 \\ 0 & -1 & 1 & 1 \\ -1 & 0 & -1 & 0 \end{pmatrix}$$

$$= \begin{pmatrix} 0 & -1 & 1 & 1 \\ 1 & 0 & 1 & 0 \\ -1 & 1 & -1 & -1 \end{pmatrix}$$

Hence, $Q_f = [U \ Q_l]$

$$= \begin{pmatrix} 2 & 3 & 4 \\ 1 & 0 & 0 & 0 & -1 & 1 & 1 \\ 0 & 1 & 0 & 1 & 0 & 1 & 0 \\ 0 & 0 & 1 & -1 & 1 & -1 & -1 \end{pmatrix}$$

$$\underbrace{\phantom{\quad\quad\quad\quad}}_{\text{links}}$$

To determine $B_f$

$$B_f = [B_t U]$$

$$B_t = -Q_l^T = \begin{pmatrix} 0 & -1 & 1 \\ 1 & 0 & -1 \\ -1 & -1 & 1 \\ -1 & 0 & 1 \end{pmatrix}$$

Hence,
$$B_f = \begin{array}{ccccccc} 2 & 3 & 4 & 1 & 5 & 6 & 7 \end{array} \\ \begin{pmatrix} 0 & -1 & 1 & 1 & 0 & 0 & 0 \\ 1 & 0 & -1 & 0 & 1 & 0 & 0 \\ -1 & -1 & 1 & 0 & 0 & 1 & 0 \\ -1 & 0 & 1 & 0 & 0 & 0 & 1 \end{pmatrix}$$

The augmented incidence matrix is given here

$$A_a = \begin{array}{c} \text{Nodes} \end{array} \quad \begin{array}{c} \text{Branches} \end{array}$$

|  | 1 | 2 | 3 | 4 | 5 | 6 | 7 |
|---|---|---|---|---|---|---|---|
| 1 | 0 | 0 | 1 | 1 | 1 | 0 | -1 |
| 2 | 0 | 1 | 0 | 0 | -1 | 1 | 1 |
| 3 | -1 | 0 | -1 | 0 | 0 | -1 | 0 |
| 4 | 1 | -1 | 0 | -1 | 0 | 0 | 0 |

The oriented graph is as shown here

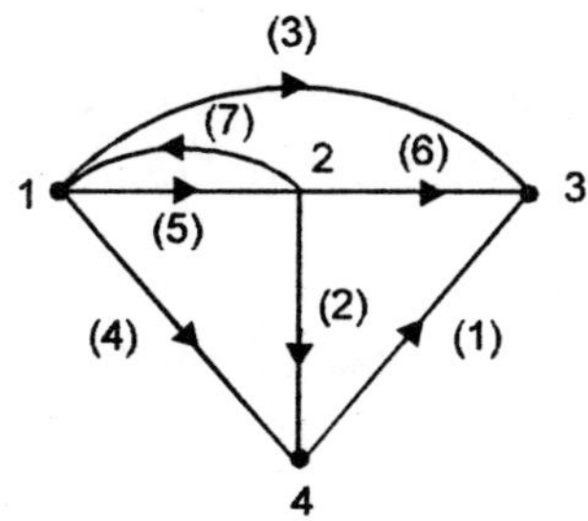

**Fig. E4.10**

Taking $2, 3, 4$ as twigs we have the requisite tree

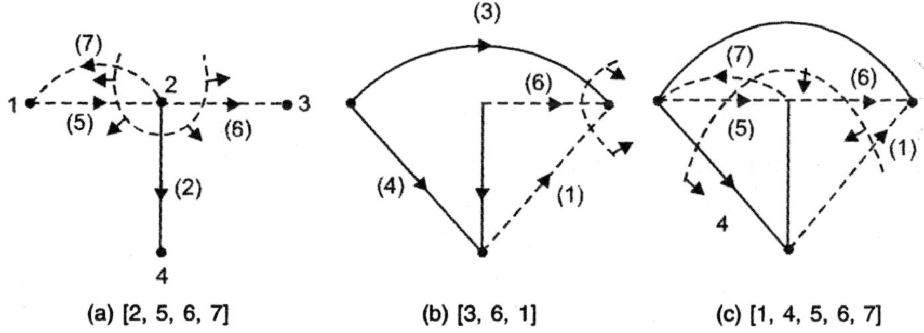

(a) [2, 5, 6, 7]    (b) [3, 6, 1]    (c) [1, 4, 5, 6, 7]

**Fig. E4.10.1**

Twigs Branches

$$Q_f = \begin{array}{c} \\ 2 \\ 3 \\ 4 \end{array} \begin{array}{c} \begin{array}{ccccccc} 2 & 3 & 4 & 1 & 5 & 6 & 7 \end{array} \\ \left( \begin{array}{ccccccc} 1 & 0 & 0 & 0 & -1 & 1 & 1 \\ 0 & 1 & 0 & 1 & 0 & 1 & 0 \\ 0 & 0 & 1 & -1 & 1 & -1 & -1 \end{array} \right) \end{array}$$

Hence this verifies with the one obtained mathematically. While drawing the curved lines for a cut-set consisting of one twig and other links connected to the same node, the direction of arrows along the curved lines should coincide with that of the twig, because in any case the entry in the twig and branch of that twig i.e. row and column corresponding to the twig has to be +1.

Now to obtain $B_f$ we redraw the tree and by connecting one link at a time and taking the direction of the link as reference i.e. taken as +1, the sign for the twigs, is assigned. Those branches which don't form part of the loop are assigned zero.

$$B_f = \begin{array}{c} \\ 1 \\ 5 \\ 6 \\ 7 \end{array} \begin{array}{c} \begin{array}{ccccccc} 2 & 3 & 4 & 1 & 5 & 6 & 7 \end{array} \\ \left( \begin{array}{ccccccc} 0 & -1 & 1 & 1 & 0 & 0 & 0 \\ 1 & 0 & -1 & 0 & 1 & 0 & 0 \\ -1 & -1 & 1 & 0 & 0 & 1 & 0 \\ -1 & 0 & 1 & 0 & 0 & 0 & 1 \end{array} \right) \end{array}$$

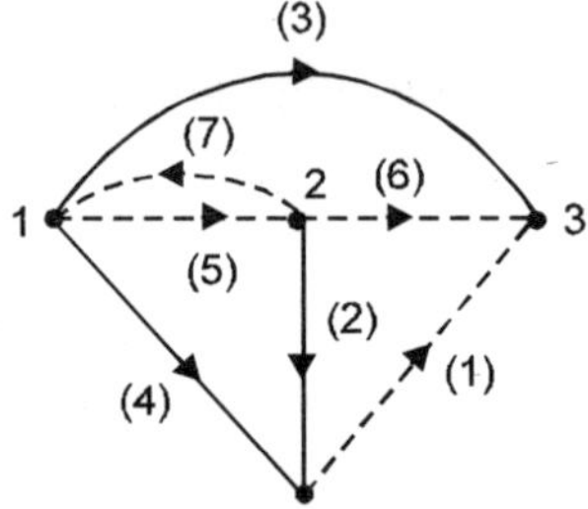


**Fig. E4.10.2**

Hence this verifies with the one obtained through mathematical manipulation.

**Example 4.11:** An oriented graph has nodes $1, 2, 3, 4$ and six branches 1 to 6. Taking node 4 as datum, the node voltages are $V_{n1}\ V_{n2}\ V_{n3}$. The branch voltages are $V_1$ to $V_6$ and expressed as

$$V_b = A^T V_n$$

Without drawing the graph, make use of the matrix $A^T$ and answer the following questions:

$$A^T = \begin{array}{c} \\ 1 \\ 2 \\ 3 \\ 4 \\ 5 \\ 6 \end{array} \begin{array}{c} \begin{array}{ccc} \text{node } 1 & 2 & 3 \end{array} \\ \left( \begin{array}{ccc} 1 & 0 & 0 \\ 0 & 1 & 0 \\ 0 & 0 & 1 \\ 1 & -1 & 0 \\ -1 & 1 & 0 \\ 0 & -1 & 1 \end{array} \right) \end{array}$$

(*i*) Which branches are in parallel and which are in series (*ii*) Are there branches corresponding to $V_{n1}\ V_{n2}\ V_{n3}$. Name them if any (*iii*) Express the currents in branches identified in (*ii*) in term of their link currents.

**Solution:** Since $A^T$ is the transpose of incidence matrix, the column represent the branches connected to the node.

Since branch 4 and 5 are connected between nodes 1 and 2 these are the branches connected in parallel.

At node 3 only there are two branches, at all other nodes we have more than two branches, hence, the two branches at node 3 constitute series branches i.e. branch 3 and 6.

(*ii*) Now $V_b = A^T V_n$

$$V_{b1} = V_{n1} \qquad V_{b2} = V_{n2} \quad \text{and} \quad V_{b3} = V_{n3}$$

Therefore branches $1, 2, 3$ correspond to the voltages $V_{n1}$, $V_{n2}$, $V_{n3}$ respectively.

(*iii*) At node 1, three branches are incident 1, 4, 5

Therefore, $\qquad I_1 + I_4 - I_5 = 0 \quad$ or $\quad I_1 = I_5 - I_4$

Similarly, $\quad I_2 - I_4 + I_5 - I_6 = 0 \quad$ or $\quad I_2 = I_4 + I_6 - I_5$

$$I_3 + I_6 = 0 \quad \text{or} \quad I_3 = -I_6$$

**PROBLEMS**

**4.1.** For the data given draw the network and draw the oriented graph and write down the incidence matrix and hence the nodal admittance matrix.

| Nodes | Admittance |
|-------|-----------|
| 1 – 2 | $2 - j\,8.0$ |
| 1 – 3 | $1 - j\,4.0$ |
| 2 – 3 | $0.666 - j\,2.664$ |
| 2 – 4 | $1 - j\,4.0$ |
| 3 – 4 | $2 - j\,8.0$ |

**4.2.** A linear graph has 5 nodes and 7 branches. The reduced incidence matrix for the graph is given as

nodes   branches

$$
A = \begin{array}{c c}
 & \begin{array}{ccccccc} 1 & 2 & 3 & 4 & 5 & 6 & 7 \end{array} \\
\begin{array}{c} 1 \\ 2 \\ 3 \\ 4 \end{array} &
\left( \begin{array}{ccccccc}
1 & 1 & 0 & 0 & 0 & 0 & 1 \\
-1 & -1 & 1 & 0 & 0 & 0 & 0 \\
0 & 0 & -1 & 1 & 0 & 0 & 0 \\
0 & 0 & 0 & -1 & -1 & -1 & 0
\end{array} \right)
\end{array}
$$

Show that the branches $(1, 3, 4, 5)$ constitute a tree (*Hint*. Determinant of $A_t \neq 0$) Also show that the branches $(1, 3, 4)$ do not constitute a tree. Obtain $B_f$ mathematically.

**4.3.** For the network shown, write the incidence matrix, the nodal admittance matrix and solve for the nodal voltages and branch currents. Take node 4 as datum node.

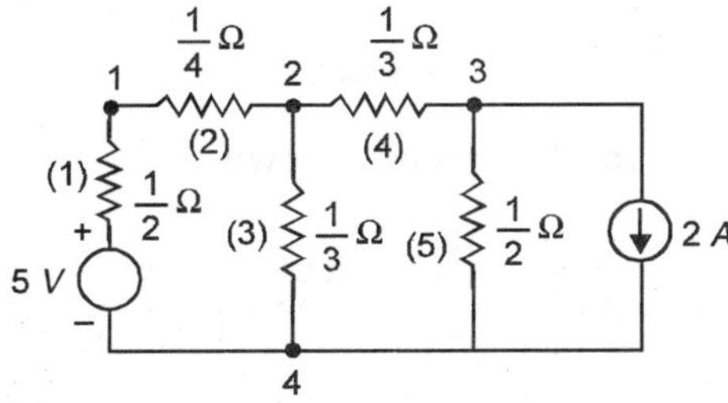


**Fig. P4.3**

**4.4.** For the network in P4.3 write down f-cut-set matrix taking 2, 3, 4 as twigs of the tree. Also obtain f-cut-set admittance matrix. Solve for the twig voltages and hence the branch voltages and currents.

**4.5.** For the network in P4.3 write down $B_f$ matrix taking 2, 3, 4 as twigs of the tree. Also obtain loop impedance matrix using $B_f$. Solve for the loop currents and the branch currents.

**4.6.** Show that the graphs shown in Fig. P4.6 are isomorphic.

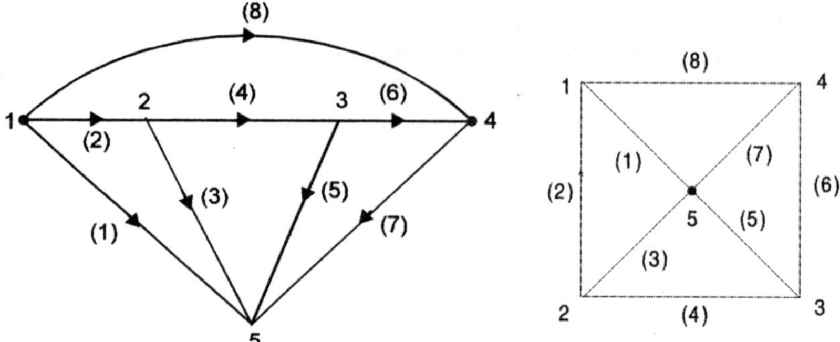

**Fig. P4.6 graph**

**4.7.** The f-cut set matrix of a network is given as

$$
\begin{array}{cc}
\text{Twigs} & \text{Links} \\
\begin{array}{cccc} 1 & 2 & 3 & 4 \end{array} & \begin{array}{ccc} 5 & 6 & 7 \end{array}
\end{array}
$$

$$
\left(
\begin{array}{cccc|ccc}
 & & & & -1 & 0 & 0 \\
 & & U & & 1 & 0 & 1 \\
 & & & & 0 & 1 & 1 \\
 & & & & 0 & 1 & 0
\end{array}
\right)
$$

Draw the oriented graph.

**4.8.** Determine the number of trees in the graphs shown here

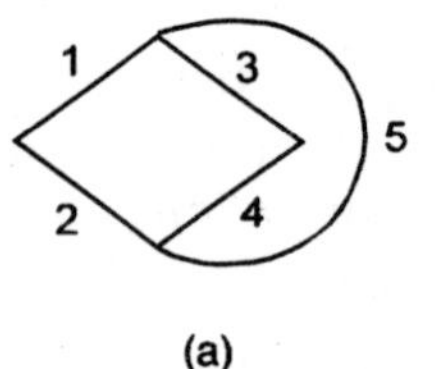

(a)

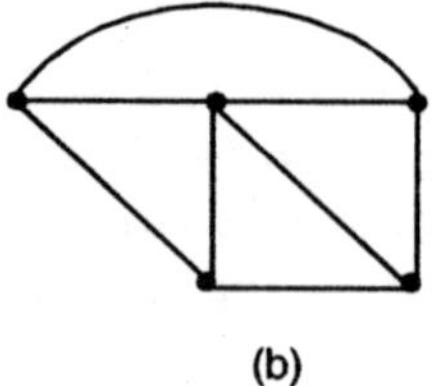

(b)

**Fig P4.8**

**4.9.** Determine the current $i_1$ in the circuit shown

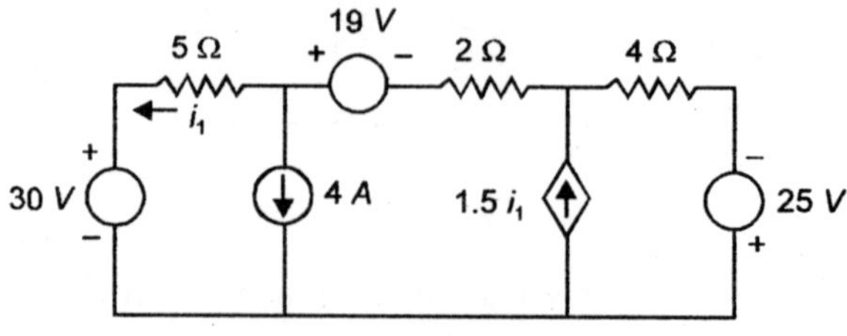

**Fig P4.9**

**4.10.** For the circuit shown in Fig. P4.10 construct a tree in which $V_1$ and $V_2$ are tree branch voltages. Using nodal approach, solve for $V_1$

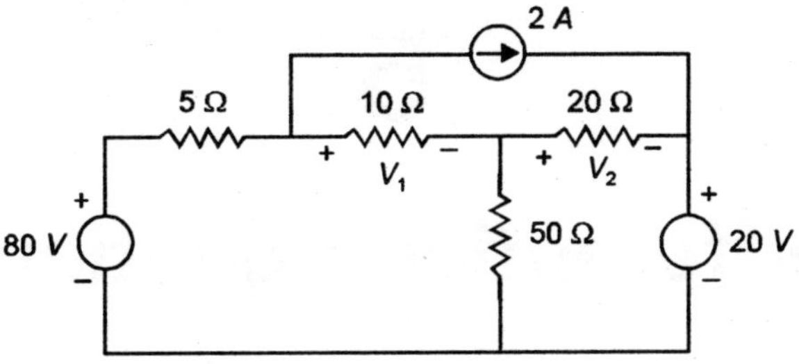

**Fig. P4.10**

**4.11.** For the network shown draw a tree and obtain $i_1$

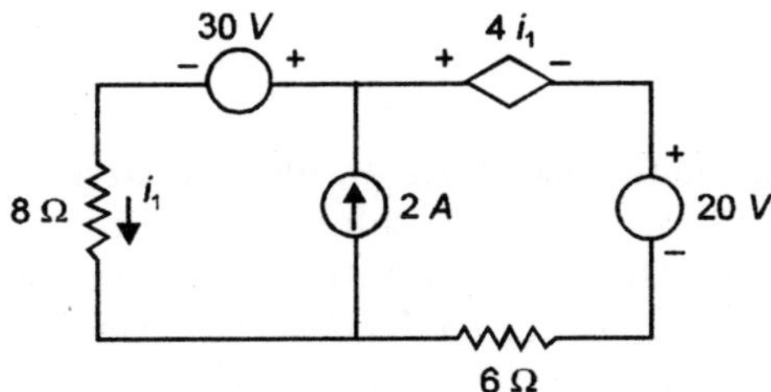

**Fig. P. 4.11**

**4.12.** For the circuit shown in Fig. P4.12 select a suitable tree and determine branch voltages and branch currents using (*i*) nodal voltages and (*ii*) mesh current methods.

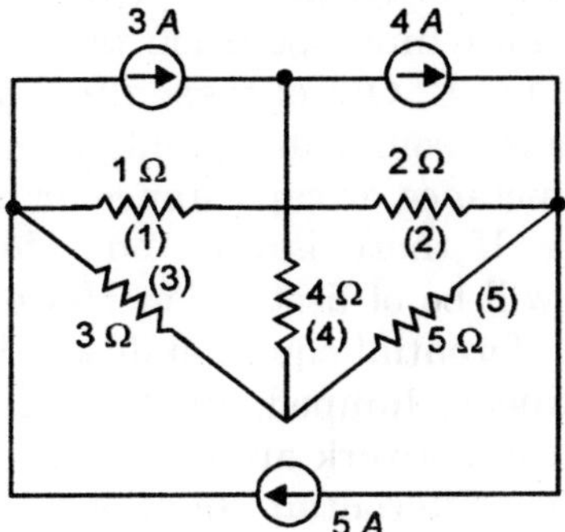

**Fig. P4.12**

# 5
# Differential Equations

## 5.1 INTRODUCTION

The solution of algebraic equations describing an electric circuit gives steady state operation of the circuit. However, to obtain transient solution of a network, the solution of a set of differential equations describing the network with time as independent variable and current or voltage or any other electrical quantity as dependent variable is required. Whenever, there is redistribution of energy in an electric network normally due to a switching operation, transients appear in the network. These transients even though appear in a circuit for a short duration their presence plays an important role in the performance of the circuit. The time for which these transients exist in a network is known as transient period and it is a period between two steady state operating conditions of the network.

As mentioned in the previous paragraph, the transients are due to redistribution of energy in the network whenever switching operation takes place. In the electric circuits we have mainly three elements $R$, $L$ and $C$. Out of these three only $L$ and $C$ have the properties of storing energy to be taken from the source and giving energy back to the source. Therefore, for transients to occur, either inductance or capacitance or both should be present and there should be a switching operation. If there is only one element present, then differential equation describing the network will be of first order. However, if both $L$ and $C$ are present and switching takes place, then differential equation describing the network will be of second order. Since the elements are linear, lumped, finite, passive and bilateral (LLFPB), the differential equations describing the network are ordinary, linear differential equations with constant co-efficients. An ordinary differential equation is one in which there is only one independent variable, in our case time $t$. As a result there is no need for partial derivatives.

A differential equation is linear if it contains only terms of the first degree of dependent variable and all its higher derivatives, e.g. if $i(t)$ is the dependent variable the equation

$$i'(t) + 2i(t) = \sin t \qquad \text{...(5.1)}$$

is a linear differential equation.

However, the equation

$$[i'(t)]^2 + 2i(t)i'(t) + 4i(t) = \sin t \qquad \text{...(5.2)}$$

is a non-linear differential equations

An implication of linear property is the superposition property which implies that if for a forcing function $f_1(t)$ the response is $i_1(t)$ and for $f_2(t)$, it is $i_2(t)$ then for a linear combination of the forcing functions $af_1(t) + bf_2(t)$, the response is also the sum of the linear combination of the responses $i_1(t)$ and $i_2(t)$ i.e. the response is $ai_1(t) + bi_2(t)$, where $a$ and $b$ are arbitrary constants.

By constant co-efficient we mean the co-efficient, in the differential equation describing the network should be independent of time $t$.

The circuit changes are assumed to occur at time $t = 0$ when a switch is operated. We assume an ideal switch which takes zero time for its operation and we define two very important times.

$$t = 0_- \text{ the instant prior to } t = 0 \text{ and}$$

$$t = 0_+ \text{ the instant immediately after switching.}$$

## 5.2  HOMOGENEOUS LINEAR DIFFERENTIAL EQUATIONS

Since we are restricting our analysis only to second order circuits the differential equation describing such a circuit will also be of second order. A typical second order differential equation with current $i$ as the dependent variable and $t$ as the independent variable will be

$$i''(t) + ai'(t) + bi(t) = f(t) \qquad \qquad ...(5.3)$$

Where $a$ and $b$ are constant co-efficients and $f(t)$ the forcing function or source of excitation. If for a certain circuit $f(t) = 0$ i.e. the excitation is absent then differential equation is known as homogeneous differential equation otherwise it is known as non-homogeneous. The complete solution of a differential equation depends upon the initial condition of the circuit and the forcing functions.

The solution of the differential equation is made up of two components:

$$i(t) = i_{CF}(t) + i_{PI}(t) \qquad \qquad ...(5.4)$$

Where $i_{CF}(t)$ is the complementary function solution also known as free response or natural response or zero-input response and is obtained from the homogeneous equation

$$i''(t) + ai'(t) + bi(t) = 0 \qquad \qquad ...(5.5)$$

Replacing $\dfrac{d}{dt} = p$  where $p$ is the heaviside operator

We have $(p^2 + ap + b)\, i(t) = 0 \qquad \qquad ...(5.6)$

Here $p^2 + ap + b = 0$ is known as the auxiliary or characteristic equation. The zeros or roots of the polynomial or the second order equation are known as natural frequencies of the network variable $i(t)$.

Depending upon the nature of roots, we get different types of force free or zero input response.

(a) When the roots are real and distinct. Let these be $p_1$ and $p_2$. The solution is

$$i(t) = C_1 e^{-p_1 t} + C_2 e^{-p_2 t} \qquad \qquad ...(5.7)$$

(b) The roots are equal and real

$$i(t) = C_1 e^{-p_0 t} + C_2 t e^{-p_0 t} \qquad \qquad ...(5.8)$$

(c) The roots are complex. Let these be

$$p_1, p_2 = -\sigma \pm j\omega$$

Hence the solution

$$i(t) = C_1 e^{(-\sigma - j\omega)t} + C_2 e^{(-\sigma + j\omega)t}$$

$$= C_1 e^{-\sigma t}(\cos \omega t - j \sin \omega t) + C_2 e^{-\sigma t}(\cos \omega t + j \sin \omega t)$$

$$= (c_1 + c_2)e^{-\sigma t} \cos \omega t + j(c_2 - c_1)e^{-\sigma t} \sin \omega t$$

$$= M_1 e^{-\sigma t} \cos \omega t + M_2 e^{-\sigma t} \sin \omega t \qquad \qquad \text{...(5.9)}$$

Where $\qquad c_1 + c_2 = M_1$ and $j(c_2 - c_1) = M_2$

The constants $M_1$ and $M_2$ can be obtained from the initial conditions.

Dividing the right hand side of equation (5.9) by $\sqrt{M_1^2 + M_2^2}$

We have $\qquad i(t) = \dfrac{M_1}{\sqrt{M_1^2 + M_2^2}} e^{-\sigma t} \cos \omega t + \dfrac{M_2}{\sqrt{M_1^2 + M_2^2}} e^{-\sigma t} \sin \omega t$

$$= e^{-\sigma t}(\sin \phi \cos \omega t + \cos \phi \sin \omega t)$$

$$= e^{-\sigma t} \sin(\omega t + \phi) \qquad \qquad \text{...(5.10)}$$

Where $\qquad \phi = \tan^{-1} \dfrac{M_1}{M_2}$

**Example 5.1:** Determine the solution for

$$i''(t) + 3i'(t) + 2i(t) = 0$$

given the initial condition

$$i(0_+) = 2 \text{ and } i'(0_+) = -1$$

**Solution:** The characteristic equation is

$$p^2 + 3p + 2 = 0$$

$$p = -1, \ p = -2$$

Hence

$$i(t) = c_1 e^{-t} + c_2 e^{-2t}$$

Now at $t = 0_+$ $\qquad i(0_+) = 2 = c_1 + c_2$

$$i'(t) = -c_1 e^{-t} - 2c_2 e^{-2t}$$

$$i'(0_+) = -1 = -c_1 - 2c_2$$

or $\qquad c_1 + 2c_2 = 1$

and $\qquad c_1 + c_2 = 2$

or $\qquad c_2 = -1$

and $\qquad c_1 = 3$

Hence the solution $\qquad 3e^{-t} - e^{-2t} = i(t)$ **Ans.**

**Example 5.2:** Solve the equation

$$i''(t) + 4i(t) + 4 = 0$$

subject to $\qquad i(0_+) = 1$ and $i'(0_+) = -2$

**Solution:** The characteristic equation is

$$p^2 + 4p + 4 = 0$$

Hence the roots $\qquad\qquad p = -2,\ -2$

Therefore, the solution is, $\qquad i(t) = c_1 e^{-2t} + c_2 t e^{-2t}$

At $t' = 0_+$ $\qquad\qquad\qquad i(0_+) = 1 = c_1 + 0$ or $c_1 = 1$

and $\qquad\qquad\qquad i'(t) = -2c_1 e^{-2t} - c_2\,[t.(-2)e^{-2t} + e^{-2t}]$

$$i'(0_+) = -2c_1 e^{-2t} - c_2\,e^{-2t} = -2c_1 - c_2 = -3$$

or $\qquad\qquad\qquad 2c_1 + c_2 = 3$ $\quad$ or $c_2 = 1$

Hence the final solution is

$$i(t) = e^{-2t}\,(1 + t)\quad\textbf{Ans.}$$

**Example 5.3:** Solve the differential equation

$$i''(t) + 2i'(t) + 5i(t) = 0$$

subject to the condition

$$i(0_+) = 1 \text{ and } i'(0_+) = 0$$

Solution: The characteristic equation is given as

$$p^2 + 2p + 5 = 0$$

The roots of the characteristic equation are

$$p_1,\ p_2 = \frac{-2 \pm \sqrt{4 - 20}}{2}$$

$$= -1 \pm j2$$

Let $\sigma = 1$ and $\omega = 2$ and using equations 5-9 the solution is

$$i(t) = M_1 e^{-t} \cos 2t + M_2 e^{-t} \sin 2t$$

At $t = 0_+$ $\qquad\qquad\qquad i(0_+) = 1$

Hence $\qquad\qquad\qquad\qquad 1 = M_1$

Also, At $t = 0_+$ $\qquad\qquad i'(0_+) = 0$

Now

$$i'(t) = M_1(-e^{-t} \cos 2t - 2e^{-t} \sin 2t) +$$
$$M_2(-e^{-t} \sin 2t + 2e^{-t} \cos 2t)$$
$$0 = M_1(-1) + M_2(2)$$
$$0 = -M_1 + 2M_2$$

or $\qquad\qquad\qquad\qquad 2M_2 = M_1 = 1$

$$M_2 = \frac{1}{2}$$

Hence the final solution is

$$i(t) = e^{-t} \cos 2t + \frac{1}{2} e^{-t} \sin 2t \qquad\textbf{Ans.}$$

## 5.3 NON-HOMOGENEOUS EQUATIONS

As mentioned earlier in equation (5.4) the complete solution of the differential equation (non-homogeneous) is given by

$$i(t) = i_{CF}(t) + i_{pI}(t) \qquad \qquad ...(5.4)$$

According to uniqueness theorem the solution $i(t)$ in equation (5.4) is the unique solution for the non homogeneous differential equation if it satisfies the specified initial conditions at $t = 0_+$

Since we already know how to find the complementary function $i_{CF}(t)$, we now have to find out the particular integral $i_{pI}(t)$. In solving for this a very reliable rule of thumb is that $i_{pI}(t)$ usually takes the same force as the forcing function if $f(t)$ can be expressed as a sum of exponential functions. To be more specific, $i_{pI}(t)$ assumes the form of $f(t)$ plus all its derivatives. For example, if $f(t) = a \sin \omega t$ then $i_{pI}(t)$ takes the form

$$i_{pI}(t) = A \sin \omega t + B \cos \omega t$$

The only unknowns that must be determined are the co-efficient $A$ and $B$ of the terms in $i_{PI}(t)$

In illustrating the method of unknown co-efficients suppose $f(t) = \alpha e^{+\beta t}$ where $\alpha$ and $\beta$ are arbitrary constant, we assume $i_{PI}(t)$ to have a similar form i.e.

$$i_{pI}(t) = A \, e^{\beta t}$$

Where $A$ is the unknown co-efficient. To determine $A$, we simply substitute the assumed solution $i_{pI}(t)$ into the differential equation (5.5).

$$A \, e^{\beta t}[\beta^2 + a\beta + b] = \alpha \, e^{\beta t} \qquad \qquad ...(5.11)$$

We see that the polynomial within the bracket is the characteristic equation with $p = \beta$, as a result the unknown co-efficient is obtained as

$$A = \frac{\alpha}{\beta^2 + a\beta + b} \qquad \qquad ...(5.12)$$

provided that the polynomial in the denominator does not equal zero.

**Example 5.4:** Determine the solution of the equation

$$i''(t) + 3i'(t) + 2i(t) = 4e^t$$

subject to the condition $i(0_+) = 1$, $i'(0_+) = -1$

**Solution:** The characteristic equation

$$p^2 + 3p + 2 = (p + 1) \, (p + 2)$$

so that the complementary function is

$$i_{CF}(t) = c_1 e^{-t} + c_2 e^{-2t}$$

Since the forcing function is $f(t) = 4e^t$

here $\qquad\qquad \alpha = 4 \text{ and } \beta = 1$

Hence $\qquad\qquad A = \dfrac{4}{1 + 3 \times 1 + 2} = \dfrac{4}{6} = \dfrac{2}{3}$

The total solution, therefore, is

$$i(t) = i_{CF} + i_{pI} = C_1 e^{-t} + C_2 e^{-2t} + \frac{2}{3} e^t$$

To evaluate $c_1$ and $c_2$ we make use of the initial conditions

At $t = 0_+$ $\qquad\qquad\qquad i(0_+) = 1$ and $i'(0_+) = -1$

$$1 = c_1 + c_2 + \frac{2}{3} \quad \text{or} \quad c_1 + c_2 = \frac{1}{3} \quad \text{or} \quad c_1 = \frac{1}{3} - c_2$$

and

$$i'(t) = -c_1 e^{-t} - 2c_2 e^{-2t} + \frac{2}{3} e^t$$

$$i'(0_+) = -1 = -c_1 - 2c_2 + \frac{2}{3}$$

$$-\frac{1}{3} + c_2 - 2c_2 + \frac{2}{3} = -1$$

$$-c_2 = -1 - \frac{1}{3} = -\frac{4}{3}$$

or

$$c_2 = \frac{4}{3} \quad \text{and hence} \quad c_1 = \frac{1}{3} - \frac{4}{3} = -1$$

Therefore, the final solution is

$$i(t) = -e^{-t} + \frac{4}{3} e^{-2t} + \frac{2}{3} e^t \quad \textbf{Ans.}$$

If $f(t)$ the forcing function is a constant $f(t) = \alpha$

We can use this as $f(t) = \alpha\, e^{j0t}$ where $\beta = 0$ and follow the procedure outlined above.

Similarly if the forcing function is either a sin or a cos function $f(t) = \alpha \sin \omega t$ or $f(t) = \alpha \cos \omega t$ we can still consider the forcing function to be of exponential form and make use of the procedure outlined above to determine the unknown co-efficients

If $f(t) = \alpha \cos \omega t$ then $i_p(t) = R_e\, i_{pI}(t)$

If $f(t) = \alpha \sin \omega t$ then $i_p(t) = I_m\, i_{pI}(t)$

i.e. if the excitations is a cos function $\alpha \cos \omega t$ or a sin function $\alpha \sin \omega t$ we can use $f(t) = \alpha\, e^{j\omega t}$ and then we take the real or imaginary part respectively of the resulting particular integral.

As mentioned earlier, in the solution of network differential equations, the complementary function is called free response whereas the particular integral is known as the forced response.

## 5.4  INITIAL CONDITIONS IN CIRCUITS

The free response or natural response of a circuit is decided by the initial conditions of the circuit. Also, to solve a differential equation of $n$th order, $n$ initial conditions are required to obtain complete solution.

### 5.4.1 Initial Conditions for the Resistor

For a purely resistive element the *v-i* relation is given as

$$v(t) = R\, i(t) \qquad\qquad ...(5.13)$$

This shows that the voltage across the resistor changes instantaneously if the current through it changes instantaneously i.e.

$$i_R(0_+) \neq i_R(0_-) \text{ and } v_R(0_+) \neq v_R(0_-) \qquad\qquad ...(5.14)$$

### 5.4.2 Initial Conditions for an Inductor

The *v-i* relationship for an inductor is

$$v(t) = L\frac{di}{dt}$$

$$i(t) = \frac{1}{L}\int_{0_-}^{t} v(\tau)d\tau + i_L(0_-) \qquad\qquad ...(5.15)$$

Where $L$ is in henrys. The initial current $i_L(0_-)$ can be regarded as an independent current source as shown in Fig. (5.1)

**Fig. 5.1 (a)** Inductor   **(b)** Inductor with initial current

The current through the inductor is continuous for all $t$ except in the case of impulse excitation and its derivatives i.e.

$$i_L(0_+) = i_L(0_-) = I_0 \qquad\qquad ...(5.16)$$

Though the current in an inductor cannot change instantaneously, the voltage across the inductor has no such restriction i.e.

$$v_L(0_+) \neq v_L(0_-) \qquad\qquad ...(5.17)$$

Initial conditions for an inductor at $t = 0_+$. If $i_L(0_-) = 0$  i.e. if at $t = 0_-$ the current in the inductor is zero, it is zero at $t = 0_+$ also as due to principle of constant flux linkage. The flux linkages can not change instantaneously and the current through inductor can not change instantaneously. The inductor therefore, acts as an open circuit at $t = 0_+$.

### 5.4.3 Initial Condition for a Capacitor

For a capacitor shown in Fig. (5.2) the *v-i* relationships are

$$i(t) = C\frac{dv}{dt} \qquad\qquad ...(5.18)$$

$$v(t) = \frac{1}{C}\int_{0_-}^{t} i(\tau)d\tau + v_c(0_-) \qquad\qquad ...(5.19)$$

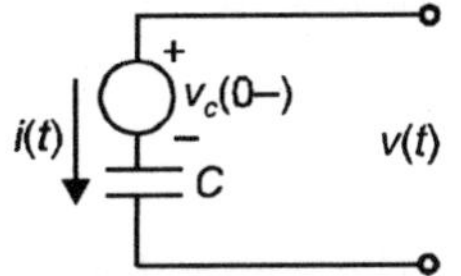

**Fig. 5.2 (a)** Capacitor　　　　　**(b)** Capacitor with initial voltage

Where $C$ is in farads. $v_c(0_-)$ is the initial voltage across the capacitor just before the switching action. It can be regarded as an independent voltage source as shown in Fig. (5.2b). In case of a capacitor voltage across its terminals can not change instantaneously i.e. $v_c(0_+) = v_c(0_-)$ for all excitations except impulses and its derivatives.

Initial conditions for capacitor at $t = 0_+$ ;

For a capacitor $v$-$i$ relationship at $t = 0_+$ is

$$v_c(0_+) = \int_{0_-}^{0_+} i(\tau)d\tau + v_c(0_-) \qquad \qquad ...(5.20)$$

If $i(t)$ does not contain an impulse or its derivatives $v_c(0_+) = v_c(0_-)$. If $q$ is the charge on the capacitor at $t = 0_-$, the initial voltage is

$$v_c(0_+) = v_c(0_-) = \frac{q}{c} \qquad \qquad ...(5.21)$$

However, if at $t = 0_-$ the change $q$ is zero, the voltage on the capacitor at $t = 0_+$ is zero i.e. $v_c(0_+) = 0$ which means when there is no energy stored in the capacitor its equivalent at $t = 0_+$ is a short circuit. This is also based on law of conservation of charge. The charge on the capacitor can not change instantaneously which otherwise requires infinite current through the capacitor which is physically not possible and hence the capacitor acts as a short circuit at $t = 0_+$. Therefore, at $t = 0_+$ we can replace the capacitor by a voltage source if an initial charge exists or by a short circuit if there is no initial charge.

**Example 5.5:** A parallel combination of $R = 2\Omega$, $L = 1H$ and $C = 1$ F is connected across a current source $I_S = 10A$ through a switch as shown in Fig. (E5.5)

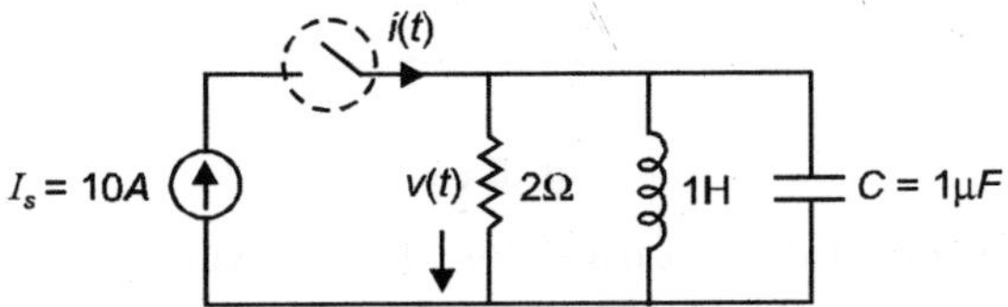

**Fig. E.5.5**

Determine $v(0_+)$, $\dfrac{dv}{dt}(0_+)$ and $\dfrac{d^2v}{dt}(0_+)$ . Given $i_L(0_+) = 0$

**Solution:** After the switch is closed the differential equation describing the network is

given as

$$I_s = \frac{v(t)}{R} + \frac{1}{L}\int_0^t v(t)dt + c\frac{dv}{dt} \qquad \qquad ...(\alpha)$$

Differentiating with respect to time we have

$$0 = \frac{1}{R}\frac{dv}{dt} + \frac{v(t)}{L} + c\frac{d^2v}{dt^2} \qquad\qquad ...(\beta)$$

At $t = 0_+$, the capacitor acts as a short circuit and inductor as an open circuit i.e. $v_c(0_+) = 0$ and $i_L(0_+) = 0$

Substituting these values in equation $(\alpha)$ we have

$$10 = \frac{0}{R} + 0 + 1\frac{dv(+)}{dt}$$

or $\qquad\qquad \left.\dfrac{dv}{dt}\right|_{0+} = 10$ V/sec.

Now using equation $(\beta)$

$$0 = \frac{1}{R}\frac{dv}{dt}(0_+) + \frac{v(0_+)}{L} + c\frac{d^2v}{dt^2}(0_+)$$

$$0 = \frac{10}{2} + 0 + 1\frac{d^2v}{dt^2}(0_+)$$

or $\qquad\qquad \dfrac{d^2v}{dt^2}(0_+) = -5$ V/sec$^2$

It is to be noted that a capacitor acts as an open circuit at $t = \infty$ and an inductor a short circuit for a step input.

**Example 5.6:** Determine the initial condition $i(0_+)$ and $i'(0_+)$ for the network shown when the switch is closed at $t = 0$.

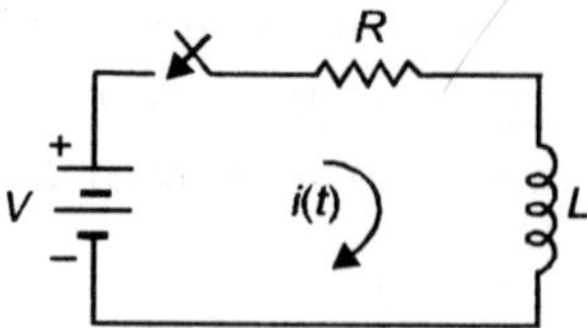


**Fig. E.5.6**

**Solution:** Writing differential equation for the network when the switch is closed

$$R\, i(t) + L\frac{di}{dt} = V$$

Since $i(0_-) = 0$, therefore, at $t = 0_+$ $i(0_+) = 0$

To find out $i'(0_+)$, we substitute $i(0_+)$ in the above equation, we have

$$R.0 + L\, i'(0_+) = V$$

or $\qquad\qquad i'(0_+) = \dfrac{V}{L} \qquad$ **Ans.**

To obtain steady state solution, since inductor acts as a short circuit for $t = \infty$, therefore current

$$i(t) = \frac{V}{R}$$

## 5.5  FIRST ORDER CIRCUITS

First order differential equations describe the first order circuits as these circuits contain only one energy storing element inductor or capacitor.

1.  Zero-Input response of First-Order Circuit (Complementary function).

Consider the circuit in Fig. (5.3). The switch $S$ is changed from position $a$ to $b$ at $t = 0$. To determine the expression for current $i(t)$ and voltage $v_R(t)$ and $v_L(t)$.

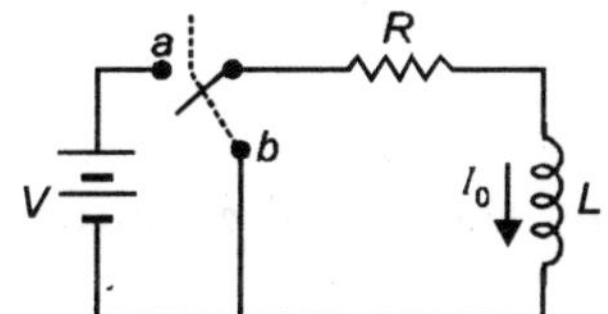

**Fig. E.5.3** (a) *R-L* circuit connected to dc source      (b)  Circuit when switch changed to position *b*.

It is presumed that switch is initially placed in position 'a' for a long time when steady state conditions are reached and current $i_L(0_-) = I_0 = \dfrac{V}{R}$

When the switch is changed to position $b$, the circuit will be as in Fig. (5.3b). The current in the circuit is initially $I_0$ and the inductor stores energy in the form of electromagnetic field which gets dissipated gradually as time passes through the resistor and finally when the total energy is dissipated the current through the circuit is zero.

Writing equation for circuit (5.3b)

$$R\,i(t) + L\frac{di}{dt} = 0 \qquad\qquad \text{...(5.22)}$$

Subject to
$$i(0_+) = \frac{V}{R}$$

$$\frac{di}{i} = -\frac{R}{L}\,dt$$

$$\log i = -\frac{R}{L}t + A$$

Where $A$ is a constant which can be determined from the given initial condition.

At $t = 0_+$
$$(i)t = \frac{V}{R}$$

$$ln\ \frac{V}{R} = 0 + A$$

Therefore the complete solution is

$$ln \; i = -\frac{R}{L}t + ln \; \frac{V}{R}$$

$$\frac{i}{V/R} = e^{-\frac{R}{L}t} \quad \text{or} \quad i = \frac{V}{R}e^{-\frac{R}{L}t} \qquad \qquad ...(5.23)$$

The solution of equation can also be obtained as follows:

The equation is homogenous as the forcing function is absent . The equation can be rewritten as

$$(Lp + R)\; i(t) = 0 \qquad \qquad ...(5.24)$$

The characteristic equation is

$$Lp + R = 0$$

or

$$p = -\frac{R}{L}$$

Therefore, the solution is $\quad i(t) = Ce^{-\frac{R}{L}t}$

Where $C$ is a constant and can be obtained using the initial condition

at

$$t = 0_+ \quad i(0_+) = \frac{V}{R}$$

$$\frac{V}{R} = C$$

Hence

$$i(t) = \frac{V}{R}e^{-\frac{R}{L}t} \qquad \qquad ...(5.25)$$

From equation (5.25) it is clear that the current response is an exponentially decaying curve. The decay characteristic is determined by the parameters $R, L$ of the circuit. Dimensionally the ratio $\frac{L}{R}$ is time and, hence the unit is in seconds. Let $\frac{L}{R} = \tau$ and let us find out the value of the current in the circuit at $t = \tau$ seconds. We have using equation (5.25)

$$i(\tau) = \frac{V}{R}e^{-1} = 0.368\frac{V}{R}$$

i.e. in one time constant $\tau$, the current reduces to 36.8% of its initial value of $\frac{V}{R}$

However, if we take $t = 4\tau$ we find that the value of the current becomes $0.0183\; I_0$ i.e. the current reduces to 2% of its initial value. The variation is shown in Fig. (5.4). We will prove now that a tangent to the curve in Fig. (5.4) intersects the time axis at $t = \tau$ i.e. if the curve is known, a tangent to the curve at $I_0$ intersects time axis at $t = \tau$

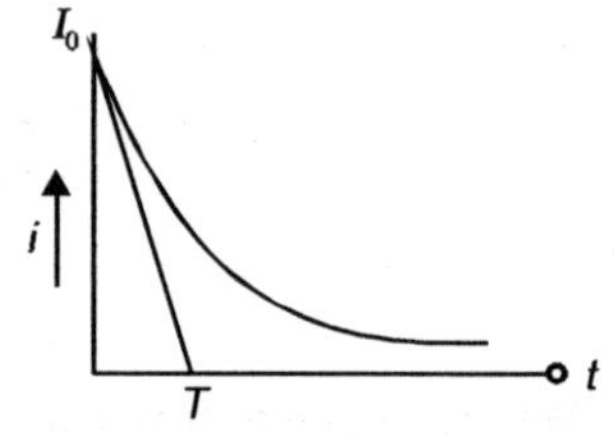

Let the tangent intersect at $t = x$, the equation to the tangent is

**Fig. 5.4** Plot of equation 5-25

$$\frac{t}{x} + \frac{i}{I_0} = 1 \qquad \qquad ...(5.26)$$

Differentiating with respect to $t$ we have

$$\frac{1}{x} + \frac{di}{dt} \cdot \frac{1}{I_0} = 0$$

or
$$\frac{di}{dt} = -\frac{I_0}{x}$$
...(5.27)

Whereas from equation (5.25)

$$\frac{di}{dt} = -\frac{V}{L}e^{-\frac{R}{L}t}$$

and at $t = 0$
$$\frac{di}{dt} = -\frac{V}{L}$$
...(5.28)

Equating equation (5.27) and (5.28) we have

$$\frac{I_0}{x} = \frac{V}{L} \qquad \text{or} \qquad x = I_0 \cdot \frac{L}{V} = \frac{V}{R}\cdot\frac{L}{V} = \frac{L}{R}$$
...(5.29)

This shows that the tangent on the curve at $I_0$ intersects the time axis at the time constant $\tau$.

Similarly RC circuit shown in Fig. (5.5) is also a first order circuit. The switch $S$ is changed from position $a$ to $b$ as in the previous case

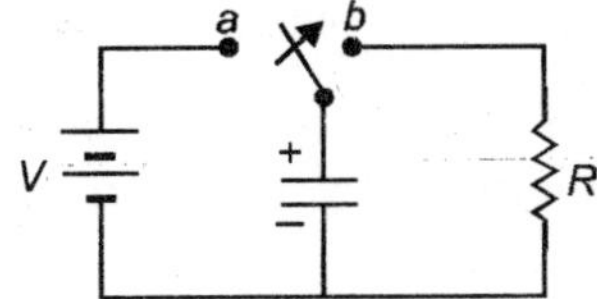
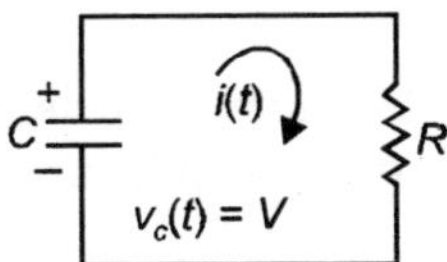


**Fig. 5.5** RC circuit (a) Capacitor charged to $V$ (b) Capacitor discharging through $R$

The equation describing the circuit (Fig. 5.5b) is

$$C\frac{dv_c}{dt} + \frac{v_c}{R} = 0$$
...(5.30)

This is a first order linear homogeneous equation. This is rewritten as

$$\left(Cp + \frac{1}{R}\right)v_c = 0$$

The characteristic equation is

$$Cp + \frac{1}{R} = 0$$

or
$$p = -\frac{1}{RC}$$

and hence the solution is

$$v_c(t) = A\,e^{-t/RC}$$

Now at $\qquad t = 0_+ \qquad v_c = V, \qquad$ hence $A = V$

and
$$V_c(t) = V\,e^{-t/RC}$$
...(5.31)

This is similar to equation (5.25) and hence time constant of the circuit $\tau = RC$ and $\dfrac{1}{RC}$ is per sec. and hence dimensionally it is frequency and hence it is known as natural frequency of the circuit expressed in radians/sec. In regard to this circuit other observations are similar to the previous *RL* circuit.

### 5.5.1  First Order Circuit with d.c. Excitation

Refer to Fig. 5.6 where we assume that the circuit is initially balanced before the switch $S$ is closed i.e. $i(0) = 0$

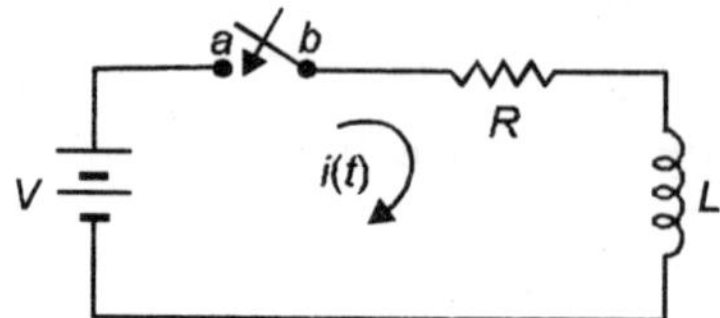

Fig. 5.6 RL circuit connected to dc source through a switch

The differential equation describing the circuit is

$$Ri + L\frac{di}{dt} = V \qquad \text{with } i(0_-) = 0 \qquad \text{...(5.32)}$$

This is a linear first order non-homogeneous equation. This also can be solved by method of variables are separable and by obtaining the complementary function and particular integral.

$$\frac{di}{dt} = \frac{V - Ri}{L} = \frac{V/R - i}{L/R}$$

$$\frac{di}{\frac{V}{R} - i} = \frac{R}{L}dt$$

or

$$-\ln\left(\frac{V}{R} - i\right) = \frac{R}{L}t + A$$

At

$$t = 0 \qquad i = 0$$

Therefore

$$-\ln\frac{V}{R} = A$$

or

$$\ln\frac{V/R}{\frac{V}{R} - i} = \frac{R}{L}t \qquad \text{or} \quad \ln\frac{V/R - i}{V/R} = -\frac{R}{L}t$$

or

$$\frac{V}{R} - i = \frac{V}{R}e^{-\frac{R}{L}t}$$

or

$$i = \frac{V}{R}\left(1 - e^{-\frac{R}{L}t}\right) \qquad \text{...(5.33)}$$

The same result can be obtained by calculating $i_{CF}$ and as $i_{pI}$ as it is a non-homogeneous equation.

To obtain $i_{CF}$

$$(Lp + R)i = 0$$

Hence

$$Lp + R = 0 \qquad \text{or } p = -\frac{R}{L}$$

Hence the solution

$$i_{CF} = Ae^{-\frac{R}{L}t}$$

To obtain $i_{pI}$

$$\text{Here} \quad f(t) = Ve^{\beta t} \quad \text{where } \beta = 0$$

$$\beta e^{0t}(R + Lp) = Ve^{0t}$$

$$\beta\left(p + \frac{R}{L}\right) = \frac{V}{L}$$

$$\beta = \frac{V}{L}\cdot\frac{L}{R} = \frac{V}{R}$$

Hence

$$i_{pI} = \frac{V}{R}$$

The complete solution

$$i(t) = \frac{V}{R} + Ae^{-\frac{R}{L}t}$$

Now at

$$t = 0 \qquad i = 0$$

$$0 = \frac{V}{R} + A \qquad \text{or } A = -\frac{V}{R}$$

Hence the complete solution is

$$i(t) = \frac{V}{R}\left(1 - e^{-\frac{R}{L}t}\right)$$

Which is same as in equation (5.33).

Fig. 5.7 shows the variation of current as a function of time. It is an exponentially rising characteristic

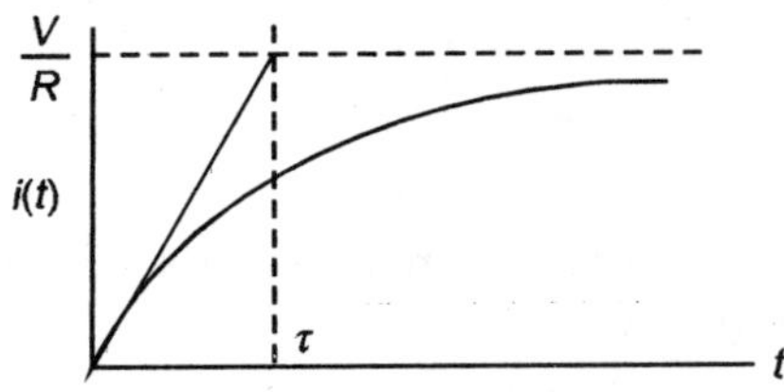

**Fig. 5.7** Plot of equation 5.33

The current increases to 68.2 percent of its final steady state value of V/R in time $t = \tau$ the time constant and reaches 98% of value in a time equal to four times the time constant.

The tangent passes through the origin. The slope of the curve is given, from equation 5.33.

$$\frac{di}{dt} = \frac{V}{R}\cdot\frac{R}{L}e^{-\frac{R}{L}t} \quad \text{and at } t = 0 \qquad \left.\frac{di}{dt}\right|_{t=0} = \frac{V}{L}$$

The equation of the tangent $\qquad y = mx$

Here it is $\qquad i = \dfrac{V}{L}t$

and at $t = \tau$ the tangent intersects the steady state $\dfrac{V}{R}$ line as at

$$t = \tau$$

$$i = \frac{V}{L} \cdot \frac{L}{R} = \frac{V}{R}$$

Therefore, if the exponential variation is given the time constant of the corresponding circuit can be obtained.

Next refer to Fig. 5.8. Initially the switch is at 'a' and after steady state conditions are reached,

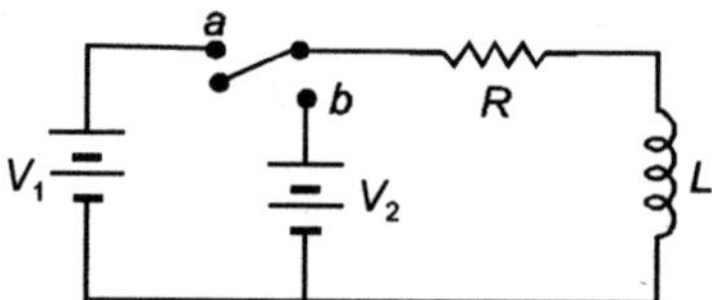

**Fig. 5.8** RL circuit connected to *dc* sources

it is required to obtain the current response after the switch is moved to '*b*' position. The net response is the sum of two responses.

(*i*)  The zero input response i.e. when $V_2$ is not there and the initial condition is that at

$t = 0 \qquad i = \dfrac{V_1}{R}$ and the response under this condition is

$$i_1(t) = \frac{V_1}{R} e^{-\frac{R}{L}t} \qquad\qquad \text{...(5.34)}$$

(*ii*) The zero state response or forced response when *dc* source $V_2$ is switched in and we know this response is

$$i_2(t) = \frac{V_2}{R}\left(1 - e^{-\frac{R}{L}t}\right) \qquad\qquad \text{...(5.35)}$$

and since it is a linear system, the net response $i(t)$ is the superposition (algebraic sum) of the two currents i.e.

$$i(t) = i_1(t) + i_2(t) \qquad\qquad \text{...(5.35)}$$

$$= \frac{V_1}{R}e^{-\frac{R}{L}t} + \frac{V_2}{R}\left(1 - e^{-\frac{R}{L}t}\right)$$

$$= \frac{1}{R}(V_1 - V_2)e^{-\frac{R}{L}t} + \frac{V_2}{R}$$

Since the first part tends to zero as $t$ tends to infinity, it is known as the transient part or transient solution and second part is a constant and hence known as steady state solution.

## 5.5.2  AC Source Switched in an RL Series Circuit

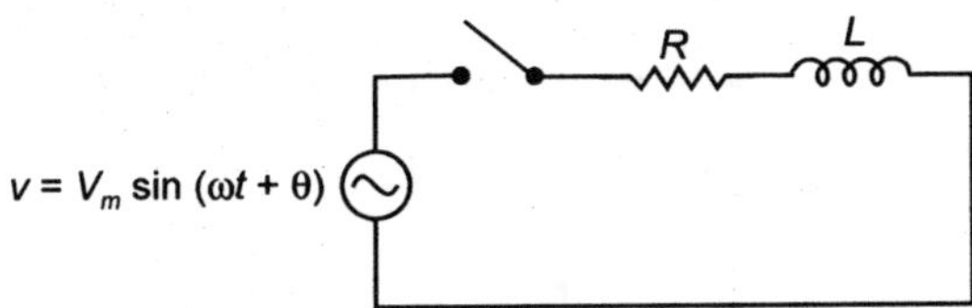


**Fig. 5.9** RL series circuit switched in ac source

The equation describing the network is

$$Ri(t) + L\frac{di}{dt} = V_m \sin(\omega t + \alpha) \text{ for } t \geq 0 \text{ when } i(0) = I_0$$

The transient solution of the equation

$$Ri(t) + L\frac{di}{dt} = 0$$

and hence

$$i_{CF}(t) = A e^{-\frac{R}{L}t}$$

and by using conventional method the steady state solution of the network

$$i_{ss} = \frac{V_m}{\sqrt{R^2 + \omega^2 L^2}} \sin(\omega t + \alpha - \theta) \text{ where } \theta = \tan^{-1}\frac{\omega L}{R}$$

Hence the complete solution

$$i(t) = i_{tr} + i_{ss} = A e^{-\frac{R}{L}t} + I_m \sin(\omega t + \alpha - \theta)$$

At $t = 0$  $i(0) = I_0$

Hence

$$I_0 = A + I_m \sin(\alpha - \theta)$$

or

$$A = I_0 - I_m \sin(\alpha - \theta)$$

The complete solution is

$$i(t) = I_0 - I_m \sin(\alpha - \theta) e^{-\frac{R}{L}t} + I_m \sin(\omega t + \alpha - \theta)$$

If $i(0) = I_0 = 0$ the initial condition

$$i(t) = I_m \sin(\omega t + \alpha - \theta) - I_m \sin(\alpha - \theta) e^{-\frac{R}{L}t} \qquad ...(5.36)$$

From equation (5.36) it is clear that when $\alpha - \theta = 0$, $\pi$....$\pi$th the transient part vanishes i.e. when $\alpha = \theta$ that is when the instant of switching in the voltage source is such that the angle through which the voltage passes coincides with the impedance angle of the circuit, even though the circuit contains inductor and switching operation takes place, there is no

transient present in the circuit. However if $\alpha - \theta = (2n + 1)\dfrac{\pi}{2}$ $n = 0, 1, 2 ...$ then the transient

will have maximum possible value. In fact it is twice the peak value. This is known as doubling effects.

**Example 5.7:** A parallel $RC$ circuit is switched into a current source (a) $i(t) = I_s$ and (b) $i(t) = I_m \sin(\omega t + \alpha)$ as shown in Fig. E 5.7. Determine the voltage across the capacitor.

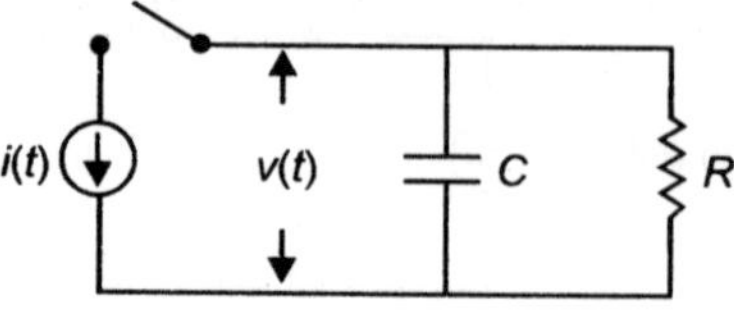

**Fig. E5.7**

**Solution:** The circuit equation is

$$C\,\frac{dv}{dt} + \frac{v}{R} = i(t) \quad \text{when} \quad v_c = V_0$$

The solution will consist of two parts, the steady state part and the transient part. The steady state part will consist of voltage drop across the resistance as the capacitor gets charged fully under steady state condition corresponding to the voltage drop across the resistance $I_s R$

The transient solution is obtained as

$$\left(Cp + \frac{1}{R}\right)v = 0 \quad \text{or} \quad p = -\frac{1}{RC}$$

Hence $\qquad\qquad v_{tr}(t) = A\,e^{-t/RC}$

The complete solution is

$$v_c(t) = A\,e^{-t/RC} + I_s R$$

Now at $\qquad\qquad t = 0 \quad v_c(0) = V_0$

$$V_0 = A + I_s R$$

Hence $\qquad\qquad A = (V_0 - I_s R)$

Hence complete solutions is

$$v(t) = (V_0 - I_s R)e^{-t/RC} + I_s R \quad \text{Ans.}$$

(b) When $i(t) = I_m \sin(\omega t + \alpha)$

By using conventional method the steady state solution of the network is

$$V_m \sin(\omega t + \alpha - \theta) \text{ for } t \geq 0$$

where

$$V_m = I_m R / \sqrt{1 + \omega^2 C^2 R^2} \quad \text{and} \quad \theta = \tan^{-1} \omega CR$$

The transient solution of course is,

$$v_{tr}(t) = A\,e^{-t/RC}$$

and, therefore,

$$v(t) = A\,e^{-t/RC} + V_m \sin(\omega t + \alpha - \theta)$$

At $\qquad\qquad t = 0 \quad v(0) = V_0$

Hence $\qquad\qquad V_0 = A + V_m \sin(\alpha - \theta)$

or
$$A = V_0 - V_m \sin(\alpha - \theta)$$

and hence the complete solution is
$$v(t) = (V_0 - V_m \sin(\alpha - \theta))\, e^{-t/RC} + V_m \sin(\omega t + \alpha - \theta)$$

## 5.6 SECOND ORDER CIRCUITS

**RLC series circuit-zero input response**: Consider the RLC series circuit shown in Fig. 5.10.

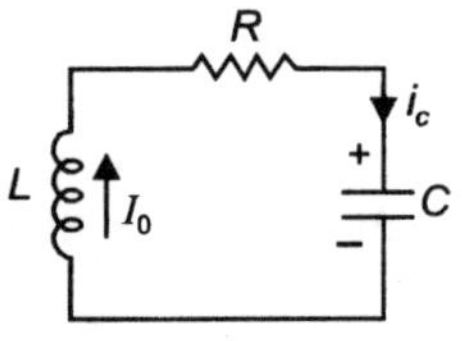

**Fig. 5.10** RLC Series Circuit

Suppose $i_L(0) = I_0$ and $v_c(0) = 0$

The circuit equation in terms of capacitor voltage $v_c(t)$ is the sum of the drops in the three elements is zero.

In terms of capacitor voltage

The drop in resistance is $Ri = RC\,\dfrac{dv}{dt}$

The drop across inductor $L\,\dfrac{di}{dt} = L\,\dfrac{d}{dt}\left(C\,\dfrac{dv}{dt}\right) = LC\,\dfrac{d^2v}{dt^2}$

Hence the equation describing the network is

$$LC\,\frac{d^2 v_C}{dt^2} + RC\,\frac{dv_C}{dt} + v_C = 0$$

Dividing the equation by $LC$ we have

$$\frac{d^2 v_C}{dt^2} + \frac{R}{L}\frac{dv_C}{dt} + \frac{v_C}{LC} = 0 \qquad (5.37)$$

Since $\quad i_L(0) = I_0 \qquad\qquad C\,\dfrac{dv_C}{dt}\bigg|_{t=0} = I_0\ $ or $\ \dfrac{dv_C}{dt} = \dfrac{I_0}{C}$

Let us define

$$\omega_n = \frac{1}{\sqrt{LC}} \quad \text{the natural frequency of the RLC circuit}$$

and
$$R_C = 2\sqrt{\frac{L}{C}} \quad \text{the critical resistance of the RLC circuit}$$

and
$$\xi = \frac{R}{R_C} = \frac{R}{2}\sqrt{\frac{C}{L}} \quad \text{the damping ratio of the circuit}$$

Writing equation (5.37) in terms of the defined parameters we have

$$\frac{d^2 v_C}{dt^2} + 2\xi\omega_n\,\frac{dv_C}{dt} + \omega_n^2 v_C = 0 \qquad \text{...(5.38)}$$

This is a second order differential equation for a second order network. The solution of the equation depends on the value of $R$ or $\xi$. The characteristic equation is
$$p^2 + 2\,\xi\omega_n p + \omega_n^2 = 0$$

The roots are

$$p_1, p_2 = -\xi\omega_n \pm \sqrt{\xi^2\omega_n^2 - \omega_n^2} \qquad \qquad ...(5.39)$$

## 1. Circuit is undamped i.e. $R = 0$ or $\xi = 0$

The roots are

$$p_1, p_2 = \pm j\omega_n$$

Hence the solution is

$$v_c(t) = A\,e^{j\omega_n t} + B\,e^{-j\omega_n t}$$

At $t = 0 \quad v_c(0) = 0 = A + B \qquad$ Hence $A = -B$

$$\frac{dv_C}{dt} = jA\omega_n\,e^{j\omega_n t} - jB\omega_n e^{-j\omega_n t}$$

$$= jA\omega_n\,(e^{j\omega_n t} + e^{-j\omega_n t})$$

At $t = 0 \quad \left.\frac{dv_C}{dt}\right|_{t=0} = \frac{I_0}{C}$

Hence $\dfrac{I_0}{C} = 2jA\omega_n \quad$ or $\quad A = \dfrac{I_0}{2j\omega_n C}$

The solution is

$$v_C(t) = \frac{I_0}{\omega_n C}\left(\frac{e^{j\omega_n t} - e^{-j\omega_n t}}{2j}\right)$$

$$= \frac{I_0}{\omega_n C}\sin\omega t \qquad \qquad ...(5.40)$$

## 2. Underdamped circuit $0 < \xi < 1$

Using the roots from equation (5.39) we have

$$v_C(t) = A\,e^{\left(-\xi\omega_n + j\omega_n\sqrt{1-\xi^2}\right)t} + B\,e^{\left(-\xi\omega_n - j\omega_n\sqrt{1-\xi^2}\right)t} \qquad \qquad ...(5.41)$$

Let $\omega_d = \omega_n\sqrt{1-\xi^2}$ damped natural frequency of RLC circuit and $\xi\omega_n = \sigma$

At $t = 0 \quad v_C(0) = 0$

Hence $A = -B$

and At $t = 0 \quad \left.\frac{dv_C}{dt}\right|_{t=0} = \frac{I_0}{C}$

$$v_C(t) = A\,e^{(-\sigma + j\omega_d)t} + B\,e^{(-\sigma - j\omega_d)t} \qquad \qquad (5.42)$$

$$\frac{dv_C}{dt} = A(-\sigma + j\omega_d)\,e^{(-\sigma + j\omega_d)t} - A(-\sigma - j\omega_d)\,e^{(-\sigma - j\omega_d)t}$$

$$= A\,e^{-\sigma t}(-\sigma + j\omega_d)\,e^{j\omega_d t} - A\,e^{-\sigma t}(-\sigma - j\omega_d)\,e^{-j\omega_d t}$$

$$\frac{I_0}{C} = A(-\sigma + j\omega_d) - A(-\sigma - j\omega_d)$$

$$= -A\sigma + jA\omega_d + A\sigma + Aj\omega_n$$

$$= 2jA\omega_d \quad \text{or} \quad A = \frac{I_0}{2j\omega_d}$$

Hence
$$v_C(t) = A\,e^{-\sigma t} \sin \omega_d t \qquad \qquad ...(5.43)$$

**3. Critically damped** i.e. $\xi = 1$

Using roots from equation (5.39) we have

$$p_1,\ p_2 = -\omega_n$$

Since the roots are real and equal

$$v_C(t) = A\,e^{-\omega_n t} + Bt\,e^{-\omega_n t}$$

$A$ and $B$ can be obtained.

**4. When the circuit is overdamped i.e. when $\xi > 1$**

The roots are

$$p_1,\ p_2 = -\xi\omega_n \pm \omega_n\sqrt{\xi^2 - 1}$$

The roots are real and distinct and the solution is

$$v_C(t) = A\,e^{p_1 t} + B\,e^{p_2 t}$$

The response characteristics have been shown in Fig. 5.11

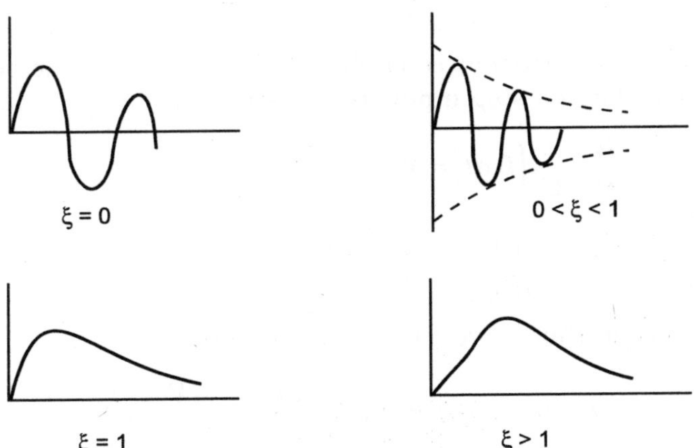


**Fig. 5.11** Response Characteristic of $v_C(t)$ for the Characteristic equations $p^2 + 2\xi\omega_n p + \omega_n^2 = 0$

**Example 5.8:** Determine the current response of the network shown in Fig. E5.8 when the switch $S$ is closed.

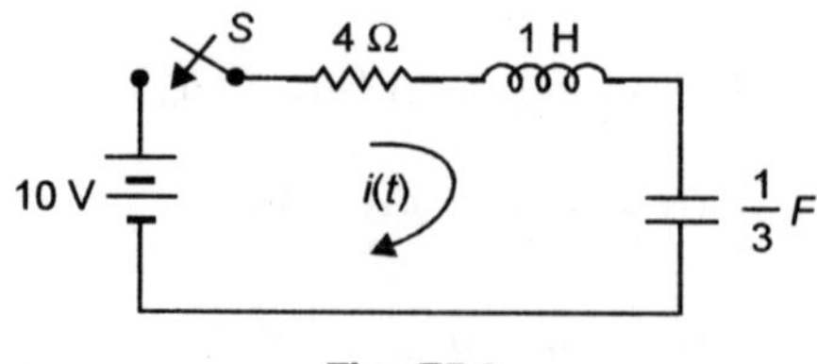

**Fig. E5.8**

**Solution:** The differential equation describing the network is given as

$$Ri + L\frac{di}{dt} + \frac{1}{C}\int idt = V \qquad\qquad \dots(1)$$

Differentiating and substituting values of the parameter, we have

or
$$L\frac{d^2i}{dt^2} + R\frac{di}{dt} + \frac{i}{C} = 0$$

or
$$\frac{d^2i}{dt^2} + 4\frac{di}{dt} + 3i = 0$$

The characteristic equation is
$$(p^2 + 4p + 3)\, i = 0$$

or
$$p^2 + 4p + 3 = 0$$

The roots are $p_1, p_2 = -1, -3$
Hence the general solution is

$$i\,(t) = A\,e^{-t} + B\,e^{-3t}$$

At $t = 0$   $i\,(0+) = 0$ as the current can't change through the inductor instantaneously as it acts as an open circuit for the step input. By substituting $i\,(0_+) = 0$ in equation 1 we find

$$R\,(0) + L\frac{di}{dt} + \frac{1}{C}\int 0\, dt = V$$

or
$$\left.\frac{di}{dt}\right|_{t=0} = \frac{V}{L} = \frac{10}{1}$$

To obtain $A$ and $B$ we thus have two initial conditions

At
$$t = 0_+ \quad i\,(0_+) = 0 \text{ and } \left.\frac{di}{dt}\right|_{t=0} = \frac{V}{L} = 10$$

$$0 = A + B \qquad\qquad A = -B$$

and
$$\frac{di}{dt} = -A\,e^{-t} - 3Be^{-3t} = -A - 3B = 10$$

or
$$B - 3B = 10 \quad \text{or} \quad B = -5 \quad \text{and} \quad A = 5$$

Hence the solution is
$$i\,(t) = 5\,(e^{-t} - e^{-3t}) \qquad \textbf{Ans.}$$

**Example 5.9:** For the network shown in Fig. determine the voltage response of the circuit when the switch $S$ is opened.

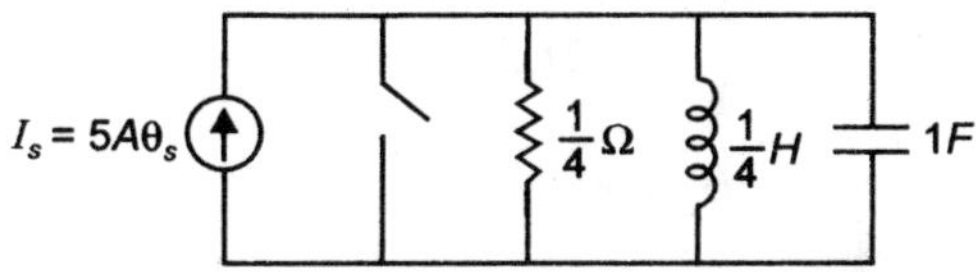



**Fig. E5.9**

**Solution:** The differential equation taking $v(t)$ as variable for the network is

$$C\frac{dv}{dt} + \frac{V}{R} + \frac{1}{L}\int V\,dt = I_s \qquad \qquad ...(1)$$

Differentiating with respect to $t$ we had

$$C\frac{d^2v}{dt^2} + \frac{1}{R}\frac{dv}{dt} + \frac{v}{L} = 0 \qquad \qquad ...(2)$$

This is a second order homogeneous equation and requires two initial conditions for its solution and only complementary solution is required.

When the switch is closed, the voltage across the capacitor at $t = 0_+$ is zero as capacitor acts as a short circuit to step inputs. Here $V_C(0_+) = 0$. Substituting this condition in equation (1) we have

$$C\frac{dv}{dt}\Big|_{t=0_+} + \frac{0}{R} + \frac{1}{L}\int 0\,dt = I_s$$

or

$$\frac{dv}{dt}\Big|_{t=0_+} = \frac{I_s}{C}$$

so the initial conditions are

At $t = 0_+$  $v_C(0_+) = 0$ and $\dfrac{dV}{dt}\Big|_{t=0_+} = \dfrac{I_s}{C}$

The characteristic equation describing equation (2) is

$$\left(Cp^2 + \frac{1}{R}p + \frac{1}{L}\right) = 0$$

Substituting the values of the parameters $R$, $L$ & $C$ we have

$$p^2 + 4p + 4 = 0$$

The roots are

$$p_1, p_2 = -2$$

Since the roots are real and equal

$$v_C(t) = A\,e^{-2t} + Bt\,e^{-2t}$$
$$\text{At } t = 0_+ \quad V_C(0_+) = 0$$
$$0 = A$$

and

$$\frac{dv}{dt} = -2A\,e^{-2t} + B\,\{t.(-2e^{-2t}) + e^{-2t}\}$$

$$\frac{5}{1} = +B \ e^{-2t} = +B$$

or
$$B = +5$$

Hence $v_C\,(t) = 5\ t\ e^{-2t}$ **Ans.**

**Example 5.10:** For the network shown determine the current response when the switch $S$ is closed.

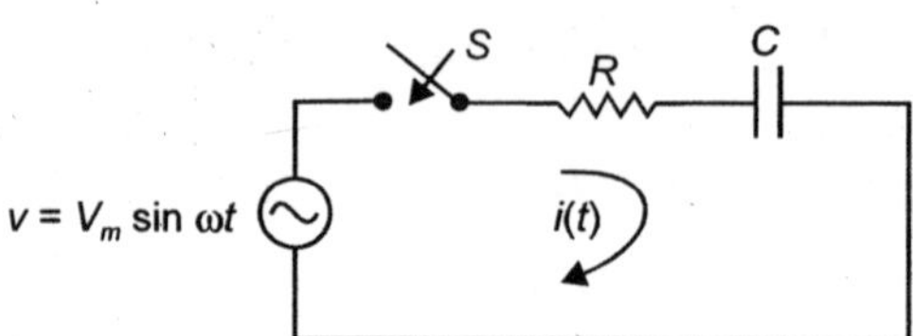


**Fig. E5.10**

**Solution:** The differential equation describing the network is

$$Ri + \frac{1}{C}\int i\,dt = V_m \sin \omega t$$

Differentiating the equation with respect to $t$ we have

$$R\frac{di}{dt} + \frac{i}{C} = V_m\,\omega \cos \omega t$$

or
$$\frac{di}{dt} + \frac{i}{RC} = \frac{V_m\,\omega}{R}\cos \omega t$$

This is a first-order non homogeneous equation which will have both complementary function and particular integral as its solution.

For complementary function, the characteristic equation is

$$\left(p + \frac{1}{RC}\right) = 0 \ \text{ or } \ p = -\frac{1}{RC}$$

Hence
$$i_{CF} = A\ e^{-t/RC}$$

Since the driving function is sinusoidal, the particular integral solution is the sum of the function and its derivative i.e.

Let
$$i_{pI} = B\ \sin \omega t + D\ \cos \omega t$$

Substituting this solution in the differential equation

$$B\omega \cos \omega t - D\omega \sin \omega t + \frac{1}{RC}(B \sin \omega t + D \cos \omega t) = \frac{V_m\omega}{R}\cos \omega t$$

Equating the co-efficients of like functions (sin or cos function)

$$B\omega + \frac{D}{RC} = \frac{V_m\omega}{R} \ \text{ and } \ \frac{B}{RC} - D\omega = 0$$

Solving for $B$ and $D$ we have

$$B = \frac{\omega^2 R C^2 V_m}{1+\omega^2 R^2 C^2} \quad \text{and} \quad D = \frac{\omega C V_m}{1+\omega^2 C^2 R^2}$$

Substituting these values in the assumed solution and after some simplification we have

$$i_{pI} = \frac{\omega^2 C^2 R V_m}{1+\omega^2 R^2 C^2}\, \sin\omega t + \frac{\omega C V_m}{1+\omega^2 C^2 R^2}\, \cos\omega t$$

$$= \frac{V_m}{R^2 + \dfrac{1}{\omega^2 C^2}} \left\{ R\,\sin\omega t + \frac{1}{\omega C}\,\cos\omega t \right\}$$

$$= \frac{V_m}{\sqrt{R^2 + \dfrac{1}{\omega^2 C^2}}} \left\{ \frac{R}{\sqrt{R^2 + 1/\omega^2 C^2}}\,\sin\omega t + \frac{1/\omega C}{\sqrt{R^2 + 1/\omega^2 C^2}}\,\cos\omega t \right\}$$

$$= \frac{V_m}{\sqrt{R^2 + 1/\omega^2 C^2}} \left\{ \sin\phi\,\sin\omega t + \cos\phi\,\cos\omega t \right\}$$

$$= \frac{V_m}{\sqrt{R^2 + 1/\omega^2 C^2}}\, \cos\left(\omega t - \phi\right)$$

where
$$\phi = \tan^{-1}\omega R C$$

Therefore, the complete solution is

$$i(t) = A\, e^{-t/RC} + \frac{V_m}{\sqrt{R^2 + 1/\omega^2 C^2}}\, \cos\left(\omega t - \phi\right)$$

$$= A\, e^{-t/RC} + I_m\, \cos\left(\omega t - \phi\right)$$

At
$$t = 0 \quad V_C(0) = 0$$

Now
$$V_C = \frac{1}{C}\int i\, dt = \frac{-1}{C}\cdot ARC e^{-t/RC} + \frac{I_m}{\omega}\,\sin\left(\omega t - \phi\right)$$

At
$$t = 0 \quad V_C = 0 = -AR - \frac{I_m}{\omega}\,\sin\phi$$

or
$$A = -\frac{I_m}{R\omega}\,\sin\phi$$

Hence the complete solution is

$$i(t) = I_m\, \cos\left(\omega t - \phi\right) - \frac{I_m}{\omega R}\,\sin\phi\, e^{-t/CR}$$

## 5.7  ANALYSIS OF TRANSFORMER

Making use of the dot convention we further analyse the operation of the two winding transformers. Let $N_1$ and $N_2$ be the No. of turns of the coils and $L_1$ and $L_2$ the self inductances of the coils, the flux linkages of the two coils are given by $N_1\phi_1$ and $N_2\phi_2$ respectively where $\phi_1$ and $\phi_2$ are the fluxes of the two coils. If both $i_1$ and $i_2$ (Fig 5.12) flow into the dots or both flow away from the dots, the sum of the flux linkages of the transformers is

$$\Sigma\phi \text{ linkages} = N_1\phi_1 + N_2\phi_2$$

However, if one of the currents say $i_1$ flows into the dot whereas $i_2$ flows away from the dot, then

$$\text{Total flux linkages} = N_1\phi_1 - N_2\phi_2$$

An important rule governing the transformer is that the sum of flux linkages is a continuous function of time.

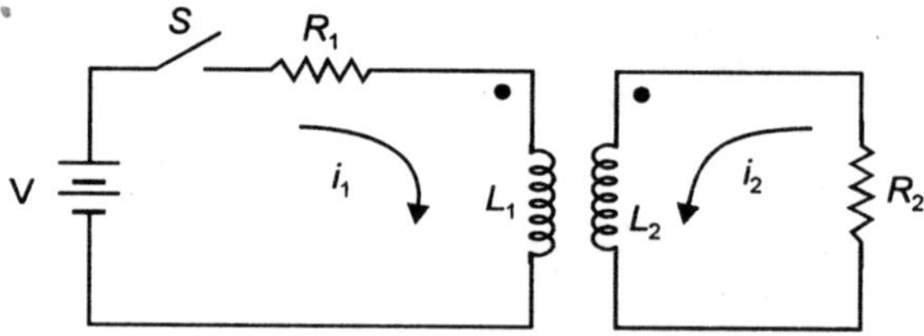


**Fig. 5.12**

The differential equations governing the transformer in Fig. 5.12 are given as

$$R_1 i_1 + L_1 \frac{di_1}{dt} + M \frac{di_2}{dt} = Vu(t)$$

and

$$R_2 i_2 + L_2 \frac{di_2}{dt} + M \frac{di_1}{dt} = 0$$

Integrating the above equations between the limits $t = 0_-$ and $t = 0_+$ we have

$$L_1(i_1 0_+) - L_1 i_1(0_-) + M i_2(0_+) - M i_2(0_-) = 0$$
$$L_2(i_2 0_+) - L_2 i_2(0_-) + M i_1(0_+) - M i_1(0_-) = 0$$

Since

$$i_1(0_-) = 0 \text{ and } i_2(0_-) = 0$$

However, their derivatives are not zeros. Similarly $\int\limits_{0_-}^{0_+} Vu(t)dt = 0$ as it is only a straight vertical line whose area is zero.

Rearranging the equations in a determinant form

$$\begin{vmatrix} L_1\{i_1(0_+) - i_1(0_-)\} & M\{i_2(0_+) - i_2(0_-)\} \\ M\{i_1(0_+) - i_1(0_-)\} & L_2\{i_2(0_+) - i_2(0_-)\} \end{vmatrix} = 0 \qquad ...(5.44)$$

Simplifying further

$$(L_1L_2 - M^2)\ [\{i_1(0_+) - i_1(0_-)\}\quad \{i_2(0_+) - i_2(0_-)\}] = 0$$

Now for $L_1L_2 > M^2$ i.e., $k < 1$, the factor $(L_1L_2 - M^2)$ is not zero and hence

$$\{i_1(0_+) - i_1(0_-)\}\quad \{i_2(0_+) - i_2(0_-)\} = 0$$

Hence,

$$i_1(0_+) = i_1(0_-)$$

and

$$i_2(0_+) = i_2(0_-)$$

This shows that for $k < 1$ which normally is the situation in case of a transformer, the current is continuous at $t = 0$ and hence the flux linkages are continuous.

However, if $k = 1$ i.e. $L_1L_2 - M^2 = 0$, then

$$i_1(0_+) - i_1(0_-) \neq 0 \qquad \text{and } i_2(0_+) - i_2(0_-) \neq 0$$

i.e. the current is not continuous at $t = 0$ for $k = 1$

In order to find out magnitude of $i_1(0_+)$ and $i_2(0_+)$ we write mesh equation for the secondary loop at $t = 0_+$

$$R_2 i_2(0_+) + L_2\frac{di_2}{dt} + M\frac{di_1}{dt} = 0$$

or

$$R_2 i_2(0_+) = -M\frac{di_1}{dt} - L_2\frac{di_2}{dt}$$

Multiplying and dividing RHS by $L_1$ we have

$$R_2 i_2(0_+) = -\frac{L_1 M}{L_1}\frac{di_1}{dt} - \frac{L_2 L_1}{L_1}\frac{di_2}{dt}$$

$$= -\frac{M}{L_1}\left(L_1\frac{di_1}{dt} + M\frac{di_2}{dt}\right)$$

$$= -\frac{M}{L_1}\left[L_1\, i_1''(0_+) + M\, i_2''(0_-)\right]$$

The mesh equation for the primary side of the transformer is

$$R_1 i_1 + L_1\frac{di_1}{dt} + M\frac{di_2}{dt} = Vu(t)$$

or $R_1 i_1(0_+) + [L_1 i_1'(0_+) - M i_2'(0_+)] = R_1 i_1(0_+) - \dfrac{L_1}{M}R_2 i_2(0_+) = V$      ...(5.45)

$T_0$ solve for $i_1(0_+)$ and $i_2(0_+)$, we need one more equation which is obtained from equation (5.44)

i.e.

$$L_1[i_1(0_+) - i_1(0_-)] + M[i_2(0_+) - i_2(0_-)] = 0$$

Since

$$i_1(0_-) = i_2(0_-) = 0 \qquad \text{We have}$$

$$L_1 i_1(0_+) + M i_2(0_+) = 0$$

or
$$i_1(0_+) = -\frac{M}{L_1} i_2(0_+)$$

and from equation (5.45)

$$i_1(0_+) = \frac{1}{R_1}\left[V + \frac{L_1}{M} R_2 i_2(0_+)\right] = -\frac{M}{L_1} i_2(0_+)$$

or
$$\frac{V}{R_1} = \left[-\frac{L_1}{M}\frac{R_2}{R_1} - \frac{M}{L_1}\right] i_2(0_+)$$

or
$$\frac{V}{R_1} = -\frac{L_1^2 R_2 + M^2 R_1}{ML_1 R_1} i_2(0_+)$$

$$= -\frac{L_1^2 R_2 + L_1 L_2 R_1}{ML_1 R_1} i_2(0_+)$$

$$\frac{V}{R_1} = -\frac{L_1^2 R_2 + L_1 L_2 R_1}{ML_1 R_1} i_2(0_+)$$

or
$$V = -\frac{L_1^2 R_2 + L_1 L_2 R_1}{ML_1} i_2(0_+)$$

$$= -\frac{L_1 R_2 + L_2 R_1}{M} i_2(0_+)$$

or
$$i_2(0_+) = \frac{-VM}{L_1 R_2 + L_2 R_1}$$

Since
$$i_1(0_+) = \frac{-M}{L_1} i_2(0_+)$$

$$= \frac{-M}{L_1}\frac{(-VM)}{L_1 R_2 + L_2 R_1}$$

$$= \frac{VL_2}{L_1 R_2 + L_2 R_1}$$

This means if it is a unity coupled transformer $i_1(0_+)$ and $i_2(0_+)$ are not identical and are given by the equations given above. However, if $k < 1$, the two currents at $t = 0+$ are continuous i.e. $\qquad i_1(0_+) = i_2(0_+) = 0$

**Example 5.11:** Consider Fig. 5.12 when Let $\quad R_1 = 2\Omega \qquad R_2 = 8\Omega$
$L_1 = 1H \quad L_2 = 4H$ and $M = 1H$ and Source Voltage $5\ u(t)$. Determine $i_1(t)$ and $i_2(t)$ assuming $i_1(0_-) = i_2(0_-) = 0$

**Solution:** The differential equations for the circuit are
$$2i_1(t) + i_1'(t) + i_2'(t) = 5\ u(t)$$

and
$$8i_2 + 4i_2'(t) + i_1'(t) = 0 \qquad\qquad ...(5.46)$$

Using Heaviside operator $\qquad p = \dfrac{d}{dt}$

$$(2 + p)\, i_1(t) + p i_2(t) = 5$$
$$p i_1(t) + (8 + 4p)\, i_2(t) = 0$$

Therefore, the equations are rewritten as

$$\begin{bmatrix} p+2 & p \\ p & 4p+8 \end{bmatrix} \begin{bmatrix} i_1(t) \\ i_2(t) \end{bmatrix} = \begin{bmatrix} 5 \\ 0 \end{bmatrix}$$

The characteristic equation is given by

$$4(p + 2)\,(p + 2) - p^2 = 0$$

$$4p^2 + 16p + 16 - p^2 = 3p^2 + 16p + 16 = 0$$
$$= 3p^2 + 12p + 4p + 16$$
$$= 3p\,(p + 4) + 4(p + 4)$$
$$= (3p + 4)\,(p + 4) = 0$$

Thus the complimentary solution of the network is

$$i_1 = k_1 e^{-4t} + k_2 e^{-\frac{4}{3}t}$$

and $\qquad i_2 = k_3 e^{-4t} + k_4 e^{-\frac{4}{3}t}$

Since there are four unknowns we need four initial conditions about the circuit.
From the circuit of Fig. 5.12 the final current $i_1(\infty)$ and $i_2(\infty)$ are (particular solution)

$$i_{1p} = \frac{V}{R_1} = \frac{5}{2} \ \text{ and } \ i_{2p} = 0$$

as there is no source in the  secondary circuit. To find out another conditions i.e. $i'_1(0_+)$ and $i'_2(0_+)$ (However, $i_1(0_+)$ and $i_2(0_+)$ are zero) we take equation (5.46)

$$i'_1(0_+) + i'_2(0_+) = 5$$
$$i'_1(0_+) + 4i'_2(0_+) = 0$$

or $\qquad i'_1(0_+) = -4i'_2(0_+)$

or $\qquad -4i'_2(0_+) + i_2(0_+) = 5$

or $\qquad i'_2(0_+) = \dfrac{-5}{3}$

and hence $\qquad i'_1(0_+) = \dfrac{20}{3}$

Now $\qquad i_1(t) = \dfrac{5}{2} + k_1 e^{-4t} + k_2 e^{-\frac{4}{3}t}$

$$i_2(t) = k_3 e^{-4t} + k_{4e}^{-\frac{4}{3}t}$$

Now
$$i'_1(t) = -4k_1 e^{-4t} - \frac{4}{3}k_{2e}^{-\frac{4}{3}t}$$

and
$$i'_2(t) = -4k_3 e^{-4t} - \frac{4}{3}k_{4e}^{-\frac{4}{3}t}$$

Now
$$i'_1(0_+) = \frac{20}{3} = -4k_1 - \frac{4}{3}k_2$$

and
$$i'_2(0_+) = -\frac{5}{3} = -4k_3 - \frac{4}{3}k_4$$

Also
$$i_1(0_+) = 0 = \frac{5}{2} + k_1 + k_2$$

and
$$i_2(0_+) = 0 = k_3 + k_4$$

It is to be noted that while using initial conditions complete solution of the current should be taken into account i.e. $i(t) = i_c + i_p$

$$-\frac{5}{3} = -4k_3 + \frac{4}{3}k_3 = -\frac{8}{3}k_3$$

or
$$k_3 = +\frac{5}{8} \text{ and } k_4 = -\frac{5}{8}$$

Similarly
$$k_1 = -k_2 - \frac{5}{2}$$

or
$$\frac{20}{3} = +4k_2 + 10 - \frac{4}{3}k_2$$

or
$$-\frac{10}{3} = \frac{8}{3}k_2 \quad \text{or} \quad k_2 = -\frac{5}{4}$$

and
$$k_1 = \frac{5}{4} - \frac{5}{2} = -\frac{5}{4}$$

Hence the complete solution

$$i_1(t) = \frac{5}{2} - \frac{5}{4}e^{-4t} - \frac{5}{4}e^{-\frac{4}{3}t}$$

$$i_2(t) = +\frac{5}{8}e^{-4t} - \frac{5}{8}e^{-\frac{4}{3}t} \quad \textbf{Ans.}$$

Let us take another example for a transformer which has unity coupling.

**Example 5.12:** Suppose the transformer has the parameters

$$L_1 = 4\text{H} \quad L_2 = 1\text{H} \quad M = 2\text{H}$$

$$R_1 = 4\Omega \quad R_2 = 1\Omega \quad \text{Voltage} = 6u(t)$$

Determine the expressions for $i_1(t)$ and $i_2(t)$ for the circuit in Fig. 5.12

**Solution:**

The equation for the primary side is

$$4i_1(t) + 4i'_1(t) + 2i'_2(t) = 6u(t)$$

For secondary side $\qquad 2i'_1(t) + i_2(t) + i'_2(t) = 0$

Using Heaviside operator $p = \dfrac{d}{dt}$ we have

$$(4 + 4p)\, i_1 + 2pi_2 = 6$$

and $\qquad 2pi_1 + (p + 1)i_2 = 0$

The characteristic equation is

$$4(1 + p)\,(p + 1) - 4p^2 = 0$$

or $\qquad 4p^2 + 8p + 4 - 4p^2 = 0$

or $\qquad 4(2p + 1) = 0 = 8(p + \dfrac{1}{2})$

Hence the complementary solution is

$$i_{1c}(t) = 8k_1 e^{-\frac{1}{2}t} \quad \text{and} \quad i_{2c}(t) = 8k_2 e^{-\frac{t}{2}}$$

Now $\qquad i_1(0_+) = \dfrac{VL_2}{L_1 R_2 + L_2 R_1} = \dfrac{6 \times 1}{4 \times 1 + 1 \times 4} = \dfrac{6}{8} = \dfrac{3}{4}$

and $\qquad i_2(0_+) = \dfrac{-VM}{L_1 R_2 + L_2 R_1} = \dfrac{-6 \times 2}{8} = -\dfrac{3}{2}$

The particular integral when $t \to \infty$ we have

$$i_{1p}(t) = \dfrac{6}{4} = \dfrac{3}{2}$$

$$i_{2p}(t) = 0$$

Hence the complete solution is

$$i_1(t) = i_{1c}(t) + i_{1p}(t)$$

$$= 8k_1 e^{-\frac{t}{2}} + \dfrac{3}{2}$$

and $\qquad i_2(t) = 8k_2 e^{-\frac{t}{2}}$

Now $\qquad i_1(0_+) = \dfrac{3}{4} = 8k_1 + \dfrac{3}{2} \qquad \text{or } 8k_1 = -\dfrac{3}{4}$

or
$$k_1 = -\frac{3}{32}$$

and
$$i_2(0_+) = -\frac{3}{2} = 8k_2 \text{ or } k_2 = -\frac{3}{16}$$

Hence
$$i_1(t) = \frac{3}{2} - \frac{3}{4}e^{-\frac{t}{2}}$$

and
$$i_2(t) = -\frac{3}{2}e^{-\frac{t}{2}} \quad \textbf{Ans.}$$

## PROBLEMS

**5.1** Explain what are transients due to in an electric circuit? Is it possible to have a switching operation and still have no transient in an electric circuit? Give reasons.

**5.2** An exponentially rising or decaying curve represents a response of an electric circuit. Show that a tangent at $t = 0$ to these curves intersects the time axis at $t = \tau$ the time constant.

**5.3** Explain what you mean by forced response and forced free response of an electric circuit.

**5.4** Determine the voltage at the terminals of a coil of resistance 10 $\Omega$ and inductance 15H at the instant when the current is 12A and (a) increasing (b) decreasing, at the rate of 6A/sec. (c) Find the stored energy under both conditions.

**5.5** A coil has a resistance of 50 $\Omega$ and a time-constant of 1.25 sec. If the coil were rewound with the same weight of wire of half the diameter, calculate (a) the inductance (b) the time constant and the stored energy when connected to a 600 V dc supply.

**5.6** In the network of fig. P5.6 the switch $S$ is closed and a steady state is reached. At $t = 0$, the switch $S$ is opened. Determine and expression for the current in the inductor, $i_2(t)$.

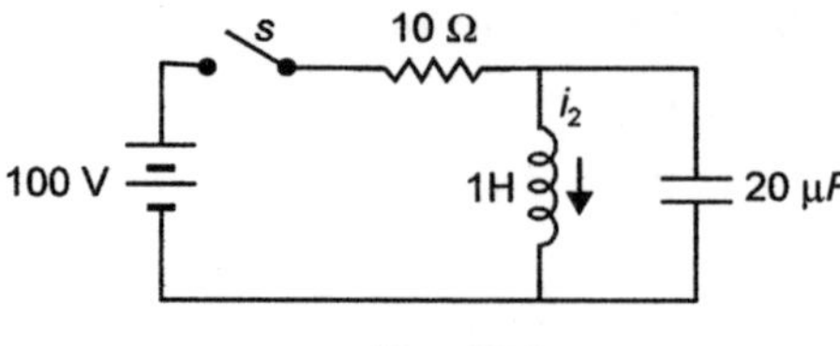


**Fig. P5.6**

**5.7** The response of a network is found to be $i = K_1 t\, e^{-\alpha t}$ $t \geq 0$ where $\alpha$ is a positive real number. Determine the time at which $i(t)$ attains a maximum value.

**5.8** Solve the following non-homogeneous differential equations for $t \geq 0$

(i) $\dfrac{d^2 i}{dt^2} + 3\dfrac{di}{dt} + 2i = 5t$

(ii) $\dfrac{d^2 v}{dt^2} + 5\dfrac{dv}{dt} + 6v = e^{-2t} + 5e^{-3t}$

**5.9** In the network shown in Fig. P5.9 the switch $S$ is closed at $t = 0$ with the capacitor initially uncharged. Determine the response $i(t)$ of the circuit.

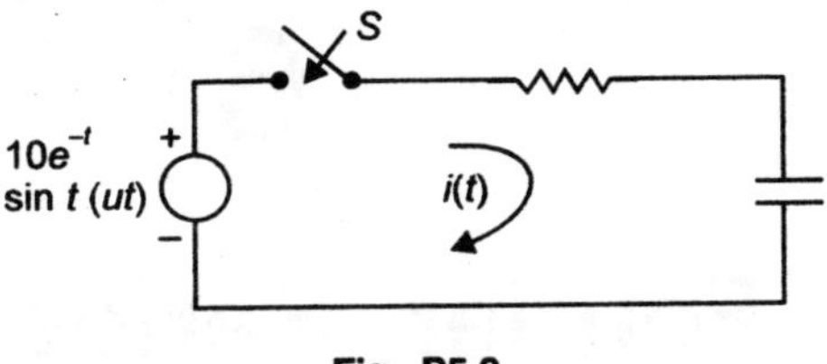


**Fig. P5.9**

**5.10** In the circuit shown, the switch is initially in position $a$ (i) For $R_d = 500\ \Omega$, determine the voltage across the field coil at the instant at which the switch is changed to position $b$ (ii) Calculate the value of $R_d$ for the voltage across the coil to be 120 V at the instant of switching.

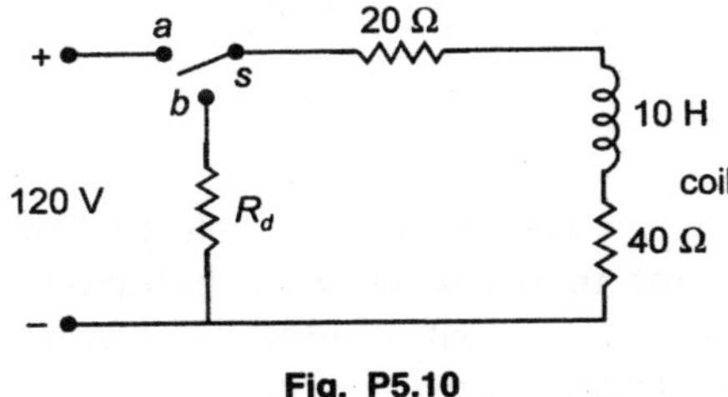


**Fig. P5.10**

**5.11** A resistance $R$ and a 2 $\mu F$ capacitor are connected in series across a 200 V *dc* supply. Across the condenser is a neon lamp that strikes at 120 V. Calculate $R$ to make the lamp strike 5 sec after the switch has been closed. If $R = 5\ M\Omega$ how long will it take the lamp to strike?

**5.12** A 50 Hz sinusoidal voltage of amplitude 400 V is applied to a series circuit of resistance 10 $\Omega$ and inductance 0.1 H. Find an expression to the value of the current at any instant after the voltage is applied, assuming the voltage is zero at the instant of application. Also calculate the value of the transient current 0.02 sec after switching on.

# 6

# Fourier Series and Fourier Transforms

## 6.1  INTRODUCTION

Sinusoidal quantities have a very important role in Engineering more so in Electrical Engineering. When the current in a capacitor or inductor or a transformer is sinusoidal, then voltage across the element is also sinusoidal. This is not true of any other wave forms. The steady state analysis of electric network with dc or sinusoidal excitations can be very easily handled using impedance concept. However, in practice we are faced with excitation other than sinusoidal both periodic as well as non-periodic. Some of these are shown in Fig. 6.1 and our objective is to find out response of linear networks when excited by such like functions.

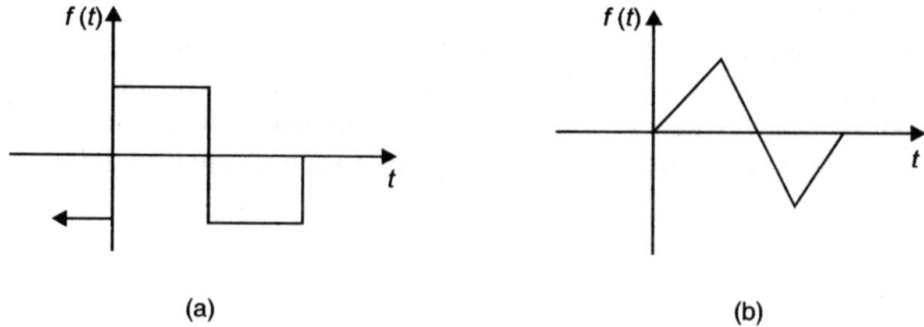


**Fig. 6.1**   (a) Periodic square wave, (b) Periodic triangular wave.

Fourier, a French mathematician, demonstrated that arbitrary periodic functions can be represented by an infinite series of sinusoids of harmonically varying frequencies and, non-periodic function wave form can be expressed by Fourier's transforms. Fourier series can be obtained of a periodic function irrespective of whether the function has Taylor's series expansion or not. It is this wide applicability of Fourier series that furnishes one reason for their importance and usefulness.

A function is said to be periodic if it repeats itself regularly over a specified time and it is expressed mathematically $f(t) = f(t + T)$ where $T$ is the time over which the function repeats itself.

Fourier series consists of a dc component and ac terms of all harmonics. Since the forced response due to each sinusoid can be obtained by impedance concept, the complete response of the linear circuit due to original function is the superposition of the responses due to various harmonics. Therefore, the main job is to evaluate the Fourier co-efficient of the infinite trigonometric series. It is to be noted that as we go for evaluation of co-efficient of higher harmonics they tend to decrease and hence the infinite series could be terminated to a good approximated value within permissible error.

## 6.2  FOURIER THEOREM

Fourier theorem which applies to all the functions that normally arise in physical and engineering application to which Fourier series can be obtained is stated as follows:

Any single valued function $f(t)$, continuous except possibly, for a finite numbers of finite discontinuities in an interval of length $2\pi$ or $T$ and having only a finite number of maxima and minima in this interval, possess a convergent Fourier series given by

$$f(t) = \frac{a_o}{2} + a_1 \cos \omega t + a_2 \cos 2\, \omega t + \ldots + b_1 \sin \omega t + b_2 \sin 2\, \omega t + \ldots$$

$$= \frac{a_o}{2} + \sum_{n=1}^{\infty} (a_n \cos n\, \omega t + b_n \sin n\, \omega t) \qquad \ldots(6.1)$$

and the co-efficient $a_n$ and $b_n$ are given by

$$a_o = \frac{2}{T} \int_0^T f(t)\, dt \qquad \ldots(6.2)$$

$$a_n = \frac{2}{T} \int_0^T f(t) \cos n\, \omega t\, dt \qquad \ldots(6.3)$$

and $$b_n = \frac{2}{T} \int_0^T f(t) \sin n\, \omega t\, dt \qquad \ldots(6.4)$$

However, if the function is given in terms of $f(\theta)$, the co-efficients are given by

$$a_o = \frac{1}{\pi} \int_0^{2\pi} f(\theta)\, d\theta \qquad \ldots(6.5)$$

$$a_n = \frac{1}{\pi} \int_0^{2\pi} f(\theta) \cos n\, \theta\, d\theta \qquad \ldots(6.6)$$

and $$b_n = \frac{1}{\pi} \int_0^{2\pi} f(\theta) \sin n\, \theta\, d\theta \qquad \ldots(6.7)$$

In order to evaluate these co-efficients following standard integral identities would frequently be used.

$$\int_0^T \sin n\omega t\, dt = 0 \quad \text{for all } n \qquad \ldots(6.8)$$

$$\int_0^T \cos n\omega t\, dt = 0 \quad \text{for } n \neq 0 \qquad \qquad ...(6.9)$$

$$\int_0^T \cos m\omega t \cos n\omega t\, dt = 0 \quad m \neq n \qquad \qquad ...(6.10)$$

$$= \frac{T}{2}, \quad m = n$$

$$\int_0^T \sin n\omega t \cos m\omega t\, dt = 0 \quad \text{for all } m \text{ and } n \qquad \qquad ...(6.11)$$

$$\int_0^T \sin n\omega t \sin m\omega t\, dt = 0 \quad m \neq n \qquad \qquad ...(6.12)$$

$$= \frac{T}{2} \quad m = n$$

In terms of angle $\theta$, these identities are given as

$$\int_0^{2\pi} \sin n\theta\, d\theta = 0 \quad \text{for all } n \qquad \qquad ...(6.13)$$

$$\int_0^{2\pi} \cos n\theta\, d\theta = 0 \quad \text{for } n \neq 0 \qquad \qquad ...(6.14)$$

$$\int_0^{2\pi} \cos m\theta \cos n\theta\, d\theta = 0 \quad \text{for } m \neq n$$

$$= \pi \quad \text{for } m = n \qquad \qquad ...(6.15)$$

$$\int_0^{2\pi} \sin n\theta \cos m\theta\, d\theta = 0 \quad \text{for all } m \text{ and } n \qquad \qquad ...(6.16)$$

$$\int_0^{2\pi} \sin n\theta \sin m\theta\, d\theta = 0 \quad \text{for } m \neq n \qquad \qquad ...(6.17)$$

$$= \pi \quad \text{for } m = n$$

It is to be noted that Fourier series can be written in a more compact form either in sin or cos term by combining the sin and cos term as follows:

Let us consider the $n$th harmonic term

$$f_n(t) = a_n \cos n\omega t + b_n \sin n\omega t$$

$$= \sqrt{a_n^2 + b_n^2}\left[\frac{a_n}{\sqrt{a_n^2 + b_n^2}} \cos n\omega t + \frac{b_n}{\sqrt{a_n^2 + b_n^2}} \sin n\omega t\right]$$

$$= A_n\left[\cos \theta_n \cos n\omega t + \sin \theta_n \sin n\omega t\right]$$

$$= A_n\left[\cos (n\omega t - \theta_n\right] \qquad \qquad ...(6.18)$$

and Fourier series in cos form can be written as

$$f(t) = A_o + \sum_{n=1}^{\infty} A_n \cos (n\omega t - \theta_n) \qquad \qquad ...(6.19)$$

where
$$\theta_n = \tan^{-1} \frac{b_n}{a_n}$$

Similarly Fourier series in sine form can be written as

$$f(t) = A_o + \sum A_n \sin(n\omega t - \phi_n) \qquad \qquad ...(6.20)$$

where
$$\phi_n = \tan^{-1} \frac{a_n}{b_n}$$

The plots of $A_n$ and $\theta_n$ or $\phi_n$ as a function of $\omega n$ or $n$ are shown here in Fig. (6.2)

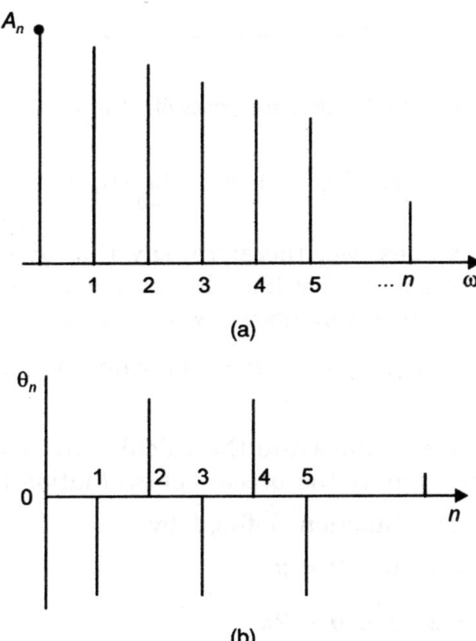

Fig. 6.2 (a) Amplitude, (b) Phase Spectra.

So our original problem that we have a non-sinusoidal but periodic source connected to a linear circuit, how to analyze it? We have been able to obtain equivalent Fourier series for the non-sinusoidal function and hence the equivalent circuit will be as shown in Fig. 6.3

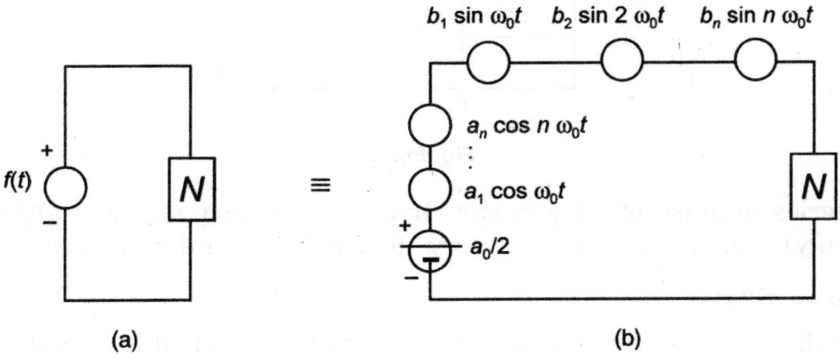

Fig. 6.3 (a) Actual circuit, (b) Equivalent circuit.

The superposition of responses due to various sources of the Fourier series gives the final response of the original source $f(t)$.

Going back to Fourier theorem a function with finite discontinuity–can have Fourier series expansion. By finite discontinuity at a point $t = t_0$ (Fig 6.4) we mean roughly speaking a finite "jump" in the value of the function. To be more specific if

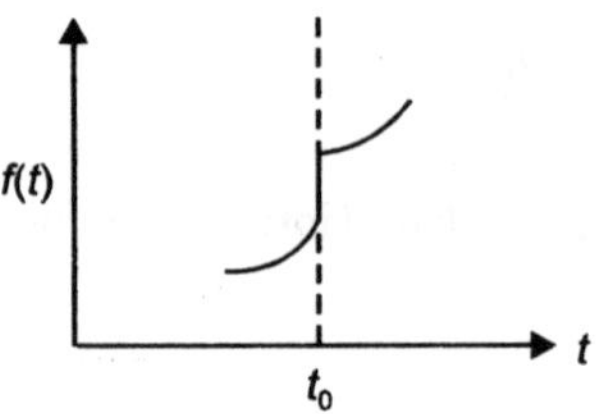

**Fig. 6.4**  A finite discontinuity of $f(t)$ at $t = t_0$

$$\underset{h \to 0}{Lt}\ f(t_0 - h) \quad \text{and} \quad \underset{h \to 0}{Lt}\ f(t_0 + h)$$

both exist but have different values, we, therefore, say that $f(t)$ has finite discontinuity at $t = t_0$. When $f(t)$ has a finite discontinuity like this at $t = t_0$, the Fourier series yields the arithmetic mean of the two limits given above when we set $t = t_0$ in the series i.e. the Fourier series converges to $\frac{1}{2}[f(t_{0+}) + f(t_{0-})]$ at every value of $t_0$ and hence at the point of discontinuity.

Let us take a few functions to illustrate the calculation of Fourier series and then we will learn a few tips to further simplify the process of evaluation to certain kind of functions.

**Example 6.1:** Consider the function defined by

$$f(\theta) = 1 \quad 0 < \theta < \pi$$

$$f(\theta) = 2 \quad \pi < \theta < 2\pi$$

i.e. the function has a discontinuity at $\theta = \pi$ at $\theta = \pi_-$ $f(\pi_-) = 1$ and at $\theta = (\pi_+)$ $f(\pi_+) = 2$ and therefore at $f(\pi) = \frac{1}{2}\ [f(\pi_-) + f(\pi_+)] = \frac{3}{2}$ as explained above.

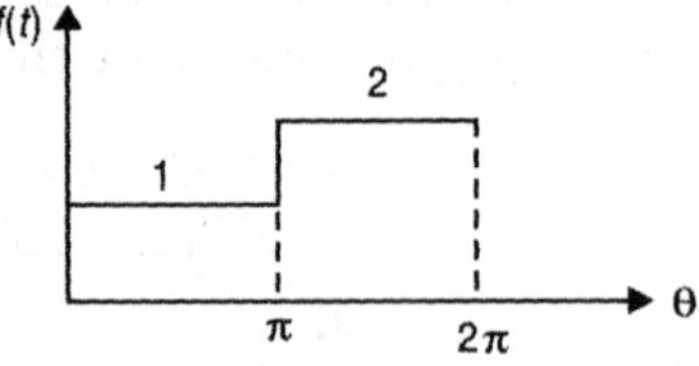

**Fig. E6.1**

Also, the series because of its periodic character (Fourier series is applicable for periodic functions only) will define $f(\theta)$ for every $\theta$ not in our interval, therefore, we should expect to get $f(o) = f(2\pi) = \frac{3}{2}$.

Now using the formulae for evaluation of Fourier co-efficients we have using equation (6.5) to (6.7)

$$a_o = \frac{1}{\pi} \int_0^{2\pi} f(\theta)\, d\theta = \frac{1}{\pi} \int_0^{\pi} 1 \cdot d\theta + \frac{1}{\pi} \int_{\pi}^{2\pi} 2 \cdot d\theta = 3$$

$$a_n = \frac{1}{\pi} \int_0^{2\pi} f(\theta) \cos n\,\theta\, d\theta$$

$$= \frac{1}{\pi} \left[ \int_0^{\pi} \cos n\theta\, d\theta + \int_{\pi}^{2\pi} 2 \cos n\theta\, d\theta \right]$$

$$= \frac{1}{\pi} \left[ \left\{ \frac{\sin n\theta}{n} \right\}_0^{\pi} + \left\{ \frac{2 \sin n\theta}{n} \right\}_{\pi}^{2\pi} \right] = 0 \text{ for } n = 1,\, 2,\, \dots$$

$$b_n = \frac{1}{\pi} \int_0^{2\pi} f(\theta) \sin n\theta\, d\theta$$

$$= \frac{1}{\pi} \left[ \int_0^{\pi} \sin n\theta\, d\theta + \int_{\pi}^{2\pi} 2 \sin n\theta\, d\theta \right]$$

$$= \frac{1}{\pi} \left[ \frac{-\cos n\pi}{n} + \frac{1}{n} - \frac{2}{n} + \frac{2 \cos n\pi}{n} \right]$$

$$= \frac{1}{\pi n} \left[ \cos n\pi - 1 \right] \text{ for } n = 1,\, 2,\, \dots\, n$$

From this it is clear that all '$a$' except $a_o$ are zeros and for $n$ = even number $b_n$ are also zeros. Therefore,

$$\frac{a_0}{2} = \frac{3}{2}$$

$$b_1 = \frac{1}{\pi}(-2) = -\frac{2}{\pi}, \quad b_3 = \frac{1}{3\pi}(-2) = -\frac{2}{3\pi}$$

and hence

$$f(\theta) = \frac{3}{2} - \frac{2}{\pi} \sin\theta - \frac{2}{3\pi} \sin 3\theta - \frac{2}{5\pi} \sin 5\theta \dots$$

Let us find out the value of function at $\theta = 0$, $\pi$ and $2\pi$. We see that

$f(o) = f(\pi) = f(2\pi) = \dfrac{3}{2}$ as envisaged earlier that the value of the function is average of the values at $\theta = \pi_-$ and $\theta = \pi_+$

Let us find out an interesting series by putting $\theta = \dfrac{\pi}{2}$ we have

$$f\left(\frac{\pi}{2}\right) = 1 = \frac{3}{2} - \frac{2}{\pi}\left(1 - \frac{1}{3} + \frac{1}{5} - \frac{1}{7} + \dots\right)$$

or

$$\frac{1}{2} = \frac{2}{\pi}\left(1 - \frac{1}{3} + \frac{1}{5} - \frac{1}{7} + \cdots\right)$$

or

$$\frac{\pi}{4} = 1 - \frac{1}{3} + \frac{1}{5} - \frac{1}{7} + \cdots$$

**Example 6.2:** Consider the functions

$$f(\theta) = -\theta, \quad -\pi \le \theta \le 0$$

$$= -\theta \quad 0 \le \theta \le \pi$$

The function is shown in Fig. E6.2

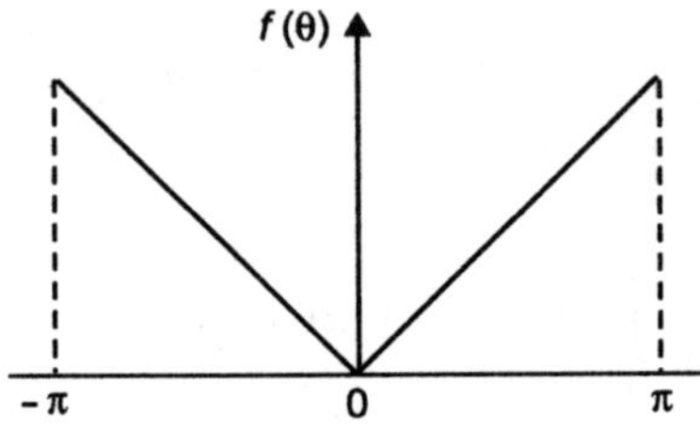


**Fig. E6.2**

There is no discontinuity in the function even though the function does not have derivative at $\theta = 0, \pm \pi, \pm 2\pi$ etc.

So the series function will be continuous everywhere. Using equation (6.5) to (6.7) we have

$$a_0 = \frac{1}{\pi}\int_{-\pi}^{\pi} f(\theta)\, d\theta = \frac{1}{\pi}\left[\int_{-\pi}^{0} -\theta\, d\theta + \int_{0}^{\pi} \theta\, d\theta\right]$$

$$= \frac{1}{\pi}\left[\left(-\frac{\theta^2}{2}\right)_{-\pi}^{0} + \left(\frac{\theta^2}{2}\right)_{0}^{\pi}\right] = \pi$$

$$a_n = \frac{1}{\pi}\left[\int_{-\pi}^{0} -\theta \cos n\theta\, d\theta + \int_{0}^{\pi} \theta \cos n\theta\, d\theta\right]$$

$$= -\frac{1}{\pi}\left[\left(\frac{\theta \sin n\theta}{n} + \frac{\cos n\theta}{n^2}\right)\right]_{-\pi}^{0} + \frac{1}{\pi}\left[\frac{\theta \sin n\theta}{n} + \frac{\cos n\theta}{n^2}\right]_{0}^{\pi}$$

$$= \frac{2}{\pi n^2}(\cos n\pi - 1)$$

Similarly,

$$b_n = \frac{1}{\pi}\left[\int_{-\pi}^{0} -\theta \sin n\theta\, d\theta + \int_{0}^{\pi} \theta \sin n\theta\, d\theta\right]$$

$$= -\frac{1}{\pi}\left[-\frac{\theta\cos n\theta}{n} + \frac{\sin n\theta}{n^2}\right]_{-\pi}^{0} + \frac{1}{\pi}\left[-\frac{\theta\cos n\theta}{n} + \frac{\sin n\theta}{n^2}\right]_{0}^{\pi}$$

$$= \frac{\cos n\pi}{n} - \frac{\cos n\pi}{n} = 0$$

Therefore,

$$f(\theta) = \frac{\pi}{2} + \sum_{n=1}^{\infty}\frac{2}{\pi n^2}(\cos n\pi - 1)\cos n\theta$$

$$= \frac{\pi}{2} - \frac{4}{\pi}\left[\cos\theta + \frac{1}{3^2}\cos 3\theta + \frac{1}{5^2}\cos 5\theta + \cdots\right]$$

Let us evaluate the value of the function at $\theta = 0$

$$f(0) = 0 = \frac{\pi}{2} - \frac{4}{\pi}\left[1 + \frac{1}{3^2} + \frac{1}{5^2} + \cdots\right]$$

or

$$\frac{\pi^2}{8} = 1 + \frac{1}{3^2} + \frac{1}{5^2} + \cdots$$

**Example 6.3:** Determine the Fourier series for the saw-tooth function shown here in Fig. E6.3.

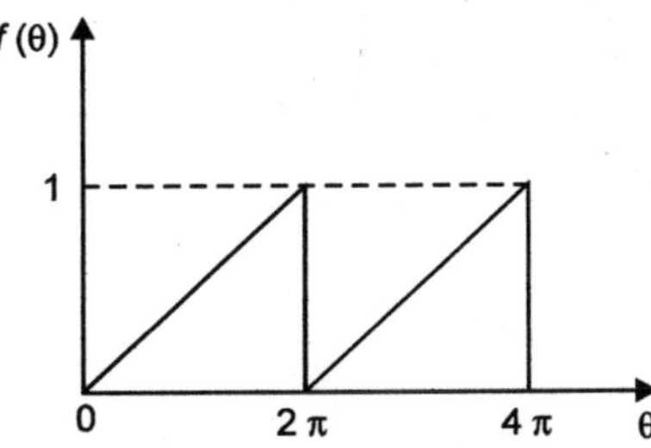


**Fig. E6.3**  Saw-tooth wave.

The function can be written analytically as

$$f(\theta) = \frac{1}{2\pi}\cdot\theta \quad 0 \le \theta \le 2\pi$$

$$a_0 = \frac{1}{\pi}\int_0^{2\pi} f(\theta)\,d\theta = \frac{1}{\pi}\int_0^{2\pi}\frac{\theta}{2\pi}\cdot d\theta = \frac{1}{\pi}\left(\frac{\theta^2}{4\pi}\right)_0^{2\pi}$$

$$= \frac{1}{\pi}\cdot\frac{4\pi^2}{4\pi} = 1$$

Therefore, $\dfrac{a_0}{2} = \dfrac{1}{2}$

$$a_n = \frac{1}{\pi}\int_0^{2\pi}\frac{\theta}{2\pi}\cos n\theta\,d\theta = \frac{1}{2\pi^2}\left[\frac{\theta\sin n\theta}{n} + \frac{\cos n\theta}{n^2}\right]_0^{2\pi}$$

$$= \frac{1}{2\pi^2}[0 + 0] = 0$$

$$b_n = \frac{1}{\pi}\int_0^{2\pi} \frac{\theta}{2\pi}\sin n\theta\, d\theta = \frac{1}{2\pi^2}\left[-\frac{\theta\cos n\theta}{n} + \frac{\sin n\theta}{n^2}\right]_0^{2\pi}$$

$$= -\frac{1}{2\pi^2}\cdot\frac{2\pi}{n} = -\frac{1}{\pi n}$$

hence the Fourier series is given as

$$f(\theta) = \frac{1}{2} - \frac{1}{\pi}\sum_{n=1}^{\infty}\frac{\sin n\theta}{n}$$

Now let us study certain kind of function where it is very easy to write Fourier series.

## 6.3 EVEN AND ODD FUNCTIONS

A function is said to be an even function when $f(\theta) = f(-\theta)$ and a function is said to be an odd one when $f(-\theta) = -f(\theta)$. Examples of even functions are $x^2$, cos $x$ etc. where as those of odd function are $x$, sin $x$ etc. Some of the even and odd functions are shown here.

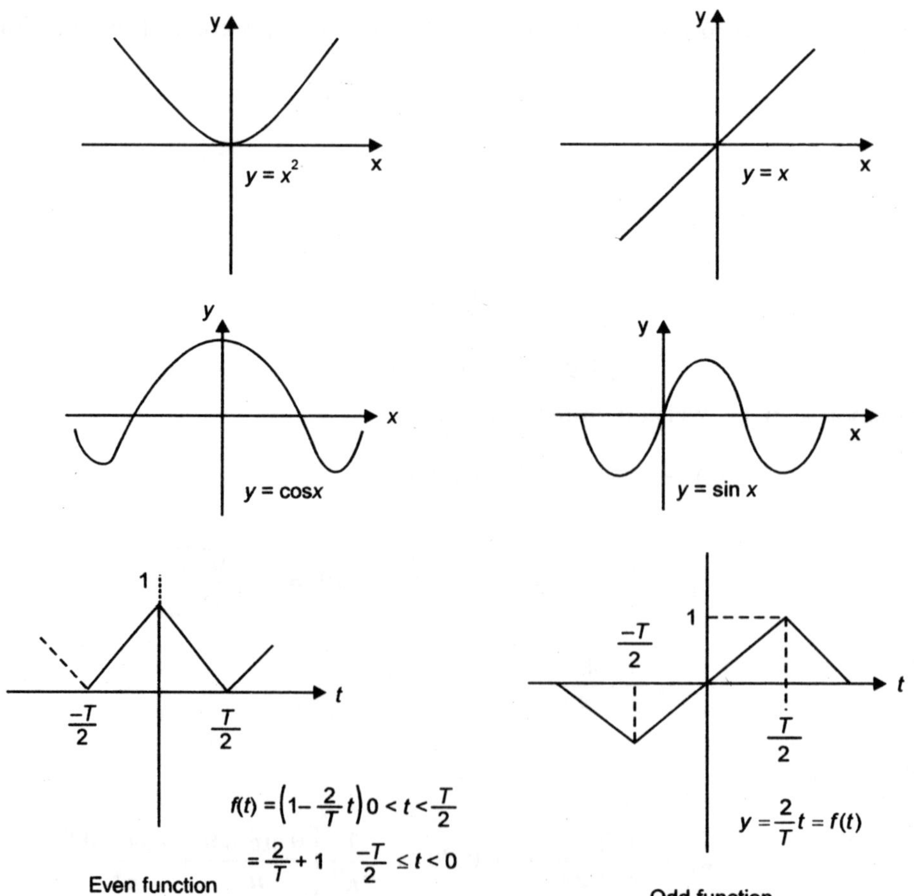


**Fig. 6.5** Examples of even and odd functions.

It is seen that the even functions are symmetrical about the ordinate whereas the odd functions are symmetrical about the origin.

If two even functions are added up the result is an even function. Similarly, if two odd functions are added up the result is an odd function. However, if an odd function is added to an even function, the result is neither an even nor an odd function, rather it may be said to have an even and an odd part.

Also if two odd functions are multiplied the result is an even function. Similarly if two even functions are multiplied the result is an even function. However, if an odd functions is multiplied to an even function the result is an odd function.

We know that

$$\int_{-t_0}^{t_0} f_e(t)\, dt = 2 \int_0^{t_0} f_e(t)\, dt \qquad \qquad ...(6.21)$$

and

$$\int_{-t_0}^{t_0} f_0(t)\, dt = 0 \qquad \qquad ...(6.22)$$

where $f_e(t)$ are even and $f_o(t)$ the odd functions. If an even function is expanded in a Fourier series over the interval $-\pi \le \theta \le \pi$ or $\dfrac{-T}{2}$ to $\dfrac{T}{2}$ the co-efficients of the series are given as

$$a_o = \frac{1}{\pi} \int_{-\pi}^{\pi} f_e(\theta)\, d\theta = \frac{2}{\pi} \int_0^{\pi} f_e(\theta)\, d\theta \qquad \qquad ...(6.23)$$

$$a_n = \frac{1}{\pi} \int_{-\pi}^{\pi} f_e(\theta) \cos n\theta\, d\theta = \frac{2}{\pi} \int_0^{\pi} f_e(\theta) \cos n\theta\, d\theta \qquad \qquad ...(6.24)$$

and

$$b_n = \frac{1}{\pi} \int_{-\pi}^{\pi} f_e(\theta) \sin n\theta\, d\theta = 0 \qquad \qquad ...(6.25)$$

i.e. the $b$ – co-efficients will be absent

In terms of time period $T$

$$a_o = \frac{4}{T} \int_0^{T/2} f_e(t)\, dt \qquad \qquad ...(6.26)$$

$$a_n = \frac{4}{T} \int_0^{T/2} f_e(t) \cos n\, wt\, dt \qquad \qquad ...(6.27)$$

and

$$b_n = 0$$

Also if the given function is an odd function it can be seen that $a_n = 0$ for all $n$ and the co-efficients will be given as

$$b_n = \frac{2}{\pi} \int_0^{\pi} f_0(\theta) \sin n\theta\, d\theta \qquad \qquad ...(6.28)$$

or
$$= \frac{4}{T} \int_0^{T/2} f_0(t) \sin n\, wt \, dt$$

## 6.4  HALF WAVE SYMMETRY

If a periodic function $f(t)$ is represented as $f\left(t \pm \dfrac{T}{2}\right) = -f(t)$ the function is said to have Half-wave symmetry. This means during the interval $T$, value of $f(t)$ during the interval $0 \leq t \leq T/2$ and $T/2 < t < T$ are equal and opposite. Similarly the value of the function $f(t)$ are also equal and opposite during the intervals $-T/2 < t < 0$ and $0 < t < T/2$. Some wave forms with half-wave symmetry are shown here in Fig. 6.6.

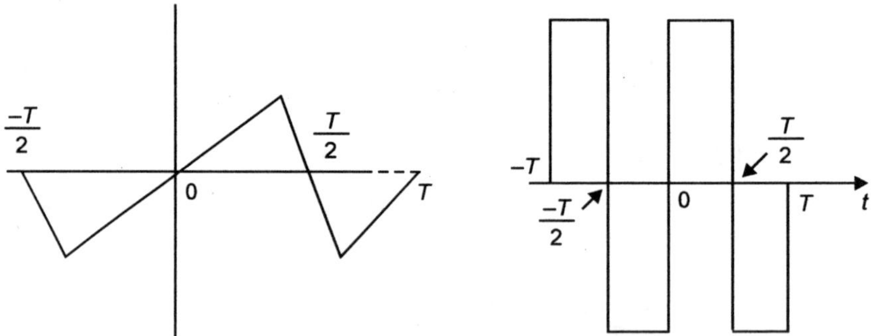

**Fig. 6.6**  Examples of Half Wave Symmetry.

It can be shown very easily that such function will have odd harmonics only in the Fourier series and even harmonics will be absent.

In general

$$a_n = \frac{1}{\pi} \int_{-\pi}^{\pi} f(\theta) \cos n\theta \, d\theta \qquad \ldots(6.29)$$

$$= \frac{1}{\pi} \left[ \int_{-\pi}^{0} f(\theta) \cos n\theta \, d\theta + \int_{0}^{\pi} f(\theta) \cos n\theta \, d\theta \right]$$

Now changing the variable $\theta$ to $(\pi + \theta)$ in the first integral we have

$$\int_{-\pi}^{0} f(\theta) \cos n\theta \, d\theta = \int_{0}^{\pi} f(\theta + \pi) \cos n(\theta + \pi) \, d\theta$$

since $\quad \theta \Leftrightarrow \pi + \theta \quad$ when $\quad \theta \to -\pi$

the lower limit becomes $\pi - \pi = 0$ an when $\theta \to 0$

the upper limit become $\pi + 0 = \pi$

$$= \int_{0}^{\pi} f(\theta + \pi) \left\{ \cos n\theta \cos n\pi - \sin n\theta \sin n\pi \right\} d\theta$$

Using $f(\theta + \pi) = -f(\theta)$ due to half have symmetry and since $\sin n\pi = 0$ for all integral value of $n$.

The above integral become

$$= -\cos n\pi \int_0^\pi f(\theta) \cos n\theta \, d\theta$$

and therefore,

$$a_n = \frac{1}{\pi}\left[ -\cos n\pi \int_0^\pi f(\theta) \cos n\theta \, d\theta + \int_0^\pi f(\theta) \cos n\theta \, d\theta \right]$$

$$= \frac{1}{\pi}(1 - \cos n\pi) \int_0^\pi f(\theta) \cos n\theta \, d\theta$$

Hence $a_n = 0$ for all even values of $n$

and
$$a_n = \frac{2}{\pi} \int_0^\pi f(\theta) \cos n\theta \, d\theta \qquad \qquad ...(6.31)$$

for $n$ odd

$$a_n = \frac{4}{T} \int_0^{T/2} f(t) \cos n\,\omega t \, dt$$

Similarly it can be shown that
$$b_n = 0 \quad \text{for } n \text{ even}$$

and
$$b_n = \frac{2}{\pi} \int_0^\pi f(\theta) \sin n\theta \, d\theta \qquad \qquad ...(6.31)$$

or
$$b_n = \frac{4}{T} \int_0^{T/2} f(t) \sin n\,\omega t \, dt \quad \text{for } n \text{ odd}$$

Thus the Fourier series for $f(t)$ with half-wave symmetry will have only odd harmonics. A typical function with half wave symmetry can be represented as

$$f(t) = a_1 \cos \omega_0 t + a_3 \cos 3\,\omega_0 t + \cdots +$$
$$b_1 \sin \omega_0 t + b_3 \sin 3\,\omega_0 t + \cdots + \qquad \qquad ...(6.32)$$

**Example 6.4:** Determine the Fourier series expansion of the function shown in Fig. E6.4

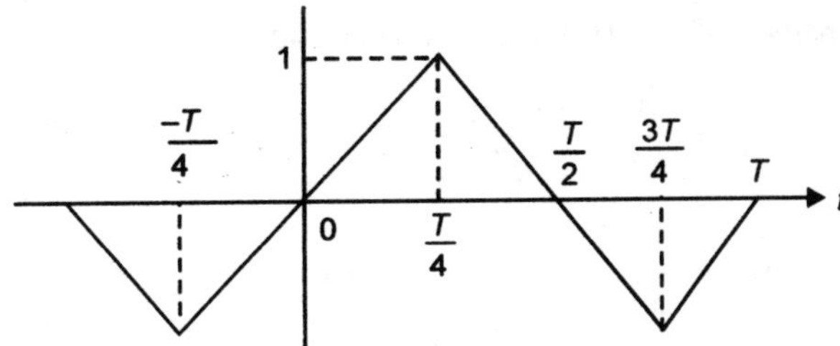

**Fig. E6.4**

The waveform is that of an odd function and also it has a half wave symmetry. Therefore, it will have only sin terms (odd function) of odd harmonics (half wave symmetry). The wave form can be written as

$$f(t) = \frac{4}{T} t \qquad\qquad 0 < t < T/4$$

$$= -\frac{4}{T} t + 2 \qquad \frac{T}{4} < t < \frac{3T}{4}$$

Using equation 6.31 (b) we have

$$b_n = \frac{4}{T} \int_0^{T/4} f(t) \sin n\,\omega t\, dt$$

$$= \frac{4}{T} \left[ \int_0^{T/4} \frac{4}{T} t \sin n\,\omega t\, dt + \int_{T/4}^{T/2} \left( 2 - \frac{4}{T} t \right) \sin n\,\omega t\, dt \right]$$

$$= \frac{4}{T} \left[ \frac{4}{T} \left\{ \left( -t\, \frac{\cos n\,\omega t}{n\,\omega} \right)_0^{T/4} + \int_0^{T/4} \frac{\cos n\,\omega t}{n\,\omega}\, dt \right\} + \right.$$

$$\left. \int_{T/4}^{T/2} 2 \sin \omega t\, dt - \frac{4}{T} \int_{T/4}^{T/2} t \sin n\,\omega t\, dt \right]$$

$$= \frac{4}{T} \left[ \frac{4}{T} \left\{ -\frac{T}{4}\, \frac{\cos n\,\omega\, T/4}{n\,\omega} + \left( \frac{\sin n\,\omega t}{n^2\,\omega^2} \right)_0^{T/4} \right\} \right] - \frac{8}{T} \left( \frac{\cos n\,\omega t}{n\,\omega} \right)_{T/4}^{T/2}$$

$$- \frac{16}{T^2} \left[ \left( -t\, \frac{\cos n\,\omega t}{n\,\omega} \right)_{T/4}^{T/2} + \left( \frac{\sin n\,\omega t}{n^2\,\omega^2} \right)_{T/4}^{T/2} \right]$$

$$= \frac{32}{T^2}\, \frac{\sin n\,\omega\, T/4}{n^2\,\omega^2} - \frac{16}{T^2}\, \frac{\sin n\,\omega\, T/2}{n^2\,\omega^2}$$

$$= \frac{8}{\pi^2 n^2}\, \sin \frac{n\,\pi}{2} - \frac{16}{T^2}\, \frac{\sin n\,\pi}{n^2\,\omega^2}$$

For all integral values of $n$, the second term is zero and hence

$$b_n = \frac{8}{\pi^2} \cdot \frac{\sin n\,\pi/2}{n^2}$$

Hence

$$f(t) = \frac{8}{\pi^2} \left[ \sin \omega t - \frac{1}{3^2} \sin 3\,\omega t + \frac{1}{5^2} \sin 5\,\omega t + \cdots \right] \quad \textbf{Ans.}$$

**Example 6.5:** Determine the Fourier series of the function given

$$f(t) = -t \qquad \frac{-T}{2} \le t \le 0$$

$$= t \qquad 0 \le t \le \frac{T}{2}$$

$$= T - t \qquad \frac{T}{2} \le t \le T$$

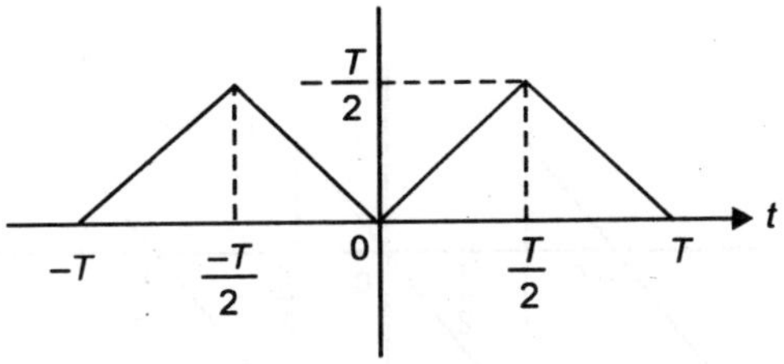

**Fig. E6.5**

Since the function has half wave symmetry, the Fourier series will have only odd harmonic terms and since it is symmetrical about the ordinate the series will have '$a$' coefficient only and $b_n = 0$

$$a_n = \frac{2}{T} \int_{-T/2}^{T/2} t \cos n\, \omega t\; dt = \frac{4}{T} \int_0^{T/2} t \cos n\, \omega t\; dt$$

$$= \left[ \left( t\, \frac{\sin n\, \omega t}{n\, \omega} \right)_0^{T/2} - \int_0^{T/2} \frac{\sin n\, \omega t}{n\, \omega}\; dt \right]$$

$$= \frac{4}{T} \left[ t\, \frac{\sin n\, \omega t}{n\, \omega} + \frac{\cos n\, \omega t}{n^2\, \omega^2} \right]_0^{T/2}$$

$$= \frac{4T^2}{4\pi^2 n^2 T} [\cos n\, \pi - 1] = \frac{T}{n^2\, \pi^2} (\cos n\, \pi - 1)$$

$$= 0 \text{ for } n \text{ even}$$

$$= \frac{-2T}{n^2\, \pi^2} \quad \text{for } n \text{ odd}$$

$$a_0 = \frac{1}{T} \int_0^T f(t)\; dt = \frac{1}{T} \cdot \frac{1}{2} \cdot T \cdot \frac{T}{2}$$

$$= \frac{T}{4}$$

Therefore, the Fourier series is

$$\frac{T}{4} - \frac{2T}{\pi^2}\left[\cos \omega t + \frac{\cos 3\,\omega t}{3^2} + \cdots\right] \textbf{ Ans.}$$

**Example 6.6:** Determine the Fourier series of the function

$$f(t) = t - \frac{T}{2} \le t \le T/2$$

and $\qquad f(t + T) = f(t)$

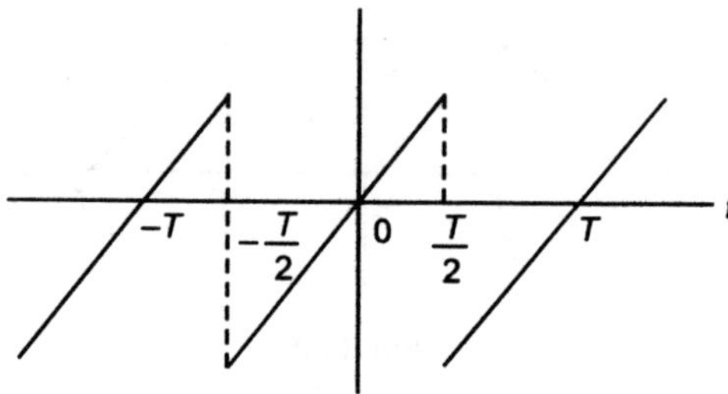

**Fig. E6.6**

Since the function is symmetric about the abscissa it is an odd function hence Fourier series will contain sin terms only Also $a_n = 0$

$$b_n = \frac{2}{T}\int\limits_{-T/2}^{T/2} f(t)\sin n\,\omega t\,dt$$

$$= \frac{4}{T}\int\limits_{0}^{T/2} t\sin n\,\omega t\,dt$$

$$= \frac{4}{T}\left[-t\,\frac{\cos n\,\omega t}{n\,\omega} + \frac{\sin n\,\omega t}{n^2\,\omega^2}\right]_0^{T/2}$$

$$= \frac{4}{T}\left[-\frac{T}{2}\,\frac{\cos n\pi}{2\pi n/T}\right] = -\frac{T}{\pi n}\cos n\pi$$

$$= -\frac{T}{\pi n} \quad \text{for } n \text{ even}$$

and $\qquad\qquad = \dfrac{T}{\pi n} \quad \text{for } n \text{ odd}$

Therefore,

$$f(t) = \frac{T}{\pi}\left[\sin \omega t - \frac{\sin 2\,\omega t}{2} + \frac{\sin 3\,\omega t}{3} \cdots\right]$$

**Example 6.7:** Determine the Fourier series of the function

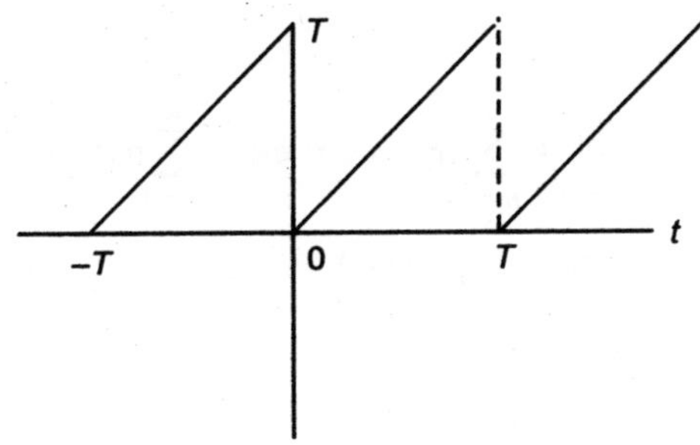

**Fig. E6.7**

$$f(t) = t \quad 0 \le t \le T$$

$$a_o = \frac{1}{2} \cdot T \cdot T \cdot \frac{1}{T} = \frac{T}{2}$$

$$a_n = \frac{2}{T} \int_0^T f(t) \cos n\omega t \, dt$$

$$= \frac{2}{T} \int_0^T t \cos n\omega t \, dt$$

$$= \frac{2}{T} \left[ t \frac{\sin n\,\omega t}{n\,\omega} + \frac{\cos n\,\omega t}{n^2\,\omega^2} \right]_0^T$$

$$= 0$$

$$b_n = \frac{2}{T} \int_0^T t \sin n\,\omega t \, dt$$

$$b_n = \frac{2}{T} \left[ -t \frac{\cos n\,\omega t}{n\,\omega} + \int \frac{\cos n\,\omega t}{n\,\omega} \, dt \right]_0^T$$

$$= \frac{2}{T} \left[ -t \frac{\cos n\,\omega t}{n\,\omega} \right]_0^T$$

$$= -\frac{2}{T} T^2 \frac{\cos 2\pi n}{2\pi n} = -\frac{T}{\pi n}$$

Therefore,

$$f(t) = \frac{T}{2} - \frac{T}{\pi} \left( \sin \omega t + \frac{\sin 2\,\omega t}{2} + \frac{\sin 3\,\omega t}{3} + \cdots \right)$$

## 6.5  EXPONENTIAL FOURIER SERIES

The Fourier series is given by

$$f(t) = \frac{a_0}{2} + \sum_{n=1}^{\infty} a_n \cos n\,\omega t + \sum_{n=1}^{\infty} b_n \sin n\,\omega t$$

We know that $\cos n\omega t = \left( \dfrac{e^{jn\omega t} + e^{-jn\omega t}}{2} \right)$

and $\qquad\qquad \sin n\omega t = \dfrac{e^{jn\omega t} - e^{-jn\omega t}}{2j}$

Therefore, $\qquad f(t) = \dfrac{a_0}{2} + \left[ \sum_{n=1}^{\infty} \left( \dfrac{a_n - jb_n}{2} \right) e^{jn\omega t} + \sum_{n=1}^{\infty} \dfrac{a_n + jb_n}{2} e^{-jn\omega t} \right]$

Let $\qquad\qquad C_n = \dfrac{a_n - jb_n}{2} \quad$ and $\quad C_{-n} = \dfrac{a_n + jb_n}{2}$

and $\qquad\qquad C_0 = \dfrac{a_0}{2} \quad$ we have

$$f(t) = C_0 + \sum_{n=1}^{\infty} C_n e^{jn\omega t} + \sum_{n=1}^{\infty} C_{-n} e^{-jn\omega t}$$

$$= \sum_{-\infty}^{\infty} C_n e^{j\omega n t} \qquad\qquad\qquad \text{...(6.33)}$$

The value of $C_n$ can be obtained as follows:

$$C_n = \frac{1}{T} \int_0^t f(t) \cos n\,\omega t\; dt - j\,\frac{1}{T} \int_0^T f(t) \sin n\,\omega t\; dt$$

$$= \frac{1}{T} \int_0^T f(t)\,(\cos n\,\omega t - j \sin n\,\omega t)\; dt$$

$$= \frac{1}{T} \int_0^T f(t)\, e^{-jn\omega t}\; dt \qquad\qquad\qquad \text{...(6.34)}$$

When equation (6.34) is substituted into (6.33) the equation represents exponential form of Fourier series

The exponential form has a number of advantages over the trigonometric form of Fourier series. Here only one integration has to be carried in contrast to three in trigonometric form for evaluation of Fourier co-efficients. Also it is usually a simpler integration.

For even functions where $b_n = 0$ and sin terms are absent $C_n$ co-efficients are real $C_n = C_{-n} = \dfrac{a_n}{2}$ whereas for odd functions since cos terms are absent i.e. $a_n = 0$ hence co-efficients are imaginary

$$C_n = C_{-n} = j \cdot \frac{b_n}{2} \, .$$

A graphical representation of the exponential co-efficients is given in the Fig. 6.7.

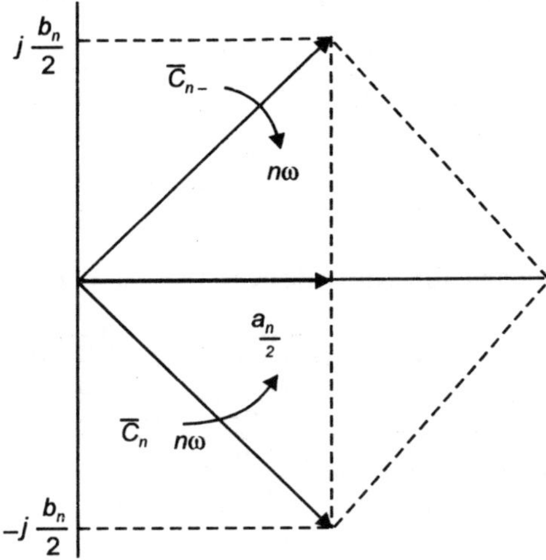

Fig. 6.7  Graphical representation of $C_n \, e^{j \, n \, \omega t} + C_{-n} \, e^{-j \, n \, \omega t}$

The phasor $C_n$ and $C_{-n}$ have been drawn when at $t = 0$ and it is seen that these are complex conjugate and the phasors representing them rotate with equal and opposite angular frequencies, the sum of these two terms will also be real and will vary sinusoidally at the rate of $n\omega$ radians per sec.

Here negative frequency means the phasor rotating clockwise where as positive angular frequency means the phasor rotating anti-clockwise. If we consider these two phasors as the two rotating magnetic field of equal magnitude but rotating in opposite directions to each other, is similar to our single phase sinusoidally varying field being decomposed to two equal and oppositely rotating magnetic field and hence the resulting field is a pulsating magnetic field due to single phase supply. The final Fourier series also contains sinusoidally or co-sinusoidally varying quantities. This means the resultant field will have fields due to various harmonics.

## 6.6  CHANGE OF INTERVAL

In most of the engineering problems the interval of the functions need not necessarily be from $-\pi$ to $\pi$. In general, say it is $-L \le \theta \le L$. In such a situation the Fourier series is given by

$$f(t) = \frac{a_0}{2} + a_1 \cos \frac{\pi t}{L} + \cdots + a_n \cos \frac{n \pi t}{L}$$

$$+ b_1 \sin \frac{\pi t}{L} + b_2 \sin \frac{2 \pi t}{L} + \cdots + b_n \sin \frac{n \pi t}{L} \qquad \text{...(6.35)}$$

where

$$a_n = \frac{1}{L} \int_{-L}^{L} f(\theta) \cos \frac{n \pi \theta}{L} \, d\theta \qquad \text{...(6.36)}$$

and
$$b_n = \frac{1}{L} \int_{-L}^{L} f(\theta) \sin \frac{n\pi\theta}{L} \, d\theta \qquad \qquad \text{...(6.37)}$$

For the half interval $0 < \theta < L$ the half range sin series for $f(\theta)$ is

$$f(\theta) = b_1 \sin \frac{\pi\theta}{L} + b_2 \frac{2\pi\theta}{L} + \cdots \qquad \qquad \text{...(6.38)}$$

where
$$b_n = \frac{2}{L} \int_{0}^{L} f(\theta) \sin \frac{n\pi\theta}{L} \, d\theta \qquad \qquad \text{...(6.39)}$$

and the half range cosine series for the interval $0 < \theta < L$ is

$$f(\theta) = \frac{a_0}{2} + a_1 \cos \frac{\pi\theta}{L} + a_2 \cos \frac{2\pi\theta}{L} + \cdots \qquad \qquad \text{...(6.40)}$$

where
$$a_n = \frac{2}{L} \int_{0}^{L} f(\theta) \cos \frac{n\pi\theta}{L} \, d\theta \qquad \qquad \text{...(6.41)}$$

**Example 6.7:** Determine Fourier series for the function
$$f(\theta) = 0 \qquad -2 < \theta < 0$$
$$f(\theta) = k \qquad 0 < \theta < 2$$

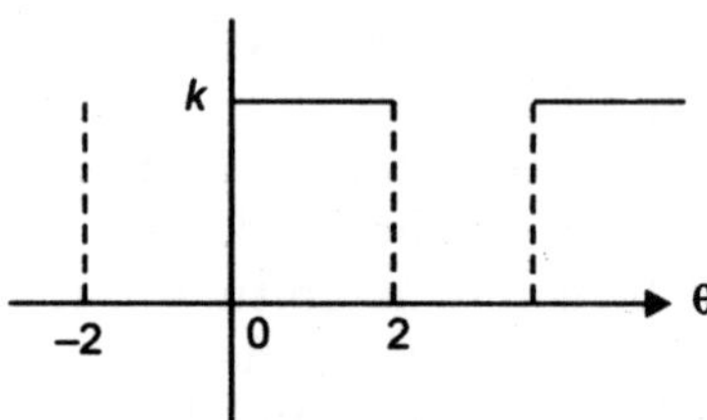


**Fig. E6.7**

$$a_0 = \frac{2 \cdot k}{2} = k$$

$$a_n = \frac{1}{2} \int_{-2}^{0} 0 \cdot \cos \frac{n\pi\theta}{2} \, d\theta + \frac{1}{2} \int_{0}^{2} k \cos \frac{n\pi\theta}{2} \, d\theta$$

$$= \left[ \frac{k}{2} \cdot \frac{2}{n\pi} \sin \frac{n\pi\theta}{2} \right]_{0}^{2} = \frac{k}{n\pi} \left( \sin \frac{n\pi\theta}{2} \right)_{0}^{2}$$

$$= 0$$

$$b_n = \frac{1}{2} \int_{0}^{2} k \sin \frac{n\pi\theta}{2} \, d\theta = -\frac{k}{2} \cdot \frac{2}{n\pi} \left[ \cos \frac{n\pi\theta}{2} \right]_{0}^{2}$$

$$= -\frac{k}{n\pi}(\cos n\pi - 1)$$

$$= \frac{k}{n\pi}(1 - \cos n\pi)$$

$$= 0 \text{ for } n \text{ even}$$

$$= \frac{2k}{n\pi} \text{ for } n \text{ odd}$$

Therefore,

$$f(\theta) = \frac{k}{2} + \frac{k}{\pi}\left(\sin\frac{\pi\theta}{2} + \frac{1}{3}\sin\frac{3\pi\theta}{2} + \frac{1}{5}\sin\frac{5\pi\theta}{2} + \cdots\right)$$

## 6.7  RMS OR EFFECTIVE VALUE OF A FUNCTION

In electric engineering we are occasionally faced with the evaluations of rms value or effective value of a given current or voltage waveform. The waveform may have a number of harmonics. Mathematically rms value of $y = f(x)$ over an interval $a \le x < b$ is defined as

$$\bar{y} = \sqrt{\frac{\int_a^b y^2\, dx}{(b-a)}} \qquad\qquad ...(6.42)$$

If the interval is over a period of $2\pi$ it is defined as

$$\bar{y} = \sqrt{\frac{\int_0^{2\pi} y^2\, dx}{2\pi}} \qquad\qquad ...(6.43)$$

Suppose $y = f(x)$ is expanded in a Fourier series

$$f(x) = \frac{a_0}{2} + \sum_{n=1}^{\infty}(a_n \cos n x + b_n \sin n x)$$

then the rms value is given by

$$\bar{y} = \sqrt{\frac{a_0^2}{4} + \frac{1}{2}\sum_{n=1}^{\infty}(a_n^2 + b_n^2)} \qquad\qquad ...(6.44)$$

Suppose the current in a circuit is given by
$$i(t) = I_1 \sin(\omega t + \alpha_1) + I_2 \sin(2\,\omega t + \alpha_2) + \ldots$$
$$I_n \sin(n\,\omega t + \alpha_n) \qquad\qquad ...(6.45)$$

The rms value of this current is given as

$$I_{rms} = \sqrt{\frac{I_1^2 + I_2^2 + \cdots + I_n^2}{2}} \qquad \qquad \text{...(6.46)}$$

Here $I_1$, $I_2$, $I_3$ are the amplitudes of the fundamental, second harmonics and other higher harmonics. Suppose the waveform contains in addition to the harmonics a.d.c. component $I_o$ also. The rms value is given by

$$I_{rms} = \sqrt{I_0^2 + \frac{I_1^2 + I_2^2 + \cdots + I_n^2}{2}} \qquad \qquad \text{...(6.47)}$$

$$= \sqrt{I_0^2 + I_1^2 + I_2^2 + \cdots I_n^2} \qquad \qquad \text{...(6.48)}$$

where $I_1$ .... $I_n$ are the rms values of individual harmonics.

**Example 6.8:** A circuit has a dc current of 10 A and ac fundamental current of 10 amp peak value. Determine the rms value of the current in the circuit.

**Solution:**

$$I_{rms} = \sqrt{10^2 + \left(\frac{10}{\sqrt{2}}\right)^2}$$

$$= \sqrt{150} = 12.45 \text{ Amp.}$$

## 6.7.1  Power

Suppose the voltage and current in a circuit are given by

$$v = v_0 + \sum_{n=1}^{\infty} V_n \sin(n\,\omega t + \phi_n) \qquad \qquad \text{...(6.49)}$$

and

$$i = I_0 + \sum_{n=1}^{\infty} I_n \sin(n\,\omega t + \theta_n) \qquad \qquad \text{...(6.50)}$$

The general expression for the average power is given by

$$p = \frac{1}{T}\int_0^T v.i\,dt$$

$$= \frac{1}{T}\int_0^T \left\{ v_0 + \sum_{n=1}^{\infty} V_n \sin(n\,\omega t + \phi_n) \right\}$$

$$\left\{ I_0 + \sum_{n=1}^{\infty} I_n \sin(n\,\omega t + \theta_n) \right\} dt \qquad \qquad \text{...(6.51)}$$

On expansion of the integral we have the following pattern of terms:

($i$) Product of two constant ($V_0\, I_0$)

($ii$) Product of a constant and a sine term

($iii$) Product of two-sine functions of same frequency

($iv$) Product of two-sine functions of different frequency

Here integration of ($ii$) and ($iv$) over a cycle will result in zero. The first term gives the value $V_0\, I_0$. The third term is given as

$$p' = \frac{1}{T} \int_0^T V_n \sin(n\,\omega t + \phi_n)\, I_n \sin(n\,\omega t + \theta_n) \qquad \text{...(6.52)}$$

$$= \frac{V_n I_n}{2} \cos(\phi_n - \theta_n)$$

Hence the expression for power becomes

$$p = V_0\, I_0 + \frac{1}{2} \sum_{n=1}^{\infty} V_n\, I_n \cos(\phi_n - \theta_n) \qquad \text{...(6.53)}$$

Therefore, the average total power is the sum total of power due to individual voltage and current of the same harmonics. No power is contributed by current and voltages of different harmonics.

The apparent power in such a circuit is given by

$$VA = \left( \sqrt{V_0^2 + \frac{V_1^2 + V_2^2 + \cdots + V_n^2}{2}} \right) \times \left( \sqrt{I_0^2 + \frac{I_1^2 + I_2^2 + \cdots + I_n^2}{2}} \right) \qquad \text{...(6.54)}$$

and hence the p.f. of the circuit is given by

$$p.f. = \frac{V_0 I_0 + \dfrac{1}{2} \sum V_n\, I_n \cos(\phi_n - \theta_n)}{VA} \qquad \text{...(6.55)}$$

**Example 6.9:** If the current in a 20 $\Omega$ resistor is given by $i = 4 + 5 \sin wt - 3 \cos 3 wt$

Determine the power consumed by the resistor

$$p = p_0 + p_1 + p_3$$

$$= 4^2 \times 20 + \left(\frac{5}{\sqrt{2}}\right)^2 \times 20 + \left(\frac{3}{\sqrt{2}}\right)^2 \times 20$$

$$= (16 + 12.5 + 4.5) \times 20$$

$$= 660 \text{ watt}$$

The effective value of current $= \sqrt{33} = 5.7$ A.

**Example 6.10:** The voltage applied to a load is

$$v = 100 \sin 337t \text{ and the current in the load is}$$

$$i = 4.2 \sin (377t + 20^\circ) + 2.3 \sin (754t - 70^\circ)$$

Determine the average power.

**Solution:** Since the power will be only corresponding to fundamental frequency as voltage has only fundamental component.

Therefore,

$$\text{Power} = \frac{100}{\sqrt{2}} \times \frac{4.2}{\sqrt{2}} \times \cos 20^\circ$$

$$p = 210 \cos 20^\circ = 197.35 \text{ watts}$$

**Example 6.11:** The voltage applied to a load is

$v = 215 \sin 377t + 140 \cos 754t + 70 \sin 1131t$ and the current in the load is $i = 15 \cos (377t + 30^\circ) + 4 \cos (754t - 60^\circ) + 3 \sin (1131t + 50^\circ)$

Determine the average power in the load and the effective value of current and voltage.

**Solution:** In order to obtain phase relation between the various harmonics and fundamental, the voltage and current expressions must be written in the same trigonometric term: Fundamental voltage can be rewritten as $215 \cos (377t - 90^\circ)$ and therefore.

The phase displacement between fundamental voltage and current is more than $90^\circ$, which is not possible, and hence it is not a well-designed problem.

The load is not a simple resistance as phase between $v$ and $i$ is not zero. Also it appears that for fundamental the voltage lags, 3rd harmonics it leads and for 5th harmonic it again lags. This is not possible in a passive RLC circuit and therefore it is not a well-designed problem.

The problem is redefined as

$$v = 215 \sin 377t + 140 \cos 754t + 70 \sin 1131t$$

$$i = 15 \sin (377t - 30^\circ) + 4 \cos (754t - 45^\circ) + 3 \sin (1131t - 60^\circ)$$

The power is 
$$\frac{215 \times 15}{2} \cos 30 + \frac{140 \times 4}{2} \cos 45^\circ + \frac{70 \times 3}{2} \cos 60^\circ$$

$$= 1396.42 + 198 + 52.5 = 164.7 \text{ watts. } \textbf{Ans.}$$

**Example 6.12:** In a two-element series network voltage and current are given as

$$v = 50 + 50 \sin 314t + 50 \sin 942t$$

$$i = 10 \sin (314t + 60^\circ) + 15 \sin (942t + 45^\circ)$$

Determine the power consumed and the network elements.

**Solution:** Here since voltage has the dc. component and current does not have, this means it is an RC series circuit.

The power consumed

$$p = \frac{50 \times 10}{2} \cos 60 + \frac{50 \times 15}{2} \cos 45^\circ$$

$$= \left(125 + \frac{375}{\sqrt{2}}\right) \text{watts}$$

$$z_1 = \frac{50}{10} = 5 = \left(R^2 + X_c^2\right)^{1/2}$$

$$z_3 = \frac{50}{15} = \frac{10}{3} = \left(R^2 + \frac{X_c^2}{9}\right)^{1/2}$$

or
$$R^2 + X_c^2 = 25$$

$$R^2 + \frac{X_c^2}{9} = \frac{100}{9}$$

or
$$\frac{8}{9} X_c^2 = \frac{125}{9} \quad \text{or} \quad X_c = 3.95 \ \Omega$$

and
$$R^2 + 3.95^2 = 25 \quad \text{or} \quad R = 3.06 \ \Omega \ \textbf{Ans.}$$

**Example 6.13:** Determine the Fourier series of the given function in complex co-efficient format

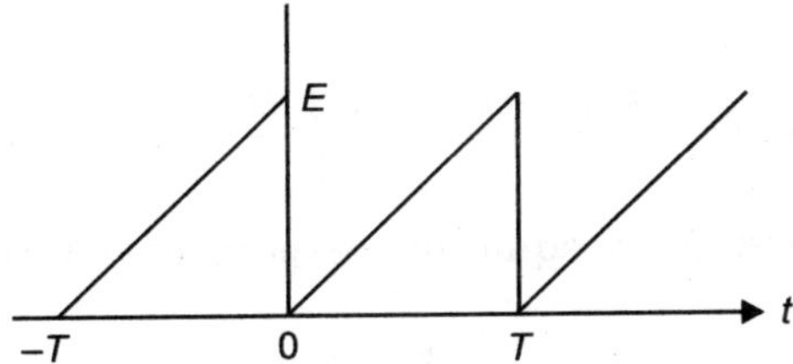


**Fig. E6.13** Saw-tooth voltage waveform.

**Soluton:** The voltage is given as $e(t) = \dfrac{(E)t}{T}$

The Fourier co-efficient

$$C_n = \frac{1}{T} \int_o^T f(t)\, e^{-jn\omega t}\, dt$$

$$= \frac{1}{T} \int_o^T \frac{E}{T}\, t\, e^{-jn\omega t}\, dt$$

$$= \frac{E}{T} \left[ -t\, \frac{e^{-jn\omega t}}{j\omega n} + \frac{e^{-j\omega nt}}{\omega^2 n^2} \right]_o^T$$

$$= -\frac{E}{T^2} \cdot T\, \frac{e^{-jn\omega t}}{j\omega n} = j\frac{E}{T}\, \frac{e^{-j\omega nt}}{\frac{2\pi}{T} \cdot n}$$

$$= j\frac{E}{2\pi n}\, e^{-j2\pi n}$$

Therefore,

$$f(t) = \sum_{-\infty}^{\infty} C_n\, e^{jn\omega_0 t} = \frac{jE}{2\pi} \sum_{-\infty}^{\infty} e^{-j2\pi n}\, e^{jn\omega_0 t}$$

$$= \frac{jE}{2\pi}\left[ \cdots\cdots \frac{e^{-3j\omega_0 t}}{3} - \frac{e^{-2j\omega_0 t}}{2} - e^{-j\omega_0 t} + e^{j\omega_0 t} + \frac{e^{2j\omega_0 t}}{2} + \frac{e^{3j\omega_0 t}}{3} + \cdots\cdots \right] + \frac{E}{2}$$

However, if we want the series in terms of sin, cos terms then $a$ and $b$ co-efficients can be found out

$$a_n = c_n + c_{n-} \quad \text{and} \quad b_n = j\,(c_n - c_{-n})$$

As the given function is an odd function of

$$t,\; a_n = 0 \quad \text{or} \quad C_n = -C_{-n}$$

or

$$b_n = j\,(cn + cn) = j2\,C_n = -\frac{E}{\pi n}$$

and

$$c_0 = a_0 = \frac{E}{2}$$

Therefore,

$$f(t) = E\left[ \frac{1}{2} - \frac{1}{\pi}\left( \sin \omega t + \frac{\sin 2\,\omega t}{2} + \frac{\sin 3\,\omega t}{3} + \cdots \right) \right] \;\textbf{Ans.}$$

The line spectra obtained from equation (exponential form) is shown in Fig. E6.13 below.

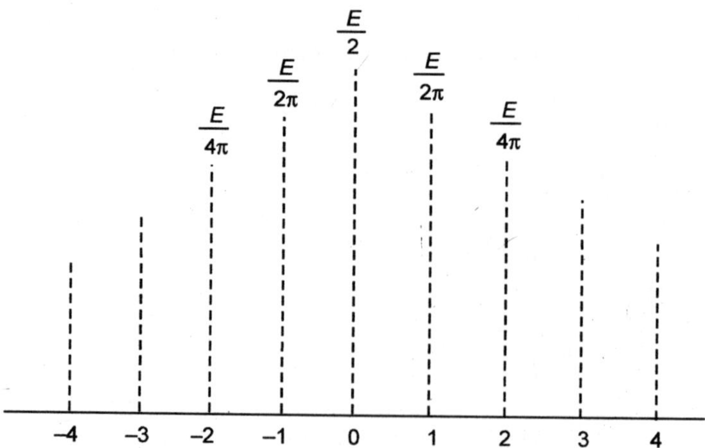

**Fig. E 6.13**   Line spectra for the waveform of the given problem function.

**Example 6.14:**  Determine Fourier series in exponential form for the function shown here.

$$C_n = \frac{1}{T} \int_{0}^{T} f(t)\, e^{-jn\omega t}\, dt$$

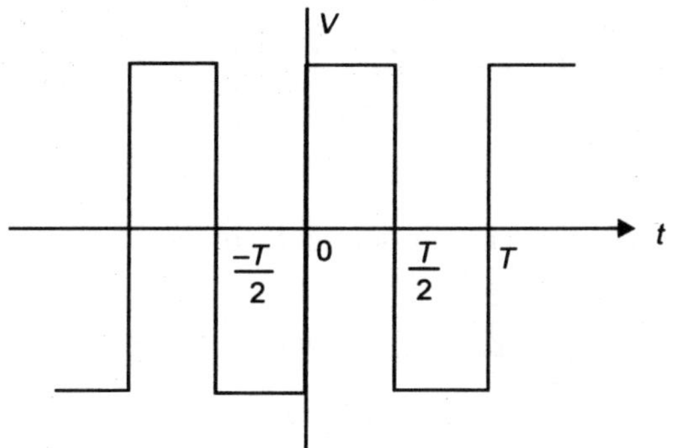

**Fig. E 6.14(a)**

$$= \frac{1}{T} \int\limits_{0}^{T/2} V \cdot e^{-jn\omega t}\, dt + \int\limits_{T/2}^{T} (-V)\, e^{-jn\omega t}\, dt$$

$$= \frac{V}{T}\left[\left(\frac{e^{-jn\omega t}}{-j\omega n}\right)_{0}^{T/2} + \left(\frac{e^{-jn\omega t}}{j\omega n}\right)_{T/2}^{T}\right]$$

$$= \frac{jV}{T}\left[\left(e^{-j\omega\frac{nT}{2}} - \frac{1}{\omega n}\right) - \left(\frac{e^{-j\omega nT}}{\omega n} - \frac{e^{-jn\omega\frac{T}{2}}}{\omega n}\right)\right]$$

$$= \frac{jV}{T}\left[\frac{2\,e^{-j\pi n}}{\omega n} - \frac{1}{\omega n} - \frac{e^{-j2\pi n}}{\omega n}\right] \quad \text{when } n \text{ is odd}$$

$$= \frac{-jV}{T}\cdot\frac{4}{\omega n} = \frac{-2jV}{\pi n}$$

When $n$ is even

$$C_n = \frac{jV}{T}\left[\frac{2}{\omega n} - \frac{1}{\omega n} - \frac{1}{\omega n}\right] = 0$$

Also $\qquad C_n = 0 \quad$ when $\quad n = 0$

The exponential form of Fourier series is given as

$$f(t) = \sum C_n\, e^{jn\omega t} = \frac{-j2V}{\pi}\sum_{-\infty}^{\infty}\frac{1}{n}\, e^{jn\omega t} \quad \text{when } n \text{ is odd}$$

Since $f(t)$ is an odd function of $t$, the co-efficient $C_n$ is, therefore, purely imaginary.

Also
$$|C_n| = |C_{-n}| = \left|\frac{2V}{\pi n}\right| = \frac{b_n}{2}$$

or
$$b_n = \frac{4V}{\pi n}$$

Therefore, Fourier series in Trigonometric term is

$$f(t) = \frac{4V}{\pi n} \sum_{n=1}^{\infty} \frac{\sin n\,\omega t}{n} \quad \text{where } n \text{ is odd}$$

The line spectra for the two formats are shown here

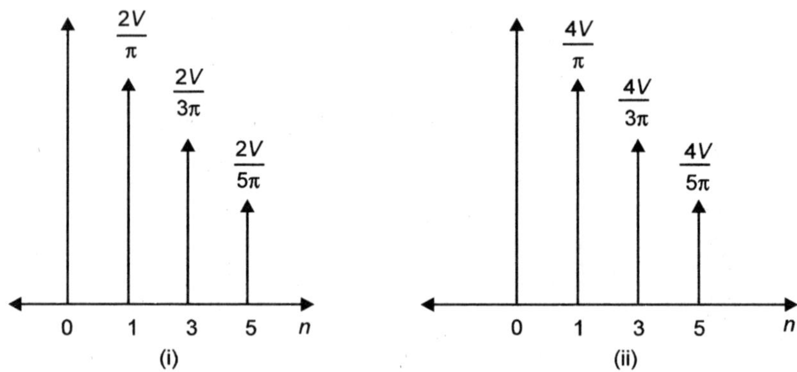

**Fig. E 6.14(b)**   (i) Line spectrum in exponential form, (ii) line spectrum in trigonometric form.

**Example 6.15:**  Determine the exponential form of Fourier series for the waveform shown here.

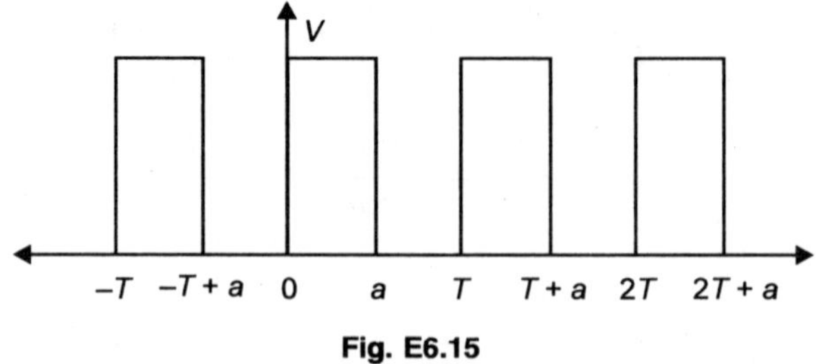

**Fig. E6.15**

**Solution:** The function is defined as
$$f(t) = v \qquad 0 < t \le a$$
$$= 0 \qquad a < t < T$$

$$C_n = \frac{1}{T} \int_0^T f(t)\, e^{-jn\omega t}\, dt = \frac{1}{T} \int_0^a V\, e^{-jn\omega t}\, dt$$

$$= \frac{V}{T} \left[ \frac{e^{-jn\omega t}}{-jn\omega} \right]_0^a$$

$$= \frac{V}{j2\pi n}\left[1 - e^{-jn\,\omega a}\right] = \frac{V}{j2\pi n}\left[1 - e^{-jn\cdot 2\pi\, a/T}\right]$$

$$= \frac{V}{j2\pi n}\left[e^{\frac{jn\,\pi a}{T}} - e^{-jn\,\pi a/T}\right]e^{-jn\,\pi a/T}$$

$$= \frac{V}{\pi n}\,e^{\frac{-jn\,\pi a}{T}}\cdot \sin\frac{n\,\pi a}{T}$$

Therefore, $\displaystyle f(t) = \sum C_n\, e^{jn\,2\pi\frac{t}{T}}$

$$= \frac{V}{\pi}\sum \frac{1}{n}\sin\frac{n\,\pi a}{T}\,e^{jn\cdot\frac{2\pi}{T}(t - a/2)}$$

The magnitude spectrum is a plot of the magnitude versus $n\,\omega$ as shown in Fig. 6.15(a)

$$\left|\overline{C}_n\right| = 2\left|C_n\right| = \frac{2}{n\pi}\left|\sin\frac{n\pi a}{T}\right| \quad n = 1,2,3,\dots$$

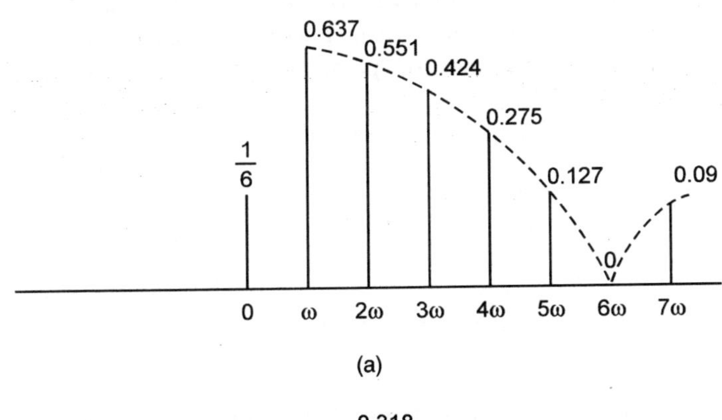


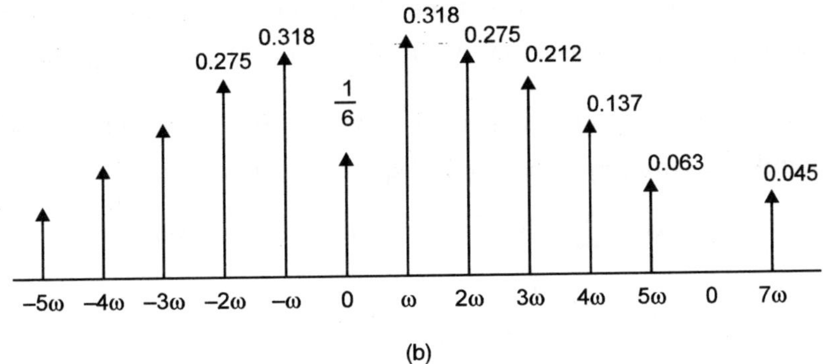


**Fig E6.15** (a) and (b) show the line spectrum of trigonometric Fourier series and exponential form of Fourier series of the given function.

For $n = 0$, we have to apply $L$ 'Hospitals' rule as $\dfrac{\sin 0}{0}$ is indeterminate

$$C_0 = \underset{n \to 0}{Lt} \frac{1}{n\pi} \sin \frac{n\pi a}{T} = \frac{1}{\pi} \underset{n \to 0}{Lt} \frac{d}{dn} \left[ \frac{\sin n\pi a/T}{n} \right] = \frac{a}{T}$$

To obtain the magnitude spectrum let us take $\dfrac{a}{T} = \dfrac{1}{6}$ as it will give suitable values of angles whose sin values are well known without referring to any table. Therefore $c_0 = \dfrac{1}{6}$

| $n$ | 0 | 1 | 2 | 3 | 4 | 5 | 6 | 7 |
|---|---|---|---|---|---|---|---|---|
| $C_n$ | $\dfrac{1}{6}$ | $\dfrac{1}{\pi}$ | $\dfrac{\sqrt{3}}{2\pi}$ | $\dfrac{2}{3\pi}$ | $\dfrac{-\sqrt{3}}{4\pi}$ | $\dfrac{-1}{5\pi}$ | $0$ | $\dfrac{-1}{7\pi}$ |
| $\overline{C}_n$ | | 0.637 | 0.551 | 0.424 | 0.275 | 0.127 | 0 | 0.091 |

## 6.8  FOURIER TRANSFORM

Fourier series is a discrete frequency representation in the frequency domain of a periodic time function. However, if a signal in time domain occurs once in some finite time interval and is non-recurring, it can be represented by a continuous frequency spectrum both in magnitude and phase angle. This is achieved by allowing the period $T$ to become infinity in the Fourier series. The line spectrum thus becomes a continuous spectrum. Therefore, starting with Fourier series expansion for periodic function we develop representation of non-recurring function.

We know that Exponential representation of Fourier series is given by

$$f(t) = \sum_{-\infty}^{\infty} C_n\, e^{jn\omega t} \qquad\qquad \text{...(6.56)}$$

where

$$C_n = \frac{1}{T} \int_{-T/2}^{T/2} f(t)\, e^{-jn\omega t}\, dt \qquad\qquad \text{...(6.57)}$$

Here $f(t)$ is a recurring function of time. However, if the function is non-recurring a few changes will have to be incorporated. As $T \to \infty$, $\omega \to 0$ as $\omega = \dfrac{2\pi}{T}$ and then $n$ becomes irrelevant and therefore $n\omega \to \omega$ and $T = \dfrac{2\pi}{\Delta\omega}$ where $\Delta\omega$ is the spacing between adjacent component of angular frequency. The above equations reduce to

$$f(t) = \sum C'\, e^{j\omega t} \qquad \omega = 0, \Delta\omega, 2\Delta\omega.. \qquad \text{...(6.58)}$$

and

$$C' = \frac{\Delta\omega}{2\pi} \int_{-T/2}^{T/2} f(t)\, e^{-j\omega t}\, dt \qquad\qquad \text{...(6.59)}$$

Therefore, substituting $C'$ in $f(t)$ we have

$$f(t) = \frac{1}{2\pi} \left[ \sum_{-\infty}^{\infty} \int_{-\infty}^{\infty} f(t)\, e^{-j\omega t}\, dt \right] e^{j\omega t}\, \Delta\omega \qquad\qquad \text{...(6.60)}$$

Changing summation sign to integral we have

$$f(t) = \frac{1}{2\pi} \int_{-\infty}^{\infty} \left[ \int_{-\infty}^{\infty} f(t)\, e^{-j\omega t}\, dt \right] e^{j\omega t}\, \Delta\omega$$

$$= \frac{1}{2\pi} \int_{-\infty}^{\infty} F(j\omega)\, e^{j\omega t}\, d\omega \qquad\qquad ...(6.61)$$

and

$$F(j\omega) = \int_{-\infty}^{\infty} f(t)\, e^{-j\omega t}\, dt \qquad\qquad ...(6.62)$$

Here $f(t)$ and $F(j\omega)$ are known as Fourier transform pairs. This derivation is subject to the Dirichlet condition i.e.

$$\int_{-\infty}^{\infty} |f(t)|\, dt < \infty \qquad\qquad ...(6.63)$$

$F(jw)$ can be written as $|F(j\omega)|\, e^{j\phi(\omega)}$. $F(j\omega)$ is thus known as the continuous amplitude spectrum and $\phi(\omega)$ is the continuous phase spectrum for a non-recurring $f(t)$. Therefore, the Fourier series spectra are the line or discrete spectra whereas Fourier transforms provide continuous spectra.

Following observations are made in regard to continuous spectra.

1. The shape of continuous amplitude and phase spectra of a non-recurring function $f(t)$ is identical with the envelope of the amplitude and phase line spectra of the same pulse but recurring.
2. All frequencies are present in the continuous spectra.

As is said earlier the sum of an even and an odd function is in general neither even nor odd. Any function $f(t)$ can be split unambiguously into an even and an odd function. The even part of a given function is the mean of the function and its reflection in the vertical axis and the odd part is the mean of the function and its negative reflection. Thus

$$E(t) = \tfrac{1}{2}\,[f(t) + f(-t)] \qquad\qquad ...(6.64)$$

$$O(t) = \tfrac{1}{2}\,[f(t) - f(-t)] \qquad\qquad ...(6.65)$$

Let $\qquad f(t) = E(t) + O(t)$

where $E$ and $O$ are in general complex. Then the Fourier transform of $f(t)$

$$F(j\omega) = \int_{-\infty}^{\infty} [E(t) + O(t)]\,(\cos \omega t - j \sin \omega t)\,dt \qquad\qquad ...(6.66)$$

$$= \int_{-\infty}^{\infty} \left[ \{E(t) \cos \omega t + O(t) \cos \omega t\} - j\,\{E(t) \sin \omega t + O(t) \sin \omega t\} \right] dt \qquad\qquad ...(6.67)$$

Since $\displaystyle\int_{-\infty}^{\infty} O(t)\, \cos \omega t\, dt$ and $\displaystyle\int_{-\infty}^{\infty} E(t)\, \sin \omega t\, dt$ vanish, therefore,

$$F(j\omega) = 2 \int_{o}^{\infty} E(t) \cos \omega t\, dt - j2 \int_{o}^{\infty} O(t) \sin \omega t\, dt \qquad\qquad ...(6.68)$$

It follows that if the function is even its transform is even and if it is odd, its transform is odd. Further conclusion can be drawn.

<table>
<tr><td>Function</td><td>Transform</td></tr>
<tr><td>Real and even</td><td>Real and even</td></tr>
<tr><td>Real and odd</td><td>imaginary and odd</td></tr>
<tr><td>Imaginary and even</td><td>imaginary and even</td></tr>
<tr><td>complex and even</td><td>complex and even</td></tr>
<tr><td>complex and odd</td><td>complex and odd</td></tr>
<tr><td>Real even plus imaginary odd</td><td>Real</td></tr>
<tr><td>Real odd plus imaginary even</td><td>Imaginary</td></tr>
<tr><td>Even</td><td>Even</td></tr>
<tr><td>Odd</td><td>Odd</td></tr>
</table>

The Fourier transform associated with the even part $E(t)$ is known as cosine transform whereas that with the odd part is known as sin transform

$$\mathcal{F}[f(t)] = \mathcal{F}_c\, E(t) - j\, \mathcal{F}_s\, O(t) \qquad \ldots(6.69)$$

## 6.9 ADDITION THEOREM

If $f(t)$ and $g(t)$ have the Fourier transform $F(j\omega)$ and $G(j\omega)$ respectively, then $f(t) + g(t)$ has the Fourier transform $F(j\omega) + G(j\omega)$. This theorem reflects the suitability of the Fourier transform for dealing with linear systems. A corollary is that $af(t)$ has the transform $aF(j\omega)$ where '$a$' is a constant

### 6.9.1  Shift Theroem

If $f(t)$ has the Fourier transform $F(j\omega)$, then $f(t-a)$ has the Fourier transform $e^{-j\omega a}\, F(j\omega)$.

$$\mathcal{F}[f(t-a) = \int_{-\infty}^{\infty} f(t-a)\, e^{-j\omega t}\, dt \qquad \ldots.(6.70)$$

Let
$$t - a = \tau \quad \text{or} \quad dt = d\tau$$

$$\mathcal{F}[f(t-a)] = \int_{-\infty-a}^{\infty-a} f(\tau)\, e^{-j\omega(\tau+a)}\, d\tau$$

$$= \int_{-\infty}^{\infty} f(\tau)\, e^{-j\omega\tau}\, e^{-j\omega a}\, d\tau$$

$$= F(j\omega)\, e^{-j\omega a} \qquad \ldots(6.71)$$

Fourier transforms are sometimes used for circuit analysis as operational methods where by the analysis is simplified as compared to classical integro-differential equation. These transforms provide important link between the time domain and frequency domain behavior of the networks. The Fourier transform of the elements (passive) of the network is very simple. However, the Fourier transform of the excitation in the network is a bit cumbersome process even for simple functions like sin $t$, $u(t)$, $\delta(t)$ etc. It turns out that none of these three has strictly speaking a Fourier transform. Of course none of them is physically possible, as for a waveform sin $t$ would have to have been switched on an infinite time ago, a step $u(t)$ would have to be maintained steady for an infinite time and an impulse $\delta(t)$ would have to be infinitely large for an infinite short time. Periodic functions fail to have Fourier transform as their infinite integral is not absolutely convergent and hence it does not satisfy Dirichlet condition.

**Example 6.16:** Determine the Fourier transform of a pulse of duration $T_D$ and amplitude $V$ as shown in Fig. E6.16(a)

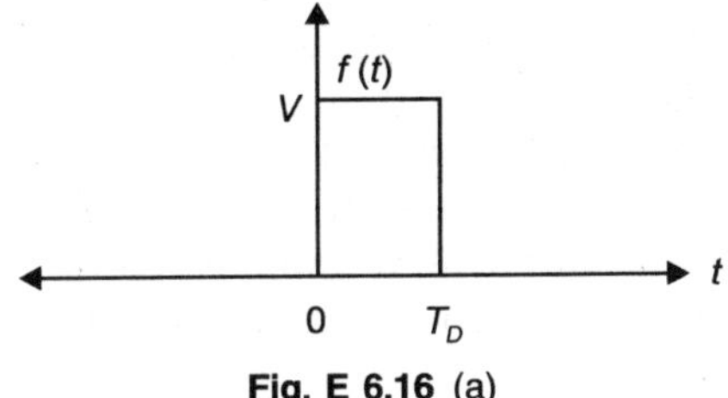


**Fig. E 6.16** (a)

**Solution:** The function $f(t)$ is defined as

$$f(t) = V \quad o < t < T_D$$
$$= o \quad t > T_D$$

The Fourier transform $F(j\omega)$ can be written as

$$F(j\omega) = \int_{-\infty}^{\infty} f(t)\, e^{-j\omega t}\, dt = \int_{o}^{T_D} v \cdot e^{-j\omega t}\, dt$$

$$= V\left[\frac{e^{-j\omega t}}{-j\omega}\right]_0^{T_D} = \frac{jV}{\omega}\left[e^{-j\omega T_D} - 1\right]$$

$$= -\frac{2V}{\omega}\left[\frac{e^{-j\omega T_D/2} - e^{j\omega T_D/2}}{2j}\right] e^{-j\omega T_D/2}$$

$$= \frac{2V}{\omega} \sin \frac{\omega T_D}{2} e^{-j\omega T_D/2}$$

$$= V T_D \frac{\sin \omega T_D/2}{\omega T_D/2} e^{-j\omega T_D/2}$$

The amplitude of $|F(j\omega)|$ is of the form $\dfrac{\sin x}{x}$ and is plotted in Fig. E 6.16(b)

The amplitude is zero when

$$\frac{\omega T_D}{2} = \pi,\ 2\,\pi,\ 3\,\pi \text{ etc.}$$

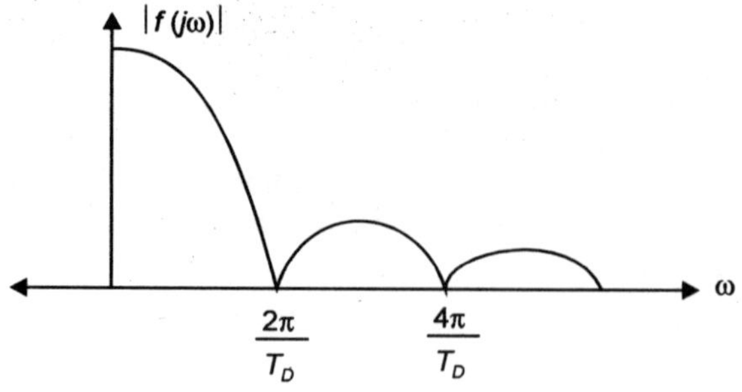

**Fig. E6.16** (b)

It is seen that the most important part of the relative frequency distribution function $F(j\omega)$ of a rectangular pulse lies in the range $0 < \omega < 2\pi/T_D$. The range widens as the pulse width $T_D$ decreases. This shows that the frequency band width required to reproduce a rectangular pulse increase inversely with the pulse duration.

Suppose
$$F(\omega) = p_{\omega_0} \qquad\qquad 0 < \omega < \omega_0$$
$$= 0 \quad \text{Elsewhere } \omega > \omega_0$$

as shown in Fig. 6.8

$$f(t) = \frac{1}{2\pi} \int_0^{\omega_0} F(\omega)\, e^{j\omega t}\, d\omega \qquad\qquad \text{...(6.72)}$$

$$= \frac{p_{\omega_0}}{2\pi} \left[\frac{e^{j\omega t}}{jt}\right]_0^{\omega_0} = \frac{1}{2\pi jt}\left[e^{j\omega_0 t} - 1\right]$$

$$= \frac{p_{\omega_0}}{\pi t} \left[\frac{e^{j\omega_0 t/2} - e^{-j\omega_0 t/2}}{2j}\right] e^{j\omega_0 t/2}$$

$$= \frac{p_{\omega_0}}{t\pi} \sin \omega_o t/2 \cdot e^{j\omega_0 t/2} \qquad\qquad \text{...(6.73)}$$

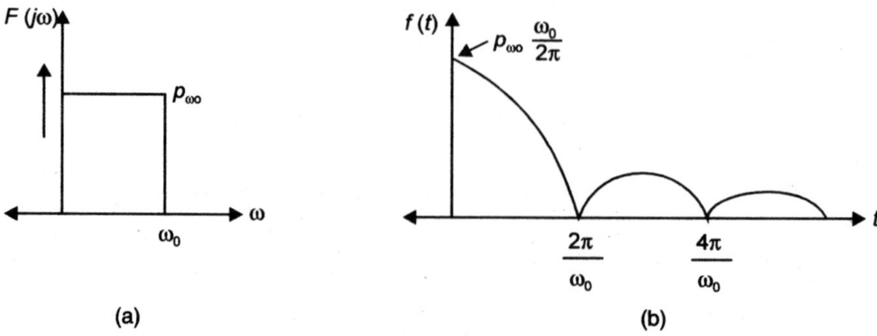

(a)                                    (b)

**Fig. 6.8**   (a) Pulse width Fourier transform and (b) Its time function.

The magnitude of the above time function is $\left| \dfrac{p_{\omega_o} \ \omega_o}{2\pi} \cdot \dfrac{\sin \omega_o t/2}{\omega_o t/2} \right|$

This shows the property of Duality for Fourier transform. If $f(t)$ is sin $c$ its transformed is a pulse whereas if Fourier transform is a pulse its inverse is a sin $c$ function in time domain.

The Fig. 6.8 shows the $f(t)$ and $p_{\omega_o}$

## 6.10  TRANSFORM OF AN IMPULSE FUNCTION δ (*T*)

Consider a pulse of duration 2 δ and height ½ δ as shown in Fig. 6.9 such that the area of the pulse is unity. Now as δ → $o$ the width of the pulse goes on decreasing whereas the height goes on increasing and in the limit the width tends to zero and the height infinity. A unit impulse is defined as

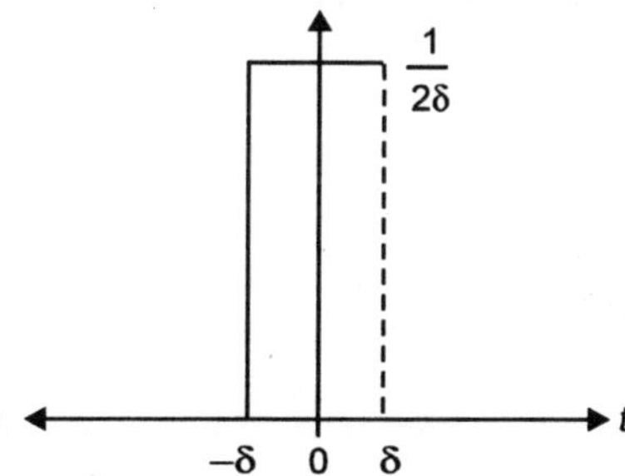

**Fig 6.9**  Impulse of duration 2 δ and height ½ δ.

$$\int_{-\infty}^{\infty} \delta(t)\, dt \ = 1$$

or $\qquad\qquad\qquad \delta(t) = 0 \quad t \neq o$ $\qquad\qquad\qquad\qquad$ ...(6.74)

It follows from the fact that since the unit impulse is the time derivative of the unit step, it must be zero for all values of $t$ except at $t = 0$ where it must be infinite. An approximation to an impulse is the charging of a pure capacitor by a d.c. battery.

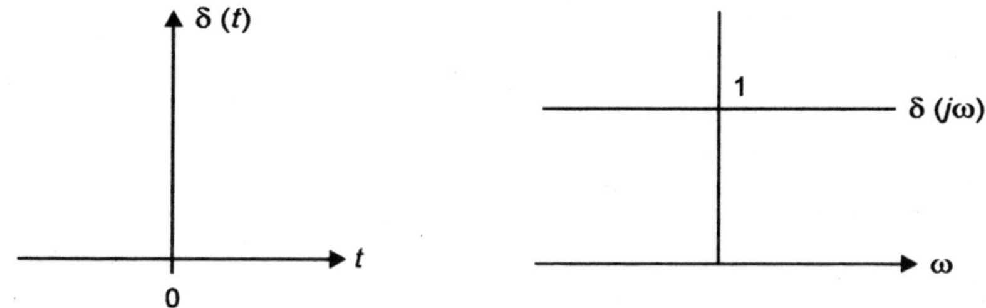

**Fig. 6.10**  Transform pair of impulse δ(*t*).

The Fourier transform of an impulse function is, therefore,

$$\delta\,(j\omega) = \int_{-\infty}^{\infty} \delta(t)\, e^{-j\omega t}\, dt$$

$\qquad\qquad\qquad\qquad\qquad\qquad\qquad\qquad\qquad\qquad\qquad\qquad$ ...(6.75)

Since $\delta(t) = 0$ except at $t = 0$, therefore

$$\delta(j\omega) = \int_{-\infty}^{\infty} \delta(t)\, dt = 1 \qquad\qquad ...(6.76)$$

as at $\qquad\qquad t = 0,\ e^{-j\omega t} = 1$

Therefore $\qquad\qquad \delta(j\omega) = 1$

It can be seen that the magnitude of $\delta(j\omega)$ is unity and hence a constant. Furthermore in the time domain we have a very narrow function an impulse $\delta(t)$ which is zero every where except at $t = 0$, in the frequency domain its Fourier transform is spread uniformly over all frequencies. This is known as reciprocal spreading i.e. if the function in time domain is compressed its frequency spectrum will be expanded and vice versa. The transform pair for an impulse function $\delta(t)$ is shown in Fig. 6.10.

## 6.11   FOURIER TRANSFORM OF A TRAIN OF IMPULSES

$$f(t) = \delta(t - nT) \quad n = 1, 2, \ldots\ldots\ n \qquad\qquad ...(6.77)$$

Consider a train of impulses of unit magnitude and separated by a time $T$. Mathematically,

$$f(t) = \sum_{n=-\infty}^{\infty} \delta(t - nT) \qquad\qquad ...(6.78)$$

Using Addition and Shifting theorem the Fourier transform becomes

$$F(jw) = \frac{1}{T} \sum_{-\infty}^{\infty} e^{jn\omega_o T} \qquad\qquad ...(6.79)$$

**Example 6.17:** Determine the Fourier transform of the exponential voltage function given by

$$v = V\, e^{-t/T} \quad t \geq o$$
$$= o \quad t < o$$

**Solution:** Fourier transform is given as

$$v(j\omega) = \int_{-\infty}^{\infty} f(t)\, e^{-j\omega t}\, dt$$

$$= \int_{-\infty}^{o} 0 \cdot e^{-j\omega t}\, dt + \int_{o}^{\infty} V \cdot e^{-t/T}\, e^{-j\omega t}\, dt$$

$$= \int_{o}^{\infty} V \cdot e^{-\left(\frac{t}{T} + j\omega\right)t}\, dt$$

$$= -V \left[ \frac{e^{-\left(\frac{1}{T} + j\omega\right)t}}{\left(\frac{1}{T} + j\omega\right)} \right]_{0}^{\infty}$$

$$= \frac{VT}{1+j\omega T}$$

$$= \frac{VT(1-j\omega T)}{1+\omega^2 T^2}$$

The magnitude is given by $\dfrac{VT}{(1+\omega^2 T^2)^{1/2}}$ and phase $-\tan^{-1}(\omega T)$ Fig. E6.17 shows the function (a) $f(t)$ (b) its amplitude Fourier transform and (c) phase angle of the transform.

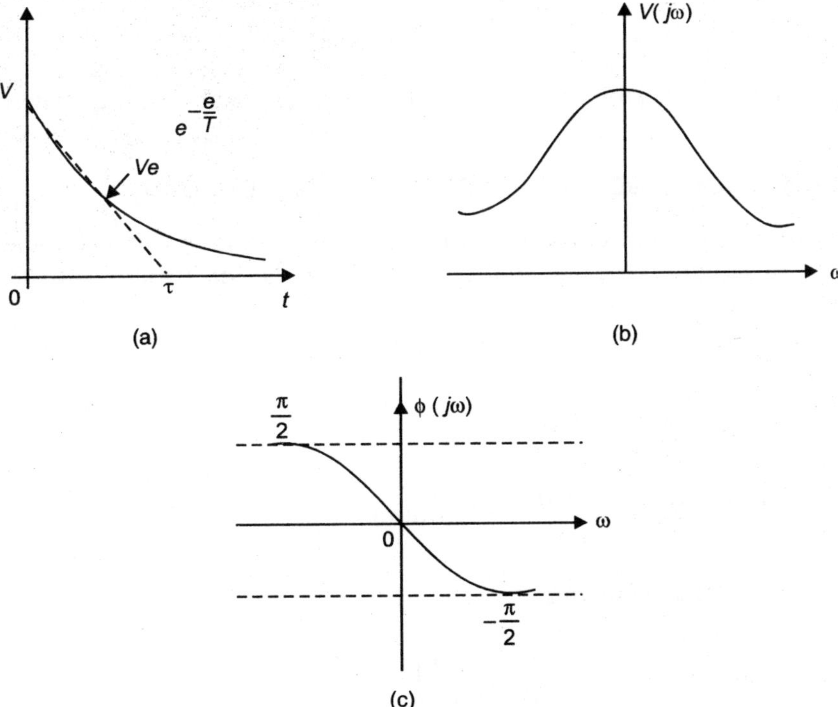


**Fig. E6.17**   (a) Function $f(t)$ (b) Amplitude Fourier of $f(t)$ (c) Phase of Fourier of $f(t)$.

**Example 6.18:**  Determine the Fourier transform of the function

$$f(t) = A\, e^{-a/t/} \quad a > o$$

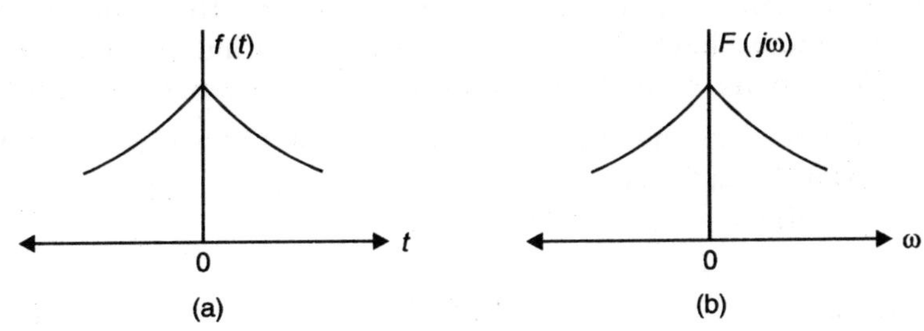


**Fig. E6.18**

As in the function, time is to be taken as its modulus therefore, the given function is as shown in Fig. E6.18 (a)

$$F(j\omega) = \int_{-\infty}^{\infty} f(t)\, e^{-j\omega t}\, dt$$

$$= \int_{-\infty}^{\infty} A\, e^{-a|t|}\, e^{-j\omega t}\, dt$$

$$= \int_{-\infty}^{0} A\, e^{at} \cdot e^{-j\omega t}\, dt + \int_{0}^{\infty} A\, e^{-at} \cdot e^{-j\omega t}\, dt$$

$$= \frac{A}{a - j\omega} + \frac{A}{a + j\omega} = \frac{2\,a\,A}{a^2 + \omega^2}$$

It may be noted that $F(j\omega)$ is also an even function and real for all $w$. The plot is shown in Fig. E 6.18 (b).

## 6.12 DIFFERENTIATION AND INTEGRATION IN TIME DOMAIN

Normally we write integro-differential equation for a given system and then make use of the Fourier operator to simplify the calculations and have better insight into the system in frequency domain.

If $\qquad F(j\omega) = \mathcal{F}[f(t)] \quad$ then

$$\mathcal{F}\left[\frac{d}{dt}\, f(t)\right] = j\omega\, F(j\omega) \qquad\qquad ...(6.80)$$

and $\qquad \mathcal{F}\left[\int f(t)dt\right] = \frac{F(j\omega)}{j\omega} - \pi F(o)\delta(\omega) \qquad\qquad ...(6.81)$

### 6.12.1 Time and Frequency Scaling

If $\;F(j\omega) = \mathcal{F}\,[f(t)]\;$ then $\;\mathcal{F}\,[f(at)] = \dfrac{1}{|a|}\, F\left(\dfrac{j\omega}{a}\right) \qquad\qquad ...(6.82)$

where $a$ is a real constant. This shows that if a function is compressed in the time domain by a factor $a$ its frequency spectrum gets expanded by the same factor. This can be understood logically also. If we make the rate of change of a function faster, it requires higher frequency components to represent it, similarly if we expand the time duration of a signal we make changes occur more slowly. Therefore, in this situation we require lower frequency components. This is known as reciprocal spreading as mentioned earlier.

**Example 6.19:** Determine the Fourier transform of the signum function.

The signum function is defined as

$$f(t) = -1 \quad -\infty < t < o$$

$$= 1 \quad o < t < \infty$$

Since the Dirichlet condition is not satisfied with this function, direct evaluation of Fourier transform of this function is not possible. Hence, the transform will be obtained by first expressing the function as a limiting case of some other function. Let us take the limiting function as $e^{-a|t|}$ with Lt $a \to o$. Therefore we express $sgn(t)$ as

$$sgn\ (t) = \underset{a \to o}{Lt}\ e^{-a|t|}\ sgn\,(t)$$

Hence Fourier transform is written as $\displaystyle\int_{-\infty}^{\infty} e^{-a|t|}\ sgn\,(t)\ e^{-j\omega t}\ dt$

$$F\,(j\omega) = \underset{a \to 0}{Lt}\left[\int_{-\infty}^{o} -e^{(a-j\omega)t}\ dt + \int_{o}^{\infty} e^{-(a+j\omega)t}\ dt\right]$$

$$= \underset{a \to 0}{Lt}\left[\frac{1}{a+j\omega} - \frac{1}{a-j\omega}\right] = \frac{2}{j\omega}$$

Fig E6.19(*a*), (*b*), (*c*) respectively show the function $f(t)$ the amplitude and phase spectra.

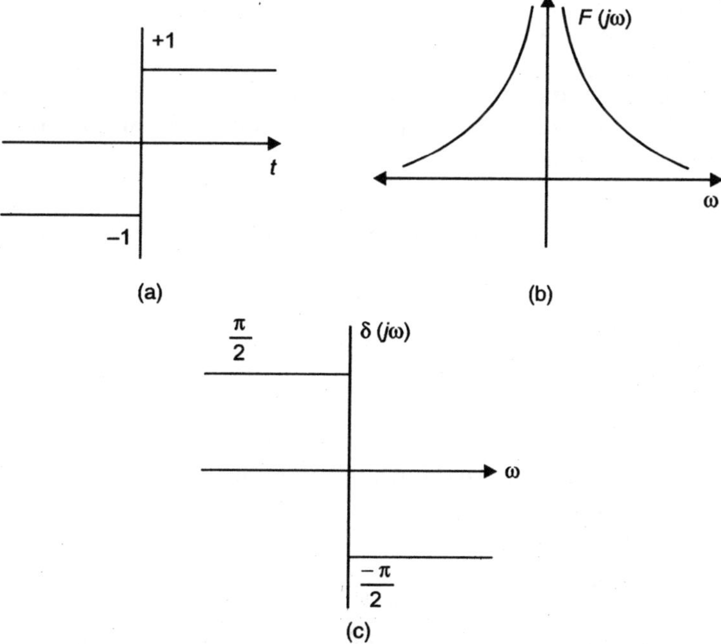

Fig. E6.19  (a) Function $f\,(t)$, (b) $[F\,(\,j\omega)]$ and (c) Phase spectra.

**Example 6.20:** Determine Fourier transform of the (a) Constant (b) unit step function as shown in Fig. E6.20 (a) and (b) respectively.

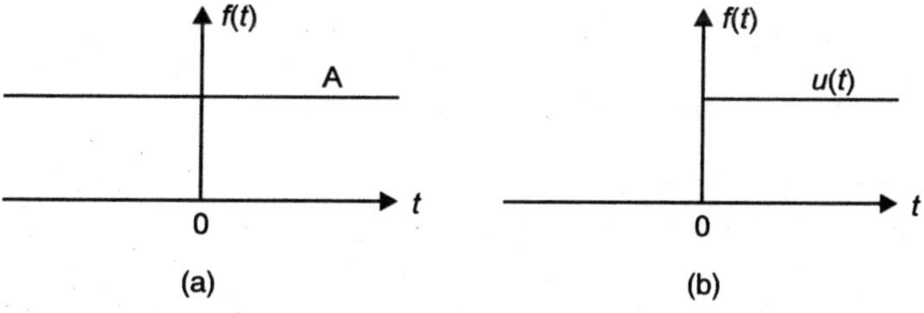

Fig. E6.20

**Solution:** (*a*) Since for this function also Dirichlet Condition is not satisfied, we again take the limiting function $e^{-a|t|}$ as $a \to o$

Therefore,

$$\mathcal{F}\,[A] = \int_{-\infty}^{\infty} A \cdot e^{-a|t|}\, e^{-j\omega_o t}\, dt$$

$$= \underset{a \to o}{Lt}\ \frac{2a\,A}{a^2 + \omega^2}$$

The value is zero for all '*a*' except at $\omega = o$ as for $\omega = o$ the function becomes indeterminate. Using L' 'Hospitals' rule and differentiate both numerator and denominator and then let $a \to o$. We have

$$\underset{a \to o}{Lt}\ \frac{2A}{2a} = \infty$$

which means the function assumes infinite value at $\omega = o$ i.e. it is in the form of an impulse. The value of the impulse function is found by integration of $F(j\omega)$ for all $\omega$ to give the area contained by the function. Thus

$$\int_{-\infty}^{\infty} \frac{2\,a\,A}{\omega^2 + a^2}\, d\omega \quad \text{Let}\quad \omega = a\,\tan\,\theta \quad \text{or}\quad d\omega = a\,\sec^2\,\theta\,d\theta$$

$$\int_{-\pi/2}^{\pi/2} \frac{2a\,A}{a^2\,\sec^2\,\theta} \cdot a\,\sec^2\,\theta\,d\theta = \int_{-\pi/2}^{\pi/2} 2\,A\,d\theta$$

$$= 2\pi\,A$$

Thus we have determined the strength of the impulse to be $2\pi$. Therefore, Fourier transform of a constant $A$ is

$$\mathcal{F}[A] = 2\pi A\delta(\omega)$$

This is shown in Fig E 6.20 (c).

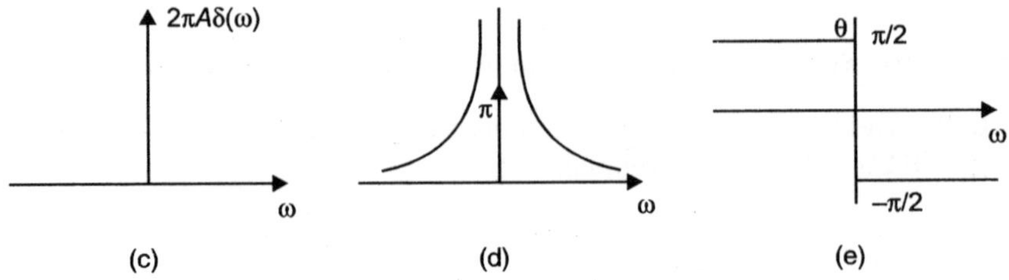


**Fig. E6.20**  (c) Transform of Fig. (a)

(*b*) Fourier transform of $u(t)$

Function $u(t)$ can be expressed as a function of *sgn* as

$$sgn\,(t) = 2\,u(t) - 1 \quad \text{or}\quad u(t) = \frac{1}{2}\,(1 + sgn\,(t))$$

Therefore, Fourier transform of  $u(t) = \mathcal{F}\left[\dfrac{1}{2}(1 + sgn(t))\right]$

$$\mathcal{F}\left[\dfrac{1}{2}\right] = \pi\,\delta\,(\omega) \quad \text{and} \quad \mathcal{F}\left[\dfrac{1}{2}\,sgn\,(t)\right] = \dfrac{1}{2}\left(\dfrac{2}{j\omega}\right) = \dfrac{1}{j\omega}$$

Hence $\qquad\qquad \mathcal{F}\,[u(t)] = \pi\,\delta\,(\omega) + \dfrac{1}{j\omega}$

The magnitude is  $\pi\,\delta\,(\omega) + \dfrac{1}{\omega}$  and the phase angle  $-\tan^{-1}\left(\dfrac{1/\omega}{o}\right)$.

The amplitude spectra and phase spectra are shown in Fig. E 6.20 (d) and (e) resectively.

**Example 6.21:** Determine the Fourier transform of the triangular wave form shown in Fig. E6.21.

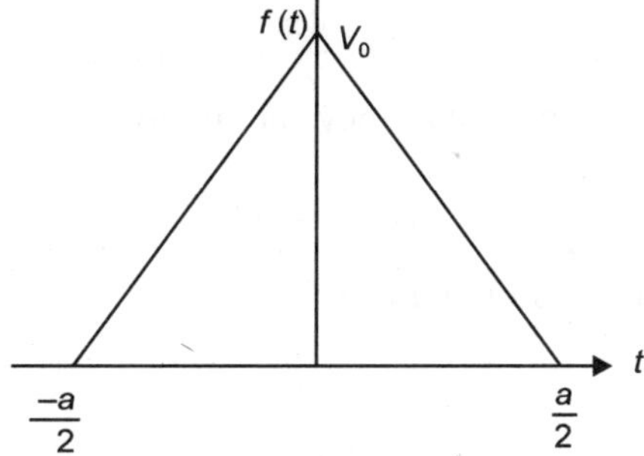


**Fig. E6.21**   Triangular waveform.

The equation describing the triangular shape is  $\dfrac{|t|}{a/2} + \dfrac{f(t)}{V_o} = 1$

Therefore  $f(t) = \left(1 - \dfrac{2|t|}{a}\right)V_0$

Since $f(t)$ satisfies the Dirichlet condition we do not require limiting function for Fourier transform of the given function.

The Fourier transform is therefore

$$F\,(j\omega) = V_0 \int\limits_{-a/2}^{a/2}\left(1 - \dfrac{2|t|}{a}\right)e^{-j\omega t}\,dt$$

$$= V_0 \int\limits_{-a/2}^{0}\left(1 + \dfrac{2t}{a}\right)e^{-j\omega t}\,dt + V_0 \int\limits_{0}^{a/2}\left(1 - \dfrac{2t}{a}\right)e^{-j\omega t}\,dt$$

Taking $e^{-j\omega t}$ as second function and using the formula for integration by part we have

$$F\,(j\omega) = \left[V_0\left(1 + \dfrac{2t}{a}\right)\dfrac{e^{-j\omega t}}{-j\omega}\right]_{-a/2}^{0} - \int\dfrac{2\,V_0}{a}\dfrac{e^{-j\omega t}}{-j\omega}\,dt$$

$$+ \left[V_0\left(1 - \dfrac{2t}{a}\right)\dfrac{e^{-j\omega t}}{-j\omega}\right]_{0}^{a/2} + \int\limits_{0}^{a/2}\dfrac{2V_0}{a}\dfrac{e^{-j\omega t}}{-j\omega}\,dt$$

$$= \frac{2V_0}{a} \left[ \frac{2}{\omega^2} - \frac{e^{-j\omega a/2}}{\omega^2} - \frac{e^{-j\omega a/2}}{\omega^2} \right]$$

$$= \frac{4V_0}{a\,\omega^2} (1 - \cos \omega\, a/2)$$

$$= \frac{4V_0}{a\,\omega^2} \cdot \frac{2 \sin^2 \omega\, a/4}{(\omega\, a/4)^2} \cdot \left( \frac{\omega a}{4} \right)^2$$

$$= \frac{8V_0}{\omega^2 a} \cdot \frac{\omega^2 a^2}{16} \frac{\sin^2 \omega a/4}{(\omega a/4)^2}$$

$$= \frac{V_0\, a}{2} \frac{\sin^2 \omega a/4}{(\omega a/4)^2}$$

The first zero of $F(j\omega)$ occurs when $\omega = \dfrac{4\pi}{a}$. Equating this to $\omega_1$ for the low pass filter (which has constant $F(j\omega)$ upto the frequency $\omega_1$ ideally) we have

$$\omega_1 = \frac{4\pi}{a} \quad \text{or} \quad \omega_1\, a = 4\pi \qquad\qquad ...(6.83)$$

The amplitude spectrum is shown here

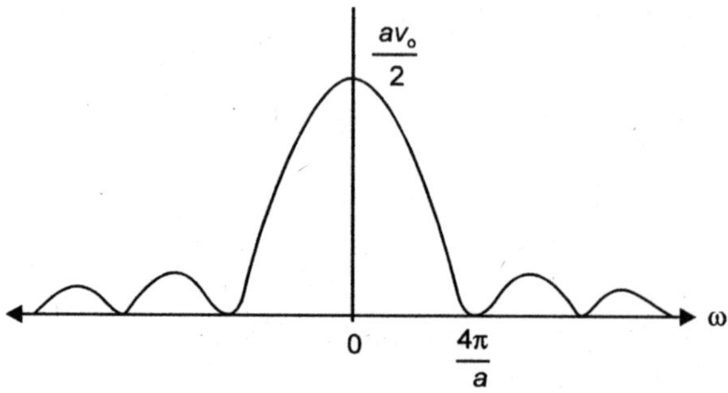


**Fig. E6.21(a)**

Similarly it has been seen in Example 6.16 for the Fourier transform of a pulse of duration $T_D$ that the product of $\omega$, and the duration of the pulse is constant. This means that

Bandwidth required × duration of pulse = constant

i.e. the bandwidth required for transmission is inversely proportional to the duration of the pulse being transmitted.

## 6.13  DUALITY

We have seen earlier that the Fourier transform of a rectangular pulse is $\dfrac{\sin x}{x}$ function where as that of $\dfrac{\sin x}{x}$ function it is a rectangular pulse. Similarly, the Fourier transform of a constant is an impulse whereas that of an impulse function it is a constant. If we look at the Fourier transform pair equations we find that they are almost symmetrical in the variable $\omega$ and $t$ except for the factor $2\pi$. This is the reason for duality that we have just observed.

An important advantage of this property is that we need not evaluate the integral in one of the equations if we have already obtained the same function by evaluating the integral in the other equation. The following relationship can be used for describing this property of duality.

If $F(\omega) = \mathcal{F}[f(t)]$ then $\mathcal{F}[F(t)] = 2\,\pi\,f\,(-\omega)$. A few examples have been shown here

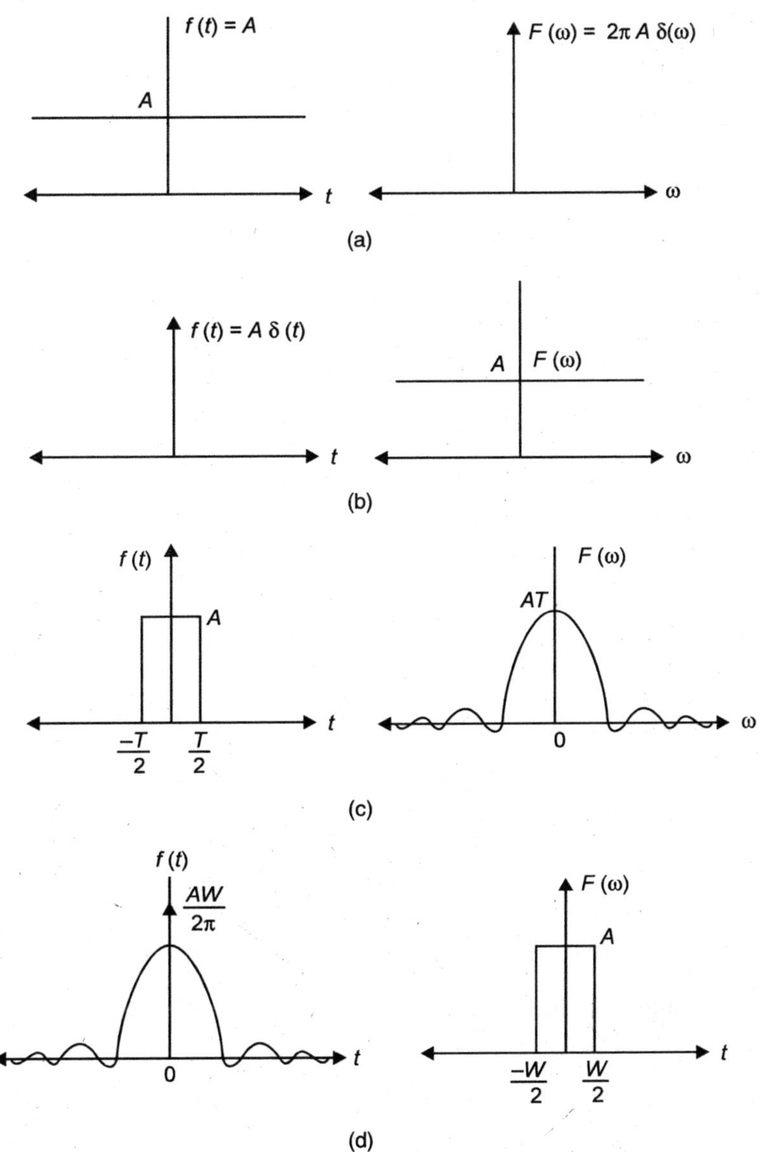

**Fig. 6.11**   Demonstration of duality.

(*a*) $f(t) = A$, (*b*) $f(t) = \delta\,(t)$, (*c*) $f(t)$ a rectangular pulse (*d*) $f(t)$ a sinc function

## 6.14 FOURIER TRANSFORM OF A SINUSOID

To obtain Fourier transform of a sinusoid we first express it in terms of exponential function as follows:

$$\sin \omega_0 t = \frac{e^{j\omega_0 t} - e^{-j\omega_0 t}}{2j}$$

and

$$\cos \omega_0 t = \frac{e^{j\omega_0 t} + e^{-j\omega_0 t}}{2}$$

We have already obtained Fourier transform for $e^{j\omega_0 t}$ and hence the Fourier transform for sinusoid can be obtained

$$F(j\omega) = \int_{-\infty}^{\infty} e^{j\omega_o t} \, e^{-j\omega t} \, dt$$

$$= \int_{-\infty}^{\infty} e^{j(\omega_o - \omega)t} \, dt = \int_{-\infty}^{\infty} e^{-j(\omega - \omega_o)t} \, dt$$

Since now $f(t)$ could be considered as a unit step and there has been a shift in frequency by $\omega_o$

$$F(j\omega) = 2\pi\,\delta\,(\omega - \omega_0)$$

Hence $\qquad \mathcal{F}\,[\sin\,\omega_0 t] = j\,\pi\,[\delta\,(\omega + \omega_0] - \delta\,(\omega - \omega_0)]$

and $\qquad \mathcal{F}\,[\cos\,\omega_0 t] = \pi\,[\delta\,(\omega + \omega_0) + \delta\,(\omega - \omega_0)]$

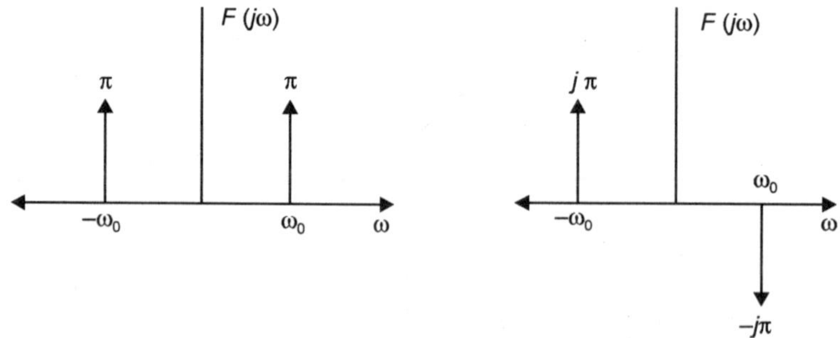


**Fig. 6.12** Fourier transform representation of cos $\omega_0 t$ and sin $\omega_0 t$.

The Fourier table gives some of the important Fourier transform pairs.

The following table 6.2 gives some important properties of Fourier transform

Table 6.2 Some Properties of Fourier transform

$$F(j\omega) = \int_{-\infty}^{\infty} f(t)\, e^{-j\omega t} \, dt$$

$$f(t) = \frac{1}{2\pi} \int_{-\infty}^{\infty} F(j\omega)\, e^{j\omega t} \, d\omega$$

### Table 6.1  Fourier Transform Pairs

| $f(t)$ | $F(j\omega)$ |
|---|---|
| $A$ | $2\pi A\,\delta(\omega)$ |
| $A\,\delta(t - t_0)$ | $A\,e^{-j\omega t_0}$ |
| $A\,u(t)$ | $A\left[\pi\delta(\omega) + \dfrac{1}{j\omega}\right]$ |
| $A\,e^{-at}\,u(t)$ | $\dfrac{A}{a + j\omega}\quad a > 0$ |
| $A\,e^{-a\lvert t\rvert}$ | $\dfrac{2aA}{a^2 + \omega^2}\quad a > 0$ |
| $A\,e^{-at}\,\sin\,\omega_0 t\,u(t)$ | $\dfrac{A\omega_0}{(a + j\omega)^2 + \omega_0^2}\quad a > 0$ |
| $A\,e^{-at}\,\cos\,\omega_0 t\,u(t)$ | $\dfrac{A(a + j\omega)}{(a + j\omega)^2 + \omega_0^2}\quad a > 0$ |
| $A\,t\,e^{-at}\,u(t)$ | $\dfrac{A}{(a + j\omega)^2}$ |
| $A\,e^{j\omega_0 t}$ | $2\pi A\,\delta(\omega - \omega_0)$ |
| $A\,\cos\,\omega_0 t$ | $A\,\pi[\delta(\omega + \omega_0) + \delta(\omega - \omega_0)]$ |
| $A\,\sin\,\omega_0 t$ | $jA\,\pi[\delta(\omega + \omega_0) - \delta(\omega - \omega_0)]$ |
| $A\,p_T(t)$ | $AT\,\dfrac{\sin\,\omega T/2}{\omega T/2}$ |

($i$) Linearity $\mathcal{F}[a\,f_1(t) + b\,f_2(t)] = a\,F_1(j\omega) + b\,F_2(j\omega)$
Simplification if

($a$) $f(t)$ is even $F(j\omega) = 2\displaystyle\int_0^\infty f(t)\,\cos\,\omega t\,dt$

($b$) $f(t)$ is odd $F(j\omega) = 2j\displaystyle\int_0^\infty f(t)\,\sin\,\omega t\,dt$

($ii$) Negative $t$, $\mathcal{F}[f(-t)] = F^*(j\omega)$

($iii$) Differentiation $\mathcal{F}\left[\dfrac{d^n}{dt^n}\,f(t)\right] = (j\omega)^n\,F(j\omega)$

($iv$) Integration $\mathcal{F}\left[\displaystyle\int_{-\infty}^{t} f(t)\,dt\right] = \dfrac{1}{j\omega}\,F(j\omega) + \pi F(o)\delta(\omega)$

($v$) Time shifting $\mathcal{F}[f(t - a)] = F(j\omega)\,e^{-j\omega a}$

($vi$) Modulation $\mathcal{F}\left[f(t)\,e^{j\omega_0 t}\right] = F[j(\omega - \omega_0)]$

$$\mathcal{F}[f(t) \cos \omega_0 t] = \frac{1}{2}[F\{j(\omega + \omega_0)\} + F\{j(\omega - \omega_0)\}]$$

$$\mathcal{F}[f(t) \sin \omega_0 t] = \frac{j}{2}[F\{j(\omega - \omega_0)\} - F\{j(\omega + \omega_0)\}]$$

(*vii*) Scaling

(*a*) Time $\mathcal{F}[f(at)] = \frac{1}{|a|} F\left(\frac{j\omega}{a}\right)$

(*b*) Magnitude $\mathcal{F}[a\, f(t)] = a\, F(j\omega)$

We now solve a few examples to illustrate further the application of Fourier transform.

**Example 6.22:** Determine the impulse response of the linear system whose transfer function is given as

$$H\,(j\omega) = \frac{3 + 2j\omega}{(j\omega)^2 + 6j\omega + 8}$$

**Solution:** The partial fraction expansion gives

$$H(j\omega) = \frac{2.5}{(4 + j\omega)} - \frac{0.5}{2 + j\omega}$$

Taking the inverse Fourier transform we have
$$h(t) = (2.5\ e^{-4t} - 0.5\ e^{-2t})\ u(t)$$

**Example 6.23:** Suppose the input to the given linear system is $v_1 = 3\ e^{-t}\ u(t)$. Determine the response of the system.

**Solution:**
$$y(t) = \mathcal{F}^{-1}[H(j\omega)\ v_1\,(j\omega)]$$

$$= \mathcal{F}^{-1}\left[\frac{3 + 2j\omega}{(4 + j\omega)(2 + j\omega)} \cdot \frac{3}{1 + j\omega}\right]$$

The partial fraction expansion is given as

$$\frac{1}{1 + j\omega} - \frac{2.5}{4 + j\omega} + \frac{1.5}{2 + j\omega}$$

Therefore taking the inverse Fourier transform
$$y(t) = e^{-t} - 2.5\ e^{-4t} + 1.5\ e^{-2t}$$

**Example 6.24:** A linear system is described by the differential equation written in $p$ operator $\left(p = \dfrac{d}{dt}\right)$ as follows:

$$(p^2 + 5p + 6)\ y(t) = v_1(t)$$

whee $y(t)$ is the output and $V_1(t)$ is the input. Determine the impulse response.

**Solution:** The transfer function is given as

$$\mathcal{F}\left[\frac{y(t)}{v_1(t)}\right] = H\,(j\omega) = \frac{1}{(j\omega)^2 + 5j\omega + 6}$$

$$= \frac{1}{(2+j\omega)(3+j\omega)}$$

Finding out partial fractions we have

$$H(j\omega) = \frac{1}{2+j\omega} - \frac{1}{3+j\omega}$$

Therefore,

$$h(t) = \left(e^{-2t} - e^{-3t}\right) u(t)$$

**Example 6.25:** Determine the output voltage across the capacitor if the excitation is a current source $i(t) = e^{-t}u(t)$

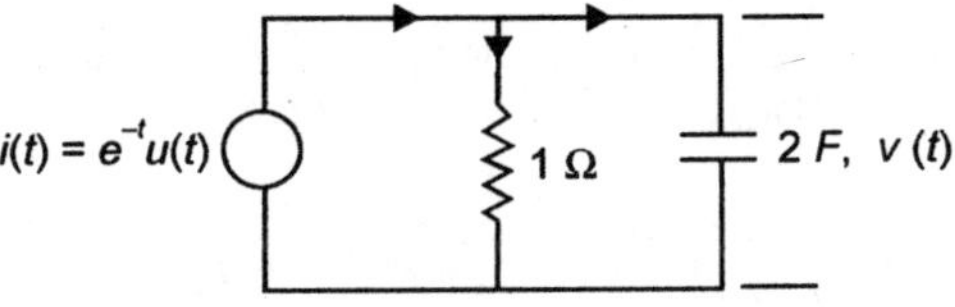


**Fig. E6.25**

$$i(t) = I_R(t) + i_c(t)$$

$$= v(t) + 2\frac{dv}{dt}$$

$$= v(t)\,[1 + 2p]$$

Therefore,

$$v(t) = \frac{i(t)}{1 + 2p}$$

$$F[v(j\omega)] = \frac{F[i(j\omega)]}{1 + 2j\omega}$$

$$F[v(j\omega)] = \frac{1}{1 + j\omega} \cdot \frac{1/2}{\frac{1}{2} + j\omega}$$

Finding out partial fraction we have

$$v(j\omega) = \frac{2}{\frac{1}{2} + j\omega} - \frac{2}{1 + j\omega}$$

Therefore,

$$v(t) = \mathcal{F}^{-1}[v(j\omega)]$$

$$= 2\,e^{-\frac{1}{2}t} - 2\,e^{-t} \ \textbf{Ans.}$$

**Example 6.26:** Determine the function $f(t)$ if the Fourier transform of the function is

$$F(j\omega) = A e^{j\,\pi/2} \qquad -\omega_o < \omega < o$$
$$= A e^{-j\,\pi/2} \qquad o < \omega < \omega_o$$

**Solution:**

$$f(t) = \frac{A}{2\pi} \int_{-\omega_o}^{0} e^{j\pi/2}\, e^{j\omega t}\, d\omega + \frac{A}{2\pi} \int_{o}^{\omega_o} e^{-j\pi/2}\, e^{j\omega t}\, d\omega$$

$$= \frac{A}{2\pi} \left[ e^{j\pi/2} \cdot \frac{e^{j\omega t}}{jt} \right]_{-\omega_o}^{o} + \frac{A}{2\pi} e^{-j\pi/2} \left[ \frac{e^{j\omega t}}{jt} \right]_{o}^{\omega_o}$$

$$= \frac{A\,j \cdot 1}{2\pi jt} \left( 1 - e^{-j\omega_o t} \right) - \frac{A\,j}{2\pi jt} \left( e^{j\omega_o t} - 1 \right)$$

$$= \frac{A}{2\pi t} \left[ 1 - e^{-j\omega_o t} - e^{j\omega_o t} + 1 \right]$$

$$= \frac{A}{2\pi t} \left[ 2 - \left( e^{j\omega_o t} + e^{-j\omega_o t} \right) \right]$$

$$= \frac{A}{2\pi t} \left[ 2 - 2\cos \omega_o t \right]$$

$$= \frac{A}{\pi t} \cdot \left( 1 - \cos \omega_0 t \right)$$

$$= \frac{A}{\pi t} \omega_0 \cdot \frac{2 \sin^2 \omega_0 t/2}{\omega_0/2}$$

$$= \frac{A\omega_0}{\pi} \cdot \frac{\sin^2 \omega_0 t/2}{(\omega_0 t)/2} \qquad \textbf{Ans.}$$

**Example 6.27:** Determine the function $f(t)$ whose Fourier transform is as shown in Fig. E6.27($a$).

From table 6.2 we see that

$$\mathcal{F}(f(t) \cos \omega_o t] = \tfrac{1}{2}( F[j(\omega + \omega_o) + F[j(\omega - \omega_o)]$$

i.e. we need find $f(t)$ of Fig. E6.27($b$) and finally multiply it with $\cos \omega_o t$.

The function is therefore,

$$F(j\omega) = 2V_o + \frac{2V_o}{2a}\omega \qquad -2a < \omega < 0$$

$$= 2V_o - \frac{2V_o}{2a}w \qquad 0 < \omega < 2a$$

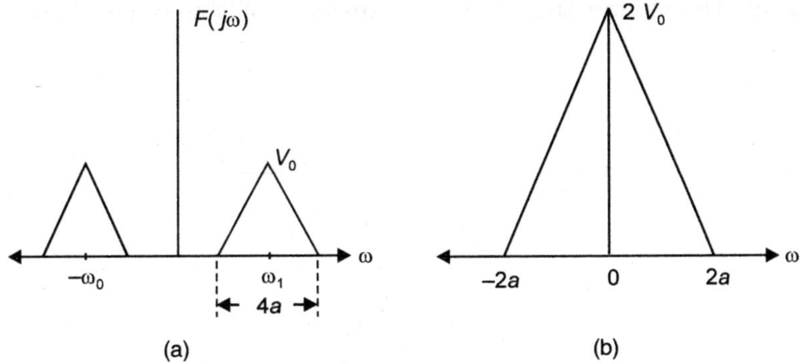



**Fig. E6.27**

$$f(t) = \frac{2}{2\pi}\left[\int_{-2a}^{0}\left(V_o + \frac{V_o}{2a}\omega\right)e^{j\omega t}d\omega + \int_{0}^{2a}\left(V_0 - \frac{V_o}{2a}\omega\right)e^{j\omega t}d\omega\right]$$

$$= \frac{V_o}{\pi}\left[\left\{\left(1+\frac{\omega}{2a}\right)\frac{e^{j\omega t}}{jt} - \int\frac{1}{2a}\cdot\frac{e^{j\omega t}}{jt}d\omega\right\}_{-2a}^{0} + \left\{\left(1-\frac{\omega}{2a}\right)\frac{e^{j\omega t}}{jt} + \int\frac{1}{2a}\frac{e^{j\omega t}}{jt}d\omega\right\}_{0}^{2a}\right]$$

$$= \frac{V_o}{\pi}\left[\left\{\frac{1}{jt}+\frac{1}{2at^2}-\frac{1}{2at^2}e^{-j2at}\right\} + \left\{-\frac{1}{2at^2}e^{j2at}-\frac{1}{jt}+\frac{1}{2at^2}\right\}\right]$$

$$= \frac{V_o}{\pi}\left[\frac{1}{at^2}-\frac{1}{2at^2}(e^{j2at}+e^{-j2at})\right]$$

$$= \frac{V_o}{\pi at^2}(1-\cos 2at)$$

$$= \frac{2V_o a}{\pi}\left\{\frac{\sin^2 at}{(at)^2}\right\}$$

Therefore the overall function is

$$f(t) = \frac{2V_o a}{\pi}\cos\omega_o t\left\{\frac{\sin^2 at}{(at)^2}\right\} \quad \textbf{Ans.}$$

The result could easily be obtained through duality. Since as in example 6.21 if the function $f(t)$ is triangular the Fourier transform is $\dfrac{\sin^2 x}{(x)^2}$ form. Now if the Fourier transform is a triangular function the time domain function $f(t)$ would be of $\dfrac{\sin^2 x}{x^2}$ form as obtained here.

**Example 6.28:** Determine the Fourier transform of the function shown here.

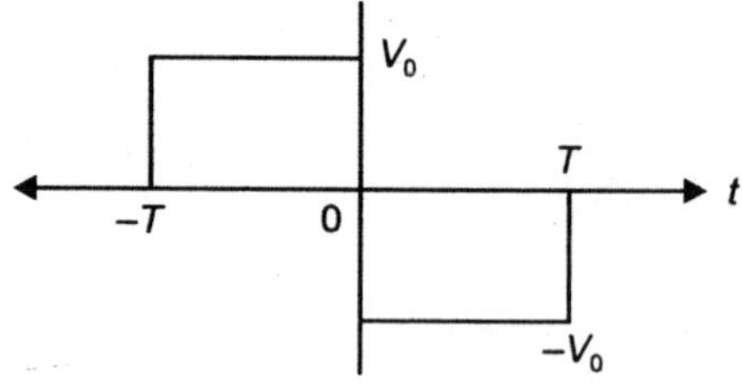

**Fig. E6.28**

The function is described as

$$f(t) = V_o, \quad -T < t < 0$$
$$= -V_o, \quad 0 < t < T$$

$$F(j\omega) = \int_{-T}^{0} V_o \, e^{-j\omega t} \, dt - \int_{0}^{T} V_o \, e^{-j\omega t} \, dt$$

$$= \left( \frac{V_o e^{-j\omega t}}{-j\omega} \right)_{-T}^{0} + \left( \frac{V_o e^{-j\omega t}}{j\omega} \right)_{0}^{T}$$

$$= V_o \left[ -\frac{1}{j\omega} + \frac{e^{j\omega T}}{j\omega} + \frac{e^{-j\omega T}}{j\omega} - \frac{1}{j\omega} \right]$$

$$= \frac{V_o}{j\omega} \left[ e^{j\omega T} + e^{-j\omega T} - 2 \right]$$

$$= \frac{2jV_o}{\omega} \left[ 1 - \cos \omega T \right]$$

$$= j\omega T^2 V_o \, \frac{\sin^2 \omega T/2}{\left( \dfrac{\omega T}{2} \right)^2} \quad \textbf{Ans.}$$

## 6.15 CAUSAL FUNCTIONS

It is well known that effects never precede their causes and so $y(t)$ the response of a physical system to an impulse applied at $t = 0$ must be zero for negative value of time $t$

$$y(t) = 0 \quad t < 0$$

Impulse response satisfying this condition is said to be "causal".

## PROBLEMS

**6.1.** Using sin series for $f(\theta) = \theta$, $\quad 0 < \theta < \pi$ show that

$$\frac{\pi^2}{6} = 1 + \frac{1}{2^2} + \frac{1}{3^2} + \cdots\cdots$$

**6.2.** From the cosine series for $f(\theta) = \theta$, $\quad 0 < \theta < \pi$ show that

$$\frac{\pi^4}{96} = 1 + \frac{1}{3^4} + \frac{1}{5^4} + \cdots\cdots$$

**6.3.** Determine the Fourier series in trigonometric form for the following functions.

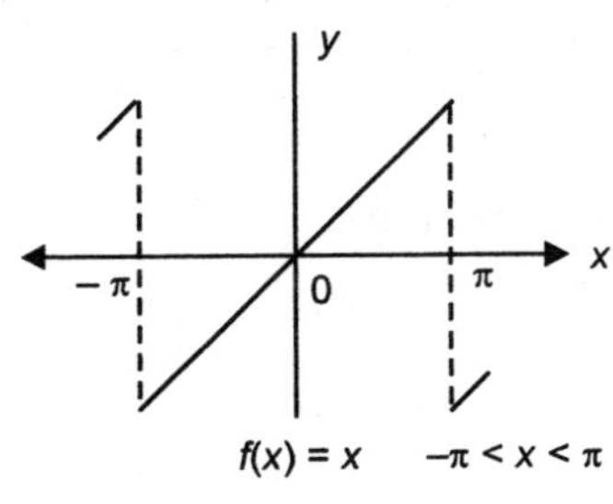


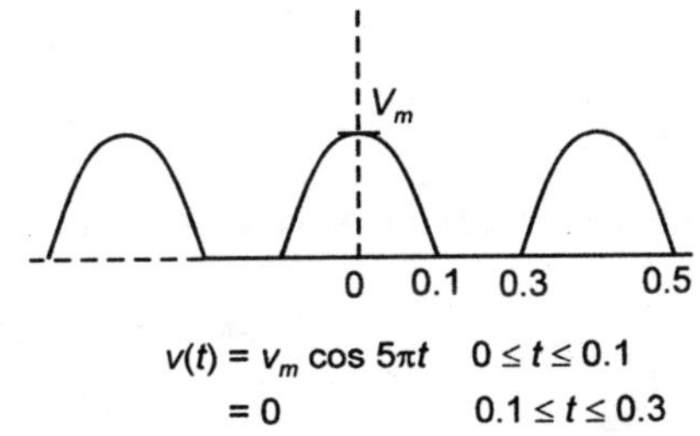


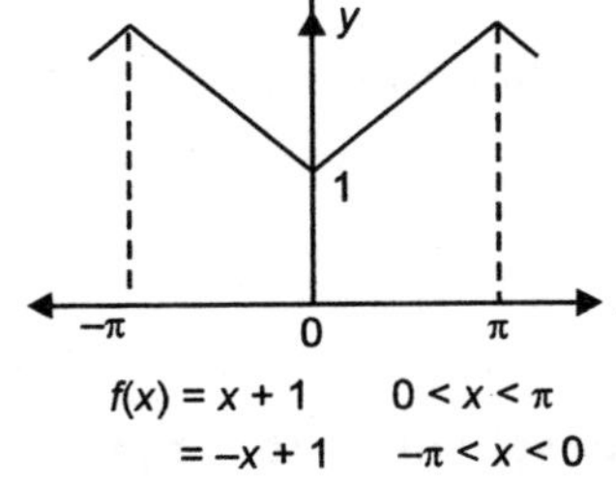


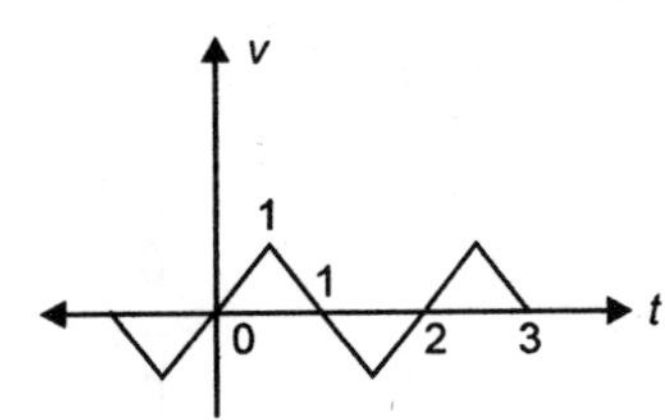


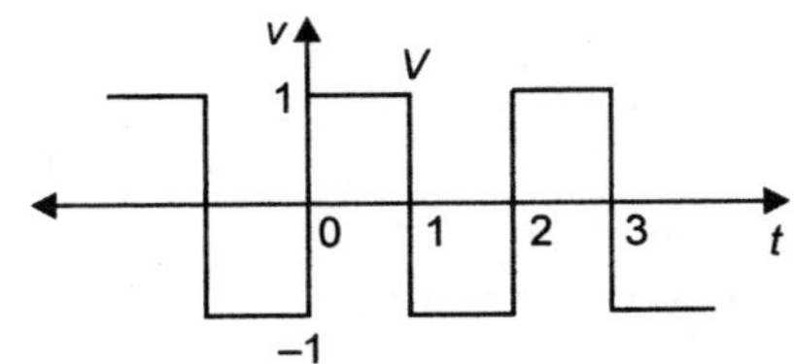


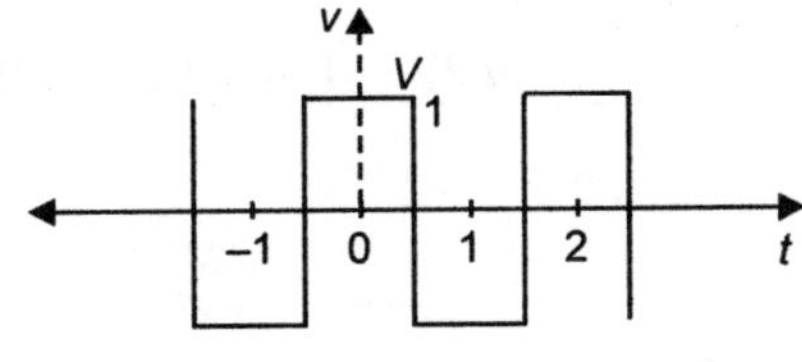


**Fig. P6.3**

**6.4.** Determine the exponential form of the Fourier series for the square wave of P6.4.

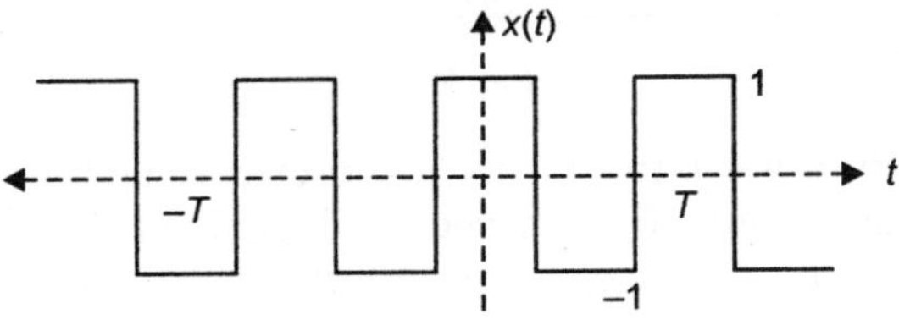


**Fig. P6.4**

**6.5.** Determine the trigonometric and exponential form of Fourier series for the following functions.

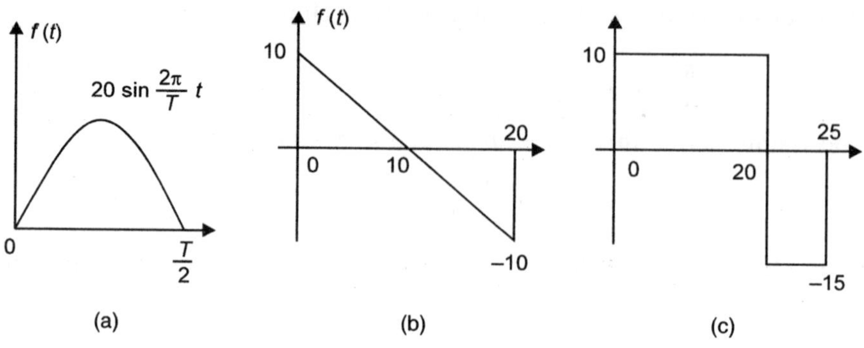

Fig. P6.5

**6.6.** Determine the Fourier series for the following functions:

$$(i) \quad f(t) = t \qquad\qquad 0 < t < \pi$$
$$\qquad f(t) = 0 \qquad\qquad +\pi < t < 2\pi$$
$$(ii) \quad f(t) = t^2 \qquad\qquad 0 < t < \pi$$
$$\qquad\quad = -t^2 \qquad\qquad \pi < t < 2\pi$$
$$(iii) \quad f(t) = e^t \qquad\qquad -\pi < t < \pi$$
$$(iv) \quad f(t) = t \sin t \qquad 0 < t < 2\pi$$

**6.7.** Determine sin and cos half range series for the following functions:

$$(i) \quad f(t) = t^2 \qquad\qquad 0 < t < \pi$$
$$(ii) \quad f(t) = e^t \qquad\qquad 0 < t < \pi$$
$$(iii) \quad f(t) = 2t - 1 \qquad 0 < t < 1$$
$$(iv) \quad f(t) = 2 \qquad\qquad -2 < t < 0$$
$$\qquad\quad = t \qquad\qquad 0 < t < 2$$

**6.8.** In Fig. P6.8, determine the steady state output voltage if the input is given as

$$u(t) = 4 + 10 \cos\left(2t + \frac{\pi}{6}\right)$$

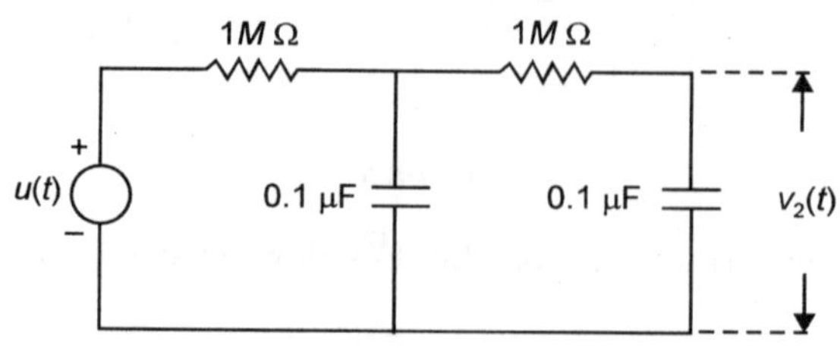

Fig. P6.8

**6.9.** Determine the steady state output voltage in circuit of Fig. P6.8 if the input is sawtooth wave with time period $T = 2\pi$ sec. and peak of 20 units.

**6.10.** The voltage and current in an electric circuit are given by

$$v = 20 + 100 \cos 377t + 50 \cos\left(1131\,t + \frac{\pi}{3}\right)$$

$$i = 5 + 7 \cos\left(377t + \frac{\pi}{10}\right) + 4 \cos\left(1131t + \frac{\pi}{4}\right)$$

Determine the average power supplied to the circuit.

**6.11.** In an electric circuit voltage and current are given as

$$v = 100 \sin\left(\omega t + \frac{\pi}{3}\right) - 50 \sin\left(3\,\omega t - \frac{\pi}{6}\right)$$

$$i = 10 \sin\left(\omega t + \frac{\pi}{3}\right) + 5 \cos\left(3\,\omega t + \frac{\pi}{3}\right)$$

Determine, the resistance, impedance, power and power factor of the circuit.

**6.12.** The current through $1 - \Omega$ resistor is given as

$$i(t) = 10 + 10 \sin\,\omega t + 5 \sin 3\omega t + 3 \sin 5\,\omega t$$

Determine rms value of current and the power dissipated in the resistance.

**6.13.** In an electric circuit voltage and currents are given by

$$v = 10 \sin t + 5 \sin 3t$$

$$i = 5 \sin 2t + 2 \sin 5t$$

Determine the average power in the circuit.

**6.14.** Fig. P6.14 shows the current waveform through $a$ 10 mH inductor. Determine expression for the voltage waveform (Fourier series) if $\omega = 500$ rac/sec.

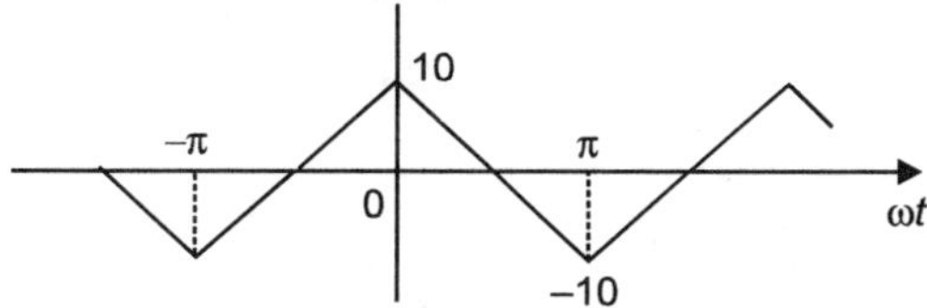


**Fig. P6.14**

**6.15.** A series RLC circuit has an applied voltage of

$$v(t) = 150 \sin 314\,t + 100 \sin 628\,t + 75 \sin 942\,t$$

If $R = 5\,\Omega$, $L = 5\,mH$ and $C = 50\,\mu F$

Determine the effective current and average power.

**6.16.** The differential equation for a linear system is given as

$$(p^2 + 4p + 13)\,y(t) = (3p + 2)\,u(t)$$

where $p = \dfrac{d}{dt}$

Determine the steady state solution for $y(t)$ if

(*i*) $u(t) = 10 \cos 3t$

(*ii*) $u(t) = 5 + 3 \cos (2t + 30º) + 4 \cos (4t - 60º)$

**6.17.** Determine the Fourier transforms of the following signals

(*i*) $f(t) = a\,[u(t) - u(t - 1)]$

(*ii*) $f(t) = A\,e^{-at}\,[u(t) - u(t - 1)]$

**6.18.** Determine the Fourier transform of the signals shown in Fig. P6.18 using suitable properties of Fourier transforms.

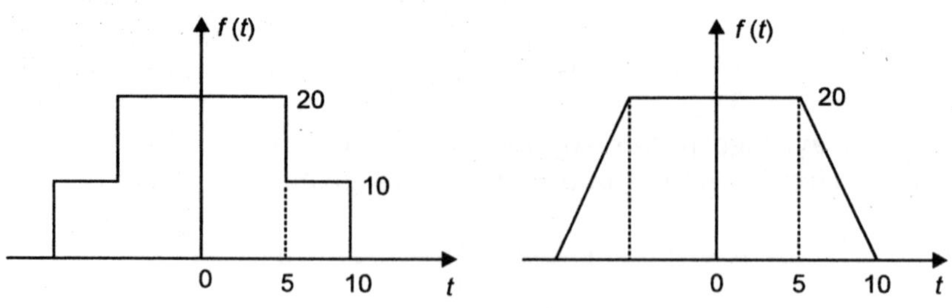

**Fig. P6.18**

**6.19.** Using the results of the previous problem, determine the signals $f(t)$ that have the Fourier transform shown in Fig. P6.19.

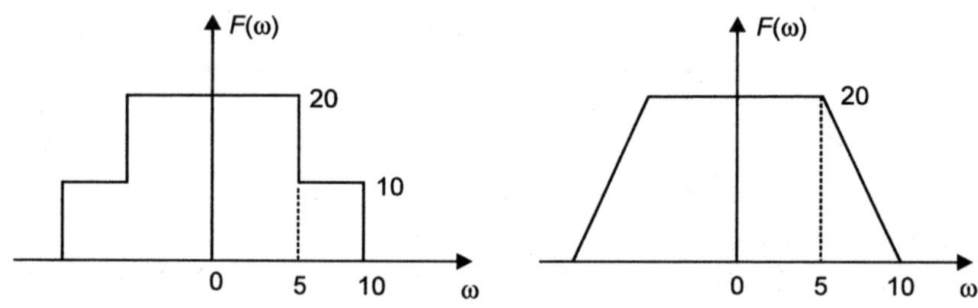

**Fig. P6.19**

**6.20.** The Fourier transform of a continuous time signal $f(t)$ is given by

$$F(\omega) = \frac{10}{j\omega + 4}$$

Determine the Fourier transform $Y(\omega)$ of each of the signals given

    (i) $y(t) = t\, f(t)$

    (ii) $y(t) = e^{-2t}\, f(t)$

    (iii) $y(t) = f(t)\, \cos 2t$

    (iv) $y(t) = f(t)\, (2t - 3)$

**6.21.** Determine $F(j2)$ if

    (i) $f(t) = 2u(t) - \delta(t - 1)$

    (ii) $f(t) = 3\, e^{-4t}\, u(t)$

# 7

# Laplace Transform

## 7.1  INTRODUCTION

We know that Fourier transorm helps us in understanding the behaviour of a system in frequency domain by expressing a signal $f(t)$ as a continuous sum of complex exponentials. Even though it is a powerful analytical tool, it suffers from the disadvantage that Fourier transform is possible only for a limited number of functions and is not available for large number of functions as it is possible only for those functions for which the Fourier integral is finite. There are some very useful functions in engineering problems e.g. a ramp or a positive exponential function etc. for which Fourier transform is not possible. However, the introduction of a convergence factor in the form of a negative real exponential into the Fourier transform leads to what is known as Laplace transform i.e.

$$\mathcal{L}[f(t)] = \mathcal{F}[e^{-at} f(t)]$$

where $\mathcal{L}$ stands for "Laplace transform of". The study of Laplace transform can be made without even mentioning Fourier transform. However, it is useful to start with Fourier transform and develop the concept of Laplace transform as it gives better insight into the relations between the two transforms and it provides a frequency domain interpretation of Laplace transform.

We start with the definition of Fourier transform i.e.

$$F(j\omega) = \int_{-\infty}^{\infty} f(t)\, e^{-j\omega t}\, dt \qquad \qquad ...(7.1)$$

As said earlier most of the $f(t)$ used in engineering problems can be made integrable (finite) if these are multiplied by a negative exponential of time $t$ e.g. $e^{-\sigma t}$ (sometimes referred as convergence factor) so that with a suitable value of $\sigma$ the overall function is made integrable (finite) and its' Fourier integral will exist for a large number of function. Thus

$$F(\sigma + j\omega) = \int_{-\infty}^{\infty} e^{-\sigma t}\, f(t)\, e^{-j\omega t}\, dt$$

$$= \int_{-\infty}^{\infty} f(t)\, e^{-(\sigma + j\omega)t}\, dt \qquad \qquad ...(7.2)$$

Since most of the time we are concerned with causal functions i.e. $f(t) = 0$ for $t < 0$
Therefore, the above equation becomes

$$F(\sigma + j\omega) = \int_{-0}^{\infty} f(t)\, e^{-(\sigma + j\omega)t}\, dt \qquad \ldots(7.3)$$

Let us define $s = \sigma + j\omega$
Therefore,

$$F(s) = \int_{0^-}^{\infty} f(t)\, e^{-st}\, dt \qquad \ldots(7.4)$$

Here $s$ is a complex frequency variable as $\omega$ appears along the
imaginary part of the frequency. The $s$-plane is shown in Fig. 7.1 and
in the lower limit $0^-$ means point before the beginning of the time and
therefore, $f(t)$ contains an impulse at $t = 0$ [initial conditions]. Laplace
transform exists for many functions which are not absolutely integrable.
The choice of $\sigma$ is very important for the convergence of Laplace transform integral. However, it may not be possible to obtain Laplace transform of a function which is a random signal as in such cases we require
the Laplace transforms of its' auto correlation functions which is an
even function of time.

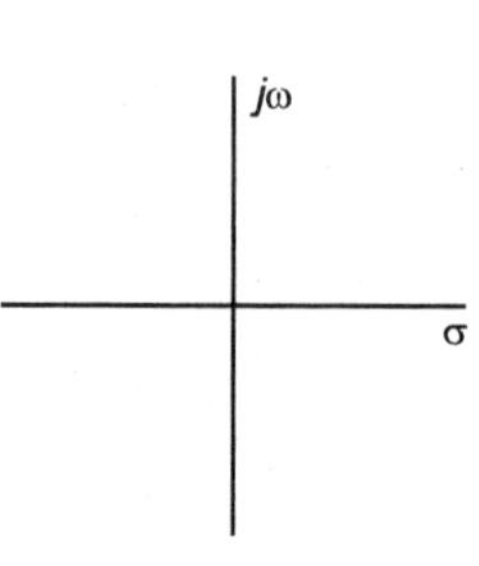

**Fig. 7.1** The $s$-plane.

For Laplace integral to be finite, the value of $\sigma$ should satisfy the
condition.

$$\underset{t \to \infty}{Lt}\ e^{-\sigma t}\, f(t) = 0 \qquad \ldots(7.5)$$

For most of the functions above condition is satisfied if a large and positive value of
$\sigma$ is selected. This will always be true with positive exponentials or with functions that
increase at a slower rate than an exponential. There are some functions like $e^{t^2}$ for which
no value of $\sigma$ can satisfy the above relation. Let us now study a few properties of Laplace
transform.

## 7.2 PROPERTIES OF LAPLACE TRANSFORM

1. **Linearity:** If $F_1(s)$ is the $\mathcal{L}\, f_1(t)$ and $F_2(s)$ that of $\mathcal{L}\, f_2(t)$, then the Laplace of $[a\, f_1(t) + b\, f_2(t)]$ is

   $a\, F_1(s) + b\, F_2(s)$ This means $\mathcal{L}$ operation is a linear operation.
2. **Frequency shift:** If $F(s)$ is $\mathcal{L}[f(t)]$ then
   $$\mathcal{L}[e^{at}\, f(t)] = F\,(s - a) \qquad \ldots(7.6)$$
   This is similar to the frequency shift in Fourier transform and is quite useful in
   determining the Laplace transform of time function multiplied by the exponential in
   time.
3. **Time and frequency scaling**
   If $F(s) = \mathcal{L}[f(t)]$ then
   $$\mathcal{L}[f(at)] = \frac{1}{a}\, F\!\left(\frac{s}{a}\right) \qquad \ldots(7.7)$$
   where $a > 0$
   This is again similar to the time and frequency scaling in Fourier transform.

4. **Differentiation w.r. to time:** This is a very useful property as it simplifies the solution of linear differential equations as compared to the classical methods where we have to follow tedious process of separately determining the complimentary function and the particular integral and then evaluate the arbitrary constant that will satisfy initial condition. However, in Laplace transform method, the initial conditions can be taken into account directly as is evident here.

If $\qquad\qquad \mathcal{L}[f(t)] = F(s)$, then

$$\mathcal{L}\left[\frac{df(t)}{dt}\right] = s\,F(s) - f(0^-) \qquad\qquad ...(7.8)$$

Here $f(0^-)$ means the initial condition of the function. Here $0^-$ means the time just before the start of analysis of a system.

*Proof:* By definition

$$\mathcal{L}[f'(t)] = \int\limits_{0^-}^{\infty} f'(t)\,e^{-st}\,dt$$

Taking $e^{-st}$ as first function and $f'(t)$ as second we have

$$e^{-st}\,f(t)\,\Big|_{0^-}^{\infty} + s\int\limits_{0^-}^{\infty} e^{-st} f(t)\,dt$$

$$= sF(s) - f(0^-)$$

A repeated application of the above gives the general property for $n$th differential as

$$\mathcal{L}\left[\frac{d^n f}{dt^n}\right] = s^n\,F(s) - s^{n-1} f(0^-) - s^{n-2} f'(0^-) - \cdots - f^{n-1}(0^-) \qquad ...(7.9)$$

5. **Integration w.r. to time**

If $\qquad\qquad \mathcal{L}[f(t)] = F(s)$

Then $\qquad\qquad \mathcal{L}\left[\int\limits_{0}^{t} f(t)\right] = \dfrac{F(s)}{s} \qquad\qquad ...(7.10)$

It is to be noted that we do not have a term like $f(0^-)$ on the right hand side of the above identity as we had in the case of differentiation. This is because we start integrating $f(t)$ at $t = 0^-$ with the result that if $f(t)$ contains an impulse at $t = 0$, this is fully taken into account.

6. **Multiplication by $t$ (Differentiation in the frequency domain):**

If $\qquad\qquad \mathcal{L}[f(t)] = F(s)$ then

$$\mathcal{L}[t\,f(t)] = -\frac{d\,F(s)}{d\,s} \qquad\qquad ...(7.11)$$

and $\qquad\qquad \mathcal{L}[t^n\,f(t)] = (-1)^n\,\frac{d^n\,F(s)}{d\,s^n} \qquad\qquad ...(7.12)$

*Proof:* Since

$$F(s) = \int\limits_{0^-}^{\infty} f(t)\,e^{-st}\,dt$$

Differentiating w.r.t $s$ we have

or
$$\frac{dF(s)}{ds} = -t \int_0^\infty f(t)\, dt\, e^{-st}$$

$$-\frac{dF(s)}{ds} = \int tf(t)\, e^{-st}\, dt = \mathcal{L}[t\, f(t)]$$

### 7. Right shift in time

Let
$$f(t) = 0 \quad \text{for} \quad t < 0$$

If
$$\mathcal{L}[f(t)] = F(s) \quad \text{then}$$

$$\mathcal{L}[f(t - t_o)] = e^{-t_o s}\, F(s) \quad \text{where} \quad t_o > 0 \qquad \text{...(7.13)}$$

*Proof:* By definition

$$\mathcal{L}[f(t - t_o)] = \int_0^\infty f(t - t_o)\, e^{-st}\, dt$$

Let
$$t - t_o = \tau$$
$$dt = d\tau$$

Hence
$$\mathcal{L}[f(t - t_o)] = \int_{-t_o}^\infty f(\tau)\, e^{-(t_o + \tau)s}\, d\tau$$

Since $f(t)$ is a causal function the integral $\displaystyle\int_{-t_o}^{0} d\tau = 0$

Therefore,
$$\int_0^\infty f(\tau)\, e^{st_0}\, e^{-s\tau}\, d\tau$$

$$= e^{-st_0} \int_0^\infty f(\tau)\, e^{-s\tau}\, d\tau$$

$$= e^{-st_0}\, F(s)$$

### 8. Convolution in Time domain:

If
$$\mathcal{L}[f_1(t)] = F_1(s) \quad \text{and} \quad \mathcal{L}f_2(t) = F_2(s)$$

then
$$F_1(s)\, F_2(s) = \mathcal{L} \int_{-\infty}^\infty f_1(\tau)\, f_2(t - \tau)\, d\tau \qquad \text{...(7.14)}$$

Since it is more convenient to evaluate Laplace transform of a function as compared to its Fourier transform, it is desirable to evaluate the convolution integral by determining the inverse Laplace of $F_1(s)\, F_2(s)$

### 9. Division by $t$ (Integration in the frequency domain)

If
$$\mathcal{L}[f(t)] = F(s) \quad \text{then}$$

$$\mathcal{L}\left[\frac{f(t)}{t}\right] = \int_s^\infty F(s)\, ds \qquad \text{...(7.15)}$$

*Proof:*

$$\int\limits_{s}^{\infty} F(s)\, ds = \int\limits_{s}^{\infty}\left[\int\limits_{0}^{\infty} f(t)\, e^{-st}\, dt\right] ds$$

$$= \int\limits_{0}^{\infty} f(t)\left[\int\limits_{s}^{\infty} e^{-st}\, ds\right] dt = \int\limits_{0}^{\infty} f(t)\left(\frac{e^{-st}}{-t}\right)_{s}^{\infty} dt$$

$$= \int\limits_{0}^{\infty} \frac{f(t)}{t}\, e^{-st}\, dt$$

$$= \mathcal{L}\left[\frac{f(t)}{t}\right]$$

## 7.3  LAPLACE TRANSFORM OF SOME TYPICAL FUNCTIONS

In engineering problems, there are some functions which are frequently used as forcing functions. While using Laplace transform for the solution of such problem we need not obtain the Laplace transform of such functions from first principles, rather these could be obtained from the standard tables. We, of course, derive these Laplace transform of some of these functions by its definition and by the use of the properties mentioned earlier. We will find that Laplace transform of most of the functions is a rational function i.e. a ratio of two polynomials in the complex frequency variable $s$. The roots of the numerator polynomial are called the zeros of the rational function and that of the denominator the poles which means at these complex frequencies the rational functions assume zero and infinite values respectively.

1. Laplace of step function

$$u(t) = 1 \quad t \geq 0$$
$$= 0 \quad t < 0$$

$$\mathcal{L}[u(t)] = \int\limits_{0}^{\infty} 1 \cdot e^{-st}\, dt = \left[\frac{e^{-st}}{-s}\right]_{0}^{\infty}$$

$$= -\frac{1}{s}\left[e^{-st}\right]_{0}^{\infty} = \frac{1}{s} \qquad \qquad ...(7.16)$$

2. An exponential function of time

$f(t) = e^{at}$. This is equivalent to $e^{at}\, u(t)$ and, therefore, using the property $\mathcal{L}\left[e^{at} f(t)\right]$ $= F(s - a)$

$$\mathcal{L}[e^{at}\, u(t)] = \frac{1}{(s - a)} \qquad \qquad ...(7.17)$$

Here $s$ is replaced by $(s - a)$

3. $\sin \omega t = \dfrac{e^{j\omega t} - e^{-j\omega t}}{2j}$ assuming that it is a causal function and using the property as in (7.17) we have

$$\mathcal{L}[\sin \omega t] = \mathcal{L}\,\frac{e^{j\omega t} - e^{-j\omega t}}{2j} = \frac{1}{2j}\left[\frac{1}{(s-j\omega)} - \frac{1}{(s+j\omega)}\right]$$

$$= \frac{\omega}{s^2 + \omega^2} \qquad\qquad \ldots(7.18)$$

4. $\cos \omega t = \dfrac{e^{j\omega t} + e^{-j\omega t}}{2}$

Therefore, $\qquad \mathcal{L}[\cos \omega t] = \mathcal{L}\,\dfrac{e^{j\omega t} + e^{-j\omega t}}{2}$

$$= \frac{1}{2}\left[\frac{1}{(s-j\omega)} + \frac{1}{(s+j\omega)}\right]$$

$$= \frac{s}{s^2 + w^2} \qquad\qquad \ldots(7.19)$$

5. Damped sin function

$$\mathcal{L}[e^{-at}\sin \omega t] = \frac{\omega}{(s+a)^2 + \omega^2} \qquad\qquad \ldots(7.20)$$

6. Damped cos function

$$\mathcal{L}[e^{-at}\cos \omega t] = \frac{s+a}{(s+a)^2 + \omega^2} \qquad\qquad \ldots(7.21)$$

7. $\sinh \omega t = \dfrac{e^{\omega t} - e^{-\omega t}}{2}$

Hence $\qquad \mathcal{L}[\sinh \omega t] = \mathcal{L}\left[\dfrac{e^{\omega t} - e^{-\omega t}}{2}\right]$

$$= \frac{1}{2}\left[\frac{1}{s-\omega} - \frac{1}{s+\omega}\right] = \frac{\omega}{s^2 - \omega^2} \qquad\qquad \ldots(7.22)$$

Similarly $\quad \mathcal{L}[\cosh \omega t] = \dfrac{s}{s^2 - \omega^2} \qquad\qquad \ldots(7.23)$

8. Damped sinh and cosh function

$$\mathcal{L}[e^{-at}\sinh \omega t] = \frac{\omega}{(s+a)^2 - \omega^2} \qquad\qquad \ldots(7.24)$$

$$\mathcal{L}[e^{-at}\cosh \omega t] = \frac{s+a}{(s+a)^2 - \omega^2} \qquad\qquad \ldots(7.25)$$

9. The unit ramp function

$$f(t) = t \quad t > 0$$
$$= 0 \quad t < 0$$

We make use of the property

$$\mathcal{L}[tf(t)] = -\frac{dF(s)}{ds}$$

Taking

$$f(t) = u(t)$$

and since

$$\mathcal{L}[u(t)] = \frac{1}{s}$$

$$\mathcal{L}[tu(t)] = -\frac{d}{ds}\left(\frac{1}{s}\right) = \frac{1}{s^2} \qquad \text{...(7.26)}$$

10. The unit Impulse function

$$\delta(t) = 1 \quad t = 0$$
$$= 0 \text{ for all other } t$$

Since impulse function is first derivative of a unit step

$$\frac{d}{dt}u(t) = \delta(t)$$

$$\mathcal{L}\left[\frac{d}{dt}u(t)\right] = s\,F(s) = s \cdot \frac{1}{s} = 1 = \mathcal{L}[\delta(t)] \qquad \text{...(7.27)}$$

Yet another possibility of obtaining the Laplace of an impulse functions is as follows: We define

$$\delta(t) = \underset{a \to 0}{Lt}\ \frac{1}{a}[u(t) - u(t - a)]$$

Therefore,

$$\mathcal{L}[\delta(t)] = \underset{a \to 0}{Lt}\ \mathcal{L}\left\{\frac{1}{a}[U(t) - U(t - a)]\right\}$$

$$= \underset{a \to 0}{Lt}\ \frac{1}{a}\left[\frac{1}{s} - \frac{1}{s}e^{-as}\right]$$

$$= \underset{a \to 0}{Lt}\ \frac{1}{a}\frac{1 - e^{-as}}{s} = \frac{1}{a}\frac{1 - \left(1 - as + \dfrac{a^2 s^2}{2} \cdots\cdots\right)}{s}$$

$$= \frac{a\,s}{a\,s} = 1$$

11. The function $\quad f(t) = t\,e^{-at}$

Since $\quad \mathcal{L}(t) = \dfrac{1}{s^2}$

$$\mathcal{L}[t\ e^{at}] = \frac{1}{(s-a)^2} \qquad\qquad ...(7.28)$$

The following table gives the Laplace transforms and hence the inverse Laplace transform of some of the typical functions normally encountered in engineering problems.

Table 7.1   Laplace transform pairs

| | $f(t)$ | $F(s)$ |
|---|---|---|
| 1. | $\delta(t-T),\ \delta(t)$ | $e^{-Ts},\ 1$ |
| 2. | $u\,(t-T),\ u(t)$ | $\dfrac{e^{-Ts}}{s},\ \dfrac{1}{s}$ |
| 3. | $e^{-at}$ | $\dfrac{1}{s+a}$ |
| 4. | $\cos\ \omega t$ | $\dfrac{s}{s^2+\omega^2}$ |
| 5. | $\sin\ \omega t$ | $\dfrac{\omega}{s^2+\omega^2}$ |
| 6. | $e^{-at}\sin\ \omega t$ | $\dfrac{\omega}{(s+a)^2+\omega^2}$ |
| 7. | $e^{-at}\cos\ \omega t$ | $\dfrac{s+a}{(s+a)^2+\omega^2}$ |
| 8. | $t$ | $1/s^2$ |
| 9. | $t^n\ e^{-at}$ | $\dfrac{n!}{(s+a)^{n+1}}$ |
| 10. | $f(t-t_0)\ u(t-t_0)$ | $e^{-t_0 s}\ F(s)$ |
| 11. | $\dfrac{d}{dt}\,f(t)$ | $sF\,(s) - f\,(0^-)$ |
| 12. | $\displaystyle\int_0^t f(t)\,dt$ | $\dfrac{F(s)}{s}$ |
| 13. | $t\,f(t),\ t^n\,f(t)$ | $-\dfrac{d\,F(s)}{ds},\ (-1)^n\,\dfrac{d^n\,F(s)}{ds^n}$ |
| 14. | $\dfrac{1}{t}\,f(t)$ | $\displaystyle\int_s^\infty F(s)\,ds$ |

**Example 7.1:** Determine the Laplace transform of the following functions:

$(i)\ t^2 \sin\ \omega t \qquad\qquad (ii)\ \dfrac{\sin\ \omega t}{t} \qquad\qquad (iii)\ t^2$

$(iv)\ t \cos\ 3\omega t \qquad\qquad (v)\ (1 - e^{-2t})\ u(t) \qquad\qquad (vi)\ t\ e^{-at}$

$(vii)\ \sin\ \omega(t-t_0) \qquad (viii)\ \sin\ \omega(t-t_0)\ u(t) \qquad (ix)\ \sin\ \omega(t-t_0)u\,(t-t_0)$

$(x)\ \sin\ \omega t\ u(t-t_0)$

**Solution:**  $\mathcal{L}[t^2 \sin \omega t]$

$(i)$  Using the relation $[t^n\, f(t)] = (-1)^n\, \dfrac{d^n\, F(s)}{ds^n}$

Since $\mathcal{L} \sin \omega t = \dfrac{\omega}{s^2 + \omega^2} = F(s)$

$$\mathcal{L} t^2 \sin \omega t = (-1)^2 \frac{d^2\, F(s)}{ds^2} = \frac{d^2}{ds^2}\left(\frac{\omega}{s^2 + \omega^2}\right)$$

$$= \frac{d}{ds} \cdot \frac{(s^2 + \omega^2)\,0 - \omega . 2s}{(s^2 + \omega^2)^2} = \frac{d}{ds}\,\frac{-2\,\omega s}{(s^2 + \omega^2)^2}$$

$$= \frac{-2\omega\,(s^2 + \omega^2) + 8\,\omega s^2}{(s^2 + \omega^2)^3} = \frac{2\omega(3s^2 - \omega^2)}{(s^2 + \omega^2)^3}$$

$(ii)$  $\left(\dfrac{\sin \omega t}{t}\right)$  using the relation  $\mathcal{L}\left(\dfrac{f(t)}{t}\right) = \displaystyle\int\limits_s^\infty F(s)\,ds$

$$= \int\limits_s^\infty \frac{\omega}{s^2 + \omega^2}\,ds = \left(\tan^{-1}\frac{s}{\omega}\right)_s^\infty$$

$$= \tan^{-1}\infty - \tan^{-1}\left(\frac{s}{\omega}\right)$$

$$= \frac{\pi}{2} - \tan^{-1}\left(\frac{s}{\omega}\right)$$

$(iii)$  $\mathcal{L}\, t^2$ using the relation $\mathcal{L}\, t^n = \dfrac{n!}{s^{n+1}}$

$$\mathcal{L}\, t^2 = \frac{2}{s^3}$$

However by definition

$$\mathcal{L}\, t^2 = \int\limits_0^\infty t^2\, e^{-st}\, dt$$

$$= t^2\,\frac{e^{-st}}{-s} - \int\limits_0^\infty 2t\,\frac{e^{-st}}{-s}\, dt$$

$$= t^2\,\frac{e^{-st}}{-s} + \frac{2}{s}\left[t \cdot \frac{e^{-st}}{-s} - \int \frac{e^{-st}}{-s}\, dt\right]$$

$$= \left[ -t^2 \frac{e^{-st}}{s} - \frac{2t}{s^2} e^{-st} - \frac{2e^{-st}}{s^3} \right]_0^\infty$$

$$= \frac{2}{s^3} \quad \textbf{Ans.}$$

(iv) $\quad \mathcal{L}\, t \cos 3\, \omega t = \mathcal{L}\,[t\, f(t)] = -\dfrac{d\,F(s)}{ds}$

Since $\quad F(s) = \dfrac{s}{s^2 + 9}$

Therefore, $\quad \mathcal{L}[t\, f(t)] = \dfrac{-d}{ds}\left( \dfrac{s}{s^2 + 9} \right)$

$$= -\frac{(s^2 + 9)\cdot 1 - s\cdot 2s}{(s^2 + 9)^2} = -\frac{s^2 + 9 - 2s^2}{(s^2 + 9)^2}$$

$$= \frac{s^2 - 9}{(s^2 + 9)^2}$$

(v) $\quad (1 - e^{-2t})\, u(t) = \dfrac{1}{s} - \dfrac{1}{s + 2} = \dfrac{2}{s\,(s + 2)}$

(vi) $\quad \mathcal{L}\, t\, e^{-at} \quad$ since $\quad \mathcal{L}\, t = \dfrac{1}{s^2}$

therefore, $\quad \mathcal{L}\, t\, e^{-at} = \mathcal{L}\, f(t)\, e^{-at} = \dfrac{1}{(s + a)^2}$

(vii) $\quad \mathcal{L} \sin \omega\, (t - t_0) = \mathcal{L}\,[\sin \omega t \cos \omega t_0 - \cos \omega t \sin \omega t_0]$

$$= \frac{\omega \cos \omega t_0}{s^2 + \omega^2} - \frac{s \sin \omega t_0}{s^2 + \omega^2}$$

(viii) $\sin \omega\, (t - t_0)\, u(t)$ The Laplace is same as for (vii) even though the function is slightly different because of $u(t)$.

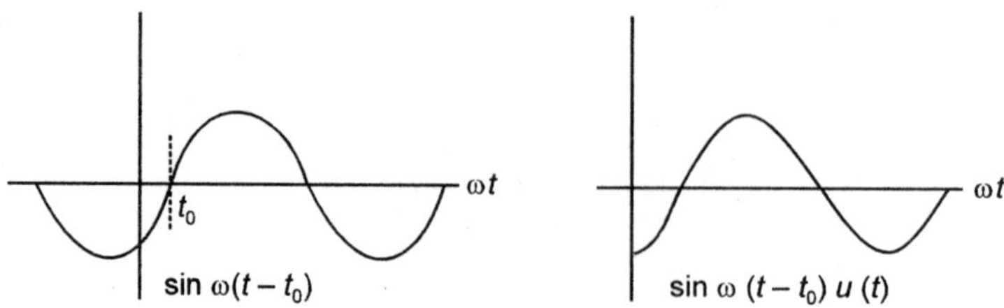


**Fig. E7.1(a)**

*(ix)* $\sin \omega (t - t_0) u(t - t_0)$. The representation of the function is

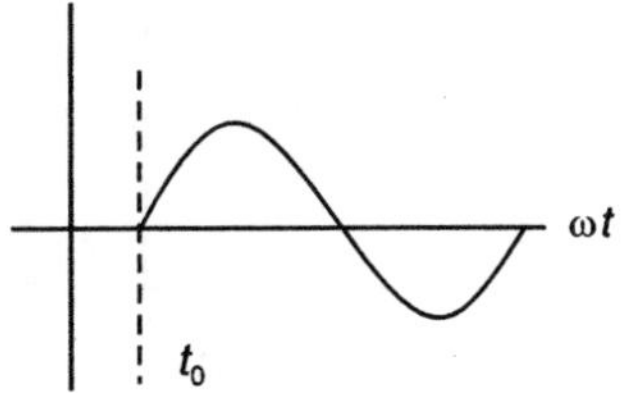

**Fig. E7.1(*b*)**

Since the origin of the sin function has been shifted to $t_0$, therefore,

$$\mathcal{L}[\sin \omega (t - t_0) u(t - t_0)] = e^{-st_0} \frac{\omega}{s^2 + \omega^2}$$

*(x)* $\sin \omega t \, u(t - t_0)$. The representation of the function is

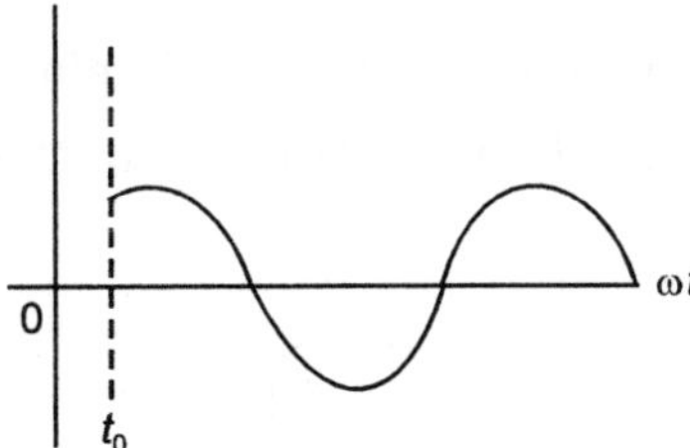

**Fig. E7.1(*c*)**

$$\mathcal{L} \sin \omega t \, u(t - t_0)$$

$$= \int_{t_0}^{\infty} \frac{e^{j\omega t} - e^{-j\omega t}}{2j} \, e^{-st} \, dt$$

$$= \frac{1}{2j} \left[ \int_{t_0}^{\infty} \left\{ e^{(-s + j\omega)t} - e^{-(s + j\omega)t} \right\} dt \right]$$

$$= \frac{1}{2j} \left[ \frac{e^{(-s + j\omega)t_0}}{s - j\omega} - \frac{e^{-(s + j\omega)t_0}}{s + j\omega} \right]$$

$$= e^{-t_0 s} \frac{\omega \cos \omega t_0 + s \sin \omega t_0}{s^2 + \omega^2}$$

## 7.4  GATE FUNCTION

A rectangular pulse of duration $\tau$ and height unity commencing at $t = t_0$ is shown in Fig. 7.2 and is represented mathematically as

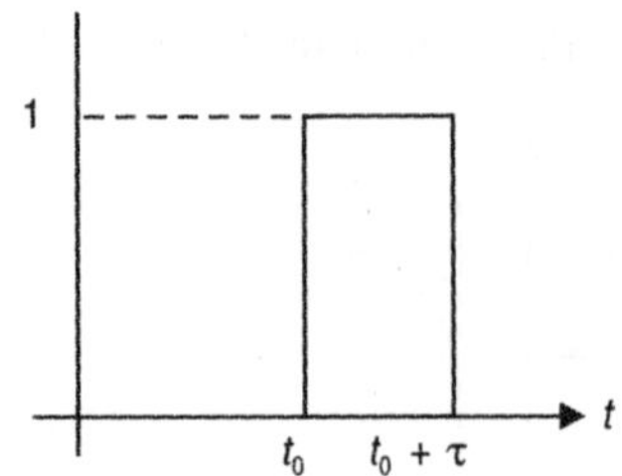

**Fig. 7.2**  Gate function.

$u(t - t_0) - u\ (t - t_0 - \tau)$ and is known as a Gate function. It is known as a Gate function because any function if multiplied by Gate function will have its natural value during the duration of the Gate function and zero elsewhere. The expression for the gate function starting from the origin is $G_0(t) = u(t) - u(t - \tau)$.

Now
$$\mathcal{L}\ G_0(t) = \frac{1}{s} - \frac{1}{s}\,e^{-s\tau} = \frac{1}{s}(1 - e^{-\tau s})  \qquad\qquad ...(7.29)$$

Gate function is quite useful for finding out Laplace transform of periodic functions.

**Example 7.2:** Determine the Laplace transform of the saw-tooth wave shown here.

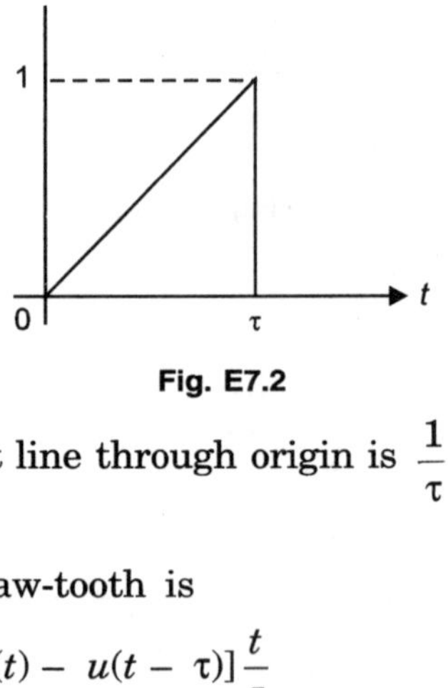

**Fig. E7.2**

Now the slope of the straight line through origin is $\dfrac{1}{\tau}$ and the equation to straight line is $\dfrac{1}{\tau}t = \dfrac{t}{\tau} = f(t)$

Therefore the equation to saw-tooth is

$$[u(t) - u(t - \tau)]\frac{t}{\tau}$$

Therefore,
$$\mathcal{L}[u(t) - u(t - \tau)]\frac{t}{\tau}$$

$$= \mathcal{L}\left[u(t)\,\frac{t}{\tau} - u(t - \tau)\,\frac{t}{\tau}\right]$$

$$= \frac{1}{\tau s^2} - \mathcal{L}\left[u(t - \tau)\left(\frac{t - \tau}{\tau}\right) - u(t - \tau)\right]$$

$$= \frac{1}{\tau s^2} - \frac{1}{\tau s^2}\,e^{-\tau s} - \frac{1}{s}\,e^{-\tau s}$$

$$= \frac{1}{\tau s^2}\left[1 - (1 + s\tau)\,e^{-\tau s}\right]$$

**Example 7.3:** Determine the Laplace transform of a single half-sine wave starting from origin.

**Solution:** The wave can be represented as

$$f(t) = \sin \frac{2\pi}{T} t\, u(t) - \sin \frac{2\pi}{T} t\, u\!\left(t - \frac{T}{2}\right)$$

$$f(t) = \sin \frac{2\pi}{T} t\, u(t) - \sin \frac{2\pi}{T} \left[\left(t - \frac{T}{2}\right) + \pi\right] u(t - T/2)$$

as $T = \pi/2$

$$= \sin \frac{2\pi}{T} t u(t) + \sin \frac{2\pi}{T}\left(t - \frac{T}{2}\right) u\!\left(t - \frac{T}{2}\right)$$

$$\mathcal{L}[f(t)] = \frac{2\pi/T}{s^2 + \left(\dfrac{2\pi}{T}\right)^2} + \frac{2\pi/T}{s^2 + \left(\dfrac{2\pi}{T}\right)^2}\, e^{-sT/2}$$

$$= \frac{2\pi T}{s^2 + \left(\dfrac{2\pi}{T}\right)^2}\left[1 + e^{-sT/2}\right] \textbf{ Ans.}$$

**Example 7.4:** Determine the Laplace transform of the periodic square wave form.

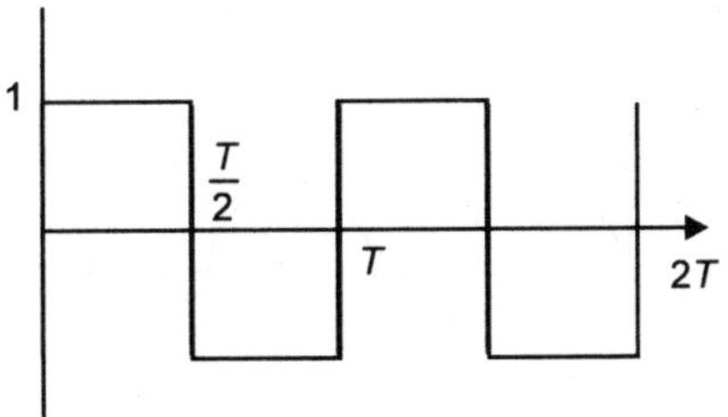


**Fig. E7.4**

The first cycle of the wave form can be expressed as

$$u(t) - u\!\left(t - \frac{T}{2}\right) - u\!\left(t - \frac{T}{2}\right) + u(t - T)$$

$$= u(t) - 2u\!\left(t - \frac{T}{2}\right) + u(t - T)$$

The Laplace transform is

$$\frac{1}{s} - \frac{2}{s} e^{-sT/2} + \frac{1}{s} e^{-sT}$$

$$= \frac{1 - 2e^{-sT/2} + e^{-sT}}{s}$$

$$= \frac{1 - e^{-sT/2} - e^{-sT/2} + e^{-sT}}{s}$$

$$= \frac{\left(1 - e^{-sT/2}\right) - e^{-sT/2}\left(1 - e^{-sT/2}\right)}{s}$$

$$F_1(s) = \frac{\left(1 - e^{-sT/2}\right)^2}{s}$$

Now since all subsequent complete cycles of the function can be obtained by shifting the first cycle by $T$, $2T$, $3T$ ... we can write the following expression for the Laplace transform of the entire function $f(t)$.

$$F(s) = F_1(s)\left[1 + e^{-sT} + e^{-2sT} + e^{-3sT} + \cdots\cdots\right]$$

The series is a geometric progression series with first term 1 and common ratio as $e^{-sT}$. Therefore the sum is $\dfrac{1}{1 - e^{-sT}} \left/ \left[\dfrac{a}{1-r}\right]\right.$ formula used.

Therefore,  $$F(s) = \frac{F_1(s)}{1 - e^{-sT}} = \frac{1 - e^{-sT/2}}{s\left(1 + e^{-sT/2}\right)} \quad \textbf{Ans.}$$

**Example 7.5:** Determine the Laplace transform of the periodic rectified half-wave as shown.

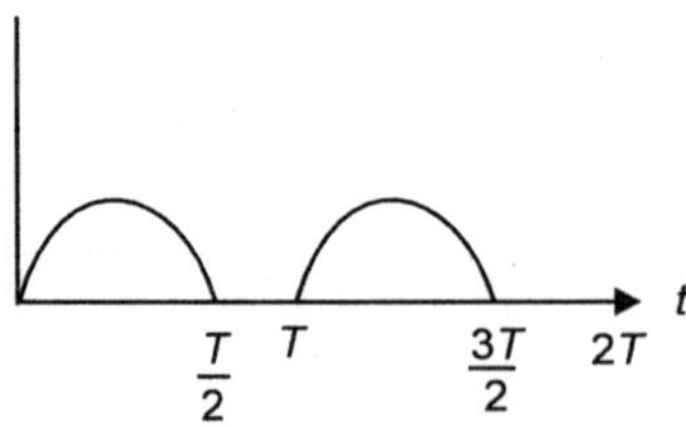


**Fig E7.5**

**Solution:** We have already found out the Laplace of a single half wave. Since these are being repeated after every time period $T$ therefore.

$$F(s) = \frac{F_1(s)}{1 - e^{-Ts}} = \frac{2\pi/T}{s^2 + \left(\dfrac{2\pi}{T}\right)^2} \cdot \frac{\left(1 + e^{-sT/2}\right)}{\left(1 - e^{-sT}\right)}$$

$$= \frac{2\pi/T}{s^2 + \left(\dfrac{2\pi}{T}\right)^2} \cdot \frac{1}{\left(1 - e^{-sT/2}\right)}$$

**Example 7.6:** Determine the Laplace transform of the function shown here.

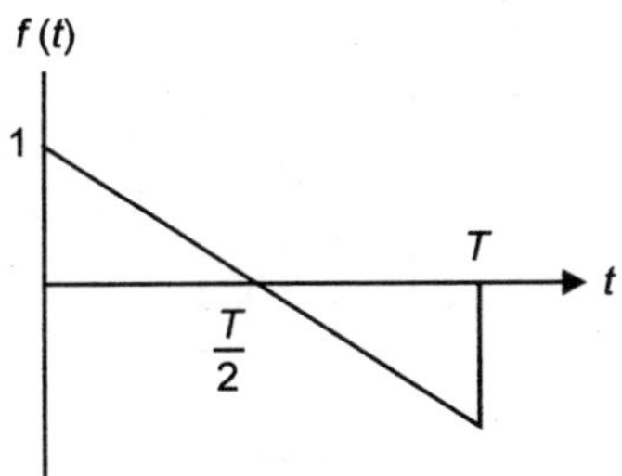



**Fig. E7.6**

using the straight line equations $\dfrac{x}{a} + \dfrac{y}{b} = 1$ we have

$$\frac{t}{T/2} + \frac{f(t)}{1} = 1$$

$$f(t) = 1 - \frac{2t}{T}$$

Since it is only one cycle using Gate function we have

$$f(t) = \left(1 - \frac{2t}{T}\right)\left[u(t) - u(t - T)\right]$$

$$= \left(1 - \frac{2t}{T}\right)u(t) - \left(1 - \frac{2t}{T}\right)u(t - T)$$

$$= \left(1 - \frac{2t}{T}\right)u(t) - \frac{1}{T}(T - 2t)\,u(t - T)$$

$$= \left(1 - \frac{2t}{T}\right)u(t) + \frac{1}{T}(2t - T)\,u(t - T)$$

$$= \left(1 - \frac{2t}{T}\right)u(t) + \frac{2}{T}(t - T)\,u(t - T) + u(t - T)$$

Therefore,

$$\mathcal{L}[f(t)] = \frac{1}{s} - \frac{2}{T s^2} + \frac{2}{T s^2}\,e^{-sT} + \frac{e^{-sT}}{s}$$

$$= \frac{1}{s}(1 + e^{-sT}) - \frac{2}{T s^2}(1 - e^{-sT})$$

**Example 7.7:** Determine the Laplace transform of the function shown here.

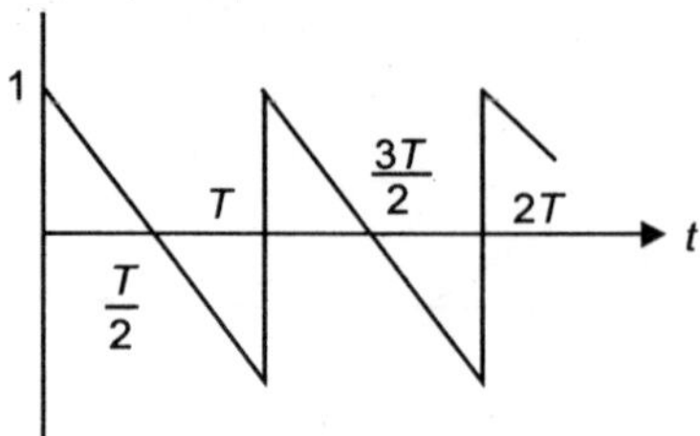

**Fig. E7.7**

**Solution:** Since it is a periodic wave form therefore the overall transform is the transform of the first wave divided by $(1 - e^{-sT})$. Therefore the Laplace transform is

$$\frac{1}{s} \cdot \frac{1 + e^{-sT}}{1 - e^{-sT}} - \frac{2}{Ts^2} \quad \textbf{Ans.}$$

**Example 7.8:** Determine the Laplace transform of the function represented by the waveform in Fig E7.8.

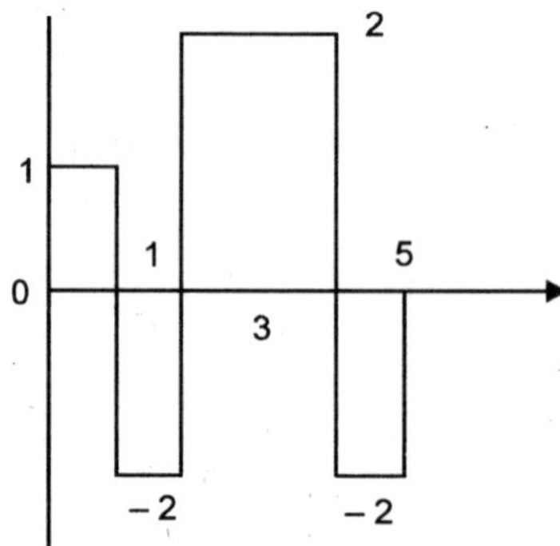

**Fig. E7.8**

$$f(t) = u(t) - u(t - 1) - 2u(t - 1) + 4u(t - 2) - 2u(t - 4) - 2u(t - 4) + 2u(t - 5)$$
$$= u(t) - 3u(t - 1) + 4u(t - 2) - 4u(t - 4) + 2u(t - 5)$$

Therefore,

$$L\,f(t) = \frac{1}{s} - \frac{3}{s}e^{-s} + \frac{4}{s}e^{-2s} - \frac{4}{s}e^{-4s} + \frac{2}{s}e^{-5s} \quad \textbf{Ans.}$$

**Example 7.9:** Determine the Laplace transform of the triangle shown here.

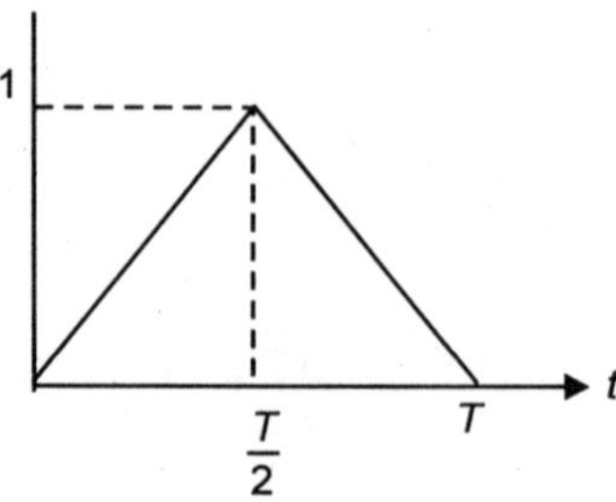

**Fig. E7.9**

The function $\qquad f(t) = \dfrac{2t}{T} \qquad\qquad 0 < t < T/2$

$$= 2\left(1 - \frac{t}{T}\right) \quad \frac{T}{2} < t < T$$

Using the definition of Laplace transform

$$\mathscr{L}\,f(t) = \int_{0}^{T/2} \frac{2t}{T}\, e^{-st}\, dt + \int_{T/2}^{T} \left(2 - \frac{2t}{T}\right) e^{-st}\, dt$$

$$= \frac{2t}{T}\frac{e^{-st}}{-s}\bigg|_{0}^{T/2} - \int \frac{2}{T}\frac{e^{-st}}{-s}\,dt + \left(2 - \frac{2t}{T}\right)\cdot\frac{e^{-st}}{-s}\bigg|_{T/2}^{T} - \int_{T/2}^{T} -\frac{2}{T}\frac{e^{-st}}{-s}\,dt$$

$$= -\frac{e^{-sT/2}}{s} + \frac{2}{Ts}\int_{0}^{T/2} e^{-st}\,dt + \left[-\frac{e^{-sT/2}}{s}\right] - \frac{2}{Ts}\int_{T/2}^{T} e^{-st}\,dt$$

$$= -\frac{e^{-sT/2}}{s} - \frac{2}{Ts^2}\left[e^{-st}\right]_{0}^{T/2} - \frac{e^{-sT/2}}{s} + \frac{2}{Ts^2}\left[e^{-st}\right]_{T/2}^{T}$$

$$= -\frac{e^{-sT/2}}{s} - \frac{2}{Ts^2}\,e^{-sT/2} + \frac{2}{Ts^2} + \frac{e^{-sT/2}}{s} + \frac{2}{Ts^2}\,e^{-sT} - \frac{2}{Ts^2}\,e^{-sT/2}$$

$$= \frac{2}{Ts^2}\left(1 - 2\,e^{-sT/2} + e^{-sT}\right)$$

Therefore,

$$\mathscr{L}\,f(t) = \left(1 - 2\,e^{-sT/2} + e^{-sT}\right)\frac{2}{Ts^2}$$

$$= \frac{2}{Ts^2}\left(1 - e^{-sT/2}\right)^2 \ \textbf{Ans.}$$

**Example 7.10:** The first derivative of a function is shown in Fig E7.10 by the impulse train. Determine the function $f(t)$.

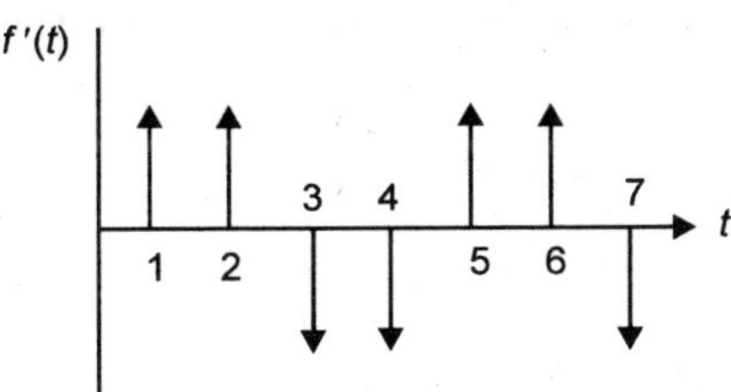


**Fig. E7.10 (a)**

**Solution:** Since the first derivative is impulse, its Laplace transform is
$$F'(s) = e^{-s} + e^{-2s} - e^{-3s} - e^{-4s} + e^{-5s} + e^{-6s} - e^{-7s}$$

Now $F'(s) = \mathscr{L}\left[\dfrac{d\,f(t)}{dt}\right] = s\,F(s)$

Therefore,

$$F(s) = \frac{F'(s)}{s} = \frac{1}{s}\left(e^{-s} + e^{-2s} - e^{-3s} - e^{-4s} + e^{-5s} + e^{-6s} - e^{-7s}\right)$$

Therefore,

$$f(t) = u(t-1) + u(t-2) - u(t-3) - u(t-4) + u(t-5) + u(t-6) - u(t-7) \quad \textbf{Ans.}$$

The wave form is shown here

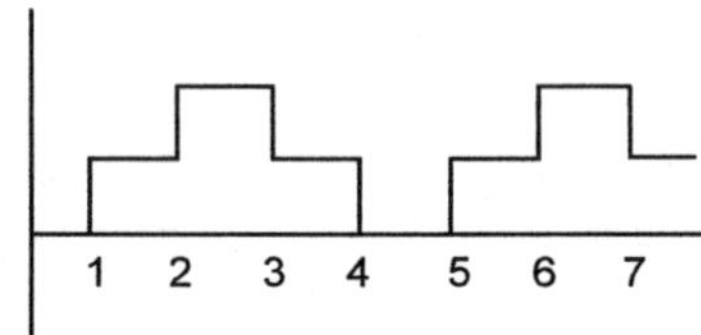

**Fig. E7.10** (*b*) Wave form of *f*(*t*) for Fig. E7.10 (*a*)

**Example 7.11:** Fig. E7.11 shows the impulse train which is second derivative of a function $z(t)$. Determine the function $z(t)$

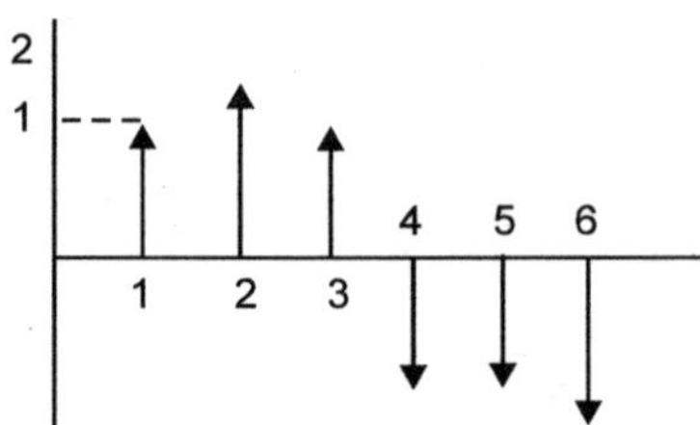

**Fig. E7.11**

The Laplace transform of the impulse train is given as

$$F(s) = e^{-s} + 2\,e^{-2s} + e^{-3s} - e^{-4s} - e^{-5s} - 2e^{-6s}$$

$$Z(s) = \frac{F(s)}{s^2} = \frac{1}{s^2}\left(e^{-s} + 2e^{-2s} + e^{-3s} - e^{-4s} - e^{-5s} - 2e^{-6s}\right)$$

$$Z(t) = (t-1)\,u(t-1) + 2\,(t-2)\,u\,(t-2) + (t-3)\,u(t-3) - (t-4)\,u(t-4)$$
$$- (t-5)\,u(t-5) - 2(t-6)\,u(t-6) \quad \textbf{Ans.}$$

**Example 7.12:** Determine the Laplace transform of the impulse function.

$$f(t) = \delta\,(t^2 - 5t + 6)$$

The function is impulse at $t = 2$ and $t = 3$ i.e.

$$f(t) = \delta\,[(t-2)\,(t-3)]$$
$$= \delta\,(t-2)\,u(t-2) + \delta(t-3)\,u(t-3)$$
$$= \delta(t-2) + \delta(t-3)$$

Therefore the Laplace transform is

$$F(s) = e^{-2s} + e^{-3s}$$

Hence $\;L\,\delta(t^2 - 5t + 6) = e^{-2s} + e^{-3s}$

## 7.5 THE CONCEPT OF COMPLEX FREQUENCY

It is seen that the solution of differential equation gives rise to terms like

$$I = I_0 \, e^{st} \qquad \qquad ...(7.30)$$

where

$$s = \sigma + j\omega$$

Since $e^{st}$ is a dimensionless quantity and so also the product of $st$ is a dimensionless quantity the units of $s$ must be (time)$^{-1}$. Here $\omega$ is interpreted as radian frequency. Since radian is dimensionless as it is a ratio of two length's, $\omega$ is effectively (time)$^{-1}$ i.e. frequency in hertz. Also it is seen that $\sigma$ and $\omega$ must have the same dimension i.e. the dimension of $\sigma$ also should be (time)$^{-1}$. Also since $\sigma$ appears as an exponential, therefore,

$$I = I_0 \, e^{\sigma t}$$

or

$$\sigma = \frac{1}{t} \ln \frac{I}{I_0} \qquad \qquad ...(7.31)$$

Since the unit for the ln of some number is the neper the units of $\sigma$ is neper per sec. The complex sum $s = \sigma + j\omega$ is therefore defined as the complex frequency. The imaginary part of the complex frequency is the radian frequency or the real frequency $f$ and the real part is neper frequency (rather than the misleading term imaginary frequency). Here $\sigma$ and $\omega$ form the real axis and imaginary axis component of $s$ respectively in an $s$-plane.

In order to understand physical significance of complex frequency, we rewrite the term

$$I = I_0 \, e^{st} = I_0 \, e^{(\sigma + j\omega)t}$$
$$= I_0 \, e^{\sigma t} \cdot e^{j\omega t}$$
$$= I_0 \, e^{\sigma t} (\cos \omega t + j \sin \omega t) \qquad \qquad ...(7.32)$$

It turns out that $\omega$ is the angular frequency which we encounter in a.c. quantities and $\sigma$ affects the overall quantity $I$ in the following manner. If $\sigma = 0$

it does not affect $I$

if $\sigma < 0$

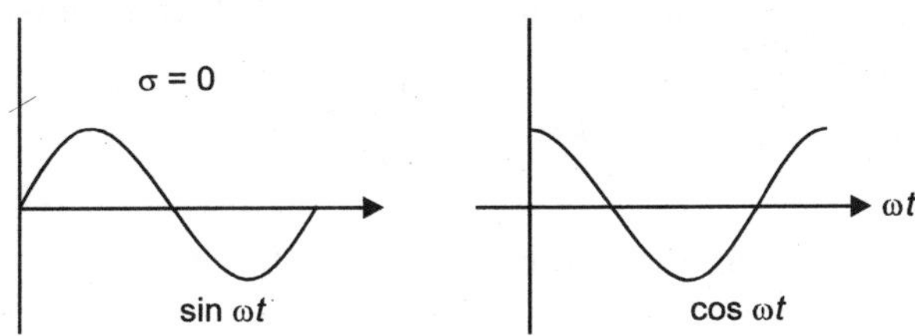


**Fig. 7.3**  (*a*) σ = 0

the quantity $I$ decreases exponentially as a function of time

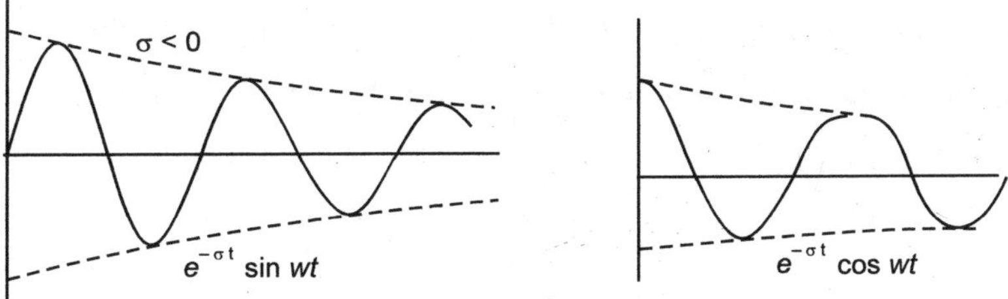


**Fig. 7.3**  (*b*) σ < 0

If $\sigma > 0$ the quantity $I$ increases exponentially as a function of $t$

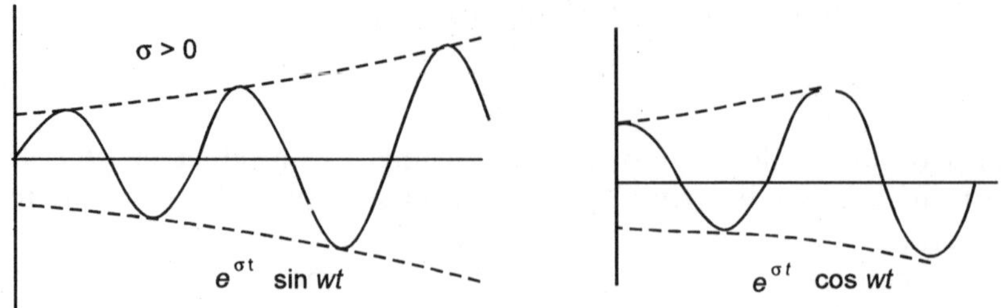



**Fig. 7.3**   (a) $\sigma = 0$ (b) $\sigma < 0$ (c) $\sigma > 0$

From above it is clear that $\omega$ decides the number of oscillation's per second of the function $I$ and $\sigma$ decides the magnitudes of these oscillations. Therefore, the role of the two kinds of frequencies is same even though the consequences are different. We combine the two frequencies and call it complex frequency.

## 7.6  PARTIAL FRACTION EXPANSION

Whenever a circuit is analyzed using Laplace transform the final form of the solution is a quotient of polynomials in $s$. Suppose the numerator and denominator of the polynomial are designated as $P(s)$ and $Q(s)$ respectively and the function.

$$I(s) = \frac{P(s)}{Q(s)} \qquad \qquad ...(7.33)$$

To find out $i(t)$ we will have to find the inverse Laplace of $I(s)$ i.e. that of $P(s)/Q(s)$. In general, however, the transform expression for $I(s)$ must be written in factored form so that inverse could be obtained.

The first step is to see that the order of $P(s)$ should be less than that of $Q(s)$. If not, $P(s)$ should be divided by $Q(s)$ to obtain an expression in the following form.

$$\frac{P(s)}{Q(s)} = A_0 + A_1 s + \cdots\cdots + \frac{P_1(s)}{Q(s)} \qquad \qquad ...(7.34)$$

Now $\dfrac{P_1(s)}{Q(s)}$ can be obtained in factored form as follows:

There are various possibilities of the roots of $Q(s)$. These could be simple roots, multiple roots or complex roots. We will now take up these possibilities and discuss the procedure for obtaining partial fraction expansion of $P_1(s)/Q(s)$.

1. Partial fraction expansion when all the roots of $Q(s)$ are simple:

If all the roots of $Q(s)$ are simple $\dfrac{P_1(s)}{Q(s)}$ could be written as

$$\frac{P_1(s)}{Q(s)} = \frac{P_1(s)}{(s + s_1)(s + s_2) \cdots\cdots (s + s_n)}$$

$$= \frac{A_1}{s + s_1} + \frac{A_2}{s + s_2} + \cdots\cdots + \frac{A_n}{s + s_n} \qquad \qquad ...(7.35)$$

where
$$A_K = (s + s_K) \left. \frac{P_1(s)}{Q(s)} \right|_{s = -s_K} \qquad \qquad ...(7.36)$$

i.e. to find $A_K$ multiply $\dfrac{P_1(s)}{Q(s)}$ by $(s + s_K)$ whereby the factor $(s + s_K)$ from the denominator of $P_1(s)/Q(s)$ will be eliminated and then substitute $s = -s_K$ in the numerator and the denominator.

For example, consider the function
$$\frac{P_1(s)}{Q(s)} = \frac{2s + 1}{(s + 1)(s + 2)(s + 3)}$$

which can be written in partial fraction form as
$$\frac{P_1(s)}{Q(s)} = \frac{A_1}{s + 1} + \frac{A_2}{s + 2} + \frac{A_3}{s + 3}$$

where
$$A_1 = (s + 1) \left. \frac{P_1(s)}{Q(s)} \right|_{s = -1}$$

$$= \left. \frac{(s + 1)(2s + 1)}{(s + 1)(s + 2)(s + 3)} \right|_{s = -1}$$

$$= \left. \frac{(2s + 1)}{(s + 2)(s + 3)} \right|_{s = -1} = \frac{-2 + 1}{1 \times 2} = -\frac{1}{2}$$

Similarly
$$A_2 = \frac{-4 + 1}{(-1)(1)} = 3$$

and
$$A_3 = \frac{-6 + 1}{(-2)(-1)} = -\frac{5}{2}$$

or
$$\frac{P_1(s)}{Q(s)} = \frac{2s + 1}{(s + 1)(s + 2)(s + 3)}$$

$$= -\frac{1}{2(s + 1)} + \frac{3}{(s + 2)} - \frac{5}{2} \cdot \frac{1}{s + 3}$$

2. **Partial fraction expansion when some roots of $Q(s)$ are of multiple order i.e. repeated.**

If $r$ of the $n$ roots of $Q(s)$ are similar the function $\dfrac{P_1(s)}{Q(s)}$ can be written as

$$\frac{P_1(s)}{Q(s)} = \frac{P_1(s)}{(s + s_1)(s + s_2) \cdots (s + s_i)^r \cdots (s + s_n)} \qquad \text{...(7.37)}$$

which is expanded as

$$\frac{P_1(s)}{Q(s)} = \frac{A_1}{s + s_1} + \frac{A_2}{s + s_2} + \cdots + \frac{A_n}{(s + s_n)}$$

$$+ \frac{A_{i_1}}{(s + s_i)} + \cdots + \frac{A_{ir}}{(s + s_i)^r} \qquad \text{...(7.38)}$$

Here first $(n - r)$ terms are simple roots whereas the remaining terms are repeated roots. The $(n - r)$ co-efficients $A_1$, $A_2$, ... $A_{n-r}$ can be obtained as discussed earlier whereas the co-efficients for the repeated roots are obtained by using the following equations.

$$A_{ir} = (s + s_i)^r \left. \frac{P_1(s)}{Q(s)} \right|_{s = -s_i} \qquad \text{...(7.39)}$$

$$A_{ir-1} = \frac{d}{ds} \left[ (s + s_i)^r \frac{P_1(s)}{Q(s)} \right]_{s = -s_i} \qquad \text{...(7.41)}$$

$$A_{ir-2} = \frac{d^2}{2! \, ds^2} \left[ (s + s_i)^r \frac{P_1(s)}{Q(s)} \right]_{s = -s_i} \qquad \text{...(7.42)}$$

$$A_{i1} = \frac{1}{(r - 1)!} \cdot \frac{d^{r-1}}{ds^{r-1}} \left[ (s + s_i)^r \frac{P_1(s)}{Q(s)} \right]_{s = -s_i} \qquad \text{...(7.43)}$$

An example will further illustrate the application of these equations.

$$\frac{P_1(s)}{Q(s)} = \frac{2s + 1}{s(s + 1)^2 (s + 2)}$$

$$= \frac{A_1}{s} + \frac{A_2}{s + 2} + \frac{A_{31}}{s + 1} + \frac{A_{32}}{(s + 1)^2}$$

$$A_1 = s \, \frac{P_1(s)}{Q(s)} = \left. \frac{2s + 1}{(s + 1)^2 (s + 2)} \right|_{s = 0} = \frac{1}{2}$$

$$A_2 = (s + 2) \frac{P_1(s)}{Q(s)} = \left. \frac{2s + 1}{s(s + 1)^2} \right|_{s = -2} = \frac{-3}{-2} = \frac{3}{2}$$

$$A_{31} = \frac{d}{ds} \left[ \frac{(s + 1)^2 (2s + 1)}{s(s + 1)^2 (s + 2)} \right]_{s = -1}$$

$$= \frac{d}{ds}\left[\frac{2s+1}{s(s+2)}\right]$$

$$= \left.\frac{s(s+2)2-(2s+1)(2s+2)}{s^2(s+2)^2}\right|_{s=-1}$$

$$= \frac{(-1)\times 1\times 2-(-2+1)(-2+2)}{1\times 1}=-2$$

$$A_{32}=\left.\frac{(s+1)^2(2s+1)}{s(s+2)(s+1)^2}\right|_{s=-1}=\frac{(2s+1)}{s(s+2)}=\frac{-1}{-1}=1$$

Therefore,

$$\frac{P_1(s)}{Q(s)}=\frac{1}{2s}+\frac{3}{2(s+2)}-\frac{2}{s+1}+\frac{1}{(s+1)^2}$$

3. Partial fraction expansion when the roots are complex conjugate. Suppose one of the complex conjugate pair of roots is given as

$$-a+jb \quad\text{and}\quad -a-jb$$

The co-efficients for these roots are obtained on the same line as the simple roots i.e.

$$A=(s+a-jb)\left.\frac{P_1(s)}{Q(s)}\right|_{s=-a+jb} \tag{7.44}$$

and

$$B=(s+a+jb)\left.\frac{P_1(s)}{Q(s)}\right|_{s=-a-jb} \tag{7.45}$$

For example suppose we have a function

$$\frac{P_1(s)}{Q(s)}=\frac{2s+1}{(s+1)(s^2+2s+5)}$$

the roots of $s^2+2s+5$ are

$$\frac{-2\pm\sqrt{4-20}}{2}=-1\pm j2$$

The co-efficients are

$$A=\left.\frac{(s+1-j2)\cdot(2s+1)}{(s+1)(s+1-j2)(s+1+j2)}\right|_{s=-1+j2}$$

$$=\frac{-2+j4+1}{(-1+j2+1)(-1+j2+1+j2)}$$

$$= \frac{-1 + j4}{(j2)(j4)} = \frac{-1 + j4}{-8} = \frac{1}{8} - j\frac{1}{2}$$

Since
$$B = A^* = \frac{1}{8} + j\frac{1}{2}$$

and
$$A_1 = \left.\frac{(s+1)(2s+1)}{(s+1)(s^2+2s+5)}\right|_{s=-1} = \frac{-2+1}{1-2+5} = -\frac{1}{4}$$

The complete expanded expression is

$$-\frac{1}{4(s+1)} + \frac{\frac{1}{8} - j/2}{s+1-j2} + \frac{\frac{1}{8} + j\frac{1}{2}}{s+1+j2}$$

$$= -\frac{1}{4(s+1)} + \frac{1}{8}\frac{1-j2}{s+1-j2} + \frac{1}{8}\frac{(1+j2)}{s+1+j2}$$

The inverse Laplace transform is

$$-\frac{1}{4}e^{-t} + \frac{1}{8}e^{(-1+j2)t} + \frac{1}{8}e^{(-1-j2)t}$$

$$= -\frac{1}{4}\left[e^{-t} - \frac{1}{2}e^{-t}e^{j2t} - \frac{1}{2}e^{-t}e^{-j2t}\right]$$

$$= -\frac{1}{4}\left[e^{-t} - e^{-t}\cos 2t\right] = -\frac{1}{2}e^{-t}\sin^2 t$$

## 7.7 OPERATIONAL REPRESENTATION (LAPLACE TRANSFORM) OF CIRCUIT ELEMENTS

Next we consider the voltage current relations in time domain of various elements in networks and then to express these in operational representation.

**Resistance:** The voltage current relation in time domain is given as

$$v_R(t) = i_R(t)\, R \qquad \qquad \text{...(7.46)}$$

and
$$i_R(t) = \frac{1}{R}v_R(t) = Gv_R(t) \qquad \qquad \text{...(7.47)}$$

The corresponding transform equations are

$$V_R(s) = I_R(s)\, R \qquad \qquad \text{...(7.48)}$$

and
$$I_R(s) = G\, V_R(s) \qquad \qquad \text{...(7.49)}$$

Now the quotient of $V_R(s)$ and $I_R(s)$ is the transform impedance (resistance) which is given here

as
$$\frac{V_R(s)}{I_R(s)} = R \qquad \qquad \text{...(7.50)}$$

whereas reciprocal of this quotient is the admittance (conductance) and is given as

$$\frac{I_R(s)}{V_R(s)} = G \qquad \qquad ...(7.51)$$

It can be seen that since resistance is not an energy storing element, it is independent of frequency.

The equivalent network representation of the resistor element in time domain and operational impedance are shown here in Fig. 7.4

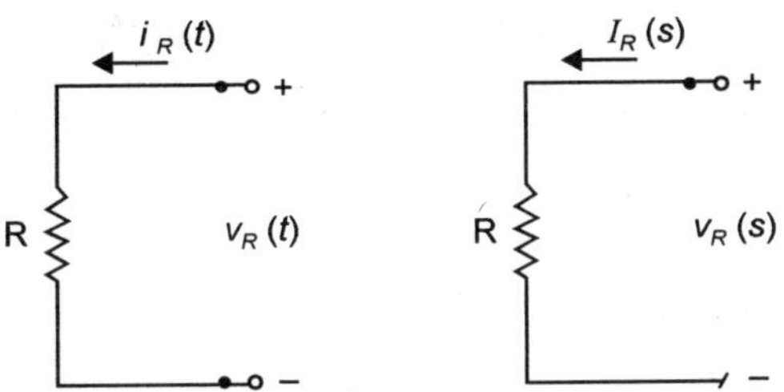

**Fig. 7.4** Equivalent network in time-domain and transform impedance.

**Inductance:** The voltage current relationship of inductance are given as

$$v_L(t) = L \frac{d_{i_L}}{dt} \qquad \qquad ...(7.52)$$

and

$$i_L(t) = \frac{1}{L} \int_{-\infty}^{t} v_L(t)\, dt \qquad \qquad ...(7.53)$$

The equivalent transform equations are given as

$$V_L(s) = L\, s\, I_L(s) - L\, i_L(0^-) \qquad \qquad ...(7.54)$$

or

$$s\, L\, I_L(s) = V_L(s) + L\, i_L(0^-) \qquad \qquad ...(7.55)$$

Here $V_L(s)$ is the transform of the voltage applied and $L\, i_L(0^-)$ is the transform voltage due to the initial current $i(0^-)$ in the inductor. Designating the sum of two transform voltage as $V_1(s)$ we have

$$\frac{V_1(s)}{I_L(s)} = sL \qquad \qquad ...(7.56)$$

The transformed impedance of the inductor is, therefore, $sL$ and is a function of complex frequency $s$.

Now taking the current equation in time-domain and writing its transform equation we have

$$I_L(s) = \frac{1}{L} \left[ \frac{V_L(s)}{s} + \frac{V_L^{-1}(0^-)}{s} \right] \qquad \qquad ...(7.57)$$

The initial value integral $v_L^{-1}(0^-)$ can be evaluated in terms of flux linkages as

$$v_L^{-1}(0^-) = \left. \int_{-\infty}^{t} v(t)\, dt \right|_{t=0} = L\, i_L(0^-) \qquad \qquad ...(7.58)$$

Therefore,

$$I_L(s) = \frac{1}{L}\left[\frac{V_L(s)}{s} + \frac{L\,i_L(0^-)}{s}\right] = \frac{V_L(s)}{sL} + \frac{i_L(0^-)}{s} \qquad \text{...(7.59)}$$

or

$$\frac{V_L}{sL} = I_L(s) - \frac{i_L(0^-)}{s} \qquad \text{...(7.60)}$$

Designating $I_L(s) - \dfrac{i_L(0^-)}{s}$ as the transform current $I_1(s)$ for the admittance $Y_L(s)$ the transform admittance becomes

$$\frac{I_1(s)}{V_L(s)} = \frac{1}{sL} \qquad \text{...(7.61)}$$

The two equivalent transform network and their time-domain, are shown in Fig. 7.5 and 7.6 respectively

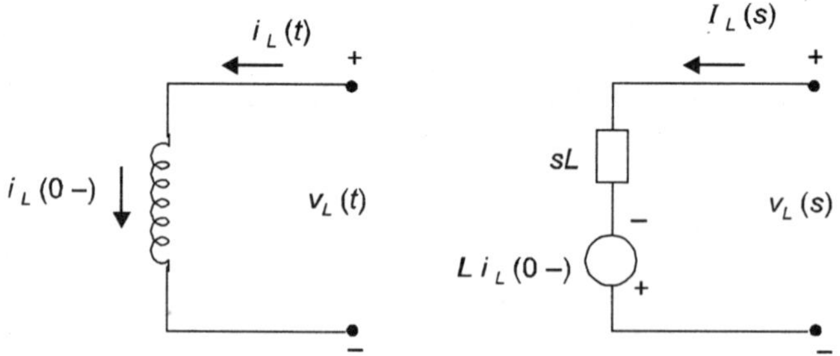

Fig. 7.5   Equivalent impedance voltage circuit.

The polarity of the voltage source is in the direction of initial current through the inductor,

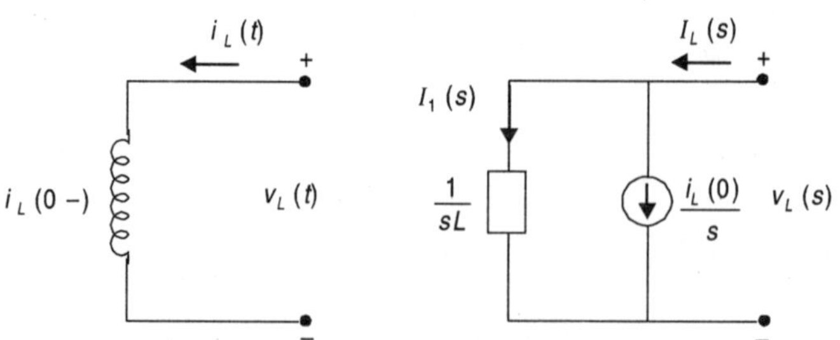

Fig. 7.6   Equivalent admittance circuit.

It is to be noted that

$$Z_L(s) = \frac{1}{Y_L(s)} = sL \qquad \text{...(7.62)}$$

**Capacitance:** The voltage current relation in time-domain for a capacitor are given as

$$v_C(t) = \frac{1}{C} \int_{-\infty}^{t} i_C(t)\, dt \qquad \qquad ...(7.63)$$

$$i_c = C\, \frac{d\, v_c(t)}{dt} \qquad \qquad ...(7.64)$$

Taking transform of the voltage equation we have

$$V_c(s) = \frac{1}{C} \left[ \frac{I_C(s)}{s} + \frac{q_C(0^-)}{s} \right] \qquad \qquad ...(7.65)$$

Here $\dfrac{q(0^-)}{C}$ is the initial voltage of the capacitor which due to the reference direction for voltage is $-\,V_0$

Therefore,

$$V_c(s) = \frac{I_C(s)}{sC} - \frac{V_0}{s} \qquad \qquad ...(7.66)$$

or

$$\frac{I_C(s)}{sC} = V_C(s) + \frac{V_0}{s} \qquad \qquad ...(7.67)$$

Here $\dfrac{V_0}{s}$ is the transform voltage due to initial voltage on the capacitor and $V_C(s)$ the transform of the applied voltage. Let the combination of the two transform voltages be designated by $V_1(s)$.

Therefore,

$$\frac{I_C(s)}{sC} = V_1(s) \qquad \qquad ...(7.68)$$

or

$$\frac{V_1(s)}{I_C(s)} = Z_C(s) = \frac{1}{sC} \qquad \qquad ...(7.69)$$

If the initial conditions are zero

$V_1(s) = V_C(s)$ and then also the transform impedance of the capacitors is $\dfrac{1}{sC}$.

Now taking the current equation we have

$$I_C(s) = C\,[sV_C(s) - V_C(0^-)] \qquad \qquad ...(7.70)$$

or

$$sC\, V_C(s) = I_C(s) + CV_C(0^-) \qquad \qquad ...(7.71)$$

Let the transform current in $Y_C(s)$ be designated as

$$I_1(s) = I_C(s) - CV_C(0^-) \qquad \qquad ...(7.72)$$

Therefore,

$$\frac{I_1(s)}{V_C(s)} = sC \qquad \qquad ...(7.73)$$

Therefore, the capacitor with an initial voltage $V_0$ has an equivalent transform diagram with an impedance $\dfrac{1}{sC}$ in series with a voltage source having transform $V_0/s$, whereas in admittance representation it consists of admittance of $Cs$ in parallel with a transform current source of $CV_0$. The equivalent transform network along with their corresponding versions in time domain are shown in Fig. 7.7.

The polarity of the voltage source is same as the $i_C(t)$ initial charge on the capacitor.

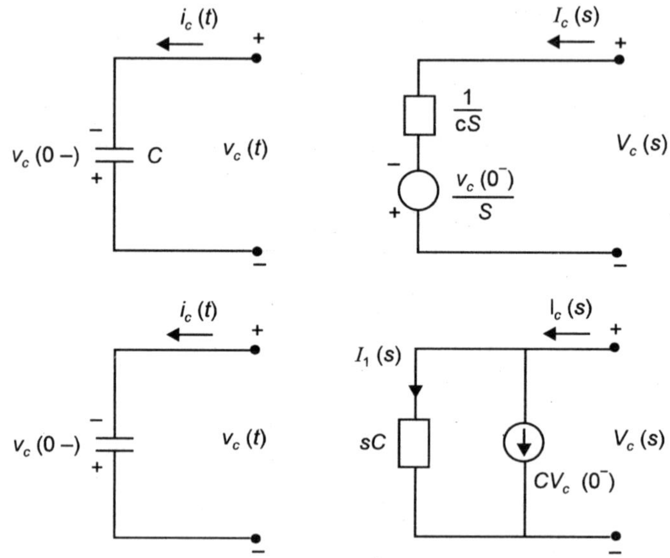

**Fig. 7.7**  Transform Impedance and admittance representation of capacitor.

It is seen that for a capacitor

$$Z_C(s) = \frac{1}{Y_C(s)} = \frac{1}{Cs} \qquad \qquad ...(7.73)$$

**Transformer:** Fig. 7.8 shows a lossless transformer with polarities marked in the time-domain.

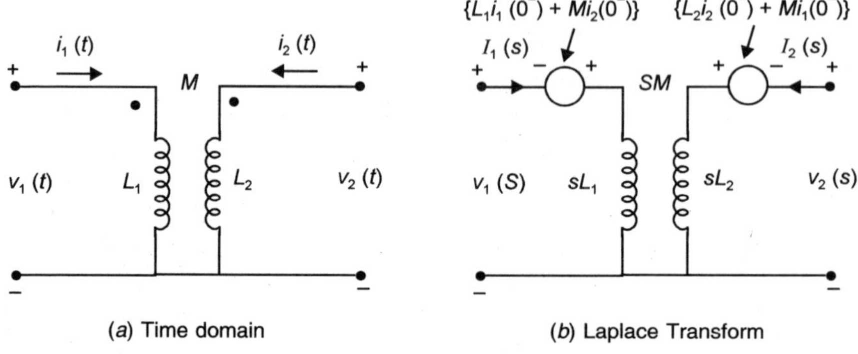

(a) Time domain  (b) Laplace Transform

**Fig. 7.8**  Equivalent circuit of a transformer.

Kirchhoff's voltage law equations can be written as

$$v_1(t) = L_1 \frac{di_1(t)}{dt} + M \frac{di_2(t)}{dt} \qquad \text{...(7.74)}$$

and
$$v_2(t) = M \frac{di_1(t)}{dt} + L_2 \frac{di_2(t)}{dt} \qquad \text{...(7.75)}$$

The corresponding transform equations are:

$$V_1(s) = L_1 s\, I_1(s) - L_1 i_1(0^-) + Ms\, I_2(s) - M i_2(0^-) \qquad \text{...(7.76)}$$

and
$$V_2(s) = Ms\, I_1(s) - M i_1(0^-) + L_2 s\, I_2(s) - L_2 i_2(0^-) \qquad \text{...(7.77)}$$

The equivalent circuit representing the transform equations is given in Fig. 7.8.

**The voltage controlled source:** Refer to Fig. 7.9 for this device.

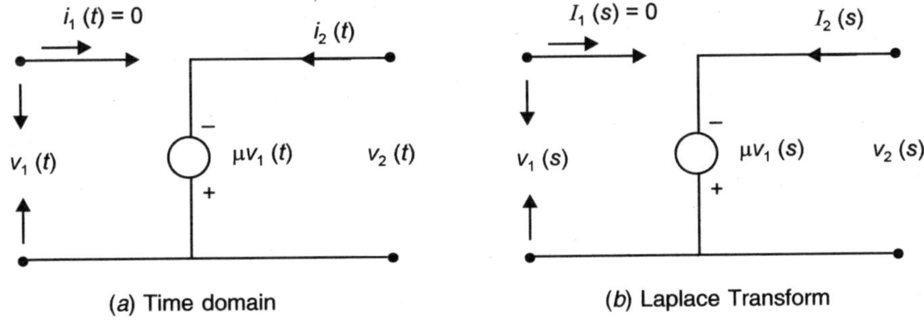


**Fig. 7.9**  Equivalent circuit of voltage controlled source.

The basic equations describing the device in time domain is
$$v_2(t) = \mu\, v_1(t) \qquad \text{...(7.78)}$$
Therefore, the transform equation is
$$v_2(s) = \mu\, v_1(s) \qquad \text{...(7.79)}$$
and equivalent transform network is shown in Fig. 7.9(b).

From the above discussion it is concluded that the initial conditions in the elements are represented by transform voltage (in case of inductor) and transform current (in case of capacitor) sources.

### 7.7.1  Initial and Final Value Theorem

The initial and final values of $f(t)$ from $F(s)$: The initial value i.e. $f(0^-)$ is given as
$$f(0^-) = \underset{t \to 0}{Lt}\, f(t) = \underset{s \to \infty}{Lt}\, s\, F(s) \qquad \text{...(7.80)}$$

and the final value i.e. $f(\infty)$ is given as
$$f(\infty) = \underset{t \to \infty}{Lt}\, f(t) = \underset{s \to 0}{Lt}\, s\, F(s) \qquad \text{...(7.81)}$$

A few examples will clarify these equations.

**Example 7.13:** Consider the transform function

$$F(s) = \frac{2s+1}{(s+1)(s+2)(s+3)} = \frac{2s+1}{s^3 + 6s^2 + 11s + 6}$$

$$f(0^-) = \underset{s \to \infty}{Lt}\ s\ F(s) = \frac{2s^2 + s}{s^3 + 6s^2 + 11s + 6}$$

$$= \underset{s \to \infty}{Lt}\ \frac{\dfrac{2}{s} + \dfrac{1}{s^2}}{1 + \dfrac{6}{s} + \dfrac{11}{s^2} + \dfrac{6}{s^3}} = 0$$

whereas the time response as from example is,

$$\underset{t \to 0}{Lt}\ -\frac{1}{2} e^{-t} + 3 e^{-2t} - \frac{5}{2} e^{-3t}$$

$$= -\frac{1}{2} + 3 - \frac{5}{2} = 0$$

For the same example final value $f(\infty)$ can also be obtained as follows:

$$f(\infty) = \underset{t \to \infty}{Lt}\ f(t) = \underset{s \to 0}{Lt}\ s\ F(s)$$

$$= \frac{2s^2 + 5}{s^3 + 6s^2 + 11s + 6} = 0$$

which is verified from the time domain expression by $t \to \infty$.

It is to be noted that the initial value theorem and final value theorems are applicable only if $f(t)$ and its derivative are Laplace transformable. However, final value theorem is not applicable if the denominator of $sF(s)$ contains any zero whose real part is zero or positive e.g.

Let $$F(s) = \frac{\omega}{s^2 + \omega^2}$$

Here $sF(s)$ has two poles on the imaginary axis and, therefore, final value theorem can't be applied even though the final value theorem gives $f(\infty) = 0$.

## 7.8 APPLICATION OF LAPLACE TRANSFORM TO ELECTRIC NETWORKS

We start with the most simple single element networks and proceed on to more complex networks and use Laplace transform technique for the solution of these networks. We first consider d.c. source connected to the network and the initial condition's are zero.

**Resistance:** Suppose a resistance is switched into a d.c. supply, it is required to find out the current response.

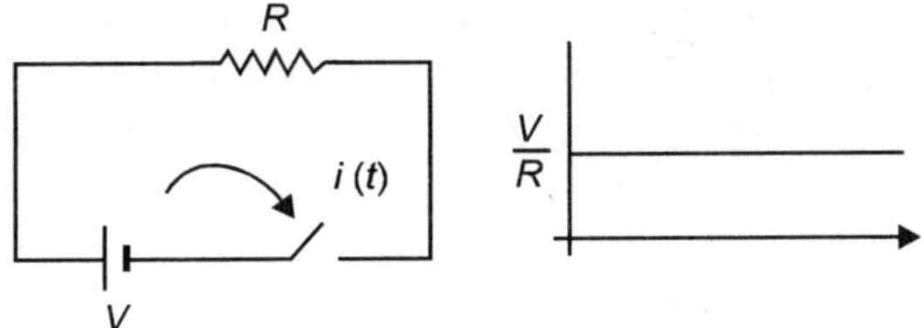

**Fig. 7.10** (a) Resistance switched into a d.c. source (b) current time response.

Since we know that resistance is insensitive to frequency, the current response is given as

$$i(t) = \frac{V}{R} \qquad \qquad ...(7.82)$$

**Inductor:** The equation giving voltage and current relation is given by

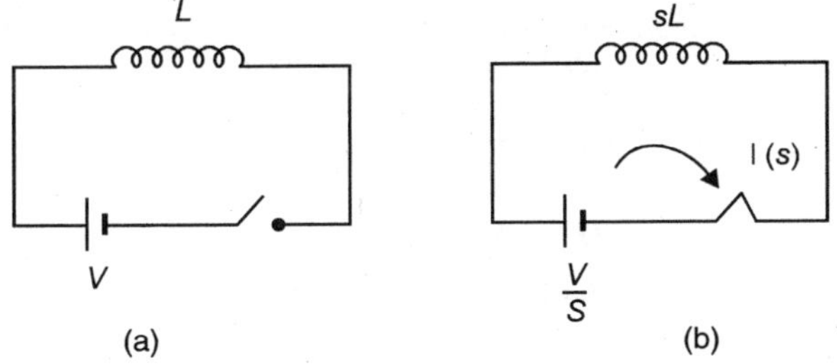

(a)                       (b)

**Fig. 7.11** (a) Pure inductor switched into a d.c. source (b) transform equivalent.

$$I(s) = \frac{V}{s} \cdot \frac{1}{sL}$$

$$= \frac{V}{L} \cdot \frac{1}{s^2} \qquad \qquad ...(7.83)$$

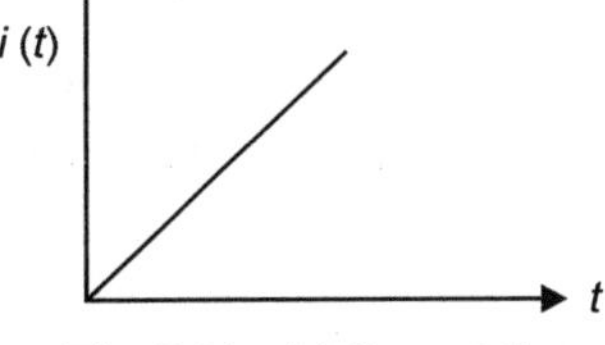

**Fig. 7.11** (c) Current time response.

Taking inverse Laplace transform

$$i(t) = \frac{V}{L}t \qquad \qquad ...(7.84)$$

This is a straight line equation passing through the origin.

Now for such a circuit.

Initial value of current $i(t)$ i.e

$$\underset{t \to 0}{i(t)} = i(0^+) = \underset{s \to \infty}{Lt}\ sI(s) = s \cdot \frac{V}{L} \cdot \frac{1}{s^2} = 0$$

which is true if $t \to 0$ in $i(t)$. The final value of the current

$$i(\infty) = \underset{s \to 0}{Lt}\ s\, I(s) = s \cdot \frac{V}{L} \cdot \frac{1}{s^2} = \infty$$

which again is true from the expression for

$$\underset{t \to \infty}{i(t)} = \frac{V}{L}t$$

This shows that an inductor acts as an open circuit at $t = 0$ when a d.c. source is switched in and it acts as a short circuit at $t = \infty$. However, it is to be noted that since inductance will have some finite resistance $R$ in actual practice, the current will settle down to a value $V/R$.

**Capacitance**

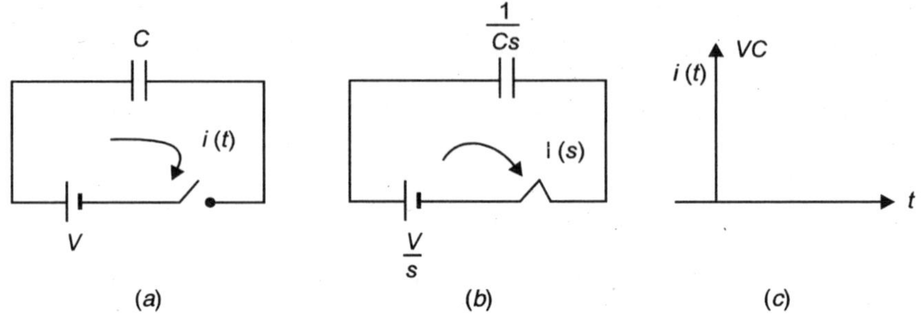

**Fig. 7.12** (a) Pure capacitor switched into a dc source  (b) Transform equivalent
(c) Time current response.

The current
$$I(s) = \frac{V}{s} \cdot \frac{1}{1/cs} = VC \qquad \qquad \text{...(7.85)}$$

$$I(t) = VC\, \delta\,(t) \qquad \qquad \text{...(7.86)}$$

i.e. it is independent of time and has finite value of an impulse of strength $VC$ at the origin. Here the initial value is given by

$$I(0^+) = \underset{s \to \infty}{Lt}\; sI(s) = \underset{s \to \infty}{Lt}\; sVC = \infty$$

it is an impulse of height $\to \infty$ and the width $\to 0$ such that the product is $VC$.

Similarly the final value $f(\infty)$

$$f(\infty) = \underset{s \to 0}{sI(s)} = 0$$

which is true as $\qquad\qquad \delta(t) = 0 \quad$ for $\quad t > 0$

This means the capacitor acts as a short circuit at $t = 0$ and open circuit as $t \to \infty$ and in fact the capacitor gets charged to voltage $V$ and the conduction stops.

**R-L series circuit:** d.c. source is switched in as shown in Fig. 7.13(a) and (b) shows the equivalent transform circuit.

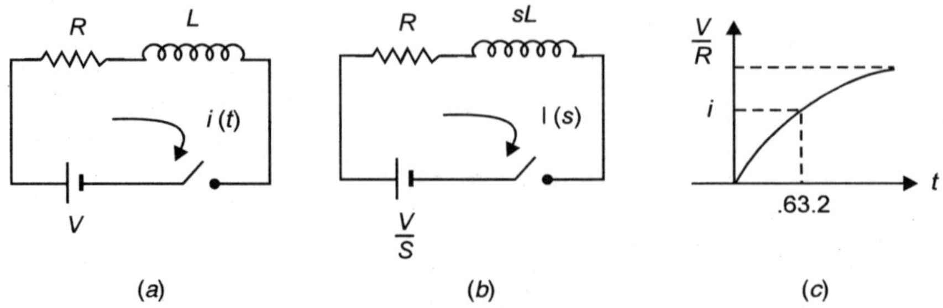

**Fig. 7.13** (a) R-L switched into d.c. source  (b) Transform equivalent
(c) Time-current response.

Assuming zero initial condition, the current $I(s)$ is given as

$$I(s) = \frac{V/s}{R + sL} = \frac{V/L}{s\,(s + R/L)} \qquad \qquad ...(7.87)$$

The current transform has poles at $s = 0$ and $s = -R/L$.

Let

$$I(s) = \frac{V/L}{s\,(s + R/L)} = \frac{A}{s} + \frac{B}{s + R/L}$$

$$A = sI(s) = \mathop{Lt}_{s \to 0} \frac{s\,V/L}{s\,(s + R/L)} = \frac{V}{R}$$

$$B = \mathop{Lt}_{s \to -R/L} (s + R/L)\,I(s) = \frac{V/L}{s} = -\frac{V}{R}$$

Therefore,

$$I(s) = \frac{V}{R}\left[\frac{1}{s} - \frac{1}{s + R/L}\right]$$

Hence

$$i(t) = \frac{V}{R}\left(1 - e^{-\frac{R}{L}t}\right) \qquad \qquad ...(7.88)$$

The initial and final values of $i(t)$ are obtained as

$$i(0^+) = \mathop{Lt}_{s \to \infty} s\,I(s) = \frac{V/L}{s + R/L} = 0$$

and

$$i(\infty) = \mathop{s\,I(s)}_{s \to 0} = \frac{V/L}{R/L} = \frac{V}{R}$$

These values can be obtained from the expression for $i(t)$ by substituting $t = 0$ and $t = \infty$ respectively and are seen to be same as obtained through the theorems.

This shows that initially at $t = 0^+$ the inductor acts as an open circuit to d.c. source and at $t = \infty$ as a short circuit.

Here $\dfrac{L}{R}$ is the time constant $\tau$ of the $RL$ series circuit and is the time in sec. to reach to 63.2% of its final steady state value.

The larger the value of $L$ as compared to resistance the slower is the build up of current in the circuit.

**R-C series circuit:** d.c. source is switched in as shown in Fig. 7.14(a) and 7.14(b) shows the equivalent transform network.

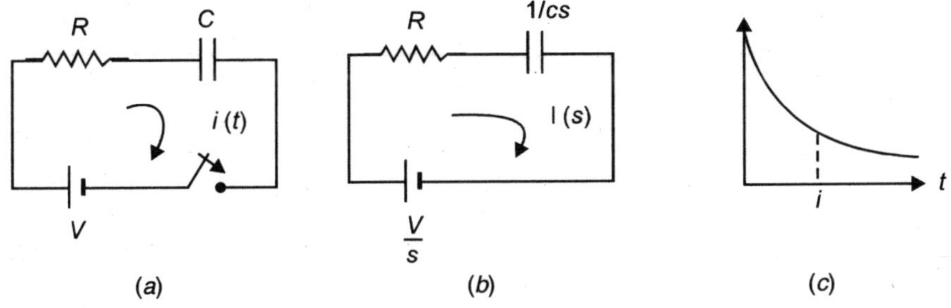


**Fig. 7.14** (a) RC circuit switch into a d.c. source (b) Transform equivalent
(c) current time response.

The current transform is given as

$$I(s) = \frac{V/s}{R + 1/cs} = \frac{(V/s)\,cs}{RCs + 1}$$

$$= \frac{V/R}{s + 1/RC} \qquad\qquad\qquad \text{...(7.89)}$$

Hence

$$i(t) = \frac{V}{R}\,e^{-t/CR} \qquad\qquad\qquad \text{...(7.90)}$$

Using initial value and final value theorems to obtain $i(0^+)$ and $i(\infty)$ we have

$$i(0^+) = \underset{s \to \infty}{Lt}\ s\ I\ (s) = \frac{s \cdot V/R}{s + 1/RC}$$

$$= \underset{s \to \infty}{Lt}\ \frac{V/R}{1 + \dfrac{1}{RCs}} = \frac{V}{R}$$

and

$$i(\infty) = \underset{s \to 0}{s\ I(s)} = \frac{s \cdot V/R}{s + 1/RC} = 0$$

which is same as when obtained by substituting $t = 0$ and $t = \infty$ respectively in the expression for $i(t)$.

Here $CR$ is known as the time constant in sec. and is the value for which current reduces to 36.8% of its initial value.

**R-L-C series:** d.c. source is switched in as shown in Fig. 7.15 $(a)$. Fig. 7.15 $(b)$ shows the equivalent transform network with zero initial conditions.

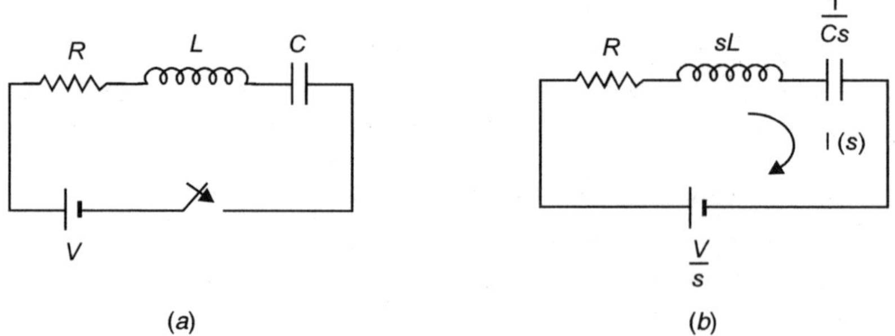

(a)        (b)

**Fig. 7.15**  $(a)$ RLC Circuit switched into a d.c. source $(b)$ Transform equivalent.

$$I(s) = \frac{V/s}{R + sL + 1/Cs}$$

$$= \frac{V}{L} \cdot \frac{1}{s^2 + \dfrac{R}{L}s + \dfrac{1}{LC}} \qquad\qquad \text{...(7.91)}$$

$$= \frac{V}{L} \cdot \frac{1}{\left\{s + \left(\dfrac{R}{2L} - \sqrt{\dfrac{R^2}{4L^2} - \dfrac{1}{LC}}\right)\right\}\left\{s + \left(\dfrac{R}{2L} + \sqrt{\dfrac{R^2}{4L^2} - \dfrac{1}{LC}}\right)\right\}}$$

Let
$$\frac{R}{2L} = a \quad \text{and} \quad \sqrt{\frac{R^2}{4L^2} - \frac{1}{LC}} = b$$

$$I(s) = \frac{V}{L} \cdot \frac{1}{(s + a - b)(s + a + b)} \qquad \qquad ...(7.92)$$

$$I(s) = \frac{V}{2bL} \left\{ \frac{1}{s + a - b} - \frac{1}{s + a + b} \right\}$$

$$i(t) = \frac{V}{2bL} \left\{ e^{-(a-b)t} - e^{-(a+b)t} \right\} \qquad \qquad ...(7.93)$$

There are three conditions based on the value of $b$.

(i) If $\dfrac{R^2}{4L^2} > \dfrac{1}{LC}$ $b$ is real.

(ii) If $\dfrac{R^2}{4L^2} = \dfrac{1}{LC}$ $b$ is zero.

(iii) If $\dfrac{R^2}{4L^2} < \dfrac{1}{LC}$ $b$ is imaginary $\qquad \qquad ...(7.94)$

**Case I:** When $b$ is real.

The expression for current will be

$$i(t) = \frac{V/L}{2\sqrt{\dfrac{R^2}{4L^2} - \dfrac{1}{LC}}} \left[ \exp\left\{ -\left( \frac{R}{2L} + \sqrt{\frac{R^2}{4L^2} - \frac{1}{LC}} \right)t \right\} \right.$$

$$\left. -\exp\left\{ -\left( \frac{R}{2L} - \sqrt{\frac{R^2}{4L^2} - \frac{1}{LC}} \right)t \right\} \right] \qquad \qquad ...(7.95)$$

and the variation of current is given in Fig. 7.16 (a)

**Case II:** When $b = 0$

The expression for current becomes

$$(t) = \frac{V}{2bL} \left( e^{-at} - e^{-at} \right) \text{ which is indeterminate} \qquad \qquad ...(7.96)$$

Therefore, differentiating $i(t)$ with respect to $b$ gives

$$i(t) = \frac{V}{2L} \cdot t \left\{ e^{-(a-b)t} + e^{-(a+b)t} \right\}$$

Now at $b = 0$

$$i(t) = \frac{V}{L} t\, e^{-at} = \frac{V}{L} t\, e^{-\frac{R}{2L}t} \qquad \qquad ...(7.97)$$

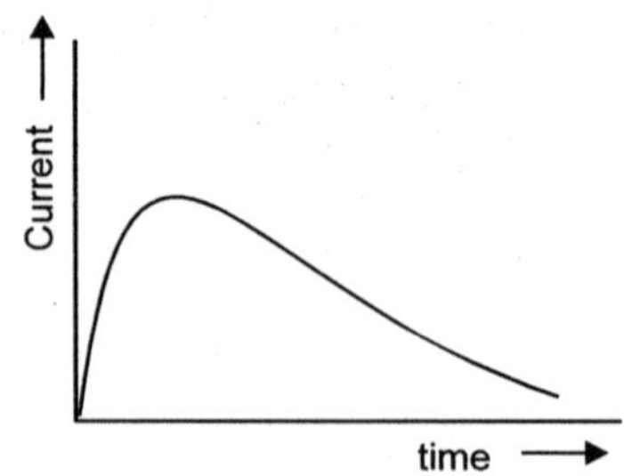

**Fig. 7.16(*a*)**   Waveform when *b* is real.

The variation of current is given in Fig. 7.16(*b*).

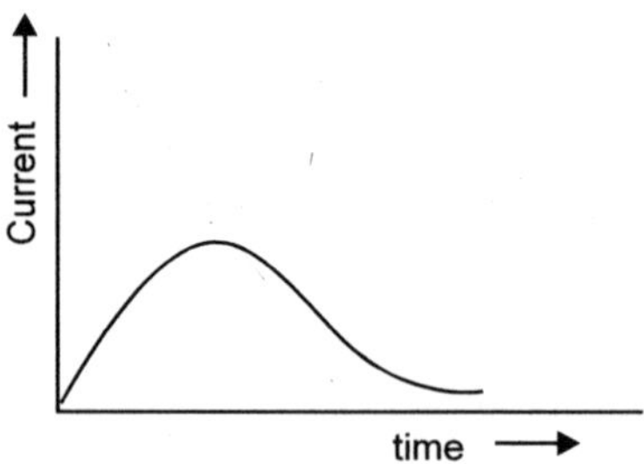

**Fig 7.16(*b*)**   Waveform when *b* = 0.

**Case III:** When *b* is imaginary.

$$i(t) = \frac{V}{2bL}\left\{e^{-at} \cdot e^{jbt} - e^{-at} \cdot e^{-jbt}\right\}$$

$$= \frac{V}{2L\sqrt{\dfrac{R^2}{4L^2} - \dfrac{1}{LC}}}\, e^{-at} \cdot 2\sin\left(\sqrt{-\dfrac{R^2}{4L^2} + \dfrac{1}{LC}}\;\right)t \qquad\qquad \ldots(7.98)$$

The wave shape of the current is shown in Fig. 7.16(*c*).

When *b* is positive or zero, the variation of current is non-oscillatory whereas it is oscillatory when *b* is imaginary. Because of the presence of the capacitance, the current in all the three cases dies down to zero value with d.c. source in the circuit.

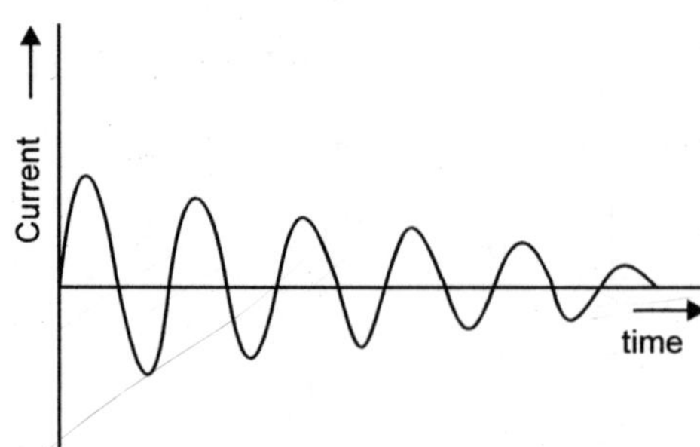

**Fig. 7.16(*c*)**   Waveform when *b* is imaginary.

**2. a.c. Source:** R-L circuit (Fig. 7.17): When switch $s$ is closed, the current in the circuit is given by

$$I(s) = \frac{V(s)}{Z(s)} = V_m \left\{ \frac{\omega \cos \phi}{s^2 + \omega^2} + \frac{s \sin \phi}{s^2 + \omega^2} \right\} \cdot \frac{1}{R + sL}$$

$$= \frac{V_m}{L} \left\{ \frac{\omega \cos \phi}{s^2 + \omega^2} + \frac{s \sin \phi}{s^2 + \omega^2} \right\} \cdot \frac{1}{s + R/L} \qquad \dots(7.99)$$

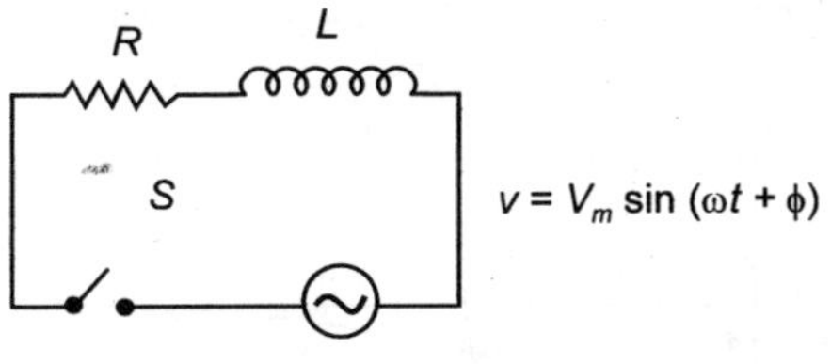

**Fig. 7.17**  R-L circuit connected to an a.c. source.

Let $\qquad \dfrac{R}{L} = a$; then

$$I(s) = \frac{V_m}{L} \left\{ \frac{\omega \cos \phi}{(s + a)(s^2 + \omega^2)} + \frac{s \sin \phi}{(s + a)(s^2 + \omega^2)} \right\}$$

Now

$$\frac{1}{(s + a)(s^2 + \omega^2)} = \frac{1}{(a^2 + \omega^2)} \left\{ \frac{1}{s + a} + \frac{a}{s^2 + \omega^2} - \frac{s}{s^2 + \omega^2} \right\}$$

and

$$\frac{s}{(s + a)(s^2 + \omega^2)} = \frac{1}{(a^2 + \omega^2)} \left\{ \frac{a s}{s^2 + \omega^2} + \frac{\omega^2}{s^2 + \omega^2} - \frac{a}{s + a} \right\}$$

Therefore,

$$\mathcal{L}^{-1}[I(s)] = i(t) = \frac{V_m}{(a^2 + \omega^2)L} \left[ \omega \cos \phi \left\{ e^{-at} + \frac{a}{\omega} \sin \omega t - \cos \omega t \right\} \right.$$

$$\left. + \sin \phi \left\{ a \cos \omega t + \omega \sin \omega t - a\, e^{-at} \right\} \right]$$

The equation can be further simplified to

$$i(t) = \frac{V_m}{L \sqrt{a^2 + \omega^2}} \left\{ \sin (\omega t + \phi - \theta) - \sin (\phi - \theta)\, e^{-at} \right\}$$

$$= \frac{V_m}{(R^2 + \omega^2 L^2)^{1/2}} \left\{ \sin (\omega t + \phi - \theta) - \sin (\phi - \theta)\, e^{-at} \right\}$$

$$\dots(7.100)$$

where $\qquad \theta = \tan^{-1} \dfrac{\omega L}{R}$

The variation of current is shown in Fig. 7.18.

The first term in the expression above is the steady state sinusoidal variation and the second term is the transient part of it which vanishes theoretically after infinite time. But practically, it vanishes very quickly after two or three cycles. The transient decay as is seen depends upon the time constant $\dfrac{1}{a} = \dfrac{L}{R}$ of the circuit. Also at $t = 0$ it can be seen that the transient component equals the steady state

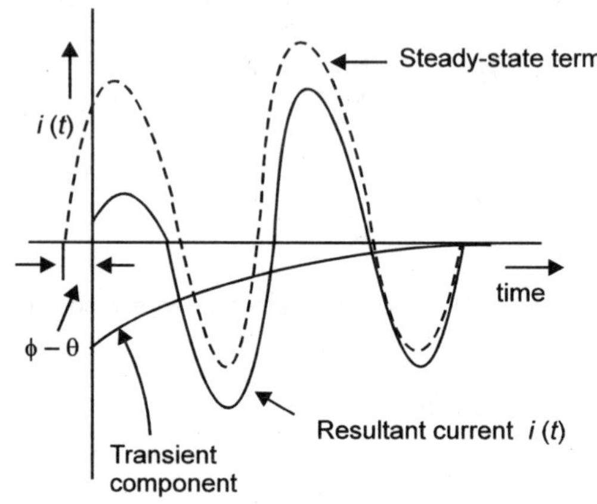


**Fig. 7.18**  Asymmetrical alternting current.

component and since the transient component is negative the net current is zero at $t = 0$. It can be seen that the transient component will be zero in case the switching on of the voltage wave is done when $\theta = \phi$, i.e. when the wave is passing through an angle $\phi = \tan^{-1} \dfrac{\omega L}{R}$. This is the situation when we have no transient even though the circuit contains inductance and there is switching operation also. On the other hand if $\phi - \theta = \pm\pi/2$, the transient term will have its maximum value and the first peak of the resulting current will be twice the peak value of the sinusoidal steady state component.

## 7.9  CONVOLUTION INTEGRAL

Convolution integral finds two very important applications in network theorem.
- (a) It enables to evaluate the response of a network to an arbitrary input in terms of the impulse response of the network.
- (b) If $F(s)$ is the Laplace transform of a function $f(t)$ and if $F(s)$ could be expressed as product of two functions $F_1(s)$ and $F_2(s)$, then Convolutions integral can be used to obtain $f(t)$ which makes the calculations of Laplace inverse much easier.

Consider a two-terminal pair network shown in Fig. 7.19.

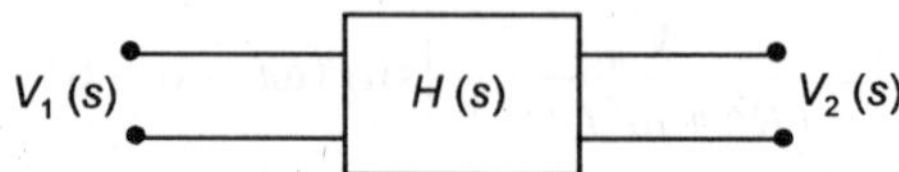


**Fig. 7.19**  Two-terminal-pair network.

Assuming initial conditions to be zero, the input the output voltage transform can be related by the equation.

$$V_2(s) = V_1(s)\ H(s) \qquad \qquad ...(7.101)$$

Here $H(s)$ is known as the transfer function of the network. If $V_1(s)$ is the transform of the impulse function $\delta(t)$, $V_1(s) = 1$ and hence

$$V_2(s) = H(s) \qquad \qquad ...(7.102)$$

or
$$V_2(t) = h(t) \qquad \qquad ...(7.103)$$

Since $h(t)$ is the impulse response of the network and is also the inverse Laplace transform of the transfer function $H(s)$, therefore, the impulse response characterizes the network. This means if the impulse response is given, the input voltage need be specified (for which the output, response is now required) so that the convolution of this voltage/current with $h(t)$ gives the output voltage. In other words any input convolved with the unit impulse response gives the output corresponding to the input. For example consider a two-port R.C. network in Fig. 7.20.

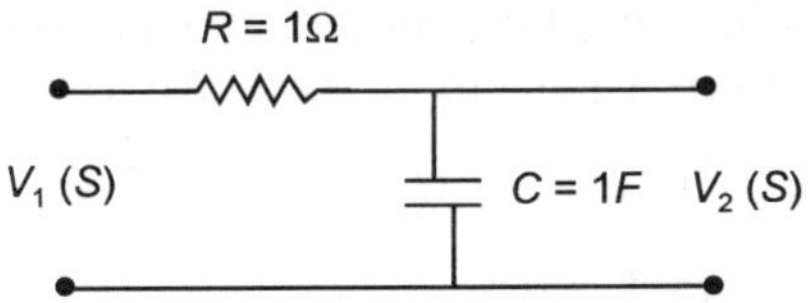


**Fig. 7.20**   R.C. two port network.

The transfer function of the network is $H(s) = \dfrac{1}{s+1}$ therefore $h(t) = e^{-t}$ which is the impulse response of the network.

To find out the output response $V_2(t)$ of this network when the input voltage is $V_1(t) = e^{-3t}$, we convolve $e^{-t}$ with $e^{-3t}$ and the result is the output response.

i.e.
$$v_2(t) = \int_0^t e^{-3(t-\tau)} \cdot e^{-\tau}\, d\tau = \int_0^t e^{-3t} \cdot e^{3\tau} \cdot e^{-\tau}\, d\tau$$

$$v_2(t) = e^{-3t} \int_0^t e^{2\tau}\, d\tau$$

$$= \frac{e^{-3t}}{2} \left[ e^{2\tau} \right]_0^t = \frac{1}{2}\left( e^{-t} - e^{-3t} \right)$$

For this particular example even though partial fraction method is more convenient yet this explains the procedure of applications of convolution integral to such problem and this can be used to advantage for more complex form of the input functions.

The application of convolution integral to obtain inverse Laplace of a function is explained as follows:

Let the two functions $f_1(t)$ and $f_2(t)$ be Laplace transformable and have the transforms $F_1(s)$ and $F_2(s)$. The product of $F_1(s)$ and $F_2(s)$ is the Laplace transform of $f(t)$ which results from the convolution of $f_1(t)$ and $f_2(t)$ and is given by the equation.

$$f(t) = \mathcal{L}^{-1}[F_1(s)\, F_2(s)] = \int_0^t f_1(\tau)\, f_2(t-\tau)\, d\tau$$

$$= \int_0^t f_2(\tau)\, f_1(t-\tau)\, d\tau \qquad \dots(7.104)$$

where $\tau$ is a dummy variable for $t$ and the integrals in these equations are known as convolution integrals and are typically denoted as

$$f(t) = f_1(t) * f_2(t) \qquad \dots(7.105)$$

using the notation we find that

$$F(s) = \mathcal{L}[\, f_1(t) * f_2(t)] = \mathcal{L}[\, f_2(t) * f_1(t)]$$
$$= F_1(s)\, F_2(s) \qquad \dots(7.106)$$

Therefore, the inverse Laplace of the product of $F_1(s)$ and $F_2(s)$ is obtained by convolving $f_1(t)$ and $f_2(t)$. The Convolution theorem of Laplace transform is restated as

$$\mathcal{L}[f_1(t) * f_2(t)] = \mathcal{L}[\, f_2(t) * f_1(t)] = F_1(s)\, F_2(s) \qquad \dots(7.107)$$

To prove this theorem we start with

$$\mathcal{L}[f_1(t) * f_2(t)] = \int_0^\infty \left[ \int_0^t f_1(t-\tau)\, f_2(\tau)\, d\tau \right] e^{-st}\, dt \qquad \dots(7.108)$$

We know that

$$u(t-\tau) = 1 \quad \text{for} \quad t \geq \tau$$

$$= 0 \quad \text{for} \quad t \leq \tau \qquad \dots(7.109)$$

We, therefore, also have the identity

$$\int_0^t f_1(t-\tau)\, f_2(\tau)\, d\tau = \int_0^\infty f_1(t-\tau)\, u(t-\tau)\, f_2(\tau)\, d\tau \qquad \dots(7.110)$$

Therefore the equation (7.108) can be written as

$$\mathcal{L}[f_1(t) * f_2(t)] = \int_0^\infty \left[ \int_0^\infty f_1(t-\tau)\, u\,(t-\tau)\, f_2(\tau)\, d\tau \right] e^{-st}\, dt \qquad \dots(7.111)$$

At this point we introduce a new variable

$$y = t - \tau \quad \text{or} \quad t = y + \tau$$

and we rewrite the equation (7.111)

$$\mathcal{L}\left[f_1(t) * f_2(t)\right] = \int_0^\infty \left[\int_0^\infty f_1(y)\, u(y)\, f_2(\tau)\, d\tau\right] e^{-s(y+\tau)}\, dy$$

$$= \int_0^\infty f_1(y)\, u(y)\, e^{-sy}\, dy \int_0^\infty f_2(\tau)\, e^{-s\tau}\, d\tau$$

$$= F_1(s)\, F_2(s) \qquad \qquad ...(7.112)$$

Thus the convolution theorem is proved.

Suppose we are required to find out Inverse Laplace of $\ F(s) = \dfrac{1}{s(s+1)}$

We let $\qquad \qquad F_1(s) = \dfrac{1}{s}\ $ and $\ F_2(s) = \dfrac{1}{s+1}$

Therefore, $\qquad \qquad f_1(t) = u(t)\ $ and $\ f_2(t) = e^{-t}$

Therefore, $\qquad \quad f(t) = \mathcal{L}^{-1}\, F(s) = \int_0^t u(t-\tau)\, e^{-\tau\, d\tau}$

$$= \int_0^t u(\tau)\, e^{-(t-\tau)}\, d\tau = e^{-t}\int_0^t u(\tau)\, e^\tau\, d\tau$$

$$= e^{-t}(e^t - 1) = 1 - e^{-t}$$

Next suppose we are required to find out inverse Laplace of $F(s) = \dfrac{1}{(s^2+1)^2}$ . Let $F_1(s)$

$= \dfrac{1}{s^2+1} = F_2(s)$

$f_1(t) = \sin\ t$ and $f_2(t) = \sin\ t$

Therefore inverse is given as

$$\mathcal{L}^{-1}[F(s)] = f(t) = \int_0^t \sin\,(t-\tau)\,\sin\,\tau\,d\tau$$

$$= \frac{1}{2}\int_0^t [\cos\,(t-2\tau) - \cos\,t]d\tau$$

$$= \frac{1}{2}\left[-\frac{1}{2}\sin\,(t-2\tau) - \tau\cos\,t\right]_0^t$$

$$= \frac{1}{2}\,(\sin\,t - t\cos\,t)$$

However if the method of partial fraction is used it would really be a tedious job to obtain the Laplace inverse of this function.

In order to understand physical significance of convolution integral let us consider the convolution process. We show here that convolution may be interpreted in terms of steps (*i*) folding (*ii*) translating (*iii*) multiplying; and (*iv*) integrating. These four steps describe the English equivalent of the German word Faltung; the convolution integral is also known as Faltung integral.

We will illustrate this with a simple function $f_1(t) = u(t)$ and $f_2(t) = e^{-t}\,u(t)$. However, as it is a graphical representation, function for which mathematical representation is tedious, can even be handled to find the response to the system to such an input signal.

The various steps associated in the convolution process of these two functions are illustrated in Fig. 7.21.

In Fig. 7.21 (a) $f_1(t)$ and $f_2(t)$ and in (b) $f_1(\tau)$ and $f_2(\tau)$ have been shown. In (c) the functions are folded along the vertical axis ($t = 0$). The function $f_1\,(t - \tau) = u(t - \tau)$ in Fig. 7.21 (d) advances $f_1(-\tau)$ by some typical value of $t$. Similarly $f_2(-\tau)$ is also advanced by the time $t$ to make it $f_2(t - \tau)$. In (e) multiplication of $f_1(t - \tau)$ and $f_2(\tau)$, as shown an integrand, is carried out. Also in (e) the multiplication of $f_1(\tau)$ and $f_2(t - \tau)$ is carried out. The integration of the cross hatched area gives a point on the curve $f(t)$ as shown in (f) for the value of $t$ selected in step (d). We now take next value of $t$ and repeat the procedure to obtain second point on $f(t)$. Carrying out the multiplication and integration steps for various values of $t$ as shown in Fig. 7.22 we obtain the response $f(t)$ as shown in (d) of Fig. 7.22.

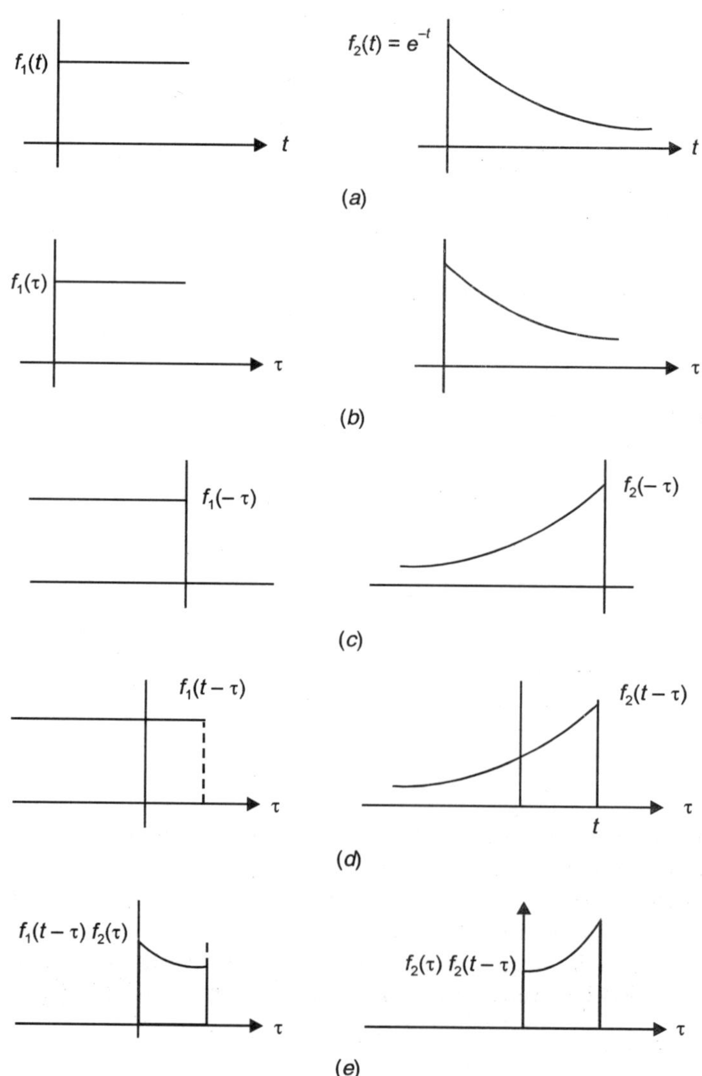

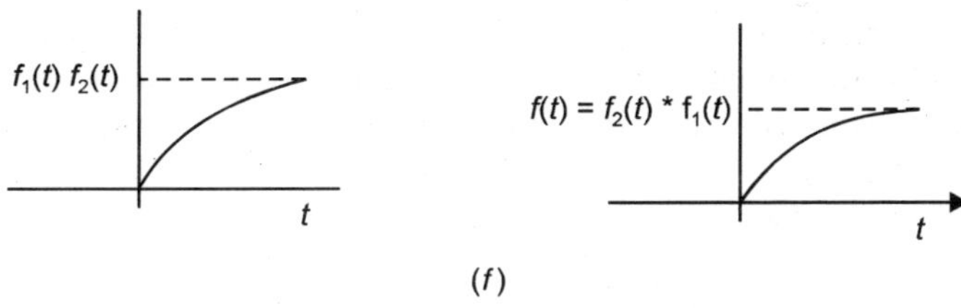

**Fig. 7.21**  Steps in the process of convolutions.

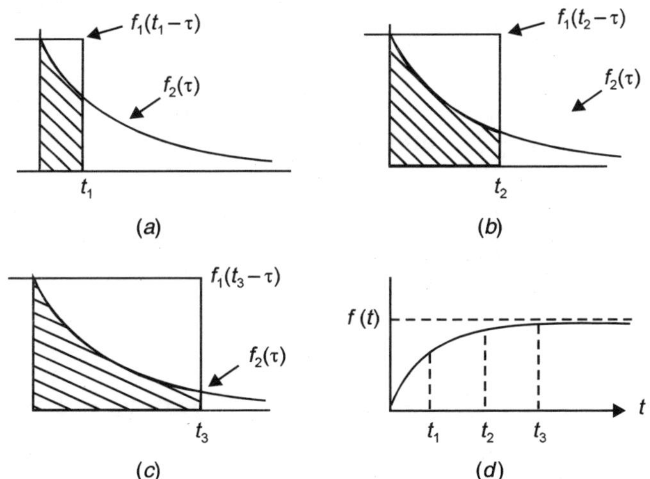

**Fig. 7.22**  Step (*e*) in Fig. 7.21 repeated for three values of *t*.

The shaded area represents a point on the curve $f(t)$ shown in part (*f*) of Fig. 7.21.

**Example 7.14:** Consider a circuit consisting of RLC elements in series. The capacitor is initially charged to a voltage $V_0$ and these are in series through a switch. If the switch is closed, determine the current response.

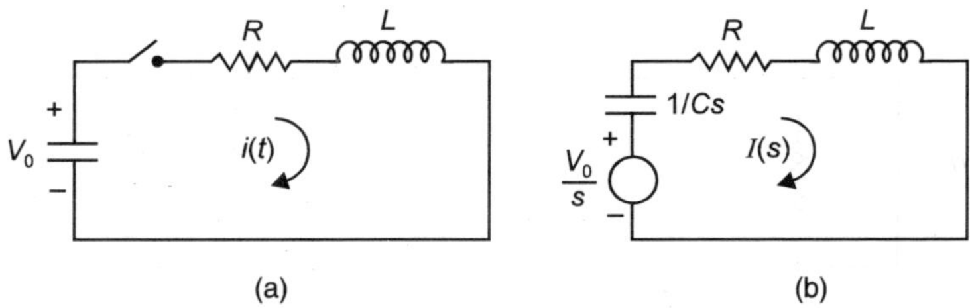

**Fig. E7.14**  (*a*) Original circuit (*b*) Transform equivalent.

We have seen from the previous example how simple it is to solve the network using Laplace transform. In the previous example we have assumed zero initial conditions and we find that the transformed network is a simple network to handle. A few steps of algebraic manipulation's and we arrive at the final results. The whole process becomes very simple, there is no $j$ term or anything that we come across even in steady state solution of the network.

Even when initial conditions are given the equivalent transform network can be drawn and straight away algebraic equations are written and the required result in the form of its transform is obtained. From this the time-domain solution can be obtained through Laplace inversion.

From the transform network

$$I(s) = \frac{V_0/s}{R + sL + 1/cs}$$

$$= \frac{V_0}{L}\frac{1}{s^2 + \dfrac{R}{L}s + \dfrac{1}{LC}} = \frac{V_0}{L}\frac{1}{(s+a)^2 + \omega^2}$$

where

$$a = \frac{R}{2L} \quad \text{and} \quad \omega = \sqrt{\frac{1}{LC} - \frac{R^2}{4L^2}}$$

$$I(s) = \frac{V_0}{L\omega} \cdot \frac{\omega}{(s+a)^2 + \omega^2}$$

Therefore

$$i(t) = \frac{V_0}{\omega L} \cdot e^{-at} \sin \omega t$$

Suppose

$$R = 2\ \Omega, L = 1H, C = \frac{1}{2}F \quad \text{and} \quad V_0 = 1 \text{ volt}$$

then

$$a = \frac{R}{2L} = \frac{2}{2} = 1$$

$$\omega = \sqrt{2 - \frac{4}{4 \times 1}} = 1$$

Therefore

$$i(t) = e^{-t} \sin t\ u(t)$$

**Example 7.15:** Determine the voltage across the parallel combination in the given circuit when it is connected across a current source $I_0\ \delta(t)$.

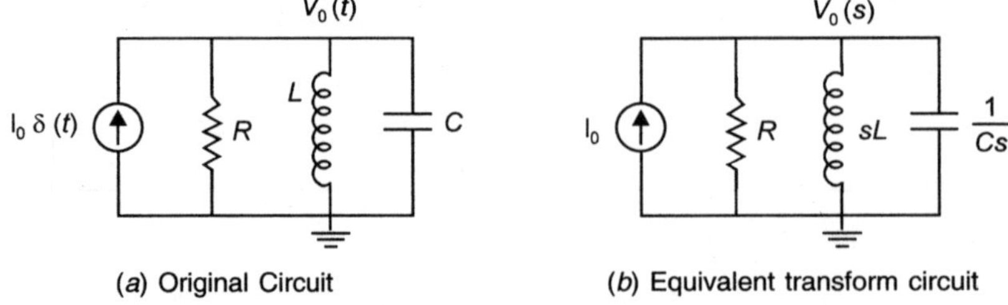



(a) Original Circuit  (b) Equivalent transform circuit

**Fig. E7.15**  Network.

Using transform network and writing nodal equation we have

$$\frac{V_0(s)}{R} + \frac{V_0(s)}{sL} + V_0 Cs = I_0$$

$$V_0(s) = \frac{RLI_0\,s}{RLCs^2 + sL + R} = \frac{I_0\,s/C}{s^2 + \dfrac{s}{RC} + \dfrac{1}{LC}}$$

$$= \frac{I_0\,s/C}{\left(s + \dfrac{1}{2RC}\right)^2 + \dfrac{1}{LC} - \dfrac{1}{4R^2C^2}} = \frac{I_0\,s/C}{(s + a)^2 + b^2}$$

where
$$a = \frac{1}{2RC} \quad \text{and} \quad b = \sqrt{\frac{1}{LC} - \frac{1}{4R^2C^2}}$$

or
$$V_0(s) = \frac{I_0}{C}\left[\frac{s + a - a}{(s + a)^2 + b^2}\right]$$

$$= \frac{I_0}{C}\left[\frac{s + a}{(s + a)^2 + b^2} - \frac{a}{b}\,\frac{b}{(s + a)^2 + b^2}\right]$$

$$V_0(t) = \frac{I_0}{C}\left[e^{-at}\cos b\,t - e^{-at}\,\frac{a}{b}\,\sin bt\right]$$

$$= \frac{I_0}{C}\,e^{-at}\left[\cos bt - \frac{a}{b}\,\sin bt\right]$$

$$= \frac{I_0}{bC}\,e^{-at}\left[b\cos bt - a\sin bt\right]\ \textbf{Ans.}$$

**Example 7.16:** Determine the currents in the loop of the network shown assuming zero initial conditions.

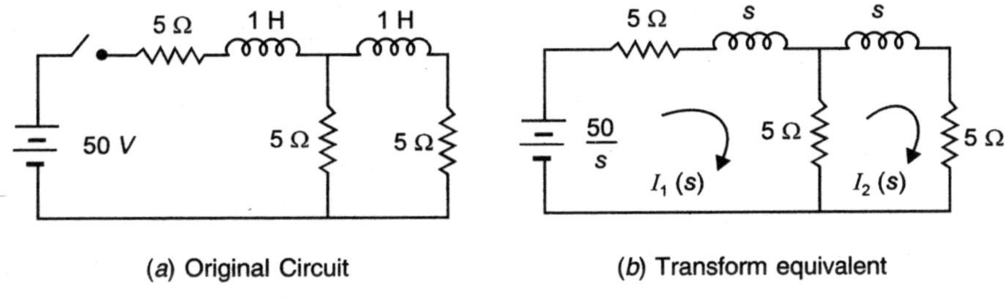


**Fig. E7.16**  Network.

The loop equations are:

$$\frac{50}{s} = 5\,I_1(s) + s\,I_1(s) + 5\,[I_1(s) - I_2(s)] = 0$$

$$= (10 + s)\, I_1(s) - 5\, I_2(s)$$

and

$$sI_2(s) + 5\, I_2(s) + 5\, [I_2(s) - I_1(s)] = 0$$

or

$$0 = -5\, I_1(s) + (10 + s)\, I_2(s)$$

In matrix form

$$\begin{pmatrix} 10 + s & -5 \\ -5 & 10 + s \end{pmatrix} \begin{pmatrix} I_1(s) \\ I_2(s) \end{pmatrix} = \begin{pmatrix} 50/s \\ 0 \end{pmatrix}$$

Using Cramer's rule

$$I_1(s) = \frac{\begin{vmatrix} 50/s & -5 \\ 0 & 10 + s \end{vmatrix}}{\begin{vmatrix} 10 + s & -5 \\ -5 & 10 + s \end{vmatrix}} = \frac{\dfrac{50}{s}(10 + s)}{(s + 10)^2 - 25}$$

$$= \frac{50\,(s + 10)}{s\,(s + 15)\,(s + 5)} = \frac{A}{s} + \frac{B}{(s + 15)} + \frac{C}{(s + 5)}$$

$$A = s\,I_1(s)\Big|_{s \to 0} = \frac{50\,(s + 10)}{(s + 15)\,(s + 5)}\Big|_{s \to 0} = \frac{20}{3}$$

$$B = (s + 15)\,I_1(s)\Big|_{s \to -15} = \frac{50(-5)}{(-15)\,(-10)} = -\frac{5}{3}$$

$$C = (s + 5)\,I_1(s)\Big|_{s \to -5} = \frac{50 \times 5}{(-5) \times 10} = -5$$

$$I_1(s) = \frac{6.666}{s} - \frac{5}{3} \cdot \frac{1}{(s + 15)} - \frac{5}{s + 5}$$

$$I_1(t) = 6.66\ u(t) - \frac{5}{3}\, e^{-15t} - 5\, e^{-5t}$$

Similarly

$$I_2(s) = \frac{\begin{vmatrix} 10 + s & 50/s \\ -5 & 0 \end{vmatrix}}{\begin{vmatrix} 10 + s & -5 \\ -5 & 10 + s \end{vmatrix}} = \frac{250/s}{(s + 15)\,(s + 5)}$$

$$= \frac{250}{s\,(s + 5)\,(s + 15)} = \frac{A}{s} + \frac{B}{s + 5} + \frac{C}{(s + 15)}$$

$$A = \underset{s \to 0}{Lt}\ s\,I_2(s) = \frac{250}{5 \times 15} = \frac{10}{3}$$

$$B = \underset{s \to -5}{Lt} (s + 5) \; I_2 (s) = \frac{250}{(-5) \times 10} = -5$$

$$C = \underset{s \to -15}{Lt} (s + 15) \; I_2(s) = \frac{250}{(-15)(-10)} = 1.67$$

Therefore,

$$I_2(s) = \frac{3 \cdot 33}{s} - \frac{5}{s + 5} + \frac{1.67}{s + 15}$$

$$i_2(t) = 3.33 \; u(t) - 5 \; e^{-5t} + 1.67 \; e^{-15t} \textbf{ Ans.}$$

**Example 7.17:** Determine the transform current $I(s)$ in the given circuit assuming initial conditions to be zero.

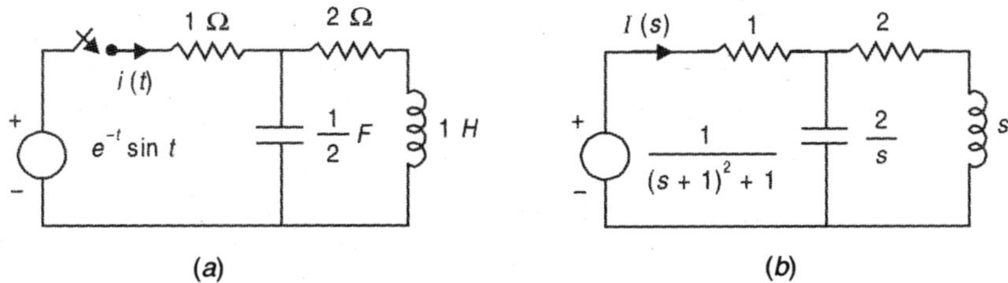



(a)       (b)

**Fig. 7.17** (a) Original circuit, (b) Transform equivalent.

Fig. 7.17 (b) shows the equivalent transform of Fig. 7.17 (a)

Current
$$I(s) = \frac{V(s)}{Z(s)}$$

$$Z(s) = \frac{(s + 2) \cdot 2/s}{s + 2 + \dfrac{2}{s}} + 1 = \frac{s^2 + 4s + 6}{s^2 + 2s + 2}$$

Therefore,
$$I(s) = \frac{s^2 + 2s + 2}{\left\{(s + 1)^2 + 1\right\} (s^2 + 4s + 6)}$$

**Example 7.18:** Solve the differential equation
$$\ddot{x} + 5\dot{x} + 6x = 0$$

Assuming
$$x(0) = 0 \quad \dot{x}(0) = 5$$

**Solution:** Taking Laplace transform of the equation

$$s^2 \, x(s) - s \, x(0) - \dot{x}(0) + 5 \, s x(s) - 5 \, x(0) + 6 \, x(s) = 0$$

or
$$s^2 \, x(s) - 5 + 5 \, s x(s) + 6 \, x(s) = 0$$

or
$$x(s) \{s^2 + 5s + 6\} = 5$$

or
$$x(s) = \frac{5}{s^2 + 5s + 6} = \frac{5}{(s+3)(s+2)} = \frac{A}{s+3} + \frac{B}{s+2}$$

$$A = \underset{s \to -3}{Lt} (s + 3) \; x(s) = -5$$

$$B = \underset{s \to -2}{Lt} \; (s + 2) \; x \; (s) = 5$$

or

$$x(s) = \frac{5}{s + 2} - \frac{5}{s + 3}$$

$$x(t) = 5 \; [e^{-2t} - e^{-3t}] \; \textbf{Ans.}$$

**Example 7.19:** Determine the transform currents in the loops of the given network.

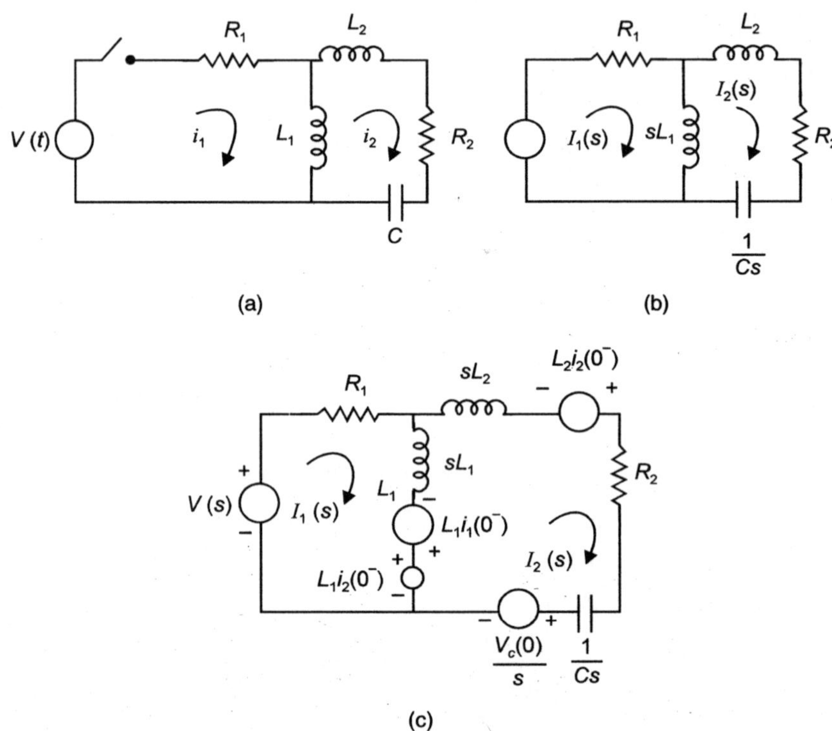



**Fig. E7.19** (a) Original Network (b) Transform Network without initial condition (c) with initial condition.

From the transform network

$$V(s) = I_1(s) \; R_1 + sL_1 I_1(s) - sL_1 I_2(s) - L_1 i_1(0^-) + L_1 i_2(0^-)$$

or

$$V(s) + L_1 \, i_1(0^-) - L_1 \, i_2 \, (0^-) = I_1(s) \, \{R_1 + sL_1\} - I_2 \, (s) \, sL_1$$

$$V_1(s) = I_1(s) \, \{R_1 + sL_1\} - I_2 \, sL_1 = V(s) + L_1 i_1(0^-) - L_1 i_2(0^-)$$

Similarly for loop 2

$$sL_2 \, I_2 \, (s) + R_2 \, I_2 \, (s) + I_2 \, (s) \; . \; 1/Cs + sL_1 \, I_2 \, (s) - sL_1 \, I_1 \, (s)$$

$$= -\frac{V_c(0)}{s} + L_1 \, i_2 \, (0^-) - L_1 \, i_1 \, (0^-) + L_2 \, i_2 \, (0^-)$$

or

$$-sL_1 \, I_1(s) + I_2(s) \, \{R_2 + sL_1 + sL_2 + 1/Cs\} = V_2(s)$$

or

$$V_2(s) = -sL_1 \, I_1(s) + \left( R_2 + sL_1 + sL_2 + \frac{1}{Cs} \right) I_2(s)$$

$$= -\frac{V_c(0)}{s} + L_1[i_2(0) - i_1(0)] + L_2 i_2(0)$$

These can be written in matrix form as

$$\begin{pmatrix} R_1 + sL_1 & -sL_1 \\ -sL_1 & R_2 + sL_1 + sL_2 + \dfrac{1}{Cs} \end{pmatrix} \begin{pmatrix} I_1(s) \\ I_2(s) \end{pmatrix} = \begin{pmatrix} V_1(s) \\ V_2(s) \end{pmatrix}$$

Hence by using Cramer's rule

$$I_1(s) = \frac{\begin{vmatrix} V_1(s) & -sL_1 \\ V_2(s) & R_2 + sL_1 + sL_2 + \dfrac{1}{Cs} \end{vmatrix}}{\Delta}$$

where

$$\Delta = \begin{vmatrix} R_1 + sL_1 & -sL_1 \\ -sL_1 & R_2 + sL_1 + sL_2 + \dfrac{1}{Cs} \end{vmatrix}$$

and

$$I_2(s) = \frac{\begin{vmatrix} R_1 + sL_1 & V_1(s) \\ -sL_1 & V_2(s) \end{vmatrix}}{\Delta}$$

**Example 7.20:** The circuit in Fig. E7.20 is operating under steady state condition when the switch is at position '*a*' and at $t = 0$, the switch is moved to position '*b*' determine the current $i(t)$.

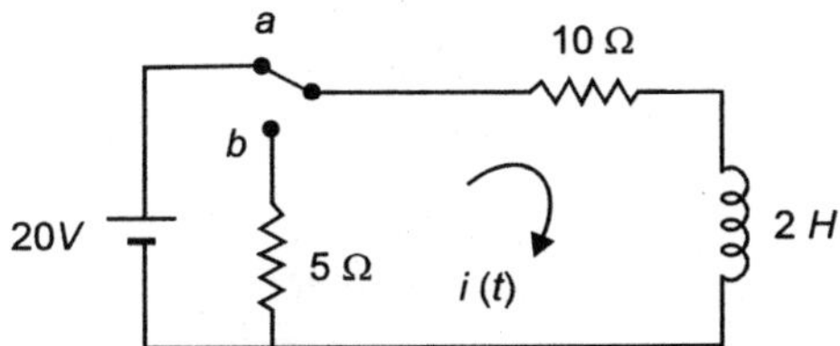



**Fig. E7.20** (a) Original circuit when switch is at position '*a*' and (b) When it is at (b).

**Solution:** When the switch is in position '*a*' the steady state current through the 2H inductor is $\dfrac{20}{10} = 2A$.

Now when the switch is thrown to position '*b*', the equivalent transform network is given as

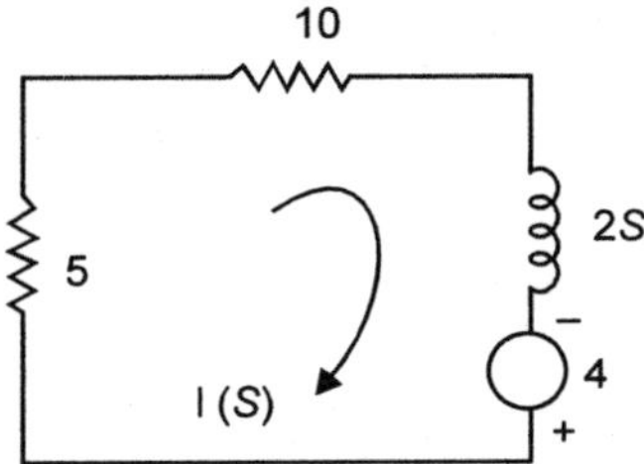



**Fig. E7.20** (b)

The loop equation

$$15\ I(s) + 2\ s\ I(s) = 4$$

$$I(s) = \frac{4}{2s + 15} = \frac{2}{s + 7.5}$$

$$i(t) = 2\ e^{-7.5t}\ A\ \textbf{Ans.}$$

**Example 7.21:** A voltage pulse of height $k$ and duration $T$ is applied to a series $R\text{-}C$ network as shown. Determine the current response of the network.

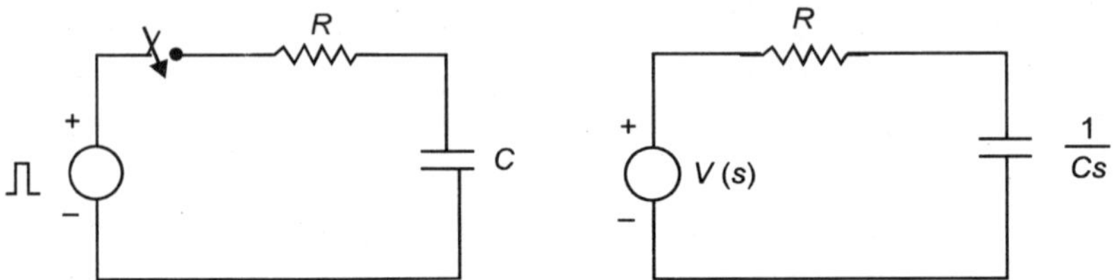

**Fig 7.21**   (*a*) Original circuit (*b*) Transform equivalent.

$$v(t) = k\ u(t) - k\ u(t - T)$$

$$v(s) = \frac{k}{s} - \frac{k}{s}\ e^{-Ts}$$

Since the circuit is initially relaxed, the transform loop equation

$$\left(R + \frac{1}{Cs}\right) I(s) = \frac{k}{s}\ (1 - e^{-Ts})$$

Therefore,
$$I(s) = \frac{k}{s} \cdot (1 - e^{-Ts})\ \frac{Cs}{RCs + 1}$$

$$= \frac{k}{s} \cdot (1 - e^{-Ts}) \cdot \frac{Cs/RC}{s + 1/RC}$$

$$= \frac{k}{R}\ \frac{1 - e^{-Ts}}{s + 1/RC}$$

$$= \frac{k}{R}\left[\frac{1}{s + 1/RC} - \frac{e^{-Ts}}{s + 1/RC}\right]$$

$$i(t) = \frac{k}{R}\left[e^{-t/CR}\ u(t) - e^{-(t-T)/CR}\ u(t - T)\right]$$

To obtain voltage across the capacitor

$$V_2(s) = I(s) \cdot \frac{1}{Cs}$$

$$= \frac{k}{RC}\left[\frac{1}{s\left(s+\dfrac{1}{RC}\right)} - \frac{e^{-Ts}}{s\left(s+\dfrac{1}{RC}\right)}\right]$$

Let

$$\frac{1}{s\left(s+\dfrac{1}{RC}\right)} = \frac{A}{s} + \frac{B}{\left(s+\dfrac{1}{RC}\right)}$$

$$A = \underset{s\to 0}{Lt}\, sV_2(s) = RC$$

$$B = \underset{s\to -\frac{1}{Rc}}{Lt}\left(s+\frac{1}{RC}\right) = -\,RC$$

Therefore,

$$V_2(s) = k\left[\frac{1}{s} - \frac{1}{s+RC}\right] - k\,e^{-Ts}\left[\frac{1}{s} - \frac{1}{s+1/RC}\right]$$

Hence

$$v_2(t) = k\left(1 - e^{-t/RC}\right)u(t) - k\left\{1 - e^{-(t-T)/RC}\right\}u(t-T)\ \textbf{Ans.}$$

**Example 7.22:** A triangular voltage pulse of duration $T$ and peak value unity is switched into a series $RL$ and initially relaxed network. Determine the current response.

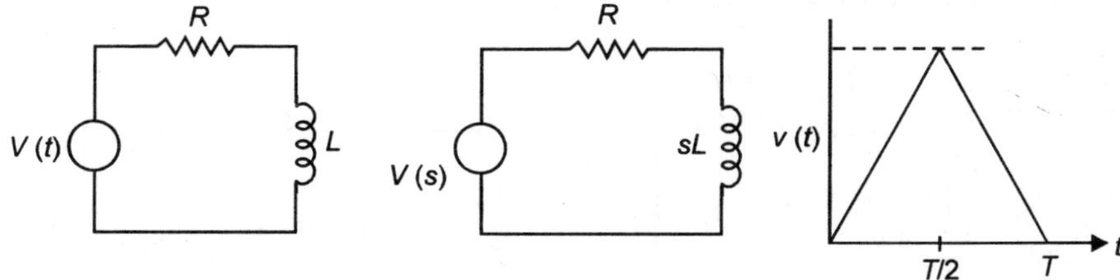



**Fig. E7.22** (*a*) Circuit (*b*) Transform equivalent (*c*) Input waveform.

As derived in Example 7.9 the transform of the voltage source is $\dfrac{2\left(1-e^{-Ts/2}\right)^2}{Ts^2}$

The current transform $I(S) = \dfrac{2}{Ts^2}\cdot\dfrac{\left(1-e^{-Ts/2}\right)^2}{R+sL}$

$$= \frac{2}{LTs^2}\,\frac{\left(1-e^{-Ts/2}\right)^2}{s+R/L}$$

Let   $R = 1$   and   $L = 1$ H then the transform current

$$I(s) = \frac{2}{T}\,\frac{1 - 2e^{-Ts/2} + e^{-Ts}}{s^2(s+1)}$$

To consider only $\dfrac{1}{s^2\,(s+1)}$

Let

$$\frac{1}{s^2\,(s+1)} = \frac{A_1}{s} + \frac{A_2}{s^2} + \frac{B}{s+1}$$

$$A_1 = \frac{d}{ds}\left[s^2 \cdot \frac{1}{(s+1)s^2}\right]_{s\to 0} = -1$$

$$A_2 = \mathop{Lt}_{s\to 0}\ s^2 \cdot \frac{1}{s^2\,(s+1)} = 1$$

$$B = \mathop{Lt}_{s\to -1}\ (s+1)\cdot \frac{1}{s^2\,(s+1)} = 1$$

Therefore,

$$I(s) = \left(1 - 2e^{-Ts/2} + e^{-Ts}\right)\left[\frac{1}{s^2} - \frac{1}{s} + \frac{1}{s+1}\right]$$

Hence

$$i(t) = \left[t - 1 + e^{-t}\right]\left[u(t) - 2u\left(t - \frac{T}{2}\right) + u(t - T)\right]\ \textbf{Ans.}$$

**Example 7.23:** In Fig. E7.23 the switch is initially at position 'a'. After steady state condition is reached when $i(0^-) = 2A$ and $V_c(0^-) = 2$ V. The switch is now thrown to position 'b'. Determine the current in the circuit.

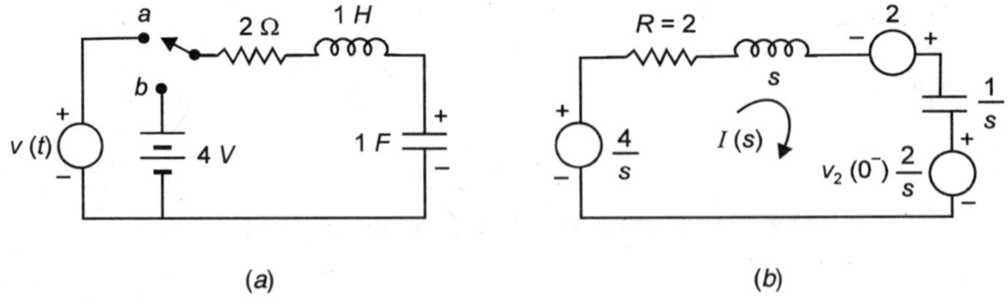


**Fig. E7.23** (a) Original circuit (b) Transform equivalent.

The transform equation of the above network is

$$\frac{4}{s} = I(s)\left\{2 + s + \frac{1}{s}\right\} - 2 + \frac{2}{s}$$

$$\frac{2}{s} + 2 = I(s)\,\frac{s^2 + 2s + 1}{s}$$

or

$$\frac{2(s+1)}{s} = I(s)\cdot\frac{(s+1)^2}{s}$$

or
$$I(s) = \frac{2}{s+1}$$

or
$$i(t) = 2\,e^{-t}\,A \quad \textbf{Ans.}$$

**Example 7.24:** In Fig. E7.24 the switch is initially closed. After steady state is reached the switch is opened. Determine the nodal voltage $V_a(t)$ and $V_b(t)$.

**Solution:** Under steady state condition $i_L(0^-) = \dfrac{5}{1} = 5A$ and

$V_c(0^-) = 5$ volt
The transform network becomes

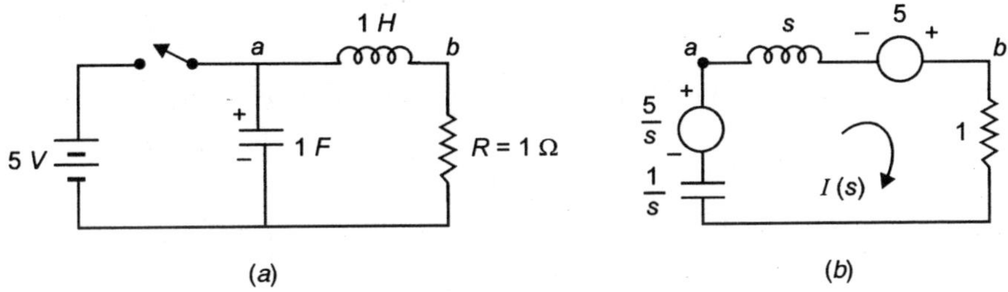

(a)           (b)

**Fig. E7.24** (a) Original circuit, (b) Transform equivalents.

Transforming the voltage sources into current sources we have the network.

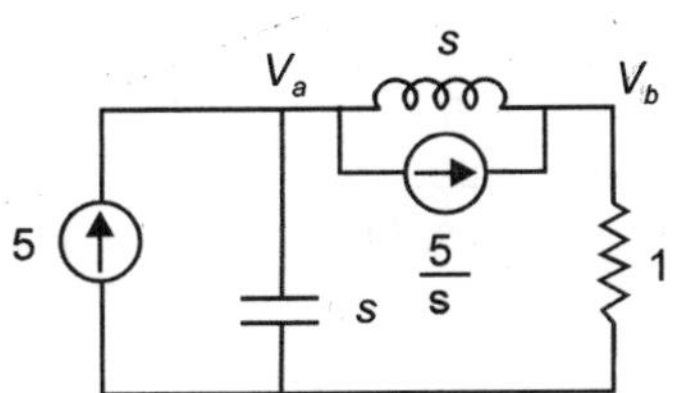

**Fig. E7.24** (c) Voltage source replaced by current source.

Writing nodal equations at node $a$ and $b$ respectively we have

$$\begin{bmatrix} \left(s + \dfrac{1}{s}\right) & -1/s \\[2mm] -\dfrac{1}{s} & 1 + \dfrac{1}{s} \end{bmatrix} \begin{bmatrix} V_a \\[2mm] V_b \end{bmatrix} = \begin{bmatrix} 5 - 5/s \\[2mm] 5/s \end{bmatrix}$$

Using Cramer's rule we have

$$V_a = \frac{\begin{vmatrix} 5 - \dfrac{5}{s} & -1/s \\[3mm] 5/s & 1 + \dfrac{1}{s} \end{vmatrix}}{\begin{vmatrix} s + \dfrac{1}{s} & -1/s \\[3mm] -1/s & 1 + \dfrac{1}{s} \end{vmatrix}} = \frac{\left(5 - \dfrac{5}{s}\right)\left(1 + \dfrac{1}{s}\right) + 5/s^2}{\left(s + \dfrac{1}{s}\right)\left(1 + \dfrac{1}{s}\right) - \dfrac{1}{s^2}}$$

$$= \frac{5\,s^2}{s^3 + s^2 + s} = \frac{5s + 5/2 - 5/2}{\left(s + \dfrac{1}{2}\right)^2 + \left(\dfrac{\sqrt{3}}{2}\right)^2}$$

$$= \frac{5\left(s + \dfrac{1}{2}\right)}{\left(s + \dfrac{1}{2}\right)^2 + \left(\dfrac{\sqrt{3}}{2}\right)^2} - \frac{\dfrac{5}{2}\cdot\dfrac{2}{\sqrt{3}}\cdot\dfrac{\sqrt{3}}{2}}{\left(s + \dfrac{1}{2}\right)^2 + \left(\dfrac{\sqrt{3}}{2}\right)^2}$$

$$V_a(t) = 5\,e^{-t/2}\,\cos\frac{\sqrt{3}}{2}\,t - \frac{5}{\sqrt{3}}\,\sin\frac{\sqrt{3}}{2}\,t\cdot e^{-t/2}$$

$$= 5\,e^{-t/2}\left[\cos\frac{\sqrt{3}}{2}\,t - \frac{1}{\sqrt{3}}\,\sin\frac{\sqrt{3}}{2}\,t\right]$$

Similarly

$$V_b(s) = \frac{\begin{vmatrix} s + \dfrac{1}{s} & 5 - \dfrac{5}{s} \\[2mm] -\dfrac{1}{s} & 5/s \end{vmatrix}}{(s^3 + s^2 + s)/s^2} = \frac{5\,(s + 1)}{s^2 + s + 1}$$

$$= 5\left[\frac{s + \dfrac{1}{2} - \dfrac{1}{2}}{s^2 + s + 1} + \frac{1}{s^2 + s + 1}\right]$$

$$= 5\left[\frac{s + \dfrac{1}{2}}{\left(s + \dfrac{1}{2}\right)^2 + \left(\dfrac{\sqrt{3}}{2}\right)^2} + \frac{\dfrac{1}{2}\cdot\dfrac{2}{\sqrt{3}}\cdot\dfrac{\sqrt{3}}{2}}{\left(s + \dfrac{1}{2}\right)^2 + \left(\dfrac{\sqrt{3}}{2}\right)^2}\right]$$

or

$$V_b(t) = 5\,e^{-t/2}\,\cos\frac{\sqrt{3}}{2}\,t + \frac{5}{\sqrt{3}}\,e^{-t/2}\,\sin\frac{\sqrt{3}}{2}\,t$$

$$= 5\,e^{-t/2}\left[\cos\frac{\sqrt{3}}{2}\,t + \frac{1}{\sqrt{3}}\,\sin\frac{\sqrt{3}}{2}\,t\right] \quad \textbf{Ans.}$$

**Example 7.25:** For the network shown determine the Thevenin's and Norton's equivalent circuit at the load terminal having load $R$ in series with $L$ as shown in the circuit.

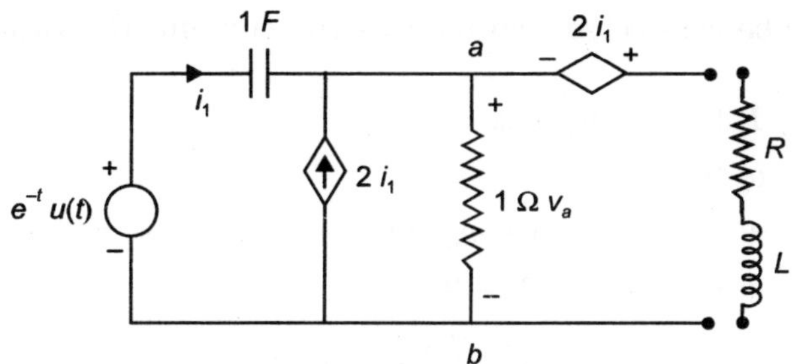

**Fig. E7.25**

**Solution:** The transformed circuit is as shown

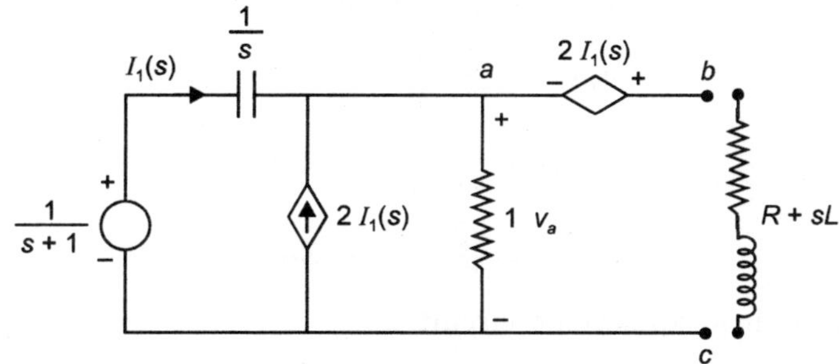

**Fig. E7.25.1**

Current through $1\,\Omega$ resistence is $I_1(s) + 2I_1(s) = 3\ I_1(s)$ downward

Hence,
$$v_a(s) = 3\ I_1(s)\ .1 = 3\ I_1(s)$$

$$V_{TH} = v_a + 2\ I_1(s) = 3\ I_1(s) + 2\ I_1(s) = 5\ I_1(s)$$

Now
$$I_1(s) = \frac{[v(s) - v_a(s)]}{1/s} = s\left[\frac{1}{s+1} - 3I_1(s)\right]$$

or
$$I_1(s) + 3I_1(s)\ s = \frac{s}{s+1}$$

or
$$I_1(s)\ [1 + 3s] = \frac{s}{s+1}$$

or
$$I_1(s) = \frac{s}{(s+1)(1+3s)}$$

$$\therefore \qquad V_{TH} = \frac{5s}{(s+1)(3s+1)}$$

Now to find $R_{TH}$ short circuit independent voltage sources and open circuiting the indepedent current sources, leaving the dependent sources as they are, we have

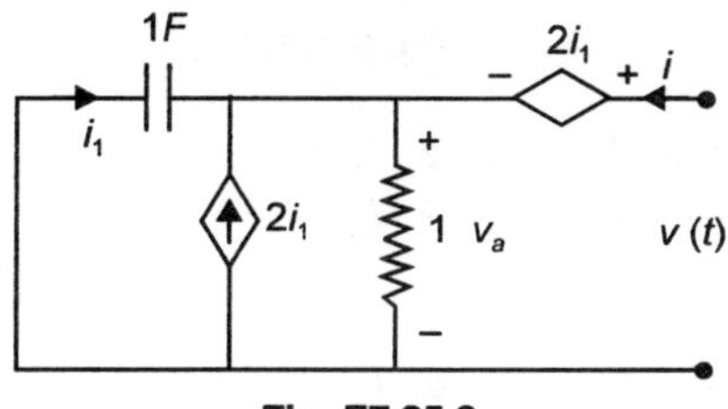

**Fig. E7.25.2**

Let a sources $v(t)$ be connected and let $i$ be the current, the ratio $v(s)$ to $i(s)$ would give $Z_{TH}(s)$

$$I_1(s) = -v_a(s).s$$

$$v_a(s) = 3I_1(s)$$

$$v(s) = 2 I_1(s) + v_a(s) = -2s\, v_a(s) + v_a(s)$$

$$v(s) = v_a(s)(1 - 2s)$$

$$I(s) = \frac{v_a(s)}{1} - 3 I_1(s)$$

$$= v_a(s) + 3s\, v_a(s)$$

$$= v_a(1 + 3s)$$

Now

$$v_a(s) = \frac{v(s)}{1 - 2s}$$

Therefore,

$$I(s) = \frac{v(s)}{1 - 2s} \cdot (1 + 3s)$$

or

$$\frac{v(s)}{I(s)} = \frac{-2s + 1}{3s + 1} = Z_{TH}$$

Hence the Thevenin's equivalent circuit is

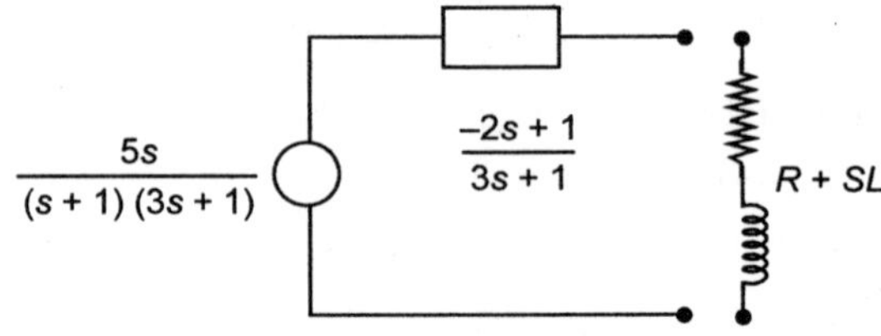

**Fig. E7.25.3**

Now to obtain Norton's equivalent the equivalent shunt impedance is same as calculated for equivalent Thevenin's, we, therefore, need calculate equivalent Norton's current source. To do that we short circuit the *bc* terminal of the circuit in Fig. 7.25.1. Therefore, we have

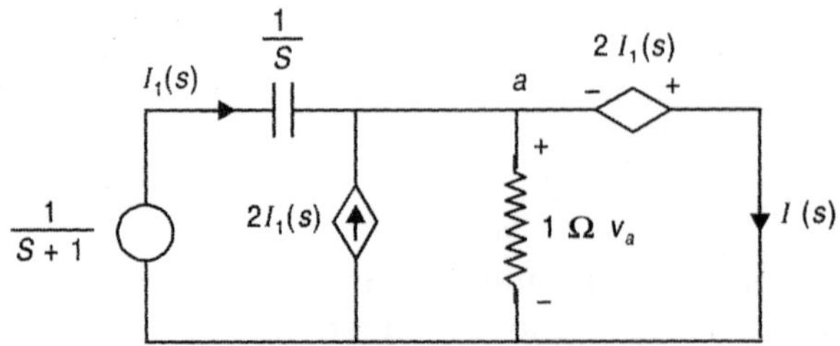

**Fig. E7.25.4**

$$-I(s) + 3 I_1(s) = -2 I_1(s)$$

or

$$-I(s) = -5 I_1(s)$$

In the third loop

$$v_a(s) = -2 I_1(s)$$

$$I_1 = s\left[\frac{1}{s+1} - v_a(s)\right] = \left[\frac{s}{s+1} - s\,v_a(s)\right]$$

$$= \frac{s}{s+1} + 2\,s\,I_1(s)$$

$$I_1(s)\,[1 - 2s] = \frac{s}{s+1}$$

or
$$I_1(s) = \frac{s}{(-2s+1)(s+1)}$$

or
$$I(s) = 5\,I_1(s) = \frac{5s}{(-2s+1)(s+1)}$$

hence Norton's equivalent circuit is

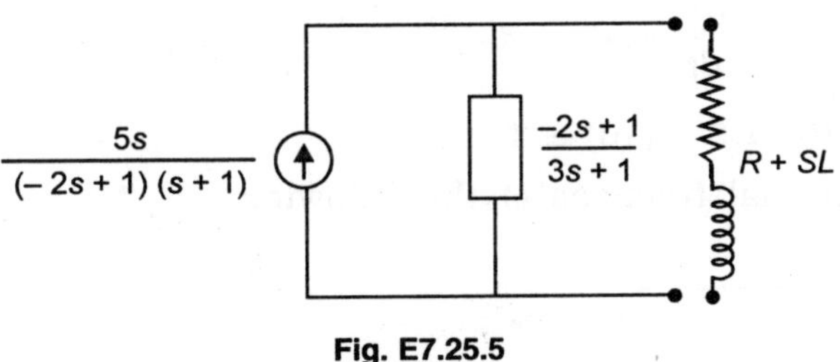


**Fig. E7.25.5**

**PROBLEMS**

**7.1.** Write an equation in time domain for the non-recurring wave form of the Fig. P7.1.

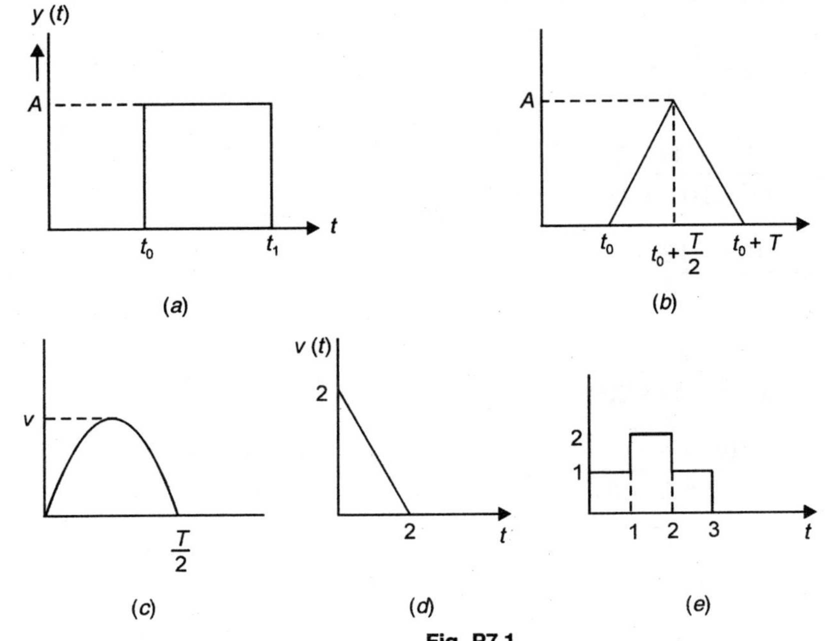


**Fig. P7.1**

**7.2.** Solve the following differential equations using Laplace transform

(a) $\dfrac{d^2y}{dt^2} + 3\dfrac{dy}{dt} + 2y = 5$

$\dot{y}(0^+) = 2 \quad y(0^+) = -1$

(b) $\dfrac{d^3y}{dt^3} + 5\dfrac{d^2y}{dt^2} + 8\dfrac{dy}{dt} = 0$

$y(0^+) = 5 \quad \text{and} \quad \dot{y}(0) = \ddot{y}(0) = 0$

(c) $\dfrac{d^2y}{dt^2} + 8\dfrac{dy}{dt} + 16y = 6e^{-2t}$

$y(0) = 1 \quad \dot{y}(0) = 0$

(d) $\dfrac{d^3y}{dt^3} + 8\dfrac{d^2y}{dt^2} + 37\dfrac{dy}{dt} + 50y = 4e^{-3t}$

$y(0) = 2, \quad \dot{y}(0) = 3 \quad \ddot{y}(0) = 1$

**7.3.** Determine the partial fractions of the following:

(a) $F(s) = \dfrac{s}{s^2\,(s+1)^2\,(s+2)}$

(b) $F(s) = \dfrac{s^2 + 7s + 8}{(s+1)\,(s+2)\,(s+3)}$

(c) $F(s) = \dfrac{(s+4)}{s\,(s+1)\,(s+3)^2\,(s+2)^3}$

(d) $F(s) = \dfrac{1}{s^2 + 4s + 13}$

(e) $F(s) = \dfrac{5s^2 + 4s + 1}{(s+3)\,(s+1)^2}$

(f) $F(s) = \dfrac{10s + 12}{s\,(s+2)^2}$

(g) $F(s) = \dfrac{20}{s\,(s^2 + 4s + 20)}$

(h) $F(s) = \dfrac{10\,(s+2)}{(s+1)^2\,(s+3)}$

(i) $F(s) = \dfrac{1}{s^3 + 2s^2 + 2s + 1}$

(*j*) $G(s) = \dfrac{(s+1)(s+3)}{s(s+2)(s+4)}$

**7.4.** Determine the inverse Laplace transform of the following:

(*a*) $\dfrac{1}{(s+1)(s+2)^2}$

(*b*) $\dfrac{1}{s^2(s^2-1)}$

(*c*) $\dfrac{80(s+1)}{s(s+3)^2(s^2+16)}$

(*d*) $\dfrac{32(s+4)}{s^2(s^2+4s+8)^2}$

(*e*) $\dfrac{1}{s^2+4s+13}$

(*f*) $\dfrac{5s^2+4s+1}{(s+3)(s+1)^2}$

(*g*) $\dfrac{s^2}{(s^2+1)^2}$

(*h*) $\dfrac{s+1}{s^2+2s}$

(*i*) $\dfrac{2s}{(s^2+4)(s+5)}$

(*j*) $\dfrac{5s}{s^2+3s+2}$

(*k*) $\dfrac{s^3+3s^2+3s+2}{s^2+2s+2}$

(*l*) $\dfrac{3\,e^{-s/3}}{s^2(s^2+2)}$

**7.5.** Obtain the Laplace transform of the following functions $f(t)$.

(*a*) $\sinh at$

(*b*) $t^2 e^{-at}$

(*c*) $1-\cos 3\,t$

(*d*) $t^2 e^{-2t}\cos 3t$

(*e*) $\dfrac{d}{dt}\left(5e^{-2t}\cos 3t\right)$

(*f*) $t\cos at$

(*g*) $\sin^2 t$

(*h*) $t^{1/2}$

(*i*) $e^{-at^2}$

(*j*) $\dfrac{1}{2a}\left[\sin at + at\cos at\right]$

(*k*) $f(t) = \sin t \quad 0 < t < \pi$
$\qquad\;\; = \;\; 0 \quad\;\;\; t > \pi$

**7.6.** In the Fig. P7.6 shown here, the capacitor $c_1$ is initially charged to a voltage $V_0$. If now the switch is closed, determine the $V(t)$ across the capacitor $c_2$.

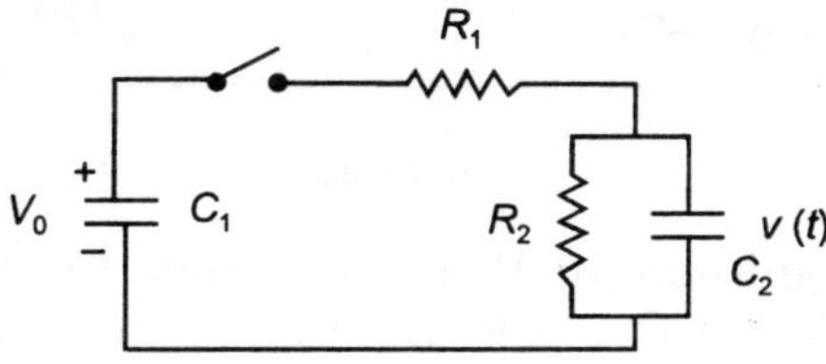


**Fig. P7.6**

**7.7.** In the Fig. P7.7 shown here, the capacitor $c_1$ is initially charged to a voltage $V_0$. If now the switch is closed determine the $v(t)$ across the capacitor $c_2$.

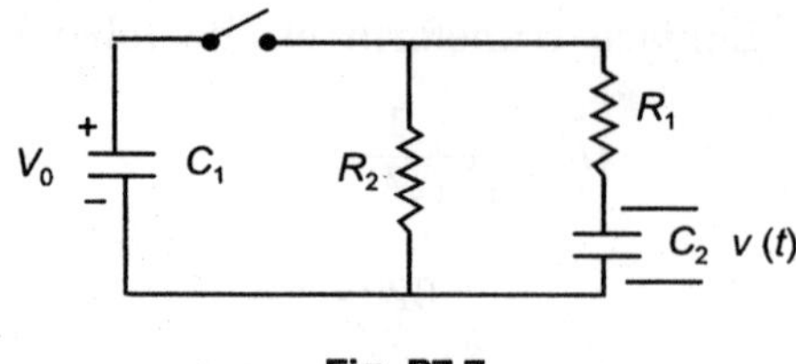

**Fig. P7.7**

**7.8** In the Fig. P7.8 shown the switch is closed at $t = 0$. Determine the current $i(t)$.

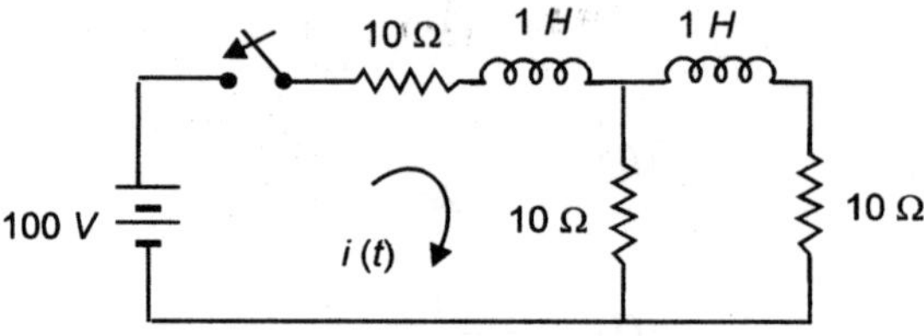

**Fig. P7.8**

**7.9** For the network shown, determine the current $i(t)$ when the switch is closed assume that the network is initially relaxed.

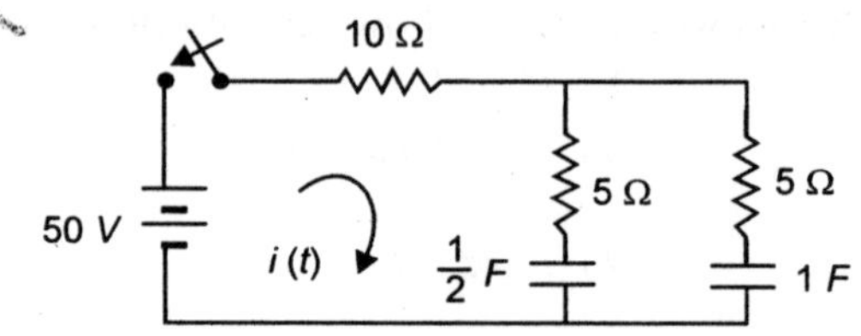

**Fig. P7.9**

**7.10** For the network shown in Fig. P7.10 determine the current $i(t)$ when both the switches are closed simultaneously. Assume that the system is initially relaxed.

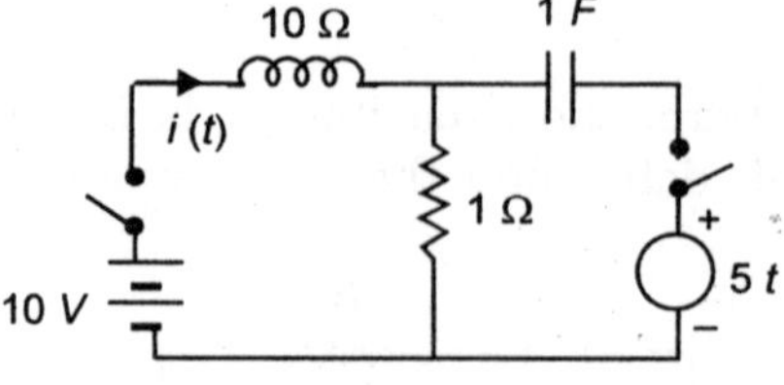

**Fig. P7.10**

**7.11.** A pulse voltage of magnitude 10 V and duration $1\,\mu$ sec is applied to a series $RC$ circuit with $R = 0.5\ \Omega$ and $C = 2\,\text{F}$. Determine the current response.

**7.12.** In the network shown the switch is initially at position '*a*'. After steady state is reached, the switch is changed to position '*b*' at $t = 0$. Determine the current $i(t)$.

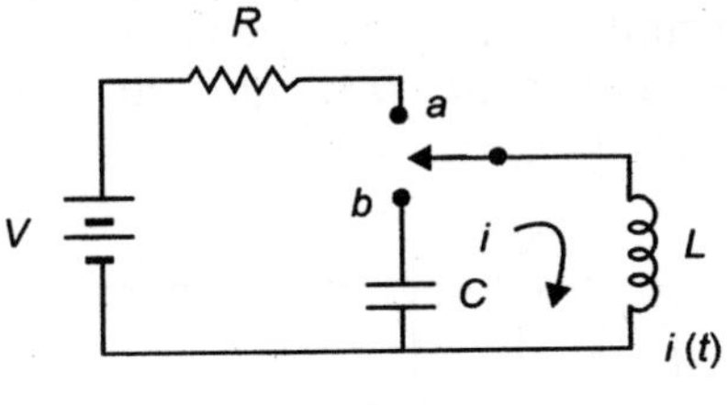

**Fig. P7.12**

**7.13.** In the network shown in Fig. P7.13 the switch is opened at $t = 0$ after steady state condition in the network has reached. Determine current $i_1(t)$ and $i_2(t)$.

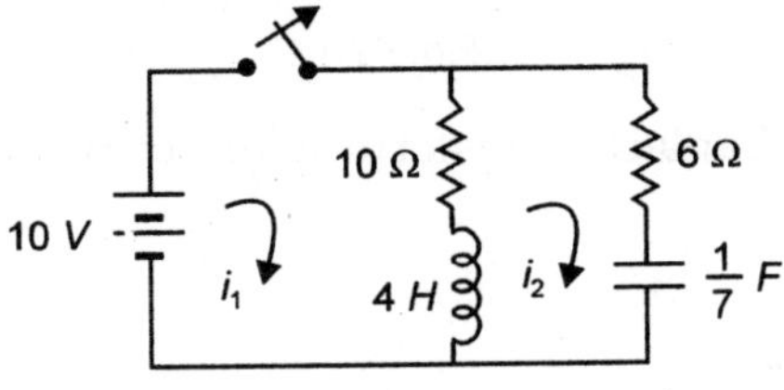

**Fig. P7.13**

**7.14.** In the circuit shown the capacitor is initially charged to a voltage $V_0$. Determine the current $i(t)$ after the switch is closed at $t = 0$.

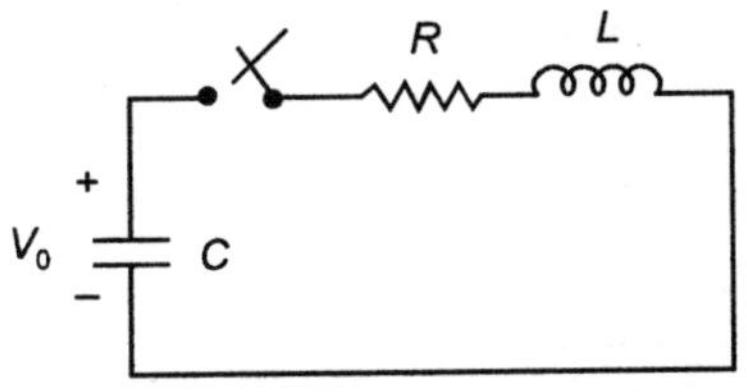

**Fig. P7.14**

**7.15.** Determine the voltage across the capacitor if a unit step voltage is switched in at $t = 0$. Assume the circuit to be initially relaxed.

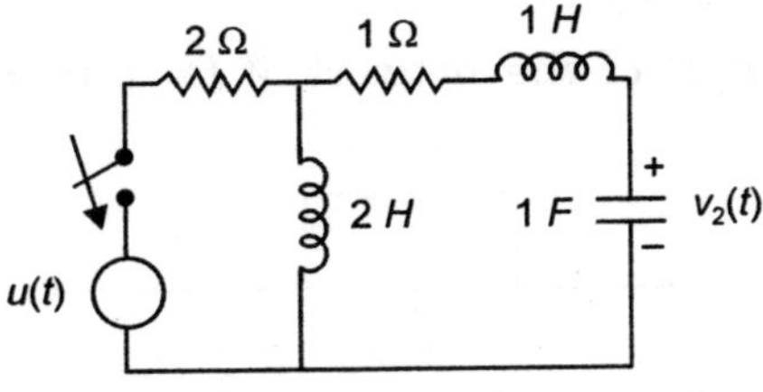

**Fig. P7.15**

**7.16.** Initially the switch is in position 'a' till steady state is reached in the network. Next it is changed over to 'b' at $t = 0$, determine the voltage across $R_2$.

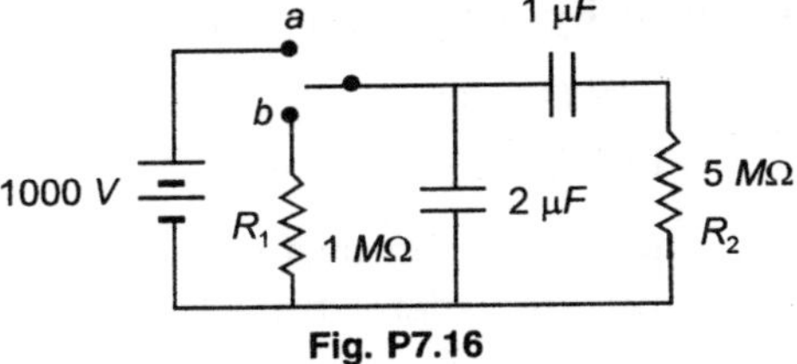

**Fig. P7.16**

**7.17** In the circuit shown initially switch is open and the circuit is under steady state condition. Determine the current $i(t)$ through 400 ohm resistance when the switch is closed. Determine $i(0^-)$ and $i(\infty)$.

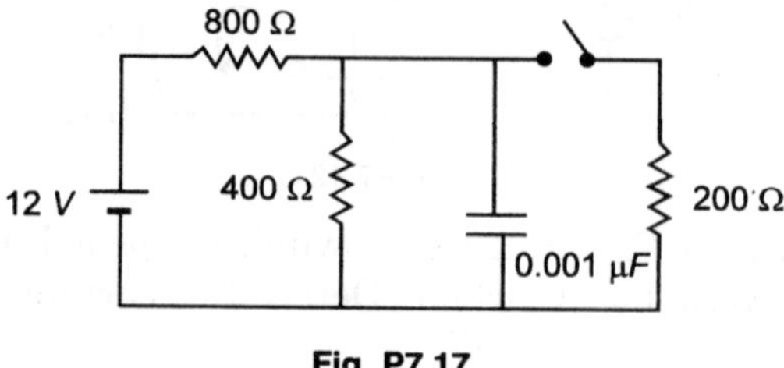

**Fig. P7.17**

**7.18** Determine the inverse Laplace transform using convolution integral for the following functions:

$$F(s) = \frac{1}{(s-a)(s-a)}$$

$$F(s) = \frac{1}{(s+1)(s+2)}$$

$$F(s) = \frac{s}{(s+1)} \cdot \frac{1}{(s^2+1)}$$

$$F(s) = \frac{1}{(s+1)} \cdot \frac{1}{\{(s+1)^2+1\}}$$

$$F(s) = \frac{1}{(s+a)} \cdot \frac{1}{(s+b)(s+c)}$$

**7.19** Determine the output $v_2(t)$ of the network if (a) $v_1(t) = u(t)$ (b) $v_1(t) = 10 \cos 2t$

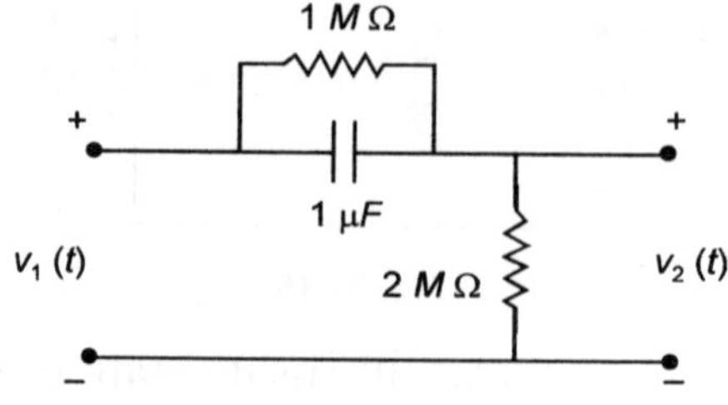

**Fig. P7.19**

**7.20** A linear system is described by a transfer function.

$$H(s) = \frac{12(s+1)}{s(s+2)(s+3)}$$

(a) Determine impulse response of the system

(b) When the input is a unit step, determine the complete response of the system.

**7.21** In the circuit shown determine the currents in the two loops (the initial and final values) using the corresponding theorem and verify the results by logic (without using any mathematical concept).

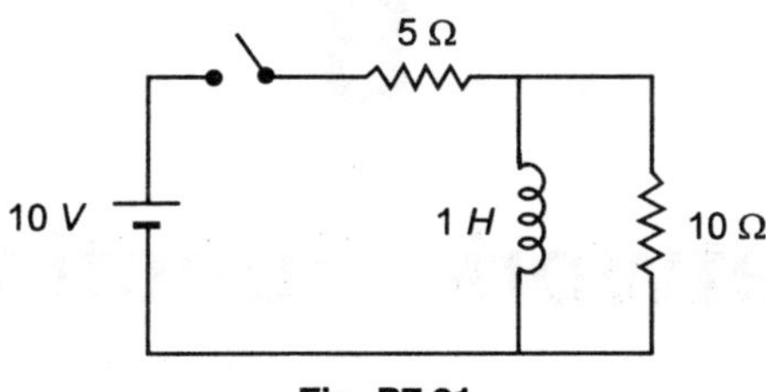


**Fig. P7.21**

**7.22** In the circuit, the switch is initially open till steady state condition is reached. It is closed at $t = 0$. Determine the loop currents and their initial and final values.

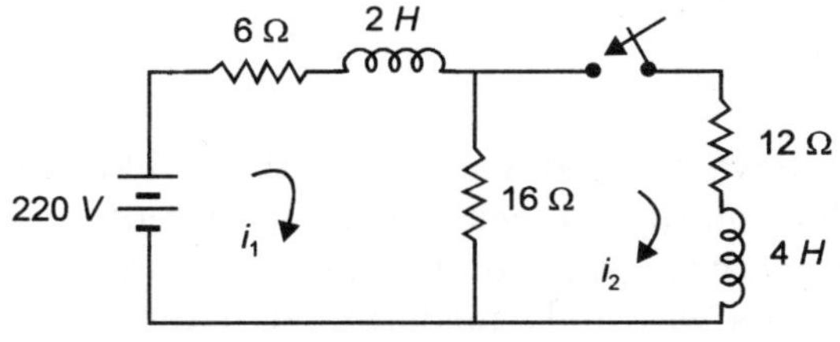


**Fig. P7.22**

**7.23** The impulse response $h(t)$ of a continuous time system is given by
$$h(t) = e^{-2t} \cos 3t \, u(t)$$
Determine its response to a unit step input.

**7.24** The transfer function of a linear system is given by
$$H(s) = \frac{10 (s + 2)}{(s + 1) (s^2 + 4)}$$

(a) Determine the impulse response of this system.

(b) Determine the response of the system when the input is $f(t) = 4 \sin 2t$.

**7.25** Determine the initial value $f(0^-)$ and final values $f(\infty)$ for the following functions:

(a) $\dfrac{s}{(s + 1) (s + 2)}$

(b) $\dfrac{s^2 + 2s + 3}{s^3 + 3s^2 + 3s + 1}$

(c) $\dfrac{(s + 3) (s + 7)}{(s + 2) (s + 5)}$

(d) $\dfrac{17 s^3 + 7 s^2 + s + 6}{s^5 + 3 s^4 + 5 s^3 + 4 s^2 + 2s}$

# 8

# Network Functions

## 8.1 INTRODUCTION

In previous chapters we have studied various methods like node analysis, loop analysis, node pair analysis etc. for solution of networks. These require setting up of simultaneous equations. These formal methods are not necessarily simple to use to solve all kinds of networks. In many situations with only moderate structural complexity solution by inspection, by Thevenin's theorem or similar short cut techniques could be used to solve the networks rather than the use of formal techniques as mentioned above. However, the formal techniques have systematic procedures and because of their generality and the availability of mathematical tools like matrices and the computing tools, the computers, these are better adaptable to network solving.

Assuming we have been able to write these network equations, we now describe the techniques for the solution of these equations and the network functions in terms of which the network behaviour is described.

## 8.2 DRIVING POINT AND TRANSFER FUNCTIONS

For a linear time invariant network with any number of independent current and voltage sources and with arbitrary initial conditions in terms of capacitor voltages and inductor currents which can also be represented by independent voltage and current sources, a set of loop or node equations can be written. It may be noted that the following analysis may be good even if network is non-passive and non-reciprocal. In matrix form these equations can be written as

$$Z_l\ I_l = E \qquad\qquad\qquad ...(8.1)$$

$$Y_n\ V_n = J_n \qquad\qquad\qquad ...(8.2)$$

On the right hand side we have the sources including the initial conditions equivalent sources e.g. $J_n$ is the sum of current sources (including Norton's equivalent of accompanied voltage sources) connected at the node with due regard to their orientations. These equations can be solved by pre-multiplying the equation by inverse of the co-efficient matrix. Thus,

$$Z_l^{-1}\ Z_l\ I_l = Z_l^{-1}\ E$$

or
$$I_l = Z_l^{-1}\ E \qquad\qquad\qquad ...(8.3)$$

Similarly

$$V_n = Y_n^{-1} J \qquad \ldots(8.4)$$

Since there is symmetry in the two sets of equations let us take nodal equations for further understanding.

The nodal equation can be written as

$$
\begin{bmatrix} V_1 \\ V_2 \\ \cdot \\ \cdot \\ \cdot \\ V_n \end{bmatrix}
=
\begin{bmatrix}
\dfrac{\Delta_{11}}{\Delta} & \dfrac{\Delta_{21}}{\Delta} & \cdots & \dfrac{\Delta_{n1}}{\Delta} \\
\dfrac{\Delta_{12}}{\Delta} & \dfrac{\Delta_{22}}{\Delta} & \cdots & \dfrac{\Delta_{n2}}{\Delta} \\
\cdot & \cdot & \cdot & \cdot \\
\cdot & \cdot & \cdot & \cdot \\
\dfrac{\Delta_{1_n}}{\Delta} & \dfrac{\Delta_{2_n}}{\Delta} & \cdots & \dfrac{\Delta_{nn}}{\Delta}
\end{bmatrix}
\begin{bmatrix} J_1 \\ J_2 \\ \cdot \\ \cdot \\ \cdot \\ J_n \end{bmatrix}
\qquad \ldots(8.5)
$$

where $\Delta$ is the determinant of the nodal admittance matrix and $\Delta_{ij}$ are the co-factors of the same matrix. The expression for one of the voltages is

$$V_j = \frac{\Delta_{1j}}{\Delta} J_1 + \frac{\Delta_{2j}}{\Delta} J_2 + \ldots + \frac{\Delta_{nj}}{\Delta} J_n \qquad \ldots(8.6)$$

This gives nodal voltages as linear combination of the equivalent nodal current sources. Consider Fig. 8.1 and assuming no initial conditions the current sources are:

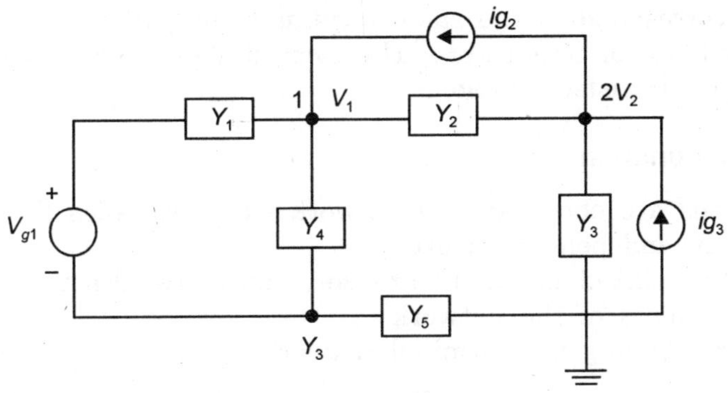


**Fig. 8.1**   Illustrating nodal equation approaches.

At node 1, $i_{g_2}$ + the Norton's equivalent of the accompanied voltage source $V_{g_1}$ which is $V_{g_1} V_1$ i.e.

$$J_1 = i_{g_2} + V_{g_1} Y_1$$

$$\text{Node 2,} \quad J_2 = i_{g_3} - i_{g_2}$$

and

$$\text{Node 3,} \quad V_{g_1} Y_1 = J_3$$

Therefore,

$$J = \begin{vmatrix} I_{g_2} + V_{g_1} Y_1 \\ I_{g_3} - I_{g_2} \\ -V_{g_1} Y_1 \end{vmatrix} \qquad \ldots(8.7)$$

Writing down the solution equation for 3 – bus system in Fig. 8.1, we have

$$\begin{bmatrix} V_1 \\ V_2 \\ V_3 \end{bmatrix} = \begin{bmatrix} \dfrac{\Delta_{11}}{\Delta} & \dfrac{\Delta_{21}}{\Delta} & \dfrac{\Delta_{31}}{\Delta} \\ \dfrac{\Delta_{12}}{\Delta} & \dfrac{\Delta_{22}}{\Delta} & \dfrac{\Delta_{32}}{\Delta} \\ \dfrac{\Delta_{13}}{\Delta} & \dfrac{\Delta_{23}}{\Delta} & \dfrac{\Delta_{33}}{\Delta} \end{bmatrix} \begin{bmatrix} I_{g2} + V_{g1}\,Y_1 \\ I_{g3} - I_{g2} \\ -V_{g1}\,Y_1 \end{bmatrix} \qquad \text{...(8.8)}$$

Writing equation say $V_2$

$$V_2 = \frac{\Delta_{12}}{\Delta}\left(I_{g2} + V_{g1}\,Y_1\right) + \frac{\Delta_{22}}{\Delta}\left(I_{g3} - I_{g2}\right) + \frac{\Delta_{32}}{\Delta}\left(-V_{g1}\,Y_1\right) \qquad \text{...(8.9)}$$

$$= \frac{\Delta_{12}\,Y_1 - \Delta_{32}\,Y_1}{\Delta}\,V_{g1} + \frac{\Delta_{12} - \Delta_{22}}{\Delta}\,I_{g2} + \frac{\Delta_{22}}{\Delta}\,I_{g3} \qquad \text{...(8.10)}$$

It can, therefore, be said that any response (or response transform if the equations have been obtained in Laplace transform) can be written as a linear combination of excitation transform. The co-efficient of this linear combination are themselves linear combinations of some functions of the Laplace operator. These functions are ratios of two determinants, the denominator being the determinant of the nodal admittance matrix and the numerator being some co-factors of the $Y$ matrix. Therefore, if the function relating the response transform with the excitation is known, response of any other excitation can be obtained.

We now define these network functions as the ratio of response transform to excitation transform. Both the response and excitation may be voltage or current. If both the excitations and response correspond to the same terminals the ratio is known as driving point function either impedance or admittance. However, if the two correspond to different terminals, it is known as transfer function.

### 8.2.1  Driving Point Function

Driving point functions are obtained of a network which satisfies the condition:
    (1) It contains no independent sources.
    (2) It is initially relaxed i.e. there are zero initial condition for the capacitors and
        inductive elements of the network.
Consider Fig. 8.2 having two-terminal networks.

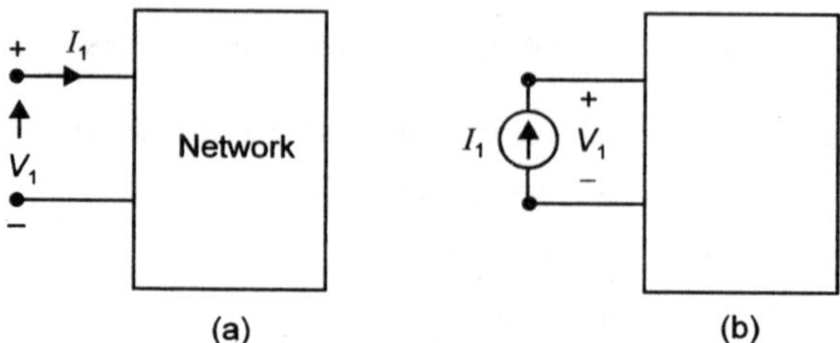


**Fig. 8.2**  Driving point functions.

Conventionally the source is connected between the two terminals and is known as the driving force for the network and the pair of terminals, the driving point of the network. Driving point impedance and driving point admittance are mathematically defined as

$$Z(s) = \frac{V_1(s)}{I_1(s)} \qquad \qquad ...(8.11)$$

and
$$Y(s) = \frac{I_1(s)}{V_1(s)} = \frac{1}{Z(s)} \qquad \qquad ...(8.12)$$

where $V_1(s)$ and $I_1(s)$ are the transforms of the terminal voltage and currents respectively. It is emphasized once again that the two conditions in regard to the networks given above are very important for the definition of the driving point quantities. Therefore, if the network contains independent sources and the initial conditions are non-zero then $Z$ and $Y$ will take on different values.

Also it is to be noted that whatever be the wave shapes of $V$ and $I$ whether sinusoidal or non-sinusoidal the relation given in equation (8.11) and (8.12) hold good.

Assume that the network does not contain any non-passive and non-reciprocal elements then driving point impedance at terminal pair 1 is given as Fig. 8.2

$$Z_{11} = \frac{V_1(s)}{I_1(s)} = \left.\frac{\Delta_{11}(s)}{\Delta(s)}\right|_y \qquad \qquad ...(8.13)$$

Since here response (numerator) is voltage and the excitation (denominator) as the current source at node 1, the notation $y$ is used to indicate that the determinant is due to nodal equations. Similarly if voltage is excitation and current is the response, then

$$Y_{11}(s) = \frac{I_1(s)}{V_1(s)} = \left.\frac{\Delta_{11}(s)}{\Delta(s)}\right|_z \qquad \qquad ...(8.14)$$

Here $Y_{11}$ is the driving point admittance and the notation $z$ means that $\Delta$ is the determinant of the loop impedance matrix. The above expressions for evaluation of $Z$ and $Y$ are applicable even when the network contains controlled source even though not always. It is, therefore, always desirable to use equation (8.11) and (8.12) which are more general as compared to (8.13) and (8.14) which are a special formulae for a certain kind of network.

### 8.2.2 Transfer Functions

Transfer functions are defined as the ratio of response to excitation at different pair of terminals and, therefore, we refer to Fig. (8.3).

Again we assume that the network does not contain independent sources and also it is initially relaxed. Also assume that there are no controlled sources.

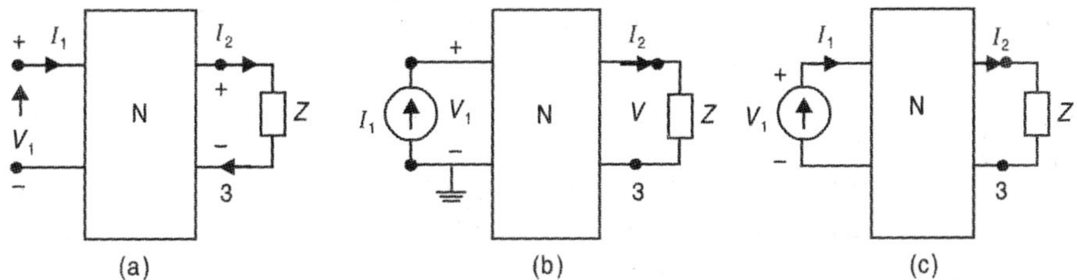

**Fig. 8.3**  Transfer functions

From equation (8.6), since there is only one source $I_1$ in Fig. 8.3 (b) above

$$V_1 = \frac{\Delta_{11}}{\Delta} I_1 \qquad \qquad \text{...(8.15)}$$

$$V_2 = (\Delta_{12}/\Delta)\, I_1 \quad \text{and} \quad V_3 = \frac{\Delta_{13}}{\Delta} I_1$$

Therefore, $\qquad \qquad V = V_2 - V_3 = \dfrac{(\Delta_{12} - \Delta_{13})}{\Delta} I_1$

We now define the following transfer functions in terms of the determinants.

Transfer impedance $\qquad \dfrac{V(s)}{I_1(s)} = \dfrac{\Delta_{12} - \Delta_{13}}{\Delta}\bigg|_y \qquad \qquad \text{...(8.16)}$

In terms of loop equations the general expressions for current $I_k$ as a linear combination of voltage sources is

$$I_k = \frac{\Delta_{1k}}{\Delta} V_1 + \frac{\Delta_{2k}}{\Delta} V_2 + \dots + \frac{\Delta_{nk}}{\Delta} V_n \qquad \qquad \text{...(8.17)}$$

Since here in Fig. 8.3 (c) there is only one source $V_1$

$$I_1 = \frac{\Delta_{11}}{\Delta} V_1\bigg|_z \quad \text{and} \quad I_2 = V_1 \frac{\Delta_{12}}{\Delta}\bigg|_z$$

Since $\qquad \qquad V = I_2 Z$

$$\frac{V(s)}{I_1(s)} = \frac{\Delta_{12} - \Delta_{13}}{\Delta}\bigg|_y = \frac{I_2 Z}{I_1} = \frac{\Delta_{12}/\Delta}{\Delta_{11}/\Delta} Z\bigg|_z = \frac{\Delta_{12}}{\Delta_{11}} Z\bigg|_z \qquad \text{...(8.18)}$$

Similarly transfer admittance

$$\frac{I_2(s)}{V_1(s)} = Y \frac{\Delta_{12} - \Delta_{13}}{\Delta_{11}}\bigg|_y = \frac{\Delta_{12}}{\Delta}\bigg|_z \qquad \qquad \text{...(8.19)}$$

Voltage gain or transfer voltage ratio

$$= \frac{V(s)}{V_1(s)} = \frac{\Delta_{12} - \Delta_{13}}{\Delta_{11}}\bigg|_y = Z \cdot \frac{\Delta_{12}}{\Delta}\bigg|_z \qquad \qquad \text{...(8.20)}$$

Current gain $\qquad \dfrac{I_2(s)}{I_1(s)} = Y \dfrac{\Delta_{12} - \Delta_{13}}{\Delta}\bigg|_y = \dfrac{\Delta_{12}}{\Delta_{11}}\bigg|_z \qquad \qquad \text{...(8.21)}$

Where $\qquad \qquad Y = \dfrac{1}{Z}$

From the above analysis following observations are made:

(1) The formulae in terms of determinants are valid for non-passive non-reciprocal network without controlled sources.

(2) The transfer impedance is not the reciprocal of transfer admittance.

When controlled sources are used the above formulae cannot be used. However, the basic definition of transfer function should be used.

**Example 8.1:** As an illustration let us find out the network function of the following network using the basic definition:

($i$) The driving point impedance is given by

$$Z(s) = \frac{V_1(s)}{I_1(s)} = R + sL + \frac{1}{Cs} = \frac{1 + RCs + LCs^2}{Cs}$$

($ii$) Here $I_1(s) = \dfrac{V_1(s)}{sL} + \dfrac{V_1(s)}{R + \dfrac{1}{Cs}} = V_1(s)\left[\dfrac{R + \dfrac{1}{Cs} + sL}{sL\left(R + \dfrac{1}{Cs}\right)}\right]$

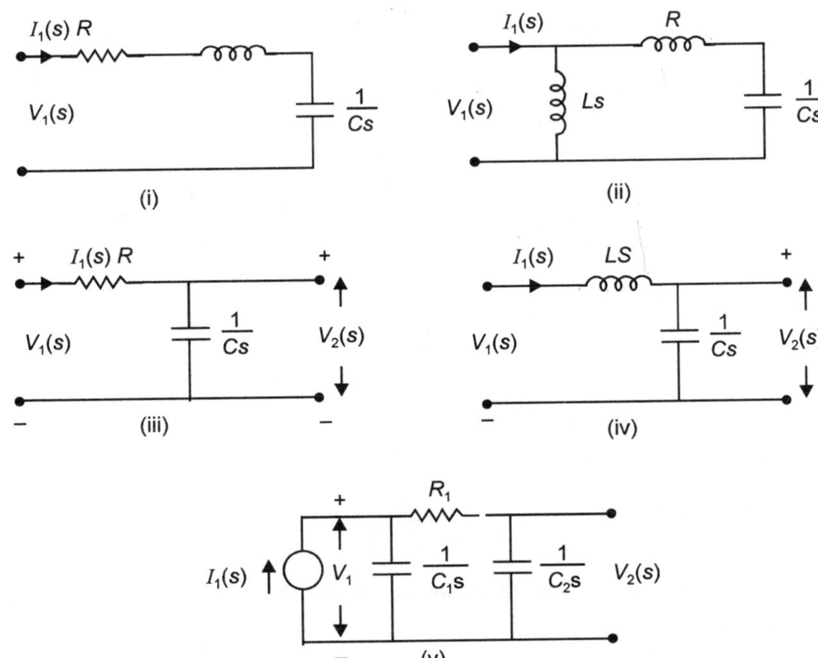



**Fig. E8.1**

or driving point impedance by definition

$$Z(s) = \frac{V_1(s)}{I_1(s)} = \frac{Rs^2 + s/C}{s^2 + \dfrac{R}{L}s + \dfrac{1}{LC}}$$

($iii$) The equations describing the network are:

$$RI_1(s) + \frac{1}{Cs}I_1(s) = V_1(s)$$

and

$$V_2(s) = I_1(s)\cdot\frac{1}{Cs}$$

The driving point impedance

$$Z_1 = \frac{V_1(s)}{I_1(s)} = R + \frac{1}{Cs} = \frac{RCs+1}{Cs}$$

The driving point admittance

$$Y_1(s) = \frac{I_1(s)}{V_1(s)} = \frac{Cs}{RCs+1} = \frac{1}{R} \cdot \frac{s}{s + \dfrac{1}{RC}}$$

The transfer voltage ratio $= \dfrac{V_2(s)}{V_1(s)} = \dfrac{\dfrac{1}{Cs}}{R + \dfrac{1}{Cs}}$

($iv$) The equations describing the network are:

$$I_1(s)\, Ls + I_1(s) . \frac{1}{Cs} = V_1(s)$$

and

$$V_2(s) = I_1(s) . \frac{1}{Cs}$$

Therefore,

$$Z_1 = \frac{V_1(s)}{I_1(s)} = sL + \frac{1}{Cs} = \frac{LCs^2 + 1}{Cs}$$

The transfer voltage ratio

$$\frac{V_2(s)}{V_1(s)} = \frac{1/Cs}{sL + 1/Cs}$$

$$= \frac{1/LC}{s^2 + 1/LC}$$

($v$) The network can be solved considering current source and voltage source (equivalent) separately.

($a$) Current source

$$I_1(s) = V_1 C_1 s + \frac{V_1 - V_2}{R_1}$$

$$V_2(s) = \frac{V_1 - V_2}{R_1} . \frac{1}{C_2 s}$$

or

$$V_2(s)\left[1 + \frac{1}{R_1 C_2 s}\right] = \frac{V_1(s)}{R_1 C_2 s}$$

or

$$\frac{V_2(s)}{V_1(s)} = \frac{1}{1 + R_1 C_2 s}$$

Using equivalent voltage source, the circuit becomes

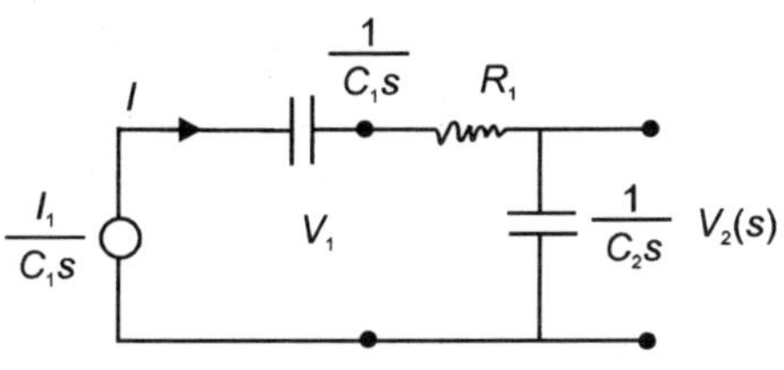

**Fig. E.8.1.1**

Current
$$I = \frac{I_1/C_1 s}{R_1 + \dfrac{1}{C_1 s} + \dfrac{1}{C_2 s}}$$

Voltage
$$V_1 = \frac{I_1/C_1 s}{R_1 + \dfrac{1}{C_1 s} + \dfrac{1}{C_2 s}} \cdot \left( R_1 + \frac{1}{C_2 s} \right)$$

and
$$V_2 = \frac{I_1/C_1 s}{R_1 + \dfrac{1}{C_1 s} + \dfrac{1}{C_2 s}} \cdot \frac{1}{C_2 s}$$

Therefore,
$$\frac{V_2(s)}{V_1(s)} = \frac{1}{1 + R_1 C_2 s}$$

We observe that all the network functions we have computed have been quotients of polynomials in $s$ having the general form

$$\frac{p(s)}{q(s)} = \frac{a_0 s^n + a_1 s^{n-1} + \cdots + a_{n-1} s + a_n}{b_0 s^m + b_1 s^{m-1} + \cdots + b_{m-1} s + b_m}$$

which is a rational function of $s$. Here $m$ and $n$ are integers.

Rewriting equations (8.6) and (8.17)

$$V_j = \frac{\Delta_{1j}}{\Delta} I_1 + \frac{\Delta_{2j}}{\Delta} I_2 + \cdots + \frac{\Delta_{nj}}{\Delta} I_n$$

and
$$I_k = \frac{\Delta'_{1k}}{\Delta'} V_1 + \frac{\Delta'_{2k}}{\Delta'} V_2 + \cdots + \frac{\Delta'_{nk}}{\Delta'} V_n$$

If in the first equation all current sources are zero except $k$th then

$$V_j = \frac{\Delta_{kj}}{\Delta} I_k$$

Therefore,
$$\frac{V_j}{I_k} = Z_{jk} = \left. \frac{\Delta_{kj}}{\Delta} \right|_y \qquad \ldots(8.22)$$

where $\Delta$ is the determinant of the nodal admittance matrix and $\Delta_{kj}$ is the co-factor of $\Delta$ with row $k$ and column $j$ deleted.

Similarly in equations (8.17) if all $V^s$ are zero except $V_j$ then

$$I_k = \frac{\Delta'_{jk}}{\Delta'} V_j$$

or

$$\frac{I_k}{V_j} = Y_{jk} = \frac{\Delta'_{jk}}{\Delta'} \qquad \qquad ...(8.23)$$

where $\Delta'$ is the determinant of the loop impedance matrix and $\Delta'_{jk}$ is a co-factor of $\Delta'$ with row $j$ and column $k$ deleted.

## 8.3  THE CALCULATION OF NETWORK FUNCTIONS

We begin with calculation of network function for a very important class of network known as ladder network because of its structure. If each element represents either a single impedance or admittance, it is known as simple ladder as shown in Fig. 8.4.

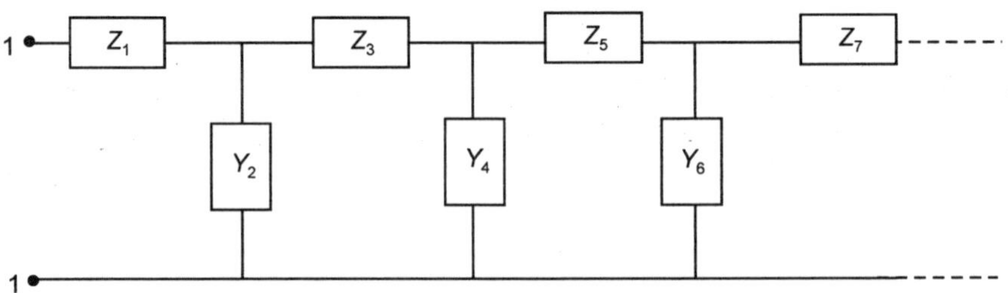

**Fig. 8.4**  Simple ladder network.

However if the ladder network contains arms that are arbitrarily complicated as shown in Fig. 8.5 it is known as a non-simple network.

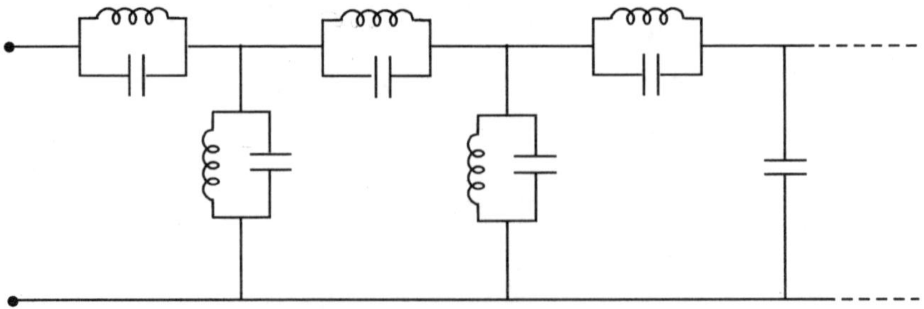

**Fig. 8.5**  A non simple ladder network.

It is to be noted that $L/C$ characterize the series arms by the impedance and the shunt branch by the admittances.

Let us find out driving point impedance of the network at terminals 11 in Fig. 8.4. To do this, we begin with $Y_6$, take its reciprocal and add to $Z_5$. Now the reciprocal of whole of this is added to $Y_4$. This pattern is continued until the process terminates. The driving point impedance is given as

$$Z = Z_1 + \cfrac{1}{Y_2 + \cfrac{1}{Z_3 + \cfrac{1}{Y_4 + \cfrac{1}{Z_5 + \cfrac{1}{Y_6}}}}} \qquad \qquad \dots(8.24)$$

This expression is known as continued fraction. Finally, of course, it is simplified to determine $Z$ for the ladder network.

**Example 8.2:** Determine the driving point impedance of the given network

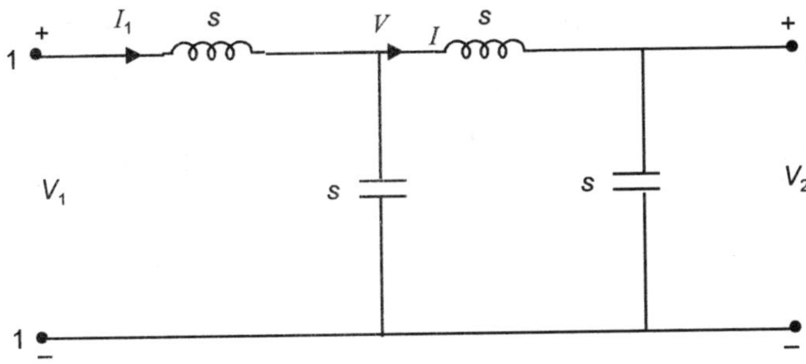

**Fig. E8.2**

$$Z = s + \cfrac{1}{s + \cfrac{1}{s + \cfrac{1}{s}}}$$

$$= s + \cfrac{1}{s + \cfrac{s}{s^2 + 1}} = s + \frac{s^2 + 1}{s^3 + s + s}$$

$$= \frac{s^4 + 3s^2 + 1}{s\left(s^2 + 2\right)}$$

In order to obtain transfer function of a given network we start from the output port using KVL and KCL equations. The first equation may be substituted into the second, second into third and so on. This process is continued till we find one equation which relates output with input and from here the requisite transfer functions can be obtained.

As an illustration, consider the previous network. Here

$$I = V_2\, s, \quad V = V_2 + I\, s$$
$$I_1 = I + V\, s = V_2\, s + (V_2 + I\, s)\, s$$
$$= V_2\, (s^2 + 2\, s)$$

Therefore, $\qquad \dfrac{V_2}{I_1} = \dfrac{1}{s^2 + 2s}$

Also
$$V_1 = V + I_1 s = V_2 + V_2 s^2 + V_2 (s^2 + 2s) s$$
$$= V_2 (s^3 + 3s^2 + 1)$$

or
$$\frac{V_2}{V_1} = \frac{1}{s^3 + 3s^2 + 1}$$

**Example 8.3:** Determine the transfer voltage ratio function of the given networks.

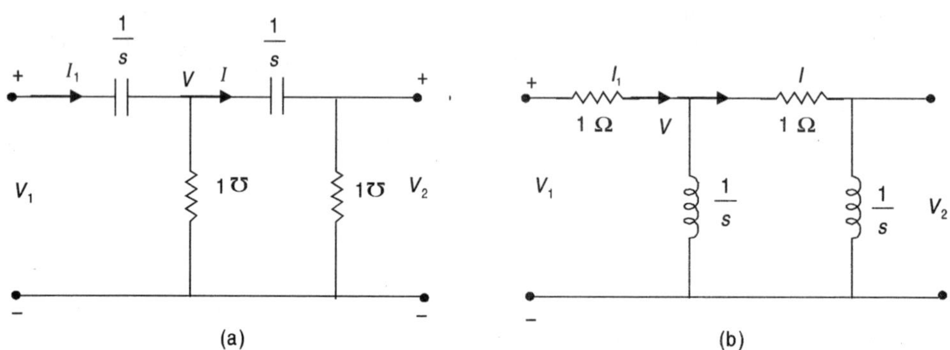

**Fig. E8.3**

Fig. (*a*):
$$I = V_2, \qquad V = V_2 + I/s = V_2 + V_2/s$$

$$I_1 = I + V \cdot 1 = 2\,V_2 + \frac{V_2}{s}$$

$$V_1 = V + I_1/s$$

$$= V_2 + \frac{V_2}{s} + \left(2V_2 + \frac{V_2}{s}\right) \cdot \frac{1}{s}$$

$$= V_2 \left[1 + \frac{3}{s} + \frac{1}{s^2}\right]$$

or
$$\frac{V_2}{V_1} = \frac{s^2}{s^2 + 3s + 1}$$

Fig. (*b*):
$$I = \frac{V_2}{s}, \qquad V = V_2 + I$$

or
$$V = V_2 + \frac{V_2}{s}$$

$$I_1 = I + \frac{V}{s} = \frac{V_2}{2} + V_2 \left(1 + \frac{1}{s}\right) \cdot \frac{1}{s}$$

Now
$$V_1 = V + I_1 = V_2 \left[ 1 + \frac{1}{s} + \frac{1}{s} + \frac{1}{s} + \frac{1}{s^2} \right]$$

$$= V_2 \left( \frac{s^2 + 3s + 1}{s^2} \right)$$

or
$$\frac{V_2}{V_1} = \frac{s^2}{s^2 + 3s + 1}$$

Next we consider the network function evaluation of general networks. The method outlined above is useful only to ladder networks. Some of the general networks are Fig. (8.6) (a) the bridged – T (b) the parallel – T or Twin – T and (c) the lattice networks. These networks can be solved by loop or nodal system of analysis and we make use of the general equations (8.22) and (8.23)

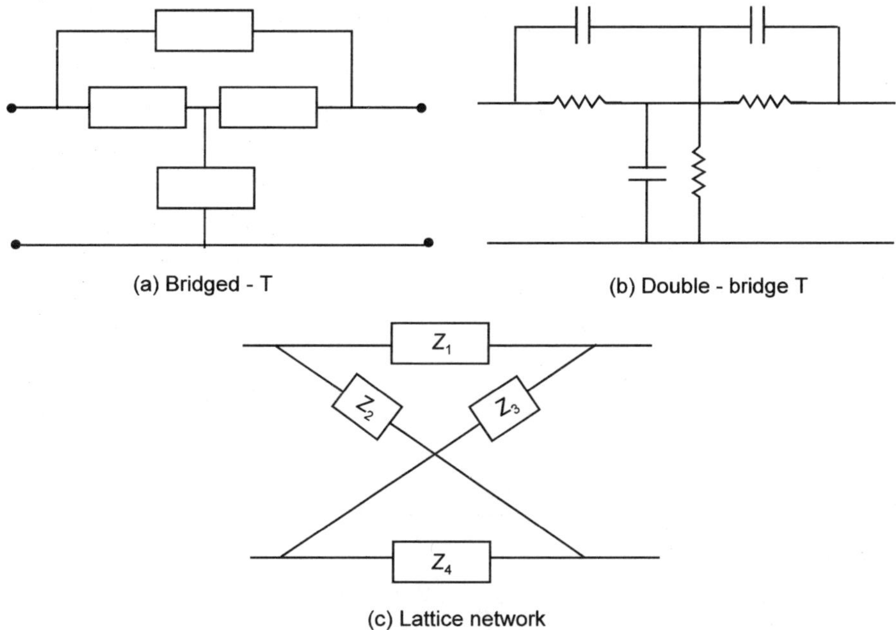

(a) Bridged - T

(b) Double - bridge T



(c) Lattice network

**Fig. 8.6**  General networks (a) Bridged – T (b) Double bridged T (c) Lattice.

**Example 8.4:** Determine the driving point admittance of the given T–bridged network.

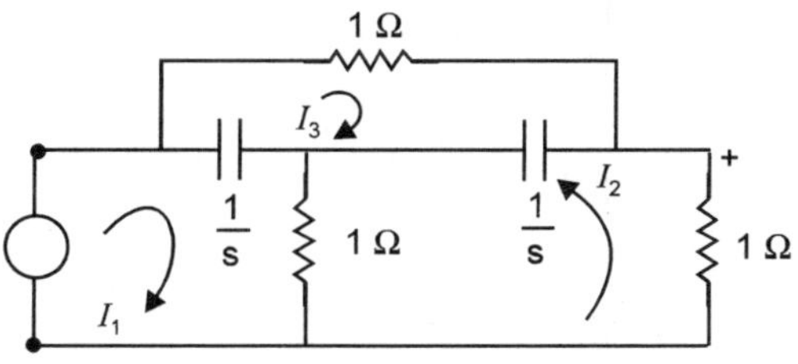


**Fig. E8.4**

**Solution:** Since driving point admittance is to be obtained, we write down the loop impedance matrix. By observation, loop impedance matrix is

$$\begin{bmatrix} \dfrac{1}{s} + 1 & +1 & -\dfrac{1}{s} \\[2ex] +1 & \dfrac{1}{s} + 2 & +\dfrac{1}{s} \\[2ex] -\dfrac{1}{s} & +\dfrac{1}{s} & \dfrac{2}{s} + 1 \end{bmatrix}$$

The determinant is

$$\left(1 + \frac{1}{s}\right)\left[\left(\frac{1}{s} + 2\right)\left(1 + \frac{2}{s}\right) - \frac{1}{s^2}\right] + \left[\left(1 + \frac{2}{s} + \frac{1}{s^2}\right)\right]$$

$$-\frac{1}{s}\left[\frac{1}{s} + \frac{1}{s}\left(2 + \frac{1}{s}\right)\right]$$

$$= \left(1 + \frac{1}{s}\right)\left[\frac{1}{s^2} + \frac{5}{s} + 2\right] - 1 - \frac{2}{s} - \frac{4}{s^2} - \frac{1}{s^3}$$

$$= \frac{2 + 5s + s^2}{s^2}$$

Now
$$Y_{11} = \frac{\Delta_{11}}{\Delta}$$

Also
$$\Delta_{11} = \left(\frac{1}{s} + 2\right)\left(1 + \frac{2}{s}\right) - \frac{1}{s^2}$$

$$= \frac{2s^2 + 5s + 1}{s^2}$$

Therefore,
$$Y_{11} = \frac{\Delta_{11}}{\Delta} = \frac{2s^2 + 5s + 1}{s^2 + 5s + 2}$$

and transfer admitance

$$Y_{12} = \frac{\Delta_{21}}{\Delta}$$

$$\Delta_{21} = \left[\left(\frac{2}{s} + 1\right) + \frac{1}{s^2}\right](-1)^{2+1} = -\frac{s^2 + 2s + 1}{s^2}$$

Therefore,
$$Y_{12} = \frac{\Delta_{21}}{\Delta} = -\frac{s^2 + 2s + 1}{s^2 + 5s + 2}$$

**Example 8.5:** Determine $G_{12}$, $Z_{12}$ and $Y_{12}$ of the given bridged–T network.

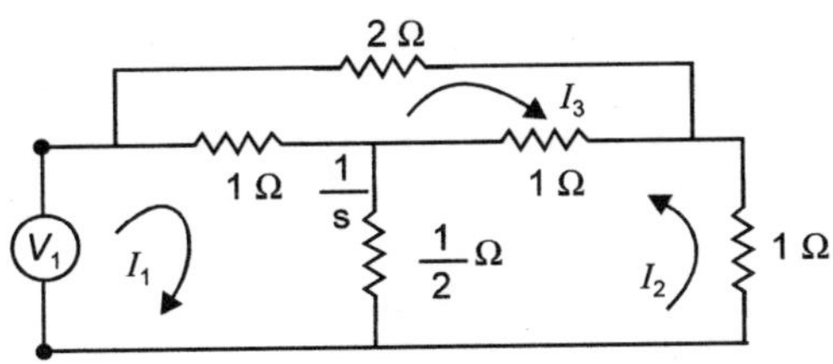

**Fig. E8.5**

The loop impedance matrix is

$$\begin{bmatrix} 1.5 & +0.5 & -1.0 \\ +0.5 & 2.5 & +1.0 \\ -1.0 & +1.0 & 4.0 \end{bmatrix}$$

The determinant $= 1.5\,(10 - 1) - 0.5\,(+2 + 1) - 1.0\,(0.5 + 2.5)$

$$= 13.5 - 1.5 - 3 = 9$$

$$\Delta_{21} = (+2 + 1) = 3\,(-1)^3 = -3$$

$$Y_{12} = \left.\frac{\Delta_{21}}{\Delta}\right|_z = \frac{-3}{9} = -\frac{1}{3} = \frac{I_2}{V_1}$$

The nodal admittance matrix

$$\begin{bmatrix} 1.5 & -1 & -0.5 \\ -1.0 & 4.0 & -1.0 \\ -0.5 & -1.0 & 2.5 \end{bmatrix}$$

The determinant

$$1.5\,(10 - 1) + 1\,(-2.5 - 0.5) - 0.5\,(1 + 2)$$

$$13.5 - 3 - 1.5 = 9$$

$$Z_{12} = \frac{\Delta_{21}}{\Delta}$$

$$\Delta_{21} = + (-\,2.5 - 0.5) = -3\,(-1)^3 = 3$$

$$Z_{12} = \left.\frac{\Delta_{21}}{\Delta}\right|_y = \frac{3}{9} = \frac{1}{3}$$

Since

$$\frac{V_2}{I_1} = Z_{12} = \frac{1}{3}$$

Since

$$V_2 = -I_2$$

Therefore,

$$\frac{I_2}{I_1} = -\frac{1}{3}$$

Now
$$\frac{V_2}{I_1} \times \frac{I_2}{V_1} = \frac{V_2}{V_1} \cdot \frac{I_2}{I_1} = -\frac{1}{9}$$

Hence,
$$G_{12} = \frac{V_2}{V_1} = \left(-\frac{1}{9}\right)(-3)$$

$$= \frac{1}{3} \textbf{ Ans.}$$

## 8.4  TWO PORT NETWORKS

Fig. 8.7 shows a two–port network. Since  a network is used as a transmission network, one of the ports usually labelled as 1 1 is known as input port and the other labelled as 2 2 is

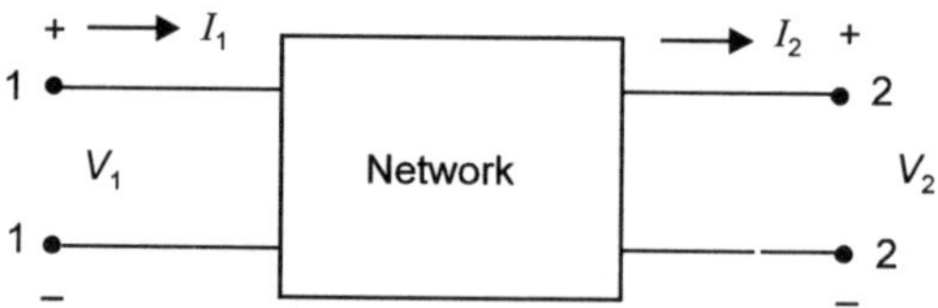



**Fig. 8.7**  Two-port network.

termed as output port. The port variables are two port voltages ($V_1$ and $V_2$) and two port currents $I_1$ and $I_2$. To correlate the port voltages and port currents of a multiport network (linear) requires as many linear equations as are the no. of ports. For a two port network two linearly independent equations are required among the four variables. The choice of independent and dependent variables depends upon the convenience in a given application.

### 8.4.1  Short Circuit Admittances

One set of equations results when the port currents are expressed in terms of the two port voltages.

$$\begin{bmatrix} I_1(s) \\ I_2(s) \end{bmatrix} = \begin{bmatrix} Y_{11} & Y_{12} \\ Y_{21} & Y_{22} \end{bmatrix} \begin{bmatrix} V_1(s) \\ V_2(s) \end{bmatrix} \qquad \text{...(8.25)}$$

Here $Y_s$ are the parameters of the network which relate the two port voltages and currents. The interpretation of the parameters is as follows.
From the equation
$$I_1(s) = Y_{11}\, V_1(s) + Y_{12}\, V_2(s)$$
It is clear that

$$Y_{11} = \left.\frac{I_1(s)}{V_1(s)}\right|_{v_2(s)=0} \qquad \text{...(8.26)}$$

i.e. $Y_{11}$ is the ratio of input current (transformed) to the transform of the input voltage when output port is shorted. Since the admittances are obtained after short circuiting one of the ports, these are known as short circuit admittances. The matrix of $Y_s$ is designated as $Y_{sc}$ and is called as short circuit admittance matrix. The terms $Y_{11}$ and $Y_{22}$ are the short circuit driving point admittances at the two ports and $Y_{21}$ and $Y_{12}$ are short circuit transfer admittances. In particular $Y_{21}$ is the forward transfer admittance i.e. the ratio of current response in port 2 to a voltage excitation in port 1 and $Y_{12}$ is the reverse transfer admittance.

The equivalent circuit representation of equation (8.25) is given in Fig. (8.8) where $Y_{12} V_2$ and $Y_{21} V_1$ are designated as voltage controlled current sources.

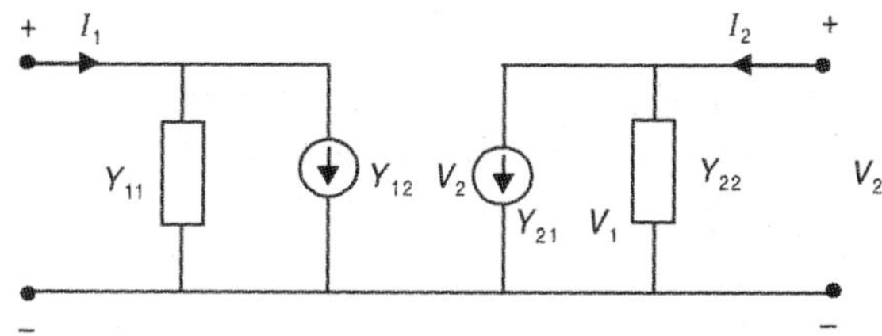

**Fig. 8.8 (a)**   Equivalent representation of equation (8.25).

Rewriting equation (8.25)

$$I_1 = Y_{11} V_1 + Y_{12} V_2$$
$$= (Y_{11} + Y_{12}) V_1 - Y_{12} (V_1 - V_2) \qquad \qquad ...(8.27)$$

and

$$I_2 = Y_{21} V_1 + Y_{22} V_2$$
$$= (Y_{21} - Y_{12}) V_1 + (Y_{22} + Y_{12}) V_2$$
$$- Y_{12} (V_2 - V_1) \qquad \qquad ...(8.28)$$

and its equivalent circuit equation (8.28) is given in Fig. 8.8(b)

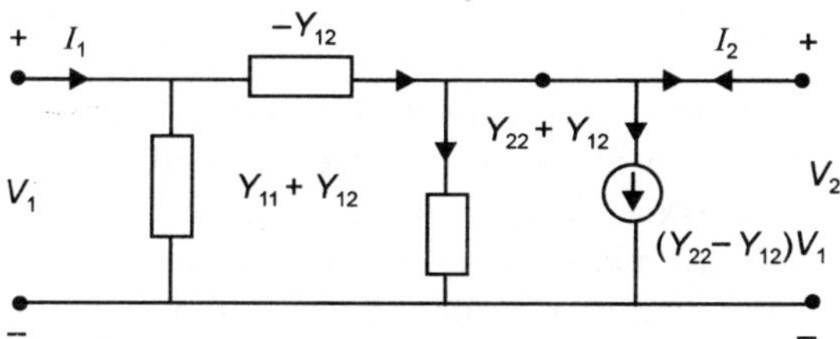

**Fig. 8.8(b)**

For reciprocity

$$\left. \frac{I_2}{V_1} \right|_{V_2 = 0} = \left. \frac{I_1}{V_2} \right|_{V_1 = 0} \qquad \qquad ...(8.29)$$

Now
$$\left. \frac{I_2}{V_1} \right|_{V_2 = 0}$$

Also
$$I_2 = Y_{21} V_1$$

when
$$V_2 = 0$$

or
$$\frac{I_2}{V_1} = Y_{21} \qquad \qquad ...(8.30)$$

Similarly,
$$\left.\frac{I_1}{V_2}\right|_{V_1=0} \quad \text{and} \quad I_1 = Y_{12}\, V_2$$

when
$$V_1 = 0$$

or
$$\frac{I_1}{V_2} = Y_{12} \qquad\qquad ...(8.31)$$

For network to be reciprocal $Y_{12} = Y_{21}$. Also a two port network is said to be symmetric if the ports can be interchanged without changing the port voltage and currents and the condition of symmetry is $Y_{11} = Y_{22}$.

### 8.4.2 Open Circuit Impedances

The port voltages can be expressed in terms of network parameters and the port currents as follows:

$$\begin{bmatrix} V_1(s) \\ V_2(s) \end{bmatrix} = \begin{bmatrix} Z_{11} & Z_{12} \\ Z_{21} & Z_{22} \end{bmatrix} \begin{bmatrix} I_1(s) \\ I_2(s) \end{bmatrix} \qquad\qquad ...(8.32)$$

Where
$$Z_{11} = \left.\frac{V_1}{I_1}\right|_{I_2=0} \qquad\qquad Z_{12} = \left.\frac{V_1}{I_2}\right|_{I_1=0}$$

$$Z_{21} = \left.\frac{V_2}{I_1}\right|_{I_2=0} \qquad\qquad Z_{22} = \left.\frac{V_2}{I_2}\right|_{I_1=0} \qquad\qquad ...(8.33)$$

Dimensionally the parameters are impedances and since these are obtained by letting one of the ports open circuited, these impedances are termed as open circuit impedances (Z-parameters for short). The matrix is designated as $Z_{oc}$ and is called open circuit impedance marix. The elements $Z_{11}$ and $Z_{22}$ are driving point impedances at the ports 1 and 2 respectively and $Z_{12}$ and $Z_{21}$ are the transfer impedances, $Z_{21}$ is the forward transfer impedance and $Z_{12}$ is the reserve transfer impedance. It is to be noted that $Y_{SC}$ and $Z_{OC}$ are inverse of each other i.e.

$$Y_{SC} = Z_{OC}^{-1} \qquad\qquad ...(8.34)$$

and
$$\det Y_{SC} = \frac{1}{\det Z_{OC}} \qquad\qquad ...(8.35)$$

The results hold good whether the network is passive or active, reciprocal or non-reciprocal. However for reciprocal network

$$Y_{12} = Y_{21} \qquad \text{and} \qquad Z_{12} = Z_{21}$$

and therefore, the two matrices $Y_{SC}$ and $Z_{OC}$ are symmetrical.

Equation (8.32) can be written as

$$V_1 = I_1\, Z_{11} + I_2\, Z_{12}$$

$$= (Z_{11} - Z_{12})\, I_1 + (I_2 + I_1)\, Z_{12} \qquad\qquad ...(8.36)$$

$$V_2 = (Z_{21} - Z_{12})\, I_1 + (Z_{22} - Z_{12})\, I_2 + Z_{12}\, (I_1 + I_2) \qquad\qquad ...(8.37)$$

The equivalent circuit is given in Fig. 8.9

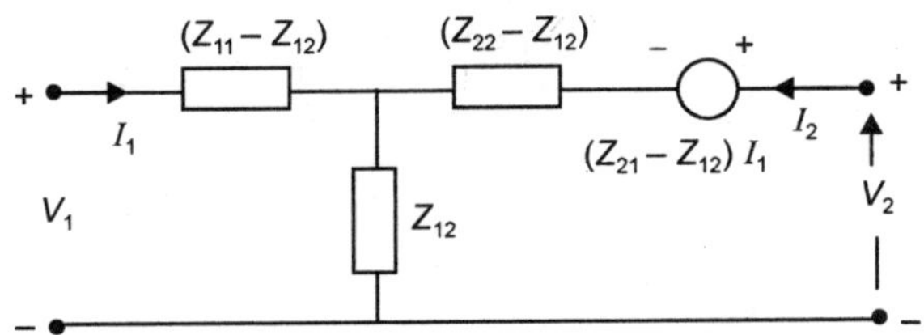



**Fig. 8.9** Equivalent with one generator of general two port network.

For reciprocity $\quad \left.\dfrac{V_1}{I_2}\right|_{V_2=0} = \left.\dfrac{V_2}{I_1}\right|_{V_1=0}$

when $\qquad \left.\dfrac{V_1}{I_2}\right|_{V_2=0}$

$$V_1 = I_1\,Z_{11} + I_2\,Z_{12}$$

$$O = Z_{21}\,I_1 + Z_{22}\,I_2$$

$$I_1 = -(Z_{22}/Z_{21})\,I_2$$

or

$$V_1 = -Z_{11}\,\frac{Z_{22}}{Z_{11}}\,I_2 + I_2\,Z_{12}$$

$$= I_2\left(\frac{Z_{12}\,Z_{21} - Z_{11}\,Z_{22}}{Z_{21}}\right)$$

or

$$\frac{V_1}{I_2} = \frac{Z_{12}\,Z_{21} - Z_{11}\,Z_{22}}{Z_{21}} \qquad\qquad \text{...(8.38)}$$

Similarly to find

$$\left.\frac{V_2}{I_1}\right|_{V_1=0}$$

$$O = I_1\,Z_{11} + I_2\,Z_{12} \qquad\qquad V_2 = Z_{21}\,I_1 + Z_{22}\,I_2$$

$$I_2 = -(Z_{11}/Z_{12})\,I_1 \qquad\qquad V_2 = Z_{21}\,I_1 - \frac{Z_{11}}{Z_{12}}\,Z_{22}\,I_1$$

or

$$\frac{V_2}{I_1} = \frac{Z_{12}\,Z_{21} - Z_{11}\,Z_{22}}{Z_{12}}$$

For reciprocity, therefore, $\quad Z_{21} = Z_{12}$

For symmetrical network

$$\left.\frac{V_1}{I_1}\right|_{I_2=0} = \left.\frac{V_2}{I_2}\right|_{I_1=0}$$

$$V_1 = I_1 Z_{11} + I_2 Z_{12}$$

Therefore,
$$\left.\frac{V_1}{I_1}\right|_{I_2=0} = Z_{11}$$

and
$$V_2 = I_1 Z_{21} + I_2 Z_{22}$$

Therefore,
$$\left.\frac{V_2}{I_2}\right|_{I_1=0} = Z_{22}$$

Therefore,
$$Z_{11} = Z_{22}$$

### 8.4.3  Hybrid Parameters

The other possibility of inter-relating the two port voltages and currents is to express a current and voltage from opposite ports in terms of the other voltage and current. Thus

$$\begin{bmatrix} V_1(s) \\ I_2(s) \end{bmatrix} = \begin{bmatrix} h_{11} & h_{12} \\ h_{21} & h_{22} \end{bmatrix} \begin{bmatrix} I_1(s) \\ V_2(s) \end{bmatrix} \qquad \dots(8.39)$$

and
$$\begin{bmatrix} I_1(s) \\ V_2(s) \end{bmatrix} = \begin{bmatrix} g_{11} & g_{12} \\ g_{21} & g_{22} \end{bmatrix} \begin{bmatrix} V_1(s) \\ I_2(s) \end{bmatrix} \qquad \dots(8.40)$$

The interpretation of the parameters is given as follows:
$$V_1\,(s) = h_{11}\ I_1\,(s) + h_{12}\ V_2\,(s)$$
$$I_2\,(s) = h_{21}\ I_1\,(s) + h_{22}\ V_2\,(s)$$

Here
$$h_{11} = \left.\frac{V_1\,(s)}{I_1\,(s)}\right|_{V_2=0} \qquad h_{12} = \left.\frac{V_1\,(s)}{V_2\,(s)}\right|_{I_1=0}$$

$$h_{21} = \left.\frac{I_2\,(s)}{I_1\,(s)}\right|_{V_2=0} \qquad h_{22} = \left.\frac{I_2\,(s)}{V_2\,(s)}\right|_{I_1=0} \qquad \dots(8.41)$$

Similarly,

$$g_{11} = \left.\frac{I_1\,(s)}{V_1\,(s)}\right|_{I_2=0} \qquad g_{12} = \left.\frac{I_1\,(s)}{I_2\,(s)}\right|_{V_1=0}$$

$$g_{21} = \left.\frac{V_2\,(s)}{V_1\,(s)}\right|_{I_2=0} \qquad g_{22} = \left.\frac{V_2\,(s)}{I_2\,(s)}\right|_{V_1=0} \qquad \dots(8.42)$$

It can be seen that $h$- and $g$- parameters are interpreted under a mixed set of terminal conditions, some of them under open circuit and some under short circuit conditions. These are, therefore, termed as hybrid $h$- and hybrid $g$-parameters. It can also be seen that $h_{11}$ and $g_{22}$ are impedances whereas $h_{22}$ and $g_{11}$ are admittances. Also, it can be seen that

$$h_{11} = \frac{1}{Y_{11}} \qquad g_{11} = \frac{1}{Z_{11}}$$

$$h_{22} = \frac{1}{Z_{22}} \quad \text{and} \quad g_{22} = \frac{1}{Y_{22}} \qquad \qquad \text{...(8.43)}$$

The transfers *gs* and *hs* are dimensionless. The quantity $g_{12}$ is the reverse short circuit current gain and $h_{21}$ is the forward short circuit current gain, $g_{21}$ is the forward open circuit voltage gain whereas $h_{21}$ is the reverse open circuit voltage gain.

Through algebraic manipulations it can be proved that

$$h_{12} = \frac{-Z_{12}}{Z_{21}} h_{21} \qquad \qquad \text{...(8.44)}$$

and

$$g_{12} = -\frac{Y_{12}}{Y_{21}} g_{21} \qquad \qquad \text{...(8.45)}$$

In case the network is reciprocal (i.e. $Z_{12} = Z_{21}$ and $Y_{12} = Y_{21}$)

$$h_{12} = -h_{21} \quad \text{and} \quad g_{12} = -g_{21} \qquad \qquad \text{...(8.46)}$$

which means that for reciprocal network the open circuit voltage gain for transmission in one direction through the two-port equals the negative of the short circuit current gain for transmission in the opposite direction.

Just as $\qquad Y_{SC} = Z_{OC}^{-1}$

Similarly

$$G = H^{-1} \qquad \text{or} \qquad \det G = \frac{1}{\det H} \qquad \qquad \text{...(8.47)}$$

From equation (8.39) the equivalent circuit to a two-port network is given in Fig. (8.10)

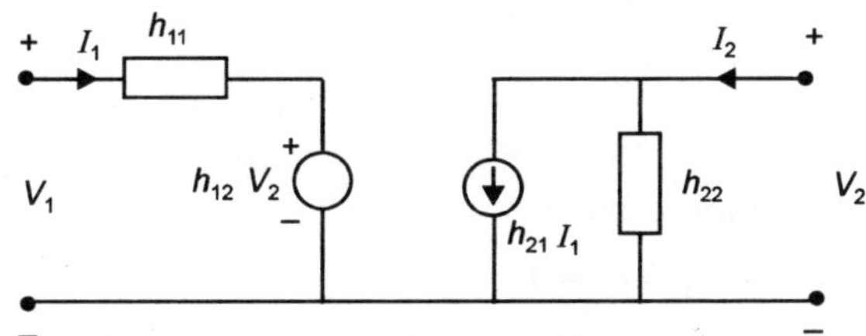

**Fig 8.10(a)**  h-parameter equivalent circuit.

For reciprocity

$$\left.\frac{I_2}{V_1}\right|_{V_2=0} = \left.\frac{I_1}{V_2}\right|_{V_1=0}$$

Now

$$V_1 = h_{11} I_1 + h_{12} V_2, \quad 0 = h_{11} I_1 + h_{12} V_2 \quad \text{when } V_1 = 0$$

$$I_2 = h_{21} I_1 + h_{22} V_2 \quad \text{or} \quad \frac{I_1}{V_2} = -\frac{h_{12}}{h_{11}}$$

Also  $I_2 = h_{21}\, I_1$  when  $V_2 = 0$  and  $V_1 = h_{11}\, I_1$

or
$$\frac{I_2}{V_1} = \frac{h_{21}}{h_{11}}$$

Hence,
$$h_{21} = -h_{12} \qquad \qquad \text{...(8.48)}$$

For the network to be symmetrical

$$\left.\frac{V_1}{I_1}\right|_{I_2=0} = \left.\frac{V_2}{I_2}\right|_{I_1=0}$$

$$O = h_{21}\, I_1 + h_{22}\, V_2$$

$$V_2 = -\, (h_{21} / h_{22}) \cdot I_1$$

Therefore,
$$V_1 = h_{11}\, I_1 - \frac{h_{21}}{h_{22}} \cdot h_{12}\, I_1$$

or
$$V_1 = \left( h_{11} - \frac{h_{21}\, h_{12}}{h_{22}} \right) I_1$$

or
$$\frac{V_1}{I_1} = \frac{h_{11}\, h_{22} - h_{21}\, h_{12}}{h_{22}} \qquad \qquad \text{...(8.49)}$$

To obtain
$$\left.\frac{V_2}{I_2}\right|_{I_1=0}$$

$$I_2 = h_{22} V_2$$

or
$$\frac{V_2}{I_2} = \frac{1}{h_{22}} \qquad \qquad \text{...(8.50)}$$

For the network to be symmetric

$$h_{11}\, h_{22} - h_{21}\, h_{12} = 1 \qquad \qquad \text{...(8.51)}$$

From equation (8.40) the equivalent circuit can be drawn as in Fig. 8.10(b)

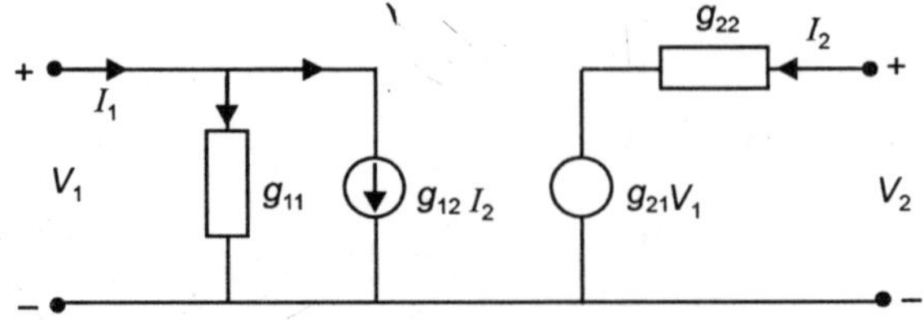

**Fig. 8.10(b)**  g-parameter equivalent.

For reciprocity
$$\left.\frac{I_2}{V_1}\right|_{V_2=0} = \left.\frac{I_1}{V_2}\right|_{V_1=0}$$

The equations are:

$$I_1 = g_{11} V_1 + g_{12} I_2$$
$$V_2 = g_{21} V_1 + g_{22} I_2$$
$$o = g_{21} V_1 + g_{22} I_2$$

or

$$\frac{I_2}{V_1} = -\frac{g_{21}}{g_{22}}$$

and

$$\frac{I_1}{V_2} = \frac{g_{12}}{g_{22}} \quad \text{when} \quad V_1 = 0$$

Therefore, $\qquad g_{12} = -g_{21}$ $\qquad\qquad\qquad$ ...(8.52)

For symmetry of the network

$$\left.\frac{V_1}{I_1}\right|_{I_2=0} = \left.\frac{V_2}{I_2}\right|_{I_1=0}$$

$$\frac{V_1}{I_1} = \frac{1}{g_{11}} \qquad\qquad\qquad ...(8.53)$$

and

$$o = g_{11} V_1 + g_{12} I_2$$
$$V_2 = g_{21} V_1 + g_{22} I_2$$

$$V_1 = -\frac{g_{12}}{g_{11}} I_2$$

or

$$V_2 = -\frac{g_{21} g_{12}}{g_{11}} I_2 + g_{22} I_2$$

$$= I_2 \left( \frac{g_{11} g_{22} - g_{21} g_{12}}{g_{11}} \right)$$

or

$$\frac{V_2}{I_2} = \frac{g_{11} g_{22} - g_{21} g_{12}}{g_{11}} \qquad\qquad ...(8.54)$$

Comparing (8.53) and (8.54), the condition for symmetry of the network is

$$g_{11} g_{22} - g_{21} g_{12} = 1 \qquad\qquad\qquad ...(8.55)$$

## 8.4.4 Chain Parameters

The voltage and current at one port can be correlated to voltage and current at the other port through a set of parameters known as chain parameters or *ABCD* parameters

$$\begin{bmatrix} V_1(s) \\ I_1(s) \end{bmatrix} = \begin{bmatrix} A & B \\ C & D \end{bmatrix} \begin{bmatrix} V_2(s) \\ I_2(s) \end{bmatrix} \qquad\qquad ...(8.56)$$

These are very well known parameters in transmission network in power systems.

These are known as chain parameters as they are the natural ones to use in a cascade or tandem or chain connection typical of a transmission network. The transmission parameters are:

$$A = \left.\frac{V_1}{V_2}\right|_{I_2 = 0} \qquad \qquad ...(8.57)$$

i.e. the reverse voltage ratio with receiving end open (output port open circuited)

$$B = \left.\frac{V_1}{I_2}\right|_{V_2 = 0} \qquad \qquad ...(8.58)$$

i.e. the transfer impedance with receiving end short circuited (output port short circuited)

$$C = \left.\frac{I_1}{V_2}\right|_{I_2 = 0} \qquad \qquad ...(8.59)$$

i.e. the transfer admittance with receiving end open

$$D = \left.\frac{I_1}{I_2}\right|_{V_2 = 0} \qquad \qquad ...(8.60)$$

i.e. the reverse current ratio with the receiving end short circuited.

If the network is reciprocal the ratio of response transform to the excitation remains same if we interchange the position of the excitation and response in the network.

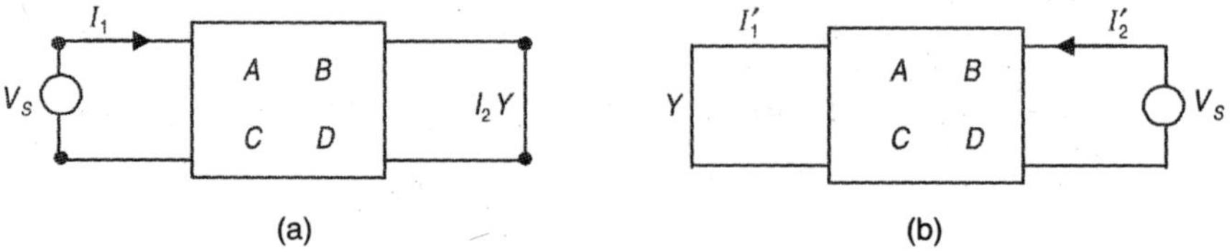

**Fig. 8.11** (a) Output port shorted (b) Input port shorted.

$$V_s = AV_2 + BI_2 \qquad \qquad o = AV_s - BI'_2$$
$$= B\,I_2 \text{ as } V_2 = 0 \qquad -I'_1 = C\,V_s - D\,I'_2$$

From Fig. 8.11 (*a*) the ratio of response to excitation is

$$\frac{I_2}{V_s} = \frac{1}{B}$$

From Fig. 8.11 (*b*)

$$I'_2 = \frac{A}{B}\,V_s$$

and
$$-I'_1 = C\,V_s - D \cdot \frac{A}{B}\,V_s = V_s\left(C - \frac{DA}{B}\right)$$

or
$$\frac{I'_1}{V_s} = \frac{AD - BC}{B}$$

Therefore the two should be equal i.e.

$$\frac{1}{B} = \frac{AD - BC}{B} \quad \text{or} \quad AD - BC = 1 \qquad \qquad ...(8.61)$$

In case the transmission network is symmetric

$$Z_{11} = \frac{V_1}{I_1}\bigg|_{I_2=0} = \frac{AV_2 + BI_2}{CV_2 + DI_2}\bigg|_{I_2=0} = \frac{A}{C}$$

Similarly, $\quad Z_{22} = \dfrac{V_2}{I_2}\bigg|_{I_1=0} \quad$ Here from the equation

$$I_1 = CV_2 - DI_2 = 0$$

or

$$\frac{V_2}{I_2} = \frac{D}{C}$$

Since in a symmetric network $\quad Z_{11} = Z_{22}$

$$\frac{A}{C} = \frac{D}{C} \quad \text{or} \quad A = D \qquad \qquad ...(8.62)$$

This shows that for a symmetric network $A = D$ and reciprocal network $AD - BC = 1$.

### 8.4.5 Inverse Transmission Parameters

Earlier we expressed $V_1$, $I_1$ in terms of $V_2$ and $I_2$ through the parameters $ABCD$. We now express $V_2$, $I_2$ in terms of $V_1$, $I_1$ through a set of parameters known as inverse transmission parameters and usually denoted by $A'$ $B'$ $C'$ and $D'$.

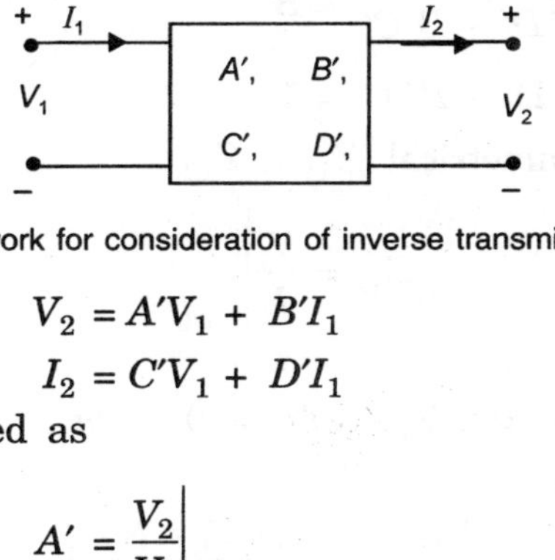

**Fig. 8.12** Two port network for consideration of inverse transmission parameters.

$$V_2 = A'V_1 + B'I_1$$
$$I_2 = C'V_1 + D'I_1 \qquad \qquad ...(8.63)$$

These parameters are defined as

$$A' = \frac{V_2}{V_1}\bigg|_{I_1=0} \qquad \qquad ...(8.64)$$

i.e. forward voltage ratio with $I_1 = 0$

$$B' = \frac{V_2}{I_1}\bigg|_{V_1=0} \qquad \qquad ...(8.65)$$

transfer impedance with $V_1 = 0$

$$C' = \frac{I_2}{V_1}\bigg|_{I_1=0} \qquad \qquad ...(8.66)$$

transfer admittance with $I_1 = 0$

and
$$D' = \frac{I_2}{I_1}\Bigg|_{V_1 = 0} \qquad \qquad ...(8.67)$$

forward current ratio with $V_1 = 0$

For condition of reciprocity we write the following equations

When $V_2 = 0$ $\qquad\qquad\qquad\qquad$ when $V_1 = 0$

$$0 = A'V_1 + B'I_1 \qquad\qquad\qquad V_2 = A'V_1 - B'I_1$$
$$I_2 = C'V_1 + D'I_1 \qquad\qquad\qquad I_2 = C'V_1 - D'I_1$$

$$V_1 = -\frac{B'}{A'}I_1 \quad \text{or} \quad I_1 = -\frac{A'}{B'}V_1 \qquad \frac{V_2}{I_1} = -B'$$

Now
$$I_2 = C'V_1 - \frac{A'}{B'}D'V_1$$

$$= \frac{B'C' - A'D'}{B'}V_1$$

or
$$\frac{V_1}{I_2} = \frac{B'}{B'C' - A'D'}$$

For network to be reciprocal $\quad \dfrac{V_1}{I_2} = \dfrac{V_2}{I_1}$

or
$$\frac{B'}{A'D' - B'C'} = B'$$

or
$$A'D' - B'C' = 1 \qquad\qquad ...(8.68)$$

For network to be symmetrical $Z_{11} = Z_{22}$

$$Z_{11} = \frac{V_1}{I_1}\Bigg|_{I_2 = 0}$$

$$I_2 = C'V_1 - D'I_1 = 0 \qquad \text{or} \qquad \frac{V_1}{I_1} = \frac{D'}{C'}$$

$$Z_{22} = \frac{V_2}{I_2}\Bigg|_{I_1 = 0} = \frac{A'V_1 - B'I_1}{C'V_1 - D'I_1}\Bigg|_{I_1 = 0} = \frac{A'}{C'}$$

Hence,
$$\frac{D'}{C'} = \frac{A'}{C'}$$

Therefore,
$$A' = D' \qquad\qquad ...(8.69)$$

**Example 8.6:** For the network shown in Fig. E8.6

(*i*) Determine the open circuit impedance parameters and draw its equivalent circuit.

(*ii*) Determine the short circuit admittance parameters and draw its equivalent circuit.

(*iii*) Determine hybrid parameters and draw its equivalent circuit.

(*iv*) Determine transmission parameters.

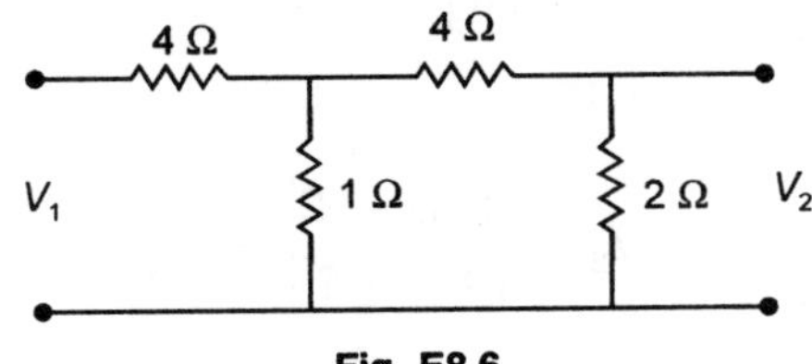

**Fig. E8.6**

(*i*) **z-parameters:** To determine z-parameters the circuits to be considered are:

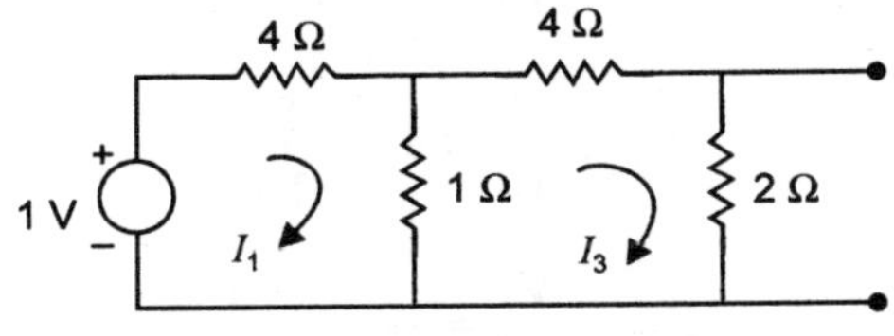

**Fig. E8.6.1**

The loop impedance matrix by inspection

$$\begin{bmatrix} 5 & -1 \\ -1 & 7 \end{bmatrix}\begin{bmatrix} I_1 \\ I_3 \end{bmatrix} = \begin{bmatrix} 1 \\ 0 \end{bmatrix}$$

The determinant is $\Delta = 35 - 1 = 34$

$$I_1 = \frac{1}{34}\begin{vmatrix} 1 & -1 \\ 0 & 7 \end{vmatrix} = \frac{7}{34}$$

$$I_3 = \frac{1}{34}\begin{vmatrix} 5 & 1 \\ -1 & 0 \end{vmatrix} = \frac{1}{34}$$

$$Z_{11} = \frac{V_1}{I_1}\bigg|_{I_2=0} = \frac{1}{7/34} = \frac{34}{7}$$

$$Z_{21} = \frac{V_2}{I_1}\bigg|_{I_2=0} = \frac{2/34}{7/34} = \frac{2}{7}$$

To calculate $Z_{12}$ and $Z_{22}$ we take the following circuit

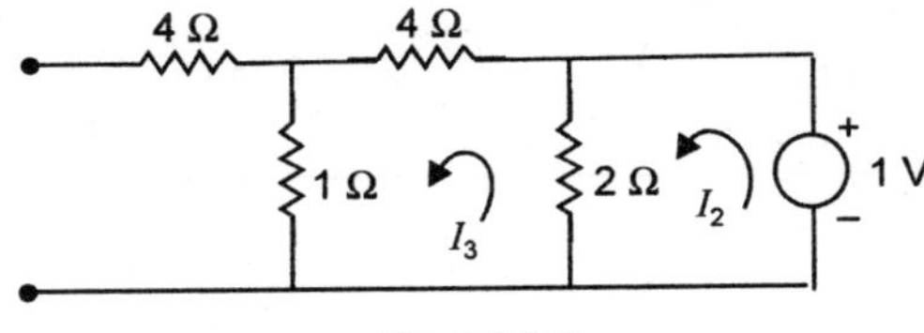

**Fig. E8.6.2**

$$\begin{bmatrix} 2 & -2 \\ -2 & 7 \end{bmatrix} \begin{bmatrix} I_2 \\ I_3 \end{bmatrix} = \begin{bmatrix} 1 \\ 0 \end{bmatrix}$$

The determinant $\Delta = 14 - 4 = 10$

$$I_2 = \frac{1}{10} \begin{vmatrix} 1 & -2 \\ 0 & 7 \end{vmatrix} = \frac{7}{10} A$$

$$I_3 = \frac{1}{10} \begin{vmatrix} 2 & 1 \\ -2 & 0 \end{vmatrix} = \frac{2}{10} A$$

$$V_1 = \frac{2}{10} \times 1 = \frac{2}{10}$$

$$Z_{22} = \frac{V_2}{I_2} = \frac{1}{7/10} = \frac{10}{7}$$

and

$$Z_{12} = \frac{V_1}{I_2} = \frac{2/10}{7/10} = \frac{2}{7}$$

Since here $Z_{12} = Z_{21} = \dfrac{2}{7}$ the network is reciprocal but is asymmetric as $Z_{11} \neq Z_{22}$
The $z$-parameter equivalent circuit.

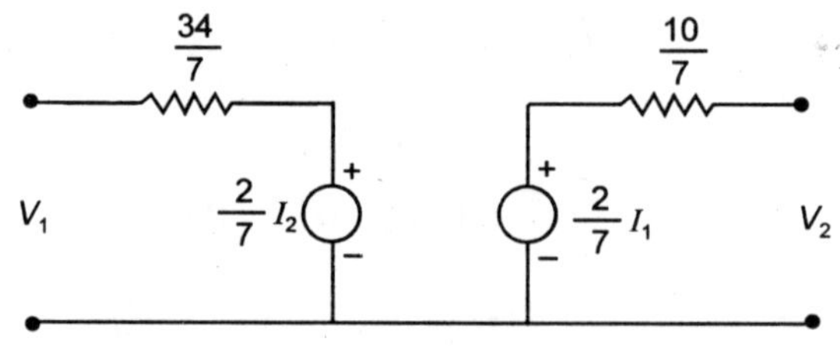

**Fig. E8.6.3**

(*ii*) **Y-Parameter's:** Since these are short circuit parameters, to obtain these we draw
the following network:

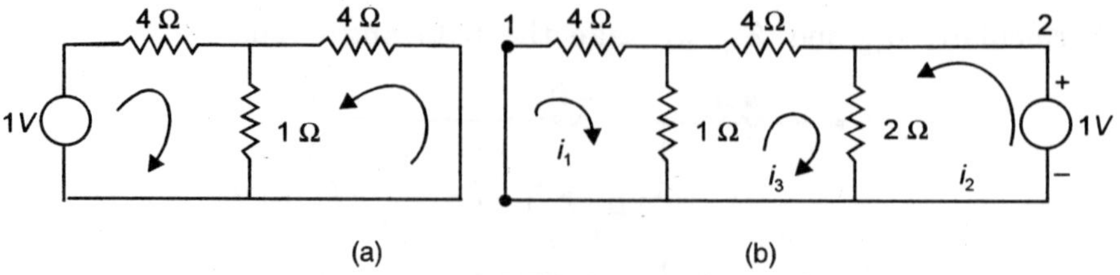

(a)          (b)

**Fig. E8.6.4**

For circuit $(a)$

$$\begin{bmatrix} 5 & +1 \\ +1 & 5 \end{bmatrix}\begin{bmatrix} I_1 \\ I_2 \end{bmatrix} = \begin{bmatrix} 1 \\ 0 \end{bmatrix}$$

Determinant $\Delta = 25 - 1 = 24$

$$I_1 = \frac{1}{24}\begin{vmatrix} 1 & 1 \\ 0 & 5 \end{vmatrix} = \frac{5}{24}$$

Similarly,

$$I_2 = -\frac{1}{24}$$

$$Y_{11} = \frac{I_1}{V_1}\bigg|_{V_2=0} = \frac{5/24}{1} = \frac{5}{24}\,\mho$$

$$Y_{21} = \frac{I_2}{V_1}\bigg|_{V_2=0} = -\frac{1}{24}\,\mho$$

For circuit $(b)$ the loop impedance matrix is,

$$\begin{bmatrix} 2 & +2 & 0 \\ +2 & 7 & -1 \\ 0 & -1 & 5 \end{bmatrix}\begin{bmatrix} I_2 \\ I_3 \\ I_1 \end{bmatrix} = \begin{bmatrix} 1 \\ 0 \\ 0 \end{bmatrix}$$

The determinant $2(35 - 1) - 2(+10) = 68 - 20 = 48$

$$I_2 = \frac{1}{48}\begin{vmatrix} 1 & +2 & 0 \\ 0 & 7 & -1 \\ 0 & -1 & 5 \end{vmatrix} = \frac{34}{48} = \frac{17}{24}\,A$$

$$I_1 = \frac{1}{48}\begin{vmatrix} 2 & +2 & 1 \\ +2 & 7 & 0 \\ 0 & -1 & 0 \end{vmatrix}$$

$$= \frac{1}{48}\left[2 \times 0 = 2 \times 0 + 1\,(-2)\right] = -\frac{1}{24}$$

Now

$$Y_{12} = \frac{I_1}{V_2}\bigg|_{V_1=0} = \frac{-1/24}{1} = -\frac{1}{24}\,\mho$$

$$Y_{22} = \frac{I_2}{V_2}\bigg|_{V_1=0} \quad \frac{17/24}{1} = \frac{17}{24}\,\mho$$

Again $Y_{12} = Y_{21}$, hence the network is reciprocal but asymmetrical as
$$Y_{11} \neq Y_{22}$$

The equivalent circuit is given as

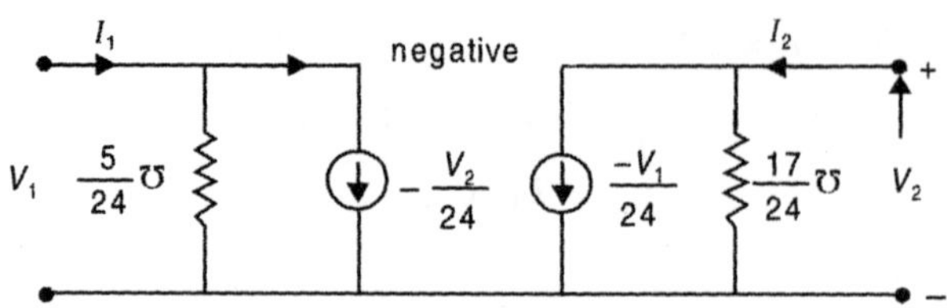


**Fig. E8.6.5**

(*iii*) **Hybrid parameters:**

$$V_1 = h_{11} I_1 + h_{12} V_2$$
$$I_2 = h_{21} I_1 + h_{22} V_2$$

$$h_{11} = \left. \frac{V_1}{I_1} \right|_{V_2=0} \qquad \text{Here} \quad V_1 = 1 \text{ volt and } I_1 = \frac{5}{24} \text{ A}$$

Hence
$$h_{11} = \frac{24}{5} \ \Omega$$

$$h_{21} = \left. \frac{I_2}{I_1} \right|_{V_2=0} = \frac{1/24}{5/24} = \frac{1}{5}$$

$$h_{22} = \left. \frac{I_2}{V_2} \right|_{I_1=0} = \frac{7/10}{1} = \frac{7}{10} \ \text{℧}$$

and
$$h_{12} = \left. \frac{V_1}{V_2} \right|_{I_1=0} = \frac{-2/10}{1} = -\frac{1}{5}$$

Here again $h_{12} = -h_{21}$ and the network is reciprocal. However, since $h_{11} h_{22} - h_{12} h_{21} \neq 1$ the network is asymmetrical.

The equivalent circuit is given as

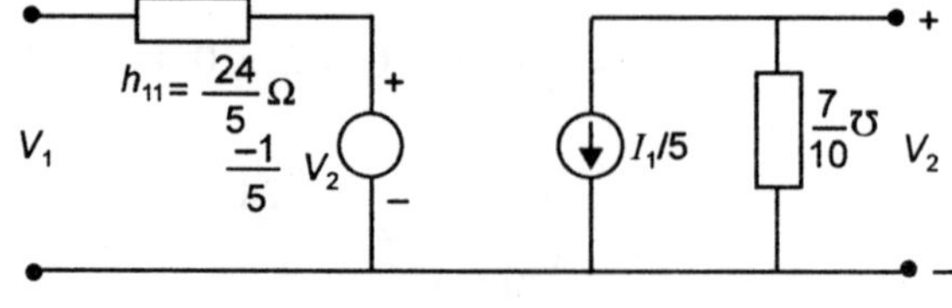


**Fig. E8.6.6**

Similarly to find out *g*-parameters

$$I_1 = g_{11} V_1 + g_{12} I_2$$
$$V_2 = g_{21} V_1 + g_{22} I_2$$

$$g_{11} = \left.\frac{I_1}{V_1}\right|_{I_2=0} = \frac{7/34}{1} = \frac{7}{34} \, \mho$$

$$g_{12} = \left.\frac{I_1}{I_2}\right|_{V_1=0} = -\,\frac{1/24}{17/24} = -\,\frac{1}{17}$$

$$g_{21} = \left.\frac{V_2}{V_1}\right|_{I_2=0} = \frac{2/34}{1} = \frac{1}{17}$$

$$g_{22} = \left.\frac{V_2}{I_2}\right|_{V_1=0} = \frac{1}{17/24} = \frac{24}{17} \, \Omega$$

The equivalent circuit is as follows:

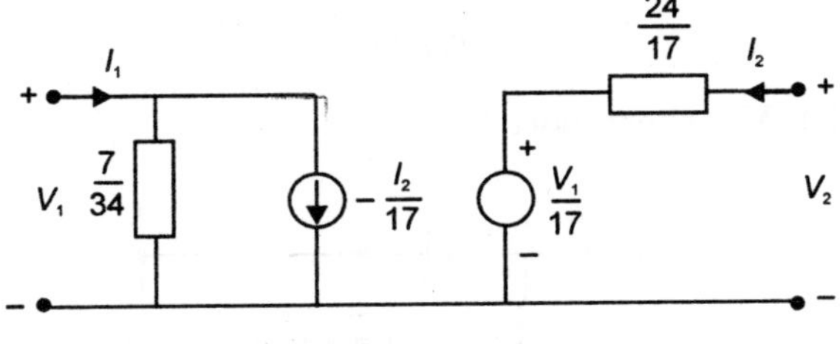


**Fig. E8.6.7**

(*iv*) ABCD parameters

$$V_1 = AV_2 + BI_2$$
$$I_1 = CV_2 + DI_2$$

From Fig. E 8.6.1

$$A = \left.\frac{V_1}{V_2}\right|_{I_2=0} = 17$$

$$B = \left.\frac{V_1}{I_2}\right|_{V_2=0} = 24$$

$$C = \left.\frac{I_1}{V_2}\right|_{I_2=0} = \frac{7}{2}$$

$$D = \left.\frac{I_1}{I_2}\right|_{V_2=0} = 5$$

Since $AD - BC = 85 - 84 = 1$ the network is reciprocal but as $A \neq D$ it is unsymmetrical.

It is to be noted that the parameter $h$, $g$ and $ABCD$ could be evaluated with the values of voltages and current as evaluated under $Z$ and $Y$ parameter as here all the four conditions $I_1 = o$, $I_2 = o$ $V_1 = o$ and $V_2 = o$ have been simulated.

$$V_1 = 17\ V_2 + 24\ I_2$$

$$I_1 = \frac{7}{2}\ V_2 + 5\ I_2$$

Similarly inverse $ABCD$ parameters can also be evaluated.

**Example 8.7:** Determine the $Y$ parameters of the network shown.

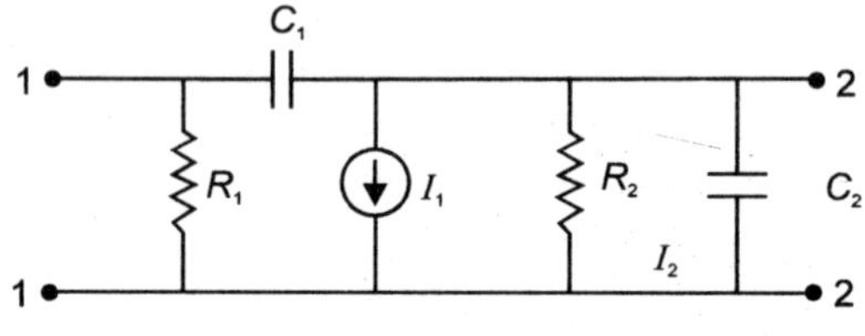

**Fig. E8.7**

To find out $Y_{11}$ and $Y_{21}$ short port 2

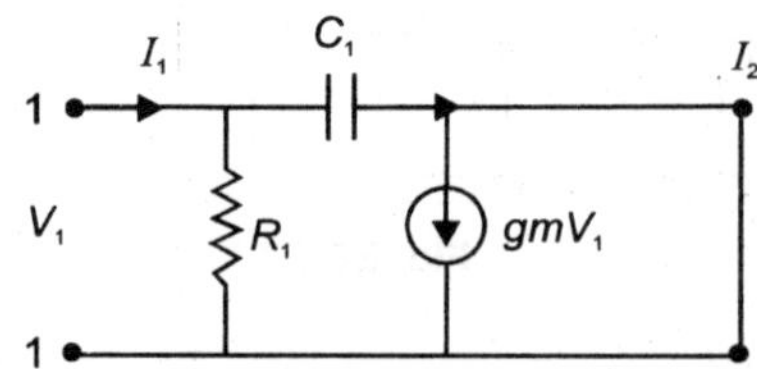

**Fig. E8.7.1**

$$I_1 = \frac{V_1}{R_1} + V_1\ sC_1 = V_1\left(\frac{1}{R_1} + sC_1\right)$$

or

$$Y_{11} = \frac{I_1}{V_1} = \frac{1}{R_1} + sC_1$$

$$Y_{21} = \frac{I_2}{V_1} = \frac{V_1\ sC_1 - g_m\ V_1}{V_1} = sC_1 - g_m$$

Similarly to find out $Y_{12}$ and $Y_{22}$ short circuit port 1.

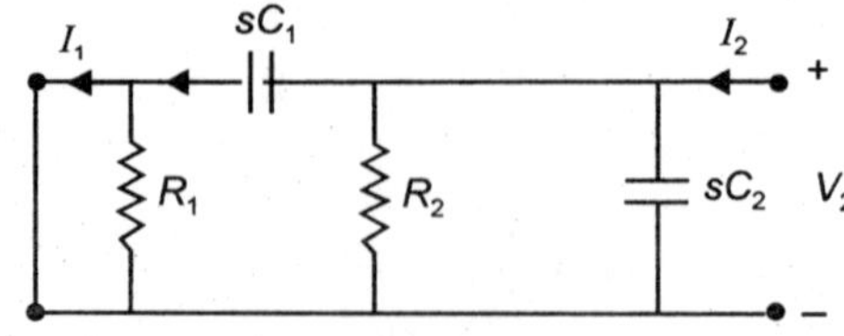

**Fig. E8.7.2**

$$I_1 = V_2 \, sC_1$$

$$Y_{12} = \left. \frac{I_1}{V_2} \right|_{V_1=0} = \frac{V_2 \, sC_1}{V_2} = sC_1$$

Since $V_1 = 0$ the dependent current source gives $g_m \, V_1 = 0$ hence it is left open.

Therefore,
$$I_2 = V_2 \left[ s(C_1 + C_2) + \frac{1}{R_2} \right]$$

or
$$Y_{22} = \frac{I_2}{V_2} = s(C_1 + C_2) + \frac{1}{R_2}$$

**Example 8.8:** Determine the *ABCD* parameters of the network shown below:

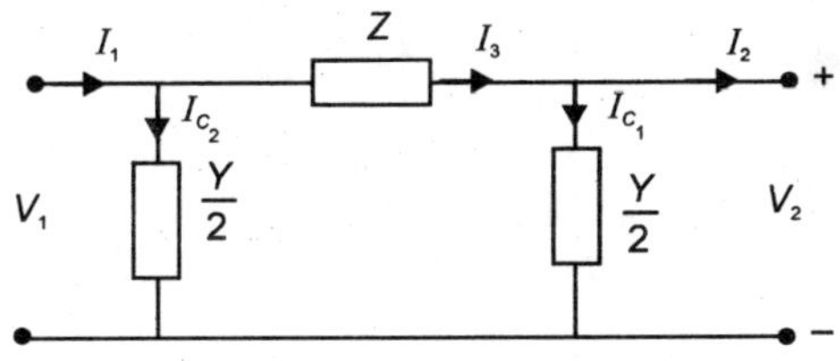


**Fig. E8.8**

**Solution:**

$$I_{C_1} = V_2 \, Y/2$$
$$I_3 = I_2 + I_{C_1} = I_2 + V_2 \, Y/2$$
$$V_1 = V_2 + I_3 \, Z = V_2 + (I_2 + V_2 \, Y/2) \, Z$$
$$V_1 = V_2 + I_2 \, Z = V_2 \, (YZ)/2$$
$$= V_2 \, (1 + YZ/2) + ZI_2$$
$$I_1 = I_3 + I_{C_2} = I_2 + V_2 \, Y/2 + V_1 \, Y/2$$
$$= V_2 \, Y_2 + I_2 + [V_2 \, (1 + YZ/2) + ZI_2] \, Y/2$$
$$= V_2 \left( Y + \frac{Y^2 Z}{4} \right) + I_2 \left( 1 + \frac{YZ}{2} \right)$$

Therefore, $A = 1 + \dfrac{YZ}{2}$ $\qquad\qquad B = Z$

$C = Y + \dfrac{Y^2 Z}{4}$ $\qquad\qquad D = 1 + \dfrac{YZ}{2}$

Since $A = D$, the network is symmetric

Now $AD - BC = \left( 1 + \dfrac{YZ}{2} \right)^2 - Z \left( Y + \dfrac{Y^2 Z}{4} \right) = 1$

Hence it is a reciprocal network.

**Example 8.9:** For the network, determine the  $z$  and  $y$  parameters if these exist

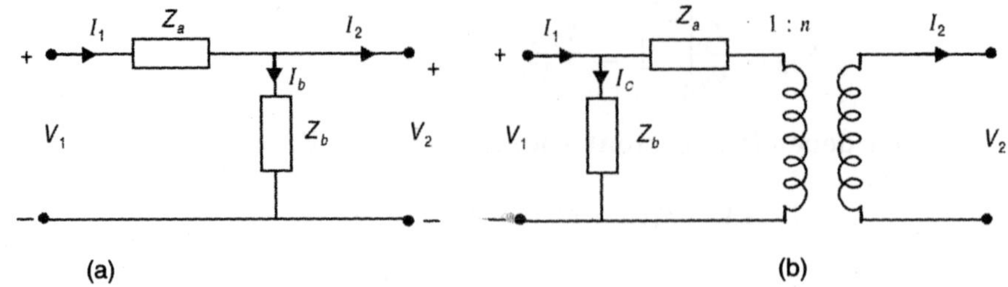

(a)                (b)

**Fig. E.8.9 (a) and (b)**

## Z-parameters

$(a)$     $I_b = V_2 / Z_b$     $\therefore\;\; I_1 = I_2 + \dfrac{V_2}{Z_b}$

$$V_1 = V_2 + I_1\,Z_a = V_2 + \left(I_2 + \dfrac{V_2}{Z_b}\right)Z_a$$

$$V_1 = V_2\left(1 + \dfrac{Z_a}{Z_b}\right) + I_2\,Z_a \qquad\qquad ...(1)$$

$$I_1 = (V_2 / Z_b) + I_2$$
$$V_2 = I_1\,Z_b - I_2\,Z_b$$

$$V_1 = (I_1\,Z_b - I_2\,Z_b)\left(1 + \dfrac{Z_a}{Z_b}\right) + I_2\,Z_a$$

$$= I_1\,Z_b - I_2\,Z_b + I_1\,Z_a - I_2\,Z_a + I_2\,Z_a$$
$$V_1 = I_1\,(Z_b + Z_a) - I_2\,Z_b$$
$$V_2 = I_1\,Z_b - I_2\,Z_b$$
$$Z_{11} = Z_a + Z_b,\;\; Z_{12} = -\,Z_b,\;\; Z_{21} = Z_b$$
$$Z_{22} = -Z_b$$

## Y-Parameters

$$I_1 = \dfrac{V_2}{Z_b} + I_2$$

Substituting for  $\dfrac{V_2}{Z_b}$  from equation (1) we have

$$I_1 = \dfrac{V_1}{Z_a + Z_b} + I_2\left(\dfrac{Z_b}{Z_a + Z_b}\right)$$

$$I_1\left(1 - \dfrac{Z_b}{Z_a + Z_b}\right) = \dfrac{V_1}{Z_a + Z_b} - \dfrac{V_2}{Z_a + Z_b}$$

$$I_1 = \frac{V_1}{Z_a} - \frac{V_2}{Z_a}$$

$$I_2 = I_1 - \frac{V_2}{Z_b} = \frac{V_1}{Z_a} - \frac{V_2}{Z_a} - \frac{V_2}{Z_b} = \frac{V_1}{Z_a} - V_2\left(\frac{1}{Z_a} + \frac{1}{Z_b}\right)$$

Therefore, $\quad Y_{11} = \dfrac{1}{Z_a}$, $\;Y_{12} = -\dfrac{1}{Z_a}$, $\;Y_{21} = \dfrac{1}{Z_a}$

and $\qquad Y_{22} = \left(\dfrac{1}{Z_a} + \dfrac{1}{Z_b}\right)$

(*b*) The equivalent circuit referred to port 1, when port 2 is open circuited, is

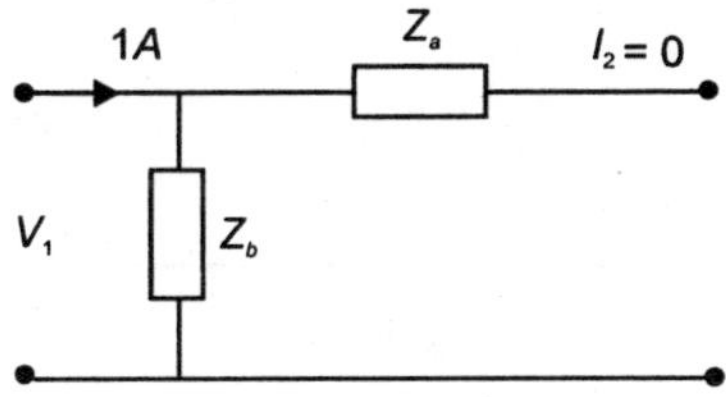



**Fig. E8.9.1**

$$\left.\frac{V_1}{I_1}\right|_{I_2=0} = Z_{11} = Z_b, \quad Z_{21} = \left.\frac{V_2}{I_1}\right|_{I_2=0} = \frac{nV_1}{I_1} = nZ_b$$

Let $\qquad I_2 = 1 \quad \text{and} \quad I_1 = 0$

$$Z_{12} = \left.\frac{V_1}{I_2}\right|_{I_1=0} = \frac{V_1}{1/n} = nV_1 = nZ_b$$

as secondary current referred to primary side is $1/n$

$$Z_{22} = \left.\frac{V_2}{I_2}\right|_{I_1=0} = Z_a + Z_b$$

when referred to the port 1 side whereas on port 2 it will be $n^2\,(Z_a + Z_b)$, *Calculation of Y-parameters:* The equivalent circuit for calculation of *Y*-parameters when referred to port 1 is as given in Fig. E 8.9.2.

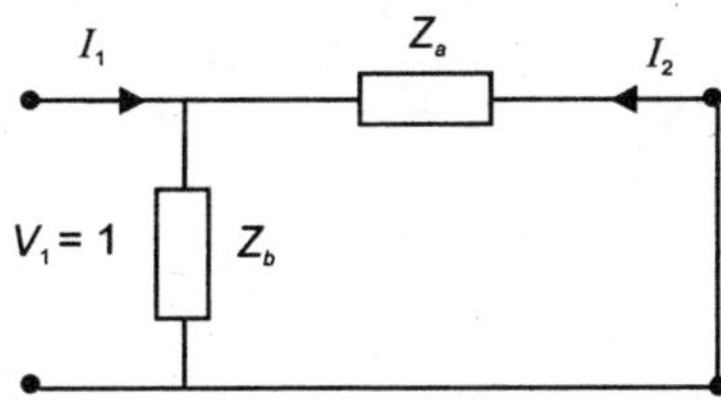



**Fig. E8.9.2**

Port II is short circuited i.e. $V_2 = 0$

$$Y_{11} = \frac{I_1}{V_1}\bigg|_{V_2=0} = \frac{1}{Z_a} + \frac{1}{Z_b}$$

Since,

$$I_2 = \frac{-I_1}{n} = -\frac{V_1}{n\,Z_a}$$

Therefore,

$$Y_{21} = -\frac{1}{n\,Z_a}$$

Let $\qquad V_1 = 0 \qquad V_2 = 1$

The equivalent circuit referred to port 2 is

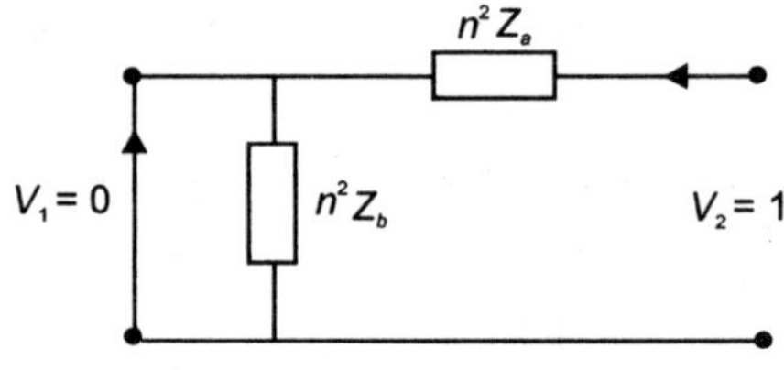


**Fig. E8.9.3**

$$I_2 = \frac{1}{n^2\,Z_a}$$

Therefore,

$$Y_{22} = \frac{1}{n^2\,Z_a}$$

Since

$$I_1 = -nI_2$$

$$Y_{12} = \frac{I_1}{V_2} = -\frac{1}{n\,Z_a}$$

**Example 8.10:** Determine the $Y$- and $Z$-parameters of the network shown

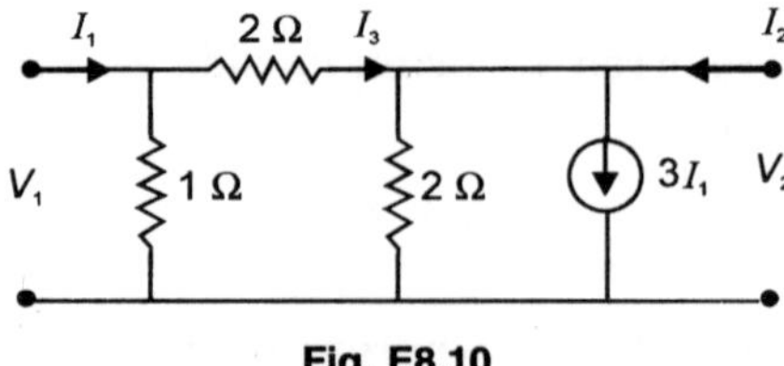


**Fig. E8.10**

Calculation of $Y_{11}$ and $Y_{21}$

$$Y_{11} = \frac{I_1}{V_1}\bigg|_{V_2=0} = \frac{1/2/3}{1} = 1.5\,\mho$$

$$I_2 - 3\,I_1 + \frac{V_1}{2} = 0$$

$$I_2 - 4.5\,V_1 + \frac{V_1}{2} = 0$$

$$I_2 = 4\ V_1$$
$$I_2 = 4 \text{ as } V_1 = 1$$
$$Y_{21} = \frac{I_2}{V_1} = \frac{4}{1} = 4\ \mho$$

Calculation of $Y_{22}$ and $Y_{12}$

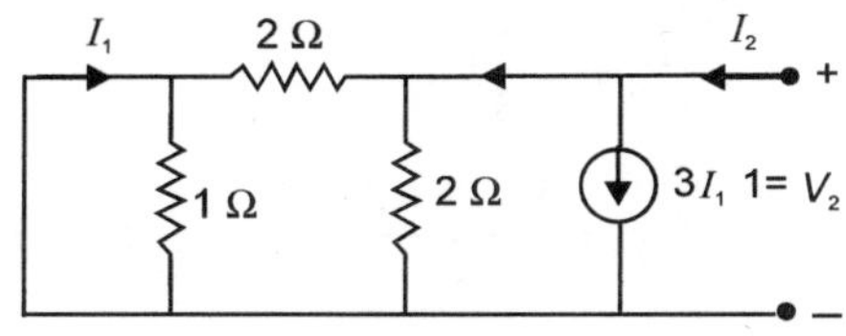


**Fig. E8.10.1**

Now
$$Y_{22} = \left.\frac{I_2}{V_2}\right|_{V_1 = 0}$$

The equivalent circuit is shown in Fig. E.8.10.1
Since 2 Ω resistance are in parallel, KCL at port 2 gives

$$I_1 = -\frac{1}{2} \text{ unit} \qquad I_2 - 3I_1 - 1 = 0, \; I_2 + \frac{3}{2} - 1 = 0$$

$$Y_{22} = \frac{I_2}{V_2} = -0.5 \qquad\qquad I_2 = -0.5$$

$$Y_{12} = \left.\frac{I_1}{V_2}\right|_{V_1 = 0} = -0.5\ \mho \textbf{ Ans.}$$

**Z-Parameters Calculations:** The equivalent circuit is

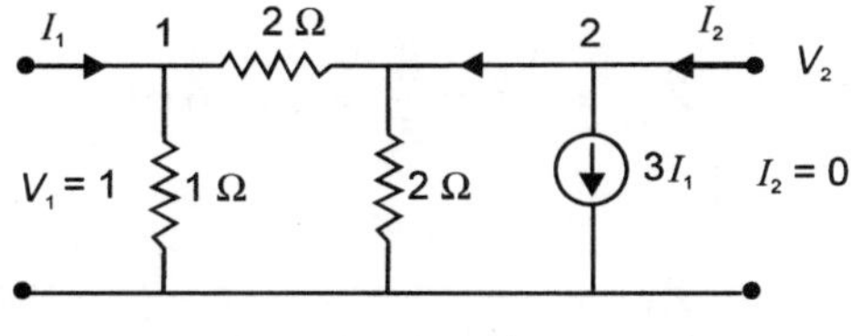


**Fig. E8.10.2**

$$Z_{11} = \left.\frac{V_1}{I_1}\right|_{I_2 = 0}$$

$$I_1 = V_1 + (V_1 - V_2)\ 0.5$$
$$= 1.5\ V_1 - 0.5\ V_2 - \text{At node I}$$
$$3I_1 - (V_1 - V_2)\ 0.5 + 0.5\ V_2 = 0 - \text{node II}$$
$$3I_1 = 0.5\ V_1 - V_2 \tag{1}$$
$$I_1 = 1.5\ V_1 - 0.5 V_2 \tag{2}$$

$$3I_1 = 0.5\ V_1 - V_2 \qquad\qquad \text{...(3)}$$

$$\frac{I_1}{3} = 0.5\ V_1 - \frac{0.5}{3}V_2 \qquad\qquad \text{...(4)}$$

Dividing equation (2) by 3. and

subtracting equations (4) from (1)

$$\frac{8I_1}{3} = -\frac{5}{6}V_2$$

or

$$Z_{21} = \frac{V_2}{I_1} = -\frac{8}{3}\times\frac{6}{5} = -3.2\ \Omega$$

From equations (1) and (2)

$$I_1 = -2.5\ V_1$$

or

$$\frac{V_1}{I_1} = -0.4\ \Omega = Z_{11}$$

To find out $Z_{12}$ and $Z_{22}$ when $I_1 = 0$

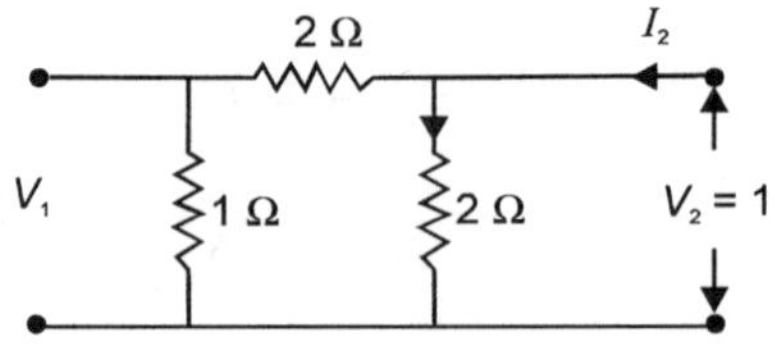



**Fig. E8.10.3**

$$V_1 = \frac{V_2}{3}, I_2 = \frac{V_2}{2} + \frac{V_2}{3} = \frac{5V_2}{6}$$

or

$$Z_{22} = \frac{V_2}{I_2} = \frac{6}{5} = 1.2\ \textbf{Ans.}$$

and

$$Z_{12} = \left.\frac{V_1}{I_2}\right|_{I_i = 0} = \frac{V_2/3}{5V_2/6} = 0.4\ \Omega\ \textbf{Ans.}$$

It is to be noted that the source is to be absent when $I_1 = 0$

**Example 8.11:** For the network shown determine the Y- and Z-parameters

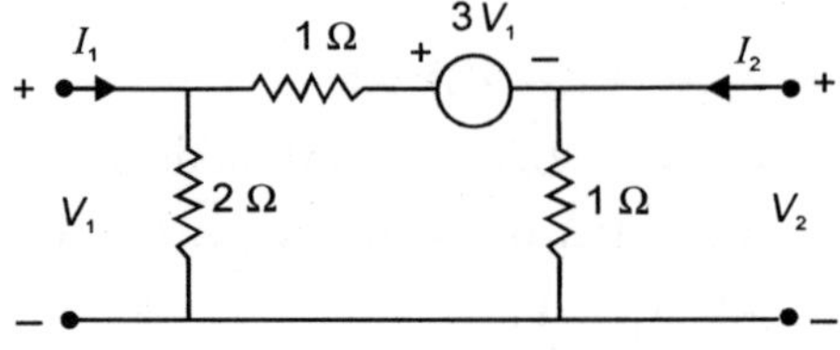



**Fig. E8.11**

*To determine $Y_{11}$ and $Y_{21}$*

$$Y_{11} = \frac{I_1}{V_1}\bigg|_{V_2=0} \qquad Y_{21} = \frac{I_2}{V_1}\bigg|_{V_2=0}$$

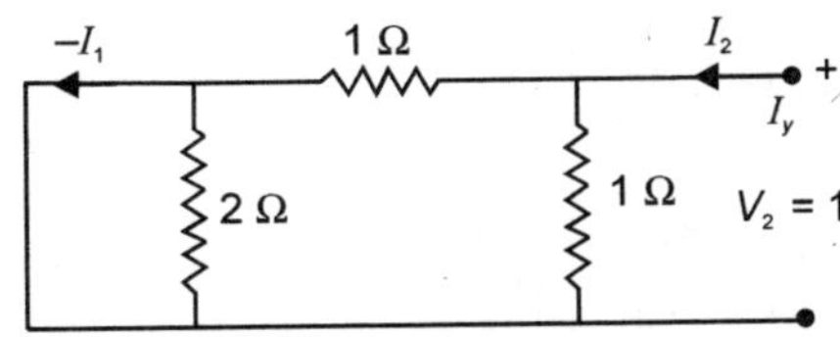


**Fig. E8.11.1**

$$I_x = \frac{V_1}{2} \qquad\qquad -I_y - 3\,V_1 + V_1 = 0$$

$$I_1 = \frac{V_1}{2} - 2V_1 \qquad\qquad I_y = -\,2V_1$$

$$= -\frac{3}{2}V_1 \quad \text{or} \quad \frac{I_1}{V_1} = -\frac{3}{2}\,\mho = Y_{11}$$

Since $\quad I_2 = -I_y = 2V_1 \qquad \therefore\ Y_{21} = \dfrac{I_2}{V_1} = \dfrac{2V_1}{V_1} = 2\,\mho$

*To determine $Y_{22}$ and $Y_{12}$*



**Fig. E 8.11.2**

$$I_1 = -V_2 \qquad\qquad I_2 = V_2 + V_2 = 2V_2$$

$$Y_{12} = \frac{I_1}{V_2} = -1.0\,\mho \qquad\qquad Y_{22} = 2\,\mho$$

*To determine $Z_{11}$ and $Z_{21}$ where* $Z_{11} = \dfrac{V_1}{I_1}\bigg|_{I_2=0} \qquad Z_{21} = \dfrac{V_2}{I_1}\bigg|_{I_2=0}$

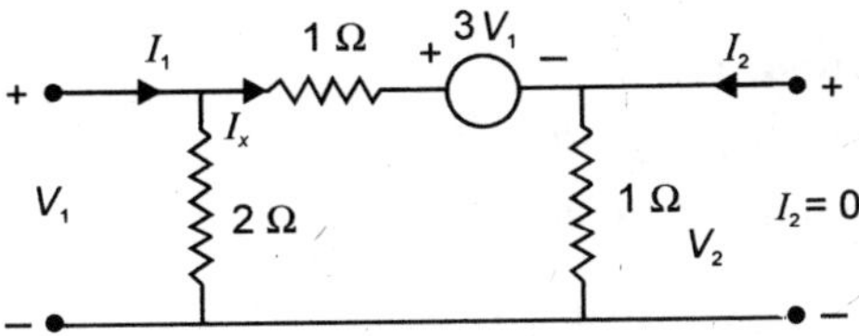


**Fig. E.8.11.3**

$$V_1 - I_x - 3V_1 - I_x = 0$$

$$I_1 = \frac{V_1}{2} - V_1 \qquad\qquad 2I_x = -2V_1$$

$$I_1 = -\frac{V_1}{2} \qquad \text{or} \quad I_x = -V_1$$

or $\quad (V_1/I_1) = -2 \ \Omega = Z_{11}$

$$V_2 = I_x = -V_1 \qquad\qquad Z_{21} = \frac{V_2}{I_1} = \frac{-V_1}{-V_1/2} = 2\,\Omega \ \textbf{Ans.}$$

## 8.5  INTERRELATION BETWEEN PARAMETERS

### 8.5.1  z-Parameters in Terms of other Parameters

The $z$-parameters are given by the expression

$$V_1 = I_1 Z_{11} + Z_{12} I_2 \qquad\qquad\qquad \text{...(8.69)}$$
$$V_2 = I_1 Z_{21} + Z_{22} I_2 \qquad\qquad\qquad \text{...(8.70)}$$

*z-parameters in terms of y-parameters*
The idea here is to express $y$-parameter equations $V = F(I)$. Then comparing the co-efficient of the two equations a relation between $z$- and $y$-parameters can be obtained.

$$I_1 = V_1 Y_{11} + V_2 Y_{12} \qquad\qquad\qquad \text{...(8.71)}$$
$$I_2 = V_1 Y_{21} + V_2 Y_{22} \qquad\qquad\qquad \text{...(8.72)}$$

Since $\qquad\qquad Z = Y^{-1}$

$$\begin{bmatrix} Z_{11} & Z_{12} \\ Z_{21} & Z_{22} \end{bmatrix} = \begin{bmatrix} Y_{11} & Y_{12} \\ Y_{21} & Y_{22} \end{bmatrix}^{-1}$$

$$= \frac{1}{\Delta} \begin{bmatrix} Y_{22} & -Y_{12} \\ -Y_{21} & Y_{11} \end{bmatrix}$$

Where $\qquad\qquad \Delta = Y_{11} Y_{22} - Y_{12} Y_{21}$

Therefore, $\qquad\qquad Z_{11} = \dfrac{Y_{22}}{\Delta}, \quad Z_{12} = \dfrac{-Y_{12}}{\Delta}$

$$Z_{21} = \frac{-Y_{21}}{\Delta} \quad \text{and} \quad Z_{22} = \frac{Y_{11}}{\Delta}$$

We know that the condition for reciprocity in terms of z-parameters is $Z_{12} = Z_{21}$ and that for symmetry is $Z_{11} = Z_{22}$. Here the condition for reciprocity is when $Y_{12} = Y_{21}$ and that for symmetry is $Y_{11} = Y_{22}$.

*z-Parameters in terms of h-Parameters*

$$V_1 = h_{11} I_1 + h_{12} V_2 \qquad\qquad\qquad \text{...(8.73)}$$
$$I_2 = h_{21} I_1 + h_{22} V_2 \qquad\qquad\qquad \text{...(8.74)}$$

From (8.74) $\qquad\qquad V_2 = \dfrac{I_2}{h_{22}} - \dfrac{h_{21}}{h_{22}} I_1$

Substituting for $V_2$ in (8.73) we have

$$V_1 = h_{11} \, I_1 + h_{12} \left( -\frac{h_{21}}{h_{22}} \, I_1 + \frac{1}{h_{22}} \, I_2 \right)$$

$$V_1 = \left( h_{11} - \frac{h_{21} \, h_{12}}{h_{22}} \right) I_1 + \frac{h_{12}}{h_{22}} \, I_2$$

Comparing the co-efficients of the sets of equation we have

$$Z_{11} = \frac{h_{11} h_{22} - h_{12} h_{21}}{h_{22}}, \quad Z_{12} = \frac{h_{12}}{h_{22}}, \quad Z_{21} = -\frac{h_{21}}{h_{22}}$$

and
$$Z_{22} = \frac{1}{h_{22}}$$

Thus the condition for symmetry is $h_{11} \, h_{22} - h_{12} \, h_{21} = 1$ and for reciprocity $h_{21} = -h_{12}$

*z-Parameters in terms of inverse hybrid parameters*
The equations relating the *g*-parameters are:

$$I_1 = g_{11} \, V_1 + g_{12} \, I_2 \qquad\qquad ...(8.75)$$

$$V_2 = g_{21} \, V_1 + g_{22} \, I_2 \qquad\qquad ...(8.76)$$

From equation (8.75)

$$V_1 = \frac{I_1}{g_{11}} - \frac{g_{12}}{g_{11}} \, I_2$$

Substituting $V_1$ in equation (8.76) we have

$$V_2 = \frac{g_{21}}{g_{11}} \, I_1 + \left( g_{22} - \frac{g_{12} \, g_{21}}{g_{11}} \right) I_2$$

Again comparing the co-efficients we have

$$Z_{11} = \frac{1}{g_{11}} \qquad\qquad Z_{12} = -\frac{g_{12}}{g_{11}}, \qquad\qquad Z_{21} = \frac{g_{21}}{g_{11}}$$

and
$$Z_{22} = \frac{g_{11} \, g_{22} - g_{12} \, g_{21}}{g_{11}}$$

*z-Parameters in term of ABCD parameters*
The transmission parameters are related through the equations

$$V_1 = AV_2 + B \, (-I_2) \qquad\qquad ...(8.77)$$

$$I_1 = CV_2 + D(-I_2) \qquad\qquad ...(8.78)$$

From (8.78) we have

$$V_2 = \frac{1}{C} \, I_1 + \frac{D}{C} \, I_2$$

## 348 Network Analysis and Synthesis

Substituting in (8.77) we have

$$V_1 = \frac{A}{C} I_1 + \left(-B + \frac{AD}{C}\right) I_2$$

$$Z_{11} = \frac{A}{C} \qquad Z_{12} = \frac{AD - BC}{C}, \qquad Z_{21} = \frac{1}{C}$$

and

$$Z_{22} = \frac{D}{C}$$

From the above it is clear that for reciprocity $AD - BC = 1$ and for symmetry $A = D$.

This completes the analysis expressing the $z$-parameters in terms of other parameters. Similarly, if some other parameter is to be expressed in terms of remaining parameters the remaining parameter equations must be rewritten in the same format as those of the parameters which are under consideration. Finally comparing the co-efficient of the two sets of equations a relation can be obtained. This is left as an exercise to be taken up by the reader. However the interrelation is given in a tabular (Table 8.1) form for ready reference. In this table the matrices appearing in each of the rows are equivalent. Also note that the equivalence involve a factor $\Delta x = x_{11} x_{22} - x_{12} x_{21}$ where $x$ is either $z$, $Y$, $T$, $T'$ $h$ or $g$. Here $T$ and $T'$ are Transmission $(A, B, C, D)$ and inverse transmission parameters.

If it is $\Delta z$, this means $\Delta z = z_{11} z_{22} - z_{12} z_{21}$ and so on.

**Table 8.1 Conversion Chart (Matrices in the same row in the table are equivalent) $\Delta x = x_{11} x_{22} - x_{12} x_{21}$**

| | [z] | | [y] | | [T] | | [T'] | | [h] | | [g] | |
|---|---|---|---|---|---|---|---|---|---|---|---|---|
| [z] | $z_{11}$ | $z_{12}$ | $\dfrac{y_{22}}{\Delta_y}$ | $-\dfrac{y_{12}}{\Delta_y}$ | $\dfrac{A}{C}$ | $\dfrac{\Delta_T}{C}$ | $\dfrac{D'}{C'}$ | $\dfrac{1}{C'}$ | $\dfrac{\Delta_h}{h_{22}}$ | $\dfrac{h_{12}}{h_{22}}$ | $\dfrac{1}{g_{11}}$ | $-\dfrac{g_{12}}{g_{11}}$ |
| | $z_{21}$ | $z_{22}$ | $-\dfrac{y_{21}}{\Delta_y}$ | $\dfrac{y_{11}}{\Delta_y}$ | $\dfrac{1}{C}$ | $\dfrac{D}{C}$ | $\dfrac{\Delta_{T'}}{C'}$ | $\dfrac{A'}{C'}$ | $-\dfrac{h_{21}}{h_{22}}$ | $\dfrac{1}{h_{22}}$ | $\dfrac{g_{21}}{g_{11}}$ | $\dfrac{\Delta_g}{g_{11}}$ |
| [y] | $\dfrac{z_{22}}{\Delta_z}$ | $-\dfrac{z_{12}}{\Delta_z}$ | $y_{11}$ | $y_{12}$ | $\dfrac{D}{B}$ | $-\dfrac{\Delta_T}{B}$ | $\dfrac{A'}{B'}$ | $-\dfrac{1}{B'}$ | $\dfrac{1}{h_{11}}$ | $-\dfrac{h_{12}}{h_{11}}$ | $\dfrac{\Delta_g}{g_{22}}$ | $\dfrac{g_{12}}{g_{22}}$ |
| | $-\dfrac{z_{21}}{\Delta_z}$ | $\dfrac{z_{11}}{\Delta_z}$ | $y_{21}$ | $y_{22}$ | $-\dfrac{1}{B}$ | $\dfrac{A}{B}$ | $-\dfrac{\Delta_{T'}}{B'}$ | $\dfrac{D'}{B'}$ | $\dfrac{h_{21}}{h_{11}}$ | $\dfrac{\Delta_h}{h_{11}}$ | $-\dfrac{g_{21}}{g_{22}}$ | $\dfrac{1}{g_{22}}$ |
| [T] | $\dfrac{z_{11}}{z_{21}}$ | $\dfrac{\Delta_z}{Z_{21}}$ | $-\dfrac{y_{22}}{y_{21}}$ | $-\dfrac{1}{y_{21}}$ | $A$ | $B$ | $\dfrac{D'}{\Delta_{T'}}$ | $\dfrac{B'}{\Delta_{T'}}$ | $-\dfrac{\Delta_h}{h_{21}}$ | $-\dfrac{h_{11}}{h_{21}}$ | $\dfrac{1}{g_{21}}$ | $\dfrac{g_{22}}{g_{21}}$ |
| | $\dfrac{1}{z_{21}}$ | $\dfrac{z_{22}}{z_{21}}$ | $-\dfrac{\Delta_y}{y_{21}}$ | $-\dfrac{y_{11}}{y_{21}}$ | $C$ | $D$ | $\dfrac{C'}{\Delta_{T'}}$ | $\dfrac{A'}{\Delta_{T'}}$ | $-\dfrac{h_{22}}{h_{21}}$ | $-\dfrac{1}{h_{21}}$ | $\dfrac{g_{11}}{g_{21}}$ | $\dfrac{\Delta_g}{g_{21}}$ |
| [T'] | $\dfrac{z_{22}}{z_{12}}$ | $\dfrac{\Delta_z}{z_{12}}$ | $-\dfrac{y_{11}}{y_{12}}$ | $-\dfrac{1}{y_{12}}$ | $\dfrac{D}{\Delta_T}$ | $\dfrac{B}{\Delta_T}$ | $A'$ | $B'$ | $\dfrac{1}{h_{12}}$ | $\dfrac{h_{11}}{h_{12}}$ | $-\dfrac{\Delta_g}{g_{12}}$ | $-\dfrac{g_{22}}{g_{12}}$ |
| | $\dfrac{1}{z_{12}}$ | $\dfrac{z_{11}}{z_{12}}$ | $-\dfrac{\Delta_y}{y_{12}}$ | $-\dfrac{y_{22}}{y_{12}}$ | $\dfrac{C}{\Delta_T}$ | $\dfrac{A}{\Delta_T}$ | $C'$ | $D'$ | $\dfrac{h_{22}}{h_{12}}$ | $\dfrac{\Delta_h}{h_{12}}$ | $-\dfrac{g_{11}}{g_{12}}$ | $\dfrac{1}{g_{12}}$ |

(Contd.)

| | | | | | | | | | | | | |
|---|---|---|---|---|---|---|---|---|---|---|---|---|
| **[h]** | $\dfrac{\Delta_z}{z_{22}}$ | $\dfrac{z_{12}}{z_{22}}$ | $\dfrac{1}{y_{11}}$ | $-\dfrac{y_{12}}{y_{11}}$ | $\dfrac{B}{D}$ | $\dfrac{\Delta_T}{D}$ | $\dfrac{B'}{A'}$ | $\dfrac{1}{A'}$ | $h_{11}$ | $h_{12}$ | $\dfrac{g_{22}}{\Delta_g}$ | $\dfrac{g_{12}}{\Delta_g}$ |
| | $-\dfrac{z_{21}}{z_{22}}$ | $\dfrac{1}{z_{22}}$ | $\dfrac{y_{21}}{z_{11}}$ | $\dfrac{\Delta_y}{y_{11}}$ | $-\dfrac{1}{D}$ | $\dfrac{C}{D}$ | $-\dfrac{\Delta_{T'}}{A'}$ | $\dfrac{C'}{A'}$ | $h_{21}$ | $h_{22}$ | $-\dfrac{g_{21}}{\Delta_g}$ | $\dfrac{g_{11}}{\Delta_g}$ |
| **[g]** | $\dfrac{1}{z_{11}}$ | $-\dfrac{z_{12}}{z_{11}}$ | $\dfrac{\Delta_y}{y_{22}}$ | $\dfrac{y_{12}}{y_{22}}$ | $\dfrac{C}{A}$ | $-\dfrac{\Delta_T}{A}$ | $\dfrac{C'}{D'}$ | $-\dfrac{1}{D'}$ | $\dfrac{h_{22}}{\Delta_h}$ | $-\dfrac{h_{12}}{\Delta_h}$ | $g_{11}$ | $g_{12}$ |
| | $\dfrac{z_{21}}{z_{11}}$ | $\dfrac{\Delta_z}{z_{11}}$ | $-\dfrac{y_{21}}{y_{22}}$ | $\dfrac{1}{y_{22}}$ | $\dfrac{1}{A}$ | $\dfrac{B}{A}$ | $\dfrac{\Delta_{T'}}{D'}$ | $\dfrac{B'}{D'}$ | $-\dfrac{h_{21}}{\Delta_h}$ | $\dfrac{h_{11}}{\Delta_h}$ | $g_{21}$ | $g_{22}$ |

Table (8.2) shows the condition for reciprocity and symmetry for various parameters. A passive reciprocal two-ports network is symmetrical in the sense that the ports may be exchanged without affecting the port voltages and currents.

### Table 8.2 Conditions for Reciprocity and Symmetry

| Parameter | Condition for passive networks | Condition for electrical symmetry |
|---|---|---|
| $z$ | $z_{12} = z_{21}$ | $z_{11} = z_{22}$ |
| $y$ | $y_{12} = y_{21}$ | $y_{11} = y_{22}$ |
| $ABCD$ | $AD - BC = 1$ | $A = D$ |
| $A'B'C'D'$ | $A'D' - B'C' = 1$ | $A' = D'$ |
| $h$ | $h_{12} = -h_{21}$ | $\Delta_h = 1$ |
| $g$ | $g_{12} = -g_{21}$ | $\Delta_g = 1$ |

## 8.6  INTERCONNECTION OF TWO PORT NETWORKS

A two-port network with a modest level of complexity can be considered to be constructed from many simpler two-port networks whose ports are interconnected in certain ways. Conversely two-port network which is to be built up can be designed by combining simple two-port structures as building blocks. From designer's point of view it is easier to design simple block and then interconnecting them than to design a single complex network. Also it is much easier to shield smaller units and thus reduce parasitic capacitances to ground.

### 8.6.1  Cascade Connection

There are various ways of interconnecting two port networks. The simpler of these is the cascade or tandem connection of two two-port networks. Two two-port networks are said to be connected in cascade when the output port of one is the input port of the other as shown in Fig. 8.13.

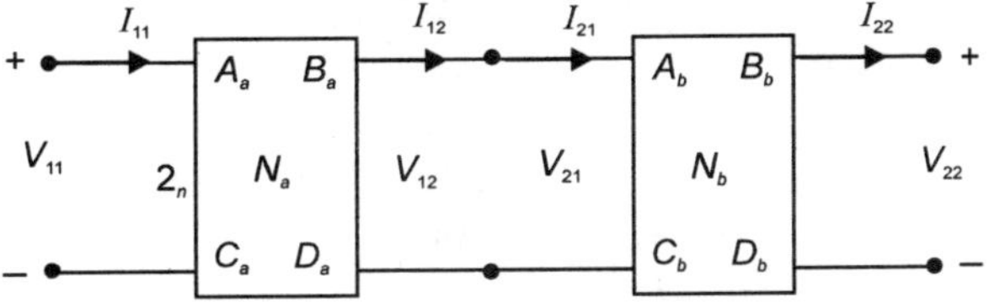


**Fig. 8.13**  Cascade Connection.

The cascade connection is most conveniently studied in terms of *ABCD* parameters of the network. The objective here is to find out the parameters of the overall network in terms of the parameters of the individual network. From the network we see that

$$\begin{bmatrix} V_{11} \\ I_{11} \end{bmatrix} = \begin{bmatrix} A_a & B_a \\ C_a & D_a \end{bmatrix} \begin{bmatrix} V_{12} \\ I_{12} \end{bmatrix} \qquad \qquad \dots(8.79)$$

and

$$\begin{bmatrix} V_{21} \\ I_{21} \end{bmatrix} = \begin{bmatrix} A_b & B_b \\ C_b & D_b \end{bmatrix} \begin{bmatrix} V_{22} \\ I_{22} \end{bmatrix} \qquad \qquad \dots(8.80)$$

Since

$$\begin{bmatrix} V_{21} \\ I_{21} \end{bmatrix} = \begin{bmatrix} V_{12} \\ I_{12} \end{bmatrix}$$

Therefore,

$$\begin{bmatrix} V_{11} \\ I_{11} \end{bmatrix} = \begin{bmatrix} A_a & B_a \\ C_a & D_a \end{bmatrix} \begin{bmatrix} A_b & B_b \\ C_b & D_b \end{bmatrix} \begin{bmatrix} V_{22} \\ I_{22} \end{bmatrix}$$

$$= \begin{bmatrix} A & B \\ C & D \end{bmatrix} \begin{bmatrix} V_{22} \\ I_{22} \end{bmatrix} \qquad \qquad \dots(8.81)$$

where

$$\begin{bmatrix} A & B \\ C & D \end{bmatrix} = \begin{bmatrix} A_a & B_a \\ C_a & D_a \end{bmatrix} \begin{bmatrix} A_b & B_b \\ C_b & D_b \end{bmatrix} \qquad \qquad \dots(8.82)$$

Thus the *ABCD* parameters of the overall two port network is equal to the product of the matrices of *ABCD* parameters of the individual networks.

If the parameters of the overall network can be expressed in terms of the similar parameters of the individual network, it is convenient to express any other set of parameters through simple algebraic manipulations. Suppose we want to express open circuit parameters of the overall network in terms of the open circuit parameters of the individual network. Express *ABCD* parameters in term of *z*-parameters and substitute them in the expression above to get the *z*-parameters for the overall network.

We know that

$$C = \frac{1}{Z_{21}} \qquad D = \frac{Z_{22}}{Z_{21}}$$

$$A = \frac{Z_{11}}{Z_{21}}$$

Now

$$Z_{12} = \frac{AD - BC}{C}$$

$$B = \frac{Z_{11} Z_{22}}{Z_{21}} - Z_{12} \qquad \qquad \dots(8.83)$$

$$\begin{bmatrix} A & B \\ C & D \end{bmatrix} = \begin{bmatrix} \dfrac{Z_{11a}}{Z_{21a}} & \dfrac{Z_{11a} Z_{22a} - Z_{12a} Z_{21a}}{Z_{21a}} \\ \dfrac{1}{Z_{21a}} & \dfrac{Z_{22a}}{Z_{21a}} \end{bmatrix} \qquad \text{[Network b]}$$

Here
$$C = \frac{Z_{11b} + Z_{22a}}{Z_{21a}\, Z_{21b}}$$

Since
$$Z_{21} = \frac{1}{C} = \frac{Z_{21a}\, Z_{21b}}{Z_{11b} + Z_{22a}} \qquad \text{...(8.85)}$$

Similarly other parameters can be obtained. It is found that

$$\begin{bmatrix} Z_{11} & Z_{12} \\ Z_{21} & Z_{22} \end{bmatrix} = \begin{bmatrix} Z_{11a} - \dfrac{Z_{12a}\, Z_{21a}}{Z_{22a} + Z_{11b}} & \dfrac{Z_{12a}\, Z_{12b}}{Z_{22a} + Z_{11b}} \\[3mm] \dfrac{Z_{21a}\, Z_{21b}}{Z_{22a} + Z_{11b}} & Z_{22b} - \dfrac{Z_{12b}\, Z_{21b}}{Z_{22a} + Z_{11b}} \end{bmatrix} \qquad \text{...(8.86)}$$

Here $Z_{11}$, $Z_{12}$, $Z_{21}$ and $Z_{22}$ are the open circuit impedances of the two cascaded two-port networks.

Many a times it is more convenient to find out overall parameters using basic concepts rather than the conversion formulae as suggested above. For example to determine $Z_{11}$ in the above network we find out

$$Z_{21} = \frac{V_2}{I_1} \quad \text{when} \quad I_2 = 0$$

Suppose a current source $I_1$ is applied.

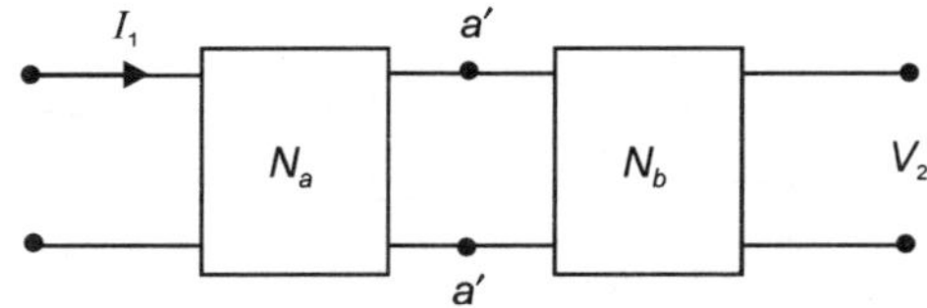


**Fig. 8.14**   Two two-port networks in cascade.

The Thevenin's equivalent voltage source will correspond to open circuit voltage across $a'a'$ which is $I_1 Z_{21a}$ and the Thevenin's equivalent series impedance will be the output impedance $Z_{22a}$.

Therefore, the Thevenin's equivalent circuit is given as in Fig. 8.14($a$).

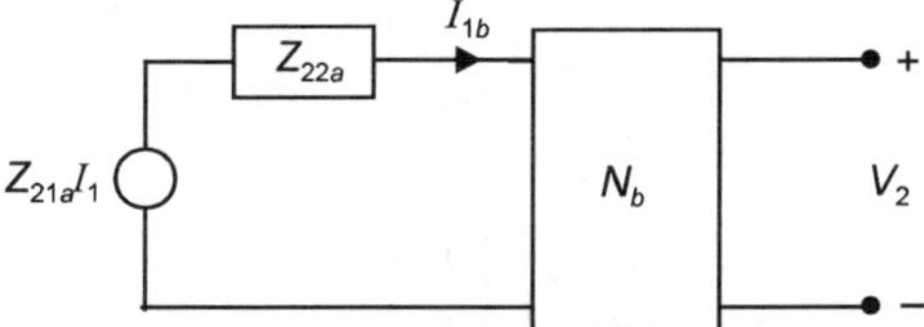


**Fig. 8.14(a)**   Thevenin's equivalent of Fig. 8.14.

Now
$$I_{1b} = \frac{Z_{21a}\, I_1}{Z_{22a} + Z_{11b}} \qquad \text{...(8.87)}$$

Hence,
$$Z_{21b} = \frac{V_2}{I_{1b}} = \frac{V_2\, (Z_{22a} + Z_{11b})}{Z_{21a}\, I_1}$$

$$= \frac{V_2}{I_1} \cdot \frac{Z_{22a} + Z_{11b}}{Z_{21a}}$$

or
$$Z_{21} = \frac{V_2}{I_1} = \frac{Z_{21a} \, Z_{21b}}{Z_{22a} + Z_{11b}} \qquad \qquad ...(8.88)$$

which is same as obtained through the conversion formulae.

It is to be noted that the zeros of the transfer impedances of the cascaded networks are the zeros of the transfer impedances of the individual two port networks i.e. zeros of $z_{12}$ are the zeros of $z_{12a}$ and $z_{12b}$. Similarly the zeros of $z_{21}$ are the zeros of $z_{21a}$ and $z_{21b}$. It is an important information from network synthesis point of view. It permits individual two ports to be designed to achieve certain transmission zeros before they are connected together. It also permits independent adjustment and tuning of elements within each two-ports to achieve desired performances without effecting adjustment of the cascaded two ports.

### 8.6.2  Series Connection

Two two-port networks are said to be connected in series if corresponding ports (input and output) are connected in series as shown in Fig. 8.15.

Here the input currents of the two two-ports networks are forced to be same. Similarly the output current of the two two-port networks is also forced to be the same. The $z$-parameter relation for the two networks are:

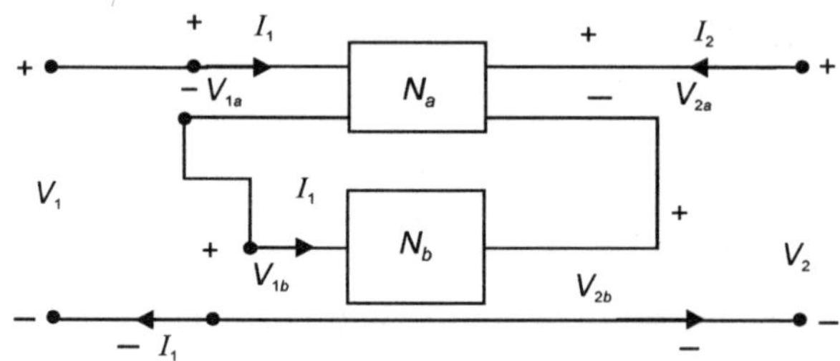

**Fig. 8.15**  Two two-ports connected in series.

$$\begin{bmatrix} V_{1a} \\ V_{2a} \end{bmatrix} = \begin{bmatrix} Z_{11a} & Z_{12a} \\ Z_{21a} & Z_{22a} \end{bmatrix} \begin{bmatrix} I_1 \\ I_2 \end{bmatrix}$$

and
$$\begin{bmatrix} V_{1b} \\ V_{2b} \end{bmatrix} = \begin{bmatrix} Z_{11b} & Z_{12b} \\ Z_{21b} & Z_{22b} \end{bmatrix} \begin{bmatrix} I_1 \\ I_2 \end{bmatrix}$$

Since
$$\begin{bmatrix} V_1 \\ V_2 \end{bmatrix} = \begin{bmatrix} V_{1a} \\ V_{2a} \end{bmatrix} + \begin{bmatrix} V_{1b} \\ V_{2b} \end{bmatrix} = \begin{bmatrix} Z_{11a} + Z_{11b} & Z_{12a} + Z_{12b} \\ Z_{21a} + Z_{21b} & Z_{22a} + Z_{22b} \end{bmatrix} \qquad ...(8.89)$$

Thus the open circuit impedance matrix of two port network connected in series is the sum of the open circuit matrices of the individual networks.

### 8.6.3 Parallel Connections

Two two-port networks are said to be connected in parallel if corresponding ports (input and output) are connected in parallel as shown in Fig. 8.16.

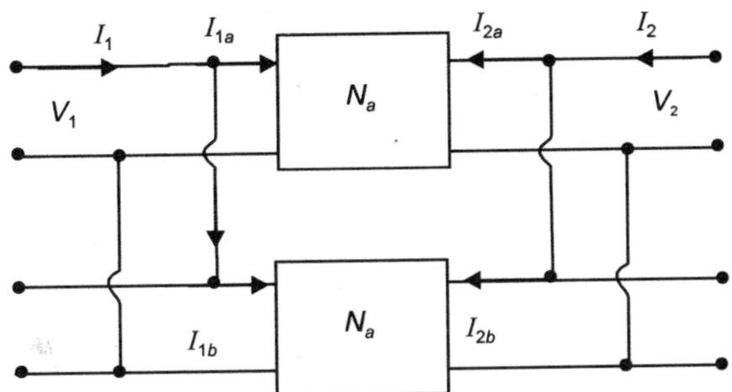

**Fig. 8.16** Two two-port networks connected in parallel.

Here the voltages of the two input ports are forced to be same as also those of the output ports. The equation in terms of short circuit admittances are given as

$$\begin{bmatrix} I_{1a} \\ I_{2a} \end{bmatrix} = \begin{bmatrix} Y_{11a} & Y_{12a} \\ Y_{21a} & Y_{22a} \end{bmatrix} \begin{bmatrix} V_1 \\ V_2 \end{bmatrix}$$

and

$$\begin{bmatrix} I_{1b} \\ I_{2b} \end{bmatrix} = \begin{bmatrix} Y_{11b} & Y_{12b} \\ Y_{21b} & Y_{22b} \end{bmatrix} \begin{bmatrix} V_1 \\ V_2 \end{bmatrix}$$

Hence,

$$\begin{bmatrix} I_1 \\ I_2 \end{bmatrix} = \begin{bmatrix} I_{1a} \\ I_{2a} \end{bmatrix} + \begin{bmatrix} I_{1b} \\ I_{2b} \end{bmatrix} = \begin{bmatrix} Y_{11a} + Y_{11b} & Y_{12a} + Y_{12b} \\ Y_{21a} + Y_{21b} & Y_{22a} + Y_{22b} \end{bmatrix} \begin{bmatrix} V_1 \\ V_2 \end{bmatrix} \qquad ...(8.90)$$

Thus the short circuit admittance matrix of two ports network connected in parallel is the sum of the short circuit admittance matrices of individual two port networks.

### 8.6.4 Series-Parallel Connection

If one port each of the two two-port networks is connected in series and the other in parallel it is convenient to find the overall parameters of the two networks using the $h$-parameters. Suppose two networks as shown in Fig. 8.17 are connected in series-parallel.

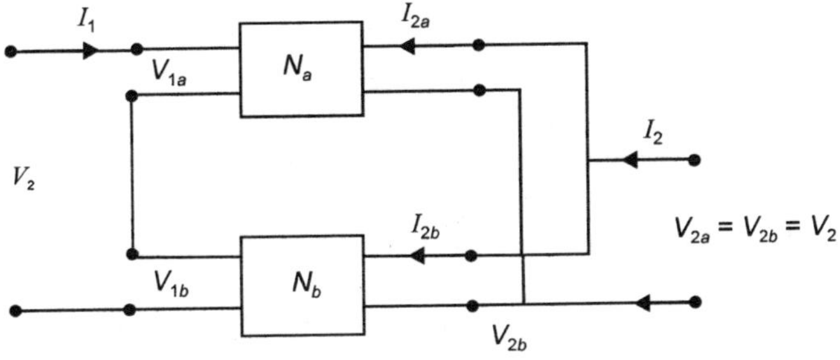

**Fig. 8.17** Two two-port networks connected in series-parallel.

For network $N_a$

$$V_{1a} = h_{11a} I_{1a} + h_{12a} V_{2a}$$

$$I_{2a} = h_{21a} I_{1a} + h_{22a} V_{2a}$$

Similarly for network $N_b$

$$V_{1b} = h_{11b} I_{1b} + h_{12b} V_{2b}$$

$$I_{2b} = h_{21b} I_{1b} + h_{22b} V_{2b}$$

Since the input ports are connected in series $V_1 = V_{1a} + V_{1b}$ and $I_{1a} = I_{1b} = I_1$ and as port-2 are in parallel

$$I_2 = I_{2a} + I_{2b} \quad \text{and} \quad V_{2a} = V_{2b} = V_2$$

Now

$$V_1 = V_{1a} + V_{1b} = h_{11a} I_{1a} + h_{12a} V_{2a} + h_{11b} I_{1b} + h_{12b} V_{2b}$$

$$= (h_{11a} + h_{11b}) I_1 + (h_{12a} + h_{12b}) V_2$$

Similarly

$$I_2 = I_{2a} + I_{2b} = h_{21a} I_{1a} + h_{22a} V_{2a} + h_{21b} I_{1b} + h_{22b} V_{2b}$$

$$= (h_{21a} + h_{21b}) I_1 + (h_{22a} + h_{22b}) V_2$$

Therefore,

$$\begin{bmatrix} h_{11} & h_{12} \\ h_{21} & h_{22} \end{bmatrix} = \begin{bmatrix} h_{11a} + h_{11b} & h_{12a} + h_{12b} \\ h_{21a} + h_{21b} & h_{22a} + h_{22b} \end{bmatrix}$$

This shows that the $h$-parameters of the networks whose input ports are connected in series and output in parallel is the algebraic sum of the respective $h$-parameters of the individual networks.

### 8.6.5 Parallel-Series Connection

If the input ports are connected in parallel and the output in series as shown in Fig. 8.18 then it is convenient to obtain $g$-parameters of the overall networks.

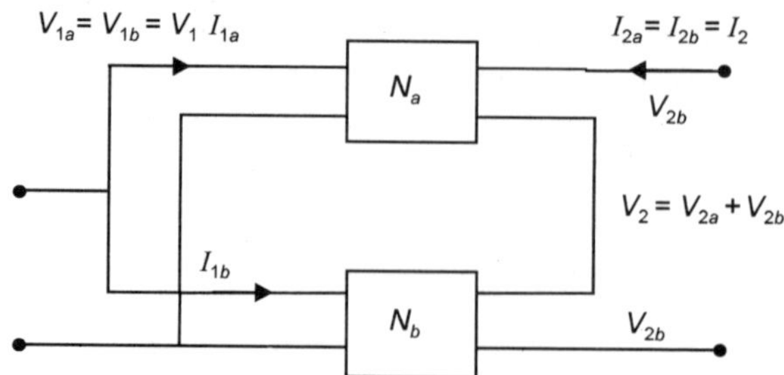


**Fig. 8.18**   Two two-port networks connected in parallel-series.

The equations relating the 'a' network in terms of $g$-parameters are given as

$$I_{1a} = g_{11a} V_{1a} + g_{12a} I_{2a}$$

$$V_{2a} = g_{21a} V_{1a} + g_{22a} I_{2a}$$

and for the network '*b*'

$$I_{1b} = g_{11b}\, V_{1b} + g_{12b}\, I_{2b}$$
$$V_{2b} = g_{21b}\, V_{1b} + g_{22b}\, I_{2b}$$

Since $\qquad V_{1a} = V_{1b} = V_1, \qquad I_1 = I_{1a} + I_{1b}$

and $\qquad V_2 = V_{2a} + V_{2b}, \qquad I_{2a} = I_{2b} = I_2$

$$I_1 = I_{1a} + I_{1b} = g_{11a}\, V_{1a} + g_{12a}\, I_{2a} + g_{11b}\, V_{1b} + g_{12b}\, I_{2b}$$
$$= (g_{11a} + g_{11b})\, V_1 + (g_{12a} + g_{12b})\, I_2 \qquad\qquad ...(8.92)$$

and $\qquad V_2 = V_{2a} + V_{2b} = g_{21a}\, V_{1a} + g_{22a}\, I_{2a} + g_{21b}\, V_{1b} + g_{21b}\, I_{2b}$

$$= (g_{21a} + g_{21b})\, V_1 + (g_{22a} + g_{22b})\, I_2 \qquad\qquad ...(8.93)$$

In a large number of two-port networks if the input ports are parallel connected and output ports series, the overall $g$-parameters of the interconnected network is the sum of the corresponding $g$-parameters of the individual networks.

From the analysis of various interconnections of the two-port networks it is observed that it is convenient to study a particular interconnection with a certain set of parameter e.g. cascade $A, B, C, D$ parameter, series connection $Z$-parameters, parallel connection $Y$-parameters and so on. In case this convenience is followed it is observed that–

(*a*) For cascade connection of any number of two port networks, the overall parameters are the multiplication of A, B, C, D matrices of individual networks.

(*b*) For series it is the sum of the corresponding $Z$-parameters of the individual networks.

(*c*) For parallel it is the sum of the corresponding $Y$-parameters of the individual networks.

(*d*) For series-parallel it is the sum of the corresponding $h$-parameters of the networks.

(*e*) For parallel-series it is the sum of the corresponding g-parameters of the individual networks.

The parallel connection is more useful and finds wider application in network synthesis as in this when two two-port circuits are connected in parallel the final circuit is also a common-terminal two port network.

A typical example is a parallel ladder network as shown here in Fig. 8.19.

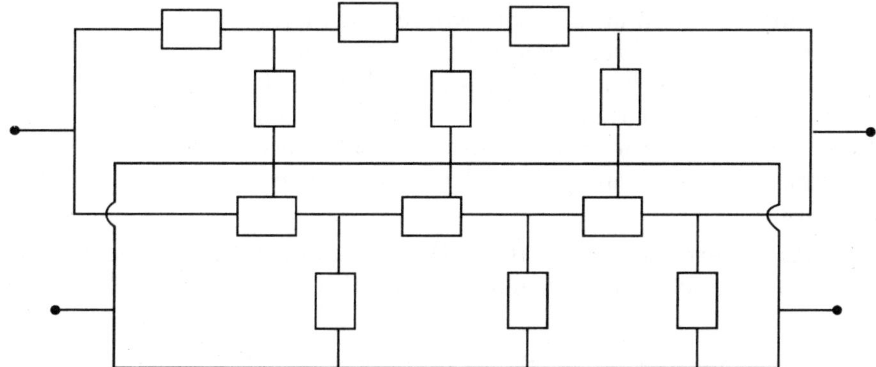

**Fig. 8.19**   Parallel ladder network.

However the series connection of two common-terminals two ports is not a common-terminal two port unless one of them is a T-network as shown in Fig. 8.20.

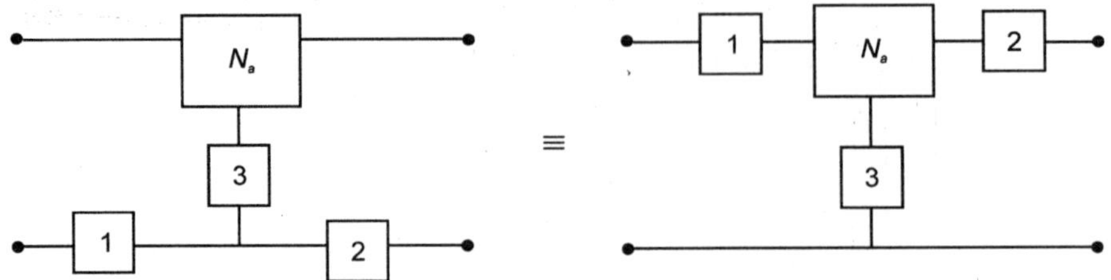

**Fig. 8.20**   T-network 1, 2, 3 and the second network Na.

## 8.6.6   Permissibility of Interconnection

Let us find out the condition required for parallel or series connection of two two-port networks. The basic requirement is that after these connections are made the port relations of the individual two port networks are not affected. To satisfy this condition for parallel connection, suppose input ports of the two-port networks are connected in parallel and the output ports are shorted on to themselves as shown in Fig. 8.21. If the voltage as measured here is zero the two networks can be connected in parallel as otherwise there will be circulating current between the two networks which will affect the port relation of the individual networks. Hence the condition that the current leaving one terminal of a port be equal to the current entering the other terminal of each individual two port is violated and the port relationships of the individual two ports are affected.

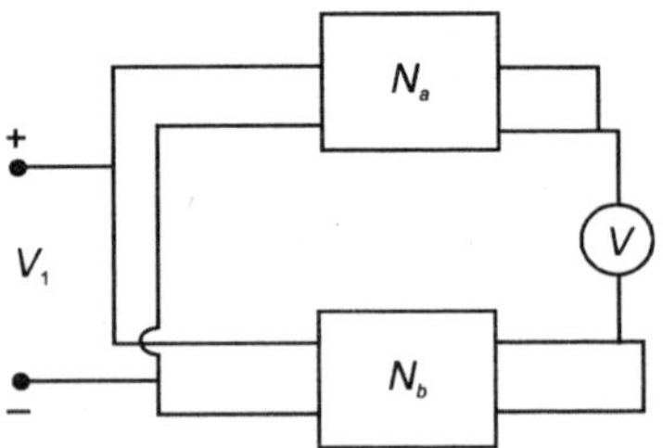

**Fig. 8.21**   Test for parallel interconnection

For series connection, the input ports are connected in series and the output ports left open as shown in Fig. 8.22. The open circuits are used here as the parameters characterizing the individual two ports and the overall two ports are the open circuit impedance parameters. If the voltage recorded as shown in Fig. 8.22 is zero and also if the voltage recorded on the input side with output port where volt meter is shown are short circuited is zero again, series connection can be made, otherwise if the voltage is non-zero, there will be circulating current and port relationship of the individual two-ports will be modified by the inter-connection and hence the addition of impedance of parameters will not be valid for the overall network.

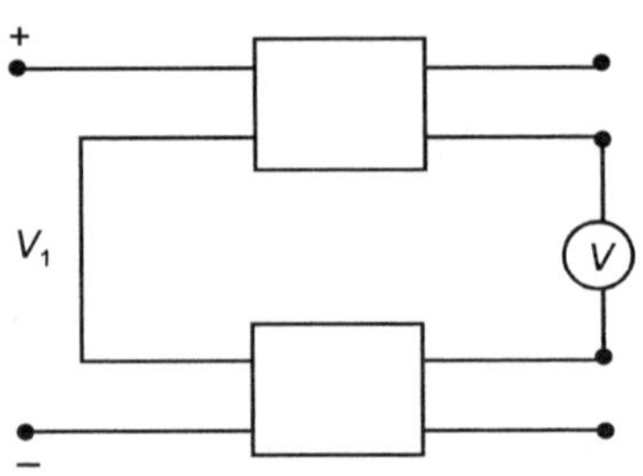

**Fig. 8.22**   For series connected two port network

**Example 8.12:** Determine the $A, B, C, D$ parameters of the networks $N_a$, $N_b$ and $N_c$ and hence determine the overall parameters if the three networks are cascade connected.

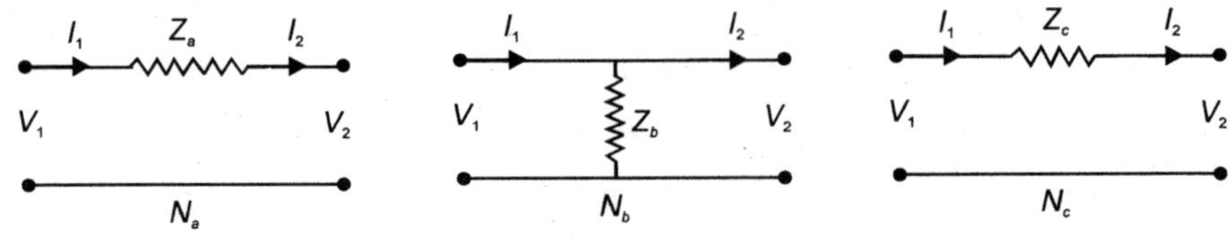

**Fig.  E8.12**

For network $N_a$

$$V_1 = V_2 + I_2 Z_a$$
$$I_1 = 0.V_2 + I_2$$

Hence $\qquad A_a = 1, \quad B_a = Z_a, \quad C_a = 0, \quad D_a = 1$

For network $N_b$

$$V_1 = V_2 + 0.I_2$$
$$I_1 = \frac{V_2}{Z_b} + I_2$$

Hence $\qquad A_b = 1, \quad B_b = 0 \quad C_b = \dfrac{1}{Z_b}, \quad D_b = 1$

Similarly for Network $N_c$

$$A_c = 1, \quad B_c = Z_c, \quad C_c = 0, \quad D_c = 1$$

Therefore the overall $A, B, C, D$ parameters are:

$$\begin{bmatrix} A & B \\ C & D \end{bmatrix} = \begin{bmatrix} 1 & Z_a \\ 0 & 1 \end{bmatrix} \begin{bmatrix} 1 & 0 \\ \dfrac{1}{Z_b} & 1 \end{bmatrix} \begin{bmatrix} 1 & Z_c \\ 0 & 1 \end{bmatrix}$$

$$= \begin{bmatrix} 1 + \dfrac{Z_a}{Z_b} & Z_c\left(1 + \dfrac{Z_a}{Z_b}\right) + Z_a \\[3mm] \dfrac{1}{Z_b} & \dfrac{Z_c}{Z_b} + 1 \end{bmatrix}$$

**Example 8.13:** Find the $Y$- and $Z$-parameters for the $RC$ ladder network shown

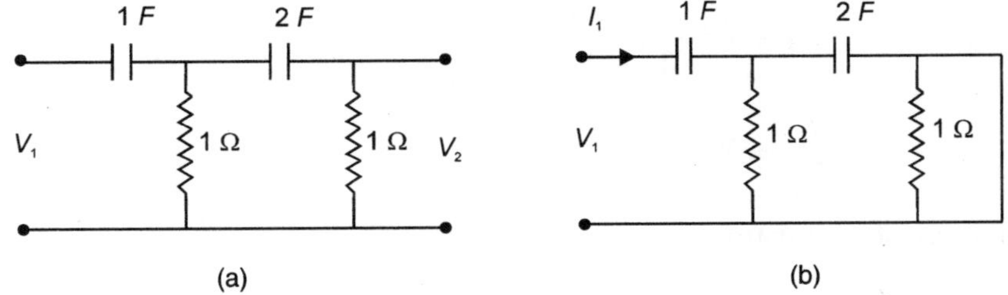


**Fig E8.13**

($a$) Calculation of $Y$-parameters

$$Y_{11} = \left.\frac{I_1}{V_1}\right|_{V_2=0} \qquad\qquad Y_{21} = \left.\frac{I_2}{V_1}\right|_{V_2=0}$$

$$I_1 = \cfrac{V_1}{\cfrac{1}{s} + \cfrac{1/2s}{1 + \cfrac{1}{2s}}} \qquad \text{or} \qquad \frac{I_1}{V_1} = \frac{s(2s+1)}{3s+1} = Y_{11}$$

Now
$$I_2 = I_1 \cdot \frac{1}{1 + \dfrac{1}{2s}} = I_1 \cdot \frac{2s}{1 + 2s}$$

$$= \frac{-V_1 \cdot 2s^2}{3s + 1}$$

or
$$\frac{I_2}{V_1} = \frac{-2s^2}{3s + 1} = Y_{21}$$

Now
$$Y_{12} = \left. \frac{I_1}{V_2} \right|_{V_1 = 0} \qquad Y_{22} = \left. \frac{I_2}{V_2} \right|_{V_1 = 0} \qquad \text{(see Fig. E8.13.1)}$$

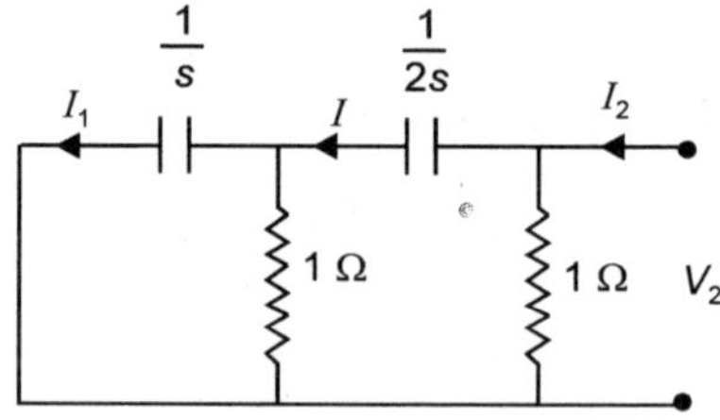

Fig. E8.13.1

$$I_1 = \frac{V_2 \cdot 2s^2}{(3s + 1)} \qquad \text{or} \quad \frac{I_1}{V_2} = \frac{-2s^2}{3s + 1} = Y_{12}$$

Now
$$I_2 = I + V_2 = \frac{2s^2 + 5s + 1}{(3s + 1)} V_2$$

where
$$I = \frac{V_2 \cdot 2s \, (s + 1)}{(3s + 1)}$$

Hence
$$\frac{I_2}{V_2} = Y_{22} = \frac{2s^2 + 5s + 1}{3s + 1}$$

(*b*) Calculation of *z*-parameters

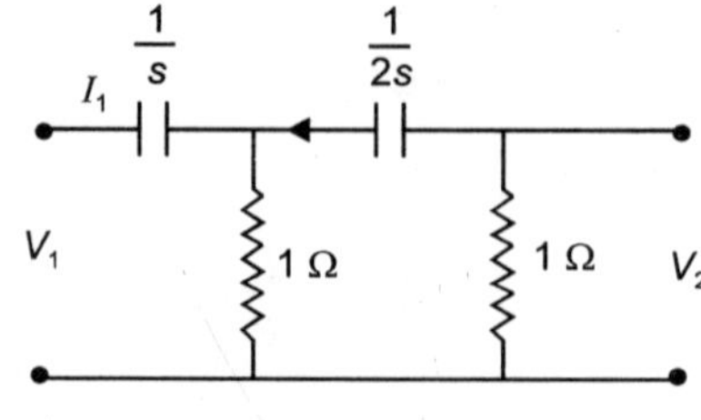

Fig. E 8.13.2

$$Z_{11} = \left. \frac{V_1}{I_1} \right|_{I_2 = 0} \qquad Z_{21} = \left. \frac{V_2}{I_1} \right|_{I_2 = 0}$$

$$I_1 = \cfrac{V_1}{\cfrac{1}{s} + \cfrac{1+2s}{1+4s}} = V_1 \, \frac{s(1+4s)}{2s^2 + 5s + 1}$$

Hence

$$\frac{V_1}{I_1} = Z_{11} = \frac{2s^2 + 5s + 1}{s(1+4s)}$$

Current

$$I = I_1 \, \cfrac{1}{1 + 1 + \cfrac{1}{2s}} = \frac{2s\,I_1}{(4s+1)}$$

Therefore,

$$V_2 = I = \frac{2s\,I_1}{4s+1}$$

$$Z_{21} = \left.\frac{V_2}{I_1}\right|_{I_2 = 0} = \frac{2s}{4s+1}$$

$$Z_{22} = \left.\frac{V_2}{I_2}\right|_{I_1 = 0}$$

$$I_2 = V_2 + \cfrac{V_2}{1 + \cfrac{1}{2s}} = V_2 \left(\frac{1+4s}{1+2s}\right)$$

or

$$\frac{V_2}{I_2} = \frac{1+2s}{1+4s} = Z_{22} \ \textbf{Ans.}$$

$$Z_{12} = \left.\frac{V_1}{I_2}\right|_{I_1 = 0}$$

Now

$$V_1 = \cfrac{V_2}{1 + \cfrac{1}{2s}} \cdot 1 = \frac{2s}{1+2s} \, V_2$$

$$Z_{12} = \frac{2s}{1+4s} \ \textbf{Ans.}$$

**Example 8.14:** Determine $A, B, C, D$ parameters of the network shown

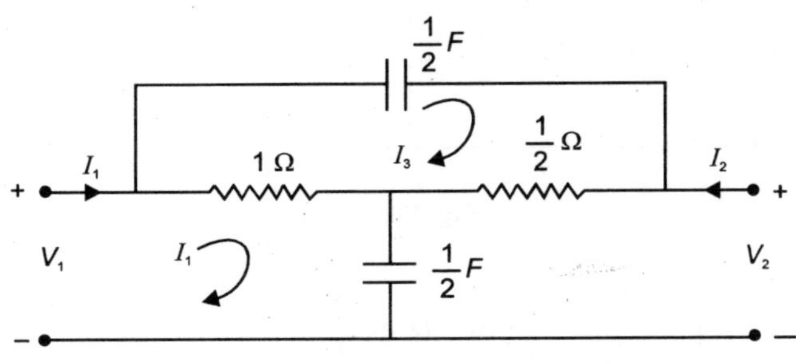


**Fig. E8.14**

**Writing loop equations**

$$\begin{bmatrix} 1 + \dfrac{2}{s} & -1 \\[2mm] -1 & \dfrac{3}{2} + \dfrac{2}{s} \end{bmatrix} \begin{bmatrix} I_1 \\[2mm] I_3 \end{bmatrix} = \begin{bmatrix} V_1 \\[2mm] 0 \end{bmatrix}$$

$$\Delta = \left(1 + \frac{2}{s}\right)\left(\frac{3}{2} + \frac{2}{s}\right) - 1 = \frac{4}{s^2} + \frac{5}{s} + \frac{1}{2}$$

$$I_1 = \frac{1}{\Delta}\begin{vmatrix} V_1 & -1 \\[2mm] 0 & \dfrac{3}{2} + \dfrac{2}{s} \end{vmatrix} = \frac{V_1\left(3 + \dfrac{4}{s}\right)}{\dfrac{8}{s^2} + \dfrac{10}{s} + 1}$$

$$I_3 = \frac{1}{\Delta}\begin{vmatrix} 1 + \dfrac{2}{s} & V_1 \\[2mm] -1 & 0 \end{vmatrix} = \frac{V_1}{\dfrac{4}{s^2} + \dfrac{5}{s} + \dfrac{1}{2}}$$

$$V_2 = \frac{V_1\left(3 + \dfrac{4}{s}\right)\cdot\dfrac{2}{s}}{\dfrac{8}{s^2} + \dfrac{10}{s} + 1} + \frac{I_3}{2} = \frac{\left(\dfrac{6}{s} + \dfrac{8}{s^2} + 1\right)V_1}{\dfrac{8}{s^2} + \dfrac{10}{s} + 1}$$

$$A = \frac{V_1}{V_2}\bigg|_{I_2 = 0} = \frac{s^2 + 10s + 8}{s^2 + 6s + 8}$$

$$C = \frac{I_1}{V_2}\bigg|_{I_2 = 0} = \frac{3 + 4/s}{\dfrac{6}{s} + \dfrac{8}{s^2} + 1} = \frac{s\left(3s + 4\right)}{s^2 + 6s + 8}$$

**To Calculate $B$ and $D$**

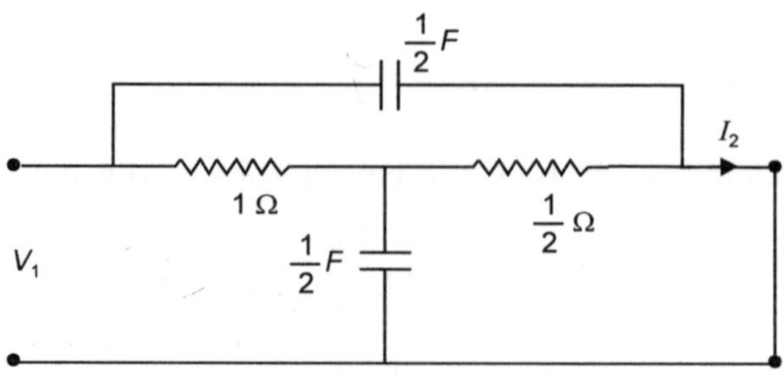


**Fig. E8.14.1**

$$B = \frac{V_1}{I_2}\bigg|_{V_2 = 0} \quad \text{and} \quad D = \frac{I_1}{I_2}\bigg|_{V_2 = 0}$$

$$I_2 = \frac{V_1}{2/s} + \frac{V_1}{\dfrac{\dfrac{1}{2}\cdot\dfrac{2}{s}}{\dfrac{1}{2}+\dfrac{2}{s}} + 1} = \frac{s^2 + 6s + 8}{2(s+6)}\cdot V_1$$

Hence
$$B = \frac{V_1}{I_2} = \frac{2(s+6)}{s^2 + 6s + 8}$$

$$I_1 = V\cdot\frac{s}{2} + \frac{V_1(s+4)}{s+6} = V_1\frac{s^2 + 8s + 8}{2(s+6)}$$

$$D = \frac{I_1}{I_2} = \frac{s^2 + 8s + 8}{s^2 + 6s + 8}\quad\textbf{Ans.}$$

**Example 8.15:** Determine the $A, B, C, D$ parameters of the two networks connected in cascade as shown in Fig. E8.15.

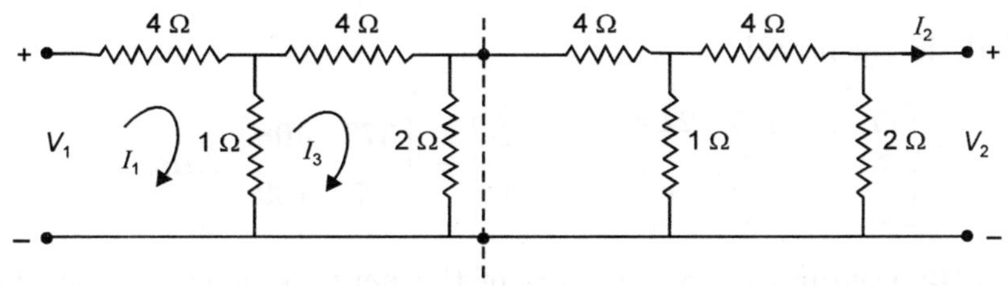

**Fig. E8.15**

**Solution:** Parameters of the single network are obtained as follows:
The loop equations are:

$$\begin{bmatrix} 5 & -1 \\ -1 & 7 \end{bmatrix}\begin{bmatrix} I_1 \\ I_3 \end{bmatrix} = \begin{bmatrix} V_1 \\ 0 \end{bmatrix}$$

The determinant $\Delta = 35 - 1 = 34$

$$I_1 = \frac{1}{\Delta}\begin{vmatrix} V_1 & -1 \\ 0 & 7 \end{vmatrix} = \frac{7V_1}{34}$$

$$I_3 = \frac{1}{34}\begin{vmatrix} 5 & V_1 \\ -1 & 0 \end{vmatrix} = \frac{V_1}{34}$$

$$V_2 = 2\,I_3 = \frac{2V_1}{34}$$

$$A = \frac{V_1}{V_2} = 17 \quad\text{and}\quad C = \frac{I_1}{V_2} = \frac{7}{2}\,\mho$$

To calculate $B$ & $D$   $V_2 = 0$

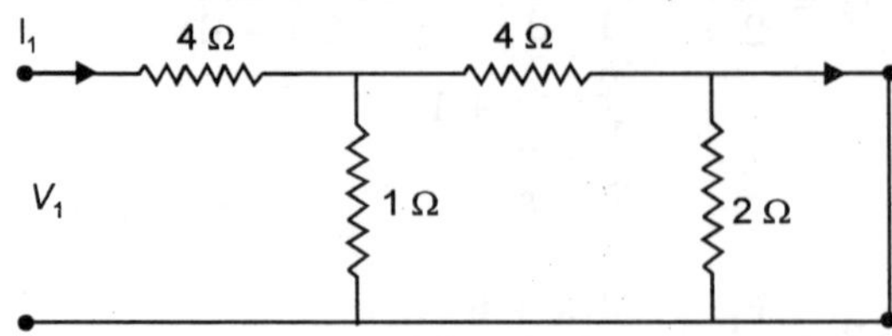

**Fig. E 8.15.1**

$$I_2 = \frac{V_1}{4 + \dfrac{4}{s}} \cdot \frac{1}{5} = \frac{V_1}{24} \quad \text{and} \quad I_1 = \frac{5V_1}{24}$$

$$B = \frac{V_1}{I_2} = 24 \quad \text{and} \quad D = \frac{I_1}{I_2} = 5$$

Since $AD - BC = 17 \times 5 - 24 \times \dfrac{7}{2} = 1$

The network is reciprocal

$$\begin{bmatrix} 17 & 24 \\ \dfrac{7}{2} & 5 \end{bmatrix} \begin{bmatrix} 17 & 24 \\ \dfrac{7}{2} & 5 \end{bmatrix} = \begin{bmatrix} A & B \\ C & D \end{bmatrix} = \begin{bmatrix} 373 & 408 \\ 77 & 109 \end{bmatrix} \textbf{ Ans.}$$

**Example 8.16:** Determine $z$-parameters of the network shown in Fig. E8.16. An identical network is connected in series with this network. Determine the $z$-parameters of the overall network and verify the result by direct computation

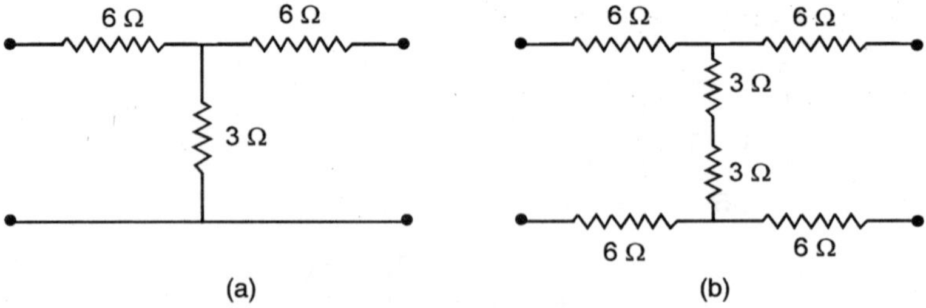

(a)                    (b)

**Fig E8.16**

$$Z_{11} = \frac{V_1}{I_1}\bigg|_{I_2=0} = 6 + 3 = 9 \ \Omega$$

$$Z_{22} = \frac{V_2}{I_2}\bigg|_{I_1=0} = 6 + 3 = 9 \ \Omega$$

$$Z_{12} = \left.\frac{V_1}{I_2}\right|_{I_1=0} \qquad \text{say} \quad I_2 = 1 \text{ Amp.}$$

$$V_1 = 3 \times 1 = 3 \text{ Volt.} \qquad \text{Hence } Z_{12} = Z_{21} = 3 \ \Omega$$

Therefore the overall $z$-parameters of the network are:

$$\begin{bmatrix} 18 & 6 \\ 6 & 18 \end{bmatrix} = \begin{bmatrix} Z_{11} & Z_{12} \\ Z_{21} & Z_{22} \end{bmatrix}$$

By direct computation Fig. E8.16(b).

$$Z_{11} = \left.\frac{V_1}{I_1}\right|_{I_2=0} = 18 \ \Omega = Z_{22}$$

$$Z_{12} = \left.\frac{V_1}{I_2}\right|_{I_1=0} \qquad \text{Let } I_2 = 1 \text{ A}$$

$$V_1 = 6 \times 1 = 6 \text{ volt}$$

Hence $\qquad Z_{12} = 6 \ \Omega = Z_{21}$ **Ans.**

**Example 8.17:** Determine the $Y$-parameters of the two $T$-networks of previous example when connected in parallel.

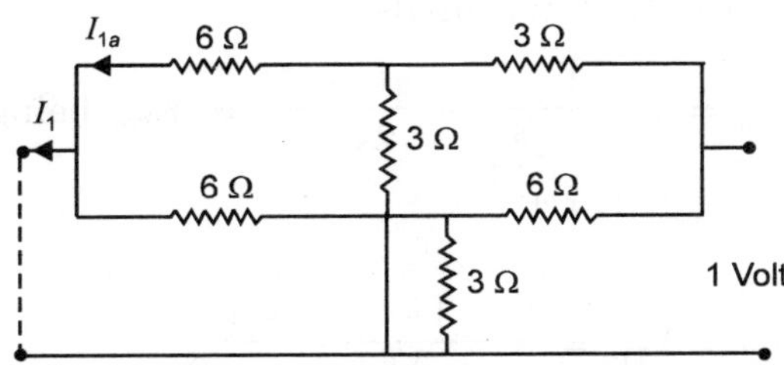


**Fig. E8.17**

**Solution:** Say the upper $T$, is network $a$, and lower $b$ network. By inspection, after short circuiting the output terminals.

$$Y_{11a} = \frac{1}{6 + \dfrac{18}{9}} = 0.125 = Y_{11b} = Y_{22a} = Y_{22b}$$

as the networks are symmetrical.

Therefore, $\qquad Y_{11} = Y_{22} = \dfrac{2}{8} = 0.25 \ \mho$

Now $\qquad Y_{12a} = \left.\dfrac{I_{1a}}{V_{2a}}\right|_{V_{1a}=0}$

For $\qquad V_{2a} = 1 \text{ volt}$

$$I_{1a} = \frac{1}{8} \times \frac{3}{9} = \frac{1}{24} A$$

Therefore, $\qquad Y_{12a} = Y_{12b} = Y_{21a} = Y_{21b} = -\dfrac{1}{24}$

Therefore the short circuit admittance parameters are:

$$\begin{bmatrix} \dfrac{1}{4} & -\dfrac{1}{12} \\[2ex] -\dfrac{1}{12} & \dfrac{1}{4} \end{bmatrix} = \begin{bmatrix} Y_{11} & Y_{12} \\[2ex] Y_{21} & Y_{22} \end{bmatrix}$$

**Example 8.18:** Determine the Y-parameters of the Twin-T network shown here

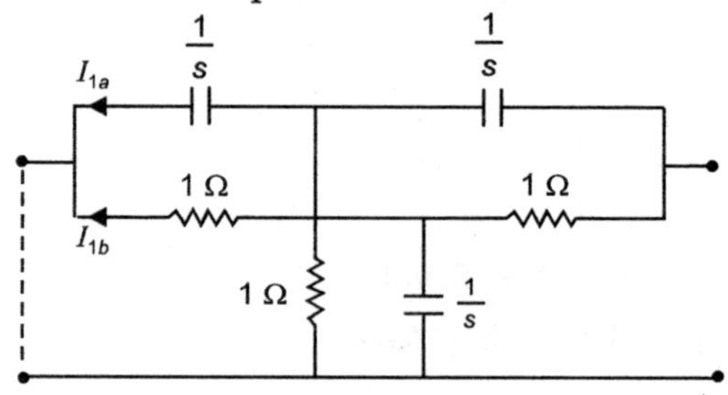

**Fig. E.8.18**

**Solution:** Shorting the output terminals

$$Y_{11a} = \frac{1}{\dfrac{1}{s} + \dfrac{1/s}{1 + \dfrac{1}{s}}} = \frac{s(s+1)}{2s+1} = Y_{22a} \text{ being symmetrical}$$

$$Y_{22b} = Y_{11b} = \frac{1}{1 + \dfrac{1/s}{1 + \dfrac{1}{s}}} = \frac{s+1}{s+2}$$

To calculate $\qquad Y_{12a} = \dfrac{I_{1a}}{V_2}\bigg|_{V_1=0} \qquad$ Let $\;\; V_2 = 1$ Volt

$$I_{1a} = \frac{1}{\dfrac{1}{s} + \dfrac{1/s}{1 + \dfrac{1}{s}}} \times \frac{1}{1 + 1/s} = \frac{s^2}{2s+1}$$

Therefore, $\qquad Y_{12a} = Y_{21a} = \dfrac{-s^2}{2s+1}$

Similarly $\qquad Y_{12b} = Y_{21b} = \dfrac{I_{1b}}{V_2}\bigg|_{V_1=0}$

$$= \frac{s+1}{s+2} \cdot \frac{1/s}{1+1/s} = \frac{-1}{s+2}$$

Here −ve sign has been assigned as the direction of $I_1$ is in opposition to conventional direction

$$Y_{11} = \frac{s(s+1)}{2s+1} + \frac{s+1}{s+2} = \frac{(s+1)(s^2+4s+1)}{(s+2)(2s+1)} = Y_{22} \textbf{ Ans.}$$

$$Y_{12} = Y_{21} = -\left( \frac{s^2}{2s+1} + \frac{1}{s+2} \right)$$

$$= -\frac{s^3 + 2s^2 + 2s + 1}{(2s+1)(s+2)} \textbf{ Ans.}$$

**Example 8.19:** Determine $Y$-parameters of the network shown in Fig. E.8.19

$$Y_{11a} = \frac{1}{\dfrac{1}{2} + \dfrac{2}{3}} = \frac{6}{7}$$

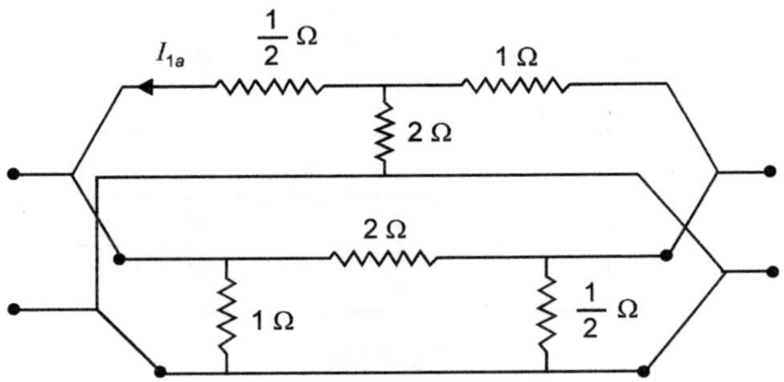


**Fig. E8.19**

$$Y_{11b} = \frac{1}{2/3} = \frac{3}{2}$$

$$Y_{22a} = \frac{1}{1 + \dfrac{1}{5/2}} = \frac{5}{7}$$

$$Y_{22b} = \frac{1}{1/5/2} = \frac{5}{2}$$

Hence

$$Y_{11} = Y_{11a} + Y_{11b} = \frac{6}{7} + \frac{3}{2} = \frac{33}{14} \; \mho$$

$$Y_{22} = \frac{5}{7} + \frac{5}{2} = \frac{45}{14} \; \mho$$

$$Y_{12a} = \left.\frac{I_{1a}}{V_2}\right|_{V_1=0} \qquad \text{Let} \quad V_2 = 1 \text{ Volt}$$

$$I_{1a} = \frac{1}{1 + \dfrac{1}{5/2}} \cdot \frac{2}{5/2} = \frac{4}{7}$$

Hence
$$Y_{12a} = -\frac{4}{7} = Y_{21a}$$

Similarly
$$Y_{12b} = -\frac{1}{2} = Y_{21b}$$

Therefore,
$$Y_{12} = -\left(\frac{4}{7} + \frac{1}{2}\right) = -\frac{15}{14} = Y_{21} \text{ \textbf{Ans.}}$$

**Example 8.20:** Two identical networks are connected series-parallel in Fig. E8.20 here. Determine the *h*-parameters of the overall network.

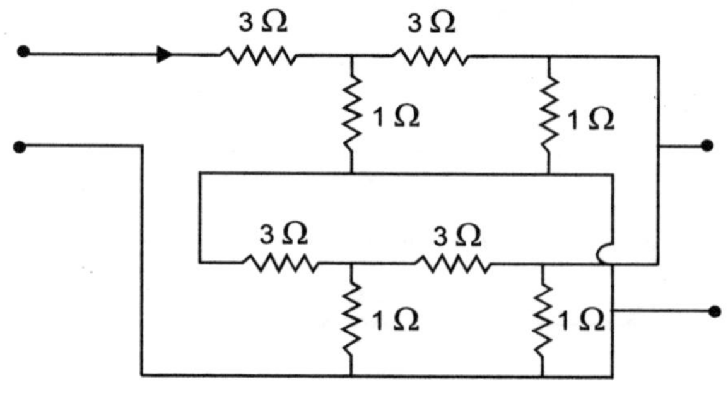

**Fig. E8.20**

The *h*-parameters of the individual networks are obtained as follows:

$$h_{11a} = \left.\frac{V_{1a}}{I_{1a}}\right|_{V_{a2}=0} = 3 + \frac{3}{4} = \frac{15}{4}$$

$$h_{22a} = \left.\frac{I_{2a}}{V_{2a}}\right|_{I_{1a}=0} = \frac{1}{4/5} = \frac{5}{4}$$

$$h_{12a} = \left.\frac{V_{1a}}{V_{2a}}\right|_{I_{1a}=0} \qquad \text{Let} \quad V_{2a} = 1, \; I_{2a} = \frac{5}{4}$$

Therefore,
$$V_{1a} = \frac{5}{4} \times \frac{1}{5} = \frac{1}{4}$$

Hence
$$h_{12a} = -\frac{1}{4} = h_{21a}$$

Therefore overall h-parameters are

$$h_{11} = \frac{15}{2}, \qquad h_{22} = \frac{5}{2}$$

$$h_{12} = h_{21} = -\frac{1}{2}$$

**Example 8.21:** If the two networks in the previous example are parallel series connected, determine the $g$-parameters.

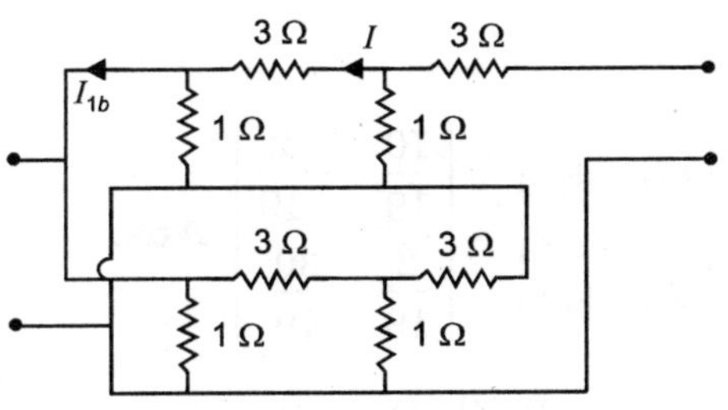

**Fig. E8.21**

$$g_{11a} = \left.\frac{I_{1a}}{V_{1a}}\right|_{I_{2a}=0} = \frac{1}{3 + \dfrac{4}{5}} = \frac{5}{19}$$

$$g_{22a} = \left.\frac{V_{2a}}{I_{2a}}\right|_{V_{1a}=0} \qquad \text{Let} \quad V_{2a} = 1 \text{ volt}$$

$$I_{2a} = \frac{1}{\dfrac{\left(3 + \dfrac{3}{4}\right) \times 1}{19/4}} = \frac{19}{15}$$

Hence

$$g_{22a} = \frac{15}{19}$$

Now

$$g_{12a} = \left.\frac{I_{1a}}{I_{2a}}\right|_{V_{1a}=0} \qquad \text{Let} \quad I_{2a} = 1 \text{ A}$$

$$I = \frac{1 \times 1}{19/4} = \frac{4}{19}$$

Therefore,

$$I_{1a} = \frac{4}{19} \times \frac{1}{4} = -\frac{1}{19}$$

Hence

$$g_{12a} = -\frac{1}{19}$$

$$g_{21a} = \left.\frac{V_{2a}}{V_{1a}}\right|_{I_{2a}=0} \qquad \text{Let} \quad V_{1a} = 1 \text{ Volt}$$

$$I_{1a} = \frac{1}{3 + \dfrac{4}{5}} = \frac{5}{19} \;,\quad I = \frac{5}{19} \times \frac{1}{5} = \frac{1}{19}$$

Therefore,
$$V_{2a} = \frac{1}{19} \times 1 = \frac{1}{19}$$

Hence
$$g_{21a} = \frac{1}{19}$$

Therefore,

$$\begin{bmatrix} g_{11} & g_{12} \\[2mm] g_{21} & g_{22} \end{bmatrix} = \begin{bmatrix} \dfrac{10}{19} & -\dfrac{2}{19} \\[3mm] \dfrac{2}{19} & \dfrac{30}{19} \end{bmatrix} \quad \textbf{Ans.}$$

**Example 8.22:** Determine $A$, $B$, $C$, $D$ parameters of the network shown

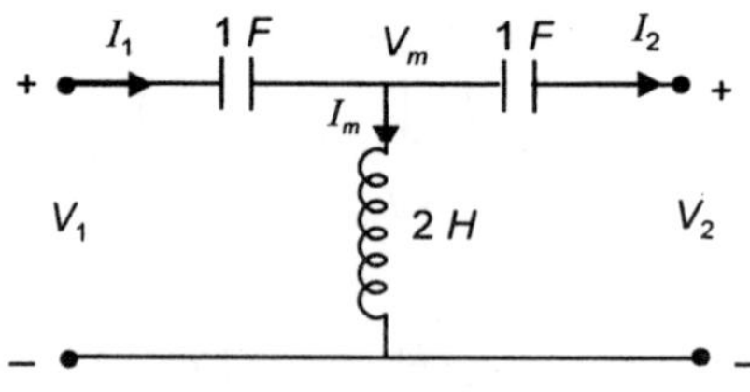


**Fig. E8.22**

$$V_m = V_2 + \frac{I_2}{s}$$

$$I_m = \frac{V_m}{2s} = \frac{V_2 + I_2/s}{2s}$$

and
$$I_1 = I_2 + \frac{V_2 + I_2/s}{2s}$$

Now
$$V_1 = V_m + I_1/s$$

$$= V_2 + \frac{I_2}{s} + \frac{I_2}{s} + \frac{sV_2 + I_2}{2s^3}$$

$$= V_2\left(1 + \frac{1}{2s^2}\right) + I_2\left(\frac{4s^2 + 1}{2s^3}\right)$$

and
$$I_1 = \frac{V_2}{2s} + I_2\left(1 + \frac{1}{2s^2}\right)$$

Hence
$$A = 1 + \frac{1}{2s^2} \qquad\qquad B = \frac{4s^2 + 1}{2s^3}$$

$$C = \frac{1}{2s} \qquad\qquad D = 1 + \frac{1}{2s^2}$$

**Example 8.23:** Fig. E8.23(*a*) and (*b*) here show two networks. Connect them in series and determine the *z*-parameters of the overall network.

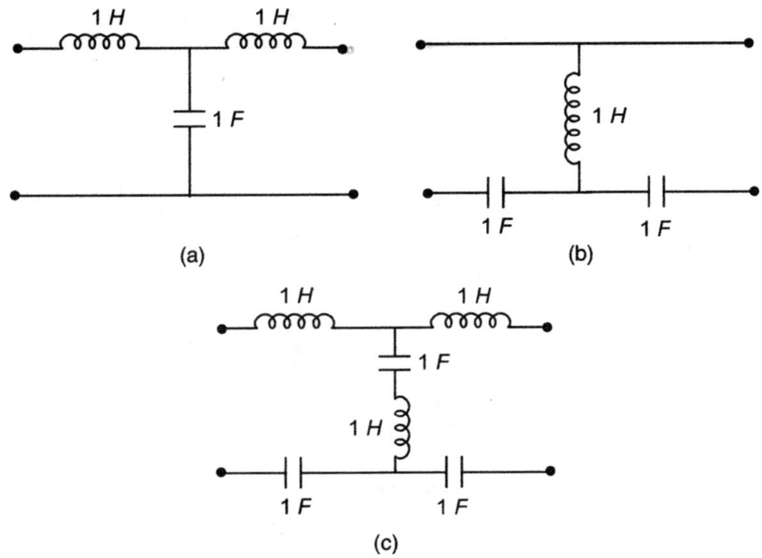

(a)  (b)

(c)

**Fig. E8.23**

$$Z_{11a} = \left( s + \frac{1}{s} \right) = Z_{22a}$$

and

$$Z_{11b} = Z_{22b} = s + \frac{1}{s}$$

$$Z_{12a} = \left. \frac{V_1}{I_2} \right|_{I_1 = 0} \qquad \text{Let} \qquad I_2 = 1\ A$$

Therefore,

$$Z_{12a} = Z_{21} = \frac{1}{s} \quad \text{and} \quad Z_{12b} = Z_{21b} = s$$

Therefore,

$$\begin{bmatrix} Z_{11} & Z_{12} \\ Z_{21} & Z_{22} \end{bmatrix} = \begin{bmatrix} 2\left( s + \dfrac{1}{s} \right) & s + \dfrac{1}{s} \\ s + \dfrac{1}{s} & 2\left( s + \dfrac{1}{s} \right) \end{bmatrix} \quad \textbf{Ans.}$$

**Example 8.24:** A model for a transistor in the common emitter connection is shown in Fig. E8.24. Determine the *h*-parameters of the transistor.

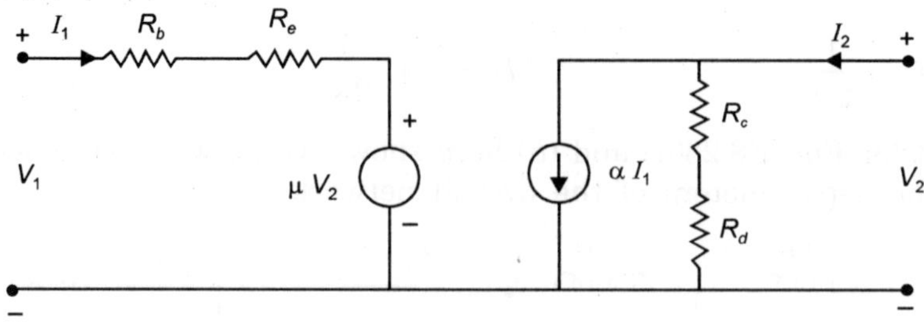

**Fig. E8.24**

From Fig. E8.24 $(a)$ $V_2 = 0$

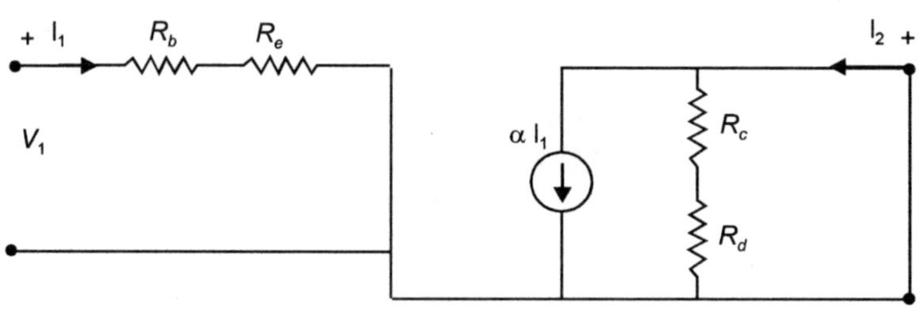

**Fig. E.8.24 (a)**

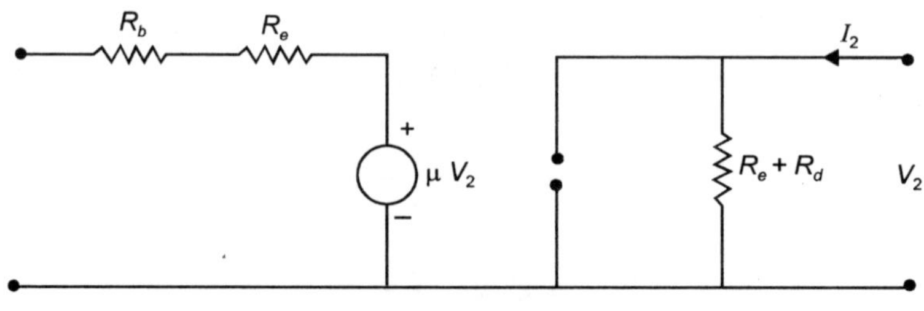

**Fig. E.8.24 (b)**

$$h_{11} = \frac{V_1}{I_1}\bigg|_{V_2 = 0} = R_b + R_e$$

Since
$$I_2 = \alpha I_1$$

Hence
$$h_{21} = \frac{I_2}{I_1}\bigg|_{V_2 = 0} = \alpha$$

From Fig 8.24 $(b)$ $I_1 = 0$

$$h_{22} = \frac{I_2}{V_2}\bigg|_{I_1 = 0} = \frac{1}{R_c + R_d}$$

and
$$h_{12} = \frac{I_2}{V_2}\bigg|_{I_1 = 0} = \frac{\mu V_2}{V_2} = \mu \;\; \textbf{Ans.}$$

**Example 8.25:** Consider the network shown in Fig. E.8.25. Determine whether the network is reciprocal

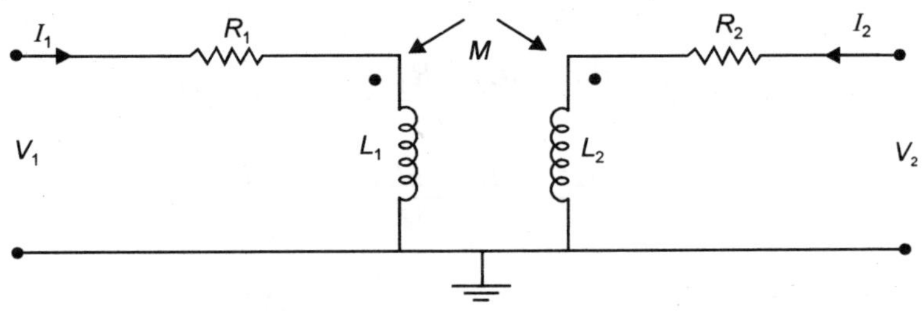

**Fig. E.8.25**

The equivalent circuit is given by

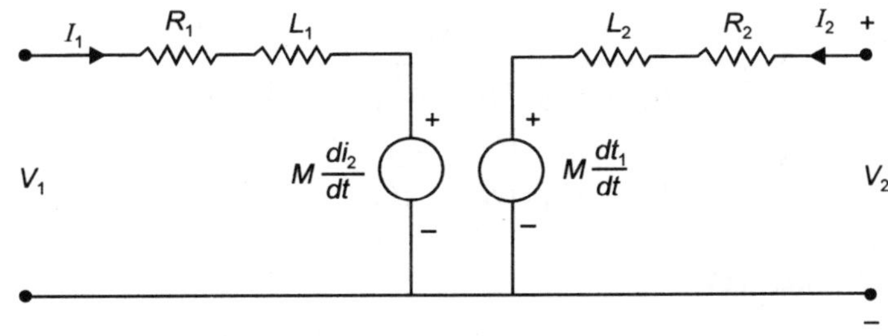

**Fig. E8.25(a)**

Writing down loop equations

$$V_1 = I_1 R_1 + sL_1 I_1 + sM I_2 = I_1 (R_1 + sL_1) + I_2 sM$$

$$V_2 = I_2 R_2 + sL_2 I_2 + sM I_1 = I_1 sM + I_2 (R_2 + sL_2)$$

Therefore, $Z_{11} = R_1 + sL_1 \qquad Z_{22} = R_2 + sL_2$

and $\qquad Z_{12} = Z_{21} = sM$

Since $Z_{12} = Z_{21}$ the network is reciprocal.

**Example 8.26:** Determine $Y$-parameters of the network shown in Fig. E8.25

$$Y_{11} = \left.\frac{I_1}{V_1}\right|_{V_2=0} \qquad \text{For } V_2 = 0 = sMI_1 + (R_2 + sL_2) I_2$$

or $\qquad I_2 = -\dfrac{sM}{R_2 + sL_2} I_1$

Now $\qquad V_1 = I_1 (R_1 + sL_1) - sM \cdot \dfrac{sM}{R_2 + sL_2} I_1$

$$= I_1 \left[ \frac{(R_1 + sL_1)(R_2 + sL_2) - s^2 M^2}{R_2 + sL_2} \right]$$

Therefore, $\qquad \dfrac{I_1}{V_1} = Y_{11} = \left[ \dfrac{R_2 + sL_2}{(R_1 + sL_1)(R_2 + sL_2) - s^2 M^2} \right]$

$$Y_{21} = \frac{I_2}{V_1}\bigg|_{V_2 = 0}$$

Now
$$V_1 = -I_2 \frac{(R_2 + sL_2)(R_1 + sL_1)}{sM} + sMI_2$$

$$= I_2 \left[ \frac{s^2M^2 - (R_2 + sL_2)(R_1 + sL_1)}{sM} \right]$$

or
$$Y_{21} = \frac{I_2}{V_1}\bigg|_{V_2 = 0} = \frac{-sM}{(R_2 + sL_2)(R_1 + sL_1) - s^2 M^2}$$

Similarly by substituting $V_1 = 0$ we can find out $Y_{22}$ and $Y_{12}$. It is found that

$$Y_{12} = \frac{-sM}{(R_2 + sL_2)(R_1 + sL_1) - s^2M^2}$$

and
$$Y_{22} = \frac{R_1 + sL_1}{(R_1 + sL_1)(R_2 + sL_2) - s^2M^2}$$

## 8.7  TWO-PORT SYMMETRY

We have studied so far the open circuit impedance and short circuit admittance parameters. We define short circuit impedances as

$$Z_{1s} = \frac{V_1}{I_1}\bigg|_{V_2 = 0} \quad \text{and} \quad Z_{2s} = \frac{V_2}{I_2}\bigg|_{V_1 = 0} \qquad \qquad ...(8.94)$$

Also open circuit impedances are defined as

$$Z_{10} = \frac{V_1}{I_1}\bigg|_{I_2 = 0} \qquad \qquad Z_{20} = \frac{V_2}{I_2}\bigg|_{I_1 = 0} \qquad \qquad ...(8.95)$$

Here the first suffix to $z$ indicates the port where measurements have been made and the second suffix represents status of the second port when short circuited $(s)$ or open circuited $(o)$.

**$ABCD$ parameters in terms of open and short circuit impedances**
We know that

$$V_1 = AV_2 - BI_2 \qquad \qquad ...(8.96)$$

$$I_1 = CV_2 - DI_2 \qquad \qquad ...(8.97)$$

Here $-$ve sign is because of direction of $I_2$

$$Z_{10} = \frac{V_1}{I_1}\bigg|_{I_2 = 0} = \frac{A}{C} \qquad \qquad ...(8.98)$$

$$Z_{20} = \frac{V_2}{I_2}\bigg|_{I_1 = 0} = \frac{D}{C} \qquad \qquad ...(8.99)$$

$$Z_{1s} = \left.\frac{V_1}{I_1}\right|_{V_2=0} = \frac{B}{D} \qquad \qquad ...(8.100)$$

$$Z_{2s} = \left.\frac{V_2}{I_2}\right|_{V_1=0} = \frac{B}{A} \qquad \qquad ...(8.101)$$

Using equations (8.98) and (8.100) we have

$$\frac{A}{C} - \frac{B}{D} = Z_{10} - Z_{1s} = \frac{AD - BC}{CD} = \frac{1}{CD}$$

and

$$Z_{20}\,(Z_{10} - Z_{1s}) = \frac{D}{C} \cdot \frac{1}{CD} = \frac{1}{C^2}$$

Hence,

$$C = \frac{1}{\sqrt{Z_{20}\,(Z_{10} - Z_{1s})}} \qquad \qquad ...(8.102)$$

Also from equation (8.99) we have

$$D = CZ_{20} = \frac{Z_{20}}{\sqrt{Z_{20}\,(Z_{10} - Z_{1s})}} \qquad \qquad ...(8.103)$$

Similarly,

$$Z_{20} - Z_{2s} = \frac{AD - BC}{CA} = \frac{1}{CA}$$

and

$$Z_{10} = \frac{A}{C}$$

Therefore,

$$A^2 = \frac{A}{C} \cdot CA = \frac{Z_{10}}{Z_{20} - Z_{2s}}$$

or

$$A = \pm\sqrt{\frac{Z_{10}}{Z_{20} - Z_{2s}}}$$

Similarly $B$ can be obtained.

## 8.8 POLES AND ZEROS OF NETWORK FUNCTION

We have seen in the previous articles that the network functions e.g. driving point imped-ance or transfer function etc. can be represented as a ratio of two polynomials $p(s)$ and $q(s)$ and is written as

$$N(s) = \frac{p(s)}{q(s)} = \frac{a_0\,s^n + a_1\,s^{n-1} + .... + a_{n-1}s + a_n}{b_0\,s^m + b_1\,s^{m-1} + .... + b_{m-1}s + b_m} \qquad \qquad ...(8.106)$$

Where $a^s$ and $b^s$ are the real positive co-efficients. After factorizing the numerator and denominator the network function $N(s)$ could be re-written as

$$N(s) = \frac{H\,(s - z_1)(s - z_2) ......... (s - z_n)}{(s - p_1)\,(s - p_2) ......... (s - p_{m)}} \qquad \qquad ...(8.107)$$

where $H = \dfrac{a_0}{b_0}$ is a constant kno wn as the scale factor. The factors in the numerator contain values like $z_1$, $z_2$ ..... $z_n$ which are the zeros of network function $N(s)$ which means the values at which the function takes its value as zero. In fact in network theory these values represent the complex frequencies at which the function vanishes. On the other hand the factors in the denominator contain values like $p_1$, $p_2$ .... $p_m$ which are the poles of the network function $N(s)$ which means the values at which the function $N(s)$ takes its value as infinity. The value $p_1$ through $p_m$ represent complex frequencies at which the function $N(s)$ blows up to infinity. The function $N(s)$ given above is said tò have $n$ zeros and $m$ poles. Poles and zeros are useful in describing the network function. The network function is completely described by its poles zeros and the scale factor.

If the poles and zeros do not repeat i.e. $p_i \neq p_j$ and $z_i \neq z_j$ the poles and zeros are said to be distinct. However, if $r$ finite poles and zeros have the same value the pole or zero is said to have repeated multiplicity $r$. When $n > m$, then $(n - m)$ poles are at $s = \infty$ and when $m > n$, $(m - n)$ number of zeros are at $s = \infty$ i.e for any rational network function the total number of zeros equals the total number of poles and, therefore, in addition to the finite poles and zeros, the poles and zeros at infinity have also to be taken into account. The symbol $\times$ is used for a pole, zero $(0)$ is used for the zero. At any frequency other than the critical frequency (corresponding to pole and zero) the network function has a finite non-zero value.

Network with all zeros in the left half are known as minimum – phase network; those with any zero in the right half plane are non-minimum phase network. The poles must lie in the left half plane including the boundary (if simple).

### 8.8.1 Physical Significance of Poles and Zeros

Consider the transfer function $G(s) = \dfrac{V_0(s)}{V_i(s)}$ which may be rewritten as

$$V_0(s) = G(s)\ V_i(s) \qquad \qquad ...(8.108)$$

Normally $V_i(t)$ is specified and if $G(s)$ of a network is known $V_0(t)$ can be obtained. The partial fraction expansion of the term $G(s)\ V_i(s)$ gives, in the denominator of each factor a pole of either $G(s)$ or $V_i(s)$, i.e. with no repeated roots in the denominator of $V_0(s)$

$$G(s)\ V_i(s) = \sum_{i=1}^{\alpha} \frac{K_i}{s - p_i} + \sum_{j=1}^{\beta} \frac{K_j}{s - p_i} \qquad \qquad ...(8.109)$$

where $\alpha$ is the number of poles of $G(s)$ and $\beta$ is the number of poles of $V_i(s)$. The inverse laplace transform of this function is

$$V_0(t) = \sum_{i=1}^{\alpha} K_i\ e^{p_i t} + \sum_{j=1}^{\beta} K_j\ e^{p_j t} \qquad \qquad ...(8.110)$$

Here the frequencies $p_i$ are the natural complex frequencies corresponding to natural response or free oscillations. It represents the natural response as it is derived out of a passive network function $G(s)$ and not out of the driving force function (voltage or current source). Similarly $p_j$ are the driving force complex frequencies corresponding to forced response or forced oscillation. It represents forced response as it is derived from the poles of the forcing function or driving function $V_i(s)$. The poles, therefore, determine the wave form of the time variation of the output response $V_o(t)$. However, the zeros determine the magnitude of each part of the response since they determine the magnitude $K_i$ and $K_j$ in the partial function expansion.

In regard to driving point impedance/admittance (usually referred as immitances) the poles and zeros have apparent significance. Since $Z(s) = V(s)/I(s)$ a pole of $Z(s)$ signifies infinite impedance i.e. zero current for a finite voltage in other words a natural voltage and, therefore, it is natural response or free oscillation and pole in $Z(s)$ is an open circuit condition. Similarly a zero of $Z(s)$ means no voltage for finite current, in other words a natural current or natural response and under this condition it represents a short circuit. Therefore, for a one terminal pair network, a pole frequency represents an open circuit and a zero frequency represents a short circuit. This can be further understood by considering single element network. For a capacitor the driving point impedance is $Z(s) = 1/Cs$. This network function has its pole at $s = 0$ as then $Z(s) = \infty$ and its zero is at $s = \infty$ as then it reduces to zero i.e. $1/Cs = 0$ i.e. it acts as a short circuit. Similarly for an inductor the driving point impedance $Z(s) = Ls$ has its pole at $s = \infty$ and zero at $s = 0$ i.e. the element behaves open circuit at $s = \infty$ and short circuit at $s = 0$.

**Example 8.27:** A circuit has an impedance

$$Z(s) = \frac{(s^2 + 4s + 8)(s + 2)}{s^2 + 2s + 5}$$

The function has zero at $s = -2$; and $-2 + j2$ and $-2 - j2$ whereas the poles are at $s = -1 + j2$ and $-1-j2$ and since the no. of zeros is three the third pole is at infinity.

**Example 8.28:** Suppose the network function is given by

$$N(s) = \frac{s^2(s + 3)}{(s + 1)(s + 2 + j2)(s + 2 - j2)}$$

This function has double zero at $s = 0$ and a zero at $s = -3$.
The poles are at $s = -1$ and $s = -2 \pm j2$.

**Example 8.29:** Determine $Z(s)$ for the given network. If this is represented by

$$Z(s) = \frac{K(s - Z_1)}{(s - p_1)(s - p_2)}$$ and determine $z_1$, $p_1$ and $p_2$ in terms of $R$, $L$ and $C$.

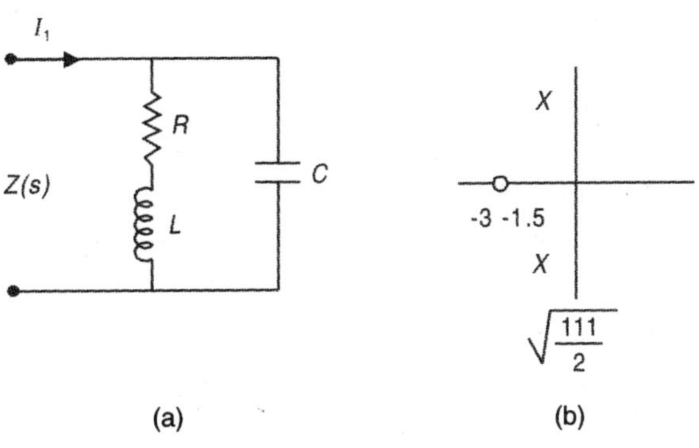


**Fig. E8.29**

**Solution:** From Fig E8.29(a)

$$Z(s) = \frac{\dfrac{1}{Cs}(R + sL)}{R + sL + 1/Cs} = \frac{\dfrac{1}{C}(s + R/L)}{s^2 + \dfrac{R}{L}s + \dfrac{1}{LC}}$$

Hence
$$z_1 = -\frac{R}{L}$$

and
$$p_1 = -\frac{R}{2L} + \sqrt{\left(\frac{R}{2L}\right)^2 - \frac{1}{LC}}$$

$$p_2 = -\frac{R}{2L} + \sqrt{\left(\frac{R}{2L}\right)^2 - \frac{1}{LC}}$$

If in the above example poles and zeros have the location as shown above in Fig. E8.29 (b) with $z(j0) = 1$, determine the value of $R$, $L$ and $C$.

Now
$$Z(j_o) = \frac{\dfrac{1}{C}\left(\dfrac{R}{L} + o\right)}{o + o + 1/LC} = 1 = \frac{R}{LC} \cdot LC = R$$

Since $Z(s)$ has zero at $s = -3 = -\dfrac{R}{L} = -\dfrac{1}{L}$ or $L = \dfrac{1}{3}H$

Also $w_n^2 = \dfrac{1}{LC} = 1.5^2 + \dfrac{111}{4}$ since the pole is at $-1.5 \pm j\sqrt{\dfrac{111}{4}}$

$$= 30$$

Therefore,
$$C = \frac{1}{10}F$$

**Example 8.30:** Suppose an impedance function is given as

$$Z(s) = \frac{15(s^3 + 2s^2 + 3s + 2)}{s^4 + 6s^3 + 8s^2}$$

Determine the poles and zeros.

For zeros
$$s^3 + 2s^2 + 3s + 2 = s^2(s + 1) + s(s + 1) + 2(s + 1)$$
$$= (s + 1)(s^2 + s + 2)$$
$$= (s + 1)(s + 0.5 + j3.5)(s + 0.5 - j3.5)$$

Poles $\quad s^4 + 6s^3 + 8s^2$
$$= s^2(s^2 + 6s + 8)$$
$$= s^2(s + 2)(s + 4)$$

The zeros are at $-1$, $-0.5 \pm j\,3.5$ and poles at $s = 0$ and $-2$, $-4$.

Since the impedance function has double pole at the origin, this function is not permissible i.e. it is not a positive real function.

### 8.8.2 Necessary Conditions for Driving Point Impedance

The necessary condition for an impedance (driving point, therefore, two terminal network) to be positive real are, after the common factors from the numerator and denominator are cancelled.

1. The co-efficient in the polynominals $p(s)$ and $q(s)$ of $N(s) = \dfrac{p(s)}{q(s)}$ must be real and positive.

2. If the poles and zeros are imaginary or complex, they should be conjugate.
3. (*a*) The real parts of poles and zero should either be zero or negative, it should not be positive.
   (*b*) If the real part is zero then that pole or zero should be simple.
4. The polynominals $p(s)$ or $q(s)$ should not have any missing terms between those of highest and lowest degree, unless all even or all odd terms are missing.
5. The degree $p(s)$ and $q(s)$ may differ at the most by one.
6. The terms of lowest degree in $p(s)$ and $q(s)$ may differ in degree by one at the most.

### 8.8.3 Necessary Conditions for Transfer Function

The necessary condition for a network function to be a transfer function after cancelling the common factors from the numerator and denominator are:

1. The co-efficients in the polynomials $p(s)$ and $q(s)$ of $N = \dfrac{p(s)}{q(s)}$ must be real, and those for $q(s)$ must be positive.
2. If the poles are imaginary or complex, they should be conjugate.
3. (*a*) The real part of poles must be zero or negative.
   (*b*) If the real part is zero then that pole must be simple. This includes the origin i.e. at the origin the pole must be simple.
4. The polynomial of $q(s)$ may not have any missing terms between that of highest and lowest degree unless all even or all odd terms are missing.
5. The polynomial $p(s)$ may have terms missing between the terms of lowest and highest degree and some of the co-efficients may be negative.
6. The degree of $p(s)$ can be as small as zero independent of the degree of $q(s)$.
7. For $G_{12}$ and $\alpha_{12}$ : The maximum degree of $p(s)$ is the degree of $q(s)$.

For $Z_{12}$ and $Y_{12}$: The maximum degree of $p(s)$ is the degree of $q(s)$ plus one.
Here $G_{12}$ is the voltage transfer function and $\alpha_{12}$ is the current transfer function.

### 8.8.4 Response of the System vis-a-vis Location of Poles and Zeros

We know that the poles and zeros of a circuit consisting of RLC elements will always lie in the left half of the $s$-plans. Let us consider a simple RLC series circuit excited by a voltage source $v(t)$. For this network the KVL equation is

$$Ri + L\,\frac{di}{dt} + \frac{1}{C}\int i\, dt = v(t) \qquad \qquad ...(8.111)$$

The corresponding homogenous equation is of the second order and is given as

$$L\,\frac{d^2 i}{d^2 t} + R\,\frac{di}{dt} + \frac{i}{C} = 0$$

or

$$\frac{d^2 i}{dt^2} + \frac{R}{L}\,\frac{di}{dt} + \frac{i}{LC} = 0$$

The roots of the equation are

$$s_1, \ s_2 = -\frac{R}{L} \pm \sqrt{\left(\frac{R}{2L}\right)^2 - \frac{1}{LC}}$$

For critical damping the quantity within the square root should be zero i.e.

$$\left(\frac{R_C}{2L}\right)^2 = \frac{1}{LC} \quad \text{or} \quad R_C = 2\sqrt{\frac{L}{C}} \qquad \qquad ...(8.112)$$

We next define the quantity $\delta = \dfrac{R}{R_C} = \dfrac{R}{2}\sqrt{\dfrac{C}{L}}$ where $\delta$ is known as damping ratio as
it is a ratio of actual resistance to the critical resistance. If the actual system resistance
is less than the critical resistance $\delta < 1$ the system is under damped and when actual
resistance equals the critical resistance i.e. $\delta = 1$ the circuit is critically damped whereas if
$\delta > 1$ the circuit is over damped.

Also we define here another term
$\omega_n = \dfrac{1}{\sqrt{LC}}$ where $\omega_n$ is the undamped natural frequency or simply the natural frequency.

Writing $\dfrac{d}{dt} = s$ and using the quantities defined above, the original differential equation can
be written as

$$(s^2 + 2\ \delta\ \omega_n\ s + \omega_n^2)\ i = 0 \qquad \qquad ...(8.113)$$

Since $i \neq 0$ we rewrite as

$$s^2 + 2\ \delta\ \omega_n\ s + \omega_n^2 = 0 \quad \text{and the roots of this equation are}$$

$$s_1, \ s_2 = -\ \delta\ \omega_n \pm \omega_n\ \sqrt{\delta^2 - 1}$$

The general solution of the equation is given as

$$i(t) = K_1 e^{-\left(-\delta\omega_n + \omega_n\sqrt{\delta^2 - 1}\right)t} + K_2\, e^{-\left(-\delta\omega_n - \omega_n\sqrt{\delta^2 - 1}\right)t} \qquad \qquad ...(8.114)$$

Here $\delta\ \omega_n$ is the damping co-efficient and is denoted by $\sigma$ in the $s$-plane. $\omega_n\ \sqrt{\delta^2 - 1}$ is the
frequency in the $s$-plane

i.e. $$\sigma = \delta\ \omega_n \quad \text{and} \quad \omega = \omega_n\ \sqrt{\delta^2 - 1} \qquad \qquad ...(8.115)$$

and $$s = \sigma + j\ \omega$$

Before we seek the solution of the above equation a look at the behaviour of the roots
of the equation as the damping ratio $\delta$ varies from zero (when $R = 0$) to infinity (when
$R \to \infty$). There are three possible forms for the roots:

| | |
|---|---|
| Case I | $\delta < 1$ the roots are complex conjugate |
| Case II | $\delta = 1$ the roots are real and repeated |
| Case III | $\delta > 1$ the roots are real |

For $\delta = 0$, $s_1, \ s_2 = \pm j\ \omega_n$

If we want to study the nature of roots when $\delta$ is varied from $o$ to $\infty$ we shall see a
loci of roots in the complex $s$-plane.

For $$\delta = 0, \ s_1, \ s_2 = j\omega_n$$

i.e. the roots are purely imaginary. For $\delta < 1$ the roots are complex conjugate as

$$s_1,\ s_2 = -\delta\,\omega_n \pm j\,\omega_n\,\sqrt{1-\delta^2}$$

i.e. $$\sigma = -\delta\,\omega_n \quad \text{and} \quad \omega = \pm\,\omega_n\,\sqrt{1-\delta^2}$$

as shown in Fig 8.23.

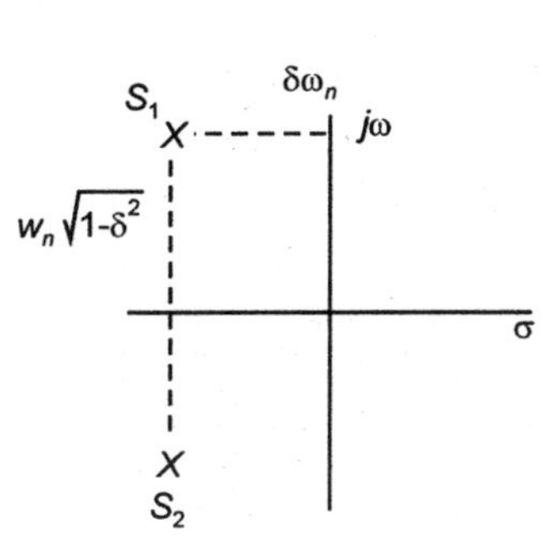

**Fig. 8.23(a)** Roots of second order equation plotted.

**Fig. 8.23(b)** Locus of the roots shown in Fig. 8.21 (a) as $0 \le \delta \le \infty$.

From these equations we have

$$\sigma^2 + \omega^2 = \delta^2\,\omega_n^2 + \omega_n^2 - \omega_n^2\,\delta^2$$
$$= \omega_n^2$$

This is an equation to a circle in the *s*-plane with center at the origin and the radius equal to $\omega_n$. Also this locus is formed by $\delta$ varying from 0 to 1. The locus is shown in Fig. 8.23 (b) Fig. 8.24 (a) shows constant damping ratio ($\delta$ = constant) contours as the straight lines ($\sigma = -\delta\omega_n$ or $\sigma = -K\omega_n$) through the origin and that contours of constant damping co-efficient ($\sigma = -\delta\,\omega_n$ or $\sigma = K_1$) are straight lines parallel to the $j\omega$ axis of the *s*-plane. Similarly lines parallel to $\sigma$-axis of the *s*-plane are lines of constant frequency $\omega$ ($\omega$ = constant or $\omega_n\,\sqrt{1-\delta^2}$ = constant)

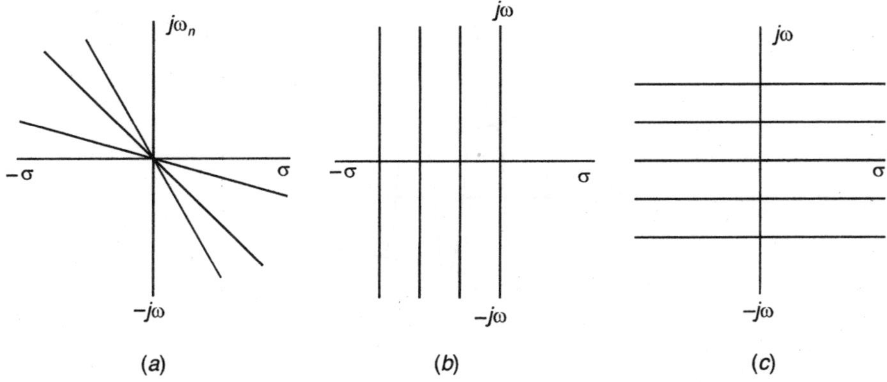

(a)  (b)  (c)

**Fig. 8.24** Contours of Constant parameter values in the *s*-plane
(a) Lines of constant $\delta$ (b) Constant damping contours (c) Constant
frequency of oscillation $\omega = \omega_n\,\sqrt{1-\delta^2}$

The location of poles in the $s$-plane can be used to explain the performance of the network in time domain in term of $\delta$ and $w_n$.

In order to understand the significance of various contours given above let us consider poles of a network as shown in Fig. 8.25.

(Zeros have been omitted for clarity sake)

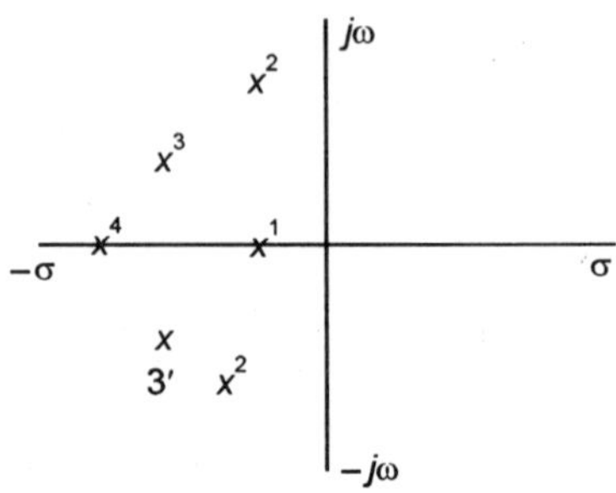


**Fig. 8.25**   Poles of a typical network.

The poles at 1 and 4 correspond to over damped case and have an exponential decay ($e^{-s_1 t}$ and $e^{-s_4 t}$) form in the time domain. The damping is greater for a pole, the greater is its distance from the origin along negative $\sigma$ axis i.e. here damping is greater due to pole $s_4$ as compared to $s_1$. Therefore the time constant (Time to decay to 63.2% of its initial value) for pole $s_1$ is higher than that for $s_4$. Typical time domain response corresponding to these poles is shown here in Fig. 8.26 (a).

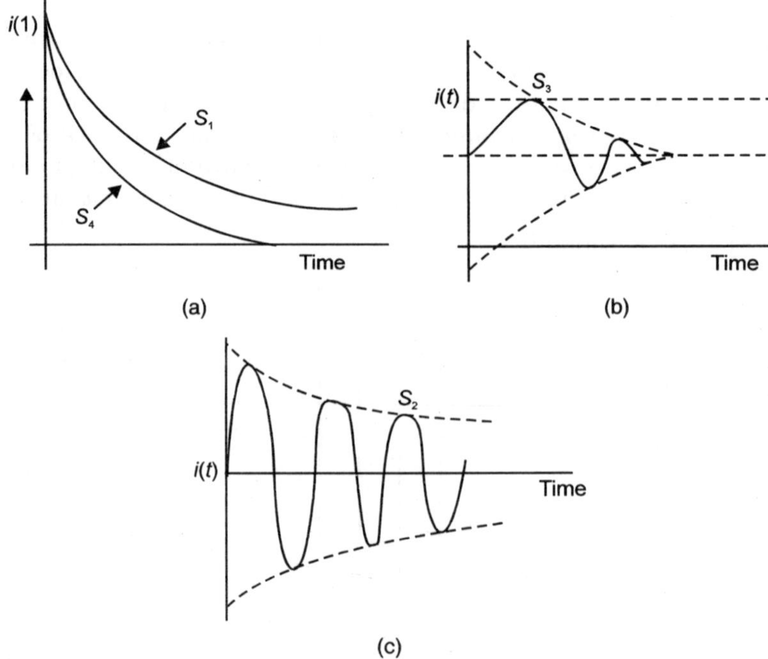


**Fig. 8.26**   Typical time-domain response.

Poles $s_2$ and $s_3$ are complex conjugate pairs of pole $s_2$ and $s_2'$ and $s_3$ and $s_3'$ respectively and these represent oscillator expressions in the time domain. The frequency of oscillation is determined by the ordinate and the damping is determine by its distance from the origin along–σ axis. The longer the ordinate the higher the frequency of oscillation and smaller its distance from the origin along – σ axis the smaller the damping. Therefore the poles $s_2$ and $s_2'$ provide higher frequency of oscillation with smaller damping whereas poles at $s_3$ have relatively smaller frequency of oscillation and larger damping as shown in Fig. 8.26(c) and (b) above. However the natural frequency $\omega_n$ of both the pairs of complex conjugate poles is more or less the same as their distances from the origin (radius) is more or less the same. A more detailed figure shows the responses of the network with different location of the poles. The location of the figure is the relative location of the complex conjugate pole (s-plane) or conjugate poles (on the $j\omega$ axis) or the real pole (on the – σ axis).

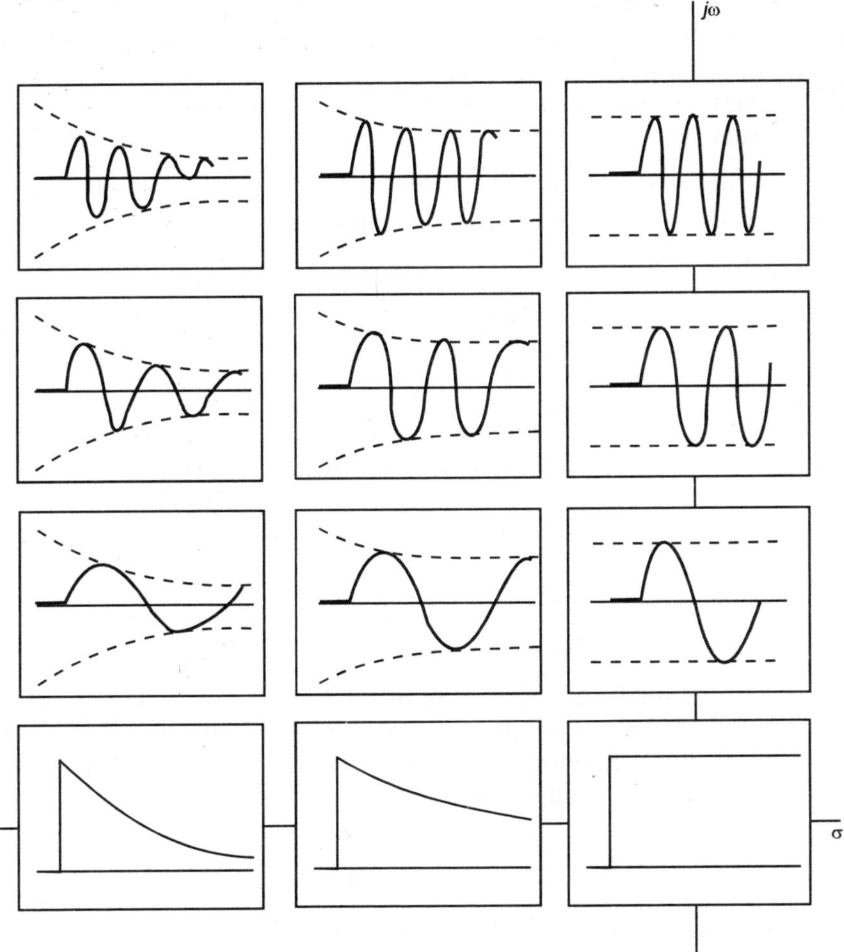

**Fig. 8.27** Response for the various *s*-plane locations.

The complete response of the network is given by the sum of the response due to individual poles i.e.

$$i(t) = K_1 e^{s_1 t} + K_2 e^{s_2 t} + K_3 e^{s_2 t} + K_4 e^{s_2' t} + K_5 e^{s_3' t} + K_6 e^{s_4 t} \qquad ...(8.116)$$

However it remains to calculate $K_1$ through $K_6$ to get the complete solution. Any of the co-efficient ($K_r$ the residue) can be obtained by using Heaviside method. Say corresponding to $r$th pole.

$$K_r = \frac{H\,(s - s_1)\,(s - s_2)\,........\,(s - s_n)}{(s - s_a)\,(s - s_b)\,..\,........\,(s - s_m)} \cdot (s - s_r)\Big|_{s = s_r} \qquad ...(8.117)$$

Substituting $s_r$ *for* $s$ in the above equation we get $K_r$ as

$$K_r = H\,\frac{(s_r - s_1)\,(s_r - s_2)\,.....\,(s_r - s_n)}{(s_r - s_a)\,(s_r - s_b)\,......\,(s_r - s_m)}$$

Here both $s_r$ and $s_n$ are known hence $K_r$ can be calculated. If all the poles and zeros are real then there is no problem, the above expression can be calculated easily. However, if the roots are complex, the terms containing $(s_r - s_n)$ are also complex which may be written in polar from as

$$(s_r - s_n) = M_{nr}\,e^{j\,\phi_{nr}} \quad \text{where} \quad M_{nr}$$

is the magnitude of the phasor and $\phi_{nr}$ is the phase angle of the same phasor as shown in Fig. (8.28)

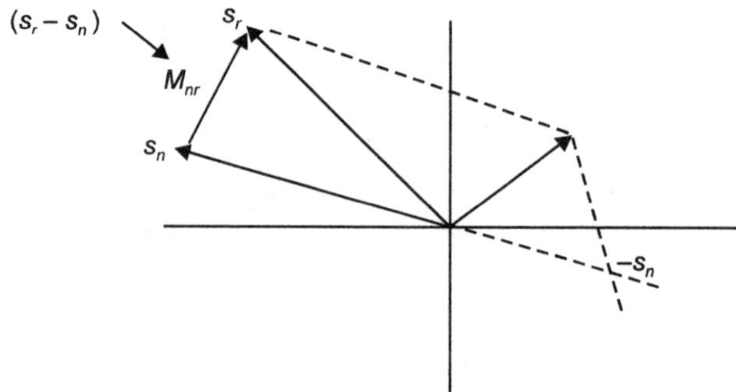


**Fig. 8.28**  Graphical calculation of $K_r$.

The $(s_r - s_n)$ is signified as a phasor directed from $s_n$ to $s_r$. The magnitude $M_{nr}$ is the distance from $s_n$ to $s_r$, the phase angle $\phi_{nr}$ is the angle of the line from $s_n$ to $s_r$ measured with respect to $\phi = 0$ line. Therefore, each factor in the above expression can be expressed in terms of $M$ and $\phi$ and the value of $K_r$ is obtained as

$$K_r = H\,\frac{M_{1r}\,M_{2r}\,.......\,M_{nr}}{M_{ar}\,M_{br}\,.......\,M_{mr}}\,e^{j\,(\phi_{1r} + \cdots \phi_{nr} - \phi_{ar} \cdots)}$$

$K_r$ is thus known in magnitude as well as phase

The value of $K_r$ can also be calculated graphically with the following steps:

(*a*) Plot the poles and zeros of $\dfrac{p(s)}{q(s)}$ to scale on the complex $s$-plane.

(*b*) Measure the distance from each of the other finite poles and zeros to a given pole $s_r$.

(*c*) Measure the angle from each of the other finite poles and zeros to a given pole $s_r$.

(*d*) Substitute these quantities into the equation above and calculate $K_r$.

The following example illustrates the procedure.

**Example 8.30:** Suppose $\dfrac{p(s)}{q(s)} = \dfrac{5s}{(s+1)(s+4)}$

Since the poles are real, a simple procedure follows:

$$K_1 = \frac{5s}{(s+1)(s+4)} (s+1)\Big|_{s=-1}$$

$$= -\frac{5}{3}$$

$$K_2 = \frac{5s}{(s+1)(s+4)} (s+4)\Big|_{s=-4}$$

$$= \frac{20}{3}$$

Therefore, $\qquad i(t) = \dfrac{20}{3} e^{-4t} - \dfrac{5}{3} e^{-t}$

However using the procedure (rigorous) as outlined here we proceed as follows:
Here $H = 5$

and $\qquad M_{01} e^{j\phi_{01}} = 1\, e^{j\,180}$

$$M_{41} e^{j\phi_{41}} = 3\, e^{j\,0}$$

$$K_1 = H\frac{M_{01}}{M_{41}} e^{j(\phi_{01} - \phi_{41})}$$

$$= 5 \times \frac{1}{3} e^{j\,180} = -\frac{5}{3}$$

$$K_2 = H\frac{M_{04}}{M_{14}} e^{j(\phi_{04} - \phi_{41})} = 5 \times \frac{4}{3} e^{j(180-180)} = \frac{20}{3}$$

Therefore, $\qquad i(t) = \dfrac{20}{3} e^{-4t} - \dfrac{5}{3} e^{-t}$

Let us consider the effect of a zero on the performance of the network, consider a pole $S_r$. If all other poles and zeros remain fixed to their position except $S_n$ which could be shifted. The closer is the zero $S_n$ to $S_r$ the smaller will be the co-efficient $K_r$. In fact when $S_n$ coincide with $S_r$, $K_r$ reduces to zero.

However, if a pole $S_m$ is moved closer to $S_r$, the co-efficient $K_r$ increases in magnitude.

Therefore, if in the design of a network, poles and zeros are to be selected following points should be considered.

(*i*) Poles should be so selected that these give a desired time response.

(*ii*) The zeros should be so selected to adjust the magnitude of the various $K$ co-efficients.

We know that all networks with R,L,C as its elements, its function will always have its poles and zeros in the left half of the $s$-plane and hence their response is stable. However active networks for example-containing controlled or dependent sources are not necessarily stable. This is illustrated by the following Example.

**Example 8.31:** Consider the network in Fig. E8.31

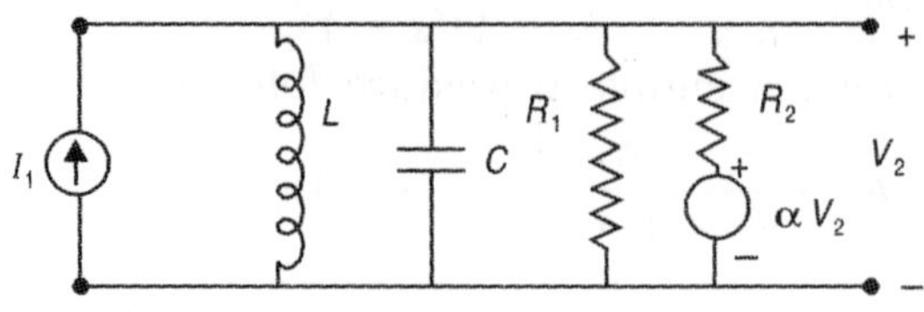

**Fig. E8.31**

From the network

$$I_1 = V_2 \left[ \frac{1}{sL} + sC + \frac{1}{R_1} \right] + \frac{V_2 - aV_2}{R_2}$$

$$= V_2 \left[ \frac{1}{sL} + sC + \frac{1}{R_1} + \frac{1}{R_2} - \frac{a}{R_2} \right]$$

or

$$\frac{V_2}{I_1} = \frac{1}{\dfrac{1}{sL} + sC + \dfrac{1}{R_1} + \dfrac{1}{R_2}(1 - a)}$$

$$= \frac{s/C}{s^2 + \dfrac{s}{C}\left[ \dfrac{1}{R_1} + \dfrac{1}{R_2}(1 - a) \right] + \dfrac{1}{LC}}$$

Let

$$R_1 = \frac{1}{2}\ \Omega,\ R_2 = 1\ \Omega,\ L = \frac{1}{2}\ H \text{ and } C = 1\ F$$

The above equation becomes

$$\frac{V_2}{I_1} = \frac{s}{s^2 + s\left[2 + (1 - a)\right] + 2}$$

$$= \frac{s}{s^2 + (3 - a)s + 2}$$

It can be seen that depending upon what value is given to $a$, the roots of the denominator will have their locations in the right half or left half of the $s$-place. Suppose $a = 0$ i.e., there is no dependent source i.e., there is no active source, it can be seen that the poles are

$$(s + 2)\ (s + 1) = 0 \quad \text{or} \quad s = -2,\ s = -1$$

which is the expected result.

In general, the roots are given by

$$s_1,\ s_2 = \frac{-(3 - a) \pm \sqrt{(3 - a)^2 - 8}}{2}$$

$$= \frac{-(3 - a) \pm \sqrt{a^2 - 6a + 1}}{2}$$

For the roots to be real $a^2 - 6a + 1 = 0$

or

$$a_1,\ a_2 = 3 \pm 2\sqrt{2}$$

This shows that the roots are located on the negative real axis at $a = 0$ and as '$a$' increases the poles move towards each other meeting at $s = -\sqrt{2}$ as shown in Fig. 8.29. Thereafter the locus is a circle for the range of values,

$3 - 2\sqrt{2} \leq a \leq 3 + 2\sqrt{2}$ as shown in Fig 8.29.

However when $a = 3$ $s_1, s_2 = \pm j \sqrt{2}$

For $a > 3 + 2\sqrt{2}$, the poles are again on the real axis but in the right half of the $s$-plane, one moving towards zero and the other towards infinity.

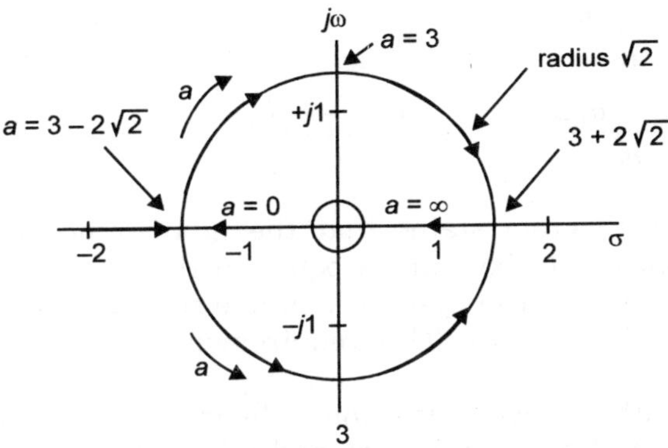


**Fig. 8.29**  Plotting poles of example 8.31.

We now define a system to be stable if the poles of the transfer functions lie on the left half of the $s$-plane and the imaginary axis and strictly stable if the poles lie in the left half of the $s$-plane only. Thus an active network is stable if its output oscillates with constant magnitude when the poles are on the imaginary axis. Similarly an active network is said to be strictly stable if its poles lie in the left half of the $s$-plane i.e. poles on the imaginary axis are excluded. This statement is equivalent to saying that a system is said to be stable if for a bounded input there is bounded output which means if a step input is given, the output term should not contain terms like, $t$, $t^2$, $e^t$, $t \sin t$ etc.

In regard to the above network, it is said to be stable if $a \leq 3$, strictly stable if $a < 3$ and when $a > 3$ the system is unstable as the output becomes unbounded with time.

Since the stability of the network is determined by the locations of poles in the $s$-plane, a criterion to determine the stability of a system under the name Routh-Hurwitz criterion is described here briefly. This criterion determines the absolute stability of the system and not the relative stability.

Suppose the denominator of the transfer function is expressed as

$$p(s) = a_0 \, s^n + a_1 \, s^{n-1} + a_2 \, s^{n-2} + \ldots\ldots + a_{n-1} + a_n$$

We separate $p(s)$ into its even and odd parts and form two rows of co-efficients. The first row consists of co-efficients $a_0 \, a_2 \ldots$ and the second row those of $a_1 \, a_3 \ldots$ .

$$
\begin{array}{lll}
a_0 & a_2 & a_4 \ldots \\
a_1 & a_3 & a_5 \ldots
\end{array}
$$

As the next step, complete the Routh-Hurwitz array of numbers. For $n = 6$ this is,

$$
\begin{array}{c|cccc}
s^6 & a_0 & a_2 & a_4 & a_6 \\
s^5 & a_1 & a_3 & a_5 &
\end{array}
$$

$$
\begin{array}{c|ccc}
s^4 & b_1 & b_3 & b_5 \\
s^3 & c_1 & c_3 & \\
s^2 & d_1 & d_3 & \\
s^1 & e_1 & & \\
s^0 & f_1 & &
\end{array}
$$

consisting of $n + 1$ rows. Here $\quad b_1 = \dfrac{a_1 a_2 - a_0 a_3}{a_1}$

$$
b_3 = \frac{a_1 a_4 - a_0 a_5}{a_1}, \quad b_5 = a_6
$$

Similarly, $c_1 = \dfrac{b_1 a_3 - a_1 b_3}{b_1} \qquad c_3 = \dfrac{b_1 a_5 - a_1 b_5}{b_1}$

and so on.

The Routh Hurwitz theorem states that the number of changes in sign of the first column (as we scan from top to bottom) is equal to the number of roots of $p(s) = 0$ with positive real parts. That means the system is stable if and only if there are no changes of sign in the first column of the array. This requirement is both necessary and sufficient for stability of the system.

While applying Routh Hurwitz criterion, following two situations may arise:

1. The first element of any row is zero and all the elements in that row are non-zero.
2. All the elements in a row are zero.

($i$) If a zero occurs in the first column of a row, the elements in the next row become infinite and hence Routh's test breaks down. To take into account such a situation the original polynomial is multiplied by a factor $(s + a)$ where $a$ is any positive real number, and carry out the usual Routh's test.
Consider the polynomial
$s^4 + s^3 + 2s^2 + 2s + 3$. Arranging as per the Routh's arrays we have

$$
\begin{array}{c|ccc}
s^4 & 1 & 2 & 3 \\
s^3 & 1 & 2 & \\
s^2 & 0 & 3 & \\
s^1 & & & \\
s^0 & & &
\end{array}
$$

This shows in the third row the first element is zero and the other element is non zero. Therefore we multiply with a factor $(s + a)$ where let $a = 3$, we have the new polynomial as

$$
(s + 3)\,(s^4 + s^3 + 2s + 2s + 3)
$$
$$
= s^5 + 4s^4 + 5s^3 + 8s^2 + 9s + 9
$$

The array is

$$
\begin{array}{c|ccc}
s^5 & 1 & 5 & 9 \\
s^4 & 4 & 8 & 9 \\
s^3 & 3 & \dfrac{27}{4} & \\
s^2 & -1 & 9 & \\
s^1 & \dfrac{27 \times 5}{4} & & \\
s^0 & 9 & &
\end{array}
$$

Since there are two changes in sign in the first column of the Routh's tabulation, two roots of the polynomial are in the right half of the $s$-plane.

2.  The second situation when all elements in a Routh's tabulation row are zero. This situation arises when there are pair of roots with opposite sign, the roots could be real or imaginary or both or conjugate roots forming a quadrate in the $s$-plane. The equation corresponding to the co-efficient just above the row of zeros is called the auxiliary equation. The order of the equation is always even and, therefore, the number of root pairs are equal in magnitude and opposite in sign. Suppose the auxiliary equation is of second order, this will have two equal and opposite roots. For a fourth order auxiliary equation, there are two pairs of equal and opposite roots. Here again Routh's Hurwitz criterion breaks down because of row of zeros. To overcome this situation, the derivative of the auxiliary equation is taken with reference to $s$ and the row of the zeros is replaced by the co-efficients of the resultant equation derived by taking the derivatives of the auxiliary equation. To illustrate further we take the following polynomial.

**Example 8.32:**

$$p(s) = s^6 + 5s^5 + 11s^4 + 25s^3 + 36s^2 + 30s + 36$$

The Routh's Hurwitz array is

| | | | | |
|---|---|---|---|---|
| $s^6$ | 1 | 11 | 36 | 36 |
| $s^5$ | 5 | 25 | 30 | |
| $s^4$ | 6 | 30 | 36 | |
| $s^3$ | 0 | 0 | 0 | |
| | 24 | 60 | 0 | |
| $s^2$ | 15 | 36 | 0 | |
| $s^1$ | $\dfrac{12}{5}$ | 0 | | |
| $s^0$ | 36 | | | |

The row corresponding to $s^3$ has zeros. Therefore, the auxiliary equation just above the row of zeros is

$$6s^4 + 30s^2 + 36$$
$$= 6\,(s^4 + 5s^2 + 6)$$

The derivative of this is

$$24s^3 + 60s$$

Therefore, the array will be as indicated below zeros of the original array. Since there is no change in sign of the elements in the first Column the system has no root in the right half of the $s$-plane and hence the system is stable.

Next consider another polynomial

$$p(s) = s^6 + s^5 - 2s^4 - 3s^3 - 7s^2 - 4s - 4$$

The Routh Hurwitz array is as follows:

| | | | | |
|---|---|---|---|---|
| $s^6$ | 1 | −2 | −7 | −4 |
| $s^5$ | 1 | −3 | −4 | 0 |
| $s^4$ | 1 | −3 | −4 | |
| $s^3$ | 0(4) | 0(−6) | 0 | |

$$s^2 \qquad -\dfrac{3}{2} \qquad -4$$

$$s^1 \qquad -\dfrac{50}{3} \qquad 0$$

$$s^0 \qquad -4$$

Therefore, the elements corresponding to row $s^3$ are zero and hence the auxiliary equation is $s^4 - 3s^2 - 4 = 0$ and its derivative with reference to $s$ is $4s^3 - 6s$ and therefore, the co-efficient against $s^3$ row are as shown above and the table is completed.

Since there is only one change in sign in the elements of first column, there is one root in the right half of the $s$-plane. The two pairs of roots can be obtained by solving the auxiliary equation.

$$s^4 - 3s^2 - 4 = 0$$

$$s^3 (s - 2) + 2s^2 (s - 2) + s (s - 2) + 2 (s - 2)$$

$$(s - 2) (s^3 + 2s^2 + s + 2) = 0$$

$$(s - 2) \{s^2 (s + 2) + 1 (s + 2)\} = 0$$

$$(s - 2) (s + 2) (s^2 + 1) = 0$$

Therefore the two pairs of roots are

$s = \pm 2$ and $s = \pm j$ and the root in the right half of the $s$-plane corresponds to $s = 2$.

**PROBLEMS**

**8.1.** For the two port $RC$ network shown here determine $G_{12}$.

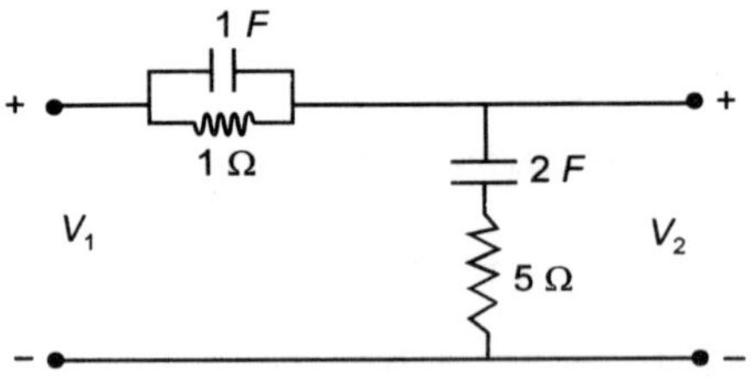

**Fig. P8.1**

**8.2.** For the resistive two-port network shown in the Fig determine (a) $G_{12}$ (b) $Z_{12}$ (c) $Y_{12}$ and (d) $\alpha_{12}$.

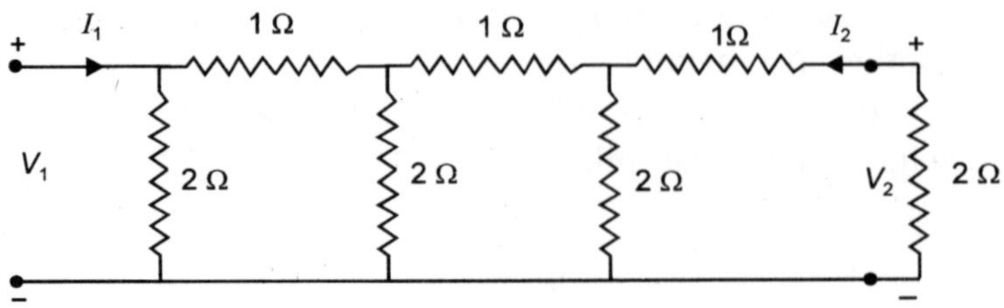

**Fig. P8.2**

**8.3.** For the network shown determine $z_{11}$ and $y_{11}$.

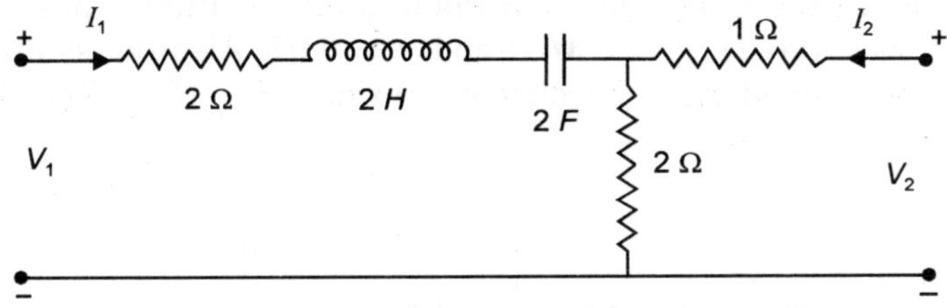

**Fig. P8.3**

**8.4.** Find the $y$ and $z$ parameters for the resistive network of the Fig. P8.4.

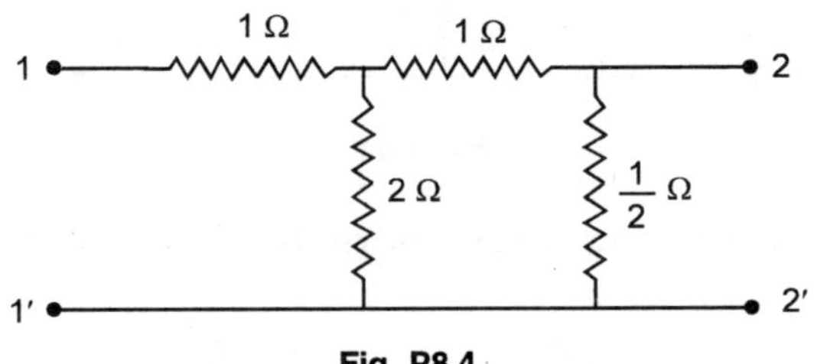

**Fig. P8.4**

**8.5.** For the network containing a current controlled current source shown in Fig. P8.5, determine the $y$ and $z$ parameters.

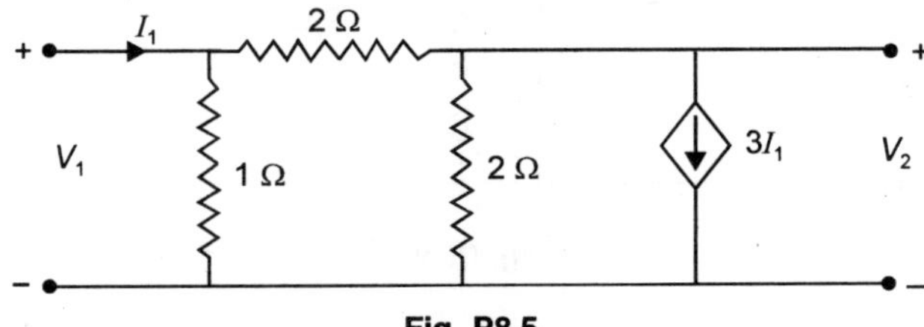

**Fig. P8.5**

**8.6.** For the network shown determine $y_{11}$ and $y_{22}$.

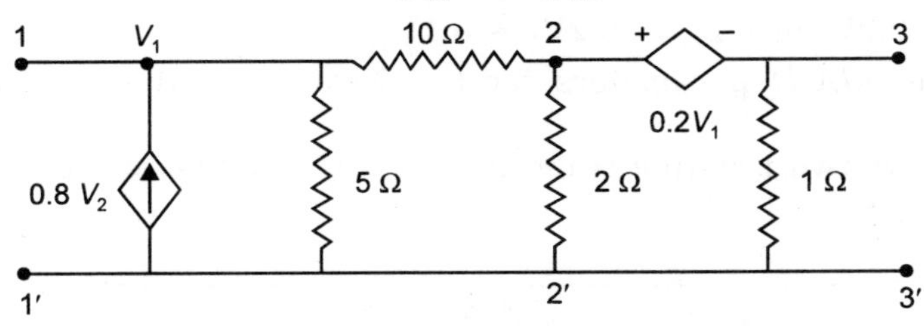

**Fig. P8.6**

**8.7.** For the network shown determine $y$ parameters and the output impedance.

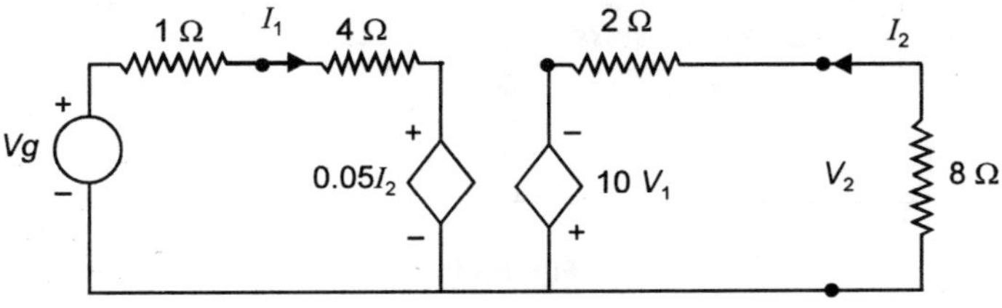

**Fig. P8.7**

**8.8.** A common base amplifier is represented by the low frequency $T$ equivalent circuit as shown in Fig. P8.8. Determine the $z$-parameters and evaluate these values for the case when $r_e = 20\ \Omega$, $r_b = 700\ \Omega$, $r_c = 1\ M\ \Omega$ and $\alpha = 0.97$. Also determine the voltage gain, current gain, input impedance and output impedance if $Z_s = 500\ \Omega$ and $Z_L = 3\ K\ \Omega$.

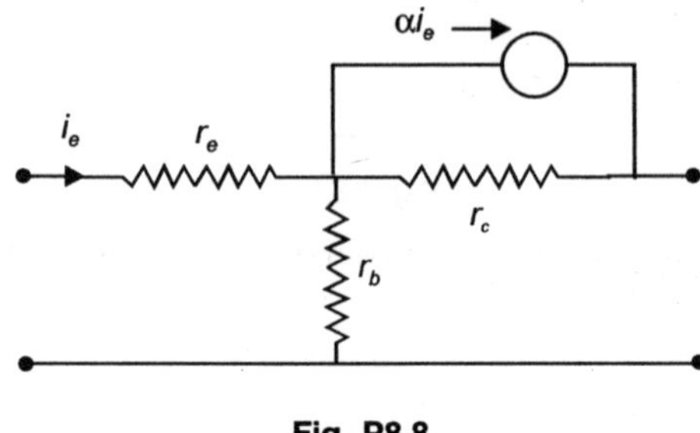

**Fig. P8.8**

**8.9.** For the network shown in Fig. P8.9 determine the $y$-parameters.

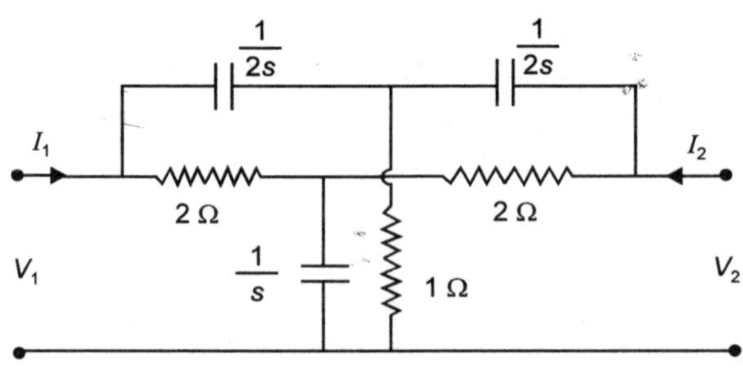

**Fig. P8.9**

**8.10.** The network equations for a two port network give the current $I_1$ and $I_2$ at the two ports as

$I_1 = 0.25V_1 - 0.2V_2$ and $I_2 = -0.2V_1 + 0.1V_2$

Determine the *ABCD* parameters for the Network and Hence write the network equation.

**8.11.** Fig. P8.11 shows two isolating transformers with passive elements. Determine the $z$-parameters.

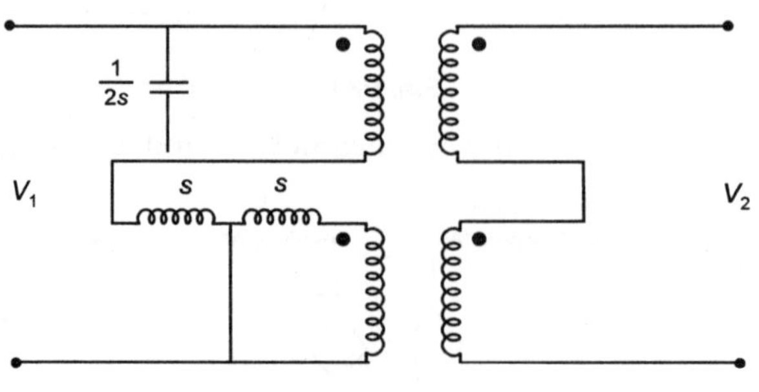

**Fig. P8.11**

**8.12.** For the network shown in Fig. P8.12 determine $y$-parameters.

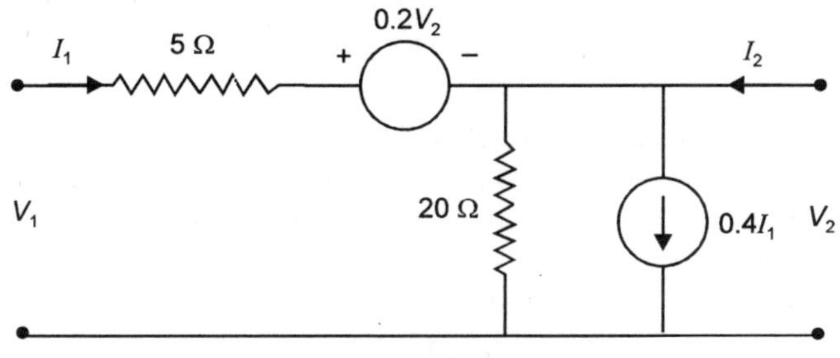

**Fig. P8.12**

**8.13.** The $z$-parameters of a certain two port network are $Z_{11} = 5\ \Omega$, $z_{12} = Z_{21} = 3\ \Omega$ and $z_{22} = 4\ \Omega$. Determine its $y$-parameters, transmission parameters, inverse transmission parameters and hybrid parameters $(h)$ and inverse hybrid parameters $(g)$.

**8.14.** A two port network is characterized by the following equations:

$$4V_1 + 2I_1 + 3I_2 = 0$$

and $$10I_1 + 10I_2 - 2V_1 - 3V_2 = 0$$

Write the equation in standard $y$-parameters form and hence determine $y$-parameters.

**8.15.** For the network shown determine the $z$-parameters.

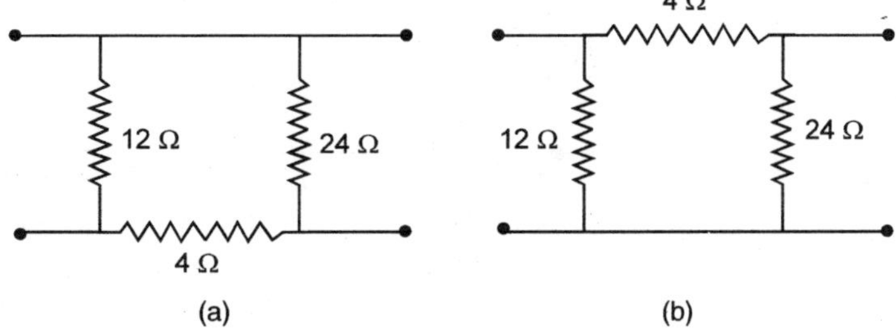

(a)  (b)

**Fig. P8.15**

**8.16.** For the network shown in Fig. P8.16 determine the $z$-parameters.

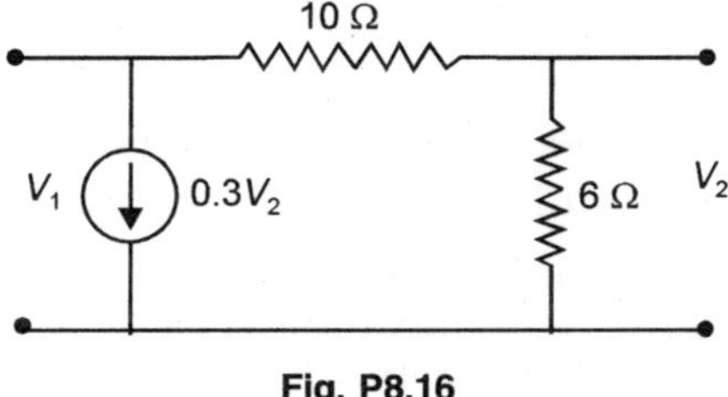

**Fig. P8.16**

**8.17.** If $y = \begin{bmatrix} 10 & -2 \\ 100 & 20 \end{bmatrix}$ all values in m-mhos determine $z$. Also

if $z = \begin{bmatrix} 8 & 6 \\ 2 & 4 \end{bmatrix}$ determine $y$.

**8.18.** Find  $h$-parameters if

$$(a)\ y = \begin{bmatrix} 10 & -2 \\ 100 & 20 \end{bmatrix} \qquad\qquad (b)\ z = \begin{bmatrix} 8 & 6 \\ 2 & 4 \end{bmatrix}$$

**8.19.** Obtain the input and output impedances of an amplifier having $h_{11} = 2\ \Omega$, $h_{12} = 1$ $h_{21} = 5$ and $h_{22} = 2\ \mho$, if it is driven by a source having an internal resistance of $4\ \Omega$ and is terminated through a load which draws maximum power from the amplifier.

**8.20.** A typical two-port network is characterized by the equation

$$2V_1 + 4I_2 = I_1 \quad \text{and} \quad V_2 + 6V_1 = 8I_2$$

Determine the values of ($i$) $y_{11}$ ($ii$) $z_{21}$ ($iii$) $h_{21}$

**8.21.** Determine the $z$, $y$, $h$-parameters and transmission parameters of the network shown.

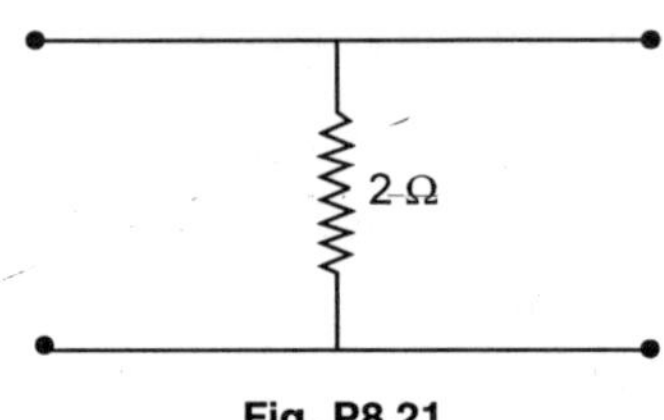

**Fig. P8.21**

**8.22.** For the network shown in Fig. P8.22 determine the overall parameters if the  $z$-parameters of the two individual networks are

$$Zn_1 = \begin{bmatrix} 1 & 3 \\ 3 & 4 \end{bmatrix} \quad Zn_2 = \begin{bmatrix} 5 & 7 \\ 7 & 8 \end{bmatrix}$$

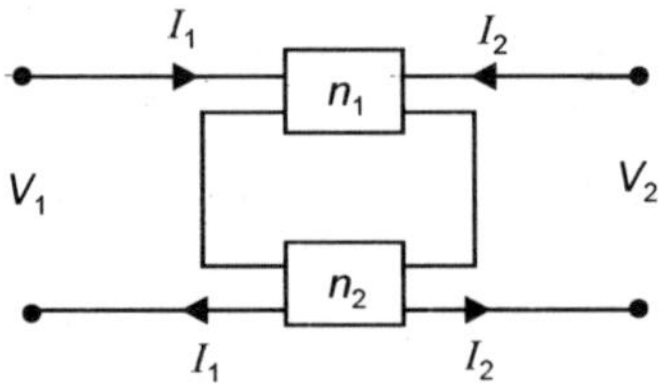

**Fig. P8.22**

**8.23.** Determine the input impedance $Z$ of a two port network, if a load resistance of $5\ \Omega$ is connected across its output port. The $z$-parameters of the network are

$$Z_{11} = 5\ \Omega,\ Z_{12} = Z_{21} = 3\ \Omega,\ Z_{22} = 2\ \Omega$$

**8.24.** Fig. P8.24 shows a two port network $N$ having the $h$-parameters as

$$h_{11} = 1\ k\Omega,\quad h_{12} = 0.0015,\quad h_{21} = 100,\quad h_{22} = 100\ \mu\mho$$

($a$) Determine the output voltage $V_2$ if load resistance is $10\ k\Omega$

($b$) Determine load resistance for maximum power transfer

($c$) Determine z-parameters of the network $N$.

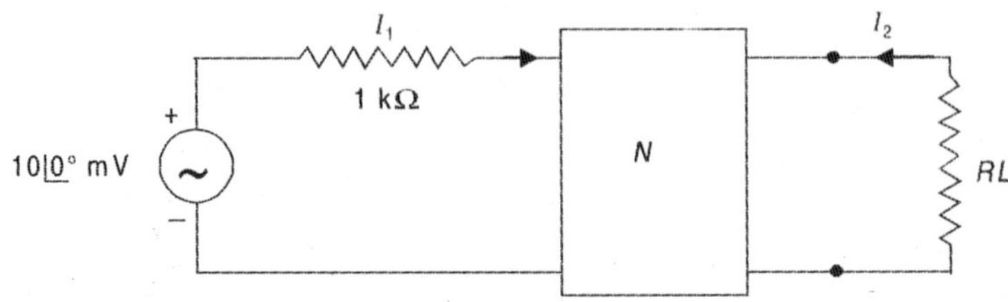

**Fig. P8.24**

**8.25.** The linear network shown in Fig. P8.25 has only resistors. If $I_1 = 10A$ and $I_2 = 15A$, $V$ is found to be 100 volt. However, if $I_1 = -10A$ and $I_2 = 5A$ $V = 0$. Determine $V$ if $I_1 = 6A$ and $I_2 = 10\,A$.

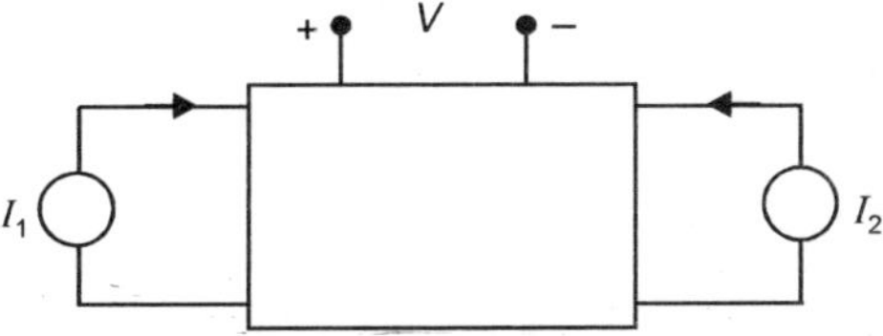

**Fig. P8.25**

**8.26.** The series combination of 2mh inductor and 10 $\Omega$ resistor is in parallel with series combination of 5 $\mu F$ capacitor and 10 $\Omega$ resistor. Determine $Z_{in}(s)$ and identify all its critical frequencies.

**8.27.** Fig P8.27 shows three pole zero configurations each representing admittance. Determine expression for each admittance in $s$-domain.

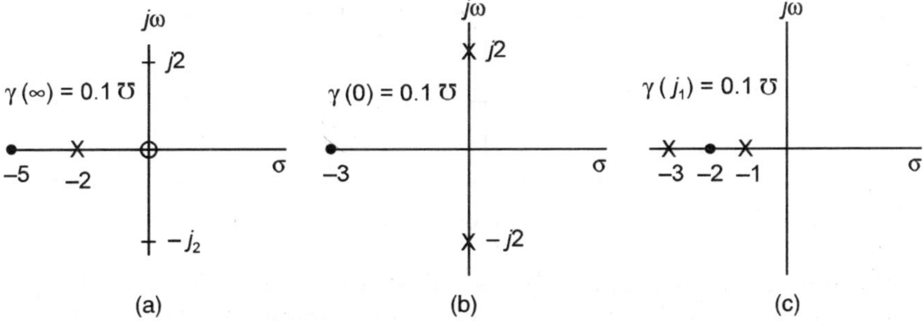

**Fig. P8.27**  Pole-zero configuration of admittances.

**8.28.** Calculate the damping ratio and the undamped natural frequencies of oscillation of a network whose transfer function is given as

$$\frac{s}{s^2 + 3s + 6}$$

**8.29.** A two terminal network consists of a coil with resistance $R$ and inductance $L$ Henries and it is shunted by a capacitor $C$. The poles and zero of the driving point impedance function $z(s)$ are poles $-\dfrac{1}{2} \pm j\dfrac{\sqrt{3}}{2}$, zero at $-1$

If $z(jo) = -1$, determine the values of $R$, $L$ and $C$.

**8.30.** Given the equation
$$s^3 + 5s^2 + ks + 1 = 0$$
Determine the range of values of $K$ for which the roots of the equation have negative real parts. Also determine the value of $K$ for which real part vanishes.

**8.31.** For the following equations use Routh Hurwitz criterion to determine $(a)$ the number of roots with positive real parts $(b)$ the number of roots with zero real parts and $(b)$ the number of roots with negative real parts.

$(i)$ $4s^3 + 7s^2 + 7s + 2 = 0$

$(ii)$ $s^3 + 3s^2 + 4s + 1 = 0$

$(iii)$ $5s^3 + s^2 + 6s + 2 = 0$

$(iv)$ $s^5 + 2s^4 + 2s^3 + 4s^2 + 11s + 10 = 0$

**8.32.** Determine whether following functions represent driving point impedance functions:

$(i)$ $\quad z(s) = \dfrac{s + 1}{(s + 2)(s + 3)}$

$(ii)$ $\quad z(s) = \dfrac{s^2 + s - 1}{s^3 + s^2 + s - 1}$

$(iii)$ $\quad z(s) = \dfrac{(s^2 + 1)^2}{(s + 1)(s - 1)}$

$(iv)$ $\quad z(s) = \dfrac{s^4 + s^2}{s^4 + 3s^3 + 2s^2 + 5s + 2}$

**8.33.** For the network shown determine the value of $K$ for which the network is stable.

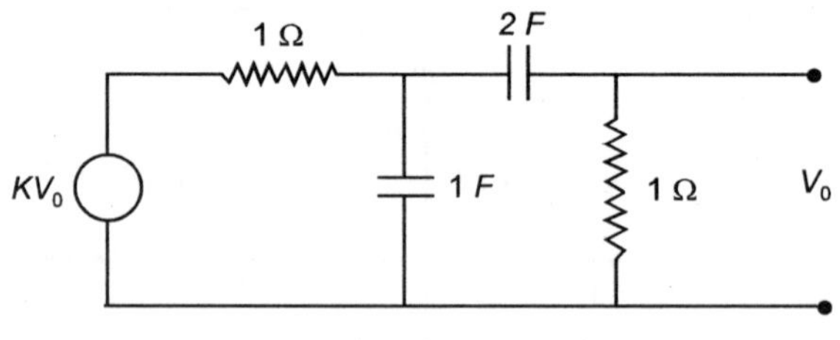


**Fig. P8.33**

**8.34.** A plaster model of an impedance shows poles at $s = -1 \pm j5$ and zeros at $s = -3 \pm j4$. If the height of the model is 8 cm at the origin; determine the height at $s = j1$, $s = 3 + j1$ and $s = \infty$.

# 9

# Resonance

## 9.1 INTRODUCTION

Whenever the natural frequency of oscillation of a system (could be electrical, mechanical or a civil structure or a hydraulic) coincides with the frequency of the driving force (a voltage source in an electric circuit or a wind force in a civil structure etc.), the two system resonate with respect to each other and the system has maximum response to a fixed magnitude of driving force. This phenomenon is known as resonance. This phenomenon may be useful under certain conditions and sometimes it may prove to be disastrous for the system.

There are many engineering applications of resonance. In a refrigerator, the reciprocating compressor is mounted on a support designed to minimize the vibrations transmitted to the cabinet. A suspension bridge in Washington showed tendencies to oscillate up and down during construction and only a few months after construction, it began to build up oscillations under a moderate wind and within an hour the multi-billion dollar bridge was reduced to pieces. This is a typical example of designing a bridge ignoring the possibility of phenomenon of resonance on the bridges.

In case of a series compensated power system if the difference between the normal power frequency and the natural frequency of oscillation of the system after compensation coincides with one of the natural torsional frequencies of the machine's shaft system (Turbine, or combination of turbine and generator), torsional oscillation may be excited which may build up sufficiently to break the shaft of these machines. This is known as subsynchronous resonance as the resonance occurs at a frequency less than the normal supply frequency.

A power transformer is normally operated at the knee-point of its B-H curve due to economic reasons. If some how the system voltage increases due to Ferranti effect or due to rejection of load, the transformer operates along the saturation region of the B-H curve where the inductance offered by the winding decreases and keeps on changing depending upon the dynamic behaviour of increase of system voltage. Saturation of transformers subjected to over voltages can produce high currents rich in harmonics and in the presence of sufficient capacitance there is risk of phenomenon of resonance taking place which may result in further increase in system voltages which may prove disastrous for the system insulation. This is known as Ferro-resonance as it is due to the non-linear behaviour of the iron core of the transformer. In such a situation it may be necessary to disconnect shunt capacitors very quickly to reduce the chances of occurrences of Ferro-resonance.

In an electric circuit with inductance and capacitor in series, there is always a frequency at which the two reactances just cancel resulting in the minimum impedance (resistive circuit) characteristic of series resonance. In electronic circuits many a times resonance condition is desired for maximum response for a given magnitude of excitation.

In general there are two types of resonance in the electric circuits:

(*a*) Series resonance,

(*b*) Parallel resonance.

We now study basic features and characteristic of these phenomena.

## 9.2  SERIES RESONANCE

Consider a series RLC circuit connected to a variable frequency voltage source as shown in Fig. 9.1.

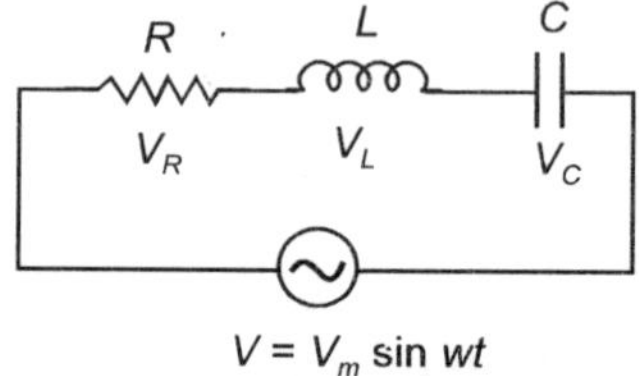



**Fig. 9.1**  Series RLC circuit with a.c. source.

The impedance of the circuit at any frequency $\omega$ is given as

$$Z = R + j\omega L - \frac{j}{\omega C}$$

$$= R + j\left(\omega L - \frac{1}{\omega C}\right) \qquad \qquad ...(9.1)$$

Since resistance is independent of frequency the circuit will have minimum impedance at some frequency.

When
$$Z = R \qquad \qquad ...(9.2)$$

or when
$$\omega_0 L - \frac{1}{\omega_0 C} = 0$$

or
$$\omega_0 = \frac{1}{\sqrt{LC}}$$

$$f_0 = \frac{1}{2\pi \sqrt{LC}} \qquad \qquad ...(9.3)$$

Here $f_0$ is the frequency of resonance i.e. if a circuit has fixed values of $R$, $L$ and $C$, resonance will take place if the supply frequency

$f_0 = \dfrac{1}{2\pi \sqrt{LC}}$ when the impedance of the circuit is purely resistive i.e. the power factor of the circuit is unity, the supply voltage and current are in phase. However, it is to be noted the phase relations between the voltage and current in the individual elements $R$, $L$ and $C$ are not same. The current in the inductor lags its voltage by 90º and in the capacitor it leads its voltage by 90º.

The variation of impedance of the three elements as a function of frequency is shown in Fig. 9.2.

Since $R$ is independent of frequency it is shown by a horizontal line $Z = R$.

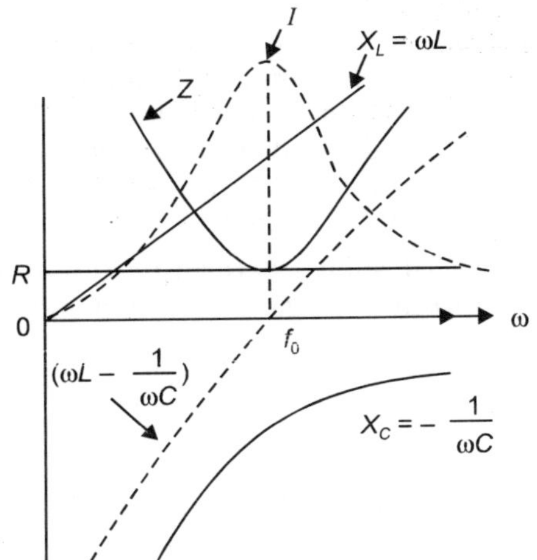

**Fig. 9.2**  $ZV_s$ $\omega$ for RLC series circuit.

Also $X_L = \omega L$ the inductive reactance is a straight line passing through the origin and inductive reactance is taken as +ve, where as $X_C = \dfrac{1}{\omega C}$ the capacitive reactance as a function of $\omega$ is a rectangular hyperbola and the reactance is taken as –ve. The net impedance is shown a positive quantity. The resonance frequency is $f_0$ where $\left(\omega L - \dfrac{1}{\omega C}\right)$ is zero and at this frequency the impedance curve has minimum value equal to $R$. The variation of current is also shown in Fig. 9.2 as a function of frequency and is maximum at $f_0$ whereas on either side the current decreases. It is to be noted that at $\omega = 0$, the current in the $RLC$ series circuit is zero as the capacitor reactance is infinite and, therefore, the graph starts from origin whereas it is again zero at $\omega = \infty$ and hence the graph should not be passing through zero rather it should have some finite value as indicated in the diagram.

Again, it can be seen that the series circuit is capacitive for all frequency $\omega < \omega_0$ and at $\omega = \omega_0$ the circuit is resistive. Fig. 9.3 shows the current response of the $RLC$ series circuit for certain voltage $V$ when one of the three parameters is changed at a time.

It is seen that the current response curve must always start from origin and the frequency of resonance can be varied by varying either $L$, or $C$ or both. The steepness of the response curve can be varied by varying the resistance of the circuit.

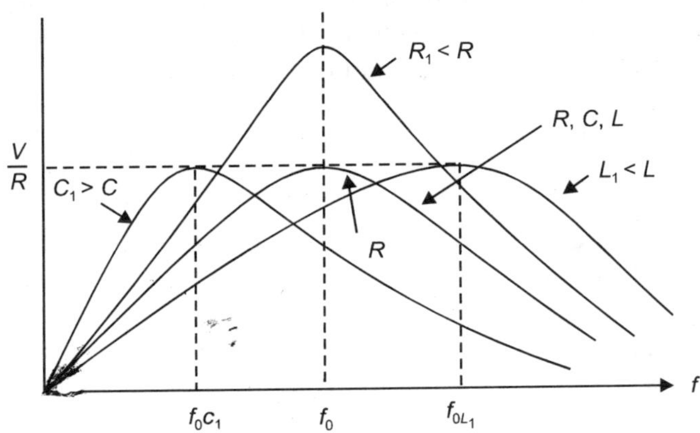

**Fig. 9.3**  Current response for $RLC$ series circuit.

There are various applications of a series resonant circuit where the frequency is fixed and either $L$ or $C$ is varied to obtain the condition of resonance. A typical example is that of tuning a radio receiver to a particular desired station that is operating at a fixed frequency. Here a circuit or $L$ and $C$ is adjusted to resonance at the operating frequency of the desired station. The capacitor $C$ (parallel plate) is variable in most portable radio receivers and the inductance of the coil is usually varied in tuning of an automobile radio receiver.

**Example 9.1:** For the Fig. E9.1 shown determine the maximum current, the frequency at which it occurs and the resulting voltage across the inductance and capacitance.

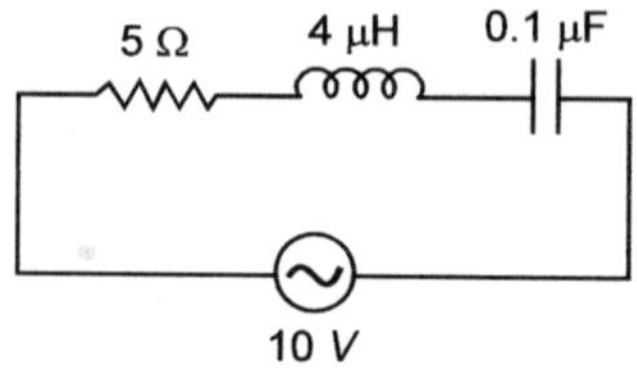


**Fig. E9.1**

**Solution:**  The frequency at which the current is maximum is

$$\omega_0 = \frac{1}{\sqrt{4 \times 10^{-3} \times 0.1 \times 10^{-6}}}$$

$$= 5 \times 10^4 \text{ rad/sec.}$$

The current at resonance is $\dfrac{V}{R} = \dfrac{10}{5} = 2 \text{ A}$

The voltage across the inductance

$$= \omega_0 LI = 5 \times 10^4 \times 4 \times 10^{-3} \times 2$$

$$= 400 \ V$$

and voltage magnitude across the capacitor is

$$= \frac{I}{\omega_0 C} = \frac{2}{5 \times 10^4 \times 0.1 \times 10^{-6}}$$

$$= 400 \ V$$

It is seen that the voltage across $L$ or $C$ under resonance is greater than the voltage across all the three elements i.e. the supply voltage.

Fig. E9.1.(a) shows phasor diagram of the $RLC$ series circuit corresponding to $\omega < \omega_0$, $\omega = \omega_0$ and $\omega > \omega_0$. We have taken current as the reference phasor as it is a series circuit and current is same through all the elements.

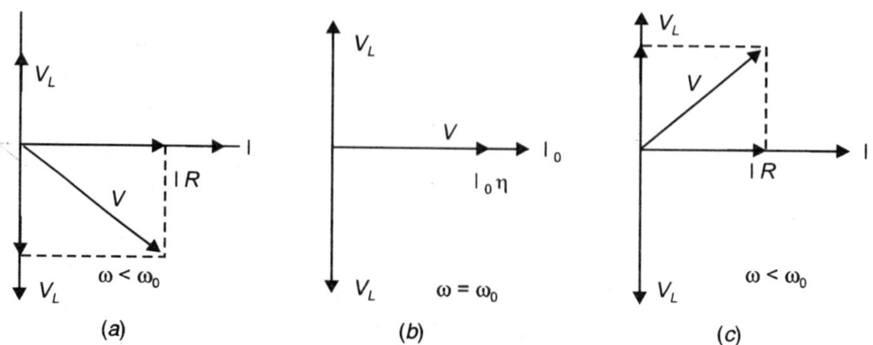


**Fig E9.1**   Phasor diagram of $RLC$ series circuit (a) $\omega < \omega_0$ (b) $\omega = \omega_0$ (c) $\omega > \omega_0$

In Fig. E9.1($a$) since $\omega < \omega_0$, the circuit is predominantly capacitive hence the net supply voltage lags behind the circuit current. However the voltage across the inductor leads the current by 90°, the voltage across the resistor is in phase with the circuit current whereas voltage across the capacitor lags behind the current by 90°. The voltage across inductor and capacitor are 180° out of phase with reference to each other. Since the circuit at this frequency is capacitive the net voltage lags behind the circuit current by some angle.

In Fig. E9.1($b$) since $\omega = \omega_0$ the circuit is resistive as the voltage drop across inductance exactly cancels the voltage across the capacitance as these are equal in magnitude but opposite in phase relation. It is to be seen that under this condition voltage across $L$ or $C$ individually may far exceed the supply voltage.

Next we find expression for voltage drops across individual elements under resonance condition.

Under resonance

$$I_0 = \frac{V}{R} \qquad \qquad \text{...(9.4)}$$

Therefore,

$$V_R = I_0 R = \frac{V}{R} \cdot R = V \text{ the supply voltage} \qquad \text{...(9.5)}$$

$$V_L = I_0 \omega L = \frac{V}{R} \omega_0 L = V \cdot \frac{\omega_0 L}{R}$$

$$= VQ \qquad \qquad \text{...(9.6)}$$

where $Q$ is known as the quality factor of the series network. Usually $Q \gg 1$, hence it is also known as voltage gain, as the voltage across the inductor is much greater than the supply voltage.

Also

$$V_C = \frac{I_0}{\omega_0 C} = V \cdot \frac{1}{\omega_0 CR}$$

$$= V Q \qquad \qquad \text{...(9.7)}$$

Hence

$$V_L = V_C \gg V$$

Therefore, extreme care must be taken when working on series circuits that may become resonant when connected to power line sources.

**Example 9.2:** In an *RLC* series circuit shown in Fig. E9.2 determine ($i$) the necessary value of capacitor ($ii$) the supply voltage to produce a voltage of 5 volts across the capacitance if resonance frequency is 5 KHz ($iii$) If the capacitance is made 1/2 of at ($i$), determine the frequency of resonance, the $Q$ of the new circuit.

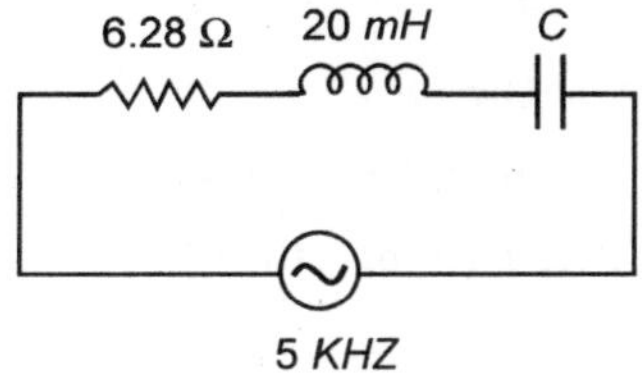


**Fig. E9.2**

**Solution:**

$$5000 = \frac{1}{2\pi \sqrt{20 \times 10^{-3} \times C \times 10^{-6}}}$$

$$C = 0.0507 \ \mu F$$

$$V_C \ = 5 = \ V \cdot \frac{1}{2\pi \times 5 \times 10^3 \times 0.0507 \times 10^{-6} \times 6.28}$$

or
$$V = 0.05 \ \text{volt}$$

(*iii*) If the capacitance is halved, the frequency of resonance would be $\sqrt{2} \ f_0$

$$= \sqrt{2} \times 5000 = 7071 \ \text{Hz}$$

and $Q$ of the coil $= \dfrac{2\pi \times 7071 \times 20 \times 10^{-3}}{6.28}$

$$= 141.4 \ \textbf{Ans.}$$

### 9.2.1   Quality Factor Q

In a practical circuit $R$ is essentially resistance of the coil since practical capacitors have very low loss in comparison to practical inductor. Hence $Q$ is a measure of the energy storage property ($L \ I^2$) in relation to the energy dissipation property ($I^2 \ R$) of a coil or a circuit. The $Q$ is, therefore, defined as

$$Q = 2\pi \cdot \frac{\text{maximum energy stored}}{\text{energy dissipated per cycle}} \qquad \text{...(9.8)}$$

In electric circuit energy is stored in the form of electromagnetic field in the inductance where as in electrostatic form of energy across a capacitance. It can be proved that at any instant at a certain frequency the sum of energy stored by the inductor and the capacitor is constant. At the extreme situation when the current through the inductor is maximum, the voltage across the capacitor is zero hence the total energy is

$$\frac{1}{2} L \left( \sqrt{2} \ I \right)^2 = L \ I^2 \qquad \text{....(9.9)}$$

where $\sqrt{2} \ I$ is the instantaneous maximum value of the current. At this since $V_C$ is zero, therefore maximum energy stored is $L \ I^2$. The power consumed per cycle is the energy per

sec divided by $f_0$ under resonance condition. Therefore $P_R = \dfrac{I^2 R}{f_0}$

Hence
$$Q = \frac{2\pi L I^2 f_0}{I^2 R} = \frac{\omega_0 L}{R} \qquad \text{...(9.10)}$$

$Q$ can also be looked as the ratio of

$$\frac{\text{Time rate of change of energy stored}}{\text{Time rate of change of energy dissipated}}$$

$$= \frac{\text{Reactive power absorbed by the inductor}}{\text{Active power consumed by the resistor}} \qquad \text{...(9.11)}$$

The great advantage of this definition of $Q$ is that it is also applicable to more complicated lumped circuits, to distributed circuits such as transmisson lines and to non-electrical circuits.

$Q$ is also a measure of the frequency selectivity of the circuit. A circuit with high $Q$ will have a very sharp current response curve as compared to one which has a low value of $Q$. To understand this let us consider Fig. 9.4. Here we find that the current response is maximum at $f_0$ and on either side of $f_0$, the current decreases sharply.

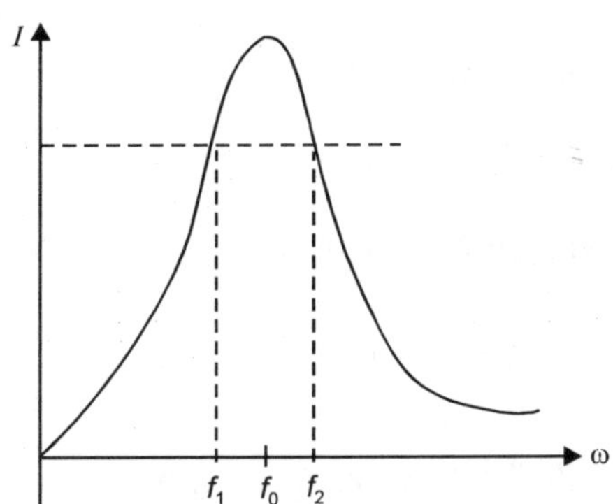

**Fig. 9.4**  Frequency selectivity.

In order to obtain quantitative analysis of this reduction in current, we specify two frequencies $f_1$ and $f_2$ at which the magnitude $(X_L - X_C)$ is equal to $R$.

Since at $f_1$ the circuit is capacitive $X_C$ is greater than $X_L$, therefore, at $f_1$,

$$X_C - X_L = R \qquad \qquad \text{...(9.12)}$$

and at $f_2$

$$X_L - X_C = R \qquad \qquad \text{...(9.13)}$$

The corresponding impedances are:

$$Z_1 = \sqrt{R^2 + R^2} \ \angle \tan^{-1} \frac{-R}{R} = \sqrt{2}\ R \ \angle -45^\circ \qquad \qquad \text{...(9.14)}$$

and

$$Z_2 = \sqrt{R^2 + R^2} \ \angle \tan^{-1} \frac{-R}{R} = \sqrt{2}\ R \ \angle 45^\circ \qquad \qquad \text{...(9.15)}$$

The current at these two frequencies are:

$$I_1 = \frac{V \ \angle 45^\circ}{\sqrt{2}\ R} \qquad \qquad \text{...(9.16)}$$

and

$$I_2 = \frac{V \ \angle -45^\circ}{\sqrt{2}\ R} \qquad \qquad \text{...(9.17)}$$

Since $\dfrac{V}{R} = I_0$ the current at resonance at $f_0$,

$$I_1 = 0.707\ I_0 \ \angle 45^\circ \qquad \qquad \text{...(9.18)}$$

and

$$I_2 = 0.707\ I_0 \ \angle -45^\circ \qquad \qquad \text{...(9.19)}$$

Frequencies $f_1$ and $f_2$ are known as half-power frequencies as at these frequencies the power dissipated by the circuit is half of that dissipated at $f_0$.

The band width of a resonant circuit is defined as the frequency range between the 70.7% current points.

i.e. 
$$BW = f_2 - f_1 \qquad \qquad \qquad ...(9.20)$$

At 
$$f_1 \quad X_C - X_L = R \qquad \qquad ...(9.21)$$

$$\frac{1}{2\pi f_1 C} - 2\pi f_1 L = R \qquad \qquad ...(9.22)$$

and at $f_2$

$$X_L - X_C = R \qquad \qquad ...(9.23)$$

$$2\pi f_2 L - \frac{1}{2\pi f_2 C} = R \qquad \qquad ...(9.24)$$

On subtracting (9.24) from (9.22) we have

$$\left( \frac{1}{\omega_1} + \frac{1}{\omega_2} \right) \frac{1}{C} - (\omega_1 + \omega_2) L = 0$$

or 
$$\frac{\omega_2 + \omega_1}{\omega_1 \omega_2} = LC (\omega_1 + \omega_2)$$

or 
$$\frac{1}{\omega_1 \omega_2} = LC = \frac{1}{\omega_0^2}$$

or 
$$\omega_0 = \sqrt{\omega_1 \omega_2} \qquad \qquad ...(9.25)$$

This means the resonance frequency is the geometric mean of the half-power frequencies.

Now on addition of (9.22) and (9.24) we have

$$\frac{1}{\omega_1 C} - \frac{1}{\omega_2 C} + \omega_2 L - \omega_1 L = 2R$$

or 
$$\frac{\omega_2 - \omega_1}{\omega_1 \omega_2 C} + (\omega_2 - \omega_1)L = 2R$$

or 
$$(\omega_2 - \omega_1)\left( \frac{1}{\omega_0^2 C} + L \right) = 2R$$

or 
$$(\omega_2 - \omega_1)\left( \frac{2}{\omega_0^2 C} \right) = 2R$$

or 
$$(\omega_2 - \omega_1) = \omega_0^2 CR$$

$$\frac{\omega_2 - \omega_1}{\omega_0} = \omega_0 CR = \frac{1}{Q}$$

$$f_2 - f_1 = \frac{f_0}{Q} \qquad \qquad ...(9.26)$$

Bandwidth is thus given by the ratio of the frequency of resonance to the quality factor and selectivity is defined as the ratio of resonant frequency to the bandwidth $f_0/(f_2 - f_1)$. This, therefore, shows that the larger the value of $Q$ the smaller is $(f_2 - f_1)$ and hence sharper is the current response.

**Example 9.3:** For the circuit shown determine the value of inductance for resonance if $Q = 50$ and $f_0 = 175$ KHZ. Also find the circuit current the voltage across the capacitor and the bandwidth of the circuit.

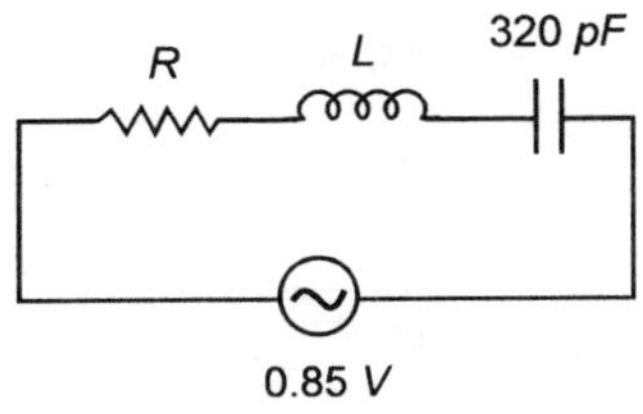



**Fig. E9.3**

**Solution:**
$$f_0 = \frac{1}{2\pi\sqrt{LC}}$$

$$175 \times 10^3 = \frac{1}{2\pi\sqrt{320 \times 10^{-12}\,L}}$$

or
$$L = 2.58 \text{ mH}$$

The reactance of the coil at resonance is
$$2\pi \times 175 \times 10^3 \times 2.58 \times 10^{-3} = 2840\ \Omega$$

Since
$$Q = \frac{\omega_0 L}{R}$$

or
$$R = \frac{\omega_0 L}{Q} = \frac{2840}{50} = 56.8\ \Omega$$

The impedance of the circuit at resonance is
$$Z = R = 56.8\ \Omega$$

Therefore, current $I_0 = \dfrac{0.85}{56.8} = 14.96$ mA

The voltage across the capacitor $= QV$
$$= 50 \times 0.85 = 42.5 \text{ volt}$$

The bandwidth of the circuit is $\dfrac{f_0}{Q} = \dfrac{175 \times 10^3}{50}$

$$\frac{f_0}{Q} = 3.5 \times 10^3 \text{ Hz}$$

$$= 3.5 \text{ KHz}$$

**Example 9.4:** For the circuit shown, determine the impedance magnitude at resonance, at 1 KHz below resonance and, at 1KHz above resonance.

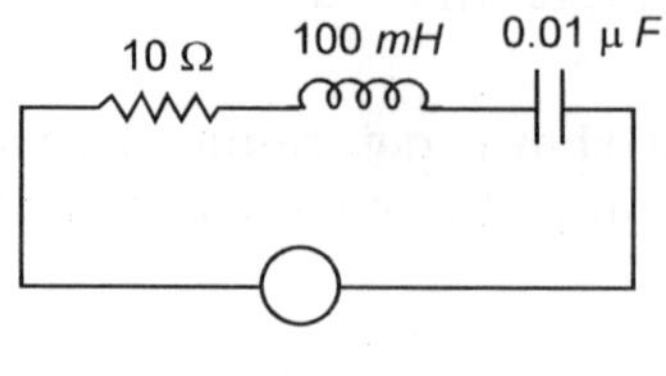


**Fig. E9.4**

**Solution:**

$$f_0 = \frac{1}{2\pi \sqrt{100 \times 10^{-3} \times 0.01 \times 10^{-6}}}$$

$$= 5.03 \text{ KHz}$$

Frequency $\qquad f_1 = 5.03 - 1 = 4.03 \text{ KHz}$

and $\qquad f_2 = 5.03 + 1 = 6.03 \text{ KHz}$

Note these $f_1$ and $f_2$ are not necessarily half power frequencies and hence $Z_1$ need not necessarily equal $Z_2$.

$$Z_1 = \sqrt{10^2 + (2.53 - 3.95) \times 10^6}$$

Since 10 is negligibly small as compared to $X_L$ and $X_C$ which are in $K\,\Omega$, therefore,

$$Z_1 = - j\ 1.42 \text{ K}\Omega$$

$$Z_2 = j3.79 - j2.64 = j1.15 \text{ K}\Omega \textbf{ Ans.}$$

Before we move to parallel resonance, let us summarize the characteristics of series resonance circuit.

1. At resonance power factor of the circuit is unity.
2. Therefore, supply voltage and current are in phase.
3. The reactive component of the input impedance is zero and hence the circuit is resistive and hence current drawn by the circuit is maximum.
4. The frequency of resonance is given by

$$f_0 = \frac{1}{2\pi \sqrt{LC}}$$

5. Even there is voltage across the individual reactive element but across both the elements the net voltage is zero as the  two voltages are equal in magnitude but 180º out of phase. The voltage across each reactive element is $QV$ where $Q$ is  the quality factor of the element and $V$ is the supply voltage.
6. The quality factor

$$Q = \frac{\omega_0 L}{R} = \frac{1}{\omega_0 CR}$$

7. For $\omega < \omega_0$ the circuit is capacitive

$\qquad \omega = \omega_0$ the circuit is resistive

$\qquad \omega > \omega_0$ the circuit is inductive.

8. The band width $f_2 - f_1 = \dfrac{f_0}{Q}$

and $f_0 = \sqrt{f_1 f_2}$

**Example 9.5:** For the circuit shown in Fig. E9.5 $R_1 = 0.5\ \Omega$, $R_2 = 1.5\ \Omega$, $R_3 = 0.5\ \Omega$ $C_1 = 6\ \mu F$ and $C_2 = 12\ \mu F$ $L_1 = 25$ mH and $L_2 = 15$ mH. Determine $(i)$ the frequency of resonance $(ii)$ $Q$ of the circuit $(iii)$ $Q$ of coil 1 and coil 2 individually

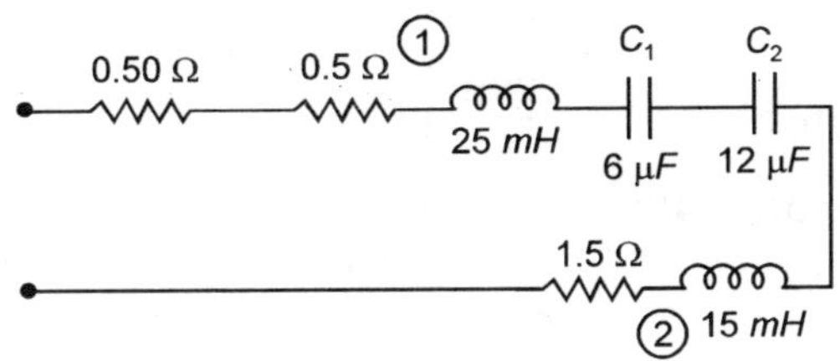


**Fig. E9.5**

**Solution:**

Total inductance of the circuit $25 + 15 = 40$ mH

Total capacitance of the circuit $\dfrac{6 \times 12}{18} = 4\mu F$

The frequency of resonance $= \dfrac{1}{2\pi \sqrt{40 \times 10^{-3} \times 4 \times 10^{-6}}}$

$$= \dfrac{10^4}{8\pi} \text{ Hz or } \omega_0 = 2.5 \times 10^3 \text{ rad/sec}$$

$Q$ of the circuit $= \dfrac{\omega_0 L_{eq}}{R_{eq}} = \dfrac{2.5 \times 10^3 \times 40 \times 10^{-3}}{2.5} = 125$

$= 40$ **Ans.**

$Q$ of coil 1 $\dfrac{\omega_0 L_1}{R_1} = \dfrac{2.5 \times 10^3 \times 25 \times 10^{-3}}{0.5} = 125$

$Q$ of coil 2 $\dfrac{\omega_0 L_2}{R_2} = \dfrac{2.5 \times 10^3 \times 15 \times 10^{-3}}{1.5} = \dfrac{2.5 \times 15}{1.5}$

$= 25$ **Ans.**

Bandwidth of the circuit $= \dfrac{f_0}{Q}$

$$= \dfrac{10^4}{8\pi \times 40} = 9.95 \text{ Hz}$$

**Example 9.6:** A coil having a 5 ohm resistor is connected in series with a 50 µF capacitor. The circuit resonates at 100 Hz $(a)$ Determine the inductance of the coil. $(b)$ If the circuit is connected across a 200 V 100 Hz a.c. source, determine the power delivered to the coil $(c)$ the voltage across the capacitor and the coil $(d)$ the bandwidth of the circuit.

**Solution:** At resonance

$$\omega_0 = \frac{1}{\sqrt{LC}}$$

$$2\pi \times 100 = \frac{1}{\sqrt{50 \times 10^{-6}\, L}}$$

or

$$L = \frac{25}{500} = 50 \text{ mH}$$

The current at resonance

$$= \frac{200}{5}\ 40\,A$$

Therefore, power dissipated $40^2 \times 5 = 8000$ watts.
$$= 8 \text{ kW.}$$

Voltage across capacitor

$$= I\,X_c$$

$$= \frac{40}{50 \times 10^{-6} \times 100 \times 6.28}$$

$$= \frac{40 \times 10^6}{5 \times 10^3 \times 6.28} = \frac{8000}{6.28}$$

The impedance of the coil $R + j\,\omega_0 L$
$$= 5 \times j\,628 \times 50 \times 10^{-3}$$
$$= 5 + j\,31.40$$

The voltage
$$V_L = 40\,(5 + j\,31.4)$$
$$\simeq 1256 \text{ volts } \textbf{Ans.}$$

The $Q_0$ of the coil
$$= \frac{31.4}{5} \simeq 6.3$$

The bandwidth
$$= \frac{f_0}{Q_0} = \frac{100}{6.3} \simeq 16 \text{ Hz}$$

## 9.3 PARALLEL RESONANCE

Consider Fig. 9.5 where $L$ and $C$ are connected in parallel. In series with these elements, resistances shown are their own resistances. Coil will have its own resistance and the capacitor may have some loss component. In general there could be external resistance in series with these elements.

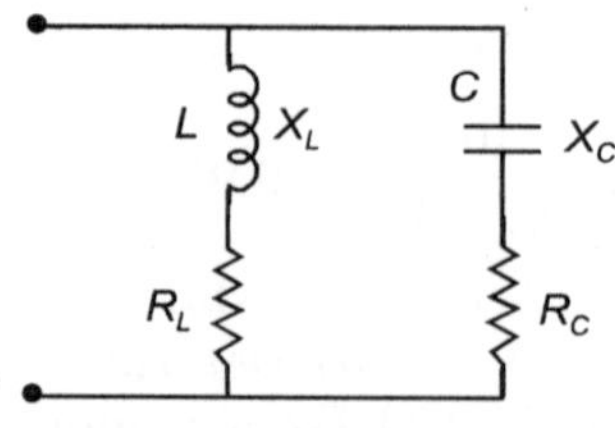


**Fig. 9.5** Parallel resonance.

The input admittance is the sum of the admittances of the two branches.

$$Y_{eq} = \left( \frac{R_c}{R_c^2 + X_c^2} + j\,\frac{X_c}{R_c^2 + X_c^2} \right) + \left( \frac{R_L}{R_L^2 + X_L^2} - j\,\frac{X_L}{R_L^2 + X_L^2} \right)$$

$$= \left( \frac{R_c}{R_c^2 + X_c^2} + \frac{R_L}{R_L^2 + X_L^2} \right) + j\left( \frac{X_c}{R_c^2 + X_c^2} - \frac{X_L}{R_L^2 + X_L^2} \right)$$

Now at resonance the imaginary part (susceptance) of admittance is zero.
Therefore,

$$\frac{X_{c_0}}{R_c^2 + X_{c_0}^2} = \frac{X_{L_0}}{R_L^2 + X_{L_0}^2} \tag{...(9.28)}$$

or
$$X_{c_0} R_L^2 + X_{c_0} X_{L_0}^2 = X_{L_0} R_c^2 + X_{L_0} X_{c_0}^2$$

$$\frac{1}{2\pi f_0 C}\, R_L^2 + \frac{1}{2\pi f_0 C}\cdot(2\pi f_0 L)^2 = 2\pi f_0 L R_c^2 + \frac{2\pi f_0 L}{(2\pi f_0 C)^2}$$

Multiplying by $2\pi f_0 C$ we have

$$R_L^2 + 4\pi^2 f_0^2 L^2 = 4\pi^2 f_0^2 R_c^2 LC + \frac{L}{C}$$

or
$$4\pi^2 f_0^2 LC\left( R_c^2 - \frac{L}{C} \right) = R_L^2 - \frac{L}{C}$$

or
$$4\pi^2 f_0^2 = \frac{1}{LC}\cdot\frac{R_L^2 - L/C}{R_c^2 - L/C}$$

or
$$f_0 = \frac{1}{2\pi\sqrt{LC}}\cdot\sqrt{\frac{R_L^2 - L/C}{R_c^2 - L/C}} \tag{...(9.29)}$$

This is the frequency of resonance of the general parallel *RLC* circuit.
The equivalent input admittance is real and is given as

$$Y_{eq_0} = G_0 = \frac{R_c}{R_c^2 + X_c^2} + \frac{R_L}{R_L^2 + X_L^2} \tag{...(9.30)}$$

Here if $R_c$ is the equivalent series resistance of the capacitance, it can be ignored as it usually is small in magnitude, the equivalent circuit then is given in Fig. 9.6.

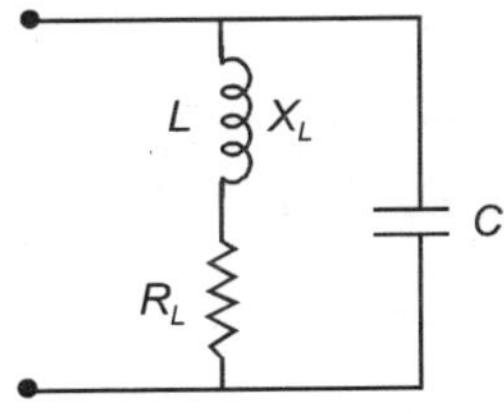


**Fig. 9.6**  Parallel LC circuit with $R_c = 0$.

The resonance frequency for the circuit is given by the following expression when $R_c = 0$ in the general circuit expression.

$$f_0 = \frac{1}{2\pi\sqrt{LC}} \cdot \sqrt{\frac{R_L^2 - L/C}{-L/C}}$$

$$= \frac{1}{2\pi\sqrt{LC}} \cdot \sqrt{1 - R_L^2 \frac{C}{L}} \qquad \qquad ...(9.31)$$

The general expression for $Y_{in}$ in this case is obtained by putting $R_c = 0$ in the expression for $Y_{eq}$ and variation of $Z_{in}$ and $I_{input}$ vs $\omega$ are drawn in Fig. 9.7 (*a*) and (*b*).

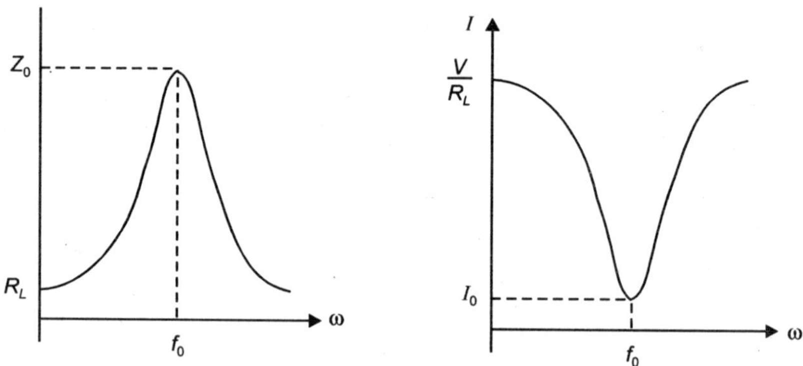


**Fig. 9.7** (*a*) $Z$ Vs $f$ (*b*) $I$ Vs $f$ for parallel resonance.

The shape of curve for current and $Y_{in}$ will be identical as $I = Y_{in} V$.

It is seen that at resonance the circuit has maximum impedance and hence the current is minimum. At this frequency since the susceptance of the circuit is zero, the power factor of the circuit is unity.

At frequency $f < f_0$ since the capacitance acts as an open circuit the net circuit is inductive and at higher frequencies $f > f_0$ the inductor acts as an open circuit the net circuit is capacitive.

If we now consider that in the general circuit if $R_L$ is also zero besides $R_c$ being zero, we have the equivalent circuit (Fig. 9.8).

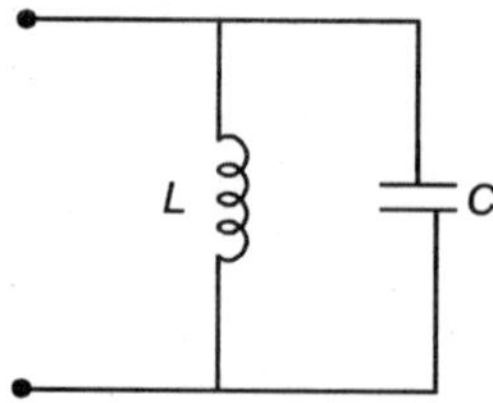


**Fig. 9.8** Parallel LC circuit.

For this the frequency of resonance is

$$f_0 = \frac{1}{2\pi\sqrt{LC}} \qquad \qquad ...(9.32)$$

and it can be seen that at the frequency the input impedance is

$$\frac{j\,X_L\,(-j\,X_c)}{j\,X_L - j\,X_c} = \frac{j\,X_L\,(-j\,X_c)}{0}$$

$$= \infty$$

and hence theoretically the current drawn from the source is zero. This means there will be current in each parallel path but the net current from the supply is zero i.e. there will be exchange of energy between the inductance and the capacitance of the circuit. Actually because of the small winding resistance of the coil some power is lost so the reactive currents are not exactly identical and there is a small current from the source to take into account this loss.

If we assume that

$$R_L = R_c = \sqrt{\frac{L}{C}} = \sqrt{\frac{\omega L}{\omega C}} = \sqrt{X_L\,X_c}$$

Therefore, $Y_{\text{in}} = \dfrac{R_c}{X_L\,X_c + X_c^2} + \dfrac{R_L}{X_L\,X_c + X_L^2} + j\left(\dfrac{X_c}{X_L X_c + X_c^2} - \dfrac{X_L}{X_L X_c + X_L^2}\right)$

The imaginary part

$$= \frac{X_L\,X_c^2 + X_c\,X_L^2 - X_L^2\,X_c - X_L\,X_c^2}{\left(X_L\,X_c + X_c^2\right)\left(X_c\,X_L + X_L^2\right)}$$

$$= 0$$

Therefore $Y_{\text{in}}$ is resistive at all frequencies that means resonance will occur at all frequencies.

Consider Fig. 9.6. The impedance to the parallel circuit is

$$Z_T = \frac{(R + jX_L)\,(-jX_c)}{R + jX_L - jX_c} \qquad \qquad ...(9.33)$$

Let

$$X_L = X_C$$

Therefore,

$$Z_T = \frac{(X_L - jR)\,X_c}{R} \qquad \qquad ...(9.34)$$

Let

$$\frac{X_L}{R} = Q \text{ the quality factor of the coil,}$$

$$Z_T = (Q - j\,1)\,X_c \qquad \qquad ...(9.35)$$

Usually $Q \gg 1$, therefore,
At resonance

$$Z_T = Q\,X_{C0} \qquad \qquad ...(9.36)$$

Since this impedance is effectively a pure resistance, for the high $Q$-circuit, resonance occurs when $X_{L0} = X_{C0}$

and

$$f_0 = \frac{1}{2\pi\,\sqrt{LC}}$$

These relations hold good only when $Q >> 1$ (should be greater than 5). Similarly bandwidth of the circuit is

$$BW = \frac{f_0}{Q}$$

**Example 9.7:** For the circuit shown in Fig. E9.7 determine the current through the parallel branches and, the supply current.
(*b*) if the frequency is made $f_0/2$ and $2f_0$ determine the currents from the source under these conditions.

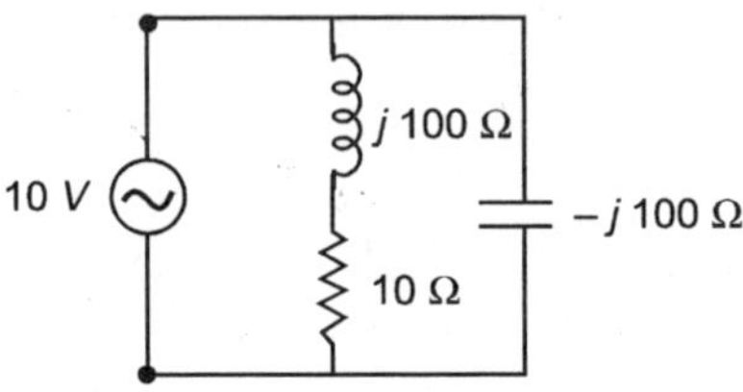


**Fig. E9.7**   Value at resonance.

**Solution:**   (*a*) $Z_0 = Q\,X_{C0} = \dfrac{100}{10} \times 100 = 1000\,\Omega$

The current   $I_0 = \dfrac{V}{Z_0} = \dfrac{10}{1000} = 10$ mA

The current through the capacitor branch

$$I_C = \frac{10}{-j\,100} = j\,100 \text{ mA}$$

$$I_L = \frac{10}{j\,100} = -j\,100 \text{ mA}$$

The current through capacitor and inductor are $Q\,I_0$. These large currents in $L$ and $C$ are called oscillating currents because they are confined within the $L$ and $C$ of the parallel resonant circuit. The supply current is 10 mA which supplies the losses of the parallel resonance circuit. The input power from the source is $10 \times 10 = 100$ mW
The $I^2R$ loss in the 10 $\Omega$ resistor is
$$(0.01)^2 \times 10 = 100 \text{ mW}$$

(b) If however, the frequency is reduced to $f_0/2$ the inductive reactance become 50 $\Omega$ and capacitor 200 $\Omega$.

The net impedance $\quad = \dfrac{(10 + j50)\,(-j\,200)}{10 + j50 - j\,200}$

$\quad = 68.8\,\angle 75^\circ$

The current $\quad I = \dfrac{10\,\angle -75^\circ}{68.8}$

$\quad = 0.145\,\angle -75^\circ$

If the frequency is doubled

$$X_L = j\,200 \quad \text{and} \quad X_C = -j50$$

$$Z_T = \frac{(10 + j200)\,(-j50)}{10 + j200 - j50}$$

$$= 66.4 \angle -89°$$

The current

$$I = \frac{10 \angle 89°}{66.4}$$

$$= 0.151 \angle 89°$$

This shows that impedance at resonance is maximum and the current is minimum. The currents through the parallel branches gets magnified at resonance and is $QI_0$. The circuit is inductive for $f < f_0$ and is capacitive for $f > f_0$.

If we load this parallel resonant circuit with some resistance $R$ the effective resistance will be less than $Q\,X_{C0}$. Let the resistance be $R' = Q'\,X_{C0}$. Therefore $Q'$ is smaller than $Q$ and hence the bandwidth of the circuit is increased when the resonant circuit is shunted with the resistance $R$.

### 9.3.1  Q of Parallel RLC Circuit

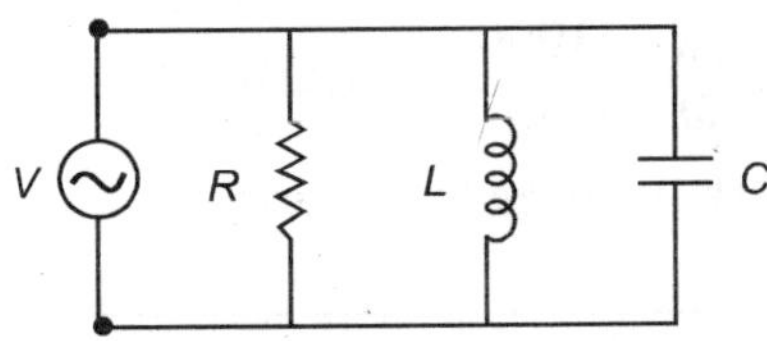

**Fig. 9.9**  Parallel RLC circuit.

At resonance there will be exchange of energy between the inductor and the capacitor. When the inductor is carrying maximum current, the voltage across capacitor is zero and hence the electrostatic energy is zero.

The time rate of change of energy through inductor is $\dfrac{V^2}{\omega_0 L}$ and the time rate of change of power is $\dfrac{V^2}{R}$.

Therefore,

$$Q_0 = \frac{V^2}{\omega_0 L} \cdot \frac{R}{V^2} = \frac{R}{\omega_0 L}$$

$$= \frac{V^2 \omega_0 c R}{V^2} = \omega_0 C R \qquad \qquad ...(9.37)$$

and bandwidth

$$= \frac{f_0}{Q}$$

**Example 9.8:** Consider Fig. E9.8 determine the original bandwidth of the circuit and the loaded circuit bandwidth.

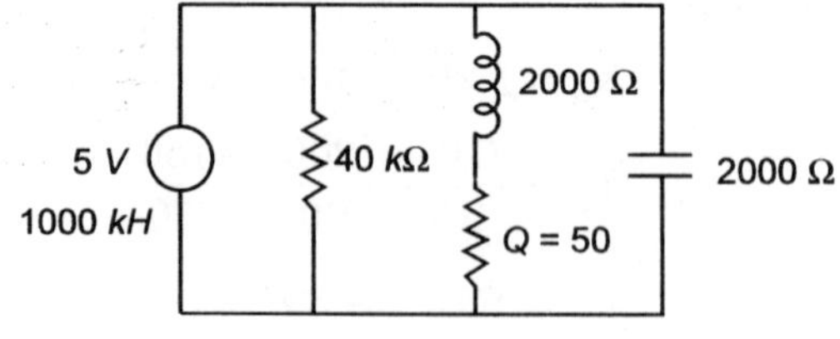

**Fig. E9.8**

The original equivalent circuit is

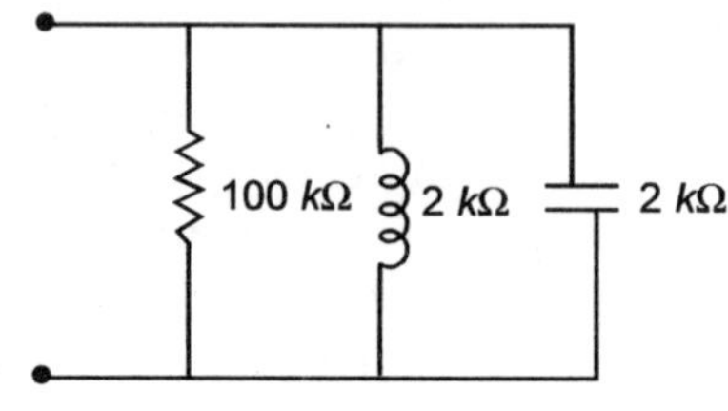

**Fig. E9.8** (*a*)

$$Z_0 = Q\, X_{C0} = 50 \times 2 = 100 \ k\Omega$$

The bandwidth 
$$= \frac{f_0}{Q} = \frac{1000}{50} = 20 \text{ kHz}$$

The equivalent circuit with loading resistance is

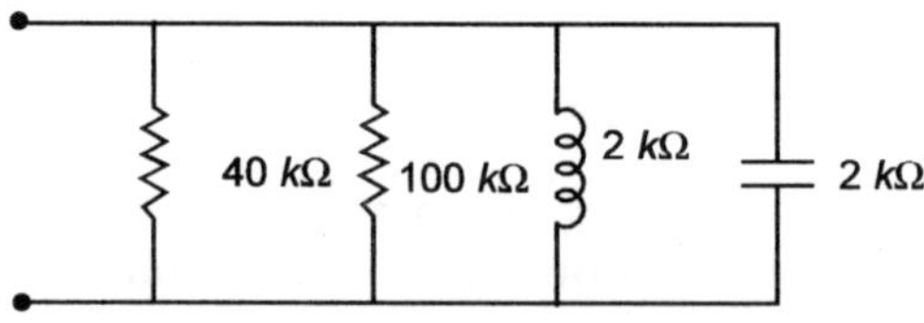

**Fig. E9.8** (*b*)

$$Q'\, X_{C0} = \frac{40 \times 100}{140} = \frac{200}{7}$$

Therefore,
$$Q' = \frac{200}{7} \times \frac{1}{2} = \frac{100}{7}$$

Therefore bandwidth 
$$= \frac{1000}{100} \times 7 = 70 \text{ kHz}$$

It is to be noted that bandwidth in case of parallel resonance is obtained when net impedance of the circuit is 70.7% of the maximum value.

For a parallel *RLC* circuit the admittance is given as

$$Y_{\text{in}} = \frac{1}{R} + j\left(\omega C - \frac{1}{\omega L}\right) \qquad\qquad \text{...(9.38)}$$

Since $Q_0 = \omega_0 \, CR$ we have

$$Y_{\text{in}} = \frac{1}{R} + j\left( \frac{\omega \, \omega_0 \, CR}{\omega_0 \, R} - \frac{\omega_0}{\omega \, \omega_0 \, L} \right)$$

$$= \frac{1}{R} + \frac{j}{R}\left( \frac{\omega \, \omega_0 \, CR}{\omega_0} - \frac{\omega_0 \, R}{\omega \, \omega_0 \, L} \right)$$

$$= \frac{1}{R}\left[ 1 + j\left( \frac{\omega}{\omega_0} Q_0 - \frac{\omega_0}{\omega} Q_0 \right) \right]$$

$$= \frac{1}{R}\left[ 1 + j \, Q_0 \left( \frac{\omega}{\omega_0} - \frac{\omega_0}{\omega} \right) \right] \qquad \qquad ...(9.39)$$

Since at resonance the equivalent admittance of the circuit is $1/R$, at 1/2 power frequencies it would be $\dfrac{\sqrt{2}}{R}$. Therefore, at $\dfrac{1}{2}$ power frequencies the imaginary quantity should be equal to unity so that

$$Y_{\text{in}} = \left| \frac{\sqrt{2}}{R} \right|$$

Therefore,

$$Q_0 \, \frac{\omega_1}{\omega_0} - \frac{\omega_0}{\omega_1} Q_0 = -1 \qquad \qquad ...(9.40)$$

or

$$\omega_1^2 - \omega_0^2 = -\frac{\omega_1 \, \omega_0}{Q_0}$$

or

$$\omega_1^2 + \frac{\omega_1 \, \omega_0}{Q_0} - \omega_0^2 = 0$$

or

$$\omega_1 = \frac{-\dfrac{\omega_0}{Q_0} + \sqrt{\dfrac{\omega_0^2}{Q_0^2} + 4 \, \omega_0^2}}{2}$$

$$= \omega_0 \left[ \sqrt{\left( \frac{1}{2Q_0} \right)^2 + 1} - \frac{1}{2 \, Q_0} \right] \qquad \qquad ...(9.41)$$

Similarly at $\omega_2$

$$Q_0 \left[ \frac{\omega_2}{\omega_0} - \frac{\omega_0}{\omega_2} \right] = 1$$

This gives on solution

$$\omega_2 = \omega_0 \left[ \sqrt{\left(\frac{1}{2\,Q_0}\right)^2 + 1} + \frac{1}{2\,Q_0} \right] \qquad \text{...(9.42)}$$

Hence bandwidth  $\Delta\omega = \omega_2 - \omega_1 = \dfrac{\omega_0}{Q_0}$

or
$$f_2 - f_1 = \frac{f_0}{Q_0}$$

Similarly using the equations (9.41) and (9.42) it can be proved that

$$\omega_1\omega_2 = \omega_0^2 \quad \text{or} \quad \omega_0 = \sqrt{\omega_1\,\omega_2} \qquad \text{...(9.43)}$$

Summarising the characteristics of a parallel resonance we have

1. At resonance the input impedance is maximum or the input admittance is minimum and the circuit is purely resistive and hence power factor is unity.
2. The circuit is capacitive for frequencies $f > f_0$. It is inductive for $f < f_0$.
3. The current through the inductor and capacitor are equal in magnitude but opposite in phase and the current is amplified by the factor $Q$, hence the current in each branch is $Q\,I_0$ where $I_0$ is current from the source.
4. Quality factor is given as  $Q_0 = \dfrac{R}{\omega_0\,L} = \omega_0\,C\,R$
5. The resonance frequency when $R_L \to 0$ or $R_C \to 0$ is given by

$$f_0 = \frac{1}{2\pi\sqrt{LC}}$$

6. Bandwidth is given as  $f_2 - f_1 = \dfrac{f_0}{Q_0}$ .
7. $f_0 = \sqrt{f_1 f_2}$ where $f_1$ and $f_2$ are the lower and upper half power frequencies respectively.

## 9.4  POLE-ZERO CONFIGURATION OF RLC SERIES CIRCUIT

For a series $RLC$ circuit.

$$Z(s) = R + sL + \frac{1}{Cs}$$

$$= \frac{L}{s}\left[ s^2 + \frac{R}{L}\,s + \frac{L}{C} \right] \qquad \text{...(9.44)}$$

The zero of $Z(s)$ are obtained by putting

$$s^2 + \frac{R}{L}\,s + \frac{L}{C} = 0$$

$$s = -\frac{R}{2L} \pm \frac{1}{2}\sqrt{\frac{R^2}{L^2} - \frac{4L}{C}}$$

$$= -\frac{R}{2L} \pm \sqrt{\left(\frac{R}{2L}\right)^2 - \frac{L}{C}} \qquad \text{...(9.45)}$$

Here
$$Z(s) = \frac{L}{s}\left[(s + \alpha - j\omega_d)(s + \alpha + j\omega_d)\right] \qquad \text{...(9.46)}$$

where $\alpha$ is the exponential damping co-efficient and $\omega_d$ is the damped natural resonant frequency and is given by

$$\omega_d = \sqrt{\omega_0^2 - \alpha^2} \qquad \text{...(9.47)}$$

The above impedance expression can also be written as

$$Z(s) = \frac{L}{s}\left(s^2 + 2\,\xi\,\omega_0\,s + \omega_0^2\right) \qquad \text{...(9.48)}$$

where
$$\omega_0 = \frac{1}{\sqrt{LC}}\,,\; 2\,\xi\,\omega_0 = \frac{R}{L} \qquad \text{...(9.49)}$$

$$\xi = \frac{R}{2\omega_0 L} = \frac{1}{2\,Q_0} \qquad \text{...(9.50)}$$

Here $\xi$ is known as damping factor and is a dimensionless quantity.

The value of $Q$ varies inversely with damping ratio $\xi$. A high value of $Q$ means a low value of $\xi$ which means a series RLC circuit with low $R$ has high $Q$.

The admittance function for the series network is written as

$$Y(s) = \frac{s}{L} \cdot \frac{1}{(s + \alpha + j\omega_d)(s + \alpha - j\omega_d)} \qquad \text{...(9.51)}$$

The plot of $Y(s)$ on s-plane is shown here in Fig. 9.10

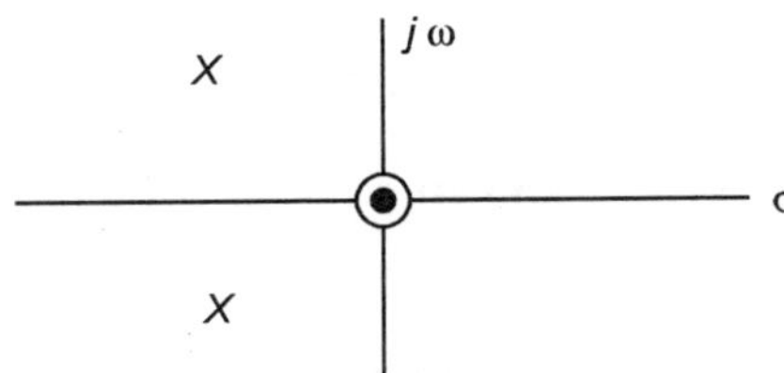


**Fig. 9.10**  The poles and zeros of $Y(s)$ for series RLC circuit.

The closer the poles $s_1$ and $s_2$ are to the $j\omega$-axis (the smaller the value of $R$) the higher the value of $Q$.

The phase angle plot of the *RLC* series circuit as a function of $\omega$ is shown here in Fig. 9.11.

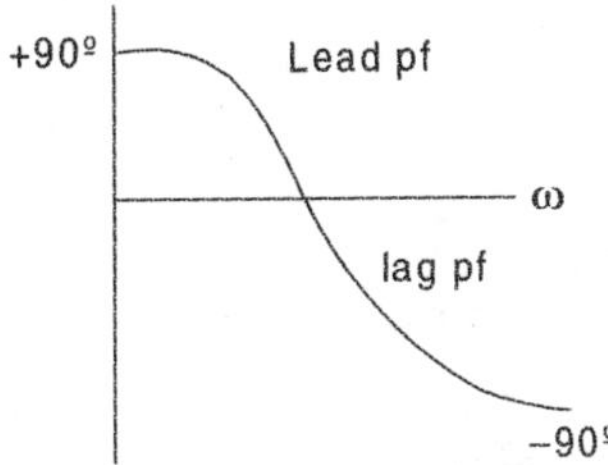


**Fig. 9.11**  Plot of power factor angle vs $\omega$ for *RLC* series circuit.

This shows that for frequencies $f < f_0$ the circuit is capacitive and $f > f_0$ the circuit is inductive.

## 9.5  PARALLEL RESONANCE CIRCUIT POLE—ZERO CONFIGURATION

Consider a circuit where $R$, $L$ and $C$ are connected in parallel. The input admittance as a function of complex frequency is

$$Y(s) = \frac{1}{R} + Cs + \frac{1}{sL} = \frac{C}{s}\left[\frac{s}{RC} + s^2 + \frac{1}{LC}\right]$$

$$= \frac{C}{s}\left(s^2 + 2\alpha s + \omega_0^2\right)$$

$$= \frac{C}{s}(s + \alpha - j\omega_d)(s + \alpha + j\omega_d) \qquad \qquad \text{...(9.52)}$$

where

$$\alpha = \frac{1}{2RC}, \; \omega_d = \sqrt{\omega_0^2 - \alpha^2}$$

and

$$\omega_0 = \frac{1}{\sqrt{LC}}$$

$\omega_d$ is the damped natural resonance frequency

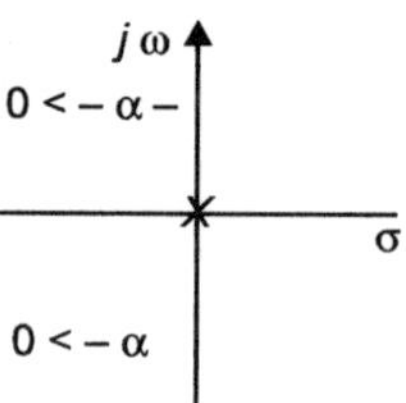


**Fig. 9.12**  The poles and zeros of  $Y(s)$  for parallel *RLC* circuit.

The poles and zeros of  $Y(s)$  are shown in Fig. 9.12

$Y(s)$  can also be written as

$$Y(s) = \frac{C}{s}\left(s^2 + 2\,\xi\omega_0 s + \omega_0^2\right) \qquad \qquad \text{...(9.53)}$$

where  $\xi$  is damping factor

Hence from these equations we have

$$\alpha = \frac{1}{2RC} = \frac{\omega_0}{2\,Q_0} \quad \text{as } Q_0 = \omega_0\,RC$$

$$\omega_d = \sqrt{\omega_0^2 - \alpha^2} = \omega_0\sqrt{1 - \left(\frac{1}{2\,Q}\right)^2} \qquad \qquad \text{...(9.54)}$$

and

$$\xi = \frac{\alpha}{\omega_0} = \frac{1}{2\,Q_0} \qquad \qquad \text{...(9.55)}$$

Again for parallel *RLC* circuit the value of $Q$ varies inversely as the damping factor $\xi$. A high value of $Q$ means a low value of $\xi$. Also it is to be noted that closer the zeros are to $j\omega$ axis, higher is the value of $RC$ and hence higher the value of $Q$.

The phase relation as a function of frequency in parallel *RLC* circuit is shown in Fig. 9.13.

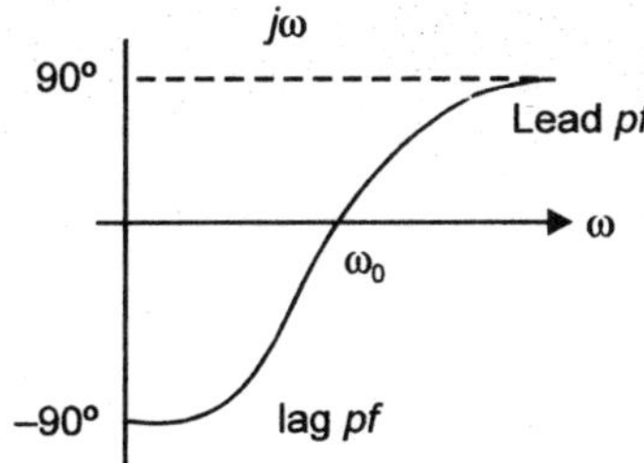



**Fig. 9.13**  Plot of power factor angle *vs* $\omega$ for parallel *RLC* circuit.

This shows that at resonance the power factor is unity. For frequencies $f < f_0$ the circuit is inductive and $f > f_0$ the circuit is capacitive.

**Example 9.9:** Show that the sum of energy stored by the inductor and the capacitor connected in series at resonance at any instant is constant and is given by $L\,I^2$.

**Solution:** Let $i$ and $v$ be the instantaneous current through inductor and the voltage across the capacitor at any time $t$.

Let
$$i = I_m \cos \omega_0 t$$

The energy stored is
$$\frac{1}{2} L i^2 = \frac{1}{2} L I_m^2 \cos^2 \omega_0 t$$

The energy stored in the capacitor
$$\frac{q^2}{2C} = \frac{1}{2C}\left[\int_0^t i\,dt\right]^2$$

$$= \frac{1}{2C} \cdot I_m^2 \left[\int_0^t \cos\omega_0 t\,dt\right]^2$$

$$= \frac{1}{2C} I_m^2 \left[\left(\frac{\sin \omega_0 t}{\omega_0}\right)_0^t\right]^2$$

$$= \frac{I_m^2}{2} L \sin^2 \omega_0 t$$

Therefore total energy

$$= \frac{1}{2} L I_m^2 \left(\cos^2 \omega_0 t + \sin^2 \omega_0 t\right)$$

$$= \frac{1}{2} I_m^2 L$$

$$= L\,I^2$$

**Example 9.10:** Show that the sum of energy stored by the inductor and the capacitor in a parallel *RLC* circuit at any instant is constant at resonant frequency and is equal to $CV^2$.

**Solution:** Let
$$v = V_m \cos \omega_0 t$$

The energy stored by the capacitor is

$$= \frac{1}{2} C V_m^2 \cos^2 \omega_0 t$$

The energy stored by the inductor is $\dfrac{1}{2} L i^2$

Now
$$v = L \frac{di}{dt}$$

$$di = \frac{V}{L} dt$$

$$i^2 = \left( \int_0^t \frac{v}{L} dt \right)^2$$

$$= \frac{V_m^2}{L^2} \cdot \frac{\sin^2 \omega_0 t}{\omega_0^2} = \frac{V_m^2}{L^2} \cdot \sin^2 \omega_0 t \, LC$$

$$i^2 = \frac{V_m^2 C}{L} \sin^2 \omega_0 t$$

Energy
$$= \frac{1}{2} L C^2 = \frac{1}{2} L \cdot \frac{V_m^2 C}{L} \sin^2 \omega_0 t$$

$$= \frac{C}{2} V_m^2 \sin^2 \omega_0 t$$

Therefore total energy $= \dfrac{1}{2} C V_m^2 = C V^2$  Hence proved.

**Example 9.11:** Determine the resonant frequency, the source current and the input impedance for the circuit shown in Fig. E9.11 for each of the following cases:

| | | |
|---|---|---|
| Case I | $R_L = 150 \ \Omega$ | $R_C = 100 \ \Omega$ |
| Case II | $R_L = 150 \ \Omega$ | $R_C = 0 \ \Omega$ |
| Case III | $R_L = 0 \ \Omega$ | $R_C = 0 \ \Omega$ |

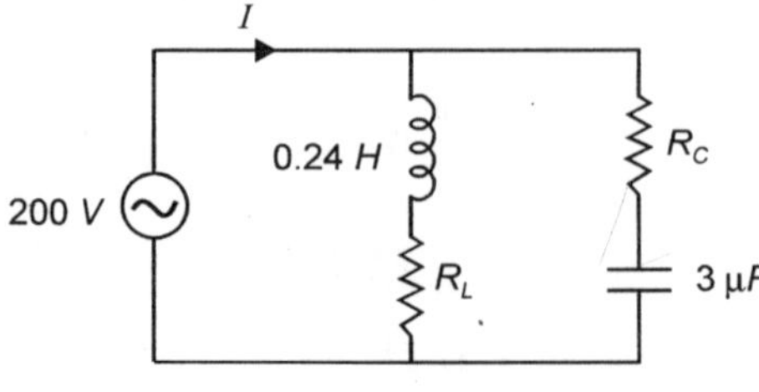


**Fig. E9.11**

**Case I.** The resonant frequency

$$f_0 = \frac{1}{2\pi\sqrt{LC}} \cdot \sqrt{\frac{R_L^2 - L/C}{R_c^2 - L/C}}$$

$$= \frac{1}{2\pi\sqrt{0.24 \times 3 \times 10^{-6}}} \cdot \sqrt{\frac{150^2 - \dfrac{0.24 \times 10^6}{3}}{100^2 - \dfrac{0.24 \times 10^6}{3}}}$$

$$= 170 \text{ Hz}$$

The reactances at this frequency

$$X_L = j2\pi \times 170 \times 0.24 = j256\,\Omega$$

$$X_C = -\frac{j\,10^6}{2\pi \times 170 \times 3} = -j312\,\Omega$$

The branch currents are

$$I_L = \frac{200\,(150 - j256)}{(150 + j256)\,(150 - j256)} = 0.34 - j0.582$$

and

$$I_C = \frac{200\,(100 + j312)}{(100 - j312)\,(100 + j312)}$$

$$= 0.186 + j\,0.582$$

The total current (source)

$$= I_L + I_C$$
$$= 0.34 - j0.582 + 0.186 + j0.582$$
$$= 0.526 \text{ A}$$

and the circuit impedance is

$$\frac{200}{0.526} = 380\ \Omega$$

**Case II.** The resonant frequency if $R_C = 0$ is given by

$$f_0 = \frac{1}{2\pi\sqrt{LC}} \cdot \sqrt{1 - R_L^2\,\frac{C}{L}}$$

$$= \frac{1}{2\pi\sqrt{0.24 \times 3 \times 10^{-6}}} \cdot \sqrt{1 - 150^2\,\frac{3 \times 10^{-6}}{0.24}}$$

$$= 159 \text{ Hz}$$

The reactances at this frequency are

$$X_L = j2\pi \times 159 \times 0.24 = j\,240\ \Omega$$

$$X_C = \frac{-j}{2\pi \times 159 \times 3 \times 10^{-6}} = -j\,334\ \Omega$$

The branch currents are:

$$I_L = \frac{200(150 - j240)}{(150 + j240)\,(150 - j240)}$$

$$= 0.374 - j0.598 \text{ A}$$

$$I_C = \frac{200}{-j334} = j0.598$$

Total source current = 0.374 A

The circuit impedance = $\dfrac{200}{0.374} = 535$ **Ans.**

**Case III.** The resonance frequency is given as

$$f_0 = \frac{1}{2\pi\sqrt{0.24 \times 3 \times 10^{-6}}}$$

$$= 188 \ \text{Hz}$$

The reactances at this frequency are

$$X_L = j2\pi \times 188 \times 0.24 = j283 \ \Omega$$

$$X_C = \frac{-j}{2\pi \times 188 \times 3 \times 10} = -j283 \ \Omega$$

The branch currents are:

$$I_L = \frac{200}{j283} = -j0.706 \ \text{A}$$

$$I_C = \frac{200}{-j283} = j0.706 \ \text{A}$$

Source current = 0

The total impedance = $\infty$ **Ans.**

**Example 9.12:** A coil having a resistance of 50 $\Omega$ and inductances 10 mH is connected in series with a capacitor and is supplied at constant voltage and variable frequency source. The maximum current is 1A at 750 Hz. Determine the bandwidth and half power frequencies.

**Solution:** $Q$ of the coil = $\dfrac{\omega_0 L}{R} = \dfrac{2\pi \times 750 \times 10 \times 10^{-3}}{50}$

$$= 0.9425$$

Bandwidth = $\dfrac{f_0}{Q} = \dfrac{750}{0.9425} = 795.8 \ \text{Hz}$

i.e.
$$f_2 - f_1 = 795.8 \ \text{Hz}$$

and
$$f_0 = \sqrt{f_1 \, f_2}$$

$$750^2 = f_1 f_2$$

$$(f_2 + f_1)^2 = 795.8^2 + 4 \times 750^2$$

$$= 1697.8^2$$

or
$$f_2 + f_1 = 1697.8$$

$$f_2 - f_1 = 795.8$$

$$2 f_2 = 2493.6$$

or
$$f_2 = 1246.8 \text{ Hz} \quad \text{and} \quad f_1 = 451 \text{ Hz} \textbf{ Ans.}$$

**Example 9.13:** For the circuit of Fig. E9.13 determine the frequency of resonance and the current drawn from the source.

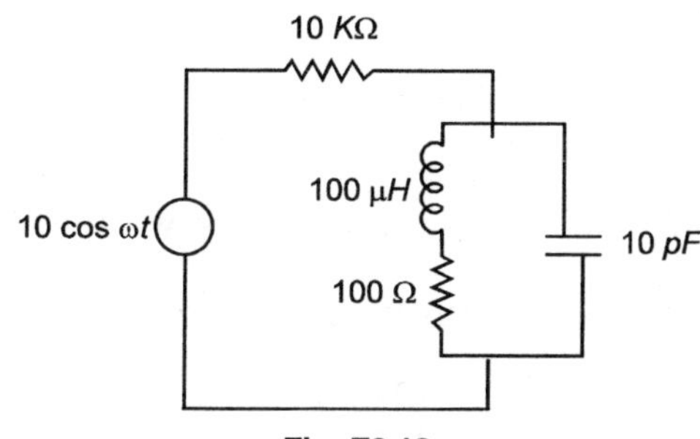


**Fig. E9.13**

$$f_0 = \frac{1}{2\pi \sqrt{LC}} \sqrt{\frac{R_L^2 - L/C}{-L/C}}$$

$$= \frac{1}{2\pi \sqrt{100 \times 10^{-6} \times 10 \times 10^{-12}}} \cdot \sqrt{\frac{100^2 - \dfrac{100 \times 10^{-6}}{10 \times 10^{-12}}}{-10 \times 10^6}}$$

$$\omega_0 \simeq \frac{10^8}{\sqrt{10}} = 31.606 \times 10^6 \text{ rad/sec.}$$

$Q$ of the coil $= \dfrac{\omega_0 L}{R} = \dfrac{31.6 \times 10^6 \times 100 \times 10^{-6}}{100} = 31.6$

Effective parallel resistance $= QXco$

$$= 31.6 \times \frac{1}{31.6 \times 10^6 \times 10 \times 10^{-12}}$$

$$= 10^5 = 100 \ k \ \Omega$$

Therefore current $= \dfrac{10}{\sqrt{2} \times 110}$

$$= 0.06427 \text{ mA} \textbf{ Ans.}$$

**Example 9.14:** Two branches of a parallel circuit have element $R_L = 6 \ \Omega \ L = 1$ mH and $R_C = 4 \ \Omega$ and $C = 20 \ \mu F$. Determine the frequency of resonance. Also determine the maximum value of $R_C$ for which resonance can occur.

**Solution:** We know that for such a circuit.

$$\omega_0 = \frac{1}{\sqrt{LC}} \cdot \sqrt{\frac{R_L^2 - L/C}{R_c^2 - L/C}}$$

$$= \sqrt{\frac{1}{1 \times 10^{-3} \times 20 \times 10^{-6}}} \cdot \sqrt{\frac{6^2 - \dfrac{10^{-3}}{20 \times 10^{-6}}}{4^2 - \dfrac{10^{-3}}{20 \times 10^{-6}}}}$$

$$= \frac{10^4}{\sqrt{200}} \sqrt{\frac{14}{34}}$$

$$= 4537 \ \text{rad/sec.}$$

In order for resonance to take place the quantity within the square root sign should be positive. Since $R_L^2 < \dfrac{L}{C}$, $R_c^2$ should also be less than $\dfrac{L}{C}$ or $R_c < \sqrt{\dfrac{L}{C}}$ or $R_c < 7.07\,\Omega$ **Ans.**

**Example 9.15:** For the circuit shown in Fig. E9.15 determine (*a*) $Q$ of the coil (*b*) Capacitance $C$ (*c*) $Q$ of the circuit (*d*) bandwidth of the circuit (*e*) maximum energy stored in the capacitor of the circuit (*f*) power dissipated in the resistor.

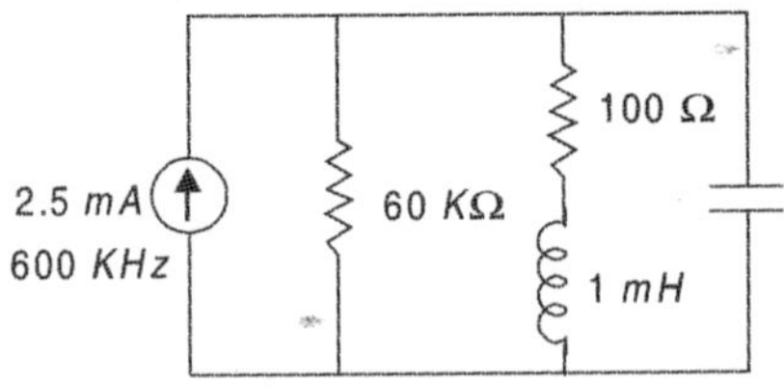



**Fig. E9.15**

**Solution:**

(*a*) $Q$ of the coil $= \dfrac{\omega_0 L}{R} = \dfrac{2\pi \times 600 \times 1000 \times 1 \times 10^{-3}}{100}$

$$= 2\pi \times 6 = 6.28 \times 6$$
$$= 37.68$$

(*b*) $\quad 600 \times 10^3 = \dfrac{1}{2\pi \sqrt{1 \times 10^{-3} C}}$

$$C = \frac{10^{-3}}{4\pi^2 \times 36 \times 10^4}$$

$$= \frac{10^{-7}}{1419.78}$$

$$= 70.43 \ \text{pF}$$

(*c*) $Q$ of the circuit

The equivalent resistance of coil and capacitor

$$= Q_c \, X_{co} = 37.68 \times \frac{10^{12}}{2\pi \times 600 \times 10^3 \times 70.43}$$

$$= 3.78 \ k\Omega$$

The equivalent resistance $\dfrac{60 \times 3.78}{63.78} = 3.56 \ k\Omega$

Therefore $Q$ of the circuit $= \omega_0 \ R_{eq} \ C$

$$= 2 \ \pi \times 600 \times 10^3 \times 3.56 \times 10^3 \times 70.43 \times 10^{-12}$$

$$= 9.45 \ \textbf{Ans.}$$

(*d*) Bandwidth of the circuit $= \dfrac{f_0}{Q} = \dfrac{600 \times 10^3}{9.45}$

$$= \quad 63.5 \ \text{kHz}$$

(*e*) The voltage across the capacitor at resonance

$$I \ R_{eq} = 2.5 \times 10^{-3} \times 3.56 \times 10^3$$

$$= 8.75 \ \text{volt}$$

Maximum energy stored by the capacitor

$$Cv^2 = 70.43 \times 10^{-12} \times 8.75^2$$

$$= 5392 \times 10^{-12} \ \text{joule}$$

(*f*) Power dissipated in the resistor

$$= 2.5 \times 2.5 \times 3.56 \times 10^3$$

$$= 22.25 \ \text{mW} \ \textbf{Ans.}$$

**Example 9.16:** In the problem of example 9.14 if $R_c = 5 \ \Omega$ and $R_L$ is variable, the circuit is fed from a sinusoidal source at a frequency $f$ such that $X_c = 6 \ \Omega$ and $X_L = 15 \ \Omega$. Determine the value of $R_L$ for which resonance can take place.

**Solution:** At resonance

$$\frac{X_c}{R_c^2 + X_c^2} - \frac{X_L}{R_L^2 + X_L^2} = \frac{5}{25 + 36} - \frac{15}{R_L^2 + 225} = 0$$

or
$$\frac{5}{61} - \frac{15}{R_L^2 + 225} = 0$$

or
$$R_L^2 + 225 = 183$$

or
$$R_L^2 = -42$$

This means $R_L$ should be negative which is not possible. Hence no value of $R_L$ will be able to bring resonance in the circuit at the given condition.

**Example 9.17:** A coil has an inductance of 1.3 mH and resonates at 600 KHz and its $Q = 30$. If the bandwidth required is 50 KHz what resistor should be connected across the coil?

**Solution:** The reactance of the coil $= 2\pi (600 \times 10^3) \times 1.3 \times 10^{-3}$

$$= 4900 \ \Omega$$

Since at resonance $X_{L0} = X_{C0}$

$$X_{c0} = 4900$$

The input resistance $= Z_{T0} = Q \ X_{c0}$

$$= 30 \times 4900$$

$$= 147000 \ \Omega$$

The required bandwidth is 50 kHz

Therefore, required $Q$ of the circuit is $\dfrac{f_0}{50}$

$$= \dfrac{600}{50} = 12$$

The equivalent input resistance required is

$$Z'_{T0} = Q' \, X_{c0} = 12 \times 4900$$
$$= 58800$$

The resistance required for shunting is say $R'$

$$58800 = \dfrac{147000 R'}{147000 + R'}$$

or $\qquad 147000 \times 58800 = 147000 \ R' - 58800 \ R'$

or $\qquad R' = \dfrac{147000 \times 58800}{88200}$

$$= 98 \ k\Omega \ \textbf{Ans.}$$

**Example 9.18:** The circuit shown is connected across a voltage source of such that voltage across capacitor is $V_c = 20\sqrt{2}\ \sin 0.5t$. Determine the instantaneous energy stored in the capacitor and inductor (*b*) $Q$ of the circuit.

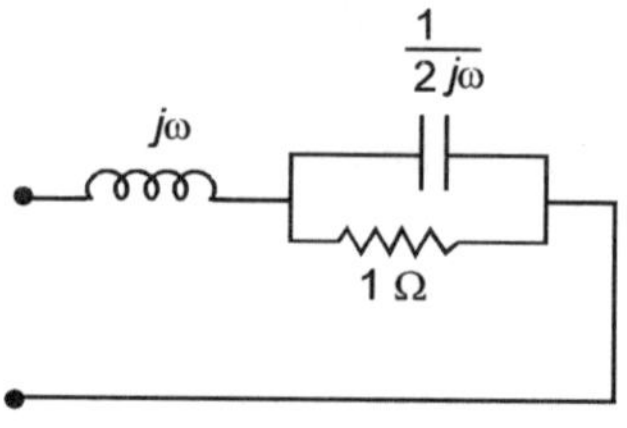



**Fig. E9.18**

**Solution:** The input impedance of the circuit is

$$j\omega + \dfrac{1 \times \dfrac{1}{2j\omega}}{1 + \dfrac{1}{2j\omega}} = j\omega + \dfrac{1}{1 + 2j\omega}$$

$$= j\omega + \dfrac{1}{1 + 4\omega^2} - j\,\dfrac{2\omega}{1 + 4\omega^2}$$

Putting the imaginary quantity to zero for resonance

$$\omega = \dfrac{2\omega}{1 + 4\omega^2}$$

or $\qquad \omega + 4\omega^3 = 2\omega$

$$1 + 4\omega^2 = 2 \quad \text{or} \quad 4\omega^2 = 1$$

$$\omega = \tfrac{1}{2} = 0.5 \ \text{rad/sec}$$

Taking capacitor voltage as reference

Since the voltage across capacitor is $20\sqrt{2}\ \sin 0.5t$ the current through capacitor is

$$C\,\dfrac{d}{dt}\left(20\sqrt{2}\ \sin 0.5t\right)$$

The current $$I_c = \frac{20}{1/j2\omega_0} = j20 \text{ A}$$

Current through resistor $= \dfrac{20}{1} = 20\ \angle 0^\circ$

Therefore, current through the inductor

$$20\ \angle 90^\circ + 20\ \angle 0^\circ$$
$$= 20\sqrt{2}\ \angle 45^\circ$$

Hence the instantaneous current through the inductor $= 20\sqrt{2}\ \sqrt{2}\ \sin(0.5\ t + 45^\circ)$

The instantaneous energy through the capacitor

$$\frac{1}{2}cv^2 = \frac{1}{2} \times 2\left(20\sqrt{2}\ \sin 0.5t\right)^2$$
$$= 800\ \sin^2 0.5t$$

Energy through the inductor

$$\frac{1}{2}L\,i^2 = \frac{1}{2} \times 1 \times [40\sin(0.5t + 45^\circ)]^2$$
$$= 800\ \sin^2(0.5t + 45^\circ)$$

$$Q \text{ of the circuit} = \frac{2\pi\ \text{Energy stored | max}}{\text{Energy dissipated}}$$

loss in the resistor $= 400$ watts

loss per cycle $= \dfrac{400}{f_0}$

Now

$$w_c(t) + w_L(t) = 800\left[1 + \frac{\sin t - \cos t}{2}\right]$$
$$= 800\left[1 + \frac{\sqrt{2}}{2}\right]$$

Therefore

$$Q_0 = \frac{0.5 \times 800}{400}\left[1 + \frac{1}{\sqrt{2}}\right]$$
$$= 1 + \frac{1}{\sqrt{2}} = 1.707 \textbf{ Ans.}$$

**Example 9.19:** Determine the current through all the branches of the given network.

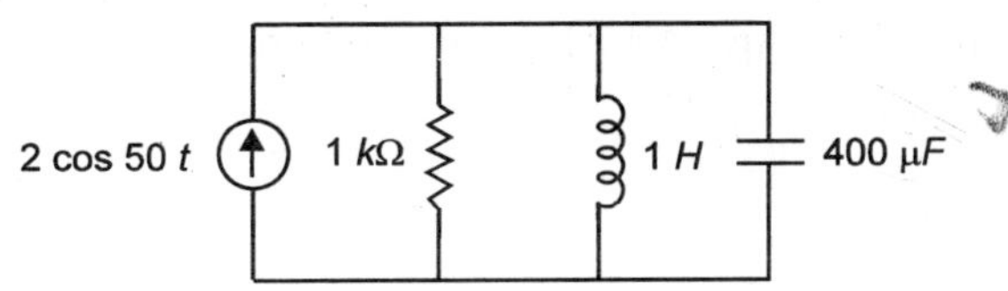


**Fig. E9.19**

**Solution:**  The natural angular frequency

$$= \frac{1}{\sqrt{1 \times 400 \times 10^{-6}}} = \frac{1000}{20} = 50 \;\; \text{rad/sec}$$

Therefore current source at $\omega = 50$ rad/sec produces resonance in the circuit.
The current $i$ supplied by the source is 2 A peak
The voltage across the source is

$$iR = 2 \times 1000 = 2000 \;\text{volts}$$
$$v(t) = 2000 \sin 50 \; t$$

The current through the inductor

$$\frac{2000}{j \times 50 \times 1} = -j40$$
$$= 40 \; \cos \; (\omega t - 90^\circ)$$
$$= 40 \; \sin \; \omega t$$

Current through capacitor

$$i_c \; (t) = \frac{v(t)}{1/j\omega_0 C} = j \; 2000 \times 50 \times 10^{-6} \times 400$$
$$= j40 \; \text{A}$$
$$= 40 \; \cos \; (50t + 90^\circ)$$
$$= -40 \; \sin \; 50t \; \textbf{Ans.}$$

## 9.6  MULTIPLE RESONANCE IN HIGH—Q CIRCUITS

Consider Fig. 9.14. The input impedance of the circuit,

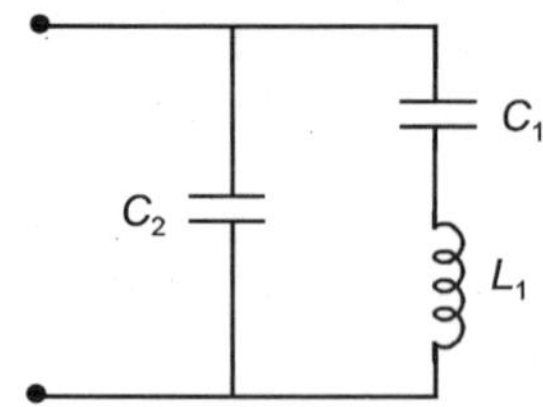

**Fig. 9.14**  Multiple resonance.

$$Z_{\text{in}} = \frac{(-j\omega C_2)\,(j\omega L_1 - j/\omega C_1)}{j\omega L_1 - j\left(\dfrac{1}{\omega C_1} + \dfrac{1}{\omega C_2}\right)}$$

For series resonance the equivalent input impedance should be zero which means

$$\omega L_1 - \frac{1}{\omega C_1} = 0$$

or

$$\omega_{0s} = \frac{1}{\sqrt{L_1 C_1}}$$

Similarly for parallel resonance the equivalent input impedance should be infinite which means

$$\omega L_1 = \frac{1}{\omega C_1} + \frac{1}{\omega C_2}$$

Let

$$C_T = \frac{C_1 C_2}{C_1 + C_2}$$

Therefore,

$$\omega L_1 = \frac{1}{\omega C_T}$$

and

$$\omega_{0p} = \frac{1}{\sqrt{L_1 C_T}}$$

As $C_T$ is smaller than $C_1$, the high impedance of the parallel resonance must occur at a higher frequency than the low impedance of the series resonance. At low frequencies (d.c approx.) the capacitor blocks the current in the circuit and hence the impedance is very high. At very high frequencies the capacitor $C_2$ acts a short circuit and hence the impedance is low.

In actual practice at series resonance the impedance of the circuit is not zero and is equal to the resistance of the coil whereas the impedance of the circuit at parallel resonance is not infinite but is a finite value less than $QX_C$ value we found for the conventional parallel resonant circuit.

Consider a circuit having an ideal coil and two capacitors as shown here.

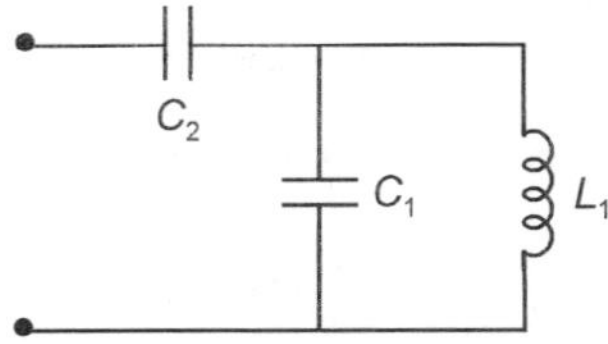


**Fig. 9.15**  Multiple resonance.

The input impedance of the circuit is

$$Z_{\text{in}} = -\frac{j}{\omega C_2} + \frac{(-j/\omega C_1)(j\omega L_1)}{j\omega L_1 - j/\omega C_1}$$

$$= \frac{\omega L_1\left(\dfrac{1}{\omega C_1} + \dfrac{1}{\omega C_2}\right) - \dfrac{1}{\omega C_1} \cdot \dfrac{1}{\omega C_2}}{j\left(\omega L_1 - \dfrac{1}{\omega c_1}\right)}$$

For series resonance the effective impedance is zero and hence

$$\omega_{0s} L_1 \frac{(\omega_{0s} C_2 + \omega_{0s} C_1)}{\omega_{0s}^2 C_1 C_2} = \frac{1}{\omega_{0s}^2 C_1 C_2}$$

or
$$\omega_{0s}^2 \, L_1 \, (C_1 + C_2) = 1$$

or
$$\omega_{0s} = \frac{1}{\sqrt{L_1(C_1 + C_2)}}$$

However, for parallel resonance the input impedance should be infinite hence the term in the denominator should be zero.

or
$$\omega_{0p} = \frac{1}{\sqrt{L_1 C_1}}$$

Here again parallel resonance occurs at a higher frequency as compared to series resonance.

As with the previous circuit the magnitude of impedance at low frequency is very high whereas at high frequencies it presents a very low impedance.

**Example 9.20:** Determine the frequencies of resonance for the circuits in Fig. 9.14 and Fig. 9.15 given.

Fig. 9.14    $C_1 = 6 \ \mu F$    $C_2 = 3 \ \mu F$    $L_1 = 200 \ mH$
Fig. 9.15    $C_1 = 3 \ \mu F$    $C_2 = 6 \ \mu F$    $L_1 = 200 \ mH$

**Solution:** For (*a*) Fig. 9.14

$$f_{0s} = \frac{1}{2\pi \sqrt{L_1 C_1}}$$

$$= \frac{1}{2\pi \sqrt{200 \times 6 \times 10^{-9}}} = 145 \ Hz$$

$$f_{0p} = \frac{1}{2\pi \sqrt{200 \times 2 \times 10^{-9}}} = 251 \ Hz$$

(*b*) $f_{0p}$ in this circuit is same as $f_{0s}$ in circuit (a) and hence (In Fig. 9.15 $C_2$ is in series with $L_1$)

$$f_{0p} = 145 \ Hz$$

$$f_{0s} = \frac{1}{2\pi \sqrt{200 \times 9 \times 10^{-9}}}$$

$$= 118 \ Hz \ \textbf{Ans.}$$

**Example 9.21:** Determine the resonant frequency, $Q_0$ and bandwidth for the circuit shown in Fig. E9.21.

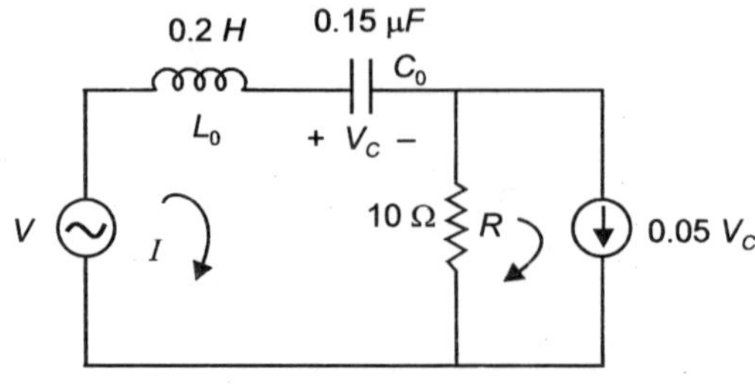


**Fig. E9.21**

**Solution:** The KVL equation is

$$j\omega L I + \frac{I}{j\omega C} + (I - 0.05\,V_c)\,R = V$$

and
$$V_c = \frac{I}{j\omega C}$$

The above equation can be rewritten as

$$j\omega L I + \frac{I}{j\omega C} + I R - 0.05\,\frac{I R}{j\omega C} = V$$

or
$$\frac{V}{I} = R + j\left\{\omega L - \frac{1}{\omega C}\,(1 - 0.05\,R)\right\}$$

At resonance the imaginary part is zero, therefore,

$$\omega_0\,L = \frac{1 - 0.05\,R}{\omega_0\,C}$$

or
$$\omega_0^2 = \frac{1 - 0.05\,R}{LC}$$

$$f_0 = \frac{1}{2\pi\,\sqrt{LC}}\cdot\sqrt{1 - 0.05 \times 10}$$

$$= \frac{0.707}{2\pi\,\sqrt{0.2 \times 0.15 \times 10^{-6}}}$$

$$= 650\ \text{Hz}$$

The quality factor is

$$Q_0 = \frac{\omega_0\,L}{R}$$

$$= \frac{4.082 \times 10^3 \times 0.2}{10}$$

$$= 96.4$$

Bandwidth
$$= \frac{f_0}{Q_0} = \frac{650}{96.4} = 6.74\ \text{Hz}\ \textbf{Ans.}$$

**Example 9.22:** Determine the resonant frequency, quality factor $Q_0$ and bandwidth of the given circuit.

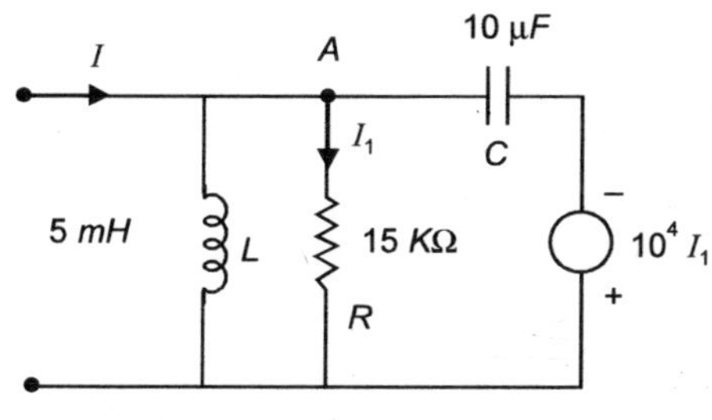


**Fig. E9.22**

**Solution:** Applying Kirchhoff current law at node $A$ we have

$$I - \frac{V_A}{j\omega L} - \frac{V_A}{R} - \frac{V_A + 10^4 I_1}{1/j\omega C} = 0$$

and

$$I_1 = \frac{V_A}{R}$$

Substituting this we have

$$I = \frac{V_A}{j\omega L} + \frac{V_A}{R} + \frac{V_A + \dfrac{10^4 V_A}{R}}{1/j\omega C}$$

$$= V_A \left[ \frac{1}{R} - \frac{j}{\omega L} + j\omega C \left( 1 + \frac{10^4}{R} \right) \right]$$

$$= V_A \left[ \frac{1}{R} + j \left\{ \omega C \left( 1 + \frac{10^4}{R} \right) - \frac{1}{\omega L} \right\} \right]$$

or

$$Y_{\text{in}} = \frac{I}{V_A} = \frac{1}{R} + j \left[ \omega C \left( 1 + \frac{10^4}{1.5 \times 10^4} \right) - \frac{1}{5 \times 10^{-3}\,\omega} \right].$$

For resonance

$$\omega_0 C \,(1 + 0.667) = \frac{10^3}{5\,\omega_0}$$

or

$$\omega_0 \times 10 \times 10^{-6} \,(1.667) = \frac{10^3}{5\,\omega_0}$$

or

$$\omega_0^2 = \frac{10^3 \times 10^5}{1.667 \times 5} = 11.99 \times 10^6$$

$$\omega_0 = 3.46 \; k \; \text{rad/sec}$$

or

$$f_0 = 551 \; \text{Hz}$$

The quality factor

$$\frac{R}{\omega_0 L} = \frac{15 \times 10^3}{3.46 \times 10^3 \times 5 \times 10^{-3}}$$

$$= \frac{15 \times 10^3}{5 \times 3.46}$$

$$= 867 \; \textbf{Ans.}$$

Hence bandwidth

$$\frac{f_0}{Q_0} = \frac{551}{867}$$

$$= 0.635 \; \text{Hz} \; \textbf{Ans.}$$

## QUESTIONS

**9.1.** Discuss various characteristics of a series RLC resonant circuit. Derive mathematical expressions in support of your discussion.

**9.2.** Define $Q$ of a coil and $Q$ of a series RLC circuit. Show that the voltage across the inductor or capacitor in a series RLC circuit is $Q$ times the supply voltage at resonance.

**9.3.** What are half-power frequencies in a series RLC resonance circuit? Derive an expression for bandwidth of the circuit.

**9.4.** On overhead lines why it is important to avoid series resonance phenomenon?

**9.5.** Explain very briefly subsynchronous resonance and Ferroresonance phenomena in power systems.

**9.6.** Explain the variation of phase between the supply voltage and circuit current when frequency is varied in a series RLC circuit.

**9.7.** Discuss various characteristics of a general parallel RLC circuit at resonance. Derive resonance frequency when (a) coil is loss free (b) capacitor is loss free (c) both are loss free. Draw the equivalent circuit when $Q$ of the circuit is high.

**9.8.** Derive an expression for $Q$ of a parallel resonance circuit from first principles.
A given parallel circuit has certain bandwidth. If it is required to obtain a lower bandwidth what measure should be taken to achieve this?

**9.9.** Explain how the tuning (the resonant frequency) can be changed in a parallel resonant circuit by varying the resistance of the coil.

**9.10.** Explain what you mean by selectivity of a circuit and explain how it can be improved.

**9.11.** "Voltage is magnified in a series resonance circuit while current is magnified in a parallel resonance circuit". Explain.

**9.12.** If multiple resonance occurs in a high $Q$ parallel circuit, explain how both a series resonance and a parallel resonance can occur.

## PROBLEMS

**9.1.** A 70 ohm resistor, a 5 mH coil and a 15 pF capacitor are in series across a 110 volt supply. Determine the resonant frequency the $Q$ of the circuit at this frequency, the voltage across the capacitor at resonance.

**9.2.** A variable inductor is connected in series with a resistor and a capacitor. The circuit is connected to a 200 volt 50 Hz supply and the maximum current obtainable by varying inductance is 0.314 A. The voltage across the capacitor then is 300 V. Determine the circuit constants.

**9.3.** A coil of resistance 50 ohm and inductance 9 H is connected in series with a capacitor and is supplied at constant voltage and variable frequency source. If the maximum current is 1A at 75 Hz determine the frequencies when the current is 0.5A.

**9.4.** An inductor in series with a variable capacitor is connected across a constant voltage at a frequencies of 1 MHz. The current is maximum when the capacitor is 500 pF and it reduces to half when it is 600 pF. Determine (a) the resistance (b) inductance and (c) $Q$ factor of the inductor.

**9.5.** A coil of 20 $\Omega$ resistance and 0.2 H inductance is connected in parallel with a capacitor of 100 $\mu$F capacitance. Find the frequency of resonance and the effective impedance at resonance.

**9.6.** A coil of resistance $R$ and inductance $L$ is connected across a capacitor $C$. Determine the impedance of the circuit at resonance. Show that the circulating current is $V\sqrt{C/L}$ so long as the resistance is negligible. If in the above circuit $R = 1\ \Omega$, $L = 10\ \mu H$ and $C = 10^4$ pF determine the current at resonance if $V$ is the supply voltage.

**9.7.** A series circuit consists of a 115 $\Omega$ resistor, a 0.024 $\mu$F capacitor and coil of inductance $L$. If the resonant frequency of the circuit is 1000 Hz determine $L$ and the bandwidth.

**9.8.** A 0.0015 $\mu$F capacitor is connected in parallel with a coil of 75 $\Omega$ and 25 mH inductance. Determine the impedance of the circuit at resonance and its bandwidth.

**9.9.** If in the above problem P9.8 a loading resistance of 0.5 M $\Omega$ is placed in parallel with the circuit, determine the new circuit impedance at resonance, the effective $Q$ and the new bandwidth.

**9.10.** Determine the loading resistor required to produce a bandwidth of 500 Hz for the circuit of problem 9.8.

**9.11.** For the network shown derive expression for series resonance frequency and parallel resonance frequencies.

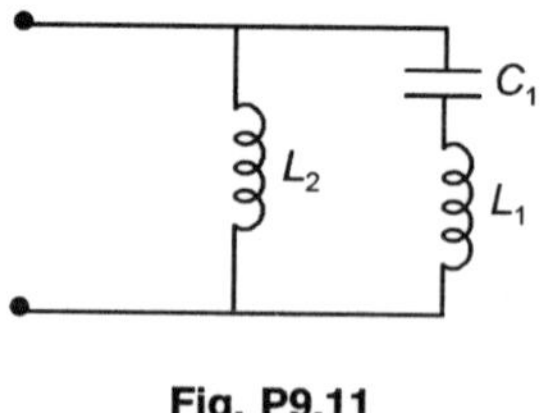

**Fig. P9.11**

If $L_1 = 50$ mH $C_1 = 1$ $\mu$F and $L_2 = 450$ mH
determine these frequencies.

**9.12.** For the network shown, derive expression for series and parallel resonance frequencies.

If $L_1 = 450$ mH $C_1 = 1$ $\mu$F and $L_2 = 50$ mH
determine these frequencies

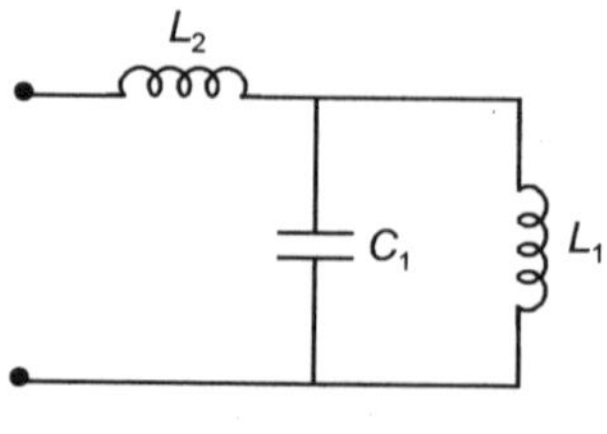

**Fig. P9.12**

**9.13.** (*a*) A resonant series $RLC$ circuit has $R = 5$ $\Omega$ and operates at 40 V. Find the power at half-power frequency (*b*) the bandwidth of a series resonant circuit is 100 Hz and the resonant frequency 1 KHz. If the circuit resistance is 10 $\Omega$ find $L$ and $C$ (*c*) Find the half power frequency of a series RLC circuit having $Q_0 = 50$ and $f_0 = 10$ KHz.

**9.14.** In the circuit shown in Fig. P9.14 $v_1(t) = 220 \cos \omega_0 t$. Determine equivalent parallel RLC circuit and calculate $\omega_0$ and $Q_0$. Also determine $v_2(t)$, $i_1(t)$, $i_2(t)$, $i_3(t)$ average power loss in the 10 k $\Omega$ resistor and the maximum energy stored in the inductor.

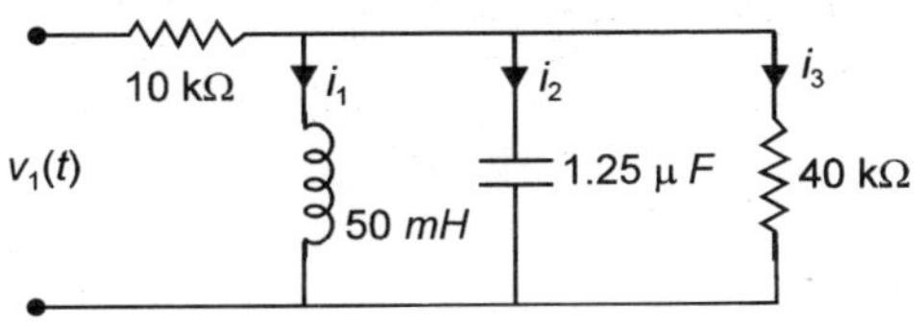

**Fig. P9.14**

**9.15.** A certain series resonant circuit has a maximum current of 50 mA and $V_L$ of 100 volts. The applied voltage is 10 V. What is $Z$? What are $X_L$, $X_C$ and $Q$ of the circuit?

**9.16.** A coil of resistance 10 ohm and inductance 0.5 H is connected in series with a capacitor $C_1$. On applying a sinusoidal voltage the current is maximum when the frequency is 50 Hz. A second capacitor $C_2$ is connected in parallel with this circuit. Determine the value of capacitance so that the combination resonates at 100 Hz. Determine the total current in each case if the applied voltage is 230 V.

# 10
# Passive Network Synthesis

## 10.1 INTRODUCTION

So far we have studied what is known as Network analysis that is, given a network and the source of excitation to find out the response of the network that may be to find out branch voltages or currents nodal voltages etc. and this solution of the network is unique. However, network synthesis implies given response for a given source of excitation, to obtain a network that would adhere to these conditions. In general the network synthesis gives various options to satisfy the excitation response requirements so it does not give a unique solution. The network synthesis problem can be stated as follows. Given the impulse response (where excitation is an impulse) to obtain the requisite network elements which gives this response. We know that the Laplace transform of the impulse response in time-domain when the system is relaxed is known as the transfer function of the network. The networks we are considering (passive networks) are known to be 'causal' and 'stable'. By causality is meant that effects never precede their causes i.e. response can't be obtained without excitation and so the response $h(t)$ of a physical system to an impulse applied at $t = 0$ must be zero for negative values of $t$ :

$$h(t) = 0 \quad t < 0 \qquad \qquad ...(10.1)$$

Impulse responses satisfying this condition are said to be causal so this is one of the conditions for physical realisability of the network. As an example the impulse response.

$$h(t) = e^{-at} \, . \, u(t) \text{ is causal whereas}$$

$$h(t) = e^{-a|t|} \text{ is not causal}$$

Under certain circumstances the impulse response could be made causal (realizable) by delaying it appropriately. For example the impulse response in Fig. 10.1($a$) is non-causal. However, if we delay it by a time $t_0$ we find that the response now becomes causal or realizable.

**Fig. 10.1**   (a) Non causal function (b) Causal function.

However in frequency domain the causality (realisability) is implied when Paley-Wiener condition is satisfied which states that $H(j\omega)$ Fourier transform of $h(t)$ is realizable if and only if

$$\int_{-\infty}^{\infty} \frac{\ln |H\,(j\omega)|}{1 + \omega^2}\, d\omega < \infty \qquad \ldots(10.2)$$

In fact before $H\,(j\omega)$ is tested for Paley Wiener condition it must be ascertained whether the function $|H(j\omega)|^2$ is integrable or not i.e. whether

$$\int_{-\infty}^{\infty} |H\,(j\omega)|\, d\omega < \infty \qquad \ldots(10.3)$$

The physical interpretation of Paley Wiener criterion is that the magnitude $|H(j\omega)|$ of a realisable network must not be zero over a finite band of frequencies or the amplitude function can't fall off to zero value abruptly or falls off to zero value faster than exponential order. For example consider an ideal low pass filler shown in Fig. 10.2 $(a)$.

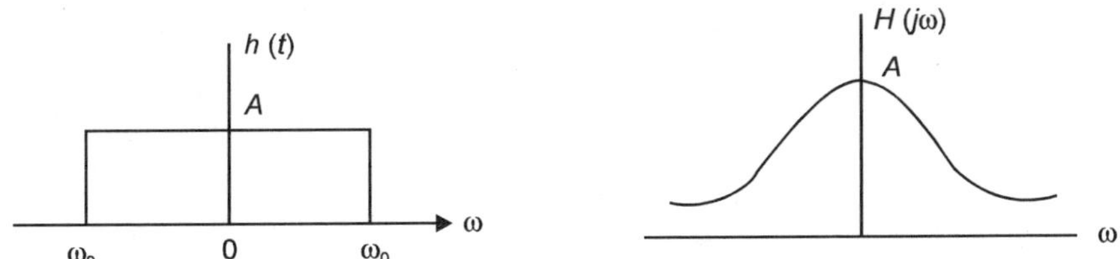


**Fig. 10.2**  (a) Ideal low pass filter.

A filter with this $H(j\omega)$ can't be realized because $|H(j\omega)|$ abruptly becomes zero beyond $\omega_0$.

Also the Gaussian shaped curve

$H(j\omega) = e^{-\omega^2}$ shown in Fig. 10.2 $(b)$ is also not realizable because

$$\ln\,(|H(j\omega)| = \omega^2 \qquad \ldots(10.4)$$

and the integral $\displaystyle\int_{-\infty}^{\infty} \frac{\omega^2}{1 + \omega^2}\, d\omega$ is not finite.

On the other hand the amplitude function

$$|H(j\omega)| = \frac{1}{\sqrt{1 + \omega^2}} \qquad \ldots(10.5)$$

represents a realizable network.

Now the second requirement for the network to be realizable is that it should be stable which means for a bounded input there should be bounded output i.e. if

$$|e(t)| < C_1 \qquad 0 \leq t \leq \infty \qquad \ldots(10.6)$$

then

$$|r(t)| < C_2 \qquad 0 \leq t \leq \infty \qquad \ldots(10.7)$$

where $C_1$ and $C_2$ are real positive, finite quantities.

The stability criterion in frequency domain requires that the system function posses poles in the left half plane or on the $j$-$\omega$ axis only. Also the poles on $j$-$\omega$ axis must be simple. Since only simple poles on $j$-$\omega$ axis are permissible, if $H(s)$ is given as

$$H(s) = \frac{a_n s^n + a_{n-1} s^{n-1} + \cdots + a_1 s + a_0}{b_m s^m + b_{m-1} s^{m-1} + \cdots + b_1 s + b_0} \qquad \qquad ...(10.8)$$

then the order of the numerator '$n$' can't exceed the order of the denominator by more than unity i.e. $(n - m) \leq 1$. If $n$ exceeded $m$ by more than unity this would mean that at $s = j\omega$ $= \infty$ and there would be a multiple pole. It is to be noted that $H(s)$ is the transfer function of the proposed network to be synthesized.

The elements required in synthesizing these networks are linear passive and time – invariant and are inherently causal i.e. in such networks the response (effect) can't precede the excitation (cause) and therefore, given the transfer function we need not bother about the causality condition. We must test for the second condition that is the stability of the network which indirectly means we are left with the checking of physical realisability of $H(s)$ in terms of the following conditions:

1. $H(s)$ should not have poles in the right half of the $s$-plane as this would mean a term of the type $e^{at}$ where $a$ is positive in the time domain and the system would be unstable.
2. $H(s)$ should not have multiple poles in the $j$-$\omega$ axis.
3. The degree of numerator of $H(s)$ should not be more than unity with reference to the denominator. This means in order to test $H(s)$ for its physical realisability we have to test the denominator polynomial of $H(s)$ in the light of the above requirements. In fact the denominator polynomials of the transfer function belong to a class of polynomials known as Hurwitz polynomials.

A polynomial $P(s)$ is said to be Hurwitz if the following conditions are satisfied:
1. $P(s)$ is real when $s$ is real.
2. The roots of $P(s)$ have real parts which are zero or negative which means these lie along $j$-$\omega$ axis or in the negative half of $s$-plane.
Suppose $P(s)$ is given as

$$P(s) = a_n s^n + a_{n-1} s^{n-1} + \cdots + a_1 s + a_0 \qquad \qquad ...(.10.9)$$

then (1) all the co-efficients $a_n \ldots a_0$ must be real and positive which means that between the highest order term in $s$ and the lowest order term, none of the co-efficients is zero unless of course the polynomial is even or odd.

(2) The second property of Hurwitz polynomial is that both odd and even parts have roots on the $j$-$\omega$ axis only. If we denote the odd and even parts of $P(s)$ by $n(s)$ and $m(s)$ respectively then

$$P(s) = n\,(s) + m\,(s) \qquad \qquad ...(10.10)$$

Then $n(s)$ and $m(s)$ both have roots on the $j$-$\omega$ axis only. Therefore, if $P(s)$ is either even or odd all its roots are on $j\omega$ axis.

(3) The continued fraction expansion of the ratio of the odd to even parts or even to odd parts of $P(s)$ yields all positive quotients.

Continued fraction expansion requires division and inversion and has the following important characteristics:

1. The continued fraction expansion of even to odd parts or vice versa of a polynomial must be finite in length i.e. the process terminates and does not continue indefinitely.
2. If the continued fraction expansion of odd to even parts or vice versa of a polynomial yields positive quotient term, then the polynomial must be Hurwitz to within a multiplicative factor $W(s)$. That is if we write

$$F(s) = W(s) \cdot F_1(s) \qquad \qquad ...(10.11)$$

then $F(s)$ is Hurwitz if $W(s)$ and $F_1(s)$ are Hurwitz.

Therefore, in order to check whether the given polynomial is Hurwitz or not following tests must be carried out:

1. All the co-efficients of the polynomial must be real and positive and none of them must be missing except if the polynomial has only even or odd order terms. This can be done by inspection of the polynomial, no calculation is required.
2. The quotients of the continued fraction expansion of even to odd parts or vice versa of the polynomial are positive (This is done by actual carrying out the continued fraction expansion).

However, it is to be noted that if $P(s)$ is completely even or completely odd then the second parts for continued fraction expansion are obtained by differentiating $P(s)$ with reference to $s$ i.e. the continued fraction to begin with has the parts $\dfrac{P(s)}{P'(s)}$. It can be seen that if $P(s)$ is completely odd then $P'(s)$ is completely even and vice versa.

We will now illustrate with the help of few examples the process of testing the polynomial whether it is Hurwitzian or not.

**Example 10.1:** Check if the polynomial
$$P(s) = 2\,s^4 + 5\,s^3 + 6\,s^2 + 2\,s + 1$$
is Hurwitzian.

**Solution:**
$$m(s) = 2\,s^4 + 6\,s^2 + 1$$
$$n(s) = 5\,s^3 + 2\,s$$

By inspection, all the co-efficients in the polynomial are positive and no power of $s$ is missing. The next test is to obtain continued fraction expansion.

$$5\,s^3 + 2s \overline{\left)\,2\,s^4 + 6\,s^2 + 1\,\right.}\,\dfrac{2s}{5}$$
$$\underline{2\,s^4 + \dfrac{4}{5}s^2}$$
$$\dfrac{26}{5}s^2 + 1 \overline{\left)\,5\,s^3 + 2\,s\,\right.}\,\dfrac{25}{26}s$$
$$\underline{5\,s^3 + \dfrac{25}{26}s}$$
$$\dfrac{27}{26}s \overline{\left)\,\dfrac{26}{5}s^2 + 1\,\right.}\,\dfrac{676}{135}s$$
$$\underline{\dfrac{26}{5}s^2}$$
$$1 \overline{\left)\,\dfrac{27}{26}s\,\right.}\,\dfrac{27}{26}s$$
$$\underline{\dfrac{27}{26}s}$$
$$\times$$

Since all the quotients are positive, the given polynomial is Hurwitzian.

**Example 10.2:** Consider the polynomial
$$P(s) = s^3 + 3 s^2 + 4 s + 12$$
and test whether it is Hurwitzian.

By inspection, all the co-efficient are positive and no power of $s$ is missing, therefore, it should be Hurwitzian

The continued fraction expansion
$$m (s) = 3 s^2 + 12$$
$$n (s) = s^3 + 4 s$$

$$\frac{n(s)}{m(s)} = 3s^2 + 12 \; \dfrac{\left. \begin{array}{l} s^3 + 4s \\ \underline{s^3 + 4s} \end{array} \right| ^{s/3}}{0}$$

It is seen that the division has been terminated abruptly by a common factor $s^3 + 4s$. The polynomial can therefore be written as

$$P(s) = (s^3 + 4s)\left(1 + \frac{3}{s}\right) = W(s)\, F_1\,(s)$$

We know that the factor $\left(1 + \dfrac{3}{s}\right)$ is Hurwitz. Since the multiplication factor $W(s)$ is also Hurwitz, $P(s) = W(s)\, F_1(s)$ is Hurwitz. The term $(s^3 + 4\,s)$ is the multiplicative factor which we referred to earlier.

**Example 10.3:** Test the polynomial
$$P(s) = s^4 + s^3 + 2s^2 + 4s + 3$$
whether it is Hurwitz or not
$$m(s) = s^4 + 2s^2 + 3$$
$$n(s) = s^3 + 4s$$

$$s^3 + 4s \,\overline{\left| s^4 + 2\,s^2 + 3 \right.}\, s$$
$$\underline{s^4 + 4\,s^2}$$
$$-2\,s^2 + 3 \,\overline{\left| s^3 + 4\,s \right.}\, -\frac{s}{2}$$
$$\underline{s^3 - \frac{3s}{2}}$$
$$\frac{11s}{2} \,\overline{\left| -2\,s^2 + 3 \right.}\, -\frac{4}{11}s$$
$$\underline{-2\,s^2}$$
$$3 \,\overline{\left| \frac{11}{2}\,s \right.}\, \frac{11}{6}\,s$$
$$\underline{\frac{11}{2}\,s}$$
$$\times$$

Since there are negative quotients the polynomial is non-Hurwitz.

**Example 10.4:** Test whether the polynomial
$$P(s) = s^7 + 2\ s^6 + 2\ s^5 + s^4 + 4\ s^3 + 8\ s^2 + 8\ s + 4$$
is Hurwitz or not
$$m(s) = 2\ s^6 + s^4 + 8\ s^2 + 4$$
$$n(s) = s^7 = 2\ s^5 + 4\ s^3 + 8\ s \cdot$$

The continued fraction expansion

$$2\ s^6 + s^4 + 8\ s^2 + 4 | s^7 + 2\ s^5 + 4\ s^3 + 8\ s | s/2$$

$$s^7 + \frac{s^5}{2} + 4\ s^3 + 2\ s$$
______

$$\frac{3}{2}\ s^5 \qquad\qquad + 6\ s | 2\ s^6 + s^4 + 8\ s^2 + 4 | \frac{4}{3}\ s$$

$$2\ s^6 + \qquad + 8\ s^2$$
______

$$s^4 + 4 \quad \left|\frac{3}{2}\ s^5 + 6 s\right| \frac{3}{2}\ s$$

$$\frac{3}{2}\ s^5 + 6 s$$
______

$$\times$$
______

Here again the division has terminated abruptly by a common factor $W(s)$ $(s^4 + 4)$ which can be factorised as follows:
$$s^4 + 4\ s^2 + 4 - 4\ s^2$$
$$(s^2 + 2)^2 - (2\ s)^2$$
$$= (s^2 + 2\ s + 2)\ (s^2 - 2\ s + 2)$$

Since here one of the co-efficients in $(s^2 - 2\ s + 2)$ is negative, the polynomial as a whole is non-Hurwitz. Also since in $s^4 + 4$, the $s^2$ term is missing hence it is non-Hurwitz.

As mentioned earlier, in case the polynomial has all even order terms or all odd terms, then we can't formulate either $m(s)$ or $n(s)$. To consider for such polynomials for Hurwitz or non-Hurwitz we take the derivative of the original polynomial which will have its order one less than the original polynomial and hence the ratio $\dfrac{m(s)}{n(s)}$ or $\dfrac{n(s)}{m(s)}$ can be tested.

**Example 10.5:** Test whether the polynomial
$$P(s) = s^5 + 3\ s^3 + s$$
is Hurwitzian or not

**Solution:** $\qquad P'(s) = 5\ s^4 + 9\ s^2 + 1$

Therefore, finding the continued fraction expansion

$$5 s^4 + 9 s^2 + 1 \left| s^5 + 3 s^3 + s \right| \frac{s}{5}$$

$$s^5 + \frac{9}{5}\ s^3 + \frac{s}{5}$$
______

$$\frac{6}{5}s^3 + \frac{4}{5}s \overline{\smash{\big)}5s^4 + 9s^2 + 1}\ \Big|\ \frac{25}{6}s$$

$$\underline{5s^4 + \frac{10}{3}s^2}$$

$$\frac{17}{3}s^2 + 1 \overline{\smash{\big)}\frac{6}{5}s^3 + \frac{4}{5}s}\ \Big|\ \frac{18}{85}s$$

$$\underline{\frac{6}{5}s^3 + \frac{18}{85}s}$$

$$\frac{10}{17}s \overline{\smash{\big)}\frac{17}{3}s^2 + 1}\ \Big|\ \frac{17}{30}s$$

$$\underline{\frac{17}{3}s^2}$$

$$1 \overline{\smash{\big)}\frac{10}{17}s}\ \Big|\ \frac{10}{17}s$$

$$\underline{\frac{10}{17}s}$$

$$\times$$

Since all the quotients are positive the polynomial is Hurwitz.

## 10.2 POSITIVE REAL FUNCTION (PRF)

We next study a very important class of functions known as positive real functions. These functions are important in the sense, that if a function is positive real, this represents a physically realizable passive driving point immittance.

Now, a function $F(s)$ is said to be positive real if it satisfies the following conditions:

1. $F(s)$ is real for real $s$ i.e. $F(\sigma)$ is real (as $s$ in general is $\sigma + j\omega$ we take only real part of $s$).
2. Real $F(s) \geq 0$ if $Re(s) \geq 0$

To understand the first requirement consider the $s$-plane and the $F(s)$ plane shown in Fig. 10.3 $(a)$ and $(b)$.

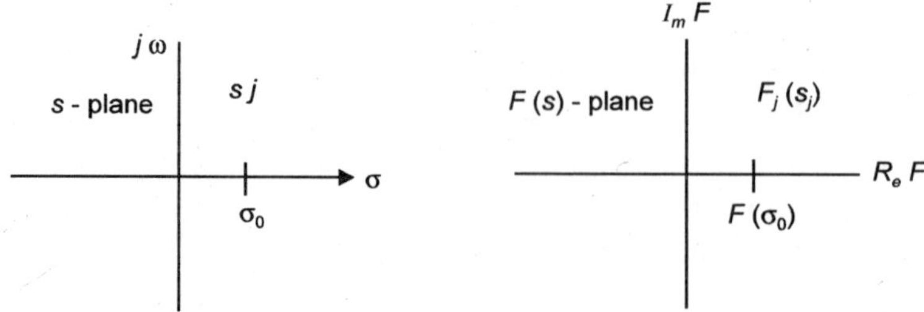


**Fig. 10.3**   (a) $s$-plane (b) $F(s)$ plane.

If $F(s)$ is positive real, a point $\sigma_0$ on the positive real axis of the $s$-plane would correspond to a point $F(\sigma_0)$ in the $F(s)$ plane and this point must be on the positive real axis as shown. Similarly a point $S_i$ in the right half of the $s$-plane must correspond to $F(S_i)$ in the right half of the $F(s)$ plane. This means for a positive real function the right half of the $s$-plane must correspond to (or map on to) the right half of the $F(s)$ plane. The real of the $s$-plane maps on to the real axis of the $F(s)$ plane. This means in order to test function for positive realness we will have to test the function for every single point in the right half of the $s$-plane which is a very tedious and time consuming job.

Therefore, we provide here an alternative set of necessary and sufficient conditions for a rational function with real co-efficient to be positive real. These are:

($a$) $F(s)$ must have no poles in the right-half of the s-plane.

($b$) $F(s)$ may have only simple poles on the $j$-$\omega$ axis with real and positive residues.

($c$) $R_e\, F(j\omega) \geq 0$   for all  $\omega$

Let us understand one by one the implication of these three conditions.

The first one requires that we must test the denominator of $F(s)$ whether it is Hurwitz or not which can be obtained through a continued fraction expansion of the polynomial as discussed earlier.

The second condition is tested by finding partial fraction expansion of $F(s)$ and checking whether the residues of the poles on the $j$-$\omega$ axis are positive and real. Therefore if $F(s)$ has a pair of poles at $s = \pm\, j\omega_1$, a partial fraction expansion gives terms of the form.

$$\frac{K_1}{s - j\omega_1} + \frac{K_1^*}{s + j\omega_1}$$

The residues of complex conjugate poles are conjugate themselves. If the residues are real which these should be for $F(s)$ to be positive real then $K_1 = K_1^*$ so that

$$\frac{K_1}{s - j\omega_1} + \frac{K_1^*}{s + j\omega_1} = \frac{2\,K_1\,s}{s^2 + \omega_1^2} \qquad \qquad ...(10.12)$$

Therefore, if $K_1$ is found to be positive, then $F(s)$ satisfies the second of the three conditions.

For the third condition we first find the real part of $F(j\omega)$ from the original function $F(s)$ suppose

$$F(s) = \frac{P(s)}{Q(s)} \qquad \qquad ...(10.13)$$

Say $M_1\, N_1$ and $M_2,\, N_2$ are the odd and even parts of $P(s)$ and $Q(s)$ respectively such that

$$F(s) = \frac{M_1(s) + N_1(s)}{M_2(s) + N_2(s)} \qquad \qquad ...(10.14)$$

Let us obtain even and odd parts of $F(s)$. To do that, we multiply numerator and denominator of $F(s)$ by $(M_2(s) - N_2(s))$ we obtain

$$\frac{M_1 + N_1}{M_2 + N_2}\; \frac{M_2 - N_2}{M_2 - N_2}$$

$$F(s) = \frac{M_1 M_2 - N_1 N_2}{M_2^2 - N_2^2} + \frac{M_2 N_1 - M_1 N_2}{M_2^2 - N_2^2} \qquad \qquad ...(10.15)$$

Here the first term is even whereas the second is odd as multiplication of even with even or odd with odd gives even function. Whereas multiplication of odd with even or vice versa gives an odd function.

If we substitute $s = j\text{-}\omega$ we find that the even part of any polynomial is real whereas the odd part is imaginary so that if $F(j\omega)$ is written as

$$F(j\omega) = R_e\,[F\,(j\omega)] + j\,I_m\,[\,F\,(j\omega)] \qquad \text{...(10.16)}$$

then
$$Re\,[F(j\omega)] = \text{Even}\,[F(s)]\,_{s\,=\,j\omega} \qquad \text{...(10.17)}$$

and
$$j\,I_m\,[F(j\omega)] = \text{Odd}\,[F(s)]\ s\,=\,j\omega \qquad \text{...(10.18)}$$

Therefore, to test for the third condition for positive realness we determine the real part of $F(j\omega)$ by finding out the even part of $F(s)$ and then substituting $s = j\omega$ we then check to see whether $R_e\,[F\,(j\omega)] \geq 0$ for all $\omega$.

The denominator of $R_e\,[F(j\omega)]$ is always a positive quantity because

$$M_2^2\,(j\omega) - N_2^2\,(j\omega) = M_2^2\,(\omega) + N_2^2\,(\omega) \geq 0$$

Here an extra $j$ or imaginary term in $N_2\,(j\omega)$ which when squared gives $-1$ and hence the denominator of $R_e\,F(j\omega)$ is the sum of two squared numbers and is always positive. Therefore, our problem is reduced to finding out whether

$$A(\omega^2) = M_1\,(j\omega)\,M_2\,(j\omega) - N_1\,(j\omega)\,N_2\,(j\omega) \geq 0 \qquad \text{...(10.19)}$$

Since $A(\omega^2)$ represents an even polynomial and may be written as

$$A(\omega^2) = A_0\,\omega^{2r} + A_2\,\omega^{2r-2} + A_4\,\omega^{2r-4} + \cdots + A_{2r-2}\,\omega^2 + A_{2r}$$

If all the co-efficients of $A\,(\omega^2)$ are positive then $A\,(\omega^2)$ is positive for all values of $\omega$ between 0 and $\infty$. However, if all the A-coefficients are not positive then a test known as Sturm's test is carried out which is explained as follows:

Here we first replace $\omega^2$ by $x$ and redesignate $A\,(\omega^2)$ by say $p_0\,(x)$. Therefore $A(\omega^2)$ is expressed as

$$A(\omega^2) = p_0(x) = a_0\,x^r + a_1\,x^{r-1} + \cdots + a_{r-1}\,x + a_r \qquad \text{...(10.20)}$$

Here $p_0(x)$ is the first of a set of functions known as Sturm's functions. The second function is obtained by differentiating the first one

Thus
$$p_1(x) = a_0\,r\,x^{r-1} + (r-1)\,a\,x^{r-2} + \cdots \qquad \text{...(10.21)}$$

Now if $p_0(x)$ is divided by $p_1(x)$, it gives a two-term quotient, the remainder is the negative of the next Sturm's function $p_2(x)$ i.e.

$$\frac{\mathrm{p}_0(x)}{p_1(x)} = \alpha_1\,x + \alpha_2 + \frac{-p_2(x)}{p_1(x)} \qquad \text{...(10.22)}$$

Here $p_2(x)$ is one degree lower than $p_1(x)$. The division is repeated to yield the next Sturm's function $p_3(x)$ i.e.

$$\frac{p_1(x)}{p_2(x)} = \alpha_3 + \alpha_4 + \frac{-p_3(x)}{p_2(x)} \qquad \text{...(10.23)}$$

The procedure is continued till ($i$) the last Sturm's function $p_r$ of degree zero is found or ($ii$) the remainder resulting from the division process is identically zero (premature termination).

After we have obtained $p_0$ ....... $p_r$, Sturm theorem states that the number of zeros of $p_0$ $(x)$ in the interval $0 < x < \infty$ is equal to $S_\infty - S_0$ where $S_\infty$ and $S_0$ are the number of sign changes in the set $(p_0.\ p_1 \ldots \ldots p_r)$ evaluated at $x = 0$ and $x = \infty$ respectively. If, however, there are no sign change i.e. $S_\infty - S_0 = 0$ then $A\,(\omega^2) \geq 0$, for all $\omega$. It is to be noted that the co-efficient $\alpha_1 \ldots \ldots \alpha_r$ are of no consequence to find $P_0$ through $P_r$ and are to be ignored.

Before we illustrate the application of Sturm theorem and to other related procedure for testing positive realness of given function (through problem) we list here the properties of positive real function.

## 10.2.1 Properties of positive real (p.r.) functions

1. If $F(s)$ is a p.r. function, then $1/F(s)$ is also positive real. This means, if a driving point impedance is positive real, then its reciprocal driving point admittance is also positive real.
2. The sum of two positive real functions is also positive real. From network point of view, if two impedances are connected in series the sum of the impedances is positive real. Similarly this holds good for two admittances. However, the difference of two positive real functions need not necessarily be positive real e.g.

$$F(s) = s - \frac{1}{s} \text{ is not positive real}$$

3. The poles and zeros of a positive real function can't have positive real parts i.e. they can't lie in the right half of the $s$-plane.
4. Simple poles with real positive residues can exist on the $j$-$\omega$ axis.
5. The poles and zeros of a positive real function are real or conjugate. These are never imaginary as that would mean that the elements constituting the network contains some imaginary element which is not true.
6. The difference between the highest power of the numerator and denominator polynomial is at most unity.
7. Similarly the difference between the lowest powers of the denominator and numerator is at the most unity. This condition, therefore, prevents the possibility of multiple poles or zeros at $s = 0$.

The passive elements to be used for synthesizing a network the $R$, $L$ and $C$ and the impedance and/or admittances of these elements are rational function of $s$ as is seen hereunder.

1. If $F(s) = sL$ where $L$ is a real positive number, it is a positive real by definition and $L$ is an inductance if $F(s)$ is an impedance.
2. $F(s) = R$ where $R$ is a real positive number, here again $F(s)$ is positive real by definition and if $F(s)$ is an impedance $R$ is the resistance.
3. $F(s) = \dfrac{k}{s}$ where $k$ is a real positive, here $F(s)$ is positive real by definition and $1/k$ is the capacitance in Farads. Here $F(s)$ is positive real because when $s$ is real $F(s)$ is real and when real part of $s > 0$, $R_e(s) = \sigma > 0$.
   Then

$$R_e\left(\frac{k}{s}\right) = \frac{k\sigma}{\sigma^2 + \omega^2} > 0$$

Therefore $F(s)$ is positive real.

Therefore, we find that the basic passive impedances are positive real functions. Similarly it is clear that the admittance

$$Y(s) = ks$$
$$Y(s) = k$$

and $$Y(s) = k/s$$
are positive real if $k$ is real and positive.

We now take a few examples for testing positive realness of the given function.

**Example 10.6:** Test whether the following polynomial is Hurwitzian or not.

$$F(s) = \frac{2s^4 + 6s^3 + 11s^2 + 10s + 5}{s^4 + 5s^3 + 8s^2 + 9s + 6}$$

**Solution:** Using Hurwitz Criterion

$$\frac{m_1(s)}{n_1(s)} = \frac{2s^4 + 11s^2 + 5}{6s^3 + 10s}$$

$$6s^3 + 10s\ \bigg|\ 2s^4 + 11s^2 + 5\ \bigg|\ \frac{s}{3}$$

$$\underline{2s^4 + \frac{10}{3}s^2\ \ \ \ \ \ }$$

$$\frac{23}{3}s^2 + 5\ \bigg|\ 6s^3 + 10s\ \bigg|\ \frac{18s}{23}$$

$$\underline{6s^3 + \frac{90}{23}s\ \ }$$

$$\frac{140}{23}s\ \bigg|\ \frac{23}{3}s^2 + 5\ \bigg|\ \frac{529}{420}$$

$$\underline{\frac{23}{3}s^2\ \ }$$

$$5\ \bigg|\ \frac{140}{23}s\ \bigg|\ \frac{28}{23}s$$

$$\underline{\frac{140}{23}s\ \ }$$

$$0$$

$$\frac{m_1(s)}{n_1(s)} = \frac{s}{3} + \cfrac{1}{\cfrac{18}{23}s + \cfrac{1}{\cfrac{529}{420}s + \cfrac{1}{\cfrac{28}{23}s}}}$$

Similarly $$\frac{m_2(s)}{n_2(s)} = \frac{s^4 + 8s^2 + 6}{5s^3 + 9s}$$

$$5s^3 + 9s\ \bigg|\ s^4 + 8\,s^2 + 6\ \bigg|\ \frac{s}{5}$$

$$\underline{s^4 + \frac{9}{5}s^2\ \ }$$

$$\frac{31}{5}s^2 + 6\ \bigg|\ 5\,s^3 + 9\ s\ \bigg|\ \frac{25}{31}s$$

$$\begin{array}{r} 5\,s^3 + \dfrac{150}{31}s \\[2pt] \underline{\phantom{5\,s^3 + \dfrac{150}{31}s}} \end{array}$$

$$\dfrac{129}{31}s \left|\; \dfrac{31}{5}s^2 + 6 \;\right| \dfrac{31^2\,s}{129 \times 5}$$

$$\dfrac{31}{5}s^2$$

$$\underline{\phantom{xxxx}}$$

$$6 \left|\; \dfrac{129}{31}s \;\right| \dfrac{129}{6 \times 31}s$$

$$\dfrac{129}{31}s$$

$$\underline{\phantom{xxxx}}$$

$$0$$

$$\frac{m_2(s)}{n_2(s)} = \frac{s}{5} + \cfrac{1}{\dfrac{25}{31}s + \cfrac{1}{\dfrac{31^2\,s}{129 \times 5} + \cfrac{1}{\dfrac{129}{31 \times 6}s}}}$$

Since all the quotients are positive the polynomial $F(s)$ is Hurwitz.

**Example 10.7:** Test whether the function

$$\frac{s + a}{s^2 + bs + c} \quad \text{is positive real}$$

Without going through the details it can be said that it is positive real if
1.  $a, b, c \geq 0$
2.  $b \geq a$

Test whether

$$\frac{s + 4}{s^2 + 5s + 3} \quad \text{is positive real}$$

Here $a = 4,\; b = 5,\; c = 3$

Therefore $a, b, c \geq 0$

and $b > a$ therefore it is positive real.

However the function

$$\frac{s + 2}{s^2 + 3}$$

is not positive real because $b \not> a$

**Example 10.8:** Test whether the function

$$A(\omega^2) = 3\omega^4 - 12\omega^2 + 4 \text{ is positive real}$$
$$P_0(x) = 3x^2 - 12x + 4$$
$$P_1(s) = 6x - 12$$

$$\frac{P_0(x)}{P_1(x)} = \frac{3x^2 - 12x + 4}{6x - 12}$$

or

$$6x - 12 \overline{)\begin{array}{l} 3x^2 - 12x + 4 \\ 3x^2 - 6x \end{array}} \quad \dfrac{x}{2} - 1$$

$$\begin{array}{r} - \ 6x + 4 \\ - \ 6x + 12 \\ \hline - \ 8 \ (-P_2) \end{array}$$

Therefore

$$P_0(x) = 3x^2 - 12x + 4$$
$$P_1(x) = 6x - 12$$
$$P_2(x) = +8$$

|         | $P_0(x)$ | $P_1(x)$ | $P_2(x)$ |          |
|---------|----------|----------|----------|----------|
| $x = 0$ | 4        | $-12$    | 8        | $s_0 = (2)$ |
| $x = \infty$ | $+$ | $+$      | $+$      | $s_\infty = 0$ |

Therefore, there are two changes in sign hence the function is not positive real.
If a function has both the polynomials as bi-quadratic as

$$F(s) = \frac{s^2 + a_1 s + a_0}{s^2 + b_1 s + b_0} \quad \text{to be positive real}$$

Again without going through the details the requirement for the function to be positive real are, calculate $a_1 b_1$ and $\left(\sqrt{a_0} - \sqrt{b_0}\right)^2$.

For $F(s)$ to be positive real

$$a_1 b_1 \geq \left(\sqrt{a_0} - \sqrt{b_0}\right)^2$$

For consider a function

$$F(s) = \frac{s^2 + 3s + 36}{s^2 + 3s + 25}$$

Since

$$a_1 b_1 = 3 \times 3 = 9$$

and

$$\sqrt{a_0} - \sqrt{b_0} = 6 - 5 = 1$$

Therefore $a_1 b_1 > 1$. Hence the function is positive real.

Let us consider a number of other helpful points by which a function might be tested quickly. Let the function $F(s)$ have poles on the $j\omega$-axis, a partial fraction expansion will show if the residues of these poles are positive real. Consider the function

$$F(s) = \frac{3s^2 + 8}{s(s^2 + 4)}$$

Here $F(s)$ has a pair of poles at $s = \pm j2$
The partial fraction expansion

$$\frac{A}{s} + \frac{Bs}{s^2 + 4} = \frac{As^2 + 4A + Bs^2}{s(s^2 + 4)}$$

Hence

$$A + B = 3$$
$$4A = 8 \quad \text{or} \quad A = 2, \ B = 1$$

and

$$F(s) = \frac{2}{s} + \frac{s}{s^2 + 4}$$

Since the residues are positive, the function is positive real.

It has already been mentioned that the driving point immittance of a time invariant passive network are positive real function. Let us consider, series and/or parallel combination of impedances and using the above property, we test these functions for positive realness. For example, consider two impedances $Z_1(s)$ and $Z_2(s)$ connected in parallel, the net impedance is

$$Z(s) = \frac{Z_1(s)\,Z_2(s)}{Z_1(s) + Z_2(s)} \qquad \qquad \text{...(10.24)}$$

Since the net impedance is also passive, therefore, $Z(s)$ must also be positive real. Consider the functions

$$F(s) = \frac{Ks}{s + a} \qquad \qquad \text{...(10.25)}$$

and

$$F(s) = \frac{K}{s + a} \qquad \qquad \text{...(10.26)}$$

where $K$ and $a$ are real and positive quantities the functions must be positive real. Thus it may be concluded that the function of the type.

$$F(s) = \frac{s + b}{s + a} = \frac{s}{s + a} + \frac{b}{s + a} \qquad \qquad \text{...(10.27)}$$

must be positive for $a, b \geq 0$

Again, let us consider the function of the type

$$F(s) = \frac{ks}{s^2 + a} \quad k, a \geq 0 \qquad \qquad \text{...(10.28)}$$

$F(s)$ could be rewritten as

$$F(s) = \frac{1}{\dfrac{s}{k} + \dfrac{a}{ks}} \qquad \qquad \text{...(10.29)}$$

Since $\dfrac{s}{k}$ and $\dfrac{a}{ks}$ are both positive real for $k, a \geq 0$ the sum of the two positive real function is also positive real. Also if a function is positive real its reciprocal is also positive real. Therefore, the given function $F(s)$ is positive real.

## 10.3 BASIC SYNTHESIS PROCEDURE

The basic philosophy of synthesis is, given a positive real function; decompose this into smaller function each one of them is again a positive real function and each function represents a single element ($R$, $L$ or $C$) or a suitable series or parallel combination of these elements. Of course, if all $Zi(s)$ were given to us, we would synthesise a network whose driving point impedance is $Z(s)$ by suitably connecting these $Zi(s)$. However, if we are given

$Z(s)$ then how would we obtain various $Zi(s)$ is the problem of synthesis. Suppose $Z(s)$ is given to be a ratio of two polynomials as

$$Z(s) = \frac{P(s)}{Q(s)} = \frac{a_n\, s^n + a_{n-1}\, s^{n-1} + \cdots + a_1 s + a_0}{b_m\, s^m + b_{m-1}\, s^{m-1} + \cdots + b_1 s + b_0} \qquad \ldots(10.30)$$

This function $Z(s)$ can be decomposed by either of the following methods or a combination of these. The methods are:

1. Removal of a pole at infinity.
2. Removal of a pole at origin.
3. Removal of conjugate imaginary poles.
4. Removal of a constant.

We will discuss these now one by one.

## 1. Removal of a pole at infinity

This means that if $Z(s)$ is such that the order of numerator is greater than that of denominator by one i.e. $n - m = 1$, then by dividing the numerator by the denominator a term, $Hs$ can be taken out of $Z(s)$ where $H$ is a positive constant and this $H$ as can be seen is

$H = \dfrac{a_n}{b_m}$ and $Z(s)$ can now be rewritten as

$Z(s) = Hs + Z_2(s)$ where $Z_2(s)$ is the remainder of the original $Z(s)$ after taking out $Hs$.

Now since $Z(s)$ is positive real and $Hs$ is positive real $Z_2(s)$ is, therefore, positive real. This is true also because the poles of $Z_2(s)$ on the imaginary axis are same as those of $Z(s)$.

Now the third condition for positive realness is that $R_e\,[Z_2\,(j\omega)] \geq 0$ for all $\omega$. Since for $Z(s)$ this condition holds good and out of $Z(s)$ an element with operational impedance $Hs$ is taken out which in effect is an inductor which is only an energy storing element and not the energy dissipating element, therefore, the above condition would hold good for the network which has all the elements of $Z(s)$ minus an inductor.

It is to be noted that if the original function is an admittance $Y(s)$ then removal of a pole at infinity would be taking out a capacitor from $Y(s)$ and is a shunt capacitor. This is indicated in the Fig. 10.4 below:

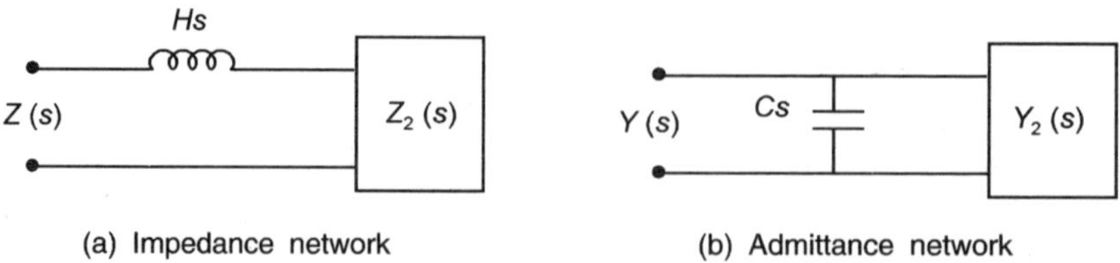


(a) Impedance network　　　　　(b) Admittance network

**Fig. 10.4** Removal of pole at infinity

## 2. Removal of a pole at origin

We can remove a pole at origin from $Z(s)$ only if it has one. This really means that removal of a pole at origin means the denominator should be such that it should be possible to take $s$ as common. This is possible if $b_0$ is zero and we arrange the numerator and denominator in the ascending order of $s$ as shown here.

$$Z(s) = \frac{a_0 + a_1 s + \cdots + a_{n-1} s^{n-1} + a_n s^n}{b_1 s + b_2 s^2 + \cdots + b_{m-1} s^{m-1} + b_m s^m} \qquad \text{...(10.32)}$$

Again dividing numerator by denominator we have

$$Z(s) = \frac{K_0}{s} + \frac{c_0 + c_1 s + \cdots + c_{n-1} s^{n-1}}{b_1 + b_2 s + \cdots + b_{m-1} s^{m-1}} \qquad \text{...(10.33)}$$

$$= Z_1(s) + Z_2(s)$$

where
$$K_0 = \frac{a_0}{b_1}$$

Since $Z_1(s) = \dfrac{K_0}{s}$ is a positive real which represents a capacitor in case of $Z(s)$, shunt inductor, in case the original function is admittance and by the same argument as in case of removal of pole at infinity, $Z_2(s)$ is also positive real and this activity of removal of pole at origin gives the following configuration Fig. 10.5

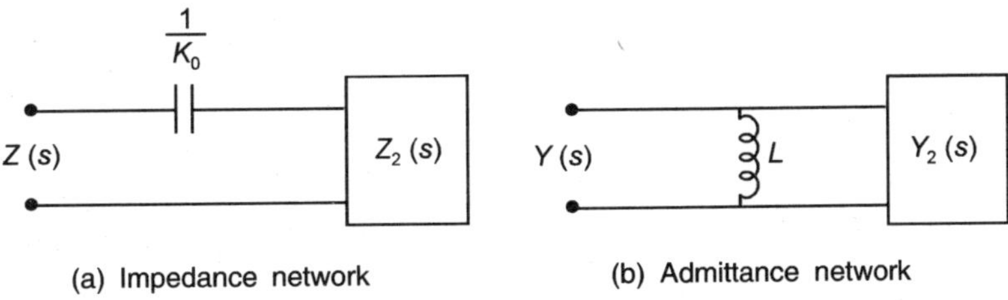


(a) Impedance network      (b) Admittance network

**Fig. 10.5** Removal of a pole at origin.

## 3. Removal of Conjugate Imaginary Poles

If $Z(s)$ has poles along $j\text{-}\omega$ axis, say poles at $s = \pm j\omega$, then $Z(s)$ will have
$(s + j\omega_1)(s - j\omega_1) = (s^2 + \omega^2)$ factors in the denominator polynomial and hence $Z(s)$ can be rewritten as

$$Z(s) = \frac{p(s)}{(s^2 + \omega^2) Q_1(s)} \qquad \text{...(10.34)}$$

and the partial fraction expansion of $Z(s)$ gives

$$Z(s) = \frac{2 K_1 s}{s^2 + \omega^2} + Z_2(s) \qquad \text{...(10.35)}$$

The factor $\dfrac{2 K_1 s}{s^2 + \omega^2}$ represents a suitable combination of $L$ and $C$ and hence it is a reactive element, therefore $Z_2(s)$ is again a positive real function.

The network represented by this expansion is a parallel combination of $L$ and $C$ and hence

$$Z(s) = \frac{2 K_1 s}{s^2 + \omega_1^2} + Z_2(s) \qquad \text{...(10.36)}$$

This combination is to be connected in series with $Z_2(s)$

Now operational impedance of $L$ and $C$ in parallel is

$$\frac{sL \cdot 1/Cs}{sL + \dfrac{1}{Cs}} = \frac{\dfrac{1}{C} \cdot s}{s^L + \dfrac{1}{LC}} = \frac{2K_1 s}{s^2 + \omega_1^2} \qquad \text{...(10.37)}$$

Therefore $C = \dfrac{1}{2K_1}$ and $L = \dfrac{2K_1}{\omega_1^2}$ $\qquad$ ...(10.38)

However, if the function given is admittance then

$$Y(s) = \frac{2K_1 s}{s^2 + \omega^2} + Y_2(s) \qquad \text{...(10.39)}$$

and $\dfrac{2K_1 s}{s^2 + \omega^2}$ represents a shunt across the terminals with $L$ and $C$ in series

$$Y_1(s) = \frac{1}{sL + 1/Cs} = \frac{Cs}{LCs^2 + 1}$$

$$= \frac{s/L}{s^2 + 1/LC} \qquad \text{...(10.40)}$$

Therefore, $L = \dfrac{1}{2K_1}$ and $C = \dfrac{2K_1}{\omega_1^2}$ $\qquad$ ...(10.41)

These two representations are shown in Fig. 10.6

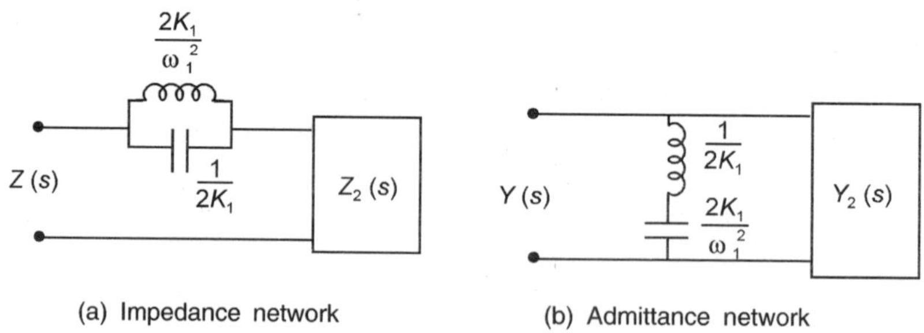



**Fig. 10.6**  Removal of conjugate imaginary poles.

**Removal of a Constant:** If we plot $R_e\,[Z(j\omega)]$ as a function of $\omega$ we get the variations as shown in Fig. 10.7.

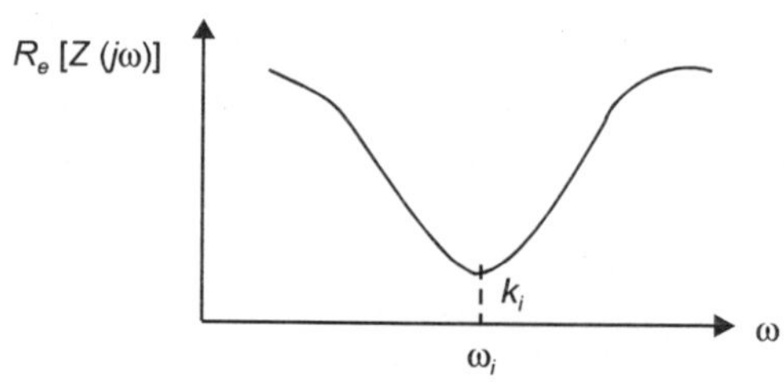



**Fig. 10.7**  Variation of $R_e\,[Z(j\omega)]$ as a function of $\omega$.

Here $R_e [Z(j\omega)]$ is found to be minimum at $\omega_i$ and is equal to $K_i$. We can take out a constant $K \le Ki$ from $Z(s)$ so that the remainder is still positive real as $R_e [Z(j\omega)]$ still will be $\ge 0$ for all $\omega$. If the original function is $Z(s)$ the constant $K$ corresponds to resistance whereas if it is $Y(s)$ then it corresponds to conductance. The representation is shown in Fig. 10.8.

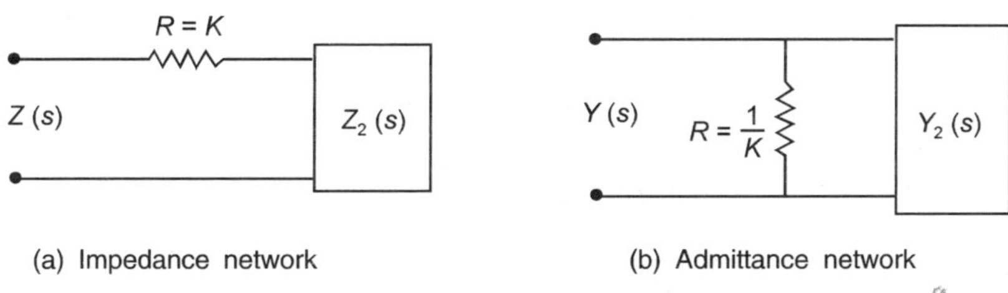


(a) Impedance network  (b) Admittance network

**Fig. 10.8**  Removal of a constant.

The four operations described above can be utilized to synthesise a network for a given impedance or admittance function network. We illustrate with a few examples.

**Example 10.9:** Synthesize the impedance function.

$$Z(s) = \frac{s^2 + 6s + 60}{s(s + 10)}$$

**Solution:** Since $b_0$ here is zero, the impedance function has a pole at origin.

$$10s + s^2 \overline{\left)\, 60 + 6s + s^2 \left(\frac{6}{s}\right.\right.}$$
$$\underline{\phantom{xx}60 + 6s\phantom{xxx}}$$
$$s^2$$

Therefore, $Z_1(s) = \dfrac{6}{s}$ and $Z_2(s) = \dfrac{s^2}{s^2 + 10s} = \dfrac{s}{s + 10}$

Now $\qquad Y_2(s) = \dfrac{1}{Z_2(s)} = \dfrac{s + 10}{s}$

$$= 1 + \frac{10}{s}$$

This is a parallel combination of conductance $1\ \mho$ and inductance $\dfrac{1}{10} H$.

The final network is

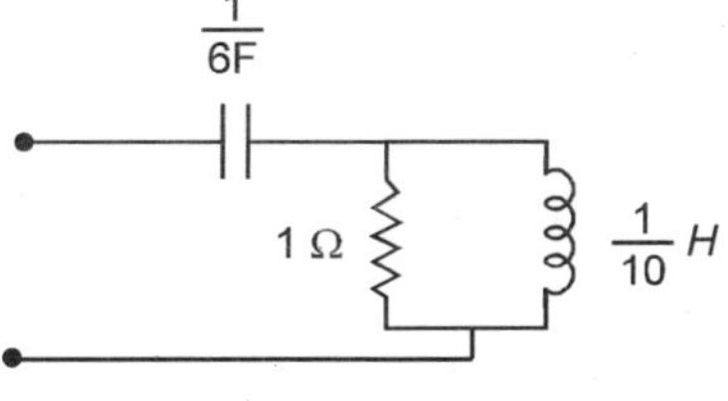


**Fig. E10.9**

**Example 10.10:** Synthesize the network which has admittance function as

$$Y(s) = \frac{5s + 3}{2s + 4} \quad \text{where } Y(s) \text{ is positive real.}$$

**Solution:** Let us synthesize by removing the constant term (here minimum).

$$R_e \left[ \frac{3 + 5s}{4 + 2s} \right]$$

$$= \frac{(3 + j5\omega)(4 - j2\omega)}{(4 + 2j\omega)(4 - j2\omega)}$$

$$= \frac{12 + 10\omega^2}{16 + 4\omega^2}$$

Here the minimum occurs at $\omega = 0$ and is equal to $\dfrac{12}{16} = \dfrac{3}{4} = Y_1(s)$

Therefore, $\qquad Y_2(s) = \dfrac{5s + 3}{2s + 4} - \dfrac{3}{4}$

$$= \frac{14s}{8(s + 2)}$$

$$Y_2(s) = \frac{(7/4)s}{s + 2}$$

or $\qquad Z_2(s) = \dfrac{s + 2}{(7/4)s} = \dfrac{4}{7} + \dfrac{8}{7s}$

i.e. it is a resistance of $\dfrac{4}{7}\,\Omega$ and capacitance of $\dfrac{7}{8}$ F. Alternatively, the constant term can be obtained as follows:

$$
\begin{array}{r}
\phantom{4 + 2s\,|}\,3/4 \\
4 + 2s\,\big|\,\overline{3 + 5s} \\
\underline{3 + \dfrac{3s}{2}} \\
\dfrac{7s}{2}
\end{array}
$$

Hence $\qquad Y(s) = \dfrac{3}{4} + \dfrac{(7/2)\,s}{4 + 2s}$

$$= Y_1 + Y_2$$

Hence $\qquad Z_2 = \dfrac{1}{Y_2} = \dfrac{2(2 + s)}{\dfrac{7}{2}s}$

$$= \frac{8}{7s} + \frac{4}{7}$$

Therefore, the circuit is

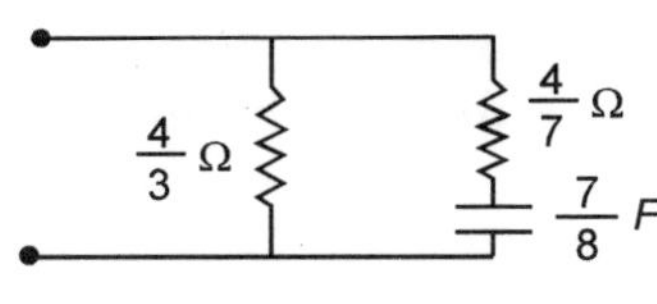

**Fig. E10.10**

**Example 10.11:** Synthesize a network which has the impedance function.

$$Z(s) = \frac{5s^3 + 2s^2 + 3s + 1}{5s^3 + 3s}$$

**Solution:** The real part of the function is a constant equal to unity. Taking out unity from $Z(s)$ we have

$$Z_1(s) = Z(s) - 1 = \frac{5s^3 + 2s^2 + 3s + 1 - 5s^3 - 3s}{5s^3 + 3s}$$

$$= \frac{2s^2 + 1}{5s^3 + 3s}$$

The reciprocal of $Z_1(s) = Y_1(s) = \dfrac{5s^3 + 3s}{2s^2 + 1}$

$$2s^2 + 1 \overline{\left) \begin{array}{l} 5s^3 + 3s \\ 5s^3 + \dfrac{3}{2}s \\ \hline \quad \dfrac{s}{2} \end{array} \right.} \dfrac{5}{2}s$$

and it has a pole at $s = \infty$ and since it is admittance, therefore, it represents a shunt capacitance of $\dfrac{5}{2}$F.

$$Y_1(s) = \frac{5s^3 + 3s}{2s^2 + 1} = \frac{5}{2}s + \frac{s/2}{2s^2 + 1}$$

$$Y_2(s) = \frac{s/2}{2s^2 + 1}$$

or

$$Z_2(s) = \frac{2s^2 + 1}{s/2} = 4\,s + \frac{2}{s}$$

These represent an inductor of 4H and a capacitor of 1/2 F in series. The network is

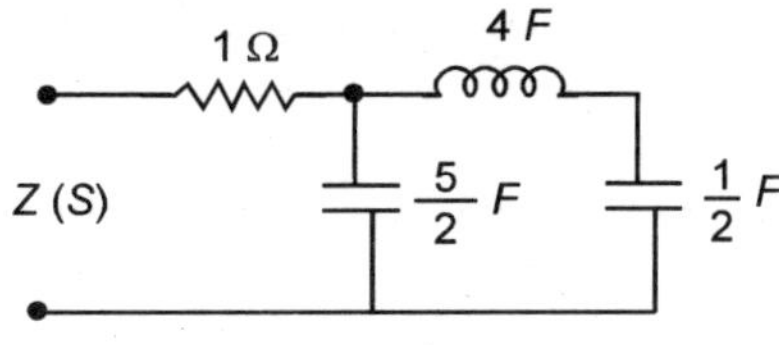


**Fig. E10.11**

The examples above illustrate special cases of driving point synthesis procedure. However, they also show basic procedure involved. We next move to the discussion of synthesizing networks with two kinds of elements either *LC* or *RC* or *RL* networks.

We first study the properties of these two element networks and then we synthesise them.

## 10.4 L-C IMMITTANCE FUNCTIONS

Let us consider the impedance of a passive one port network.

$$Z(s) = \frac{M_1(s) + N_1(s)}{M_2(s) + N_2(s)} \qquad \qquad ...(10.42)$$

The average power dissipated by the one port network having reactance elements $L$ and $C$ is zero which mean that the real part of $Z(j\omega)$ is zero i.e.

$$R_e\,[Z(j\omega)] = E_v\,[Z\,(j\omega)] = 0 \qquad \qquad ...(10.43)$$

where
$$E_v\,[Z\,(j\omega)] = \frac{M_1 M_2 - N_1 N_2}{M_2^2 - N_2^2} \qquad \qquad ...(10.44)$$

which means $M_1(s)\,M_2(s) - N_1(s)\,N_2(s) = 0$. For this situation to arise either of the following cases must hold

    (*a*) $M_1 = 0 = N_2$

    (*b*) $M_2 = 0 = N_1$

From equation 10.42

For (*a*)
$$Z(s) = \frac{N_1}{M_2} \qquad \qquad ...(10.45)$$

and (*b*)
$$Z(s) = \frac{M_1}{N_2} \qquad \qquad ...(10.46)$$

This means that the driving point immittance
(1) is a ratio of even to odd or odd to even polynomials.
(2) has poles and zeros of $Z(s)$ or $Y(s)$ on the imaginary axis.
(3) the poles and zeros interlace on the $j\omega$ axis i.e. the poles and zeros alternate on the $j$-$\omega$ axis. This is known as separation property. The separation property is given by $0 \leq \omega_0 < \omega_1 < \omega_2 < \omega_3 < \omega_4 \ldots < \infty$ where these frequencies in $Z(s)$ are given as

$$Z(s) = \frac{K\,(s^2 + \omega_1^2)\,(s^2 + \omega_3^2) \cdots (s^2 + \omega_{2n-1}^2)}{s\,(s^2 + \omega_2^2)\,(s^2 + \omega_4^2) \cdots (s^2 + \omega_{2m-2}^2)} \qquad \qquad ...(10.47)$$

Here frequencies $\omega_1\,\omega_3 \ldots \omega_{2n-1}$ are known as internal zeros and $\omega_2, \omega_4 \ldots \omega_{2m-2}$ as internal poles. The critical frequencies at $s = 0$ and $s = \infty$ are called external critical frequencies. Expanding $Z(s)$ into partial fraction we have

$$Z(s) = \frac{k_0}{s} + \frac{2k_2\,s}{s^2 + \omega_2^2} + \frac{2k_4\,s}{s^2 + \omega_4^2} + \cdots + k_\infty\,s \qquad \qquad ...(10.48)$$

Let $s = j\omega$ we see that $Z(j\omega)$ has zero real part and can thus be rewritten as a pure reactance $j\,X\,(\omega)$.

$$Z(j\omega) = j\left[-\frac{k_0}{\omega} + \frac{2\,k_2\,\omega}{\omega_2^2 - \omega^2} + \cdots + k_\infty\,\omega\right] \qquad \ldots(10.49)$$

$$= j\,X\,(\omega)$$

Differentiating w.r. to $\omega$ we have

$$\frac{d\,X\,(\omega)}{d\omega} = \frac{k_0}{\omega^2} + k_\infty + \frac{2k_2\,(\omega^2 + \omega_2^2)}{(\omega_2^2 - \omega^2)^2} + \cdots \qquad \ldots(10.50)$$

Since all the residues $ki$ are positive it is found that for an $LC$ function

$$\frac{d\,X\,(\omega)}{d\omega} \geq 0 \qquad \ldots(10.51)$$

Similarly for an admittance function it can be shown that $\dfrac{d\,B(\omega)}{d\omega} \geq 0$ where $B(\omega)$ is the suspectance of the $LC$ function

$$\ldots(10.52)$$

If we consider a function

$$Z(s) = \frac{ks\,(s^2 + \omega_3^2)}{(s^2 + \omega_2^2)\,(s^2 + \omega_4^2)} \qquad \ldots(10.53)$$

and if we let $s = j\omega$ we have

$$Z(j\omega) = j\,X\,(\omega) = j\,\frac{k\omega\,(-\omega^2 + \omega_3^2)}{(-\omega^2 + \omega_2^2)\,(-\omega^2 + \omega_4^2)} \qquad \ldots(10.54)$$

Fig. 10.9 shows the curve of $X\,(\omega)$ vs $\omega$ At $\omega = 0$ $X\,(\omega) = 0$ and as $\omega$ increases the plot is shown in Fig. 10.9. The zeros will occur at $\omega = 0$ and $\omega_3$ and poles at $\omega_2$ and $\omega_4$.

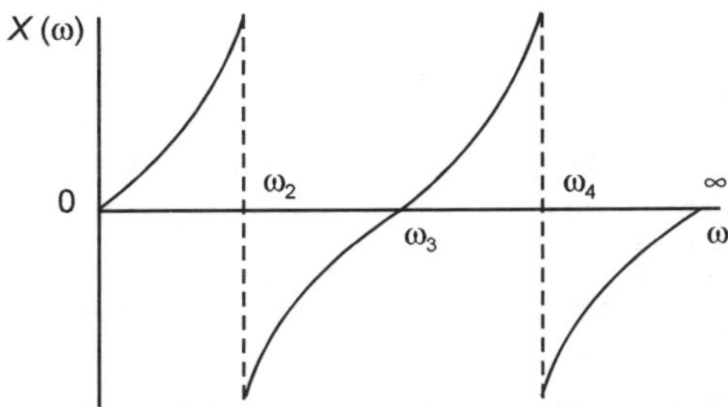


**Fig. 10.9**   Poles and zeros configuration of $x(\omega)$

(4) The highest powers of numerator and denominator must differ by unity; the lowest powers also differ by unity.

(5) With this we observe that at $s = 0$ and at $s = \infty$, there is always a critical frequency whether zero or a pole.

**Example 10.12:** Test whether the following functions are $LC$ immittance function.

$$\text{(1) } Z(s) = \frac{ks(s^2 + 6)}{(s^2 + 2)(s^2 + 4)}$$

$$\text{(2) } Z(s) = \frac{s^5 + 3s^3 + 5s}{3s^4 + 6s^2}$$

$$\text{(3) } Z(s) = \frac{k(s^2 + 4)(s^2 + 9)}{(s^2 + 2)(s^2 + 6)}$$

$$\text{(4) } Z(s) = \frac{k(s^2 + 4)(s^2 + 9)}{s(s^2 + 6)}$$

**Solution:**
(1) The first one is not $LC$ immitance because the separation property does not hold good.
(2) The poles and zeros do not occur along $j\text{-}\omega$ axis and hence it is not $LC$ immitance function.
(3) The powers of $s$ are same for numerator and denominator and since these do not differ by one, it is not an $LC$ immittance function.
(4) It satisfies all the five requirements mentioned above, hence it is an $LC$ impedance function.

As is said earlier, the network synthesis does not provide a unique solution to a given problem. Given p.r. $F(s)$, following methods can be used to synthesize the network and each one gives different configuration. Mainly there are two methods:

1. Foster form which uses partial fraction expansion of the given $F(s)$.
2. Cauer form which utilizes continued fraction expansion of the function $F(s)$. By virtue of the technique the network comes out to be of ladder formation. Again Foster method is divided into two sections.

Foster form I provides series network realization $Z(s)$ and Foster form II the parallel network $Y(s)$.

### 10.4.1 Foster Form I

Since we know that the poles and zeros of an $LC$ immittance function lie on the $j\text{-}\omega$ axis, the partial fraction of the $LC$ immittance function will, in general, be of the form.

$$F(s) = \frac{k_0}{s} + \frac{2k_2 s}{s^2 + \omega_2^2} + \frac{2k_4 s}{s^2 + \omega_4^2} + \cdots + k_\infty s \qquad \text{...(10.55)}$$

While synthesizing such a network each term in the partial fraction is associated with an element or a pair of elements and then these are to be connected in series. For example if $F(s)$ is $Z(s)$ then the term $\dfrac{k_0}{s}$ represents a capacitor of $\dfrac{1}{k_0}$ farads, the term $k_\infty s$ an inductor of $k_\infty$ henry and the term $\dfrac{2ks}{s^2 + \omega^2}$ represents a parallel combination of $L$ and $C$ where $C$ is $\dfrac{1}{2k_2}$ farads and the inductor is of $\dfrac{2k_2}{\omega_2^2}$ and the elements are connected as shown in Fig. 10.10.

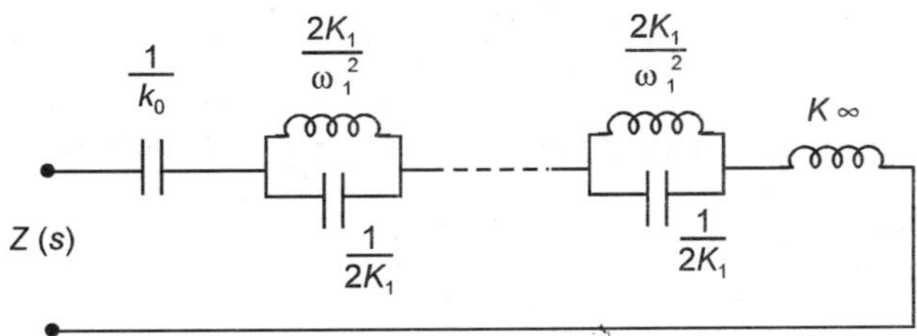

**Fig. 10.10**  First Foster form of *LC* impedance function.

It is to be noted that since all the poles of *LC* immittance function lie on $j$-$\omega$ axis we can remove all the poles simultaneously using partial fraction. If $Z(s)$ has no pole at the origin (and so it has a zero at origin) the first term in the partial fraction will be missing ($C_0$ will be absent). Similarly if there is a zero rather than a pole at infinity then the inductor will be absent.

## 10.4.2 Foster Form II

However if the function $F(s)$ is given in admittance from $Y(s)$, then the partial fraction expansion of $Y(s)$ gives a circuit consisting of parallel branches.

Let

$$Y(s) = \frac{k_0}{s} + \frac{2k_2 s}{s^2 + \omega_2^2} + \cdots + k_\infty s \qquad \qquad \text{...(10.56)}$$

Each term here represents an admittance of an element or a pair of elements to be connected across a two-port network.

$\dfrac{k_0}{s}$ represents an inductor element of $\dfrac{1}{k_0}$ henry and $k_\infty$ represents capacitor of $k_\infty$ farad. Similarly if $L$ and $C$ are in series, their admittance is

$$Y = \frac{1}{z} = \frac{1}{sL + 1/Cs} = \frac{Cs}{s^2 + LC + 1} = \frac{s/L}{s^2 + \omega^2}$$

$$= \frac{2ks}{s^2 + \omega^2}$$

where
$$2\,k = \frac{1}{L}$$

and
$$L = \frac{1}{2k} \quad \text{and} \quad C = \frac{2k}{\omega^2} \qquad \qquad \text{...(10.57)}$$

Therefore the network takes the shape as in Fig. 10.11.

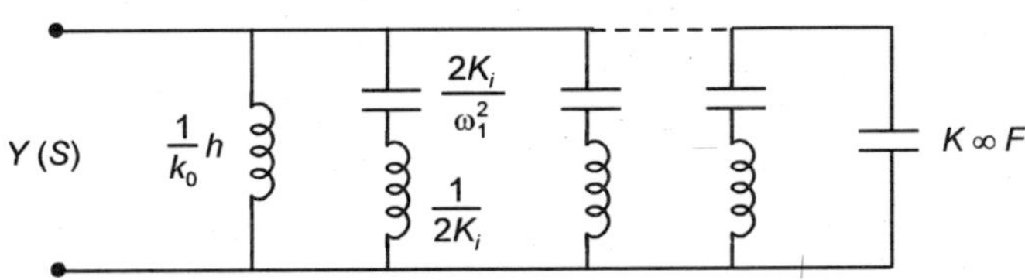

**Fig. 10.11**  Foster Form II representation.

Let us illustrate the application of Foster I and Foster II forms of network realization with some examples.

**Example 10.13:** Synthesise the function $Z(s)$ using first Foster form of realization.

$$Z(s) = \frac{s\,(s^2 + 10)}{(s^2 + 4)\,(s^2 + 16)}$$

**Solution:** The partial fractions are found as follows:

$$Z(s) = \frac{s\,(s^2 + 10)}{(s^2 + 4)\,(s^2 + 16)} = \frac{2k_2\,s}{s^2 + 4} + \frac{2k_4\,s}{s^2 + 16}$$

Now

$$2\,k_2 s = \mathop{\mathrm{Lt}}_{s^2 \to -4}\ \frac{(s^2 + 4)\,s\,(s^2 + 10)}{(s^2 + 4)\,(s^2 + 16)}$$

or

$$2\,k_2 = \frac{1}{2}$$

Similarly

$$2\,k_4 = \mathop{\mathrm{Lt}}_{s^2 \to -16}\ \frac{s^2 + 10}{s^2 + 4} = \frac{-16 + 10}{-16 + 4} = \frac{1}{2}$$

Hence

$$Z(s) = \frac{s/2}{s^2 + 4} + \frac{s/2}{s^2 + 16}$$

Therefore

$$C_1 = 2\ F \quad L_1 = \frac{1}{8}H$$

$$C_2 = 2\ F \quad L_2 = \frac{1}{32}H$$

Therefore, the network becomes

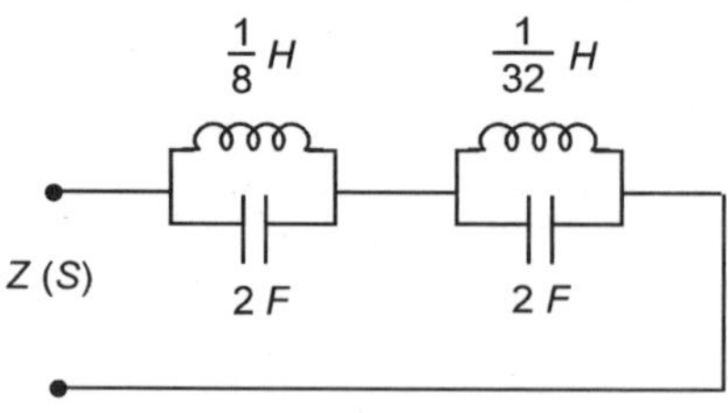


**Fig. E10.13**

**Example 10.14:** Obtain the second Foster form network using same $Z(s)$ as in Example 10.13.

The

$$Y(s) = \frac{1}{Z(s)} = \frac{(s^2 + 4)(s^2 + 16)}{s(s^2 + 10)}$$

$$
\begin{array}{r}
s \\
s^3 + 10s\ \overline{\smash{\big)}\ s^4 + 20s^2 + 64} \\
\underline{s^4 + 10s^2} \\
10s^2 + 64
\end{array}
$$

Therefore
$$Y(s) = s + \frac{10s^2 + 64}{s^3 + 10s}$$

$$= s + \frac{A}{s} + \frac{2k_2 s}{s^2 + 10}$$

$$A = \underset{s \to 0}{\text{Lt}}\ s F(s) = \frac{(s^2 + 4)(s^2 + 16)}{s^2 + 10} = \frac{32}{5}$$

$$2 k_2 s = \underset{s^2 \to -10}{\text{Lt}}\ \frac{(s^2 + 10)(s^2 + 4)(s^2 + 16)}{s(s^2 + 10)}$$

or
$$2 k_2 = \frac{(-10 + 4)(-10 + 16)}{-10}$$

$$= \frac{18}{5}$$

The elements are capacitor 1F

An inductor
$$H = \frac{5}{32} H$$

An inductor of $\dfrac{5}{18} H$ in series with $\dfrac{18/5}{10} = \dfrac{9}{25} F$ capacitor

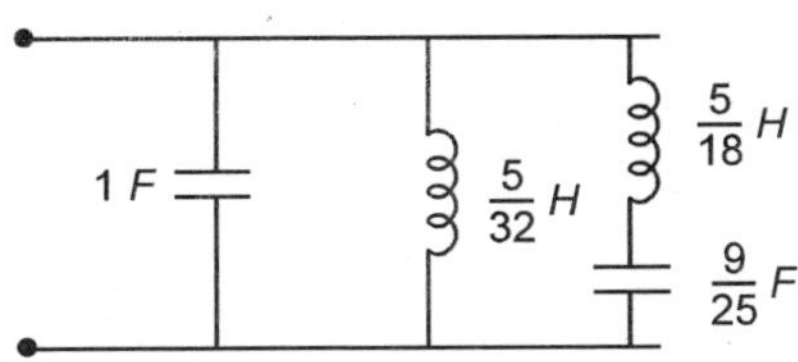


**Fig. E10.14**

**Example 10.15:** Synthesise impedance function

$$Z(s) = \frac{(s^2 + 2)(s^2 + 4)}{s(s^2 + 3)(s^2 + 5)}$$

$$= \frac{k_0}{s} + \frac{2k_2 s}{s^2 + 3} + \frac{2k_4 s}{s^2 + 5}$$

$$k_0 = \underset{s \to 0}{\text{Lt}}\ sZ(s) = \frac{2 \times 4}{15} = \frac{8}{15}$$

$$2 k_2 = \underset{s^2 \to -3}{\text{Lt}}\ \frac{(s^2 + 2)(s^2 + 4)}{s^2(s^2 + 5)} = \frac{-1 \times 1}{-3(2)} = \frac{1}{6}$$

Similarly  $k_4 = \dfrac{3}{20}$

$$C_0 = \frac{15}{8}\,\text{F}, \qquad C_2 = 6\,\text{F} \quad L_2 = \frac{1}{18}\text{H}$$

$$C_4 = \frac{10}{3}\,\text{F} \qquad L_4 = \frac{3}{50}\text{H}$$

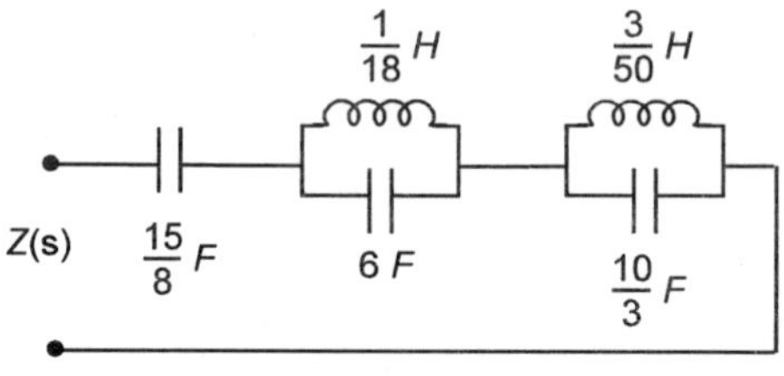

**Fig. E10.15**

**Example 10.16:** Synthesise the Foster II form network when its admittance function is given as

$$Y(s) = \frac{s(s^2 + 3)(s^2 + 5)}{(s^2 + 2)(s^2 + 4)}$$

Since the numerator is of higher power of $s$ we divide it by the denominator.

$$s^4 + 6\,s^2 + 8\,\bigl|\,s^5 + 8s^3 + 15s\,\bigl|\,s$$
$$\underline{\quad\quad\quad\quad\;\; s^5 + 6s^3 + 8s\quad}$$
$$2s^3 + 7s$$

Therefore
$$Y(s) = s + \frac{2s^3 + 7s}{(s^2 + 2)\,(s^2 + 4)}$$

$$= s + \frac{2k_2\,s}{s^2 + 2} + \frac{2k_4\,s}{s^2 + 4}$$

using the procedure as outlined in the previous example we find

$$2\,k_2 = \frac{3}{2}$$

and
$$2\,k_4 = \frac{1}{2}$$

Therefore
$$Y(s) = s + \frac{(3/2)s}{s^2 + 2} + \frac{s/2}{s^2 + 4}$$

$$C_0 = 1\,\text{F} \qquad C_2 = \frac{3}{4}\text{F} \;\text{ and }\; L_2 = \frac{2}{3}\text{H}$$

$$C_4 = \frac{1}{8}\,\text{F} \qquad L_4 = 2\,\text{H}$$

Therefore the network is

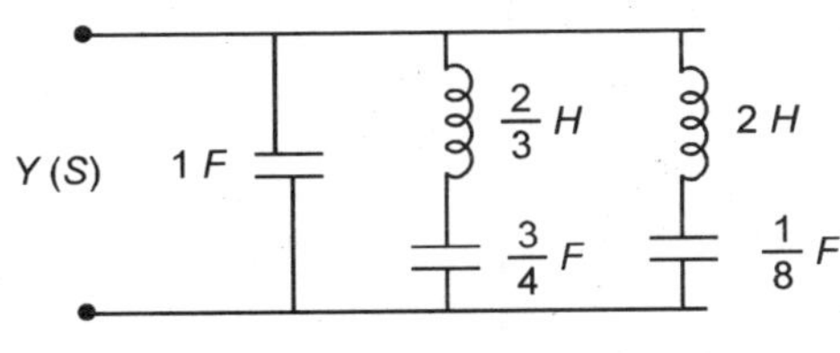



**Fig. E10.16**

It is found that the number of elements in both forms of Foster realization is same and it equals the number of internal critical frequencies plus one. In other words the total number of elements is equal to the degree of highest power of $s$ in $F(s)$ whether $Z(s)$ or $Y(s)$.

### 10.4.3 First Cauer Form

As mentioned earlier continued fraction expansion is used in Cauer method. In first Cauer form we arrange the polynomials in the numerator and denominator in the descending power of $s$ and we eliminate during every step a pole at infinity ($s = \infty$). If the original function is $Z(s)$ and the order of the numerator is $2n$ and that of the denominator is $2n - 1$, we eliminate a pole at infinity by dividing numerator by the denominator, thereby we get the quotient as $L_1 s$. Therefore, the remainder function $Z_2(s)$ is still $LC$ immittance and is given as

$$Z_2(s) = Z(s) - L_1 s \qquad \qquad ...(10.58)$$

Now the denominator of $Z_2(s)$ is of order $2n - 1$ whereas the numerator becomes $(2n - 2)$ as the difference in power must be one. Since in Cauer I we always eliminate a pole at infinity, therefore, we invert $Z_2(s)$ to obtain

$$Y_2(s) = \frac{1}{z_2(s)}$$

Now of course the numerator will be of power $2n - 1$ and the denominator $(2n - 2)$. Again divide this numerator of $Y_2(s)$ by its denominator we have $C_2 s$ which will form shunt element of the two terminal network where as $L_1 s$ forms the series element. Next we again invert $Y_4(s)$ so that we can have a pole at infinity. So every time we divide numerator (of, in general, $2r$ power) by the denominator (of in general $2r - 1$ power) to eliminate a pole at infinity. The process is continued till the remainder is zero. In fact, this process is known as continued fraction expansion and is given as follows:

$$Z(s) = L_1 s + \cfrac{1}{Y_2(s) + \cfrac{1}{Z_3(s) + \cfrac{1}{Y_4(s) + \cfrac{1}{z_2(s)}}}} \qquad \qquad ...(10.59)$$

Since the expansion for $Z(s)$ looks like a ladder, the networks so synthesized are known as ladder networks.

Whatever be the form of the original function whether $Z(s)$ or $Y(s)$, in first form of Cauer a pole is eliminated at infinity ($s = \infty$) at every step. Therefore, if the numerator of the original function has power of $s$ smaller than that of its denominator, it should be inverted and the continued fraction expansion carried out. The final network synthesized is a ladder network whose series elements are inductors and shunt elements, capacitors shown here in Fig. 10.12

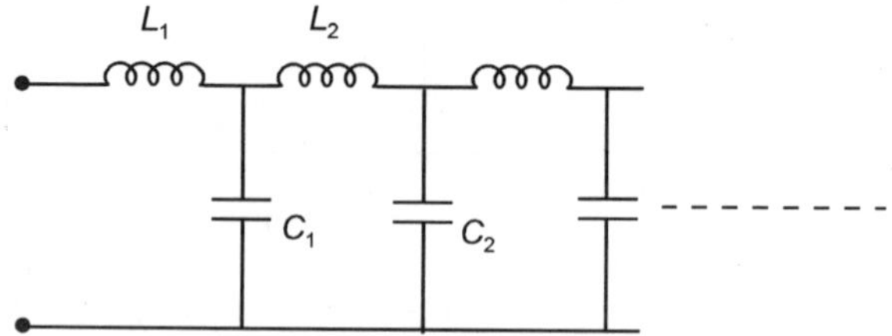



**Fig. 10.12**  First Cauer form of *LC* network.

**Example 10.17:** Synthesise the impedance function

$$Z(s) = \frac{s(s^2 + 3)(s^2 + 5)}{(s^2 + 2)(s^2 + 4)} \quad \text{using I form of Cauer network}$$

$$= \frac{s^5 + 8s^3 + 15s}{s^4 + 6s^2 + 8}$$

Since the numerator has 5th power and denominator 4th power we remove a pole at infinity by dividing numerator by the denominator.

$$s^4 + 6s^2 + 8 \overline{\smash{\big)}\ s^5 + 8s^3 + 15s}\ \big(\ s \to Z_1$$
$$\underline{s^5 + 6s^3 + 8s}$$

$$2s^3 + 7s \overline{\smash{\big)}\ s^4 + 6s^2 + 8}\ \Big(\ \frac{s}{2} \to Y_1$$
$$\underline{s^4 + \frac{7}{2}s^2}$$

$$\frac{5}{2}s^2 + 8 \overline{\smash{\big)}\ 2s^3 + 7s}\ \Big(\ \frac{4}{5}s \to Z_2$$
$$\underline{2s^3 + \frac{32}{5}s}$$

$$\frac{3}{5}s \overline{\smash{\big)}\ \frac{5}{2}s^2 + 8}\ \bigg/\ \frac{25}{6}s \to Y_2$$
$$\underline{\frac{5}{2}s^2}$$

$$8 \overline{\smash{\big)}\ \frac{3s}{5}}\ \Big(\ \frac{3}{40}s \to Z_3$$
$$\underline{\frac{3}{5}s}$$
$$\times$$

The elements are

$$L_1 = 1\text{H} \quad C_1 = \frac{1}{2}\text{F}$$

$$L_2 = \frac{4}{5}\text{H} \quad C_2 = \frac{25}{6}F$$

$$L_3 = \frac{3}{40}\text{H}$$

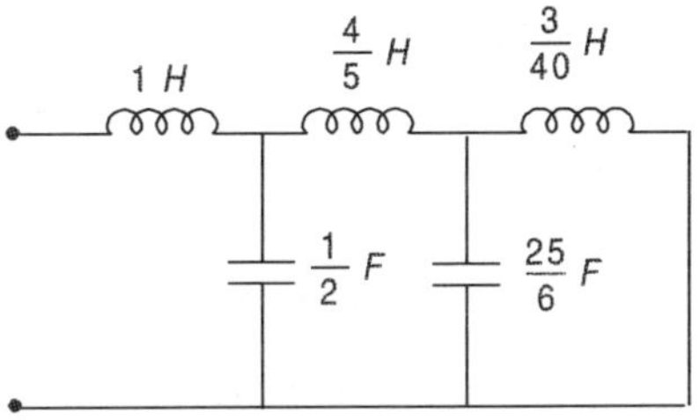

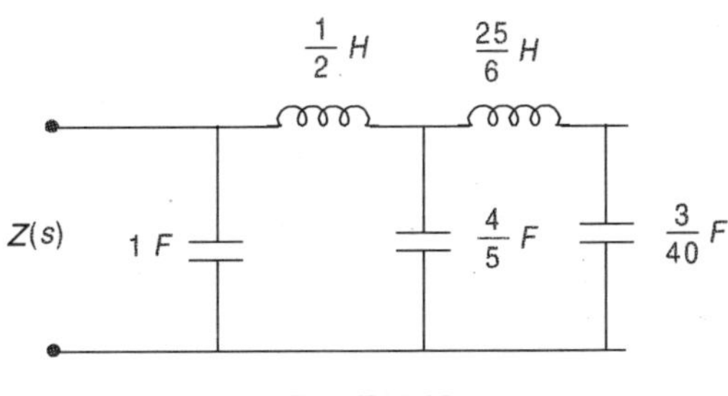

**Fig. E10.17**

**Example 10.18:** Synthesise the impedance function

$$Z(s) = \frac{(s^2 + 2)\,(s^2 + 4)}{s\,(s^2 + 3)\,(s^2 + 5)}$$ using first form of Cauer network.

**Solution:** Since here we remove poles at infinity in which case the power of $s$ of the numerator should be greater than that of denominator by one. Therefore, we invert $Z(s)$ and we have

$$Y(s) = \frac{s\,(s^2 + 3)\,(s^2 + 5)}{(s^2 + 2)\,(s^2 + 4)}$$

If we follow the continued fraction expansion we will have the same quotients as in the previous example. However, their significance is just dual i.e., wherever it is $Z(s)$, it would be $Y(s)$ now. Therefore, the network would be

**Fig. E10.18**

## 10.4.4 II Cauer Form

Here we arrange the given function polynomials (numerator as well denominator) in the ascending power of $s$ and remove a pole at origin ($s = 0$) successively. Since the lowest degree of numerator and denominator of an $LC$ admittance must differ by one, it follows that there must be a zero or a pole at $s = 0$. After arranging the numerator and denominator in the ascending power of $s$ we divide the lowest power of the numerator by the lowest power of the denominator, then we invert the remainder and divide again. With this we will have an alternate form of ladder network which will have series capacitors and shunt reactors. Series capacitor impedance has a pole at origin where as a shunt reactor admittance has a pole at origin. A general network is shown here in Fig. 10.13.

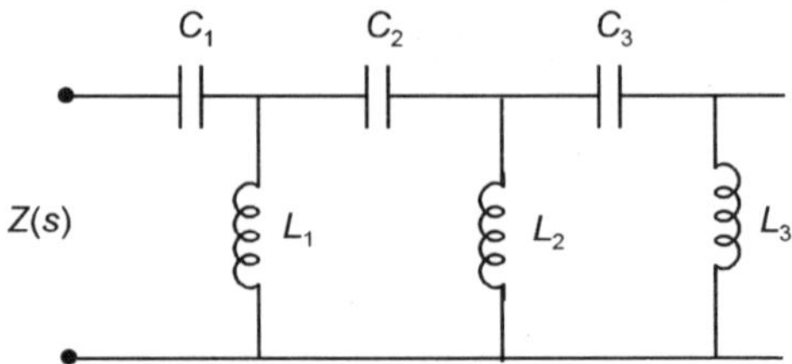

**Fig. 10.13**   Second Cauer form of *LC* network.

The following example explains the procedure.

**Example 10.19:** Synthesise the impedance function

$$Z(s) = \frac{s(s^2 + 3)(s^2 + 5)}{(s^2 + 2)(s^2 + 4)}$$

**Solution:** Using II form of Cauer network

$$Z(s) = \frac{15s + 8s^3 + s^5}{8 + 6s^2 + s^4}$$

Since here we have to remove pole at the origin, we take the reciprocal of $Z(s)$.

$$Y(s) = \frac{1}{Z(s)} = \frac{8 + 6s^2 + s^4}{15s + 8s^3 + s^5}$$

So the first quotientt will indicate shunt inductive susceptance.

$$15s + 8s^3 + s^5 \overline{\smash{\big)}\, 8 + 6s^2 + s^4 \,\left(\dfrac{8}{15s}\right)}$$
$$8 + \frac{64}{15}s^2 + \frac{8}{15}s^4$$
$$-\qquad\qquad -$$
$$\frac{26}{15}s^2 + \frac{7}{15}s^4 \overline{\smash{\big)}\, 15s + 8s^3 + s^5 \,\left(\dfrac{225}{26s}\right)}$$
$$15s + \frac{105}{26}s^3$$
$$-\qquad -$$
$$\frac{103}{26}s^3 + s^5 \overline{\smash{\big)}\, \frac{26}{15}s^2 + \frac{7}{15}s^4 \,\left(\dfrac{676}{1545\,s}\right)}$$
$$\frac{26}{15}s^2 + \frac{676}{1545}s^4$$
$$-\qquad\quad -$$
$$\frac{3}{103}s^4$$

$$\frac{3}{103}s^4 \left|\frac{103}{26}s^3 + s^5\right| \frac{103 \times 103}{78\,s}$$

$$\frac{103}{26}s^3$$

$$s^5 \left|\frac{3}{103}s^4\right| \frac{3}{103\,s}$$

$$\frac{3}{103}s^4$$

$$\times$$

The network is shown here

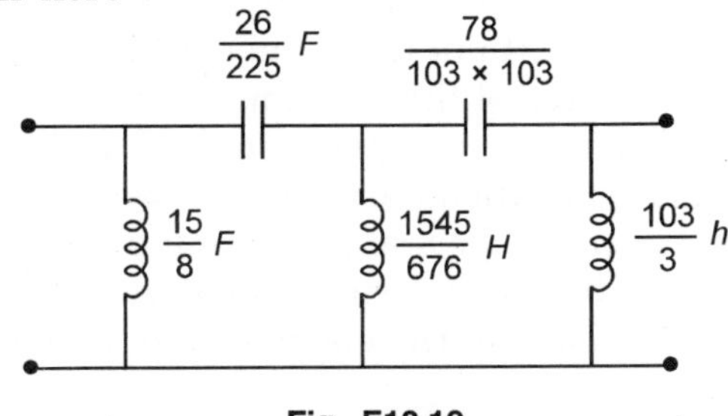


**Fig. E10.19**

**Example 10.20:** Synthesise the *LC* impedance function

$$Z(s) = \frac{(s^2 + 1)(s^2 + 3)}{s(s^2 + 2)} \quad \text{in II Cauer form}$$

**Solution:** Arranging in the ascending order

$$Z(s) = \frac{3 + 4s^2 + s^4}{2s + s^3}$$

$$2s + s^3 \left|3 + 4s^2 + s^4\right| \frac{3}{2s}$$

$$3 + \frac{3}{2}s^2$$

$$\frac{5}{2}s^2 + s^4 \left|2s + s^3\right| \frac{4}{5\,s}$$

$$2s + \frac{4}{5}s^3$$

$$\frac{s^3}{5} \left|\frac{5}{2}s^2 + s^4\right| \frac{25}{2s}$$

$$\frac{5}{2}s^2$$

$$s^4 \left|\frac{s^3}{5}\right| \frac{1}{5s}$$

$$\frac{s^3}{5}$$

$$\times$$

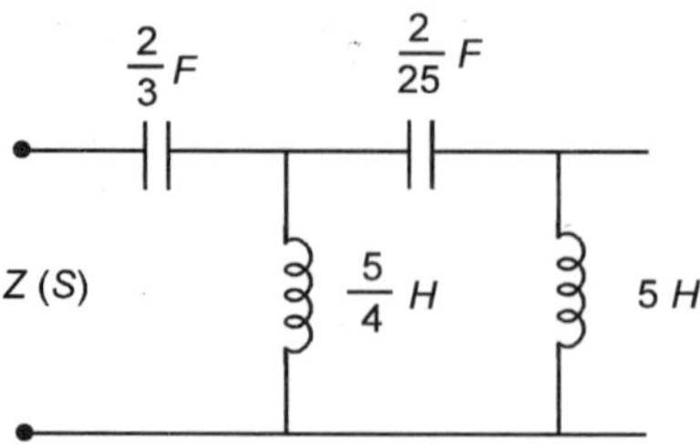

**Fig. E10.20**  IInd form of Cauer network for example 10.20.

It can be seen that the number of elements in Foster form or Cauer's form of network are the same and equals the highest power of $s$ in the function $Z(s)$ or $Y(s)$. Note that in general the number of elements is one greater than the number of internal critical frequencies defined earlier as being all the poles and zeros of the function excluding those at $s = 0$ and $s = \infty$. Also it is seen that the Foster and Cauer form give the minimum number of elements for a specified immittance $LC$ function. These realizations are also known as canonical forms.

Also it is seen that the networks for a specified immittance function are all equivalent, whether obtained through Foster or Cauer formulations. However, the network to be selected depends upon practical consideration like the element size, compensation for parasitic effects, cost etc.

## 10.5 RC IMPEDANCE FUNCTION

Let us find out the impedance of a parallel $RC$ network and the admittance of a series $RL$ network.

Parallel $RC$ network gives

$$Z(s) = \frac{R \cdot 1/Cs}{R + \dfrac{1}{Cs}} = \frac{1/C}{s + \dfrac{1}{RC}} \qquad \ldots(10.60)$$

whereas the admittance of a series $RL$ network

$$Y(s) = \frac{1}{R + sL} = \frac{1/L}{s + R/L} \qquad \ldots(10.61)$$

The nature of the two expressions is same and it shows that these functions have poles along the negative real axis in the $s$-plane. Assuming that all driving point functions that can be realized with two kinds of elements can be synthesized in Foster form, it is possible to derive all the properties pertaining to $RC$ or $RL$ driving point immittance.

If we replace the inductances in the series Foster form of $LC$ immittances by the resistances, the circuit configurations is given as in Fig. 10.14.

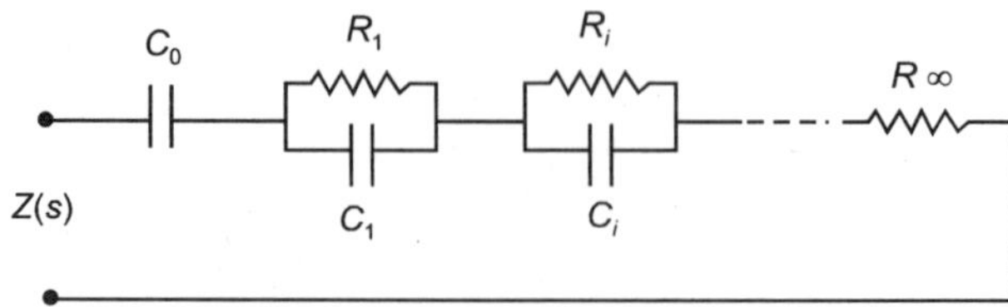

**Fig. 10.14**  The First Foster form of RC impedance network.

From this Fig. we can write down

$$Z(s) = \frac{k_0}{s} + \frac{k_1}{s + \sigma_1} + \cdots \frac{k_i}{s + \sigma_i} + \cdots + k_\infty \qquad \ldots(10.62)$$

where

$$C_0 = \frac{1}{k_0}, \ C_1 = \frac{1}{k_1}, \ R_1 = \frac{k_1}{\sigma_1}$$

$$C_i = \frac{1}{k_i}, \ R_i = \frac{k_i}{\sigma_i}, \ R_\infty = \frac{1}{k_\infty}$$

If we compare the partial fraction terms having $RC$ parallel combination with the one derived above, it is essential that the constants $k_i$ and $\sigma_i$ must be positive and real.

However, if the function $F(s)$ has to represent $Y_{RL}(s)$ rather than $Z_{Rc}(s)$, the configuration of the $RL$ admittance network will be as shown in Fig. 10.15.

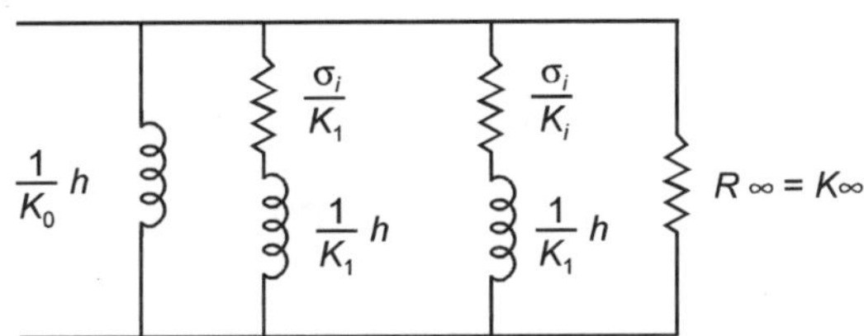

**Fig. 10.15**  First Foster form of $RL$ admittance.

where the correspondence between the elements and the partial fraction is given as

$$L_0 = \frac{1}{k_0} \quad L_1 = \frac{1}{k_1} \quad R_1 = \frac{\sigma_1}{k_1}$$

$$L_i = \frac{1}{k_i} \quad R_i = \frac{\sigma_i}{k_i} \quad R_\infty = K_\infty$$

We see that an $RC$ impedance $Z_{RC}(s)$ also can be realized as an $RL$ admittance $Y_{RL}(s)$. All the properties of $RC$ impedances are same as those of $RL$ admittances. It is, therefore, important to specify whether a function is to be realized as an $RC$ impedance or an $RL$ admittance.

Let us now study properties of $RC$ impedance function.

From the analysis so far, of course two important properties have already been mentioned, these are:

1. The poles of an $RC$ driving point impedance are on the negative real axis. Also it can be shown using Foster II configuration of $RC$ ($RC$ admittance) network that the poles of an $RC$ admittance functions are also on the negative real axis. therefore, it can be said that the zeros of $RC$ driving point impedances are also on the negative real axis.
2. The residues of the poles $K_i$ are all real and positive. However, we would show that this property does not apply to $RC$ admittance.

Now that the poles and zeros of $RC$ impedances are on the $-\sigma$ axis, let us find out the relative locations of the critical frequency, for which, we find out the slope of $Z(s)$ at $s = \sigma$ i.e.

$$\frac{dz(s)}{d\sigma} = -\frac{k_0}{\sigma^2} + \frac{-k_1}{(\sigma + \sigma_1)^2} + \cdots + \frac{-k_i}{(\sigma + \sigma_i)^2} \qquad \ldots(10.63)$$

It is clear that $\dfrac{dZ(\sigma)}{d\sigma} \leq 0$ i.e. it is negative. Let us study the behaviour of $Z(s)$ as $\sigma$ varies. The extreme conditions are:

When $\sigma = \omega = 0$ and $\sigma = \omega = \infty$. If we look at the network (general) of first Foster form $\sigma = \omega = 0$ means this is a steady state d.c. and here capacitor $C_0$ acts as an open circuit, therefore,

$Z(0) = \infty$ and there is a pole of $Z(s)$ at $\sigma = 0$. However, if $C_0$ is absent then at $\sigma = 0$, the impedance $Z(0) = R_1 + R_2 + \cdots + R_\infty$ as the capacitances act as open circuit.

At $\sigma = \infty$, all the capacitance will act as short circuits and hence $Z(\infty) = 0$ if $R_\infty$ is missing, otherwise $Z(\infty) = R_\infty$. To summarise these statement we have

$$Z(0) = \begin{cases} \infty & C_0 \text{ present} \\ \displaystyle\sum_{i=1}^{\infty} R_i & C_0 \text{ absent} \end{cases} \qquad \ldots(10.64)$$

$$Z(\infty) = \begin{cases} 0 & R_\infty \text{ absent} \\ R_\infty & R_\infty \text{ present} \end{cases} \qquad \ldots(10.65)$$

This shows that $Z(0) \geq Z(\infty)$

To find out relative location of critical frequency we have just seen that for $RC$ impedance function in Foster I form there is a pole at origin and a zero at infinity. Therefore, if $Z(s)$ is given as

$$Z(s) = \frac{(s + \sigma_2)(s + \sigma_4)}{(s + \sigma_1)(s + \sigma_3)} \qquad \ldots(10.66)$$

then $\sigma_1$, must be near the origin as it is a pole, the singularity farther from the origin must be a zero i.e. $-\sigma_4$. Let us plot (Fig. 10.16)

$$Z(\sigma) = \frac{(\sigma + \sigma_2)(\sigma + \sigma_4)}{(\sigma + \sigma_1)(\sigma + \sigma_3)} V_s \ \sigma$$

Beginning at $\sigma = 0 \quad Z(0) = \dfrac{\sigma_2 \sigma_4}{\sigma_1 \sigma_3} > 1$

$Z(0)$ = a positive quantity greater than unity as $\sigma_1 < \sigma_2$ and $\sigma_3 < \sigma_4$

Since the slope of $Z(\sigma)$ is always negative as $-\sigma$ decreases (or $\sigma$ increases), $Z(\sigma)$ must increase until at $s = \sigma_1$, $Z(-\sigma_1) = \infty$. At $\sigma = -\sigma_1$ $Z(\sigma)$ changes sign and is negative because between $-\sigma_1$ and $-\sigma_2$, the term in the denominator $(\sigma + \sigma_1)$ becomes negative whereas the other three factors are positive and it continues until the next critical frequency $(-\sigma_2)$ is reached where the function is zero. Since $Z(\sigma)$ increase for decreasing $-\sigma$, the third critical frequency must be a pole

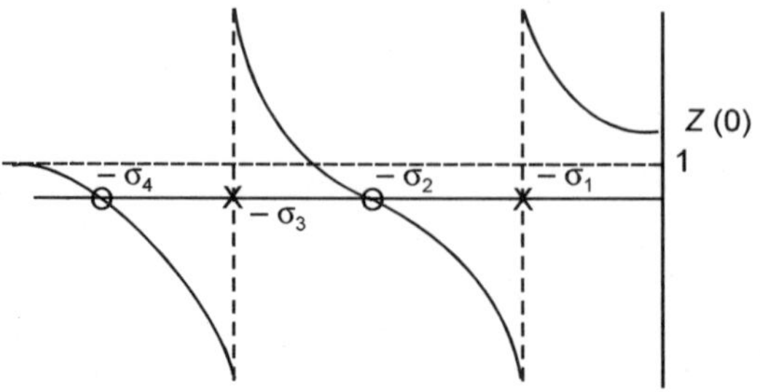


**Fig. 10.16** Plot of $Z(0)$ *Vs* $\sigma$ for *RC* network in Foster I form.

at $s = -\sigma_3$ As $Z(\sigma)$ changes sign at $-\sigma_3$, the final critical frequency must be the zero at $s = -\sigma_4$. Beyond $\sigma = -\sigma_4$ the curve becomes equal to $Z(\infty) = 1$. From this discussion we find that the poles and zeros of an *RC* impedance must alternate so that for the case under consideration

$$0 \le \sigma_1 < \sigma_2 < \sigma_3 < \sigma_4 < \infty$$

We, therefore, summarize the properties of *RC* impedance (First Foster form) as

1. Poles and zeros lie on the negative real axis and they alternate.
2. The critical frequency nearest to the origin or at the origin must be a pole whereas the critical frequency nearest to infinity or at infinity must be a zero. Since there can be no pole at infinity, the degree of numerator can't exceed that of the denominator.
3. The residues of the poles must be positive and real.
4. The impedance $Z(0) > Z(\infty)$.

We started the *RC* impedance synthesis procedure with an assumption that this could be realized with Foster first form of configuration. Here one thing is very important that before partial fraction expansion of a $Z(s)$ is obtained, we must remove the minimum real part of $Z(j\omega)$ from the original function $Z(s)$. It can be shown that $[R_e\, Z(j\omega)] = Z(\infty)$ so that we have to remove $Z(\infty)$ as a resistor in the partial fraction expansion. In cases where the numerator is of lower degree than the denominator $Z(\infty) = 0$ and hence there is no need to remove this part. However, if both are of the same degree then $Z(\infty)$ can be obtained by dividing the numerator by the denominator, the quotient is then $Z(\infty)$ and partial fraction of the remainder is then obtained.

**Example 10.21:** Synthesise the *RC* impedance using first Foster form

$$Z(s) = \frac{4(s+2)(s+6)}{s(s+4)}$$

**Solution:** Since the numerator and denominator are of the same degree, we remove $Z(\infty)$ from the function $Z(s)$ which is obtained by dividing numerator by the denominator and the quotient is $Z(\infty)$ or dividing each factor of numerator and denominator by $s$ and then let $s \to \infty$ we have

$$Z(\infty) = \operatorname*{Lt}_{s \to \infty} \frac{4\left(1 + \dfrac{2}{s}\right)\left(1 + \dfrac{6}{s}\right)}{1 + 4/s} = 4$$

Therefore, $Z_1(s) = \dfrac{4s^2 + 32s + 48 - 4s^2 - 16s}{s(s+4)}$

$$= \frac{16(s+3)}{s(s+4)}$$

$$= \frac{A}{s} + \frac{B}{s+4}$$

$$A = \operatorname*{Lt}_{s \to 0} \frac{16(s+3)}{s+4} = \frac{16 \times 3}{4} = 12$$

$$B = \mathop{\mathrm{Lt}}_{s \to -4} \frac{16\,(s+3)}{s} = \frac{16\,(-1)}{-4} = 4$$

Therefore,
$$Z(s) = \frac{12}{s} + \frac{4}{s+4} + 4$$

The network is

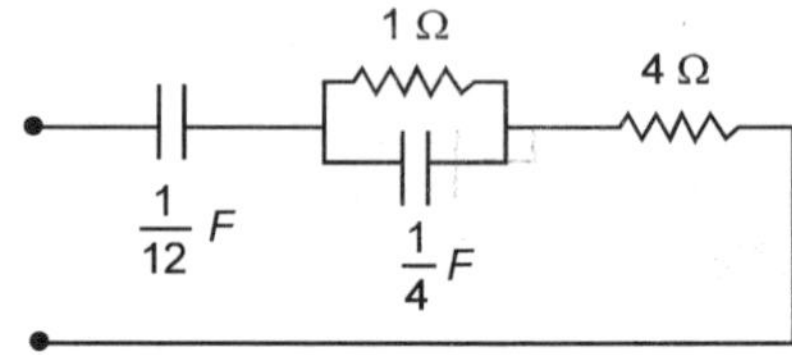



**Fig. E10.21**   *RC* impedance network.

**Example 10.22:** Synthesise the *RL* admittance using second Foster form (Admittance form)

$$Y(s) = \frac{4\,(s+2)\,(s+6)}{s\,(s+4)}$$

**Solution:** Since *RC* impedance form and *RL* admittance have the similar term in partial fraction expansion, the network will have

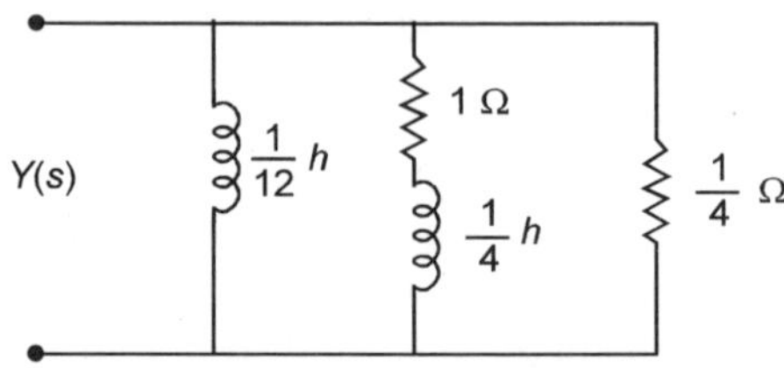



**Fig. E10.22**   *RL* admittance.

$$L_0 = \frac{1}{12}H \quad L_1 = \frac{1}{4}H \quad R_1 = 1\ \Omega \text{ and } R_\infty = \frac{1}{4}\ \Omega$$

**Example 10.23:** Synthesise the driving point impedance function using Foster first form of realization.

$$Z(s) = \frac{(s+1)\,(s+3)\,(s+5)}{s\,(s+2)\,(s+4)\,(s+6)}$$

**Solution:** Since the degree of numerator in *s* is lower than that of denominator there is no need to find out $Z(\infty)$ as it turns out to be zero.

Finding the partial fraction

$$Z(s) = \frac{A}{s} + \frac{k_1}{s+2} + \frac{k_2}{s+4} + \frac{k_3}{s+6}$$

$$A = \mathop{\mathrm{Lt}}_{s \to 0} sZ(s) = \frac{1 \times 3 \times 5}{2 \times 4 \times 6} = \frac{5}{16}$$

$$K_1 = \underset{s \to -2}{\text{Lt}} (s + 2)\, Z(s) = \frac{(-1)\,(1)\,(3)}{(-2)\,(2)\,(4)} = \frac{3}{16}$$

$$K_2 = \underset{s \to -4}{\text{Lt}} (s + 4)\, Z(s) = \frac{(-3)\,(-1)\,(1)}{(-4)\,(-2)\,(2)} = \frac{3}{16}$$

$$K_3 = \underset{s \to -6}{\text{Lt}} (s + 6)\, Z(s) = \frac{(-5)\,(-3)\,(-1)}{(-6)\,(-4)\,(-2)} = \frac{5}{16}$$

Therefore, the network has the values

$$C_0 = \frac{16}{5}F \quad C_1 = \frac{16}{3}F \quad R_1 = \frac{3}{32}\,\Omega$$

$$C_2 = \frac{16}{3}F \quad R_2 = \frac{3}{64}\,\Omega \quad C_3 = \frac{16}{5}F \quad R_3 = \frac{5}{96}\,\Omega$$

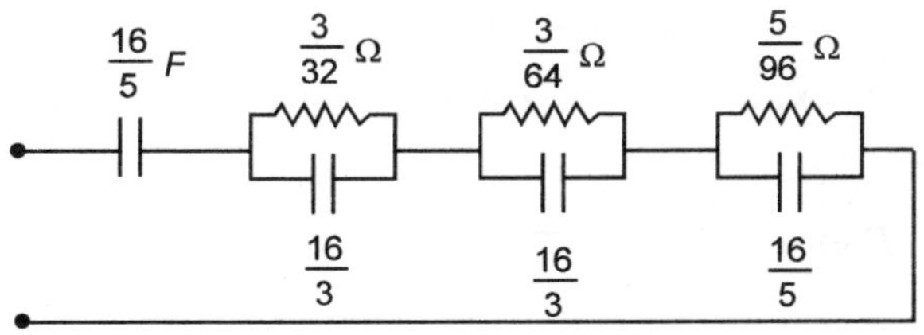

**Fig. E10.23**   First Foster form network.

## 10.5.1 Cauer Forms of RC Networks

The Cauer form as mentioned earlier are obtained by continued fraction expansion. In first Cauer form the expansion is about infinity. If the degree in $s$ of the numerator equals that of the denominator then a constant can be obtained from the functions $Z(s)$ by letting $s \to \infty$. However, if the degree of denominator is higher than that of numerator, no constant term can be removed from $Z(s)$. However, under Cauer I form there is no necessity to go through this process. The check is inbuilt in the process and hence it is seen that the whole synthesis process can be resolved by a continued fraction expansion where the quotients represents the elements of a ladder network.

The continued fraction is given by

$$F(s) = \alpha_1 + \cfrac{1}{\alpha_2 s + \cfrac{1}{\alpha_3 + \cfrac{1}{\alpha_4 s + \cfrac{1}{\alpha_5 + \cfrac{1}{\alpha_{r-1} + \cfrac{1}{\alpha_r s}}}}}} \qquad \qquad ...(10.67)$$

In case $\alpha_1 = 0$ i.e. $Z(\infty) = 0$ the first resistance element is absent. The network is shown in Fig. 10.17.

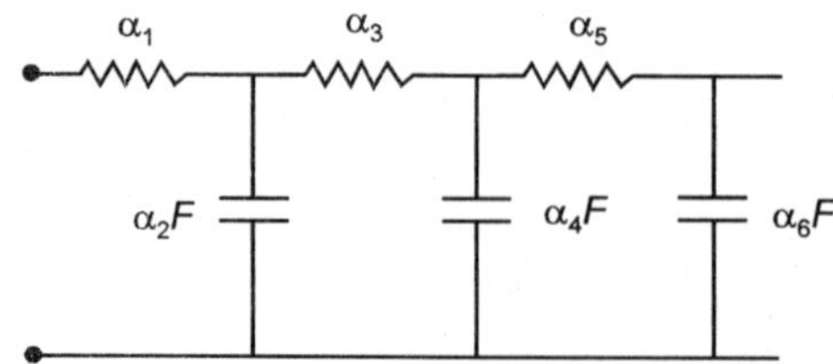

**Fig. 10.17**  Ist Cauer form of RC impedance.

**10.5.2**  In the IInd Cauer form, the continued fraction expansion is about the origin. Here we arrange the polynomials numerator and denominator of the impedance function in the ascending powers of $s$ as given here

$$F(s) = \frac{a_0 + a_1 s + a_2 s^2 + \cdots\cdots + a_{n-1} s^{n-1} + a_n s^n}{b_0 + b_1 s + b_2 s^2 + \cdots\cdots + b_{m-1} s^{m-1} + b_m s^m} \qquad \text{...(10.68)}$$

where $\qquad\qquad m = n \quad \text{or} \quad n = m - 1$

and the continued fraction expansion for a general situation is given as here

$$F(s) = \cfrac{1}{C_1 s} + \cfrac{1}{G_1 + \cfrac{1}{\cfrac{1}{C_2 s} + \cfrac{1}{G_2 + \vdots}}} \qquad \text{...(10.69)}$$
$$\cfrac{1}{C_{r-1} s} + G_r$$

If $F(s)$ has a pole at origin which means $b_0 = 0$ then the first quotient will be $\dfrac{1}{C_1}$ and then the continued fraction expansion is continued till the remainder is zero. However, if the impedance function $F(s) = Z(s)$ does not have a pole at origin, in that case the function is inverted right at the first step and the continued fraction is carried out till the remainder is zero. The first quotient will be $G_1$ and this will form the shunt branch of the network shown here in Fig. 10.18.

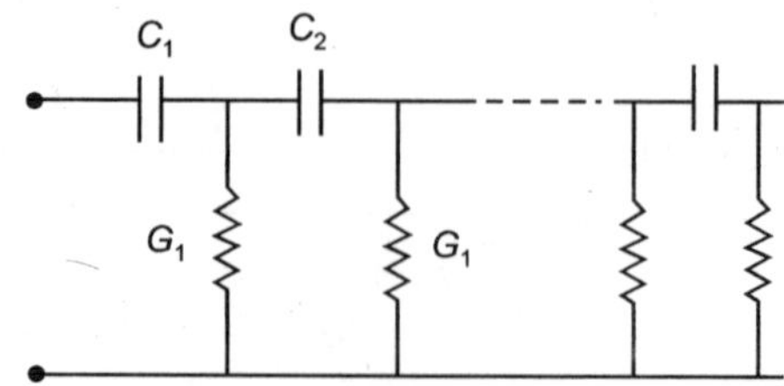

**Fig. 10.18**  Second Cauer form of RC impedance.

**Example 10.24:**  Synthesise the function using Cauer form of realization

**Solution** Ist Cauer form:

$$F(s) = Z(s) = \frac{3(s+2)(s+4)}{s(s+3)}$$

Since the degree of s in the numerator and the denominator is same the constant term $Z(\infty)$ can be obtained.

$$Z(\infty) = \underset{s \to \infty}{\text{Lt}} \frac{3(s+2)(s+4)}{s(s+3)} = \underset{s \to \infty}{\text{Lt}} \frac{3(1+2/s)(1+4/s)}{1(1+3/s)}$$

$$= 3$$

Therefore, the first element (series) resistance $R_1 = 3 \ \Omega$

$$Z_1(s) = Z(s) - 3 = \frac{3s^2 + 18s + 24}{s^2 + 3s} - 3$$

$$= \frac{9s + 24}{s^2 + 3s}$$

Hence
$$Y_1(s) = \frac{s^2 + 3s}{9s + 24}$$

$$
9s + 24 \ \big|\ s^2 + 3s \ \big|\ \frac{s}{9}
$$
$$
\qquad s^2 + \frac{24}{9}s
$$
$$
\qquad\quad -\ -
$$
$$
\overline{\qquad\qquad\qquad}
$$
$$
\frac{s}{3}\ \big|\ 9s + 24 \ \big|\ 27
$$
$$
\qquad 9s
$$
$$
\overline{\qquad\qquad\qquad}
$$
$$
24\ \big|\ \frac{s}{3}\ \big|\ \frac{s}{72}
$$
$$
\qquad \frac{s}{3}
$$
$$
\overline{\qquad\qquad}
$$
$$
\qquad \times
$$

Therefore, the elements are

$$R_1 = 3 \ \Omega \quad C_1 = \frac{1}{9}F \quad R_2 = 27 \ \Omega \quad C_2 = \frac{1}{72}F$$

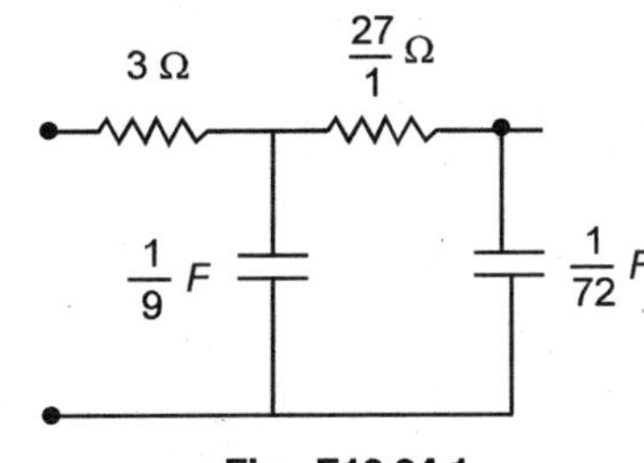


**Fig. E10.24.1**

However, if $F(s)$ is taken as $Y_{RL}(s)$, then the procedure is exactly same as for 1st Cauer only thing is the quotients will be

$$G_1 = 3 \ \mho \quad L_1 = \frac{1}{9}H \quad G_2 = 27 \ \mho \quad L_2 = \frac{1}{72}H$$

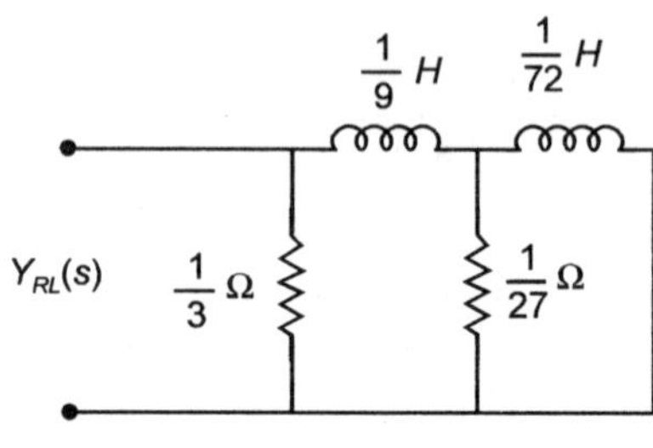

**Fig. E10.24.2**

Cauer II

$$Z(s) \;=\; \frac{3(s+2)(s+4)}{s(s+3)}$$

Arranging in the ascending order of $s$ we have

$$\frac{24 + 18s + 3s^2}{3s + s^2}$$

Since $Z(s)$ has a pole at the origin as $b_0 = 0$, therefore, we straightaway carry out continued fraction expansion.

$$3s + s^2 \,\overline{\left)\, 24 + 18s + 3s^2 \,\right.}\; \dfrac{8}{s} \longrightarrow Z$$

$$\underline{24 + 8s}$$

$$10s + 3s^2 \,\overline{\left)\, 3s + s^2 \,\right.}\; \dfrac{3}{10} \longrightarrow Y$$

$$\underline{3s + \dfrac{9}{10}s^2}$$

$$\dfrac{s^2}{10} \,\overline{\left)\, 10s + 3s^2 \,\right.}\; \dfrac{100}{s} \longrightarrow Z$$

$$\underline{10s}$$

$$3s^2 \,\overline{\left)\, \dfrac{s^2}{10} \,\right.}\; \dfrac{1}{30} \longrightarrow Y$$

$$\underline{\dfrac{s^2}{10}}$$
$$\times$$

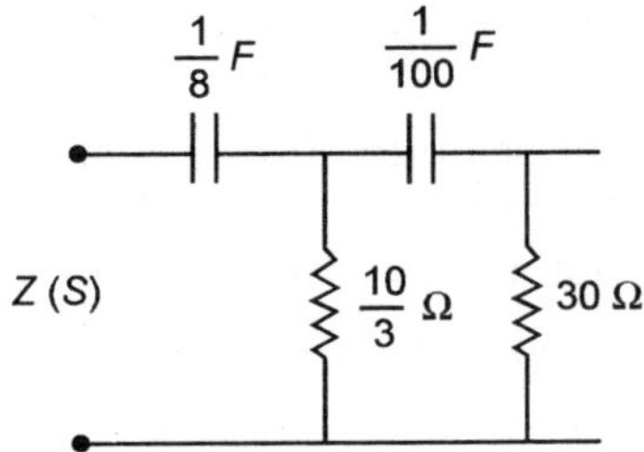

**Fig. E10.24.3**  Cauer II form.

However, if $F(s)$ denotes an admittance then

$$L_1 = \frac{1}{8}\,\text{H} \qquad R_1 = \frac{3}{10}\,\Omega \qquad L_2 = \frac{1}{100}\,\text{H} \qquad R_2 = \frac{1}{30}\,\Omega$$

and the network is as shown

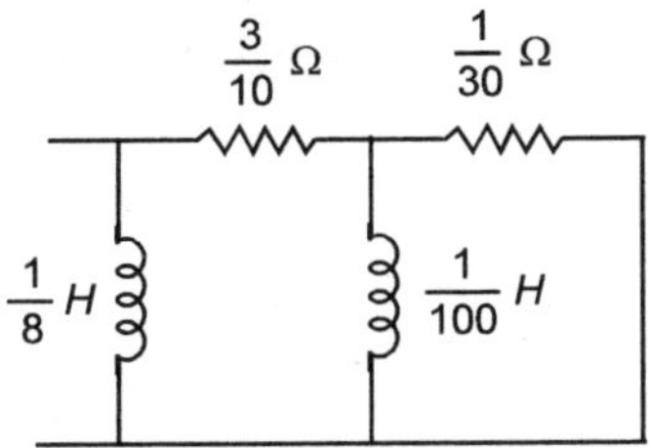

**Fig. E10.24.4**  $Y_{RL}$ realization.

## 10.6  RL IMPEDANCE OR RC ADMITTANCE FUNCTIONS

**Properties**

The impedance of a parallel combination of $R$ and $L$ and the admittance of a series combination of $R$ and $C$ are given as

$$Z_{RL}(s) = \frac{R \cdot sL}{R + sL} = R\,\frac{s}{s + \dfrac{R}{L}} \qquad\qquad \text{...(10.70)}$$

and

$$Y_{RC}(s) = \frac{1}{R + \dfrac{1}{Cs}} = \frac{Cs}{RCs + 1} = \frac{1}{R} \cdot \frac{s}{s + \dfrac{1}{RC}} \qquad\qquad \text{...(10.71)}$$

So it is seen that these have identical expressions. Therefore, the immitance that represents a series Foster $RL$ impedance or a parallel Foster $RC$ admittance is written, in general as

$$F(s) = K_0 + \frac{K_1 s}{s + \sigma_1} + \cdots \frac{K_i s}{s + \sigma_i} + \cdots + K_\infty s \qquad\qquad \text{...(10.72)}$$

Comparing this with the $F(s)$ for a series Foster $RC$ impedance or a parallel Foster $RL$ admittance it is found that whereas the latter does not have $s$ in the partial fraction term, the former is associated with $s$, in order to give the *RL* tank circuit a resistor in parallel with an inductor.

The properties of *RL* series impedance Foster I or *RC* admittance Foster II can be derived in a manner similar to the one's for *RC* impedance Foster I and *RL* admittance Foster II form. The procedure is not repeated here but conclusions are given as under:

1. Poles and zeros of an *RL* impedance or *RC* admittance are located on the negative real axis and are interlaced.
2. The singularity nearest to the origin is a zero whereas the singularity near or at $s = \infty$ is a pole. The fact that there can't be a zero at $s = \infty$ the degree of denominator can't be greater than that of numerator of $Z_{RL}(s)$.
3. The residues of the poles of $Z_{RL}(s)$ are real and negative. However, the residues of $\dfrac{Z_{RL}(s)}{s}$ are real and positive.
4. The slope of $Z_{RL}(s)$ is positive and $Z_{RL}(\infty) > Z_{RL}(0)$

The plot of $Z_{RL}(s)$ as a function of $\sigma$ is shown in Fig. 10.19 and the corresponding pole zero configuration is shown in Fig. 10.20.

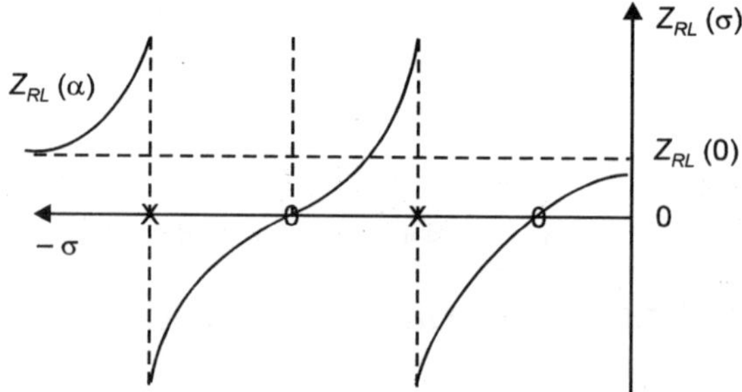


**Fig. 10.19** A typical plot of $Z_{RL}(\sigma)$ or $Y_{RC}(\sigma)$ Vs $\sigma$

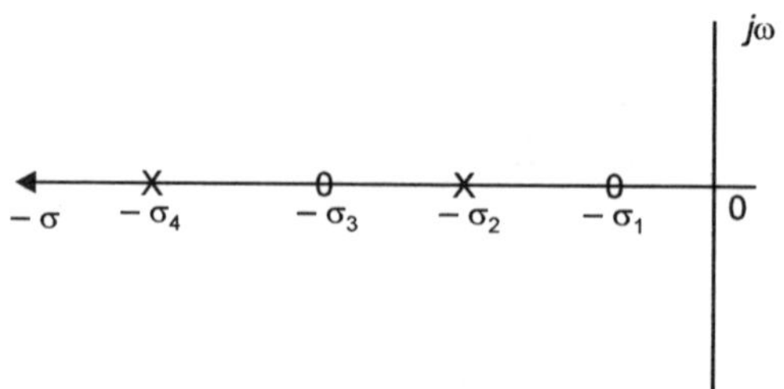


**Fig. 10.20** Pole Zero configuration of $Z_{RL}(\sigma)$ and $Y_{RC}(\sigma)$

## Synthesis of RL Impedance (Foster I Form) and RC Admittance (Foster II Form)

Based on III property listed above the partial fraction expansion of $Z_{RL}(s)$ gives terms of the form

$$\frac{-k_i}{s + \sigma i}$$

However, the $R$ and $L$ in parallel do not involve negative sign and there is no reason for concern. In order to obtain Foster form of $RL$ impedance we expand $(Z_{RL}/s)$ rather than $Z_{RL}(s)$ into partial fraction. With the minor manipulation it is found that the residues turn out to be positive. Thus

$$\frac{Z_{RL}(s)}{s} = \frac{K_0}{s} + \frac{K_1}{s+\sigma_1} + \cdots + \frac{K_i}{s+\sigma_i} + \cdots + K_\infty \qquad \qquad \text{...(10.70)}$$

where all $K^s$ are positive and real. Multiplying both sides of the above equation by $s$ we have

$$Z_{RL}(s) = K_0 + \frac{K_1 s}{s+\sigma_1} + \cdots + \frac{K_i s}{s+\sigma_i} + \cdots + K_\infty s$$

which can be used to synthesise the network. For $RL$ impedance ($R$ and $L$ in parallel)

$$R_0 = K_0, \quad L_\infty = K_\infty, \quad R_1 = K_1, \quad L_1 = \frac{K_1}{\sigma_1} \quad \text{and } R_i = K_i \text{ and } L_i = \frac{K_i}{\sigma_i}$$

and the general format of the network is shown in Fig. 10.21.

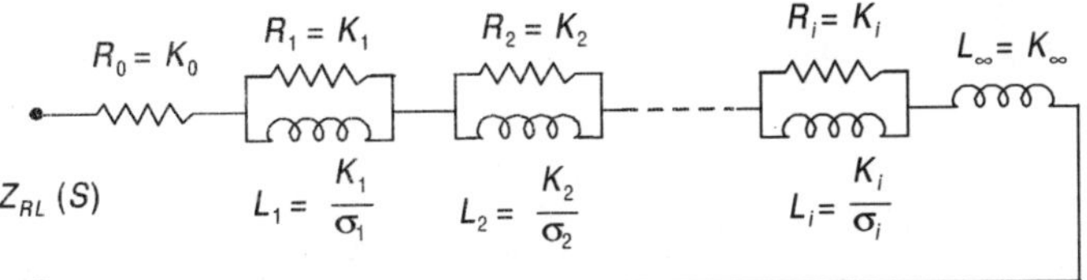


**Fig. 10.21**  Series Foster network for $RL$ impedance function.

However, if the original function is $Y_{RC}(s)$ then

$$R_0 = \frac{1}{K_0}, \quad C_\infty = K_\infty, \quad R_1 = \frac{1}{K_1}, \quad C_1 = \frac{K_1}{\sigma_1}, \quad R_i = \frac{1}{K_i}, \quad C_i = \frac{K_i}{\sigma_i}$$

and the general format of the network is given as in Fig. 10.22.

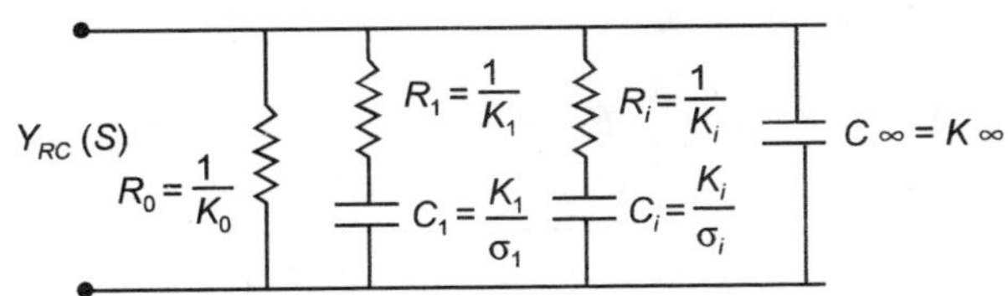


**Fig. 10.22**  Parallel Foster network for $RC$ admittance function.

**Example 10.25:** Synthesise the network in the two Foster forms (Series $RL$ impedance and parallel $RC$ admittance) of the given function.

$$F(s) = \frac{(s+1)(s+5)}{(s+3)(s+7)}$$

**Solution:** Since near origin the function has a zero ($s = -1$), therefore, the residues are going to be negative. So to avoid we find partial fraction expansion of

$$\frac{Z_{RL}(s)}{s} = \frac{F(s)}{s} = \frac{(s+1)(s+5)}{s(s+3)(s+7)}$$

$$= \frac{K_0}{s} + \frac{K_1}{s+3} + \frac{K_2}{s+7}$$

$$K_0 \underset{s \to 0}{Lt} = \frac{s(s+1)(s+5)}{s(s+3)(5+7)} = \frac{5}{21}$$

$$K_1 \underset{s \to -3}{Lt} = \frac{(s+3)(s+1)(s+5)}{s(s+3)(s+7)} = \frac{(-2)(2)}{(-3)(4)} = \frac{1}{3}$$

$$K_2 \underset{s \to -7}{Lt} = \frac{(-6)(-2)}{(-7)(-4)} = \frac{3}{7}$$

Therefore, $\quad R_0 = \dfrac{5}{21}\,\Omega \qquad R_1 = \dfrac{1}{3}\Omega \qquad L_1 = \dfrac{1}{9}\mathrm{H}$

$$R_2 = \frac{3}{7}\Omega \qquad L_2 = \frac{3}{49}\,\mathrm{H}$$

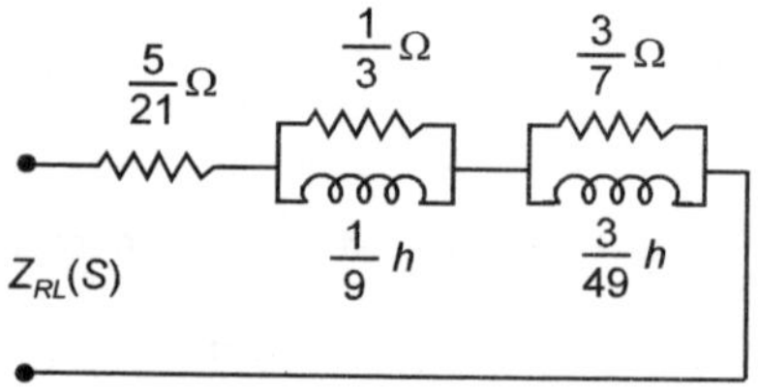

$\frac{5}{21}\Omega \qquad \frac{1}{3}\Omega \qquad \frac{3}{7}\Omega$

$Z_{RL}(S) \qquad \frac{1}{9}\,h \qquad \frac{3}{49}\,h$

**Fig. E10.25.1**

However, for Foster, II, the residues remain same but their significance is

$$R_0 = \frac{21}{5}\,\Omega \qquad R_1 = \frac{1}{K_1} = 3\ \Omega \qquad C_1 = \frac{1}{9}F$$

$$R_2 = \frac{7}{3}\,\Omega \qquad C_2 = \frac{3}{49}F$$

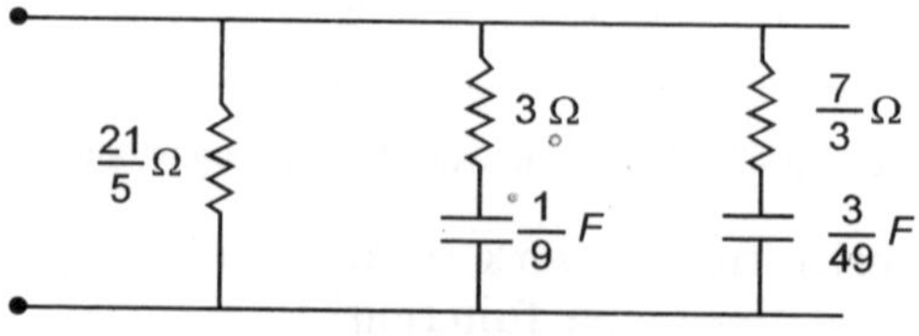

$\frac{21}{5}\Omega \qquad 3\,\Omega \qquad \frac{7}{3}\Omega$

$\frac{1}{9}F \qquad \frac{3}{49}F$

**Fig. E10.25.2**

## 10.6.1  Cauer Forms of RL Impedance and RC Admittance

In order to synthesise an *RL* impedance in ladder form (Cauer form) we make use of the fact that min $[Z_{RL}(j\omega)] = Z(0)$ thereby we remove $Z(0)$. After removing $Z(o)$ from $Z(s)$ we obtain $Z_1(s)$, we invert it and divide it, this will give us the admittance of inductance *i.e.* a pole at $s = o$, we remove this and again invert and divide. We continue this process by continued fraction expansion after arranging the numerator and the denominator terms in the ascending power of $s$, till the remainder is zero. The quotient with constant term forms the series resistances and the quotients with $\dfrac{1}{Ks}$ term constitute the shunt inductances of the Ist Cauer form.

However, keeping the same continued fraction expansion process, the constant terms in the quotients represent shunt resistances and $\dfrac{1}{Ks}$ term represents the series capacitances of the IInd Cauer format of network.

The number of element in all these networks is equal to number of internal critical frequencies plus 1. The critical frequencies at $s = o$ and $s = \infty$ are not to be counted as these are external critical frequencies.

**Example 10.26:** Synthesise the network in Cauer I and II form (*RL* impedance and *RC* admittance) of the given function.

$$Z_{RL}(s) = \frac{2(s+1)(s+3)}{(s+2)(s+4)}$$

**Solution:** Arranging the polynomials in the ascending power of $s$ we have

$$8 + 6s + s^2 \;\Big|\; 3 + 4s + s^2 \;\Big|\; \dfrac{3}{8}$$
$$\underline{3 + \dfrac{9}{4}s + \dfrac{3}{8}s^2}$$
$$\dfrac{7}{4}s + \dfrac{5}{8}s^2 \;\Big|\; 8 + 6s + s^2 \;\Big|\; \dfrac{32}{7s}$$
$$\underline{8 + \dfrac{20}{7}s}$$
$$\dfrac{22}{7}s + s^2 \;\Big|\; \dfrac{7}{4}s + \dfrac{5}{8}s^2 \;\Big|\; \dfrac{49}{88}$$
$$\underline{\dfrac{7}{4}s + \dfrac{49}{88}s^2}$$
$$\dfrac{3}{44}s^2 \;\Big|\; \dfrac{22}{7}s + s^2 \;\Big|\; \dfrac{22 \times 44}{21s}$$
$$\underline{\dfrac{22}{7}s}$$
$$s^2 \;\Big|\; \dfrac{3}{44}s^2 \;\Big|\; \dfrac{3}{44}$$
$$\underline{\dfrac{3}{44}s^2}$$
$$\times$$

The I Cauer circuit $Z_{RL}(s)$ is given as

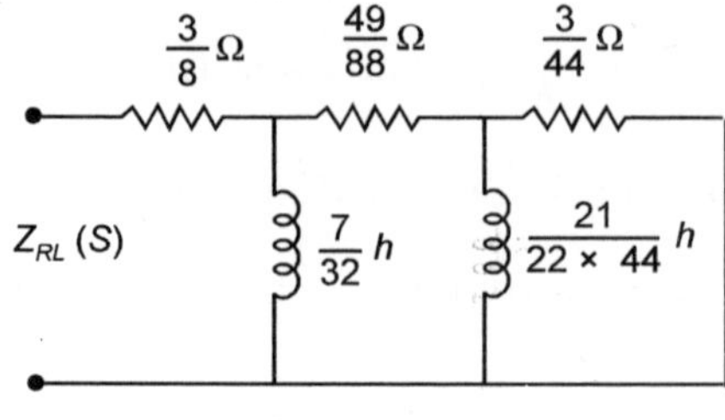

**Fig. E10.26.1**

However, the II Cauer circuit is

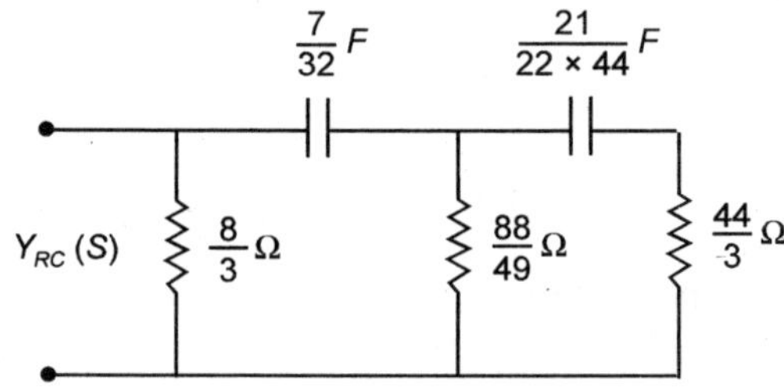

**Fig. E10.26.2** II Cauer form of the network.

**Example 10.27:** Synthesise all the four forms of RC driving point function.

$$Z_{RC}\,(s) = \frac{2\,(s+2)(s+4)}{(s+1)(s+3)}$$

Foster I

Since the degree of numerator and the denominator of $Z_{RC}(s)$ is same, we take out the constant term $Z(\infty) = 2$.

Therefore,

$$Z_2(s) = \frac{2s^2 + 12s + 16}{s^2 + 4s + 3} - 2$$

$$= \frac{4s + 10}{s^2 + 4s + 3}$$

Therefore,

$$Z_{RC}(s) = 2 + \frac{K_1}{s+1} + \frac{K_2}{s+3}$$

$$K_1 = \operatorname*{Lt}_{s \to -1} \frac{4s + 10}{s + 3} = \frac{6}{2} = 3$$

$$K_2 = \operatorname*{Lt}_{s \to -3} \frac{4s + 10}{s + 1} = 1$$

Therefore,

$$Z_{RC} = 2 + \frac{3}{s+1} + \frac{1}{s+3}$$

The elements are
$$R_o = 2\ \Omega \qquad C_1 = \frac{1}{K_1} = \frac{1}{3}\ \text{F}$$

$$R_1 = \frac{K_1}{\sigma_1} = \frac{3}{1} = 3\ \Omega \qquad C_2 = \frac{1}{K_2} = 1\ \text{F}$$

$$R_2 = \frac{K_2}{\sigma_2} = \frac{1}{3}$$

Therefore, the circuit is

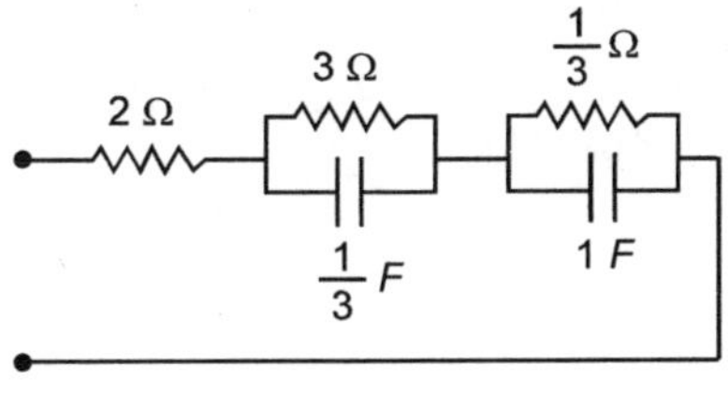


**Fig. E10.27.1**

**Foster II**

Here since it is $Y_{RC}(s)$ which requires that the partial fraction should be taken for $(Y_{RC}/s)$ rather than $Y_{RC}(s)$ as in the latter case the residues of partial fraction would turn out to be negative.

$$\frac{Y_{RC}(s)}{s} = \frac{(s+1)(s+3)}{2s\,(s+2)(s+4)}$$

$$= \frac{K_1}{s} + \frac{K_2}{s+2} + \frac{K_3}{s+4}$$

$$K_1 = \underset{s\to 0}{\text{Lt}}\ \frac{(s+1)(s+3)}{2\,(s+2)(s+4)} = \frac{3}{16}$$

$$K_2 = \underset{s\to -2}{\text{Lt}}\ \frac{(s+1)(s+3)}{2s\,(s+4)} = \frac{1}{8}$$

$$K_3 = \underset{s\to -4}{\text{Lt}}\ \frac{(s+1)(s+3)}{2s\,(s+2)} = \frac{3}{16}$$

Therefore,
$$Y_{RC}(s) = \frac{3}{16s} + \frac{s/8}{s+2} + \frac{3s/16}{s+4}$$

$$R_0 = \frac{1}{K_0} = \frac{16}{3}\ \Omega \qquad R_1 = \frac{1}{K_1} = 8\ \Omega \qquad C_1 = \frac{1}{16}\ \text{F}$$

$$R_2 = \frac{16}{3}\ \Omega \qquad C_2 = \frac{3}{64}\ \text{F}$$

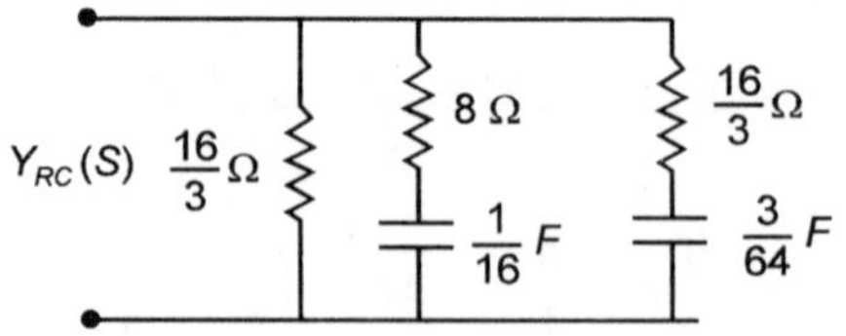

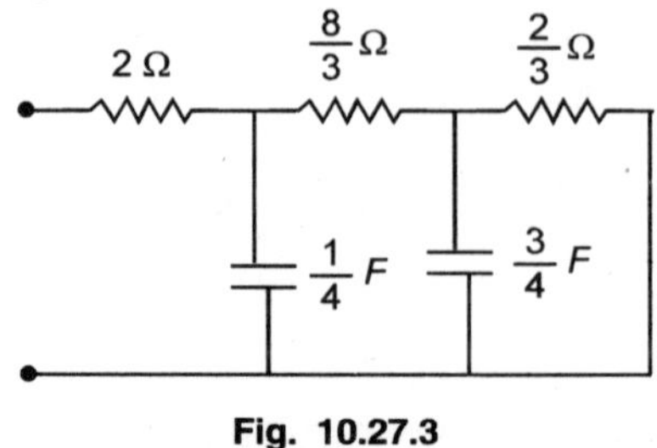

**Fig. E10.27.2**

II Foster form network of example.

Cauer I

Here we arrange the numerator and denominator in the descending power of $s$ and divide numerator by the denominator.

$$Z_{RC}(s) = \frac{2s^2 + 12s + 16}{s^2 + 4s + 3}$$

$$
s^2 + 4s + 3 \,\big|\, 2s^2 + 12s + 16 \,\big|\, 2
$$

$$
\underline{2s^2 + 8s + 6}
$$

$$
\overline{\phantom{-}} \quad \overline{\phantom{-}} \quad \overline{\phantom{-}}
$$

$$
4s + 10 \,\big|\, s^2 + 4s + 3 \,\Big|\, \frac{s}{4}
$$

$$
\underline{s^2 + \frac{5}{2}s}
$$

$$
\frac{3}{2}s + 3 \,\big|\, 4s + 10 \,\Big|\, \frac{8}{3}
$$

$$
\underline{4s + 8}
$$

$$
2 \,\Big|\, \frac{3}{2}s + 3 \,\Big|\, \frac{3}{4}s
$$

$$
\underline{\frac{3}{2}s}
$$

$$
3 \,\big|\, 2 \,\big|\, \tfrac{2}{3}
$$

$$
\underline{2}
$$

$$
\times
$$

The first quotient represents series resistance $2\,\Omega$, $\dfrac{s}{4}$ represent a shunt admittance $Cs$ therefore, $C = \dfrac{1}{4}$ F. Similarly series resistance $\dfrac{8}{3}\,\Omega$ and shunt capacitor of $\dfrac{3}{4}$ F and resistance $\dfrac{2}{3}\,\Omega$.

Therefore, the network is given as

**Fig. 10.27.3**

**Cauer II**

Here we arrange the numerator and denominator in the ascending order *i.e.*

$$Z_{RC}(s) = \frac{16 + 12s + 2s^2}{3 + 4s + s^2}$$

Since $b_o \neq 0$ in the denominator the first element will be shunt resistance, therefore, invert the impedance function

$$Y_{RC}(s) = \frac{3 + 4s + s^2}{16 + 12s + 2s^2}$$

$$16 + 12s + 2s^2 \left|\; 3 + 4s + s^2 \;\right| \frac{3}{16} \longrightarrow G$$

$$\underline{3 + \frac{9}{4}s + \frac{3}{8}s^2}$$

$$\frac{7}{4}s + \frac{5}{8}s^2 \left|\; 16 + 12s + 2s^2 \;\right| \frac{64}{7s} \longrightarrow \frac{1}{C}$$

$$\underline{16 + \frac{40}{7}s}$$

$$\frac{44s}{7} + 2s^2 \left|\; \frac{7}{4}s + \frac{5}{8}s^2 \;\right| \frac{49}{176} \longrightarrow G$$

$$\underline{\frac{7}{4}s + \frac{49}{88}s^2}$$

$$\frac{6}{88}s^2 \left|\; \frac{44}{7}s + 2s^2 \;\right| \frac{44 \times 88}{42s} \longrightarrow \frac{1}{C}$$

$$\underline{\frac{44}{7}s}$$

$$2s^2 \left|\; \frac{6}{88}s^2 \;\right| \frac{3}{88} \longrightarrow G$$

$$\underline{\frac{6}{88}s^2}$$

$$\times$$

The II Cauer network is

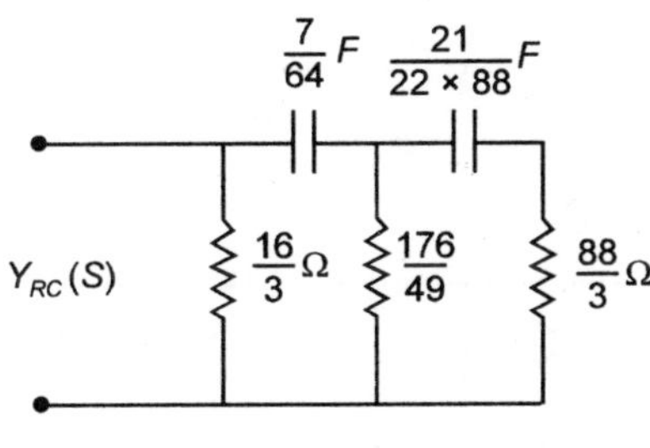


**Fig. E10.27.4**

**Example 10.28:** Determine the II Foster and Cauer I and II of the $Z_{RC}(s)$

$$Z_{RC} = \frac{(s+1)(s+3)(s+5)}{s(s+2)(s+4)(s+6)}$$

**Solution:** Foster II requires $Y_{RC}(s)$ which $= \dfrac{1}{Z_{RC}(s)}$

$$Y_{RC}(s) = \frac{s(s+2)(s+4)(s+6)}{(s+1)(s+3)(s+5)}$$

In Foster II for $RC$ circuit and Foster I $RL$ circuit we make use of the function $\dfrac{Y_{RC}(s)}{s}$ and $\dfrac{Y_{RL}(s)}{s}$ respectively as this provides positive real residues.

$$\frac{Y_{RC}(s)}{s} = \frac{(s+2)(s+4)(s+6)}{(s+1)(s+3)(s+5)}$$

$$= K_0 + \frac{K_1}{s+1} + \frac{K_2}{s+3} + \frac{K_3}{s+5}$$

$$K_1 = \operatorname*{Lt}_{s \to -1} \frac{(s+2)(s+4)(s+6)}{(s+3)(s+5)} = \frac{15}{8}$$

$$K_2 = \operatorname*{Lt}_{s \to -3} \frac{(s+2)(s+4)(s+6)}{(s+1)(s+5)} = \frac{3}{4}$$

$$K_3 = \operatorname*{Lt}_{s \to -5} \frac{(s+2)(s+4)(s+6)}{(s+1)(s+3)} = \frac{3}{8}$$

Therefore,

$$Y_{RC}(s) = \frac{(15/8)s}{s+1} + \frac{(3/4)s}{s+3} + \frac{(3/8)s}{s+5}$$

$$R_1 = \frac{1}{K_1} = \frac{8}{15}\,\Omega \qquad C_1 = \frac{15}{8}\,F$$

$$R_2 = \frac{4}{3}\,\Omega \qquad\qquad C_2 = \frac{1}{4}\,F$$

$$R_3 = \frac{8}{3}\,\Omega \qquad\qquad C_3 = \frac{3}{40}\,F$$

$$K_0 = \operatorname*{Lt}_{s \to \infty} \frac{Y_{RC}(s)}{s} = 1 = C_\infty$$

The network comes out to be

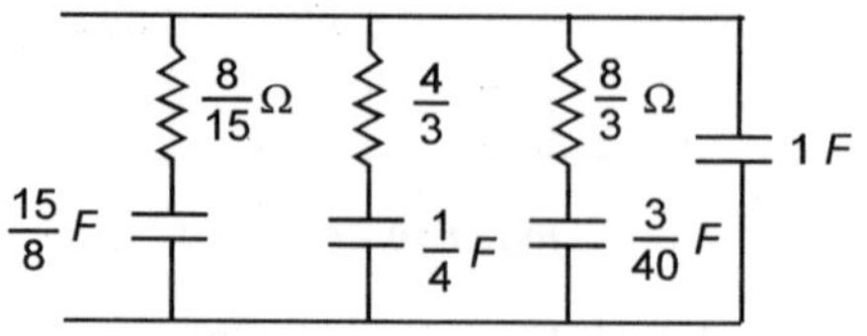

**Fig. E10.28.1**  Foster II form.

## Cauer I

$$Z_{RC}(s) = \frac{s^3 + 9s^2 + 23s + 15}{s^4 + 12s^3 + 44s^2 + 48s}$$

Since the degree of denominator in $s$ is greater than that of numerator hence the first element will be a shunt capacitor. Therefore, inverting $Z_{RC}(s)$ we have

$$s^3 + 9s^2 + 23s + 15 \,\big|\, s^4 + 12s^3 + 44s^2 + 48s \,\big|\, s$$

$$\underline{s^4 + 9s^3 + 23s^2 + 15s}$$

$$3s^3 + 21s^2 + 33s \,\big|\, s^3 + 9s^2 + 23s + 15 \,\Big|\, \frac{1}{3}$$

$$\underline{s^3 + 7s^2 + 11s}$$

$$2s^2 + 12s + 15 \,\big|\, 3s^2 + 21s^2 + 33s \,\Big|\, \frac{3}{2}s$$

$$\underline{3s^3 + 18s^2 + \frac{45}{2}s}$$

$$3s^2 + \frac{21}{2}s \,\big|\, 2s^2 + 12s + 15 \,\Big|\, \frac{2}{3}$$

$$\underline{2s^2 + 7s}$$

$$5s + 15 \,\big|\, 3s^2 + \frac{21}{2}s \,\Big|\, \frac{3}{5}s$$

$$\underline{3s^2 + 9s}$$

$$\frac{3}{2}s \,\big|\, 5s + 15 \,\Big|\, \frac{10}{3}$$

$$\underline{5s}$$

$$15 \,\Big|\, \frac{3}{2}s \,\Big|\, \frac{s}{10}$$

$$\underline{\frac{3}{2}s}$$

$$\times$$

Therefore, the network is

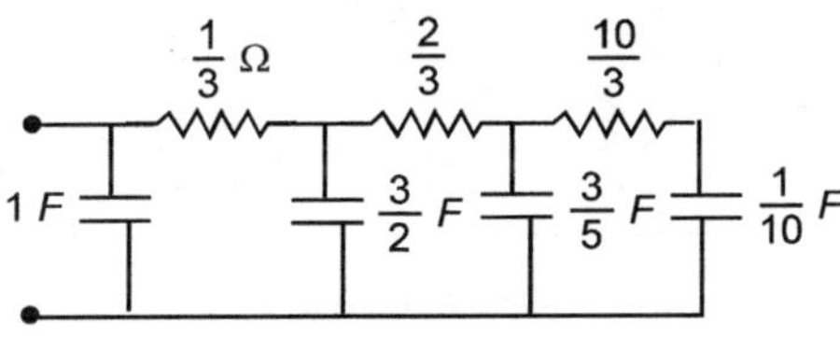



**Fig. E10.28.2**

## Cauer II

$$Z_{RC}(s) = \frac{s^3 + 9s^2 + 23s + 15}{s^4 + 12s^3 + 44s^2 + 48s}$$

Since $bo = 0$ therefore, arranging numerator and denominator in the ascending powers of $s$ and divide the numerator by the denominator we will have a quotient $\dfrac{1}{K_1 s}$ which represent a series capacitor $K_1$ and therefore, we have to retain in the impedance form only.

$$48s + 44s^2 + 12s^3 + s^4 \left|\; 15 + 23s + 9s^2 + s^3 \;\right| \dfrac{15}{48s}$$

$$15 + \dfrac{55}{4}s + \dfrac{15}{4}s^2 + \dfrac{5}{16}s^3$$

$$\overline{\dfrac{37}{4}s + \dfrac{21}{4}s^2 + \dfrac{11}{16}s^3}$$

If we continue the process we have the following continued fraction:

$$\dfrac{15}{48s} + \cfrac{1}{5.19 + \cfrac{1}{\cfrac{0.55}{s} + \cfrac{1}{28.25 + \cfrac{1}{\cfrac{0.13}{s} + \cfrac{1}{6.87 + \cfrac{1}{\cfrac{0.0067}{s}}}}}}}$$

or

$$Z_{RC}(s) = \dfrac{1}{3.2s} + \cfrac{1}{\cfrac{1}{0.192} + \cfrac{1}{\cfrac{1}{1.8s} + \cfrac{1}{\cfrac{1}{0.035} + \cfrac{1}{\cfrac{1}{7.7s} + \cfrac{1}{\cfrac{1}{0.0015} + \cfrac{1}{\cfrac{1}{149s}}}}}}}$$

and the network is given as

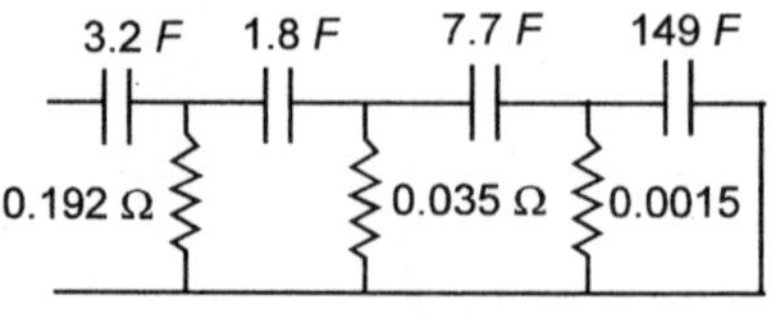


**Fig. E10.28.3**

**Example 10.29:** With suitable reasoning justify whether the given impedance functions are *LC*, *RC* or *RL*.

1.  $Z(s) = \dfrac{s^3 + 2s}{s^4 + 4s^2 + 3}$

2. $Z(s) = \dfrac{s^2 + 6s + 8}{s^2 + 4s + 3}$

3. $Z(s) = \dfrac{s^2 + 4s + 3}{s^2 + 6s + 8}$

4. $Z(s) = \dfrac{s^2 + 5s + 6}{s^2 + s}$

5. $Z(s) = \dfrac{s^4 + 5s^2 + 6}{s^3 + s}$

1. $Z(s) = \dfrac{s(s^2 + 2)}{(s^2 + 1)(s^2 + 3)}$

It is an *LC* impedance as (1) it is a ratio of odd to even powers of $s$ (2) Poles and zeros are on $j\omega$ axis and are interlaced (poles $\pm j\ \pm j\sqrt{3}$ and zeros $\pm j\sqrt{2}$ (3). The highest and lowest powers in the numerator and denominator differ by one.

2. $Z(s) = \dfrac{(s+2)(s+4)}{(s+1)(s+3)}$

It is an *RC* impedance function ($i$) the internal critical frequency near to the origin is a pole $s = -1$ ($ii$) the critical frequency nearest to infinity is a zero at $s = -4$ ($iii$) the poles and zero lie on the negative real axis and are interlaced ($iv$) $Z(o) = \dfrac{8}{3}$ where as $Z(\infty) = 1$. Therefore, $Z(o) > Z(\infty)$.

3. $Z(s) = \dfrac{(s+1)(s+3)}{(s+2)(s+4)}$

It is an *RL* impedance as ($i$) the critical frequency near to origin is a zero $s = -1$ whereas near infinity it is pole $s = -4$. ($ii$) The poles and zeros lie on negative real axis and are interlaced. ($iii$) $Z_{RL}(\infty) = 1$ and $Z_{RL}(o) = \dfrac{3}{8}$ therefore, $Z_{RL}(\infty) > Z_{RL}(o)$.

4. $Z(s) = \dfrac{(s+2)(s+3)}{s(s+1)}$

Since poles zeros do not alternate it is none of the functions given

5. $Z(s) = \dfrac{s^4 + 5s^2 + 6}{s(s^2 + 1)}$

It does not satisfy any of the element combination as the critical frequencies are not interlaced.

**Example 10.30:** The input impedance for the network shown is

$$Z_{in} = \dfrac{2s^2 + 2}{s^3 + 2s^2 + 2s + 2}$$

If $Z_o$ is an *LC* network, determine the expression for $Z_o$ and synthesise $Z_o$ in Foster series form.

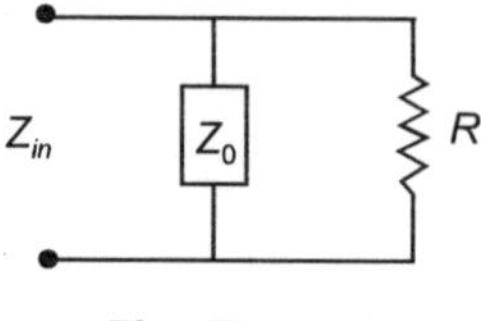

**Fig. E10.30.1**

Since $Z_o$ is an *LC* network, therefore $Z_{in}$ at $s = o$ should be a resistance $Z_{in}$ at $s = o$ is equal to 1

Now
$$\frac{1}{Z_{in}} = \frac{1}{Z_o} + 1$$

or
$$\frac{1}{Z_o} = \frac{1}{Z_{in}} - 1 = \frac{s^3 + 2s^2 + 2s + 2}{2s^2 + 2} - 1 = \frac{s(s^2 + 2)}{2(s^2 + 1)}$$

$$Z_o = \frac{2(s^2 + 1)}{s(s^2 + 2)}$$

$$= \frac{K_1}{s} + \frac{K_2 s}{s^2 + 2}$$

$$K_1 = \underset{s \to 0}{\text{Lt}} \; \frac{2(s^2 + 1)}{s^2 + 2} = 1$$

and
$$sK_2 = \underset{s^2 \to -2}{\text{Lt}} \; \frac{2(s^2 + 1)}{s}$$

or
$$K_2 = \underset{s^2 \to -2}{\text{Lt}} \; \frac{2(s^2 + 1)}{s^2} = 1$$

Therefore,

$$Z_o = \frac{1}{s} + \frac{s}{s^2 + 2}$$

Therefore
$$C_o = 1 \text{ F} \qquad C_2 = K_2 = 1 \text{ F} \qquad L_2 = \frac{1}{2} \text{ H}$$

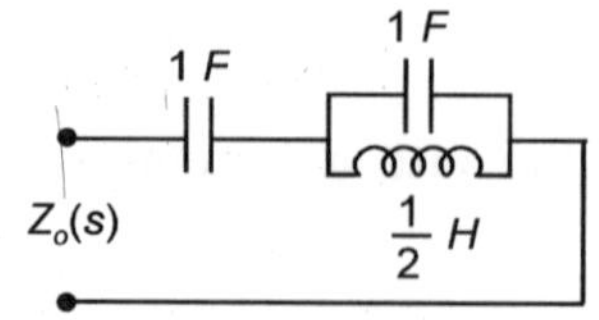

**Fig. E10.30.2**

**Example 10.31:** Synthesise $\dfrac{s(s^2 + 2)}{(s^2 + 1)(s^2 + 3)} = Z(s)$ in all the four formats.

As mentioned earlier it is an *LC* function. Since degree of numerator is lower than that of denominator.

$$Z(s) = \frac{K_1 s}{s^2 + 1} + \frac{K_2 s}{s^2 + 3}$$

$$K_1 s = \operatorname*{Lt}_{s^2 \to -1} \frac{s\,(s^2 + 2)}{(s^2 + 3)}$$

or

$$K_1 = \operatorname*{Lt}_{s^2 \to -1} \frac{s^2 + 2}{s^2 + 3} = \frac{1}{2}$$

Similarly

$$K_2 s = \operatorname*{Lt}_{s^2 \to -3} \frac{s\,(s^2 + 2)}{s^2 + 1}$$

or

$$K_2 = \operatorname*{Lt}_{s^2 \to -3} \frac{s^2 + 2}{s^2 + 1} = \frac{1}{2}$$

or

$$Z(s) = \frac{s/2}{s^2 + 1} + \frac{s/2}{s^2 + 3}$$

Therefore, $\quad C_1 = 2 \text{ F} \qquad L_1 = \dfrac{1}{2}\text{H}$

$$C_2 = 2 \text{ F} \qquad L_2 = \frac{1}{6}\text{H}$$

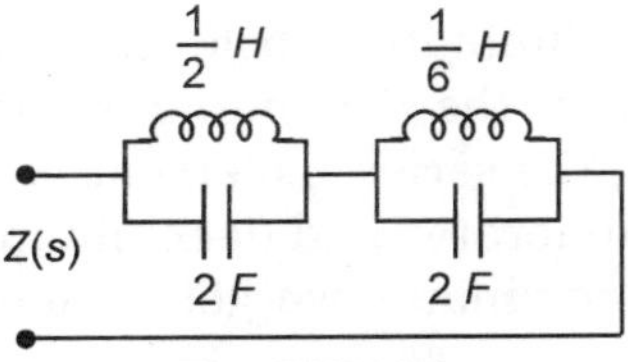


**Fig. E10.31.1**

To find out II Foster we rewrite

$$Y(s) = \frac{1}{Z(s)} = \frac{(s^2 + 1)(s^2 + 3)}{s\,(s^2 + 2)}$$

To find the partial fraction, since degree of numerator is greater than that of denominator, we divide

$$s^3 + 2s\,\big|\,\overline{s^4 + 4s^2 + 3}\,\big|\,s$$
$$\phantom{s^3 + 2s\,\big|\,}s^4 + 2s^2$$
$$\phantom{s^3 + 2s\,\big|\,}\overline{\phantom{xxxx}-\ \ -\phantom{xxx}}$$
$$\phantom{s^3 + 2s\,\big|\,}2s^2 + 3$$

$$Y(s) = s + \frac{2s^2 + 3}{s\,(s^2 + 2)} = s + Y_1(s)$$

$$Y_1(s) = \frac{K_1}{s} + \frac{K_2 s}{s^2 + 2}$$

$$K_1 = \operatorname*{Lt}_{s \to 0} sY_1(s) = \frac{2s^2 + 3}{s^2 + 2} = \frac{3}{2}$$

$$K_2 = \operatorname*{Lt}_{s^2 \to -2} (s^2 + 2)Y_1(s) = \frac{2s^2 + 3}{s}$$

or 

$$K_2 = \operatorname*{Lt}_{s^2 \to -2} \frac{2s^2 + 3}{s^2} = \frac{1}{2}$$

$$Y(s) = s + \frac{3}{2s} + \frac{s/2}{s^2 + 2}$$

Therefore, the elements are $C_o = 1\text{F}$, $L_\infty = \dfrac{2}{3}\text{H}$, $L_1 = 2h$, $C_2 = \dfrac{1}{4}\text{F}$

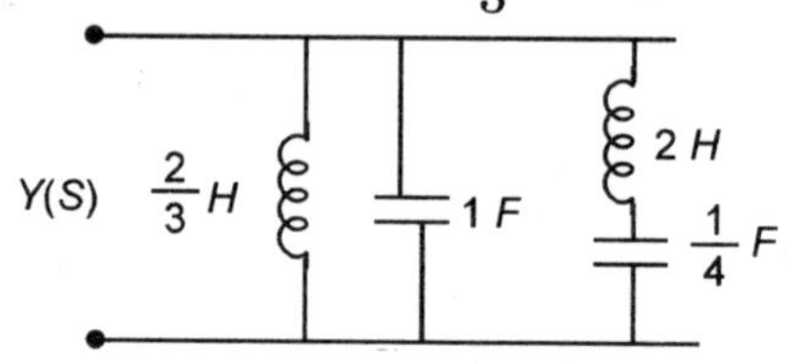


**Fig. E10.31.2**

Cauer I

$$Z(s) = \frac{s(s^2 + 2)}{(s^2 + 1)(s^2 + 3)}$$

Since in Cauer I we always eliminate a pole at infinity which means, the numerator must be one degree in $s$ higher than the denominator. Here since the degree of numerator is one lower than the denominator, the series $L$ element of Cauer I is absent. Hence we invert $Z(s)$ to get $Y(s)$ and arrange the numerator and denominator in the descending order of s and dividing the numerator by the denominator we get a pole at $s = \infty$.

$$Y(s) = \frac{s^4 + 4s^2 + 3}{s^3 + 2s}$$

$$s^3 + 2s \,\Big|\, s^4 + 4s^2 + 3 \,\Big|\, s \qquad\qquad -Y$$
$$\underline{s^4 + 2s^2}$$
$$\overline{\phantom{xxxx}}$$

$$2s^2 + 3 \,\Big|\, s^3 + 2s \,\Big|\, \frac{s}{2} \qquad\qquad -Z$$
$$\underline{s^3 + \frac{3}{2}s}$$
$$\overline{\phantom{xxxx}}$$

$$\frac{s}{2} \,\Big|\, 2s^2 + 3 \,\Big|\, 4s \qquad\qquad -Y$$
$$\underline{2s^2}$$
$$\overline{\phantom{xxx}}$$

$$3 \,\Big|\, \frac{s}{2} \,\Big|\, \frac{s}{6} \qquad\qquad -Z$$
$$\underline{\frac{s}{2}}$$
$$\times$$

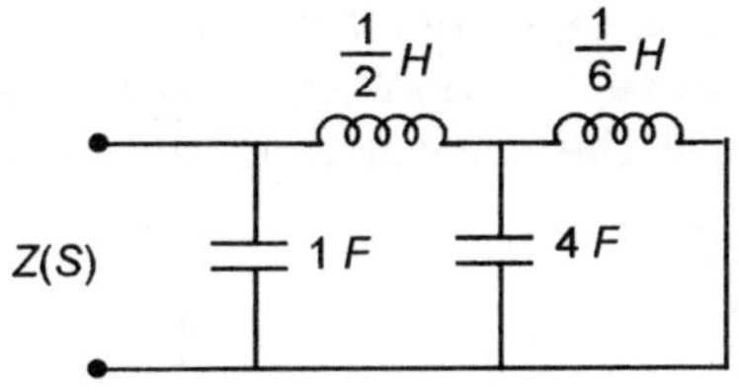

**Fig. E10.31.3**  Causer I Form.

## Cauer II

We remove here a pole at origin after arranging the polynomials in the ascending powers of $s$. We find that $Z(s)$ has a zero at origin, therefore, $Y(s)$ will have a pole at origin. Inverting $Z(s)$ and arranging in ascending order we have

$$Y(s) = \frac{3 + 4s^2 + s^4}{2s + s^3}$$

$$
2s + s^3 \;\Big|\; 3 + 4s^2 + s^4 \;\Big|\; \frac{3}{2s} \qquad\qquad -Y
$$

$$
\underline{3 + \frac{3}{2}s^2}
$$

$$
\frac{5}{2}s^2 + s^4 \;\Big|\; 2s + s^3 \;\Big|\; \frac{4}{5s} \qquad\qquad -Z
$$

$$
\underline{2s + \frac{4}{5}s^3}
$$

$$
\frac{s^3}{5} \;\Big|\; \frac{5}{2}s^2 + s^4 \;\Big|\; \frac{25}{2s} \qquad\qquad -Y
$$

$$
\underline{\frac{5}{2}s^2}
$$

$$
s^4 \;\Big|\; \frac{s^3}{5} \;\Big|\; \frac{1}{5s} \qquad\qquad -Z
$$

$$
\underline{\frac{s^3}{5}}
$$

$$
\times
$$

**Fig. E10.31.4**

It is to be noted that in $LC$ network in Cauer I the quotients will be $Ks$ and in Cauer II $\dfrac{K}{s}$ as in the latter we remove a pole at the origin where as in the former we remove a pole at infinity.

**Example 10.32:** The following pole zero configuration has three diagrams. Pick up the one which represents *RL* impedance function and synthesise in all the four formats.

**Solution:** For *RL* impedance the critical frequency near or at origin is a zero.

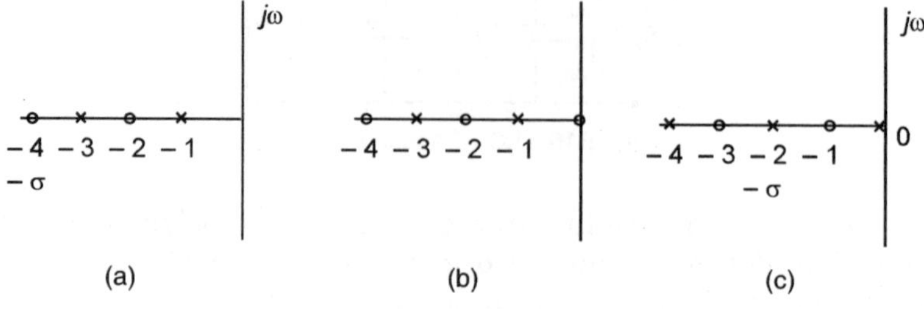

(a)  (b)  (c)

**Fig. E10.32.1**

Therefore, Fig. (b) represents the poles zeros of an *RL* impedance.

$$Z_{RL}(s) = \frac{s(s+2)(s+4)}{(s+1)(s+3)}$$

Since $Z_{RL}$ and $Y_{RL}$ may have negative residues these are expanded as $\dfrac{Z_{RL}}{s}$ rather than $Z_{RL}(s)$.

$$\frac{Z_{RL}(s)}{s} = \frac{(s+2)(s+4)}{(s+1)(s+3)} = 1 + \frac{2s+5}{(s+1)(s+3)}$$

$$\frac{Z_{RL}(s)}{s} = 1 + \frac{K_1}{s+1} + \frac{K_2}{s+3}$$

$$K_1 = \operatorname*{Lt}_{s \to -1} \frac{2s+5}{s+3} = \frac{3}{2}$$

$$K_2 = \operatorname*{Lt}_{s \to -3} \frac{2s+5}{s+1} = \frac{1}{2}$$

Therefore,
$$\frac{Z_{RL}(s)}{s} = 1 + \frac{(3/2)}{s+1} + \frac{1/2}{s+3}$$

or
$$Z_{RL} = s + \frac{(3/2)s}{s+1} + \frac{(1/2)s}{s+3}$$

Therefore, the components are:

$$L_\infty = 1\ \text{H} \qquad L_1 = \frac{3}{2}\text{H} \qquad R_1 = \frac{3}{2}\,\Omega$$

$$L_2 = \frac{1}{6}\text{H} \qquad R_2 = \frac{1}{2}\,\Omega$$

The circuit Foster I is shown here

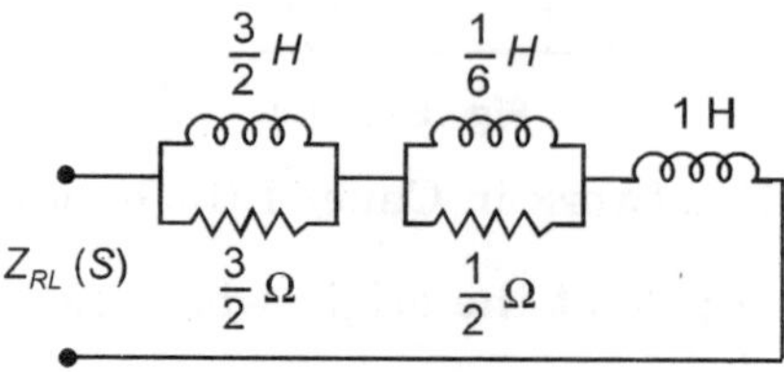


**Fig. E10.32.2**

Foster II

$$Y_{RL}(s) = \frac{(s+1)(s+3)}{s(s+2)(s+4)}$$

$$= \frac{K_0}{s} + \frac{K_1}{s+2} + \frac{K_2}{s+4}$$

$$K_0 = \operatorname*{Lt}_{s \to 0} \frac{(s+1)(s+3)}{(s+2)(s+4)} = \frac{3}{8}$$

$$K_1 = \operatorname*{Lt}_{s \to -2} \frac{(s+1)(s+3)}{s(s+4)} = \frac{1}{4}$$

$$K_2 = \operatorname*{Lt}_{s \to -4} \frac{(s+1)(s+3)}{s(s+2)} = \frac{3}{8}$$

$$Y_{RL}(s) = \frac{3/8}{s} + \frac{1/4}{s+2} + \frac{3/8}{s+4}$$

The elements are $L_0 = \dfrac{8}{3}H$

$$L_1 = \frac{1}{K_1} = 4H \qquad R_1 = \frac{\sigma_1}{K_1} = 8\ \Omega$$

$$L_2 = \frac{1}{K_2} = \frac{8}{3}H \qquad R_2 = \frac{\sigma_2}{K_2} = \frac{4}{3/8} = \frac{32}{3}\ \Omega$$

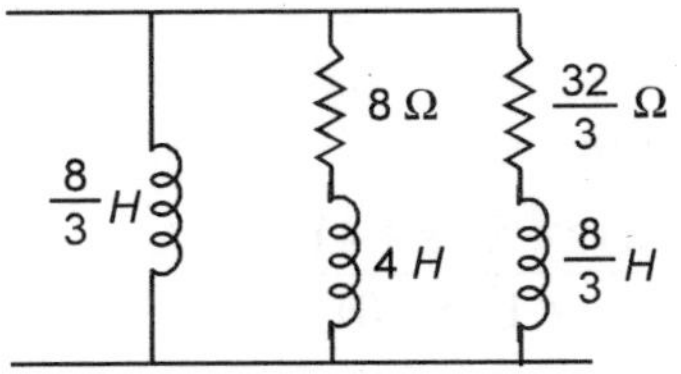

**Fig. E10.32.3**

Cauer I

$$Z_{RL}(s) = \frac{s(s+2)(s+4)}{(s+1)(s+3)}$$

In Cauer I we always eliminate pole at infinity *i.e.* we begin with a function having numerator one degree in $s$ higher than the denominator. Since here $Z_{RL}(s)$ has the requisite specification we arrange the numerator and denominator in the descending power in $s$ and divide the numerator by the denominator. We have

$$s^2 + 4s + 3 \left|\ s^3 + 6s^2 + 8s\ \right|\ s \qquad\qquad -Z$$

$$\underline{s^3 + 4s^2 + 3s}$$

$$2s^2 + 5s \left|\ s^2 + 4s + 3\ \right|\ \frac{1}{2} \qquad\qquad -Y$$

$$s^2 + \frac{5}{2}s$$

$$\frac{3}{2}s + 3 \left|\ 2s^2 + 5s\ \right|\ \frac{4}{3}s \qquad\qquad -Z$$

$$2s^2 + 4s$$

$$s \left|\ \frac{3}{2}s + 3\ \right|\ \frac{3}{2} \qquad -Y$$

$$\frac{3}{2}s$$

$$3 \left|\ s\ \right|\ \frac{1}{3} \qquad -Z$$

$$\frac{s}{\times}$$

The network is given as follows:

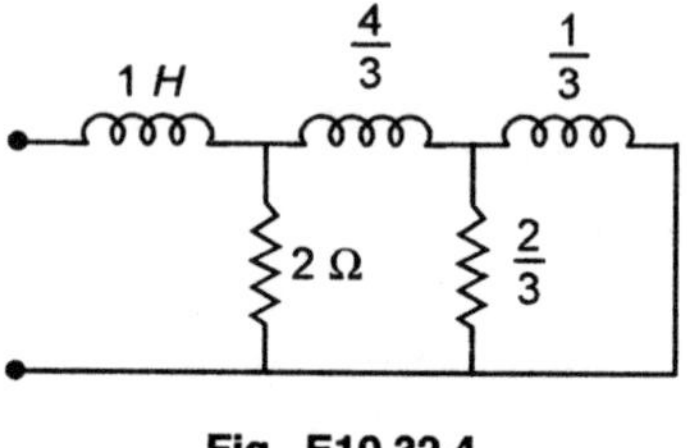



**Fig. E10.32.4**

Cauer II

Here we remove pole at origin.

$$Z_{RL} = \frac{s(s+2)(s+4)}{(s+1)(s+3)}$$

Since the order of $s$ of the numerator is greater than that of the denominator by one, pole at origin is not possible. Therefore, we invert it.

$$Y_{RL}(s) = \frac{(s+1)(s+3)}{s(s+2)(s+4)}$$

Arranging in the ascending power of $s$ we have

$$8s + 6s^2 + s^3 \left|\ 3 + 4s + s^2\ \right|\ \frac{3}{8}s$$

$$3 + \frac{9}{4}s + \frac{3}{8}s^2$$

$$\frac{7}{4}s + \frac{5}{8}s^2 \;\Big|\; 8s + 6s^2 + s^3 \;\Big|\; \frac{32}{7}$$

$$8s + \frac{20}{7}s^2$$

$$\frac{22}{7}s^2 + s^3 \;\Big|\; \frac{7}{4}s + \frac{5}{8}s^2 \;\Big|\; \frac{49}{88s}$$

$$\frac{7}{4}s + \frac{49}{88}s^2$$

$$\frac{3}{44}s^2 \;\Big|\; \frac{22}{7}s^2 + s^3 \;\Big|\; \frac{22 \times 44}{21}$$

$$\frac{22}{7}s^2$$

$$s^3 \;\Big|\; \frac{3}{44}s^2 \;\Big|\; \frac{3}{44s}$$

$$\frac{3}{44}s^2$$

$$\times$$

The network is shown here as

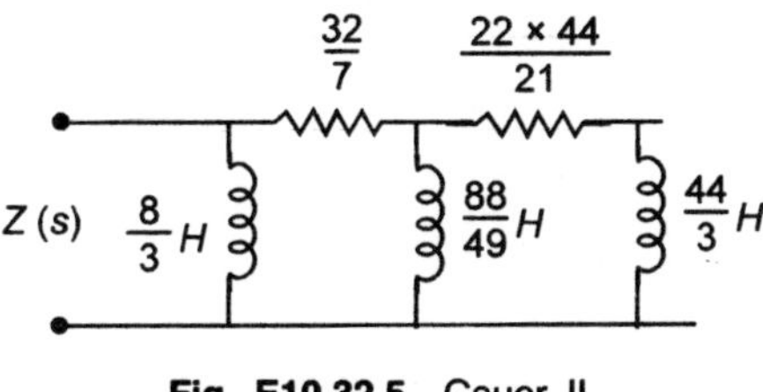


**Fig. E10.32.5**  Cauer II.

It is seen that since there are 4 internal critical frequencies, the number of elements required is 4 + 1 = 5 in all the formats whether Foster or Cauer.

## PROBLEMS

**10.1.** Test whether the following polynomials are Hurwitz.

(a) $s^7 + 3s^5 + 2s^3 + s$

(b) $s^4 + 7s^3 + 6s^2 + 21s + 8$

(c) $s^7 + 2s^6 + 2s^5 + s^4 + 4s^3 + 8s^2 + 8s + 4$

(d) $s^4 + 3s^3 + 9s^2 - s + 20$

(e) $s^2 + 4s + 7$

(f) $s^3 + 4s^2 + 5s + 2$

(g) $s^5 + s^3 + s$

(h) $s^4 + 3s^2 + 6s + 20$

(i) $-2s^2 - 4s - 12$

**10.2.** Test whether the following functions are positive real.

(a) $\dfrac{s^2 - s - 6}{s^2 + 2s - 2}$

(b) $\dfrac{s^2 + 4}{s^3 + 3s^2 + 3s + 1}$

(c) $\dfrac{s^2 + 1}{s^3 + 4s}$

(d) $\dfrac{2s^2 + 2s + 4}{(s+1)(s^2+2)}$

(e) $\dfrac{s^2 + 6s + 5}{s^2 + 9s + 14}$

(f) $\dfrac{5s^2 + s}{s^2 + 1}$

(g) $\dfrac{3s^2 + 5}{s^3 + s}$

(h) $\dfrac{6 - s}{s + 1}$

(i) $\dfrac{s^2 + 8s + 5}{s + 3}$

**10.3.** Identify the functions whether these are *LC*, *LR* or *CR* functions

(a) $Y(s) = \dfrac{2(s+1)(s+3)}{(s+2)(s+4)}$

(b) $Y(s) = \dfrac{4(s+1)(s+5)}{s(s+3)}$

(c) $Y(s) = \dfrac{s(s+2)(s+6)}{(s+1)(s+4)}$

(d) $Y(s) = \dfrac{(s+2)(s+6)}{s(s+4)}$

(e) $Z(s) = \dfrac{s+2}{s(s+4)}$

(f) $Z(s) = \dfrac{s^2 + 2s + 8}{s(s+3)}$

(g) $Z(s) = \dfrac{3(s+1)(s+4)}{(s+3)}$

(h) $Z(s) = \dfrac{s(s^2+4)(s^2+16)}{(s^2+9)(s^2+25)}$

(i) $Z(s) = \dfrac{(s^2+1)(s^2+6)}{s(s^2+4)}$

**10.4.** Synthesise the *LC* driving point impedance

$$Z(s) = \dfrac{6s^4 + 42s^2 + 48}{s^5 + 18s^3 + 48s}$$

in the form shown in the Fig. P10.4.

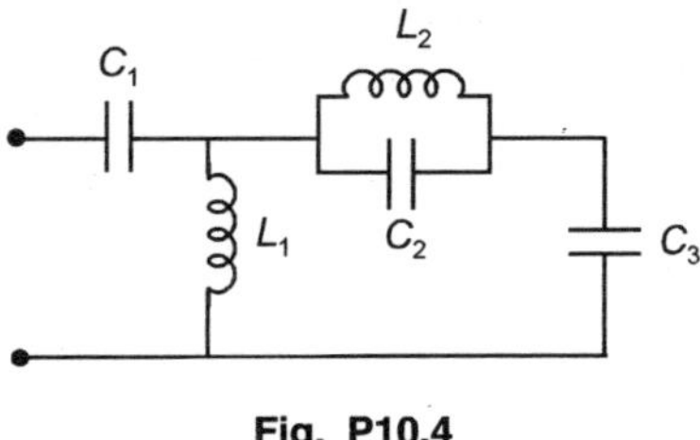

**Fig. P10.4**

Determine the values of the elements.

**10.5** An impedance function has the pole zero pattern shown in Fig. P10.5 below:

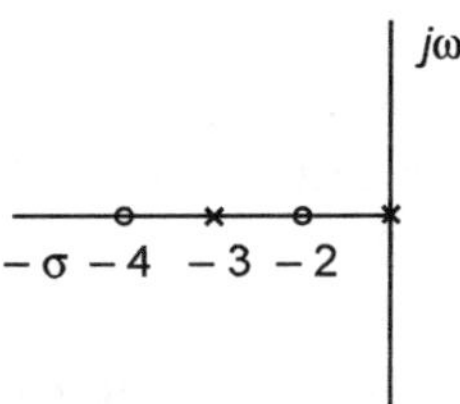


**Fig. P10.5**

If $Z(-3) = 3$ Synthesise the impedance in a Foster and Cauer form.

**10.6** For a network synthesise $Y$ as an $LC$ admittance if it is related as

$$\frac{1}{2+Y} = \frac{s(s^2+3)}{2s^3+s^2+5s+1}$$

**10.7.** There is an $LC$ network with the same driving point impedance as the network shown in the Fig. P10.7. The alternate network should contain two elements only. Determine the network.

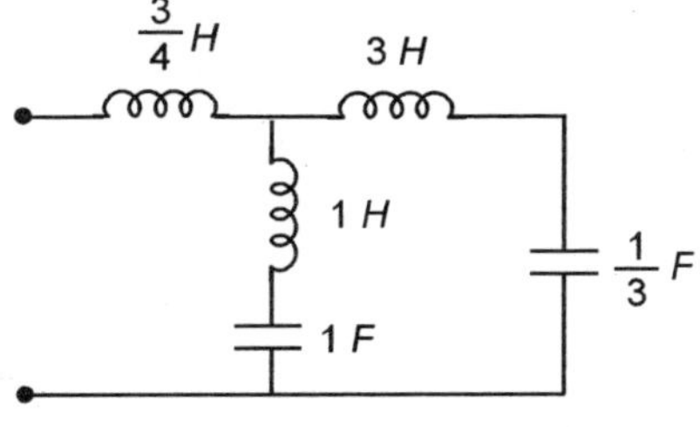


**Fig. P10.7**

**10.8.** Find the network for the following in Foster and Cauer form

$$Z(s) = \frac{2(s+1)(s+4)}{s(s+2)}$$

**10.9.** Synthesise the following function in Cauer form.

$$(a)\quad Z(s) = \frac{s^3+s^2+2s+1}{s^4+s^3+3s^2+s+1}$$

(b)  $Z(s) = \dfrac{4s^3 + 3s^2 + 4s + 2}{2s^2 + s}$

(c)  $Z(s) = \dfrac{s\,(s^2 + 4)}{(s^2 + 2)(s^2 + 9)}$

(d)  $Y(s) = \dfrac{(s^2 + 1)(s^2 + 4)}{s\,(s^2 + 2)}$

**10.10.** Poles and zeros are given for a function $Z(s)$ as

| | | |
|---|---|---|
| poles | 0, | −2 |
| zeros | −1, | −3 |

and $Z(\infty) = 4$. Determine $Z(s)$ and synthesise it.

**10.11.** An impedance function has poles and zeros as follows.

| | | |
|---|---|---|
| poles | −1 | −3 |
| zeros | −2 | −4 |

and the impedance has its value $Z(o) = 8$. Determine $Z(s)$ and Synthesise the network.

**10.12.** Synthesise in Foster I form

$$Y(s) = \frac{(s+3)(s+5)(s+7)}{s\,(s+4)(s+6)}$$

# 11

# State Variable Analysis

## 11.1  INTRODUCTION

In chapter II we have studied how to seek a solution of an electric network by using KCL, KVL and $v$-$i$ relationships. For a network we write down a set of integro-differential equations and the solution of these equations gives the required parameters of the network. These equations have been solved using classical methods which require homogenous solution (undriven circuit i.e. zero input) plus the particular solution (with inputs). These have also been solved by transforming these equations into algebraic ones with the use of Laplace transform or Fourier transform and then manipulation of these algebraic equations for certain response and finally corresponding inverse transform is obtained which gives the solution in time domain. Both these transforms analyses based on what is known as frequency domain due mainly to the interpretation of the Laplace transform variable $s$ as complex frequency. These methods provide a unique solution to an $n$th order differential equation if $n$ boundary conditions of the system described by the $n$th order differential equation are given. In the simplest case these can be the initial values of the output and its first $(n-1)$ derivatives.

However, it would be quite useful if we can write these equations in time domain that too first order differential equations rather than $n$th order equation and obtain the solution in time domain only without bringing any transformation in between the process of analysis. In fact with the advent of computers and the knowledge of matrices has revolutioned the process of analysing a system howsoever complex it could be.

From an engineering view point important groups of systems consisting of mainly electromechanical, electrical and certain chemical system, it is possible to relate the system dynamics directly to the energy concepts. Any transient or dynamic changes of a system selected from this group are caused basically by the redistribution of energy within the system. The energy is transformed from one form to another (in electric circuit from electromagnetic form to electrostatic form and vice versa) with accompanying losses (in electric system as ohmic loss in the resistance). The fact that no energy transfer can take place instantaneously as this would require infinite energy rates which is physically impossible. However, all energy changes take place with finite flow rates and time is thus required for the redistribution process. The number of independent energy storage elements thus form the number of independent variables in the new mathematical formulation which we will study here. These independent variables are known as state variables. A circuit consisting of resistance elements and with voltage sources, has energy dissipating elements and hence for such a network the response at any time is a function of the source voltage at that time, such a system is known as a static system. However, if the circuit has inductance and or capacitor in a circuit, the system response at any time not only depends upon the source voltage at that time but on the previous values of energy stored by the inductor and or capacitor also. From the instant $t_0$ when the circuit is operated, to time $t$ when the response is desired, there is continuous, redistribution of energy amongst the various inductive and capacitive elements and the sources. So such a system is a dynamic system.

It is to be noted that in general, there can be a large number of sets of state variables for a particular system. However, it is always desirable to select those variables as state variable which have some physical significance. We are interested only in seeking the solution of the network from time to time, but we are monitoring various parameters of the network so that these do not go beyond pre-specified values. For example, in an electric network we should see to it that the voltage across no capacitor should exceed corresponding voltage as this would otherwise damage the insulation of the capacitor. Similarly current through an inductor should not exceed a predetermined value as it may overheat the coil and again damage the coil. Similarly in a control system, there could be a situation when the output of a system is stable yet some of the system element may have a tendency to exceed their specified ratings. Also, it may sometimes be necessary and advantageous to provide a feedback proportional to some of the internal variables of a system rather than the output alone for the purpose of stabilizing and improving the performance of the system. Therefore, we should select such variables which have physical significance so that better insight into the operation of the system is possible. The state variable formulation has the following advantages:

1. The system equations expressed in state space form are well suited to direct programming on both analog and digital computers.
2. When the system is complex i.e. it has higher dimensionality or dependent variables, the normal form combined with the compact notation features of matrix and vector algebra makes it possible to utilize a very convenient and elegant mathematical language.
3. The state variables give information about the internal behaviour of the system besides the input output behaviour. This helps in operating the system with better performance.
4. The analysis is applicable to both linear as well as non-linear systems, time-invariant as well as time variant system.

In this chapter we will study the state variables analysis as applied to linear time-invariant systems only.

Before we proceed further to develop mathematical models using state variables analysis, we first formally define the concept of state.

## 11.2 THE CONCEPT OF STATE

The state of a system is defined as the minimum set of variables known as state variables which contain sufficient information about the past history of the system so that it is possible to compute future states of the system if the future inputs and equations describing the system are known. For an electric network the number of state variables equals the total number of energy storing elements (capacitors and inductors) minus all-capacitor loops and all inductor cut-sets i.e. if $n$ is the number of energy storing element, $n_1$ is the number of all-capacitor loops and $n_2$ the number of all inductor cut sets, the number of state variables for the given network is $(n - n_1 - n_2)$. As mentioned earlier the number of possible sets of state variables can be infinite but we usually select the one which have physical significance for the reason cited earlier.

For a linear time-invariant system the general form of the state equations is given as

$$\dot{X}(t) = Ax(t) + Bu(t) \qquad \qquad \text{...(11.1)}$$
$$Y(t) = Cx(t) + Du(t) \qquad \qquad \text{...(11.2)}$$

Here equation (11.1) is a vector differential state equation whereas (11.2) is the output equation, $x(t)$ is the state vector, $u(t)$ the input and $y(t)$ the output vector. $A$, $B$, $C$ and $D$ are the compatible matrices.

The following example of a simple electric network will illustrate the choice of state variable.

**Example 11.1:** For the network shown write the state equations.

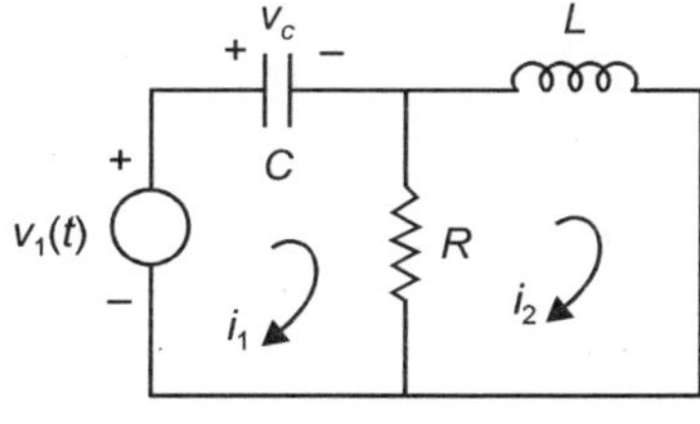

**Fig. E11.1**

**Solution:** Writing KVL equation for the two loops

$$\frac{1}{C} \int i_1 \, dt + (i_1 - i_2)R = v_1(t) \qquad \qquad \text{...(1)}$$

and

$$L \frac{d i_2}{dt} + (i_2 - i_1)R = 0 \qquad \qquad \text{...(2)}$$

From the first equation after differentiation, we have

$$\frac{i_1}{C} + \left( \frac{d i_1}{dt} - \frac{d i_2}{dt} \right) R = \dot{v}_1(t) \qquad \qquad \text{...(3)}$$

and

$$L \dot{i}_2 (t) + (i_2 - i_1) R = 0$$

or

$$\dot{i}_2 (t) = -i_2 \cdot \frac{R}{L} + i_1 \cdot \frac{R}{L} \qquad \qquad \text{...(4)}$$

Substituting $\dot{i}_2 (t)$ in equation (3) we have

$$\frac{d i_1}{dt} + i_2 \frac{R}{L} - i_1 \frac{R}{L} = \frac{\dot{v}_1}{R} - \frac{i_1}{RC}$$

$$\frac{d i_1}{dt} = i_1 \left( \frac{R}{L} - \frac{1}{RC} \right) - i_2 \cdot \frac{R}{L} + \frac{\dot{v}_1}{R} \qquad \qquad \text{...(5)}$$

and

$$\frac{d i_2}{dt} = i_1 \frac{R}{L} - i_2 \frac{R}{L} \qquad \qquad \text{...(6)}$$

This is one set of state equations where $i_1$ and $i_2$ are taken as the state variables. However, the term $(\dot{v}_1/R)$ is not desirable. Hence taking $V_C$ and $i_2(i_L)$ as the state variables we derive again the state equations. We need only two state variables as there are two energy storing elements.

Since $\qquad v_c = \dfrac{1}{C} \int i_1 \, dt \quad$ we rewrite the equation

$$\frac{d\,v_c}{dt} = \frac{i_1}{C} \qquad\qquad\qquad \ldots(7)$$

$$v_c + (i_1 - i_2)\,R = v_1\,(t) \qquad\qquad \ldots(8)$$

and $\qquad\qquad i_2 + (i_2 - i_1)\,\dfrac{R}{L} = 0 \qquad\qquad\qquad \ldots(9)$

Now $i_1$ should be eliminated from (8)
From equation (8) we have

$$i_1 = \frac{v_1(t) - v_c + i_2\,R}{R}$$

$$\frac{dv_c}{dt} = \frac{1}{C}\,\frac{v_1(t) - v_c + i_2\,R}{R} = -\frac{v_c}{RC} + \frac{i_2}{C} + \frac{v_1}{RC} \qquad \ldots(10)$$

and $\qquad \dfrac{di_2}{dt} = (i_1 - i_2)\,\dfrac{R}{L} = \dfrac{v_1 - v_c + i_2\,R}{R}\cdot\dfrac{R}{L} - i_2\,\dfrac{R}{L}$

$$= \frac{v_1}{L} - \frac{v_c}{L} \qquad\qquad\qquad \ldots(11)$$

Therefore, the state equations are

$$\begin{bmatrix} \dfrac{dv_c}{dt} \\[2ex] \dfrac{di_2}{dt} \end{bmatrix} = \begin{bmatrix} -\dfrac{1}{RC} & \dfrac{1}{C} \\[2ex] -\dfrac{1}{L} & 0 \end{bmatrix}\begin{bmatrix} v_c \\[2ex] i_2 \end{bmatrix} + \begin{bmatrix} \dfrac{1}{RC} \\[2ex] \dfrac{1}{L} \end{bmatrix} v_1(t) \qquad \ldots(12)$$

Thus by taking the state variables corresponding to $V_C$ and $i_2\,(i_L)$ we obtain the state equations in this normal form.

It is to be noted that sometimes in the literature the phrase. "The order of complexity of a network" is used to mean the number of state equations that can be written for a network or the number of state variables of the network which as we know equals the number of all the energy storing elements in the network minus the number of all capacitor loops and all-inductor cut sets.

**Example 11.2:** Write state equation for the network shown.

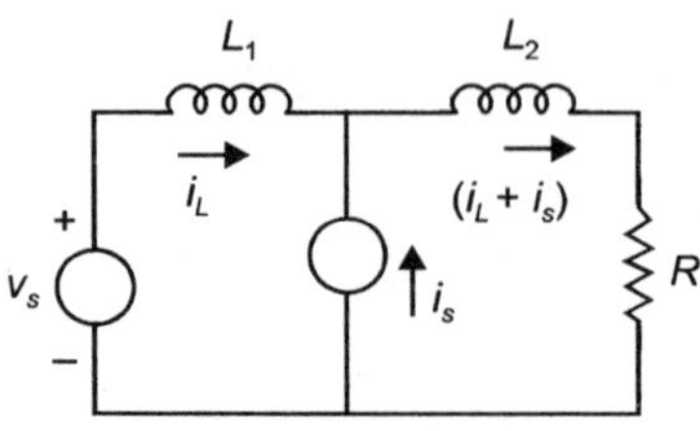


**Fig. E11.2**

**Solution:** There are two energy storing elements. However, there is an all – inductor cut set containing $L_1$, $L_2$ and $i_s$, therefore, the order of complexity is $2 - 1 = 1$ i.e. we need write only one state equation.

Let the current through $L_1$ is $i_L$, the chosen state variable. The current through the resistance branch is $(i_L + i_s)$.

Applying KVL to the outer loop consisting of $V_s$, $L_1$, $L_2$ and $R$ we have

$$v_s = L_1 \frac{di_L}{dt} + L_2 \frac{di_L}{dt} + L_2 \frac{di_s}{dt} + R\,(i_L + i_s)$$

or
$$(L_1 + L_2) \frac{di_L}{dt} = -R\,(i_L + i_s) - L_2 \frac{di_s}{dt} + v_s$$

or
$$\frac{di_L}{dt} = -\frac{R}{L_1 + L_2}\,(i_L + i_s) - \frac{L_2}{L_1 + L_2}\frac{di_s}{dt} + \frac{v_s}{L_1 + L_2} \quad \textbf{Ans.}$$

**Example 11.3:** Write state equations for the network shown.

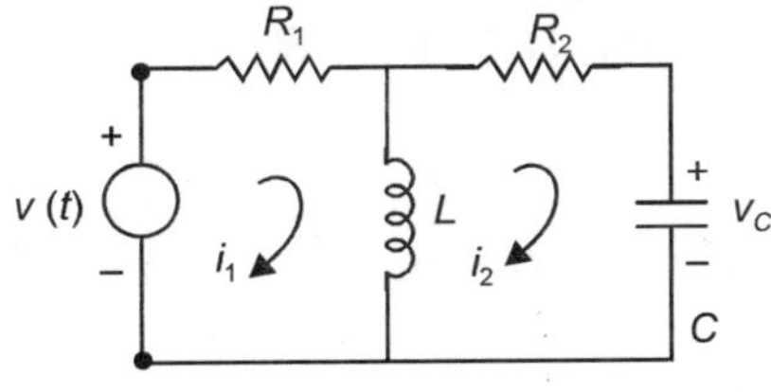


**Fig. E.11.3**

**Solution:** Since there are two energy storing elements and there are no all-capacitor loops or all-inductor cut sets, therefore, the order of complexity of the network is 2 and we need write two state equations with state variables $i_L$ and $v_c$

Writing KVL equation we have

$$L \frac{d}{dt}(i_1 - i_2) + R_1 i_1 = v(t)$$

and
$$L \frac{d}{dt}(i_2 - i_1) + R_2 i_2 + v_c = 0$$

since
$$i_1 - i_2 = i_L \text{ and } i_2 = C\frac{dv_c}{dt}$$

$$L \frac{di_L}{dt} + R_1 i_1 = v(t)$$

$$-L \frac{di_L}{dt} + R_2 i_2 + v_c = 0$$

Now
$$i_2 = C\frac{dv_c}{dt} \text{ and } i_1 = i_L + C\frac{dv_c}{dt}$$

$$L \frac{di_L}{dt} + R_1\left(i_L + C\frac{dv_c}{dt}\right) = v(t)$$

and
$$L \frac{di_L}{dt} = v_c + R_2 i_2 = v_c + R_2 C\frac{dv_c}{dt}$$

$$v(t) - R_1 \left( i_L + C\,\frac{dv_c}{dt} \right) = v_c + R_2\,C\,\frac{dv_c}{dt}$$

$$v(t) - v_c - R_1\,i_L = (R_2\,C + R_1\,C)\,\frac{dv_c}{dt}$$

or

$$\frac{dv_c}{dt} = -\frac{R_1\,i_L}{(R_1 + R_2)C} - \frac{v_c}{(R_1 + R_2)C} + \frac{v(t)}{(R_1 + R_2)C}$$

$$L\,\frac{di_L}{dt} = v(t) - R_1\,i_1 = v(t) - R_1 \left( i_L + C\,\frac{dv_c}{dt} \right)$$

$$= v(t) - R_1\,i_L - C\,R_1 \left[ \frac{-R_1\,i_L}{(R_1 + R_2)\,C} - \frac{v_c}{C\,(R_1 + R_2)} + \frac{v(t)}{(R_1 + R_2)\,C} \right]$$

$$\frac{di_L}{dt} = \frac{v(t)}{L} - \frac{R_1}{L}\,i_L + \frac{R_1^2\,i_L}{(R_1 + R_2)\,L} + \frac{v_c\,R_1}{(R_1 + R_2)L} - \frac{v(t)}{L}\cdot\frac{R_1}{R_1 + R_2}$$

$$= -i_L\,\frac{R_1}{L}\left( 1 - \frac{R_1}{R_1 + R_2} \right) + \frac{v_c\,R_1}{(R_1 + R_2)L} + \frac{v(t)}{L}\left( 1 - \frac{R_1}{R_1 + R_2} \right)$$

$$= -i_L\,\frac{R_1\,R_2}{L(R_1 + R_2)} + \frac{v_c\,R_1}{(R_1 + R_2)L} + \frac{v(t)}{L}\cdot\frac{R_2}{R_1 + R_2}$$

Therefore, the state equation are

$$\begin{bmatrix} \dfrac{di_L}{dt} \\[2mm] \dfrac{dv_c}{dt} \end{bmatrix} = \begin{bmatrix} \dfrac{-R_1\,R_2}{L\,(R_1 + R_2)} & \dfrac{R_1}{L\,(R_1 + R_2)} \\[3mm] -\dfrac{R_1}{C\,(R_1 + R_2)} & -\dfrac{1}{C\,(R_1 + R_2)} \end{bmatrix} \begin{bmatrix} i_L \\[2mm] v_c \end{bmatrix} + \begin{bmatrix} \dfrac{R_2}{L\,(R_1 + R_2)} \\[3mm] \dfrac{1}{C\,(R_1 + R_2)} \end{bmatrix} v(t)$$

Hence the state equations are obtained

## 11.3  STATE EQUATIONS

It can be seen that after writing KVL equations and some algebraic manipulations it is possible to write state equations for a network.

However, the following methods can be used to write state equations in a systematic way.

1. Equivalent source method.
2. Network topological method.

### 11.3.1  Equivalent Source Method

Here the capacitor and inductors are replaced by the equivalent voltage and current sources respectively. The circuit is reduced to thereby as a resistive one. The current through the capacitors and voltage across the inductors are obtained using superposition principle. To begin with we assume that there are no degeneracies i.e. there are no all-capacitor loops or all-inductor cut sets present in the network. The following example illustrates the application of Equivalent Source Method.

**Example 11.4:** Write the state equations using Equivalent Source Method for the network shown.

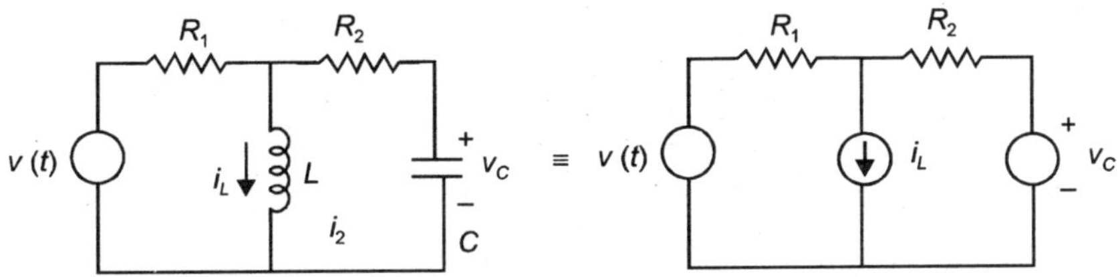

**Fig. E11.4.1**

Using superposition we get the network as in Fig. (*a*) (*b*) and (*c*), the sources $v(t)$, $v_c$ and $i_L$ have been retained respectively and the other sources replaced by the respective internal resistances.

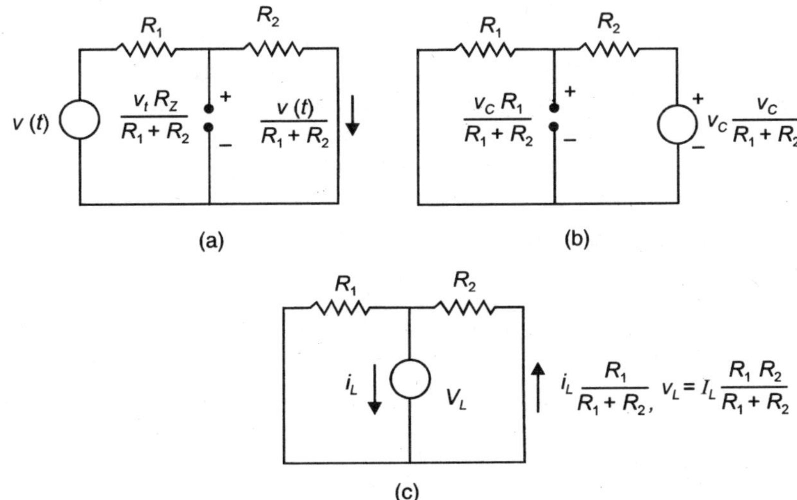

**Fig. E11.4.2**

The voltage across inductor is sum of three voltages as shown in Fig. (*a*), (*b*) and (*c*) Therefore,

$$L\,\frac{di_L}{dt} = -i_L\,\frac{R_1\,R_2}{R_1 + R_2} + \frac{v_c\,R_1}{R_1 + R_2} + \frac{v(t)\,R_2}{R_1 + R_2}$$

The current through $C$ is the algebraic sum of the current in the three circuits.

$$C\,\frac{dv_c}{dt} = -i_L\,\frac{R_1}{R_1 + R_2} - \frac{v_c}{R_1 + R_2} + \frac{v(t)}{R_1 + R_2}$$

or

$$\frac{dv_c}{dt} = -\frac{R_1}{(R_1 + R_2)C}\,i_L - \frac{v_c}{(R_1 + R_2)\,C} + \frac{v(t)}{C(R_1 + R_2)}$$

and

$$\frac{di_L}{dt} = -\frac{R_1 R_2 i_L}{L(R_1 + R_2)} + \frac{v_c R_1}{L(R_1 + R_2)} + \frac{v(t)}{(R_1 + R_2)} \cdot \frac{R_2}{L}$$

Therefore, the state equation is given as

$$
\begin{bmatrix} \dfrac{di_L}{dt} \\[2mm] \dfrac{dv_c}{dt} \end{bmatrix}
=
\begin{bmatrix} -\dfrac{R_1 R_2}{L(R_1 + R_2)} & \dfrac{R_1}{L(R_1 + R_2)} \\[3mm] -\dfrac{R_1}{(R_1 + R_2)C} & -\dfrac{1}{(R_1 + R_2)C} \end{bmatrix}
\begin{bmatrix} i_L \\[2mm] v_c \end{bmatrix}
+
\begin{bmatrix} \dfrac{R_2}{L(R_1 + R_2)} \\[3mm] \dfrac{1}{C(R_1 + R_2)} \end{bmatrix} v(t)
$$

If, however, the network has all-capacitor loops or all-inductor cut set, the method has to be modified in terms of selection of state variables. In an all-capacitor loops it is not desirable to assign state variables to all the capacitors because KVL has to be satisfied. Therefore, the voltage across one of the capacitors in the loop will be a linear combination of the remaining capacitors whose voltages have been chosen as state variables. Such a capacitor is known as excess element. Similarly in an all-inductor cut-set, we choose state variables for all the inductors of the cut set except the one, current is taken as the linear combination of the remaining inductors where state variable has been chosen as per KCL. Here this inductor is known as excess element.

In order to obtain equivalent source elements, the excess capacitor is replaced by current source and excess inductor by a voltage source. The current through the excess capacitor is expressed in terms of state variable voltages and source voltage (if one is present in the loop) and the voltage across an inductor is expressed in terms of state variable currents and current source (if one is present in the cut set). Let us take an example to illustrate this procedure.

**Example 11.5:** Write state equations for the network shown here.

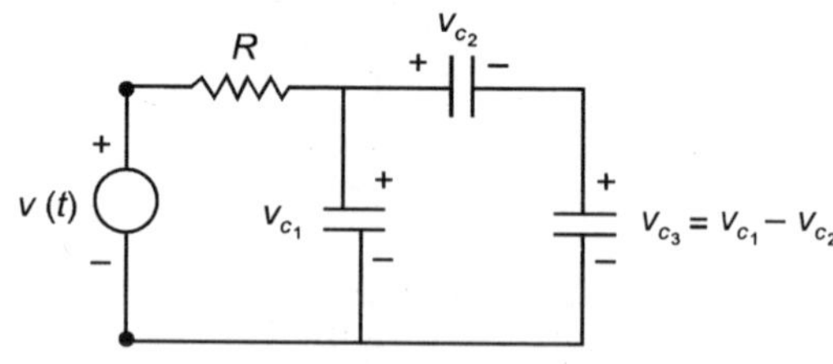

**Fig. E11.5.1**

Since there are three energy storing elements and there is an all-capacitor loop we need select only two state variables say $v_{c_1}$ and $v_{c_2}$ and from the Fig. $v_{c_3} = v_{c_1} - v_{c_2}$ and this being an excess element (capacitor) we represent it by a current source.

The equivalent source network is shown here.

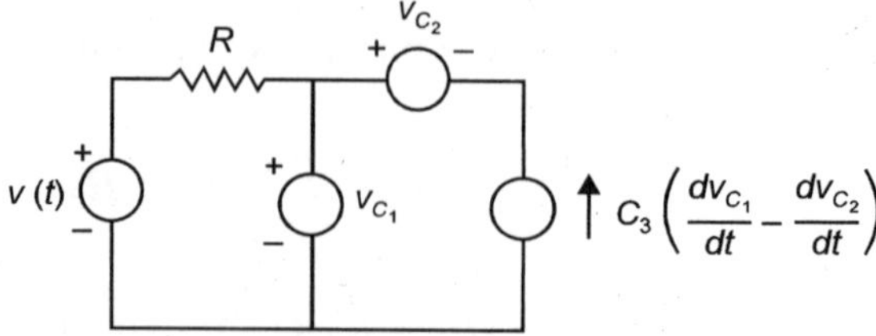

**Fig. E11.5.2**

For application of superposition principle, following circuits are required.

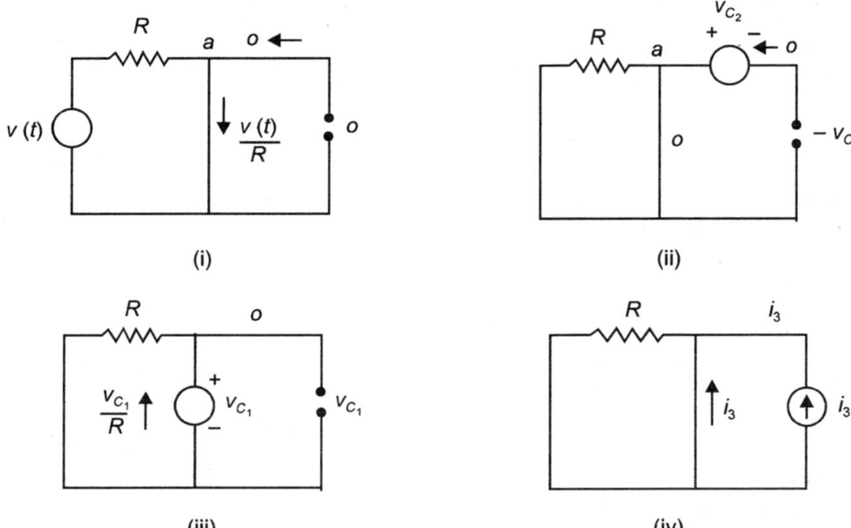

(i)   (ii)

(iii)   (iv)

**Fig. E11.5.3**

The current through $C_1$ is the algebraic sum of currents in the four circuits (i) – (iv) taking current entering the node a as +ve.

$$C_1 \frac{dv_{c_1}}{dt} = \frac{v(t)}{R} - C_3 \frac{dv_{c_1}}{dt} + C_3 \frac{dv_{c_2}}{dt} - \frac{v_{c_1}}{R} \qquad ...(1)$$

The current through $C_2$

$$C_2 \frac{dv_{c_2}}{dt} = C_3 \frac{dv_{c_1}}{dt} - C_3 \frac{dv_{c_2}}{dt}$$

$$\frac{dv_{c_2}}{dt} = \frac{C_3}{C_2 + C_3} \frac{dv_{c_1}}{dt}$$

Substituting for $C_3 \dfrac{dv_{c_2}}{dt}$ in (1) we have

$$(C_1 + C_3) \frac{dv_{c_1}}{dt} = -\frac{v_{c_1}}{R} + \frac{C_3^2}{C_2 + C_3} \frac{dv_{c_1}}{dt} + \frac{v(t)}{R}$$

or $\qquad \left\{ (C_1 + C_3) - \dfrac{C_3^2}{C_2 + C_3} \right\} \dfrac{dv_{c_1}}{dt} = -\dfrac{v_{c_1}}{R} + \dfrac{v(t)}{R} \qquad ...(2)$

and $\qquad \dfrac{dv_{c_2}}{dt} = \dfrac{C_3}{C_2 + C_3} \dfrac{dv_{c_1}}{dt} = K_1 \dfrac{dv_{c_1}}{dt}$

We rewrite equation (2) let the quantity in the bracket be $K$ then

$$K \frac{dv_{c_1}}{dt} = -\frac{v_{c_1}}{R} + \frac{v(t)}{R}$$

or

$$\frac{dv_{c_1}}{dt} = -\frac{v_{c_1}}{KR} + \frac{v(t)}{KR}$$

Therefore,

$$\frac{dv_{c_2}}{dt} = -\frac{K_1}{KR}v_{c_1} + \frac{K_1}{KR} v(t)$$

$$\begin{bmatrix} \dfrac{dv_{c_1}}{dt} \\[3ex] \dfrac{dv_{c_2}}{dt} \end{bmatrix} = \begin{bmatrix} -\dfrac{1}{KR} & 0 \\[3ex] -\dfrac{K_1}{KR} & 0 \end{bmatrix} \begin{bmatrix} v_{c_1} \\[3ex] v_{c_2} \end{bmatrix} + \begin{bmatrix} \dfrac{1}{KR} \\[3ex] \dfrac{K_1}{KR} \end{bmatrix} v(t)$$

## 11.3.2  Network Topological Method

The Equivalent source method even though simple, can't be used on digital computer as it is difficult to formulate algorithm for the procedure.

(1) Network graph theory can be applied to advantage for writing state equations as this can be used on digital computer. (1) Here a tree is selected with the priority of twigs as follows: Independent voltage source, capacitors, inductors and independent current source i.e. take all the independent voltage sources, maximum number of capacitors, minimum number of inductors and no independent current sources as far as possible. Place controlled voltage sources in the tree and controlled current sources in the co-tree.

(2) In the diagram all the capacitors in the tree are assigned voltage (state variable) with polarity markings and the inductor in the co-tree are assigned current with direction (with arrow). These voltages and currents form the state variables of the network. However, if a resistance branch is to be taken as a twig or a link, its' voltage or current should be expressed in term of the state variables. If these are not possible, assign a new voltage or current variable to that resistive twig or link.

(3) Write capacitor current $C\dfrac{dv}{dt}$ for each capacitor twig equal to the algebraic sum of link currents at the node (or supernode) on either end of the capacitor, using KCL. The supernode is identified as the set of all tree twigs connected to that terminal of the capacitor. It is to be noted with caution that no new variable should be introduced during the process.

(4) Write inductor voltage $L\dfrac{di}{dt}$ for each inductor by using KVL round the loop containing that inductor in terms of the twig voltages constituting the loop. Again no new variable is to be introduced.

(5) If in step 2, new voltage variables (other than state voltage variables) were assigned to resistors (twig), set $(v_R/R)$ equal to the algebraic sum of link currents using KCL. However, if new current variables (other than state current variables) were assigned to resistor (link), set $i_R R$ equal to an algebraic sum of twig voltages around the loop containing the resistor link. Now, solve these resistor equations simultenously to obtain expressions to each $v_R$ and $i_R$ in terms of state variables and the source quantities.

(6) While writing equations in step 3 and 4, use of $v_R$ and $i_R$ for various resistor branches must have been made. Replace them by the corresponding values as obtained in step 5. Write the state equations in the normal form. There is one variation in graph theory procedure in connection with independent sources. Whereas in graph theory we consider independent sources as accompanied i.e. a voltage source with series branch and a current source with shunt branch was considered as a single element. Here we take them as separate element.

The following example illustrates the application of the above procedure using graph theory for writing state equations.

**Example 11.6:** Write the state equations using graph theory for the network shown.

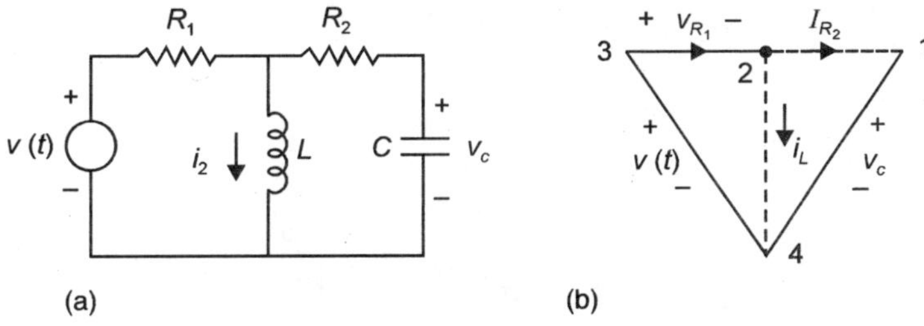


**Fig. E11.6**

**Solution:** In Fig. b, as mentioned earlier all the independent voltage sources, maximum number of capacitors are to be chosen as twigs of the tree. The inductor is taken as link of the co-tree. There are two resistances in the original circuit $R_1$ is taken as twig and hence voltage $V_{R1}$ is assigned to it as this can't be expressed in term of $V_C$ and $I_L$ which are the state variables. The other resistance $R_2$ is taken as link and current $I_{R2}$ is assigned in the direction shown.

We first write KCL equation for the twigs at suitable nodes.

At node 1
$$C\frac{dv_c}{dt} = I_{R_2}$$

At node 2
$$\frac{v_{R1}}{R_1} = i_L + i_{R_2}$$

Loop containing nodes 2,3,4

$$L\frac{di_L}{dt} + v_{R_1} = v(t)$$

Loop containing nodes 1,2,4

$$L\frac{di_L}{dt} = I_{R_2} R_2 + v_c$$

Therefore, we have two equations
$$I_{R_2} R2 + v_c = v(t) - v_{R1}$$
$$i_L R_1 + I_{R_2} R_1 = v_{R1}$$

Solving for $V_{R_1}$ and $I_{R_2}$ we have
$$I_{R_2}(R_1 + R_2) = v(t) - v_c - i_L R_1$$

$$I_{R_2} = \frac{1}{R_1 + R_2}\left[v(t) - v_c - i_L R_1\right]$$

Therefore, the state equations are

$$\frac{dv_c}{dt} = -\frac{v_c}{(R_1 + R_2)C} - \frac{R_1}{(R_1 + R_2)C}\,i_L + \frac{v(t)}{(R_1 + R_2)C}$$

and

$$L\frac{di_L}{dt} = \frac{R_2}{R_1 + R_2}\left[v(t) - v_c - i_L R_1\right] + v_c$$

$$= v_c\left(1 - \frac{R_2}{R_1 + R_2}\right) - \frac{R_1 R_2}{R_1 + R_2}\,i_L + \frac{R_2}{R_1 + R_2}\,v(t)$$

$$= v_c \cdot \frac{R_1}{R_1 + R_2} - \frac{R_1 R_2}{R_1 + R_2}\,i_L + \frac{R_2}{R_1 + R_2}\,v(t)$$

or

$$\frac{di_L}{dt} = v_c\,\frac{R_1}{L\,(R_1 + R_2)} - \frac{R_1 \cdot R_2}{L\,(R_1 + R_2)}\,i_L + \frac{R_2}{L\,(R_1 + R_2)}\,v(t)$$

Therefore, the state equations are

$$\begin{bmatrix} \dfrac{dv_c}{dt} \\[2ex] \dfrac{di_L}{dt} \end{bmatrix} = \begin{bmatrix} -\dfrac{1}{C\,(R_1 + R_2)} & \dfrac{-R_1}{(R_1 + R_2)C} \\[2ex] \dfrac{R_1}{L\,(R_1 + R_2)} & \dfrac{-R_1 R_2}{L\,(R_1 + R_2)} \end{bmatrix} \begin{bmatrix} v_c \\[2ex] i_L \end{bmatrix} + \begin{bmatrix} \dfrac{1}{C\,(R_1 + R_2)} \\[2ex] \dfrac{R_2}{L\,(R_1 + R_2)} \end{bmatrix} v(t)$$

**Example 11.7:** Write state equations for the network shown using graph theory.

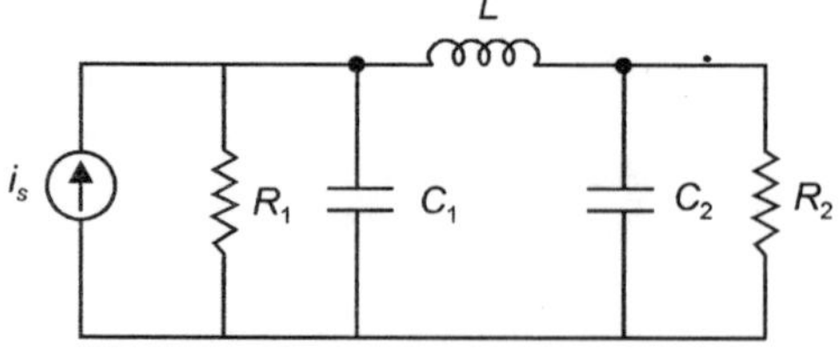

**Fig. E11.7.1**

**Solution:** For drawing the tree we take capacitor as twigs, inductor and current source as link. Since $R_1$ and $R_2$ are connected across $C_1$ and $C_2$ respectively, these are to be taken as links.

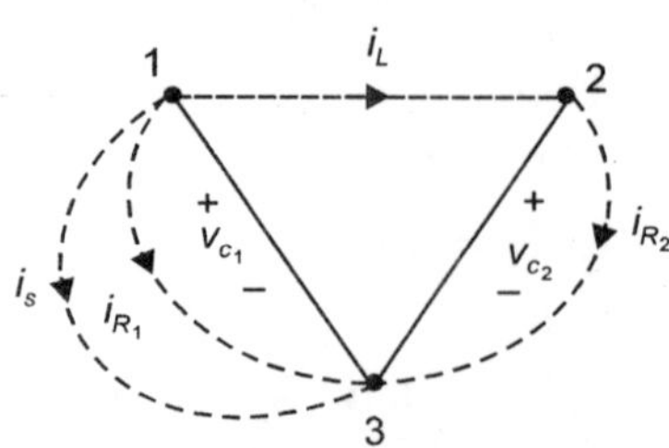

**Fig. E11.7.2

Using KCL, we write current equation at node 1

$$C_1 \frac{dv_{c_1}}{dt} = i_s - i_{R1} - i_L$$

$$= -\frac{v_{c_1}}{R_1} - i_L + i_s$$

At node 2
$$C_2 \frac{dv_{c_2}}{dt} = i_L - i_{R2} = i_L - \frac{v_{c_2}}{R_2}$$

$$= 0\, v_{c1} - \frac{v_{c_1}}{R_2} + i_L$$

Considering the loop 1,2,3 containing the link (inductor current $i_L$)

$$L \frac{di_L}{dt} + v_{c_2} = v_{c1}$$

or
$$L \frac{di_L}{dt} = v_{c1} - v_{c2}$$

or
$$\frac{di_L}{dt} = \frac{v_{c_1}}{L} - \frac{v_{c_2}}{L} + 0 \cdot i_L$$

Therefore, the state equations are

$$
\begin{bmatrix} \dfrac{dv_{c_1}}{dt} \\[2ex] \dfrac{dv_{c_2}}{dt} \\[2ex] \dfrac{di_L}{dt} \end{bmatrix}
=
\begin{bmatrix}
-\dfrac{1}{C_1 R_1} & 0 & -\dfrac{1}{C_1} \\[2ex]
0 & -\dfrac{1}{C_2 R_2} & \dfrac{1}{C_2} \\[2ex]
\dfrac{1}{L} & -\dfrac{1}{L} & 0
\end{bmatrix}
\begin{bmatrix} v_{c_1} \\[2ex] v_{c_2} \\[2ex] i_L \end{bmatrix}
+
\begin{bmatrix} 1 \\[2ex] 0 \\[2ex] 0 \end{bmatrix}
\begin{bmatrix} i_s \end{bmatrix}
$$

Since there are three energy storing elements and there is neither an all-capacitor loop nor an all-inductor cut set, the number of state variables are three and hence three state equations have been obtained.

If the network contains an all-capacitor loop, or an all-inductor cut set, this is known as excess element situation. Here we leave one capacitor out of the two and one inductor from the co-tree. So one capacitor becomes link where current is to specified and is given as capacitor times the time derivative of the voltage across it which is given as algebraic sum of voltages around the loop containing the excess capacitor. Similarly, the excess inductor becomes a twig where the voltage is given by inductance times the derivative of the algebraic sum of current at the node at either terminal of the inductor. Let us demonstrate this with an example.

**Example 11.8:** Write state equations for the network shown using graph theory.

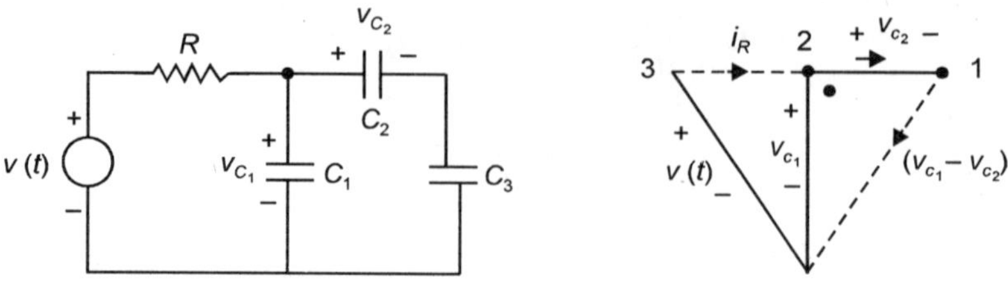

**Fig. E11.8**

There are three capacitors forming the loop, hence there are two state variables and we need only two equations. Capacitor 3 is made a link.

At node 1

$$C_2 \frac{dv_{c_2}}{dt} = C_3 \frac{dv_{c_1}}{dt} - C_3 \frac{dv_{c_2}}{dt}$$

At node 2

$$C_1 \frac{dv_{c_1}}{dt} + C_2 \frac{dv_{c_2}}{dt} = i_R = \frac{v_R}{R} = \frac{v(t) - v_{c_1}}{R}$$

$$C_1 \frac{dv_{c_1}}{dt} + C_3 \frac{dv_{c_1}}{dt} - C_3 \frac{dv_{c_2}}{dt} = \frac{v(t)}{R} - \frac{v_{c_1}}{R}$$

$$(C_1 + C_3) \frac{dv_{c_1}}{dt} - \frac{C_3^2}{C_2 + C_3} \frac{dv_{c_1}}{dt} = \frac{v(t)}{R} - \frac{v_{c_1}}{R}$$

$$\frac{dv_{c_1}}{dt} \left[ C_1 + C_3 - \frac{C_3^2}{C_2 + C_3} \right] = \frac{v(t)}{R} - \frac{v_{c_1}}{R}$$

or

$$K \frac{dv_{c_1}}{dt} = \frac{v(t)}{R} - \frac{v_{c_1}}{R}$$

and

$$\frac{dv_{c_2}}{dt} = \frac{C_3}{C_2 + C_3} \frac{dv_{c_1}}{dt} = K_1 \left\{ -\frac{v_{c_1}}{KR} + \frac{v(t)}{KR} \right\}$$

Therefore, the state equations are:

$$\begin{bmatrix} \dfrac{dv_{C_1}}{dt} \\[2ex] \dfrac{dv_{C_2}}{dt} \end{bmatrix} = \begin{bmatrix} -\dfrac{1}{KR} & 0 \\[2ex] -\dfrac{K_1}{KR} & 0 \end{bmatrix} \begin{bmatrix} v_{C_1} \\[2ex] v_{C_2} \end{bmatrix} + \begin{bmatrix} \dfrac{1}{KR} \\[2ex] \dfrac{K_1}{KR} \end{bmatrix} \big[ v(t) \big]$$

**Example 11.9:** Write state equations for the network shown using graph theory approach.

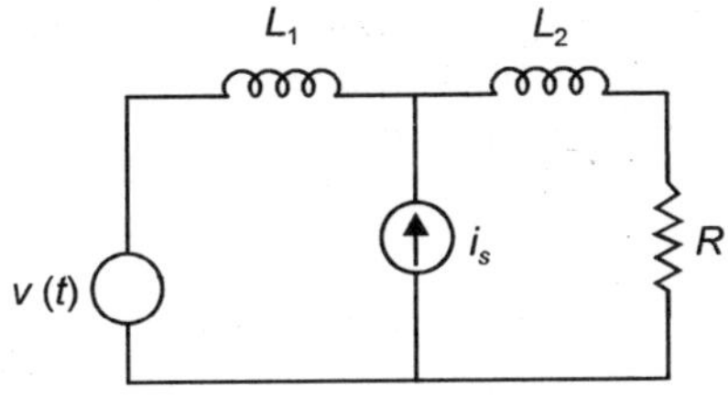 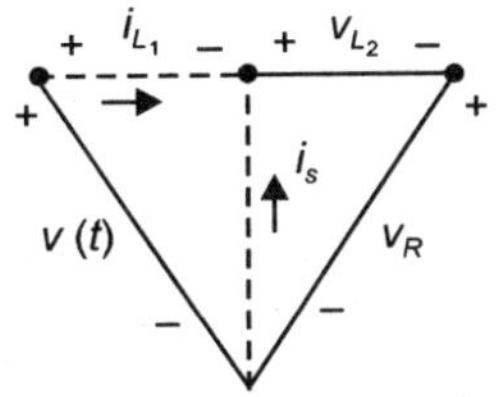

**Fig. E11.9**

**Solution:** Since $L_1$ $L_2$ and $i_s$ constitute an all-inductor cut-set, there will be one state equation. Also one of the two inductors form a twig and the other a link.

Current through $L_2$ branch $= i_{L_1} + i_s = \dfrac{v_R}{R}$

Therefore $V_R = (i_{L_1} + i_s)\, R$

Also

$$L_1 \frac{di_{L_1}}{dt} + v_{L_2} + v_R = v(t)$$

$$L_1 \frac{di_{L_1}}{dt} + L_2\left(\frac{di_{L_1}}{dt} + \frac{di_s}{dt}\right) + \left(i_{L_1} + i_s\right) R = v(t)$$

$$\frac{di_{L_1}}{dt} = \frac{1}{L_1 + L_2}\left[-i_{L_1} R - L_2 \frac{di_s}{dt} - i_s R + v(t)\right]$$

$$= -\frac{R}{L_1 + L_2} i_{L_1} - \frac{L_2}{L_1 + L_2} \frac{di_s}{dt} - \frac{R}{L_1 + L_2} i_s + \frac{v(t)}{L_1 + L_2} \qquad \textbf{Ans.}$$

**Example 11.10:** Write state equation for the network using graph theory approach.

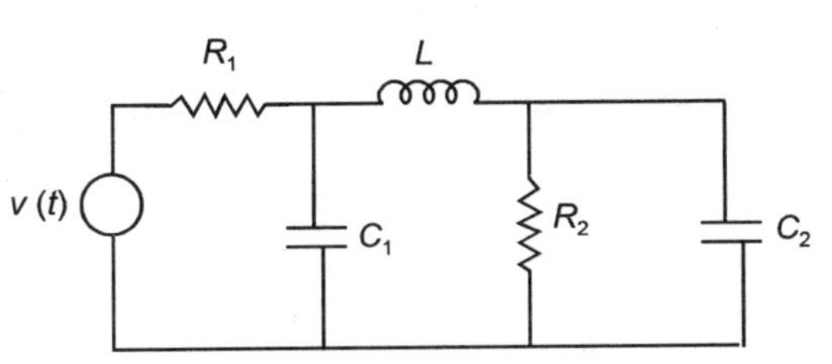 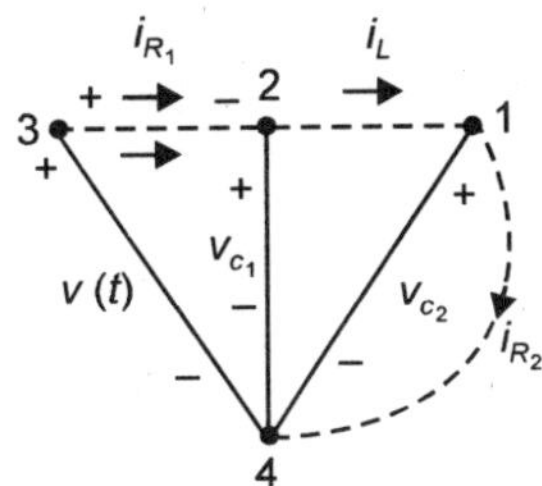

**Fig. E11.10**

**Solution:** $V_{C_1}$ $V_{C_2}$ and $i_L$ are the state variales. At node 1

$$C_2 \frac{dv_{c_2}}{dt} = i_L - i_{R2} = i_L - \frac{v_{c_2}}{R_2}$$

At node 2

$$C_1 \frac{dv_{c_1}}{dt} = i_{R1} - i_L = \frac{v(t) - v_{c_1}}{R_1} - i_L$$

$$\frac{dv_{c_1}}{dt} = -\frac{v_{c_1}}{R_1 C_1} - \frac{i_L}{C_1} + \frac{v(t)}{R_1 C_1}$$

$$\frac{dv_{c_2}}{dt} = -\frac{v_{c_2}}{R_2 C_2} + \frac{i_L}{C_2}$$

$$\frac{di_L}{dt} = \frac{v_{c_1}}{L} - \frac{v_{c_2}}{L}$$

The equations are rewritten as

$$\begin{bmatrix} \dfrac{dv_{c_1}}{dt} \\[2ex] \dfrac{dv_{c_2}}{dt} \\[2ex] \dfrac{di_L}{dt} \end{bmatrix} = \begin{bmatrix} -\dfrac{1}{R_1 C_1} & 0 & -\dfrac{1}{C_1} \\[2ex] 0 & -\dfrac{1}{R_2 C_2} & \dfrac{1}{C_2} \\[2ex] \dfrac{1}{L} & -\dfrac{1}{L} & 0 \end{bmatrix} \begin{bmatrix} v_{c_1} \\[2ex] v_{c_2} \\[2ex] i_L \end{bmatrix} + \begin{bmatrix} \dfrac{1}{R_1 C_1} \\[2ex] 0 \\[2ex] 0 \end{bmatrix} \begin{bmatrix} v(t) \end{bmatrix}$$

**Example 11.11:** Write state equations for the network shown using graph theory approach

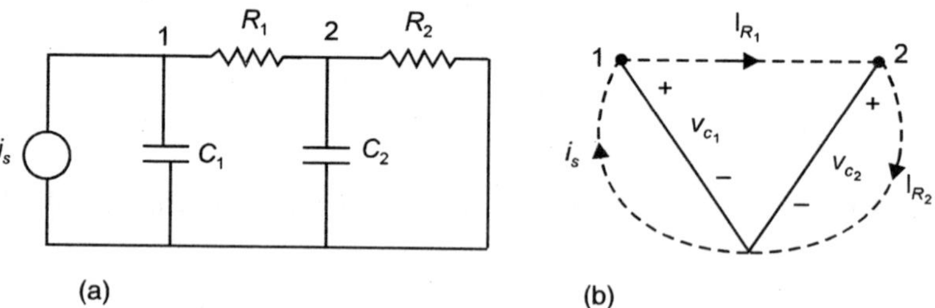


**Fig. E11.11(a)**

At node 1

$$C_1 \frac{dv_{c_1}}{dt} = i_s - \frac{v_{c_1} - v_{c_2}}{R_1}$$

or

$$\frac{dv_{c_1}}{dt} = -\frac{v_{c_1}}{R_1 C_1} + \frac{v_{c_2}}{R_1 C_1} + \frac{i_s}{C_1}$$

At node 2

$$C_2 \frac{dv_{c_2}}{dt} = \frac{v_{c_1} - v_{c_2}}{R_1} - \frac{v_{c_2}}{R_2}$$

$$\frac{dv_{c_2}}{dt} = \frac{v_{c_1}}{R_1 C_1} - \frac{R_1 + R_2}{R_1 R_2 C_2} v_{c_2}$$

Therefore the equations are:

$$\begin{bmatrix} \dfrac{dv_{c_1}}{dt} \\[3mm] \dfrac{dv_{c_2}}{dt} \end{bmatrix} = \begin{bmatrix} -\dfrac{1}{R_1 C_1} & \dfrac{1}{R_1 C_1} \\[3mm] \dfrac{1}{R_1 C_2} & -\dfrac{R_1 + R_2}{R_1 R_2 C_2} \end{bmatrix} \begin{bmatrix} v_{c_1} \\[3mm] v_{c_2} \end{bmatrix} + \begin{bmatrix} \dfrac{1}{C_1} \\[3mm] 0 \end{bmatrix} [i_s]$$

So far we have considered independent current and voltage sources. However, if a network contains a controlled source, it would be represented by a twig or a link depending upon which variable is specified. If voltage of the branch is specified it should be a twig. If current is specified it should be a link. Let us illustrate with an example.

**Example 11.12:** Write state equations for the network shown.

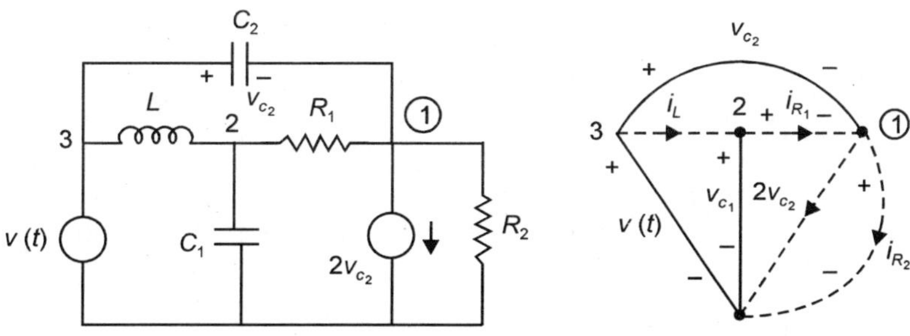

**Fig. E11.12**

Here the controlling quantity is voltage across a capacitor and the controlled quantity is current. Therefore, the controlled branch is taken as a link whereas the controlling branch a twig as the voltage is across the capacitor.

At node 2

$$C_1 \frac{dv_{c_1}}{dt} = i_L - i_{R_1} = i_L - \frac{v_{R_1}}{R_1}$$

Now

$$v_{R1} - v_{c2} + v(t) - v_{c1} = 0$$

or

$$v_{R1} = v_{c1} + v_{c2} - v(t)$$

Therefore,

$$C_1 \frac{dv_{c_1}}{dt} = i_L - \frac{v_{c_1}}{R_1} - \frac{v_{c_2}}{R_1} + \frac{v(t)}{R_1}$$

or

$$\frac{dv_{c_1}}{dt} = -\frac{v_{c_1}}{R_1 C_1} - \frac{v_{c_2}}{R_1 C_1} + \frac{i_L}{C_1} + \frac{v(t)}{R_1 C_1}$$

At node 1

$$C_2 \frac{dv_{c_2}}{dt} = 2\, v_{c_2} + i_{R_2} - i_{R_1}$$

$$= 2v_{c_2} + \frac{v_{R_2}}{R_2} - \frac{v_{c_1}}{R_1} - \frac{v_{c_2}}{R_1} + \frac{v(t)}{R_1}$$

$$= 2v_{c_2} + \frac{v(t)}{R_2} - \frac{v_{c_2}}{R_2} - \frac{v_{c_1}}{R_1} - \frac{v_{c_2}}{R_1} + \frac{v(t)}{R_1}$$

or

$$\frac{dv_{c_2}}{dt} = \frac{v_{c_1}}{R_1 C_2} - \frac{v_{c_2}}{C_2}\left(\frac{1}{R_1} + \frac{1}{R_2} - 2\right) + \frac{R_1 + R_2}{R_1 R_2 C_2} v(t)$$

Between nodes 2 and 3

$$L\frac{di_L}{dt} = v(t) - v_{c_1}$$

or

$$\frac{di_L}{dt} = -\frac{v_{c_1}}{L} + \frac{v(t)}{L}$$

The equations are

$$\begin{bmatrix} \dfrac{dv_{c_1}}{dt} \\[2ex] \dfrac{dv_{c_2}}{dt} \\[2ex] \dfrac{di_L}{dt} \end{bmatrix} = \begin{bmatrix} -\dfrac{1}{R_1 C_1} & -\dfrac{1}{R_1 C_1} & \dfrac{1}{C_1} \\[2ex] -\dfrac{1}{R_1 C_2} & \dfrac{R_1 + R_2 - 2R_1 R_2}{R_1 R_2 C_2} & 0 \\[2ex] -\dfrac{1}{L} & 0 & 0 \end{bmatrix} \begin{bmatrix} v_{c_1} \\[2ex] v_{c_2} \\[2ex] i_L \end{bmatrix} + \begin{bmatrix} 1/R_1 \\[2ex] \dfrac{R_1 + R_2}{R_1 R_2 C_2} \\[2ex] \dfrac{1}{L} \end{bmatrix} \begin{bmatrix} v(t) \end{bmatrix} \quad \textbf{Ans.}$$

## 11.4  STATE MODEL OR STATE SPACE DESCRIPTION

The state equations for a continuous time system with $m$ inputs can be written in a compact form as

$$\dot{x} = Ax + Bu \quad \text{where} \qquad \qquad \qquad \text{...(11.3)}$$
$$x = [x_1, x_2, \ \dots\dots \ x_n]^T \quad \text{is an } n\text{-dimensional state vector}$$

$$A = \begin{bmatrix} a_{11} & a_{12} & \dots\dots & a_{1n} \\ \vdots & & & \\ a_{n1} & a_{n2} & \dots\dots & a_{nn} \end{bmatrix}$$

is an $n \times n$ system matrix

$$
B = \begin{bmatrix} b_{11} & b_{12} & \cdots\cdots & b_{1m} \\ & \vdots & & \\ & \vdots & & \\ b_{n1} & b_{n2} & \cdots\cdots & b_{nm} \end{bmatrix}
$$

is an $n \times m$ input matrix or the distribution matrix and $u = [u_1, u_{2\ldots} \ldots u_m]^T$ is the $m$ dimensional input vector and its components are the various inputs to the system. Equation (11.3) is also known as the vector differential equation. The elements of the matrices $A$ and $B$ are all real numbers and constants for linear time-invariant system. However, for time-variant system these are function of time.

The output of the system can be expressed as a linear combination of the state and the input as

$$Y = Cx + Du \qquad\qquad \ldots(11.4)$$

where $Y = [Y_1, Y_2, \ldots\ldots Y_p]^T$ is the $p$-dimensional output vector.
Here

$$
C = \begin{bmatrix} C_{11} & \cdots\cdots\cdots & C_{1n} \\ & \vdots & \\ & \vdots & \\ & \vdots & \\ & \vdots & \\ C_{p1} & \cdots\cdots\cdots & C_{pn} \end{bmatrix}
$$

is a $p \times n$ output matrix and

$$
D = \begin{bmatrix} d_{11} & \cdots\cdots\cdots & d_{1m} \\ & \vdots & \\ & \vdots & \\ & \vdots & \\ & \vdots & \\ d_{p1} & \cdots\cdots\cdots & d_{pm} \end{bmatrix}
$$

is a $p \times m$ transmission matrix

The elements of the $C$ and $D$ matrices are real and constant for linear time-invariant system. The equations (11.3) and (11.4) constitute the state space description or state model of the linear time-invariant system. These equations also represent time-varying system if $A$, $B$, $C$ and $D$ matrices are functions of time.

Before we seek solution of these state equations it is important to mention that we assume here that the initial time is Zero i.e. $t = 0$ without loss of generality because we have assumed a stationary process i.e. the process, the parameters $A$, $B$ of which are real and constant. For a stationary process the change of state does not depend upon the initial time but only on the control interval i.e. the length of time during which the force (input drive) is applied.

## 11.5  SOLUTION OF STATE EQUATIONS

The vector differential equation can be solved on the same lines as the first order scalar differential equation. The scalar differential equation is given as

$$\dot{x} = ax + bu \qquad x(0) = x_0 \qquad \qquad \text{...(11.5)}$$

This equation has solution in two parts. The first is known as complementary solution where the equation is assumed to be homogeneous i.e. it is given as

$$\dot{x} = ax \qquad \qquad \text{...(11.6)}$$

i.e. the solution where the input to the system is zero. The second part is known as the particular solution to the entire equation when the input is there i.e.

$\dot{x} = ax + bu$. However, this solution is independent of the initial condition. The complete solution is the sum of the two solutions. This follows from the superposition principle that the total response of the system to $u(t)$ is the sum of the responses to zero and $u$ inputs taken together. In other words the complete response is the sum of the zero-input and zero-state (relaxed) response.

Taking Laplace transform of the scalar homogeneous equation we have,

$$sx(s) - x(0) = ax(s) \qquad \qquad \text{...(11.7)}$$

$$(s - a)\, x(s) = x(0) \qquad \qquad \text{...(11.8)}$$

$$x(s) = (s - a)^{-1} x(0) \qquad \qquad \text{...(11.9)}$$

or
$$x(t) = e^{at} \cdot x(0) = e^{at} \cdot x_0 \qquad \qquad \text{...(11.10)}$$

The particular solution using convolution integral is found to be

$$x_p = e^{at} \int_0^t e^{-a\tau}\, bu(\tau)\, d\tau \qquad \qquad \text{...(11.11)}$$

and thus the complete solution is given to be

$$x(t) = e^{at} x_0 + e^{at} \int_0^t e^{-a\tau}\, bu(\tau)\, d\tau \qquad \qquad \text{...(11.12)}$$

Reverting to the homogeneous vector differential equation.

$$\dot{X} = AX \qquad X(0) = X_0 \qquad \qquad \text{...(11.13)}$$

Taking Laplace transform on both sides

$$sX(s) - X(0) = AX(s) \qquad \qquad \text{...(11.14)}$$

or
$$(Is - A)\, X(s) = X(0) \qquad \qquad \text{...(11.15)}$$

or
$$X(s) = (sI - A)^{-1} X(0) \qquad \qquad \text{...(11.16)}$$

or
$$X(t) = \mathscr{L}^{-1} \left[ (sI - A)\, x(0) \right] \qquad \qquad \text{...(11.17)}$$

Note that

$$[sI - A]^{-1} = \frac{1}{s}\left[ I - \frac{A}{s} \right]^{-1} \qquad \qquad \text{...(11.18)}$$

$$= \frac{1}{s}\left[ I + \frac{A}{s} + \frac{A^2}{s^2} + \cdots\cdots \right]$$

$$= \frac{I}{s} + \frac{A}{s^2} + \frac{A^2}{s^3} + \cdots \qquad \qquad ...(11.19)$$

$$\mathcal{L}^{-1}[sI - A]^{-1} = I + At + \frac{A^2}{2!}t^2 + \cdots = e^{At}$$

Therefore, $\qquad\qquad\qquad X(t) = \mathcal{L}^{-1}[sI - A]\,x(0) \qquad\qquad ...(11.20)$

$$= e^{At} \cdot x(0)$$

$e^{At}$ is a matrix of order $n \times n$ and because of its similarity to $e^{at}$ we name it matrix exponential. Also from the above equation the matrix $e^{At}$ could be viewed as a transformation (linear) through which the initial state $x(0)$ is being transformed into the updated state $x(t)$. The same thing can be viewed as the one where the network initial state $x(0)$ at $t = 0$ is driven to a state $x(t)$ at time $t$. This transition in state is carried out by matrix exponential $e^{At}$ and is, therefore, known as state transition matrix. This matrix is sometimes known as fundamental matrix and is denoted by $\phi(t)$

Let us study a few properties of matrix exponential

## 11.5.1  Properties of State Transition Matrix

Let

(1) $\qquad\qquad\qquad\qquad \phi(t) = e^{At} \qquad\qquad\qquad\qquad\qquad ...(11.21)$

$$\phi(0) = e^{A.0} = I \text{ (Identity matrix)}$$

The differential of $\phi(t)$

(2) $\qquad\qquad\qquad\qquad \dfrac{d\,\phi(t)}{dt} = \dfrac{d}{dt}(e^{At}) = e^{At}.A \qquad\qquad ...(11.22)$

This is again an $n \times m$ matrix

(3) $\qquad\qquad e^{A(t_1 + t_2)} = e^{At_1} \cdot e^{At_2} = e^{At_2} \cdot e^{At_1} \qquad\qquad ...(11.23)$

This means the two matrices $e^{At_1}$ and $e^{At_2}$ are commutative.

Let $\qquad\qquad\qquad\qquad t_1 = -t_2 = t$

$$e^{At} \cdot e^{-At} = e^{0.t} = I$$

(4) $\qquad\qquad\qquad\qquad \phi(t)\,\phi(-t) = I \qquad\qquad\qquad\qquad ...(11.24)$

This means the inverse of the matrix $e^{At}$ is obtained by replacing $t$ with $-t$.

(5) Another interesting feature of the homogenous solution is found by studying the process at $t = 0$, $t = t_1$ and $t = t_2$. The corresponding states are $x_0$, $x_1$ and $x_2$ respectively.

$$x_1 = e^{At_1}x_0 \text{ and } x_2 = e^{At_2}x_0 = e^{A(t_2 - t_1)} \cdot e^{At_1}x_0$$

or $\qquad\qquad\qquad x_2 = e^{A(t_2 - t_1)} \cdot x_1 \qquad\qquad\qquad\qquad ...(11.25)$

This means the state of the system at any time $t_2$ can be obtained by a transformation of the state at any other time $t_1$ using the matrix exponential as transformation matrix. It is to be noted that it is not necessary that $t_2$ be greater than $t_1$, which means the transformation also works backward in time.

(6) The other property is

$$[\phi(t)]^n = [e^{At}]^n = e^{Ant} = \phi(nt) \qquad \qquad ...(11.26)$$

Let us now consider the complete vector differential equation.

$$\dot{\overline{X}} = \overline{A}\,\overline{X} + \overline{B}\,\overline{u} \qquad \qquad ...(11.27)$$

Taking Laplace transform on both the sides we have

$$s\overline{X}(s) - \overline{X}(0) = \overline{A}\,\overline{X}(s) + \overline{B}\,\overline{u}(s) \qquad \qquad ...(11.28)$$

or
$$\left(sI - \overline{A}\right)\overline{X}(s) = \overline{X}(0) + \overline{B}\,\overline{u}(s) \qquad \qquad ...(11.29)$$

or
$$X(s) = \left(sI - \overline{A}\right)^{-1}\left[\overline{X}(0) + \overline{B}\,\overline{u}\,(s)\right] \qquad \qquad ...(11.30)$$

since
$$\mathcal{L}^{-1}\,[s\,I - A]^{-1} = e^{At}$$

or
$$[s\,I - A]^{-1} = \mathcal{L}\,[e^{At}] \qquad \qquad ...(11.31)$$

or
$$X(s) = \mathcal{L}\!\left[e^{At}\right]x(0) + \mathcal{L}\!\left[e^{At}\right]B\,u(s) \qquad \qquad ...(11.32)$$

Taking inverse Laplace transform of the equation using convolution integral we have

$$X(t) = e^{At}\,x(0) + \int_0^t e^{A(t-\tau)}\,B\,u(\tau)\,d\tau \qquad \qquad ...(11.33)$$

So far we have considered the initial time to be zero. However, if the initial time given is $t_0$, then the solution becomes

$$X(t) = e^{A(t-t_0)}\,X(t_0) + \int_{t_0}^t e^{A(t-\tau)}\,B\,u(\tau)\,d\tau \qquad \qquad ...(11.34)$$

This solution is known as the state transition equation. If we replace matrix exponential $e^{At}$ by the fundamental matrix $\phi(t)$, the solution can be given by

$$X(t) = \phi(t - t_0)\,X(t_0) + \int_{t_0}^t \phi(t - \tau)\,B\,u(\tau)\,d\tau \qquad \qquad ...(11.35)$$

**Example 11.13:** Determine the state transition matrix and solution of the system described by the vector differential equation.

$$\begin{bmatrix} \dot{x}_1 \\ \dot{x}_2 \end{bmatrix} = \begin{bmatrix} 1 & 2 \\ -3 & -4 \end{bmatrix}\begin{bmatrix} x_1 \\ x_2 \end{bmatrix} + \begin{bmatrix} 0 \\ 1 \end{bmatrix}u(t)$$

where
$$u(t) = 1 \quad t \geq 0$$
$$= 0 \quad t < 0$$

Also compute the inverse of the state transition matrix. Assume the system to be initially relaxed i.e. $x_1(0) = x_2(0) = 0$

**Solution:** State transition matrix $\phi\,(t) = e^{At} = \mathcal{L}^{-1}\,[sI - A]^{-1}$

$$= \left\{ \begin{bmatrix} s & 0 \\ 0 & s \end{bmatrix} - \begin{bmatrix} 1 & 2 \\ -3 & -4 \end{bmatrix} \right\}^{-1}$$

$$[s\,I - A] = \begin{bmatrix} s-1 & -2 \\ 3 & s+4 \end{bmatrix}$$

**Now**

$$[s\,I - A]^{-1} = \frac{Adj\,[s\,I - A]}{|s\,I - A|} = \frac{\begin{bmatrix} s+4 & -3 \\ 2 & s-1 \end{bmatrix}}{(s-1)(s+4)+6}$$

$$= \frac{1}{(s+2)(s+1)} \cdot \begin{bmatrix} s+4 & 2 \\ -3 & s-1 \end{bmatrix}$$

$$= \begin{bmatrix} \dfrac{s+4}{(s+1)(s+2)} & \dfrac{2}{(s+1)(s+2)} \\[3mm] \dfrac{-3}{(s+1)(s+2)} & \dfrac{s-1}{(s+1)(s+2)} \end{bmatrix}$$

Expanding into partial fraction and taking inverse Laplace for each term of the matrix, we have

$$e^{At} = \begin{bmatrix} 3e^{-t} - 2e^{-2t} & 2e^{-t} - 2e^{-2t} \\ -3e^{-t} + 3e^{-2t} & -2e^{-t} + 3e^{-2t} \end{bmatrix} \quad \textbf{Ans.}$$

It can be verified that at $t = 0 \qquad \phi(t) = I$

$$\begin{bmatrix} 3-2 & 2-2 \\ -3+3 & -2+3 \end{bmatrix} = \begin{bmatrix} 1 & 0 \\ 0 & 1 \end{bmatrix}$$

To obtain the inverse of the transition matrix we have to replace $t = -t$ i.e.
$$\phi^{-1}(t) = \phi(-t)$$

**Therefore,**

$$\phi^{-1}(t) = \phi(-t) = \begin{bmatrix} 3e^{t} - 2e^{2t} & 2e^{t} - 2e^{2t} \\ -3e^{t} + 3e^{2t} & -2e^{t} + 3e^{2t} \end{bmatrix}$$

The complete response is given as

$$X(t) = e^{At}\,X(0) + \int_{0}^{t} e^{A(t-\tau)}\,B\,u(\tau)\,d\tau$$

The first term on the right side has already been calculated. Here $x(0) = \begin{bmatrix} 0 \\ 0 \end{bmatrix}$

Taking the second term and replacing $t$ by $(t - \tau)$ in the matrix $e^{At}$ we have

$$\int_0^t \begin{bmatrix} 3e^{-(t-\tau)} - 2e^{-2t+2\tau} & 2e^{-t+\tau} - 2e^{-2t+2\tau} \\ -3e^{-t+\tau} + 3e^{-2t+2\tau} & -2e^{-t+\tau} + 3e^{-2t+2\tau} \end{bmatrix} \begin{bmatrix} 0 \\ 1 \end{bmatrix} [1]\, d\tau$$

$$= \int_0^t \begin{bmatrix} 2e^{-t}e^{\tau} & -2e^{-2t}e^{2\tau} \\ -2e^{-t}e^{\tau} & +3e^{-2t}e^{2\tau} \end{bmatrix} d\tau$$

$$= \begin{bmatrix} 2e^{-t}e^{\tau} - e^{-2t}e^{2\tau} \\ -2e^{-t}e^{\tau} + \dfrac{3}{2}e^{-2t}e^{2\tau} \end{bmatrix}_0^t$$

$$= \begin{bmatrix} 1 \\ -2 + \dfrac{3}{2} \end{bmatrix} - \begin{bmatrix} 2e^{-t} - e^{-2t} \\ -2e^{-t} + \dfrac{3}{2}e^{-2t} \end{bmatrix}$$

$$= \begin{bmatrix} 1 - 2e^{-t} + e^{-2t} \\ -\dfrac{1}{2} + 2e^{-t} - \dfrac{3}{2}e^{-2t} \end{bmatrix}$$

Therefore, the complete solution is given as

$$\begin{bmatrix} x_1 \\ x_2 \end{bmatrix} = \begin{bmatrix} 3e^{-t} - 2e^{-2t} & 2e^{-t} - 2e^{-2t} \\ -3e^{-t} + 3e^{-2t} & -2e^{-t} + 3e^{-2t} \end{bmatrix} \begin{bmatrix} x_1(0) \\ x_2(0) \end{bmatrix} + \begin{bmatrix} 1 - 2e^{-t} + e^{-2t} \\ -\dfrac{1}{2} + 2e^{-t} - \dfrac{3}{2}e^{-2t} \end{bmatrix}$$

Since $x_1(0) = x_2(0) = 0$

$$\begin{bmatrix} x_1 \\ x_2 \end{bmatrix} = \begin{bmatrix} 1 - 2e^{-t} + e^{-2t} \\ -\dfrac{1}{2} + 2e^{-t} - \dfrac{3}{2}e^{-2t} \end{bmatrix}$$

## 11.6  EVALUATION OF MATRIX EXPONENTIAL $e^{At}$

The fundamental matrix or matrix exponential can be evaluated by the following methods:

1. Laplace transform method.
2. Leverrier algorithm (or Sourian – Frame algorithm).
3. Use of Sylvester's interpolation formulae.
4. Use of Diagonal or Jordan forms.

### 11.6.1 Laplace Transform Method

The approach utilized is the application of equation.

$$e^{At} = \mathcal{L}^{-1} \{[s\,I - A]^{-1}\} \qquad \text{...(11.6)}$$

which involves first the inversion of matrix $[sI - A]$ and then taking the inverse Laplace transform of the inverted matrix. This inverse Laplace transform step requires obtaining the inverse Laplace of each element of the matrix $[sI - A]^{-1}$. Since each element of the inverse matrix is a rational function $[p(s)/Q(s)]$ instead of being a real number, the inverse Laplace on a digital computer cannot be carried out. The various steps associated with this method have already been explained in the previous example.

### 11.6.2 Leverrier Algorithm

As mentioned, the Laplace transform method is rather cumbersome by long hand calculation if the dimension of the matrices exceed $3 \times 3$. In such situations Leverrier algorithm is always suggested as the algorithm is easily programmable on digital computers.

The algorithm is explained as follows:

$$\text{Let} \qquad [s\,I - A]^{-1} = \frac{a\,dj\,[s\,I - A]}{|s\,I - A|} \qquad \text{...(11.37)}$$

$$= \frac{P_{n-1}\,s^{n-1} + P_{n-2}\,s^{n-2} + \cdots + P_1 s + P_0}{s^n + \alpha_{n-1}\,s^{n-1} + \cdots + \alpha_1 s + a_0} \qquad \text{...(11.38)}$$

where $P_i$ are $n \times n$ matrices of real numbers and $\alpha_i$ are scalars again real number for time-invarient linear system under consideration. These are obtained as follows:

$$\text{Let} \qquad P_{n-1} = I \qquad \text{...(11.39)}$$

$$\alpha_{n-1} = -\,t_r\,[A] \qquad \text{...(11.40)}$$

$$P_{n-2} = P_{n-1}\,A + \alpha_{n-1}\,I \qquad \text{...(11.41)}$$

$$\alpha_{n-2} = -\,\frac{1}{2} t_r\,(P_{n-2}\,A) \qquad \text{...(11.42)}$$

$$P_i = P_{i+1}\,A + \alpha_{i+1}\,I \qquad \text{...(11.43)}$$

$$\text{and} \qquad \alpha_i = -\frac{1}{n-i}\,t_r\,(P_i\,A) \qquad \text{...(11.44)}$$

where $tr(A)$ means trace of matrix $A$ and equals the sum of the diagonal elements of $A$.

The calculations carried out as outlined in the algorithm can be verified by the check.

$$P_0 A + \alpha_0\,I = 0 \qquad \text{...(11.45)}$$

**Example 11.14:** Consider a $3 \times 3$ matrix and obtain the state transition matrix using Leverrier algorithm.

$$A = \begin{pmatrix} -1 & 2 & -8 \\ 0 & -2 & 4 \\ 0 & 0 & -4 \end{pmatrix}$$

**Here**
$$n = 3$$

**Therefore**
$$P_2 = I = \begin{pmatrix} 1 & 0 & 0 \\ 0 & 1 & 0 \\ 0 & 0 & 1 \end{pmatrix}$$

$$\alpha_2 = -\,t_r\,(A) = 7$$

$$P_1 = P_2 A + \alpha_2 I = \begin{pmatrix} -1 & 2 & -8 \\ 0 & -2 & 4 \\ 0 & 0 & -4 \end{pmatrix} + 7 \begin{pmatrix} 1 & 0 & 0 \\ 0 & 1 & 0 \\ 0 & 0 & 1 \end{pmatrix}$$

$$= \begin{pmatrix} 6 & 2 & -8 \\ 0 & 5 & 4 \\ 0 & 0 & 3 \end{pmatrix}$$

$$\alpha_1 = -\frac{1}{2} t_r\,(P_1 A) = -\frac{1}{2} t_r \begin{pmatrix} 6 & 2 & -8 \\ 0 & 5 & 4 \\ 0 & 0 & 3 \end{pmatrix}\begin{pmatrix} -1 & 2 & -8 \\ 0 & -2 & 4 \\ 0 & 0 & -4 \end{pmatrix}$$

$$= -\frac{1}{2} t_r \begin{pmatrix} -6 & 8 & -8 \\ 0 & -10 & +4 \\ 0 & 0 & -12 \end{pmatrix} = -\frac{1}{2}(-28) = 14$$

$$P_0 = P_1 A + \alpha_1 I$$

$$= \begin{pmatrix} -6 & 8 & -8 \\ 0 & -10 & +4 \\ 0 & 0 & -12 \end{pmatrix} + \begin{pmatrix} 14 & 0 & 0 \\ 0 & 14 & 0 \\ 0 & 0 & 14 \end{pmatrix}$$

$$= \begin{pmatrix} 8 & 8 & -8 \\ 0 & 4 & +4 \\ 0 & 0 & 2 \end{pmatrix}$$

$$\alpha_0 = -\frac{1}{3}\,t_r\,[P_0 A] = -\frac{1}{3}\,t_r \begin{pmatrix} 8 & 8 & -8 \\ 0 & 4 & +4 \\ 0 & 0 & 2 \end{pmatrix}\begin{pmatrix} -1 & 2 & -8 \\ 0 & -2 & 4 \\ 0 & 0 & -4 \end{pmatrix}$$

$$= -\frac{1}{3}\, t_r \begin{pmatrix} -8 & 0 & 0 \\ 0 & -8 & 0 \\ 0 & 0 & -8 \end{pmatrix}$$

Let us verify our calculation substituting the values in the relation.

$$P_0 A + \alpha_0\, I = \begin{pmatrix} -8 & 0 & 0 \\ 0 & -8 & 0 \\ 0 & 0 & -8 \end{pmatrix} + 8 \begin{pmatrix} 1 & 0 & 0 \\ 0 & 1 & 0 \\ 0 & 0 & 1 \end{pmatrix} = [0]$$

Hence the calculations are alright.

The $Adj\,[s\,I - A] = P_2\, s^2 + P_1\, s + P_0$

$$= \begin{pmatrix} s^2 & 0 & 0 \\ 0 & s^2 & 0 \\ 0 & 0 & s^2 \end{pmatrix} + \begin{pmatrix} 6 & 2 & -8 \\ 0 & 5 & 4 \\ 0 & 0 & 3 \end{pmatrix} s + \begin{pmatrix} 8 & 8 & -8 \\ 0 & 4 & 4 \\ 0 & 0 & 2 \end{pmatrix}$$

$$\begin{pmatrix} s^2 + 6s + 8 & 2s + 8 & -8s - 8 \\ 0 & s^2 + 5s + 4 & 4s + 4 \\ 0 & 0 & s^2 + 3s + 2 \end{pmatrix}$$

and the
$$|s\,I - A| = s^3 + \alpha_2\, s^2 + \alpha_1\, s + \alpha_0$$
$$= s^3 + 7s^2 + 14s + 8$$

using Remainder theorem to find roots of the equation we find one of the roots as $s = -2$.
$$(s + 2)\,(s^2 + 5s + 4) = (s + 2)\,(s + 1)\,(s + 4)$$

Therefore,

$$[s\,I - A]^{-1} = \begin{pmatrix} \dfrac{1}{s + 1} & \dfrac{2}{(s + 1)\,(s + 2)} & \dfrac{-8}{(s + 2)\,(s + 4)} \\[2ex] 0 & \dfrac{1}{(s + 2)} & \dfrac{4}{(s + 2)\,(s + 4)} \\[2ex] 0 & 0 & \dfrac{1}{(s + 4)} \end{pmatrix}$$

Therefore,
$$e^{At} = \mathcal{L}^{-1}\left\{[s\,I - A]^{-1}\right\}$$

$$
\begin{pmatrix}
e^{-t} & 2e^{-t} - 2e^{-2t} & -4(e^{-2t} - e^{-4t}) \\
0 & e^{-2t} & 2(e^{-2t} - e^{-4t}) \\
0 & 0 & e^{-4t}
\end{pmatrix}
$$

Here the inverse of $[sI - A]$ was obtained through a recursive algorithm and is easy as compared to the Laplace transform method. However, in this method the inverse Laplace transform is obtained in the same way as in Laplace transform method and hence is tedious and time consuming and can't be adopted so easily on digital computers. The methods that follow can be used on digital computers easily.

### 11.6.3  Use of Sylvester's Interpolations Method

According to Sylvester's Interpolation method the state transition matrix can be expressed as

$$e^{At} = a_0 I + a_1 A + a_2 A^2 + \ldots + a_{n-1} A^{n-1} \qquad \ldots(11.46)$$

where $a^s$ are scalar functions of time $t$ which can be evaluated by solving a set of simultaneous linear equations.

$$
\begin{pmatrix}
1 & \lambda_1 & \lambda_1^2 & \cdots & \lambda_1^{n-1} \\
1 & \lambda_2 & \lambda_2^2 & \cdots & \lambda_2^{n-1} \\
\vdots & & & & \\
1 & \lambda_n & \lambda_n^2 & \cdots & \lambda_n^{n-1}
\end{pmatrix}
\begin{pmatrix}
a_0 \\
a_1 \\
\vdots \\
a_{n-1}
\end{pmatrix}
=
\begin{pmatrix}
e^{\lambda_1 t} \\
e^{\lambda_2 t} \\
\vdots \\
e^{\lambda_n t}
\end{pmatrix}
\qquad \ldots(11.47)
$$

where $\lambda^s$ are the eigen values of the matrix which are obtained using Cayley-Hamilton theorem which states that every square matrix $A$ satisfies its own characteristic equation.

$$|\lambda I - A| = 0$$

The solution of the equation gives different eigen values of the matrix $A$.

Let us take an example to illustrate the procedure.

**Example 11.15:** Evaluate state transition matrix if matrix $A$ is given as follows, using Sylveder's interpolation method.

$$A = \begin{pmatrix} 1 & 2 \\ -3 & -4 \end{pmatrix}$$

The eigenvalues are obtained by solving.

$$\lambda I - A = \begin{pmatrix} \lambda & 0 \\ 0 & \lambda \end{pmatrix} - \begin{pmatrix} 1 & 2 \\ -3 & -4 \end{pmatrix} = 0$$

$$= \begin{pmatrix} \lambda - 1 & -2 \\ +3 & \lambda + 4 \end{pmatrix} = (\lambda - 1)(\lambda + 4) + 6 = 0$$

or
$$\lambda^2 + 3\,\lambda - 4 + 6 = 0$$

or
$$\lambda^2 + 3\,\lambda + 2 = 0$$

or
$$\lambda_1 = -1$$
$$\lambda_2 = -2$$

Therefore, to obtain $a_0\,a$, we write the equation.

$$\begin{pmatrix} 1 & -1 \\ 1 & -2 \end{pmatrix} \begin{pmatrix} a_0 \\ a_1 \end{pmatrix} = \begin{pmatrix} e^{-t} \\ e^{-2t} \end{pmatrix}$$

$$a_0 - a_1 = e^{-t}$$
$$a_0 - 2a_1 = e^{-2t}$$

Hence
$$a_1 = e^{-t} - e^{-2t}$$
and
$$a_0 = 2\,e^{-t} - e^{-2t}$$

Therefore,

$$e^{At} = \begin{bmatrix} 2e^{-t} - e^{-2t} & 0 \\ 0 & +2e^{-t} - e^{-2t} \end{bmatrix} + \left( e^{-t} - e^{-2t} \right) \begin{bmatrix} 1 & 2 \\ -3 & -4 \end{bmatrix}$$

$$= \begin{bmatrix} 3e^{-t} - 2e^{-2t} & 2(e^{-t} - e^{-2t}) \\ -3(e^{-t} - e^{-2t}) & -2e^{-t} + 3e^{-2t} \end{bmatrix}$$

But putting $t = 0$ it can be verified that $e^{At}$ becomes an identity matrix and hence the calculations are alright.

It is seen that as compared to the previous two methods, the method suggested by Sylvester can easily be adopted on digital computer as standard program for evaluation of eigenvalues and solution of linear simultaneous equations (matrix inversion) are readily available.

The procedure outlined requires all the eigenvalues to be distinct. However, if $A$ matrix has repeated (multiple) eigenvalues, a small modification as suggested below is required.

If $A$ matrix has multiple eigenvalues the matrix on the left-hand side of equation (11.47) will have equal number of identical rows and hence this matrix becomes singular. Therefore, these rows need modification with the first row corresponding to repeated eigenvalue of this group as in the above equation, the following rows will be replaced by successive derivatives with respect to the eigenvalue. For example matrix $A$ is of order $6 \times 6$ and $\lambda_2$ has multiplicity of 3, the equation corresponding to equation (11.47) will be

$$\begin{bmatrix} 1 & \lambda_1 & \lambda_1^2 & \lambda_1^3 & \lambda_1^4 & \lambda_1^5 \\ 1 & \lambda_2 & \lambda_2^2 & \lambda_2^3 & \lambda_2^4 & \lambda_2^5 \\ 0 & 1 & 2\lambda_2 & 3\lambda_2^2 & 4\lambda_2^3 & 5\lambda_2^4 \\ 0 & 0 & 2 & 6\lambda_2 & 12\lambda_2^2 & 20\lambda_2^3 \\ 1 & \lambda_5 & \lambda_5^2 & \lambda_5^3 & \lambda_5^4 & \lambda_5^5 \\ 1 & \lambda_6 & \lambda_6^2 & \lambda_6^3 & \lambda_6^4 & \lambda_6^5 \end{bmatrix} \begin{bmatrix} a_0 \\ a_1 \\ a_2 \\ a_3 \\ a_4 \\ a_5 \end{bmatrix} = \begin{bmatrix} e^{\lambda_1 t} \\ e^{\lambda_2 t} \\ t\,e^{\lambda_2 t} \\ t^2 e^{\lambda_2 t} \\ e^{\lambda_5 t} \\ e^{\lambda_6 t} \end{bmatrix}$$

To illustrate the procedure further consider the following example.

**Example 11.16:** Given

$$A = \begin{pmatrix} 0 & 1 \\ -9 & -6 \end{pmatrix}$$

Determine $e^{At}$

**Solution:**
$$|\lambda I - A| = \begin{pmatrix} \lambda & -1 \\ 9 & \lambda + 6 \end{pmatrix} = 0$$

$$\lambda^2 + 6\lambda + 9 = 0$$

or
$$\lambda_1 = \lambda_2 = -3$$

$$\begin{pmatrix} 1 & \lambda_1 \\ 0 & 1 \end{pmatrix} \begin{pmatrix} a_0 \\ a_1 \end{pmatrix} = \begin{pmatrix} e^{-3t} \\ t\, e^{-3t} \end{pmatrix}$$

Therefore, the set of equations

$$\begin{pmatrix} 1 & -3 \\ 0 & 1 \end{pmatrix} \begin{pmatrix} a_0 \\ a_1 \end{pmatrix} = \begin{pmatrix} e^{-3t} \\ t\, e^{-3t} \end{pmatrix}$$

$$a_0 - 3a_1 = e^{-3t}$$

$$a_1 = t\, e^{-3t} \text{ and } a_0 = e^{-3t} + 3\, t\, e^{-3t}$$

Now
$$e^{At} = a_0\, I + a_1\, A$$

$$= \begin{bmatrix} 3t\, e^{-3t} + e^{-3t} & 0 \\ 0 & 3t\, e^{-3t} + e^{-3t} \end{bmatrix} + t\, e^{-3t} \begin{bmatrix} 0 & 1 \\ -9 & -6 \end{bmatrix}$$

$$= \begin{bmatrix} (3t + 1)\, e^{-3t} & t\, e^{-3t} \\ -9\, t\, e^{-3t} & (3t - 5)\, e^{-3t} \end{bmatrix} \textbf{Ans.}$$

It is to be noted that even if $A$ has complex numbers as its elements, there is no cause for anxiety as standard computer program for inversion of complex matrices are also available.

### 11.6.4  Use of Diagonal or Jordan Form

If we have $A$ as the diagonal matrix or if we are able to transform $A$ into a diagonal matrix which otherwise is a normal matrix it becomes very simple to evaluate $e^{At}$.

Consider a diagonal matrix $D$

$$D = \begin{pmatrix} \lambda_1 & 0 & 0 & 0 & 0 \\ 0 & \lambda_2 & 0 & 0 & 0 \\ 0 & 0 & \lambda_3 & 0 & 0 \\ 0 & 0 & 0 & \lambda_4 & 0 \\ 0 & 0 & 0 & 0 & \lambda_5 \end{pmatrix} \qquad \text{...(11.49)}$$

Since $(s\,I - D)$ is a diagonal matrix it is evident that its inverse will also be a diagonal matrix and element in the diagonal will correspond to $\dfrac{1}{s - \lambda_i}$ and then $\mathcal{L}^{-1}[(s\,I - D)^{-1}]$ will be

$$= \begin{pmatrix} e^{\lambda_1 t} & 0 & 0 & 0 & 0 \\ 0 & e^{\lambda_2 t} & 0 & 0 & 0 \\ 0 & 0 & e^{\lambda_3 t} & 0 & 0 \\ 0 & 0 & 0 & e^{\lambda_4 t} & 0 \\ 0 & 0 & 0 & 0 & e^{\lambda_5 t} \end{pmatrix} \qquad \text{...(11.50)}$$

In our evaluation for $e^{At}$ if we somehow can obtain $A$ in the diagonal form we know that

$$\mathcal{L}^{-1}[(sI - A)^{-1}] = e^{At} \qquad \text{...(11.51)}$$

The physical significance of Diagonal matrix is that it represents a set of decoupled first order differential equations i.e. the state variables do not affect each other and every equation will have a state variable on the right hand side and its derivature on the left hand side. However, in a physical system it is almost never feasible. There is always effect of one state variable on some of the other state variables if not on all state variables.

Here there are two possibilities. One, $A$ has all distinct eigenvalues. Two, it may have some repeated eigenvalues.

We first consider the case when $A$ has all distinct eigenvalues.

Let $\lambda_i$ be an eigenvalue of $A$ and $m_i$ its corresponding eigenvector, then from the theory of eigenvalue and eigen vector, the following relations are obtained.

$$Am_1 = \lambda_1 m_1$$
$$Am_2 = \lambda_2 m_2$$
$$Am_i = \lambda_i m_i \qquad \text{...(11.52)}$$

These equations can be rearranged to obtain

$$AM = MD \qquad \text{...(11.53)}$$

where $D$ is a diagonal matrix defined above in equation (11.49)

and
$$M = [m_1, m_2, \dots\ m_i] \qquad \text{...(11.54)}$$

i.e. columns of $M$ are the eigenvectors of $A$.

Since $A$ has distinct eigenvalues, $M$ is a non-singular matrix. Therefore, its inverse exists.

Therefore,
$$D = M^{-1}\,AM \qquad \text{...(11.55)}$$

and
$$(s\,I - D) = M^{-1}\,(s\,I - A)\,M \qquad \text{...(11.56)}$$

or
$$(s\,I - D)^{-1} = M^{-1}\,(s\,I - A)^{-1}\,M \qquad \text{...(11.57)}$$

or
$$(s\,I - A)^{-1} = M\,(s\,I - D)^{-1}\,M^{-1} \qquad \text{...(11.58)}$$

Taking Laplace transform we have

$$e^{At} = M\,e^{Dt}\,M^{-1} \qquad \text{...(11.59)}$$

Therefore, to find out $e^{At}$ following steps are required:

1. Obtain eigen-values of $A$ and corresponding to each eigenvalue obtain eigen vector.
2. The columns of $M$ are the eigen vectors of various eigenvalues obtain $D = M^{-1}AM$

and hence obtain

$$e^{At} = M e^{Dt} M^{-1}$$

The following example illustrates the procedure:

**Example 11.17:** Given

$$A = \begin{pmatrix} 1 & 2 \\ -3 & -4 \end{pmatrix}$$

Determine $e^{At}$ using Jordan form.

**Solution:** The eigen values as in previous example are $\lambda_1 = -1$ and $\lambda_2 = -2$
Eigen vector corresponding to $\lambda_1 = -1$ is obtained

$$[\lambda_1 I - A] \begin{bmatrix} x_1 \\ x_2 \end{bmatrix} = 0$$

$$\left\{ \begin{pmatrix} -1 & 0 \\ 0 & -1 \end{pmatrix} - \begin{pmatrix} 1 & 2 \\ -3 & -4 \end{pmatrix} \right\} \begin{pmatrix} x_1 \\ x_2 \end{pmatrix} = 0$$

or

$$\begin{pmatrix} -2 & -2 \\ 3 & 3 \end{pmatrix} \begin{pmatrix} x_1 \\ x_2 \end{pmatrix} = 0$$

$$-2\,x_1 - 2\,x_2 = 0$$
$$3\,x_1 + 3\,x_2 = 0$$

Let
$$x_1 = 1 \qquad x_2 = -1$$

Therefore
$$m_1 = \begin{pmatrix} 1 \\ -1 \end{pmatrix}$$

Similarly for $\lambda_2 = -2$, the eigen vector is

$$\begin{pmatrix} -2 & 0 \\ 0 & -2 \end{pmatrix} - \begin{pmatrix} 1 & 2 \\ -3 & -4 \end{pmatrix} \begin{pmatrix} x_1 \\ x_2 \end{pmatrix} = 0$$

or

$$\begin{pmatrix} -3 & -2 \\ 3 & 2 \end{pmatrix} \begin{pmatrix} x_1 \\ x_2 \end{pmatrix} = 0$$

$$-3\,x_1 - 2\,x_2 = 0$$
$$3\,x_1 + 2\,x_2 = 0$$

Let
$$x_1 = 1 \qquad x_2 = -3/2$$

$$m_2 = \begin{pmatrix} 1 \\ -3/2 \end{pmatrix}$$

Therefore,

$$M = \begin{pmatrix} 1 & 1 \\ -1 & -3/2 \end{pmatrix}$$

$$M^{-1} = \frac{Adj \begin{bmatrix} 1 & 1 \\ -1 & -3/2 \end{bmatrix}}{\Delta M}$$

Since $\quad\quad \Delta M = -\dfrac{3}{2} + 1 = -\dfrac{1}{2}$

Therefore, $\quad M^{-1} = -2 \begin{pmatrix} -3/2 & -1 \\ 1 & 1 \end{pmatrix} = \begin{pmatrix} 3 & 2 \\ -2 & -2 \end{pmatrix}$

Therefore, $\quad\quad D = M^{-1}\,A\,M$

$$= \begin{pmatrix} 3 & 2 \\ -2 & -2 \end{pmatrix} \begin{pmatrix} 1 & 2 \\ -3 & -4 \end{pmatrix} \begin{pmatrix} 1 & 1 \\ -1 & -3/2 \end{pmatrix}$$

$$= \begin{pmatrix} -3 & -2 \\ 4 & 4 \end{pmatrix} \begin{pmatrix} 1 & 1 \\ -1 & -3/2 \end{pmatrix}$$

$$= \begin{pmatrix} -1 & 0 \\ 0 & -2 \end{pmatrix}$$

Therefore $\quad\quad e^{Dt} = \begin{pmatrix} e^{-t} & 0 \\ 0 & e^{-2t} \end{pmatrix}$

$$e^{At} = M\,e^{Dt}\,M^{-1}$$

and

$$= \begin{pmatrix} 1 & 1 \\ -1 & -3/2 \end{pmatrix} \begin{pmatrix} e^{-t} & 0 \\ 0 & e^{-2t} \end{pmatrix} \begin{pmatrix} 3 & 2 \\ -2 & -2 \end{pmatrix}$$

$$= \begin{pmatrix} e^{-t} & e^{-2t} \\ -e^{-t} & -\dfrac{3}{2}e^{-2t} \end{pmatrix} \begin{pmatrix} 3 & 2 \\ -2 & -2 \end{pmatrix} = \begin{pmatrix} 3e^{-t} - 2e^{-2t} & 2e^{-t} - 2e^{-2t} \\ -3e^{-t} + 3e^{-2t} & -2e^{-t} + 3e^{-2t} \end{pmatrix} \textbf{Ans.}$$

### 11.6.5 Series Evaluation of $e^{At}$

We know that

$$e^{At} = I + At + \frac{A^2 t^2}{2!} + \frac{A^3 t^3}{3!} + \cdots \quad\quad ...(11.60)$$

Depending upon the accuracy required, the series can be terminated for evaluation purpose.

Suppose $\quad\quad A = \begin{pmatrix} 1 & 2 \\ -3 & -4 \end{pmatrix}$

Evaluate $e^{At}$ upto 3rd term
To do that we have to evaluate $A^2$

$$= \begin{pmatrix} 1 & 2 \\ -3 & -4 \end{pmatrix} \begin{pmatrix} 1 & 2 \\ -3 & -4 \end{pmatrix}$$

$$= \begin{pmatrix} -5 & -6 \\ 9 & 10 \end{pmatrix}$$

Therefore,
$$\frac{A^2 t^2}{2} = \begin{pmatrix} -2 \cdot 5 t^2 & -3t^2 \\ 4.5t^2 & 5t^2 \end{pmatrix}$$

and
$$At = \begin{pmatrix} t & 2t \\ -3t & -4t \end{pmatrix}$$

Hence

$$e^{At} = \begin{pmatrix} 1 + t - 2.5t^2 & 2t - 3t^2 \\ -3t + 4.5t^2 & 1 - 4t + 5t^2 \end{pmatrix}$$

## 11.7  SOLUTION OF DIFFERENTIAL EQUATION USING STATE VARIABLES

Consider an $n$th order differential equation.

$$\frac{d^n x}{dt^n} + a_1 \frac{d^{n-1}x}{dt^{n-1}} + a_2 \frac{d^{n-2}x}{dt^{n-2}} + \cdots + a_{n-1}\frac{dx}{dt} + a_n x = u(t) \qquad \text{...(11.61)}$$

As said earlier to solve $n$th order differntial equation we should know initial value of $x$ and the $(n-1)$ derivatives besides the input $u(t)$.

Let us take $x(t),\ \dot{x}(t),\ \ldots\ \dot{x}^{(n-1)}(t)$
as the state variables

Let us define
$$x_1 = x(t)$$
$$\dot{x}_1 = x_2$$
$$\dot{x}_2 = x_3$$
$$\vdots$$
$$\dot{x}_{n-1} = x_n \qquad \text{...(11.62)}$$

and from the original $n$th order differential equation.

$$\dot{x}_n = -a_n x_1 - a_{n-1} x_2 \ldots \ldots -a_1 x_n + u(t) \qquad \text{...(11.63)}$$

These equations can be written in the following matrix form:

$$\dot{x} = Ax + Bu \qquad \text{...(11.64)}$$

where
$$x = [x_1,\ x_2,\ \ldots\ x_n]^T$$
and

$$A = \begin{bmatrix} 0 & 1 & 0 & 0 & \cdots\cdots & 0 \\ 0 & 0 & 1 & 0 & \cdots\cdots & 0 \\ \vdots & & & & & \\ 0 & 0 & 0 & 0 & \cdots\cdots & 1 \\ -a_n & -a_{n-1} & -a_{n-2} & \cdots\cdots & & -a_1 \end{bmatrix} \qquad \text{...(11.65)}$$

$$B = [0\ 0\ 0\ \ldots\ldots\ 0\ 1]^T$$

and
$$u = [u] = [1]$$

**Example 11.18:** Consider the system being represented by the third order differential equation.

$$\dddot{x} + 9\ddot{x} + 26\dot{x} + 24x = 12u(t)$$

where $x$ is the output and $u(t)$ is the input. Obtain the state space representation of the system.

**Solution:** We select the state variables as

$$x_1 = x$$
$$x_2 = \dot{x}_1 = \dot{x}$$
$$x_3 = \dot{x}_2 = \ddot{x}$$
$$\dot{x}_3 = \dddot{x} = -24x_1 - 26x_2 - 9x_3 + 12u(t)$$

Therefore, in the vector matrix differential form

$$\begin{pmatrix} \dot{x}_1 \\ \dot{x}_2 \\ \dot{x}_3 \end{pmatrix} = \begin{pmatrix} 0 & 1 & 0 \\ 0 & 0 & 1 \\ -24 & -26 & -9 \end{pmatrix} \begin{pmatrix} x_1 \\ x_2 \\ x_3 \end{pmatrix} + \begin{pmatrix} 0 \\ 0 \\ 12 \end{pmatrix} [u(t)]$$

The block diagram representation of the state equation is shown in Fig. E11.18.

The block diagram also suggests the solution of the equation on analog computer using state variable approach. The block diagram makes use of integrators, potentiometers and summators.

It can be shown that the choice of state variables is not unique. We consider the same system represented by the third order equation example 11.18. Let us assume that $x(t)$ is the output and $u(t)$ the input.

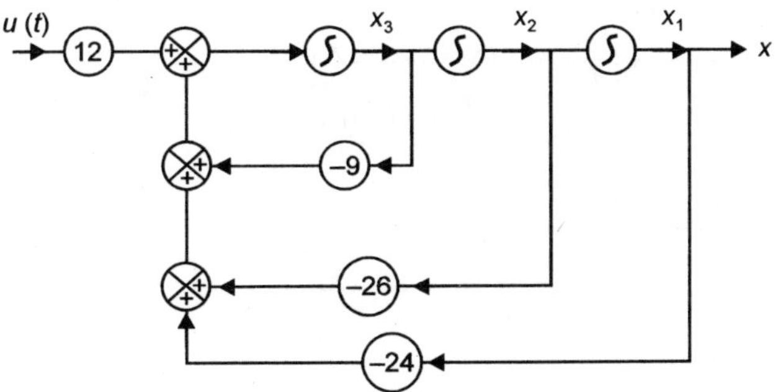


**Fig. E11.18**

Therefore,

$$\frac{x(s)}{u(s)} = \frac{12}{(s+2)(s+3)(s+4)}$$

$$= \frac{A}{(s+2)} + \frac{B}{(s+3)} + \frac{C}{(s+4)}$$

$$= \frac{6}{(s+2)} + \frac{-12}{(s+3)} + \frac{6}{(s+4)}$$

$$x(s) = \frac{6\,u(s)}{(s+2)} + \frac{-12u(s)}{(s+3)} + \frac{6u(s)}{(s+4)}$$

These state equations can be written for each of these transfer functions and the output obtained as the sum. The main advantage of this formulation is that the matrix $A$ is diagonal if the eigenvalues are distinct. In this case as eigenvalues are : $-2, -3, -4$ therefore the matrix $A$ would be diagonal.

Let

$$x(s) = x_1(s) + x_2(s) + x_3(s)$$

$$x_1(s) = \frac{6u(s)}{s+2}$$

$$(s+2)\,(x_1(s)) = 6u(s)$$

$$sx_1(s) + 2\,x_1(s) = 6u(s)$$

$$\dot{x}_1 = -2x_1 + 6u(t)$$

Similarly

$$\dot{x}_2 = -3x_2 - 12u(t)$$

and

$$\dot{x}_3 = -4x_3 + 6u(t)$$

Therefore, the equations are:

$$\begin{pmatrix} \dot{x}_1 \\ \dot{x}_2 \\ \dot{x}_3 \end{pmatrix} = \begin{pmatrix} -2 & 0 & 0 \\ 0 & -3 & 0 \\ 0 & 0 & -4 \end{pmatrix} \begin{pmatrix} x_1 \\ x_2 \\ x_3 \end{pmatrix} + \begin{pmatrix} 6 \\ -12 \\ 6 \end{pmatrix} [u(t)]$$

$$x = [1\ 1\ 1]\,[x_1\ x_2\ x_3]^T$$

and the block diagram is as shown

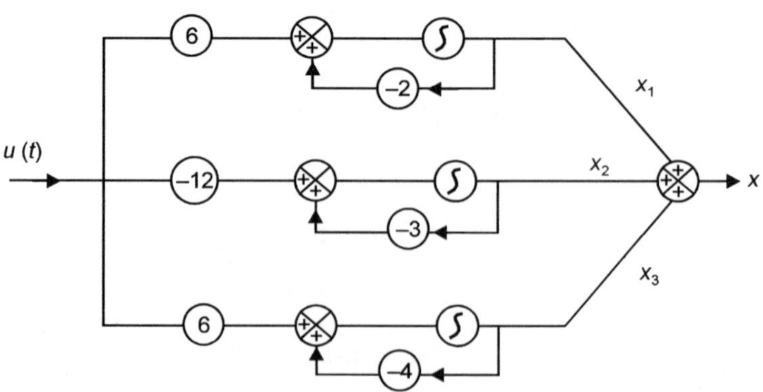


**Fig. E11.18.1**  Block diagram representation of example 11.18

It is seen that there is no unique selection of state variables. It has a very large number of possibilities. With every selection of state variables the most important thing is that the eigenvalues of all these 'A' matrices remain unchanged.

We now study another method of transforming $A$ matrix into a diagonal matrix especially when $A$ is given in the form of the example under consideration. This is known as companion form. To transform this into a diagonal matrix we use a matrix known as Vander Monde matrix.

$$M = V = \begin{pmatrix} 1 & 1 & \cdots\cdots & 1 \\ \lambda_1 & \lambda_2 & \cdots\cdots & \lambda_n \\ \lambda_1^2 & \lambda_2^2 & \cdots\cdots & \lambda_n^2 \\ \vdots & & & \\ \lambda_1^{n-1} & \lambda_2^{n-1} & \cdots\cdots & \lambda_n^{n-1} \end{pmatrix} \qquad ...(11.66)$$

where $\lambda^s$ are the eigenvalues of the companion matrix. Then diagonal matrix $D$ is given as

$$D = M^{-1} A M$$

Consider matrix $\qquad A = \begin{pmatrix} 0 & 1 \\ -2 & -3 \end{pmatrix}$ and obtain $e^{At}$

This is of the companion form. The eigenvalues are:

$$|\lambda I - A| = 0 = \begin{vmatrix} \lambda & 1 \\ -2 & \lambda + 3 \end{vmatrix} = 0$$

$$\lambda(\lambda + 3) + 2 = 0 \quad \text{or} \quad \lambda^2 + 3\lambda + 2 = 0$$

or $\qquad\qquad\qquad \lambda_1 = -1, \ \lambda_2 = -2$

The modal matrix or Vander Monde matrix would be

$$M = \begin{pmatrix} 1 & 1 \\ -1 & -2 \end{pmatrix}$$

Hence $\qquad M^{-1} = -\begin{pmatrix} -2 & -1 \\ 1 & 1 \end{pmatrix} = \begin{pmatrix} 2 & 1 \\ -1 & -1 \end{pmatrix}$

Therefore, $\qquad D = M^{-1} A M = \begin{pmatrix} -1 & 0 \\ 0 & -2 \end{pmatrix}$

In fact it is not necessary to make all these calculations one can straight away write down once the matrix $A$ is of the companion form.

$$e^{Dt} = \begin{pmatrix} e^{-t} & 0 \\ 0 & e^{-2t} \end{pmatrix}$$

Therefore, $\qquad e^{At} = M\, e^{Dt}\, M^{-1}$

$$e^{At} = \begin{pmatrix} 1 & 1 \\ -1 & -2 \end{pmatrix} \begin{pmatrix} e^{-t} & 0 \\ 0 & e^{-2t} \end{pmatrix} \begin{pmatrix} 2 & 1 \\ -1 & -1 \end{pmatrix}$$

$$\begin{pmatrix} 2e^{-t} - e^{-2t} & e^{-t} - e^{-2t} \\ -2e^{-t} + 2e^{-2t} & -e^{-t} + 2e^{-2t} \end{pmatrix} \quad \textbf{Ans.}$$

**Example 11.19:** A network is described by the following state equations:

$$\dot{x} = \begin{pmatrix} -1 & 2 & -8 \\ 0 & -2 & 4 \\ 0 & 0 & -4 \end{pmatrix} \begin{pmatrix} x_1 \\ x_2 \\ x_3 \end{pmatrix} + \begin{pmatrix} 0 \\ 0 \\ 1 \end{pmatrix} u(t)$$

and 

$$x(0) = [0 \ 0 \ 0]^T$$

Determine the time response of the network when excited by a unity step at $t = 0+$

**Solution:** The time response is given by

$$x(t) = e^{At}\, X(0) + \int_0^t e^{A(t-\tau)}\, Bu(\tau)\, d\tau$$

$e^{At}$ has already been evaluated in example 11.14 and is given to be

$$e^{At} = \begin{pmatrix} e^{-t} & 2e^{-t} - 2e^{-2t} & -4(e^{-2t} - e^{-4t}) \\ 0 & e^{-2t} & 2(e^{-2t} - e^{-4t}) \\ 0 & 0 & e^{-4t} \end{pmatrix}$$

Since $X(0)$ is given to be a null vector
Therefore,

$$x(t) = \int_0^t e^{-A(t-\tau)}\, Bu(\tau)\, d\tau$$

Now the integrand is

$$\begin{pmatrix} e^{-(t-\tau)} & 2e^{-(t-\tau)} - 2e^{-2(t-\tau)} & -4\left[e^{-2(t-\tau)} - e^{-4(t-\tau)}\right] \\ 0 & e^{-2(t-\tau)} & 2\left(e^{-2(t-\tau)} - e^{-4(t-\tau)}\right) \\ 0 & 0 & e^{-4(t-\tau)} \end{pmatrix} \begin{pmatrix} 0 \\ 0 \\ 1 \end{pmatrix} [1]$$

$$= \int_0^t \begin{pmatrix} -4\left(e^{-2(t-\tau)} - e^{-4(t-\tau)}\right) \\ 2\left(e^{-2(t-\tau)} - e^{-4(t-\tau)}\right) \\ e^{-4(t-\tau)} \end{pmatrix} d\tau$$

$$x_1 = \int_0^t \left(-4\, e^{-2t}\, e^{2\tau} + 4\, e^{-4t}\, e^{4\tau}\right) d\tau$$

$$= \left[ -\frac{4}{2}\, e^{-2t}\, e^{2\tau} + \frac{4}{4}\, e^{-4t}\, e^{4\tau} \right]_0^t$$

$$= 2\, e^{-2t} - e^{-4t} - 1$$

$$x_2 = \int_0^t \left(2e^{-2t}\,e^{2\tau} - 2e^{-4t}\,e^{4\tau}\right) d\tau$$

$$= \frac{1}{2} - e^{-2t} + \frac{e^{-4t}}{2}$$

$$x_3 = \int_0^t e^{-4(t-\tau)}\,d\tau = \left[\frac{1}{4}\,e^{-4t}\cdot e^{4\tau}\right]_0^t$$

$$= \frac{1}{4}\left[1 - e^{-4t}\right]$$

Therefore,

$$\begin{bmatrix} x_1 \\ x_2 \\ x_3 \end{bmatrix} = \begin{bmatrix} 2e^{-2t} - e^{-4t} - 1 \\ \dfrac{1}{2} - e^{-2t} + \dfrac{e^{-4t}}{2} \\ \dfrac{1}{4} - \dfrac{e^{-4t}}{4} \end{bmatrix} \quad \textbf{Ans.}$$

**Example 11.20:** A system is represented by the state equation

$$\dot{x} = A\,x$$

If for $\qquad\qquad x(0) = \begin{bmatrix} 1 \\ -2 \end{bmatrix}$ the response is

$$x(t) = \begin{bmatrix} e^{-2t} \\ -2e^{-2t} \end{bmatrix}$$

and for $\qquad\qquad x(0) = \begin{bmatrix} 1 \\ -1 \end{bmatrix}$ the response is

$$x(t) = \begin{bmatrix} e^{-t} \\ -e^{-t} \end{bmatrix}$$

Determine the system marix and the state transition matrix.

Suppose $e^{At}$ has four components $\begin{bmatrix} A_{11} & A_{12} \\ A_{21} & A_{22} \end{bmatrix}$

Therefore,

$$\left.\begin{array}{l} A_{11} - 2A_{12} = e^{-2t} \\ A_{21} - 2A_{22} = -2e^{-2t} \end{array}\right\} \text{Condition I}$$

and

$$\left.\begin{array}{l} A_{11} - A_{12} = e^{-t} \\ A_{21} - A_{22} = -e^{-t} \end{array}\right\} \begin{array}{l} A_{11} = A_{12} + e^{-t} \\ \text{Condition II} \end{array}$$

$$A_{12} + e^{-t} - 2A_{12} = e^{-2t}$$

$$-A_{12} = e^{-2t} - e^{-t}$$

$$A_{12} = e^{-t} - e^{-2t}$$

$$A_{11} = e^{-t} - e^{-2t} + e^{-t} = 2\,e^{-t} - e^{-2t}$$

Now

$$A_{21} = A_{22} - e^{-t}$$

$$A_{22} - e^{-t} - 2\,A_{22} = -2\,e^{-2t}$$

$$-A_{22} = e^{-t} - 2\,e^{-2t}$$

$$A_{22} = -e^{-t} + 2\,e^{-2t}$$

$$A_{21} = -2\,e^{-t} + 2\,e^{-2t}$$

Therefore,

$$e^{At} = \begin{bmatrix} 2e^{-t} - e^{-2t} & e^{-t} - e^{-2t} \\ -2e^{-t} + 2e^{-2t} & -e^{-t} + 2e^{-2t} \end{bmatrix}$$

$$\alpha[e^{At}] = \begin{bmatrix} \dfrac{2}{s+1} - \dfrac{1}{s+2} & \dfrac{1}{s+1} - \dfrac{1}{s+2} \\ \dfrac{2}{s+2} - \dfrac{2}{s+1} & \dfrac{2}{s+2} - \dfrac{1}{s+1} \end{bmatrix}$$

$$[sI - A]^{-1} = \begin{bmatrix} \dfrac{s+3}{(s+1)(s+2)} & \dfrac{1}{(s+1)(s+2)} \\ \dfrac{-2}{(s+1)(s+2)} & \dfrac{s}{(s+1)(s+2)} \end{bmatrix}$$

$$[sI - A] = \begin{bmatrix} s & -1 \\ +2 & s+3 \end{bmatrix}$$

Therefore,

$$A = \begin{bmatrix} 0 & 1 \\ -2 & -3 \end{bmatrix} \textbf{ Ans.}$$

## PROBLEMS

**11.1.** For the network shown obtain the state equations in terms of physical state variables.

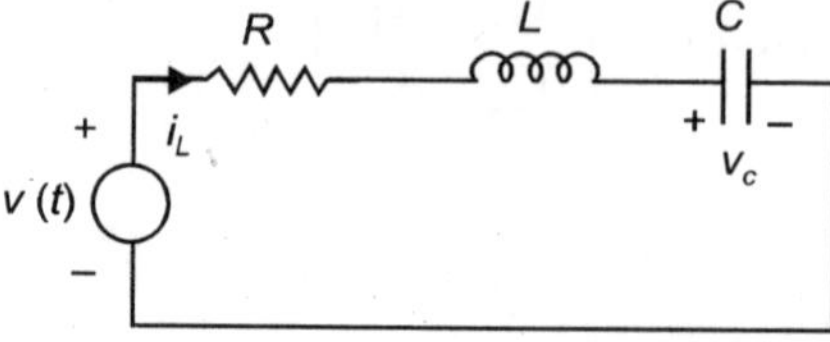


**Fig. P11.1**

**11.2.** For the network shown obtain the state equations in terms of $I_{L1}$, $I_{L2}$ and $v_c$ as state variables.

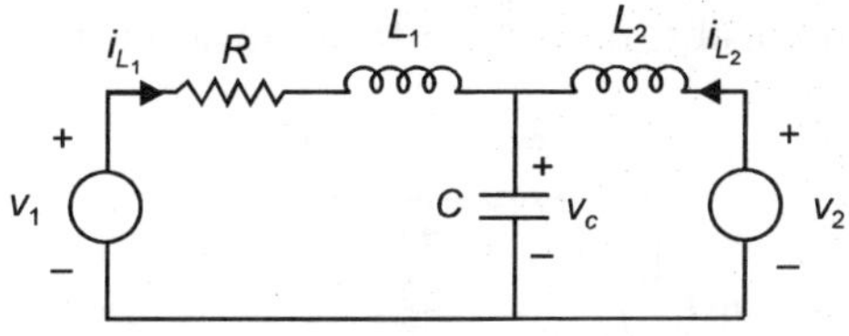

**Fig. P11.2**

**11.3.** For the network shown determine the state equations in terms of $i_L$ and $v_c$ as state variables.

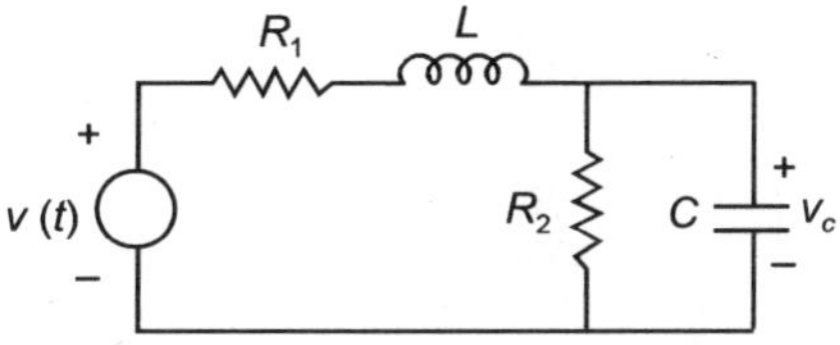

**Fig. P11.3**

**11.4.** For the network shown obtain the state equations in terms of $v_{c1}$ and $v_{c2}$ as state variables.

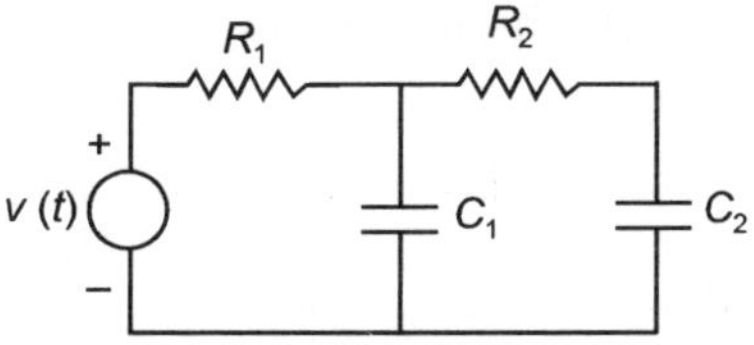

**Fig. P11.4**

**11.5.** For the network shown obtain the state equation using topological method.

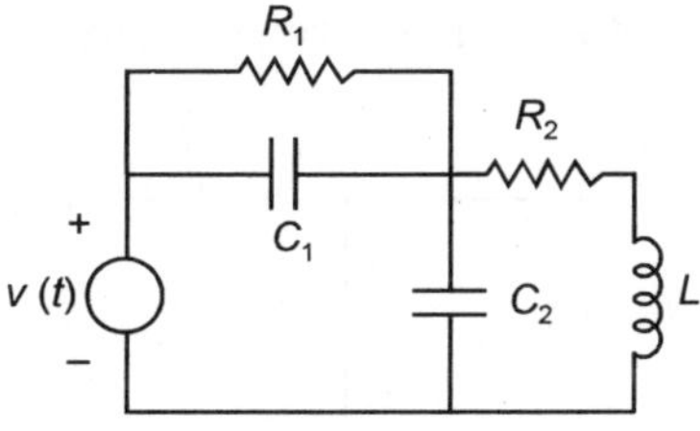

**Fig. P11.5**

**11.6.** For the network shown obtain the state equation using topological method.

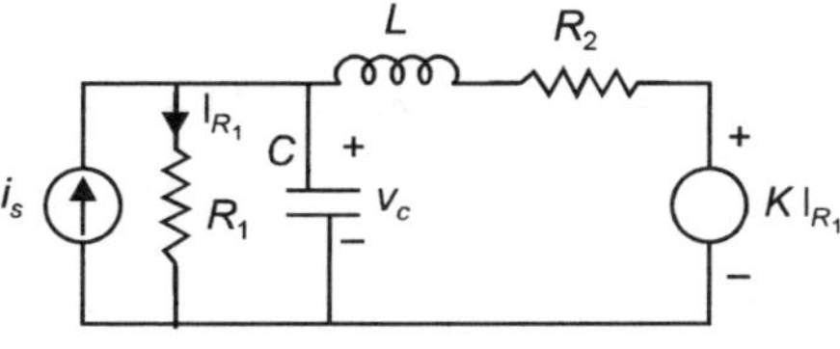

**Fig. P11.6**

**11.7.** For the network shown obtain the state equations using equivalent source method.

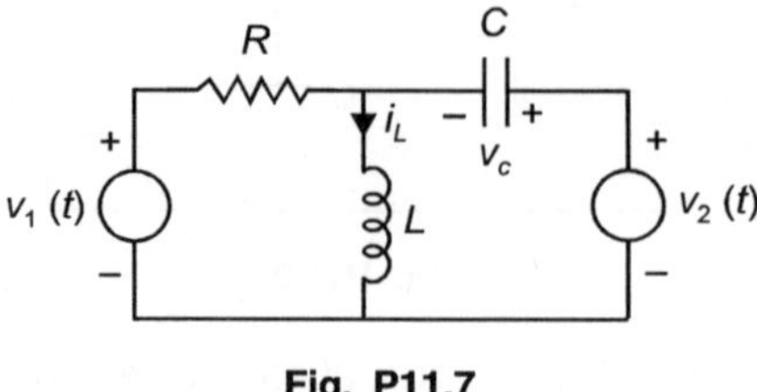

**Fig. P11.7**

**11.8.** For the network shown obtain the state equations taking $v_{c1}$, $vc_2$, and $i_L$ as state variables.

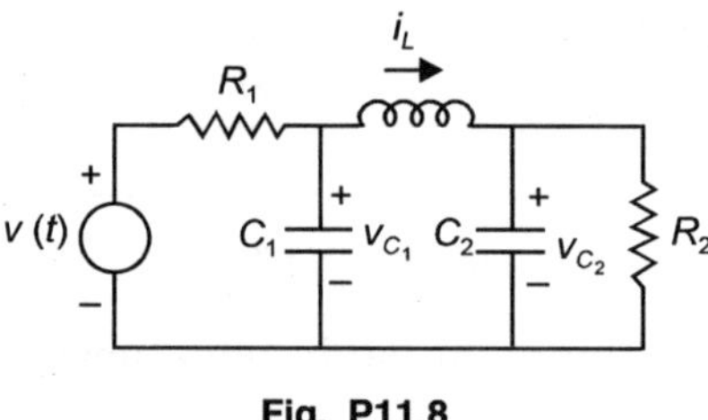

**Fig. P11.8**

**11.9.** For the network shown obtain the state equations taking $i_L$ and $v_c$ as state variables.

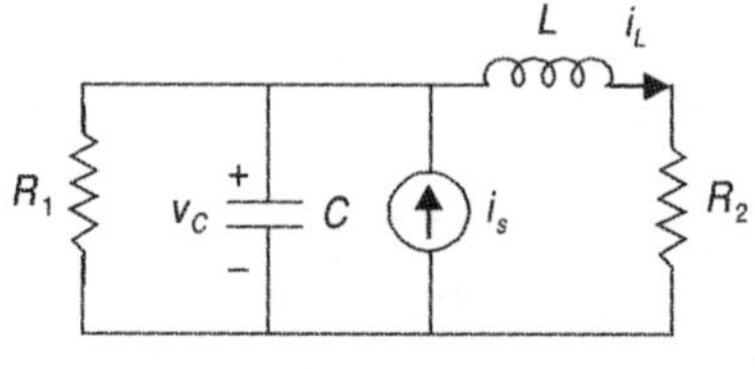

**Fig. P11.9**

**11.10.** For the network shown obtain the state equations.

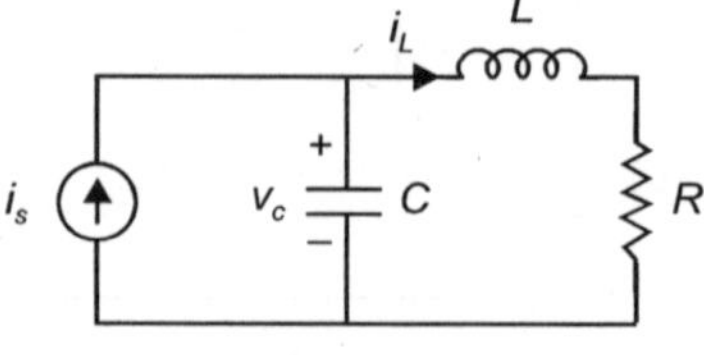

**Fig. P11.10**

**11.11.** For the network shown obtain the state equations.

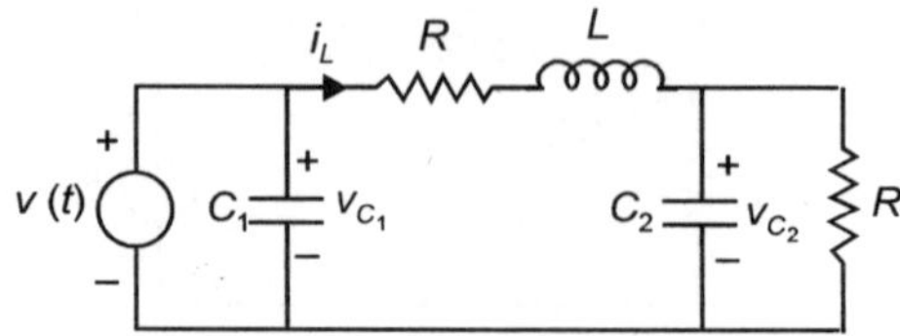

**Fig. P11.11**

**11.12.** For the network shown obtain the state equations.

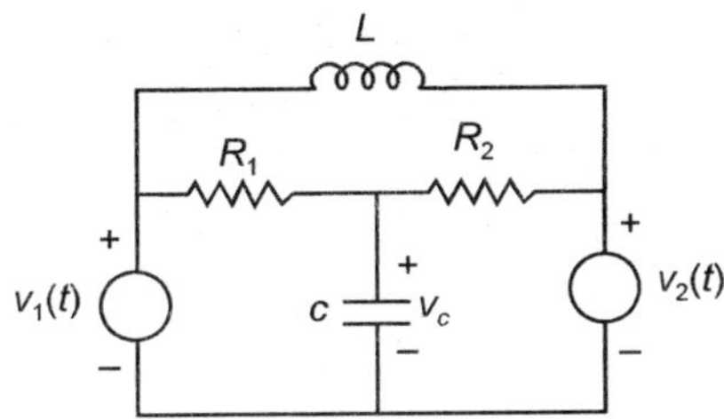



**Fig. P11.12**

**11.13.** A network is described by the following state equations:

$$\begin{bmatrix} \dot{x}_1 \\ \dot{x}_2 \end{bmatrix} = \begin{bmatrix} 1 & 2 \\ -3 & -4 \end{bmatrix} \begin{bmatrix} x_1 \\ x_2 \end{bmatrix} + \begin{bmatrix} 0 \\ 1 \end{bmatrix} u(t)$$

obtain $[x_1 \ x_2]$ if $x_1(0) = 1$   $x_2(0) = 1$

**11.14.** A network is described by the following state equations:

$$\dot{x} = \begin{bmatrix} 0 & 1 \\ 0 & -0.4 \end{bmatrix} \begin{bmatrix} x_1 \\ x_2 \end{bmatrix} + \begin{bmatrix} 0 \\ 2 \end{bmatrix} u$$

If $x_1(0) = 1$ and $x_2(0) = 0$. Obtain the complete solution of the network.

**11.15.** Determine $e^{At}$ for the following $A$–matrices using ($i$) series evaluation ($ii$) Laplace transform ($iii$) Leverrier algorithm ($iv$) Sylvelsters interpolation formulae.

$$(a) \begin{bmatrix} 0 & 2 \\ -1 & -3 \end{bmatrix} \qquad (b) \begin{bmatrix} -1 & 1 \\ 0 & 1 \end{bmatrix} \qquad (c) \begin{bmatrix} -1 & 1 & 0 \\ 0 & -4 & 1 \\ 0 & 0 & -8 \end{bmatrix}$$

**11.16.** A system is described by a second order differential equation.

$\ddot{x} + 5\dot{x} + 6x = u(t)$

Obtain state equations with two different sets of state variables.

**11.17.** A system is described by a third order differential equation.

$\dddot{x} + 7\ddot{x} + 14\dot{x} + 8x = u(t)$

Develop state space representation where the system matrix  is a diagonal one.

**11.18.** A system is represented by the state equation.

$$\dot{x} = Ax$$

If for $\qquad\qquad x(0) = \begin{bmatrix} 1 \\ -3/2 \end{bmatrix}$ the response is

$$x(t) = \begin{bmatrix} e^{-2t} \\ -\dfrac{3}{2} e^{-2t} \end{bmatrix}$$

and for
$$x(0) = \begin{bmatrix} 1 \\ -1 \end{bmatrix} \text{ the response is}$$

$$x(t) = \begin{bmatrix} e^{-t} \\ -e^{-t} \end{bmatrix}$$

Determine the system matrix and the state transition matrix.

# 12

# Attenuators and Classical Filters

## 12.1 INTRODUCTION

Whenever there is a network between the source and the load, the signals like voltage current and power get attenuated as these reach the load end from the source end via the network. These attenuating networks sometimes we have to use, as there is no option e.g. if the source (power plant) is far away from the load, a transmission network should be used to transport the power. In this process there is attenuation in current, voltage (under normal loading) and power. At times even though the source and load are very near to each other still we have to use a network to attenuate the voltage/current e.g. attenuators (potential dividers for voltage and current limiters for current) are used in the laboratory to obtain small values of voltage and current for some specific purposes. Resistance attenuators are also used as volume and mixing controls in broadcasting station.

Since our study is limited only to resistance attenuators we will henceforth not even mention the general attenuating networks having other than resistance elements like the transmission lines where all the four parameters $R$, $L$, $C$ and $G$ are prevalent.

A resistance attenuator is a network consisting of resistor and is designed to reduce by known amounts the voltage, current or power between the source (input) and the load (output) terminals when the network is properly terminated by a resistance. Fig. 12.1 shows the general arrangement of an attenuating network.

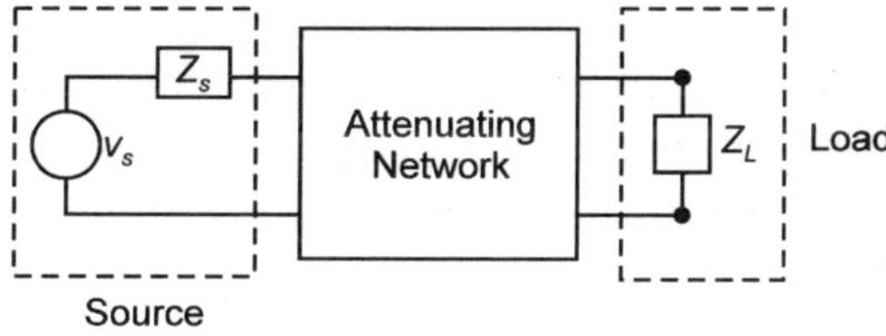


**Fig. 12.1** General arrangement of attenuating network.

When the attenuator network is symmetrical the design specifications are attenuation constant and the characteristic resistance. In order to match impedances between two networks, resistance attenuators are used only if insertion loss can be tolerated. In such cases the attenuation network has to be unsymmetrical and the image impedances rather than the characteristic impedances (resistances) must be specified along with the attenuation constant.

A purely resistive attenuator has a constant loss characteristic and is independent of frequency. An attenuator of constant attenuation is known as a 'pad'.

Since the two important parameters for the design of attenuators, image impedance for unsymmetrical attenuator, characteristic impedance (resistance) for symmetrical attenuator and the attenuation constant for both types of attenuators we study the significance of these terms.

## 12.2  IMAGE AND CHARACTERISTIC IMPEDANCES

Consider a single two terminal pair nework in Fig. 12.2 where the two image impedances $Z_{I1}$ and $Z_{I2}$ are the two values of impedances such that when the terminal marked 2 are terminated in $Z_{I2}$ the input impedance at terminals marked '1' is $Z_{I1}$ and when terminal marked '1' are terminated with $Z_{I1}$ the input impedance at terminals marked '2' is $Z_{I2}$.

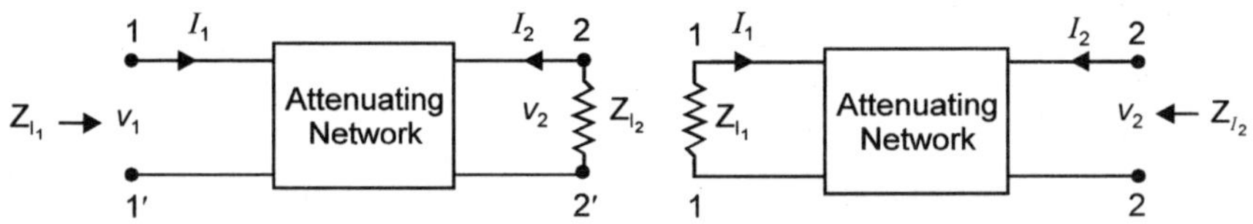

**Fig. 12.2**  Network terminals by image impedances.

Using *ABCD* parameters the two-terminal pair impedances and admittances ($Z_{11}$ $Z_{12}$ . $Y_{11}$ $Y_{12}$ . . .) and certain algebraic manipulation it can be shown that

$$Z_{I1} = \sqrt{Z_{0c1}\, Z_{sc1}} \qquad\qquad ...(12.1)$$

where $Z_{0c1}$ is the input impedance (open circuit) measured at terminals marked '11' when the terminals marked '22' are kept open and $Z_{sc1}$ is the short circuit impedance as measured at terminals '11' when terminals marked '2' are short circuited.

Similarly

$$Z_{I2} = \sqrt{Z_{0c_2}\, Z_{sc_2}} \qquad\qquad ...(12.2)$$

However, if the network is symmetrical

$$Z_{I1} = Z_{I2} = Z_0 = \sqrt{Z_{0c_1}\, Z_{sc1}} = \sqrt{Z_{oc_2}\, Z_{sc_2}} \qquad\qquad ...(12.3)$$

where $Z_0$ is the charactristic impedance (resistance) of the attenuating network. Therefore, the characteristic resistance of a network is defined as that resistance with which a network must be terminated so that the input and terminating resistances are equal. The following example illustrates the concept of $Z_0$.

**Example 12.1:** For the T-network shown obtain $Z_0$ and input impedance if $Z_L = Z_0$

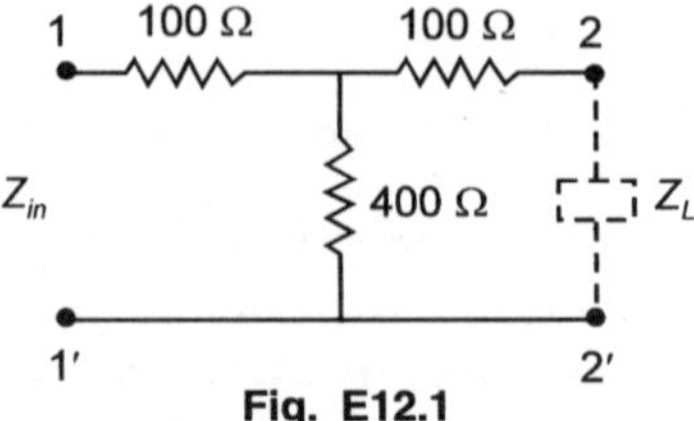

**Fig. E12.1**

**Solution:** Since T-network is a symmetrical network

$$Z_{I1} = Z_{I2} = Z_0 = \sqrt{Z_{oc}\, Z_{sc}}$$

$$Z_{oc} = 100 + 400 = 500\ \Omega$$

$$Z_{sc} = 100 + \frac{100 \times 400}{500} = 180\ \Omega$$

Therefore,
$$Z_0 = \sqrt{500 \times 180}$$
$$= 300\ \Omega$$

If now
$$Z_L = Z_0 = 300\ \Omega$$

$$Z_{in} = \frac{(300 + 100) \times 400}{400 + 400} + 100$$

$$= 300\ \Omega$$

This proves the statement that for a symmetrical network the characteristic resistance is that resistance ($Z_0 = 300\ \Omega$ here) with which if the network is terminated ($Z_L = 300\ \Omega$) then the input impedance ($Z_{in} = 300\ \Omega$) equals the characteristic resistance.

Since it is a symmetrical network, if input terminals '11' are terminating through $Z_0$ and the output impedance is measured at terminals '22' we find that it equals again $Z_0$.

Sometimes it may be necessary to obtain a higher level of attenuation and this may call for connecting more number of symmetrical networks in cascade. Let us now illustrate with an example the effect of cascading similar networks on the characteristic impedance and the input impedance of the network.

**Example 12.2:** Consider two T-networks connected in cascade. Determine $Z_0$ and $Z_{\text{input}}$ when $Z_L = Z_0$.

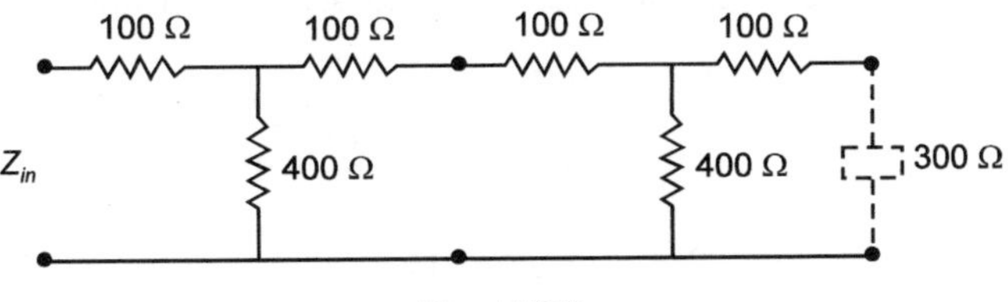


**Fig. E12.2**

**Solution:** Again
$$Z_{0c} = 100 + \frac{600 \times 400}{1000}$$
$$= 340\ \Omega$$

$$Z_{sc} = \left[ \left\{ \frac{(100 \times 400)}{500} + 200 \right\} \| 400 \right] + 100$$

$$= \frac{280 \times 400}{680} + 100$$

$$= 164.7 + 100$$

$$= 264.7\ \Omega$$

Therefore,
$$Z_0 = Z_{oc} \, Z_{sc} = \sqrt{340 \times 264.7} = 300 \ \Omega$$

From Example 12.1 it is clear that a network can be connected between a matched source and the load ($Z_s = Z_L$) to produce attenuation without affecting the matching. In example 12.2 it is shown that when a second identical network is connected in cascade for effecting more attenuation the input impedance (resistance here) does not change. In fact, we may connect any number of similar networks in cascade, $Z_{sc}$ and $Z_{oc}$ will turn towards the characteristic impedance and thus these do not change the input impedance, provided of course, that the network is terminated through $R_0$ (or $Z_0$). This property of attenuation network is effectively used in filter circuits. This gives us an alternate definition of characteristic resistance as one which equals the input impedance of an infinite number of similar symmetrical network connected in cascade.

As mentioned earlier if the attenuation network is asymmetrical this will have two different characteristic impedances known as image impedances. The values are different depending on which end is used as the input. We illustrate with an example.

**Example 12.3:** Fig E12.3 shows an asymmetrical resistance attenuator network. Determine $Z_{I1}$, $Z_{I2}$ and the input impedance if the load end is terminated through a resistance $Z_{I2}$ and the output impedance when the source end is terminated through $Z_{I1}$

**Solution:** Since
$$Z_{I1} = \sqrt{Z_{oc1} \, Z_{sc1}}$$

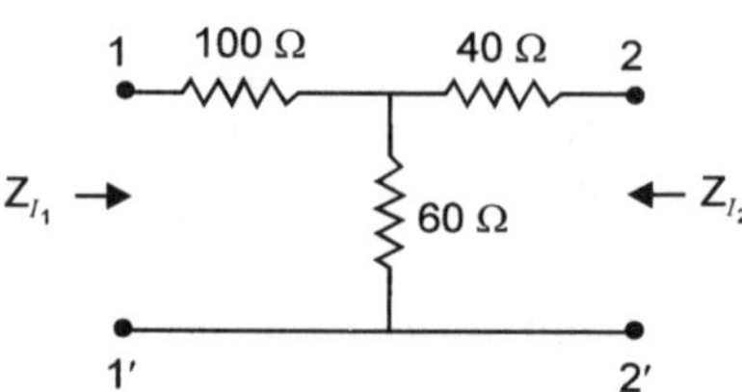


**Fig. E12.3**  Asymmetrical resistance attenuator network.

$$Z_{0c1} = 160 \ \Omega$$

$$Z_{sc1} = 100 + \frac{40 \times 60}{100} = 124\,\Omega$$

Therefore,
$$Z_{I1} = \sqrt{160 \times 124}$$
$$= 140.8 \ \Omega$$

Similarly,
$$Z_{0c2} = 100 \ \Omega$$

$$Z_{sc2} = 40 + \frac{100 \times 60}{160}$$
$$= 77.5 \ \Omega$$

$$Z_{I2} = \sqrt{100 \times 77.5} = 88 \ \Omega$$

$$Z_{\text{input 1}} = 100 + \frac{60 \times 128}{188} = 140.8\,\Omega = Z_{I1}$$

Similarly
$$Z_{\text{output}} = 40 + \frac{240.8 \times 60}{300.8}$$

$$= 40 + 48 = 88 \ \Omega = Z_{I_2}$$

This shows that the image impedance at terminals '1' is the same as the input impedance at terminals '1' when the terminals '2' are terminated by its image impedance $Z_{I2}$ and vice versa.

As mentioned earlier an attenuation network is characterized by its characteristic impedance (Symmetrical network) or image impedances (asymmetrical network) and the propagation function from input to output.

### 12.2.1  Propagation Function

Consider the general transmission network between the source and the load as shown in Fig. 12.3.

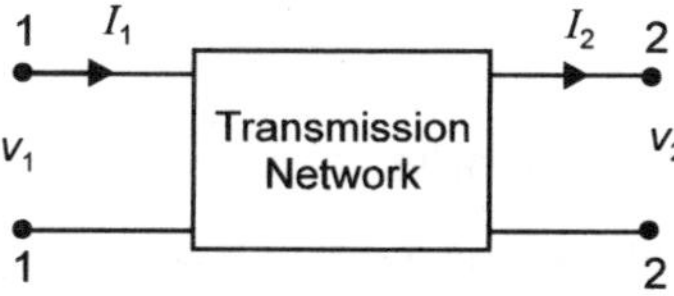


**Fig. 12.3**  Transmission  network.

We know transmission network is a symmetric network and the voltage at the two ends are related as

$$V_1 = V_2 \cos h\gamma + I_2 Z_0 \sin h\gamma \qquad \qquad ...(12.4)$$

$$I_1 = V_2 \frac{\sin h\gamma}{Z_0} + I_2 \cos h\gamma \qquad \qquad ...(12.5)$$

Now using the relation $V_2 = I_2 Z_0$
we have

$$\frac{V_1}{V_2} = \cos h\gamma + \sin h\gamma = e^\gamma = \frac{I_1}{I_2} \qquad \qquad ...(12.6)$$

Here $\gamma$ is known as a propagation function and is given as

$$\gamma = \alpha + j\beta \qquad \qquad ...(12.7)$$

where $\alpha$ is known as the attenuation constant or function and $\beta$ the phase function. For resistance networks $\beta = 0$ and hence $\gamma$ is a real number and it means that when a signal is propagated from input terminals to output terminals through a resistive network, its magnitude reduces by a factor $e^\alpha$ where $\alpha$ is the attenuation constant of the network. Phase function $\beta$ means there is a change in phase of the quantity (V and current) as it propagates from input to output terminals.

If there are $(n - 1)$ number of circuits with propagation function $\gamma_1, \gamma_2 \ldots \gamma_{n-1}$ connected in cascade, the ratio of the input voltage $V_1$ at the first circuit and $V_n$ the output voltage at the output terminals of $(n - 1)$th circuit is given as

$$\frac{V_1}{V_n} = e^{(\gamma_1 + \gamma_2 + \cdots + \gamma_{n-1})} = e^\gamma \qquad \qquad ...(12.8)$$

This means the propagation function of cascaded networks is the sum of the propagation of individual networks.

Also it is seen from equation (12.4) that $\cosh\gamma = A$ and is the ratio of input voltage to output voltage when there is no load at the output terminals.

Since
$$\cos\, h\gamma = A \qquad\qquad\qquad\text{...(12.9)}$$

$$\sinh\, \frac{\gamma}{2} = \sqrt{\frac{A-1}{2}} \qquad\qquad\qquad\text{...(12.10)}$$

$$\cosh\, \frac{\gamma}{2} = \sqrt{\frac{A+1}{2}} \qquad\qquad\qquad\text{...(12.11)}$$

$$\tanh\, \frac{\gamma}{2} = \sqrt{\frac{A-1}{A+1}} \qquad\qquad\qquad\text{...(12.12)}$$

### 12.2.2 Balanced and Unbalanced Networks

We know that a symmetric network is one which has input and output impedance equal or $A = D$. A network may also be classified as a balanced or unbalanced network. Fig. 12.4 (a) shows an unbalanced whereas 12.4 (b) the balanced network. Fig. 12.4 (a) is said to be unbalanced because the impedance between one input terminal and one output terminal (lower ones) is zero whereas the impedance between one input and one output terminal (top) is $Z_1$. This is a completely unbalanced network. The lower input and output terminals in this case are usually at ground potential. A balanced symmetrical network is shown in Fig. 12.4 (b). Here the impedance is divided equally between the upper and lower terminals.

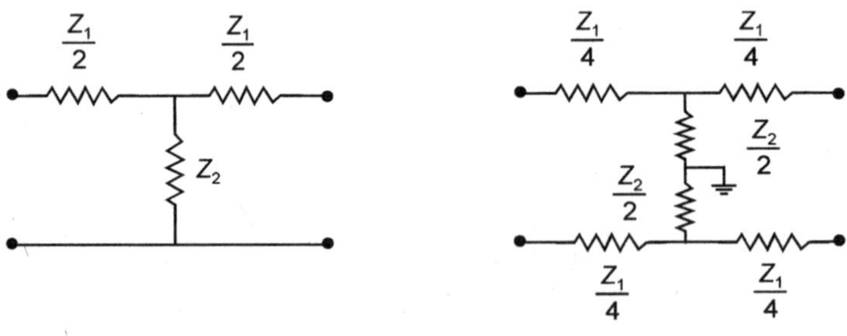

<table>
<tr><td>(a)  T-unbalanced.</td><td>(b)  H-network balanced.</td></tr>
</table>

**Fig. 12.4**  Symmetrical network.

However, it is to be noted that the characteristic impedance and propagation function for both the networks are same.

## 12.3  LATTICE ATTENUATION

The Lattice is the general type of symmetrical and balanced network. Any symmetrical balanced or unbalanced network can be transformed into an equivalent Lattice. However, the inverse of the statement is not true. We, therefore, derive design equation for the lattice network and the same equation would be used for $T-$, $\pi-$ or bridged networks after finding equivalent Lattice network using what is known as bisection theorem.

Fig. 12.5 shows a Lattice structure terminated through characteristic resistance $R_0$.

Fig. 12.5 (a) is essentially a bridge network where opposite sides are equal 11′ terminals are input and 22′ are the output where $R_0$ the characteristic resistance is shown connected.

We know that $R_0 = \sqrt{R_{sc}\,R_{0c}}$ where $R_{sc}$ is the input resistance when output terminals are short circuited and $R_{oc}$ the input resistance when the output terminals 22′ are open circuited.

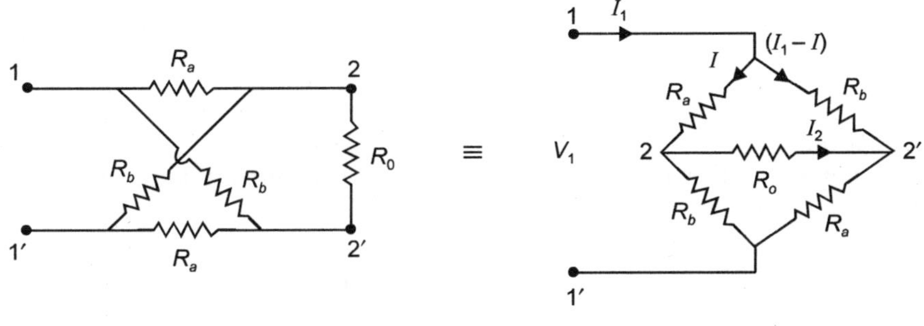

(a) Lattice structure         (b) Equivalent of Fig. 12.5(a)

**Fig. 12.5**

$$R_{sc} = \frac{2R_a R_b}{R_a + R_b} \qquad \qquad ...(12.13)$$

and
$$R_{oc} = \frac{R_a + R_b}{2} \qquad \qquad ...(12.14)$$

Hence
$$R_0 = \sqrt{R_a R_b} \qquad \qquad ...(12.15)$$

To find out propagation constant or attenuation constant here as for resistive network, phase constant is zero, we write the following equation from Fig. 12.5 (b). Since the attenuator is terminated by its characteristic resistance $R_0$ and since it is a symmetric network.

$$V_1 = I_1 R_0 = (I_1 - I)\,R_b - I_2 R_0 + (I - I_2)\,R_b \qquad ...(12.16)$$
$$= I_1 R_b - I_2\,(R_0 + R_b)$$

or
$$I_1(R_0 - R_b) = -I_2\,(R_0 + R_b) \qquad ...(12.17)$$

or
$$\frac{I_1}{I_2} = \frac{R_0 + R_b}{R_b - R_0} = n \qquad ...(12.18)$$

or
$$R_b = R_0\,\frac{n+1}{n-1} \qquad ...(12.19)$$

and
$$R_a = R_0\,\frac{n-1}{n+1} \qquad ...(12.20)$$

Now
$$\frac{I_1}{I_2} = \frac{V_1}{V_2} = e^{\alpha} = \frac{1 + R_a/R_0}{1 - R_a/R_0} \qquad ...(12.21)$$

We obtain this relation from Fig. 12.5 (b) as follows:

$$I_1 R_0 = I R_a + I_2 R_0 + (I_1 - I + I_2)\,R_a$$

Hence

$$\frac{I_1}{I_2} = e^{\alpha} = \frac{1 + R_a/R_0}{1 - R_a/R_0}$$

Since

$$R_0 = \sqrt{R_a\, R_b}$$

Hence

$$\frac{I_1}{I_2} = e^{\alpha} = \frac{1 + \sqrt{R_a/R_b}}{1 - \sqrt{R_a/R_b}} \qquad \text{...(12.22)}$$

Therefore

$$\alpha = \ln \frac{V_1}{V_2} = 2.303 \log_{10} \frac{V_1}{V_2} \qquad \text{...(12.23)}$$

where $\alpha$ is in nepers.

If $P_1$ is the power at the input terminals, $P_2$ at the output terminals, attenuation is defined as

Attenuation

$$= \log_{10} = \frac{P_1}{P_2} \text{ Bels} \qquad \text{...(12.24)}$$

However, a smaller unit decibel is normally used for attenuation hence

Attenuation in

$$db = 10 \log_{10} \frac{P_1}{P_2} \qquad \text{...(12.25)}$$

Since

$$P_1 = \frac{V_1^2}{R_0} \quad \text{and} \quad P_2 = \frac{V_2^2}{R_0}$$

Therefore, Attenuation

$$= 10 \log_{10} \left(\frac{V_1}{V_2}\right)^2$$

$$= 20 \log_{10} \frac{V_1}{V_2} \, dbs \qquad \text{...(12.26)}$$

Since

$$\frac{V_1}{V_2} = e^{\alpha} = n$$

Therefore, Attenuation in $\quad db = 20 \log_{10} n$

or

$$n = \text{anti log} \left(\frac{db}{20}\right) \qquad \text{...(12.27)}$$

Since attenuation in neper is given by

$$\alpha = 2.303 \log_{10} \frac{V_1}{V_2}$$

and that in

$$dB = 20 \log_{10} \frac{V_1}{V_2}$$

Therefore,

$$\frac{\text{neper}}{2.303} = \frac{dB}{20}$$

or

$$dB = \frac{20}{2.303} \text{ neper}$$

$$= 8.686 \text{ neper} \qquad \text{...(12.28)}$$

or
$$\text{neper} = \frac{dB}{8.686} = 0.115 \; dB.$$

Therefore, 1 neper of attenuation equals 8.686 dB and 1 dB means 0.115 neper.

The reason for using logarithmic for attenuation is that it is easier to calculate the overall attenuation of cascaded network, if the attenuation for individual networks is known, as then it is the sum of attenuation of individual networks.

Having obtained the design equation for Lattice attenuation we now state bisection theorem and without going through its proof we apply this to find equivalent lattice networks for $\pi$, $T$ and bridge-T networks and make use of the design equation to find the desired values of resistances required for a given attenuation.

### 12.3.1 Bisection Theorem

A network is said to have been bisected when the open circuited and short circuited input impedances of the two bisected networks are equivalent. Also the square root of the product of these impedances is the characteristic impedance of the original whole network.

Bisection theorem states that any symmetrical balanced or unbalanced network can be transformed into an equivalent lattice. The series arm of the lattice is equal to the short circuited resistance of the bisected network and the diagonal arm $R_b$ is the open-circuited impedance of the bisected network. The transformation is always possible when all the elements of $T - \pi$ or bridged $- T$ are of the same kind i.e. all resistors or all inductors etc, otherwise the transformation may not be physically realisable because of the need for negative resistances etc.

### 12.3.2 T-type Attenuator

T-type attenuator is a symmetrical but unbalanced circuit and is shown

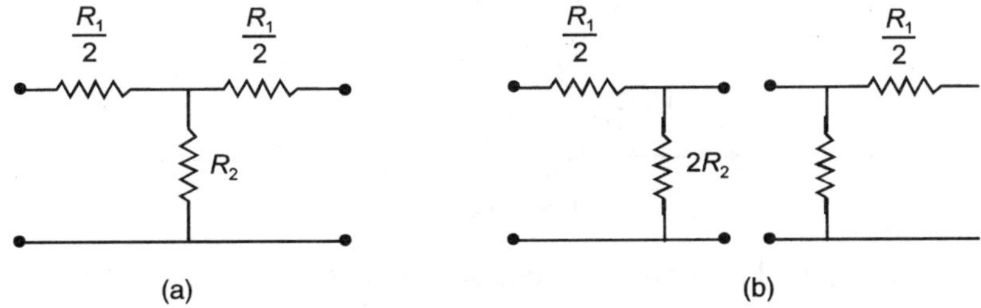

**Fig. 12.6**   (a) T-type attenuator, (b) Bisected of (a).

in Fig. 12.6 (a) and its bisected network is shown in Fig. 12.6 (b). Let $Z_{hsc}$ and $Z_{hoc}$ be the half section short circuit and open circuit impedance of the bisected network. As mentioned earlier the series arm of the lattice network equals the short circuit input resistance of the bisected network whereas the diagonal arm the $Z_{hoc}$.

Also using equations (12.19) and (12.20) we have

$$R_a = \frac{R_1}{2} = Z_{hsc} = R_0 \frac{n-1}{n+1} \qquad \qquad \text{...(12.29)}$$

and
$$R_b = \frac{R_1}{2} + 2R_2 = Z_{hoc} = R_0 \frac{n+1}{n-1} \qquad \qquad \text{...(12.30)}$$

Therefore the design equations are:

$$R_1 = 2R_0 \frac{n-1}{n+1} \qquad \qquad \qquad \text{...(12.31)}$$

and

$$2\,R_2 = R_0 \left[ \frac{n+1}{n-1} - \frac{n-1}{n+1} \right]$$

or

$$R_2 = \frac{R_0}{2} \left[ \frac{4n}{n^2-1} \right] = \frac{2n\,R_0}{n^2-1} \qquad \qquad \text{...(12.32)}$$

These are the design equations for the T-type attenuator i.e. given the characteristic resistance of the desired attenuator and its attenuation the resistances required can be obtained. Following example illustrates the procedure for Lattice and T-type attenuators.

**Example 12.4:** Design a lattice and a T-type attenuator if the characteristic resistance is 200 ohm and the attenuation 20dB.

**Solution:** Since attenuation is 20dB.

$$n = \text{anti log}_{10}\ \frac{dB}{20} = \text{anti log}\ \frac{20}{20} = 10$$

*(i) For Lattice attenuator*

$$R_a = R_0\ \frac{n-1}{n+1} = 200 \times \frac{9}{11} = 163.6\,\Omega$$

and

$$R_b = 200 \times \frac{11}{9} = 244.5\,\Omega$$

The lattice attenuators is shown here

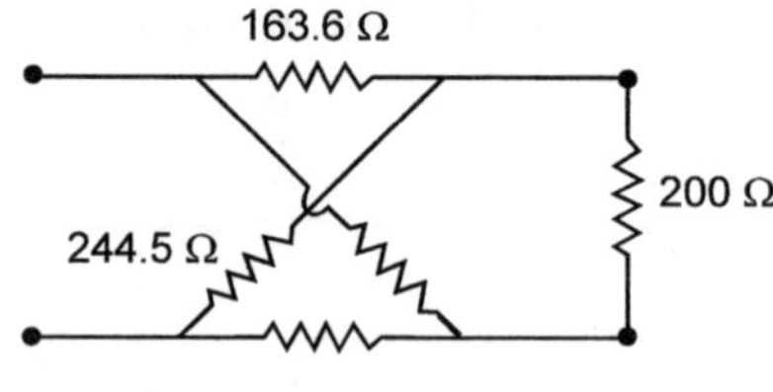


**Fig. E12.4.1**

*(ii) T-type attenuator:* The value of resistances are:

$$R_1 = 2\,R_0\ \frac{n-1}{n+1} = 2 \times 200 \times \frac{9}{11}$$
$$= 327.2\ \Omega$$

and

$$R_2 = \frac{R_0}{2} \cdot \frac{4n}{n^2-1} = \frac{200 \times 20}{99} = 40.4\,\Omega$$

The T-type attenuator is

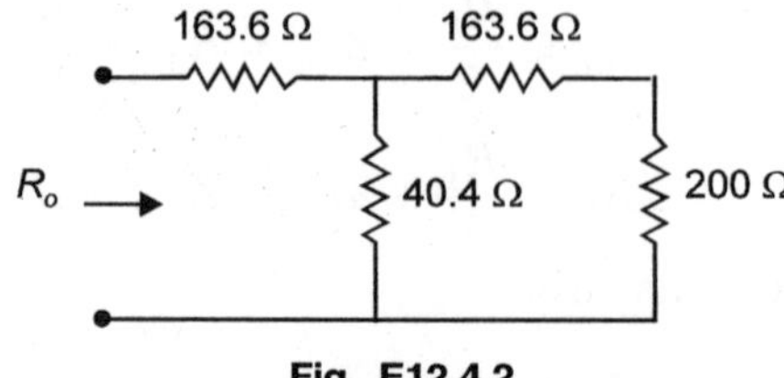


**Fig. E12.4.2**

Next we consider $\pi$-type attenuator

### 12.3.3 $\pi$-type Attenuator

Using bisection theorem we first bisect the $\pi$-network (Fig. 12.7(a)).

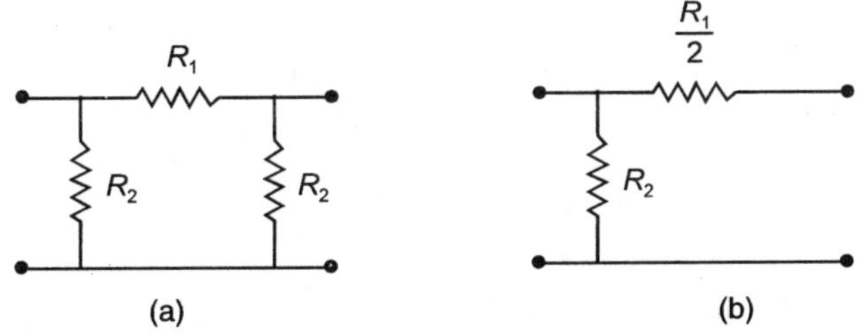



**Fig. 12.7**   (a) $\pi$ – network (b) half section of bisected network (a).

$$R_a = R_{hsc} = \frac{(R_1/2) \cdot R_2}{\dfrac{R_1}{2} + R_2} = \frac{R_1 R_2}{R_1 + 2R_2} = R_0 \frac{n-1}{n+1} \qquad ...(12.33)$$

and 
$$R_b = R_2 = R_0 \frac{n+1}{n-1} \qquad ...(12.34)$$

To obtain $R_1$ we eliminate $R_2$ from these equations and substitute value of $R_2$ from (12.34) into (12.33)

$$\frac{R_1 R_0 \dfrac{n+1}{n-1}}{R_1 + 2R_0 \dfrac{n+1}{n-1}} = R_0 \frac{n-1}{n+1} \qquad ...(12.35)$$

$$R_1 \left(\frac{n+1}{n-1}\right)(n+1) = \left(R_1 + 2R_0 \frac{n+1}{n-1}\right)(n-1)$$

$$R_1 \left[\frac{(n+1)^2}{n-1} - (n-1)\right] = 2 R_0(n+1)$$

$$R_1 \frac{4n}{n-1} = 2 R_0(n+1)$$

or 
$$R_1 = \frac{R_0}{2}\left(\frac{n^2-1}{n}\right) \qquad ...(12.36)$$

The $\pi$-type attenuator is shown here

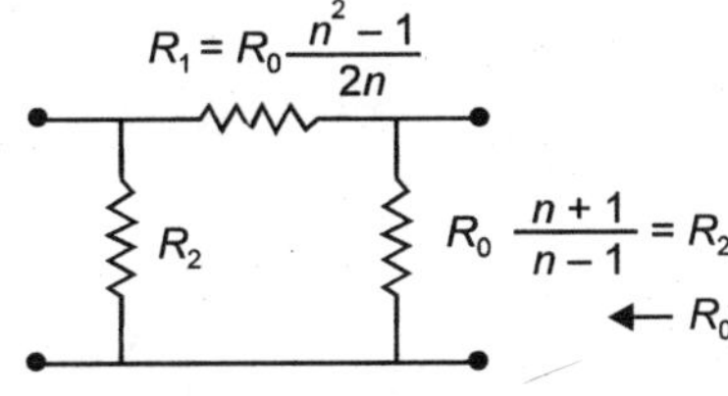



**Fig. 12.8**   $\pi$-type attenuator

**Example 12.5:** For the specification given as in example 12.4 obtain a $\pi$-type attenuator.

$$R_2 = R_0 \frac{n+1}{n-1} = 200 \times \frac{11}{9} = 244.5\,\Omega$$

$$R_1 = 200 \times \frac{100-1}{2 \times 10} = 10 \times 99 = 990\ \Omega$$

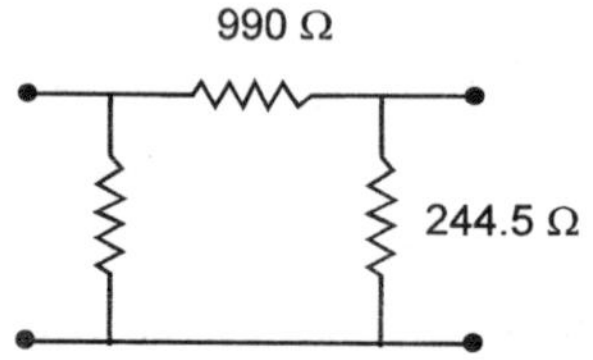

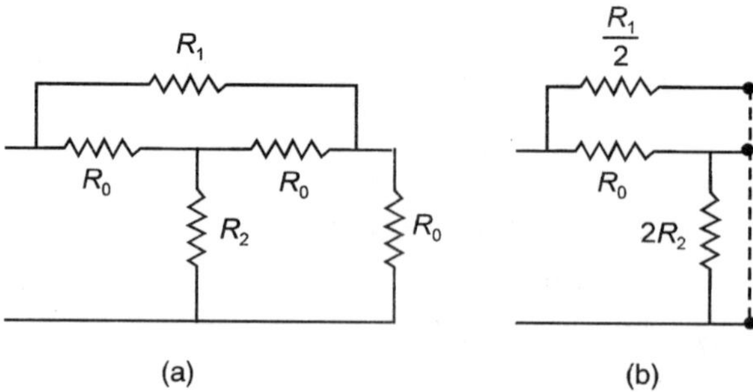


**Fig. E12.5.1**

### 12.3.4  Bridged T - type Attenuator

Bridged T circuit is also a symmetrical network, hence bisection principle is used to bisect the network.

$$R_a = R_{hsc} = R_0 \frac{n-1}{n+1} = \frac{R_0\,R_1/2}{R_0 + R_1/2} \qquad \text{...(12.37)}$$

or
$$\frac{n-1}{n+1} = \frac{R_1}{R_1 + 2R_0}$$

or
$$(n+1)\,R_1 = (n-1)\,(R_1 + 2\,R_0)$$

(a)                    (b)

**Fig. 12.9**  (a) Bridged T network (b) Bisected half section of (a)

$$(n + 1 - n + 1)\,R_1 = 2(n-1)\,R_0$$

or
$$R_1 = (n-1)\,R_0 \qquad \text{...(12.38)}$$

Now

$$R_b = R_{hoc} = R_0 \frac{n+1}{n-1} = R_0 + 2\,R_2$$

or
$$2\,R_2 = R_0\left[\frac{n+1}{n-1} - 1\right] = \frac{2R_0}{n-1}$$

or
$$R_2 = \frac{R_0}{n-1} \qquad \qquad ...(12.39)$$

It is to be noted that if T- or $\pi$-type attenuators are to be made variable, all the three resistances must be varied simultaneously whereas in bridged T-type attenuator only two resistances used be made variable.

**Example 12.6:** Taking the specification of previous example the bridged T-attenuator will have the following resistances:

$$R_1 = (n - 1) \, R_0 = 9 \times 200 = 1800 \; \Omega$$

$$R_2 = \frac{200}{9} = 22.2 \; \Omega$$

The attenuator is shown here

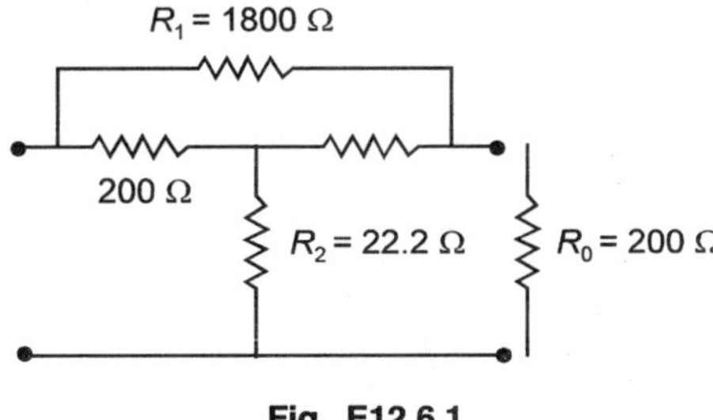

**Fig. E12.6.1**

So far we have considered symmetrical unbalanced attenuators. Balanced version of these networks can be designed by placing half of the series resistances in each line and using the same formulae. The balanced version of the above network is given here Fig. E12.6.2. Next we consider the asymmetrical attenuator networks. These are:

1. L-type
2. T-type

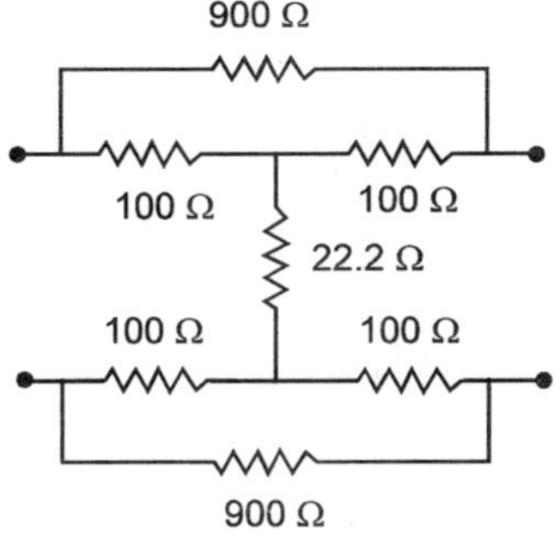

**Fig. E12.6.2**   Balanced version of Fig. E12.6.1

## 12.3.5 L-type Attenuator

Normally an asymmerical attenuator network is specified by the image impedances (resistances) and attenuation. L-type attenuator can be designed by a single or same image impedance for input and output circuit and attenuation. It can be designed by two different image impedances for the input and output terminals. This attenuator requires only two variable resistances but maintains a constant resistance in only one direction. The two values of resistances are denoted by $R_{I1}$ and $R_{I2}$ respectively in Fig. 12.10.

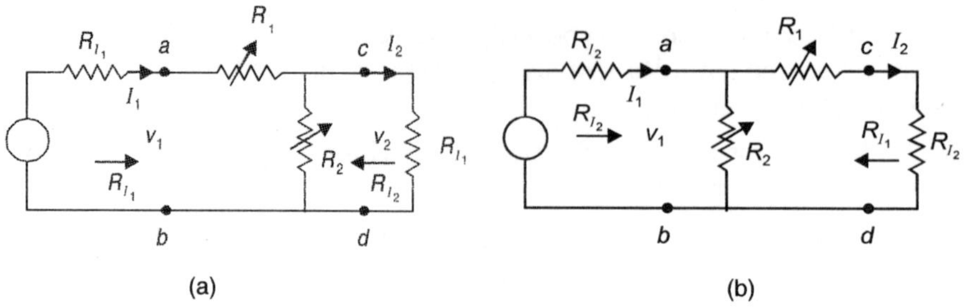

**Fig. 12.10**  Two versions of L-type attenuators.

Considering Fig. 12.10 (a) and using KVL for the loop containing $R_{I1}$ we have

$$I_2 R_{I1} + (I_2 - I_1) R_2 = 0 \qquad \text{...(12.40)}$$

or
$$I_2 (R_{I1} + R_2) = I_1 R_2$$

or
$$\frac{I_1}{I_2} = n = \frac{R_{I1} + R_2}{R_2} \qquad \text{...(12.41)}$$

or
$$R_2 = \frac{R_{I1}}{n-1} \qquad \text{...(12.42)}$$

Now the input impedance between point $a$ $b$

$$R_{I1} = R_1 + \frac{R_2 R_{I1}}{R_2 + R_{I1}} \qquad \text{...(12.43)}$$

$$R_1 = R_{I1} - \frac{R_2 R_{I1}}{R_2 + R_{I1}} = R_{I1}\left(1 - \frac{R_2}{R_2 + R_{I1}}\right)$$

Since
$$\frac{R_2}{R_2 + R_{I1}} = \frac{1}{n}$$

Therefore
$$R_1 = R_{I1}\left(1 - \frac{1}{n}\right) = \frac{n-1}{n} R_{I1} \qquad \text{...(12.44)}$$

These are the design equations for L-type attenuator when the L-shape faces the input terminals.

If it faces the output terminals as shown in Fig. 12.10(b) the analysis is as follows:
Writing down the KVL equation for the loop not containing voltage source.

$$I_2 (R_1 + R_{I2}) + (I_2 - I_1) R_2 = 0 \qquad \text{...(12.45)}$$

or
$$I_2 (R_1 + R_{I2} + R_2) = I_1 R_2$$

or
$$\frac{I_1}{I_2} = n = \frac{R_1 + R_{I2} + R_2}{R_2} \qquad \text{...(12.46)}$$

$$R_1 = nR_2 - R_2 - R_{I2} \qquad \text{...(12.47)}$$

Also the input impedance between $a\ b$ towards the load side is

$$R_{I2} = \frac{R_2\,(R_1 + R_{I2})}{R_2 + R_1 + R_{I2}} \qquad \text{...(12.48)}$$

Substituting for $R_1$ in this expression, we have

$$R_{I2} = \frac{R_2\,(n\,R_2 - R_2 - R_{I2} + R_{I2})}{R_2 + n\,R_2 - R_2 - R_{I2} + R_{I2}}$$

$$= \frac{R_2^2\,(n-1)}{n\,R_2}$$

or $\qquad\qquad R_2 = \dfrac{n}{n-1}\,R_{I2} \qquad\qquad$ ...(12.49)

Therefore, $\qquad R_1 = n\,R_2 - R_2 - R_{I2}$

$$= R_2\,(n-1) - R_{I2}$$

$$= \frac{n}{n-1}\,(n-1)R_{I2} - R_{I2}$$

$$= (n-1)\,R_{I2} \qquad \text{...(12.50)}$$

We have considered L-type attenuator where the source impedance, load impedance and the image impedance looked into the network from input terminals are all three equal. In case, the input source impedance is different from the load impedance, however, the image impedances are equal to the impedance corresponding to input terminals and output terminal i.e. if $R_{i1}$ is the source resistance, the image resistance of the network looked from the input terminals is $R_{i1}$. Similarly if load resistance is $R_{i2}$, the image resistance looked from the output terminals is $R_{i2}$ as shown in Fig. 12.11.

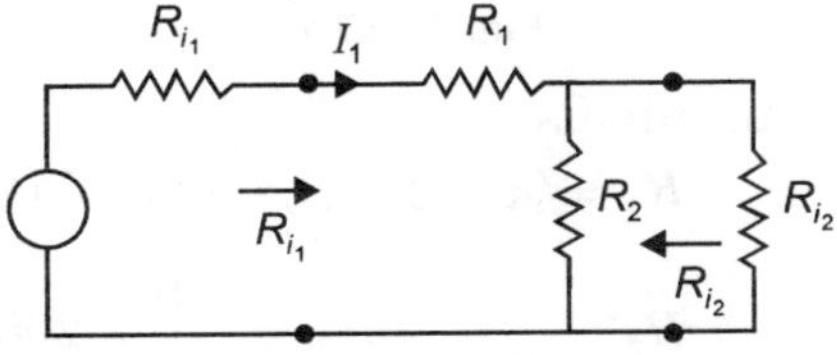


**Fig. 12.11**   Minimum loss network (L-Type)

This network is also known as minimum loss network and is used to provide matching between unequal source and load resistances with minimum attenuation for a resistive network.

Since there are two unknowns we can write down one equation each for the image impedances.

$$R_{i1} = R_1 + \frac{R_2\,R_{i2}}{R_2 + R_{i2}} \qquad \text{...(12.51)}$$

and $\qquad\qquad R_{i2} = \dfrac{R_2\,(R_{i1} + R_1)}{R_1 + R_2 + R_{i1}} \qquad\qquad$ ...(12.52)

After algebraic manipulations, we have

$$R_1 = \sqrt{R_{i1}\,(R_{i1} - R_{i2})} \qquad\qquad ...(12.53)$$

and
$$R_2 = \sqrt{\dfrac{R_{i1}\,R_{i2}^2}{R_{i1} - R_{i2}}} \qquad\qquad ...(12.54)$$

From the expression for $R_1$ and $R_2$, it is clear that such an L-type attenuator can be realized when

$$R_{i1} > R_{i2}$$

and the L-shape should be towards the source end.

**Example 12.7:** Design an L-section attenuator which has input resistance, load resistance and the image resistance looked from the input terminals as 200 $\Omega$ and the attenuation is 10 when the L-section face (*a*) Input terminals (*b*) Output terminals.

**Solution:** (*a*) L-section faces input terminals

$$R_1 = \dfrac{n-1}{n} \times 200 = \dfrac{10-1}{10} \times 200 = 180\ \Omega$$

and
$$R_2 = \dfrac{R_{i1}}{n-1} = \dfrac{200}{9} = 22.22\ \Omega$$

The L-type attenuator is shown here

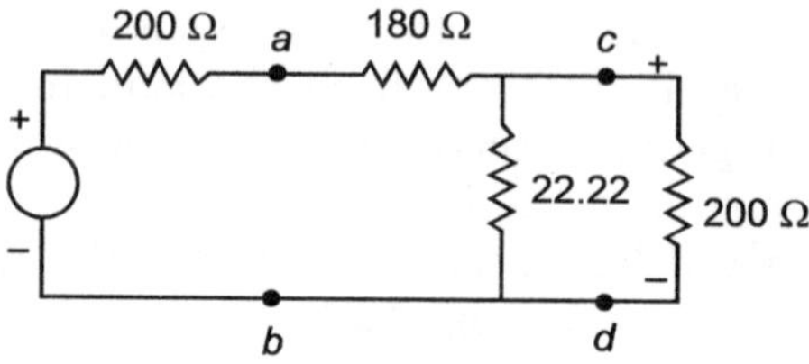



**Fig. E12.7.1**

(*b*) L-section faces output terminals

$$R_1 = (n-1)\,R_{i2} = (10-1) \times 200 = 1800\ \Omega$$

$$R_2 = \dfrac{n}{n-1}\,R_{I2} = \dfrac{10}{9} \times 200 = 222.22\ \Omega$$

The attenuator is shown here

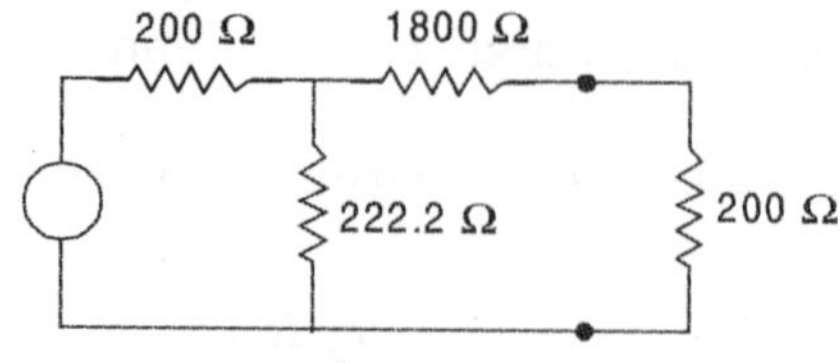



**Fig. E12.7.2**

**Example 12.8:** Design a minimum loss (L-section) resistive network to match a source impedance of 1200 $\Omega$ to a load impedance of 900 ohms. Also calculate the transmission loss of this network.

**Solution:** Since the load impedance and source impedance are different.

$$R_1 = \sqrt{R_{I1}(R_{I1} - R_{I2})} = \sqrt{1200 \times (1200 - 900)}$$

$$= \sqrt{1200 \times 300}$$

$$= 600 \ \Omega$$

$$R_2 = \sqrt{\frac{1200 \times 900^2}{1200 - 900}} = \sqrt{1200 \times 900 \times 3}$$

$$= 1800 \ \Omega$$

The attenuator is shown here

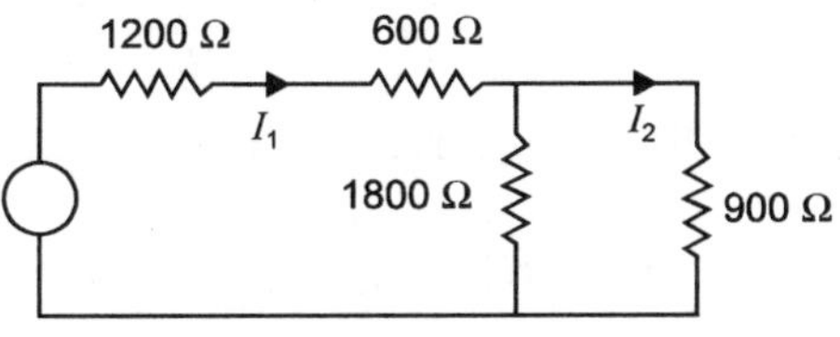


**Fig. E12.8**

Now
$$I_2 = I_1 \frac{1800}{2700}$$

or
$$\frac{I_1}{I_2} = \frac{2700}{1800} = 1.5$$

Input power $\quad = I_1^2 \ R_{i1}$

Output power $\quad = I_2^2 \ R_{i2}$

Therefore, Transmission loss $= 10 \ \log_{10} \dfrac{P_1}{P_2}$

$$= 10 \ \log_{10} \frac{I_1^2 \ R_{i1}}{I_2^2 \ R_{i2}}$$

$$= 10 \ \log_{10} \left( 1.5^2 \times \frac{1200}{900} \right)$$

$$= 10 \ \log_{10} 3$$

$$= 4.77 \text{ dbs } \textbf{Ans.}$$

### 12.3.6  Asymmetrical T-type Attenuator

Fig. 12.12 shows an asymmetrical T-type attenuator which has input and output image
impedances as $R_{i1}$ and $R_{i2}$ respectively.

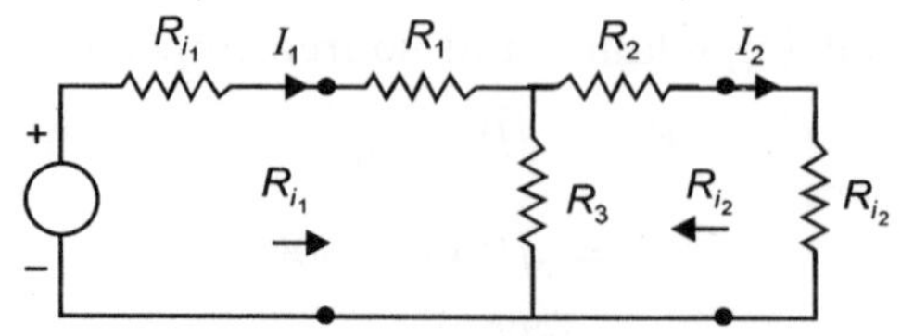

**Fig. 12.12** Asymmetrical T-type.

Applying KVL for the loop not containing the source.

$$I_2\,(R_2 + R_{i2}) + (I_2 - I_1)\,R_3 = 0 \qquad \text{...(12.55)}$$

or

$$I_2\,(R_2 + R_3 + R_{i2}) = I_1\,R_3$$

or

$$\frac{I_1}{I_2} = \frac{R_2 + R_3 + R_{i2}}{R_3} = n \qquad \text{...(12.56)}$$

Now writing down equation for image resistances we have

$$R_{i1} = R_1 + \frac{(R_2 + R_{i2})\,R_3}{R_2 + R_3 + R_{i2}} \qquad \text{...(12.57)}$$

Similarly,

$$R_{i2} = R_2 + \frac{(R_1 + R_{i1})\,R_3}{R_1 + R_3 + R_{i1}} \qquad \text{...(12.58)}$$

Solution of these three equations (12.56) to (12.58) provides

$$R_3 = \frac{2n\,\sqrt{R_{i1}\,R_{i2}}}{n^2 - 1} \qquad \text{...(12.58)}$$

$$R_1 = \frac{n^2 + 1}{n^2 - 1}\,R_{i1} - R_3 \qquad \text{...(12.59)}$$

and

$$R_2 = \frac{n^2 + 1}{n^2 - 1}\,R_{i2} - R_3 \qquad \text{...(12.60)}$$

**Example 12.9:** Design an asymmetrical T-type attenuator with 20 db attenuation and having source and load impedances as 200 ohms and 800 ohms respectivley.

**Solution:** Given $R_{i1} = 200\ \Omega$ and $R_{i2} = 800\ \Omega$

$$n = \text{anti}\,\log_{10}\frac{db}{20} = \text{antilog}_{10}\,1 = 10$$

$$R_3 = \frac{2 \times 10\,\sqrt{200 \times 800}}{100 - 1} = \frac{20 \times 400}{99} = 80.8\ \Omega \quad \textbf{Ans.}$$

$$R_1 = \frac{101}{99} \times 200 - 80.8 = 123.2\ \Omega \quad \textbf{Ans.}$$

$$R_2 = \frac{101}{90} \times 800 - 80.8 = 735.36\ \Omega \ \textbf{Ans.}$$

### 12.3.7  Ladder Type Attenuator

In order to provide attenuation in steps, a number of symmetrical T or $\pi$ networks are connected in cascade. The resultant network is known as ladder-type attenuator. Fig. 12.13 shows a ladder type attenuator with three identical and symmetrical $\pi$ section connected in cascade. Here the output terminals are terminated with resistance $R_0$ which is the characteristic resistance of the network. The source end also has resistance $R_0$ so that it is a matched circuit. One of the input terminals is connected to any intermediate points depending upon the attenuation required. The net attenuation is the sum of the attenuation of individual sections between the input and output terminals, However, at any of these numbered points when looked in either directions, the resistance equals the characteristic resistance $R_0$.

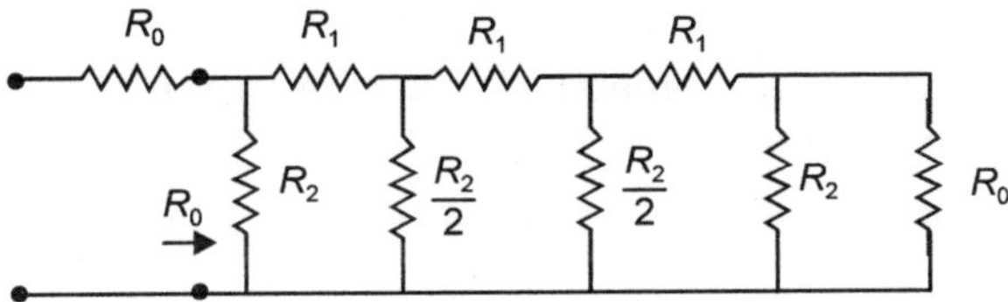

**Fig. 12.13**  Ladder type attenuator.

**Example 12.10:** Design a ladder network for a load resistance of 200 $\Omega$ and an attenuation of 5 dbs per step.

**Solution:** $n = \text{anti}\log_{10}\dfrac{db}{20} = \text{anti}\log_{10}\dfrac{5}{20} = 1.778$

$$R_1 = R_0\,\frac{n^2 - 1}{2n} = 200 \times \frac{2.16}{2 \times 1.778} = 122\ \Omega$$

$$R_2 = R_0\,\frac{n+1}{n-1} = 200 \times \frac{2.778}{0.778} = 714\ \Omega$$

Therefore ladder network is

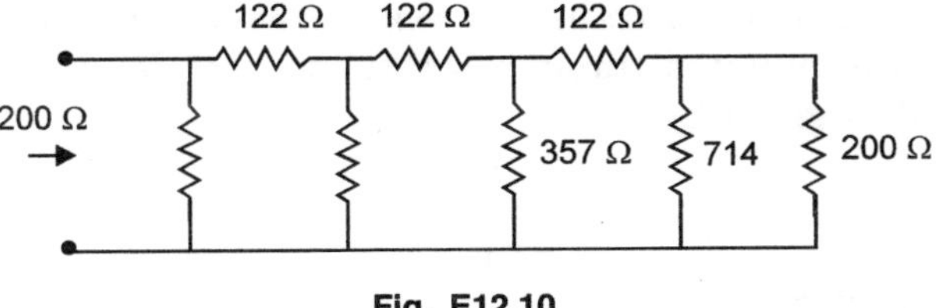

**Fig. E12.10**

### 12.3.8  $\pi$-type asymmetrical Attenuator

Fig 12.14 shows a $\pi$-type attenuator where $R_{I1}$ and $R_{I2}$ are the input and output image resistances respectively.

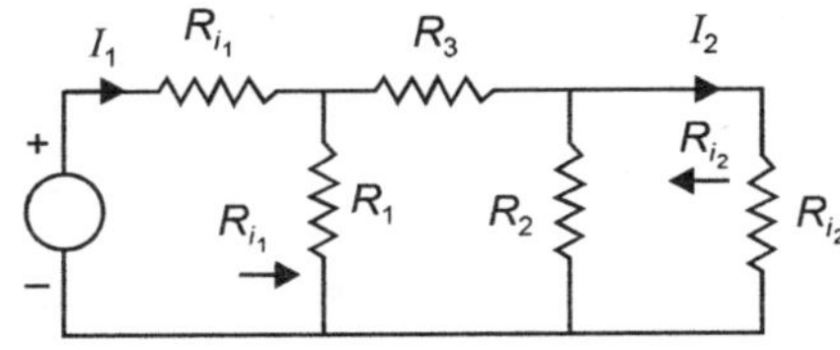

**Fig. 12.14**  $\pi$-type asymmetrical attenuator.

After writing down equations for $I_1$ and $I_2$ we obtain the ratio $\dfrac{I_1}{I_2} = n$ and hence write equation for $R_{I1}$ and $R_{I2}$ and simplifying we obtain the following relations:

$$G_3 = \frac{2n \cdot \sqrt{G_{i1}\, G_{i2}}}{n^2 - 1} \qquad \qquad \text{...(12.61)}$$

$$G_1 = G_{i1} \frac{n^2 + 1}{n^2 - 1} - G_3 \qquad \qquad \text{...(12.62)}$$

and
$$G_2 = G_{i2} \frac{n^2 + 1}{n^2 - 1} - G_3 \qquad \qquad \text{...(12.63)}$$

**Example 12.11:** Design an asymmetrical $\pi$-attenuator to operate between 200 $\Omega$ and 500 $\Omega$ and having a loss of 20 dbs.

**Solution:** Since attenuation is 20 dbs, $n = 10$
Let

$$G_{i1} = \frac{1}{200}\, \mho \qquad G_{i2} = \frac{1}{500}\, \mho$$

$$G_3 = \frac{2 \times 10 \sqrt{(1/200)(1/500)}}{99} = \frac{2 \times 10}{99 \times 100}\, \mho \quad \text{or} \quad R_3 = 495 \ \Omega$$

$$G_1 = \frac{1}{200} \times \frac{101}{99} - 2.02 \times 10^{-3} = 2.08 \times 10^{-3}\, \mho \quad \text{or} \quad R_1 = 480.7 \ \Omega$$

$$G_2 = \frac{1}{50} \times \frac{101}{99} - 2.02 \times 10^{-3} = 18.38 \times 10^{-3}\, \mho \quad \text{or} \quad R_2 = 54.4 \ \Omega$$

## 12.4  INSERTION LOSS

Let us consider the maximum power transfer problem. We know that if a source $V_1$ volts and internal resistance $R_1$ is terminated through a resistance of $R_1$, the power delivered by the source is maximum and is given by

$$P_{2\ \text{max}} = \frac{V_1^2}{4R_1} \qquad \qquad \text{...(12.64)}$$

However, as shown in Fig. 12.15 (a).

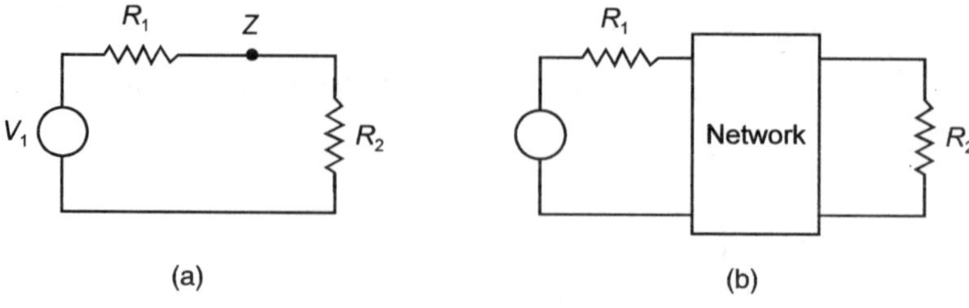


**Fig. 12.15**   (a) without network (b) with network.

if the source is terminated through a resistane $R_2 \neq R_1$, the power transfer is given as

$$P_{20} = \left(\frac{V_1}{R_1 + R_2}\right)^2 R_2 \qquad \qquad ...(12.65)$$

Now, if a network is inserted in between the source and the load $(R_2$ resistance) the power delivered at the load is

$$P_2 = \frac{V_2^2}{R_2} \qquad \qquad ...(12.66)$$

where $V_2$ is the terminal voltage across the load after the network insertion and this $P_2 < P_{20}$, as a part of the power is lost in the network inserted. The difference $(P_{20} - P_2)$ i.e. the power before insertion and power after insertion as received at the load is known as insertion loss and the effect of inserting two-port network as in Fig. 12.15(b) is characterized by the ratio of $\dfrac{P_{20}}{P_2}$ i.e. the ratio of power before to after insertion is given as

$$\frac{P_{20}}{P_2} = e^{2\alpha} \qquad \qquad ...(12.67)$$

and is known as insertion power ratio and $\alpha$ is the insertion loss in nepers. In terms of network variables

$$\frac{P_{20}}{P_2} = \frac{V_1^2 \, R_2^2}{(R_1 + R_2)^2} \cdot \frac{1}{V_2^2} = e^{2\alpha} \qquad \qquad ...(12.68)$$

or

$$\alpha = 10 \, \log_{10} \left[\frac{V_1 R_2}{V_2 (R_1 + R_2)}\right]^2$$

$$= 20 \, \log_{10} \left(\frac{V_1}{V_2} \cdot \frac{R_2}{R_1 + R_2}\right) db \qquad \qquad ...(12.69)$$

If we are considering harmonically varying voltages, then

$$\alpha = 20 \, \log_{10} \left[\frac{V_1 \, (j\omega)}{V_2 \, (j\omega)} \cdot \frac{R_2}{R_1 + R_2}\right] db \qquad \qquad ...(12.70)$$

This is an important result from the view point of design of filters or equalizers for transmission systems such as telephone lines.

Now, we can view the power difference or insertion loss as "reflected power" from the load and lost in the network between the source and the load.

i.e.

$$P_R = P_{2\mathrm{max}} - P_2$$

we have taken reference power as $P_{2\mathrm{max}}$ rather than $P_{20}$ as, in general, with the network inserted, maximum power transfer may not be possible whereas if the networks were not inserted it was possible to transer $P_{2\mathrm{max}}$ power by making $R_2 = R_1$

Here $P_2$ is the transmitted power into the load.

Therefore we define here

$$(\text{reflection co-efficient}) = \rho^2 = \frac{P_R}{P_{2\,\text{max}}} = \frac{P_{2\,\text{max}} - P_2}{P_{2\,\text{max}}}$$

$$= 1 - \frac{P_2}{P_{2\,\text{max}}} \qquad \qquad ...(12.71)$$

we have taken here $\rho^2$ rather than $\rho$ as the reflection is with reference to voltage and current. Therefore, reflection of power would be designated by square of reflection co-efficient.

Substituting for $P_2$ and $P_{2\text{max}}$ we have

$$\rho^2 = 1 - \frac{V_2^2}{R_2\,V_1^2} \cdot 4R_1 = 1 - \frac{4R_1}{R_2}\left(\frac{V_2}{V_1}\right)^2 \qquad \qquad ...(12.72)$$

Since $P_2$ is the transmitted power, the transmission co-efficient is defined as

$$[t(j\omega)]^2 = \frac{P_2}{P_{2\,\text{max}}} = \frac{4R_1}{R_2}\cdot\left(\frac{V_2}{V_1}\right)^2 \qquad \qquad ...(12.73)$$

Therefore if we combine these two equations (12.72) and (12.73) we have

$$\rho^2 + t^2 = 1 \qquad \qquad ...(12.74)$$

Which means

$$\frac{\text{Reflected power + transmitted power}}{\text{Available power}} = 1 \qquad \qquad ...(12.75)$$

### 12.4.1  Transmission Loss

When the loss in the network itself is of interest, transmission loss is normally obtained. It is a comparison of the power transmitted to the network from the source and the transmitted power from the network to the load. The ratio in db or neper is given as

$$\text{Transmission loss in dbs} = 10\,\log_{10}\frac{P_1}{P_2} \qquad \qquad ...(12.76)$$

where $P_1$ is the power input to the network and $P_2$ the power from the network to the load.

$$\text{Transmission loss in nepers} = \frac{1}{2}\ln\frac{P_1}{P_2} \qquad \qquad ...(12.77)$$

### 12.5  MATCHED SYSTEM

Consider a matched system as shown in Fig. 12.16 where the source resistance, the characteristic resistance of the attenuating network and the load resistance are equal and each equals say $R_0$.

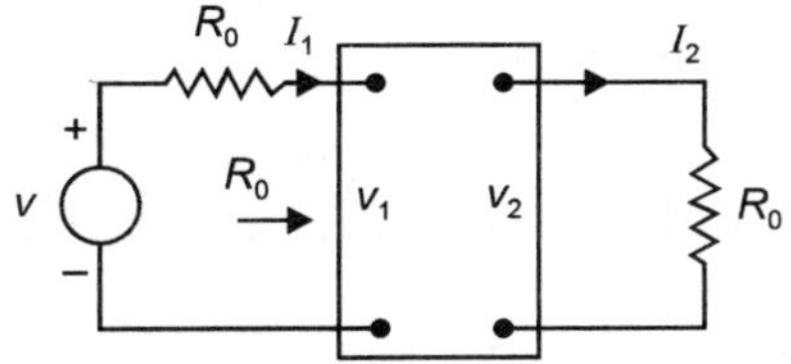

**Fig. 12.16**  Matched network.

The power input to the network
$$P_1 = I_1^2 \, R_0 \qquad \qquad \text{...(12.78)}$$

Power output from the network or

power fed to the load
$$P_2 = I_2^2 \, R_0 \qquad \qquad \text{...(12.79)}$$

$$\text{Transmission loss} = 10 \, \log_{10} \frac{P_1}{P_2} = 20 \, \log_{10} \frac{I_1}{I_2} \qquad \qquad \text{...(12.80)}$$

To find out insertion loss let us find out power to the load when the attenuating network is not connected in the circuit (Fig. 12.17)

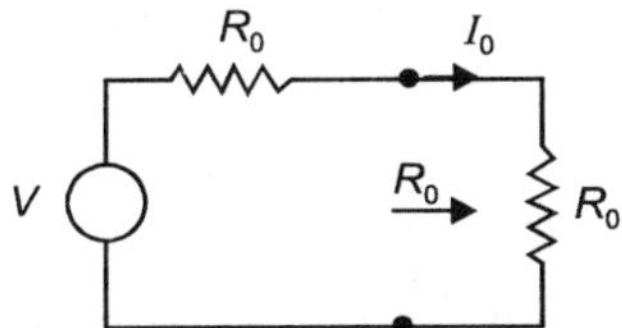

**Fig. 12.17**  Load connected to the source.

The power to the load $= I_0^2 \, R_0$

In fact $I_0 = I_1$ as the input resistance is same whether the network is connected or not and hence $I_1 = I_0$ and here

$$P_0 = I_1^2 \, R_0 \qquad \qquad \text{...(12.81)}$$

where $P_0$ is the power to the load when the attenuating network is inserted between the source and the load.

$$\text{Hence Insertion loss} \qquad = 10 \, \log_{10} \frac{P_0}{P_2} \qquad \qquad \text{...(12.82)}$$

$$= 10 \, \log_{10} \frac{I_1^2 \, R_0}{I_2^2 \, R_0} \qquad \qquad \text{...(12.83)}$$

$$= 20 \, \log_{10} \frac{I_1}{I_2} \qquad \qquad \text{...(12.84)}$$

Therefore, for a symmetrical network terminated at each end in its characteristic impedance.

Transmission loss = Insertion loss = Attenuation

$$= 20 \, \log_{10} \frac{I_1}{I_2} = 20 \, \log_{10} \frac{V_1}{V_2} \text{ in dbs} \qquad \qquad \text{...(12.85)}$$

**Example 12.12:** Determine the, transmission loss of the attenuator shown in Fig. E12.12. Also determine the insertion loss.

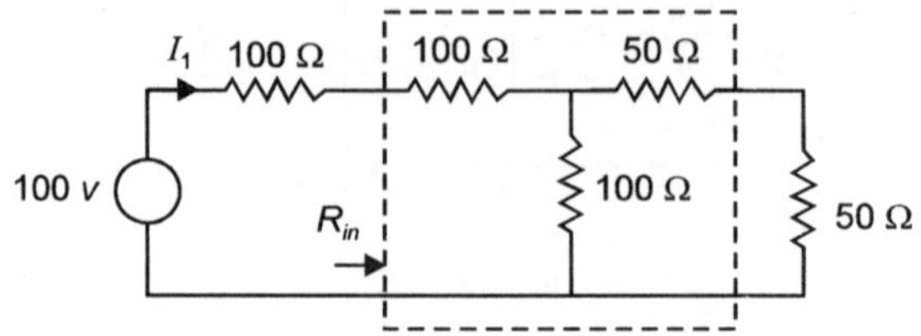


**Fig. E12.12.1**

**Solution:**

$$R_{in} = 100 + 100 \parallel 100$$
$$= 150 \ \Omega$$

Therefore total resistance

$$= 100 + 150 = 250 \ \Omega$$

Current

$$I_1 = \frac{100}{250} = 0.4 \text{ A}$$

Current

$$I_2 = 0.4 \times \frac{100}{200} = 0.2 \text{ A}$$

Power input to attenuator

$$= I_1^2 \, R_{in}$$
$$= 0.4^2 \times 150$$
$$P_1 = 24 \text{ watt}$$

Power input to the load or output from the attenuator

$$= I_2^2 \, R_L$$
$$= 0.2^2 \times 50$$
$$P_2 = 2 \text{ watt}$$

Therefore, transmission loss in db 10 $\log_{10} \dfrac{P_1}{P_2}$

$$= 10 \, \log_{10} \frac{24}{2}$$
$$= 10 \, \log_{10} 12$$
$$= 10.8 \text{ dbs}$$

Transmission loss in neper

$$= \frac{1}{2} \, \log \frac{P_1}{P_2}$$

$$= \frac{1}{2} \, \log_{10} 12$$

$$= 1.24 \text{ nepers}$$

(*b*) **Insertion loss:** As insertion loss is due to the difference in the power received by the load before the attenuating network is inserted to the power received after the network is inserted. We have already calculated the power received by the load after the attenuating network is inserted and is found to be $P_2 = 2$ watt. Now to find out power before the network is inserted we remove the network and the circuit is then as follows:

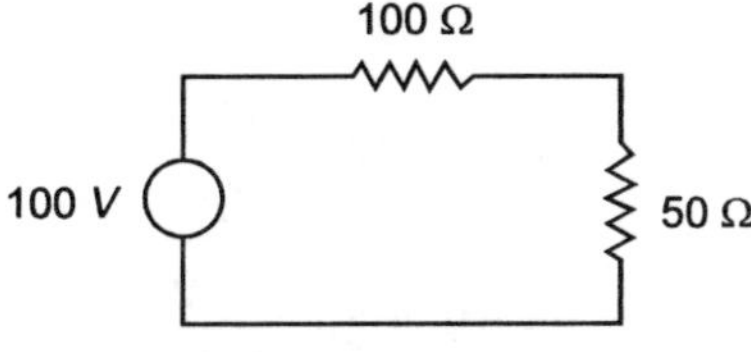

**Fig. E12.12.2**

The power is given as $\left(\dfrac{100}{150}\right)^2 \times 50 = 22.2$ watt

Therefore insertion loss in db

$$= 10 \, \log_{10} \frac{22.2}{2} = 10.45 \text{ dbs}$$

and insertion loss in neper

$$= \frac{1}{2} \, ln \, \frac{22.2}{2}$$

$$= \frac{1}{2} \, ln \, 11.1 = 1.2 \text{ neper}$$

**Example 12.13:** For the circuit shown in Fig. E12.13 determine the transmission loss and insertion loss.

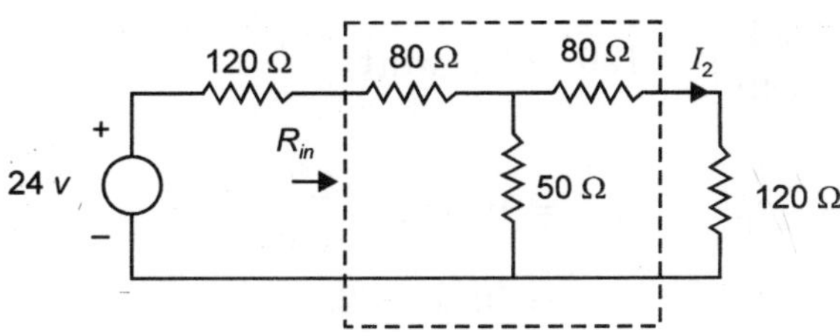

**Fig. E12.13.1**

**Solution:**

$$R_{in} = 80 + \frac{200 \times 50}{250} = 120 \ \Omega$$

Therefore,

$$I_1 = \frac{24}{120 + 120} = 0.1 \text{ A}$$

$$I_2 = 0.1 \times \frac{50}{250} = 0.02 \text{ A}$$

Power input to the attenuator

$$= I_1^2 \, R_{in}$$

$$P_1 = 0.1^2 \times 120 = 1.2 \text{ watt}$$

Power output from the attenuator or power loss in the load

$$P_2 = 0.02^2 \times 120 = 0.048 \text{ watt}$$

Therefore transmission loss

$$= 10 \log_{10} \frac{P_1}{P_2}$$

$$= 10 \log_{10} \frac{1.2}{0.048}$$

$$= 10 \log_{10} 25 = 13.98 \text{ dbs.}$$

Since the attenuating network is symmetrical and $R_{in} = R_L$. Therefore, its characteristic resistance is also equal to $R_{in} = R_L = R_0 = 120 \ \Omega$.

Hence Transmission loss = Insertion loss. This can be verified further. We have already calculated power received by the load when the attenuator is connected. If it is not inserted the circuit is as follows:

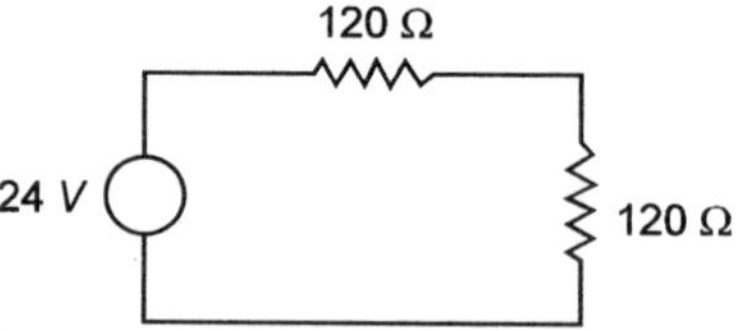

**Fig. E12.13.2**

Therefore, power loss in the load before the attenuator is inserted as

$$\left(\frac{24}{240}\right)^2 \times 120 = 1.2 \text{ watt.}$$

Hence insertion loss

$$= 10 \log_{10} \frac{1.2}{0.048}$$

$$= 10 \log_{10} 25$$

$$= 13.98 \text{ dbs } \textbf{Ans.}$$

## 12.6  CLASSICAL FILTERS: INTRODUCTION

Filters are frequency selective networks that attenuate signals at some frequency and allow others to pass with little or no attenuation. However, in practical filters there is always some loss of energy in the pass band or transmitting range and never complete attenuation in the rejection range of the filters. Attenuator networks as mentioned earlier use only one type of element namely a resistor whereas for filters normally reactive elements viz inductor and capacitors are used. ·

Electric filters find enormous application in Electronics and Electrical equipments. It is difficult to conceive a modern electronic device or a system which does not employ an electric filter.

The theory of filters was first advanced by Campbell and Wagner. The design of these filters is based on image parameters which basically depended on image or interactive impedances.

## 12.7  CLASSIFICATION OF FILTERS

Filters are classified based on their selection or rejection of frequency range as follows (see Fig. 12.18).

*Low pass:* These transmit all frequencies from zero to a cut-off frequency and attenuate all higher frequencies Fig. 12.18 (a)

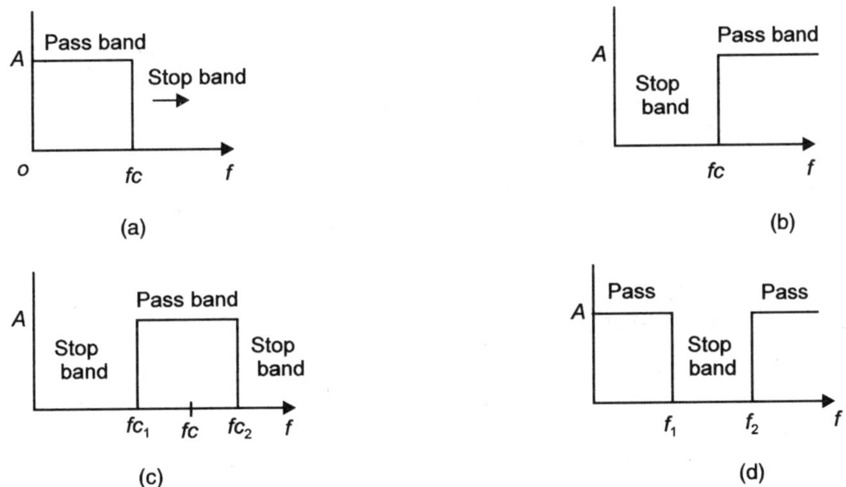


**Fig. 12.18**  (a) Low pass filter (b) High pass filter
(c) Band pass filter (d) Band stop filter.

*High pass filters:* These filters transmit signals of all frequencies above a certain cut-off frequency and attenuate or reject all frequencies below the same cut-off frequency. Fig. 12.18 (b).

*Band pass filters:* These filters transmit signals of all frequencies above a fixed minimum $f_{c_1}$ and below a fixed maximum $f_{C2}$, the band being bounded by $f_{c_1}$ and $f_{c_2}$ as in Fig. 12.18 (c).

*Band stop or Band elimination filters:* These filters transmit signals of all frequencies from 0 to $f_1$ and from $f_2$ to infinity and reject all signals of frequencies between $f_1$ and $f_2$ Fig. 12.18 (d).

Fig. 12.18 shows typical characteristics of various types of ideal filters where the filter transmits a certain band of frequencies without attenuation and causes infinite attenuation for all other frequencies and offers constant impedances during the transmission band. However, such a filter is not realizable in practice as losses in the inductor and capacitor would cause attenuation during the pass band and a finite rejection band is feasible. One can approach (not realize ideally) an ideal filter characteristic by using several specially designed sections in cascade. However, economy demands that these be designed and fabricated as simply as possible and still have a reasonably good characteristic. Therefore, only the simpler and more commonly used networks will be considered here e.g. the symmetrical-$T$ and

$\pi$-networks. In addition, the bisected $T$ and $\pi$ sections generally known as L-section find applications as terminating networks. These will be considered after the impedance characteristics (characteristic impedance or interactive impedance and the propagation constant) of the symmetrical structures ($T$ or $\pi$) have been investigated.

## 12.8 CHARACTERISTIC IMPEDANCE

We have seen in the previous chapter on attenuators that for symmetrical $T$ or $\pi$ network terminated by its characteristic impedance has both its image impedance equal and each is equal to the terminated impedance i.e. its characteristic impedance. Consider $T$-network shown in Fig. 12.19.

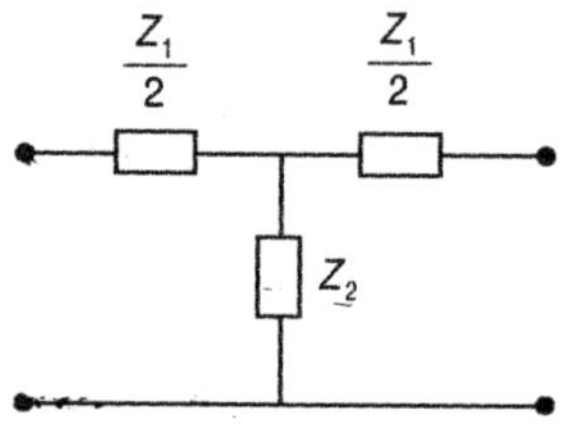

**Fig. 12.19**   T-network

$Z_0$ the characteristic impedance is given as

$$Z_0 = \sqrt{Z_{oc}\, Z_{sc}} \quad \text{where} \qquad \qquad \text{...(12.86)}$$

$$Z_{oc} = \frac{Z_1}{2} + Z_2 \quad \text{and} \quad Z_{sc} = \frac{Z_1}{2} + \frac{Z_1 Z_2 / 2}{\dfrac{Z_1}{2} + Z_2} \qquad \text{...(12.87)}$$

Therefore,

$$Z_0 = \sqrt{Z_1 Z_2 \left(1 + \frac{Z_1}{4 Z_2}\right)} \qquad \qquad \text{...(12.88)}$$

Also from equation (12.31) we have

$$Z_1 = 2\, Z_0\, \frac{n-1}{n+1}$$

where

$$n = e^{\gamma}$$

Therefore,

$$Z_1 = 2 \sqrt{Z_1 Z_2 \left(1 + \frac{Z_1}{4 Z_2}\right)} \cdot \frac{e^{\gamma} - 1}{e^{\gamma} + 1}$$

$$= 2 \sqrt{Z_1 Z_2 \left(1 + \frac{Z_1}{4 Z_2}\right)} \cdot \frac{e^{\gamma/2} - e^{-\gamma/2}}{e^{\gamma/2} + e^{-\gamma/2}}$$

$$= 2 \sqrt{Z_1 Z_2 \left(1 + \frac{Z_1}{4 Z_2}\right)} \tanh \gamma/2 \qquad \text{...(12.89)}$$

$$\text{or} \qquad \tanh \gamma/2 = \frac{Z_1}{2} \cdot \frac{1}{\sqrt{Z_1 Z_2 \left(1 + \dfrac{Z_1}{4Z_2}\right)}}$$

$$= \frac{1}{\sqrt{\dfrac{4Z_2}{Z_1}\left(1 + \dfrac{Z_1}{4Z_2}\right)}} = \frac{1}{\sqrt{\left(\dfrac{4Z_2}{Z_1} + 1\right)}} \qquad ...(12.90)$$

$$\text{or} \qquad \sqrt{\frac{\cosh \gamma - 1}{\cosh \gamma + 1}} = \frac{1}{\sqrt{\dfrac{4Z_2}{Z_1} + 1}}$$

$$\text{or} \qquad \frac{\cosh \gamma + 1}{\cosh \gamma - 1} = \frac{4Z_2}{Z_1} + 1$$

$$\text{or} \qquad \frac{2}{\cosh \gamma - 1} = \frac{4Z_2}{Z_1}$$

$$\text{or} \qquad \frac{2}{2 \sinh^2 \gamma/2} = \frac{4Z_2}{Z_1}$$

$$\text{or} \qquad \sinh \gamma/2 = \sqrt{\frac{Z_1}{4Z_2}} \qquad ...(12.91)$$

Ideally a filter offers zero attenuation during pass band frequencies and infinite attenuation during stop band frequencies. However, $\gamma$ being a function of frequency as $Z_1$, $Z_2$ are function of frequency the attenuation is not exactly zero over the complete pass band of frequencies.

It is to be noted that for $T$ and $\pi$ networks $Z_1$ and $Z_2$ are of opposite types e.g. inductance and capacitors so that $Z_1/4Z_2$ is a negative quantity and hence $\sqrt{\dfrac{Z_1}{4Z_2}}$ becomes an imaginary quantity.

Therefore,

$$\sinh \left(\frac{\alpha}{2} + j\frac{\beta}{2}\right) = \sqrt{\frac{Z_1}{4Z_2}} \qquad ...(12.92)$$

$$\sinh \frac{\alpha}{2} \cos \frac{\beta}{2} + j \cosh \frac{\alpha}{2} \sin \frac{\beta}{2} = \sqrt{\frac{Z_1}{4Z_2}}$$

Since the right hand quantity is an imaginary number, therefore,

$$\sinh \frac{\alpha}{2} \cos \frac{\beta}{2} = 0 \quad \text{and}$$

$$\cosh \frac{\alpha}{2} \sin \frac{\beta}{2} = \sqrt{\frac{Z_1}{4Z_2}} \qquad \qquad ...(12.93)$$

The two conditions of operation of the filter viz the pass band and stop band are mathematically obtained from these equations.

Since $\qquad \qquad \sinh \dfrac{\alpha}{2} \cos \dfrac{\beta}{2} = 0$

Either $\qquad \qquad \sinh \dfrac{\alpha}{2} = 0 \quad \text{or} \quad \cos \dfrac{\beta}{2} = 0 \qquad \qquad ...(12.94)$

1. $\qquad \qquad \sinh \dfrac{\alpha}{2} = 0.$ Therefore, $\dfrac{\alpha}{2} = 0$ or $\alpha = 0$

Therefore, $\qquad \qquad \sin \dfrac{\beta}{2} = \left| \sqrt{\dfrac{Z_1}{4Z_2}} \right| \quad \text{as } \cosh \dfrac{\alpha}{2} = 1 \qquad \qquad ...(12.95)$

$$\alpha = 0 \text{ represents pass band region}$$

since $Z_1/4Z_2$ is negative, in order to satisfy the condition

$$\sin \frac{\beta}{2} = \left| \sqrt{\frac{Z_1}{4Z_2}} \right|$$

the requirement is $\qquad -1 < \dfrac{Z_1}{4Z_2} < 0 \qquad \qquad ...(12.95a)$

Therefore $\qquad \qquad \beta = 2 \sin^{-1} \sqrt{\left| \dfrac{Z_1}{4Z_2} \right|} \qquad \qquad ...(12.96)$

2. When $\qquad \qquad \cos \dfrac{\beta}{2} = 0$

$$\frac{\beta}{2} = \pm \frac{\pi}{2} \quad \text{and} \quad \sin \frac{\beta}{2} = \pm 1$$

Here $\qquad \alpha \neq 0$ and $\cosh \dfrac{\alpha}{2} = \sqrt{\dfrac{Z_1}{4Z_2}} \qquad \qquad ...(12.97)$

In general $\qquad \dfrac{\beta}{2} = (2n-1)\dfrac{\pi}{2}$ or $\beta = (2n-1)\pi$

where $n$ is an integer

since $\alpha \neq 0$ it represents stop band operation of the filter and the attenuation constant

$$\alpha = 2 \cosh^{-1} \sqrt{\left| \frac{Z_1}{4Z_2} \right|}$$

since $\cosh \gamma \geq 1$ i.e. $\alpha \geq 0$ the stop band is characterized by

$$\frac{Z_1}{4Z_2} \leq -1 \qquad \qquad ...(12.97a)$$

We have two inequality conditions, one for the pass band and the other for the stop band. The transition period (cut off freuency point) is described by the common part of the two inequalities i.e.

$$\frac{Z_1}{4Z_2} = -1 \qquad\qquad ...(12.98)$$

Also it is to be noted that at transition (cut off frequency) the characteristic impedance becomes reactive. Since characteristic impedance for T-network is given as

$$Z_{0T} = \sqrt{Z_1 Z_2}\ \sqrt{1 + \frac{Z_1}{4Z_2}} \qquad\qquad ...(12.99)$$

And if $Z_1$, $Z_2$ are reactances of opposite sign $\sqrt{Z_1 Z_2}$ is a real number and $\sqrt{1 + \dfrac{Z_1}{4Z_2}}$ is imaginary when $\left|\dfrac{Z_1}{4Z_2}\right| > 1$. Therefore, the characteristic impedance becomes reactive only when $\dfrac{Z_1}{4Z_2} = -1$.

The characteristic impedance of the $\pi$-network is obtained as follows:

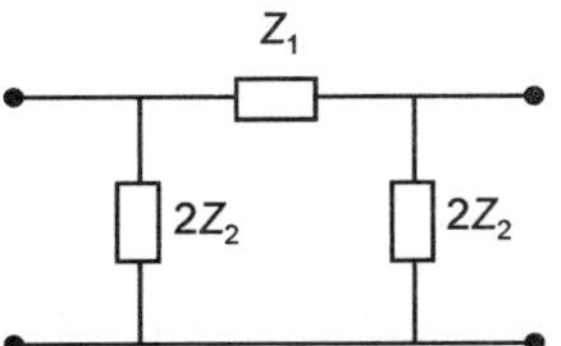

**Fig. 12.20**  $\pi$-network.

$$Z_{0\pi} = \sqrt{Z_{0C}\ Z_{SC}} = \sqrt{\frac{2Z_2\,(Z_1 + 2Z_2)}{Z_1 + 4Z_2} \cdot \frac{Z_1 \cdot 2Z_2}{Z_1 + 2Z_2}}$$

$$= \sqrt{Z_1 Z_2\ \frac{1}{(Z_1 + 4Z_2)} \cdot \frac{1}{4Z_2}}$$

$$= \sqrt{Z_1 Z_2 \left(1 + \frac{Z_1}{4Z_2}\right)^{-1}}$$

$$= \sqrt{\frac{Z_1 Z_2}{1 + \dfrac{Z_1}{4Z_2}}} \qquad\qquad ...(12.100)$$

It can be seen that $Z_{0T}\, Z_{0\pi} = Z_1 Z_2 = \dfrac{L}{C} = k^2$ and therefore, sometimes $\sqrt{Z_1 Z_2}$ is known as nominal characteristic impedance of the filter.

Since the nature of characteristic impedance of the $\pi$-network is similar as for *T*-network, the previous discussion in terms of cut off frequency and propagation hold good for $\pi$-network also,

1. Attenuation constant

$$\alpha = 0 \text{ for pass band}$$

$$= 2 \cosh^{-1} \sqrt{\frac{Z_1}{4Z_2}} \text{ in stop band} \qquad \text{...(12.101)}$$

2. Phase

$$\beta = 2 \sin^{-1} \sqrt{\left|\frac{Z_1}{4Z_2}\right|} \text{ in pass band}$$

and

$$\beta = \pi \text{ in stop band} \qquad \text{...(12.102)}$$

3. Characteristic impedance

$$Z_{0T} = \sqrt{Z_1 Z_2 \left(1 + \frac{Z_1}{4Z_2}\right)} \text{ for } T \text{ network}$$

$$Z_{0\pi} = \sqrt{Z_1 Z_2 \Big/ \left(1 + \frac{Z_1}{4Z_2}\right)} \text{ for } \pi \text{ network} \qquad \text{...(12.103)}$$

4. Cut-off frequency

$$Z_1 + 4\,Z_2 = 0$$

and

$$\frac{Z_1}{4Z_2} = 0 \quad \text{or} \quad Z_1 = 0 \qquad \text{...(12.104)}$$

The configuration of Low pass filter with $T$-section and $\pi$-section using reactance elements is shown in Fig. 12.21.

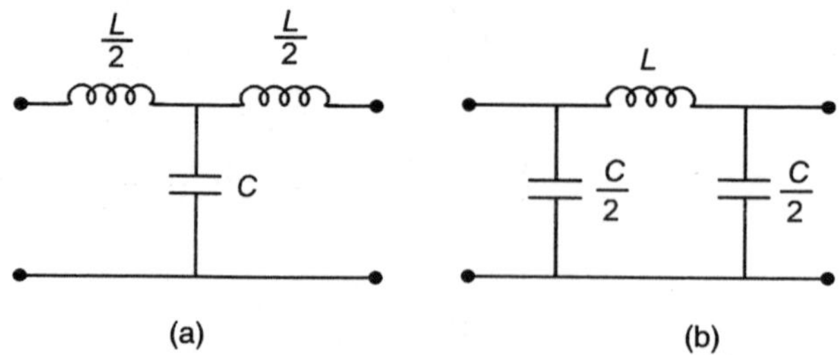

**Fig. 12.21** Low pass filter (a) T–section (b) $\pi$–section.

It is seen that the total series impedance and shunt impedance in each case is $Z_1 = j\omega L$, $Z_2 = \dfrac{1}{j\omega C}$ respectively.

Hence, the product $Z_1 Z_2 = \dfrac{L}{C}$ which means for each section the product is constant independent of frequency. Such an arrangement is known as a constant-$k$ filter.

## 12.9  LOW PASS CONSTANT-K FILTER

After substituting for $Z_1 = j\omega L$ and $Z_2 = \dfrac{1}{j\omega C}$, various characteristics for T- and $\pi$-section for low pass filter can be obtained as follows:

1. *Cut-off frequency:* From equation 12.104 the two frequencies are when $Z_1 = 0$  $j\omega L = 0$ since $L$ is not Zero $\omega = 0$ or $\omega_c = 0$.

and
$$\frac{Z_1}{4Z_2} = -1 \quad \text{or} \quad \frac{j\omega_c L \; j\omega_c C}{4} = -1$$

or
$$\omega_c^2 LC = 4$$

or
$$\omega_c = \frac{2}{\sqrt{LC}}$$

or
$$f_c = \frac{1}{\pi\sqrt{LC}} \qquad\qquad ...(12.105)$$

Thus the pass band for constant-k low pass filter extends between 0 Hz to $\dfrac{1}{\pi\sqrt{LC}}$ Hz

2. *Attenuation constant*: From equation (12.101) $\alpha = 0$ in pass band and

$$\alpha = 2 \cosh^{-1} \sqrt{\left|\frac{Z_1}{4Z_2}\right|}$$

$$= 2 \cosh^{-1} \sqrt{\left|\frac{j\omega L \cdot j\omega C}{4}\right|}$$

$$= 2 \cosh^{-1} \frac{f}{f_C} \qquad\qquad ...(12.106)$$

neper in stop band

3. *Phase Constant* : From equation (12.102) in pass band

$$\beta = 2 \sin^{-1} \sqrt{\left|\frac{Z_1}{4Z_2}\right|}$$

$$= 2 \sin^{-1} \frac{f}{f_C}$$

and in stop band $\qquad\qquad \beta = \pi \qquad\qquad\qquad\qquad ...(12.107)$

4. *Characteristic impedance T-Section:*

$$Z_{OT} = \sqrt{Z_1 Z_2 \left(1 + \frac{Z_1}{4Z_2}\right)}$$

$$= \sqrt{\frac{L}{C}\left(1 - \frac{\omega^2 LC}{4}\right)}$$

$$= \sqrt{\frac{L}{C}}\left\{1 - \left(\frac{f}{f_c}\right)^2\right\}^{1/2} \qquad ...(12.108)$$

It is seen that during pass band $f < f_c$, therefore, the filter acts as resistive whereas during stop band $f > f_c$ and hence $Z_{OT}$ is imaginary and the filter acts as reactive element. Similarly the characteristic impedance for $\pi$-section is given as

$$Z_{0\pi} = \frac{\sqrt{L/C}}{\left\{1 - (f/f_c)^2\right\}^{1/2}} \qquad ...(12.109)$$

This low pass filter acts resistive during pass band ($f < f_c$) and reactive during stop band ($f > f_c$). When the characteristic impedance is resistive (pass band) the network absorbs real power from the source and since the network is loss less (reactive elements) all of this power is transmitted to the load without attenuation. However, when the characteristic impedance $Z_0$ is reactive (stop band) and the filter is terminated by $Z_0$, the output voltage and current may have any values and still satisfy the condition that the real power is zero. Hence, the ratio of input to output voltage and the ratio of input to output current may vary in the attenuation range. This accounts for variable attenuation in this range.

Fig. 12.22 show the variation of normalized characteristic impedance and attenuation constant $\alpha$, phase constant $\beta$ as a function of frequency of the prototype low pass filter. The normalized characteristic impedance is given by

$$R_K = \frac{\text{Actual characteristic impedance}}{\sqrt{L/C}} \qquad ...(12.110)$$

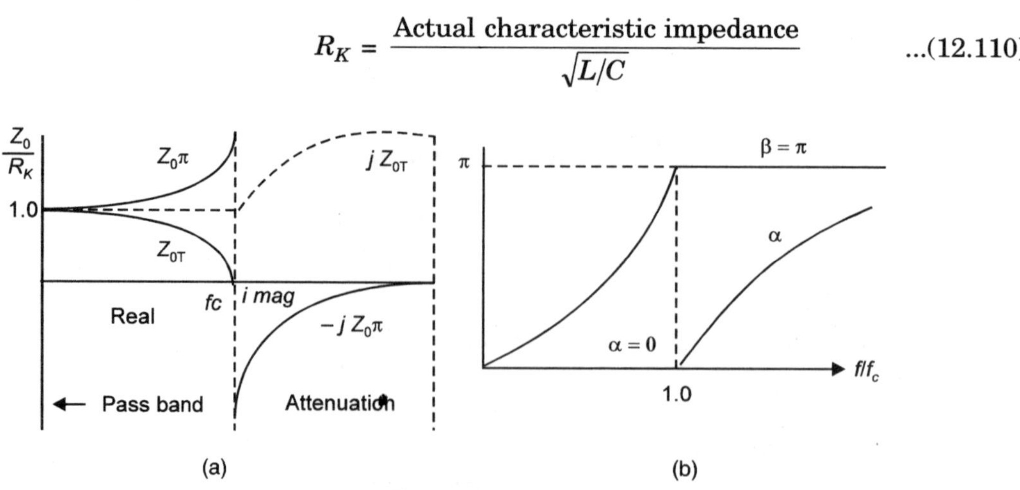


**Fig. 12.22**

(a) Normalized characteristic impedance of prototype low-pass filter
(b) Attenuation and phase function of a prototype low-pass filter.

## 12.10 HIGH PASS, CONSTANT-K FILTER

Fig. 12.23 shows a general, arrangement of high pass constant-k filter. It is seen that as compared to low-pass, the location of inductor and capacitor has been interchanged. Thus it becomes a high pass filter. Here low frequency signals are shunted by the low impedance of the inductance whereas high frequency signals are transmitted to the load.

For symmetric $T$ and $\pi$ network filters the general expression for various quantities remain same (in term of $Z_1$ and $Z_2$) only variation is that $Z_1$ now equals $\dfrac{1}{j\omega C}$ and $Z_2 = j\omega L$

1. *Nominal characteristic impedance:*

$$Z = \sqrt{Z_1 Z_2} = \sqrt{\frac{L}{C}} \qquad \qquad ...(12.111)$$

2. *Cut-off frequency:*

$$Z_1 = 0 = \frac{1}{j\omega C}$$

For finite $C$, $\omega \to \infty$ or $f_C \to \infty$

The other cut-off frequency is when

$$Z_1 + 4\,Z_2 = 0$$

$$\frac{1}{j\omega_c C} + 4\,j\omega_c L = 0$$

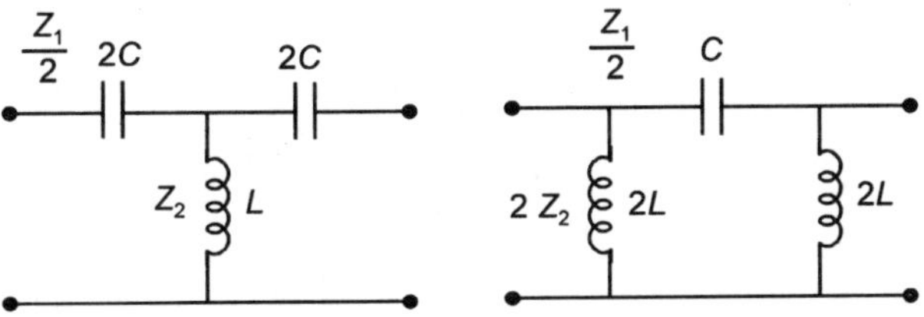


**Fig. 12.23**  High pass, constant-k type filters
(a) T-section   (b) π-section.

or

$$4\omega_c^2\,LC = 1$$

or

$$\omega_c = \frac{1}{2\sqrt{LC}}$$

or

$$f_c = \frac{1}{4\pi\sqrt{LC}}$$

Therefore, the pass band for high pass constant-k type filter lies between

$$f_c = \frac{1}{4\pi\sqrt{LC}} \text{ to } \infty \qquad \qquad ...(12.112)$$

and stop band is from

$$0 \text{ to } \frac{1}{4\pi\sqrt{LC}} \qquad \qquad ...(12.113)$$

3. *Attenuation Constant:*
   In pass band

$$\alpha = 0$$

   whereas in stop band

$$\alpha = 2\,\cosh^{-1}\sqrt{\left|\frac{Z_1}{4Z_2}\right|} = 2\,\cosh^{-1}\left(\frac{f_c}{f}\right) \qquad ...(12.114)$$

4. *Phase Constant:*
   In pass band

$$\beta = 2\,\sin^{-1}\sqrt{\left|\frac{Z_1}{4Z_2}\right|} = 2\sin^{-1}\left(\frac{f_c}{f}\right)$$

   In stop band

$$\beta = \pi \qquad \qquad ...(12.115)$$

5. *Characteristic impedance:*

$$Z_{OT} = \sqrt{Z_1 Z_2 \left(1 + \frac{Z_1}{4Z_2}\right)}$$

$$= \sqrt{\frac{L}{C}\left(1 + \frac{1/j\omega C}{4j\omega L}\right)}$$

$$= \sqrt{\frac{L}{C}\left(1 - \frac{1}{4}\omega^2 LC\right)}$$

$$= \sqrt{\frac{L}{C}\left\{1 - \left(\frac{f_c}{f}\right)^2\right\}} \qquad \qquad ...(12.116)$$

Similarly

$$Z_{0\pi} = \frac{\sqrt{L/C}}{\sqrt{1 - \left(\frac{f_c}{f}\right)^2}} \qquad \qquad ...(12.117)$$

It is seen that when $f < f_c$ the characteristic impedance is reactive and hence it is stop band.

However, if $f > f_c$ the characteristic impedance is a real number i.e. it is resistive and this corresponds to pass band.

Fig. 12.24 shows the variation of normalized characteristic impedance and attenuation constant $\alpha$, phase constant $\beta$ as a function of frequency of a high pass, constant-k filter.

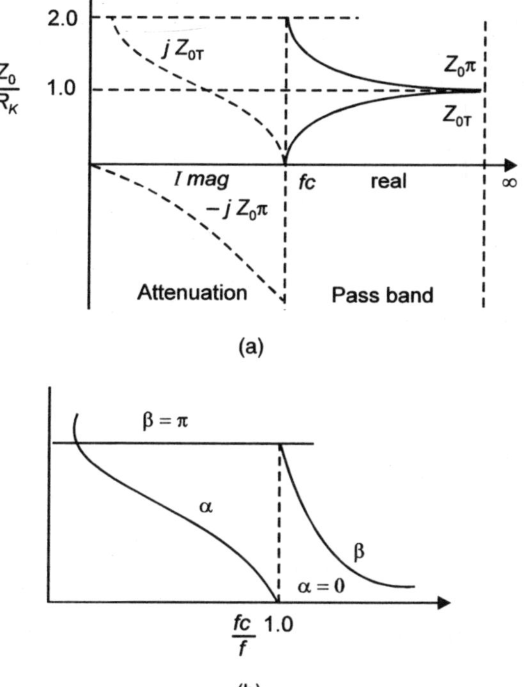



**Fig. 12.24**

(a) Variation of normalized characteristic impedance
(b) Variation of $\alpha$ and $\beta$, as a function of frequency

## 12.11 BAND PASS CONSTANT-K TYPE FILTER

The basic principle involved in low pass and high pass filters also applies to band pass filter. This filter has transmission band between two specified frequencies $\omega_1$ and $\omega_2$ and no transmission elsewhere. A band pass filter is developed by the series connection of a low pass filter and a high pass filter where cut-off frequency of the low pass filter is higher than that of the high pass filter. Thus, the overlap allows only a band of frequencies to be transmitted as shown in Fig. 12.25.

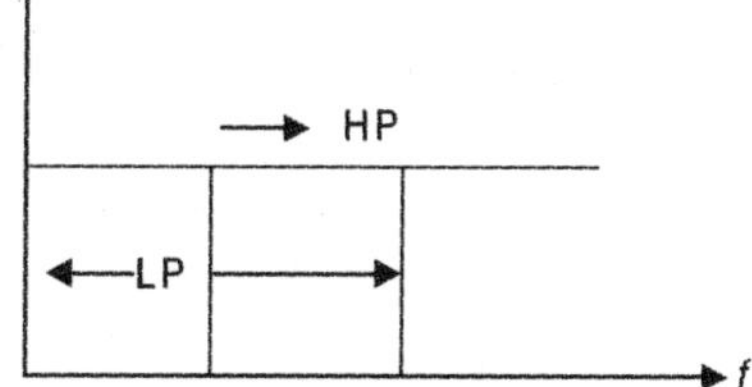



**Fig. 12.25** Series connection of LP and HP filter.

Fig. 12.26 shows a band pass, constant-k filter with a series resonant series arm and an anti-resonant shunt arm. The parameters of the series arm and the shunt arm are so selected that they resonate at the same frequency.

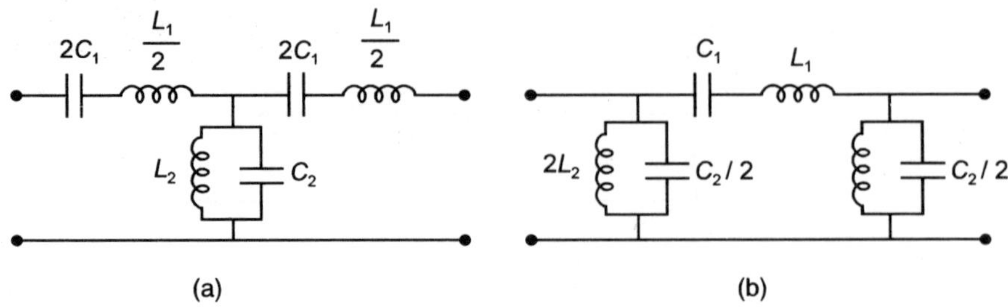

**Fig. 12.26**  General Configuration of Band pass (a) T-type constant-k filter
(b) $\pi$–type constant-k filter

Therefore,

$$L_1 C_1 = L_2 C_2 = \frac{1}{\omega_c^2} \qquad \text{...(12.118)}$$

Series arm impedance

$$Z_1 = j\omega L_1 + \frac{1}{j\omega C_1} = j\,\frac{\omega^2 L_1 C_1 - 1}{\omega C_1} \qquad \text{...(12.119)}$$

Similarly shunt arm impedance

$$Z_2 = \frac{j\omega L_2 \cdot 1/j\omega C_2}{j\omega L_2 + 1/j\omega C_2} = \frac{j\omega L_2}{1 - \omega^2 L_2 C_2} \qquad \text{...(12.120)}$$

Therefore,

$$Z_1 Z_2 = j\,\frac{\omega^2 L_1 C_1 - 1}{\omega C_1} \cdot \frac{j\omega L_2}{1 - \omega^2 L_2 C_2}$$

$$= \frac{L_2}{C_1} \cdot \frac{1 - \omega^2 L_1 C_1}{1 - \omega^2 L_2 C_2} \qquad \text{...(12.121)}$$

Since

$$L_1 C_1 = L_2 C_2$$

Therefore,

$$Z_1 Z_2 = \frac{L_2}{C_1} = R_K^2$$

or

$$R_k = \sqrt{\frac{L_2}{C_1}} = \sqrt{\frac{L_1}{C_2}} \qquad \text{...(12.122)}$$

The characteristic impedance

$$Z_{OT} = \sqrt{Z_1 Z_2 \left(1 + \frac{Z_1}{4 Z_2}\right)}$$

Now

$$\frac{Z_1}{4 Z_2} = j\,\frac{\omega^2 L_1 C_1 - 1}{\omega C_1} \cdot \frac{1 - \omega^2 L_2 C_2}{4 j\omega L_2}$$

$$= \frac{(\omega^2 L_1 C_1 - 1)(1 - \omega^2 L_2 C_2)}{4\omega^2 L_2 C_1}$$

$$1 + \frac{Z_1}{4Z_2} = \frac{4\omega^2 L_2 C_1 + \left(\dfrac{\omega^2}{\omega_0^2} - 1\right)\left(1 - \dfrac{\omega^2}{\omega_0^2}\right)}{4\omega^2 L_2 C_1}$$

Therefore,

$$Z_{OT} = \sqrt{\frac{L_2}{C_1} \cdot \frac{4\omega^2 L_2 C_1 + \left(\dfrac{\omega^2}{\omega_0^2} - 1\right)\left(1 - \dfrac{\omega^2}{\omega_0^2}\right)}{4\omega^2 L_2 C_1}} \qquad ...(12.123)$$

Now cut-off frequencies occur when

$$Z_1 + 4Z_2 = 0 \quad \text{or} \quad Z_1 = -4Z_2$$

Multiplying both sides by $Z_1$ we have

$$Z_1^2 = -4Z_1 Z_2 = -4R_k^2$$

$$Z_1 = \pm j \, 2 \, R_k \qquad ...(12.124)$$

This equation defines two cut-off frequencies when

$$Z_1 = -j \, 2 \, R_k \text{ at frequency } f_1$$

and
$$Z_1 = +j2 \, R \text{ at frequency } f_2 \qquad ...(12.125)$$

which means impedance $Z_1$ of series arm at $f_1$ is negative of impedance $Z_1$ at $f_2$

$$\omega_1 L_1 - \frac{1}{\omega_1 C_1} = -\left(\omega_2 L_1 - \frac{1}{\omega_2 C_1}\right)$$

or
$$\frac{\omega_1^2 L_1 C_1 - 1}{\omega_1 C_1} = -\frac{\omega_2^2 L_1 C_1 - 1}{\omega_2 C_1} \qquad ...(12.126)$$

or
$$\left(\frac{\omega_1}{\omega_0}\right)^2 - 1 = \frac{\omega_1}{\omega_2}\left[\left(\frac{\omega_2}{\omega_0}\right)^2 - 1\right]$$

$$\omega_1^2 \omega_2 - \omega_0^2 \omega_2 = -\omega_1 \omega_2^2 + \omega_1 \omega_0^2$$

or
$$\omega_1^2 \omega_2 + \omega_1 \omega_2^2 = \omega_1 \omega_0^2 + \omega_2 \omega_0^2$$

$$\omega_1 \omega_2 = \omega_0^2$$

or
$$f_0 = \sqrt{f_1 f_2} \qquad ...(12.127)$$

Hence the resonant frequency $f_0$ of the components of series arm and the shunt arm should be such that it is the geometric mean value of the two cut-off frequencies $f_1$ and $f_2$.

To obtain expressions for various parameters $L_1$, $C_1$, $L_2$ and $C_2$ following procedure is adopted. We know that

$$j\omega_1 L_1 + \frac{1}{j\omega_1 C_1} = -2\, jR_k \qquad\qquad ...(12.128)$$

or
$$\omega_1 L_1 - \frac{1}{\omega_1 C_1} = -2R_k \quad \text{Now} \quad \frac{1}{C_1} = L_1 \omega_0^2$$

$$\omega_1 L_1 - \frac{\omega_0^2}{\omega_1} L_1 = -2R_k$$

or
$$\frac{\omega_1^2 - \omega_0^2}{\omega_1} L_1 = -2R_k$$

or
$$\frac{\omega_1^2 - \omega_1 \omega_2}{\omega_1} L_1 = -2R_k$$

or
$$L_1 = \frac{2R_k}{\omega_2 - \omega_1} = \frac{R_k}{\pi(f_2 - f_1)} \qquad\qquad ...(12.129)$$

Now
$$C_1 = \frac{1}{L_1 \omega_0^2} = \frac{\pi(f_2 - f_1)}{R_k \omega_1 \omega_2} = \frac{f_2 - f_1}{4\pi f_1 f_2 R_k}$$

Also since
$$R_k = \sqrt{\frac{L_2}{C_1}} = \sqrt{\frac{L_1}{C_2}} \qquad\qquad ...(12.130)$$

or
$$C_1 R_k^2 = L_2$$

$$L_2 = \frac{f_2 - f_1}{4\pi f_1 f_2 R_k} \cdot R_k^2 = \frac{(f_2 - f_1) R_k}{4\pi f_1 f_2} \qquad\qquad ...(12.131)$$

Also
$$C_2 R_k^2 = L_1$$

or
$$C_2 = \frac{L_1}{R_k^2} = \frac{R_k}{\pi(f_2 - f_1) R_k^2}$$

$$C_2 = \frac{1}{\pi(f_2 - f_1) R_k} \qquad\qquad ...(12.132)$$

Fig. 12.27 shows variation of characteristic impedance, attenuation $\alpha$ and phase shift $\beta$ as a function of frequency for a band pass constant–k filter.

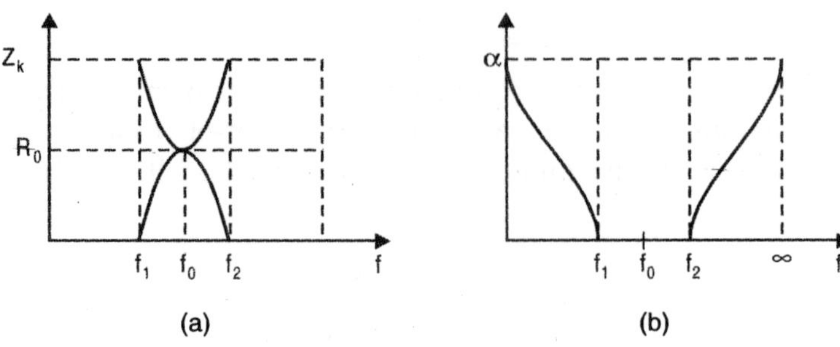

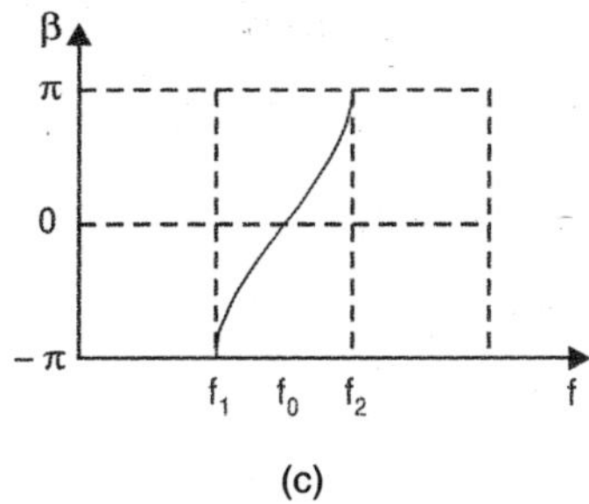



**Fig. 12.27** Variation of (a) Impedance characteristic (b) attenuation
(c) phase shift as a function of frequency for a band pass, constant–k filter.

## 12.12 BAND STOP OR BAND ELIMINATION CONSTANT–k TYPE FILTER

In order to develop a band stop filter the series and shunt arm of the band pass are interchanged and we thus have the general configuration of band stop constant–k type filter as
shown in Fig. 12.28.

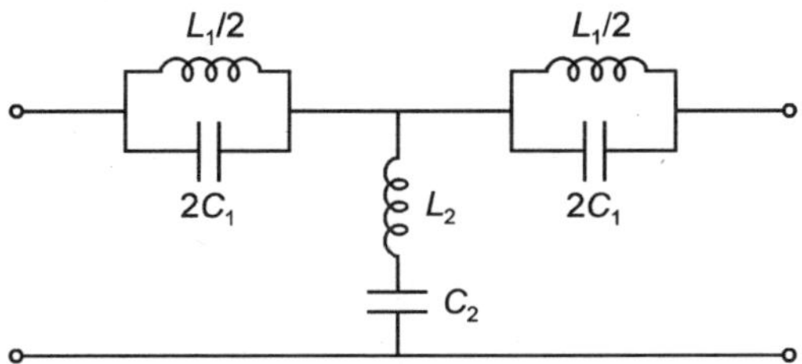



**Fig 12.28** General Configuration of band stop constant–k type filter.

Here again the components of the filter for series and shunt arm are so selected that
their corresponding resonance frequencies are same. This frequency is known as the resonant
frequency or centre frequency or rms frequency of the filter.

Therefore,

$$L_1 C_1 = L_2 C_2 = \frac{1}{\omega_0^2} \qquad \qquad ...(12.133)$$

$$Z_1 = \frac{j\omega L_1 \cdot 1/j\omega C_1}{j\omega L_1 + \dfrac{1}{j\omega C_1}} = j\,\frac{\omega L_1}{1 - \omega^2 L_1 C_1} \qquad \qquad ...(12.134)$$

and

$$Z_2 = j\,\frac{\omega^2 L_2 C_2 - 1}{\omega C_2} \qquad \qquad ...(12.135)$$

and

$$Z_1 Z_2 = \frac{L_1}{C_2} = \frac{L_2}{C_1} = R_k^2$$

Therefore,

$$R_k = \sqrt{\frac{L_1}{C_2}} = \sqrt{\frac{L_2}{C_1}} \qquad \qquad ...(12.136)$$

1. Cut off frequencies are obtained from the expression

$$\frac{Z_1}{4Z_2} = -1 \quad \text{or} \quad Z_1 = -4\,Z_2$$

$$Z_1^2 = -4Z_1Z_2 = -4\,R_k^2$$

or
$$Z_1 = \pm\,2jR_k \qquad\qquad \text{...(12.137)}$$

This relation is similar to the one obtained for the band pass filter and it means that $Z_1$ at $f_1$ is negative of $Z_1$ at $f_2$, the two cut off frequencies.

Therefore,

$$j\,\frac{\omega_1 L_1}{1 - \omega_1^2\,L_1C_1} = -j\,\frac{\omega_2\,L_1}{1 - \omega_2^2\,L_1C_1} \qquad\qquad \text{...(12.138)}$$

Since
$$L_1C_1 = \frac{1}{\omega_0^2}$$

$$\frac{1}{1 - \left(\dfrac{\omega_1}{\omega_0}\right)^2} = \frac{\omega_2}{\omega_1} \cdot \frac{1}{\left(\dfrac{\omega_2}{\omega_0}\right)^2 - 1}$$

or
$$1 - \left(\frac{\omega_1}{\omega_0}\right)^2 = \frac{\omega_1}{\omega_2}\left\{\left(\frac{\omega_2}{\omega_0}\right)^2 - 1\right\}$$

$$\omega_2\omega_0^2 - \omega_1^2\omega_2 = \omega_1\omega_2^2 - \omega_1\omega_0^2$$

or
$$\omega_0^2\,(\omega_2 - \omega_1) = \omega_1\omega_2\,(\omega_2 - \omega_1),$$

or
$$\omega_0^2 = \omega_1\omega_2 \quad \text{or} \quad f_0 = \sqrt{f_1 f_2} \qquad\qquad \text{...(12.139)}$$

Hence the resonant frequency $f_0$ of the components of series arm and shunt arm of the filter circuit should be such that it is equal to the geometric mean value of the two cut-off frequencies $f_1$ and $f_2$.

Normally for filter design the nominal characteristic impedance and the cut-off frequencies would be given. It would be required to obtain the parameters of the filter circuit $L_1$, $C_1$, $L_2$ and $C_2$ to achieve the given requirements.

Since
$$Z_1 = \pm\,j2R_k$$

$$j\,\frac{\omega_1 L_1}{1 - \omega_1^2\,L_1C_1} = +\,j2R_k \qquad\qquad \text{...(12.140)}$$

or
$$L_1 = 2R_k\,\frac{(1 - \omega_1^2\,L_1C_1)}{\omega_1}$$

$$= \frac{R_k}{\pi\,f_1}\left\{1 - \left(\frac{\omega_1}{\omega_0}\right)^2\right\}$$

$$= \frac{R_k}{\pi f_1} \cdot \frac{\omega_0^2 - \omega_1^2}{\omega_0^2}$$

$$L_1 = \frac{R_k}{\pi\, f_1 f_2}\, (f_2 - f_1) \qquad\qquad ...(12.141)$$

Now
$$C_1 = \frac{1}{L_1 \omega_0^2} = \frac{1}{4\pi^2 L_1 f_1 f_2} = \frac{\pi f_1 f_2}{4\pi^2 R_k\, (f_2 - f_1)} \cdot \frac{1}{f_1 f_2}$$

$$C_1 = \frac{1}{4\pi\, (f_2 - f_1) R_k} \qquad\qquad ...(12.142)$$

Also we know that

$$\frac{L_2}{C_1} = R_k^2 = \frac{L_1}{C_2}$$

or
$$L_2 = C_1 R_k^2$$

$$L_2 = \frac{R_k}{4\pi\, (f_2 - f_1)} \qquad\qquad ...(12.143)$$

and
$$C_2 = \frac{L_1}{R_k^2} = \frac{f_2 - f_1}{\pi f_1 f_2 R_k} \qquad\qquad ...(12.144)$$

Let us solve a few problems to illustrate the application of various formulae derived in connection with various types of filters.

**Example 12.14:** Design a low-pass constant-k type T-section and $\pi$-section filters with $f_c = 3$ KHz and nominal characteristic impedance 500 ohms. Also determine the frequency at which the filter offers attenuation of 20dbs. Determine $\beta$ for $f = 2$ KHz and $f = 10$ KHz.

**Solution:** We normally terminate the filter with nominal characteristic impedance rather than the actual characteristic impedance as otherwise it would require a very large number of circuits to be connected in tandem at the terminating load point. To avoid that we terminate with nominal characteristic impedance which is within the engineering accuracy requirements.

Solution. We know that (Equation 12.20)

$$f_0 = \frac{1}{\pi \sqrt{LC}}$$

and
$$R_k = \sqrt{\frac{L}{C}}$$

$$f_0 R_k = \frac{1}{\pi \sqrt{LC}} \cdot \sqrt{\frac{L}{C}} = \frac{1}{\pi C}$$

or
$$C = \frac{1}{\pi f_0 R_k} = \frac{1}{\pi \times 3 \times 10^3 \times 500}\, F = 0.212\,\mu F$$

and
$$L = CR^2{}_k = \frac{500 \times 500}{\pi \times 3 \times 10^3 \times 500} = 53 \text{ mH}$$

These are the total series inductance and capacitance. Therefore for T-section $\dfrac{L_1}{2} = 26.5$ mH and $C = 0.212$ μF

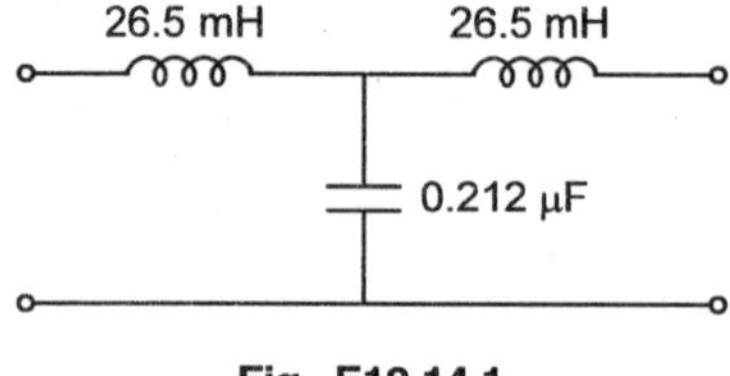

Fig. E12.14.1

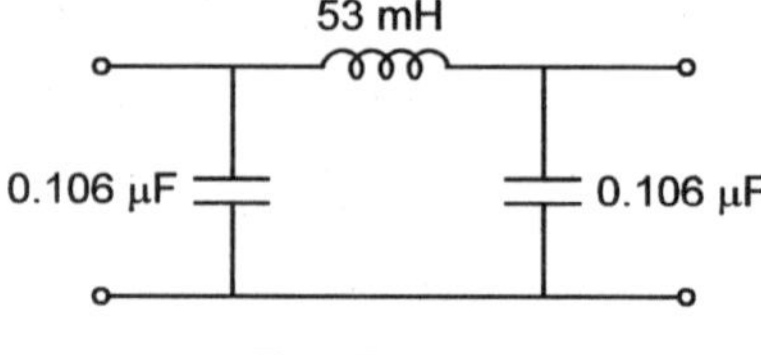

Fig. E12.14.2

whereas for the π–section $L = 53$ mH and $\dfrac{C}{2} = 0.106$ μF

Now attenuation $\alpha = 2 \cos h^{-1} \dfrac{f}{f_c}$ nepers (Eqn. (12.21)

where the required attenuation is in db. Therefore, attenuation in neper (required)

$$2 \cosh^{-1} \frac{f}{f_c} = \frac{20}{8.686} = 2.3025$$

or
$$1.1512 = \cosh^{-1} \frac{f}{f_c}$$

or
$$\frac{f}{f_c} = \cosh(1.1512) = 1.739$$

$$f = 5.217 \text{ KHz}$$

The phase constant corresponding to 2 KHz is given as

$$\beta = 2 \sin^{-1} \frac{f}{f_c} = 2 \sin^{-1} \frac{2}{3} = 83.6°$$

or
$$\beta = 1.46 \text{ radian}$$

and for $f = 10$ KHz, since it is much beyond the pass band, phase constant is $\pi$ radian. **Ans.**

**Example 12.15:** Design a high pass, constant-k type filter with T-section and π-section, when the cut off frequency is 8 KHz and the nominal characteristic impedance is 500 Ω. Also determine the attenuation and phase constant for frequencies (*i*) 5 KHz (*ii*) 20 KHz.

**Solution:** We know that for high pass, constant-k filter (Equation 12.27)

$$R_k = \sqrt{\frac{L}{C}} \quad \text{and} \quad f_0 = \frac{1}{4\pi\sqrt{LC}}$$

$$R_k f_0 = \sqrt{\frac{L}{C}} \cdot \frac{1}{4\pi\sqrt{LC}} = \frac{1}{4\pi C}$$

or
$$C = \frac{1}{4\pi R_k f_0}$$

$$= \frac{1}{4\pi \times 500 \times 8 \times 10^3} = 0.02 \ \mu F$$

$$L = C R_k^2 = 0.02 \times 500^2 \times 10^{-6} = 5 \ mH$$

For T-section, the circuit is as follows in Fig E12.15 (a)

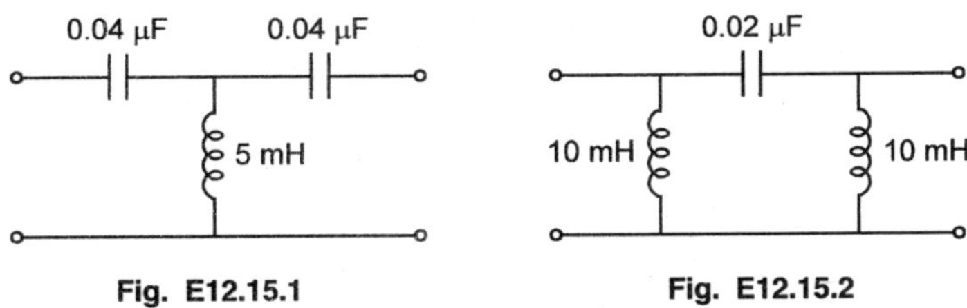


**Fig. E12.15.1**          **Fig. E12.15.2**

and for $\pi$-section the circuit is shown in Fig. E12.2.2
(*i*) Attenuation and phase constant at 5 KHz, since it is a stop band frequency for the high pass filter, attenuation

$$\alpha = 2 \cosh^{-1} \frac{f_c}{f} = 2 \cos h^{-1} \frac{8}{5}$$

$$\simeq 2 \ nepers$$

phase constant $\beta = \pi$ radian.

(*ii*) Attenuation and phase constant at 20 KHz since it is a pass band $\alpha = 0$

and
$$\beta = 2 \sin^{-1} \frac{f_c}{f} = 2 \sin^{-1} \frac{8}{20}$$

$$= 0.822 \ radian \ \textbf{Ans.}$$

**Example 12.16:** Design a band pass, constant-k filter with cut-off frequency of 4 KHz and 10 KHz and nominal characteristic impedance of 500 $\Omega$.

**Solution:** Given $R_k = 500 \ \Omega$   $f_1 = 4 \ KHz$   $f_2 = 10 \ KHz$
To obtain $L_1, C_1, L_2, C_2$ the series and shunt arm parameters.
Using the formulae for various parameters (equation 12.43 to (12.46) we have

$$L_1 = \frac{R_k}{\pi(f_2 - f_1)} = \frac{500}{\pi(10 - 4) \times 10^3}$$

$$= 26.54 \ mH$$

$$C_1 = \frac{f_2 - f_1}{4\pi f_1 f_2 R_k} = \frac{6}{4\pi \times 4 \times 10 \times 10^3 \times 500}$$

$$= 0.023 \ \mu F$$

$$L_2 = \frac{f_2 - f_1}{4 f_1 f_2} R_k = \frac{6 \times 10^3 \times 500}{4 \times 4 \times 10 \times 10^6}$$

$$= 18.75 \text{ mH}$$

$$C_2 = \frac{1}{\pi \left(f_2 - f_1\right) R_k} = \frac{1}{\pi \times 6 \times 10^3 \times 500}$$

$$= 0.106 \ \mu\text{F}$$

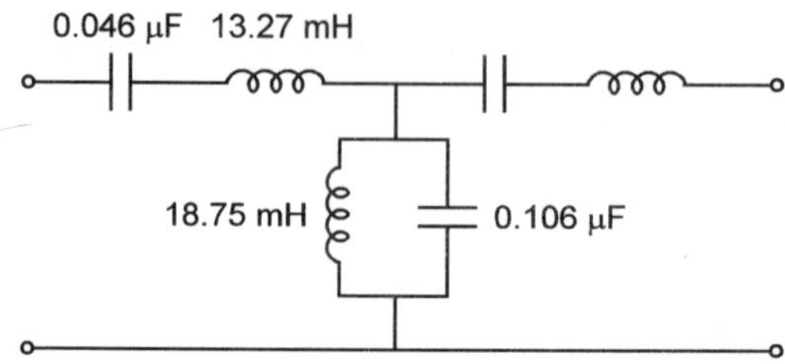



**Fig. E12.16**

**Example 12.17:** Design a band stop, constant-k filter with cut off frequencies of 4 KHz and 10 KHz and nominal characteristic impedance of 500 $\Omega$.

**Solution:** Given
$$R_k = 500 \ \Omega$$

$$f_1 = 4 \text{ KHz} \qquad f_2 = 10 \text{ KHz}$$

To obtain $L_1 \ C_1$, $L_2 \ C_2$ the series and shunt arm parameters (Equation 12.55 to 12.58)

$$L_1 = \frac{f_2 - f_1}{\pi f_1 f_2} R_k = \frac{6 \times 500}{\pi \times 4 \times 10 \times 10^3}$$

$$= 23.88 \text{ mH}$$

$$C_1 = \frac{1}{4\pi \times 6 \times 10^3 \times 500} = 0.026 \ \mu\text{F}$$

$$L_2 = \frac{R_k}{4\pi \left(f_2 - f_1\right)} = \frac{500}{4\pi \times 6 \times 10^3} = 6.6 \text{ mH}$$

$$C_2 = \frac{f_2 - f_1}{\pi f_1 f_2 R_k} = \frac{6}{\pi \times 4 \times 10 \times 10^3 \times 500}$$

$$= 0.095 \ \mu\text{F}$$

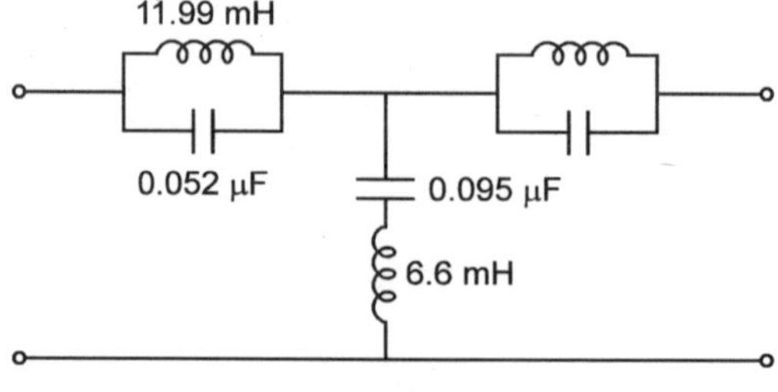



**Fig. E12.17**

From the foregoing analysis of various types of constant-k filters following observations are made:

1. The characteristic impedance of the filter circuit does not remain constant over the pass band and is a function of frequency i.e. it varies with frequency. As a result it is not possible to terminate the filter with a suitable circuit. We need to have a very large number of circuits to be connected in cascade at the load terminal to equal to the varying characteristic impedance. For economic reasons, therefore, it is terminated by a resistance equal to nominal characteristic impedance.

2. The attenuation near cut-off is not very high i.e. this filter does not provide sharp cut-off. Again, in order to have sharp cut-off we need to connect large number of circuits in cascade which is uneconomical.

## 12.13 m-DERIVED FILTERS

The m-derived filters to be discussed next improve the performance of the constant-k filters. The design starts with the constant-k filter sections, the series impedance is multiplied by a factor $m$, the shunt impedance is divided by the same factor and an additional reactance of the opposite sign is added either in series or in parallel to produce an m-derived filter. Fig. 12.29 shows a typical m-derived filter.

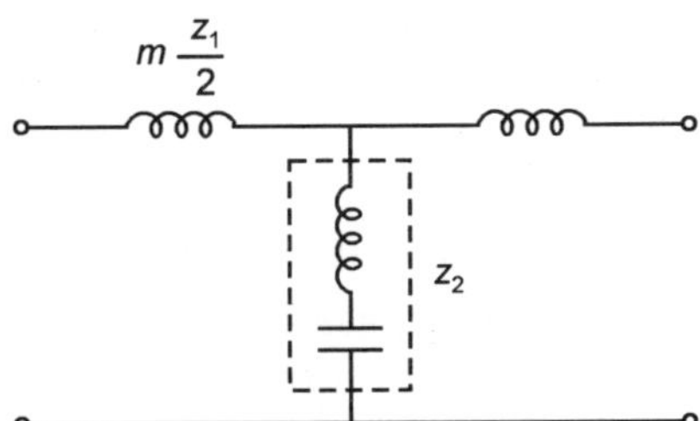

**Fig. 12.29** A typical m-derived filter.

The frequency of resonance of the shunt arm is so chosen that it is slightly higher than that of the k-type filter. Let this frequency be represented as $f_\infty > f_c$. At this freuency $(f_\infty)$ the impedance of the shunt branch is zero hence it is a short circuit i.e. the attenuation is infinite. This produces very sharp cut-off but the attenuation falls off beyond $f_\infty$. The decrease in attenuation is a disadvantage that can be overcome by cascading the m-derived section with a constant-k filter having a rising attenuation characteristic beyond cut-off. No reflection occurs because both sections have the same characteristic impedance. Fig. 12.30 shows the attenuation characteristic of a constant-k low pass filter cascaded with an m-derived filter.

It is to be noted that value of $m$ is so chosen that the characteristic **impedance of the** constant-k and m-derived is same.

Now characteristic impedance of the constant k is

$$Z_{OT} = \sqrt{Z_1 Z_2 \left(1 + \frac{Z_1}{4Z_2}\right)} = \sqrt{Z_1 Z_2 + \frac{Z_1^2}{4}}$$

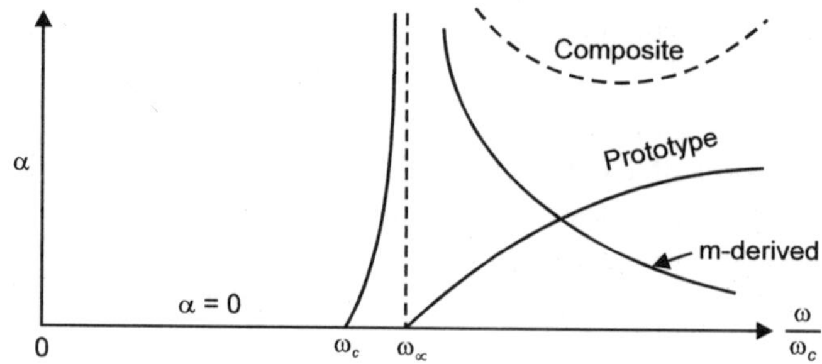



**Fig. 12.30** Attenuation characteristic of low-pass constant-k filter cascaded with m-derived section ($m = 0.6$).

For the m-derived

$$Z'_{OT} = \sqrt{m\,Z_1\,Z'_2 + \frac{m^2 Z_1^2}{4}} = Z_{OT} \qquad\qquad \ldots(12.145)$$

or

$$mZ_1\,Z'_2 + m^2\frac{Z_1^2}{4} = Z_1 Z_2 + \frac{Z_1^2}{4}$$

$$Z'_2 = \frac{Z_2}{m} + \frac{1-m^2}{m}\cdot\frac{Z_1}{4}$$

From the expression it is clear that the shunt arm of the m-derived filter consists of the impedance $Z_2/m$ and $\dfrac{1-m^2}{m}\cdot\dfrac{Z_1}{4}$ connected in series where $Z_1$ and $Z_2$ are the series and shunt impedance of the constant-k type low pass filter and m is a parameter suitably selected so that it provides sharp cut-off near the cut-off frequency. Since $Z_2$ is a capacitive reactance, $\dfrac{1-m^2}{4m}Z_1$, should be an inductive reactance so that series resonance may take place. This means $(1-m^2)$ should be a positive quantity which means $m$ should be less than one i.e.

$$0 < m < 1$$

Therefore, m-derived T-section is represented for low pass filter as in Fig. 12.31.

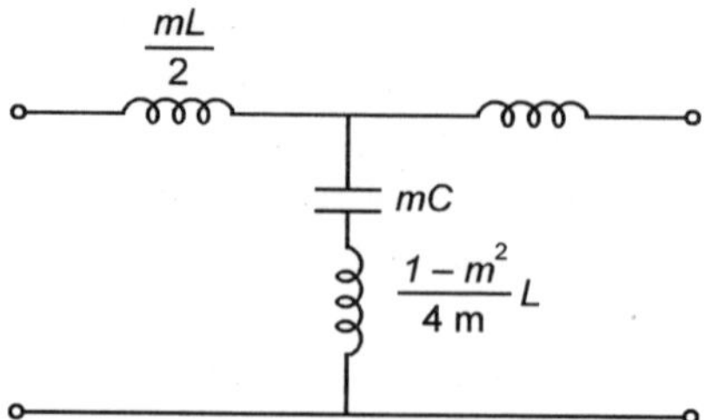



**Fig. 12.31** m-derived T-section filter.

To calculate m we take the condition where series resonance of the shunt arm takes place, we equate the reactances of the two elements at $f_\infty$.

$$\frac{1}{2\pi f_\infty mC} = \frac{1-m^2}{4m} L \cdot 2\pi f_\infty$$

or

$$\frac{1}{f_\infty^2} = (1-m^2) LC \pi^2 = (1-m^2) \cdot \frac{1}{f_c^2}$$

$$\left(\frac{f_c}{f_\infty}\right)^2 = 1 - m^2$$

or

$$m^2 = 1 - \left(\frac{f_c}{f_\infty}\right)^2$$

$$m = \sqrt{1 - \left(\frac{f_c}{f_\infty}\right)^2} \qquad \qquad ...(12.146)$$

With the calculation of $m$, all parameters of the m-derived filter can be obtained. However, if we select a suitable value of $m$, $f_\infty$ can be obtained from the same relations.

### 12.13.1 m-Derived $\pi$–section LPF

The shunt arm is taken as $m$ times the value of the impedance of the constant-k filter and obtain the value of series arm if the characteristic impedance remains same as that of the constant-k filter.

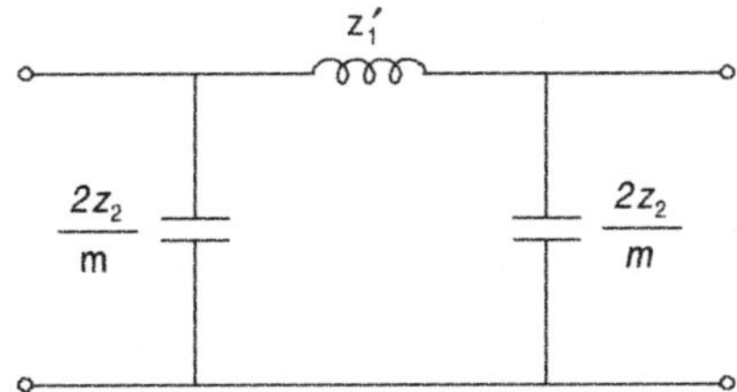


**Fig. 12.32**  m-derived  $\pi$–section LPF.

Equating the two characteristic impedances we have

$$\frac{Z_1 Z_2}{1 + \dfrac{Z_1}{4Z_2}} = \frac{Z_1' \cdot Z_2/m}{1 + \dfrac{Z_1'}{4Z_2/m}} \qquad \qquad ...(12.147)$$

$$\frac{Z_1 Z_2 \cdot 4Z_2}{Z_1 + 4Z_2} = \frac{Z_1' \dfrac{Z_2}{m} \cdot (4Z_2/m)}{4\dfrac{Z_2}{m} + Z_1'}$$

$$\frac{Z_1}{Z_1 + 4Z_2} = \frac{Z_1'/m^2}{4\dfrac{Z_2}{m} + Z_1'}$$

$$\frac{Z_1 + 4Z_2}{Z_1} = \frac{(4Z_2/m) + Z_1'}{Z_1'/m^2}$$

or
$$1 + \frac{4Z_2}{Z_1} = \frac{4Z_2}{m} \cdot \frac{m^2}{Z_1'} + m^2$$

or
$$1 + \frac{4Z_2}{Z_1} - m^2 = \frac{4m\, Z_2}{Z_1'}$$

$$\frac{1}{4m\,Z_2}\left[(1 - m^2) + \frac{4\,Z_2}{Z_1}\right] = \frac{1}{Z_1'}$$

or
$$Z_1' = \frac{4\,m\,Z_2}{(1 - m^2) + \dfrac{4\,Z_2}{Z_1}} = \frac{4m\,Z_1\,Z_2}{(1 - m^2)Z_1 + 4Z_2}$$

Multiplying and dividing r.h.s by $\dfrac{m}{1 - m^2}$ we have

$$Z_1' = \frac{m\,Z_1 \cdot \dfrac{4m}{1 - m^2} \cdot Z_2}{m\,Z_1 + \dfrac{4m}{1 - m^2}\, Z_2} \qquad \qquad ...(12.148)$$

Hence the series branch is a parallel combination of $mZ_1$ and $\dfrac{4m}{1 - m^2} \cdot Z_2$ impedances. The circuit is shown in Fig. 12.33.

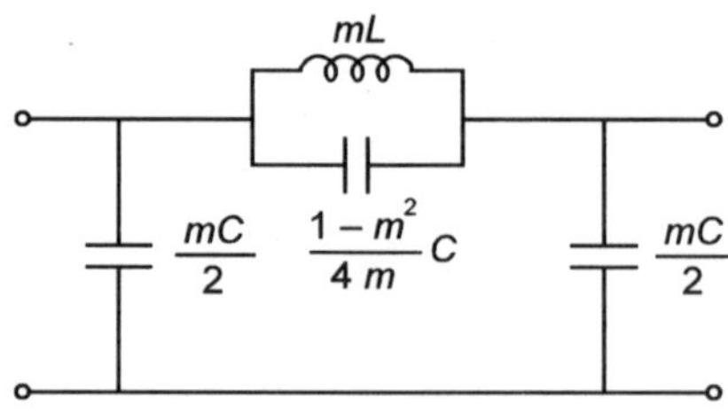


**Fig. 12.33**

Here the parameters of the series arm are so chosen that the parallel circuit resonates at $f_\infty$ when the series circuit works as an open circuit and hence infinite attenuation. At resonance

$$2\pi mL f_\infty = \frac{1}{2\pi f_\infty \dfrac{1 - m^2}{4m} C}$$

$$4\pi^2 mLC f_\infty^2 = \frac{4m}{1 - m^2}$$

$$\frac{f_\infty^2}{f_c^2} = \frac{1}{1 - m^2} \quad \text{or} \quad f_\infty = \frac{f_c}{\sqrt{1 - m^2}} \qquad \qquad ...(12.149)$$

Since $m < 1$, $f_\infty > f_c$ which is a desirable feature.

From the equation (12.63)

$$m = \sqrt{1 - \left(\frac{f_c}{f_\infty}\right)^2} \qquad \qquad ...(12.150)$$

This is the same relation as that derived for the low pass $T$–section filter.

The attenuation for m-derived low pass filter in the stop band is given as

$$\alpha = 2\,\cosh^{-1} \sqrt{\frac{Z_1}{4Z_2}} \quad f_c < f < f_\infty$$

$$= 2\,\sinh^{-1} \sqrt{\frac{Z_1}{4Z_2}} \quad f > f_\infty \qquad \qquad ...(12.151)$$

Now $Z_1 = j\omega L$ and $Z_2 = \dfrac{1}{j\omega C}$ for the low pass constant-k filter. Using the m-derived circuit we have (Fig. 12.31).

$$\frac{Z_1}{4Z_2} = \frac{m\,\omega L}{4\left[\dfrac{1}{m\,\omega C} - \dfrac{1 - m^2}{4m}\,\omega L\right]} \qquad \qquad ...(12.152)$$

$$\alpha = 2\,\cosh^{-1} \frac{mf/f_c}{\sqrt{1 - \left(\dfrac{f}{f_\infty}\right)^2}} \quad f_c < f < f_\infty$$

$$= 2\,\sinh^{-1} \frac{mf/f_c}{\sqrt{1 - \left(\dfrac{f}{f_\infty}\right)^2}} \quad f > f_\infty \qquad \qquad ...(12.153)$$

From these expressions it is clear that there is substantial increase in the sharpness of cut-off for the m-derived section as compared to the constant-k type filter.

The phase constant during the pass band is given by

$$\beta = 2\,\sin^{-1} \sqrt{\left|\frac{Z_1}{4Z_2}\right|}$$

Substituting the value of $z_1$ and $z_2$ for m-derived section we have

$$\beta = 2\,\sin^{-1} \frac{mf/f_c}{\sqrt{1 - \left(\dfrac{f}{f_c}\right)^2 (1 - m^2)}} \qquad \qquad ...(12.154)$$

During attenuation band, however, upto $f_\infty$ phase constant $\beta$ has the value $\pi$ and for $f > f_\infty$ the value drops to zero as the shunt arm goes inductive above series resonance.

The factor $m$ varies between zero and one and can be selected to achieve different responses. Fig. 12.34 shows the attenuation responses of low pass m-derived filter section for various values of $m$. For $m = 1$ it represents response of low pass constant-k filter. A value of $m$ less than one produces a response that has a frequency of infinite attenuation $f_\infty$ which corresponds to the frequency of resonance of shunt arm. As m reduces $f_\infty$ also reduces and produces a steeper cut-off characteristic.

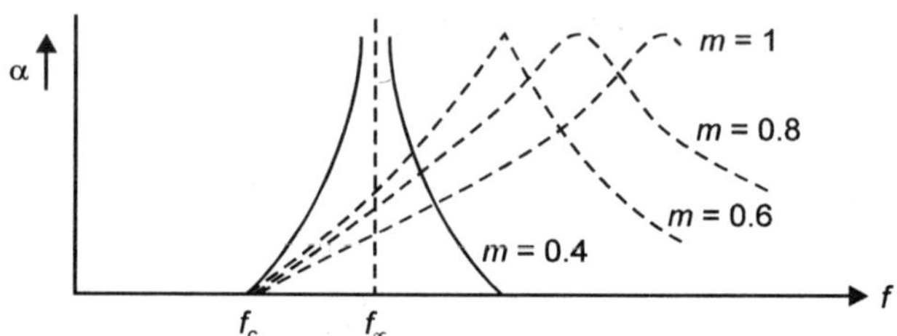


**Fig. 12.34**   Attenuation characteristic of LP m-derived filter.

Response for high pass m-derived filters are similar to, but mirror image of those shown above.

## 12.14 m-DERIVED HIGH PASS FILTER T-SECTION

The impedance of the shunt arm of the low pass filter is

$$\frac{Z_2}{m} + \frac{1-m^2}{4m}Z_1$$

In case of high pass filter

$$Z_1 = \frac{1}{j\omega C} \quad \text{and} \quad Z_2 = j\omega L$$

Therefore, the shunt arm consists of a series inductance of $\dfrac{L}{m}$ henry and a capacitance of $\dfrac{4mC}{1-m^2}$. Hence the circuit is as shown here

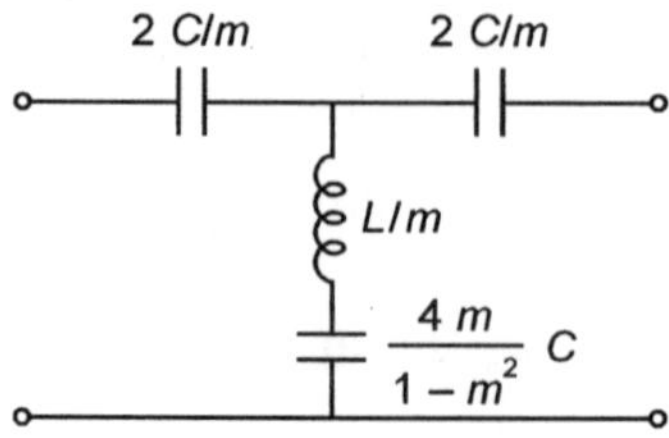


**Fig. 12.35**   m-derived high pass T-section filter.

The shunt arm resonates at a frequency $f_\infty$ where the shunt arm offers zero reactance and hence transmission at this frequency is zero or attenuation is infinity.

Therefore, at $f_\infty$

$$2\pi f_\infty \frac{L}{m} = \frac{1}{2\pi f_\infty \dfrac{4mC}{1-m^2}}$$

or
$$f_\infty^2 = \frac{1}{16\pi^2 \dfrac{1}{1-m^2} LC} \qquad \text{...(12.155)}$$

Now for a high pass filter

$$f_c = \frac{1}{4\pi \sqrt{LC}} \qquad \text{...(12.156)}$$

Therefore,
$$f_\infty^2 = f_c^2 (1-m^2)$$

or
$$f_\infty = f_c \sqrt{1-m^2}$$

or
$$f_c^2 - f_\infty^2 = m^2 f_c^2$$

or
$$m = \sqrt{1 - \left(\frac{f_\infty}{f_c}\right)^2} \qquad \text{...(12.157)}$$

Thus, if the values of $f_c$, $f_\infty$ and the nominal characteristic impedance $R_k$ are known, values of various parameters for the high pass filter can be obtained.

**π–section:** Here if we see Fig. 12.33 for a low-pass π–section, we notice that the series arm impedance is given by a parallel combination of mL and $\dfrac{1-m^2}{4m}C$, therfore for the high pass filter the respective element will be a capacitance of $\dfrac{C}{m}$ and inductor of $\dfrac{4m}{1-m^2}L$ in parallel. Therefore, the circuit is represented as shown in Fig. 12.36.

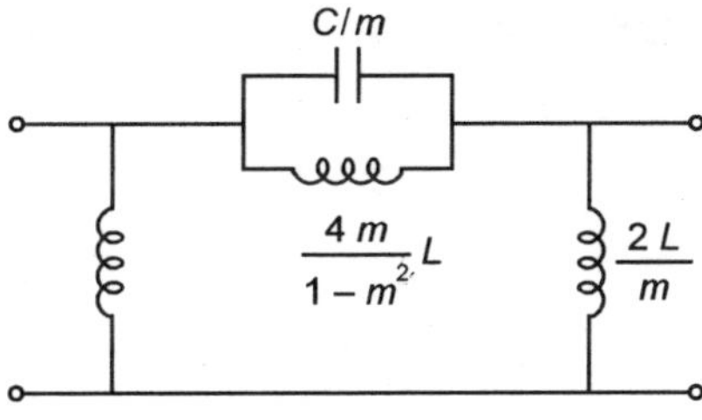


**Fig. 12.36** m-derived high pass π–section filter.

Here the resonance occurs in the series arm at frequency $f_\infty$ when the series arm acts as an open circuit being a parallel circuit hence at this frequency no transmissions of signal is possible hence the attenuation is infinite.

Therefore, at $f_\infty$

$$\frac{1}{2\pi f_\infty C/m} = 2\pi f_\infty \cdot \frac{4m}{1-m^2}L \qquad \text{...(12.158)}$$

or

$$\frac{1}{f_\infty^2} = 16\pi^2 LC \cdot \frac{1}{1-m^2}$$

or

$$\left(\frac{f_c}{f_\infty}\right)^2 = \frac{1}{1-m^2}$$

or

$$\left(\frac{f_\infty}{f_c}\right)^2 = 1 - m^2$$

or

$$m = \sqrt{1 - \left(\frac{f_\infty}{f_c}\right)^2} \qquad \text{...(12.159)}$$

The value of $m$ comes out to be same as for a High pass T-section. Having known $f_c$, $f_\infty$ and $R_k$ the parameters of the high pass $\pi$-section can be obtained.

## 12.15 m-DERIVED BAND PASS FILTER

We know that if $Z_1$ is the total series arm impedance and $Z_2$ the shunt arm impedance of a constant-k T-section network then the m-derived circuit has series arm of $\dfrac{Z_1}{m}$ and the shunt arm $\dfrac{Z_2}{m} + \dfrac{1-m^2}{4m} Z_1$. By treating the entire series arm impedance of the constant-k band pass filter as $Z_1$ and the shunt arm $Z_2$ (Fig. 12.26) we obtain the m-derived band pass filter as shown in Fig. 12.37.

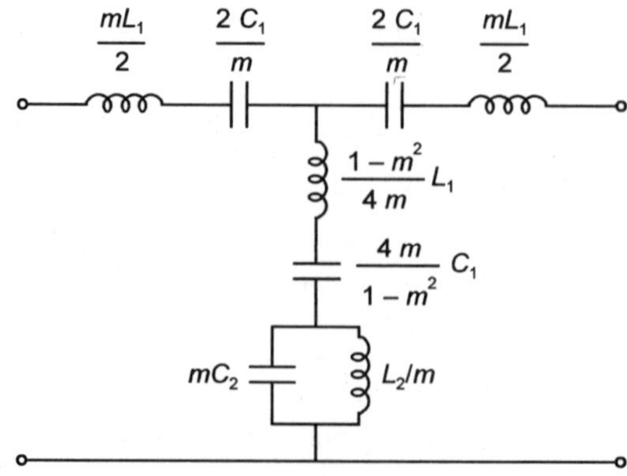


**Fig. 12.37** General configuration of m-derived band pass filter.

The shunt arm consists of a series resonant circuit and an anti-resonance parallel circuit. The anti-resonance frequency of the arm as a whole should be $f_0$ where the circuit (shunt arm) resonates at a frequency below $f_0$ and again a frequency above $f_0$. At these frequencies the shunt arm acts as a short circuit and thus at these frequencies the attenuation is infinite. These frequencies of high attenuation $f_\infty$ appear on each side of the pass band and the m-derived section could be used to have a sharp cut-off.

At resonance for the shunt arm, the net impedance should be zero.

$$j\omega_\infty L_1 \frac{1-m^2}{4m} + \frac{1}{j\omega_\infty C_1}\frac{1-m^2}{4m} + \frac{j\omega_\infty \dfrac{L_2}{m}\cdot\dfrac{1}{j\omega_\infty mC_2}}{j\omega_\infty \dfrac{L_2}{m} + \dfrac{1}{j\omega_\infty mC_2}} = 0 \qquad \text{...(12.160)}$$

or

$$\frac{1-m^2}{4}\left(\omega_\infty^2 L_1 C_1 - 1\right) = \frac{\omega_\infty^2 L_2 C_1}{\omega_\infty^2 L_2 C_2 - 1} \qquad \text{...(12.161)}$$

Now for a BPF

$$L_1 C_1 = L_2 C_2 = \frac{1}{\omega_0^2} \text{ and } f_1 f_2 = f_0^2$$

Also from equation (12.131) and (12.130), we have

$$L_2 = \frac{f_2 - f_1}{4\pi f_1 f_2} R_k \quad \text{and} \quad C_1 = \frac{f_2 - f_1}{4\pi f_1 f_2 R_k}$$

Therefore,

$$L_2 C_1 = \left(\frac{f_2 - f_1}{4\pi f_1 f_2}\right)^2 \qquad \text{...(12.162)}$$

Substituting these values in equation (12.161) we have

$$\frac{1-m^2}{4}\left[\left(\frac{f_\infty}{f_0}\right)^2 - 1\right]^2 = 4\pi^2 f_\infty^2 \left(\frac{f_2 - f_1}{4\pi f_1 f_2}\right)^2$$

$$(1-m^2)\left(f_\infty^2 - f_0^2\right)^2 = f_\infty^2 (f_2 - f_1)^2$$

or

$$\left(f_\infty^2 - f_1 f_2\right)^2 = \frac{f_\infty^2 (f_2 - f_1)^2}{1 - m^2}$$

$$f_\infty^2 - f_1 f_2 = \frac{f_\infty (f_2 - f_1)}{\sqrt{1 - m^2}}$$

$$f_\infty^2 - \frac{f_2 - f_1}{\sqrt{1 - m^2}} f_\infty - f_1 f_2 = 0 \qquad \text{...(12.163)}$$

$$f_\infty = \frac{f_2 - f_1}{2\sqrt{1 - m^2}} \pm \sqrt{\frac{(f_2 - f_1)^2}{4(1 - m)^2} + f_1 f_2} \qquad \text{...(12.164)}$$

Since the term under the radical sign is greater than the first term, one of the frequencies would turn out to be negative which has no physical relevance hence, reversing the two terms we obtain.

$$f_{\infty_1} = \sqrt{\frac{(f_2 - f_1)^2}{4(1 - m)^2} + f_1 f_2} - \frac{f_2 - f_1}{2\sqrt{1 - m^2}} \qquad \text{...(12.165)}$$

and
$$f_{\infty_2} = \sqrt{\frac{(f_2 - f_1)^2}{4(1 - m^2)} + f_1 f_2} + \frac{f_2 - f_1}{2\sqrt{1 - m^2}} \qquad \text{...(12.166)}$$

Therefore,
$$f_{\infty_2} - f_{\infty_1} = \frac{f_2 - f_1}{\sqrt{1 - m^2}} \qquad \text{...(12.167)}$$

or
$$m = \sqrt{1 - \left(\frac{f_2 - f_1}{f_{\infty_2} - f_{\infty_1}}\right)^2} \qquad \text{...(12.168)}$$

We know that for a general quadratic equation $ax^2 + bx + c = 0$ the product of the two roots equals $c/a$

Therefore, here $f_{\infty_2} f_{\infty_1} = f_1 f_2 = f_0^2$

or
$$f_0 = \sqrt{f_{\infty_1} f_{\infty_2}} \qquad \text{...(12.169)}$$

In order to obtain various parameters of the m-derived band pass filter, the values of the parameters $L_1, L_2, C_1, C_2$ are obtained using the expressions for the corresponding constant-k filter and then using the value $m$ as obtained from the above expression, the actual values of the various parameters can be obtained. It is to be noted that if the value of $m$ is not available, an optimum value of $m = 0.6$ can be used.

**Example 12.18:** A low pass constant-k filter with cut-off frequency $f_C = 36$ KHz is required to produce a maximum attenuation at 60 KHz when used with 500 $\Omega$ termination. Design a suitable m-derived (*i*) T-section (*ii*) π-section filter.

**Solution:** For a low pass constant-k filter

$$f_c = \frac{1}{\pi\sqrt{LC}} \quad \text{and} \quad R_k = \sqrt{\frac{L}{C}}$$

Therefore,
$$R_k f_c = \frac{1}{\pi C} \quad \text{or} \quad C = \frac{1}{\pi R_k f_c}$$

$$C = \frac{1}{\pi \times 500 \times 36 \times 10^3} = 0.0177 \ \mu F$$

$$L = C R_k^2 = 0.0177 \times 500^2 \times 10^{-6}$$

$$= 4.425 \ mH$$

For the m-derived low pass T-section

$$m = \sqrt{1 - \left(\frac{f_C}{f_\infty}\right)^2} = \sqrt{1 - \left(\frac{36}{60}\right)^2}$$

$$= 0.8$$

Therefore series inductance $= 0.8 \times 4.425 = 3.54$ mH

$$\text{shunt inductance} = \frac{1 - m^2}{4m} L = \frac{1 - 0.64}{4 \times 0.8} \times 4.425$$

$$= 0.4978 \text{ mH}$$

$$\text{shunt capacitor} = mC = 0.8 \times 0.0177$$

$$= 0.01416 \text{ } \mu\text{F}$$

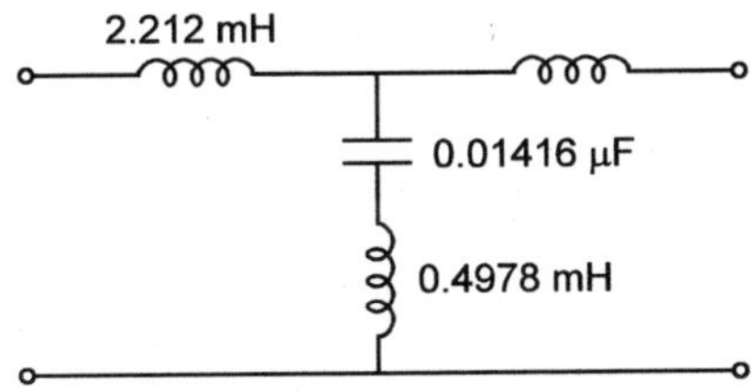


**Fig. E12.18.1**  Constant-k-type.

*π–section* For π-section the series inductance is again 4.425 mH and shunt capacitance is 0.0177 µF. The value of $m$ is same as that for the T-section i.e. $m = 0.8$.

Series arm is a parallel combination of mL and $\dfrac{1 - m^2}{4m} C$

$$mL = 0.8 \times 4.425 = 3.54 \text{ mH}$$

and

$$\frac{1 - m^2}{4m} C = \frac{0.36}{3.2} \times 0.0177 = 0.002 \text{ } \mu\text{F}$$

The shunt branch each has a capacitance $\dfrac{mc}{2}$

$$\frac{0.8 \times 0.0177}{2} = 0.007 \text{ } \mu\text{F}$$

The circuit is as shown

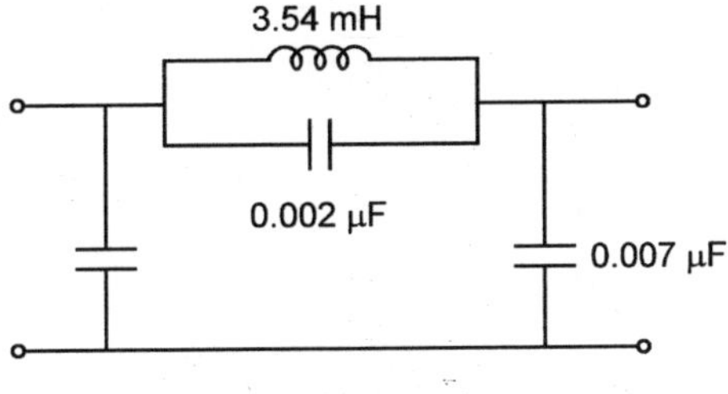


**Fig. E12.18.2**

**Example 12.19:** A high pass constant-k filter with cut-off frequency 40 KHz, is required to produce a maximum attenuation at 36 KHz when used with terminated resistance of 500 ohm. Design a suitable m-derived (*i*) T-section (*ii*) π-section.

**Solution:** T-section.

For a high pass constant-k filter

$$f_c = \frac{1}{4\pi\sqrt{LC}} \quad \text{and} \quad R_k = \sqrt{\frac{L}{C}}$$

$$f_c R_k = \frac{1}{4\pi C} \quad \text{or} \quad C = \frac{1}{4\pi f_c R_k}$$

$$C = \frac{1}{4\pi \times 40 \times 10^3 \times 500} = 3.98 \times 10^{-9} \text{ F}$$

$$= 3.98 \text{ nF}$$

$$L = C R_k^2 = 3.98 \times 10^{-9} \times 500 \times 500$$

$$= 0.995 \text{ mH.}$$

Also

$$m = \sqrt{1 - \left(\frac{f_\infty}{f_c}\right)^2} = \sqrt{1 - \left(\frac{36}{40}\right)^2} = 0.436$$

For T-section

Series arm

$$\frac{2C}{m} = \frac{2 \times 3.98}{0.436} = 18.25 \text{ nF}$$

For shunt arm

$$\frac{L}{m} = \frac{0.995}{0.436} = 2.28 \text{ mH}$$

$$\frac{4m}{1-m^2} C = \frac{4 \times 0.436}{1 - 0.19} \times 3.98 = 2.15 \text{ nF}$$

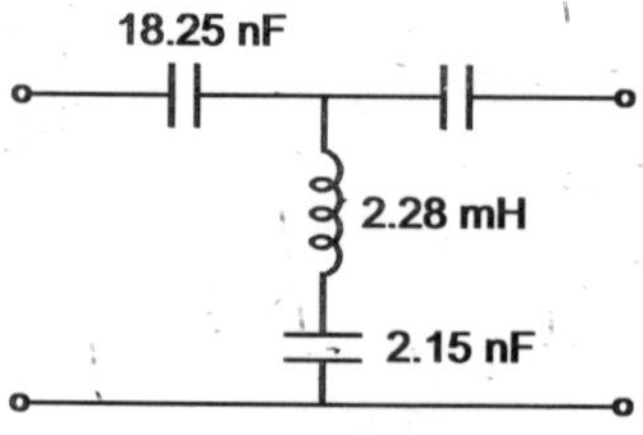


**Fig. E12.19.1**

π-section

Series arm

$$\frac{C}{m} = 9.125 \text{ nF in parallel}$$

with

$$\frac{4m}{1-m^2} L = \frac{4 \times 0.436}{0.81} \times 0.995 = 2.15 \text{ mH}$$

Shunt arm
$$\frac{2L}{m} = \frac{2 \times 0.995}{0.436} = 4.56 \text{ mH}$$

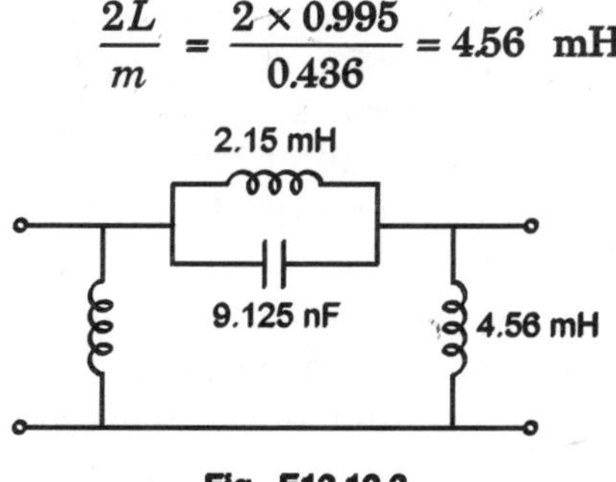

**Fig. E12.19.2**

**Example 12.20:** Design a band pass m-derived filter with cut-off frequencies of 4 KHz and 10 KHz and nominal characteristic impedance of 500 $\Omega$ Assume m = 0.6 and determine the frequency at which maximum attenuation is achieved.

**Solution:** First of all, the parameters of band pass filter, constant-k type are obtained as calculated in example 12.16 and then using $m = 0.6$ various parameters corresponding to Fig. 12.37 are obtained. To find out the frequencies where maximum attenuation is achieved we substitute the values of $f_1 = 4$ KHz $f_2 = 10$ KHz and $m = 0.6$ in the equation 12.163 and we have

$$f_\infty^2 - \frac{f_2 - f_1}{\sqrt{1 - m^2}}\, f_\infty - f_1 f_2 = 0$$
$$f_\infty^2 - 7.5 f_\infty - 40 = 0$$

$$f_\infty = \frac{7.5 \pm \sqrt{56.26 + 160}}{2} = \frac{7.5 \pm 14.7}{2}$$

Since one of the frequencies comes out to be negative we interchange the number and we have

$$\frac{14.7 \pm 7.5}{2} = 11.1 \text{ KHz and } 3.6 \text{ KHz}$$

These are the frequencies one of which is greater than $f_2$ and the other is less than $f_1$. The product of $f_{\infty_1}$ and $f_{\infty_2}$ is 39.96 which is approx. $f_1 f_2 = 40$.

## 12.16  COMPOSITE FILTERS

It is to be noted that the characteristic impedance of m-derived filters also changes with frequency as $m$ is varied. This is shown in Fig. 12.38.

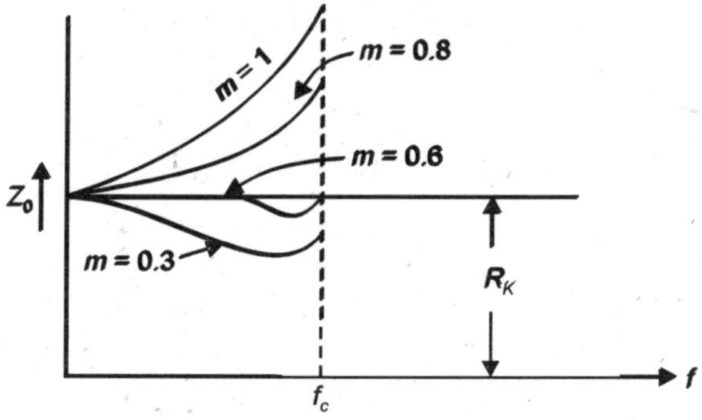


**Fig. 12.38  Characteristic** impedance of **m-derived filters.**

It is seen that characteristic impedance for the low-pass m-derived π-section filter with $m = 0.6$ remains constant over a band of 85% of the band pass region and is equal to the terminating resistance of the load. Therefore, a ½ section-π (L-section) m-derived with $m = 0.6$ can be used to improve the matching of filter which consists of a large section of filters in cascade. The m-derived filter can thus be used to improve the performance of constant-k type filters in the following ways:

1. It can provide steeper cut-off by using a low value of $m$.
2. It can provide maximum attenuation at a particular frequency $(f_\infty)$.
3. It can improve the matching of the filter to the source and the load by providing ½ π-section with m = 0.6.

Any or all of these advantages can be achieved by using several sections of m-derived or a combination of constant-k and m-derived filter section terminated at either end with a ½ section π-filters with m = 0.6. Such a circuit is shown in Fig. 12.39 and the whole arrangement is known as composite filter.

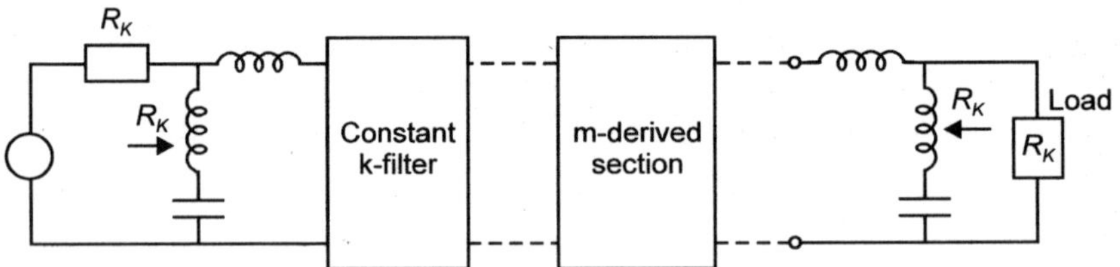

**Fig. 12.39**  A Composite Filter.

A composite filter, therefore, consists of the following components:

1. One or more constant-k sections to produce cut-off or transition between the pass band and the attenuation band at a specified frequency $f_c$.
2. One or more m-derived sections to give infinite attenuation at a frequency $f_\infty$ near to $f_c$.
3. Two-terminating m-derived half sections with m = 0.6 to provide matching with the source and the load.

The following example illustrates the design of a composite filter.

**Example 12.21:** Design a composite LP filter with the following specifications.
$$f_c = 4 \text{ KHz} \qquad f_\infty = 4.1 \text{ KHz} \qquad R_k = 500 \ \Omega.$$

**Solution:** The elements of the constant-k filter are obtained as follows:

$$f_c = \frac{1}{\pi\sqrt{LC}} \quad \text{and} \quad R_k = \sqrt{\frac{L}{C}}$$

or

$$\pi f_c R_k = \frac{1}{C} \quad \text{or} \quad C = \frac{1}{\pi R_k f_c}$$

$$C = \frac{1}{\pi \times 500 \times 4 \times 10^3} = 0.159 \ \mu\text{F}$$

$$L = C R_k^2 = 0.159 \times 500^2 \times 10^{-6} = 39.75 \text{ mH}$$

The value of $m$ to provide infinite attenuation at $f_\infty = 4.1$ KHz is given as

$$m = \sqrt{1 - \left(\frac{f_C}{f_\infty}\right)^2} = \sqrt{1 - \left(\frac{4}{4.1}\right)^2} = 0.22$$

The components of the m-derived T-section are:

$$\frac{mL}{2} = \frac{0.22 \times 39.75}{2} = 4.375 \text{ mH}$$

$$mc = 0.22 \times 0.159 = 0.035 \text{ } \mu F$$

$$\frac{1-m^2}{4m}L = \frac{1-0.22^2}{4 \times 0.22} \times 39.75 \text{ mH}$$

$$= 42.98 \text{ mH}$$

The ½ section T m-derived at each end will have elements with $m = 0.6$.
Full section T with m = 0.6

$$mL = 0.6 \times 39.75 = 23.85 \text{ mH}$$

$$mC = 0.6 \times 0.159 = 0.0954 \text{ mF}$$

$$\frac{1-m^2}{4m}L = \frac{1-0.6^2}{4 \times 0.6} \times 39.75 = 10.6 \text{ mH}$$

Half section values are

$$\frac{mL}{2} = 11.92 \text{ mH}$$

$$\frac{mC}{2} = 0.0477 \text{ } \mu F$$

$$\frac{1-m^2}{2m}L = 21.2 \text{ mH}$$

Therefore, the composite filter having one constant-k, one m-derived and one half section T at each end is shown here

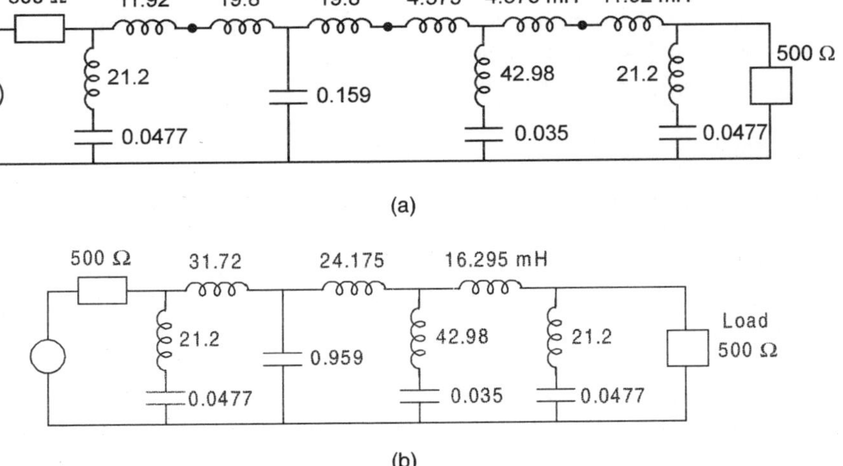


**Fig. E12.21**

In Fig. E12.21 (a) junction (terminals) of the various filters have been identified where the impedances are matched. Therefore, these junctions can be connected together without introducing reflections and the individual sections can be combined. In this process the individual sections lose their identity as T-sections and Fig. E12.21 (b) shows the composite filter which meets all the specification set forth in the statement of the problem. If a $\pi$-section is desired, similar procedure can be followed.

**PROBLEMS**

**12.1.** For a T-type attenuator if $R_1 = R_2 = 60\ \Omega$ and $R_3 = 90\ \Omega$, determine the characteristic resistance.

**12.2.** For a T-type attenuator if $R_1 = 60\ \Omega$ $R_2 = 240\ \Omega$ and $R_3 = 60\ \Omega$ determine the image resistances.

**12.3.** For a $\pi$-type attenuator $R_1 = 500\ \Omega$, $R_2 = R_3 = 750\ \Omega$ determine characteristic resistance. If the network is terminated by its characteristic resistance, determine its input resistance.

**12.4.** A symmetrical T-network has $\dfrac{R_1}{2} = 200\ \Omega$ and $R_2 = 62.5\ \Omega$. Determine its characteristic resistance using bisection theorem. If the network is terminated by its characteristic resistance, determine its input resistance. Also, determine the loss in dbs.

**12.5.** Design a bridged-T attenuator for attenuation (*i*) 0 dbs (*ii*) 10db when the load resistance $R_0 = 200\ \Omega$

**12.6.** In the circuit of Fig. below $R_1 = R_2 = 120\ \Omega$ and $R_3 = R_4 = 150\ \Omega$, determine $R_0$

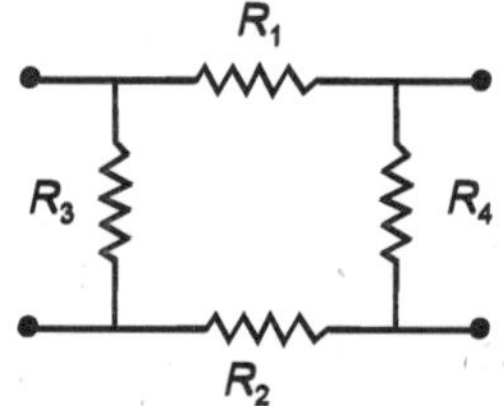


**Fig. P12.6**

Also, determine (a) the value of source and load impedance for matched condition and (b) the insertion loss in db when matched.

**12.7.** The arms for a lattice network are given by

$$R_a = R_0 \tanh \frac{\alpha}{2} \quad \text{and} \quad R_b = R_0 \coth \frac{\alpha}{2}$$

where $\alpha$ is expressed is nepers. Using bisection principle design equations for T-section and $\pi$-section attenuators.

**12.8.** In an asymmetrical T-type attenuator the series arms have resistances $500\ \Omega$ and $140\ \Omega$ whereas shunt arm has $40\ \Omega$. The network is supplied by a 10 V source with internal resistance of $160\ \Omega$. Determine the insertion loss when the load resistance is $40\ \Omega$.

**12.9.** For the matched network of the Fig. P12.9 determine (a) the transmission loss of each network and (b) the overall attenuation in dbs.

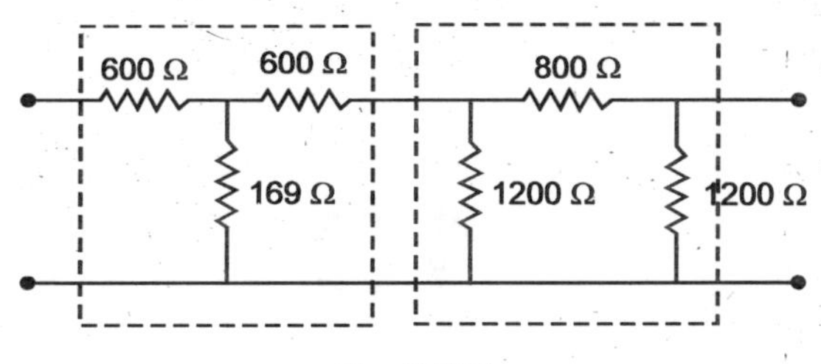

**Fig. P12.9**

**12.10.** A generator with internal resistance of 200 $\Omega$ is to be connected to a load of 150 $\Omega$ through an attenuator. Determine the parameters of the network that will provide matching with minimum attenuation. Also calculate the insertion loss provided by the network.

**12.11.** Design a bridged-T network when the internal resistance of the source and the resistance of the load equal 300 $\Omega$ and the insertion loss of 14 db.

**12.12.** Design a symmetrical T-type and $\pi$-type attenuator for 6 db attenuation and 100 ohm matched termination.

**12.13.** Design a low pass constant-k (*i*) T-section and (*ii*) $\pi$-section filter with $f_c$ = 6 KHz and $R_0$ = 500 ohm. Compute $\alpha$ and $\beta$ for the filters for $f$ = 10 KHz. Also determine the frequency at which the attenuation is 10 dbs.

**12.14.** Design a high pass constant-k (*i*) T-section and (*ii*) $\pi$-section filter with $R_k$ = 500 $\Omega$, $f_c$ = 4 KHz, one sharp cut-off section at $f_\infty$ = 3.8 KHz and terminating half section with $m$ = 0.6. Combine the elements to form a composite filter.

**12.15.** Design a band pass constant-k filter with $f_{c1}$ = 2 KHz and $f_{c2}$ = 3 KHz and $R_k$ = 500 $\Omega$.

**12.16.** For the given low pass constant-k filter determine the nominal characteristic impedance and the cut-off frequency.

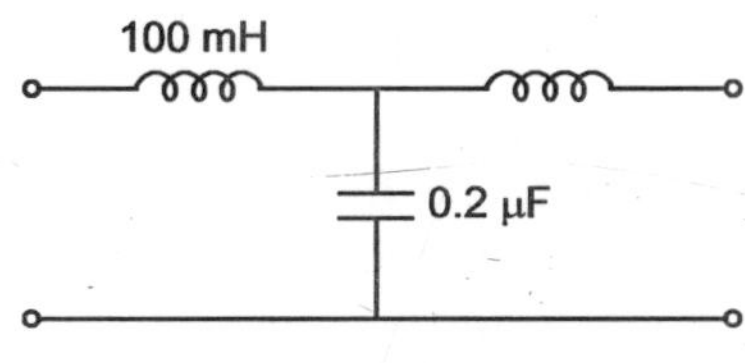

**Fig. P12.16**

Determine the attenuation and phase at $f$ = 2KHz

**12.17.** Design an m-derived low pass filter with T-section having cut-off frequency 7.2 KHz and infinite attenuation at $f_\infty$ = 7.5 KHz and $R_k$ = 500 $\Omega$.

**12.18.** A high pass constant-k filter with $f_c$ = 20 KHz is required to produce a maximum attenuation at 18 KHz when used with terminating resistance of 500 ohm. Design a suitable m-derived (*i*) T-section (*ii*) $\pi$-section.

**12.19.** Design a band-pass m-derived filter with specification as in P12.3. Assume m = 0.6 and determine the frequencies at which maximum attenuation is achieved.

**12.20.** Design a band stop constant-k filter with cut-off frequencies of 8 KHz and 12 KHz and nominal characteristic impedance of 500 $\Omega$.

**12.21.** In a band pass constant-k filter the ratio of the capacitances in the series and shunt arm is 1 : 100 and the resonant frequency of both the arms is 10 KHz, determine the bandwidth of the filter.

# 13

# Modern Filter Theory and Active RC Filters

## 13.1  INTRODUCTION

The use of active elements eliminates two basic requirement from RLC networks, those of passivity and reciprocity. Therefore, active networks not only realize network functions which can otherwise be realized by passive RLC circuit but also can be used to realize any driving point or transfer characteristic not achievable by passive networks.

In the previous chapter we studied design of passive filter where an inductor is an integral part of all types of filters. Inductor creates some of the problems and because of the other advantages of active elements passive filter remains a good academic exercise than a practical utility. Some of the limitations of passive filters are:

1. The use of inductor as a network element is not desirable especially at low frequencies (less than 1 KHz) as at these frequencies practical inductors of reasonable Q tend to become bulky, and expensive. The dissipative losses start increasing when an inductor is miniaturized. Roughly Q of a low frequency inductor is proportional to square of the scaling factor. This means an inductor having Q of 100 when reduced to 12.5% of the original size will have approximately a Q of 25.
2. While cascading different sections of filters a buffer or isolation amplifier is required to prevent loading of the circuit.
3. There is a need for an external amplifier to the required gain.

The use of linear active circuits has come about in a big way because of rapidly growing area of integrated circuits where many network elements and their functional interconnections are fabricated on a single chip. The main advantages of integrated micro circuits are:

(1) reduction in size and weight and increased equipment density.
(2) Reliability of the system is enhanced and there is reduction in cost. This is achieved by eliminating discrete joints and assembly operations.
(3) The functional performance is improved by register and compact distribution of circuit elements.
(4) Faster operation due to absence of parasites and decreased propagation delay.
(5) Reduction in power consumption.

However, the integrated circuits have the following disadvantages:

1. Integrated this film resistors and capacitors are limited to moderate values and their accuracies are poor.
2. Miniaturized inductors are of poor quality and can't be used for many applications.

Therefore, the practical solution to these problems is obtained by designing network using linear active element with resistors and capacitors as these are easily integrable.

The classical filter design is basically based on image parameters which depends upon image or iterative impedances, the modern filter design is based on the selection of the filter transfer function which satisfies certain specifications and then realisation of this function by synthesis techniques.

The modern filter design requires the following steps:

1. Selection of filter specification.
2. Selection of a suitable transfer function which satisfies the given specification (approximation).
3. Realisation of the transfer function through calculation of filter component values.
4. Fabrication and testing of the filter to see how accurate the approximation is to the real response required.

## 13.2  FILTER SPECIFICATIONS

Fig. 13.1 shows the characteristics of an ideal low pass filter.

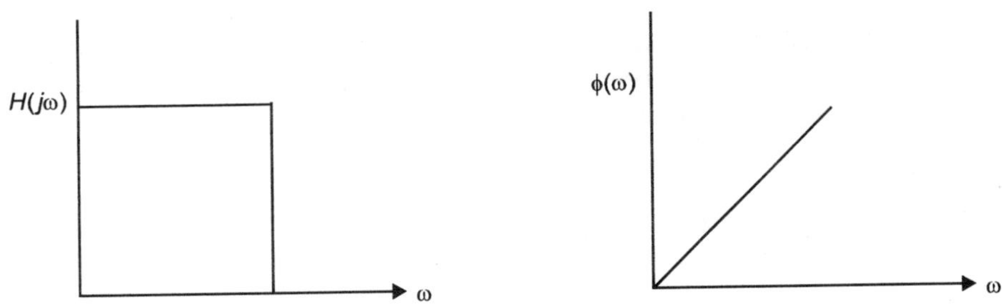


**Fig. 13.1**  Characteristic of an ideal LP filter.

The magnitude is a brick wall characteristic whereas the phase is a desirable linear characteristic. We will limit our discussion to magnitude characteristic only. Here the filter is required to allow signals of all frequencies upto $\omega_c$ without any attenuation (as it is a flat or brick-wall characteristic) and it should cut-off all signals beyond $\omega_C$ i.e. provide infinite attenuation. It is impossible to have practical filter which will have these characteristics. From Fourier transform theory we know that any band limited frequency response results in a non-causal time function as the time response of such a frequency response has the term $\sin \omega t/\omega t$ which exists even for $t < 0$ i.e. it is non-causal function and a non-causal function can not be realized with passive elements.

In the light of this we have to modify our specification so that these could be achieved by the filter circuits. Fig. 13.2 shows the specification for a realizable low-pass filter.

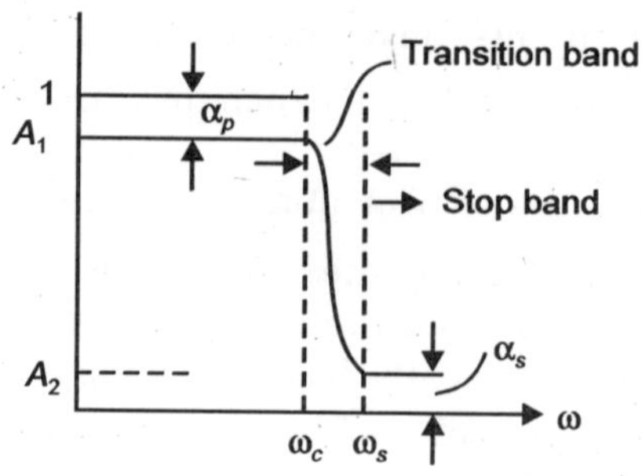



**Fig. 13.2**   Characteristic for a practical LP filter.

Here $\omega_c$ is the critical frequency of the low-pass filter and $\omega_s$ the start of the stop band region. Usually $\omega_c = \omega_s$ theoretically but here we have assumed that the function decreases $A_1$ to $A_2$ during $\omega_c < \omega < \omega_s$.

Here $\alpha_p$ is the maximum permissible pass band attenuation and $\alpha_s$ the minimum pass band attenuation i.e. the function $H(j\omega)$ does not reduce to zero but it may have value between 0 and $A_2$. $\omega_c$ need not necessarily coincide with $\omega_s$ and hence we have allowed a band of frequencies $\omega_C < \omega < \omega_S$ for the function to reduce to a value $A_2$ or less, so everywhere we have seen to it that we don't have sharp cut-off or discontinuities in the function $|H(j\omega)|$

Usually
$$\alpha \le \alpha_p \quad \omega \le \omega_c \qquad \qquad \qquad \text{...(13.1)}$$

and
$$\alpha \ge \alpha_s \quad \omega \ge \omega_s \qquad \qquad \qquad \text{...(13.2)}$$

where $\alpha$ is the attenuation in $dB$

The problem of approximation requires the selection of a suitable function whose frequency response fits into the response shown in Fig. 13.2.

Suppose $F(\omega)$ is the required response shown in Fig. 13.2 and $T(\omega)$ the function which approximates $F(\omega)$. The objective is to minimize

$$\epsilon(\omega) = \max \; |F(\omega) - T(\omega)| \qquad \qquad \qquad \text{...(13.3)}$$

over the range $\omega_1 \le \omega \le \omega_2$

The other criterion could be to minimize the mean square error i.e. $T(\omega)$ is so selected that

$$\epsilon(\omega) = \int_{\omega_1}^{\omega_2} |F(\omega) - T(\omega)|^2 \; d\omega \qquad \qquad \qquad \text{...(13.4)}$$

Yet another method to approximation to select $T(\omega)$ in such a way that at least at one point $\omega = \omega_0$, $T(\omega_0) = F(\omega_0)$.

Suppose we expand $F(\omega)$ and $T(\omega)$ around $\omega_0$ using Taylor's series expansion.

$$F(\omega) = F(\omega_0) + (\omega - \omega_0)\, f'(\omega_0) + \frac{(\omega - \omega_0)^2 \, f''(\omega_0)}{2!} + \cdots \qquad \qquad \text{...(13.5)}$$

and

$$T(\omega) = T(\omega_0) + (\omega - \omega_0)\, T'(\omega_0) + \frac{(\omega - \omega_0)}{2!} \, T''(\omega_0) + \cdots \qquad \qquad \text{...(13.6)}$$

In order that $F(\omega_0) = T(\omega_0)$ and since $(\omega - \omega_0)$ or its higher power can't be zero, the derivative of $F(\omega)$ and $T(\omega)$ at $\omega = \omega_0$ should be zero. This is known as Taylor's series approximation.

If we represent the $|F(j\omega)|^2$ as a ratio of two polynomials

$$|F(j\omega)|^2 = K_0 \frac{1 + a_1\omega^2 + a_2\omega^4 + ... + a_m\omega^{2m}}{1 + b_1\omega^2 + b_2\omega^4 + ... + b_n\omega^{2n}} \qquad ...(13.7)$$

For derivatives to be zero at $\omega = 0$, the requirement would be that $m < n$ and

$$a_1 = b_1, a_2 = b_2 ... a_m = b_m$$

$$b_m + 1 = b_m + 2 ... = b_n - 1 = 0 \text{ and } b_n \neq 0$$

The resulting function is known as maximally flat magnitude function.

In this function all the numerator and denominator terms will be same except for an extra term in the denominator equal to $b_n\,\omega^{2n}$

**Example 13.1:** Determine the values of $a$, $b$ and $c$ for the given function $|F(j\omega)|$ to be a maximally flat magnitude function.

$$F(s) = \frac{s + a}{s^2 + bs + c}$$

$$|F(j\omega)|^2 = \frac{a + j\omega}{(c - \omega^2) + jb\omega}$$

$$|F(j\omega)|^2 = \frac{a^2 + \omega^2}{c^2 + (b^2 - 2c)\omega^2 + \omega^4}$$

Now for $|F(j\omega)|^2$ to be maximally flat magnitude function, the co-efficient of various terms in $\omega^2$ in the numerator and denominator should be same i.e. $a^2 = c^2 \qquad b^2 - 2c = 1$ and $b_n = 1$

Let $a = 2 = c \qquad b^2 - 4 = 1$

$$b^2 = 5 \quad \text{or} \quad b = \sqrt{5}$$

Therefore the polynomial is

$$|F(j\omega)|^2 = \frac{4 + \omega^2}{4 + \omega^2 + \omega^4} \text{ **Ans.**}$$

A good approximation for a brick-wall kind of function is usually taken as

$$T(j\omega) = \frac{1}{1 + t^2 k_n^2(\omega)} \qquad ...(13.8)$$

where $k_n(\omega)$ is such that

$$0 \le k_n^2(\omega) << 1 \qquad 0 \le \omega \le \omega_c \qquad ...(13.9)$$

$$k_n^2(\omega) >> 1 \qquad \omega > \omega_s \qquad ...(13.10)$$

With such an approximation $T^2(j\omega) \approx 1$ in the pass band and approximately zero in the stop band.

Since $\alpha$ the attenuation is a positive quantity, from the above equation.

$$\alpha = 10 \log \left[1 + k_n^2(\omega)\, t^2\right] \qquad ...(13.11)$$

where $t$ is a constant which determines the pass band and/or stop band attenuation.

We now consider two very important methods of approximation known as Butterworth approximation and Chebychev approximation for low pass filters.

## 13.3 BUTTERWORTH APPROXIMATION

In this approximation

$$K_n(\omega) = w^n \qquad \qquad ...(13.12)$$

and is a special form of Taylor series expansion as

$$T(\omega) = 1 \text{ at } \omega = 0 \qquad \qquad ...(13.13)$$

and all the derivatives at $\omega = 0$ also vanish and $T(\omega) = F(\omega)$ at $\omega = 0$ and if we take $\omega_c = 1$ then

$$T(j\omega) = \frac{1}{\sqrt{2}} \qquad \qquad ...(13.14)$$

and

$$|T(j\omega)|^2 = \frac{1}{1 + t^2 \left( \dfrac{\omega}{\omega_c} \right)^{2n}} \qquad \qquad ...(13.15)$$

and

$$\alpha(\omega) = 10 \log \left[ 1 + t^2 \left( \frac{\omega}{\omega_c} \right)^{2n} \right] \qquad \qquad ...(13.16)$$

Fig. 13.3 shows the frequency response of the Butterworth Filter for various values of $n$

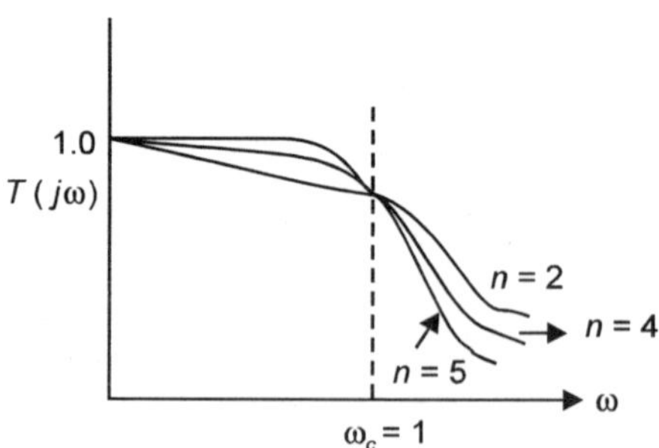


**Fig. 13.3** Frequency response of Butterworth LP filter.

All the curves pass through the same point at $\omega = \omega_C$ and this point is determined by $\alpha_p(\omega)$.

Also, it is seen that larger the value of $n$ better is the response of the filter in terms of ideal response. The point through which all the curves pass is the 3 db point of the filter and it occurs at $\omega = \omega_C = 1$, of course we have assumed $t = 1$ here.

Attenuation upto pass band is given by

$$\alpha_p = 10 \log \left[ 1 + t^2 \left( \frac{\omega_c}{\omega_c} \right)^{2n} \right] \qquad \qquad ...(13.17)$$

and beyond stop band it is given as

$$\alpha_s = 10 \log \left[ 1 + t^2 \left( \frac{\omega_s}{\omega_c} \right)^{2n} \right] \qquad ...(13.18)$$

The factor $\dfrac{\omega_c}{\omega_s}$ is called the selectivity parameter and is denoted by $K$ i.e.

$$K = \frac{\omega_c}{\omega_s} \qquad ...(13.19)$$

The equation (13.17) can be rewritten

$$1 + t^2 \left( \frac{\omega_c}{\omega_c} \right)^{2n} = 10^{0.1\alpha_p} \qquad ...(13.20)$$

or

$$t^2 = 10^{0.1\alpha_p} - 1 \qquad ...(13.21)$$

and from equation (13.18) we have

$$\frac{t^2}{K^{2n}} = 10^{0.1\alpha_s} - 1 \qquad ...(13.22)$$

Dividing the equation (13.21) by equation (13.22) we have

$$K^{2n} = \frac{10^{0.1\alpha_p} - 1}{10^{0.1\alpha_s} - 1} = K_1^2 \text{ Say} \qquad ...(13.23)$$

Then

$$n = \frac{\log K_1}{\log K} \qquad ...(13.24)$$

when $n$ is the order of the filter. While evaluating $n$ if the value comes in fraction the next higher integral should be taken for designing the filter. In the above derivations we have eliminated $t$. However, if

$$n = \frac{\log K_1}{\log K}$$ then value of $t$ could be same whether obtained from equation (13.17)

or (13.18) whereas if $n \neq \dfrac{\log K_1}{\log K}$ then $t$ can be selected to satisfy either equation (13.17) or (13.18).

In case of equation (13.17)

$$t = \sqrt{10^{0.1\alpha_p} - 1} \qquad ...(13.25)$$

whereas in case of equation (13.18)

$$t = \sqrt{K^{2n} (10^{0.1\alpha_s} - 1)} \qquad ...(13.26)$$

We have already seen that when $t = 1$ and $\omega = \omega_C = 1$ the attenuation $\alpha_p = 3$ *dB*. We now take $t = 1$ and $\omega_C = 1$ the normalized values for further investigations.

So we have now, using Butterworth approximation, the Butterworth polynomial or the transfer function for a *LP* filter as

$$|T(j\omega)|^2 = \frac{1}{|H(j\omega)|^2} \qquad \qquad ...(13.27)$$

This is an all pole transfer function as there is no zero. The general nature of $|H(j\omega)|^2$ is

$$|H(j\omega)|^2 = 1 + t^2 \left(\frac{\omega}{\omega_C}\right)^{2n} \qquad \qquad ...(13.28)$$

In particular when $t = 1$ and $\omega_C = 1$
We have

$$|H(j\omega)|^2 = H(s)\,H(-s)\big|_{s-j\omega} = 1 + (-s^2)^n \qquad ...(13.29)$$

This represents an equation to a circle with center at the origin in the *s*-plane having $n$ roots. The radius of the circle is unity if $t = 1$. However, if $t \neq 1$ the radius of the circle is

$$\frac{1}{\sqrt[n]{t}} \quad \text{units}$$

This circle where all the poles lie is known as Butterworth circle.

The roots in the Butterworth circle appear in quadrantal, so half of these lie in the RHP and other half in the LHP. We have to consider those which lie in the LHP as the transfer function so obtained is realizable as its response is stable.

Since only $n$ roots are to be considered for the transfer function; depending upon whether $n$ is even or odd, two sets of roots are possible.

(*i*) When $n$ is even, the equation (13.29) reduces to

$$s^{2n} = -1 = e^{j(2K - 1)\pi} \qquad \qquad ...(13.30)$$

The $2n$ roots are:

$$S_K = e^{\dfrac{j(2K - 1)\pi}{2n}} \qquad \qquad ...(13.31)$$

$$\text{or} \qquad = \cos\frac{2K - 1}{2n}\pi + j\sin\frac{2K - 1}{2n}\pi$$

$$K = 1, 2, \ldots 2n \qquad \qquad ...(13.32)$$

$$= \cos\theta + j\sin\theta$$

(*ii*) When $n$ is odd equation (13.29) reduces to

$$S^{2n} = 1 = e^{j2K\pi} \qquad \qquad ...(13.33)$$

$$\text{or} \qquad S_k = e^{j\dfrac{K\pi}{n}} \qquad \qquad ...(13.34)$$

$$= \cos\frac{K\pi}{n} + j\sin\frac{K\pi}{n},$$

$$K = 0, 1, 2 \ldots (2n - 1) \qquad \qquad ...(13.35)$$

$$= \cos \phi + j \sin \phi$$

Here one of the roots will be $(s + 1)$

For $n$ even if one of the roots is $-\cos \theta - j \sin \theta$, the other will be $-\cos \theta + j \sin \theta$. Therefore, for the transfer function having one pair of conjugate roots the transfer function is

$$\frac{1}{(s + \cos\theta + j \sin\theta)\,(s + \cos\theta - j\sin\theta)} \qquad \text{...(13.36)}$$

$$= \frac{1}{s^2 + 2s \cos\theta + 1} \qquad \text{...(13.37)}$$

Therefore, for $n$ roots the transfer function would be

$$T(s) = \frac{1}{\displaystyle\prod_{K=1}^{n/2} (s^2 + 2s \cos\theta_k + 1)} \qquad \text{...(13.38)}$$

Here it is $\dfrac{n}{2}$ as there will be total $n$ number of poles which are conjugate, hence $\dfrac{n}{2}$

pairs of conjugate poles.

However, if $n$ is odd.

$$T(s) = \frac{1}{(s + 1)\displaystyle\prod_{K=1}^{(n-1)/2} (s^2 + 2s \cos \phi_k + 1)}$$

For example if $n = 1$

$$T(s) = \frac{1}{s + 1}$$

If $n = 2$

$$T(s) = \frac{1}{s^2 + 2s \cos 45^\circ + 1} = \frac{1}{s^2 + \sqrt{2}s + 1}$$

For $n = 3$

$$T(s) = \frac{1}{(s + 1)\left(s^2 + 2s \cos \dfrac{\pi}{3} + 1\right)}$$

$$= \frac{1}{(s + 1)(s^2 + s + 1)}$$

In these transfer functions using Butterworth approximation, there is no pole on $j\omega$ axis.

These transfer functions have been obtained assuming radius to be unity and $t = 1$. However, if $t \neq 1$, the radius is not of unity but

$$r = \frac{1}{\sqrt[n]{t}} \qquad \qquad ...(13.39)$$

and the poles will be $r\, e^{j\theta}$ rather than $e^{j\theta}$

Also, if $r$ is the radius of Butterworth circle given by the above equation (13.39), the transfer function for $n$ even will be

$$T(s) = \frac{1}{\displaystyle\prod_{K=1}^{n/2} (s^2 + 2\, s r \cos \theta_k + r^2)} \qquad \qquad ...(13.40)$$

Similarly for $n$ odd the transfer function will be modified. Here half of the co-efficient of $s$, $r \cos \theta_k$ is known as the damping factor.

Now to calculate the transfer function for different values of $n$ we first obtain $\theta$ for each $n$ and then it is simple to obtain the transfer function.

For $n$ even,
$$\theta = \frac{2K-1}{2n}\pi = \frac{K}{n}\pi - \frac{\pi}{2n}$$

and for $n$ odd $\quad \phi = \dfrac{K}{n}\pi$

| $n$ | 2 | 4 | 6 | 8 | 10 |
|---|---|---|---|---|---|
| $\theta$ | $\dfrac{\pi}{4}$ | $\left(\dfrac{K\pi}{4} - \dfrac{\pi}{8}\right)$ | $\left(\dfrac{K\pi}{6} - \dfrac{\pi}{12}\right)$ | $\left(\dfrac{K\pi}{8} - \dfrac{\pi}{16}\right)$ | $\left(\dfrac{K\pi}{10} - \dfrac{\pi}{20}\right)$ |
| $k$ | 1 | 1,2 | 1,2,3 | 1,2,3,4 | 1,2,3,4,5 |
| $\cos \theta$ | | 0.9238 | 0.966 | 0.9801 | 0.9877 |
| | | 0.3827 | 0.707 | 0.83 | 0.891 |
| | | | | | 0.707 |
| | | | 0.2588 | 0.555 | 0.454 |
| | | | | 0.195 | 0.1564 |

For $n$ odd

| $n$ | 1 | 3 | 5 | 7 | 9 |
|---|---|---|---|---|---|
| $\phi$ | — | $\dfrac{K\pi}{3}$ | $\dfrac{K\pi}{5}$ | $\dfrac{K\pi}{7}$ | $\dfrac{K\pi}{9}$ |
| $k$ | | 1 | 1,2 | 1,2,3 | 1,2,3,4 |
| $\cos \phi$ | | $\dfrac{1}{2}$ | 0.809 | 0.9 | 0.9347 |
| | | | 0.309 | 0.623 | 0.766 |
| | | | | 0.222 | 0.5 |
| | | | | | 0.1736 |

**Table 13.1  Butterworth Function for low pass filter**
**Butterworth circle radius 1 p.u.**

| $n$ | Factors (Denominator) |
|---|---|
| 1 | $(s + 1)$ |
| 2 | $\left(s^2 + \sqrt{2}s + 1\right)$ |
| 3 | $(s + 1)\,(s^2 + s + 1)$ |
| 4 | $(s^2 + 1.8476s + 1)\,(s^2 + 0.7654s + 1)$ |
| 5 | $(s + 1)\,(s^2 + 1.618s + 1)\,(s^2 + 0.618s + 1)$ |
| 6 | $(s^2 + 1.932s + 1)\,\left(s^2 + \sqrt{2}\,s + 1\right)\,(s^2 + 0.5176s + 1)$ |
| 7 | $(s + 1)\,(s^2 + 1.8s + 1)\,(s^2 + 1.246s + 1)\,(s^2 + 0.444s + 1)$ |
| 8 | $(s^2 + 1.9614s + 1)\,(s^2 + 1.66s + 1)\,(s^2 + 1.10s + 1)\,(s^2 + 0.39s + 1)$ |
| 9 | $(s + 1)\,(s^2 + 1.8694s + 1)\,(s^2 + 1.532s + 1)\,(s^2 + s + 1)\,(s^2 + 0.3472s + 1)$ |
| 10 | $(s^2 + 1.975s + 1)\,(s^2 + 1.782\ s + 1)\,\left(s^2 + \sqrt{2}\,s + 1\right)\,(s^2 + 0.908s + 1)\,(s^2 + 0.3128s + 1)$ |

From the formulation of this table it is easy to see that Butterworth low pass filter transfer function can be obtained for any value of $n$.

## 13.4  CHEBYCHEV APPROXIMATIONS

Butterworth polynomial is exact only at $\omega = 0$ over the range $0 < \omega \le 1$ and decreases between the range.

A better approximation would be one where the response oscillates over the complete range $0 \le \omega \le 1$. This is given by Chebychev polynomial.

Chebychev polynomials are the linearily independent solutions of the differential equations.

$$(1 - \omega^2)\,\ddot{y} - \omega\dot{y} - n^2 y = 0 \qquad \qquad ...(13.41)$$

One of the solutions is

$$T_n(\omega) = \cos\,(n\,\cos^{-1}\,\omega) \qquad \omega < 1$$

and
$$= \cosh\,(n\,\cosh^{-1}\,\omega) \qquad \omega \ge 1 \qquad ...(13.42)$$

Here $T_n(\omega)$ is called the Chebychev polynomial

For $n = 0$, from both the equations

$$T_0(\omega) = 1$$

and
$$T_1(\omega) = \cos\,(\cos^{-1}\,\omega) = \omega$$

and
$$= \cosh\,(\cosh^{-1}\,\omega) = \omega \qquad ...(13.43)$$

Let $\omega = \cos\,\phi$

Therefore, for the range $\omega < 1$

$$T_{n+1}(\omega) = \cos\,(n + 1)\,\phi \qquad \qquad ...(13.44)$$

$$= \cos\,n\,\phi\,\cos\,\phi - \sin\,n\,\phi\,\sin\,\phi$$

$$= 2\,\cos\,n\,\phi\,\cos\,\phi - \sin\,n\,\phi\,\sin\,\phi - \cos\,n\,\phi\,\cos\,\phi$$

$$= 2\,\omega\,\cos\,n\,\phi - \cos\,(n - 1)\,\phi$$

$$= 2\,\omega\,T_n - T_{n-1}\,(\omega) \qquad \qquad ...(13.45)$$

Since $T_0(\omega)$ and $T_1(\omega)$ are known the above equation gives recursive relationship for obtaining Chebychev polynomials for various order (various values of $n$)

e.g.
$$T_2(\omega) = 2\,\omega\,T_1(\omega) - T_0(\omega) \qquad \qquad ...(13.46)$$
$$= 2\,\omega^2 - 1$$

Following Table 13.2 gives Chebychev polynomials of different orders.

### Table 13.2 Chebychev Polynomials of different order

| $n$ | $T_n(\omega)$ |
|---|---|
| 0 | 1 |
| 1 | $\omega$ |
| 2 | $2\,\omega^2 - 1$ |
| 3 | $4\,\omega^3 - 3$ |
| 4 | $8\,\omega^4 - 8\,\omega^2 + 1$ |
| 5 | $16\,\omega^5 - 20\,\omega^3 + 5\,\omega$ |
| 6 | $32\,\omega^6 - 48\,\omega^4 + 18\,\omega^2 - 1$ |
| 7 | $64\,\omega^7 - 112\,\omega^5 + 56\,\omega^2 - 7\,\omega$ |
| 8 | $128\,\omega^8 - 256\,\omega^6 + 160\,\omega^4 - 32\,\omega^2 + 1$ |
| 9 | $256\,\omega^9 - 576\,\omega^7 + 432\,\omega^5 - 120\,\omega^3 + 9\,\omega$ |
| 10 | $512\,\omega^{10} - 1280\,\omega^8 + 1120\,\omega^6 - 400\,\omega^4 + 50\,\omega^2 - 1$ |

With regard to the table above, following observations are made:
1.  The co-efficient of the highest power of $\omega$ is $2^{n-1}$.
2.  The zero of the polynomials are all located in the interval $|\omega| \leq 1$.
3.  For $\omega = 0$ $\quad T_n = 0 = \cos(n\,\cos^{-1} 0)$

$$= \cos\frac{n\pi}{2} = (-1)^{n/2}$$

$$\text{for } n \text{ even}$$

for $n$ odd $T_n(0) = 0$

Similarly

$$T_n(1) = \cos(n\,\cos^{-1} 1) = 1 \qquad \qquad ...(13.47)$$

This means $T_n(\omega)$ is less than equal to one for frequencies $\omega < 1$

However, beyond $|\omega| \leq 1$ i.e. for $|\omega| \geq 1$,

$T_n(\omega)$ increases rapidly for increasing values of $|\omega|$ as

$$T_n(\omega) = \cosh(n\,\cosh^{-1}\omega)$$

$$= \cosh n\,\phi = \frac{e^{n\phi} + e^{-n\phi}}{2} \qquad \qquad ...(13.48)$$

and as $n\,\omega$ increases $n\,\phi$ increases and hence $T_n(\omega)$ increases.

Therefore, for Chebychev approximation $K_n(\omega)$ is written as $T_n(\omega)$ and thus we have

$$|T(j\omega)|^2 = \frac{1}{1 + t^2\,T_n^2\!\left(\dfrac{\omega}{\omega_C}\right)^2} \qquad \qquad ...(13.49)$$

We know that when $n$ is odd $T_n(0) = 0$ and, therefore, $|T(0)|^2 = 1$ and when $n$ is even

$$T_n(0) = \pm 1 \qquad \qquad ...(13.50)$$

therefore,

$$|T_n(0)|^2 = \frac{1}{1+t^2} \qquad \qquad ...(13.51)$$

and the response oscillates between

$\dfrac{1}{1+t^2}$ and 1 between $0 \leq |\omega| \leq 1$

and the number of oscillations equals $n$, the order of the polynomial. The variation is shown in Fig. 13.4 where two values of $n$ are taken $n = 4$ and $n = 5$

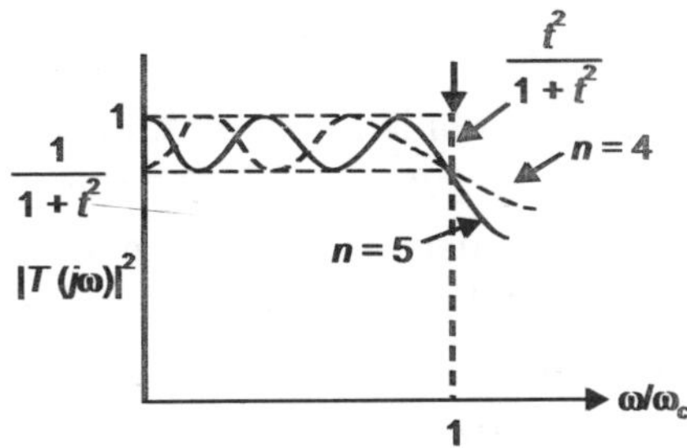

**Fig. 13.4  Chebychev approximation for Low-pass filter.**

It is seen that when $n$ is odd, the transfer function $|T(j\omega)|^2$ at $\omega = 0$ has a value unity, whereas it has a value $\dfrac{1}{1+t^2}$ when $n$ is even. However, at $\omega = \omega_C$ the function has the value

$\dfrac{1}{1+t^2}$ irrespective of the order $(n)$ of the filter.

The ripple size (peak to peak) is $= 1 - \dfrac{1}{1+t^2} = \dfrac{t^2}{1+t^2}$ $\qquad \qquad$ ...(13.52)

The attenuation is given as

$$\alpha_p = 10 \log \left[ 1 + t^2 \, T_n^2 \left( \frac{\omega_C}{\omega_C} \right) \right] = 10 \log (1 + t^2) \qquad \qquad ...(13.53)$$

Also

$$\alpha_s = 10 \log \left[ 1 + t^2 \, T_n^2 \left( \frac{\omega_S}{\omega_C} \right) \right] \qquad \qquad ...(13.54)$$

The solution of these equations give $t$ and $n$. The value of $n$ is given by

$$n \geq \frac{\cosh^{-1}(1/K_1)}{\cosh^{-1}(1/K)} \qquad \qquad ...(13.55)$$

where $K_1$ and $K$ are given as in equation (13.19) and (13.23).

It can be shown that the roots of the transfer function of Chebychev polynomial lie on an ellipse given as

$$\frac{\sigma_K^2}{\sinh^2 \gamma} + \frac{\omega_K^2}{\cosh^2 \gamma} = 1 \qquad \qquad ...(13.56)$$

where $\sigma_k = \sinh \gamma \sin \dfrac{2k+1}{2n} \pi$   $k = 0, 1, \ldots (2n-1)$ ...(13.57)

$$\omega_k = \cosh \gamma \cos \dfrac{2k+1}{2n} \pi \qquad \text{...(13.58)}$$

and
$$\gamma = \dfrac{1}{n} \sinh^{-1}(1/t) \qquad \text{...(13.59)}$$

The roots are $\qquad S_k = \sigma_k + j\omega_k$

From the given specification of a LPF it is easy to find out $t$ and $n$. Knowing $n$ we can write down $T_n(\omega)$ and then Left hand roots of the polynomial
$$1 + t^2 \, T_n^2(\omega) = 0 \qquad \text{...(13.60)}$$
are the poles of the filter.

Sometimes an approximation to $T_n(\omega)$ is made by taking only the highest power of $\omega$ i.e.
$$T_n(\omega) \simeq t^2 \, 2^{n-1} \, \omega^n \quad \text{for} \quad \omega >> 1 \qquad \text{...(13.61)}$$

Following few examples would illustrate how to obtain Butterworth and Chebychev polynomials given the specifications of the LP filter.

**Example 13.2:** The specification for a LP filter are:
$$\alpha_p \le 1 \ dB \quad \text{for} \quad f \le 2 \text{ MHz}$$
$$\alpha_s \ge 40 \ dB \quad \text{for} \quad f \ge 8 \text{ MHz}$$

Determine $n$ and $t$ for Butterworth and Chebychev polynomials.

**Solution:** The selectivity and discrimination parameters for this filter are:
$$k = \dfrac{2}{8} = \dfrac{1}{4}$$

$$k_1 = \left[ \dfrac{10^{0.1}-1}{10^4 - 1} \right]^{1/2} \simeq \dfrac{(0.2589)^{1/2}}{100} \simeq 0.5088 \times 10^{-2}$$

The order of the Butterworth polynomial
$$n \ge \dfrac{\log K_1}{\log K} = \dfrac{-2.2933}{0.602}$$
$$\ge 3.8$$

Therefore, $\qquad n = 4$

Chebychev polynomial $\qquad \dfrac{\cosh^{-1}\left(1/0.5088 \times 10^{-2}\right)}{\cosh^{-1} 4}$

$$n \simeq 2.89$$
$$n = 3$$
$$t = \sqrt{10^{0.1}-1} = 0.5089$$

For Chebychev
$$1 = 10 \log (1 + t^2)$$

$$1 + t^2 = 1.2589$$

or
$$t = 0.5088$$

**Example 13.3:** The specification for a LP filter are given as follows:

$$\alpha_p \leq 0.5 \text{ dB} \qquad \omega \leq 3 \text{ MHz}$$
$$\alpha_s \geq 30 \text{ dB} \qquad \omega > 4 \text{ MHz}$$

Determine the order of the Butterworth and Chebychev polynomials

**Solution:** The selectivity and discrimination parameters for the filter are:

$$k = \frac{\omega_C}{\omega_S} = \frac{3}{4} = 0.75$$

and
$$k_1 = \left[\frac{10^{0.1} - 1}{10^6 - 1}\right]^{1/2} = 0.5088 \times 10^{-3}$$

$n$ for Butterworth filter

$$n \geq \frac{\log K_1}{\log K} = \frac{\log 0.5088 \times 10^{-3}}{\log 0.75}$$

$$= 26.36$$

or
$$n = 27$$

For Chebychev filter

$$n \geq \frac{\cosh^{-1} 1/K_1}{\cos h^{-1} 1/K}$$

$$\geq \frac{\cosh^{-1} \dfrac{10^3}{0.5088}}{\cosh^{-1} 4/3} = 10.4$$

$$n = 11$$

It is found that to realize a filter of particular specification the order of Butterworth filter is higher as compared to Chebychev filter.

It is also found that for the same order of the filters, Chebychev filter has narrower transition band as compared to Butterworth filter.

**Example 13.4:** Determine the value of $t$ for a ripple of (*i*) 1 dB (*ii*) 2 dB (*iii*) 3 dB.

**Solution:**

(*i*)
$$dB = 10 \log (1 + t^2)$$
$$1 = 10 \log (1 + t_1^2)$$

or $\quad 1 + t_1^2 = 1.2589 \quad$ or $\quad t_1^2 = 0.2589$

$$t_1 = 0.5088$$

(*ii*)
$$2 = 10 \log (1 + t_2^2)$$

$$1 + t_2^2 = 1.58489 \quad \text{or} \quad t_2 = 0.7647$$

$(iii)$
$$3 = 10 \log (1 + t_3^2)$$
$$t_3^2 = 0.99526$$
$$t_3 = 0.9976$$
$$\simeq 1.0$$

**Example 13.5:** Determine Chebychev polynomial for $n = 2$ and for a ripple of 2 dB.

**Solution:**

For 2dB,   $t = 0.7647$

$$T_2(\omega) = 2\,\omega^2 - 1$$

Therefore, Chebychev polynomial

$$= 1 + t^2\,T_2^2(\omega)$$
$$= 1 + 0.585\,(2\,\omega^2 - 1)^2$$
$$= 0.585\,(4\omega^4 - 4\,\omega^2) + 1.585$$
$$= \omega^4 - \omega^2 + 0.677$$

Substituting $\dfrac{s}{j}$ for $\omega$ we have

$$s^4 + s^2 + 0.677$$

The roots (LHS) of the polynomial are obtained as follows:

$$s^2 = \frac{-1 \pm \sqrt{1 - 2.708}}{2}$$
$$= -0.5 + j\,0.653$$

Let the roots be $s = a + jb$

$$a^2 - b^2 + 2\,jab = -0.5 + j\,0.653$$
$$a^2 - b^2 = -0.5$$

$$ab = 0.3265 \quad \text{or} \quad b = \frac{0.3265}{a}$$

$$a^2 - \frac{0.1066}{a^2} = -0.5$$

$$a^4 + 0.5\,a^2 - 0.1066 = 0$$

$$a^2 = \frac{-0.5 \pm 0.822}{2} = \frac{-0.5 + 0.822}{2} = 0.16$$

$$a = 0.4$$

$$s = 0.4 \pm j\,0.81625 \ \textbf{Ans.}$$

Alternative method of obtaining the roots is

$$r = \frac{1}{n}\sinh^{-1}\frac{1}{t} = \frac{1}{2}\sinh^{-1} 1.3077 = 0.5415$$

Now
$$\sinh \gamma = 0.5684$$

and
$$\cosh \gamma = 1.15$$

$$\sigma = 0.5684 \sin 45°$$
$$= 0.4$$

and
$$\omega_k = \cosh \gamma \cos \frac{\pi}{4}$$

$$= 1.15 \cdot \frac{1}{\sqrt{2}} = 0.813 \text{ Ans.}$$

**Example 13.6:** Show that

(i) $B_2(s) \cdot B_2(-s) = 1 + \omega^4$

(ii) $B_3(s) \cdot B_3(-s) = 1 + \omega^6$

where B's are Butterworth functions

**Solution:** We know that

$$B_2(s) = \left(s^2 + \sqrt{2}\,s + 1\right)$$

Therefore,
$$B_2(-s) = \left(s^2 - \sqrt{2}\,s + 1\right)$$

Hence
$$B_2(s)\, B_2(-s) = s^4 + 1$$

Let
$$s = j\omega \quad \text{Hence } B_2(s)\, B_2(-s) = 1 + \omega^4$$

(ii) Now for $n = 3$, one of the roots is $(s + 1)$ and since $n$ is odd

$$\phi = \frac{K\pi}{n} = \frac{\pi}{3}$$

Therefore, the polynomial is $s^2 + 2\,s \cos \dfrac{\pi}{3} + 1$

$$= (s^2 + s + 1)$$

Hence total polynomial
$$B_3(s) = (s + 1)(s^2 + s + 1)$$
$$B_3(-s) = (1 - s)(s^2 - s + 1)$$

Hence
$$B_3(s)\, B_3(-s) = (1 - s^2)(s^4 + s^2 + 1)$$

Let $s = j\omega$, the polynomial reduces to $(1 + \omega^2)(\omega^4 - \omega^2 + 1) = 1 + \omega^6$

## 13.5  COMPARISON BETWEEN BUTTERWORTH AND CHEBYCHEV APPROX

Let us further study and compare the behaviour of Butterworth and Chebychev approximation at higher frequencies i.e. $\omega \gg 1$. We know that

$$|T_B(\omega)|^2 \simeq \frac{1}{C^2\,\omega^{2n}} \qquad \qquad \text{...(13.62)}$$

$$|T_C(\omega)|^2 \simeq \frac{1}{t^2\,2^{2(n-1)}\,\omega^{2n}} \qquad \qquad \text{...(13.63)}$$

Taking log for both the polynomials

$$10 \log |T_B(\omega)|^2 = -(20 \log C + 20\,n \log \omega)\ \text{dB} \qquad \qquad \text{...(13.64)}$$

$$10 \log |T_C(\omega)|^2 = -[20 \log t + 20\,n \log \omega + 60\,(2n - 2)]\ dB \qquad \qquad \text{...(13.65)}$$

Both the responses have the same attenuation slopes 20 $n$ log $\omega$. The intercept of Butterworth depends upon $c$ where as that for Chebychev it depends upon both $t$ and $n$. However because of the factor $3(2n - 2)$, the Chebychev response has larger attenuation in the stop band.

However, it is observed that Butterworth gives a fairly linear phase response and is better than the Chebychev response. Normally the filters ideal phase response is one that is linear over the frequency range.

In case of Chebychev approximation, the higher the order of the polynomial more non-linear is the phase response. If linear phase response is the requirement neither Butterworth nor Chebychev approximation is suitable.

It is also observed that for a given order $n$ of Chebychev polynomial larger the value of $t$ (larger ripple in pass-band) the better is the performance of the filter in the stop-band. Also as mentioned in the previous paragraph that higher the order of the polynomial better the attenuation in the stop band. Therefore, it can be concluded that a smaller pass band ripple and sharper cut-off characteristics can't be achieved simultaneously. If the pass band ripple is small, the attenuation in the stop band is lower and vice versa.

## 13.6  SOME SPECIAL FILTERS

The other approximations are Inverse Chebychev Elliptic Filter and Bessel filter. The function under Inverse Chebychev is given as

$$T_{IC} = \frac{t^2\, T_n^2\,(1/\omega)}{1 + t^2\, T_n^2\,(1/\omega)} \qquad \qquad ...(13.66)$$

This is a low pass function with monotonic pass band and equiripple stop band as shown in Fig. 13.5.

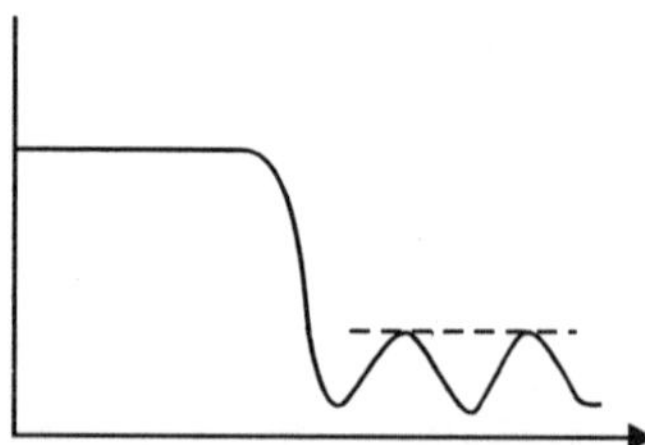

**Fig. 13.5**  Inverse Chebychev response.

Here one has to see that poles of $T_{IC}$ are in the pass band and zeros in the stop band. The response of elliptic filter is shown in Fig. 13.6.

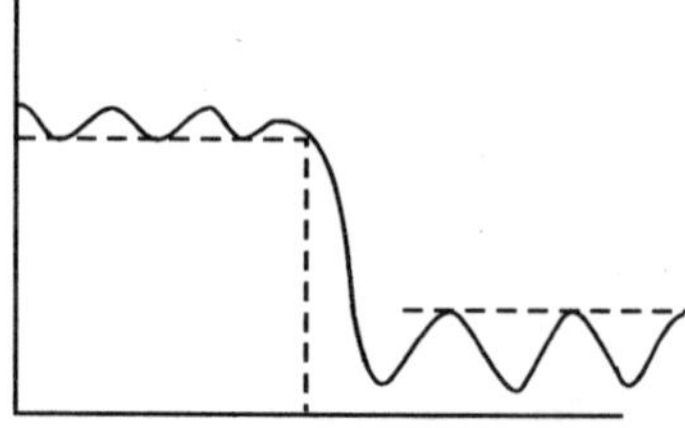

**Fig. 13.6**  Elliptic filter response.

Elliptic filters are optimal in the sense that for the same response the order of elliptic filters is always lower.

Bessel polynomial is a better approximation for linear phase filter.

The approximation techniques discussed so far result in a rational function for transfer voltage ratio. For realisation of these functions for the design of actual filters we make use of co-efficient matching technique. Before we take up design brief study of some of the active elements and active circuits would be desirable.

## 13.7 ACTIVE ELEMENTS AND CIRCUITS

There are many active elements used in filters, the most important one is the operational amplifier. Recently a few other elements like the gyrator, the negative impedance converter are in use in filter design and fabrication. The most basic active element is the controlled source. The controlled source is a unidirectional, non-autonomous active two-port having two pairs of terminals, one controlled and the other controlling. One of the terminal variables at the controlled port (output port) depends upon one of the variables at the controlling port (input port). Because of the unidirectional property the controlling terminal (input) variables are independent of the controlled terminal variables.

Following are four types of controlled sources. We call these as transducers (Fig. 13.7).

### 13.7.1  Voltage to Current Transducer (VCT)

This is an ideal voltage controlled current source i.e. the output current is dependent upon the input voltage (controlling port). VCT is characterized by the following relations Fig. 13.7(a)

$$\begin{bmatrix} I_1 \\ I_2 \end{bmatrix} = \begin{bmatrix} 0 & 0 \\ g & 0 \end{bmatrix} \begin{bmatrix} V_1 \\ V_2 \end{bmatrix} \qquad ...(13.67)$$

which indicates that the device has an infinite input and output impedance (or zero admittance), a forward admittance of $g$ and no reverse transmission.

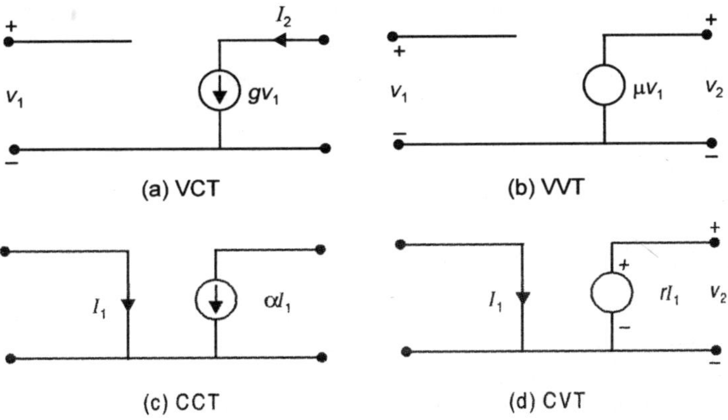


**Fig. 13.7**  Controlled sources.

VCT has infinite power gain as it can supply unlimited power with zero-input power.

### 13.7.2  Voltage to Voltage Transducer (VVT)

This is represented by the following relation (Fig. 13.7b)

$$\begin{pmatrix} I_1 \\ V_2 \end{pmatrix} = \begin{pmatrix} 0 & 0 \\ \mu & 0 \end{pmatrix} \begin{pmatrix} V_1 \\ I_2 \end{pmatrix} \qquad \qquad ...(13.68)$$

Therefore, VVT has zero input admittance and zero output impedance. The transmission is again in the forward direction with voltage transfer ratio (or voltage gain) of $\mu$. VVT is also known as voltage amplifier.

### 13.7.3  Current to Current, Transducer CCT : (Fig. 13.7 c)

Here the output current depends upon input current. The relations are:

$$\begin{pmatrix} V_1 \\ I_2 \end{pmatrix} = \begin{pmatrix} 0 & 0 \\ \alpha & 0 \end{pmatrix} \begin{pmatrix} I_1 \\ V_2 \end{pmatrix} \qquad \qquad ...(13.69)$$

The input impedance and output admittance are zero. The device has forward current transfer ratio $\alpha$. CCT consumes no input power whereas it can supply any amount of power, hence it has infinite power gain. It is also known as current amplifier.

### 13.7.4  Current to Voltage Transducer CVT Fig. 13.7 (d)

It is an ideal controlled source where output voltage is the controlled quantity and input current the controlling quantity. The relation is given as

$$\begin{pmatrix} V_1 \\ V_2 \end{pmatrix} = \begin{pmatrix} 0 & 0 \\ r & 0 \end{pmatrix} \begin{pmatrix} I_1 \\ I_2 \end{pmatrix} \qquad \qquad ...(13.70)$$

This device has zero input and output impedances, a forward transfer impedance of $r$ and no reverse transmission is possible. The device can supply unlimited power with zero input power. Hence it has infinite power gain.

The devices VCT, VVT, CCT and CVT are also known as voltage controlled current source (e.g. field effect transistor), voltage controlled voltage source (e.g. an operational amplifier) current controlled current source (a transistor) and current controlled voltage source respectively.

It can be observed from the transmission parameter matrices that we can obtain a VVT by cascading a VCT and a CVT.

$$\begin{pmatrix} 0 & 0 \\ g & 0 \end{pmatrix} \begin{pmatrix} 0 & 0 \\ r & 0 \end{pmatrix} = \begin{pmatrix} 0 & 0 \\ gr & 0 \end{pmatrix} \qquad \qquad ...(13.71)$$

Here $gr$ is a unitless real number corresponding to gain $\mu$ of a VVT.

Similarly we can also obtain a CCT by cascading a CVT and a VCT.

### 13.7.5  Operational Amplifier

An ideal operational amplifier has the following characteristics:

   (*i*) The input impedance is infinity which means it does not draw any current from the voltage source at input terminals.

   (*ii*) The output impedance is zero i.e., the output voltage is independent of the current drawn from the operational amplifier.

 (*iii*) It has infinite bandwidth which means it will amplify signals of any frequency with equal gain.

(*iv*) The open loop gain is infinite. An operational amplifier is also considered as a dual input infinite gain VVT as shown in Fig. 13.8.

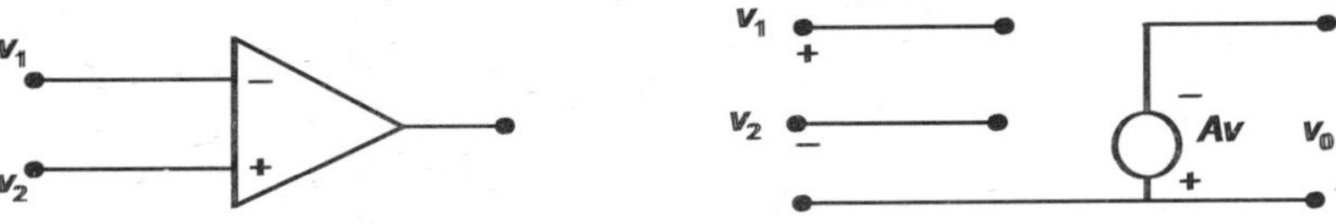

**Fig. 13.8**  Operational amplifier as a VVT.

If $A$ is the gain of the operational amplifier the output voltage $V_0$ is given as

$$V_0 = A \ (V_2 - V_1) \qquad\qquad ...(13.72)$$

For an ideal operational amplifier $A$ is infinite, hence the differential voltage $(V_2 - V_1)$ is zero and if one of the terminals of operational amplifier either 1 or 2 in feedback configuration is grounded, the other is a virtual ground. The ideal operational amplifier has zero offset which means for zero output, the input voltage is zero. When both input terminals are used, it is called a differential input amplifier.

Following circuits along with some discription of each indicates a few applications of an operational amplifier.

Fig. 13.9 (a) indicates operational amplifier as an inverting circuit, Since the feedback is applied

(*a*) Inverting amplifier  

(*b*) Non-inverting amplifier  

(*c*) Inverting  

(*d*) Integrating amplifier  

(*e*) differentiating circuit

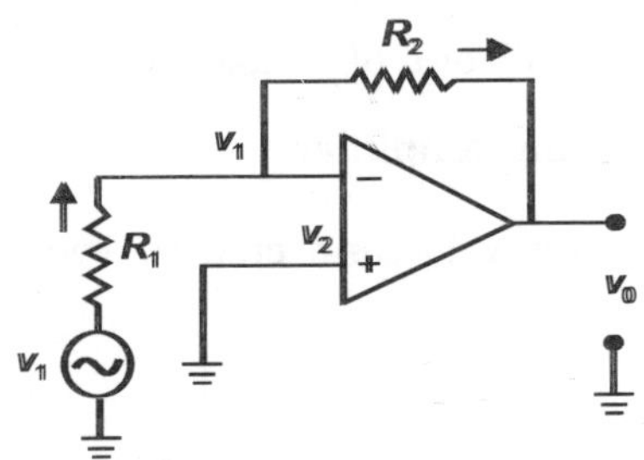

(a) Inverting amplifier

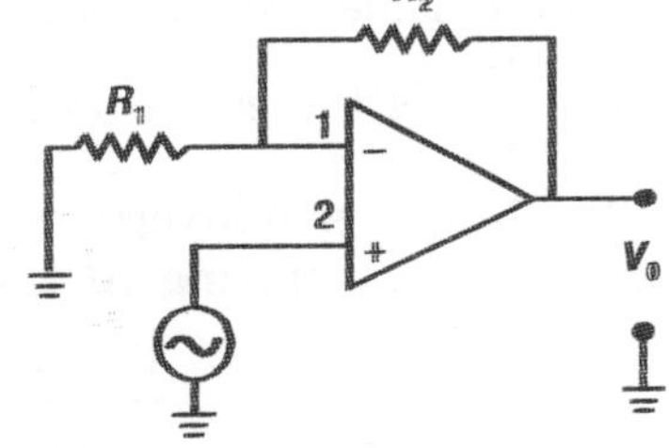

(b) Non-inverting amplifier

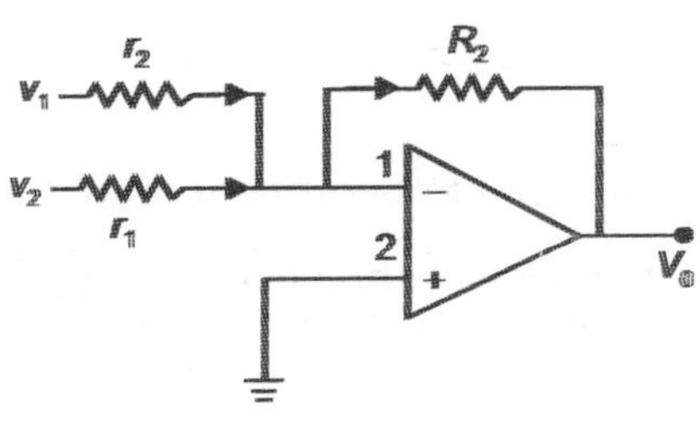

(c) Inverting summor

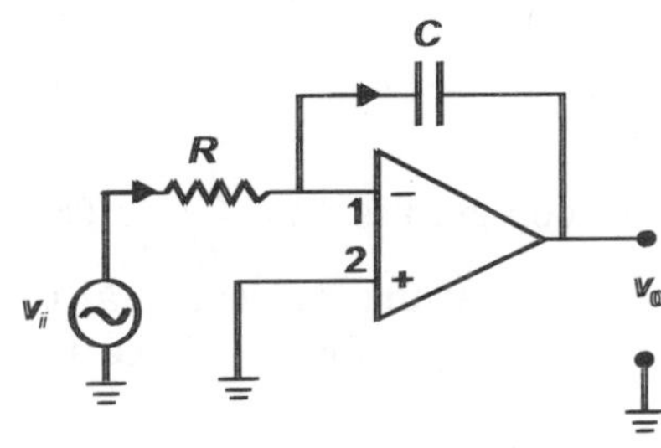

(d) Integrating amplifier

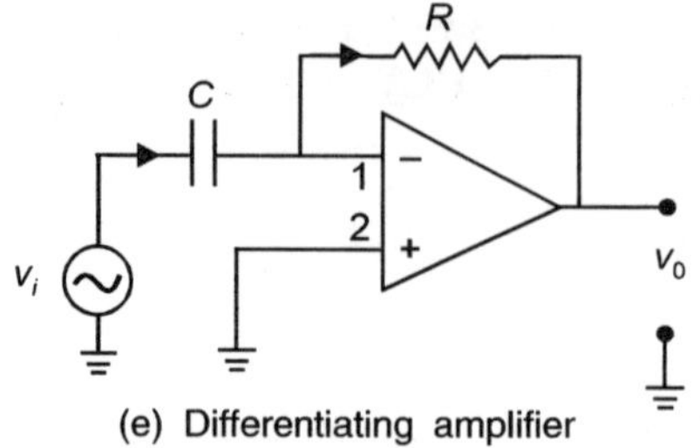



(e) Differentiating amplifier

**Fig. 13.9** Application of operational amplifier.

at the inverting terminal (terminal with −ve polarity), it is referred to as negative feedback. The supply is connected to terminal 1 through resistance $R_1$. Using ideal operational amplifier characteristics

$$V_2 - V_1 = \frac{V_0}{A}$$

For an ideal operational amplifier $A \to \infty$, $V_2 - V_1 = 0$ or $V_2 = V_1$ and therefore, terminal 1 also attains zero potential i.e. it attains virtual ground.

Applying KCL at node 1 we have

$$\frac{V_i - V_1}{R_1} = \frac{V_1 - V_0}{R_2} \qquad \qquad ...(13.73)$$

Since $$V_1 = 0$$

We have $$V_0 = -\frac{R_2}{R_1} V_i \qquad \qquad ...(13.74)$$

Here −ve sign means that the output voltage is $180^\circ$ out of phase with reference to input voltage, $\dfrac{R_2}{R_1}$ is the closed loop gain of the inverting amplifier.

Fig. 13.9 (b) is a non-inverting amplifier as the supply is connected to +ve terminal of operational amplifier. Applying KCL at node 1

$$\frac{V_i - V_0}{R_2} = -\frac{V_i}{R_1} \qquad \qquad ...(13.75)$$

or $$\frac{V_0}{V_i} = \left(1 + \frac{R_2}{R_1}\right) \qquad \qquad ...(13.76)$$

The closed loop gain is $\left(1 + \dfrac{R_2}{R_1}\right)$. Thus the closed loop gain of operational amplifier depends only on circuit parameters external to it and is independent of the open loop gain. Fig. 13.9 (c) is an inverting summer. Since terminal 2 is grounded, terminal 1 is at virtual ground i.e. $V_1 = 0$

Hence, using KCL at node 1, we have

$$\frac{v_1}{r_1} + \frac{v_2}{r_2} + \cdots + \frac{v_n}{r_n} = -\frac{V_0}{R_2} \qquad \qquad ...(13.77)$$

or
$$V_0 = -R_2 \left[ \frac{v_1}{r_1} + \frac{v_2}{r_2} + \cdots + \frac{v_n}{r_n} \right] \qquad \text{...(13.78)}$$

However, if $\qquad r_1 = r_2 = \ldots = R$ then

$$V_0 = -\frac{R_2}{R} \left[ v_1 + v_2 + \cdots + v_n \right] \qquad \text{...(13.79)}$$

Fig. 13.9 (d) is an integrating circuit. Again since terminal 2 is grounded $V_2 = V_1 = 0$. Applying KCL at node 1

$$\frac{v_i}{R} = -C \frac{dV_0}{dt} \qquad \text{...(13.80)}$$

or
$$V_0 = -\frac{1}{RC} \int V_i \, dt + V_c(0) \qquad \text{...(13.81)}$$

Therefore, the output is proportional to the integral of the input. $V_0(0)$ is the initial voltage of the capacitor.

Fig. 13.9 (e) is a differentiating circuit. Applying KCL at node 1.

$$C \frac{dv_i}{dt} = -\frac{v_0}{R} \qquad \text{...(13.82)}$$

or
$$V_0 = -\frac{C}{R} \frac{dv_i}{dt} \qquad \text{...(13.83)}$$

that is, the output voltage is proportional to differential of the input voltage.

For sinusoidal signals the gain of both these circuits (d) and (e) is unity (0 dB) when $\omega = 1/RC$. If the frequency is doubled gain for an integrating circuit is halved whereas that for the differentiating circuit is doubled i.e. gain increases by –6 dB for doubling frequency in case of integrator whereas for differentiator it increases by 6 dB. Since the gain of the integrating circuit decreases with increase in frequency, this circuit has an inherent noise suppressing property i.e. it has the property of suppressing high frequency unwanted signals which may appear at the input terminals.

In chapter 1 we have studied CNIC and VNIC i.e. the current negative impedance converter and voltage negative impedance conveter. These can be realized by using controlled sources as shown in Fig. 13.10.

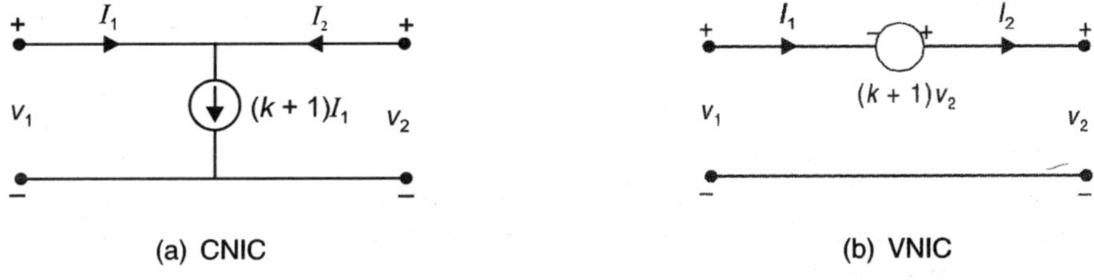


**Fig. 13.10**  Controlled source realisation of CNIC and VNIC.

The CNIC is realised using a single CCT and VNIC is realized using a single VVT and thus clearly point out the relations between the negative impedance converters and the transducers.

The equation describing these circuits are:

$$I_2 = (K + 1)\, I_1 - I_1 = KI_1 \qquad\qquad ...(13.84)\ \text{CNIC}$$
$$V_1 = V_2$$
$$V_1 = -\,(K + 1)\, V_2 + V_2 = -\,KV_2 \qquad\qquad ...(13.85)\ \text{VNIC}$$
$$I_1 = I_2$$

CNIC and VNIC can also be realized using operational amplifier as shown in Fig. 13.11.

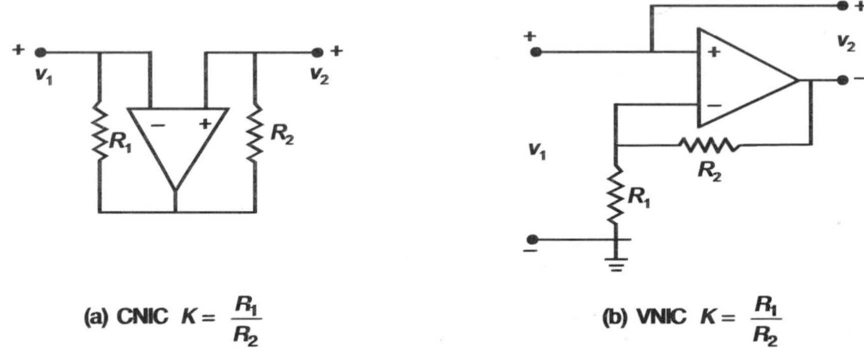

(a) CNIC  $K = \dfrac{R_1}{R_2}$
(b) VNIC  $K = \dfrac{R_1}{R_2}$

**Fig. 13.11**  Operational amplifier realisation of CNIC and VNIC.

NICs are non-reciprocal devices and are potentially unstable networks.

### 13.7.6  Nullator and Norators

There are certain networks whose terminal behaviour can't be expressed in terms of any of the known quantities like $z, y, h, g$. Such networks are usually called pathological networks. One such element is a nullator. It is a 2-terminal element defined by the equation

$$V = I = 0 \qquad\qquad ...(13.86)$$

and is shown in Fig. 13.12 (a)

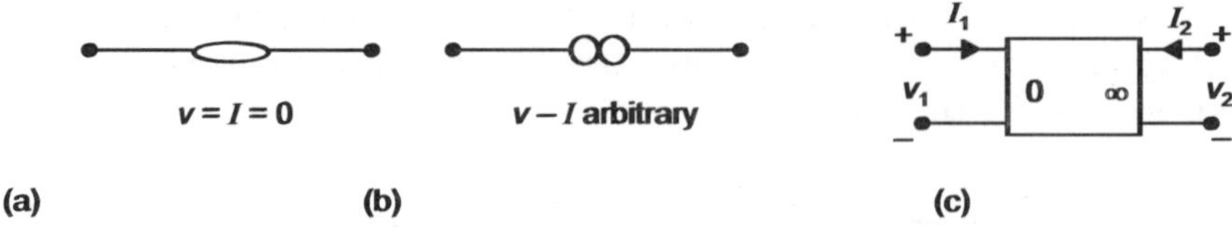

$v = I = 0$        $v - I$ arbitrary

(a)                    (b)                    (c)

**Fig. 13.12**  (a) Nullator (b) Norator (c) Nullor.

From the relation it is clear that a nullator is both an open circuit and a short circuit simultaneously. This is a bilateral and lossless element.

Norator is a 2–terminal element for which the terminal variables $V$ and $I$ are completely arbitrary. As a circuit element it is indicated as shown in Fig. 13.12 (b). By definition, it is a non-reciprocal 2 terminal element.

It is shown in the literature that the nullator and norator can not be obtained from any physically realizable network components. However, there exists a pathological two-port which is obtainable by a limiting process. Such a two port is the nullor. One pair of terminals of this device acts as a short circuit and open circuit at the same time whereas at the other pair of terminals $V$ and $I$ can assume arbitrary values. The nullor is described by a transmission matrix which is a null-matrix i.e.

$$\begin{pmatrix} V_1 \\ I_1 \end{pmatrix} = \begin{pmatrix} 0 & 0 \\ 0 & 0 \end{pmatrix} \begin{pmatrix} V_2 \\ -I_2 \end{pmatrix} \qquad \qquad ...(13.87)$$

The nullator and the norator also form a basic set of active elements. It is to be noted that in a physical system the voltages and currents are uniquely and definitely determined, there is nothing arbitrary about these and hence in a physically realizable circuit, the nullator and the norator must come in a pair and not as isolated elements.

We first consider the representation of controlled sources using nullators and norators. It should be noted that the dual of a nullator is a nullator and dual of a norator is a norator. Fig. 13.13 shows series and parallel combination of nullator and norator.

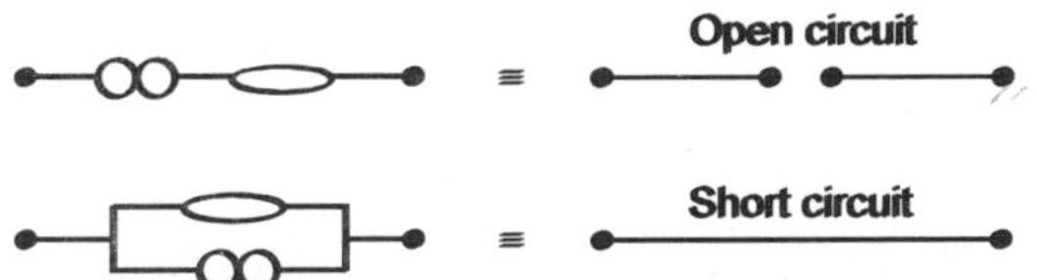


**Fig. 13.13**  Series parallel combination.

Fig. 13.14 shows basic representation of each of the four types of controlled sources. Each circuit employs a single nullator and a single norator.

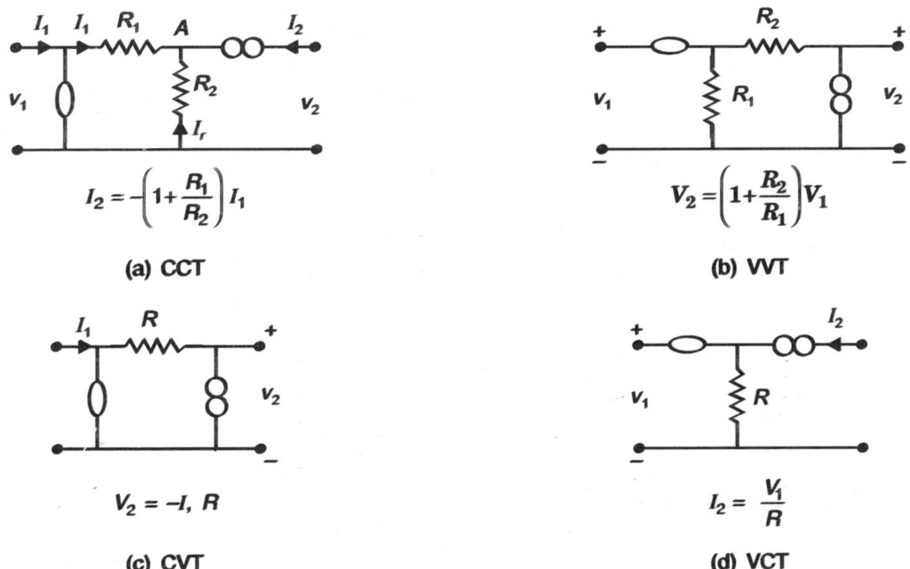


**Fig. 13.14**  Four basic nullator-norator models of controlled sources
(a) CCT (b) VVT (c) CVT and (d) VCT.

It can be seen that CCT Fig. 13.14 (a) is dual of VVT Fig. 13.14 (b) and vice versa and CVT Fig. 13.14 (c) is dual of VCT Fig. 13.14 (d).

Let us analyse circuit. (a) Since the current through nullator is zero, the input current $I_1$ flows through $R_1$ producing a drop $I_1 R_1$. In the loop containing nullator, this voltage drop

$I_1 R_1$ will flow a current $I_r$ in the resistor $R_2$ in the direction as shown in Fig. 13.14 (a) as voltage across nullator is zero. Applying KCL at node A.

We have
$$I_1 + I_2 + I_r = 0$$

or
$$I_2 = -I_1 - I_r = -I_1 - I_1 \cdot \frac{R_1}{R_2}$$

$$= -I_1 \left(1 + \frac{R_1}{R_2}\right) \qquad \qquad ...(13.88)$$

This shows that the circuit works as a current controlled current transducer (source). For circuit 13.14 (b) $R_1$, $R_2$ act as potential divider for voltage $V_2$.

Hence
$$V_1 = \frac{V_2}{R_1 + R_2} \cdot R_1 \quad \text{or} \quad V_2 = V_1 \frac{R_1 + R_2}{R_1}$$

$$= \left(1 + \frac{R_2}{R_1}\right) V_1 \qquad \qquad ...(13.89)$$

In both these cases, the gain is more than unity. The gain can be made unity if for CCT $R_1 \to 0$ and $R_2 \to \infty$ and for VVT $R_1 \to \infty$ and $R_2 \to 0$ and the circuit will be as shown here in Fig. 13.15.

**Fig. 13.15**  Unity gain (a) CCT (b) VVT.

The other circuits of Fig. 13.14 (c) and (d) are simple to analyse.

Using cascade connection of Fig. 13.15 (a) and Fig (b) various circuit for controlled sources can be constructed e.g. to obtain CVT we cascade CCT and VVT of Fig. (13.15) we obtain Fig. 13.16 (a). Similarly to obtain VCT we cascade VVT and CCT of Fig. 13.15 and we obtain Fig. 13.16 (b).

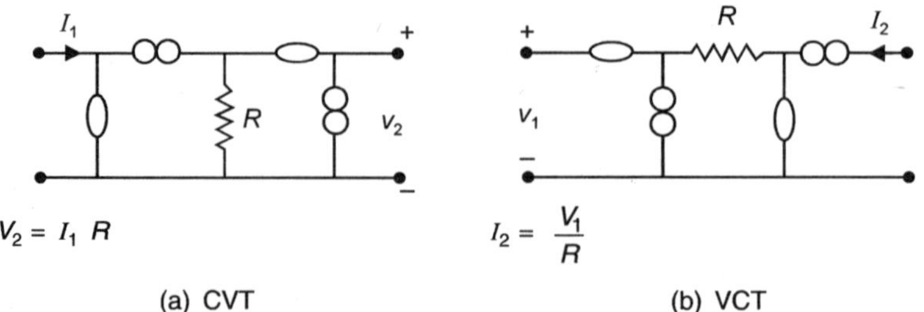


**Fig. 13.16**  (a) CVT (b) VCT.

In order to obtain CCT using two nullators and two norators we connect CVT and VCT of Fig. 13.14 in cascade as shown in Fig. 13.17 (a)

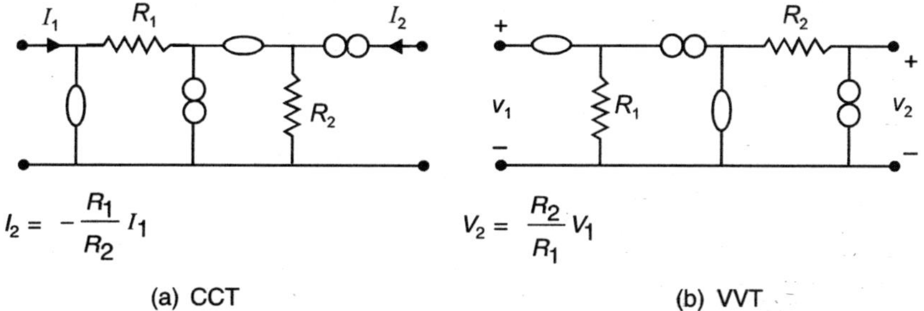

$$I_2 = -\frac{R_1}{R_2} I_1$$

(a) CCT

$$V_2 = \frac{R_2}{R_1} V_1$$

(b) VVT

**Fig. 13.17**  (a) CCT (b) VVT

Similarly VVT can be obtained by cascade connecting VCT and CVT of Fig. 13.14 and we obtain the Fig. 13.17 (b).

Following procedure suggests a method of coverting a nullator norator circuit into a transistor circuit.

The first step is to group together all the nullators and norators into pairs of a nullator and norator having a common junction. Each pair is replaced by a transistor as follows:

(*i*) The junction of the pair corresponds to the emitter.

(*ii*) The terminal of the nullator which is not connected at the junction (free terminal) corresponds to the base.

(*iii*) The free terminal of the norator corresponds to the collector.

To illustrate this let us take a modified CCT circuit where we have two nullators and two norators as shown in Fig. 13.18. Since a series combination of a nullator and a norator is an open circuit,

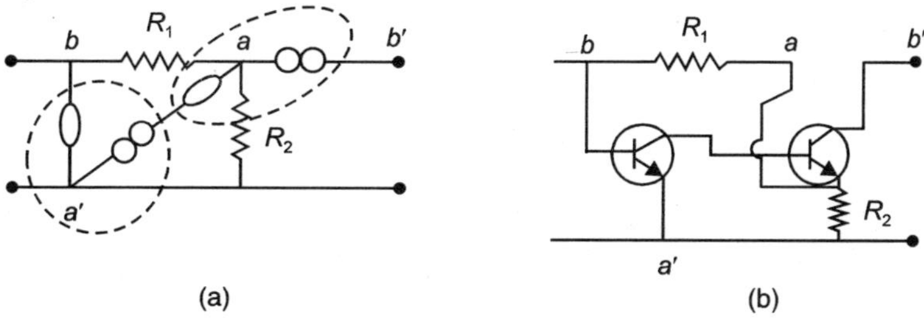

(a)

(b)

**Fig. 13.18**  (a) Nullator-norator model of CCT (b) Equivalent transistor model of CCT

the circuit is same as shown in Fig. 13.13. The two pairs are enclosed by dotted lines. We have to replace these by two transistor circuits. The junctions are at $a$ and $a'$. Following the steps mentioned above the equivalent transistor version of the nullator-norator circuit is obtained as shown in Fig. 13.18 (b).

Thus it is seen that a nullator norator model can always be reduced to its equivalent transistor model and hence physical realisation of an otherwise hypothetical configuration. Fig. 13.19 shows a nullator norator model of a CNIC. As there is a nullator between terminals 1 and 2,

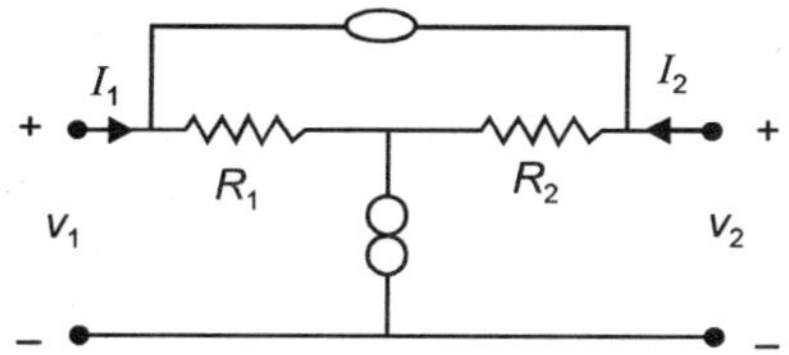

**Fig. 13.19**  Nullator-norator model of a CNIC.

voltage $V_1 = V_2$. Also, since current through the nullator is zero,

$$I_1 R_1 = I_2 R_2$$

or $$I_2 = \frac{R_1}{R_2} I_1 \text{ and } V_1 = V_2 \qquad \qquad \text{...(13.90)}$$

To obtain its equivalent transistor version we add a series combination of a nullator and a norator as shown in Fig. 13.20.

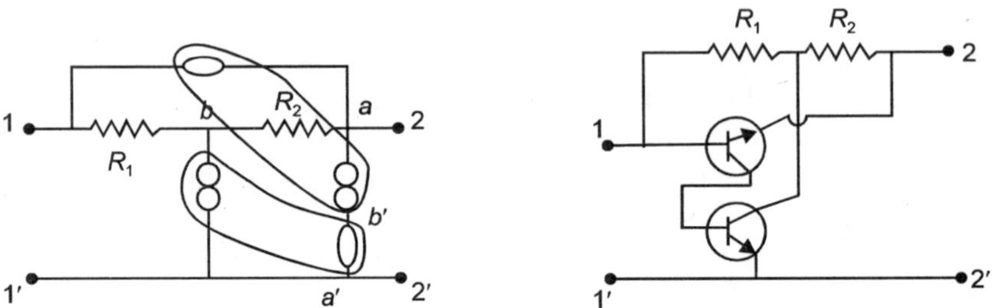

**Fig. 13.20**  CNIC realiasation of transistor circuit.

This acts as an open circuit and we have two pairs of nullator-norator each having a common point and hence we can use the procedure outlined above and we obtain the equivalent transistor version of CNIC.

We could have taken a series combination of nullator and norator on the left side of the nullator in Fig. 13.20 and we would get a different version of equivalent transistor circuit. Final selection of the circuit depends upon biasing, stability and other practical considerations.

Fig. 13.21 shows a nullor model of a VNIC

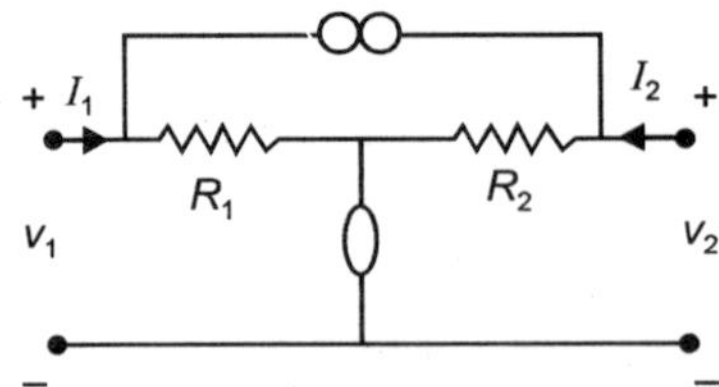

**Fig. 13.21**  Nullor model of a VNIC.

Since the nullator does not draw any current

$$I_1 = -I_2 \qquad \qquad \text{...(13.91)}$$

$$\frac{V_1}{R_1} = -\frac{V_2}{R_2} \qquad \qquad \text{...(13.92)}$$

or $$V_1 = -\frac{R_1}{R_2} V_2 \qquad \qquad \text{...(13.93)}$$

The modified VNIC and its equivalent transistorized version are shown in Fig. 13.22.

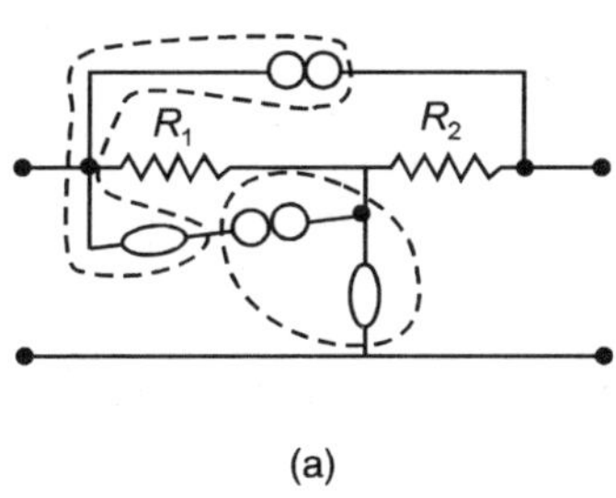

(a)

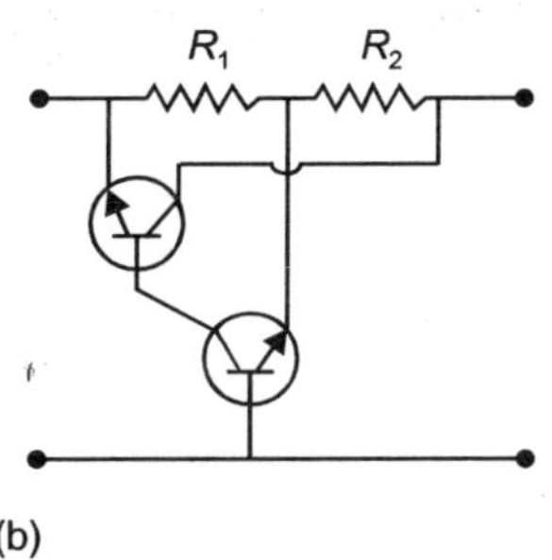

(b)

**Fig. 13.22**   (a) Modified VNIC   (b) Transistorised version of VNIC.

After studying briefly some of the active elements, we now turn to the design of active filters where we shall use operational amplifier as the active element and we make use of the co-efficient matching technique as an efficient method of filter design.

## 13.8  FILTER DESIGN BY CO-EFFICIENT MATCHING

In this method we begin with a network of given topology which is then analysed to determine the type of transfer function it can realize. Design equations are then obtained by equating like co-efficients.

The active element used is usually a non-inverting type operational amplifier of gains greater than unity. Since higher order filters can be realized by cascading second order stages, design methods in general for second order transfer function are considered. Sallen and key have done excellent work in this direction and have tabulated many such second order networks.

### 13.8.1  Second Order Low Pass Filter

Fig. 13.23 shows an active *RC* configuration for realizing LP filter originally proposed by Sallen and Key.

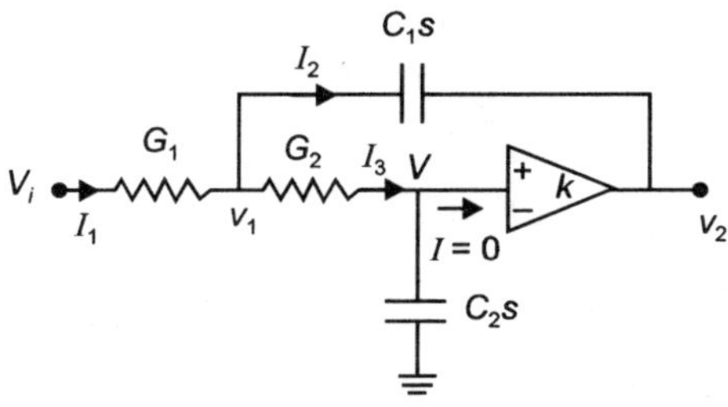

**Fig. 13.23**   Sallen-key low pass RC filter.

Applying KCL at node $V_1$ and $V$ we have

$$(V_i - V_1)\, G_1 = (V_1 - V_2)\, C_1 s + V_1\, \frac{G_2 C_2 s}{G_2 + C_2 s} \qquad \text{...(13.94)}$$

$$\left(V_1 - \frac{V_2}{K}\right) G_2 = \frac{V_2}{K}\, C_2 s \qquad \text{...(13.95)}$$

$$V_1 G_2 = \frac{G_2 + C_2 s}{K} \cdot V_2$$

or
$$V_1 = \frac{G_2 + C_2 s}{K G_2} V_2 \qquad \qquad \text{...(13.96)}$$

$$G_1 V_i = -V_2 C_1 s + V_1 \left\{ \frac{G_2 C_2 s}{G_2 + C_2 s} + C_1 s + G_1 \right\} \qquad \text{...(13.97)}$$

$$= -V_2 C_1 s + \frac{G_2 C_2 s + C_1 G_2 s + C_1 C_2 s^2 + G_1 G_2 + G_1 C_2 s}{(G_2 + C_2 s) K G_2}$$

$$\cdot (G_2 + C_2 s) V_2$$

$$= \frac{-K G_2 C_1 s + G_2 C_2 s + C_1 G_2 s + G_1 C_2 s + C_1 C_2 s^2 + G_1 G_2}{K G_2} V_2$$

or
$$\frac{V_2}{V_i} = \frac{K G_1 G_2}{C_1 C_2 s^2 + \{G_2 C_2 + G_1 C_2 + (1 - K) G_2 C_1\} s + G_1 G_2}$$

$$= \frac{K \cdot G_1 G_2 / C_1 C_2}{s^2 + \left\{ \dfrac{G_2}{C_1} + \dfrac{G_1}{C_1} + \dfrac{G_2}{C_2} (1 - K) \right\} s + \dfrac{G_1 G_2}{C_1 C_2}} \qquad \text{...(13.98)}$$

However, the general transfer function for a second order low pass filter is

$$T(s) = \frac{H \, \omega_0^2}{s^2 + \dfrac{\omega_0}{Q} s + \omega_0^2} \qquad \text{...(13.99)}$$

Where $K$, $\omega_0$ and $Q$ are the d.c. gain, resonant frequency (peak response frequency) and quality factor of the filter circuit. On comparing the co-efficients we have

$$H = K, \quad \omega_0^2 = \frac{G_1 G_2}{C_1 C_2} \quad \text{or} \quad \omega_0 = \sqrt{\frac{G_1 G_2}{C_1 C_2}} \qquad \text{...(13.100)}$$

$$Q = \frac{\sqrt{\dfrac{G_1 G_2}{C_1 C_2}}}{\dfrac{G_2}{C_1} + \dfrac{G_1}{C_1} + \dfrac{G_2}{C_2} (1 - K)} \qquad \text{...(13.101)}$$

It is to be noted that the number of unknowns is larger than the number of equations and hence it is desirable to assign suitable values to some unknowns (equal to number of unknowns – number of equations). However, it is to be kept in mind that in active filters built in thin films it is desirable to keep elemental spread (the ratio of maximum to minimum value) as small as possible. Therefore, in the present situation, there are four variables

$(G_1, G_2, C_1, C_2)$ and two equations. Let us select equal value elements $G_1 = G_2 = G$ and $C_1 = C_2 = C$. We then have

$$\omega_0 = \frac{G}{C} \qquad \qquad ...(13.102)$$

and

$$Q = \frac{G/C}{\dfrac{3G}{C} - \dfrac{G}{C}K} \qquad \qquad ...(13.103)$$

$$= \frac{1}{3 - K} \qquad \qquad ...(13.104)$$

These form the design equations for this second order filter or we may choose
$$C_1 = C_2 = 1 \text{ and solve the equation for } G_1 \text{ and } G_2$$
The following example illustrates the procedure.

**Example 13.7:** Design a low-pass second order Butterworth filter.

**Solution:** We know that Butterworth approximation for a second order filter is

$$H(s) = \frac{K}{s^2 + \sqrt{2}\, s + 1}$$

Comparing co-efficients with equation (13.98) and (13.99) we have

$$\omega_0 = \sqrt{\frac{G_1 G_2}{C_1 C_2}} = 1 \text{ (Say)}$$

and

$$\frac{\omega_0}{Q} = \sqrt{2} \quad \text{or} \quad Q = \frac{\omega_0}{\sqrt{2}} = \frac{1}{\sqrt{2}}$$

say

$$C_1 = C_2 = 1 \text{ F}$$

Now, therefore,

$$G_1 G_2 = 1$$

and

$$Q = \frac{\sqrt{G_1 G_2}}{G_1 + G_2 + G_2 (1 - K)}$$

Suppose

$$K = 2$$

Then

$$Q = \frac{\sqrt{G_1 G_2}}{G_1} = \frac{1}{G_1} = \frac{1}{\sqrt{2}}$$

$$G_1 = \sqrt{2}$$

and since

$$G_1 G_2 = 1$$

$$G_2 = \frac{1}{\sqrt{2}}$$

The filter is sketched here in Fig. E13.7.

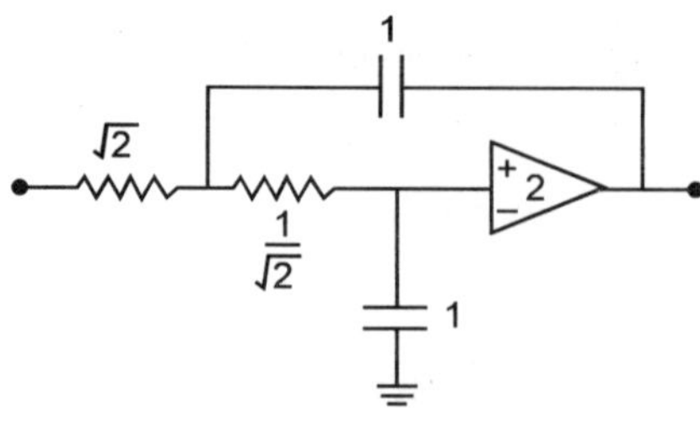

**Fig. E13.7**

It is to be noted that these values are normalized values. $C = 1$ does not mean 1 F, it is 1 if $\omega_0 = 1$. However, in practice $\omega_0$ is much larger than unity.

We have considered here a finite positive gain amplifier, one could consider a negative gain and infinite gain amplifier. But in many practical filter design situations the positive gain realisation is preferred to the other two realisations.

### 13.8.2  Second Order High Pass Filter

The configuration for the filter is also suggested by Sallen and key and is shown in Fig. 13.24.

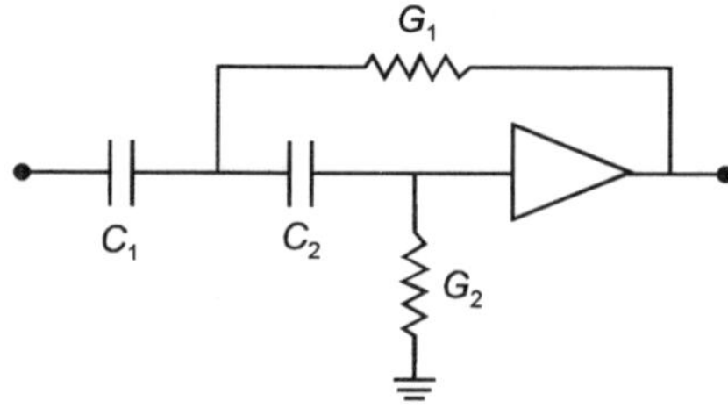

**Fig. 13.24**  Sallen and key second order High pass filter.

One way of designing this filter would be to develop equation in a manner similar to the one carried out in the case of low-pass filter. An alternative procedure would be to make an LP – HP transformation where

$$s \rightarrow 1/s$$

i.e. in the transfer function for an LP filter replace $s$ by $\dfrac{1}{s}$ e.g. if the transfer function for LP filter is

$$\frac{K}{s^2 + \sqrt{2}\,s + 1} \qquad \qquad \text{...(13.105)}$$

we replace $s$ by $\dfrac{1}{s}$

we have
$$\frac{K\,s^2}{s^2 + \sqrt{2}\,s + 1} \qquad \qquad \text{...(13.106)}$$

and this is the transfer function for an HP filter.

This transformation for active $RC$ networks can be carried out by what is known as $RC$ : $CR$ transformation which is achieved by replacing each resistance $R_i$ in the original network by a capacitance of value $1/R_i$ and replacing each capacitance $C_i$ in the original network by a resistance of value $1/C_i$ . The topological graph of the two networks would remain identical. This means if we have to design a High pass filter we will obtain its transfer function which will be similar to one given in (13.106) and we will convert this into LP transfer function as shown in (13.105). Using the transformation $s \rightarrow 1/s$ and we achieve the filter as shown in Fig. E13.7 Using $RC : CR$ transformation we will have the desired high pass filter.

Suppose the LP filter circuit is known and is the one given in Fig. E13.7. The high pass filter will have its parameters

$$C_1 = \sqrt{2} \qquad\qquad C_2 = \frac{1}{\sqrt{2}}$$

$$R_1 = 1 \qquad\qquad R_2 = 1$$

and the circuit will be as shown in Fig. 13.25

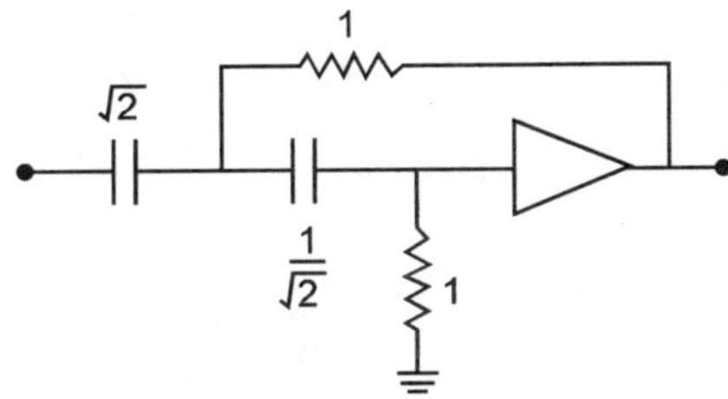

**Fig. 13.25**   A typical high pass   filter.

**Example 13.8:** Suppose the transfer function of high pass filter is found to be

$$\frac{K s^2}{s^2 + 5s + 6}$$

Transforming using transformation $s \rightarrow \dfrac{1}{s}$

$$\frac{K/s^2}{\dfrac{1}{s^2} + \dfrac{5}{s} + 6} = \frac{K}{6s^2 + 5s + 1}$$

or

$$\frac{K/6}{s^2 + \dfrac{5}{6}s + \dfrac{1}{6}}$$

Now this is the transfer function of a LP filter
With $\qquad\qquad K = 2$ we have

$$\omega_0^2 = \frac{G_1 G_2}{C_1 C_2}$$

and

$$Q = \frac{\sqrt{(G_1 G_2)/C_1 C_2}}{\dfrac{G_2}{C_1} + \dfrac{G_1}{C_1} - \dfrac{G_2}{C_2}}$$

Let $$C_1 = C_2 = 1$$

Therefore, $$\omega_0^2 = G_1\,G_2 = \frac{1}{6}$$

$$Q = \frac{\sqrt{G_1 G_2}}{G_1} = \sqrt{\frac{G_2}{G_1}}$$

Now $$\frac{\omega_0}{Q} = \frac{5}{6} = \frac{\sqrt{G_1 G_2}}{\sqrt{G_2/G_1}} = G_1$$

and $$G_2 = \frac{1}{6 G_1} = \frac{1 \times 6}{6 \times 5} = \frac{1}{5}$$

For LP filter $$G_1 = \frac{5}{6} \quad G_2 = \frac{1}{5} \quad C_1 = 1, \quad C_2 = 1$$

For HP filter using $RC : CR$ transformation

$$C_1 \to \frac{1}{R_1} = \frac{5}{6} \qquad\qquad C_2 \to \frac{1}{R_2} = \frac{1}{5}$$

$$R_1 \to \frac{1}{C_1} = 1 \qquad\qquad R_2 = \frac{1}{C_2} = 1$$

Hence the hp filter is as shown in Fig. E13.8

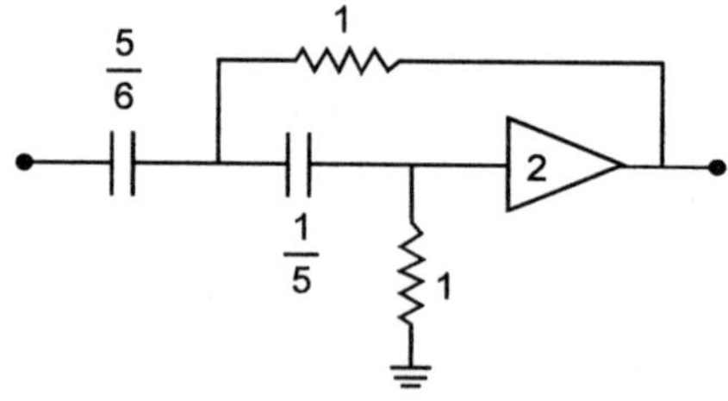

**Fig. E13.8**

### 13.8.3  Second Order Band-pass Filters

The configuration for the second order Band pass filter is shown in Fig. 13.26

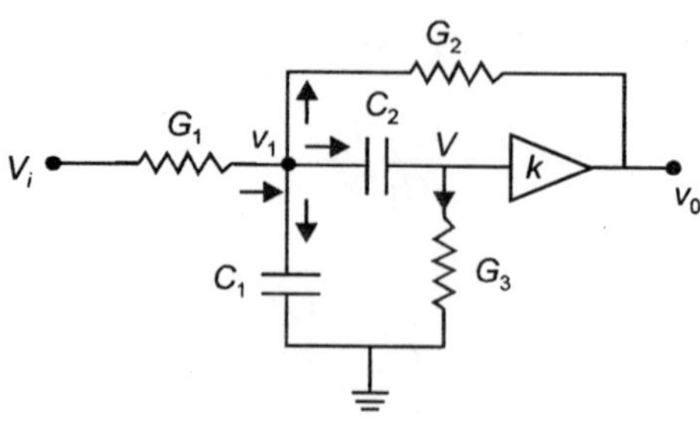

**Fig. 13.26**  Second order Band-pass filter.

Applying KCL at $V_1$ and $V$ we have

$$(V_i - V_1)\, G_1 = (V_1 - V_0)\, G_2 + \left(V_1 - \frac{V_0}{K}\right) C_2 s + V_1\, C_1 s \qquad \text{...(13.107)}$$

and

$$\left(V_1 - \frac{V_0}{K}\right) C_2 s = \frac{V_0}{K}\, G_3 \qquad \text{...(13.108)}$$

$$V_1\, C_2\, s = (G_3 + C_2 s)\, \frac{V_0}{K}$$

or

$$V_1 = \frac{G_3 + C_2 s}{C_2 s} \cdot \frac{V_0}{K}$$

$$V_i\, G_1 = (G_1 + G_2 + C_2 s + C_1 s) V_1 - \left(G_2 + \frac{C_2 s}{K}\right) V_0$$

$$= \left[G_1 + G_2 + (C_1 + C_2)s\right] \frac{G_3 + C_2 s}{C_2 s} \cdot \frac{V_0}{K} - \left(G_2 + \frac{C_2 s}{K}\right) V_0$$

$$= (G_1 + G_2 + C_1 s + C_2 s)(G_3 + C_2 s)\, V_0 - G_2\, C_2\, sK\, V_0 - C_2^2 s^2\, V_0$$

$$V_i\, G_1\, C_2\, Ks = \left[G_1 G_3 + G_1 C_2 s + G_2 C_2 s + G_3 C_1 s \right.$$

$$\left. + C_1 C_2 s^2 + G_3 C_2 s - G_2 C_2 sK + G_2 G_3\right] V_0$$

$$= \left[C_1 C_2 s^2 + \{G_1 C_2 + G_2 C_2 (1 - K) + G_3 C_1 + G_3 C_2\}\right.$$

$$\left. s + G_1 G_3 + G_2 G_3)\right] V_0$$

or

$$\frac{V_0}{V_i} = \frac{G_1 C_2\, Ks / C_1 C_2}{s^2 + \left\{\dfrac{G_1}{C_1} + \dfrac{G_2}{C_1}(1 - K) + \dfrac{G_3}{C_2} + \dfrac{G_3}{C_1}\right\} s + \dfrac{G_1 G_3 + G_2 G_3}{C_1 C_2}}$$

$$= \frac{K(G_1/C_1)s}{s^2 + \left\{\dfrac{G_1}{C_1} + \dfrac{G_2}{C_1}(1 - K) + \dfrac{G_3}{C_2} + \dfrac{G_3}{C_1}\right\} s + \dfrac{G_3(G_1 + G_2)}{C_1 C_2}} \qquad \text{...(13.109)}$$

It is to be noted that no such network transformation exists to modify an active RC low-pass filter to an active RC band pass filter and vice versa. The usual low-pass to band-pass transformations

$$s \rightarrow s + \frac{1}{s} \qquad \text{...(13.110)}$$

can be carried out only on the network function. However, a band-pass filter can be realized by cascading (with proper isolation) a low pass filter with a high pass filter.

The general transfer function for a band pass filter is

$$\frac{V_0}{V_i} = \frac{Hs}{s^2 + \dfrac{\omega_0}{Q} s + \omega_0^2} \qquad \text{...(13.111)}$$

Comparing the co-efficients of the expression (13.111) with the (13.109) above, we have

$$H = K \frac{G_1}{C_1} \qquad \omega_0^2 = \frac{G_3(G_1 + G_2)}{C_1 C_2}$$

and

$$\frac{\omega_0}{Q} = \frac{G_1}{C_1} + \frac{G_2}{C_1}(1 - k) + G_3 \frac{C_1 + C_2}{C_1 C_2}$$

Suppose, the transfer function of a band pass filter is given as

$$\frac{H s}{s^2 + 2s + 16} \qquad\qquad ...(13.112)$$

Realise this band pass filter.

Since the number of unknowns is more than the number of equations we assume the values of some of the unknowns and obtain the values of the remaining unknowns

We let $\qquad\qquad G_1 = G_2 = 1$

and $\qquad\qquad C_1 = C_2 = 1$

We have $\qquad\qquad H = K$  as  $G_1 = C_1 = 1$

$\qquad\qquad\qquad \omega_0^2 = 16 = 2G_3$  or  $G_3 = 8$

Also $\qquad\qquad 2 = \dfrac{\omega_0}{Q} = 1 + (1 - k) + 16$

or $\qquad\qquad 2 = 2 - k + 16$

or $\qquad\qquad K = 16$

Thus the band pass filter is as shown

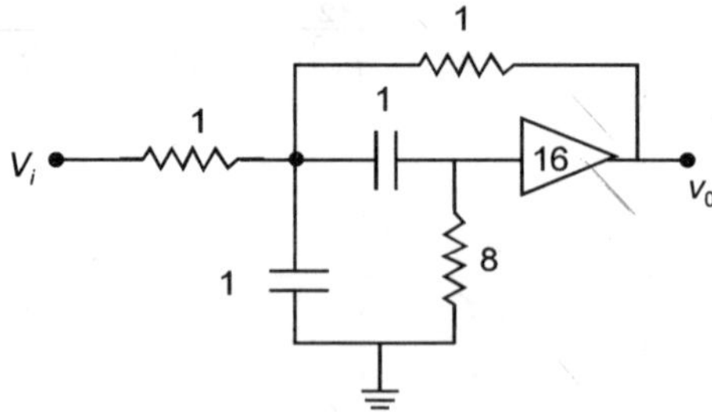


**Fig. 13.27** Band pass filter for the given problem.

We close here the design of active *RC* filters. In fact, there are a large number of configurations available in the literature for these filters and these configurations can be analysed to obtain voltage gain transfer function and then the co-efficients of the transfer function are compared with the given approximated (Butterworth, Chebychev or any other approximation) transfer function. Since, in general the number of unknowns are greater than the number of equations, some of these unknowns are assigned suitable values, the remaining are obtained from the given relations. We next move to sensitivity consideration of *RC* active filters.

## 13.9  SENSITIVITY CONSIDERATIONS

We now study the variation in response of a filter when one or more number of elements undergo minor variation in its/their values i.e. we study how sensitive the response of a filter is when one or more than one elements undergo minor variations in its/their values. This is known as sensitivity considerations. The elements like $R$ and $C$ undergo minor variations in their values due to environmental condition (Temperature humidity etc.) and again, many a times we want to replace one resistor or a capacitor as it has gone faulty. Even though we replace by the same specification resistor or capacitor, however, no resistor or capacitor will be exactly identical because of manufacturing problems and the real value of the element will be different. Now, how the performance of the circuit is affected by the replacement, is the sensitivity consideration.

The sensitivity function $S_k^{T(s)}$ of the network function $T(s, k)$ due to the variation of the parameter $k$ is defined as

$$S_k^{T(s)} = \frac{d\left(\ln T(s,k)\right)}{d\left(\ln k\right)} = \frac{dT(s,k)/T(s,k)}{dk/k} \qquad ...(13.113)$$

This means $S_k^{T(s)}$ is equal to the percentage or per unity change in $T(s, k)$ divided by the percentage or per unity change in $k$, assuming the changes are small increments e.g. if $S_k^{T(s)} = 0.5$ then a 1 percentage change in the parameter $k$ will cause a 0.5% change in the network function. Therefore, from the design view point, it is desirable to keep $S_k$ as small as possible. Ideally it should be zero i.e. we should have insensitive network, insensitive to minor variation in the element values.

From the definition of the sensitivity function the following two identities follow:

$$S_{1/k}^{T} = -S_k^{T} \qquad \text{Also } S_k^{1/T} = -S_k^{T} \qquad ...(13.114)$$

$$S_{k1}^{T} = S_{k2}^{T}\, S_{k1}^{k_2} \qquad ...(13.115)$$

Besides studying sensitivity of $T$ we have to study sensitivity of $\omega_0$, $Q$ etc. which are functions of circuit parameters.

Before we take up sensitivity of some of the active filters, we develop expressions for $S$ which are useful.

Let $F$ be a polynomial in $x$, we obtain $S_x^F$ for different formulation of $F$

1.  When $F$ is a constant, independent of $x$ i.e. $F = C$

$$\frac{dF}{dx} = 0 \qquad \text{Hence } S_x^F = 0 \qquad ...(13.116)$$

2.  When $F = C\,x$

Now
$$S_x^F = \frac{dF/F}{dx/x}$$

$$= \frac{C\,dx/C\,x}{dx/x} = 1 \qquad ...(13.117)$$

In general
$$S_x^{KF(x)} = S_x^{F(x)} \qquad ...(13.118)$$

where $k$ is a constant

3.  $F = x^n$

$$S_x^F = \frac{dF/F}{dx/x} = \frac{n\,x^{n-1}dx/x^n}{dx/x}$$

$$= n \qquad\qquad\text{...(13.119)}$$

4.  Let $F = x^2$

$$S_{x^2}^F = \frac{dF/F}{dx^2/x^2} = \frac{1/x^2}{1/x^2} = 1$$

$$S_x^F = \frac{dF/F}{dx/x} = \frac{2x\,dx/x^2}{dx/x}$$

$$= 2$$

Hence $\qquad\qquad S_x^F = 2S_{x^2}^F \qquad\qquad\text{...(13.120)}$

5.  $F = F_1\,F_2$

$$S_x^F = \frac{dF/F}{dx/x} = \frac{\left(F_2\,\partial F_1 + F_1\,\partial F_2\right)/F}{dx/x}$$

$$= \frac{\partial F_1/F_1}{dx/x} + \frac{\partial F_2/F_2}{dx/x}$$

$$= S_x^{F_1} + S_x^{F_2} \qquad\qquad\text{...(13.121)}$$

6.  $F = A + Bk$ where A and B are constant

$$S_k^F = \frac{dF/F}{dk/k} = \frac{B\,dk/(A+Bk)}{dk/k}$$

$$= \frac{Bk}{F} \qquad\qquad\text{...(13.122)}$$

Let us illustrate the application of some of the expressions derived above.

**Example 13.9:** Consider the impedance of a parallel RLC circuit.

$$Z(s) = \frac{s/c}{s^2 + \dfrac{1}{RC}s + \dfrac{1}{LC}}$$

$$= \frac{s/c}{s^2 + \dfrac{\omega_0}{Q}s + \omega_0^2}$$

where $\qquad \omega_0^2 = \dfrac{1}{LC} \quad$ and $\quad \dfrac{\omega_0}{Q} = \dfrac{1}{RC}$

Determine various sensitivities.

**Solution:** $\qquad\qquad \omega_0 = L^{-1/2}\,C^{-1/2}$

Hence $\qquad\qquad S_L^{\omega_0} = -\dfrac{1}{2}$

$$S_C^{\omega_0} = -\frac{1}{2}$$

To find $S_R^Q$ we should be able to express $Q$ as a function of $R$

Now $\dfrac{Q}{\omega_0} = RC$  or  $Q = \omega_0\, RC$

$$= \frac{RC}{\sqrt{LC}} = R \cdot L^{-1/2} C^{1/2}$$

Therefore, $$S_R^Q = 1$$

$$S_L^Q = -\frac{1}{2} \text{ and } S_C^Q = \frac{1}{2}$$

Now to find $S_R^Z$ we rewrite the impedance expression as

$$Z^{-1} = \frac{c}{s}\left(s^2 + \frac{1}{LC}\right) + R^{-1}$$

$$Y = A + x$$

$$S_x^y = \frac{x}{y} = \frac{R^{-1}}{Z^{-1}} = R^{-1}Z$$

$$S_{R^{-1}}^{Z^{-1}} = \frac{s/CR}{s^2 + \dfrac{s}{CR} + \dfrac{1}{LC}}$$

Since superscript and subscript both are inverse, therefore,

$$S_{R^{-1}}^{Z^{-1}} = S_R^Z = \frac{s/CR}{s^2 + \dfrac{s}{CR} + \dfrac{1}{LC}}$$

Similarly $S_L^Z$ or $S_C^Z$ can be obtained.

### 13.9.1  Multi-parameter Variation

If function $F$ is a function of more than one variable $x_j$, $j = 1, \ldots n$, the sensitivity due to each variable is

$$\frac{\partial F/F}{\partial x_i/x_i}$$

Therefore, total change in $F$ if there are small variations in all variables is given as

$$\frac{dF}{F} = \sum_{i=1}^{n} S_{x_i}^{F} \frac{\partial x_i}{x_i} \qquad \qquad ...(13.123)$$

i.e. the total per unit change in $F$ is equal to the sum of the products of sensitivity of individual variable and the per unit change in that corresponding variable taken over all the variables.

### 13.9.2  Pole-zero Sensitivities

In certain situations one or more number of poles (zeros) may be of great importance. The values or location of these poles depend upon circuit parameters (in circuit theory) or system parameters (in control or other system). Sometimes these poles (zero) may be occupying critical locations. A small variation in the parameter of the system may change the pole (zero) configuration such that the system may become unstable or it may affect some other performance aspect of the circuit or the system. The measure of the displacement due to an incremental change in the network parameter would thus be useful. An estimate of the change in the pole (zero) locations is given by the pole (zero) sensitivity which is defined as follows:

Let $s = s_j$ be a pole (zero) of $F(s, k)$ when $k$ takes its nominal value. The pole zero sensitivity of $F(s, k)$ is then defined as

$$S_k^{s_j} \triangleq \left. \frac{d s_j}{d k/k} \right|_{s = s_j} \qquad \qquad ...(13.124)$$

This expression could be considered on the root sensitivity of a numerator polynomial $N(s, k)$ if $s = s_j$ is a zero of $F(s, k)$. Similarly if $s = s_j$ is a pole then the same expression could be considered as the root sensitivity of denominator polynomial $D(s, k)$.

It is to be noted that pole and zero sensitivities are numbers. If the root is real, the corresponding sensitivity is real, otherwise it is a complex number.

Also, it is to be noted that in case of other sensitivities, if one parameter is changed, one performance index changes, whereas here as can be seen in the sensitivity definition when $k$ changes all the roots $s_j$ undergo variation, that is why in the superscript we have $s_j$ when subscript is $k$. We try to find out here in root sensitivity the change in value of pole or zero for a unity change in the parameter $k$, in contrast to general sensitivity where we find the unity change in $F(s, k)$ for a unity change in $k$. We rewrite here

$$S_k^{s_j} = \left. \frac{d s_j}{d k/k} \right|_{s = s_j}$$

$$= k \left. \frac{d s_j}{d k} \right|_{s = s_j} \qquad \qquad ...(13.125)$$

$$= k(-1) \left. \frac{dF/dk}{dF/ds} \right|_{s = s_j} \qquad \qquad ...(13.126)$$

This expression can be used for evaluating root sensitivity, once it is calculated, the displacement in the root ($\Delta s_j$) can be evaluated using the relation

$$\Delta s_j = S_k^{s_j} \frac{\Delta k}{k} \qquad \qquad ...(13.127)$$

where $\dfrac{\Delta k}{k}$ is the per unit change in parameter $k$.

We illustrate the procedure with a simple example.

**Example 13.10:** Suppose the polynomial is

$$F(s, k) = s^2 + ks + 6$$

where $k = 5$ is the nominal value and say the variation is 0.05 per unit increase i.e. $(\Delta k/k) = 0.05$. Determine the root sensitivity and the displacement of the roots.

**Solution:** Using (13.126) we have

$$S_K^{s_j} = k \frac{ds}{dk} = k(-1) \frac{dF}{dk} \cdot \frac{ds}{dF}\bigg|_{s=s_j}$$

From the expression for the polynomial

$$\frac{dF}{dk} = s \quad \text{and} \quad \frac{dF}{ds} = 2s + k$$

Therefore,

$$S_k^{s_j} = 5(-1) \frac{s}{2s+5}\bigg|_{s=s_j}$$

There are two roots $s_1 = -2$, $s_2 = -3$

$$S_1^{new} = S_1^{old} + \Delta s_j \quad j = 1$$

$$S_5^{s_1} = 5\,(-1)\,\frac{-2}{-4+5} = 10$$

$$S_5^{s_2} = 5\,(-1)\,\frac{-3}{-6+5} = -15$$

$$\Delta S_j = S_k^{S_j} \frac{\Delta k}{k}$$

$$\Delta S_1 = 10 \times 0.05 = 0.5$$

$$\Delta s_2 = -15 \times 0.05 = -0.75$$

Therefore,

$$S_{1new} = -2 + 0.5 = -1.5$$

$$S_{2new} = -3 - 0.75 = -3.75 \textbf{ Ans.}$$

**Example 13.11:** Consider another polynomial

$$F(s_1\, k) = s^2 + ks + 1$$

where $k$ has a nominal value of 1.5 and assume that $k$ is changed (decreased) 0.0666 per unit. The roots of the original polynomial are:

$$s_1,\, s_2 = \frac{-1.5 \pm \sqrt{1.5^2 - 4}}{2}$$

$$= \frac{-1.5 \pm j\,1.3228}{2}$$

$$= -0.75 \pm j\,0.6614$$

Now

$$S_k^{s_j} = -k \frac{ds}{dk} = -k \frac{dF/dk}{dF/ds}$$

$$\frac{dF}{dk} = s \quad \text{and} \quad \frac{dF}{ds} = 2s + k$$

$$S_k^{s_j} = -k \, \frac{s}{2s + k}$$

$$S_k^{s_1} = -1.5 \, \frac{-0.75 + j0.6614}{-1.5 + j1.3228 + 1.5}$$

$$= -1.5 \, \frac{-0.75 + j0.6614}{j1.3228}$$

$$= j1.134 \, (-0.75 + j0.6614)$$

$$= -j0.85 - 0.75 = -0.75 - j0.85$$

$$S_k^{s_2} = -0.75 + j0.85$$

$$\Delta s_1 = 0.0666 \, (-0.75 - j0.85)$$

$$= -0.05 - j0.0567$$

and

$$\Delta s_2 = -0.05 + j0.0567$$

$$S_1' = -0.75 + j0.6614 + 0.05 + j0.0567$$

$$= -0.70 + j0.7181$$

and

$$S_2' = -0.75 - j0.6614 + 0.05 - j0.0567$$

$$= -0.7 - j0.7181 \ \textbf{Ans.}$$

## 13.10  SENSITIVITY CONSIDERATIONS FOR RC ACTIVE FILTERS

Consider the Sallen-key configuration of low-pass filter for which the transfer function is found to be equation (13.98).

$$T(s) = \frac{K\,G_1 G_2/C_1 C_2}{s^2 + \left\{\dfrac{G_2}{C_1} + \dfrac{G_1}{C_1} + \dfrac{G_2}{C_2}\,(1 - K)\right\}s + \dfrac{G_1 G_2}{C_1 C_2}}$$

$$= \frac{K\,\omega_0^2}{s^2 + \dfrac{\omega_0}{Q}\,s + \omega_0^2}$$

where

$$\omega_0 = \sqrt{\frac{G_1 G_2}{C_1 C_2}} = \frac{1}{\sqrt{R_1 R_2 C_1 C_2}}$$

Since

$$\omega_0 = R_1^{-1/2}\,R_2^{-1/2}\,C_1^{-1/2}\,C_2^{-1/2} \qquad \qquad \text{...(13.128)}$$

Therefore,

$$S_{R1}^{\omega_0} = S_{R2}^{\omega_0} = S_{C1}^{\omega_0} = S_{C2}^{\omega_0} = -\frac{1}{2} \qquad \qquad \text{...(13.129)}$$

Hence

$$S_{R1}^{\omega_0} + S_{R2}^{\omega_0} = -1 \qquad \qquad \text{...(13.130)}$$

and
$$S_{C1}^{\omega_0} = S_{C2}^{\omega_0} = -1 \qquad \qquad \text{...(13.131)}$$

Now
$$\frac{\omega_0}{Q} = \frac{1}{R_1 C_1} + \frac{1}{R_2 C_1} + \frac{1}{R_2 C_2}(1 - K) \qquad \text{...(13.132)}$$

To find out sensitivity of $Q$ with respect to $R_1$, $R_2$ $C_1$ and $C_2$, it is convenient to find sensitivity of $\dfrac{\omega_0}{Q}$ rather than to find out directly and since $S_{R,C}^{\omega_0}$ is known we can find out $S_{R,C}^{Q}$ easily.

Now
$$S_{R_1 C}^{\omega_0/Q} = S_{R_1 C}^{\omega_0} - S_{R_1 C}^{Q}$$

Hence
$$S_{R_1}^{\omega_0} - S_{R_1}^{Q} = -\frac{1}{2} - S_{R_1}^{Q} = -S_{1/R_1}^{\omega_0}$$

The polynomial $\dfrac{\omega_0}{Q}$ could be rewritten as

$$\frac{\omega_0}{Q} = A + B \, \frac{1}{R_1}$$

Hence sensitivity is $\dfrac{B \cdot 1/R_1}{A + B/R_1}$ using equation (13.122)

Therefore,
$$-\frac{1}{2} - S_{R_1}^{Q} = - \frac{1/R_1 C_1}{\dfrac{1}{R_1 C_1} + \dfrac{1}{R_2 C_1} + \dfrac{1}{R_2 C_2}(1 - K)}$$

Hence
$$S_{R_1}^{Q} = -\frac{1}{2} + \frac{1/R_1 C_1}{\dfrac{1}{R_1 C_1} + \dfrac{1}{R_2 C_1} + \dfrac{1}{R_2 C_2}(1 - K)}$$

$$= -\frac{1}{2} + \frac{Q}{R_1 C_1}(R_1 C_1 R_2 C_2)^{1/2} = -\frac{1}{2} + Q\sqrt{\frac{R_2 C_2}{R_1 C_1}}$$

To find out $S_{R_2}^{Q}$ we make use of the relation (13.114)

$$S_{R_2}^{\omega_0/Q} = S_{R_2}^{\omega_0} - S_{R_2}^{Q} = -\frac{1}{2} - S_{R_2}^{Q} = -\frac{1}{2} + S_{1/R_2}^{Q}$$

Now
$$\frac{\omega_0}{Q} = A + \frac{1}{R_2}\left\{\frac{1}{C_1} + \frac{1}{C_2} - \frac{K}{C_2}\right\}$$

$$= A + B \cdot \frac{1}{R_2}$$

Hence, sensitivity is

$$= \frac{B \cdot 1/R_2}{A + B \cdot \dfrac{1}{R_2}}$$

$$= \frac{\left(\dfrac{1}{C_1} + \dfrac{1}{C_2} - \dfrac{K}{C_2}\right) \cdot \dfrac{1}{R_2}}{\dfrac{1}{R_1 C_1} + \dfrac{1}{R_2 C_1} + \dfrac{1}{R_2 C_2}(1 - K)}$$

$$S_{R_2}^Q = -\frac{1}{2} + \frac{\left\{\dfrac{1}{C_1} + \dfrac{1}{C_2}(1 - K)\right\}\dfrac{1}{R_2}}{\dfrac{1}{R_1 C_1} + \dfrac{1}{R_2 C_1} + \dfrac{1}{R_2 C_2}(1 - K)}$$

$$= -\frac{1}{2} + 1 - Q\sqrt{\frac{R_2 C_2}{R_1 C_1}} = \frac{1}{2} - Q\sqrt{\frac{R_2 C_2}{R_1 C_1}}$$

Hence
$$S_{R_1}^Q + S_{R_2}^Q = 0$$

Following the same procedure we can find out $S_{C_1}^Q$ and $S_{C_2}^Q$ and it can be shown that

$$S_{C_1}^Q + S_{C_2}^Q = 0$$

### 13.10.1 Sensitivities of some Particular Cases of Sallen-key Filter

**Case I.** Let $G_1 = G_2 = G$ and $C_1$ and $C_2 = C$ i.e. the resistors are of equal values and so are the capacitors separately. The transfer function is given as

$$T(s) = \frac{K\, G^2/C^2}{s^2 + (3 - K)\dfrac{G}{C}s + \dfrac{G^2}{C^2}} \qquad\qquad ...(13.133)$$

Hence
$$\omega_0^2 = \frac{1}{R^2 C^2} \quad \text{or} \quad \omega_0 = R^{-1}C^{-1}$$

and
$$\frac{\omega_0}{Q} = (3 - K)\frac{1}{RC} \quad \text{and} \quad Q = \frac{1}{3 - K}$$

Here $S_R^{\omega_0} = -1$ as there are two $R^s$ in the circuit and if each $R$ is varied by the same amount then $\omega_0$ varies by the same amount and

$$S_{R_1 = R}^{\omega_0} + S_{R_2 = R}^{\omega_0} = -1 \qquad\qquad ...(13.134)$$

Similarly
$$S_{C_1}^{\omega_0} + S_{C_2}^{\omega_0} = -1 \qquad\qquad ...(13.135)$$

Substituting in the expression for $S_{R_1}^Q$, $R_1 = R_2 = R$ and $C_1 = C_2 = C$ we have

$$S_{R_1}^Q = \frac{1/RC}{\dfrac{2}{RC} + \dfrac{1}{RC}(1 - K)} - \frac{1}{2} = \frac{1}{2 + 1 - K} - \frac{1}{2}$$

$$= \frac{1}{3 - K} - \frac{1}{2} = Q - \frac{1}{2}$$

Therefore,
$$S_{R_2}^Q = \frac{1}{2} - Q \qquad \qquad ...(13.136)$$

Similarly it can be shown that

$$S_{C_1}^Q = -S_{C_2}^Q = 2Q - \frac{1}{2} \qquad \qquad ...(13.137)$$

Now
$$S_k^Q = \frac{K\,dQ}{Q\,dK}$$

Also
$$Q = \frac{1}{3-K} \quad \text{or} \quad dQ = \frac{(3-K)\cdot 0 - 1\,(-1)}{(3-K)^2}\,dK$$

$$= \frac{dK}{(3-K)^2}$$

or
$$S_k^Q = \frac{K}{Q}\cdot\frac{1}{(3-K)^2} = KQ$$

$$= KQ$$

$$= Q\left(3 - \frac{1}{Q}\right) = 3Q - 1 \qquad \qquad ...(13.138)$$

and
$$K = 3 - \frac{1}{Q}$$

This is a very important result. Assume $S_k^Q$ equal to 3 Q which implies that for a $Q$ of 25, the gain must be kept within 0.1 percent to keep the percent variation of $Q$ within 7.5 percent. As a result active filters in general require a stricter tolerance for high $Q$ responses and, therefore, we face a serious problem of designing high $Q$ active $RC$ filters.

**Case II.** When $K = 1$ and $R_1 = R_2 = R$

$$T(s) = \frac{1/R^2 C_1 C_2}{s^2 + \dfrac{2}{RC_1}s + \dfrac{1}{R^2 C_1 C_2}} \qquad \qquad ...(13.139)$$

$$\omega_0^2 = \frac{1}{R^2 C_1 C_2} \quad \text{and} \quad \frac{\omega_0}{Q} = \frac{2}{RC_1}$$

$$\omega_0 = R^{-1} C_1^{-1/2} C_2^{-1/2} \quad \text{and} \quad Q = \frac{\omega_0 RC_1}{2}$$

$$= \frac{1}{R\sqrt{C_1 C_2}}\cdot\frac{RC_1}{2}$$

$$S_{R_1}^{\omega_0} = S_{R_2}^{\omega_0} = S_R^{\omega_0} = -\frac{1}{2} \qquad \qquad ...(13.140)$$

$$S_{C_1}^{\omega_0} = S_{C_1}^{\omega_0} = -\frac{1}{2} \qquad \qquad ...(13.141)$$

Since
$$Q = \frac{1}{2}\sqrt{\frac{C_1}{C_2}} = \frac{1}{2}C_1^{1/2}\,C_2^{-1/2}$$

Hence $S_{R_1}^Q = 0 = S_{R_2}^Q$ as $Q$ is independent of $R_1$ and $R_2$.

$$S_{C_1}^Q = \frac{1}{2} = -S_{C_2}^Q \qquad \qquad ...(13.142)$$

The sensitivities in this case are much lower ($\simeq 0.5$) as compared to the Case I where these were proportional to $Q$. This is an advantage. However, the ratio of $C_1$ to $C_2$ is quite large as $2\,Q = \sqrt{\dfrac{C_1}{C_2}}$ or $\dfrac{C_1}{C_2} = 4Q^2$

For higher values of $Q$, the values of $C_1$ and $C_2$ are widely different.

**Case III.** Let $\dfrac{R_2}{R_1} = \dfrac{Q}{\sqrt{3}}$ and $C_1 = Q\sqrt{3}$ and $C_2 = 1$

To determine the sensitivities

Since
$$\omega_0^2 = \frac{1}{R_1 R_2 C_1 C_2} = \frac{1}{R_1^2\,\dfrac{Q}{\sqrt{3}}\cdot Q\sqrt{3}} = \frac{1}{Q^2 R_1^2}$$

$$R_1 = \frac{1}{\omega_0 Q}, \; R_2 = \frac{1}{\omega_0\sqrt{3}}$$

$$\frac{\omega_0}{Q} = \frac{1}{R_1 C_1} + \frac{1}{R_2 C_1} + \frac{1}{R_2 C_2}(1-K)$$

or
$$= \frac{\omega_0 Q}{Q\sqrt{3}} + \frac{\omega_0\sqrt{3}}{Q\sqrt{3}} + \frac{\omega_0\sqrt{3}}{1}(1-K)$$

or
$$1 + 3\,(1-K) = 0$$

or
$$4 = 3\,K \text{ or } K = \frac{4}{3}$$

$$R_1 = \frac{1}{\omega_0 Q} \quad R_2 = \frac{1}{\sqrt{3}\,\omega_0}$$

Since
$$\omega_0^2 = \frac{1}{R_1 R_2 C_1 C_2}$$

$$S_{R_1, R_2, C_1, C_2}^{\omega_0} = -\frac{1}{2} \qquad \qquad ...(13.143)$$

$$S_{R_1}^Q = \frac{1/R_1 C_1}{\dfrac{1}{R_1 C_1} + \dfrac{1}{R_2 C_1} + \dfrac{1}{R_2 C_2}(1-K)} - \frac{1}{2}$$

$$= Q\left(\frac{R_2 C_2}{R_1 C_1}\right)^{1/2} - \frac{1}{2}$$

$$= Q\left(\frac{Q}{\sqrt{3}} \cdot \frac{1}{\sqrt{3}\,Q}\right)^{1/2} - \frac{1}{2}$$

$$= \frac{Q}{\sqrt{3}} - \frac{1}{2} = -S_{R_2}^Q \qquad \qquad ...(13.144)$$

Similarly,

$$S_{C_1}^Q = Q\left(\frac{C_2}{C_1}\right)^{1/2} \cdot \frac{R_1 + R_2}{(R_1 R_2)^{1/2}} - \frac{1}{2}$$

$$= Q\left(\frac{1}{Q\sqrt{3}}\right)^{1/2} \frac{\dfrac{1}{\omega_0 Q} + \dfrac{1}{\sqrt{3}\,\omega_0}}{\left(\dfrac{1}{\sqrt{3}\,\omega_0^2 Q}\right)^{1/2}} - \frac{1}{2}$$

$$= Q\left(\frac{1}{Q\sqrt{3}}\right)^{1/2} \cdot \frac{\sqrt{3} + Q}{\sqrt{3}\,\omega_0 Q}\, \omega_o \left(\sqrt{3}\,Q\right)^{1/2} - \frac{1}{2}$$

$$= \frac{\sqrt{3} + Q}{\sqrt{3}} - \frac{1}{2}$$

$$= 1 + \frac{Q}{\sqrt{3}} - \frac{1}{2} = \frac{Q}{\sqrt{3}} + \frac{1}{2} \qquad \qquad ...(13.145)$$

$$= -S_{C_2}^Q$$

### 13.10.2 Negative Gain Low Pass Filter

Fig. 13.28 shows a second order low pass filter with negative gain.

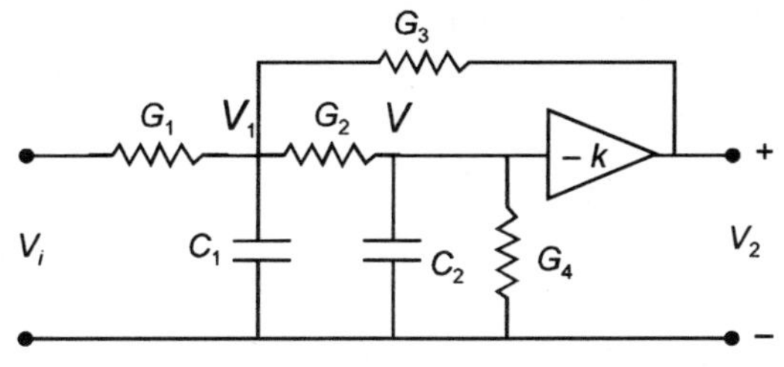



**Fig. 13.28**  Second order LP filter with −ve feedback.

Applying KCL at $V_1$ and $V$ nodes and after some algebraic manipulations; the transfer function is given as

$$T(s) = \cfrac{-K\,G_1\,G_2/C_1\,C_2}{s^2 + \left\{\dfrac{1}{C_1}(G_1 + G_2 + G_3) + \dfrac{1}{C_2}(G_2 + G_4)\right\}s}$$

$$+ \frac{G_1\,G_2}{C_1\,C_2}\left[1 + (1+K)\frac{G_3}{G_1} + \frac{G_3}{G_1}\left(1 + \frac{G_3}{G_2} + \frac{G_1}{G_2}\right)\right] \qquad ...(13.146)$$

Let $G_1 = G_2 = G_3 = G_4 = G$ and $C_1 = C_2 = C$ we have

$$T(s) = \frac{-K(G/C)^2}{s^2 + 5\dfrac{G}{C}s + (K+5)\,G^2/C^2} \qquad ...(13.147)$$

$$= \frac{-K\left(\dfrac{1}{RC}\right)^2}{s^2 + \dfrac{5}{RC}s + (K+5)\dfrac{1}{R^2\,C^2}} \qquad ...(13.148)$$

Therefore, $\quad \omega_0^2 = \dfrac{K+5}{R^2\,C^2} \quad$ or $\quad \omega_0 = \dfrac{\sqrt{K+5}}{RC}$

$$\omega_0 \propto \frac{1}{RC}$$

Also $\qquad \dfrac{\omega_0}{Q} = \dfrac{5}{RC} \quad$ or $\quad Q = \dfrac{\omega_0\,RC}{5} = \dfrac{\sqrt{K+5}}{5}$

Hence the design equations are:

$$RC = \frac{5Q}{\omega_0} \quad \text{and} \quad K = 25\,Q^2 - 5$$

$$S_{R_1}^{\omega_0} = S_{R_2}^{\omega_0} = -\frac{1}{2} = S_{C_1}^{\omega_0} = S_{C_2}^{\omega_0} \qquad ...(13.149)$$

The closed loop gain of the amplifier for a reasonable value of $Q$ is very large e.g. for $Q = 5$ results in $K = 620$. This is a common feature of all negative gain circuits. It is found that the sensitivities of $\omega_0$ and $Q$ with reference to the circuit parameters are all less than or equal to ½.

### 13.10.3 Infinite Gain Low-pass Filter

Fig. 13.29 shows an infinite gain second order low pass filter with multiple feedback.

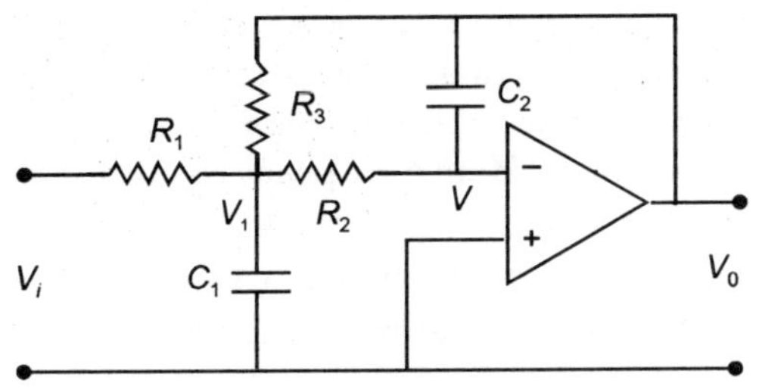

**Fig. 13.29** Second order LP filter with infinite gain

Writing KCL equation at $V_1$ and $V$ we have

$$\frac{V_i - V_1}{R_1} = \frac{V_1 - V_0}{R_3} + \frac{V_1}{R_2} + V_1 C_1 s \qquad \text{...(13.150)}$$

and

$$\frac{V_1}{R_2} = -V_0 \, C_2 \, s \qquad \text{...(13.151)}$$

or

$$V_1 = -V_0 \, C_2 \, R_2 \, s$$

Substituting for $V_1$ we have

$$\frac{V_i}{R_1} = V_1 \left[ \frac{1}{R_1} + \frac{1}{R_3} + \frac{1}{R_2} + C_1 s \right] - \frac{V_0}{R_3}$$

$$= -V_0 \, C_2 \, R_2 \, s \left[ \frac{1}{R_1} + \frac{1}{R_3} + \frac{1}{R_2} + C_1 s \right] - \frac{V_0}{R_3}$$

$$V_i = -V_0 \left[ R_2 C_2 s + R_1 C_2 s + \frac{R_1 R_2}{R_3} C_2 s + R_2 R_1 C_1 C_2 s^2 + \frac{R_1}{R_3} \right] \qquad \text{...(13.152)}$$

$$\frac{V_0}{V_i} = \frac{1/R_1 R_2 C_1 C_2}{s^2 + \left( \dfrac{R_1 C_2}{R_1 R_2 C_1 C_2} + \dfrac{R_2 C_2}{R_1 R_2 C_1 C_2} + \dfrac{R_1 R_2 C_2}{R_1 R_2 C_1 C_2} \right) s + \dfrac{1}{R_2 R_3 C_1 C_2}}$$

$$= \frac{1/R_1 R_2 C_1 C_2}{s^2 + \dfrac{1}{C_1} \left( \dfrac{1}{R_2} + \dfrac{1}{R_1} + \dfrac{1}{R_3} \right) s + \dfrac{1}{R_2 R_3 C_1 C_2}} \qquad \text{...(13.153)}$$

If we select $R_1 = R_2 = R_3 = R$ we have

$$\omega_0 = \frac{1}{R \sqrt{C_1 C_2}} \quad \text{and} \quad Q = \frac{1}{3} \sqrt{\frac{C_1}{C_2}}$$

$$S_{R_2, R_3}^{\omega_0} = -\frac{1}{2} = S_{C_1, C_2}^{\omega_0} \qquad \text{...(13.154)}$$

It can be seen that the sensitivities of $\omega_0$ and $Q$ with reference to the circuit elements are less than or equal to 1/2 but the element spread is $9Q^2$ i.e. $9Q^2 = \dfrac{C_1}{C_2}$.

If we consider sensitivity as a basis of comparison between positive, negative and infinite feedback, negative and infinite feedback are superior. However, if product of gain and sensitivity is the consideraion, positive gain low pass filter is superior. Also the element spread is smaller in positive feedback. Hence in practical filters designs, positive gain circuit is preferred to the other two circuits.

**Example 13.12:** Following Fig. E13.12 shows a first order low pass filter. Derive transfer function and obtain the value of parameters if this circuit were to represent Butterworth low pass first order filter.

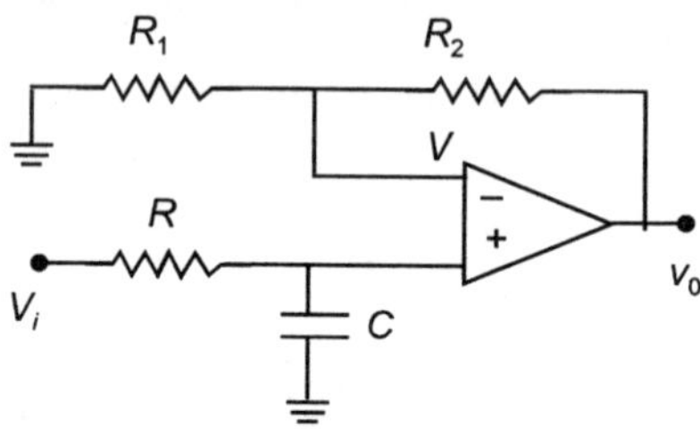

**Fig. E13.12**

**Solution:** By the use of voltage divider rule the voltage.

$$V = \frac{V_0}{R_1 + R_2} R_1$$

or
$$\frac{V_0}{V} = \frac{R_1 + R_2}{R_1} = 1 + \frac{R_2}{R_1} = A \text{ (say)}$$

Now applying KCL at node $V$ we have

$$\frac{V_i - V}{R} = VCs$$

or
$$V_i = V(1 + RCs)$$

$$= \frac{V_0}{A}(1 + RCs)$$

or
$$\frac{V_0}{V_i} = \frac{A}{1 + RCs} = \frac{A/RC}{s + 1/RC}$$

Here
$$\omega_0 = \frac{1}{RC}$$

If we select a suitable value of $C$, $R$ can be calculated. Similarly if gain $A$ is given or, is assumed some arbitrary value and if either $R_1$ or $R_2$ is assumed the value of $R_2$ or $R_1$ can be obtaind.

**Example 13.13:** Suppose $F$ is represented by

$$F = \frac{s + K}{s^2 + Ks + K}$$

Determine $K$ so that $F$ is maximally flat magnitude

$$|F\ (j\omega)|^2 = \frac{K^2 + \omega^2}{(K - \omega^2)^2 + K^2\omega^2}$$

$$= \frac{K^2 + \omega^2}{\omega^4 - 2K\omega^2 + K^2 + K^2\omega^2}$$

$$= \frac{K^2 + \omega^2}{K^2 + (K^2 - 2K)\omega^2 + \omega^4}$$

Equating co-efficients of like power in the numerator and denominator $K^2 = K^2$

$$K^2 - 2K = 1 \quad \text{or} \quad K^2 - 2K - 1 = 0$$

$$K = \frac{2 \pm \sqrt{4 + 4}}{2}$$

$$K = 1 + \sqrt{2}\ = 2.414\ \textbf{Ans.}$$

## PROBLEMS

**13.1.** Show that the transfer function of the second order active low pass filter shown in Fig. P13.1 is given by

$$T(s) = \frac{\omega_0^2}{s^2 + \dfrac{\omega_0}{Q}s + \omega_0^2}$$

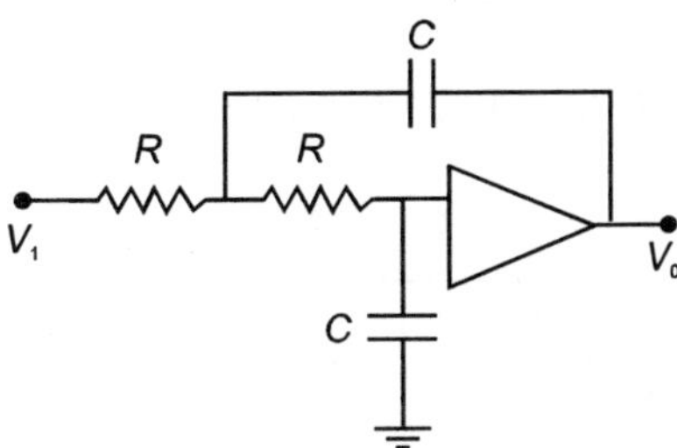

**Fig. P13.1**

Determine the value of $\omega_0$ and $Q$.

**13.2.** Fig. P13.2 shows a 3rd order low pass filter. Show that the transfer function is given by

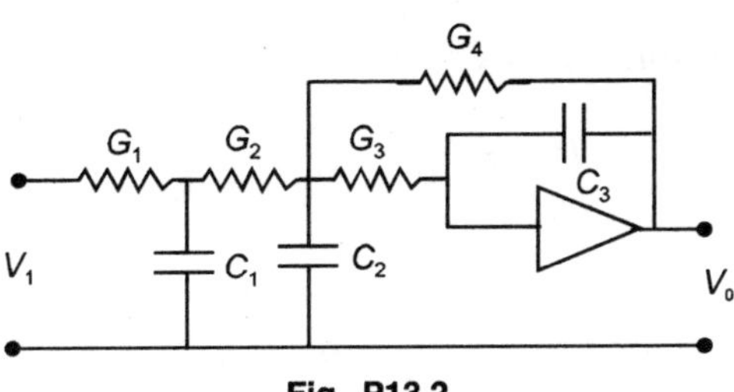

**Fig. P13.2**

$$\frac{V_0}{V_i} = \cfrac{G_1 G_2 G_3 / C_1 C_2 C_3}{s^3 + \left(\dfrac{G_1 + G_2}{C_1} + \dfrac{G_2 + G_3 + G_4}{C_2}\right) s^2}$$

$$+ \left(\frac{G_1 G_2 + G_1 G_3 + G_1 G_4 + G_2 G_3 + G_2 G_4}{C_1 C_2} + \frac{G_3 G_4}{C_2 C_3}\right) s$$

$$+ \frac{G_1 G_3 G_4 + G_2 G_3 G_4}{C_1 C_2 C_3}$$

Let this be represented by $\dfrac{K}{s^3 + 2s^2 + 2s + 1}$ and if $G_1 = G_2 = G_3 = G_4 = 1$.

Determine $K$ and other parameters.

**13.3.** The configuration of the second order band pass Butterworth filter is shown in Fig. P13.3.

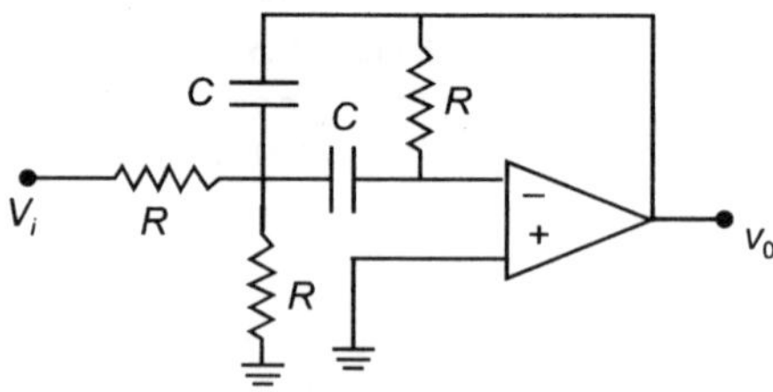


**Fig. P13.3**

Determine $\omega_0$ for the filter.

**13.4.** Fig. P13.4 shows a low pass filter. Using suitable transformation obtain the corresponding high pass filter.

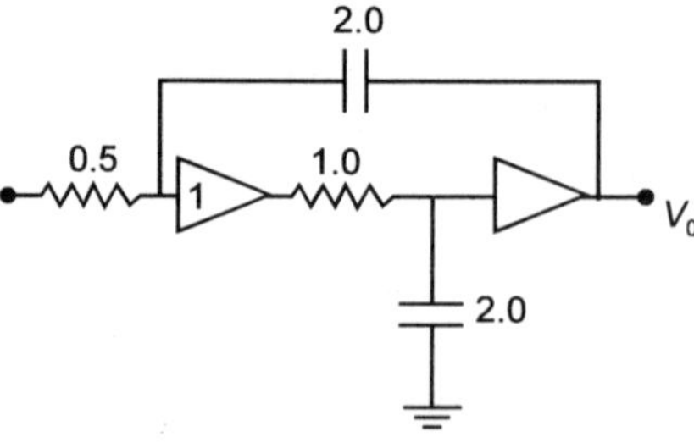


**Fig. P13.4**

**13.5.** A typical inverting amplifier is characterized by the following transfer function.

$$T(s) = \frac{-K R_2}{R_1 + R_2 + K R_1}$$

Show that

$$S_k^T = \frac{R_1 + R_2}{R_1 + R_2 + K R_1}$$

$$S^T_{R_1} = -\frac{(K+1)R_1}{R_1 + R_2 + KR_1}$$

$$S^T_{R_2} = \frac{(K+1)R_1}{R_1 + R_2 + KR_1}$$

**13.6.** Following Fig. P13.6 shows an attenuator. Compute the sensitivities of the transfer function with respect to each resistor.

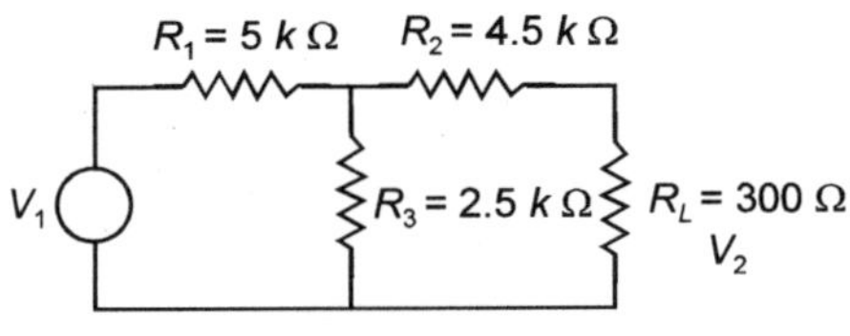


**Fig. P13.6**

Also, determine the tolerance limit of each resistor if the maximum permissible variation in $T(s)$ is 4%.

**13.7.** A Butterworth low pass filter with 3 dB loss at $f = 20$ MHz has 60 dB loss at $f = 80$ MHz. Determine the order of the filter. For the same requirement using Chebychev filter with 1 dB ripple (ripple band upto 20 MHz), determine $n$, the order of the filter.

**13.8.** Fig. P13.8 shows an active RC filter. Determine the transfer function $T(s)$. If the transfer function can be represented by

$$T(s) = \frac{Hs}{s^2 + 0.2s + 3}$$

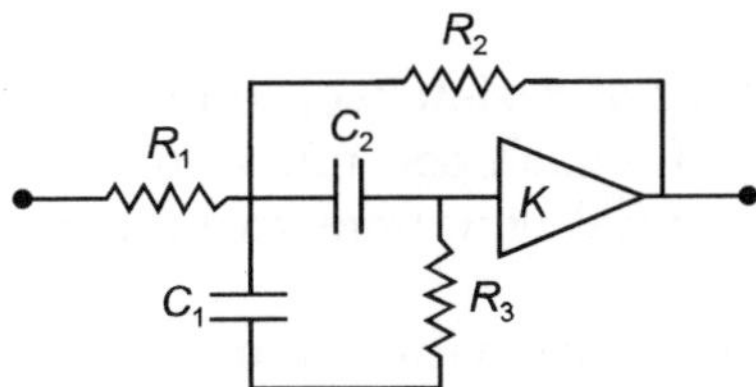


**Fig. P13.8**

Determine $R_3$ and $K$ if $R_1 = R_2 = C_1 = C_2 = 1$

# 14

# Analogous Systems

## 14.1 INTRODUCTION

Interdisciplinary interaction is the need of the hour. Today, an electrical engineer may work in a big organization jointly with a civil engineer on some hydraulic or structural problem or with a mechanical engineer. Therefore, he should be able to draw analogies between electrical system with other physical systems. Analysing and synthesing electrical system is much easier as compared to other physical systems. Hence the concept of analogous system plays an important role in bringing various engineers to work together for a final objective which it may be a combination of various disciplines. The study of water distribution system is identical to load flow solution in an electric network. Even though the variables are different, they will be described by the same set of equations. Therefore, such systems whose differential equations are of identical form (variables could be different) are called analogous systems.

An electrical system consisting of resistance, inductance and capacitance is analogous to a mechanical system consisting of a dash pot, mass and spring respectively. Dual electrical circuits governed by similar differential equations are also a special form of analogous system.

Consider Fig. 14.1 showing dual electrical circuits Fig. 14.1 (a) is dual of Fig. 14.1 (b) and vice versa. The behaviour of circuit 14.1 (a) is given by the differential equation using KVL.

$$Ri + L\frac{di}{dt} + \frac{1}{C}\left[\int i\,dt + q(0)\right] = V(t) \qquad \qquad ...(14.1)$$

For Fig. 14.1 (b) using KCL

$$GV' + C'\frac{dV'}{dt} + \frac{1}{L'}\left[\int V(t)\,dt + \phi(0)\right] = i(t) \qquad \qquad ...(14.2)$$

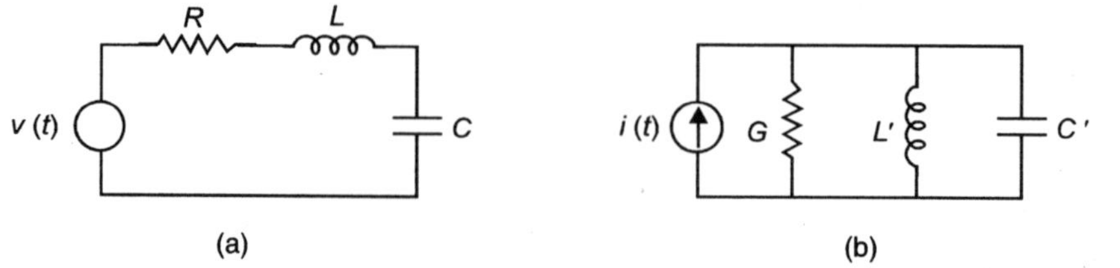

(a)

(b)

**Fig. 14.1** Dual electrical circuits.

Since the two differential equations are identical, the two networks are analogous. The analogy or equivalence between the various elements is

$$R \leftrightarrow G \qquad\qquad L \leftrightarrow C' \qquad\qquad C \leftrightarrow L'$$

$$u(t) \leftrightarrow i'(t) \qquad\qquad i(t) \leftrightarrow V'(t) \qquad\qquad q \leftrightarrow \phi$$

It is much easier to work with electric circuits for the following reasons:

(*i*) Each component of the physical (mechanical, hydraulic etc.) system is replaced by its electrical analog and these electrical analogs are interconnected as per the requirement of the physical system. Since the symbols in electric circuits are simple and the interconnection is also very simple, the whole task is very simple and the physical system is easily replaced by its electrical analog.

(*ii*) With the help of network theorems and other mathematical techniques available, one can study behaviour of the system more comprehensively. It is very simple to study the sensitivity of the system to the changes in circuit parameters like $R$, $L$, $C$ or any other electrical parameter.

(*iii*) The experimentation is very simple. All these features prove quite invaluable in the analysis and modeling of systems.

However, a physical system can be modeled in a number of ways depending upon the specific problem to be dealt with and the desired accuracy. The model would also depend upon the region of study whether small region (small perturbation) or large region (large perturbation). A mathematical model is linear if the differential equation describing it has co-efficients which are constant or are function of independent variables. If the co-efficients are function of time (independent variables), the mathematical model is linear time-varying. However, if it has constant co-efficient, the model is said to be linear time – invariant. We will study here linear systems only.

## 14.2  LINEAR MECHANICAL SYSTEMS

Three basic elements the mass, the spring and the damper form the linear mechancial systems. These appear as co-officients of the differential equation representing the system. These represent three essential forces present in a mechanical system.

Mechanical systems are again classified into two categories

(*i*) Translational system (linear motion)
(*ii*) Rotational system (circular motion)

Three elements for translational and rotational mechanical systems with their relevant properties and conventional symbols are shown in Fig. 14.2.

### 14.2.1  Translational System

The three forces which resist the motion are:

(*i*) *Inertia force*: According to Newton's second law of motion, the rate of change of momentum of a body equals the force acting on the body i.e.

$$F = M\,\frac{dv}{dt} = M\,\frac{d^2x}{dt^2} \qquad\qquad ...(14.3)$$

The Mass Element

(*ii*)  **The spring element**

$$F = \frac{1}{K}(x_1 - x_2) = Kx$$

$$= \frac{1}{K}\int_{-\infty}^{t}(v_1 - v_2)\,dt$$

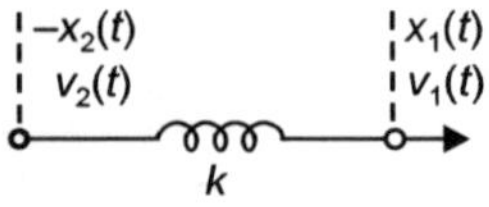

$$= K\int_{-\infty}^{t} v\,dt \qquad \qquad ...(14.4)$$

(*iii*)  **The damper element**

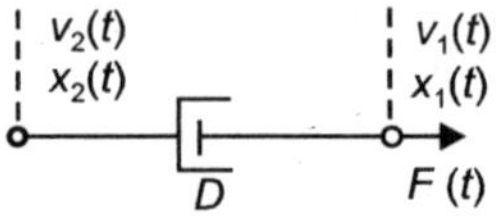

$$F = D\,(v_1 - v_2)$$

$$= D\,(\dot{x}_1 - \dot{x}_2) = f\dot{x} \qquad \qquad ...(14.5)$$

For rotational system the elements are:

(*iv*)  **The inertia element**

$$T = I\frac{d\omega}{dt} = I\frac{d^2\theta}{dt^2} \qquad \qquad ...(14.6)$$

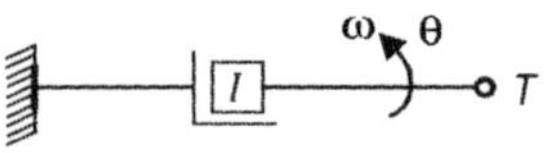

(*v*)  **The torsional spring element.**

$$T = \frac{1}{K}(\theta_1 - \theta_2)$$

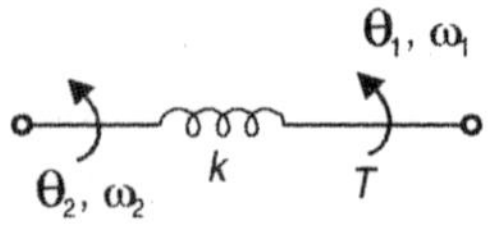

$$= \frac{1}{K}\int_{-\infty}^{t}(\omega_1 - \omega_2)\,dt$$

$$= \frac{1}{K}\,\theta \qquad \qquad ...(14.7)$$

(*vi*)  **The damper element**

$$T = D\,(\omega_1 - \omega_2)$$

$$= D\left(\dot{\theta}_1 - \dot{\theta}_2\right)$$

$$= D\,\dot{\theta} \qquad \qquad ...(14.8)$$

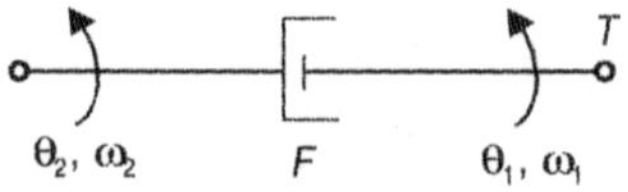

**Fig. 14.2**  The three basic elements of mechanical system.

where $v$ and $x$ are the instantaneous velocities and displacement at any time $t$. The mass $M$ of the body is the co-efficient of the force equation.

The mass element $m$ represents a particle of mass concentrated at the centre of the physical mass. It has two points or terminals, one is free which is applied force $F$ and the other terminal represents a reference with respect to which the motional variable of the free terminal ($x$) is measured.

**Damping force:** Whenever, a body comes into motion, a force called damping force comes into action and is directly proportional to velocity of the body. The damping could be due to fluid friction or due to electromagnetic forces depending upon the medium in which the moving body is in motion. The damping force

$$F_D = D\,(v_1 - v_2) = D\,(\dot{x}_1 - \dot{x}_2) = f\ddot{x} \qquad \ldots(14.9)$$

In mechanical system it signifies viscous force between piston and cylinder. The damping force always acts in opposition to the direction of motion of the body.

**Spring force:** Whenever a spring is stretched or compressed restoring forces are developed within the spring and the spring comes back to the normal state, when the external force is removed. The restoring force of a spring is proportional to its displacement with respect to the reference point.

$$F = \frac{1}{K}\,(x_1 - x_2) = \frac{1}{K}x = \frac{1}{K}\left[\int_0^t v\,di + x(0)\right] \qquad \ldots(14.10)$$

where $K$ is the compliance of the spring and $\dfrac{1}{K}$ the stiffness.

### 14.2.2 Rotational System

As in case of translational motion there are three forces to oppose the applied force, there are three torques in rotational motion which oppose the applied torque.

*Inertia torque*: Here again as per Newton's second law of motion the rate of change of angular momentum equals the applied torque.

$$T_I = I_\theta\,\frac{d\omega}{dt} = I_\theta\,\frac{d^2\theta}{dt^2} \qquad \ldots(14.11)$$

Here $I$ is moment of insertia in kg m$^2$ and $\theta$ is in radians.

*Damping force*: As the rotating mass begins to rotate, damping comes into action and is proportional to the angular velocity

$$T_D = D_\theta\,\omega = D_\theta\,\frac{d\theta}{dt} \qquad \ldots(14.12)$$

when $D_\theta$ is the rotational damping co-efficient.

*Spring torque or Control torque*: When a spring is tightened or loosened about an axis, restoring torque is developed in the spring and this torque is proportional to the angular displacement $\theta$ i.e.

$$T_C = \frac{1}{K}\theta = \frac{1}{K}\left[\int_0^t \omega\,dt + \theta(0)\right] \qquad \ldots(14.13)$$

where $K$ is rotational compliance of the spring.

If we compare the set of equations corresponding to translational system and rotational system, we find that the two systems are identical mathematically. The equivalence is

$$F \leftrightarrow T \qquad\qquad x \leftrightarrow \theta \qquad\qquad v \leftrightarrow \omega$$

$$M \leftrightarrow I \qquad\qquad D \leftrightarrow D_\theta \qquad\qquad K \leftrightarrow K_\theta$$

Therefore, within mechanical system we have analogous system just as we have one within electrical system.

## 14.3  D'ALEMBERT'S PRINCIPLE

When a body is subjected to external forces, its static equilibrium i.e. the equilibrium of various forces acting on the body at any instant of time is given by D'Alembert's principle which states that:

For anybody, the algebraic sum of externally applied forces and the forces resisting the motion in any given direction is zero.

To apply D'Alembert's principle consider Fig. 14.3 (a) for translational system and Fig. 14.3 (b) for rotational system. For 14.3 (a) the external force is $f$ and the opposing forces are:

$$M\,\frac{d^2x}{dt^2} \qquad\qquad D\,\frac{dx}{dt} \qquad\qquad \frac{1}{K}\left[\int_0^t v\,dt + x(0)\right]$$

Inertia          Damping          Spring force

By D'Alembert's principle

$$M\,\frac{d^2x}{dt^2} + D\,\frac{dx}{dt} + \frac{1}{K}\left[\int_0^t \frac{dx}{dt}\,dt + x(0)\right] = f \qquad\qquad \ldots(14.14)$$

or in terms of velocity as variable

$$M\,\frac{dv}{dt} + Dv + \frac{1}{K}\left[\int_0^t v\,dt + x(0)\right] = f \qquad\qquad \ldots(14.15)$$

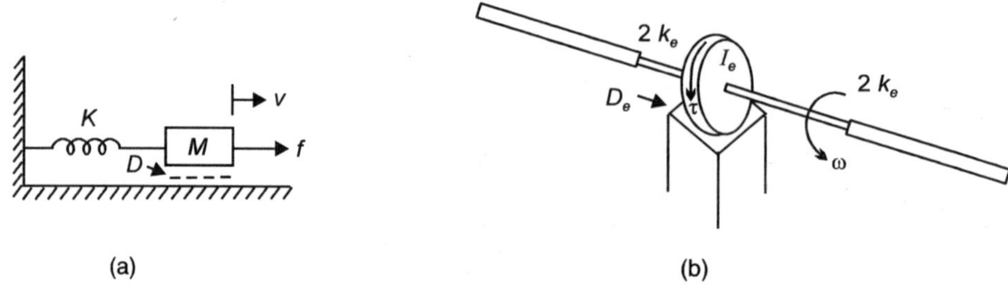


**Fig. 14.3**  (a) Translational mechanical system
(b) A rotational mechanical system.

D'Alembert's principle when applied to rotating bodies is slightly modified and is given as

For any body, the algebraic sum of externally applied torques and the torques opposing the rotation about an axis is zero.

If $T$ is the external torque applied, applying D'Alembert's principle to the system shown in Fig. 14.3 (b) gives

$$I \frac{d\omega}{dt} + D_\theta\, \omega + \frac{1}{K_\theta}\left[\int_0^t \omega\, dt + \theta(0)\right] = T \qquad\qquad ...(14.16)$$

Equation (14.15) and (14.16) are mathematically identical and hence the two systems shown in Fig. 14.3 (a) and (b) are analogous to each other.

## 14.4 FORCE-VOLTAGE ANALOGY

In order to obtain Force voltage analogy we compare the co-efficient of the differential equations describing the mechanical and electrical system.

Thus comparing equation (14.1) with equation (14.15), we have two mathematically identical systems where force $f$ in mechanical system is analogous to voltage in the RLC series circuit and hence, this is known as Force-voltage analogy. Table 14.1 shows analogous quantities in the two systems.

**Table 14.1  Analogies in Force-voltage analogous system**

| Mechanical System | Electrical System |
| --- | --- |
| Force $f$ | Voltage $V$ |
| Velocity $v$ | Current $i$ |
| Displacement $x$ | Charge $q$ |
| Mass $M$ | Inductance $L$ |
| Damping co-efficient $D$ | Resistance $R$ |
| Compliance $K$ | Capacitance $C$ |

Given a mechanical system, its' $f$-$v$ electrical analog can be drawn using the rule given hereunder.

Each junction in the mechanical system corresponds to a closed loop which consists of electrical excitation sources and passive elements analogous to the mechanical driving sources and passive elements connected to the junction. All points on a rigid mass are considered the same junction.

**Example 14.1:** Determine the electrical analog of the mechanical system shown in Fig. E14.1

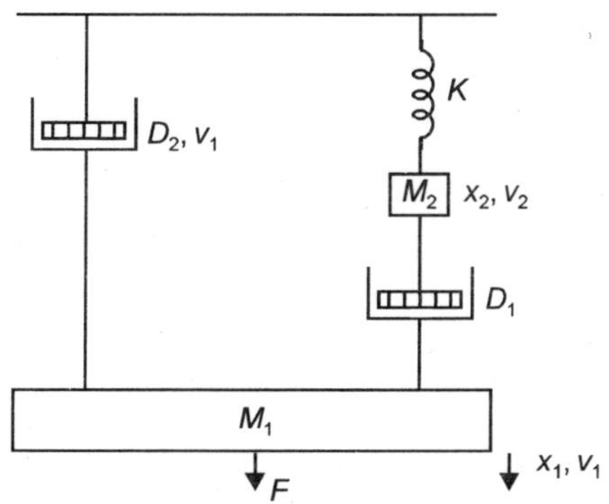


**Fig. E14.1.1**

The force acting on mass $M_1$ is $F$. The forces opposing the external force $F$ are:

1. Inertial force $M_1 \dfrac{dv_1}{dt}$
2. Damping force $D_1(v_1 - v_2)$ and
3. $D_2\, v_1$

Applying D'Alembert's principle, we have

$$M_1 \frac{dv_1}{dt} + D_1\,(v_1 - v_2) + D_2\,v_1 = F$$

Considering equilibrium of mass $M_2$ we have
The external force $F = 0$
The opposing forces are:

$$M_2 \frac{dv_2}{dt}\,,\ D_1\,(v_2 - v_1)$$

and spring force
$$\frac{1}{K}\left[\int_0^t v_2\ dt + x_2(0)\right]$$

Hence, applying D'Alembert's principle

$$M_2 \frac{dv_2}{dt} + D_1\,(v_2 - v_1) + \frac{1}{K}\left[\int_0^t v_2\ dt + x_2(0)\right] = 0$$

The electrically analogous equations are:

$$L_1 \frac{di_1}{dt} + R_1\,(i_1 - i_2) + R_2\,i_1 = V$$

and
$$L_2 \frac{di_2}{dt} + R_1\,(i_2 - i_1) + \frac{1}{k}\left[\int i_2\ dt + q_2(0)\right] = 0$$

The network representing these two equations simultaneously is drawn hereunder.

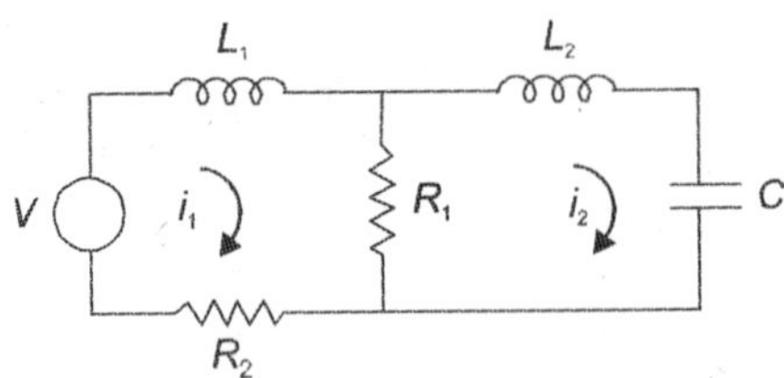


**Fig. E14.1.2**

Since there are two distances $x_1$ and $x_2$ hence two junctions and two loops in the electrical network.

## 14.5  FORCE-CURRENT ANALOGY

Comparing equation (14.15) this time with equation (14.2) of the electrical circuit, we find as far as the external force is concerned, these are $f$ and $I$ hence we call this as the force-current analogy and hence the other analogies in these two analogous system are given in Table 14.2.

**Table 14.2  Analogies Between**

| *Mechanical System* | *Electrical System* |
| --- | --- |
| $f$ | $i$ |
| velocity  $v$ | voltage  $e$ |
| Displacement  $x$ | Flux linkage  $\phi$ |
| Mass  $M$ | Capacitance  $C$ |
| Damping Co-efficient  $D$ | Conductance  $G$ |
| Compliance  $k$ | Inductance  $L$ |

**Example 14.2:** Determine the electrical analog of the system shown in the previous example.

**Solution:** The equations for the mechanical systems have already been obtained, these are reproduced for convenience

$$M_1 \frac{dv_1}{dt} + D_1(v_1 - v_2) + D_2 v_1 = F$$

and

$$M_2 \frac{dv_2}{dt} + D_1(v_2 - v_1) + \frac{1}{K}\left[\int v_2\, dt + x_2(0)\right] = 0$$

The analogous equations in electrical systems are:

$$C_1 \frac{de_1}{dt} + G_1(e_1 - e_2) + G_2 e_1 = i$$

$$C_2 \frac{de_2}{dt} + G_1(e_2 - e_1) + \frac{1}{L}\left[\int_0^t e_2\, dt + \phi(0)\right] = 0$$

Since there are two distances $x_1$ and $x_2$, there should be two nodes with voltage $e_1$ and $e_2$.

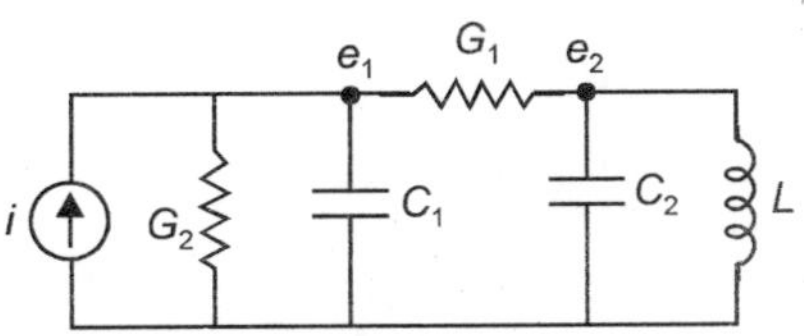


**Fig. E14.2**

However, if a mechanical system is given its' electrical analog using force-current analogy can be drawn as explained under.

Each junction in the mechanical system corresponds to a node which joins electrical source (current source here) to passive elements analogous to the mechanical driving source and to passive elements connected to the junctions. All points on the rigid mass are considered the same junction. One terminal of the capacitance analogous to a mass is always connected to the ground. This is because the velocity or displacement of a mass is referred to the stationary ground. Two masses rigidly connected are analogous to an electrical circuit having two capacitances connected across the same node and the ground. The electrical circuits for mechanical systems based on force-voltage and force-current are dual of each other.

## 14.6  MECHANICAL COUPLING DEVICES

Various mechanical coupling devices such as friction wheels (gears) and levers also have their electrical analogs.

### 14.6.1  Friction Wheel

Fig. 14.4 shows a pair of non-slipping friction wheels. The point of contact on the driving wheel is $c_1$ and $c_2$ on the driven wheel. The linear velocity of the two wheels must be same as they do not slip with respect to each other and both experience the same force magnitude though opposite in direction (action and reaction forces).

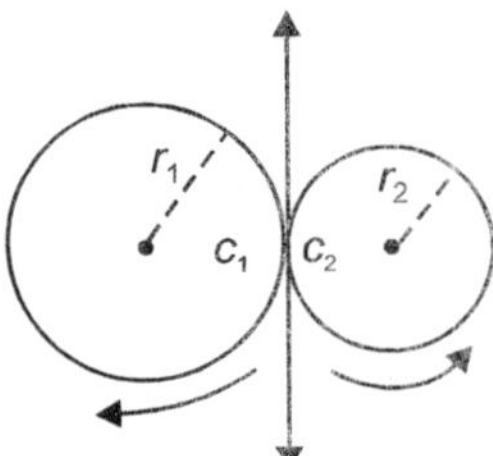

**Fig. 14.4**  Friction Wheel

If $F$ is the force, the torques

$$T_1 = F r_1 \qquad \qquad ...(14.17)$$

and

$$T_2 = F r_2 \qquad \qquad ...(14.18)$$

Therefore,

$$\frac{T_1}{T_2} = \frac{r_1}{r_2} \qquad \qquad ...(14.19)$$

Since, the radii of both the wheels are different, they will have different-angular velocities as the linear velocity is same.

$$\omega_1 r_1 = \omega_2 r_2 \qquad \qquad ...(14.20)$$

Hence

$$\frac{\omega_1}{\omega_2} = \frac{r_2}{r_1} \qquad \qquad ...(14.21)$$

These two equations representing the operations of the friction wheel or gear are identical to the voltage and current relation related to the turn's ratio of ideal transformers. In an ideal transformer the voltage drops and losses are assumed to be absent and the voltage and current relation are given as

$$\frac{V_1}{V_2} = \frac{N_1}{N_2} \qquad \qquad ...(14.22)$$

$$\frac{I_1}{I_2} = \frac{N_2}{N_1} \qquad \qquad ...(14.23)$$

Therefore, there is equivalence between torques and voltages and angular speed and currents and the radii are analogous to number of turns in the two windings of the transformer. Therefore, electrical analog is shown in Fig. 14.5.

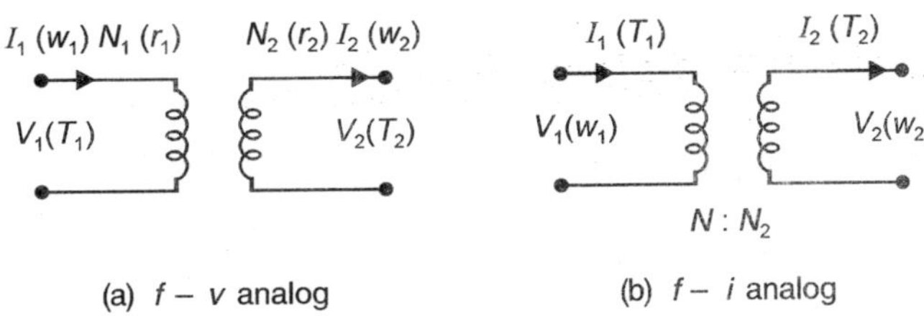

(a) $f - v$ analog  (b) $f - i$ analog

**Fig. 14.5**  Electric analog ($r_2 : r_1$) ideal transformer.

However, if we choose Torque-current analogy
Then  $T_1 \leftrightarrow I_1$  and  $T_2 \leftrightarrow I_2$
In that case, the turns ratio would be

$$\frac{N_1}{N_2} = \frac{r_2}{r_1} \qquad\qquad ...(14.24)$$

and the electrical analog would be as shown in Fig. 14.5 (b).

The only difference between the actual transformer and analog transformer is that whereas the latter can be operated on constant voltage (d.c. voltage or constant force or torque), the former can't be used on d.c. voltages. If in a mechanical system constant torque is applied on one of the wheels, then electrical analog is obtained and the solution is sought either using differential equations method or Laplace transform method. However, if a varying torque especially sinusoidally varying torque is applied to one of the wheels then electrical analog is obtained and solution is sought using impedance concept as illustrated in the following example.

**Example 14.3:** Two friction wheels with radii $r_1$ and $r_2$ are mounted on shafts with moment of inertia $I_{\theta 1}$ and $I_{\theta 2}$ respectively. The friction wheels are coupled and the assembly is supported on bearings with damping co-efficients $D_{\theta 1}$ and $D_{\theta 2}$. If a sinusoidally varying ($\omega_0$) torque is applied to the shaft of wheel 1, determine the steady state angular velocity of wheel 1. The shaft of wheel 2 is connected to ground through a spring with compliance $K$.

**Solution:** Using the torque-voltage analogy the electrical analog of the mechanical system is shown in Fig. E14.3.

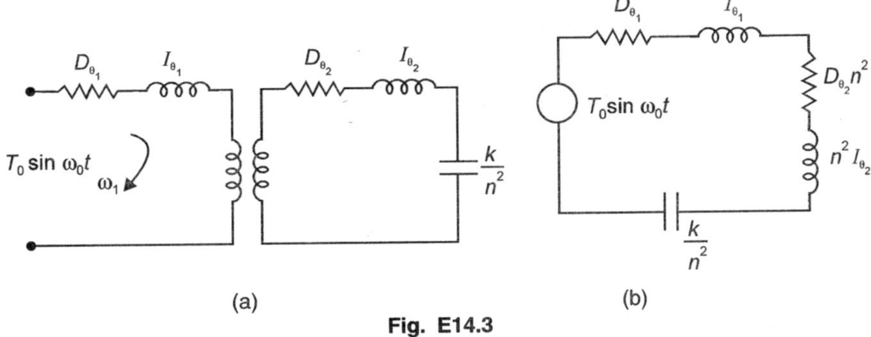

(a)  (b)

**Fig. E14.3**

If the transformation ratio $n = \dfrac{r_1}{r_2}$ the equivalent circuit referred to primary of the analog transformer (wheel 1) is as shown in Fig. E14.3 (b) and hence the steady state angular speed (current in analog transformer) is given by

$$\omega_1 = \frac{T_0 \, \sin\,(\omega_0 t - \phi)}{\left(D_{\theta_1} + n^2 \, D_{\theta_2}\right)^2 + \left\{\omega_0 \left(I_{\theta_1} + n^2 \, I_{\theta_2}\right) - \dfrac{n^2}{K\,\omega_0}\right\}^2}$$

where

$$\phi = \tan^{-1} \frac{\omega_0 \left(I_{\theta_1} + n^2 \, I_{\theta_2}\right) - \dfrac{n^2}{K\,\omega_0}}{D_{\theta_1} + n^2 \, D_{\theta_2}}$$

However, if a constant torque $T_0$ is applied then the angular speed of wheel 1 will be

$$\omega(t) = \mathcal{L}^{-1} \left[ \frac{T_0(s)}{\left(D_{\theta_1} + n^2 \, D_{\theta_2}\right) + s\left(I_{\theta_1} + n^2 \, I_{\theta_2}\right) + \dfrac{n^2}{Ks}} \right]$$

Gear trains have the same analog as the friction wheels. The only difference is that whereas in friction wheels we have the ratio $r_1 : r_2$ in gears we have $T_1 : T_2$ where $T_1$, $T_2$ are the number of teeth on the two gears.

Whenever mechanical matching between the motor and load is required, gear system is used. Sometimes a load is required to run at a lower speed as compared to the motor, a gear system with teeth ratio $T_1 : T_2$ would be required where $T_1 > T_2$. In gas turbine power plant the synchronous generator is required to operate at 3000 rpm whereas gas turbine is more efficient at speed more than 5000 rpm. Hence a gear with suitable ratio would be required to run the synchronous generator at its required speed. Thus, gear trains act as matching devices in mechanical system as transformer do in electrical system (voltage transformation or current transformation).

### 14.6.2  Lever

A lever is a mechanical device used to lift heavy weights by applying a relatively smaller effort. Fig. 14.6 shows such a lever which rests on a rigid fulcrum $p$. The weight is attached at one end of the lever and the effort is applied at the other end. The lever is assumed to be massless but rigid. The length between the fulcrum and the effort end is greater than that between the fulcrum and the weight end.

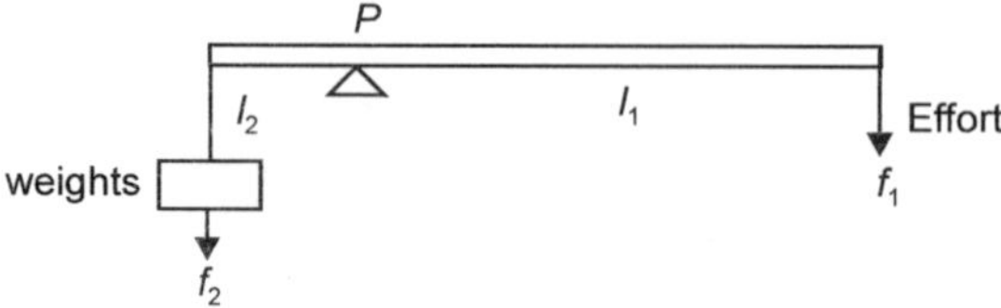


**Fig. 14.6**  A lever system.

Suppose force $f_1$ is applied to lift weight of equivalent force $f_2$ at a velocity $u_2$. At equilibrium

$$f_1 l_1 = f_2 l_2 \qquad\qquad ...(14.25)$$

$$\frac{f_1}{f_2} = \frac{l_2}{l_1} \qquad \qquad ...(14.26)$$

If $\omega$ is the angular velocity of the lever about the fulcrum, the linear velocities of the two ends of the lever are:

$$u_1 = l_1\omega \quad \text{and} \quad u_2 = l_2\omega \qquad \qquad ...(14.28)$$

or
$$\frac{u_1}{u_2} = \frac{l_1}{l_2} \qquad \qquad ...(14.29)$$

If $f_1$ and $f_2$ correspond to voltages $V_1$ and $V_2$ of the transformer and $l_2 \leftrightarrow N_1$ and $l_1 \leftrightarrow N_2$, the analogous ideal transformer is shown in Fig. 14.7 (a).

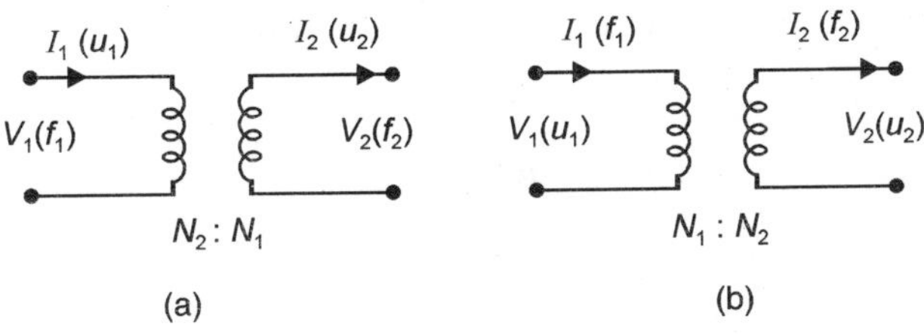


**Fig. 14.7**   Electrical analog of a lever.

The current
$$\frac{I_1}{I_2} = \frac{u_1}{u_2} = \frac{N_2}{N_1} \qquad \qquad ...(14.30)$$

However, if velocities are taken analogous to the voltages then

$$\frac{V_1}{V_2} = \frac{u_1}{u_2} = \frac{l_1}{l_2} = \frac{N_1}{N_2} \qquad \qquad ...(14.31)$$

and
$$\frac{I_1}{I_2} = \frac{f_1}{f_2} = \frac{N_2}{N_1} \qquad \qquad ...(14.32)$$

and the analogous ideal transformer is shown as in Fig. 14.7 (b).

It is to be noted that in all these analog transformers, the input and output quantities are powers as in the case of an electric transformer e.g. in case of friction wheel, the input mechanical quantities are $T_0$ and $\omega$ the product of which is power, in case of lever, the input quantities are $f_1$ and $u$ the product again is power.

**Example 14.4:** Determine the electrical analog of the given mechanical system using force-current analogy. The lever is assumed to be mass-less and rigid and the junctions can move vertically only.

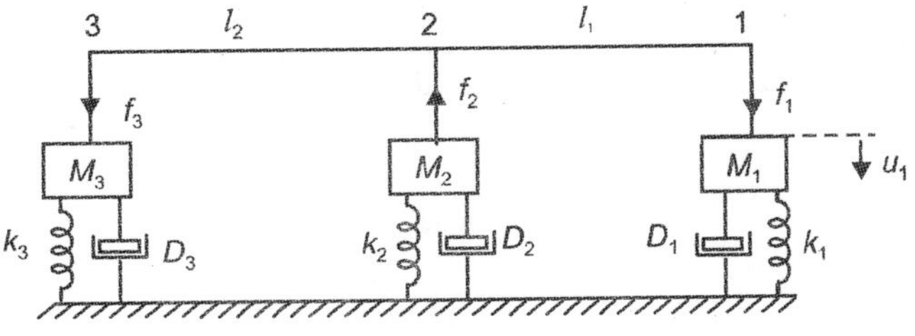


**Fig. E14.4.1**

**Solution:** Since the bar is neither simply supported nor pivoted on a fixed fulcrum, the system is not a simple lever.

In order to study the dynamics of the system when force $f_1$ is applied at junction 1, we consider junction 2 and 3 to be fixed one at a time and we apply superposition principle taking one junction fixed at one time.

Assume junction 3 to be fixed. Taking moments at equilibrium

$$\frac{f_1}{f_2} = \frac{l_2}{l_1 + l_2}$$

and

$$\frac{u_1}{u_2} = \frac{l_1 + l_2}{l_2}$$

which shows that the turns ratio is $N_1 : N_2 :: (l_1 + l_2) : l_2$ in force-current analogy.

Now assume junction 2 to be fixed. At equilibrium

$$f_1 l_1 = f_3 l_2$$

and

$$\frac{u_1}{l_1} = \frac{u_3}{l_2} \qquad\qquad \text{(same angular velocity)}$$

or

$$\frac{f_1}{f_3} = \frac{l_2}{l_1}$$

and

$$\frac{u_1}{u_3} = \frac{l_1}{l_2}$$

Again using $f - i$ analogy

$$\frac{f_1}{f_3} = \frac{I_1}{I_3} = \frac{l_2}{l_1} = \frac{N_3}{N_1}$$

and

$$\frac{u_1}{u_3} = \frac{V_1}{V_3} = \frac{N_1}{N_3} = \frac{l_1}{l_2}$$

Therefore, the turns ratio $N_1 : N_3 :: l_1 : l_2$

In each of these two secondaries (one due to junction 2 and other due to junction 3), three passive elements (a mass, a dashpot and a spring) are connected to a common node. Fig. E14.4.2 shows the $f - i$ analogous of the mechanical system

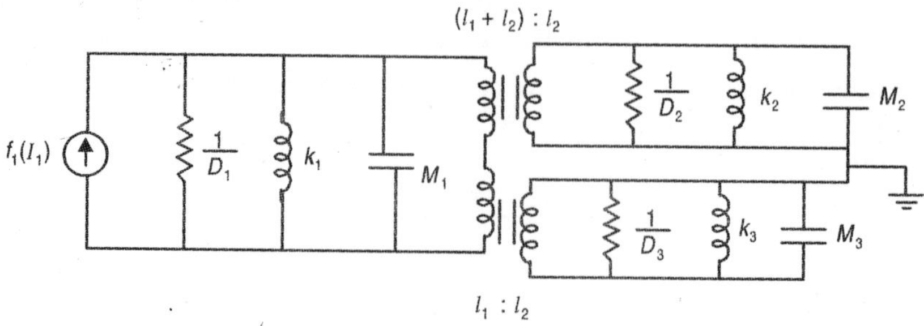


**Fig. E14.4.2**

## 14.7  ELECTROMECHANICAL SYSTEM

Here both electrical and mechanical systems are an integral part of the system. These are often known as electromechanical transducers as they convert electric energy into mechanical energy and vice versa such as an electric motor, microphone, loud speakers etc.

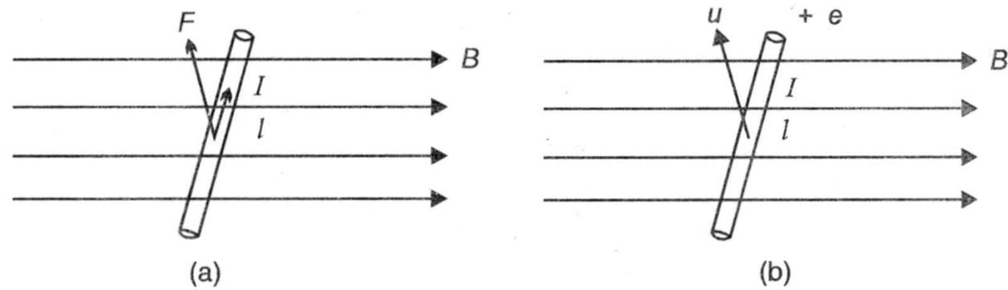

**Fig. 14.8**  (a) Motoring action, (b) Generating action.

Consider a conductor of effective length $l$ carrying a current I lying perpendicular in a uniform magnetic field of density-B as shown in Fig. 14.8 (a).

The conductor experiences a force which is given by Flemings left hand rule and in this case, the direction of force is upward as shown in Fig. 14.8 (a) and the magnitude of force $f$ is given as

$$f = B\ Il \qquad \qquad ...(14.33)$$

or

$$\frac{I}{f} = \frac{1}{Bl} \qquad \qquad ...(14.34)$$

However, if conductor of length $l$ is moved perpendicular to the field $B$ with a velocity $u$, the voltage is induced (current induced) in the conductor with polarity given by Flemings right hand rule and is as shown in Fig. 14.8 (b). The magnitude of the voltage induced is given by

$$e = Blu \qquad \qquad ...(14.35)$$

or

$$\frac{e}{u} = Bl \qquad \qquad ...(14.36)$$

The equations are analogous to the ones for an ideal transformer with turns ratio $N_1 : N_2 :: Bl : 1$. One side of the transformer has electrical quantities $e$ and $i$ and the secondary side has mechanical quantities $u$ and $f$.

$$\frac{I_1}{I_2} = \frac{I}{f} = \frac{N_2}{N_1} = \frac{1}{Bl} \qquad \qquad ...(14.37)$$

Similarly

$$\frac{V_1}{V_2} = \frac{e}{u} = \frac{N_1}{N_2} = \frac{Bl}{1} \qquad \qquad ...(14.38)$$

and the analogous circuit is shown in Fig. 14.9.

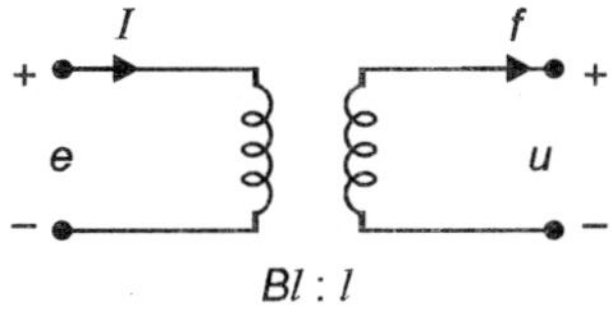

**Fig. 14.9**  Ideal transformer analogous to electro mechanical system.

Since here force and current and voltage and velocity are analogous to each other it is an $f - i$ analogous representation. In an electromechancal system, if we want to solve the system for an electrical quantity e.g. $e$ or $I$ we refer the transformer to primary side i.e. all the mechanical quantities on the secondary side are referred to the primary side and solve the system for the requisite electrical quantity. However, if some mechancial quantity is required, then refer the primary quantities on to the secondary side and solve the system for the requisite quantity. This is illustrated with the following example:

**Example 14.5:** Fig. E14.5 shows an electromechanical relay. It has an electromagnet with a coil of $n$ number of turns, the length of each turn is say $l$ and a current $I$ is passed through the coil. As a result the magnet gets magnetized and it attracts an armature of mass $m$ to which the moving contact of the relay is attached. The armature has spring with compliance $K$ and friction $D$ as shown in Fig. E14.5. The coil has $R$ and $L$ as the resistance and inductance respectively.

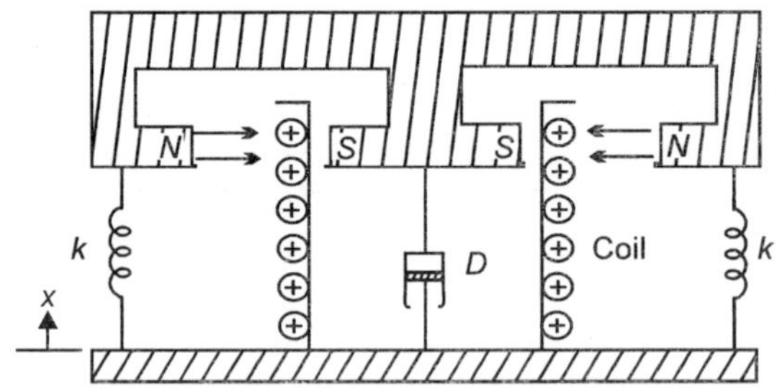

**Fig. E14.5**   An electromechanical relay

Fig. E14.5.1 shows the electrical analog of this system (the relay)

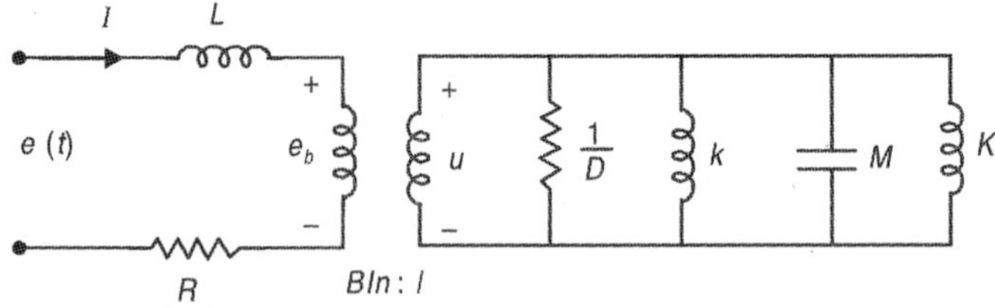

**Fig. E14.5.1**   Electrical analog of electromechanical relay.

For the primary side the KVL equation gives

$$RI + L\,\frac{dI}{dt} + e_b(t) = e(t)$$

For the secondary side the KCL equation gives

$$M\,\frac{du}{dt} + Du + \frac{2}{K}\left[\int_0^t u\,dt + x(0)\right] = f$$

Here $e$ is the voltage applied to the coil and $e_b = lBnu$ and $f = BlnI$
Substituting these values for $e_b$ and $f$ in the above equation we have

$$RI + L\,\frac{dI}{dt} + Blnu = e(t)$$

$$M \frac{du}{dt} + Du + \frac{2}{K}\left[\int_0^t u\,dt + x(0)\right] = BlnI$$

Here $x(0)$ is the initial displacement of the armature from the reference point before the supply $e$ is given to the coil. There are two unknowns $i$ and $u$ and two equations in terms of these unknowns and other known parameters of the relay and hence can be solved uniquely.

The equivalent circuits referred to primary side and referred to secondary side are shown in Fig E14.5.2 (a) and (b) respectively.

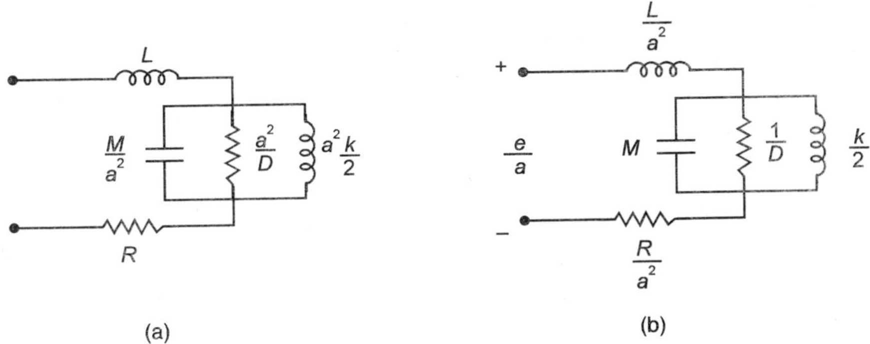

(a)  (b)

**Fig. E14.5.2**

Here 'a' is the ratio of transformation i.e.

$$a = \frac{Bln}{1} = Bln$$

Next, let us consider a few simple mechanical systems and draw their electrical analog just by inspection using force-voltage analogy. Consider the system in Fig. 14.10 (a).

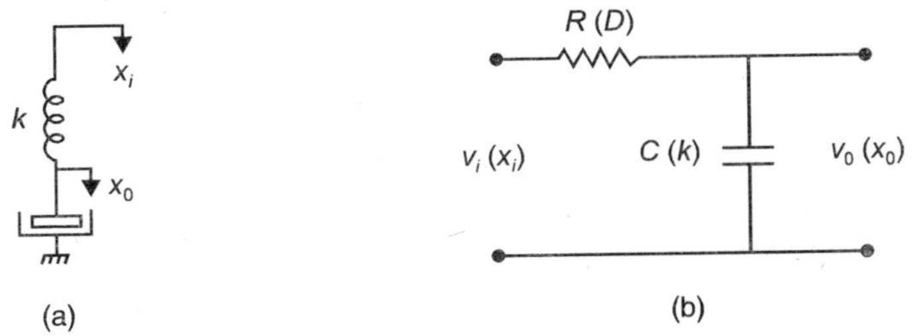

(a)  (b)

**Fig. 14.10**  (a) Mechanical system,  (b) *f–v* electrical analog.

In Fig. 14.10 (a) a step displacement $x_i$ is given to the spring, in series with which is connected a dash pot with resistivity $D$. The displacement of the other end of the spring is not $x_i$ but $x_0$. In $f$–$v$ analogy the spring has equivalence with capacitance and dash pot with resistance and displacement is equivalent to voltage. The input $x_i$ $(v_i)$ is given to series combination of $K(C)$ and $D(R)$ and the output $x_0$ $(v_0)$ is taken between the spring dash-pot common point and the overall reference. So we get the analog as shown in Fig. 14.10 (b).

It is to be noted that the transfer functions for both the mechanical and its electrical analog are similar. Since for electrical analog the transfer function is

$$\frac{V_0(s)}{V_i(s)} = \frac{1}{RCs+1}$$

For the mechanical system this would be

$$\frac{x_0(s)}{x_1(s)} = \frac{1}{DKs+1}$$

Consider systems shown in Fig. 14.11 (a) and Fig. 14.12 (a) and using the same reasoning as in the previous examples, their electrical analogs can be drawn as shown in Fig. 14.11 (b) and Fig. 14.12 (b) respectively.

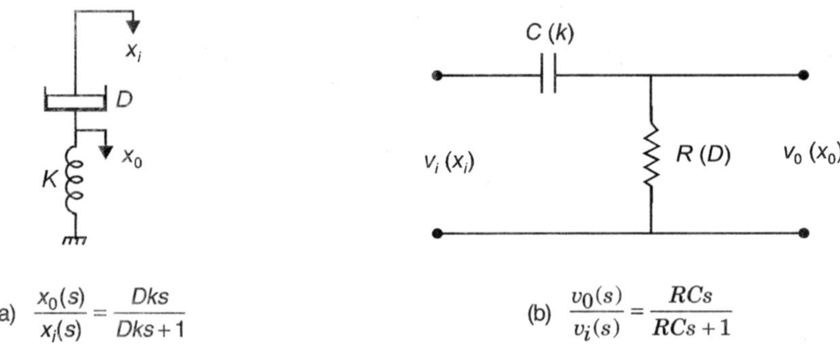

(a) $\dfrac{x_0(s)}{x_i(s)} = \dfrac{Dks}{Dks+1}$

(b) $\dfrac{v_0(s)}{v_i(s)} = \dfrac{RCs}{RCs+1}$

**Fig. 14.11**  (a) Mechanical System, (b) $f - v$ electrical analog.

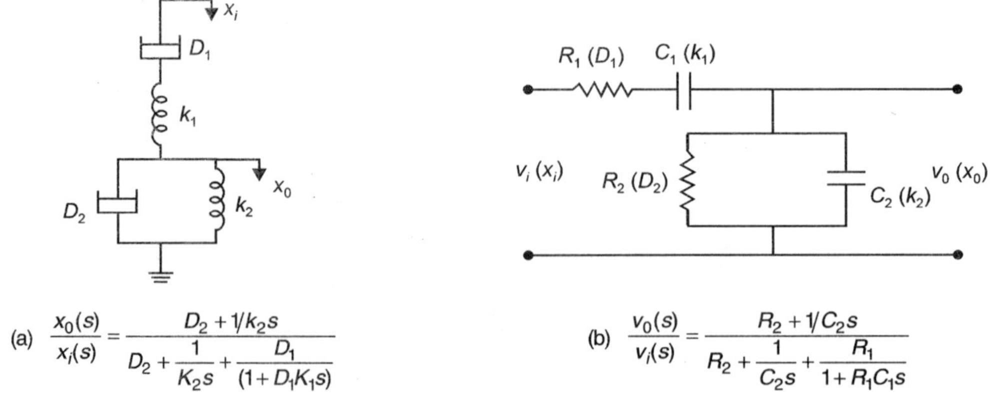

(a) $\dfrac{x_0(s)}{x_i(s)} = \dfrac{D_2 + 1/k_2 s}{D_2 + \dfrac{1}{K_2 s} + \dfrac{D_1}{(1+D_1 K_1 s)}}$

(b) $\dfrac{v_0(s)}{v_i(s)} = \dfrac{R_2 + 1/C_2 s}{R_2 + \dfrac{1}{C_2 s} + \dfrac{R_1}{1+R_1 C_1 s}}$

**Fig. 14.12**  (a) Mechanical Systems,  (b) $f - v$ electrical analog.

The transfer function for both the systems are indicated alongwith the respective diagrams

## PROBLEMS

**14.1.** Draw the electric analogous circuit of the mechanical system shown in figure.

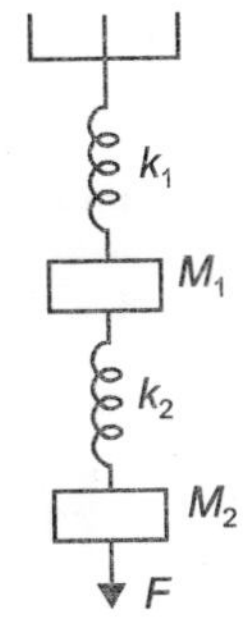

**Fig. P14.1**

**14.2.** Draw elecrical analog for the mechanical system shown using force-voltage and force current analogy.

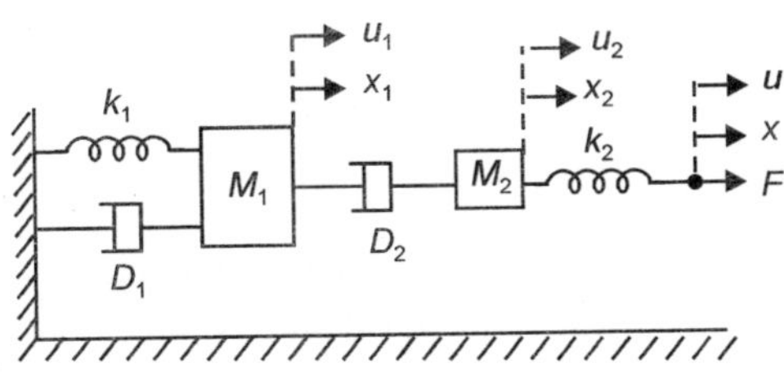

**Fig. P14.2**

**14.3.** For the mechanical system shown draw the electrical analog using $f - v$ and $f - i$ analysis.

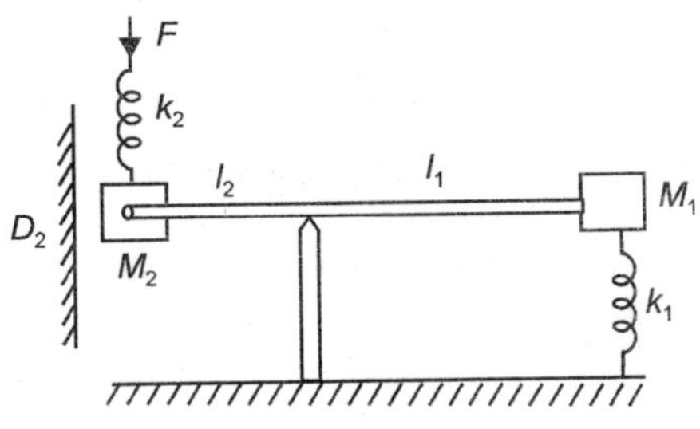

**Fig. P14.3**

**14.4.** Write the governing equations for the mechanical systems, the analogous equations for the electrical systems for the following.

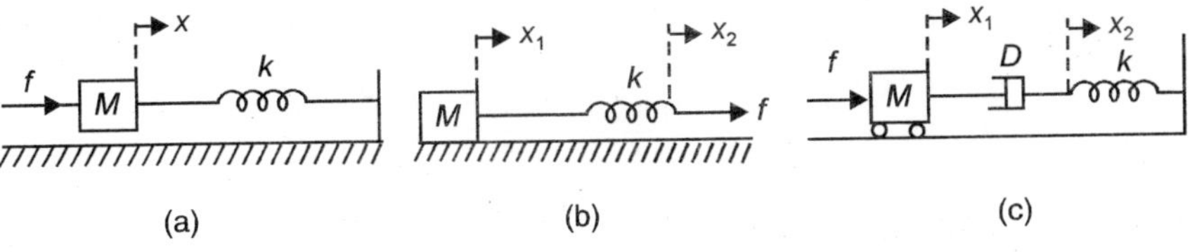

(a)        (b)        (c)

**Fig. P14.4**

# 15

# Frequency Response and Bode Plot

## 15.1 INTRODUCTION

We study here the network functions which describe networks in the sinusoidal steady state. Let $H(s)$ be any network function, a driving point or transfer function. The general network function is given as

$$H(s) = \frac{a_0 s^n + a_1 s^{n-1} + \ldots\ldots + a_n}{b_0 s^m + b_1 s^{m-1} + \ldots\ldots + b_m} \qquad \ldots(15.1)$$

Replacing $s = j\omega$ we have

$$H(j\omega) = \frac{a_0 (j\omega)^n + a_1 (j\omega)^{n-1} + \ldots\ldots + a_n}{b_0 (j\omega)^m + b_1 (j\omega)^{m-1} + \ldots\ldots + b_m} \qquad \ldots(15.2)$$

Now $H(j\omega)$ can be written in rectangular co-ordinates as

$$H(j\omega) = R(\omega) + j\, X(\omega) \qquad \ldots(15.3)$$

or in polar co-ordinates

$$H(j\omega) = M(\omega)e^{j\phi(\omega)} \qquad \ldots(15.4)$$

Here $R(\omega)$ is the real part and $X(\omega)$ the imaginary part of $H(j\omega)$. $M(\omega)$ is the magnitude of the network function and is an even function in $\omega$ and $\phi(\omega)$ represents the phase response of the net-work function and is an odd function of $\omega$.

Equation (15.2) can be rewritten in terms of zeros and poles as follows

$$H(j\omega) = \frac{a_0 (j\omega - z_1)(j\omega - z_2)\ldots..(j\omega - z_n)}{b_0 (j\omega - p_1)(j\omega - p_2)\ldots..(j\omega - p_m)} \qquad (15.5)$$

We will study the relationship between the zeroes and poles of a system function and its' steady state sinusoidal response. In other words we will investigate the effect of zero and pole position on the behaviour of $H(s)$ along $j\omega$ axis.

## 15.2 AMPLITUDE AND PHASE RESPONSE

The amplitude and phase response of a system provides valuable information in the analysis and design of electric, electronic networks and especially transmission circuits.

It is used to make use of the frequency range from 0 to $\infty$ rather than $-\infty$ to $\infty$ as generators of sine waves produce only positive frequencies. Also, the information regarding negative frequency is redundant as amplitude and real parts are even functions of frequencies whereas phase and imaginary parts are odd functions of frequencies. This is well illustrated in Fig. 15.1 for a low pass filter.

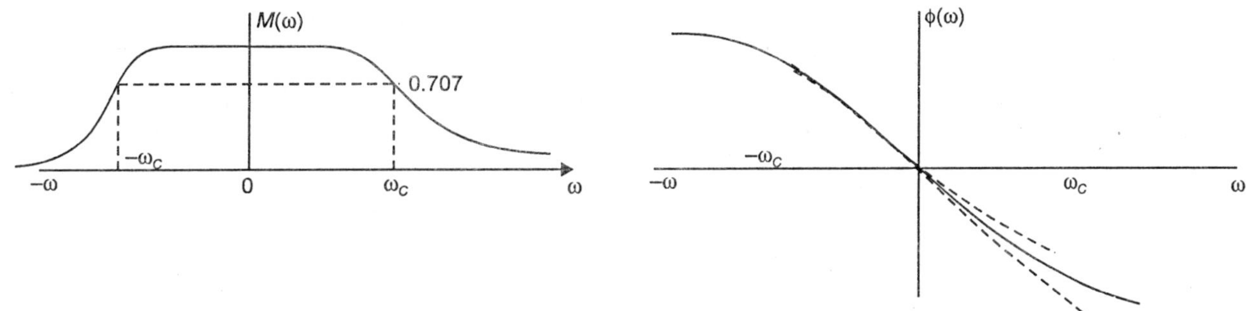

Fig. 15.1  Amplitude and phase response of a low-pass filter.

The cut-off frequency of the filter is indicated in the Fig. 15.1 as $\omega_c$ which is at the amplitude value of 0.707 of $M_{\max}(\omega)$. This frequency is known as half power frequency as sequare of 0.707 is 0.5 and square of $M(\omega)$ if it is voltage or current results in power and hence this frequency is known as half power frequency. In terms of decibels the half power point is that frequency at which $20 \log|H(\omega_c)|$ is down 3 $db$ from $20 \log|H_{\max}(\omega)|$. The characteristics in Fig. 15.1 show that the system describing these characteristics will not 'pass' frequency greater than $\omega_c$. It is found that if the phase response of $\phi(\omega)$ of a system is linear, then minimum pulse distortion will result. We see in Fig. 15.1(b) that the phase is approximately linear over the range $-\omega_c \leq \omega \leq \omega_c$, and if all the significant harmonic terms are less than $\omega_c$ then the system will produce minimum phase distortion. With this example we see the importance of an amplitude phase description of a system. Therefore, in this chapter our attention will be mainly towards methods to obtain amplitude and phase response curves both analytically and graphically.

Consider low pass filter circuit shown in Fig. 15.2

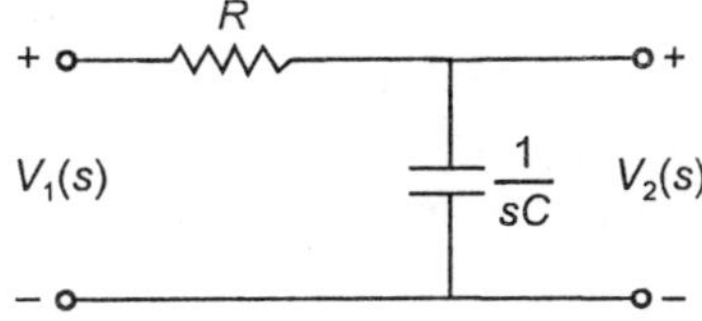

Fig. 15.2  Low pass filter or R.C. integrator circuit.

The voltage ratio transfer function of the network is

$$H(s) = \frac{V_2(s)}{V_1(s)} = \frac{\dfrac{1}{RC}}{s + \dfrac{1}{RC}} \qquad \qquad ...(15.6)$$

Let $s = j\omega$, we have

$$H(j\omega) = \frac{\dfrac{1}{RC}}{j\omega + \dfrac{1}{RC}} \qquad\qquad ...(15.7)$$

In polar form $H(j\omega)$ becomes

$$H(j\omega) = \frac{\dfrac{1}{RC}}{\sqrt{\omega^2 + \dfrac{1}{R^2C^2}}}\; e^{-j\,\tan^{-1}\omega RC} = M(\omega)e^{j\phi(\omega)} \qquad\qquad ...(15.8)$$

The asymptotic (extreme) values of the amplitude and phase response can be obtained as follows:

When $\quad \omega \to 0, \quad |H(0)| = \dfrac{\dfrac{1}{RC}}{\dfrac{1}{RC}} = 1$

and when $\quad \omega \to \infty, \quad |H(\infty)| = 0$

and at a typical frequency $\omega = \dfrac{1}{RC}$

$$\left| H\!\left(\frac{1}{RC}\right) \right| = \frac{\dfrac{1}{RC}}{\sqrt{2}\cdot\dfrac{1}{RC}} = \frac{1}{\sqrt{2}}$$

To obtain phase values

At $\omega = 0 \qquad\qquad\qquad \phi(0) = e^{-j\,\tan^{-1}\omega RC} = e^{-j\,\tan^{-1}0} = 0°$

At $\omega = \infty \qquad\qquad\qquad \phi(\infty) = e^{-j\,\tan^{-1}\omega RC} = e^{-j\,\tan^{-1}\infty} = -\dfrac{\pi}{2}$

At $\omega = \dfrac{1}{RC} \qquad\qquad \phi\!\left(\dfrac{1}{RC}\right) = e^{-j\,\tan^{-1}1} = -45°$

By obtaining magnitude and phase of the transfer function for various values of $\omega$, the complete characteristics can be plotted and are found to be as shown in Fig. 15.3.

It is observed that at cut off frequency, the network function either driving point or transfer function reduces to 0.707 times the corresponding maximum value of function.

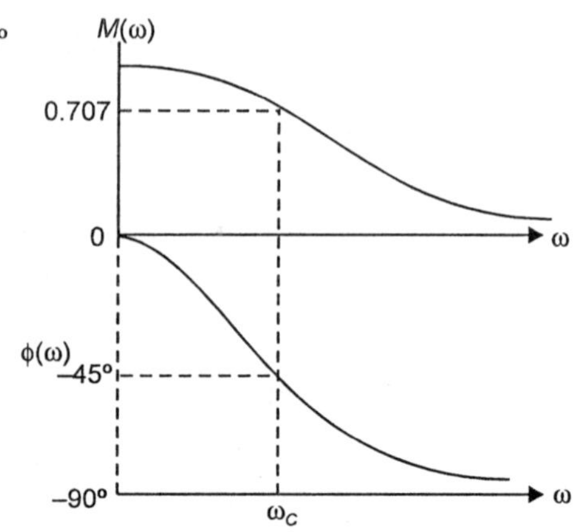

**Fig. 15.3**  Amplitude and phase response of circuit in Fig. 15.2.

## 15.3 PLOTS FROM S-PLANE PHASORS

We now consider the method to obtain amplitude and phase response from the pole zero diagram of a function is the s-plane. Consider the system function as

$$H(s) = \frac{A(s - z_0)(s - z_1)}{(s - p_0)(s - p_1)(s - p_2)} \qquad \ldots(15.9)$$

Substituting $s = j\omega$ we have

$$H(j\omega) = \frac{A(j\omega - z_0)(j\omega - z_1)}{(j\omega - p_0)(j\omega - p_1)(j\omega - p_2)} \qquad \ldots(15.10)$$

Here each factor $(j\omega - z_i)$ or $(j\omega - p_i)$ corresponds to a vector from the zero $z_i$ or pole $p_j$ directed to any point $j\omega$ on the imaginary axis. Therefore, if we express these factors in polar form

$$j\omega - z_i = N_i\, e^{j\theta i} \text{ and } j\omega - p_j = M_j e^{j\phi j}$$

Hence $H(j\omega)$ can be rewritten as

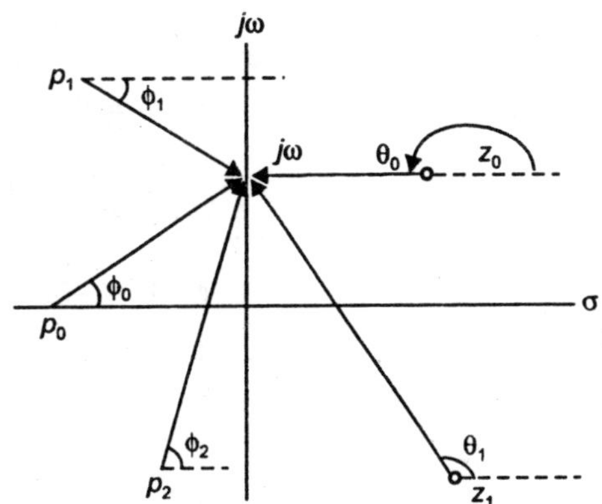


$$H(j\omega) = \frac{AN_0 N_1}{M_0 M_1 M_2} e^{j(\theta_0 + \theta_1 - \phi_0 - \phi_1 - \phi_2)} \qquad \ldots(15.11)$$

This mathematical expression is shown in the s-plane as shown in Fig. 15.4.

It is to be noted that $\theta_0$ is 180° and $\phi_1$ is negative. Therefore, the amplitude response $M(\omega)$ in general is given by the following equation.

**Fig. 15.4** Evaluation of amplitude and phase of a system function from pole zero diagram.

$$M(\omega) = \frac{\displaystyle\prod_{i=0}^{n} \text{Vector magnitude from the zeros to the point on } j\omega \text{ axis}}{\displaystyle\prod_{j=0}^{m} \text{Vector magnitude from the poles to the point on } j\omega \text{ axis}} \qquad \ldots(15.11A)$$

Similarly, the phase response is given as

$$\phi(\omega) = \sum_{i=0}^{n} \text{angles of the vectors from the zeros to the } j\omega\text{-axis}$$

$$-\sum_{j=0}^{n} \text{angles of the vectors from the poles to the } j\omega\text{-axis} \qquad \ldots(15.11B)$$

It is to be noted that these relationships for amplitude and phase are point to point relationships along $j\omega$ axis for different values of $\omega$, which means to plot amplitude and phase graphs we must draw vectors from the zeros and poles to every point on the $j\omega$ axis.

Consider the network function

$$H(s) = \frac{5s}{s^2 + 6s + 25} = \frac{5s}{(s+3+j4)(s+3-j4)} \qquad \text{...(15.12)}$$

Let us find out the amplitude and phase for $H(j5)$. From the poles and zeros of $H(s)$ we draw vectors to the point $\omega = 5$ as shown in Fig. 15.5.

From the pole zero plot it is clear that

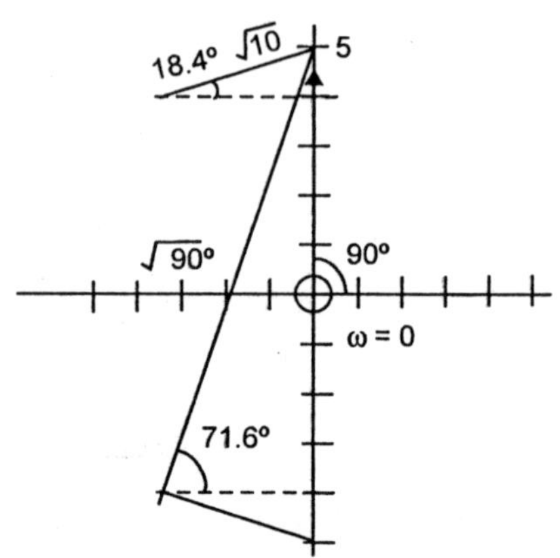


$$M(j5) = \frac{5 \times 5}{\sqrt{10} \times \sqrt{90}} = \frac{25}{30} = \frac{5}{6}$$

and

$$\phi(j5) = 90 - 18.4 - 71.6 = 0$$

Now at

$$\omega = 0$$

$$M(j0) = \frac{5(0)}{0 + 0 + 25} = 0$$

$$\phi(j0) = -53° + 53° + 90° = 90°$$

**Fig. 15.5**  Pole zero plot of the function $H(s)$

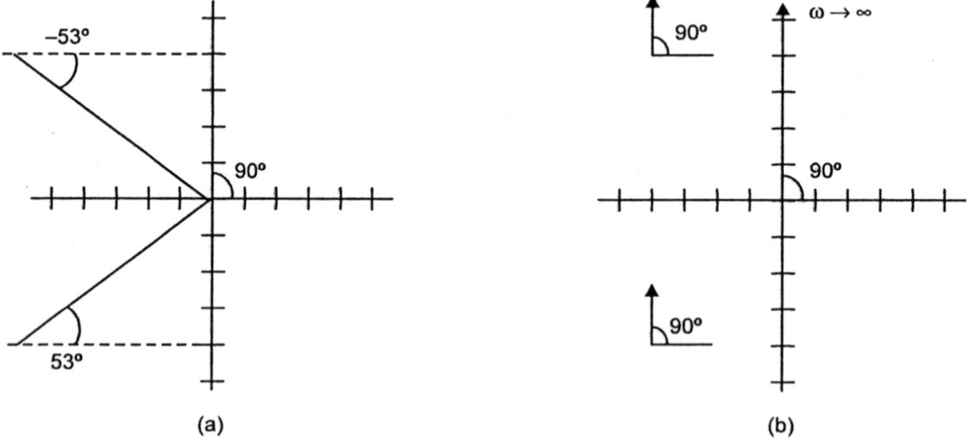


**Fig. 15.6**  Amplitude and phase at $\omega \to 0$ and $\omega \to \infty$

When $\omega \to \infty$  Fig. 15.6 (b)

We see from Fig. 15.6 (b) that the vector lengths from zero and poles are all equal and hence let this be '$a$'.

Therefore,  $M(\infty) = \dfrac{5a}{a \times a} = \dfrac{5}{a}$ and as a $\to \infty$, $M(\infty) \to 0$

However, phase $\phi (\infty) = 90 + 90 + 90 = 270$ or $-90°$

We now find out amplitude and phase analytically. Rewritting the function

$$M(s) = \frac{5s}{s^2 + 6s + 25}$$

When $\qquad \omega \to 0 \quad |H(0)| = \dfrac{5(j0)}{(j0)^2 + (j0) + 25} = 0$

and when $\qquad \omega \to \quad 0$ say $\Delta a$

$$\phi(j\Delta a) = \frac{5\,j\Delta a}{25} = \frac{5\,\Delta a j}{25} = \frac{\Delta a j}{5}$$

Hence $\qquad \phi(j\Delta a) = 90°$

Similarly when $\omega \to \infty$ say $\omega_n$

$$H(j\omega_n) = \frac{5(j\omega_n)}{-\omega_n^2}$$

or $\qquad |H(j\omega_n)| = \dfrac{5}{\omega_n}$ and as $\omega_n \to \infty$

$$|H(j\omega_n)| \to 0$$

Similarly to find out amplitude and phase for $\omega = 5$

$$H(j5) = \frac{5 \times j5}{-25 + 30j + 25} = \frac{25j}{+j30}$$

$$|H(j5)| = \frac{j25}{j30} = \frac{5}{6}$$

Hence $\qquad \phi(j5) = 0°$

So the results for amplitude and phase at $\omega = 0$, $\omega \to \infty$ and $\omega = 5$ are verified analytically. Results for a few more values of $\omega$ can be obtained and the plots can be drawn as shown in Fig. 15.7.

$$
\begin{aligned}
\omega &= 1\,|H(j1)| &&= 0.2 \text{ and} & \phi(j1) &= 76° \\
\omega &= 2\,|H(j2)| &&= 0.413 \text{ and} & \phi(j2) &= 60.3 \\
\omega &= 5\,|H(j5)| &&= 0.83 \text{ and} & \phi(j5) &= 0° \\
\omega &= 10\,|H(j10)| &&= 0.52 \text{ and} & \phi(j10) &= -51.4°
\end{aligned}
$$

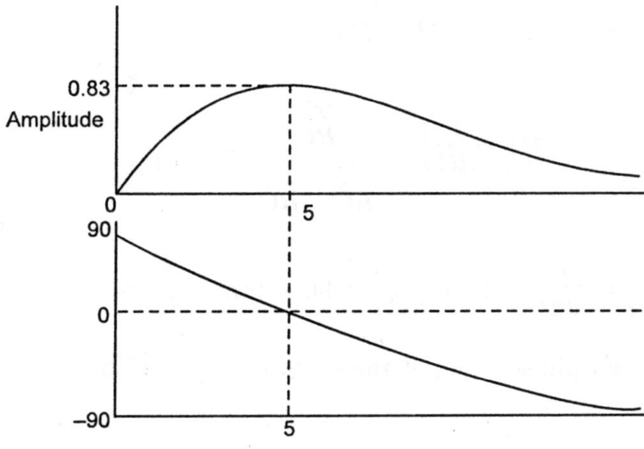


**Fig. 15.7** Amplitude and phase response for $H(s)$ of Eq. 15.12.

If we interchange the location of $R$ and $C$ in Fig. 15.2 we obtain the following as shown in Fig. 15.8.

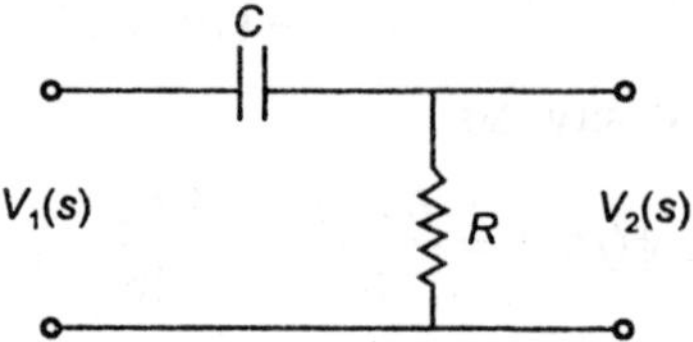

**Fig. 15.8**  High pass filler circuit Or R-C differentiator circuit.

The transfer function of the above network is given as

$$H(s) = \frac{V_2(s)}{V_1(s)} = \frac{s}{s + \dfrac{1}{RC}} \qquad \text{...(15.13)}$$

or
$$H(j\omega) = \frac{j\omega}{j\omega + \dfrac{1}{RC}} \qquad \text{...(15.14)}$$

Again for $\qquad \omega \to 0 \quad |H(j\omega)| = 0$ and $\phi(j\omega) = \tan^{-1}\dfrac{j\omega}{\dfrac{1}{RC}}$

$$= \tan^{-1} j\omega RC = 90°$$

For $\omega \to \infty$ Dividing numerator and denominator of (15.13) by $s$ we have

$$H(s) = \frac{1}{1 + \dfrac{1}{RCs}} = 1$$
$$\text{Lt } s \to \infty$$

and phase $\qquad \phi(j\omega) = 0$

For $\qquad\qquad \omega = \dfrac{1}{RC}$

$$H\left(j\frac{1}{RC}\right) = \frac{\dfrac{j}{RC}}{\dfrac{j}{RC} + \dfrac{1}{RC}} = \frac{j}{j+1}$$

Hence $\left|H\left(\dfrac{1}{RC}\right)\right| = \dfrac{1}{\sqrt{2}}$ and $\phi\left(\dfrac{1}{RC}\right) = 90 - \tan^{-1}\dfrac{1}{\sqrt{2}} = 45°$

The amplitude and phase plots of the circuit in Fig. 15.8 are shown in Fig. 15.9

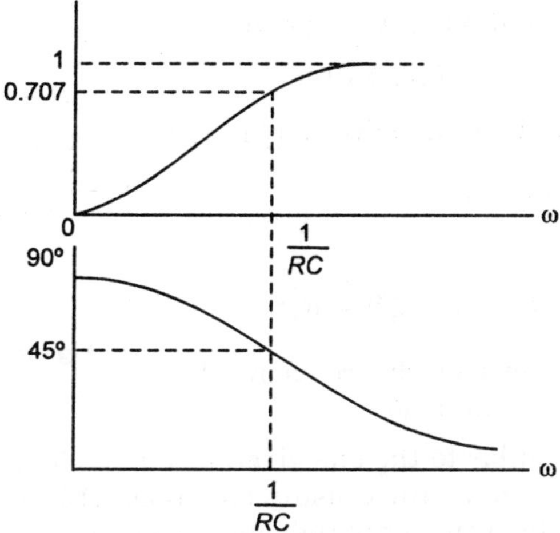

**Fig. 15.9**  Amplitude and phase plots of Network in Fig. 15.8.

Comparing the results of Fig. 15.2 (Integrator Circuit) and Fig. 15.8 (Differentiator circuit) we find that in the network of Fig. 15.2 the output lags behind the input for all $\omega$ whereas in the network of Fig. 15.8 the phase of the output voltage leads the input for all $\omega$. In control system theory these networks are called lag networks and lead networks respectively. The magnitude variation of the two networks as $\omega$ varies are inverse. For a constant input voltage, the output voltage of the network of Fig. 15.2 decreases with increase in $\omega$ and that for Fig. 15.8 it increases with increase in $\omega$.

Next let us consider a series RLC circuit having the driving point impedance

$$Z(s) = R + sL + \frac{1}{sC}$$

and the admittance

$$Y(s) = \frac{1}{L}\left[\frac{s}{s^2 + \dfrac{R}{L}s + \dfrac{1}{LC}}\right] \qquad \text{...(15.15)}$$

or

$$Y(s) = \frac{1}{L}\left[\frac{s}{(s - s_a)(s - sa^*)}\right] \qquad \text{...(15.16)}$$

where $sa,\ sa^* = -\dfrac{R}{2L} \pm j\sqrt{\dfrac{1}{LC} - \left(\dfrac{R}{2L}\right)^2} = -\xi\,\omega_n \pm j\omega_n\sqrt{1 - \xi^2}$ $\qquad \text{...(15.17)}$

where $\omega_n$ is the natural frequency of the network and $\xi$ is the damping ratio. We consider here only the case of Oscillation in the circuit when $\xi < 1$. For this the poles are located on a circle of radius $\omega_n$ and a line from the pole to the origin makes an angle $\theta$

$= \cos^{-1}\xi$ with the negative real axis. A single zero is located at the origin as shown in Fig. 15.10.

Let $\qquad \sigma = -\xi\omega_n$ and the imaginary part is

$$\omega = \omega_n \sqrt{1-\xi^2}$$

and $\sigma^2 + \omega^2 = \xi^2\omega_n^2 + \omega_n^2\,(1 - \xi^2) = \omega_n^2$

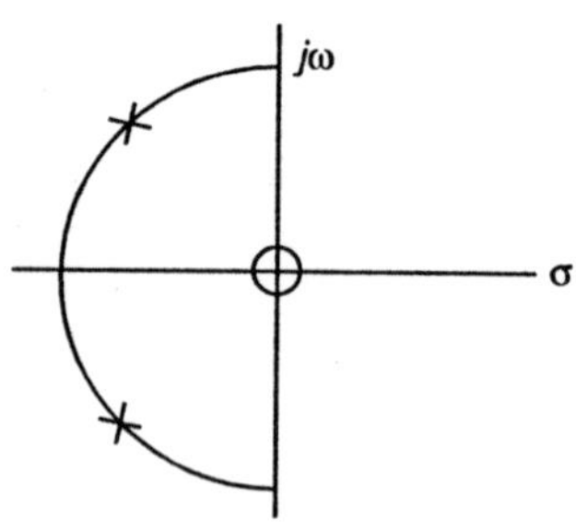

**Fig. 15.10**  The poles and zeros of $Y(S)$ given by Eqn. 15.17.

This shows that the locus of the roots in the complex s-plane is a circle of radius $\omega_n$

This pole zero plot is similar to the one given in Fig. 15.5 and the amplitude and phase relations as a function of $\omega$ are obtained using equations (15.11A) and (15.11B), Fig. 15.11 shows four values of $\omega$ for which the amplitude and phase have been evaluated and approximate values have been shown in Fig. 15.11.

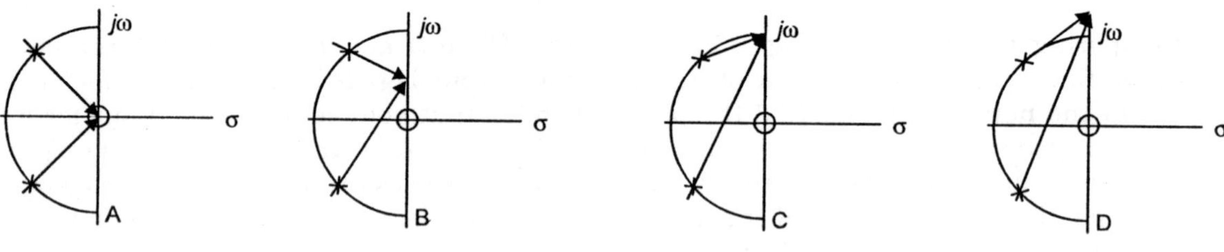

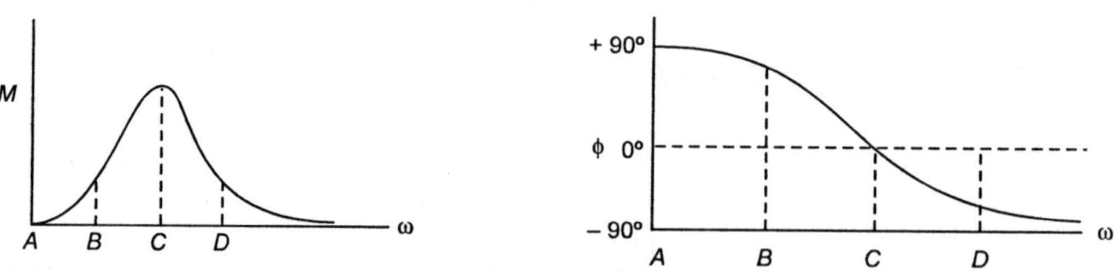

**Fig. 15.11**  Phasors which determine the frequency response shown for four different values of $\omega$.

The magnitude of $Y(j\omega)$ is given as

$$|Y(j\omega)| = \frac{1}{\sqrt{\left[R^2 + \left(\omega L - \dfrac{1}{\omega c}\right)^2\right]}} \qquad\qquad ...(15.18)$$

and for resonance to take place $|Y(j\omega)|$ should be maximum at that frequency and then

$$|Y(j\omega)| = \frac{1}{R} \text{ and for. this}$$

$$\omega L = \frac{1}{\omega C} \quad \text{or} \quad \omega = \frac{1}{\sqrt{LC}} = \omega_n$$

i.e., resonance occurs at $\omega = \omega_n$ and not at the point opposite the pole on the $j\omega$-axis. The phase of the circuit at this frequency is zero.

We define Q of the circuit (RLC series circuit)

$$\text{as} \qquad Q = \frac{\omega_n L}{R} = \frac{1}{2}\frac{\omega_n}{R/2L} = \frac{1}{2}\frac{\omega_n}{\xi \omega_n}$$

$$= \frac{1}{2\xi} \qquad\qquad ...(15.19)$$

$$= \frac{1}{2\cos\theta} \qquad\qquad ...(15.20)$$

where $\theta$ is the angle subtended by the line joining the origin and one of the conjugate poles with the $-\sigma$ axis.

From the above analysis following conclusions can be drawn.

1. The closer the poles $sa$ and $sa^*$ are to the $j\omega$ axis the higher the $Q$.
2. The value of $Q$ varies inversly with damping ratio $\xi$. A high value of $Q$ means a low value of damping ratio. A series RLC circuit with low $R$ thus has high $Q$.

Next let us consider the effect of poles and zeros on the $j\omega$ axis upon the frequency response.

Consider the function

$$H(s) = \frac{s^2 + 1.21}{s^2 + 1.44} = \frac{(s + j\,1.1)(s - j\,1.1)}{(s + j\,1.2)(s - j\,1.2)} \qquad ...(15.21)$$

The pole zero diagram is shown in Fig. 15.12

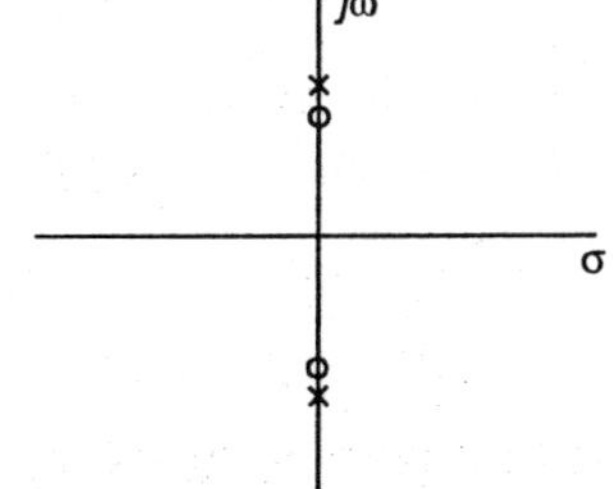

Fig. 15.12   Pole zero for equation (15.21).

At $\omega = 1.1$, the vector from zero to the frequency itself (i.e., $\omega = 1.1$) is of zero magnitude. Therefore, at a zero on the $j\omega$ axis the amplitude response is zero. At $\omega = 1.2$ the vector from the pole to the frequency itself i.e., $\omega = 1.2$ is of zero magnitude. The amplitude response is, therefore, infinite at a pole as seen from equation (15.21) or equation (15.11A)

Next we consider the phase response. When $\omega < 1.1$ it is apparent from the pole zero plot that the phase is zero. However, when $\omega > 1.1$ but $< 1.2$, the vector from the zero at $\omega = 1.1$ is now pointing upwards while that from the other poles and zeros are oriented in the same direction as for $\omega < 1.1$. We see that at a zero on the $j\omega$-axis the phase response has a step discontinuity of 180° for increasing frequency.

Similarly at a pole on the $j\omega$ axis the phase response is discontinuous by $-180°$. These observations have been illustrated in Fig. 15.13.

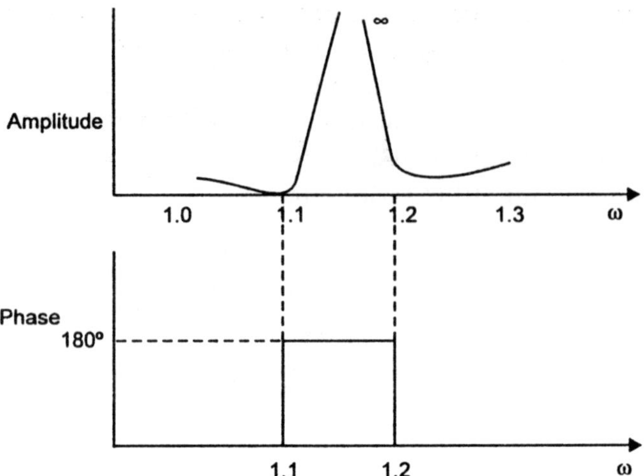

Fig. 15.13 Amplitude and phase for $H(s)$ in Fig. 15.12.

If we extend these ideas we find that if we have a zero at $z = -\sigma \pm j\omega_i$ where $\sigma$ is very small as compared to $\omega_i$ then we will have a dip in the amplitude characteristic and a rapid change of phase near $\omega = \omega_i$ as shown in Fig. 15.14. Similarly, if there is a pole at $p = -\sigma \pm j\omega_i$, with $\sigma$ very small, then the amplitude will be peaked and the phase will decrease rapidly near $\omega = \omega_i$ as shown in Fig. 15.15. A different situation would occur when we have poles and zeros far away from the $j\omega$-axis i.e. when $\sigma$ is large as compared to the frequency range of interest. Under these circumstances we see that these poles and zeros contribute little to the shaping of the amplitude and phase response curves. Their only effect is to scale up or down the overall amplitude response.

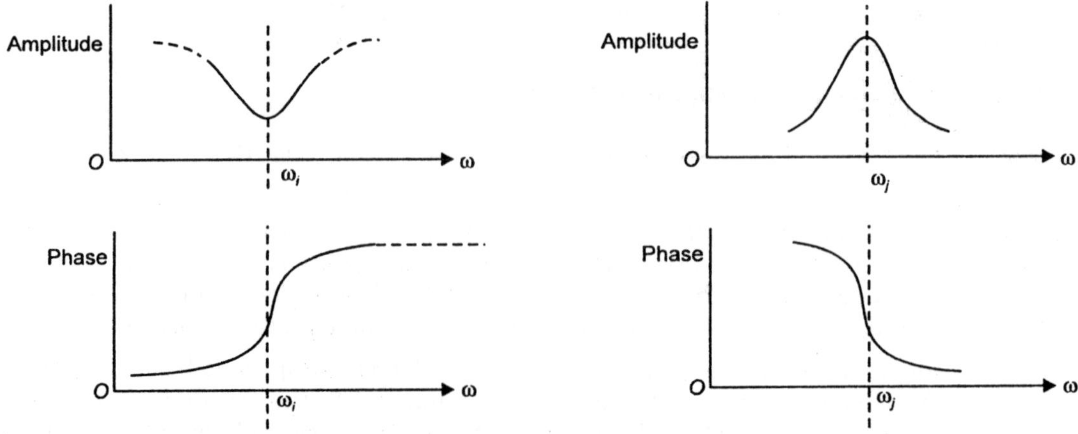

Fig. 15.14 Effect of zero very near to $j\omega$ axis.    Fig. 15.15 Effect of pole very near to $j\omega$ axis.

For a system to be stable it is essential that there should not be any pole in the right half of the $s$-plane. However, transfer functions may have zeros in the right half of the $s$-plane. Consider the pole zero configuration in Fig. 15.16(a) and (b). Both the configurations have the same poles. However, the difference is in the configuration of zeros. In Fig. 15.16(a) these are in the left half of the *s*-plane in, Fig. 15.16(b) these are the mirror image of the left plane into the right half of *s*-plane along the $j\omega$ axis (The mirror is the $j\omega$-axis)

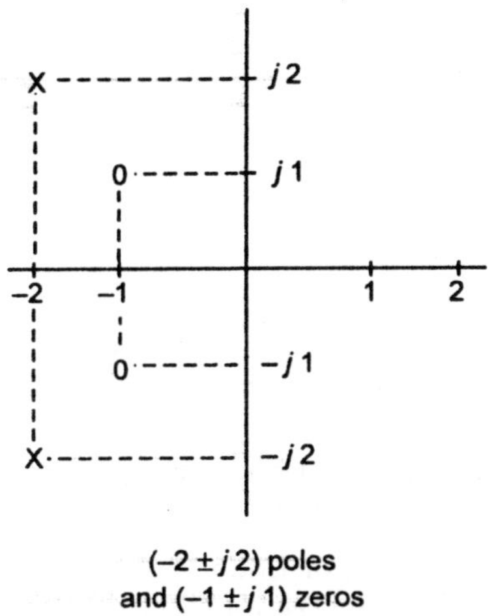


(−2 ± *j* 2) poles
and (−1 ± *j* 1) zeros

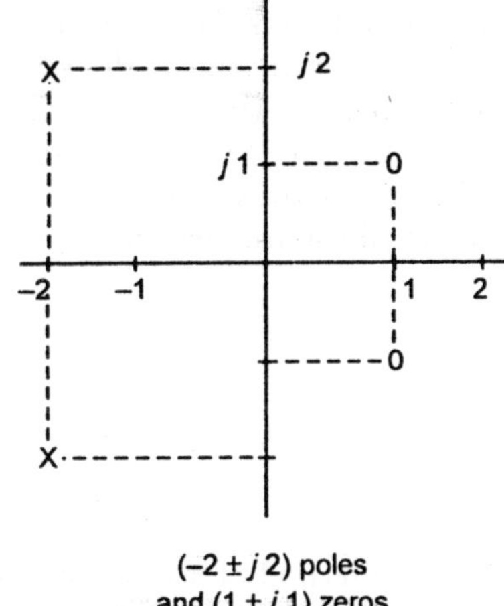


(−2 ± *j* 2) poles
and (1 ± *j* 1) zeros

**Fig. 15.16**  (a) Minimum phase function.        (b) Non-minimum phase function.

If we compare the amplitude response of the two configurations we find that these are same as the lengths of the vectors are identical for both situations. However, the absolute magnitude of phase of (b) is greater than the phase of (a) for all frequencies. This is due to the fact that the zeros in the right half plane contribute more phase shift on an absolute magnitude basis than counterpart in the left of plane.

We draw the following conclusions:

1. A system function with zeros in the left half of *s*-plane or on the $j\omega$-axis only is called a minimum phase function.

2. If the function has one or more zeros in the right half of *s*-plane, it is a non minimum phase function.

Fig 15.17 shows the phase response of the minimum and non minimum phase functions.

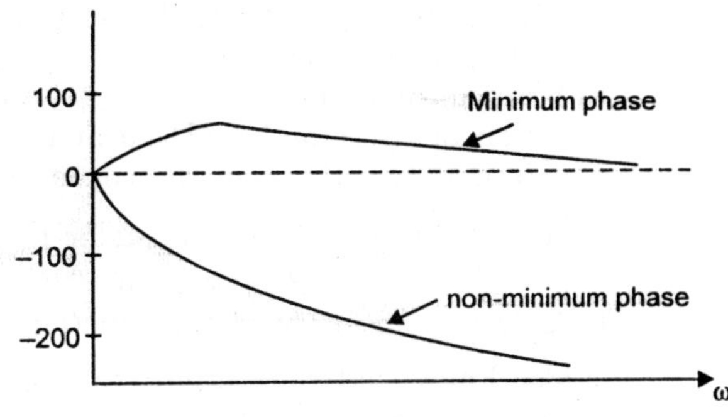


**Fig. 15.17**   Phase relative for minimum and non-minimum phase functions.

Let us consider a situation when the zeros in the right half of the *s*-plane are the mirror image of the poles in the left half of the *s*-plane as shown in Fig. 15.18.

From Fig. 15.18 it is clear that the vector drawn from a zero to any point $\omega_i$ on the $j\omega$ axis is identical in magnitude with the vector drawn from its mirror image (a pole) to $\omega_i$. Therefore, the amplitude response of such a function must be constant for all frequencies as shown in Fig. 15.19(a).

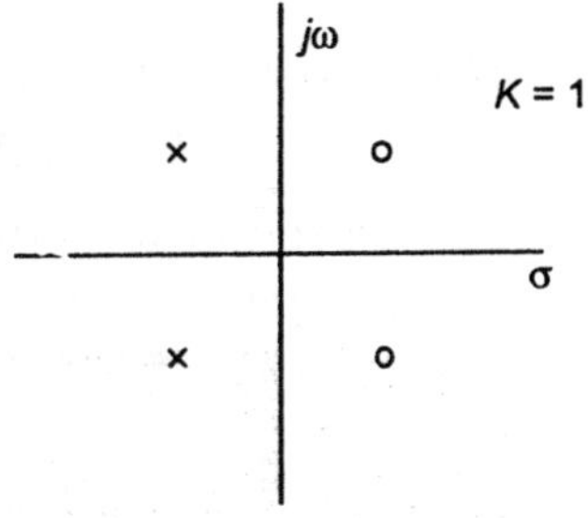


**Fig. 15.18**  All pass functions.

However, the phase response is not constant as shown in Fig. 15.19(b).

Hence an all-pass function is one which has its zeros in the right of the s-plane and poles are the mirror image of the zeros in the left half of the s-plane. These networks are normally used to correct phase distortion in transmission lines.

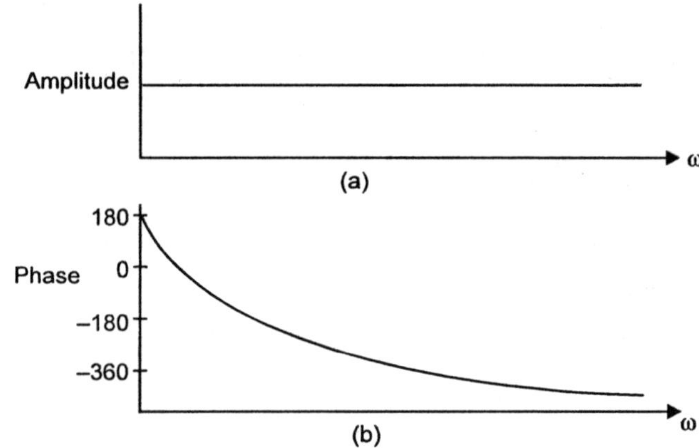


**Fig. 15.19**  (a) Amplitude and (b) phase response of all-pass function in Fig. 15.18.

## 15.4 BODE PLOT

Another useful method of studying frequency response is the use of Bode Plot which requires two graphs; one with the magnitude of $H(j\omega)$ plotted in decibel vs frequency on a logarithmic scale and the other with the phase shift of $H(j\omega)$ in degrees vs. frequency on a logarithmic scale. These graphs are often referred to as Bode Plot, the corner plot or the logarithmic plots of $H(s)$.

This form of plot has the following advantages

   (i)  The product factors in the expression for $H(j\omega)$ become additive terms since logarithms are used.

  (ii)  The shape of the corner plot for the most commonly used funcitons in servo systems makes it possible to represent approximately the exact function plot by straight line asymptotes.

 (iii)  Since the corner plots are easy to construct, the data necessary for the construction of polar plot and the magnitude in decibels vs phase shift plot can be obtained directly from the corner plot.

Consider the open loop transfer function of

$$H(j\omega) = M(\omega)e^{\,j\theta\,(\omega)} \qquad \qquad ...(15.22)$$

where $M$ is magnitude of the function.

We also know that the amplitude response is

$$M(\omega) = |H(j\omega)| = \frac{|N(j\omega)|}{|D(j\omega)|} \qquad \qquad ...(15.23)$$

If we express the amplitude in decibels we have

$$20 \log M(\omega) = 20 \log |N(j\omega)| - 20 \log |D(j\omega)| \qquad \qquad ...(15.24)$$

In factored form both $N(s)$ and $D(s)$ are made up of four kinds of factors.

   (i)  a constant $K$

  (ii)  a root at the origin, $s$ or $s^n$

 (iii)  a simple real root $s + \alpha$ or $(1 + j\omega T)^{\frac{1}{2}}$

(iv) a complex pair of roots $\left(s^2 + 2\alpha s + \alpha^2 + \beta^2\right)^{\pm 1}$ or $\left(1 + 2j\delta\mu - \mu^2\right)^{\pm 1}$

If these factors are in the numerator, their magnitude in decibels carry positive signs. However if these factors belong to the denominator, their magnitudes in decibels carry negative signs. We start with case (i)

(i) The $K$ factor: For the constant $K$, the *db* loss or gain is

$$20 \log K = K_2$$

The constant $K_2$ is negative if $|K| < 1$ or positive if $|K| > 1$. The phase response is zero or 180° depending upon whether $K$ is positive or negative. The Bode Plot showing the magnitude in decibels and phase of a constant are shown in Fig. 15.20.

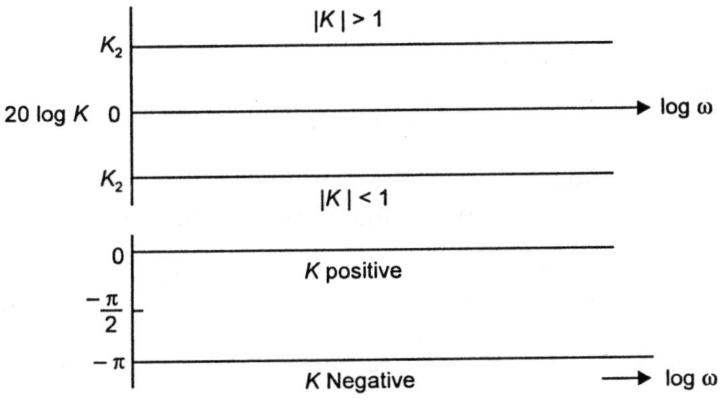


**Fig. 15.20**  Magnitude in dbs and phase of a constant.

(ii) Poles or zeros at the origin $(j\omega)^{\pm n}$

The magnitude of $(j\omega)^{\pm n}$ in decibels is

$$\pm 20n \log \omega \; db$$

which is the equation of a straight line in either semi-logarithmic coordinates or rectangular coordinates. The slope of these straight lines is given by

$$\frac{d\,20\log\left|(j\omega)^{\pm n}\right|}{d\,\log \omega} = \pm 20\,ndb \qquad \qquad \dots(15.25)$$

Hence in rectangular co-ordinates a unit change in $\log_{10}\omega$ will correspond to a change of $\pm 20n\,db$. Also a unit change in $\log_{10}\omega$ is equivalent to a change of 1 to 10, 10 to 100 etc. in the logarithmic scale. Thus the slopes of these straight lines are described by $20n$ decibels per decade of frequency.

The number of decades between any two frequencies is defined by

$$\text{Number of decades} = \frac{\log \dfrac{\omega_2}{\omega_1}}{\log 10} \qquad \qquad \dots(15.26)$$

Sometimes the unit octave is used to represent the separation of two frequencies.

Frequencies $\omega_2$ and $\omega_1$ are said to be separated by an octave if $\dfrac{\omega_2}{\omega_1} = 2$

The number of octaves between two frequencies is defined by

$$\text{Number of octaves} = \frac{\log \dfrac{\omega_2}{\omega_1}}{\log 2} \qquad \qquad ...(15.27)$$

For one decade of frequency $\dfrac{\omega_2}{\omega_1} = 10$, the relation between octaves and decades is

$$\text{No. of octaves} = \frac{\log \dfrac{\omega_2}{\omega_1}}{\log 2} = \frac{\log 10}{\log 2} = \frac{1}{0.301} \text{ decade}$$

Hence 20 *db*/decade = 20 × 0.301 = 6 *db*/octave.

For a single pole at the origin the slope of the magnitude curve is −20 *db*/decade of frequency or −6 *db*/octave of frequency.

The phase shift of $(j\omega)^{\pm n}$ is

$$\pm n \times 90° \text{ or } \pm n\frac{\pi}{2} \text{ radians}$$

The plot for magnitude in *db* and phase shift in degrees for the function $(j\omega)^{\pm n}$ are shown in Fig. 15.21.

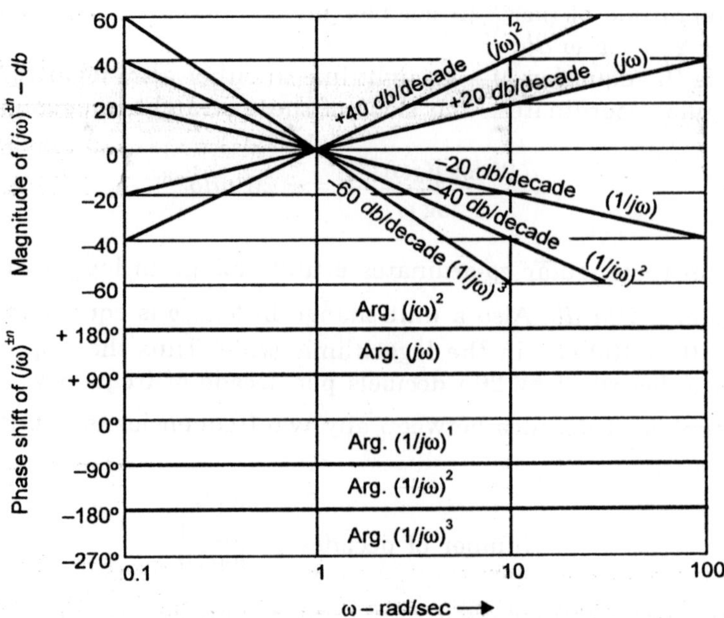


**Fig. 15.21**  Bode plot of magnitude and phase of the factor $(j\omega)^{\pm n}$.

(iii) a simple zero $(1 + j\omega T) = (1 + \omega^2 T^2)^{\frac{1}{2}}$ $\qquad\qquad$ ...(15.28)

To find out a linear asymptotic approximation we obtain the magnitude of the simple zero when $\omega T \ll 1$ and $\omega T \gg 1$.

Taking log of (15.28) we have

$20 \log |H(j\omega)| = 20 \log (1 + \omega^2 T^2)^{\frac{1}{2}}$

When $\omega T \ll 1$

$20 \log |H(j\omega)| = 20 \log 1 = 0$ $\qquad\qquad$ ...(15.29)

and when $\omega T \gg 1$

$20 \log |H(j\omega)| = 20 \log \omega T = 20 \log \omega + 20 \log T$ $\qquad$ ...(15.30)

which represents a straight line with slope of 20 *db*/decade of frequency (or 6 *db*/octave) and the intercept at $\omega = 1$, is 20 log $T$. The intersection of the low frequency and high frequency is obtained by equating equation (15.29) and (15.30) i.e.,

$20 \log \omega T = 0$

Which means $\omega T = 1$ or $\omega = \dfrac{1}{T}$ $\qquad\qquad$ ...(15.31)

The frequency obtained in equation (15.31) is known as the positive corner frequency of the plot. The actual magnitude deviates very slightly from the straight line asymptote as shown in Fig. 15.22

The phase of the single zero is given as

$$\phi(j\omega) = \tan^{-1} \omega T \qquad\qquad \text{...(15.32)}$$

and is plotted in Fig. 15.23. Table 15.1 shows the magnitude and phase shift of the zero $(1 + j\omega T)$. Table 15.2 gives a comparison of actual magnitude and the straight line asymptotes at some significant frequencies.

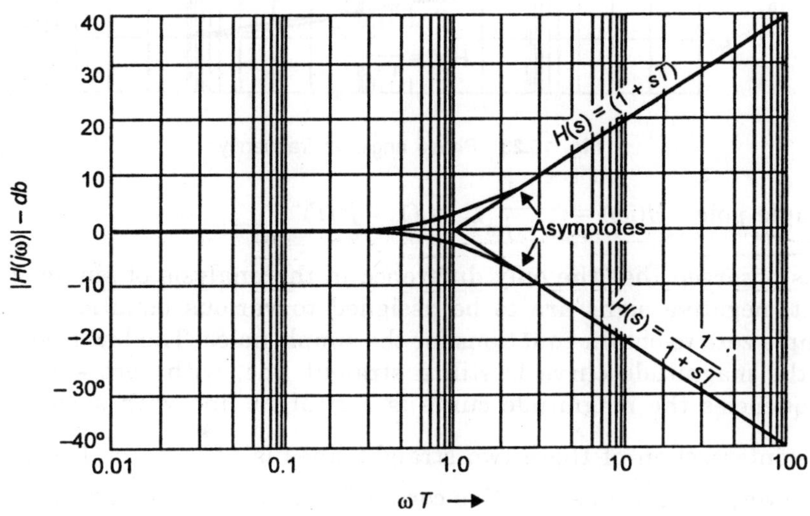


**Fig. 15.22**  Magnitude versus frequency $H(s) = 1 + sT$ and $H(s) = 1/1 + sT$.

**Table 15.1**

| $\omega T$ | $log_{10}\omega T$ | $\|1 + j\omega T\|$ | $\|1 + j\omega T\| db$ | $Arg(1 + j\omega T)$ |
|---|---|---|---|---|
| 0.01 | −2 | 1 | 0 | 0.5° |
| 0.1 | −1 | 1.04 | 0.043 | 5.7° |
| 0.5 | −0.3 | 1.12 | 1 | 26.6° |
| 0.76 | −0.12 | 1.26 | 2 | 37.4° |
| 1.0 | 0 | 1.41 | 3 | 45.0° |
| 1.31 | 0.117 | 1.65 | 4.3 | 52.7° |
| 1.73 | 0.238 | 2.0 | 6.0 | 60.0° |
| 2.0 | 0.3 | 2.23 | 7.0 | 63.4° |
| 5.0 | 0.7 | 5.1 | 14.2 | 78.7° |
| 10.0 | 1.0 | 10.4 | 20.3 | 84.3° |

**Table 15.2**

| $\omega T$ | | Magnitude of $(1 + j\omega T)$ in db | Asymptotic values of magnitude (db) | Error (db) |
|---|---|---|---|---|
| 0.1 | One decade below corner freq. | 0.3 | 0 | +0.3 |
| 0.5 | One octave below corner freq. | 1.0 | 0 | +1 |
| 0.76 | At the corner freq. | 2 | 0 | +2 |
| 1.0 | At the corner freq. | 3 | 0 | +3 |
| 1.31 | | 4.3 | 2.3 | +2 |
| 2.0 | One octave above corner freq. | 7 | 6 | +1 |
| 10 | One decade above corner freq. | 20.3 | 20 | +0.3 |

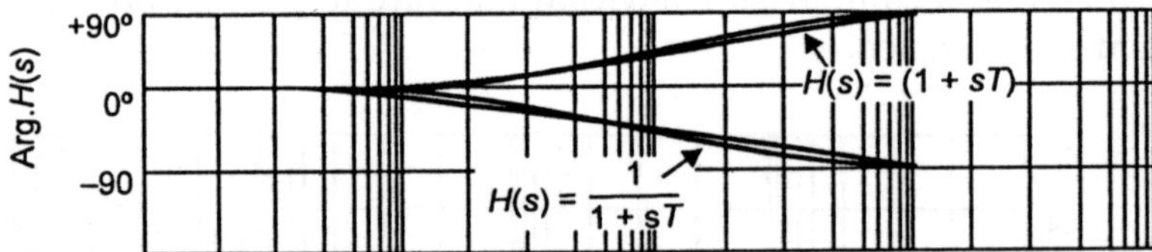


**Fig. 15.23**   Phase angle vs frequency.

(iv) Simple pole   $H(j\omega) = \dfrac{1}{1 + j\omega T} = (1 + j\omega T)^{-1}$

It is observed that the only difference in the analysis of simple zero and pole is that, negative signs are to be assigned to various equations corresponding to simple zero to obtain equations for the simple poles. The low frequency asymptote of the magnitude curve is still a straight line, with zero slope and for higher frequencies the magnitude curve is a straight line with slope of −20db/decade.

The intersection of these two straight lines is at $\omega = \dfrac{1}{T}$. The corner frequency however, is known as negative corner frequency. The intercept is −20 log T. The phase shift is negative of what we have for the simple zero and is given in Table 15.1. The magnitude and phase angle plots are given in Fig. 15.22 and 15.23 respectively. The error curve between the actual magnitude and those obtained from asymptotic lines is given in Fig. 15.24.

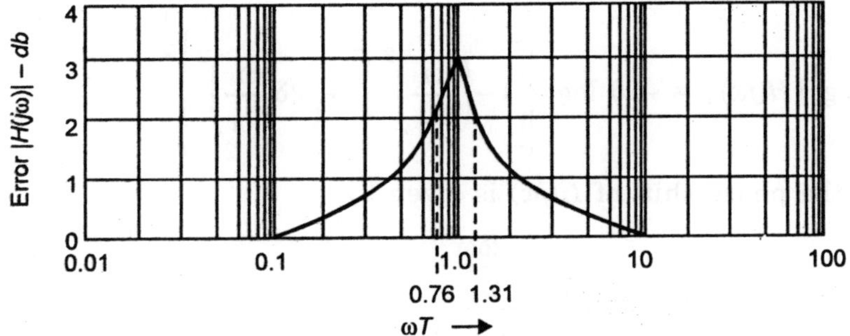

**Fig. 15.24** Error in magnitude versus frequency $H(s)$ $1 + sT$ and $\dfrac{1}{1+sT}$.

It is to be noted that the magnitude eror curve is symmetrical with respect to the corner frequency $\dfrac{1}{T}$. Also, the error is $3db$ at the corner frequency and $1db$ at one octave above and below the corner frequency. At one decade above and below the corner frequency the error is about $0.3\ db$.

(iv) Quadratic poles: Consider the second order transfer function

$$H(s) = \frac{\omega_n^2}{s^2 + 2\delta\omega_n s + \omega_n^2} = \frac{1}{\dfrac{s^2}{\omega_n^2} + \dfrac{2\delta}{\omega_n}s + 1} \qquad \text{...(15.33)}$$

Substituting $s = j\omega$

$$H(j\omega) = \frac{1}{-\left(\dfrac{\omega}{\omega_n}\right)^2 + j2\delta\dfrac{\omega}{\omega_n} + 1} \qquad \text{...(15.34)}$$

Let $\dfrac{\omega}{\omega_n} = \mu$ we have

$$H(j\omega) = \frac{1}{1 + j2\,\delta\,\mu - \mu^2} \qquad \text{...(15.35)}$$

The magnitude of the transfer function is

$$|H(j\omega)| = \frac{1}{\left\{\left[1 - \left(\dfrac{\omega}{\omega_n}\right)^2\right]^2 + \left(2\delta\dfrac{\omega}{\omega_n}\right)^2\right\}^{\frac{1}{2}}} \qquad \text{...(15.36)}$$

The magnitude in decibels is

$$20 \log |H(j\omega)| = -20 \log \left\{ \left[ 1 - \left( \frac{\omega}{\omega_n} \right)^2 \right]^2 + \left( 2\delta \frac{\omega}{\omega_n} \right)^2 \right\}^{\frac{1}{2}} \qquad ...(15.37)$$

and the phase shift of $H(j\omega)$ is given by

$$\phi\,(j\omega) = -\tan^{-1} \frac{2\delta\omega}{\dfrac{\omega_n}{1 - \left( \dfrac{\omega}{\omega_n} \right)^2}} \qquad ...(15.38)$$

Again at low frequencies i.e., $\dfrac{\omega}{\omega_n} < < 1$ equation (15.37) reduces to

$$20 \log |H(j\omega)| = -20 \log 1 = 0 \; db.$$

which means at low frequencies the asymptote for a second order system is a straight line with zero slope.

At high frequencies when $\dfrac{\omega}{\omega_n} > > 1$

the magnitude of $H(j\omega)$ in decibels becomes

$$20 \log |H(j\omega)| = -20 \log \sqrt{ \left[ 1 - \left( \frac{\omega}{\omega_n} \right)^2 \right]^2 + \left( 2\delta \frac{\omega}{\omega_n} \right)^2 }$$

$$\simeq -20 \log \sqrt{ \left( \frac{\omega}{\omega_n} \right)^4 }$$

$$\simeq -40 \log \frac{\omega}{\omega_n} \; db \qquad ...(15.39)$$

From this it is clear that the slope of the magnitude curve in decibels at higher frequencies is −40 db/decade and is a straight line.

The intersection of the two asymptotes is found by equating

$$-40 \log \frac{\omega}{\omega_n} = 0 \quad \text{or} \quad \frac{\omega}{\omega_n} = 1$$

or $\qquad\qquad\qquad \omega = \omega_n$

Hence the frequency $\omega = \omega_n$ is the corner frequency for a second order factor. In this case since magnitude and phase are not only function of corner frequencies, they also depend upon the damping ratio $\delta$, the actual magnitude plot in this case differs from the asymptotic lines. The actual magnitude plot and the phase shift plots are given in Fig. 15.25 and

Fig. 15.26 respectively. Usually if a transfer function of the quadratic form is given, first the values of $\delta$ and $\omega_n$ are determined, then the sets of equations given are used to plot the magnitude and phase shift vs frequency curves.

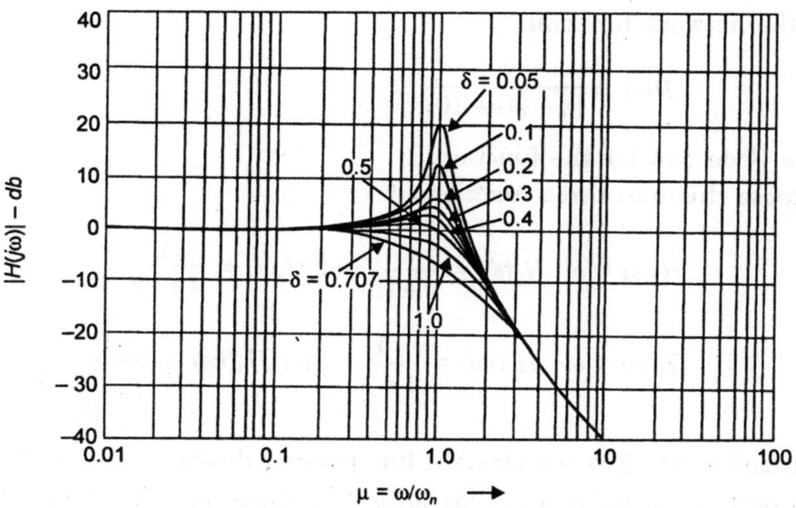

Fig. 15.25  Magnitude versus frequency.

$$H(s) = \frac{1}{1 + 2\delta(s / \omega_n) + (s / \omega_n)^2}$$

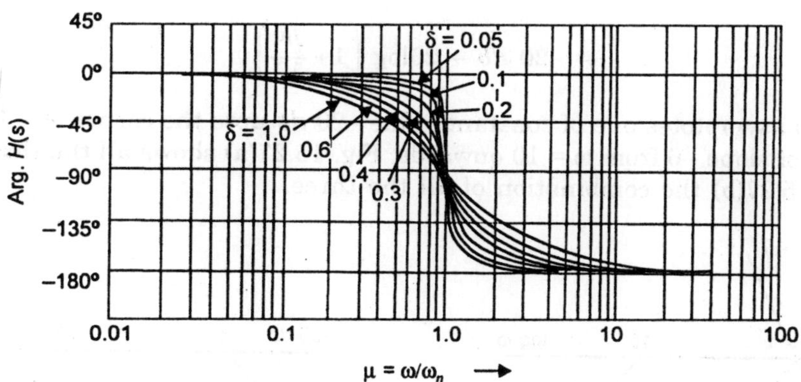

Fig. 15.26  Phase angle versus frequency.

$$H(s) = \frac{1}{1 + 2\delta(s / \omega_n) + (s / \omega_n)^2}$$

We see from Fig. 15.25 that only for $\delta \simeq 0.6$ is the straight line approximation close to the actual one.

In Fig. 15.26 we find that the phase varies from $0°$ at low frequencies to $-180°$ at high frequencies with the value $-90°$ at the break frequency for all values of $\delta$. The phase charactristics change abruptly near the break frequency, the more abrupt, the smaller the value of δ.

We have so far considered the magnitude and phase plots for a pole, for a zero, all the equations are to be multiplied by −1.

Consider the network function.

$$H(s) = \frac{s}{(s+1)(s+10)}$$

*Sketch the Bode plot for the function.*
We decompose this into three functions,

$$H_1(s) = s \quad H_2(s) = \frac{1}{s+1} \quad \text{and} \quad H_3(s) = \frac{1}{s+10}$$

or $\qquad H_1(\omega) = \omega \quad H_2(j\omega) = \dfrac{1}{1+j\omega} \quad \text{and} \quad H_3(j\omega) = \dfrac{1}{1+\dfrac{j\omega}{10}}, \dfrac{1}{10}$

$20 \log |H_1(\omega)| = 20 \log \omega$ is a straight line passing through $\omega = 1$ at $0db$ with slope 6.

$20 \log |H_2(\omega)| = -20 \log (1 + j\omega)$. At large frequency upto $\omega = 1$ the slope is $0db$ and a straight line of slope −6 from $\omega = 1$ onwards.

$$20 \log H_3(j\omega 1) = 20 \log \frac{1}{10} \cdot \frac{1}{1 + \dfrac{j\omega}{10}}$$

$$= -20\ db - 20\log \left(1 + \frac{j\omega}{10}\right)$$

It has two asymptotes one of constant value −20 *db* upto the corner $\omega = 10$ and another a straight line of slope −6 from $\omega = 10$ onwards. Fig. 15.27(a) shows all the three components whereas Fig. 15.27(b) the combination of all the three.

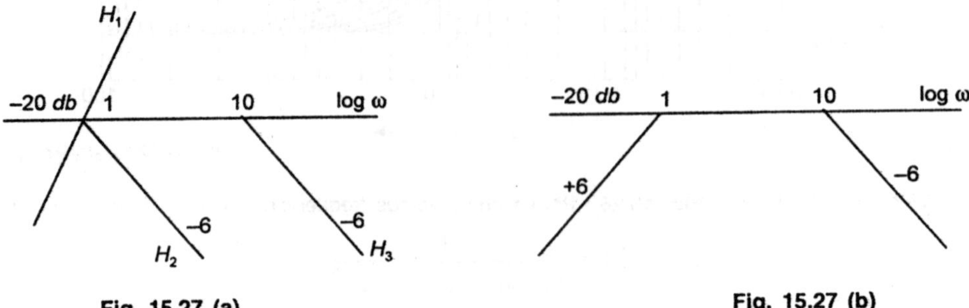


Fig. 15.27 (a)                    Fig. 15.27 (b)

Let us consider a function.

$$H(s) = 20 \left(1 + \frac{s}{100}\right). \text{ Sketch its Bode plot.}$$

$$20 \log \ |H(j\omega)| = 20 \log 20 + 20 \log \ (1 + j\omega T)$$

where $\quad T = \dfrac{1}{100}$ or $\omega = 100$ which is the corner frequency. The straight lines corresponding to two factors are shown in Fig. 15.28(a) and the combination is shown in Fig. 15.28(b).

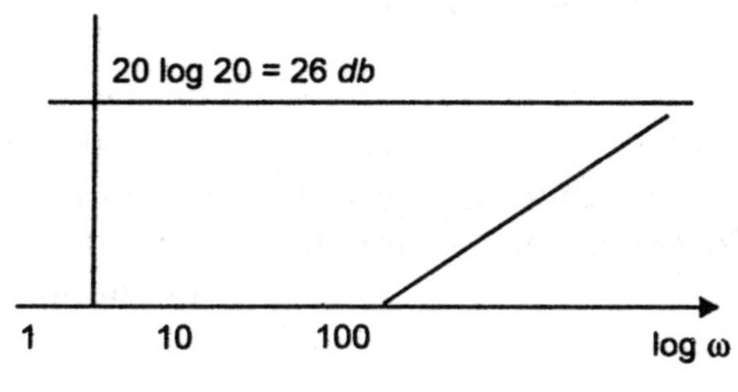


**Fig. 15.28(a)**

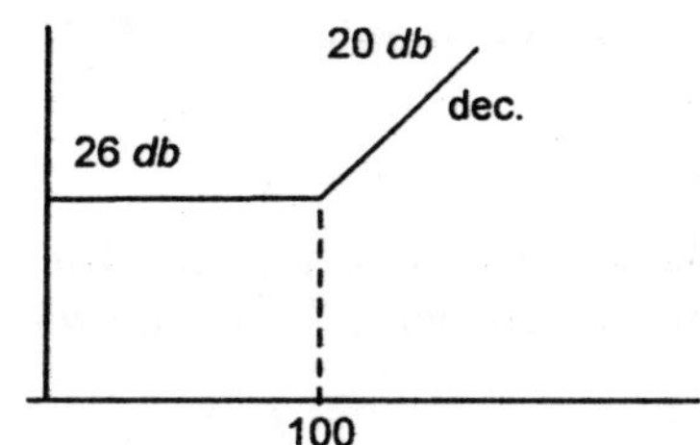


**Fig. 15.28(b)**

The Bode phase plot is the angle corresponding to the zero at $\omega = 100$. We, therefore, have $0°$ below $\omega = 10$, $90°$ above $\omega = 1000$ and $45°$ at $\omega = 100$, and a $45°$/decade slope for $10 < \omega < 1000$. The final plot is shown, in Fig. 15.29.

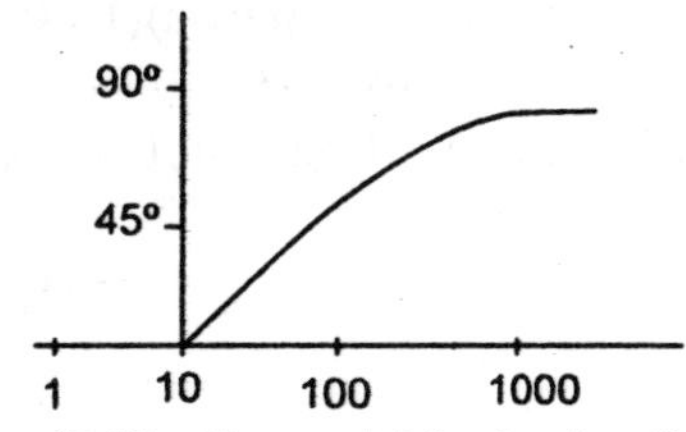


**Fig. 15.29**   Phase plot for the function

$$20 \left( 1 + \frac{s}{100} \right).$$

**PROBLEMS**

**15.1** Obtain the system function whose Bode plot is shown in Fig. p15.1

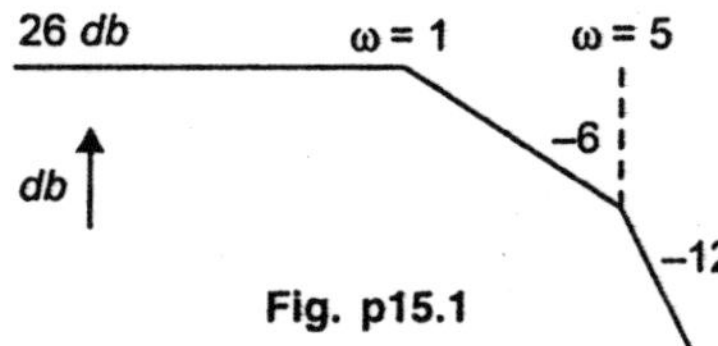


**Fig. p15.1**

**15.2** For the following system functions determine the crossover or corner frequencies.

(i)  $\quad H(s) = \dfrac{10}{s(1 + 0.5s)(1 + 0.1s)}$

(ii)  $H(s) = \dfrac{75(1 + 0.2s)}{s(s^2 + 16s + 100)}$

**15.3**  For the Bode plot shown in Fig. p15.3 obtain the system function and the values of $\omega_1$ and $\omega_2$

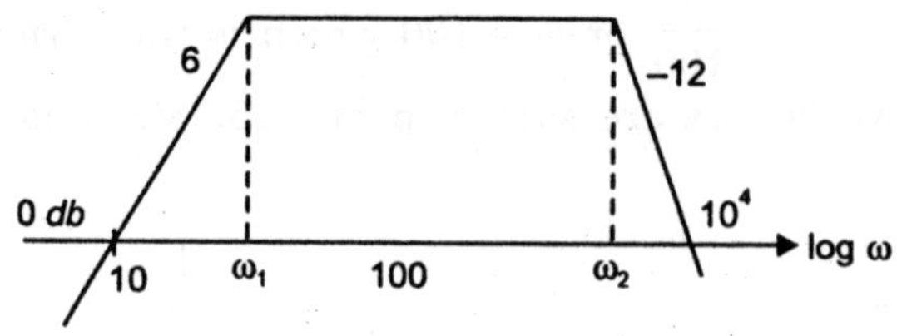


**Fig. p15.3**

**15.4**  Sketch the Bode plot for the following transfer function and determine in each case, the system gain $k$ for the gain crossover frequency to be 5 rad/sec.

(a)  $H(s) = \dfrac{Ks^2}{(1+0.2s)(1+0.02s)}$

(b)  $H(s) = \dfrac{Ke^{-0.1s}}{s(1+s)(1+0.1s)}$

**15.5**  For the Bode plot shown in Fig. p15.5 determine the system function

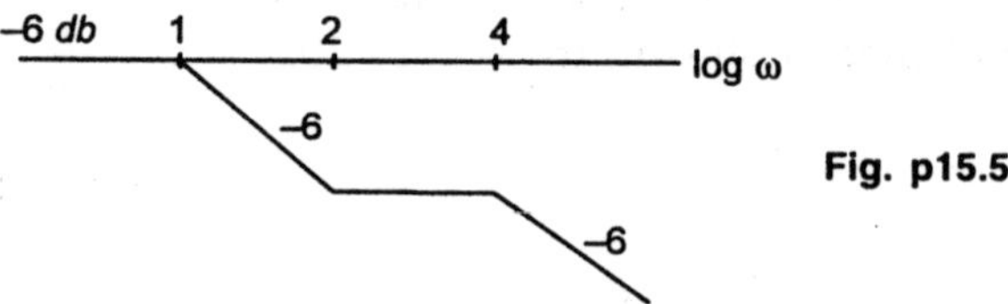


**Fig. p15.5**

**15.6**  Determine $H(j\omega)$ in *db* at $\omega = 10$ rad/sec if $H(s) =$

(a) $150/(s + 7)$    (b) $520s$    (c) $\dfrac{1}{(s+16)}$  and find $|H(j\omega)|$ if $H(j\omega)$ in *db* is

(d) $37db$    (e) $-65.4db$

# 16
# z-Transforms

## 16.1 INTRODUCTION

The method of Laplace has several advantages for the analysis of linear continuous-time systems as compared to the classical techniques of solving the integro-differential equations. An important advantage of Laplace transform method is that initial conditions can also be included in the expression for the Laplace transform of the output and the complete solution is obtained in one step. Similar advantages can be obtained in the analysis of discrete time systems using z-transform which changes the difference equations into algebraic equations in the transform domain.

Discrete time signals have values only at discrete instants of time. As a result these signals are represented by sequence of the form $x(n)$ where $n$ is an integer which may vary between $-\infty$ to $\infty$ . Such sequence of signals may be obtained either by sampling a continuous time signal or simply as a sequence of numbers that can be used for modeling signals internal to a computer.

In order to obtain a sequence of discrete time signals from a continuous time signal, the sampling rate is important. The sampling rate should be such that there is no loss of information contained in the continuous signal. This is possible only if the continuous time signal is band-limited i.e. it does not contain any component above a certain frequency, which can be achieved by passing the continuous signal through a low-pass filter.

The process of sampling can be considered as multiplication of the given continuous time signal with a periodic train of unit impulses with period $T$ i.e.

$$p(t) = \sum_{n=-\infty}^{\infty} \delta(t-nT) \qquad \qquad ...(16.1)$$

Fig. 16.1 (a) shows a typical band limited ($\omega_b$) signal and 16.1 (b) the desired amplitude spectrum of the sampled signal obtained from the band limited in (a). From Fig 16.1(b) it is clear that for proper sampling (to contain complete information in the original signal) the frequency of sampling i.e. $f_s = \dfrac{1}{T}$ should be such that

$$\omega_s > 2\omega_b$$

However, if $\omega_s < 2\omega_b$, there will be overlap between the successive samples and the information will be distorted. The sampling theorem is, therefore, stated as follows:

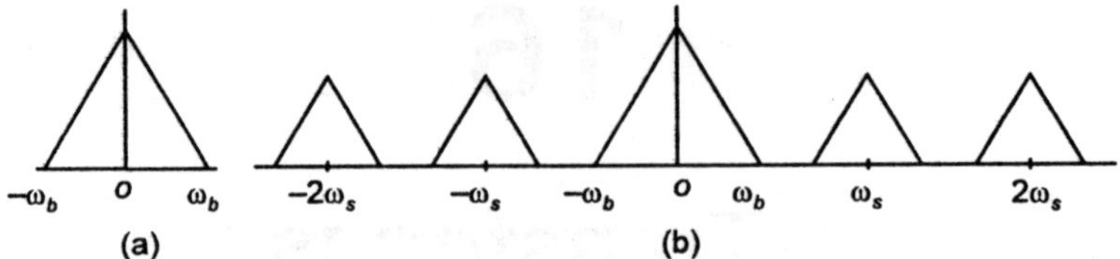

**Fig. 16.1**

A band limited signal having no frequency components above $\omega_b$ is completely described by uniformly spaced samples if the sampling frequency $\omega_s$ is greater than twice $\omega_b$.

If the frequency $\omega_s$ equals $2\omega_b$ it is called as the Nyquist rate.

However, it is to be noted that no signal is strictly band limited in practice (sharp cut offs are not possible). As a result it is not possible to sample a signal and then reconstruct the original signal from the sample without error. However, in most of the situations, the energy content of the signal is negligible beyond a certain frequency. Hence, this frequency may be considered as the bandwidth. Therefore, it is desirable to choose the sampling frequency higher than the Nyquist rate.

Regardless of the origin of these sequence signals, these may be expressed as

$$x = \{\text{- - - -}, \ x(-2), \ x(-1), \ x(0), \ x(1), \ x(2),\text{- - -}\} \qquad \text{...16.2}$$

## 16.2 DEFINITION OF z-TRANSFORM

If $x(n)$ is a sequence where $n$ may have value $-\infty$ to $\infty$, its z-transform is given as

$$X(z) = \sum_{n=-\infty}^{\infty} x(n) \, z^{-n} \qquad \text{...16.3}$$

Here $X(z)$ is known as the z-transform of the sequence $x(n)$

**Example 16.1:** Fig. E. 16.1 shows a limited sequence of some numbers. Determine its z-transform.

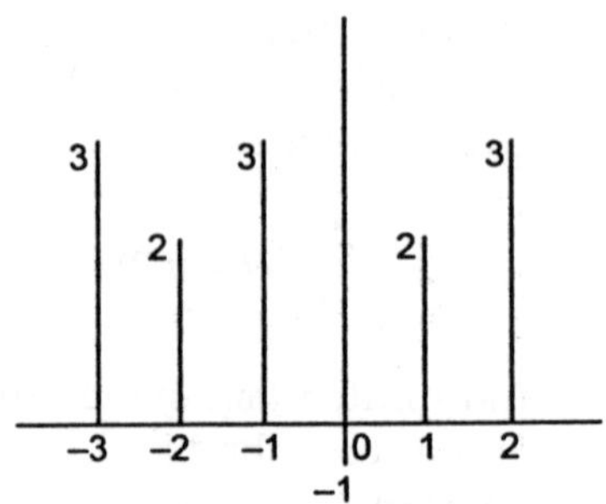

**Fig. E 16.1**

**Solution:** Using equation 16.3 the z-transform of the given sequence is

$$X(z) = 3z^3 + 2z^2 + 3z - 1 + 2z^{-1} + 3z^{-2} \qquad \text{...16.4}$$

From equation 16.4 it is clear that the exponent of $z$ is a "position marker" of the sequence. For example $z^2$ means it is a second element from the origin to the left. Similarly $z^{-k}$ means it is a $k$ th element of the sequence from the origin to the right.

Now suppose we multiply equation 16. 4 by $z$ we have

$$zX(Z) = 3z^4 + 2z^3 + 3z^2 - z + 2 + 3z^{-1} \qquad \text{...16.5}$$

Since the power of $z$ in each element of the sequence has increased by 1 which means the sequence has moved to left by one position as shown in Fig. E.16.2.

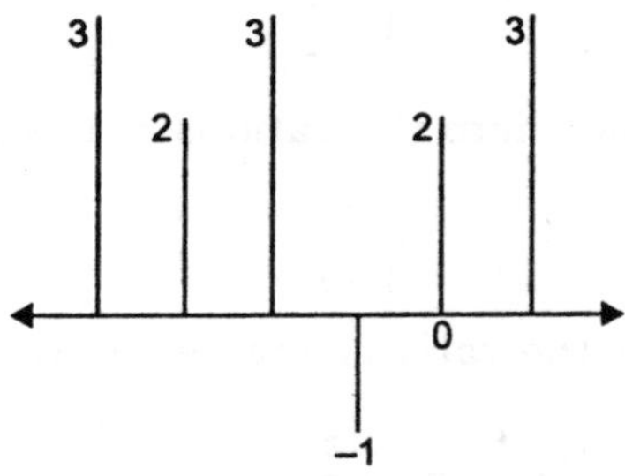

**Fig. E 16.2**  Effect of multiplying by $z$

Therefore, we may say that the result of multiplication by $z$ advances the sequence by one. Similarly multiplication of the sequence by $z^{-1}$ would move the sequence backward in time by one unit.

Equation 16.3 gives expression for bilateral $z$-transform. However, for causal functions the summation over the lower limits-$\infty$ to 0 does not exist and hence for a unilateral $z$-transform or $z$-transform for causal sequence the equation 16.3 becomes.

$$x(z) = \sum_{n=0}^{\infty} x(n)\, z^{-n} \qquad \text{...(16.6)}$$

## 16.3  SOME $z$-TRANSFORM PAIRS

### 1.  The constant sequence

Consider the constant sequence obtained by sampling a step function as shown in Fig. 16.2

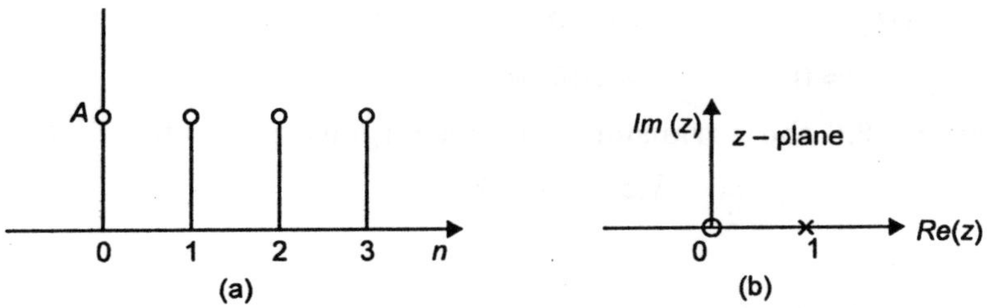

**Fig. 16.2**  (a) constant sequence (b) its $z$-transform

The sequence can be written as $x(n) = A$, $n = 0, 1, 2, \ldots\ldots$
Using equation 16.6 we have

$$x(z) = A + Az^{-1} + Az^{-2} + \cdots\cdots$$

$$= A\left[1 + \frac{1}{z} + \frac{1}{z^2} + --------\right]$$

If $z^{-1} < 1$ or $z > 1$
the sum of the infinite series is

$$X(z) = \frac{A}{1 - \dfrac{1}{z}}$$

(for the sum of an infinite series where the ratio is $< 1$ is given by

$$\frac{a}{1-r}$$

where '$a$' is the first term and $r$ the ratio of the geometric series)

Hence
$$X(z) = \frac{Az}{z-1} \qquad\qquad \ldots(16.7)$$

This shows that the $z$-transform of a constant series has a zero at $z = 0$ and a pole at $z = 1$.

However, in case of Laplace transform the Laplace transform of a step function has pole at $s = 0$.

For the constant sequence it has been fond that the series converges only if $|z| > 1$ which means the region of convergence for this case is outside the unit circle in the $z$-plane.

Also, it is to be noted that the pole of Laplace transform of the continuous time signal in the $s$-plane maps into the pole of the $z$-transform of the constant sequence in the $z$-plane. We would see later that this is true for other cases as well. The transformation between Laplace transform and $z$-transform is given by

$$z = e^{sT}$$

It is found, when $s = 0$, $z = 1$

### 2.  The Unit Impulse

The unit impulse in discrete time is expressed as

$$x(k) = 1 \qquad\qquad k = 0$$
$$= 0 \qquad\qquad \text{elsewhere} \qquad\qquad \ldots(16.8)$$

Using equation 16.6 the $z$-transform of unit impulse is given as

$$X(z) = x(k)\, z^{-n}$$
$$= 1\, z^{0} = 1$$

### 3.  The Exponential Sequence

Consider the sequence

$$x(n) = Ar^{n}$$

Such a sequence can be obtained by sampling an exponential function of time of the type

$$x(t) = Ae^{\alpha t}$$

where
$$r = e^{\alpha t}$$

Again, using equation 16.6 we have the $z$-transform of $x(n)$ as

$$X(z) = \sum_{n=0}^{\infty} A r^n \, z^{-n}$$

$$= \sum_{n=0}^{\infty} A \, (r \, z^{-1})^n$$

$$= A\left[1 + \frac{r}{z} + \frac{r^2}{z^2} + - - - - - -\right]$$

If $\left|\dfrac{r}{z}\right| \ll 1$

or $|z| \gg |r|$

$$X(z) = \frac{A}{1 - \dfrac{r}{z}} = \frac{Az}{z - r} \qquad \qquad ...(16.9)$$

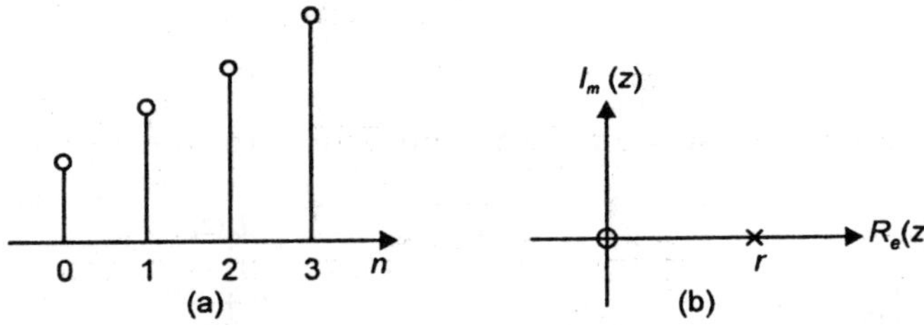


**Fig. 16.3** (a) The exponential sequence (b) Pole zero plot of (a)

Fig. 16.3 shows the exponential sequence and its pole-zero plot in the $z$-plane. For this sequence the convergence is obtained when $|z| > |r|$ i.e. it is obtained outside a circle of radius $|r|$ in the z-plane.

The Laplace transform of the continuous time function $x(t) = Ae^{\alpha t}$ is $\dfrac{A}{s - \alpha}$ and hence the pole of the $z$-transform of the sequence obtained from the continuous time function is located at $z = r$ which is the map of the pole at $s = \alpha$ in the s-plane of the Laplace transform. This is as per the mapping represented by

$$z = e^{sT} \qquad \qquad ...(16.10)$$

Further, a pole outside the unit circle i.e. $|z| > 1$ in the $z$-plane will represent the $z$-transform of a sequence that increases as $n$ is increased (see fig 16.3 as $n$ increases the magnitude of sequence element increases). This is true in case of Laplace transform also. A pole in the right half of the $s$-plane represents Laplace transform of a continuous time function that increases with time and eventually approaches infinity.

## 4. Sinusoidal Sequence

(a) $x(n) = A \cos \beta n$

This can be written as

$$x(n) = A \cdot \frac{e^{j\beta n} + e^{-j\beta n}}{2}$$

Let

$$r = e^{j\beta}$$

Hence, using the result of the previous article where if

$$x(n) = Ar^n$$

$$x(z) = \frac{Az}{z - r}$$

where

$$r = e^{\alpha T}$$

Hence

$$x(z) = \frac{1}{2}\left[\frac{Az}{z - e^{j\beta}} + \frac{Az}{z - e^{-j\beta}}\right]$$

$$= \frac{Az\,(z - \cos\beta)}{z^2 - 2z\cos\beta + 1} \qquad \qquad ...(16.11)$$

when $|z| > 1$

as $|r| = 1 = |e^{j\beta}|$

Fig 16.4 shows the cosine sequence and the pole zero plot for the $z$-transform

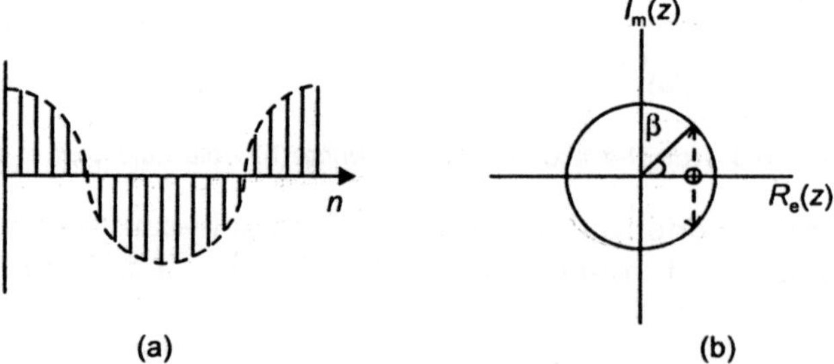

(a)         (b)

**Fig. 16.4** (a) The cosine sequence $x(n) = A \cos \beta n$ (b) Pole-Zero of $z$-transform of (a)

Similarly the $z$-transform of the sine sequence is evaluated as follows:

$$\mathscr{Z}\,[A \sin \beta n] = \mathscr{Z}\left[A \cdot \frac{e^{j\beta n} - e^{-j\beta n}}{2j}\right]$$

$$= \frac{A}{2j}\left[\frac{z}{z - e^{j\beta}} - \frac{z}{z - e^{-j\beta}}\right]$$

$$= \frac{Az \sin \beta}{z^2 - 2z\cos\beta + 1} \qquad \qquad ...(16.12)$$

for $|z| > 1$

Fig. 16.5 shows the sine sequence and its z-transform

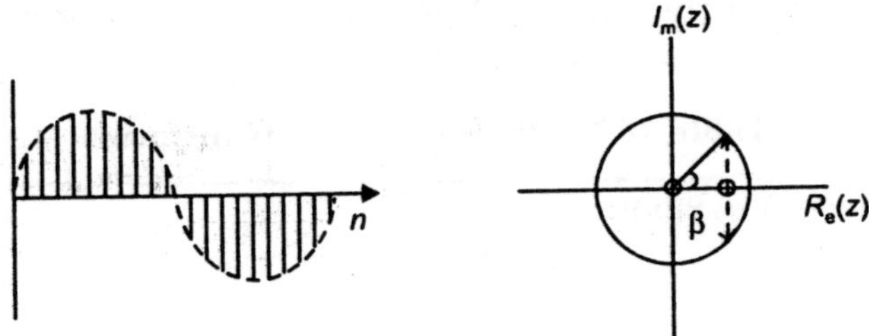

**Fig. 16.5** (a) sin sequence $x(n) = A \sin \beta n$ (b) Pole zero plot of z-transform of (a)

In case of Laplace transform of continuous time sine function the poles lie along the $j\omega$ axis whereas for z-transform the poles lie on a unit circle. This mapping is suggested by the equation.

$$z = e^{sT}$$

**Example 16.2:** Determine the z-transform of the sequence $x(n) = Ae^{-\alpha n} \cos \beta\, n$

**Solution:**

$$x(n) = A\, e^{-\alpha n} \cos \beta\, n$$

$$= A\, e^{-\alpha n}\, \frac{e^{j\beta n} + e^{-j\beta n}}{2}$$

$$= A\, \frac{e^{(-\alpha + j\beta)n} + e^{-(\alpha + j\beta)n}}{2}$$

$$\mathscr{Z}\,[x(n)] = \frac{A}{2}\left[\frac{z}{z - e^{(-\alpha + j\beta)}} + \frac{z}{z - e^{-(\alpha + j\beta)}}\right]$$

$$= \frac{A}{2}\, \frac{z^2 - z\, e^{-(\alpha + j\beta)} + z^2 - z\, e^{-(\alpha + j\beta)}}{z^2 - z\, e^{-(\alpha + j\beta)} - z\, e^{(-\alpha + j\beta)} + e^{-2\alpha}}$$

$$e^{-\alpha}\, e^{-j\beta} = e^{-\alpha}\, (\cos \beta - j \sin\beta)$$

$$e^{-\alpha}\, e^{j\beta} = e^{-\alpha}\, (\cos \beta + j \sin\beta)$$

$$\mathscr{Z}\,[x(n)] = \frac{A}{2}\, \frac{2z^2 - 2z\, e^{-\alpha} \cos\beta}{z^2 - 2z\, e^{-\alpha} \cos \beta + e^{-2\alpha}}$$

$$= \frac{A\,(z^2 - z\, e^{-\alpha} \cos \beta)}{z^2 - 2z\, e^{-\alpha} \cos\beta + e^{-2\alpha}}$$

$$= \frac{Az\,(z - e^{-\alpha}\,\cos\beta)}{z^2 - 2z\,e^{-\alpha}\,\cos\beta + e^{-2\alpha}} \qquad\qquad (16.13)$$

Table 16.1 gives some of the important properties of $z$-transform which can be used for obtaining $z$-transform of other sequence functions.

Table 16.1  Properties of z-transform

| | | |
|---|---|---|
| 1. | Linearity $\mathcal{X}\,[a_1 x_1\,(n)\,a_2\,x_2(n)]$ | |
| | $\qquad = a_1 x_1(z) + a_2 x_2\,(z)$ | ...(16.14) |
| 2. | Shift in time $\mathcal{X}\,[x(n+1)$ | |
| | $\qquad = z[x(z) - x(0)]$ | |
| | $\mathcal{X}\,[x(n+k) = z^k\,x(z) - zk\,x(0) - z^{k-1}\,x(1)\ .....zx(k-1)$ | ...(16.15) |
| 3. | Multiplication by $n$ | |
| | $\qquad \mathcal{X}\,[nx\,(n)] = -z\,\dfrac{d}{dz}\,x\,(z)$ | |
| | or $\qquad \mathcal{X}\,[n^m\,x\,(n) = (-z)^m\,\dfrac{d^m\,x\,(z)}{dz^m}$ | ...(16.16) |
| 4. | Multiplication by $r^n$ | |
| | $\qquad = \mathcal{X}\,[r^n\,x(n)] = x\left[\dfrac{z}{r}\right]$ | ...(16.17) |
| 5. | Convolution $\mathcal{X}\left[\Sigma x_1(k)\,x_2\,(n-k)\right]$ | |
| | $\qquad = x_1\,(\ )\,x_2\,(z)$ | ...(16.18) |
| 6. | Initial Value $\qquad x(0) = \underset{z\to\infty}{\mathrm{Lt}}\,\left[x(z)\right]$ | ...(16.19) |
| 7. | Final Value $\qquad x(\infty) = \underset{z\to 1}{\mathrm{Lt}}\,\left[(z-1)\,x(z)\right]$ | ...(16.20) |

We know that convolution property is very useful for finding out the response of a discrete system to an arbitrary input. If $h(n)$ represents the impulse response sequence of such a system and the input sequence is given by $x(k)$, then according to convolution property, the $z$-transform of the output sequence is obtained as

$$Y(z) = H(z)\,x(z) \qquad\qquad ...(16.21)$$

Suppose the impulse response of a system is a sequence $\{8, 4, 2, 1\}$ and if the input sequence is $\{2, 1\}$

then

$$x(z) = 2 + z^{-1}$$
$$H(z) = 8 + 4z^{-1} + 2z^{-2} + z^{-3}$$

and the $z$-transform of the output is given by

$$H(z)\,x(z) = (8 + 4z^{-1} + 2z^{-2} + z^{-3})\,(2 + z^{-1})$$
$$= 16 + 16z^{-1} + 8z^{-2} + 4z^{-3} + z^{-4}$$

Hence the output sequence is

$$\{16\ 16,\ 8,\ 4,\ 1\}$$

Conversely, if the input and output are given and it is desired to find what impulse response the system must have, we should divide the *z*-transform of the output by the *z*-transform of the input and expand the result as a polynomial in $z^{-1}$. The co-efficients is the result.

Let us solve some problems to illustrate the application of the *z*-transform properties given in table 16.2.

**Example 16.3:** Determine the solution of the given second order difference equation.

$$y(k + 2) + 3y(k + 1) + 2y(k) = u(k)$$

with initial conditions

$$y(0) = 3 \text{ and } y(1) = 2$$

**Solution:** Taking the *z*-transform of both sides of the equation and using the shifting property of *z*-transform, we have

$$z^2 y(z) - z^2 y(0) - zy(1) + 3zy(z)$$

$$- 3zy(0) + 2y(z) = u(z)$$

or

$$\{z^2 + 3z + 2\} y(z) = z^2 y(0) + zy(1) + 3zy(0) + u(z)$$

or

$$= 3z^2 + 2z + 9z + u(z)$$

$$= 3z^2 + 11z + U(z)$$

Hence

$$Y(z) = \frac{3z^2 + 11z + U(z)}{z^2 + 3z + 2}$$

Hence $y(k)$ can be obtained by taking the inverse *z*-transform of the above expression.

**Example 16.4:** Determine the *z*-transform of the ramp sequence $x(n) = An$ using the third property of *z*-transform given in table 16.1.

**Solution:** The *z*-transform of sequence of a step function of magnitude $A$ is from equation 16.7

$$\mathscr{Z}[A] = \frac{Az}{z-1}$$

Hence

$$\mathscr{Z}[An] = -z \frac{d}{dz}\left[\frac{Az}{z-1}\right]$$

$$= -zA\left[\frac{(z-1) - z.1}{(z-1)^2}\right]$$

$$= -Az \frac{-1}{(z-1)^2}$$

$$= \frac{Az}{(z-1)^2} \qquad \qquad ...(16.22)$$

**Example 16.5:** Determine the $z$-transform of the sequence $x(n) = An\, r^n$ using property 4 of the $z$-transform in Table 16.1.

**Solution.** From equation 16.22

$$\mathscr{X}\,[An] = \frac{Az}{(z-1)^2}$$

and using property 4

$$\mathscr{X}\,[r^n\, x(n)] = x\!\left(\frac{z}{r}\right)$$

i.e. replace $z$ by $\dfrac{z}{r}$

$$\mathscr{X}\,(An\, r^n] = \frac{A\,z/r}{\left(\dfrac{z}{r}-1\right)^2}$$

$$= \frac{A\,z\,r}{(z-r)^2}$$

**Example 16.6:** Determine the $z$-transform of the sequence $x(n) = Ar^n \cos \beta n$ using property 4 of the $z$-transform in Table 16.1.

**Solution:** From equation 16.11

$$\mathscr{X}\,[A \cos \beta\, n)] = \frac{Az\,(z-\cos\beta)}{z^2 - 2z \cos\beta + 1}$$

Using property 4 the

$$\mathscr{X}\,[A\, r^n \cos \beta\, n)] = \frac{A\,\dfrac{z}{r}\left(\dfrac{z}{r}-\cos\beta\right)}{\dfrac{z^2}{r^2} - 2\dfrac{z}{r}\cos\beta + 1}$$

$$= \frac{A\,(z - r \cos\beta)}{z^2 - 2\,zr \cos\beta + r^2}$$

Similarly $\mathscr{X}\,[A\, r^n \sin \beta\, n]$ can be obtained by using equation 16.12 and property 4 of the $z$-transform and we then have

$$\mathscr{X}\,[A\, r^n \sin \beta\, n) = \frac{Azr \sin \beta}{z^2 - 2zr \cos \beta + r^2}$$

Table 16.2 gives some of the $z$-transform pairs which can be used to obtain $z$-transform of the function without deriving them when using $z$-transform for the solution of discrete time system.

**Table 16.2  Some *z*-transforms pairs**

| | | |
|---|---|---|
| 1. | $\delta(n)$ | $1$ |
| 2. | A (Step) | $\dfrac{Az}{z-1}$ |
| 3. | $A\,r^n$ (exponential) | $\dfrac{Az}{z-r}$ |
| 4. | $A\,n$ (ramp) | $\dfrac{Az}{(z-1)^2}$ |
| 5. | $A\cos\beta\,n$ | $\dfrac{Az(z-\cos\beta)}{z^2-2z\cos\beta+1}$ |
| 6. | $A\sin\beta\,n$ | $\dfrac{Az\sin\beta}{z^2-2z\cos\beta+1}$ |
| 7. | $A\,n\,r^n$ | $\dfrac{Arz}{(z-r)^2}$ |
| 8. | $A\,n^2$ | $\dfrac{Az(z+1)}{(z-1)^3}$ |
| 9. | $A\,r^n\cos\beta\,n$ | $\dfrac{Az(z-r\cos\beta)}{z^2-2zr\cos\beta+r^2}$ |
| 10. | $A\,r^n\sin\beta\,n$ | $\dfrac{Azr\sin\beta}{z^2-2zr\cos\beta+r^2}$ |

## 16.4 INVERSION OF *z*-TRANSFORM

In the analysis of discrete time systems, the use of *z*-transform requires the inversion of *z*-transform to obtain the corresponding sequence. Following methods are commonly employed for the purpose:

### 16.4.1 Partial Fraction Expansion

The idea behind this procedure is same as that in the Laplace transform i.e. we obtain poles of $x(z)$ by factoring the denominator polynomial. Since the co-efficients of the polynomial here are all real, these poles will either be real or conjugate pairs if complex. Next we obtain the partial fraction expansion of $x(z)$ in order to identify the sequence for each term in the expansion i.e. we obtain the inverse *z*-transform. If we look in table 16.2 we find that most of the transforms have $z$ as multiplier in the numerator. It will, therefore, he desirable to obtain partial fraction expansion of $\dfrac{x(z)}{z}$ rather than $x(z)$.

The procedure to be followed for partial fraction expansion is exactly identical to the one followed in Laplace transformation. After the residues for various poles are obtained, each can be multiplied by $z$ to conform to the standard form in Table 16.2. The inverse transform of each term can then be obtained from the tables 16.2 and 16.3.

**Table 16.3  Partial fraction equivalents for causal sequence.**

| $X(z)$ | $x(n)$ |
|---|---|
| $\dfrac{Az}{z-r}$ | $A\,r^n\ n \geq 0$ |
| $\dfrac{(C+jD)\,z}{z-r\,e^{j\phi}} + \dfrac{(C-jD)\,z}{z-r\,e^{-j\phi}}$ | $2r^n\,(C\cos n\phi - D\sin n\phi)\ \ n \geq 0$ |
| $\dfrac{Az}{(z-r)^m}$ | $\dfrac{An\,(n-1)\ldots(n-m+2)}{(m-1)!\ r^{m-1}}\,r^n\ \ n \geq m-1$ |

The following examples will further illustrate the procedure:-

**Example 16.7:** Determine the inverse of the $z$-transform

$$x\,(z) = \frac{2z^2 + 3z}{(z-1)\,(z-0.5)}$$

**Solution:**

$$\frac{x(z)}{z} = \frac{2z+3}{(z-1)(z-0.5)}$$

$$= \frac{A}{z-1} + \frac{B}{z-0.5}$$

$$A = \operatorname*{Lt}_{z \to 1}\left[(z-1)\,\frac{x(z)}{z}\right] = \operatorname*{Lt}_{z \to 1}\frac{2z+3}{z+0.5} = \frac{5}{0.5}$$

$$= 10$$

$$B = \operatorname*{Lt}_{z \to 0.5}\frac{2z+3}{z-1} = \frac{1+3}{-0.5} = -8$$

Hence

$$x(z) = \frac{10z}{z-1} - \frac{8z}{z-0.5}$$

$$= 10 - A\,r^n$$

$$= 10 - 8(0.5)^n$$

**Example 16.8:** Determine the inverse of the following $z$-transform.

$$x(z) = \frac{2z}{z^2 + 1} + \frac{2z}{z^2 - 1}$$

**Solution:**

$$\frac{x(z)}{z} = \frac{2}{z^2 + 1} + \frac{2}{z^2 - 1}$$

$$= \frac{2}{(z+j)\,(z-j)} + \frac{2}{(z+1)\,(z-1)}$$

For the first term

$$\frac{A}{z+j} + \frac{B}{z-j}$$

$$A = \operatorname*{Lt}_{z \to -j} \frac{2}{z-j} = \frac{2}{-j2} = j$$

and hence $B = -j$

For the second term

$$\frac{A}{z+1} + \frac{B}{z-1}$$

$$A = \operatorname*{Lt}_{z \to -1} \frac{2}{z-1} = -1$$

and hence $B = 1$

$$\frac{x(z)}{z} = \frac{j}{z+j} - \frac{j}{z-j} + \frac{1}{z-1} - \frac{1}{z+1}$$

Hence

$$x(z) = \frac{jz}{z+j} - \frac{jz}{z-j} + \frac{z}{z-1} - \frac{z}{z+1}$$

$$= j\left[\frac{z}{z+j} - \frac{z}{z-j}\right] + \frac{z}{z-1} - \frac{z}{z+1}$$

$$x(n) = \frac{2\,jj\left[e^{-n\pi/2} - e^{n\pi/2}\right]}{2j} + (-1)^n - 1$$

$$= 2\sin\frac{n\pi}{2} + (-1)^n - 1 \quad \textbf{Ans.}$$

### 16.4.2 The Method of Long Division

In this method we divide the numerator polynomial by the denominator to obtain an expansion in the form of an infinite series with negative powers of $z$ (for a causal sequence). The resulting power series is

$$x(z) = \sum_{n=0}^{\infty} a_n z^{-n}$$

We know that the powers of $z$ are indicators as the position markers of series, therefore,

$$x_n = a_n \quad n = 0, 1, 2, \cdots$$

The following examples illustrate the method.

**Example 16.9:** Determine the inverse $z$-transform of the function.

$$x(z) = \frac{2z^2 - 4z}{z^3 - 3z^2 + 5z - 4}$$

**Solution:** The long division is performed as follows:

$$z^3 - 3z^2 + 5z - 4 \;\Big|\; \begin{array}{l} 2z^2 - 4z \\[4pt] 2z^2 - 6z + 10 - \dfrac{8}{z} \end{array} \;\Big|\; \dfrac{2}{z}$$

$$\begin{array}{l} 2z - 10 + \dfrac{8}{z} \\[8pt] 2z - 6 + \dfrac{10}{z} - \dfrac{8}{z^2} \end{array} \;\Big|\; \dfrac{2}{z^2}$$

$$\begin{array}{l} -4 - \dfrac{2}{z} + \dfrac{8}{z^2} \\[8pt] -4 + \dfrac{12}{z} - \dfrac{20}{z^2} + \dfrac{16}{z^3} \end{array} \;\Big|\; \dfrac{-4}{z^3}$$

$$\begin{array}{l} -\dfrac{14}{z} + \dfrac{28}{z^2} - \dfrac{16}{z^3} \\[8pt] -\dfrac{14}{z} + \dfrac{42}{z^2} - \dfrac{70}{z^3} + \dfrac{56}{z^4} \end{array} \;\Big|\; -\dfrac{14}{z^4}$$

$$-\dfrac{14}{z^2} + \dfrac{54}{z^3} - \dfrac{56}{z^4}$$

The process can continue indefinitely. However, we close it here and we find $x(n)$ for a few $n$.

$$x(0) = 0 \text{ as } z \text{ is absent}$$

i.e. the constant term is absent

$$x(1) = 2, \quad x(2) = 2$$
$$x(3) = -4 \quad x(4) = -14 \cdot \cdot \cdot \cdot \cdot \cdot$$

This method does not give the inverse transform in a closed form. It is not very convenient if the value of $x(n)$ is desired for large $n$. However, this difficulty is overcome by the partial fraction method discussed earlier.

**Example 16.10:** Determine the inverse $z$-transform of the function.

$$x(z) = \frac{2z^2 - 3z}{z^3 - 3z^2 + 6z - 4}$$

**Solution:** Using partial fraction expansion method.
The denominator

$$z^3 - 3z^2 + 6z - 4 \text{ has a zero at } z = 1, \text{ hence}$$
$$z^2(z - 1) - 2z(z - 1) + 4(z - 1)$$

Hence the denominator is

$$(z - 1)(z^2 - 2z + 4)$$

The roots of $z^2 - 2z + 4$ are

$$\frac{2 \pm \sqrt{4 - 16}}{2} = 1 \pm j\sqrt{3}$$

Hence
$$\frac{x(z)}{z} = \frac{2z - 3}{(z - 1)(z - 1 - j\sqrt{3}\,(z - 1 + j\sqrt{3})}$$

$$= \frac{A}{z - 1} + \frac{B}{z - 1 - j\sqrt{3}} + \frac{C}{z - 1 + j\sqrt{3}}$$

$$A = \operatorname*{Lt}_{z \to 1} \frac{2z - 3}{z^2 - 2z + 4} = \frac{2 - 3}{1 - 2 + 4} = -\frac{1}{3}$$

$$B = \operatorname*{Lt}_{z \to 1 + j\sqrt{3}} \frac{2z - 3}{(z - 1)(z - 1 + j\sqrt{3})}$$

$$= \frac{2 + j2\sqrt{3} - 3}{(1 + j\sqrt{3} - 1)(1 + j\sqrt{3} - 1 + j\sqrt{3})}$$

$$= \frac{-1 + j2\sqrt{3}}{j\sqrt{3} \cdot 2j\sqrt{3}} = \frac{-1 + j2\sqrt{3}}{-6}$$

$$= \frac{1}{6} - j\frac{1}{\sqrt{3}}$$

Similarly
$$C = \frac{1}{6} + j\frac{1}{\sqrt{3}}$$

$$\frac{x(z)}{z} = -\frac{1}{3} \cdot \frac{1}{z - 1} + \frac{\dfrac{1}{6} - j\dfrac{1}{\sqrt{3}}}{z - 2\,e^{j\pi/3}} + \frac{\dfrac{1}{6} + j\dfrac{1}{\sqrt{3}}}{z - 2\,e^{-j\pi/3}}$$

Now using the relation:

If
$$X(z) = \frac{(C + jD)z}{z - r\,e^{j\phi}} + \frac{(C - jD)z}{z - r\,e^{-j\phi}}$$

then $x(n) = \quad 2r^n\,[C \cos n\phi - D \sin n\phi]$

$$n \geq 0$$

Here $r = 2$, $C = \dfrac{1}{6}$ $D = \dfrac{1}{\sqrt{3}}$ and $\phi = -\dfrac{\pi}{3}$

Hence,
$$x(n) = -\frac{1}{3}(1)^n + 2.2^n\left[\frac{1}{6}\cos\frac{n\pi}{3} + \frac{1}{\sqrt{3}}\sin\frac{n\pi}{3}\right]$$

$$= \frac{1}{-3}(1)^n + 2^{n+1}\left[\frac{1}{6}\cos\frac{n\pi}{3} + \frac{1}{\sqrt{3}}\sin\frac{n\pi}{3}\right]$$

**Example 16.11:** Determine the inverse transform of

$$X(z) = \frac{3z^4 - 10z^3 - 5.75z^2}{(z-1)(z-0.5)(z^2+0.25)}$$

**Solution:**

$$\frac{X(z)}{z} = \frac{3z^3 - 10z^2 - 5.75z}{(z-1)(z-0.5)(z+j0.5)(z-j0.5)}$$

$$= \frac{A}{z-1} + \frac{B}{(z-0.5)} + \frac{C}{z+j0.5} + \frac{D}{(z-j0.5)}$$

$$A \underset{z\to 1}{\text{Lt}} = \frac{3-10-5.75}{0.5(1+j0.5)(1-j0.5)}$$

$$= \frac{-12.75}{0.5(1+0.5^2)} = -\frac{12.75}{0.5(1.25)} = -20.4$$

$$B = \underset{z\to 0.5}{\text{Lt}} = \frac{3\times 0.5^3 - 10(0.5)^2 - 5.75(0.5)}{0.5(0.5+j0.5)(0.5-j0.5)}$$

$$= \frac{0.375 - 2.5 - 2.875}{0.25} = 20$$

$$D = \underset{z\to j0.5}{\text{Lt}} = \frac{3(j0.5)^3 - 10(j0.5)^2 - 5.75(j0.5)}{(j0.5-1)(j0.5-0.5)(j1.0)}$$

$$= \frac{2.5 - j3.25}{0.75 + j0.25}$$

$$= \frac{2.5(1-j1.3)(3-j1)}{0.25(3+j1)(3-j1)} = 1.7 - j4.9$$

Hence $C = 1.7 + j4.9$

Hence

$$\frac{X(z)}{z} = -\frac{20.4}{z-1} + \frac{20}{z-0.5} + \frac{1.7+j4.9}{z+j0.5} + \frac{1.7-j4.9}{z-j0.5}$$

$$X(z) = \frac{20.4z}{z-1} + \frac{20z}{z-0.5} + \frac{(1.7+j4.9)z}{z+0.5\,e^{j\pi/2}} + \frac{1.7-j4.9}{z-e^{-j\pi/2}}$$

Here     $C = 1.7$     $D = 4.9$

$\qquad\quad r = 0.5 \qquad\quad = -\pi/2$

Hence $X(t) = (0.5)^n \left[ 3.4 \cos\dfrac{n\pi}{2} + 9.8 \sin\dfrac{n\pi}{2} \right] + 20.4 - 20\,(0.5)^n$ **Ans.**

## 16.5  SOLUTION OF DIFFERENCE EQUATIONS

In case of Laplace transform we convert the linear differential equation into a set of algebraic equations and solve for the network. Similarly for $z$-transform the difference equations are converted into algebraic equations in $z$ variables by using the shifting property of the $z$-transform as given in Table 16.1. This means multiplication by $z$ is equivalent to a forward shift. The complete solution however, can be obtained by finding the inverse $z$-transform of the function.

We take an example to illustrate further the procedure involved.

**Example 16.12:** A second order difference equations is given as

$$y(n + 2) - 0.6\,y(n + 1) + 0.25\,y(n) = u(n)$$

with initial conditions

$$y(0) = 3 \quad y(1) = 2$$

Determine $y(n)$ the output sequence if the input sequence is

$$u(n) = 8(0.5)^n$$

**Solution:** The $z$-transform of

$$u(n) \text{ is } \frac{8z}{z - 0.5}$$

$$z^2 y(z) - z^2\,y(0) - zy(1) - 0.6\,zy(z) + 0.6zy(0) + 0.25y(z) = \frac{8z}{z - 0.5}$$

Substituting values for the initial conditions we have

$$z^2 y(z) - 3z^2 - 2z - 0.6\,zy(z) + 1.8z + 0.25\,y(z) = \frac{8z}{z - 0.5}$$

$$(z^2 - 0.6z + 0.25)\,y(z) = \frac{8z}{z - 0.5} + 3z^2 - 0.2z$$

$$(z^2 - 0.6z + 0.25)\,y(z) = \frac{8z + (z - 0.5)(3z^2 - 0.2z)}{z - 0.5}$$

$$Y(z) = \frac{8z + 3z^3 - 0.2\,z^2 - 1.5z^2 + 0.1z}{(z - 0.5)(z^2 - 0.6z + 0.25)}$$

or
$$\frac{Y(z)}{z} = \frac{8.1 - 1.7z + 3z^2}{(z - 0.5)(z^2 - 0.6z + 0.25)}$$

Now roots of quadratic polynomial in the denominator are

$$\frac{0.6 \pm \sqrt{0.36 - 1}}{z} = 0.3 \pm j0.4$$

$$= 0.5 \, e^{j0.927}$$

$$\frac{Y(z)}{z} = \frac{A}{z - 0.5} + \frac{B}{z - 0.3 - i0.4} + \frac{C}{z - 0.3 + j0.4}$$

$$A = \underset{z \to 0.5}{\text{Lt}} = \frac{8.1 - 1.7(0.5) + 0.75}{0.25 - 0.3 + 0.25}$$

$$= \frac{8.1 - 0.85 + 0.75}{0.2} = \frac{8.85 - 0.85}{0.2} = 40$$

$$B = \frac{8.1 - 1.7(0.3 + j0.4) + 3(0.3 + j0.4)^2}{(0.3 + j0.4 - 0.5)(0.3 + j0.4 - 0.3 + j0.4)}$$

$$= \frac{8.1 - 0.51 - j0.68 + 3(0.09 - 0.16 + j0.24)}{(-0.2 + j0.4)(j0.8)}$$

$$= \frac{40}{z - 0.5} + \frac{18.45 - j9.22}{z - 0.5 \, e^{j0.927}} + \frac{18.45 + j9.22}{z - 0.5 \, e^{-j0.927}}$$

or
$$Y(z) = \frac{40z}{z - 0.5} + \frac{(18.45 - j9.22)z}{z - 0.5 \, e^{j0.927}} + \frac{(18.45 + j\,9.22)z}{z - 0.5 \, e^{-j0.927}}$$

Hence

$Y(n) = 40(0.5)^n + (0.5)^n (36.9 \cos 0.927n - 18.45 \sin 0.927n)$ **Ans.**

**Example 16.13:** Solve the homogeneous difference equation

$$y(n) - 3\,y(n - 1) + 2\,y(n - 2) = 0$$

**Solution:** The corresponding characteristic equation is,

$$\lambda^2 - 3\lambda + 2 = 0 \qquad \text{or} \qquad (\lambda - 2)(\lambda - 1) = 0$$

or
$$\lambda = 2 \qquad \text{and} \qquad \lambda = 1$$

Therefore, the solution is,

$$y(n) = C_1(2)^n + C_2(1)^n = C_1(2)^n + C_2$$

The unknowns $C_1$ and $C_2$ can be evaluated if the initial conditions are known.

Next we solve a homogeneous equation with repeated roots. If $\lambda_1$ is a root of multiplicity $K$, the term corresponding to $\lambda_1$, in the solution will then have the form as

$$C_1\lambda_1^n + C_2 n\, \lambda_1^n + C_3 n^2\, \lambda_1^n + \dots + C_k n^{k-1}\, \lambda_1^n$$

**Example 16.14:** Solve the homogeneous difference equation

$$y(n) - y(n-1) - 5\,y(n-2) - 3\,y(n-3) = 0$$

**Solution:** The corresponding characteristic equation is,

$$\lambda^3 - \lambda^2 - 5\lambda - 3 = 0$$

One of the roots is $\lambda = -1$     Hence

$$\lambda^2\,(\lambda+1) - 2\lambda(\lambda+1) - 3(\lambda+1) = 0$$

or $$(\lambda+1)\,(\lambda^2 - 2\lambda - 3) = 0$$

or $$(\lambda+1)(\lambda+1)(\lambda-3) = 0$$

or $$(\lambda+1)^2(\lambda-3) = 0$$

Here $\lambda = -1$ is a root of multiplicity 2

Hence the solution is

$$C_1(-1)^n + C_2 n(-1)^n + C_3(2)^n$$

or $$(C_1 + C_2)(-1)^n + C_3(2)^n \qquad \textbf{Ans.}$$

**Example 16.15:** Solve the difference equation

$$y(n) + 3y(n-1) + 2y(n-2) = u(n)$$

with initial conditions $y(0) = 1$ and $y(-1) = 0$

Using (i) classical technique (ii) z-transform

**Solution:** The characteristic equation is,

$$\lambda^2 + 3\lambda + 2 = 0$$

or $$(\lambda+1)(\lambda+2) = 0$$

Therefore, the roots of the characteristic equation are

$$\lambda_1 = -1 \qquad \text{and} \qquad \lambda_2 = -2$$

and the homogeneous solution is

$$y_c(n) = C_1(-1)^n + C_2(-2)^n$$

The forcing function $u(n)$ is a discrete time unit step function and it can be written as

$$u(n) = n^0$$

Therefore, the particular solution is

$$y_p(n) = D_0$$

Substituting the particular solution in the equation,

$$A_0 + 3A_0 + 2A_0 = 1$$

or $$A_0 = \frac{1}{6}$$

Therefore, the complete solution is $y = y_c + y_p$ and is given as

$$y(n) = C_1(-1)^n + C_2(-2)^n + \frac{1}{6}$$

Now at $\qquad n = 0 \qquad y(0) = 1$

or $\qquad 1 = C_1 + C_2 + \frac{1}{6} \qquad$ or $\qquad C_1 + C_2 = \frac{5}{6}$

and at $\qquad n = -1 \qquad y(-1) = 0$

$$y(-1) = C_1(-1)^{-1} + C_2(-2)^{-1} + \frac{1}{6}$$

$$0 = -C_1 - \frac{1}{2} C_2 + \frac{1}{6}$$

or $\qquad C_1 + \frac{C_2}{2} = \frac{1}{6}$

and $\qquad C_1 + C_2 = \frac{5}{6}$

or $\qquad \dfrac{C_2}{2} = \dfrac{2}{3}$ or $\quad C_2 = \dfrac{4}{3}$

and hence $\qquad C_1 = \dfrac{5}{6} - \dfrac{8}{6} = -\dfrac{3}{6} = -\dfrac{1}{2}$

The complete solution is, therefore,

$$y(n) = -\frac{1}{2}(-1)^n + \frac{4}{3}(-2)^n + \frac{1}{6} \qquad \textbf{Ans.}$$

(ii) Using $z$-transform

Equation 16.15 gives $z$-transform of a sequence which has a shift to the left of the axis $t = 0$. However, the sequence given in the problem has a shift to the right of the axis $t = 0$. The $z$-transform of such a sequence

$$\mathscr{Z} f(n - k) = z^{-k} F(z) + z^{-k+1} f(-1) + z^{-k+2} f(-2) + \ldots$$
$$+ z^{-1} f(-k + 1) + f(-k) \ldots \ldots \qquad (16.23)$$

However, if $\qquad f(n) = 0 \qquad$ for $\qquad n < 0$

$$\mathscr{Z} f(n - k) = z^{-k} F(z) \ldots \ldots \qquad (16.24)$$

Since in the given problem $y(-1) = 0$, we make use of equation (16.24) for taking $z$-transform of the given difference equation.

$$y(z) + 3z^{-1}y(z) + 2z^{-2}y(z) = \frac{z}{z-1}$$

$$\left(1 + \frac{3}{z} + \frac{2}{z^2}\right)y(z) = \frac{z}{z-1}$$

$$\left(\frac{z^2 + 3z + 2}{z^2}\right)y(z) = \frac{z}{z-1}$$

or
$$y(z) = \frac{z^3}{(z-1)(z+1)(z+2)}$$

or
$$\frac{y(z)}{z} = \frac{z^2}{(z-1)(z+1)(z+2)}$$

$$= \frac{A}{z-1} + \frac{B}{z+1} + \frac{C}{z+2}$$

$$A = \operatorname*{Lt}_{z\to 1}\frac{y(z)}{z}(z-1) = \operatorname*{Lt}_{z\to 1}\frac{z^2}{(z+1)(z+2)} = \frac{1}{2\times 3} = \frac{1}{6}$$

$$B = \operatorname*{Lt}_{z\to -1}\frac{y(z)}{z}(z+1) = \operatorname*{Lt}_{z\to -1}\frac{z^2}{(z-1)(z+2)} = \frac{1}{(-2)\,(1)} = -\frac{1}{2}$$

$$C = \operatorname*{Lt}_{z\to -2}\frac{y(z)}{z}(z+2) = \operatorname*{Lt}_{z\to -2}\frac{z^2}{(z-1)(z+1)} = \frac{4}{(-3)(-1)} = \frac{4}{3}$$

$$\frac{y(z)}{z} = \frac{1/6}{z-1} - \frac{1/2}{z+1} + \frac{4/3}{z+2}$$

Hence
$$y(n) = \frac{1}{6} - \frac{1}{2}(-1)^n + \frac{4}{3}(-2)^n \qquad \textbf{Ans}$$

## 16.6 DISCRETE-TIME CONVOLUTION

We have already discussed convolution integral for continuous time signal under article 7.9 of the book. Here we have found that for such systems convolution integral enables to evaluate the response of a network to an arbitrary input in terms of the impulse response of the network. In other words any input convolved with the unit impulse response gives the output corresponding to the input.

Following the same procedure we develop the convolution technique for discrete time systems.

In discrete time system a unit impulse is defined as
$$\delta(n) = 1 \text{ for } n = 0$$

$$= 0 \text{ for } n \neq 0 \qquad \qquad ...(16.25)$$

Similarly a delta founction of magnitude $m_j$ and occuring at $j$th instant is given as

$$m_j\, \delta(n - j) = m_j \quad \text{for} \quad n = j$$
$$= 0 \quad \text{for} \quad n \neq j \qquad \qquad ...(16.26)$$

Suppose the impulse response due to $\delta(n)$ of a system is given as $h(n)$, then its response to $m_j\, \delta(n - j)$ will be $m_j\, h(n - j)$. Suppose we want to find out the response of the same system when the input sequence is $v(j)$. In continuous time signals we take integral of the product of the impulse response and the new input signal. Here since, it is a discrete time signal we take summation over the product of the new input signal (sequence here) and impulse response signal i.e. the response $y(n)$ due to $v(j)$ is given as

$$y(n) = \sum_{j=0}^{\infty} v(j)h(n - j) \qquad \qquad ...(16.27)$$

Here equation (16-27) is known as discrete time convolution or the convolution sum between the input sequence $v(j)$ and the system impulse response $h(n)$ and is symbolically represented as

$$y(n) = v(j) * h(n) \qquad \qquad ...(16.28)$$

**Example 16.16:**

The impulse response of a discrete time system is given by

$$h(n) = \frac{1}{2^n}$$

Determine the response of the system when the input is a sequence of signals $v(j)$ described by

$$v(j) = 1 \text{ for } j = 0, 1, 2, 3, 4 \text{ and}$$
$$= 0 \text{ for } j \geq 5$$

**Solution:**

Using equation (16.27), the response is given as

$$y(n) = \sum_{j=0}^{\infty} v(j)h(n - j)$$

Here $j$ varies from 0 to 4. Expanding the above equation for these values of $j$

$$y(n) = v(0)\, h(n) + v(1)h(n - 1) + v(2)h(n - 2) + v(3)h(n - 3) + v(4)h(n - 4)$$
$$= h(n) + h(n - 1) + h(n - 2) + h(n - 3) + h(n - 4)$$

Starting with $n = 0$, the response $y(n)$ for the sequence output can be calculated. Since we are considering causal systems

$$y(n) = 0 \text{ for } n < 0$$

For $\qquad\qquad n = 0, \quad y(0) = v(0)\, h(0) + v(1)\, h(-1) \,.....$

$$= 1\frac{1}{2^0} = 1 \qquad \text{as } h(-1) \text{ etc are zeros}$$

$$n = 1, \quad y(1) = h(1) + h(0) = \frac{1}{2} + 1 = \frac{3}{2}$$

$$n = 2, \quad y(2) = h(2) + h(1) + h(0) = \frac{1}{4} + \frac{1}{2} + 1 = \frac{7}{4}$$

$$n = 3, \quad y(3) = h(3) + h(2) + h(1) + h(0) = \frac{1}{2^3} + \frac{1}{2^2} + \frac{1}{2} + 1$$

$$n = 5, \quad y(n) = h(5) + h(4) + h(3) + h(2) + h(1)$$

$$= \frac{1}{2^5} + \frac{1}{2^4} + \frac{1}{2^3} + \frac{1}{2^2} + \frac{1}{2}$$

$$n = 6, \quad y(6) = h(6) + h(5) + h(4) + h(3) + h(2)$$

$$= \frac{1}{2^6} + \frac{1}{2^5} + \frac{1}{2^4} + \frac{1}{2^3} + \frac{1}{2^2}$$

and so on.

Fig 16.6 shows the plots of the sequences $h(n)$, $v(j)$ and $y(n)$

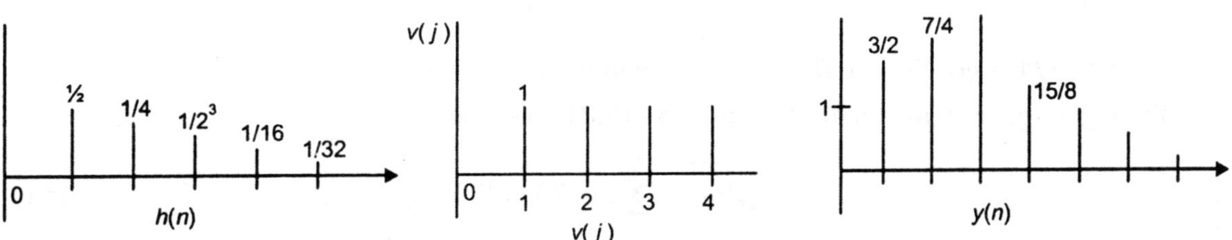



**Fig 16.6** Plots of sequences (i) $h(n)$  (ii) $v(j)$ (iii) $y(n)$

It is to be noted that the impulse response of a system is defined only for a relaxed system i.e., a system without any initial condition and hence the response obtained using convolution technique is the zero state response. If the system has initial conditions, the response due to these conditions alone should be obtained by solving the system homogeneous difference/differential equation. The response is known as zero-input response. The sum of the zero state and zero-input response is the net result.

## 16.7 THE PULSE TRANSFER FUNCTION

The system characteristics of a continuous time open loop system as shown is Fig. 16.7(a) is represented as

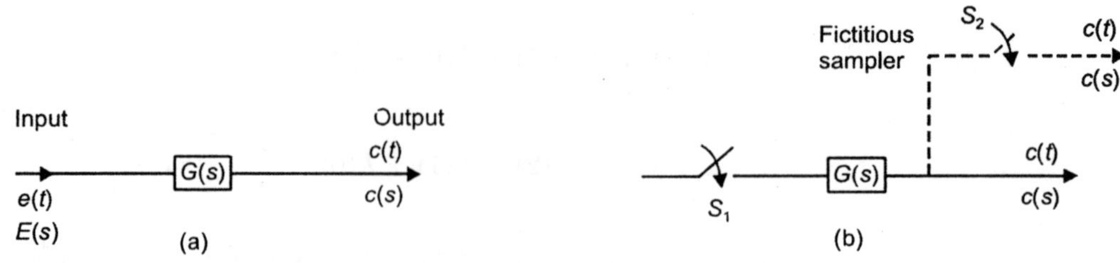



**Fig 16.7** (a) Continuous system (b) Sampled system or discrete system

$$G(s) = \frac{C(s)}{E(s)} \qquad \qquad ...(16.29)$$

For a discrete time system Fig. 16.7 (b) shows a network $G$ which is connected to a sampler $s_1$ with sampling period $T$. Suppose that in Fig. 16.7(b) a unit step function is applied at the input terminals at $t = 0$ and the sampling switch $s_1$ is closed and opened only once for a very short duration $T$, then input to the network $G$ is just a unit impulse occuring at $t = 0$. The output $c(t)$ when input is an impulse, is the impulse response of $G(s)$ or

$$c(t) = g(t) \text{ Impulse response of } G \qquad \qquad ...(16.30)$$

If a fictitious sampler $s_2$ with the same sampling period as that of $s_1$ i.e. $T$ is connected at the output terminal, the output of the switch $s_2$ is

$$C^*(t) = g^*(t) = \sum_{n=0}^{\infty} C(nT)\, \delta\, (t - nT) \qquad \qquad ...(16.31)$$

Here $C(nT) = g(nT)$ is defined as the weighting sequence of $G$

Taking Laplace transform of equation (16.31) we get

$$G^*(s) = \sum_{n=0}^{\infty} g(nT)\, e^{-nTs} \qquad \qquad ...(16.32)$$

Which is the pulsed transfer function of the system $G$.

The transformation between Laplace transform and $z$-transform is given by

$$z = e^{sT} \qquad \qquad ...(16.33)$$

Substituting this relation in equation (16.32) we have

$$G^*(s) = \sum_{n=0}^{\infty} g(nT)\, z^{-n} \qquad \qquad ...(16.34)$$

Since the right hand side of equation(16.34) is the $z$-transform of weighting sequence of $G$ hence the left hand side of equation (16.34) represents $G(z)$. Hence equation (16.34) is rewritten as

$$G(z) = \sum_{n=0}^{\infty} g(nT)\, z^{-n} \qquad \qquad ...(16.35)$$

and this is known as pulse transfer function of the open loop systems $G$. In terms of $z$ transform

$$G(z) = \frac{C(z)}{E(z)} \qquad ...(16.36)$$

Here the $z$ transform of the output $c(z)$ specifies the values of the output $c(t)$ at the sampling instants.

Alternative if we start with equation 16.27 which is reproduced here

$$y(n) = \sum_{j=0}^{\infty} v(j)\, h(n-j) \qquad ...(16.27)$$

The $z$ transform of $y(n)$ is

$$Y(z) = \sum_{n=0}^{\infty} y(n) z^{-n} \qquad ...(16.37)$$

Hence

$$Y(z) = \sum_{n=0}^{\infty}\left[\sum_{j=0}^{\infty} v(j) h(n-j)\right] z^{-n} \qquad ...(16.38)$$

$$= \sum_{n=0}^{\infty} z^{-n}\left[\sum_{j=0}^{\infty} v(j) h(n-j)\right]$$

Interchanging the order of summation

$$Y(z) = \sum_{j=0}^{\infty} v(j) \sum_{n=0}^{\infty} h(n-j) z^{-n} \qquad ...(16.39)$$

Using the shifting property for sequence on the right side of $t = 0$ axis, the second summation in equation (16.39) can be rewritten as

$$\sum_{n=0}^{\infty} z^{-n} h(n-j) = \mathcal{Z} h(n-j) = z^{-j}\, \mathcal{Z} h(n) = z^{-j} H(z)$$

Where $H(z)$ is $z$ transform of $h(n)$, the impulse reponse of the system $H$
Substituting this result in equation (16.39) we have

$$Y(z) = \sum_{j=0}^{\infty} v(j) z^{-j} H(z)$$

$$= V(z)\, H(z) \qquad ...(16.40)$$

From equation (16.40) it can be seen that the convolution of two discrete time sequences results in the multiplication of their $z$-transforms.

Using equation (16.40) we have

$$H(z) = \frac{Y(z)}{V(z)} \qquad ...(16.41)$$

Here $H(z)$ which is identical to $G(z)$ is the $z$-transfer fucntion of the discrete time system and is the ratio of the $z$-transform of the output to the $z$-transform of the input when the system is initially relaxed.

The discreate time system is usually expressed in terms of difference equations for which $z$-transform method is used for its solution.

Suppose the discrete time system is described by a general $r$th order difference equation.

$$y(n) + a_1\, y(n-1) + a_2\, y(n-2) + \ldots + ar Y\,(n-r)$$
$$= b_0\, v(n) + b_1 v(n-1) + \ldots + b_m v(n-m) \qquad\qquad \ldots(16.42)$$

Where $m \le r$. Taking $z$-transform of both sides of equation (16.42) after using the shifting properly

$$Z\, f(n-k) = z^{-n}\, F(z)$$

we get

$$[1 + a_1\, z^{-1} + a_2\, z^{-2} + \ldots + a_r\, z^{-r}]\, Y(z) = [b_0 + b_1\, z^{-1} + \ldots + b_m z^{-m}]\, V(z)$$

or

$$H(z) = \frac{Y(z)}{V(z)} = \frac{b_0 + b_1 z^{-1} + \ldots + b_m z^{-m}}{1 + a_1 z^{-1} + a_2 z^{-2} + \ldots + a_r z^{-r}} \qquad\qquad \ldots(16.43)$$

Equation (16.43) thus gives a very straight forward method of obtaining $z$-transfer function from a given difference equation describing a system.

**Example 16.17:**   The $z$- transfer function of a system is given as

$$H(z) = \frac{1}{z^2 - \dfrac{1}{2} z + \dfrac{1}{18}}$$

Determine it's (i) delta response and (ii) discrete unit step response assuming the system to be initially relaxed

**Solution:**

(i) Delta response under this condition

$$H(z) = Y(z) = \frac{1}{z^2 - \dfrac{z}{2} + \dfrac{1}{8}}$$

To obtain $h(n)$ we take inverse transform of $H(z)$

Now
$$H(z) = \frac{1}{\left(z - \dfrac{1}{3}\right)\left(z - \dfrac{1}{6}\right)}$$

Using partial fraction expansion we have

$$H(z) = \frac{A}{z - 1/3} + \frac{B}{z - 1/6}$$

$$A = \underset{z \to 1/3}{Lt}\, H(z)\left[z - \frac{1}{3}\right] = \underset{z \to 1/3}{Lt}\, \frac{1}{z - 1/6} = 6$$

$$B = \underset{z \to 1/6}{Lt}\, H(z)\left[z - \frac{1}{6}\right] = \underset{z \to 1/6}{Lt}\, \frac{1}{z - 1/3} = -6$$

Hence
$$H(z) = \frac{6}{z - 1/3} - \frac{6}{z - 1/6}$$

Since the numaretor is not having the term '$z$' in the expression for $H(z)$, the sequence is to be shifted to the right of $t = 0$ axis by one and hence

$$h(n) = 6\left(\frac{1}{3}\right)^{n-1} - 6\left(\frac{1}{6}\right)^{n-1} \qquad \textbf{Ans.}$$

**(ii) Discrete unit step**

Now
$$Y(z) = V(z)H(z)$$

For discrete unit step
$$V(z) = \frac{z}{z-1}$$

Hence
$$Y(z) = \frac{z}{z-1} \cdot \frac{1}{\left(z-\frac{1}{3}\right)\left(z-\frac{1}{6}\right)}$$

or
$$\frac{Y(z)}{z} = \frac{A}{z-1} + \frac{B}{z-1/3} + \frac{C}{z-1/6}$$

$$A = \underset{z \to 1}{Lt}\; \frac{\frac{Y(z)}{z}(z-1)}{} = \frac{1}{\left(z-\frac{1}{3}\right)\left(z-\frac{1}{6}\right)} = \frac{1}{\frac{2}{3}\times\frac{5}{6}} = \frac{9}{5}$$

$$B = \underset{z \to 1/3}{Lt}\; \frac{\frac{Y(z)}{z}\left(z-\frac{1}{3}\right)}{} = \frac{1}{(z-1)\left(z-\frac{1}{6}\right)} = \frac{1}{\left(-\frac{2}{3}\right)\times\frac{5}{6}} = -9$$
$$\scriptstyle Lt-z\to 1/3$$

$$C = \underset{z \to 1/6}{Lt}\; \frac{\frac{Y(z)}{z}(z-1/6)}{} = \frac{1}{(z-1)\left(z-\frac{1}{3}\right)} = \frac{1}{\left(-\frac{5}{6}\right)\times\left(\frac{1}{6}\right)} = +\frac{36}{5}$$

or
$$Y(z) = \frac{\frac{9}{5}z}{z-1} + \frac{9z}{z-\frac{1}{3}} + \frac{\frac{36}{5}z}{z-\frac{1}{6}}$$

or
$$y(n) = \frac{9}{5}u(n) - 9\left(\frac{1}{3}\right)^{n} + \frac{36}{5}\left(\frac{1}{6}\right)^{n} \qquad \textbf{Ans.}$$

**·Example 16.18:**

Determine the unit step response in the previous example with initial conditions
$$y(0) = 0 \text{ and } y(1) = 1$$

**Solution:**

In order to include the initial condition we write the difference equation from the z-transfer function of the system

$$Y(z) = \frac{1}{z^2 - \dfrac{z}{2} + \dfrac{1}{18}} V(z)$$

or

$$\left(z^2 - \frac{z}{2} + 18\right) Y(z) = V(z) = \frac{z}{z-1}$$

Taking inverse z-transform on both the sides

$$y(n+2) - \frac{1}{2} y(n+1) + \frac{1}{18} y(n) = u(n)$$

Using equation (16.15) the z-transform with non-zero initial conditions we get

$$z^2 y(z) - z^2\, y(0) - zy(1) - \frac{1}{2}\, [zy(z) - zy(0)] + \frac{1}{18} y(z) = \frac{z}{z-1}$$

Rearranging

$$\left(z^2 - \frac{1}{2}z + \frac{1}{18}\right) y(z) = z^2 y(0) + zy(1) - \frac{1}{2} zy(0) + \frac{z}{z-1}$$

$$= z + \frac{z}{z-1}$$

$$Y(z) = \frac{z}{z^2 - \dfrac{1}{2}z + \dfrac{1}{18}} + \frac{z}{(z-1)\left(z^2 - \dfrac{1}{2}z + \dfrac{1}{18}\right)}$$

$$= Y_1(z) + Y_2(z)$$

It can be seen that $Y_1(z)$ is the response due to initial conditions alone whereas $Y_2(z)$ is due to a forcing function only when the initial conditions are zero.

$$\frac{Y_1(z)}{z} = \frac{1}{\left(z - \dfrac{1}{3}\right)\left(z - \dfrac{1}{6}\right)} = \frac{A}{z - \dfrac{1}{3}} + \frac{B}{z - \dfrac{1}{6}}$$

$$A = \operatorname*{Lt}_{z \to \frac{1}{3}} \frac{Y_1(z)}{z}\left(z - \frac{1}{3}\right) = \operatorname*{Lt}_{z \to \frac{1}{3}} \frac{1}{\dfrac{1}{3} - \dfrac{1}{6}} = 6$$

and

$$B = \operatorname*{Lt}_{z \to \frac{1}{6}} \frac{Y_1(z)}{z}\left(z - \frac{1}{6}\right) = \frac{1}{z - \dfrac{1}{3}} = -6$$

Hence
$$Y_1(z) = \frac{6z}{z - \dfrac{1}{3}} - \frac{6z}{z - \dfrac{1}{6}}$$

Hence
$$Y_1(n) = 6\left(\frac{1}{3}\right)^n - 6\left(\frac{1}{6}\right)^n$$

However, the inverse of $Y_2(z)$ is already known from the previous example and is given as

$$Y_2(n) = \frac{9}{5}u(n) - 9\left(\frac{1}{3}\right)^n + \frac{36}{5}\left(\frac{1}{6}\right)^n$$

$$Y(n) = Y_1(n) + Y_2(n) = \frac{9}{5}u(n) - 3\left(\frac{1}{3}\right)^n + \frac{6}{5}\left(\frac{1}{6}\right)^n \qquad \textbf{Ans.}$$

## 16.8  INITIAL VALUE

If
$$Y(z) = Z\,[y(n)] \text{ then}$$
$$y(0) = \underset{z \to \infty}{Lt}\, Y(z) \qquad\qquad ...(16.44)$$

This result is used for evaluating the $y(0)$ directly from $Y(z)$ without evaluating the inverse $z$-transform.

The proof is given here

By definition
$$Y(z) = y(0) + y(1)z^{-1} + y(2)z^{-2} + \,......$$

Hence as $z \to \infty$, all terms except $y(0)$ vanish in limit and hence equation (16.44) is satisfied.

### 16.8.1 Final Value

If
$$Y(z) = Z[y(n)] \text{ then}$$
$$\underset{n \to \infty}{Lt}\, y(n) = \underset{z \to 1}{Lt}\, [(z-1)Y(z)] \qquad\qquad ...(16.45)$$

This property of evaluation of final value holds good provided that all the poles of $Y(z)$ are inside the unit circle in the $z$-plane, with the possible exception of a simple pole at $z = 1$

The proof is given here:

Consider the $z$-transform of $y(n + 1) - y(n)$ using the shifting property of equation (16.15) we have

$$Z\,[y(n + 1) - y(n)] = [z\,y(z) - z\,y(0)] - Y(z)$$

$$= \underset{k \to \infty}{Lt} \sum_{n=0}^{k} [y(n+1) - y(n)]z^{-n}$$

which is rearranged as

$$(z - 1)\ Y(z) - y(0) = \underset{k \to \infty}{Lt} \sum_{n=0}^{k} [y(n+1) - y(n)]z^{-n}$$

Taking the limit as $z \to 1$ on both sides, we have

$$\underset{z \to \infty}{Lt}\ (z-1)Y(z) = y(0) + [y(1) - y(0)] + [y(2) - y(1)]$$

$$+ \ ..... \ + [y(k) - y(k-1)] = \underset{k \to \infty}{Lt}\ y(k)$$

It is to be noted that the final value will be zero if all poles of $Y(z)$ are inside the unit circle whereas it will be infinite if any pole is outside the unit circle. If the poles are located on the unit circle the value will be indeterminate with the exception of a simple pole at $z = 1$ when the final value will be a constant. Even with all these limitation, the theorem is extensively used in the analysis of discrete time control systems.

**Example 16.19:**

Determine the initial and final values of

$$y(n) = \frac{2(z-0.5)}{z(z-1)(z-0.9)}$$

**Solution:**

(i)

$$y(0) = \underset{z \to \infty}{Lt}\ Y(z) = \frac{2(z-0.5)}{(z-1)(z-0.9)} = \frac{2\left(1 - \dfrac{0.5}{z}\right)\dfrac{1}{z}}{\left(1-\dfrac{1}{z}\right)\left(1-\dfrac{0.9}{z}\right)}$$

$$= 0 \quad \textbf{Ans.}$$

Final value

$$y(\infty) = \underset{z \to 1}{Lt}\ [(z-1)\ Y(z)] = \underset{z \to 1}{Lt} = \frac{(z-1).2(z-0.5)}{(z-1)(z-0.9)}$$

$$= \frac{2(1-0.5)}{(1-0.9)} = \frac{1}{0.1} = 10 \quad \textbf{Ans.}$$

**Example 16.20:**

Determine the initial and final values of $x\ (n)$ if

$$x(z) = \frac{10z(z-0.2)(z-0.6)}{(z-1)(z^2 - 0.6z + 0.4)}$$

**Solution:**

Now

$$x(0) = \underset{z \to \infty}{Lt}\ x(z) = \underset{z \to \infty}{Lt} = \frac{10\left(1-\dfrac{0.2}{2}\right)\left(1-\dfrac{0.6}{z}\right)}{\left(1-\dfrac{1}{z}\right)\left(1-\dfrac{0.6}{z}+\dfrac{0.4}{z^2}\right)} = 10$$

$$x\ (\infty) = \underset{z \to 1}{Lt}\ [(z-1)\ x(z)] = \frac{10z(z-0.2)(z-0.6)}{z^2 - 0.6z + 0.4}$$

$$= \frac{10 \times 0.8 \times 0.4}{1 - 0.6 + 0.4} = 4 \quad \textbf{Ans.}$$

## 16.9 STABILITY OF LINEAR DISCRETE TIME SYSTEMS

We know that in linear continuous time system, the system is said to be stable if and only if all the poles of its transfer function are located in the left half of the $s$-plane as with this situation all the components of the complete response of the system will either decay with time ($e^{-at}$ terms) or remain bounded for all possible bounded inputs. Similarly, a discrete time system is said to be stable if and only if all the poles of its transfer function are strictly inside the unit circle in the $z$-plane as only in that case all the components of the complete response will either decay with time or remain bounded for all  possible bounded input sequences.

The stability criteria for a linear discrete time system are summerised as follows:

A discrete time system is stable if and only if

(i)  All the poles of its transfer function lie strictly inside the unit circle of the $z$-plane.

(ii) Its impulse response $h(n)$ satisfies the following properties

$$\mathop{Lt}_{n \to \infty} |h(n)| = 0$$

$$\sum_{n=0}^{\infty} |h(n)|^p < \infty \quad for \ \ p \geq 1$$

(iii) The magnitude of every root of its characteristic polynomial is less than unity.

$$\boxed{\textbf{PROBLEMS}}$$

**16.1**  The impulse response of a discrete time system is given by [8, 4, 2, 1]. If the input sequence is [2, 1] determine output sequence.

$$\textbf{Ans. } [16, 16, 8, 4, 1]$$

**16.2**  Fig P16.2 shows finite number of discrete time signals. Determine its $z$-transform

**Ans.** $Y(z) = z^4 - 2z^3 + 2z^2 + z + 1 - 2z^{-1} + 3z^{-2}$

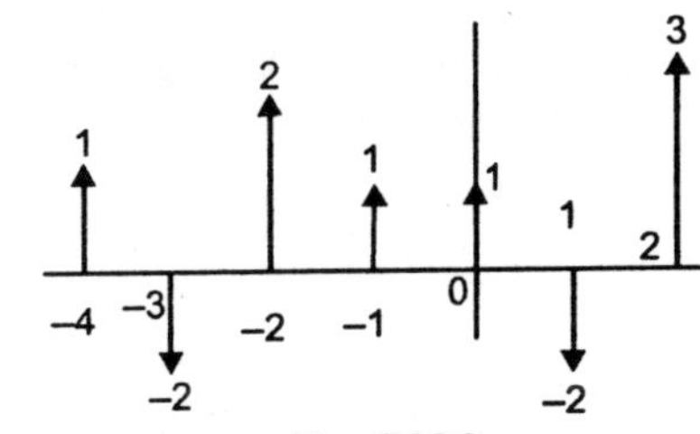

Fig. P16.2

**16.3**  Determine the $z$-transform of the sequence $x(n) = A\ e^{-\alpha n} \sin \beta n$

$$\textbf{Ans. } X(z) = \frac{A\, z \sin\beta}{z^2 - 2ze^{-\alpha} \cos\beta + e^{-2\alpha}}$$

**16.4**  Determine the $z$-transforms of the following sequences

(i)  $x(n) = An$                               (ii) $x(n) = Ar^n$

(iii)  $x(n) = An\,r^n$

(iv)  $x(n) = (n + 1)^2$

(v)  $x(n) = (n - 1)n$

**Ans.** (i) $\dfrac{Az}{(z-1)^2}$  (ii) $\dfrac{Az}{z-r}\,|z|>|r|$  (iii) $\dfrac{Arz}{(z-r)^2}$

(iv) $\dfrac{z(z^2+z+1)}{(z-1)^3}$  (v) $\dfrac{2z}{(z-1)^3}$

**16.5** Determine the inverse of the following $z$-transforms. Assume the sequence to be casual

(i)  $Y(z) = \dfrac{3z^2 - 5z}{z^3 - 4z^2 + 6z - 4}$

(ii)  $Y(z) = \dfrac{3z^4 - 10z^3 - 5.75z^2}{(z-1)(z-0.5)(z^2+0.25)}$

(iii)  $Y(z) = \dfrac{2z^2 + 3z}{(z-1)(z-0.5)}$

(iv)  $Y(z) = \dfrac{2z}{z^2+1} + \dfrac{2z}{z^2-1}$

(v)  $Y(z) = \dfrac{z^3 + 5z^2 - 7z}{(z-1)(z^2+1)}$

**Ans.** (i) $Y(n) = 0.5\,(2)^n - \left(\sqrt{2}\right)^n \left[0.5\cos\dfrac{n\pi}{4} - 2.5\sin\dfrac{n\pi}{4}\right]$

(ii)  $Y(n) = 0.5^n\left(3.4\cos\dfrac{n\pi}{2} + 9.8\sin\dfrac{n\pi}{2}\right) - 20.4 + 20(0.5)^n$

(iii)  $Y(n) = 10 - 8\,(0.5)^n$

(iv)  $2\sin\dfrac{n\pi}{2} - 1 + (-1)^n$

(v)  $1.5\cos\dfrac{n\pi}{2} + 6.5\sin\dfrac{n\pi}{2} - 0.5$

**16.6** Determine the impulse response sequence of a discrete time system with transfer function

$$H(z) = \dfrac{8z}{(z-0.5)(z-0.7)}$$

**Ans.** $h(n) = 40(0.7)^n - 40(0.5)^n$

**16.7** The impulse response sequence of a discrete time system is given by

$$h(n) = 20(0.6)^n\cos(0.4n)$$

Determine transfer function of the system (ii) the response of the system when the input $u(n) = 1$

**Ans.** (i) $\dfrac{2z(z - 0.6\cos 0.4)}{z^2 - 1.2z\cos 0.4 + 0.36}$

(ii)  $3.512 - (0.6)^n\,(1.125\cos 0.4n - 1.834\sin 0.4n)$

**16.8** Solve the difference equation given below

$$y(n) + 0.6y(n-1) + 0.08y(n-2) = 4$$

subject to the initial condition $y(0) = 2$ and $y(1) = 5$

$$\textbf{Ans. } \frac{50}{21} + \frac{259}{21}(-0.2)^n - \frac{267}{21}(-0.4)^h$$

**16.9** Assuming the system to be relaxed, determine its transfer function when it is represented by the following difference equation

$$y(n+2) - 0.6y(n+1) + 0.4y(n) = 2u(n+1) + 3u(n)$$

$$\textbf{Ans. } \frac{2z+3}{z^2 - 0.6z + 0.4}$$

**16.10** Assuming the system to be initially relaxed detemine its transfer function when it is represented by the followig difference equation

$$y(n) - y(n-1) + 0.5y(n-2) = 2u(n-1) + 3u(n-2)$$

Also determine $y(n)$ if $u(n) = 1$

$$\textbf{Ans. (i) } \frac{2z+3}{z^2 - z + 0.5} \qquad \text{(ii) } 10 - 2^{n/2}\left\{10\cos\frac{n\pi}{4} + 6\sin\frac{n\pi}{4}\right\}$$

**16.11** Determine the initial and final values of $y(n)$ if

(i) $Y(z) = \dfrac{20z(z-0.2)(z-0.5)}{(z-1)(z^2 - 0.6z + 0.4)}$

(ii) $Y(z) = \dfrac{3z^2 + 5z + 7}{(z-1)^2(z^2 - 1.2z + 0.8)}$

(iii) $Y(z) = \dfrac{30z^4 + 21z^3 + 13z^2 + 15z}{(z-0.5)(z-0.4)(z^2 - 1.4z + 0.65)}$

(iv) $Y(z) = \dfrac{2z^3 - 4z^2 + 5z}{(z-1)(z^2 - 1.2z + 0.6)}$

$$\textbf{Ans. (i) } y(0) = 20,\ y(\infty) = 10 \quad \text{(ii) } y(0) = 7.5,\ y(\infty) = \infty$$
$$\text{(iii) } y(0) = 30,\ y(\infty) = 0 \quad \text{(iv) } y(0) = 2,\ y(\infty) = 7.5$$

**16.12** The *z*-transform of a causal $x(n)$ is given by $x(z) = \dfrac{3z}{4z^2 - 2z + 1}$ Determine the *z*-transform of $y(n) = n\,x(n)$

$$\textbf{Ans. } \frac{12z^3 - 3z}{(4z^2 - 2z + 1)^2}$$

# APPENDIX A

# Matrices

## A.1  INTRODUCTION

Matrices play a very important role in the analysis of complex engineering multi-dimensional problems. With the advent of computers, matrices have found a much better place for the analysis of such systems. It is impossible to visualize beyond three dimensional problem. However, matrices help us to understand the problem beyond three dimensions. In a multi-dimensional process, it is very simple to understand the effect of a particular parameter on the overall process with the help of matrices. Therefore, matrices is a very important mathematical tool for solving linearised model of a complex physical process.

## A.2  DEFINITIONS

*Matrix:* A matrix is an ordered rectangular array of elements which may be real numbers, complex numbers, functions or operators. The matrix

$$
A = \begin{bmatrix}
a_{11} & a_{12} & \cdots\cdots & a_{1n} \\
a_{21} & a_{22} & \cdots\cdots & a_{2n} \\
\vdots & & & \\
a_{m1} & a_{m2} & \cdots\cdots & a_{mn}
\end{bmatrix}
\qquad\qquad ...(A.1)
$$

is a rectangular array of $mn$ elements.

In the number $a_{ij}$ representing a typical element, the first subscript (here $i$) denotes the row and the second subscript (here $j$) denotes the column occupied by the element.

Such an array is normally enclosed in square brackets as shown and not by vertical line. A matrix is one entity. The matrix defined above has $m$ rows and $n$ columns and is said to be a rectangular matrix of order $m \times n$. However, when $m = n$ i.e. the number of rows equals the number of columns, the matrix is said to be a square matrix of order $n$.

*Transposed matrix:* The transpose of a matrix $A$ denoted by $A^T$ is a matrix formed by interchanging the rows and columns of $A$. Thus if $A = [a_{ij}]$ then $A^T = [a_{ji}]$. For example if $A$ is the matrix of $m \times n$ order, $A^T$ will be

$$A^T = \begin{bmatrix} a_{11} & a_{21} & a_{m1} \\ a_{12} & a_{22} & a_{m2} \\ \vdots & \vdots & \vdots \\ a_{1n} & a_{2n} & a_{mn} \end{bmatrix} \qquad ...(A.2)$$

and hence $A^T$ is an $n \times m$ matrix.

A matrix of order $m \times 1$ has one column and is known as a column matrix or column vector. e.g. a column vector is written as

$$a = \begin{bmatrix} a_1 \\ a_2 \\ \vdots \\ a_m \end{bmatrix} \qquad ...(A.3)$$

This could also be written as

$$[a_1 \quad a_2 \quad ........... \quad a_m]^T$$

The transpose sign can be omitted if braces are used instead of square brackets, as given here

$$\{a_1 \quad a_2 \quad ........... \quad a_m\} \qquad ...(A.4)$$

A matrix of order $1 \times m$ has one row only and hence it is known as row matrix or row vector and is written as

$$a = [a_1 \quad a_2 \quad ........... \quad a_m] \qquad ...(A.5)$$

*Diagonal matrix:* The diagonal matrix is a square matrix whose elements off the main diagonal are all zeros ($a_{ij} = 0$ for $i \neq j$) e.g.

$$A = \begin{bmatrix} a_{11} & 0 & & \\ 0 & a_{22} & ......... & 0 \\ 0 & 0 & ......... & a_{nn} \end{bmatrix} \qquad ...(A.6)$$

**Unit or Indentity matrix:** A unit matrix usually denoted by either $I$ or $U$ is a diagonal matrix whose diagonal elements are all equal to unity e.g.

$$I = \begin{bmatrix} 1 & 0 & 0 & ......... & 0 \\ 0 & 1 & 0 & ......... & 0 \\ \vdots & & & & \\ 0 & 0 & & ......... & 1 \end{bmatrix} \qquad ...(A.7)$$

**Null matrix:** A null matrix usually denoted as 0 is matrix whose elements are all equal to zero.

$$\begin{bmatrix} 0 & 0 \\ 0 & 0 \end{bmatrix} \qquad \qquad ...(A.8)$$

**Scalar matrix:** A diagonal matrix whose diagonal elements are all equal is called a scalar matrix. Thus a scalar matrix must be of the form.

$$S = kI = \begin{bmatrix} k & 0 & \cdots\cdots & 0 \\ 0 & k & \cdots\cdots & 0 \\ \vdots & & & \\ 0 & 0 & 0 & k \end{bmatrix} \qquad ...(A.9)$$

**Real matrix:** A matrix whose elements are real is called a real matrix.

**Complex matrix:** A matrix whose some of the elements may be complex is called a complex matrix.

**Sparse matrix:** A matrix which has very few non-zero elements and most of the elements are zero is a sparse matrix.

**Upper triangular matrix:** A matrix whose elements below the diagonal are zero.

**Lower triangular matrix:** A matrix whose elements above the diagonal are zeros.

**Symmetric matrix:** It is a square matrix whose upper and lower triangular matrices are mirror image along the diagonal or it is a matrix whose transpose is the matrix itself i.e. $A = A^T$ or $a_{ij} = a_{ji}$ for all $i$ and $j$.

**Skew Symmetric matrix:** A square matrix is skew symmetric if it is equal to its 'negative transpose i.e.

$$A = -A^T \qquad\qquad ...(A.10)$$

e.g. the matrix

$$A = \begin{bmatrix} 0 & 2 & -3 \\ -2 & 0 & -1 \\ 3 & 1 & 0 \end{bmatrix}$$

is a skew symmetric

**Hermitian Matrix:** A matrix $A$ is Hermitian if its' elements satisfy the condition $a_{ij} = a^*{ji}$ where * indicates conjugate operation e.g.

$$A = \begin{bmatrix} 2 & 1-2j & 1+j4 \\ 1+2j & 4 & 0 \\ 1-j4 & 0 & 1 \end{bmatrix} \qquad ...(A.11)$$

$$A^T = \begin{bmatrix} 2 & 1+2j & 1-j4 \\ 1-2j & 4 & 0 \\ 1+j4 & 0 & 1 \end{bmatrix} \qquad ...(A.12)$$

$$A^{T*} = \begin{bmatrix} 2 & 1-2j & 1+j4 \\ 1+j2 & 4 & 0 \\ 1-j4 & 0 & 1 \end{bmatrix} = A \qquad \ldots(A.13)$$

**Skew Hermitian matrix:** A matrix $A$ is a Skew-Hermitian if its' elements satisfy the condition.

$$aij = -aj*i \quad \text{or} \quad A = -A^{T*} \qquad \ldots(A.14)$$

e.g.

$$A = \begin{bmatrix} j2 & -1+j2 & 1+j4 \\ 1+j2 & j4 & 0 \\ -1+j4 & 0 & j1 \end{bmatrix} \qquad \ldots(A.15)$$

$$A^{T} = \begin{bmatrix} j2 & 1+j2 & -1+j4 \\ -1+j2 & j4 & 0 \\ 1+j4 & 0 & j1 \end{bmatrix} \qquad \ldots(A.16)$$

$$A^{T*} = \begin{bmatrix} -j2 & 1-j2 & -1-j4 \\ -1-j2 & -j4 & 0 \\ 1-j4 & 0 & -j1 \end{bmatrix} \qquad \ldots(A.17)$$

$$-A^{T*} = \begin{bmatrix} j2 & -1+j2 & 1+j4 \\ 1+j2 & j4 & 0 \\ -1+j4 & 0 & j1 \end{bmatrix} = A \qquad \ldots(A.18)$$

Hence $A$ is a skew hermitian matrix.

It is to be noted that any square matrix can be written as the sum of a symmetric matrix and a skew-symmetric matrix. If $A$ is a square matrix, $A_s$ and $A_{sk}$ the desired symmetric and skew-symmetric matrices respectively then

$$A_s = \frac{A + A^{T}}{2} \qquad \ldots(A.19)$$

and

$$A_{sk} = \frac{A - A^{T}}{2} \qquad \ldots(A.20)$$

## A.3 DETERMINANT OF A MATRIX

It is a single value and determinant is denoted by replacing the square bracket by vertical lines as shown here.

$$|A| = \begin{vmatrix} a_{11} & \cdots\cdots\cdots & a_{1n} \\ a_{21} & \cdots\cdots\cdots & a_{2n} \\ \vdots & & \\ \vdots & & \\ a_{n1} & \cdots\cdots\cdots & a_{nn} \end{vmatrix} \qquad\qquad ...(A.21)$$

The single value is obtained by pursuing the following steps:

If the row and column containing an element $a_{ij}$ in a square matrix $A$ are deleted, the determinant of the remaining square array is called the minor of $a_{ij}$ and is denoted by $M_{ij}$. The co-factor of $a_{ij}$ denoted by $A_{ij}$ is then defined by the relations.

$$A_{ij} = (-1)^{i+j} M_{ij} \qquad\qquad ...(A.22)$$

Thus if the sum of the row and column indices of an element is even, the co-factor and the minor of that element are identical, otherwise they differ in sign.

Since the co-factor of $a_{ij}$ is the co-efficient of $a_{ij}$ in the expansion of $|A|$ we write here

$$\det A = |A| = \sum_{K=1}^{n} a_{ik} A_{ik} \qquad\qquad ...(A.23)$$

$$= \sum_{K=1}^{n} a_{kj} A_{kj} \qquad\qquad ...(A.24)$$

for any relevant value of $i$ or $j$. These relations indicate that the determinant of a square matrix is equal to the sum of the products of the element of any single row or column of that matrix by their co-factors.

From the procedure for evaluation of determinant of a square matrix $A$ outlined above a few observations about the determinants are made which would further facilitate calculations:

1. If all elements of any row or column are zero, its' determinant is zero.
2. If two rows or columns of a square matrix are interchanged, the sign of its determinant is changed.
3. If all the rows and columns of a matrix are interchanged, the value of its determinant remains unchanged.
4. If all elements of one row or one column of a square matrix are multiplied by $k$, the determinant is multiplied by $k$.
5. If to the elements of any row or column are added $k$ times corresponding elements of any other row or column the determinant remains unchanged.
6. If corresponding element of two rows or two columns are equal or in a constant ratio, then the determinant is zero.
7. If each element in one row or one column is expressed as the sum of two terms, then the determinant is equal to the sum of two determinants, in each of which one of the two terms is deleted in each element of that row or column. It is to be noted that determinant of product of two matrices equals the product of the determinant of individual matrices.

$$\det |AB| = \det |A| \det |B|$$

**Singular matrix:** A matrix whose determinant $|A| = 0$ is a singular matrix.

**Non-Singular matrix:** A matrix whose determinant is not zero is a non-singular matrix.

## A.4  RANK OF MATRIX

Rank of a matrix $A$ is the order of the largest square submatrix of $A$ formed by deleting certain rows and/or columns of $A$ whose determinant does not vanish. As a corollary of this definition it is found that if a matrix is of rank $r$ and a set of $r$ rows containing a non-singular submatrix of order $r$ is selected, then any other row in the matrix is a linear combination of these $r$ rows. This also implies that a square matrix is singular if and only if one of its row/columns is a linear combinations of the others rows/columns.

**Equality of matrices:** Two matrices $A$ and $B$ are said to be equal if they have the same order and the corresponding elements of the two matrices are equal.

## A.5  BASIC OPERATIONS

**Addition and Subtraction:** For matrices of the same order, the addition or subtraction of the two matrices requires addition or subtraction respectively of corresponding elements of the two matrices.

$$A + B = B + A \text{ and } A + (B + C) = (A + B) + C$$

These properties are known as commutative and associative properties of matrices for addition.

**Multiplication by a scalar:** If $k$ is a scalar and $A = [a_{ij}]$ is a matrix, then the product of $k$ and $A$ is another matrix whose each element $a_{ij}$ is $k$ times that of the original matrix.

**Multiplication of two matrices:** The product $P$ of the two matrices $A$ and $B$ i.e.

$$P = AB \text{ is a matrix for which the a typical element is}$$

$$P_{ij} = \sum_{K=1}^{n} a_{ik} b_{kj} \qquad \qquad …(A.26)$$

where for every $i$ and $j$, $k$ ranges over the column of order of $A$ and the row order of $B$. Thus, the entry in the $(i, j)$ position of product matrix $P$ is obtained by multiplying the corresponding element of the $i$th row of the matrix $A$ and the $j$th column of matrix $B$ and algebraic sum of the product gives the element $p_{ij}$. Thus multiplication is defined only when the number of columns in the first matrix $A$ is equal to the number of rows in the second matrix $B$ i.e. if $A$ and $B$ are of order $n \times m$ and $m \times l$ the product is possible and then it is said that the product of $A$ and $B$ is comformable. The order of matrix $p$ will then be $n \times l$. However, if the product is to be taken as $BA$, it is found that $(m \times l)\ (n \times m)$ is not compatible and hence in general $AB \neq BA$. In fact even when, there is compatibility of product of the two matrices, the inequality may still hold good.

**Example:** Find the product $AB$ of the matrices

$$A = \begin{bmatrix} 2 & -1 \\ 1 & 5 \\ -2 & 4 \end{bmatrix}_{3 \times 2} \qquad B = \begin{bmatrix} 1 & 0 & 2 \\ -3 & 1 & 0 \end{bmatrix}_{2 \times 3}$$

$$AB = \begin{bmatrix} 2 \times 1 + (-1)\,(-3) & 2 \times 0 + (1)\,(-1) & 2 \times 2 + (-1)\,(0) \\ 1 \times 1 + (5)\,(-3) & 1 \times 0 + 5 \times 1 & 2 \times 1 + 5 \times 0 \\ (-2 \times 1) + 4(-3) & -2 \times 0 + 4 \times 1 & -2 \times 2 + 4 \times 0 \end{bmatrix}_{3 \times 3}$$

$$\begin{bmatrix} 5 & -1 & 4 \\ -14 & 5 & 2 \\ -14 & 4 & -4 \end{bmatrix}_{3 \times 3}$$

Though the matrix multiplication is not commutative in general, it is associative and distributive over addition. Therefore, if the products $AB$, $BC$ and $AD$ are defined, then

$$(AB)C = A(BC) \text{ and } A(B + D) = AB + AD \qquad \text{...(A.27)}$$

## A.6  ADJOINT MATRIX

The adjoint of a square matrix $A$ is found by replacing each element $a_{ij}$ of matrix $A$ by its co-factor $A_{ij}$ and then transposing e.g. if $A$ is given as

$$A = \begin{bmatrix} 3 & -2 & 1 \\ -2 & 6 & 4 \\ 1 & 4 & 8 \end{bmatrix}$$

The *Adj* is obtained as follows:

The co-factors along the first row are

$$A_{11} = \begin{vmatrix} 6 & 4 \\ 4 & 8 \end{vmatrix} \qquad A_{12} = -\begin{vmatrix} -2 & 4 \\ 1 & 8 \end{vmatrix} \qquad A_{13} = \begin{vmatrix} -2 & 6 \\ 1 & 4 \end{vmatrix}$$

$$= 32 \qquad\qquad\qquad = 20 \qquad\qquad\qquad = -14$$

Second row:

$$A_{21} = -\begin{vmatrix} -2 & 1 \\ 4 & 8 \end{vmatrix} \qquad A_{22} = \begin{vmatrix} 3 & 1 \\ 1 & 8 \end{vmatrix} \qquad A_{23} = -\begin{vmatrix} 3 & -2 \\ 1 & 4 \end{vmatrix}$$

$$= 20 \qquad\qquad\qquad = 23 \qquad\qquad\qquad = -14$$

Third row:

$$A_{31} = \begin{vmatrix} -2 & 1 \\ 6 & 4 \end{vmatrix} \qquad A_{32} = -\begin{vmatrix} 3 & 1 \\ -2 & 4 \end{vmatrix} \qquad A_{33} = \begin{vmatrix} 3 & -2 \\ -2 & 6 \end{vmatrix}$$

$$= -14 \qquad\qquad\qquad = -14 \qquad\qquad\qquad = 14$$

Therefore, the co-factor matrix is

$$\begin{bmatrix} 32 & 20 & -14 \\ 20 & 23 & -14 \\ -14 & -14 & 14 \end{bmatrix}$$

and its transpose

$$Adj\ A = \begin{bmatrix} 32 & 20 & -14 \\ 20 & 23 & -14 \\ -14 & -14 & 14 \end{bmatrix}$$

Since $A$ is a symmetric matrix *Adj A* also happens to be symmetric matrix.

## A.7 MATRIX INVERSION

The inverse of matrix $A$ is written as $A^{-1}$ and is defined by the relation.

$$AA^{-1} = A^{-1} A = I \qquad \qquad ...(A.28)$$

However, $A^{-1}$ is determined as

$$A^{-1} = \frac{Adj\ A}{\det A} \qquad \qquad ...(A.29)$$

If we are required to obtain $A^{-1}$ of the matrix $A$, as given in the previous illustration, we have already obtained *Adj A*, we now obtain det $A$.

$$\det A = \sum_{k=1}^{3} a_{ik}\ Aik \quad \text{and} \quad i = 1$$

We have already obtained $A_{11}$, $A_{12}$, $A_{13}$, therefore,

$$\det A = 3 \times 32 - 2 \times 20 + 1 \times (-14)$$
$$= 96 - 40 - 14$$
$$= 42$$

Hence
$$A^{-1} = \begin{bmatrix} 32 & 20 & -14 \\ 20 & 23 & -14 \\ -14 & -14 & 14 \end{bmatrix} \cdot \frac{1}{42}$$

It can be verified that $AA^{-1} = I$.

It is to be noted that transpose of product of two matrices is the product of their transposes taken in reverse order i.e.

$$[AB]^T = B^T A^T \qquad \qquad ...(A.30)$$

For a singular matrix where det $A = 0$ the inverse of such a matrix $A$ does not exist.

Also, inverse of the product of two non-singular square matrices of the same order is the product of their inverses taken in reverse order e.g.

$$[A\ B]^{-1} = B^{-1}\ A^{-1} \qquad \qquad ...(A.31)$$

Also
$$[A^T]^{-1} = [A^{-1}]^T \qquad \qquad ...(A.32)$$

If matrix $A$ is not a square matrix, its inverse does not exist. In order to obtain the rank of a matrix, the following elementary operations are preferred to reduce the original matrix to an identity matrix. The highest order of the identity matrix so reduced is the rank of the matrix.

The elementary operations are:

1. The interchange of two rows or columns.
2. The multiplication of the elements of a row or column by a non zero number.
3. The addition or subtraction to the elements of a row or column, of $k$ times the corresponding elements of another row or column.

Consider for example a matrix

$$A = \begin{bmatrix} 1 & 2 & 3 & 4 \\ 0 & 1 & -1 & 0 \\ 1 & 0 & 1 & 2 \\ 1 & 1 & 0 & 2 \end{bmatrix}$$

Subtracting row 3 from row 4

$$\begin{bmatrix} 1 & 2 & 3 & 4 \\ 0 & 1 & -1 & 0 \\ 1 & 0 & 1 & 2 \\ 0 & 1 & -1 & 0 \end{bmatrix}$$

Since 2nd row and 4th row are identical matrix $A$ has det $A = 0$
Subtracting row 2 from row 4 we have

$$\begin{bmatrix} 1 & 2 & 3 & 4 \\ 0 & 1 & -1 & 0 \\ 1 & 0 & 1 & 2 \\ 0 & 0 & 0 & 0 \end{bmatrix}$$

Divide column 4 by 2

$$\begin{bmatrix} 1 & 2 & 3 & 2 \\ 0 & 1 & -1 & 0 \\ 1 & 0 & 1 & 1 \\ 0 & 0 & 0 & 0 \end{bmatrix}$$

Subtract column 4 from 2

$$\begin{bmatrix} 1 & 0 & 3 & 2 \\ 0 & 1 & -1 & 0 \\ 1 & -1 & 1 & 1 \\ 0 & 0 & 0 & 0 \end{bmatrix}$$

Add column 1 and 4 and subtract from 3

$$\begin{bmatrix} 1 & 0 & 0 & 2 \\ 0 & 1 & -1 & 0 \\ 1 & -1 & -1 & 1 \\ 0 & 0 & 0 & 0 \end{bmatrix}$$

Subtract row 1 from row 3

$$\begin{bmatrix} 1 & 0 & 0 & 2 \\ 0 & 1 & -1 & 0 \\ 1 & -1 & -1 & -1 \\ 0 & 0 & 0 & 0 \end{bmatrix}$$

Multiply row 3 by $-1$

$$\begin{bmatrix} 1 & 0 & 0 & 2 \\ 0 & 1 & -1 & 0 \\ 0 & 1 & 1 & 1 \\ 0 & 0 & 0 & 0 \end{bmatrix}$$

Subtract row 2 from 3

$$\begin{bmatrix} 1 & 0 & 0 & 2 \\ 0 & 1 & -1 & 0 \\ 0 & 0 & -2 & 1 \\ 0 & 0 & 0 & 0 \end{bmatrix}$$

Multiply column 1 by 2 and subtract from column 4.

$$\begin{bmatrix} 1 & 0 & 0 & 0 \\ 0 & 1 & -1 & 0 \\ 0 & 0 & -2 & 1 \\ 0 & 0 & 0 & 0 \end{bmatrix}$$

Interchange column 3 and 4 we have

$$\begin{bmatrix} 1 & 0 & 0 & 0 \\ 0 & 1 & 0 & -1 \\ 0 & 0 & 1 & -2 \\ 0 & 0 & 0 & 0 \end{bmatrix}$$

Therefore, the rank of matrix is 3.

While finding out inverse of a diagonal matrix, one need not go through the long process of finding out *Adj* of the matrix and the determinant of the matrix. The inverse of a diagonal matrix is a diagonal matrix with elements reciprocal of the original matrix e.g. if

$$A = \begin{bmatrix} 1 & 0 & 0 \\ 0 & 2 & 0 \\ 0 & 0 & 3 \end{bmatrix}$$

$$A^{-1} = \begin{bmatrix} 1 & 0 & 0 \\ 0 & 1/2 & 0 \\ 0 & 0 & 1/3 \end{bmatrix}$$

If one of the elements of the diagonal matrix is zero, the inverse of such a matrix does not exist as the determinant of such a matrix is zero and the matrix is singular.

One of the most important applications of matrices is the solution of linear simultaneous equations. The set of linear simultaneous equation can be written in a compact form using matrices.

$$AX = B$$

where $A$ is a co-efficient matrix

$X$ a column vector of unknown variables and $B$ is a known column vector. The solution vector $x$ is obtained by pre-multiplying both sides of the equation by $A^{-1}$ i.e.

$$A^{-1} A X = A^{-1} . B$$

or
$$X = A^{-1} . B$$

Therefore, for solution to exist $A^{-1}$ should exist. If $B = 0$, the solution is trivial when $X = 0$ is a null vector.

If $m$ is the number of equations and $n$ the number of unknowns, following situation may arise.

$m = n$ In this case the equations have a unique solution.

$m < n$ Here we solve $m$ variables in terms of the remaining $(n - m)$ variables. For one set of assumed values of $(n - m)$ variables, we have values of the $m$ variables. This means we can have more than one solution of the set of equations. This set of equation represents an underdetermined system.

However, if $m > n$, there will be more equations than the number of unknowns. Hence it is an over determined system. If the rank of the co-efficient matrix equals number of unknowns, the equation will have a unique solution otherwise it will not have a unique solution.

**APPENDIX B**

# Spice

## B.1  INTRODUCTION

The enormous complexity of the modern integrated circuits can be designed and analysed using computer aided analysis which also provides information about circuit performance which is almost impossible to obtain with the laboratory meters. Computer aided analysis provides sensitivity analysis which evaluates the effect of variation of the various elements on the performances of the circuit and also determines the permissible limits to tolerance on every component being used in the circuit. Also it is possible to have optimum design of electronic circuits in terms of circuit parameters and the performance of the circuits.

The abbreviation SPICE stands for simulation program with Integrated circuits emphasis. SPICE is a general purpose program which simulates electronic circuits and it can provide solution to time-domain response, a small signal frequency responses etc. SPICE contains models for common circuit elements passive as well as active and is capable of simulating most electronic circuits.

The input syntax for SPICE is a free format style; it does not require that data be entered in fixed column locations. It even assumes reasonable default values for unspecified circuit parameters. Also, it performs sufficient amount of error checking (Bad data detection and correction) to ensure that a circuit has been correctly entered.

A circuit is described by statements that are stored in a file called the circuit file. The circuit file is read by the SPICE simulator. Each statement is self contained and independent; the statements do not interact with each other.

There are mainly two versions of SPICE, the mainframe and the PC versions. Their methods of computation may differ but their features are almost identical. For classroom studies of course PC version is most desirable. One of these PC versions is SPICE (Microsim) which is widely used. Other versions are available in reference (1, 2). Here we will be referring only to Input, Output, formats for the SPICE software for various aspect of circuit analysis. We will not go into various techniques of analysis of electric/electronic circuits as standard programs for the same are available as in Microsim.

## B.2  CIRCUIT DESCRIPTION

The details of a circuit are fed into the computer through a file called circuit file which is normally typed from a keyboard. The file contains information about the various components (elements) and their values, the sources and the comments for what to calculate and what to provide as output. It is similar to main program in engineering problem and the subroutine to execute this program are available in program like Microsim. The circuit file is the input file to the SPICE program  which after executing the commands produces the results in another file called the output file.

## B.2.1 Element Values

The element values are written in standard floating point notations with optional scale and unit suffixes. Some of the scale suffixes are

| | | | | |
|---|---|---|---|---|
| P | = | 1 E–12 | Pico | $10^{-12}$ |
| N | = | I E–9 | Nano | $10^{-9}$ |
| μ | = | 1 E–6 | Micro | $10^{-6}$ |
| M | = | 1 E–3 | Milli | $10^{-3}$ |
| K | = | 1 E 3 | Kilo | $10^{3}$ |
| MEG | = | 1 E6 | Mega | $10^{6}$ |

whereas some of the unit suffixes are

| | | |
|---|---|---|
| V | = | Volt |
| A | = | Ampere |
| Hz | = | Hertz |
| OHM | = | Ohm (Ω) |
| H | = | Henry |
| F | = | Farad |
| DEG | = | Degrees   etc. |

The first suffix is always the scale suffix and the unit suffixes are always ignored by PSPICE. If the value of a capacitor is $10\,\mu F$ it is written as $10\,\mu$ or $10\,\mu F$. In the absence of scale and unit suffices the units of current voltage inductance, capacitance and frequency are, by default, amperes volts, henry, farad and Hertz respectively. PSPICE ignores unit suffix and hence the following are equivalent:

| | | | |
|---|---|---|---|
| 20 E–3 | 20.0 E–3 | 20 M | 20 MV |
| 20 MA | 20 M OHM | 20 MH | |

## B.2.2 Nodes

An element is always connected between two nodes and hence its location is identified by the node numbers. Node $O$ is always taken as ground node. All nodes must be connected to at least two elements and hence should appear at least twice. Node numbers must be integers from 0 to 9999. PSPICE allows any alpha numeric string upto 131 characters long. Some of the node names shown in Table B.1 are reserved and cannot be used.

**Table B.1**   Some reserved node names

| Reserved node Name | Value | Description |
|---|---|---|
| 0 | 0 Volt | Analog ground |
| SD–Hl | 1 | Digital high level |
| SD–LO | 0 | Digital low level |
| SD–X | X | Digital Unknown level |

The node numbers to which an element is connected are specified after the name of the element.

## B.2.3 Circuit Elements

Circuit elements are identified by names. A name must start with a letter symbol corresponding to the element but after that it may contains either letters or numerals. Names can be upto eight characters long. Table B.2 shows symbols of circuit elements and sources.

**Table B.2**  Symbols of Circuits Elements and Sources

| *First letter* | *Circuit elements & Sources* |
|:---:|:---|
| R | Resistance |
| L | Inductance |
| C | Capacitance |
| D | Diode |
| E | Voltage Controlled Voltage Source |
| F | Current Controlled Current Source |
| G | Voltage Controlled Current Source |
| H | Current Controlled Voltage Source |
| I | Independent Current Source |
| K | Mutual Inductor (Transformer) |
| S | Voltage Controlled Switch |
| T | Transmission Line |
| V | Independent Voltage Source |

The format for specifying passive element is

(element name)   (positive node)   (negative node)   (value)

The specification

R4                    3                    2                    800

means there is a resistor R4 in the circuit connected between the +ve terminal 3 and negative terminal 2 and its value is 800 ohms.

The format for sources is

(source name) (positive node) (negative node) (source model)

The statement

VS                    1                    0                    25V

means Vs is a voltage source connected between positive terminal 1 and negative terminal 0 and its value is 25 volts.

Similarly

IS        0              3              DC              50 MA

means there is a current source, is connected between +vc terminal 0 and negative terminal 3 (i.e. the current flows from node 0 to node 3). It is *dc* source of 50 mA.

The output variables are usually branch voltages and branch currents. The voltage of node 3 with respect to node 2 is written as $V(3, 2)$. Similarly voltage of node 4 with respect to node 0 is written as $V(4, 0)$ or simply $V(4)$.

In a *dc* circuit if current through a voltage source or through a branch is required, a dummy voltage source with 0 Volt is normally connected in series with the circuit and it is used as an ammeter to measure the current of that source or branch e.g. I(*Vz*) where *Vz* = 0 Volt.

## B.2.4 Format of Circuit Files

A circuit file which can be read by SPICE the format is given as follows:

> Title

Circuit Description
Analysis Description
Output Description
END (end of file statement)

The first line is the title which describes the type of circuit or any other comment. The last line must be the END statement so that the computer terminates the program once it has completed the computations and outputted the result. The order of the remaining lines is not important and does not affect the results of simulations.

If a PSPICE statement is more than one line, the statement can continue on the next by putting a plus sign in the first column of the next line. The PSPICE statements should be typed in uppercase preferably. In electric circuits subscripts are normally assigned to symbols for voltage current and circuit elements. However, in SPICE symbols are represented without subscripts and subscript is also used as uppercase letter along with the symbol e.g. *Vs* is printed as *VS*, $R_1$ as *R1* etc.

## B.2.5 Format of Output Files

The results of simulations by SPICE are stored in an output file. It is possible to control the type and amount of output by various commands. The output falls into four categories:

1. A description of the circuit itself.
2. Prints and plots by PRINT and PLOT statements. These include the output from the .DC, .AC, and .TRANS analysis.
3. These include some of the results or some information about the run e.g. time of computation, the memory requirement of the total simulation or a part of the simulation etc.
4. Print or plot of some of the output from .OP, .TF, .SENS etc.

Here    .OP means operating point
     .TF the transfer function
     .SENS means sensitivity analysis

**Example B.1:** The circuit of Fig. EB.1 is to be simulated on   PSPICE.

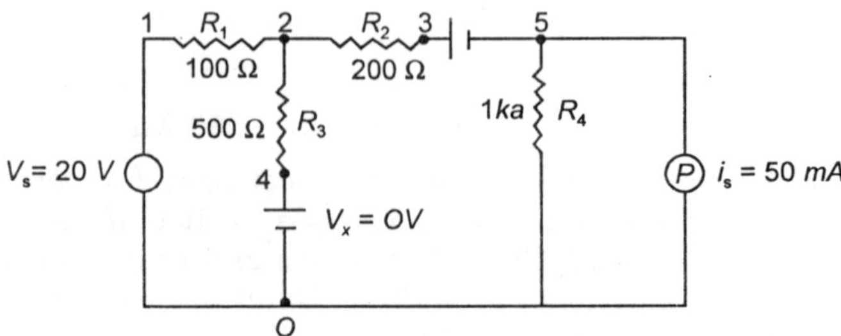


**Fig. EB.1**

Write statement for the circuit file.

**Solution:** The circuit file contains the following statements:

Example B.1 simple d.c. circuit (Title)

| VS | 1 | 0 | DC | 20 V |
| IS | 0 | 5 | DC | 50 MA |
| R1 | 1 | 2 | 100 | |
| R2 | 2 | 3 | 200 | |
| R3 | 2 | 4 | 500 | |
| R4 | 5 | 0 | 1 K OHM | |
| VX | 4 | 0 | DC OV | |
| VY | 3 | 5 | DC OV | |
| END | | | | |

**Example B.2:** In example EB.1 if Vs is varied between 10 to 40 volts in steps of 10 Volts. Write suitable input data. Print voltage V(5) and current $I_{R2}$ and $I_{R3}$. Write the statements.

**Solution:** The remaining circuit file is as in previous example. The sweep voltage format is given as

.DC SWNAME SSTART SEND SINC where SWNAME = the sweep variable name, either voltage or current source.

SSTART  =  The sweep start value
SEND    =  The sweep end value
SINC    =  The sweep increment

Here in the given problem

DC.    VS    10    40    10

For print, the statement is

.PRINT  DC    V(5,0)    I(VY)    I(VX)
.END,

If, however, the voltage *Vs* and *Is* are to be changed to arbitrary values, the statement is to be given as follows:

DC    VS    LIST    (values)

Suppose we want to analyse circuit in Fig. EB-1 for

Vs = 5V, 12V, 20V and current
Is = 20mA, 50mA and 60 mA

The statement will be

DC VS LIST 5V 12V 20V IS 20MA 50 MA 60 MA

## B.3  DEPENDENT SOURCES

Fig B.1(a) shows a voltage controlled voltage source

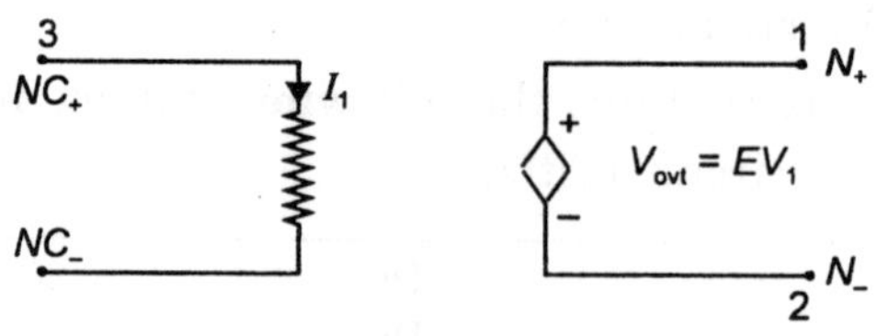

**Fig B.1 (a)**

Fig. B.1 Voltage controlled voltage source and its input format is ($E$ is the symbol for voltage controlled voltage source)

| E(name) | N+ | N– | NC+ | NC– | [(Voltage gain) value] |
|---------|----|----|-----|-----|------------------------|
| EAB     | 1  | 2  | 3   | 4   | 1.0                    |

Here N+ and N– are the positive and negative output nodes respectively and NC+ and NC– are the positive and negative nodes, respectively of the controlling voltage. So the controlling voltage is $V_1$ and controlled voltage $V_{OUT}$ of $EV_1$.

Fig. B.1 (b) shows a voltage controlled current source and its input format is

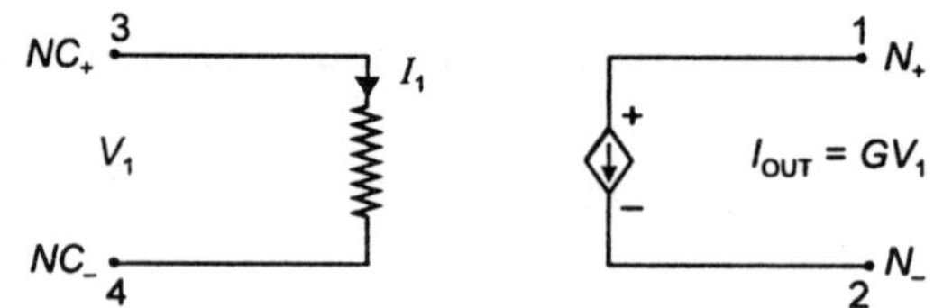

**Fig. B.1 (b)**

G(name)  N+  N–  NC+  NC–  [(transconductance) value]

Here N+ and N– are the positive and negative output nodes respectively, NC+ and NC– are the positive and negative nodes respectively of the controlling voltage. $G$ is the symbol for a voltage controlled current source. A typical statement could be

| GAB | 1 | 2 | 3 | 4 | 1.0 |
|-----|---|---|---|---|-----|

Fig. B.1 (c) shows a current controlled current source. Here $F$ is the symbol for the device.

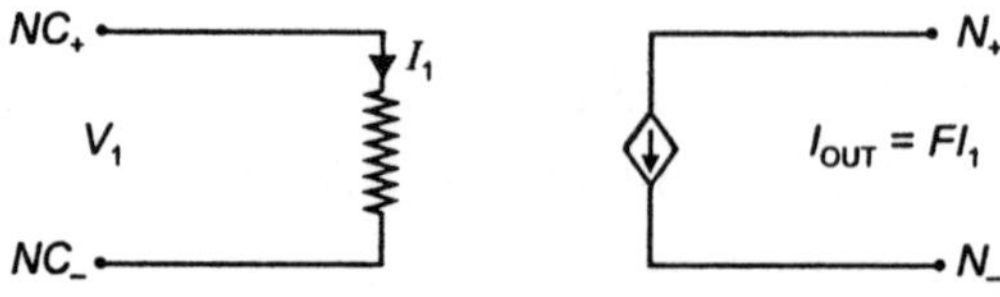

**Fig. B.1 (c)**

The input format is

F(name) N+ N– VN [(current gain) value]

N+ and N– are the positive and negative nodes respectively of the current source, VN is a voltage source through which the controlling current flows. The controlling current is assumed to flow from the positive node of VN through the voltage source VN to the negative node of VN. The current through the controlling voltage source (VN) determines the output current.

A typical statement could be

FAB        1            2                VIN            10

Fig. B.1 (d) shows a current controlled voltage source and $H$ is the symbol for the device.

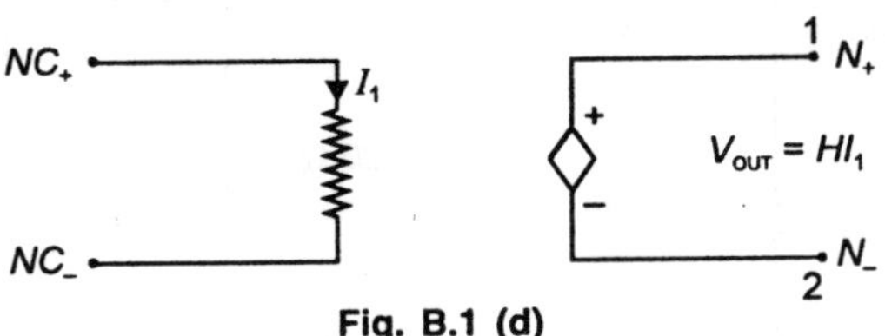

**Fig. B.1 (d)**

The input format is

H(name)    N+    N–    VN       [(transresistance) value]

N+ and N– are the positive and negative nodes, respectively of the voltage source. VN is a voltage source through which the controlling current flows and its application is similar to that for a current controlled current source.

A typical statement is

HAB        1            2                VIN            10

Some of the accepted functions in PSPICE are

| *Functions* | *Meaning* |
|---|---|
| ABS ($x$) | $\lvert x \rvert$ Absolute value |
| SGN ($x$) | Signum ($x$) |
| EXP ($x$) | $e^x$ |
| DB ($x$) | 20 log ($x$) |
| LOG ($x$) | ln $x$ |
| LOG 10 ($x$) | log $x$ (base 10) |
| PWR ($x$, $y$) | $\lvert x \rvert^y$ |
| SQRT ($x$) | $\sqrt{x}$ |
| SIN ($x$) | sin ($x$), $x$ in rad. |
| COS ($x$) | Cos $x$ |
| TAN ($x$) | tan $x$ |
| ARCTAN ($x$) | $\tan^{-1} x$ |
| $d(y)$ | $\dfrac{dy}{dx}$ |
| $s(y)$ | $\int y\,dx$ |
| AVG ($x$) | Running average of $x$ |
| RMS ($x$) | Running RMS of $x$ |

**Example B.3:** Fig. E B.3 shows a typical d.c. circuit with controlled d.c. source. Prepare circuit input file in PSPICE.

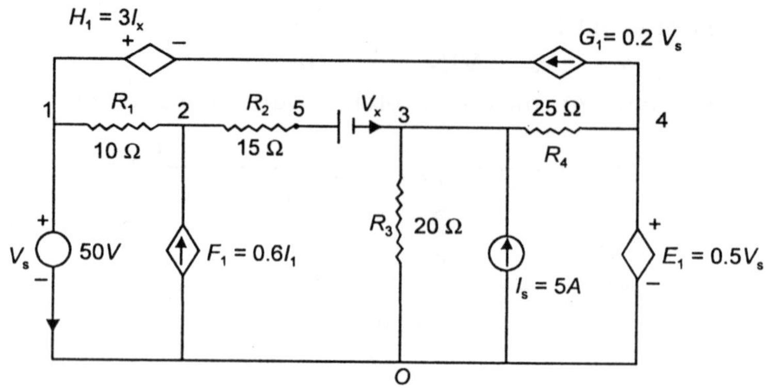


**Fig. EB.3**

**Solution:** Title Example B.3 DC circuit with controlled d.c. sources

| | | | | | | |
|---|---|---|---|---|---|---|
| VS | 1 | 0 | DC | 50V | | |
| IS | 0 | 3 | DC | 5A | | |
| R1 | 1 | 2 | 10 | | | |
| R2 | 2 | 5 | 15 | | | |
| R3 | 3 | 0 | 20 | | | |
| R4 | 3 | 4 | 25 | | | |
| VX | 5 | 3 | DC | 0V | | |
| E1 | 4 | 0 | 1 | 0 | 05 | VVT |
| G1 | 4 | 3 | 1 | 0 | 0.2 | VCT |
| F1 | 0 | 2 | VS | | 0.6 | CCT |
| H1 | 1 | 3 | VX | 3 | CVT | |
| .END | | | | | | |

**Example B.4:** Write PSPice statement for the following problems. Assume that the first node is the positive terminal and the second node is the negative terminal:

(a) A voltage source $V_{out}$ is connected between nodes 4 and 5 and is controlled by a voltage source $V_1$ and has a voltage gain of 25. The controlling voltage is connected between 8 and 10. The source is expressed as

$$V_{out} = 25\,V_1$$

(b) A current source $I_{out}$ is connected between nodes 4 and 5 and is controlled by a current source $I_1$ and has a current gain of 8. The voltage through which the controlling current flows is $Vc$. The current source is given by $I_{out} = 8\,I_1$.

(c) A current source $I_{out}$ is connected between nodes 4 and 5 and is controlled by a voltage source $V_1$ between nodes 8 and 9. The transconductance is 0.06 siemens. The current source is given by

$$I_{out} = 0.06\,V_1$$

(d) A voltage source $V_{out}$ is connected between nodes 4 and 5 and is controlled by a current source $I_1$ and has a transresistance of 100 ohms. The voltage through which the controlling current flows is $Vc$. The voltage source is expressed as

$$V_{out} = 100\,I_1$$

**Solution:**

(*a*) It is a VCT and the general format is

| E(name) | N+ | N– | NC+ | NC– | gain |
|---------|----|----|-----|-----|------|
| EAB | 4 | 5 | 8 | 10 | 25 |

(*b*) It is a CCT and the general format is

| F(name) | N+ | N– | VN | gain |
|---------|----|----|----|------|
| FAB | 4 | 5 | VC | 8 |

(*c*) It is a VCT and the general format is

| G(name) | N+ | N– | NC+ | NC– | G |
|---------|----|----|-----|-----|------|
| GAB | 4 | 5 | 8 | 9 | 0.06 |

(*d*) It is CVT and the general format is

| H(name) | N+ | N– | VN | R |
|---------|----|----|----|------|
| HAB | 4 | 5 | VC | 100 |

## B.4 AC CIRCUIT ANALYSIS

SPICE is an ideal software tool for simulating a circuit and for studying the behaviour of voltages, current and power flow under steady state and transient conditions. In a.c. circuit the variables are sinusoidal quantities and are represented by complex numbers. The following suffixes are used to represent a.c. quantities:

| *Suffix* | *Meaning* |
|----------|-----------|
| *M* | Peak Magnitude |
| *DB* | Peak Magnitude in db |
| *P* | Phase in radians |
| *G* | Group delay $\dfrac{\Delta \text{ phase}}{\Delta \text{ frequency}}$ |
| *R* | Real part |
| *I* | Imaginary part |

The voltage variable in a.c. circuit is expressed in various situations as listed below:

| *Variables* | *Meaning* |
|-------------|-----------|
| VM(5) | Magnitude of voltages at node 5 with respect to ground. |
| VM(4,2) | Magnitude of voltage at node 4 with respect to node 2. |
| VDB(R1) | DB magnitude of voltage across resistor $R_1$ where the first node (as defined in the circuit file) is assumed to be +ve with respect to second node. |
| VP(D1) | Phase of anode voltage of diode $D_1$ with respect to cathode. |
| VR(2, 3) | Real part of voltage at node 2 with respect to node 3. |
| VI(2, 3) | Imaginary part of the voltage at node 2 with respect to node 3. |

The statements for currents in ac analysis are similar to those for the dc circuit. However, only the currents through the elements listed below are available. For all other elements a zero valued voltage source must be placed in series with the device of interest.

Currents through elements for a.c. analysis.

| First letter | Element |
|---|---|
| C | Capacitor |
| I | Independent current source |
| L | Inductor |
| R | Resistor |
| T | Transmission line |
| V | Independent voltages source |

The current variable in a.c. circuit is expressed in various situations as listed below:

| Variable | Meaning |
|---|---|
| IM (R1) | Magnitude of current through resistor R1 |
| IR (R1) | Real part of current through R1 |
| II (R1) | Imaginary part of current through R1 |
| IM (VIN) | Magnitude of current through source Vin |
| IR (VIN) | Real part of current through source Vin |
| II (VIN) | Imaginary part of current through source Vin. |

The general format for the voltage/current sources is as follows:

V (name)     N+     N–     [AC (magnitude value) (phase value)]
I (name)     N+     N–     [AC (magnitude value) (phase value)]

Some typical statements

VAC          3      4      AC      5V

AC voltage source between nodes 3 and 4 and magnitude 5V with no phase delay.

VACP         3      4      AC      5      30 Deg

AC voltage source between nodes 3 and 4 with magnitude 5 volts and phase delay 30°, $5\underline{|30°}$

IAC          3      4      AC      5
IACP        ·3      4      AC      5      30 Deg

**Example 4.** For the circuit shown write the statements for the circuit file and print V (3).

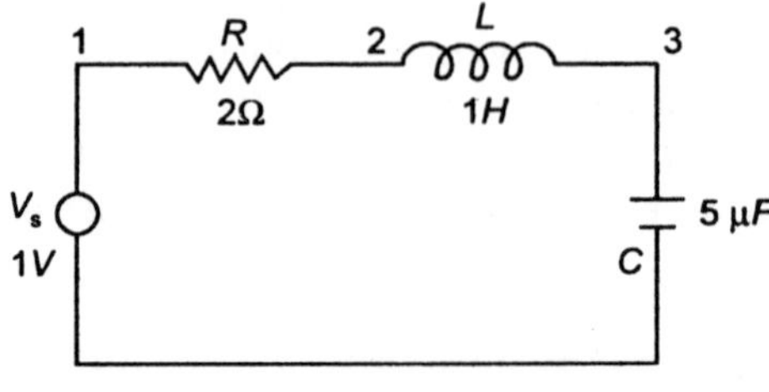


**Fig. EB-4**

**Solution.** The circuit file contain the following statements

```
VS   1   0    AC   1V
R    1   2    2
L    2   3    1H
C    3   0    5UF
```
.PRINT AC VM (3) VP (3)
.End.

**Example B.5.** For the circuit shown write the statements for the circuit file Print $I_1$ and $V_5$.

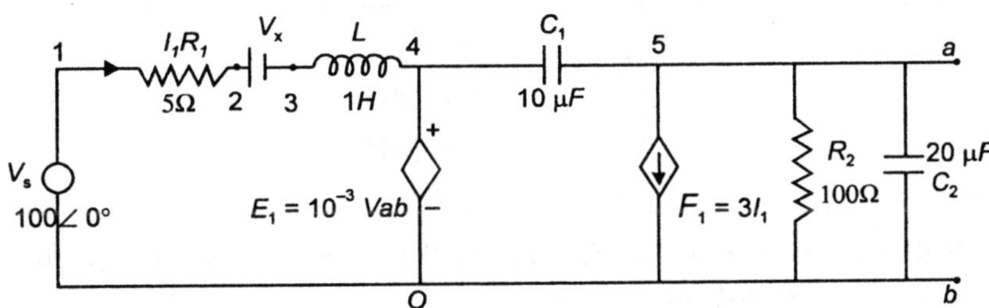


**Fig. EB.5**   AC circuit with controlled sources

**Solution.** Statements for circuit file

```
VS    1    0      Ac      100 V
R1    1    2      5
R2    5    0      100
L     3    4      1H
C1    4    5      10 UF
C2    5    0      20 UF
VX    2    3      AC      0V
E1    4    0      5       0       1M
F1    5    0      VX      3
```
.PRINT AC   IM (R1)   IP ($R_1$)
.PRINT AC   VR (5)    VI (5)    VM (5)    VP (5)

.END

## B-4.1 Magnetic Elements

A transformer is an example of mutual coupling between two coils or windings the primary and the secondary. The symbol for mutual coupling is K. The general format for a mutual inductor is

K (name)    L ((First-inductor) name L ((2nd

INDUCTOR) name) ((Coupling) value)

Fig. B.2 shows (a) Positively coupled and (b) oppositely coupled circuits

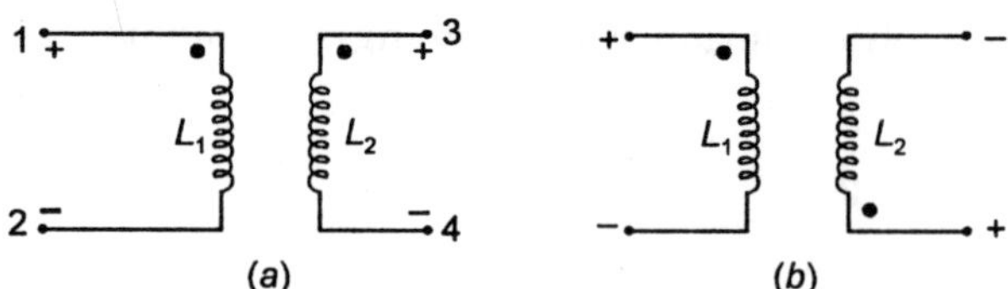

**Fig. B.2**

PSPICE assumes a dot in the first node of each inductor. Say $K = 0.8$ then the statements for the two circuits for circuit file are

(a)
| L1 | 1 | 2 | 0.6 MH (mH) |
|---|---|---|---|
| L2 | 3 | 4 | 0.6 MH |
| KXFMR | L1 | L2 | 0.8 |

(b)
| L1 | 1 | 2 | 0.6 MH |
|---|---|---|---|
| L2 | 4 | 3 | 0.6 MH |
| KXFMR | L1 | L2 | 0.8 |

**Example B.6.** For the circuit of Fig. EB.6. Write down statements for the circuit file and print statement for the output current

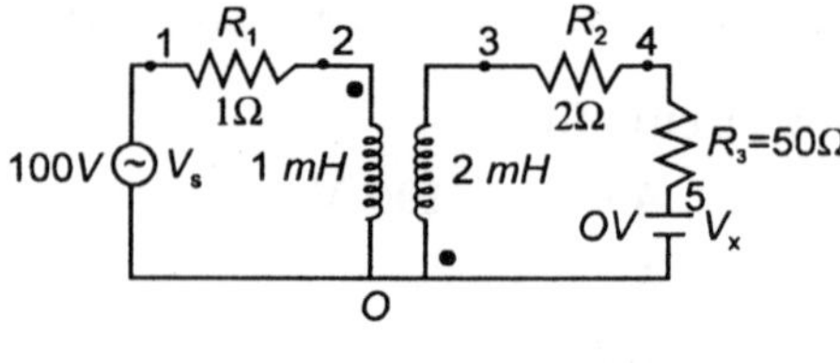

**Fig. EB.6**

**Solution.** It is to be noted that the primary and the secondary Windings have a common node. PSpice will give an error message if the common node is made as there is no dc path from the nodes of the secondary to the ground. $V_x$ is inserted to measure output current. The circuit file has the following statements.

| VS | 1 | 0 | AC | 100 |
|---|---|---|---|---|
| R1 | 1 | 2 | 1 | |
| L1 | 2 | 0 | 1 MH | |
| R2 | 3 | 4 | 2 | |
| L2 | 0 | 3 | 2 MH | |
| R3 | 4 | 5 | 50 | |
| VX | 5 | 0 | DC | 0 |
| .PRINT | AC | IM (VX) | IP (VX) | |
| .END | | | | |

## B.4.2 Transient Analysis

Whenever a circuit contains an energy storing element viz. an inductor or a capacitor or both and switching operation takes place transients are produced irrespective of whether the source is ac or dc. Spice allows simulating transient behaviours by assuming initial conditions to circuit elements.

For transient analysis the capacitor and inductor are inputted in the circuit file as follows.

*Capacitor*

C (NAME)    N+    N–    CNAME    CVALUE    IC = VO

Here CNAME is the model name, CVALUE the nominal value of the capacitor, IC the initial value of voltage Vo across capacitor.

*Inductor*

L (Name)    N+    N–    LNAME    LVALUE    IC = IO

If CNAME is omitted CVALUE is the capacitance in farads and CVALUE can be positive or negative but must not be zero. Similarly, if LNAME is omitted LVALUE is the inductance in henrys and LVALUE can be positive or negative but must not be zero.

Some capacitor and inductor statements.

| | | | | |
|---|---|---|---|---|
| C1 | 5 | 4 | 5 UF | |
| CLOAD | 10 | 12 | 5 PF | IC = 5V |
| L1 | 5 | 4 | 5 MH | |
| LLOAD | 10 | 12 | 5 MH | IC = 1 MA |

## B.5. MODELING OF TRANSIENT SOURCES

### 1. Exponential source

The waveform parameters of an exponential waveform are shown in Fig. B.5.1. The format for the wave form is

EXP        (V1 V2 TRD TRC TFD TFC)

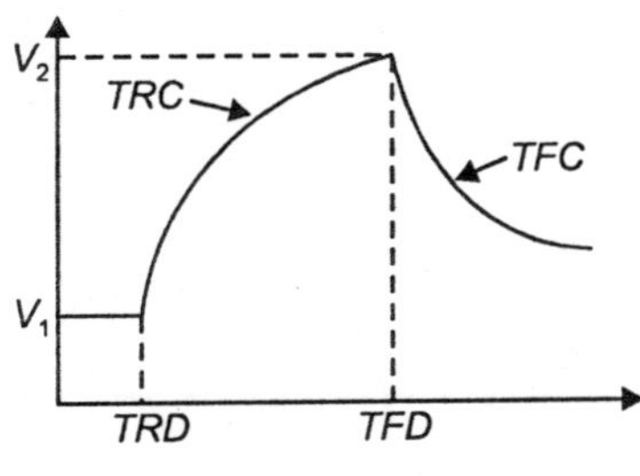


**Fig. B.3**

Here TRD is the rise delay time, TRC the rise time constant TFD is the fall delay time and TFC is the fall time constant.

It is to be noted that the values of an EXP waveform as well as the values of other time-dependent waveforms at intermediate time points are determined by PSPICE by means of linear interpolation

For $V_1$ = 1 Volt    $V_2$ = 5 V    TRD = 2 nS    TRC = 15 nS
TFD = 30 nS  and  TFC = 20 nS

EXP (1    5    2 NS    15 NS    30 NS    20 NS)

## 2. Pulse Source

Fig B.4 shows the waveform and parameters of a pulse waveform. The symbol of a pulse source is PULSE and the general format is

PULSE  (V1  V2  TD  TR  TF  PW  PER)

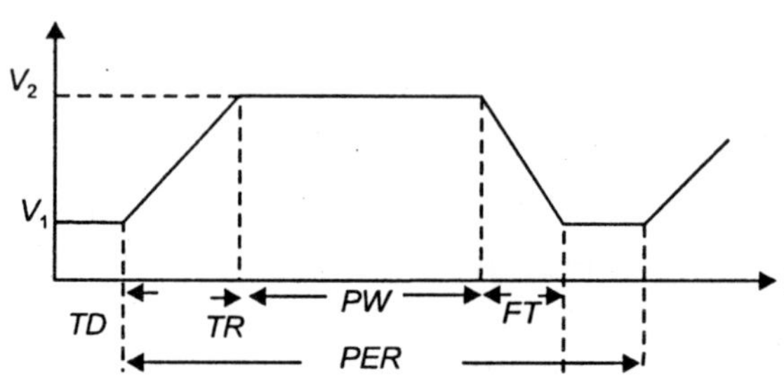


**Fig. B.4**

Here TD is delay time, TR rise time, PW pulse width in sec FT fall time and PER the period of the pulse.

If $V_1$ = 1V    $V_2$ = 5 V   TD = 2 nS   TR = 3 nS
TF = 3 nS   PW = 50 NS and PER = 90 NS, the statement is

PULSE    (1    5    2NS    3NS    3NS    50 NS  90 NS)

## 3. Sinusoidal Source

The symbol of a sinusoidal source is  SIN and the general form is

SIN (VO  VA   FREQ  TD  ALP   THETA)

where V.O is the offset-value, VA the peak voltage FREQ the frequency in HZ, TD the delay time ALPHA damping factor, THETA the phase delay. These quantities have been shown in Fig. B.5.

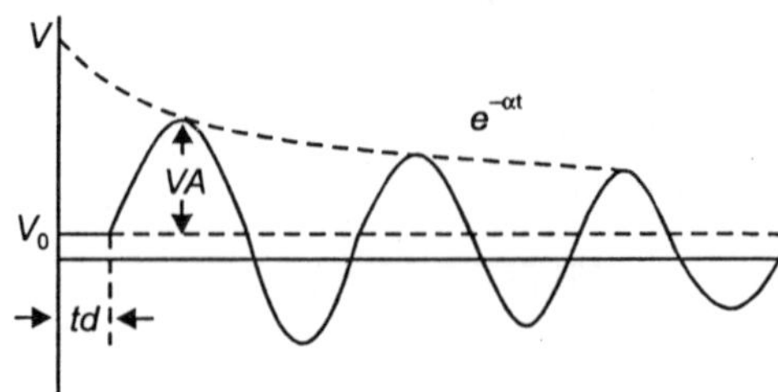


**Fig B.5**

If $V_0$ = O,   VA = 10 V,   f = 10 KHZ and $\theta$ = 30° the statement is

SIN (0  10V  10 KHZ   30 DEG)

These statements for various voltages sources have been given. However, if we have to specify current sources, similar formats can be used.

# References


1. Rashid, M.H., SPICE for circuits and electronics using PSPICE, Prentice Hall of India 2nd Edition 2002.
2. PSpice Manual, Irvine Calif. Microsim Corporation 1992.

# Multiple Choice Questions

1. A higher value of $Q$ is characterized by
   (1) narrow band of frequency  (2) sharp response
   (3) poor selectivity
   Tick out the correct one
   (a) 1 and 2        (b) 1 and 3        (c) 2 and 3        (d) 1, 2 and 3.

2. At lower half power frequency the total reactance of the series RLC circuit is
   (a) $-R$                              (b) $\sqrt{2}\ R\ \angle 45°$
   (c) $\sqrt{2}\ R\ \angle -45°$        (d) None of the above.

3. If in a series RLC circuit

$$Q = \frac{1}{R}\sqrt{\frac{L}{C}}, \text{ the circuit is under damped if}$$

   (a) $Q > \dfrac{1}{2}$    (b) $Q = 1$    (c) $Q = \dfrac{1}{2}$    (d) $Q < \dfrac{1}{2}$

4. The pole-zero configuration of input impedance on s-plane for parallel resonance is

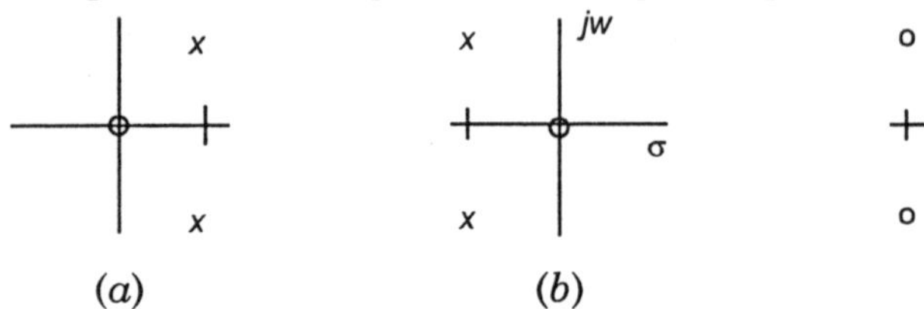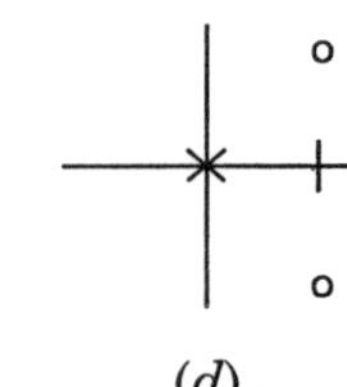

|        (a)        |        (b)        |        (c)        |        (d)        |

5. The pole zero configuration of input admittance of a parallel resonant circuit on s-plane is

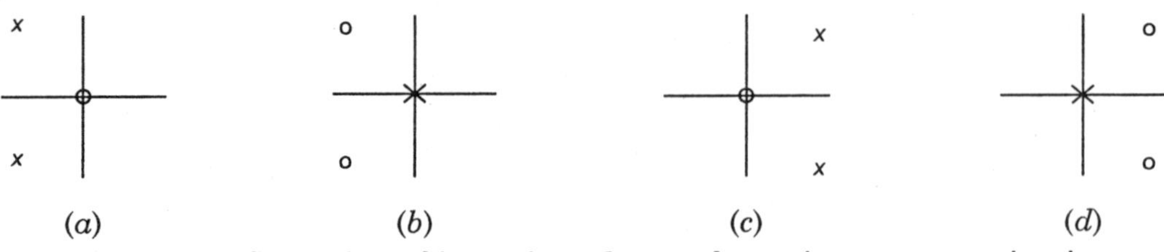

|        (a)        |        (b)        |        (c)        |        (d)        |

6. The pole-zero configuration of input impedance of a series resonant circuit on s-plane is

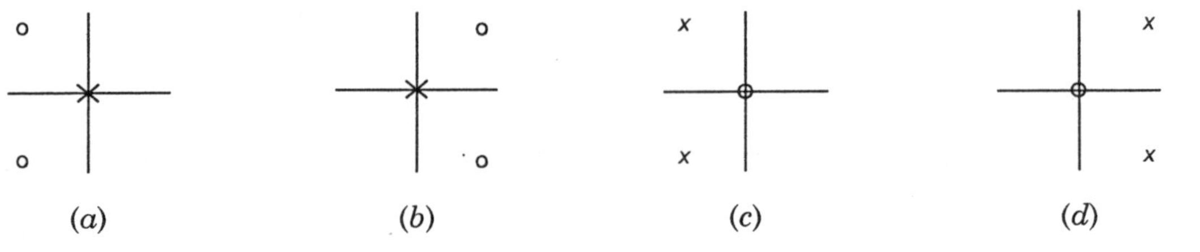

|        (a)        |        (b)        |        (c)        |        (d)        |

7. The current response as a function of frequency in a series resonant circuit is given by

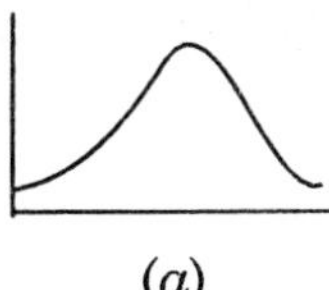 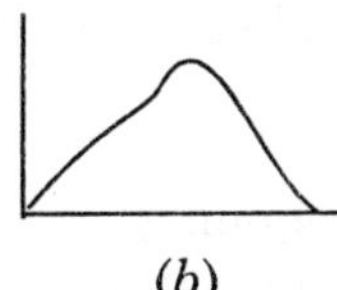 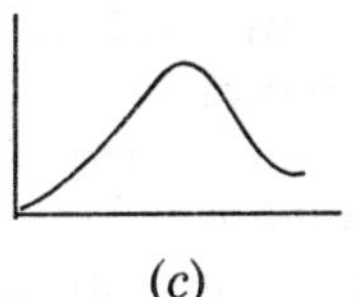 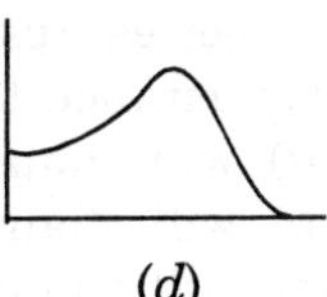

      (a)              (b)              (c)              (d)

8. At lower half power frequency the current in a series RLC circuit has the phase relation with respect to supply voltage.
   (a) 45° lead    (b) 45° lag    (c) 90° lead    (d) 90° lag
9. If a high $Q$ parallel resonant circuit is loaded
   (1) The circuit impedance reduces
   (2) The resonant frequency reduces
   (3) The bandwidth reduces
   Tick out the correct one
   (a) 1 and 2    (b) 2 and 3    (c) only 1    (d) 1, 2 and 3.
10. Pure inductive circuit takes power from the *ac* line when
    (a) both applied voltage and current rise.
    (b) both applied voltage and current decrease.
    (c) applied voltage decreases but current increases.
    (d) none of the above is necessary.
11. In an *ac* circuit having $R$, $L$ and $C$ in series and operating on lagging $p\,f$, increase in frequency will
    (a) reduce the current.        (b) increase the current.
    (c) (a) and (b) are possible.    (d) have no effect on current drawn.
12. In a circuit voltage and current are given by
    $$V = 10 \sin (\omega t + 30°)$$
    and              $$i = 10 \sin (\omega t - 30°)$$
    The power consumed in the circuit is
    (a) 100 watts    (b) 50 watts    (c) 25 watts    (d) 12.5 watts
13. A series circuit is given by $z = 5 - j12$, its susceptance is
    (a) 5/13        (b) 5/169        (c) 12/13        (d) 12/169
14. The current at a given point in a certain circuit is given as a function of time as
    $$i(t) = -3 + t$$
    The total charge passing a point between $t = 99$ sec and $t = 102$ sec is
    (a) 112 C    (b) 242.5 C    (c) 292.5 C    (d) 245.6 C
15. In a pure LC circuit (parallel) under resonance condition, current drawn from the supply main is
    (a) Very large    (b) $V\sqrt{LC}$    (c) $V/\sqrt{LC}$    (d) Zero
16. A drawn wire of resistance 5 Ω is further drawn so that its diameter becomes one fifth. Its resistance will now be (volume remaining same).
    (a) 625        (b) 125        (c) 25        (d) None of the above
17. A 500 watt 220 V bulb is supplied with 110 volts. Power consumption by the bulb will be

    (*a*) 250 watt                        (*b*) 125 watt
    (*c*) slightly more than 125 watt.     (*d*) slightly less than 125 watt.

**18.** In series resonance circuit, increasing inductance to twice its value and reducing capacitance to half its value
    (*a*) will change the resonance frequency.
    (*b*) will change the impedance at resonance frequency.
    (*c*) will change the maximum value of current.
    (*d*) will increase the selectivity of the circuit.

**19.** Voltage across capacitor in RLC series circuit is maximum
    (*a*) at resonance                     (*b*) just before resonance
    (*c*) just after resonance            (*d*) much after resonance

**20.** For the circuit shown in figure the value of current $I$ is
    (*a*) 10 A

    (*b*) 15 A

    (*c*) 20 A

    (*d*) 25 A

**21.** The maximum power that can be distributed in the load in the circuit shown is
    (*a*) 3 watts
    (*b*) 6 watts
    (*c*) 6.75 w
    (*d*) 13.5 watt

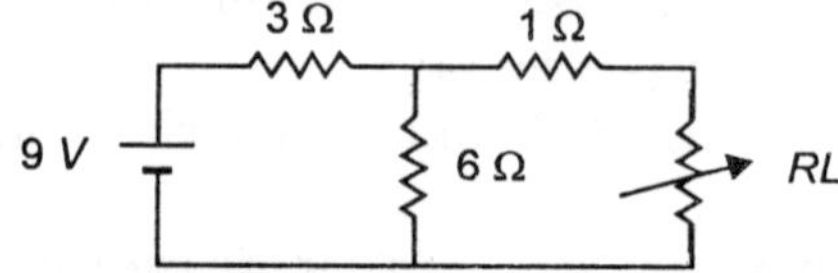

**22.** In delta connected circuit when one resister is open, the power will be
    (*a*) zero                           (*b*) reduced to 1/3
    (*c*) reduced by 1/3              (*d*) unchanged

**23.** In a network the sum of currents entering a node is $5\angle 60°$. The sum of currents leaving the node is
    (*a*) $5\angle 60°$       (*b*) $5\angle{-}60°$       (*c*) $5\angle 240°$       (*d*) 15A

**24.** A resistance of 10 ohm is connected in one branch of a network. The current in this branch is 2A. If this 10 $\Omega$ resister is replaced by a 20 ohms resistance, the current in this branch
    (*a*) may be more or less than 2A     (*b*) will be more than 2A
    (*c*) will be less than 2A                (*d*) none of these

**25.** If $V = a + jb$ and $I = (c + jd)$, the power is given by
    (*a*) $ac - bd$       (*b*) $ac + bd$       (*c*) $bc - ad$       (*d*) $bc + ad$

**26.** For a perfect transformer the voltage ratio $\dfrac{v_1}{v_2}$ is given as

    (*a*) $\dfrac{L_1}{L_2}$          (*b*) $\dfrac{L_2}{L_1}$          (*c*) $\sqrt{\dfrac{L_1}{L_2}}$          (*d*) none of the above.

**27.** An ideal transformer is
    (*i*) an abstract concept
    (*ii*) one where the voltage and current relations are derived from Faraday's law and Amperes' law respectively

(*iii*) is characterized by a single parameter $n$, the turns ratio

Tick out the correct option

(a) (*i*) and (*iii*)　　　　(b) (*i*) and (*ii*)
(c) (*ii*) and (*iii*)　　　　(d) (*i*), (*ii*) and (*iii*)

28. When a gyrator with gyration resistance $r$ is terminated through a resistor $R$, the equivalent element at the input terminals is

(a) $r^2 R$　　　(b) $r R$　　　(c) $\dfrac{r^2}{R}$　　　(d) $\dfrac{R}{r^2}$

29. When a gyrator with gyration resistance $r$ is terminated through a capacitor $C$ the equivalent element at the input terminal is

(a) a capacitor with value $r C$　　　　(b) a capacitor with value $r^2 C$
(c) an inductor with value $r C$　　　　(d) none of the above

30. When a negative converter with conversion ratio $k$ is terminated through an impedance $Z_L$, the equivalent element at the input terminals is

(a) $kZ_L$　　　(b) $k^2 Z_L$　　　(c) $- k Z_L$　　　(d) $- \dfrac{K^2}{Z_L}$

31. The $h$-parameter matrix of the NIC is

$$\begin{bmatrix} 0 & -k \\ -1/k & 0 \end{bmatrix} \qquad \begin{bmatrix} 0 & k \\ 1/k & 0 \end{bmatrix} \qquad \begin{bmatrix} k & 0 \\ 0 & 1/k \end{bmatrix} \qquad \begin{bmatrix} -k & 0 \\ 0 & -1/k \end{bmatrix}$$

　　　　(a)　　　　　　　　(b)　　　　　　　　(c)　　　　　　　　(d)

32. The input impedance of the given circuit is

(a) $\dfrac{s+2}{s+1}$　　　(b) $\dfrac{s}{s+1}$

(c) $\dfrac{s+1}{s+2}$　　　(d) none of the above.

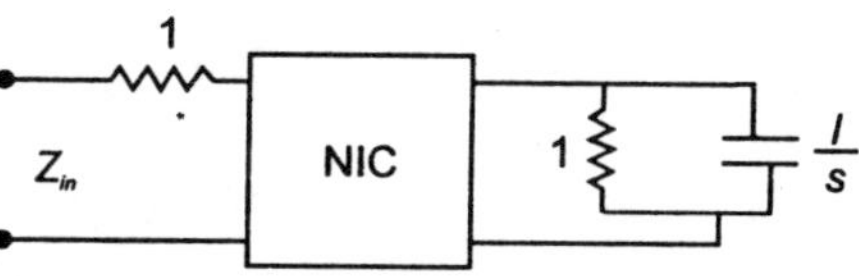

33. If a voltage source $V$ is connected in series with a capacitor $C$

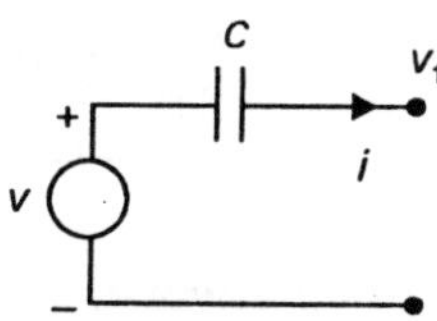

as shown in the fig its equivalent current source is given as

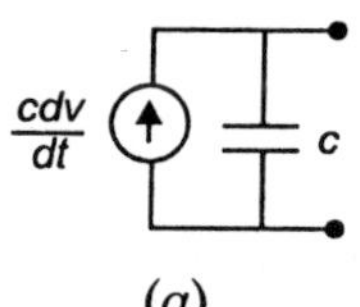

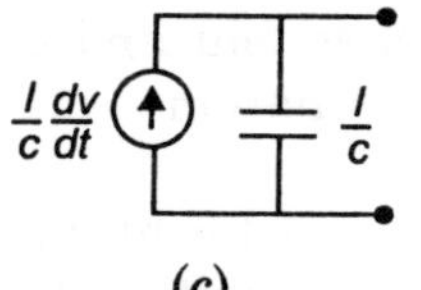

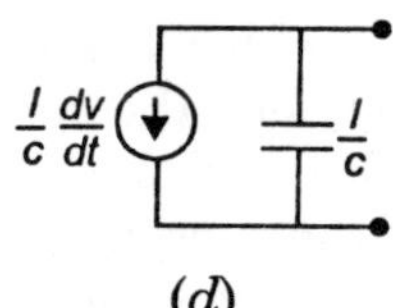

　　　　(a)　　　　　　　　(b)　　　　　　　　(c)　　　　　　　　(d)

34. KCL is a consequences of law of conservation of

(a) energy　　　(b) charge　　　(c) flux　　　(d) all the above

35. If $A$ is the incidence matrix of $N_1$ and $B$ the loop matrix of network $N_2$, the condition for the two networks to be dual of each other is that

(a) ran of [A] > rank of [B]  (b) rank of [A] < rank of [B]
(c) ran of [A] = rank of [B]  (d) none of the above is necessary

**36.** The current in the 1 Ω resistor is

(a) 5 A

(b) 10 A

(c) 15 A

(d) zero

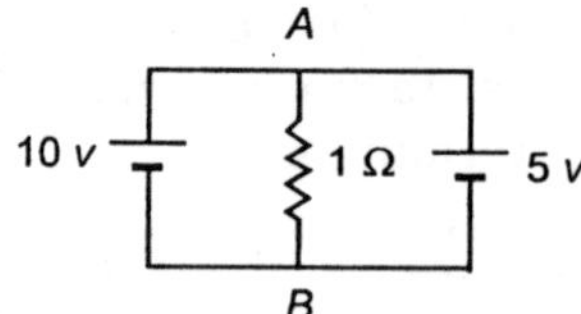


**37.** The current in a 5 Ω resistor branch in a linear network is 5 *A*. If this branch is replaced by a resistor of 10 Ω, the current in this branch will be
(a) 5 A                          (b) 10 A
(c) less than 5 A                (d) none of the above

**38.** The potential of the point *A* in the given network

(a) 6 V

(b) 7 V

(c) 8 V

(d) none of the above

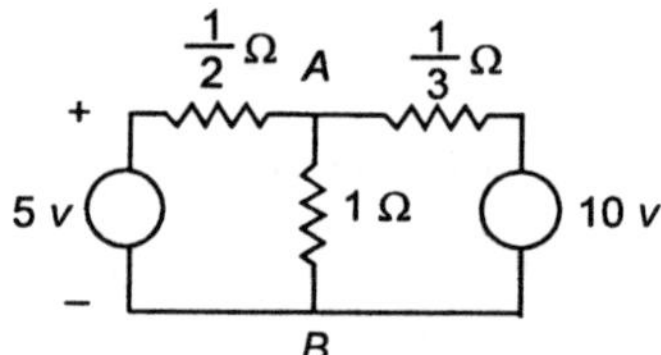


**39.** The current through 30 Ω branch in the given circuit is

(a) 2.5 A

(b) 2.25 A

(c) 2 A

(d) 10 A

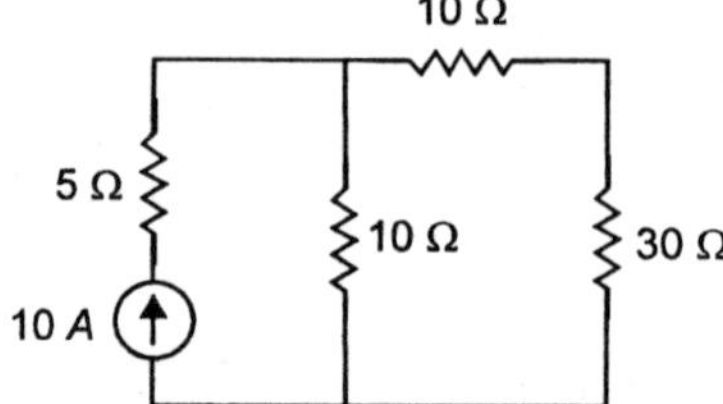


**40.** The current through 8 Ω branch is
(a) 1 A
(b) 0.5 A
(c) 1.5 A
(d) none of the above

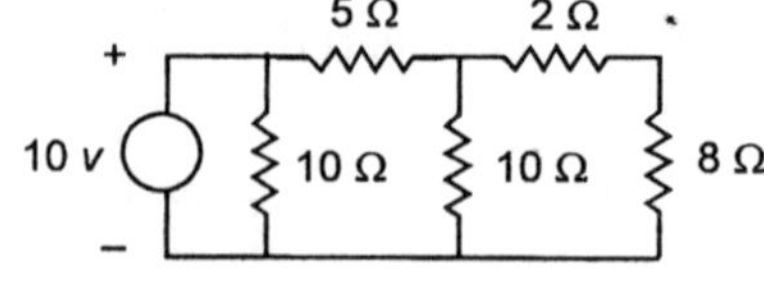


**41.** Reciprocity theorem is applicable to a network.
(a) which contains *R, L* and *C* as elements
(b) initially relaxed system
(c) which has both independent and dependent sources
Tick out the correct combination
(a) 1 and 2     (b) 1 and 3     (c) 2 and 3     (d) 1, 2 and 3

**42.** If the current in the 7 Ω resistor branch is 0.5 *A* as shown in the figure and now if the source is connected in series with 7 Ω branch and the terminals *AB* are shorted, the current in the 5 Ω resistor is
(a) 1 A
(b) 0.5 A
(c) 0.75 A
(d) none of the above

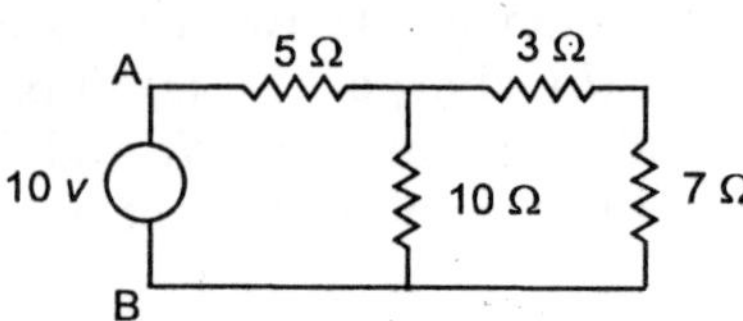

**43.** Tellegens' theorem is applicable to
   (*i*) a linear network only          (*ii*) time invariant only
(*iii*) passive and active elements
Tick out the correct combination
(*a*) 1 and 2     (*b*) 2 and 3       (*c*) 1 and 3
(*d*) none of the given combinations

**44.** The voltage across 5 A source in the given circuit is
   (*a*) 25 volt
   (*b*) 15 volt
   (*c*) 17.5 volt
   (*d*) 20 volt

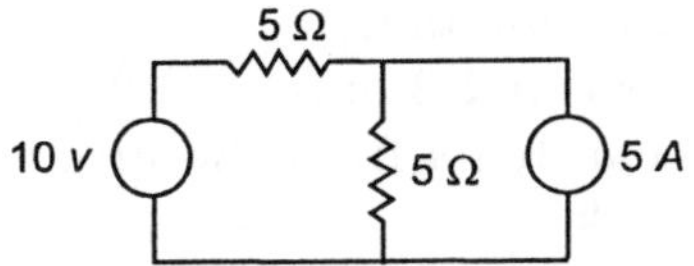


**45.** If $b$ is the number of branches and $n$ the number of nodes in a connected graph, the number of links corresponding to any tree of the graph is
(*a*) $n + 1 - b$   (*b*) $b - n + 1$   (*c*) $b - n - 1$   (*d*) $n - b - 1$

**46.** For a given network the incidence matrix is given as

$$
\begin{array}{ccccccc}
1 & 2 & 3 & 4 & 5 & 6 & 7
\end{array}
$$
$$
\begin{bmatrix}
1 & 0 & 0 & 1 & 0 & -1 & 1 \\
-1 & -1 & 1 & 0 & 0 & 0 & 0 \\
0 & 1 & 0 & -1 & 1 & 0 & 0
\end{bmatrix}
$$

The series branches in the graph are:
(*a*) 3 and 4   (*b*) 6 and 7   (*c*) 2 and 3   (*d*) none of the above.

**47.** In the previous problem the parallel branches are:
(*a*) 1 and 2   (*b*) 2 and 3   (*c*) 6 and 7   (*d*) none of the above.

**48.** For a given network the incidence matrix is given as

$$
\begin{array}{cccccc}
1 & 2 & 3 & 4 & 5 & 6
\end{array}
$$
$$
A = \begin{bmatrix}
1 & 0 & 0 & 1 & -1 & 0 \\
0 & 1 & 0 & -1 & 1 & -1 \\
0 & 0 & 1 & 0 & 0 & 1
\end{bmatrix}
$$

The series branches are:
(*a*) 3 and 4   (*b*) 3 and 5   (*c*) 3 and 6   (*d*) none of these.

**49.** In the above problem the parallel branches are:
(*a*) 3 and 5   (*b*) 4 and 5   (*c*) 3 and 6   (*d*) none of these.
**Hint:** Since at third node there are two unity entries, the two branches corresponding to these entries are in series i.e. branches 3 and 6 are in series.
Since the entries in column 4 and 5 are identical in row 1 and 2, therefore, branches 4 and 5 are in parallel.

**50.** For a given network the incidence matrix is given as in Q. 48. If $i_2 = 2A$ $i_4 = 4A$ $i_5 = 2A$, the current $i_6$ is given as
(*a*) 4 A          (*b*) 2 A          (*c*) zero A       (*d*) 6 A

**51.** $B_t$ and $B_l$ are sub-matrices of $B_f$
   (*i*) $B_t$ is an identity matrix
   (*ii*) $B_l$ is a rectangular matrix

(*iii*) The rank of $B_f$ is $b - n - 1$
Pick up the correct combination
(*a*) 1 and 2      (*b*) 2 and 3      (*c*) 1, 2, and 3      (*d*) none of the combination

**52.** $Q_t$ and $Q_l$ are the sub-matrices of $Q_f$
  (*i*) $Q_t$ is an identity matrix
  (*ii*) $Q_l$ may or may not be a rectangular matrix
 (*iii*) The rank of $Q_f$ is $(n - 1)$
Pick up the correct combination
(*a*) 1 and 2 only    (*b*) 1 and 3 only    (*c*) 2 and 3 only    (*d*) 1, 2 and 3

**53.** Arrange the topological duals from the given list
A  Loop       1  Link
B  Twig       2  node
C  Mesh       3  cut set

| | A | B | C |
|---|---|---|---|
| (*a*) | 1 | 2 | 3 |
| (*b*) | 1 | 3 | 2 |
| (*c*) | 3 | 1 | 2 |
| (*d*) | 3 | 2 | 1 |

**54.** From the given list pick up the correct combination.
A  $Z_{11}$       1  $Y_{11}/\Delta_y$
B  $Z_{22}$       2  $-Y_{12}/\Delta_y$
C  $Z_{12}$       3  $-Y_{21}/\Delta_y$
D  $Z_{21}$       4  $Y_{22}/\Delta_y$

| | A | B | C | D |
|---|---|---|---|---|
| (*a*) | 1 | 2 | 3 | 4 |
| (*b*) | 2 | 3 | 4 | 1 |
| (*c*) | 4 | 1 | 2 | 3 |
| (*d*) | 4 | 3 | 2 | 1 |

**55.** Tick out the correct relation
(*a*) $Z_{11} = Y_{11}/\Delta_z$
(*b*) $\quad\quad = Y_{22}/\Delta_y$
(*c*) $\quad\quad = Y_{11}/\Delta_y$
(*d*) $\quad\quad = Y_{22}/\Delta_y$

**56.** The $Z_{11}$ and $Z_{22}$ parameters of the given network are
(*a*) 8 Ω, 7.75 Ω
(*b*) 13 Ω, 9 Ω
(*c*) 12 Ω, 8.5 Ω
(*d*) none of the above

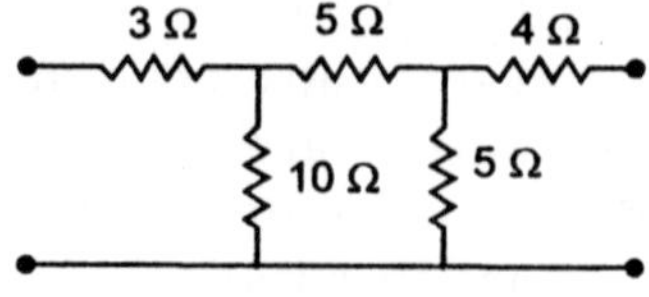

**57.** For the network shown, the parameters $h_{11}$ and $h_{21}$ are:
(*a*) 5 Ω, and $-2/3$
(*b*) 3.4 Ω and $-2/5$
(*c*) 3.4 Ω and $-3/5$
(*d*) none of the above

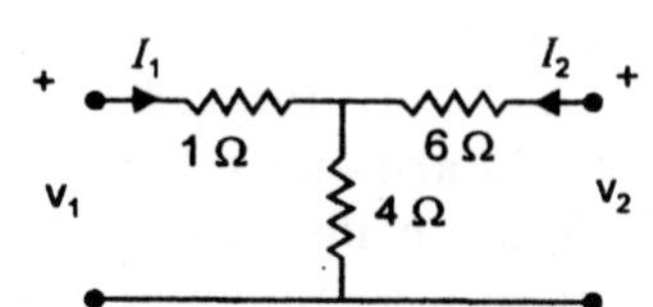

**58.** An inductor with inductance $L$ and initial current $i_0$ is shown here

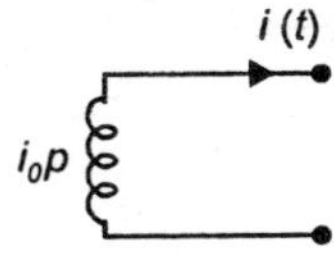

The correct impedance diagram for this circuit is

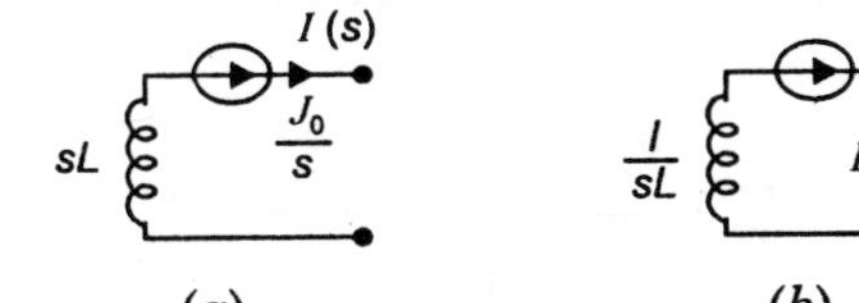

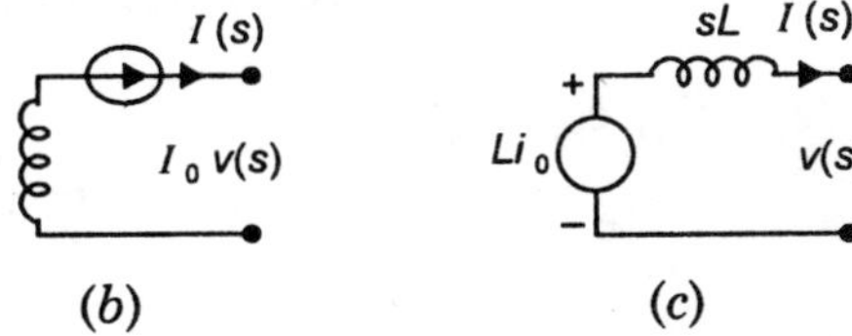

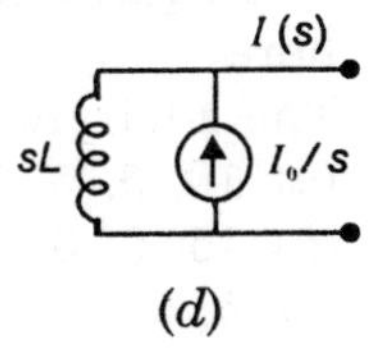

$(a)$        $(b)$        $(c)$        $(d)$

**59.** A capacitor with capacitance $c$ and initial voltage $v_c(0)$ is shown here

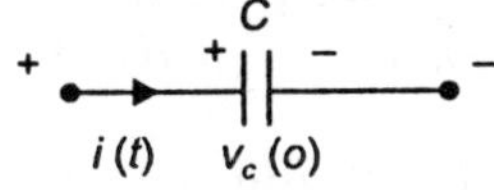

The correct admittance diagram for this circuit is

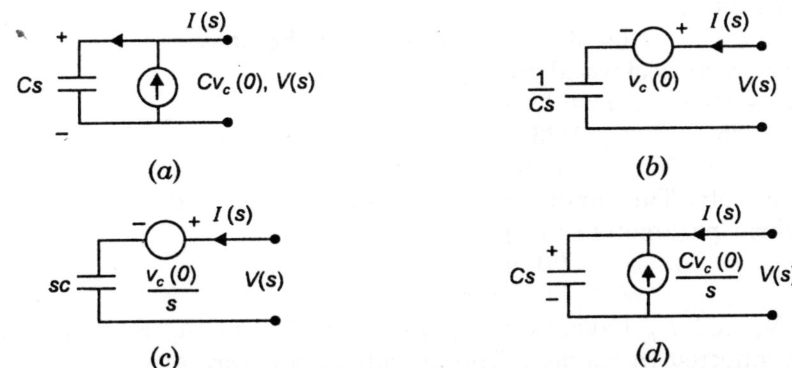

$(a)$        $(b)$

$(c)$        $(d)$

**60.** For Q. 59, the correct impedance diagram for the circuit is as shown in the above Figs a, b, c, d.

**61.** Substitution theorem is applicable for a network which has $(i)$ a unique solution $(ii)$ one or two non-linear elements $(iii)$ one non-linear or time varying element.
Choose the correct combination
$(a)$ 1 and 2     $(b)$ 1 and 3     $(c)$ 2 and 3     $(d)$ 1, 2 and 3

**62.** The equivalent circuit for a two-port network using z-parameters is

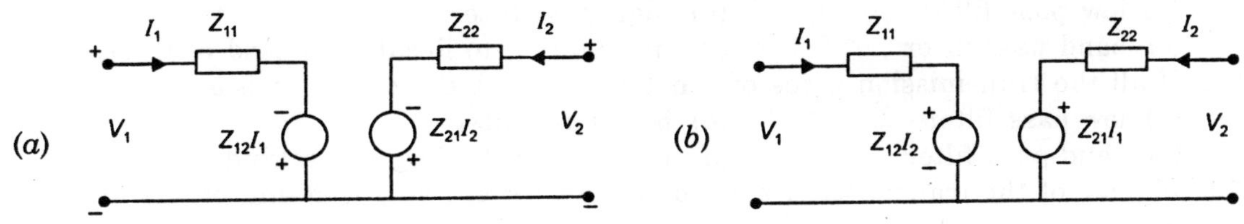

$(a)$        $(b)$

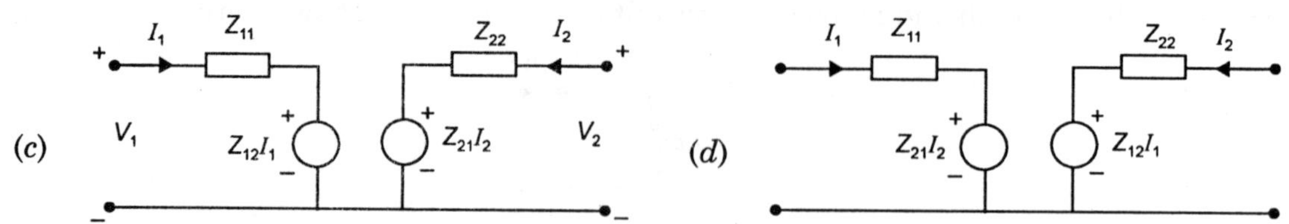


**63.** The equivalent circuit of a two-port reciprocal network using z-parameters is shown here

The z-parameters are ($z_{11}$, $z_{12}$, $z_{21}$, $z_{22}$)

(a) 10, 5, 5, 15

(b) 15, 5, 5, 20

(c) 5, 5, 5, 10

(d) 10, 10, 5, 15



**64.** The condition for reciprocity of a two-port network having different parameters are:

(i) $h_{12} = -h_{21}$          (ii) $g_{12} = -g_{21}$          (iii) $A = D$

Choose the correct combination

(a) 1 and 2      (b) 1 and 3      (c) 2 and 3      (d) 1, 2 and 3

**65.** The maximum value of the transmission parameter $A$ for a passive, reciprocal, linear two-port network is

(a) 1              (b) 2              (c) 3              (d) none of the above

**66.** For the passive linear reciprocal two port network

$A = 1$              $B = 10$              $D = 1$, the value of $C$ is

(a) 1              (b) 1/5              (c) 1/25          (d) none of the above

**67.** Two networks $N_a$ and $N_b$ have hybrid parameters $h_{11a}$, $h_{12a}$, $h_{21a}$, $h_{22a}$ and $h_{11b}$, $h_{12b}$, $h_{21b}$, $h_{22b}$ respectively. The input side is series connected and the output is parallel. The overall hybrid parameters are given as

(a) $[h_a] + [h_b]$                    (b) $[h_a] - [h_b]$

(c) $[h_a] [h_b]$                    (d) none of these

**68.** Two network $N_a$ and $N_b$ have hybrid parameters $h_a$ and $h_b$ respectively. The two networks are connected in cascade. The overall parameters are:

(a) $[h_a] + [h_b]$                    (b) $[h_a] - [h_b]$

(c) $[h_a] [h_b]$                    (d) none of these

**69.** Pick up the incorrect statement

(a) Two odd functions when added up result in an odd function.

(b) Two even functions when added up result in an even number.

(c) Two odd function multiplied together result in an odd function.

(d) Two even functions when multiplied together result in an even function.

**70.** If all the transmission zeros of a network are at infinity, it is a

(a) low pass filter                    (b) high pass filter

(c) band pass filter                    (d) no criterion to decide the kind of filter.

**71.** If all the transmission zeros of a network are at the origin, it is a

(a) low-pass filter                    (b) high-pass filter

(c) band-pass filter                    (d) no criterion to decide the kind of filter

**72.** If some of the transmission zeros of a network are at origin and others at infinity, it is a

(a) low pass filter                    (b) high pass filter

(c) band pass filter                    (d) no criterion to decide the kind of filter.

**73.** Examine the following pairs

| | Filter | | Transmission zeros |
|---|---|---|---|
| A | Low pass | 1. | infinity |
| B | High pass | 2. | origin |
| C | Band pass | 3. | some at infinity some at origin |

Arrange in the proper pairs

| | A | B | C |
|---|---|---|---|
| (a) | 1 | 3 | 2 |
| (b) | 1 | 2 | 3 |
| (c) | 2 | 1 | 3 |
| (d) | 3 | 1 | 2 |

**74.** A high-pass filter circuit is basically
(a) a differentiating circuit with low time constant
(b) a differentiating circuit with large time constant
(c) an integrating circuit with low time constant
(d) an integrating circuit with high time-constant

**75.** The ABCD parameters of an ideal transformer with $n : 1$ ratio are given as

(a) $\begin{bmatrix} n & 0 \\ 0 & -1/n \end{bmatrix}$  (b) $\begin{bmatrix} 0 & n \\ 1/n & 0 \end{bmatrix}$  (c) $\begin{bmatrix} n & 0 \\ 0 & 1/n \end{bmatrix}$  (d) $\begin{bmatrix} 0 & n \\ -1/n & 0 \end{bmatrix}$

**76.** An ideal transformer with ratio $n : 1$ is terminated through a capacitance $C$ at port 2. At port 1 it will appear as
(a) an inductor of value $n^2 C$  (b) a capacitance of value $n^2 C$
(c) an inductor of value $c/n^2$  (d) a capacitance of value $c/n^2$

**77.** A two-port network with short circuited admittance $Y_{11}$, $Y_{12}$ $Y_{21}$ $Y_{22}$ is terminated through a resistance $R$ at port 2. The overall $Y_{21}$ of the network is

(a) $\dfrac{Y_{21}}{Y_{22} + 1/R}$  (b) $Y_{21} + \dfrac{1}{R}$  (c) $\dfrac{Y_{21}/R}{Y_{22} + 1/R}$  (d) $\dfrac{Y_{21} + 1/R}{Y_{22}}$

**78.** A two-port network with open circuit impedances $Z_{11}$ $Z_{12}$ $Z_{21}$ and $Z_{22}$ is terminated through an impedance $Z_L$. The overall $Z_{21}$ is given as

(a) $Z_{21} = Z_{21} + Z_L$  (b) $Z_{21} = \dfrac{(Z_{21} + Z_L) Z_L}{Z_{22}}$

(c) $Z_{21} = \dfrac{Z_{21} Z_L}{Z_{21} + Z_L}$  (d) $Z_{21} = \dfrac{Z_{21} Z_L}{Z_{22} + Z_L}$

**79.** When a gyrator is connected in tandem with a passive reciprocal network, the overall two-port network acts as a
(a) passive reciprocal network  (b) active reciprocal network
(c) passive non-reciprocal network  (d) active non-reciprocal network

**80.** The following are the filter circuits

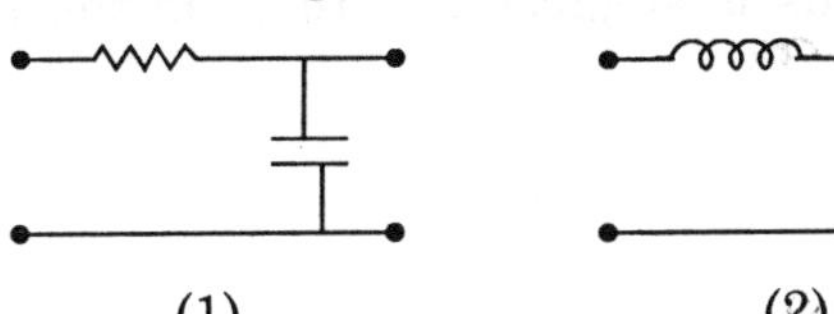

(1)　　　　　　　　(2)

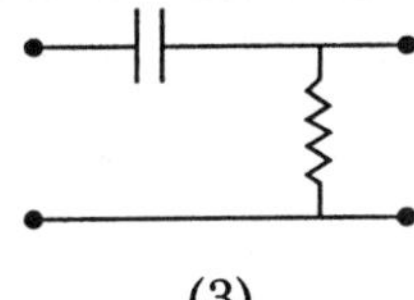

(3)

The low-pass filter circuits are:
(a) 1 alone      (b) 1 and 2      (c) 1 and 3      (d) 1, 2 and 3

**81.** The *h*-parameters of a negative impedance converter with $k$ as conversion factor are:

(a) $\begin{bmatrix} k & 0 \\ 0 & 1/k \end{bmatrix}$
(b) $\begin{bmatrix} 1/k & 0 \\ 0 & k \end{bmatrix}$
(c) $\begin{bmatrix} 0 & k \\ 1/k & 0 \end{bmatrix}$
(d) $\begin{bmatrix} 0 & 1/k \\ 0 & k \end{bmatrix}$

**82.** The value of the integral

$$\int_{-\infty}^{\infty} e^{5t}\, \delta(t-5)\, dt \text{ is}$$

(a) 1          (b) $(e^5 - 1)$      (c) $e^{25}$          (d) zero

**83.** An integrating circuit with the following options is suggested with $T$ as the pulse duration

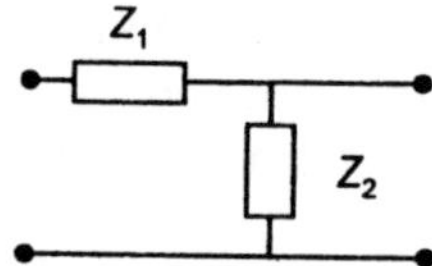

Tick out the correct arrangement
(a) $Z_1 = R$         $Z_2 = C$         $RC \ll T$
(b) $Z_1 = C$         $Z_2 = R$         $RC \ll T$
(c) $Z_1 = L$         $Z_2 = R$         $\dfrac{L}{R} \gg T$
(d) $Z_1 = R$         $Z_2 = L$         $\dfrac{L}{R} \gg T$

**84.** A differentiating circuit with the following options is suggested with $T$ as the pulse duration

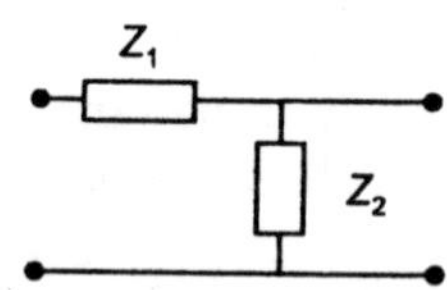

Tick out the correct arrangement
(a) $Z_1 = C$         $Z_2 = R$         $RC \ll T$
(b) $Z_1 = R$         $Z_2 = C$         $RC \ll T$
(c) $Z_1 = L$         $Z_2 = R$         $\dfrac{L}{R} \gg T$
(d) $Z_1 = R$         $Z_2 = L$         $\dfrac{L}{R} \gg T$

**85.** A differentiating circuit with the following options is suggested with $T$ as the pulse duration.

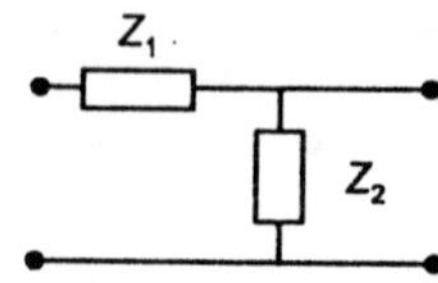

Tick out the correct arrangement

(a) $Z_1 = C$      $Z_2 = R$      $RC \gg T$

(b) $Z_1 = R$      $Z_2 = C$      $RC \gg T$

(c) $Z_1 = L$      $Z_2 = R$      $\dfrac{L}{R} \ll T$

(d) $Z_1 = R$      $Z_2 = L$      $\dfrac{L}{R} \ll T$

**86.** For the two port network shown, select the correct statement

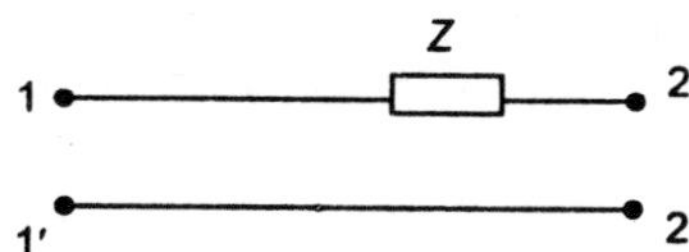

(a) It does not have $z$-parameters

(b) It has $z$-parameters

(c) It does not have $y$-parameters

(d) It does not have ABCD parameters

**87.** The convolution of a function $f(t)$ with the unit impulse function $\delta(t)$ is

  (a) $\delta(t)$      (b) $f(t)\,\delta(t)$      (c) $f(t)$      (d) $f(\tau)\,\delta(t)$

**88.** In the given figure the current I leads the voltage V if

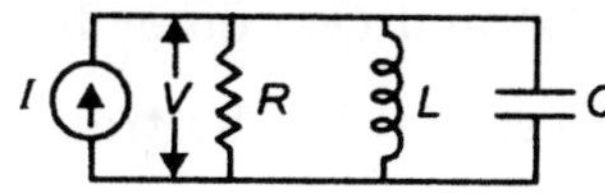

(a) $\omega L < \dfrac{1}{\omega C}$              (b) $\omega L > \dfrac{1}{\omega C}$

(c) $\omega L = \dfrac{1}{\omega C}$              (d) $R = \left[\omega L + \dfrac{1}{\omega C}\right]$

**89.** The impulse response of the first order system described by the differential equation.

$$\frac{dy}{dt} + 4y = u(t) \text{ is}$$

(a) $4\,\delta(t)$      (b) $e^{-2t}\,u(t)$      (c) $e^{-4t}$      (d) none of the above

**90.** The impulse response of a system whose transfer function is given as

$$\frac{1}{(s+1)(s+2)} \text{ is}$$

(a) $\dfrac{1}{2}e^{-t} - e^{-2t}$           (b) $e^{-t} - \dfrac{1}{2}e^{-2t}$

(c) $e^{-t} - e^{-2t}$              (d) none of the above

**91.** The impulse response of a circuit is given by

$$h(t) = \frac{1}{L}\, e^{-\frac{R}{L}t}\, u(t)$$

It's step response is given as

(a) $\left(1 - e^{-\frac{R}{L}T}\right) u(t)$

(b) $\dfrac{1}{R}\left(1 - e^{-\frac{R}{L}t}\right) u(t)$

(c) $\dfrac{L}{R}(1 - e^{-\frac{R}{L}t})\, u(t)$

(d) none of the above

**92.** Read the following statements:
1. The transform of a convolution is the product of the transforms.
2. The transform of a product is the convolution of the transform.
3. The convolution of two functions is the transform of the product of their transforms.
   Tick out the correct combination
   (a) 1 and 2    (b) 1 and 3    (c) 2 and 3    (d) 1, 2 and 3

**93.** A periodic funciton $f(t)$ with time period $T$ repeats itself after half time period $T/2$. The Fourier series of $f(t)$ would contain
(a) cosine terms only
(b) odd harmonic terms only
(c) even harmonic terms only
(d) sine terms only

**94.** For a transfer function $H(s) = P(s)\,/\,Q(s)$ where $P(s)$ and $Q(s)$ are polynomials of $s$
(a) The degree of $P(s)$ and $Q(s)$ are same.
(b) The degree of $P(s)$ is always greater than that of $Q(s)$.
(c) The degree of $P(s)$ is independent of the degree of $Q(s)$.
(d) The maximum degree of $P(s)$ and $Q(s)$ differ by one.

**95.** The d.c. gain of a system represented by the transfer function $\dfrac{25}{(s + 2)\,(s + 3)}$ is
(a) 25    (b) 25/6    (c) 5    (d) 10

**96.** Which of the following statement is true for a delayed step function $u(t - T)$?
(a) It has a finite Fourier series.
(b) It has an infinite Fourier series.
(c) It does not have Fourier series.
(d) Its Laplace transform is 1/s.

**97.** The Fourier series expansion of a periodic function with half wave symmetry contains only
(a) sine terms    (b) cosine terms    (c) odd harmonics    (d) even harmonic

**98.** The final value of the function

$$F(s) = \frac{3s + 2}{s^4 + 6s^3 + 10s^2 + s} \quad \text{is}$$

(a) 1    (b) 2    (c) zero    (d) infinity

**99.** A capacitor $C$ is connected across a coil with resistance $R$ and inductance $L$. The effective impedance of the circuit at resonance is

(a) $\dfrac{C}{RL}$    (b) $\dfrac{L}{RC}$    (c) $\dfrac{R}{LC}$    (d) $\dfrac{LC}{R}$

**100.** The current magnification of the circuit in P. 99 at resonance is

(a) $\dfrac{C}{RL}$    (b) $\dfrac{1}{R}\sqrt{\dfrac{C}{L}}$    (c) $\dfrac{1}{R}\sqrt{\dfrac{L}{C}}$    (d) $\dfrac{RC}{L}$

**101.** Two coils are wound on a common magnetic core. The sign of mutual inductance $M$ for finding out effective inductance of each coil is positive if
(a) Two coils are wound in the same sense.

(b) Fluxes produced by the two coils are equal.

(c) Fluxes produced by the coils act in the same direction.

(d) Fluxes produced by the two coils act in opposition.

**102.** The magnitude of the function:

$$F(s) = \frac{s^2 - 6s + 8}{s^2 + 6s + 8}$$

(a) increases with increase in frequency

(b) decreases with decrease in frequency

(c) has some relations with the frequency

(d) is independent of the frequency.

**103.** A 500 ohm resistance attenuator is to have an attenuation of 20 dbs. The resistances of the series and shunt arms of the lattice network, respectively are:

(a) 400 $\Omega$, 500 $\Omega$ approx.    (b) 410 $\Omega$, 510 $\Omega$ approx.

(c) 410 $\Omega$, 610 $\Omega$ approx.    (d) none of the above.

**104.** A network $N$ with impedance matrix

$$\begin{bmatrix} z_{11} & z_{12} \\ z_{21} & z_{22} \end{bmatrix} \text{ is followed by an}$$

ideal transformed with 1 : a ratio, the overall impedance matrix is

(a) $\begin{bmatrix} a\,z_{11} & z_{12} \\ z_{21} & a^2\,z_{22} \end{bmatrix}$
    (b) $\begin{bmatrix} z_{11} & a\,z_{12} \\ a\,z_{21} & z_{22} \end{bmatrix}$

(c) $\begin{bmatrix} z_{11} & a\,z_{12} \\ a\,z_{21} & a^2\,z_{22} \end{bmatrix}$
    (d) $\begin{bmatrix} a^2\,z_{11} & a\,z_{12} \\ a\,z_{21} & a^2\,z_{22} \end{bmatrix}$

**105.** A network $N$ with impedance matrix

$$\begin{bmatrix} z_{11} & z_{12} \\ z_{21} & z_{22} \end{bmatrix} \text{ is preceded by an}$$

ideal transformer with ratio 1: a, the overall impedance matrix is

(a) $\begin{bmatrix} a\,z_{11} & z_{12} \\ z_{21} & a\,z_{22} \end{bmatrix}$
    (b) $\begin{bmatrix} z_{11}/a^2 & z_{12}/a \\ z_{21}/a & z_{22} \end{bmatrix}$

(c) $\begin{bmatrix} a^2\,z_{11} & a\,z_{12} \\ a\,z_{21} & z_{22} \end{bmatrix}$
    (d) $\begin{bmatrix} z_{11}/a & a\,z_{12} \\ a\,z_{21} & z_{22} \end{bmatrix}$

**106.** A network $N$ with short circuit admittance matrix

$$\begin{bmatrix} Y_{11} & Y_{12} \\ Y_{21} & Y_{22} \end{bmatrix} \text{ is followed by}$$

and ideal transformer with 1 : $a$ ratio, the overall admittance matrix is

(a) $\begin{bmatrix} Y_{11} & a Y_{12} \\ a Y_{21} & Y_{22} \end{bmatrix}$
    (b) $\begin{bmatrix} Y_{11}/a & a Y_{12} \\ a Y_{21} & a Y_{22} \end{bmatrix}$

(c) $\begin{bmatrix} Y_{11} & Y_{12}/a \\ Y_{21}/a & Y_{22}/a^2 \end{bmatrix}$  (d) $\begin{bmatrix} Y_{11} & Y_{12}/a \\ Y_{21}/a & aY_{22} \end{bmatrix}$

**107.** A network $N$ with short circuit admittance matrix

$$\begin{bmatrix} Y_{11} & Y_{12} \\ Y_{21} & Y_{22} \end{bmatrix}$$ is preceded by

an ideal transformer with $1 : a$ ratio the overall admittance matrix is

(a) $\begin{bmatrix} Y_{11} & Y_{12}/a \\ Y_{21}/a & a^2 Y_{22} \end{bmatrix}$  (b) $\begin{bmatrix} a^2 Y_{11} & aY_{12} \\ aY_{21} & Y_{22} \end{bmatrix}$

(c) $\begin{bmatrix} Y_{11} & aY_{12} \\ aY_{21} & a^2 Y_{22} \end{bmatrix}$  (d) $\begin{bmatrix} Y_{11}/a & aY_{12} \\ aY_{21} & Y_{22}/a \end{bmatrix}$

**108.** A lattice network has its series and shunt arms 50 Ω and 60 Ω. Its equivalent $T$ network has total series resistance and shunt resistances respectively.
(a) 100 Ω, 55 Ω  (b) 100 Ω, 50 Ω
(c) 100 Ω, 5 Ω  (d) none of the above.

**109.** The equivalent $\pi$ network in the question 108 has total series resistance and one shunt branch resistance
(a) 100 Ω, 60 Ω  (b) 50 Ω, 60 Ω
(c) 600 Ω, 60 Ω  (d) none of the above.

**110.** With regard to Fourier Transform tick out the correct combination

| | Function | | Transform |
| --- | --- | --- | --- |
| A | Real and even | 1. | Real and even |
| B | Real and odd | 2. | Real and odd |
| C | Real even plus imaginary odd | 3. | Complex and even |
| | | 4. | Real |
| D | Complex and even | 5. | Imaginary and odd |

| | A | B | C | D |
| --- | --- | --- | --- | --- |
| (a) | 1 | 2 | 3 | 4 |
| (b) | 1 | 5 | 3 | 4 |
| (c) | 1 | 5 | 4 | 3 |
| (d) | 2 | 5 | 4 | 3 |

**111.** In a series RLC circuit excited by a voltage $e = E\sin \omega t$, where $LC < 1/\omega^2$
(a) current lags the applied voltage
(b) current leads the applied voltage
(c) current is in phase with the applied voltage
(d) voltages across $L$ and $C$ are equal

**112.**

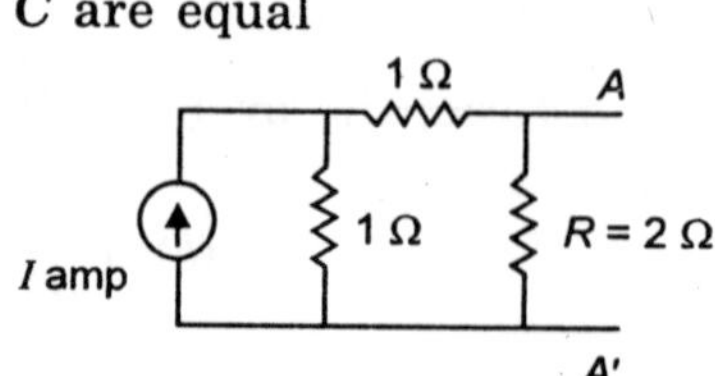

In the circuit shown in the above figure, I amp. of current flows through the resistance *R*. If a battery of emf 2 volts and internal resistance of one ohm is connected between the terminals *A–A'* (with the +ve terminal connected to *A'*, the current through *R* would be

(a) 2 amps    (b) 1.66 amps   (c) 1 amp    (d) 1.5 amps

**113.**

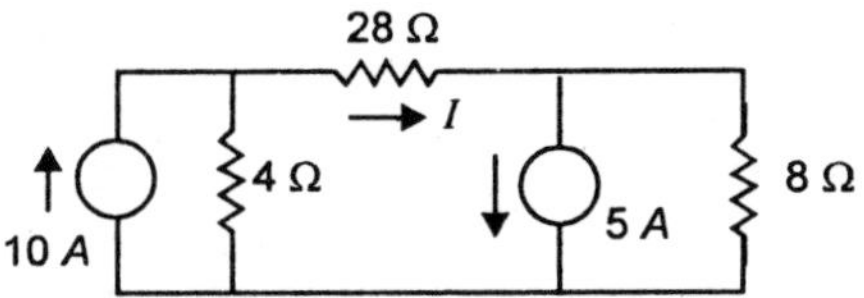

In the circuit shown in the above figure, the current I will be

(a) 1A          (b) 2A          (c) 4A          (d) 8A

**114.**

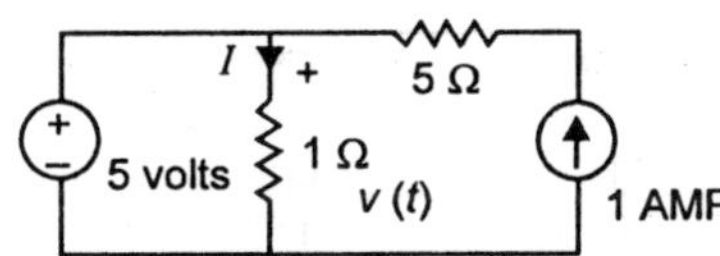

The circuit shown in the above figure is linear and time-invariant. The sources are ideal. The voltage across the one ohm resistor and the current through it will be

(a) −5V and −5A          (b) 1V and 1A
(c) 1V and 6A            (d) 5V and 5A

**115.**

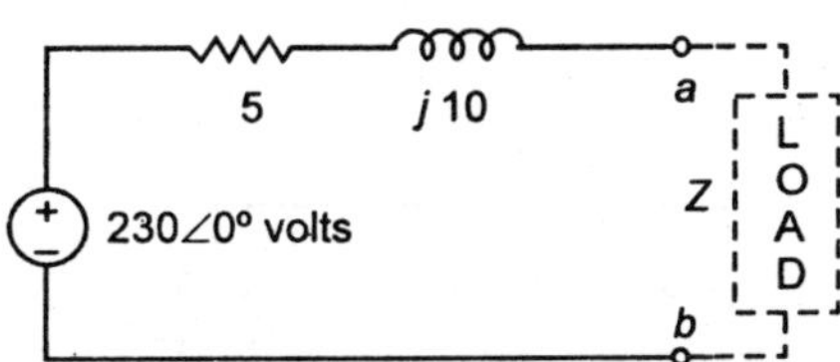

The generator with internal impedance of $5 + j10$ has a voltage $230 \angle 0°$ across it's terminals *ab* as shown in the above figure load is connected across terminals *ab*. For maximum power to be transmitted by the generator to the load, the impedance of the load to be connected across *ab* should be

(a) $+j10$       (b) $5 - j10$       (c) $-5 - j10$       (d) 15

**116.** The transfer function of an electrical low-pass RC network is

(a) $\dfrac{RCs}{1 + RCs}$    (b) $\dfrac{1}{1 + RCs}$    (c) $\dfrac{RC}{1 + RCs}$    (d) $\dfrac{s}{1 + RCs}$

**117.**

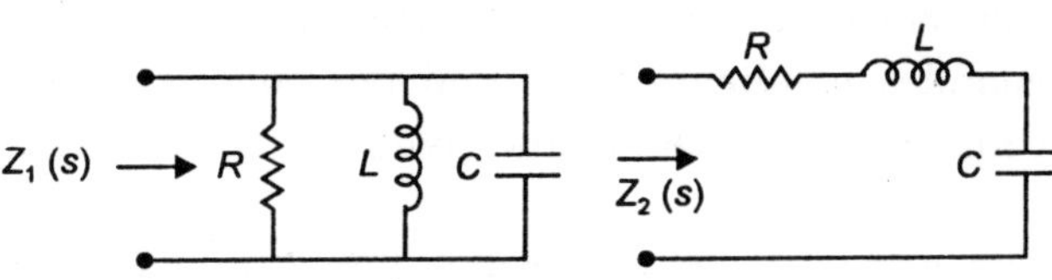

$z_1(s)$ and $z_2(s)$ are the driving point impedances of the networks shown in the above figures. Match list 1 with list 2 and select the correct answer using the codes given below the lists:

<table>
<tr><td align="center">*List 1*<br>(Stipulated condition<br>for the impedance)</td><td align="center">*List 2*<br>(Value of the<br>impedance)</td></tr>
<tr><td>A. $\lim\limits_{s \to 0} z_1(s)$</td><td>1. $R$</td></tr>
<tr><td>B. $\lim\limits_{s \to \infty} z_1(s)$</td><td>2. $sL$</td></tr>
<tr><td>C. $\lim\limits_{s \to 0} z_2(s)$</td><td>3. $1/sL$</td></tr>
<tr><td>D. $z_2(s)$ at<br>$s = j\,\omega_0 = j/\sqrt{LC}$</td><td>4. $1/sC$</td></tr>
</table>

*Codes:*

| | A | B | C | D |
|---|---|---|---|---|
| (a) | 2 | 4 | 4 | 1 |
| (b) | 1 | 2 | 3 | 4 |
| (c) | 2 | 4 | 3 | 1 |
| (d) | 1 | 3 | 4 | 2 |

**118.**

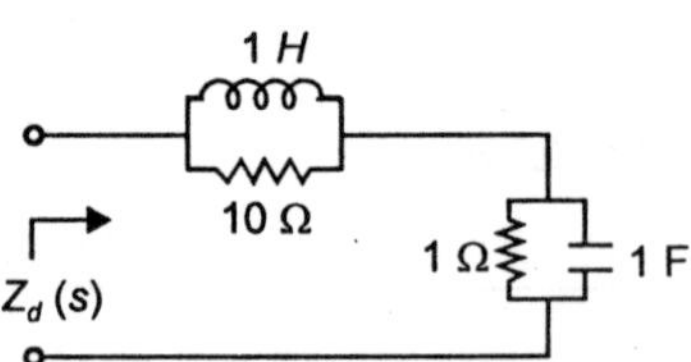


For the network shown in the above figure, which of the following statements are true?

1. $\lim\limits_{s \to 0} z_d(s) = 1$ ohm

   $\lim\limits_{s \to \infty} z_d(s) = 10$ ohm

2. $\lim\limits_{s \to 0} z_d(s) = 10$ ohm

   $\lim\limits_{s \to \infty} z_d(s) = 1$ ohm

3. $z_d(s)$ has two real poles and two complex conjugate zeros.

4. $z_d(s)$ has two complex conjugate poles and two complex conjugate zeros.

Select the correct answer using the codes given below:

*Codes:*

(a) 1 and 3     (b) 1 and 4     (c) 2 and 3     (d) 2 and 4

**119.**

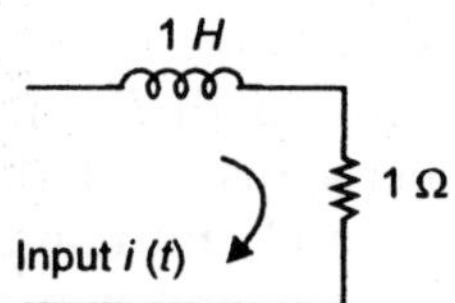



Consider the unenergised circuit shown in the above figure. Match list 1 with list 2 and select the correct answer using the codes given below the lists:

| List 1 | List 2 |
|---|---|

A. $i(t)$ for unit impulse input

$$1. \quad \mathcal{L}^{-1} \frac{1}{(s+1)(s^2+1)}$$

B. $i(t)$ for unit step input

$$2. \quad \int_0^t e^{-(t-\tau)} \, d\tau$$

C. $i(t)$ for unit ramp input

$$3. \quad e^{-t}$$

D. $i(t)$ for input of the form $\sin t$

$$4. \quad \int_0^t e^{-(t-\tau)} \tau \, d\tau$$

*Codes:*

| | A | B | C | D |
|---|---|---|---|---|
| (a) | 2 | 3 | 1 | 4 |
| (b) | 3 | 2 | 1 | 4 |
| (c) | 3 | 2 | 4 | 1 |
| (d) | 2 | 3 | 4 | 1 |

**120.**

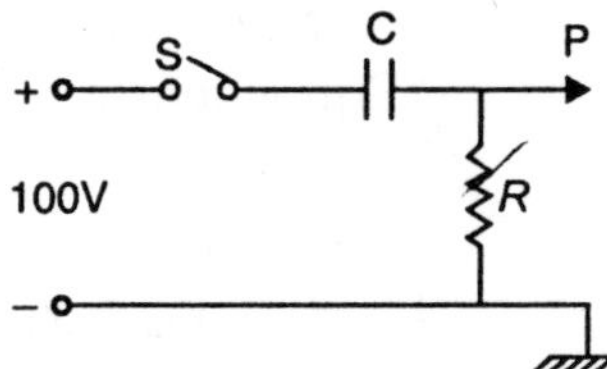



In the network shown in the above figure, $C = 5\ \mu F$ and $R = 2M\Omega$ The potential at $P$, at the instant of closing the switch $S$ and 10 seconds after closing $S$ will respectively be

(a) 0V and 63.2V      (b) 100V and 63.2V
(c) 0V and 36.8V      (d) 100V and 36.8V

**121.** The system described by

$$\dot{x}_1 = -x_1 + x_2$$
$$\dot{x}_2 = 2x_1 - 3x_2$$

will be

(a) stable      (b) unstable but not oscillatory
(c) oscillatory      (d) marginally stable

**122.** A linear series $RC$ circuit (initially relaxed so that no initial energy is stored in $C$), excited by a periodic square wave voltage source will give
(a) square wave voltage drops across both $R$ and $C$.
(b) non-square wave voltage drops across both $R$ and $C$
(c) square wave voltage drop across $R$ and non-square wave voltage drop across $C$
(d) square wave voltage drop across $C$ and non-square wave voltage drop across $R$.

**123.**

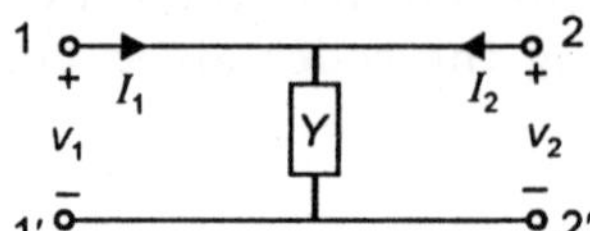

For the linear time-invariant two-port network shown in the above figure, the admittance matrix $[Y]$ representation will be

(a) $\begin{bmatrix} Y & -Y \\ -Y & Y \end{bmatrix}$ 
        (b) Null matrix

(c) $\begin{bmatrix} Y & \\ & Y \end{bmatrix}$ 
        (d) Indeterminate

**124.** A 2-port network is terminated in a load $Z_L$ at its output port, the input impedance of the terminated two-port network is

(a) $\dfrac{\Delta Z + Z_{22}\, Z_L}{Z_{11} + Z_L}$ 
        (b) $\dfrac{Z_{22}}{\Delta Z}$

(c) $\dfrac{\Delta Z + Z_{11}\, Z_L}{Z_{22} + Z_L}$ 
        (d) $\dfrac{Z_{11}}{\Delta Z}$

**125.** Two-port parameters and their equivalents are given in the two lists shown below. Match the two lists and select the correct answer using the codes given below the lists:

| | *List 1*<br>(Parameters) | | *List 2*<br>(Equivalents) |
|---|---|---|---|
| A | $z_{21}$ | 1. | $-y_{21}/y_{22}$ |
| B | $h_{21}$ | 2. | $y_{21}/y_{11}$ |
| C | $g_{21}$ | 3. | $-(y_{11}\, y_{22} - y_{12}\, y_{21})/y_{21}$ |
| D | $C$ | 4. | $-\dfrac{y_{21}}{y_{11}\, Y_{22} - y_{12}\, y_{21}}$ |

*Codes:*

| | A | B | C | D |
|---|---|---|---|---|
| (a) | 3 | 4 | 2 | 1 |
| (b) | 2 | 1 | 4 | 3 |
| (c) | 4 | 2 | 1 | 3 |
| (d) | 4 | 1 | 3 | 2 |

**126.**

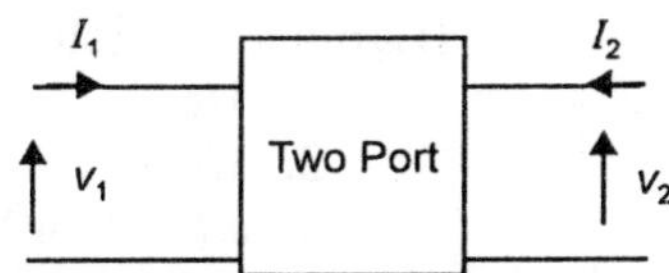



Consider the following statements regarding the two-port network shown in the above figure. The transfer functions of the network would include

(a) $V_2/V_1$, $I_2/I_1$, $V_2/I_1$ and $I_2/V_1$    (b) $V_2/I_2$ and $V_1/I_1$

(c) $V_2/V_1$ and $I_2/I_1$    (d) $V_2/I_1$ and $I_2/V_1$

**127.**

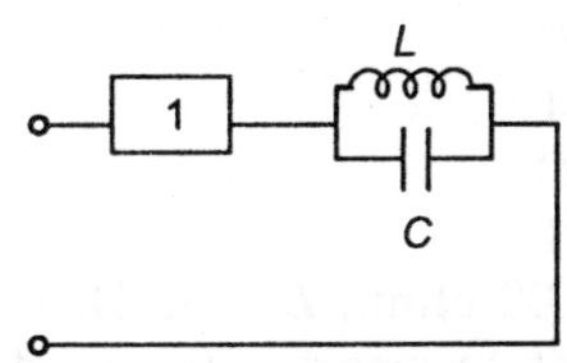



A reactance network in the first Foster form has poles at $\omega = $ zero and $\omega = $ infinity. The element in box 1 in the network is

(a) an inductor    (b) a capacitor

(c) a parallel LC circuit    (d) a series LC circuit

**128.** Consider the following statements regarding positive real function $F(s)$:

1. $F(s)$ is real when $s$ is real.
2. $F(s) \geq 0$, when $Re(s) \geq 0$.
3. The poles and zeros of $F(s)$ are in the right half of the s-plane.

Of these statements

(a) 1 and 2 are correct    (b) 1 and 3 are correct

(c) 2 and 3 are correct    (d) 1, 2 and 3 are correct

**129.**

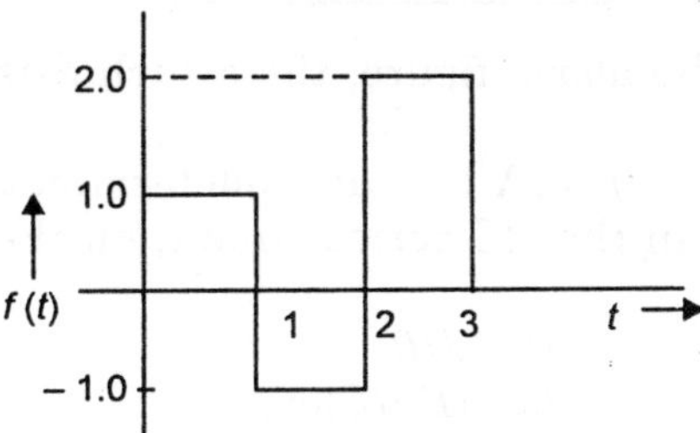



Laplace transform of $f(t)$ shown in the above figure is

(a) $F(s) = \dfrac{1}{s} - \dfrac{2}{s}e^{-s} + \dfrac{3\,e^{-s}}{s}$      (b) $F(s) = \dfrac{1}{s} - \dfrac{2e^{-s}}{s} + \dfrac{3\,e^{-2s}}{s} - \dfrac{2e^{-3s}}{s}$

(c) $F(s) = \dfrac{1}{s} - \dfrac{e^{-s}}{s} + \dfrac{2\,e^{-2s}}{s} - \dfrac{2e^{-3s}}{s}$      (d) $= \dfrac{1}{s} + \dfrac{2\,e^{-s}}{s} - \dfrac{3e^{-s}}{s}$

**130.** Match list 1 with list 2 and select the correct answer using the codes given below the lists:

|  List 1 | List 2 |
| --- | --- |
| (Time functions) | (Laplace transforms) |

| | List 1 (Time functions) | | | List 2 (Laplace transforms) |
| --- | --- | --- | --- | --- |
| A | 1 | | 1. | $1/s$ |
| B | $t$ | | 2. | $1/s^2$ |
| C | $\sin \omega t$ | | 3. | $s/(s^2 + \omega^2)$ |
| D | $\cos \omega t$ | | 4. | $\omega/(s^2 + \omega^2)$ |

*Codes:*

| | A | B | C | D |
| --- | --- | --- | --- | --- |
| (a) | A | B | C | D |
| | 1 | 2 | 3 | 4 |
| (b) | A | B | C | D |
| | 2 | 1 | 3 | 4 |
| (c) | A | B | C | D |
| | 1 | 2 | 4 | 3 |
| (d) | A | B | C | D |
| | 2 | 1 | 4 | 3 |

**131.** A series $RL$ circuit with $R = 100$ ohm., $L = 50$ H, is supplied by a d.c. source of 100V. The time taken for the current to rise to 70 p.c. of its steady state value is
(a) 0.3 sec     (b) 0.6 sec
(c) 2.4 sec     (d) 70 p.c. of time required to reach steady state

**132.** A coil with a certain number of turns has a specified time constant. If the number of turns is doubled, its time constant would
(a) remain unaffected          (b) become doubled
(c) become four-fold          (d) get halved

**133.**

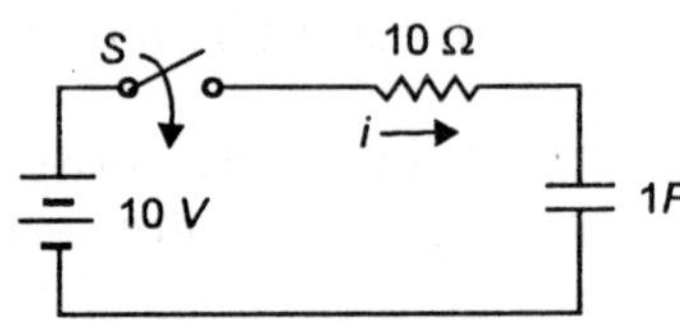


In the circuit shown in the above figure, the switch $S$ is closed at $t = 0$. The value of current at $t = 0+$ will be
(a) zero     (b) −1A     (c) +1A     (d) indeterminate

**134.** The steady-state current in the $RC$ series circuit, on the application of a step voltage of magnitude $E$ will be
(a) zero                    (b) $E/R$
(c) $(E/R)$ exp-$t/RC$          (d) $(E/RC)$exp-$t$

**135.** Consider the following statements:
In a network at resonance
1. the immittance is maximum
2. the power factor is unity irrespective of the network
3. the $Q$ of a series $RLC$ resonant circuit is independent of $R$
Of these statements
(a) 1 and 3 are correct          (b) 1 and 2 are correct
(c) 2 and 3 are correct          (d) 1 alone is correct

**136.** A minimum-phase network is one whose transfer function has
(a) zeros in the left-hand plane and poles in the right-hand plane.
(b) zeros and poles in the left-hand plane.

(c) zeros in the right-hand plane but poles in the left-hand plane.

(d) arbitrary distribution of zeros and poles in the s-plane

**137.** Three identical impedances are connected in delta to a 3-phase supply of 400 V. The line current is 34.65 A and the total power taken is 14.4 kW. The resistance of the load in each phase (in ohms) is

(a) 20      (b) 16      (c) 12      (d) 10

**138.** Readings of 1154 and 577 Watts were obtained when the two-wattmeter method was used on a balanced load. The delta connected load impedance for a system voltage of 100 V will be

(a) $15 \angle \pm 30°$    (b) $15 \angle +30°$    (c) $15 \angle -30°$    (d) $15 \angle +90°$

**139.** When two coils having self-inductances of $L_1$ and $L_2$ are coupled through a mutual inductance $M$, the coefficient of coupling, $K$ is given by

(a) $K = \dfrac{M}{\sqrt{2 L_1 L_2}}$        (b) $K = \dfrac{M}{\sqrt{L_1 L_2}}$

(c) $K = \dfrac{2 M}{\sqrt{L_1 L_2}}$        (d) $K = \dfrac{L_1 L_2}{M}$

**140.** The overall inductance of two coils connected in series, with mutual inductance aiding self-inductance is $L_1$; With mutual inductance opposing self-inductance, the overall inductance is $L_2$. The mutual inductance $M$ is given by

(a) $L_1 + L_2$    (b) $L_1 - L_2$    (c) $\dfrac{1}{4}(L_1 - L_2)$   (d) $\dfrac{1}{2}(L_1 + L_2)$

**141.** Consider the following statements:

The coefficient of coupling between two coils depends upon

1. orientation of the coils.      2. core material.
3. number of turns on the two coils.
4. self-inductances of the two coils.

Of these statements

(a) 1, 2, and 3 are correct      (b) 1 and 2 are correct

(c) 3 and 4 are correct      (d) 1, 2 and 4 are correct

**142.** Four networks are shown below in Figures (1), (2), (3) and (4)

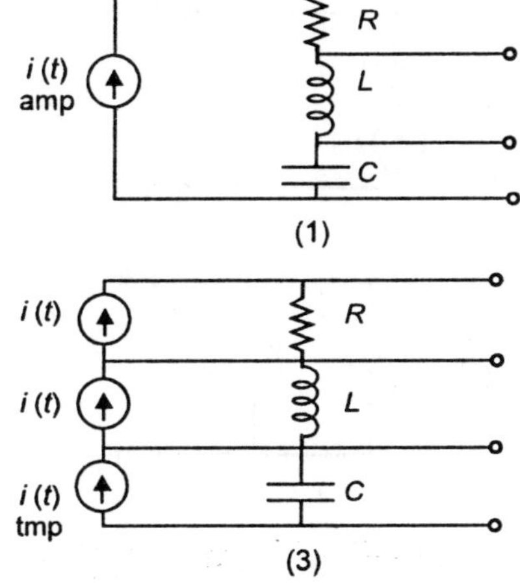


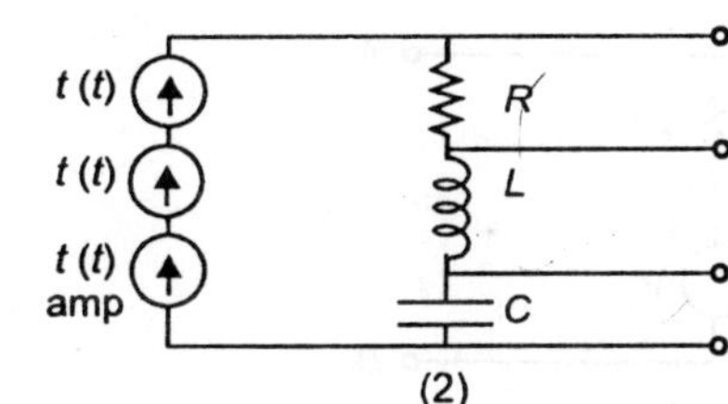


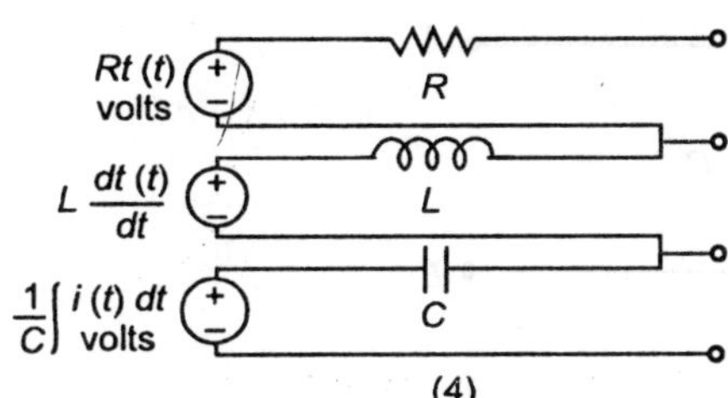

Of these networks,

(*a*) all the four networks are equivalent

(*b*) no two networks are equivalent

(*c*) networks shown in Figs. (2), (3) and (4) are equivalent

(*d*) Networks shown in figures (3) and (4) are equivalent

**143.** The number of 2 µF, 400 V capacitors needed to obtain a capacitance value of 1.5 µF rated for 1600 V is

(*a*) 12　　　　　(*b*) 8　　　　　(*c*) 6　　　　　(*d*) 4

**144.** A connected planar network has 4 nodes and 5 elements. The number of meshes in its dual network is

(*a*) 4　　　　　(*b*) 3　　　　　(*c*) 2　　　　　(*d*) 1

**145.** The value of the current $I$ flowing in the 1 ohm resistor in the circuit, shown in the given figure will be

(*a*) 10 A

(*b*) 6 A

(*c*) 5 A

(*d*) zero

**146.** In the circuit shown in the given figure, the current $I$ through $R_L$ is

(*a*) 2 A

(*b*) zero

(*c*) –2 A

(*d*) –6 A

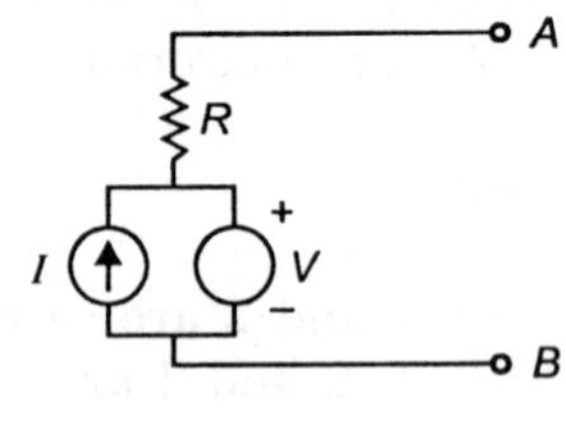

**147.** A simple equivalent circuit of the 2-terminal network shown in figure I, is

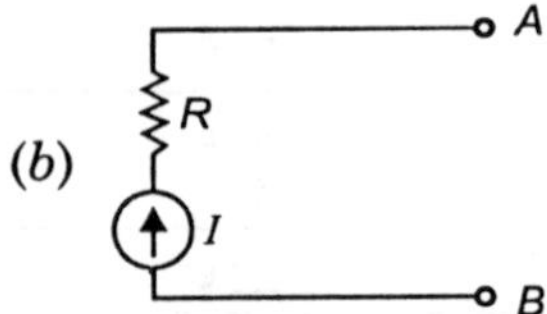

Fig. I

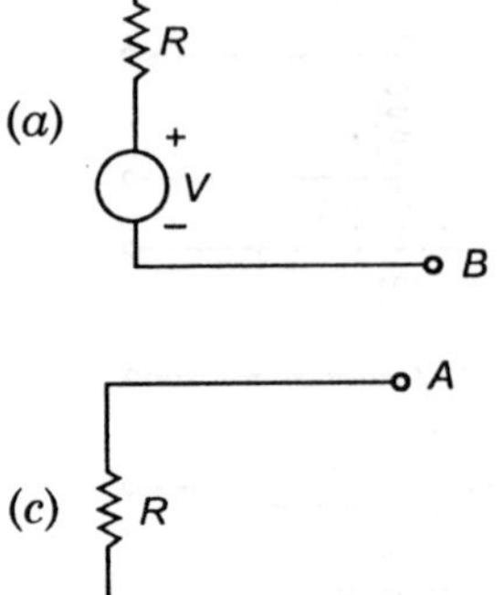

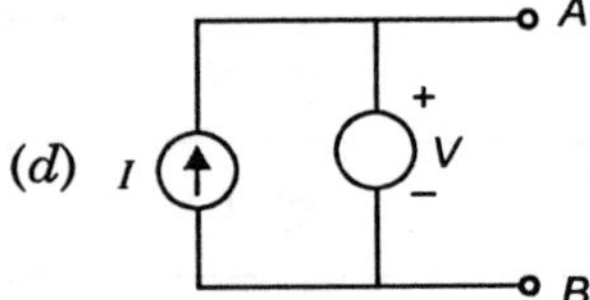

**148.** A voltage source with an internal resistance $R_s$, supplies power to a load $R_L$. The power delivered to the load varies with $R_L$ as

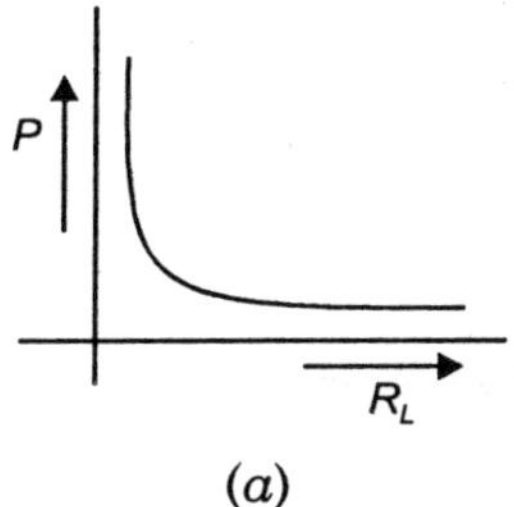

(a)

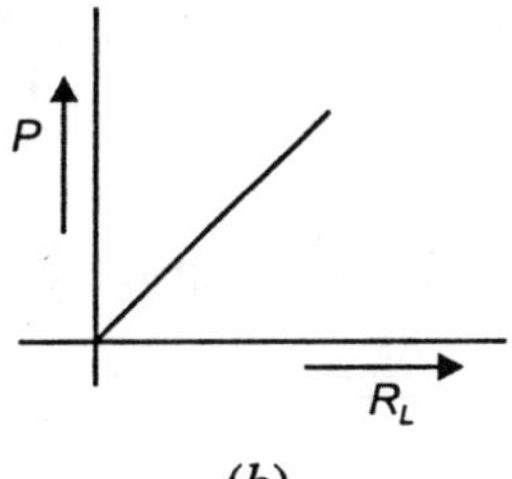

(b)

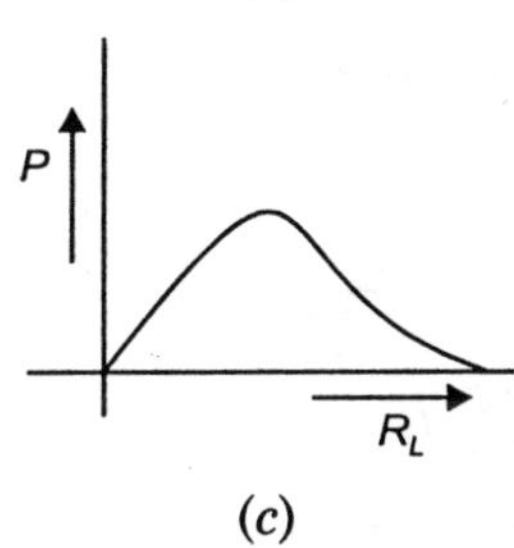

(c)

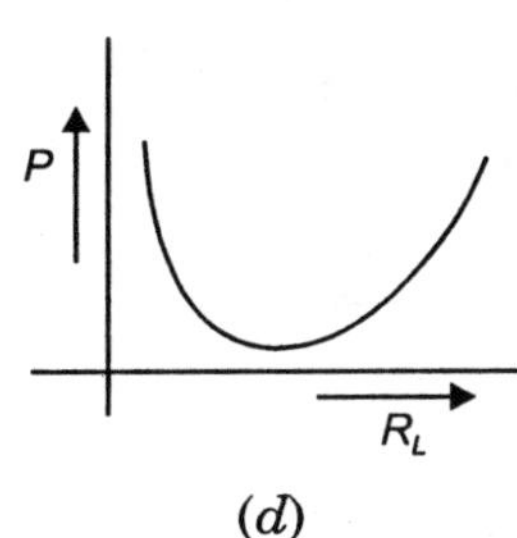

(d)

**149.** Consider the following statements:
The transfer impedances and admittances of a network remain constant when the position of excitation and response are interchanged if the network
1. is linear.
2. consists of bilateral elements.
3. has high impedance or admittance as the case may be.
4. is resonant.
Of these statements
(a) 1 and 2 are correct
(b) 1, 3 and 4 are correct
(c) 2 and 4 are correct
(d) 1, 2, 3 and 4 are correct

**150.** The Thevenin impedance across the terminals $AB$ of the given network is

(a) $\dfrac{10}{3}$

(b) $\dfrac{20}{9}$

(c) $\dfrac{13}{4}$

(d) $\dfrac{11}{5}$

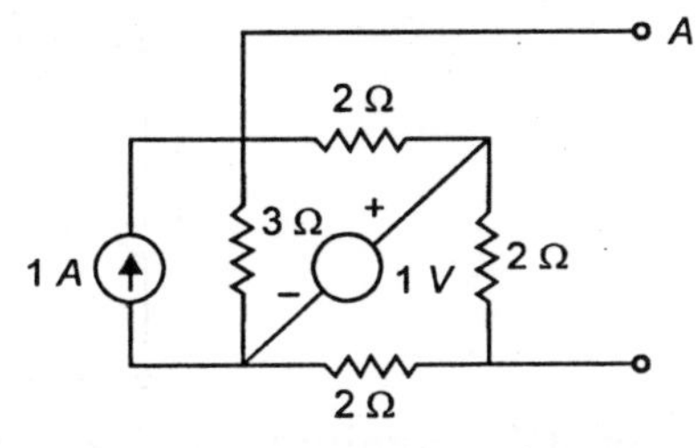

**151.** The response of an initially relaxed linear circuit to a signal $V_s$ is $e^{-2t} u(t)$. If the signal is changed to $\left(V_s + 2\,\dfrac{dV_s}{dt}\right)$, the response would be

(a) $-4\,e^{-2t}\,u(t)$
(b) $-3\,e^{-2t}\,u(t)$
(c) $4\,e^{-2t}\,u(t)$
(d) $5\,e^{-2t}\,u(t)$

**152.** An inductor with inductance $L$ and initial current $I_0$ is shown as

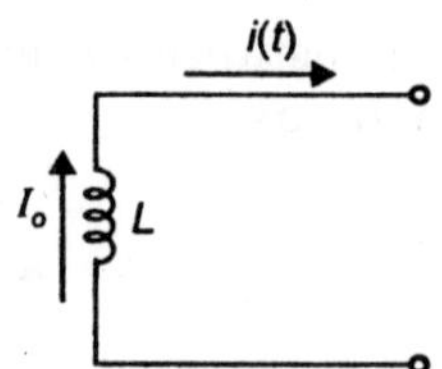

The correct admittance diagram for it is

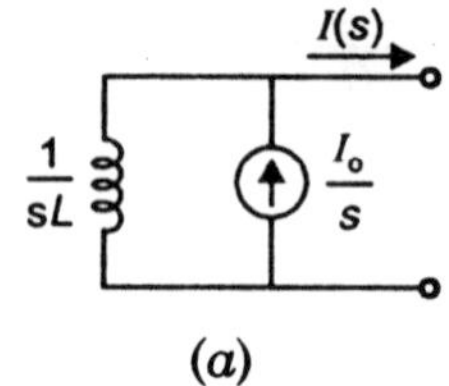

(a)

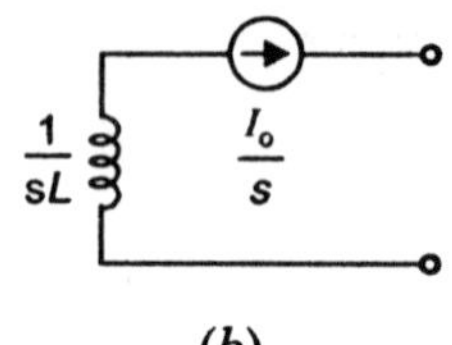

(b)

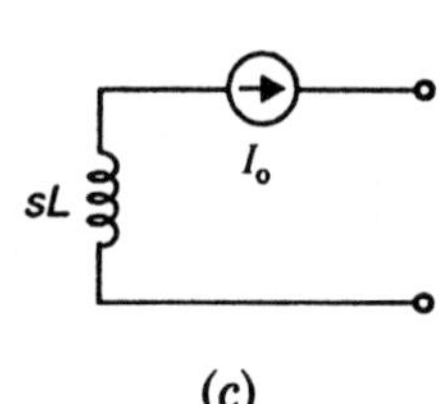

(c)

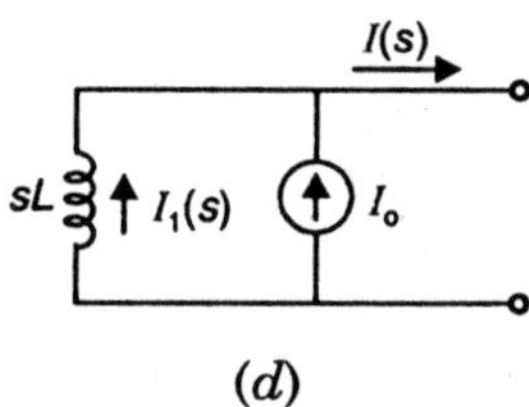

(d)

**153.** For the network shown in the given figure $Z(0) = 3\ \Omega$ and $Z(\infty) = 2\ \Omega$ The values of $R_1$ and $R_2$ will respectively be

(a) $2\ \Omega,\ 1\ \Omega$

(b) $1\ \Omega,\ 2\ \Omega$

(c) $3\ \Omega,\ 2\ \Omega$

(d) $2\ \Omega,\ 3\ \Omega$

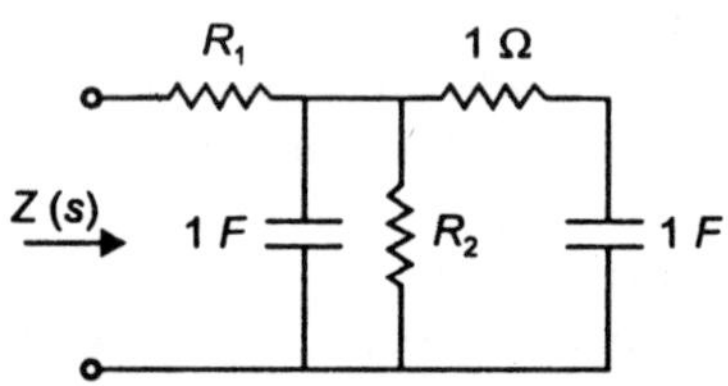

**154.** In the circuit shown in the given figure, the voltmeter indicates 30 V. The reading of the ammeter will be

(a) $20$ A

(b) $10\sqrt{2}$ A

(c) $10$ A

(d) zero

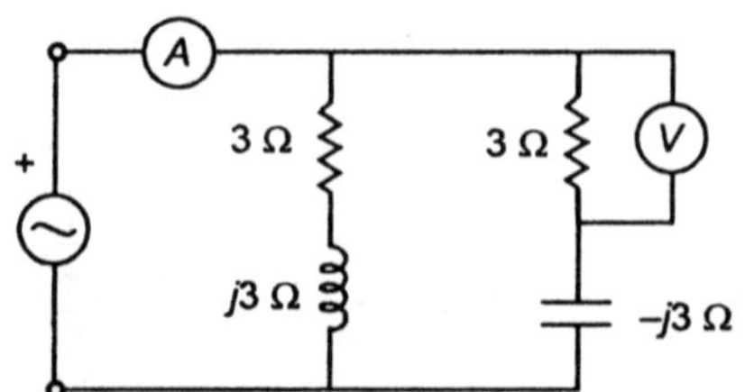

**155.** If a pulse voltage $v(t)$ of 4V magnitude and 2 secs, duration is applied to a pure inductor of 1H, with zero initial current, the current (in amps) drawn at $t = 3$ secs, will be

(a) zero

(b) 2

(c) 4

(d) 8

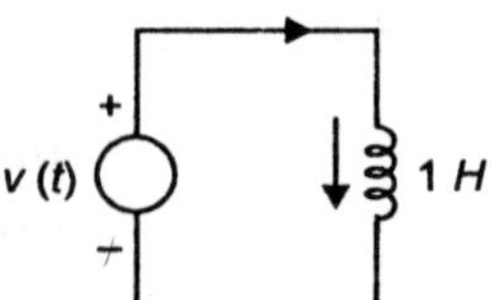

**156.** A signal is described by $s(t) = r(t - a) - r(t - b)$, $(a < b)$, where $r(t)$ is a unit ramp function starting at $t = 0$. The signal $s(t)$ is represented as

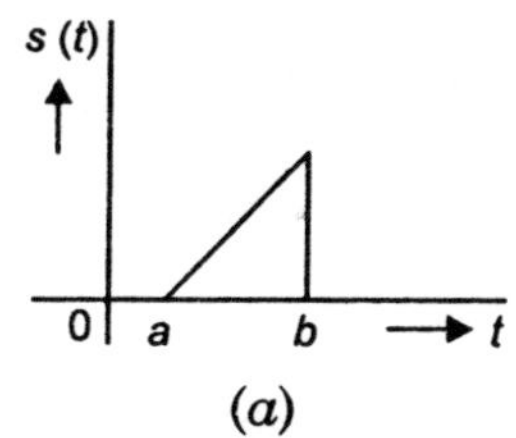

(a)

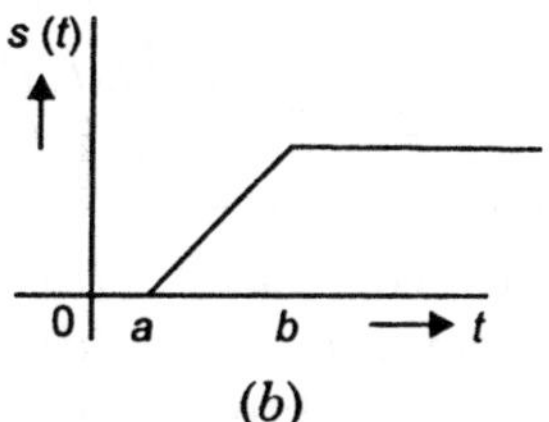

(b)

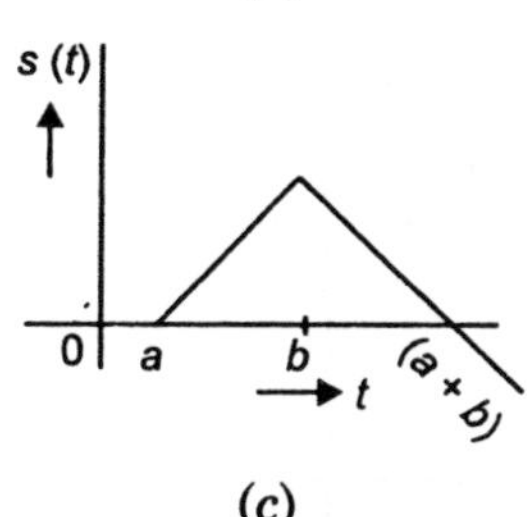

(c)

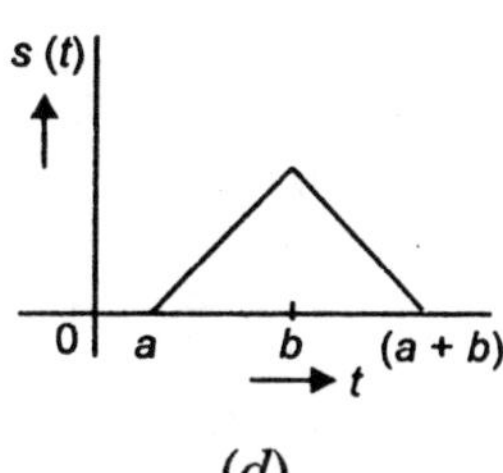

(d)

**157.** A unit impulse input to a linear network has a response $R(t)$ and a unit step input to the same network has a response $S(t)$. The response $R(t)$

(a) equals $\dfrac{dS}{dt}(t)$

(b) equals the integral of $S(t)$

(c) is the reciprocal of $S(t)$

(d) has no relation with $S(t)$

**158.** The values of $i(0^+)$ and $\dfrac{di}{dt}(0^+)$ in the circuit shown in the given figure will be respectively

(a) 10 A, 0

(b) 0, 10 A/s

(c) 0, 100 A/s

(d) 10 A, 10 A/s

**159.** The current waveform in a pure resistor of 10 $\Omega$ is shown in the given figure. Power dissipated in the resistor is

(a) 7.29 W

(b) 52.4 W

(c) 135 W

(d) 270 W

**160.** The open-circuit voltage ratio $\dfrac{V_2(s)}{V_1(s)}$ of the network shown in the given figure is

(a) $1 + 2s^2$

(b) $\dfrac{1}{1 + 2s^2}$

(c) $1 + 2s$

(d) $\dfrac{1}{1 + 2s}$

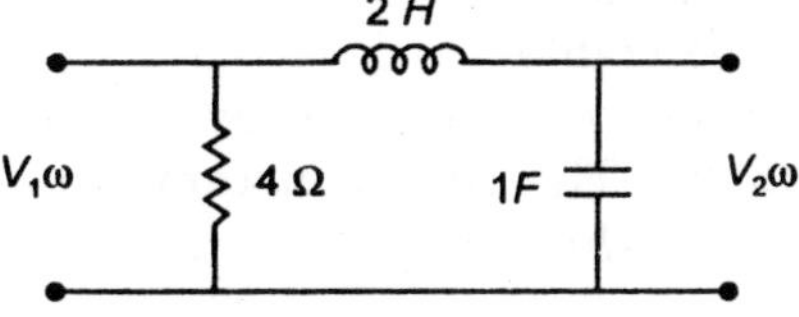

**161.** Consider the following statements:
A unit impulse $\delta(t)$ is mathematically defined as

1. $\delta(t) = 0, \; t \neq 0$  　　2. $\displaystyle\int_{0+}^{\infty} \delta(t)\, dt = 1$  　　3. $\displaystyle\int_{-\infty}^{\infty} \delta(t)\, dt = 1$

Of these statements
(a) 1, 2 and 3 are correct  　　(b) 1 and 2 are correct
(c) 2 and 3 are correct  　　(d) 1 and 3 are correct.

**162.** The short-circuit admittance matrix of the network shown in the given figure is:

(a) $\begin{pmatrix} \dfrac{1}{Z} & \dfrac{-1}{Z} \\[2ex] \dfrac{-1}{Z} & \dfrac{1}{Z} \end{pmatrix}$  　　(b) $\begin{pmatrix} \dfrac{1}{Z} & \dfrac{-1}{Z} \\[2ex] \dfrac{1}{Z} & \dfrac{1}{Z} \end{pmatrix}$

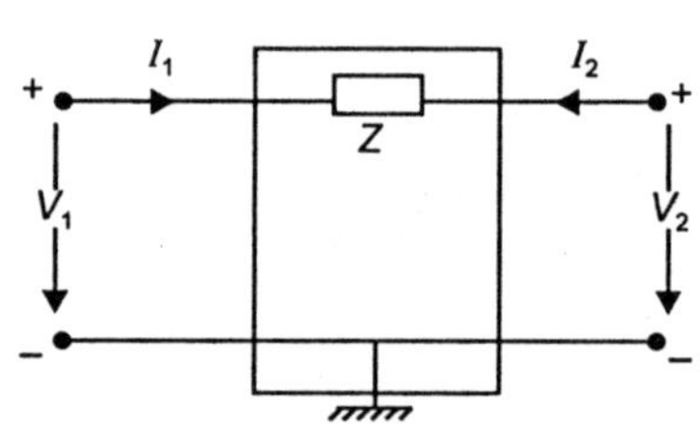

(c) $\begin{pmatrix} \dfrac{1}{Z} & \dfrac{1}{Z} \\[2ex] -1 & 1 \end{pmatrix}$  　　(d) $\begin{pmatrix} \dfrac{1}{Z} & 1 \\[2ex] 1 & \dfrac{1}{Z} \end{pmatrix}$

**163.** Consider the following statements:
For a bilateral network,
1. 　$A = D$
2. 　$z_{12} = z_{21}$
3. 　$h_{12} = -h_{21}$
Of these statements
(a) 1, 2 and 3 are correct  　　(b) 1 and 2 are correct
(c) 1 and 3 are correct  　　(d) 2 and 3 are correct.

**164.** Two two-port networks $\alpha$ and $\beta$ having $A\,B\,C\,D$ parameters as
$$A_\alpha = 4 = D_\alpha \qquad A_\beta = 3 = D_\beta$$
$$B_\alpha = 5, \; C_\alpha = 3 \quad \text{and} \quad B_\beta = 4 \text{ and } C_\beta = 2$$
are connected in cascade in the order of $\alpha$, $\beta$. The equivalent '$A$' parameter of the combination is
(a) 17  　　(b) 22  　　(c) 24  　　(d) 31

**165.** Given $F(s) = \dfrac{s+2}{s\,(s+1)}$, the initial and final values of $f(t)$ will be respectively.

(a) 1, 2  　　(b) 2, 1  　　(c) 1, 1  　　(c) 2, 2

**166.** The pole-zero configuration of an impedance function is given in the figure.
The network is
(a) *R-L* realizable
(b) *R-C* realizable
(c) *L-C* realizable
(d) *R-L-C* realizable

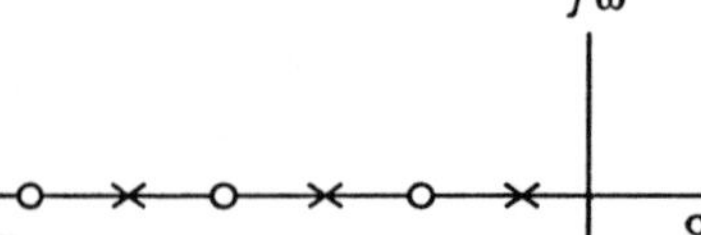

**167.** At a certain current, the energy stored in an iron-cored coil is 1000 $J$ and its copper loss is 2000 $W$. The time constant (in seconds) of the coil is
(a) 0.25  　　(b) 0.5  　　(c) 1.0  　　(d) 2.0

**168.** If an *R-L* circuit having impedance angle $\phi$ is switched in when the applied sinusoidal voltage wave is passing through an angle $\theta$, there will be no switching transient if
(a) $\theta - \phi = 0$    (b) $\theta + \phi = 0$    (c) $\theta - \phi = 90°$ (d) $\theta + \phi = 90°$

**169.** The voltage ratio transfer function for the network shown in the given figure under sinusoidal steady conditions is $G(j\omega)$

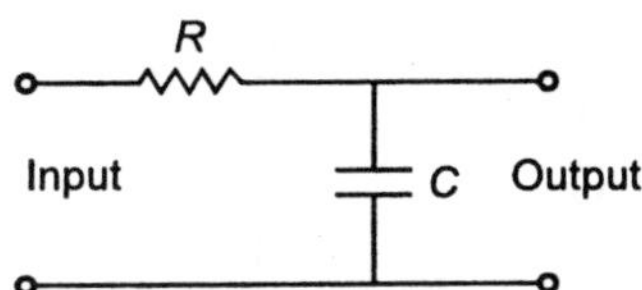


Match List I with List II and select the correct answer using the codes given below the lists:

| List I | List II |
|---|---|
| (Frequency) | (G (jω)) |
| 1.    0 | A.   1 $\angle 0°$ |
| 2.    $\dfrac{1}{RC}$ | B.   0 $\angle -90°$ |
| 3.    $\infty$ | C.   0.707 $\angle -45°$ |

*Codes:*

| | A | B | C |
|---|---|---|---|
| (a) | 1 | 3 | 2 |
| (b) | 1 | 2 | 3 |
| (c) | 2 | 1 | 3 |
| (d) | 3 | 2 | 1 |

**170.** The response of a series *RLC* circuit fed from a fixed rms voltage and variable frequency source is represented graphically in the given figure. Match List I with List II and select the correct answer using the codes given below the lists:

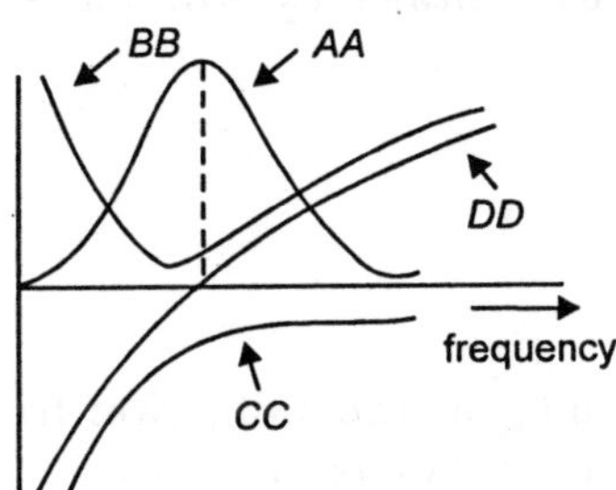


| List I | List II |
|---|---|
| A.   *AA* | 1. Current |
| B.   *BB* | 2. Impedance |
| C.   *CC* | 3. Capacitive reactance |
| D.   *DD* | 4. Net reactance |

5. Inductive reactance

*Codes:*

| | A | B | C | D |
|------|---|---|---|---|
| (a) | 2 | 1 | 3 | 5 |
| (b) | A | B | C | D |
| | 1 | 2 | 3 | 5 |
| (c) | A | B | C | D |
| | 1 | 2 | 3 | 4 |
| (d) | A | B | C | D |
| | 1 | 2 | 4 | 3 |

**171.** For the expansion of $f(\alpha x)$ in Fourier series

$$a_0 + a_1 \cos \alpha x + \dots + a_n \cos n\alpha x + \dots$$
$$+ b_1 \sin \alpha x + \dots + b_n \sin n\alpha x + \dots$$

if $f(\alpha x) = f(-\alpha x)$, then

(a)  $a_n = 0$ for all $n$ including $n = 0$
(b)  $b_n = 0$ for all $n$
(c)  $a_0 = 0$
(d)  $a_n = 0$ for all $n$ except $n = 0$

**172.** A coil having a resistance of 5 ohms and inductance of 0.1 H is connected in series with a condenser of capacitance 50μ F. A constant alternating voltage of 200 volts is applied to the circuit. The voltage across the coil at resonance is
(a) 200 volts   (b) 1788 volts   (c) 1800 volts   (d) 2000 volts

**173.** Two coupled coils connected in series have an equivalent inductance of 16 mH or 8 mH depending on the inter-connection. Then the mutual inductance $M$ between the coils is
(a) 12 mH     (b) $8\sqrt{2}$ mH     (c) 4 mH     (d) 2 mH

**174.** Two coupled coils with $L_1 = L_2 = 0.6\,H$ have a coupling coefficient of $K = 0.8$. The turns ratio $\dfrac{N_1}{N_2}$ is

(a) 4     (b) 2     (c) 1     (d) 0.5

**175.** In the circuit shown in the given figure the switch $S$ is closed at $t = 0$. The induced voltage $v_2$ will have a maximum value of

(a) 0.6 V

(b) 1 V

(c) 3.78 V

(d) 6 V

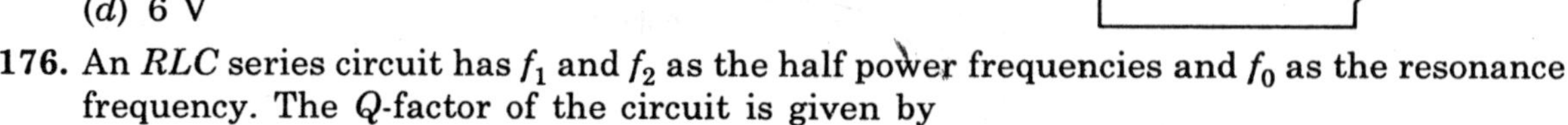


**176.** An $RLC$ series circuit has $f_1$ and $f_2$ as the half power frequencies and $f_0$ as the resonance frequency. The $Q$-factor of the circuit is given by

(a) $\dfrac{f_1 + f_2}{2f_0}$   (b) $\dfrac{f_1 - f_2}{f_2 - f_0}$   (c) $\dfrac{f_0}{f_2 - f_1}$   (d) $\dfrac{f_2 - f_1}{f_0}$

**177.** The condensers of 20 μF and 40 μF capacitance are connected in series across a 90 V supply. After charging, they are removed from the supply and are connected in parallel with positive terminals connected together and similarly the negative terminals. Then the voltage across them will be

(a)  90 V          (b)  60 V          (c)  40 V          (d)  20 V

**178.** If $R_g$ in the circuit shown in the given figure is variable between 20 $\Omega$ and 80 $\Omega$, then the maximum power transferred to the load $R_L$ will be

(a)  15 W

(b)  13.33 W

(c)  6.67 W

(d)  2.4 W

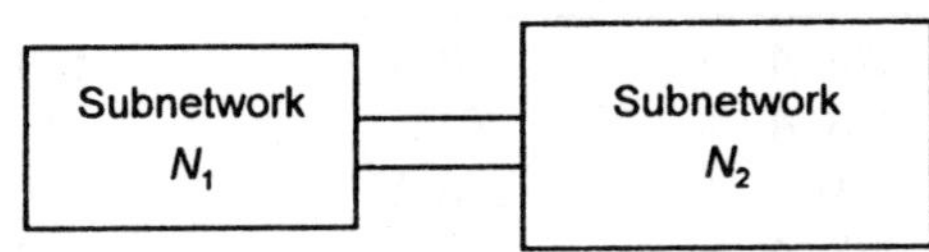


**179.** A network is composed of two sub-networks $N_1$ and $N_2$ as shown in the given figure.

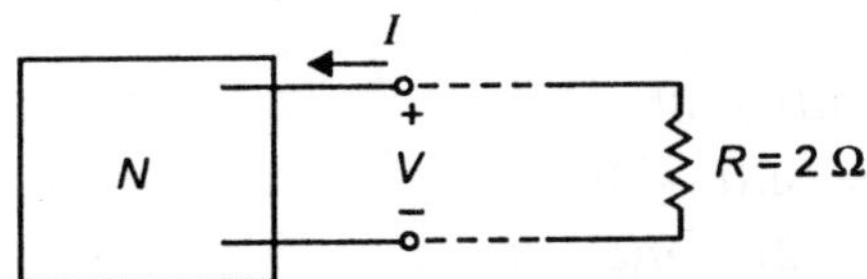


If the sub-network $N_1$ contains only linear, bilateral, time-invariant elements, then it can be replaced by its Thevenin equivalent even if the sub-network $N_2$ contains
(a) a two-terminal element which is non-linear
(b) a non-linear inductance mutually coupled to an element in $N_1$
(c) an element which is linear, but mutually coupled to some element in $N_1$
(d) a dependent source the value of which depends upon the voltage or current in some element in $N_1$

**180.** Under conditions of maximum power transfer from an a.c. source to a variable load
(a) the load impedance must also be inductive, if the generator impedance is inductive
(b) the sum of the source and load impedances is zero
(c) the sum of the source reactance and load reactance is zero
(d) the load impedance has the same phase angle as the generator impedance

**181.** The $V - I$ relation for the network shown in the given box is: $V = 4I - 9$.



If now a resistor $R = 2\ \Omega$ is connected across it, then the value of $I$ will be
(a) −4.5 A        (b) −1.5 A        (c) 1.5 A          (d) 4.5 A

**182.** The reciprocal of a network function is
(a) an immittance function, if the original function is an immitance function.
(b) a transfer function, if the original function is a transfer function.
(c) never an immittance function.
(d) never a transfer function.

**183.** An $LC$-driving point impedance function is

(a) $\dfrac{s^3 + s^2 + s + 1}{s^2 + 2s + 5}$

(b) $\dfrac{s^4 + 2s^2 + 1}{s^2 + 5}$

(c) $\dfrac{s^4 + 1}{s^3 + 2}$

(d) $\dfrac{2(s^2 + 1)}{s}$

**184.** In the network shown in the given figure, there is no initial current through $Ls$ and no initial voltage across the $C$. The switch '$S$' is closed at $t = 0$. The current $i_{L1}$ in inductor $L_1$ and the voltage $V_c$ across $C$ at $t = 0$ and $t = \infty$ will be

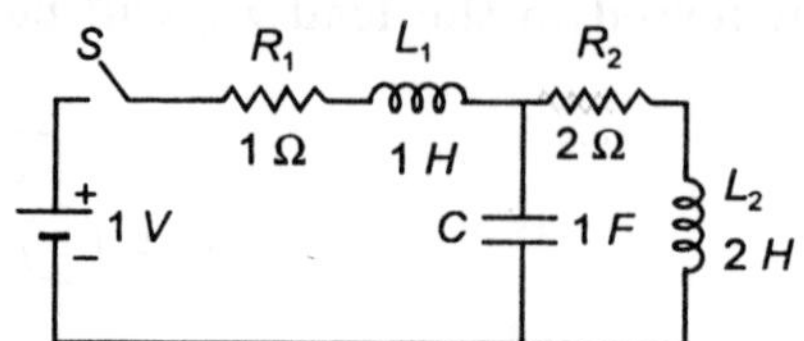

|       | $i_{L1}(0^+)$ | $i_{L1}(\infty)$ | $V_c(0^+)$ | $V_c(\infty)$ |
|-------|-----------|-----------|-----------|-----------|
| (a)   | 1/3 A     | 1/3 A     | 2/3 V     | 2/3 V     |
| (b)   | 0         | 1/3 A     | 0 V       | 1 V       |
| (c)   | 1/3 A     | 0         | 2/3 V     | 0         |
| (d)   | 0         | 1/3 A     | 0         | 2/3 V     |

**185.** A first order linear system is initially relaxed. For a unit step signal $u(t)$, the response is $v1(t) = (1 - e^{-3t})$ for $t > 0$. If a signal $3\,u(t) + \delta(t)$ is applied to the same initially relaxed system, the response will be
(a) $(3 - 6\,e^{-3t})\,u(t)$      (b) $(3 - 3\,e^{-3t})\,u(t)$
(c) $3\,u(t)$      (d) $(3 + 3e^{-3t})\,u(t)$

**186.** Consider the voltage waveform shown in the given figure

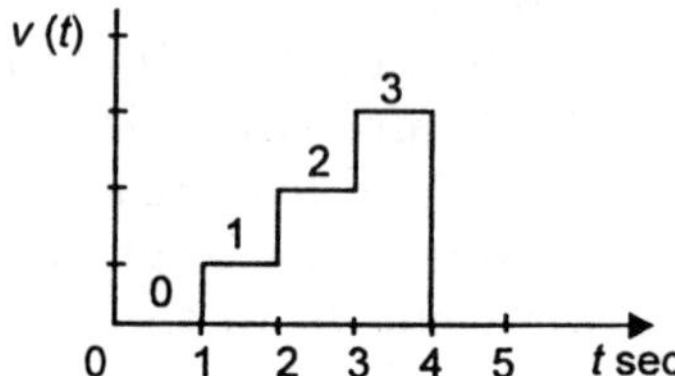

The equation for $v(t)$ is
(a) $u(t - 1) + u(t - 2) + u(t - 3)$
(b) $u(t - 1) + 2u(t - 2) + 3u(t - 3)$
(c) $u(t) + u(t - 1) + u(t - 2) + u(t - 4)$
(d) $u(t - 1) + u(t - 2) + u(t - 3) - 3u(t - 4)$

**187.** When a number of 2-port networks are connected in cascade, the individual
(a) $Z_{oc}$ matrices are added
(b) $Y_{sc}$ matrices are added
(c) chain matrices are multiplied
(d) $H$-matrices are multiplied

**188.** In the delta equivalent of the given star-connected impedance $Z_{QR}$ is equal to

(a) $40\ \Omega$
(b) $(20 + j10)\ \Omega$
(c) $\left(5 + j\dfrac{10}{3}\right)\Omega$
(d) $(10 + j\,30)\ \Omega$

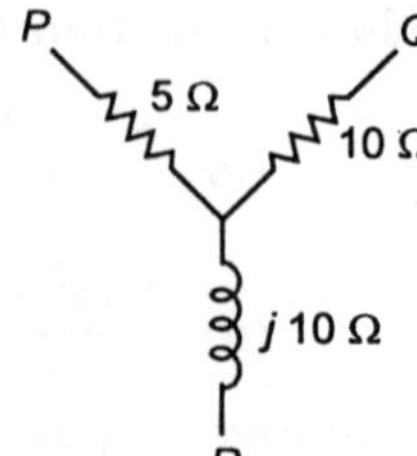

**189.** A two-port network is defined by the relations
$$I_1 = 2V_1 + V_2$$
$$I_2 = 2V_1 + 3V_2$$
Then $z_{12}$ is
(a) $-2\ \Omega$       (b) $-1\ \Omega$       (c) $-1/2\ \Omega$       (d) $-1/4\ \Omega$

**190.** With the usual notation, a two-port resistive network satisfies the condition
$$A = D = \frac{3}{2}\ \ B = \frac{4}{3}\ C$$
The $Z_{11}$ of the network is
(a) 5/3       (b) 4/3       (c) 2/3       (d) 1/3

**191.** The driving point impedance of the infinite ladder network shown in the figure given is

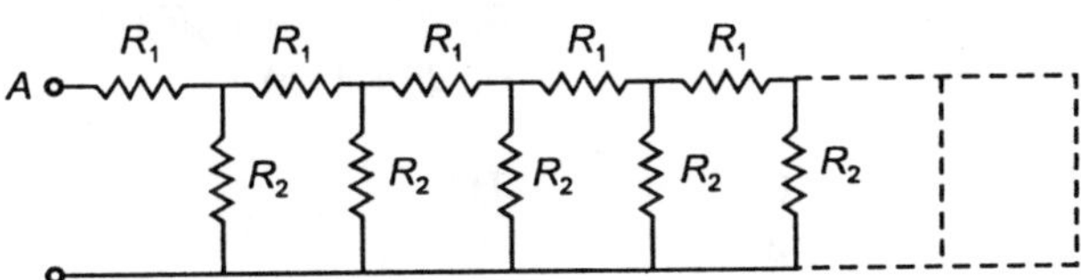

(Given: $R_1 = 2\Omega$ and $R_2 = 1.5\ \Omega$)
(a) $3\ \Omega$                  (b) $3.5\ \Omega$

(c) $\dfrac{3}{3.5}\ \Omega$          (d) $\ln\left(1 + \dfrac{3}{3.5}\right)\Omega$

**192.** A network whose impedance function is
$$\frac{4\,(s^2 + 1)\,(s^2 + 9)}{s\,(s^2 + 4)}$$
is synthesized. It consists of '$n$' $LC$ tank circuits in series with an inductance and/or capacitance. The value of '$n$' is
(a) zero       (b) 1       (c) 2       (d) 3

**193.** $F(s) = \dfrac{(s+1)\,(s+3)}{s\,(s+2)}$ represents an

(a) $RC$ impedance            (b) $RC$ admittance
(c) $RC$ impedance and an $RL$ admittance       (d) $RL$ admittance

**194.** The realization of the reactance function
$$Z(s) = \frac{4\,(s^2 + 1)\,(s^2 + 9)}{s\,(s^2 + 4)}$$

requires a minimum of
(a) 4 inductors and 4 capacitors       (b) 3 inductors and 1 capacitor
(c) 1 inductor, 1 capacitor and 1 resistor       (d) 2 inductors and 2 capacitors

**195.** In a two-port network, the condition for reciprocity in terms of '$h$'-parameters is
(a) $h_{12} = h_{21}$    (b) $h_{11} = h_{22}$    (c) $h_{11} = -h_{22}$    (d) $h_{12} = -h_{21}$

**196.** Match List I with List II and select the correct answer using the codes given below the lists:

| List I | List II |
|---|---|
| (Time function) | (Laplace transforms) |

A. $\mathscr{L}\,[af_1(t) + bf_2(t)]$     1. $aF_1(s) + bF_2(s)$

B. $\mathscr{L}\,[e^{-at} f(t)]$     2. $sF(s) + f(0)$

C. $\mathscr{L}\left[\dfrac{df(t)}{dt}\right]$     3. $\dfrac{1}{s}F(s)$

D. $\mathscr{L}\left[\displaystyle\int_0^t f(X)\,dx\right]$     4. $sF(s) - f(0^+)$

5. $aF_1(s) + a_2F_2(s)$
6. $F(s + a)$

Codes:

| | A | B | C | D |
|---|---|---|---|---|
| (a) | 6 | 2 | 3 | 4 |
| (b) | 1 | 6 | 4 | 3 |
| (c) | 2 | 5 | 3 | 4 |
| (d) | 1 | 6 | 3 | 4 |

**197.** In the given circuit, the value of $R$ that will give critical damping is

(a) $1\ \Omega$

(b) $2\ \Omega$

(c) $4\ \Omega$

(d) $10\ \Omega$

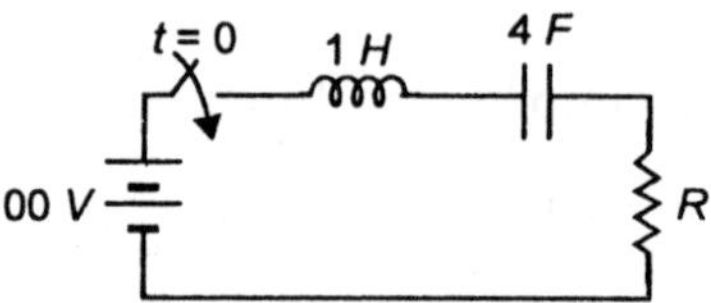


**198.** A transient current in a network is:
$$i(t) = 2\ e^{-t} - e^{-5t},\ t \geq 0.$$
The pole-zero configuration of $I(s)$ is

| | Poles | Zeros |
|---|---|---|
| (a) | 1, 5 | 9 |
| (b) | −1, −5 | −9 |
| (c) | 2, −1 | −1, −5 |
| (d) | 2, −1, | 1, 5 |

**199.** For the circuit given in the figure $V_0 = 2V$ and the inductor is initially relaxed. The switch $S$ is closed at $t = 0$. The value of $v$ at $t = 0^+$ is

(a) $3\ V$

(b) $2\ V$

(c) $0.5\ V$

(d) $0.25\ V$

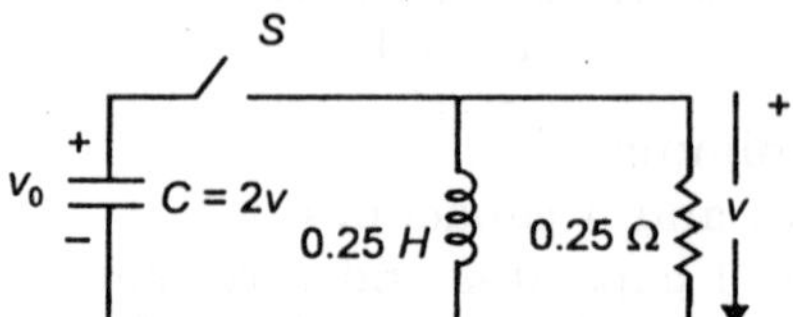


**200.** In the circuit shown in the given figure $S$ is open for a long time and steady state is reached. $S$ is closed at $t = 0$. The current $I$ at $t = 0^+$ is

(a) 4 A

(b) 3 A

(c) 2 A

(d) 1 A

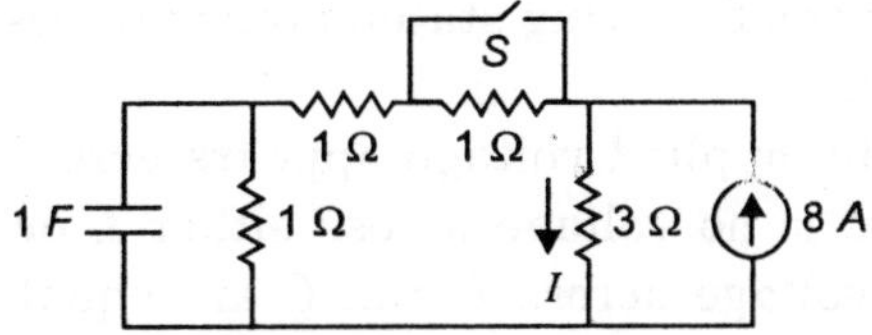

**201.** The current read by the ammeter *A* in the a.c. circuit shown in the given figure is

(a) 9 A

(b) 5 A

(c) 3 A

(d) 1 A

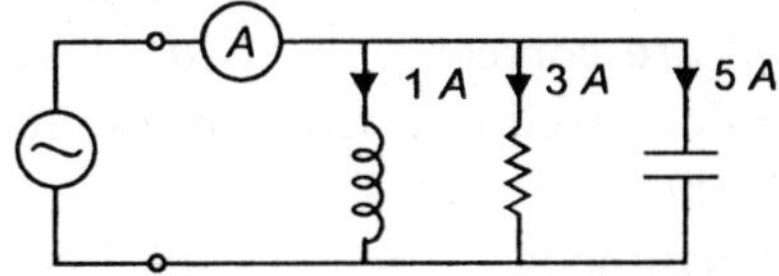

**202.** Match List I with List II and select the correct answer using the codes given below **the** lists:

| *List I* | *List II* |
|---|---|
| (Location of poles on s-plane) | (Type of response) |

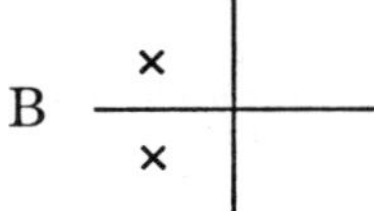

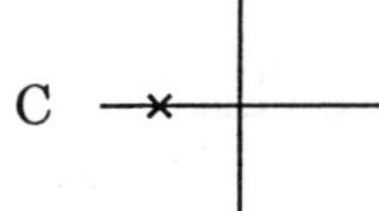
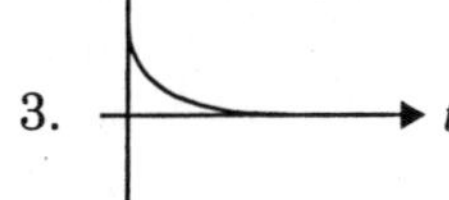

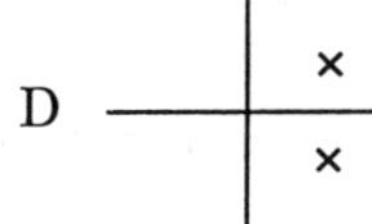

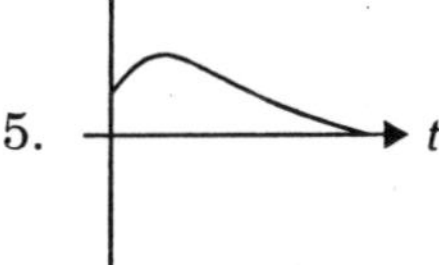

*Codes:*

|     | A | B | C | D |
|-----|---|---|---|---|
| (a) | 4 | 1 | 2 | 3 |
| (b) | 5 | 1 | 4 | 3 |
| (c) | 2 | 3 | 5 | 4 |
| (d) | 2 | 1 | 3 | 4 |

**203.** Consider the following statements with respect to a series *RLC* circuit under resonance condition:
(*a*) All the applied voltage appears across *R*.
(*b*) There is no voltage across either *L* or *C*.
(*c*) The voltage across *L* and *C* are equal and equal to their maximum values.
Of these statements
  (*a*) 1 alone is correct        (*b*) 2 alone is correct
  (*c*) 1 and 3 are correct     (*d*) 1 and 2 are correct

**204.** The power delivered to a three-phase load can be measured by the use of 2 wattmeters only when the
  (*a*) load is balanced         (*b*) load is unbalanced
  (*c*) 3-phase load is connected to the source through 3 wires
  (*d*) 3-phase load is connected to the source through 4 wires.

**205.** A 3-phase star-connected symmetrical load consumes *P* watts of power from a balanced supply. If the same load is connected in delta to the same supply, the power consumption will be
  (*a*) $P$           (*b*) $\sqrt{3}\,P$
  (*c*) $3\,P$       (*d*) not determinable from the given data.

**206.** The resonant frequency of the series circuit shown in the given figure is:

  (*a*) $\dfrac{1}{4\pi}$ Hz          (*b*) $\dfrac{1}{4\pi\sqrt{2}}$ Hz

  (*c*) $\dfrac{1}{2\pi\sqrt{2}}$ Hz     (*d*) $\dfrac{1}{4\pi\sqrt{3}}$ Hz

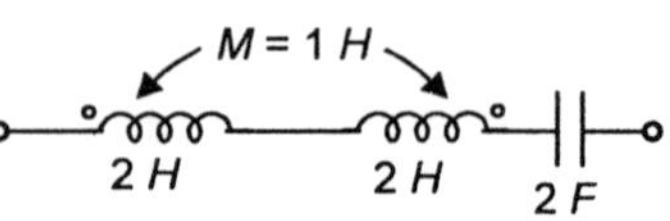


**207.** For the ideal transformer shown in the given figure
  (*a*) $V_1 = nv_2$,       $i_2 = -ni_1$
  (*b*) $V_2 = nv_1$,       $i_2 = ni_1$

  (*c*) $V_1 = nv_2$,       $i_1 = \dfrac{1}{n}i_2$

  (*d*) $V_1 = nv_2$,       $i_2 = \dfrac{-1}{n}i_1$

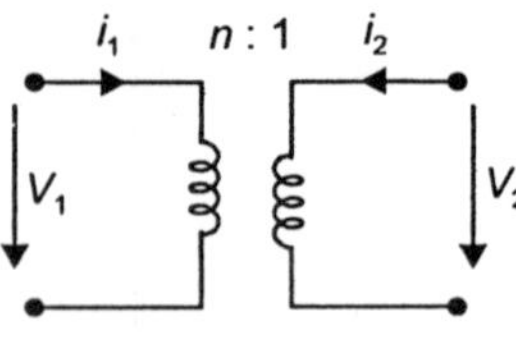


**208.** Consider the circuit shown in the given figure. For maximum power transfer to the load, the primary to secondary turns ratio must be
  (*a*) 9 : 1
  (*b*) 3 : 1
  (*c*) 1 : 3
  (*d*) 1 : 9



**209.** The transfer function, $T(s) = \dfrac{s^2}{s^2 + as + b}$ belongs to an active
  (*a*) Low-pass filter       (*b*) High-pass filter
  (*c*) Band-pass filter      (*d*) Band-reject filter

**210.** In the circuit shown in the given figure, if the power consumed by the 5 ohm resistor is 10 W, then the power factor of the circuit is

(a) 0.8

(b) 0.6

(c) 0.5

(d) zero

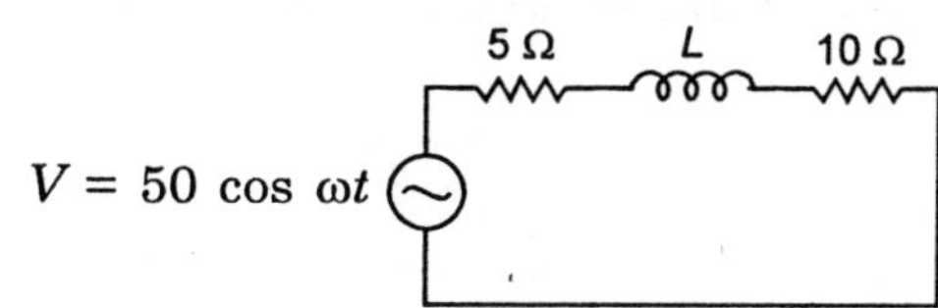


**211.** Consider the two circuits I and II shown in the following figures:

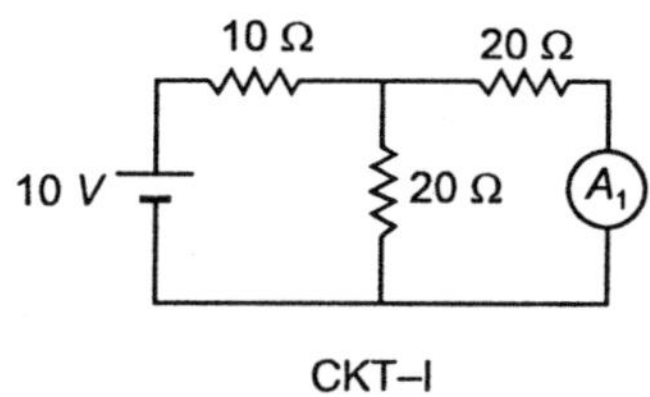


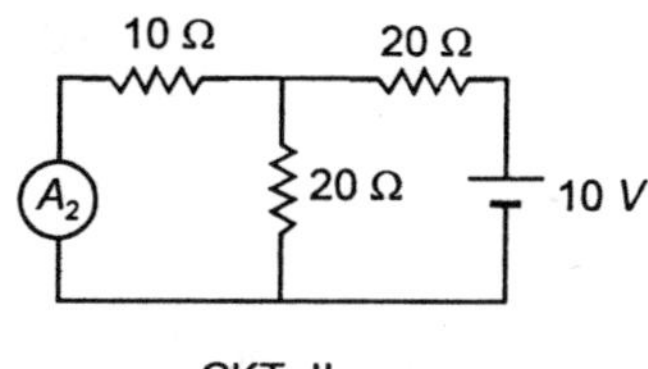


Which one of the following statements regarding the current flowing through the ammeters $A_1$ and $A_2$ is correct?

(a) The currents in $A_1$ and $A_2$ are of the same value and equal 0.25 A.

(b) The currents in $A_1$ and $A_2$ are respectively 0.25 A and 2.5 A.

(c) The currents in both the ammeters is of the same value and equals 2.5 A.

(d) The currents in $A_1$ and $A_2$ are respectively 2.5 A and 0.25 A.

**212.** In the circuit shown in the given figure, if $I_1 = 1.5\ A$, then $I_2$ will be

(a) 0.5 A

(b) 1.0 A

(c) 1.5 A

(c) 3.0 A

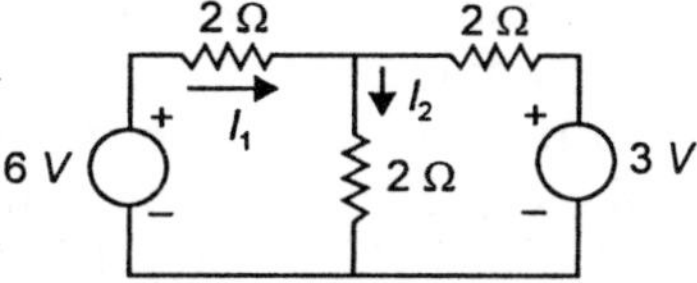


**213.** An input $f(t)$ is applied to a linear network giving a response $g(t)$. The function $f(t)$ is delayed by 1 time unit. If the network is time-invariant, then the response will be

(a) $g'(t)$      (b) $g(t-1)$      (c) $g(t)$      (d) $1 - g(t)$

**214.** For the circuit shown in the given figure, the Thevenin equivalent impedance across terminal $CD$ is given by

(a) $Z_{TH} = \dfrac{Z_4\left[\dfrac{Z_1 Z_2}{Z_1 + Z_2} + Z_3\right]}{Z_4 + Z_3 + \dfrac{Z_1 Z_2}{Z_1 + Z_2}}$

(b) $Z_{TH} = \dfrac{Z_4\left[\dfrac{Z_2 Z_3}{Z_3 + Z_2} + Z_1\right]}{Z_1 + Z_4 + \dfrac{Z_3 Z_2}{Z_2 + Z_3}}$

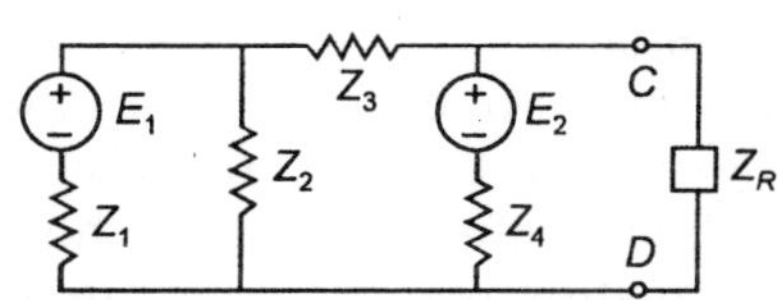


(c) $Z_{TH} = \dfrac{Z_1 + Z_2 + Z_3 Z_2}{Z_1 + Z_2 + Z_4}$      (d) $Z_{TH} = \dfrac{Z_3\,(Z_1 + Z_2 + Z_4)}{Z_1 + Z_2 + Z_3 + Z_4}$

**215.** A sinusoidal voltage source 50 $\angle 0°$ with an internal impedance $10 + j20$ is connected to a load $(Z_L)$ which is variable in both resistance and reactance. Match List I with List II for maximum power transfer and select the correct answer using the codes given below the lists:

| *List I* | *List II* |
|---|---|
| A.  Load impedance | 1.  62.5 |
| B.  Total impedance | 2.  $10 + j20$ |
| C.  Current | 3.  2.5 |
| D.  Maximum power | 4.  20 |
|  | 5.  $10 - j20$ |
|  | 6.  $2.5 - j2.5$ |

*Codes:*

| | A | B | C | D |
|---|---|---|---|---|
| (a) | 5 | 2 | 3 | 4 |
| (b) | 5 | 4 | 3 | 1 |
| (c) | 2 | 4 | 6 | 1 |
| (d) | 2 | 5 | 6 | 4 |

**216.** A Hurwitz polynomial has
(a) zeros only in the left half of the s-plane
(b) poles only in the left half of the s-plane
(c) zeros anywhere in the s-plane
(d) poles on the $j\omega$ axis only

**217.** For an all-pass function zeros are in the
(a) right half plane (RHP) and poles in the left half plane (LHP).
(b) LHP and poles in LHP.    (c) LHP and poles in RHP.
(d) RHP and poles in RHP.

**218.** Consider the following statements:
1.  Transfer impedance is the reciprocal of transfer admittance.
2.  One can derive transfer impedance of a network if its driving-point impedance and admittance are known.
3.  Driving-point impedance is the ratio of the Laplace transform of voltage and current time functions at the input.
Of these statements
(a) 1, 2 and 3 are correct    (b) 1 and 2 are correct
(c) 2 and 3 are correct    (d) 3 alone is correct

**219.** A current of the waveform shown in the given figure passes through a pure inductance of 3 mH. The instantaneous power, in Watts, during $0 < t < 2$ ms is

(a) 25000 $t$

(b) 50000 $t$

(c) 75000 $t$

(d) 100000 $t$

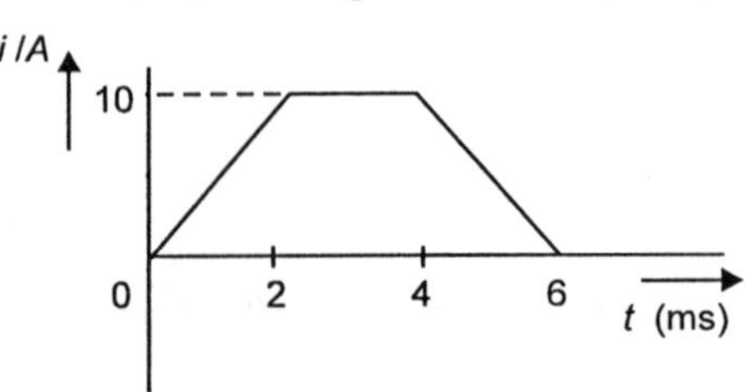

**220.** Consider the following statements:

If the energy sources are connected to the unenergised system as shown in figures $A$ and $B$, at $t = 0$, then

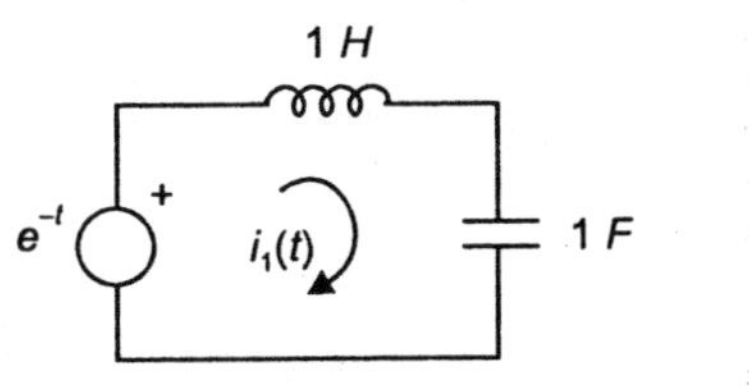


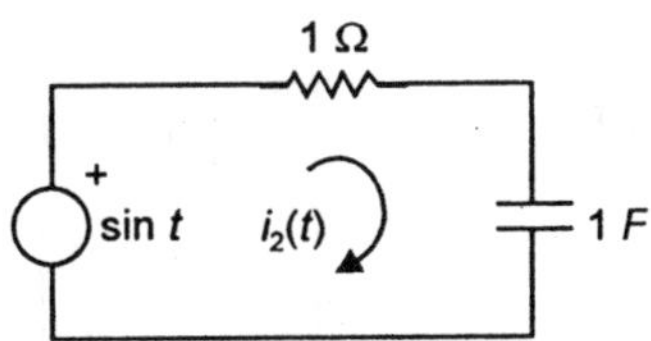


1. $i_1(t)$ and $i_2(t)$ are equal.
2. the circuit in Fig. A has one free frequency and two forced frequencies.
3. the circuit in Fig. B has one free frequency and two forced frequencies.
4. the circuit in Fig. A has two free frequencies and one forced frequency.

Of these statements

(a) 1, 2 and 3 are correct          (b) 1, 3 and 4 are correct

(c) 1 and 3 are correct             (d) 4 alone is correct

**221.** The driving-point impedance of an $RC$ network is given by

$$Z(s) = \frac{2s^2 + 7s + 3}{s^2 + 3s + 1}$$

Its canonical realization will be

(a) 6 elements  (b) 5 elements  (c) 4 elements  (d) 3 elements

**222.** A step voltage is applied to an underdamped series $RLC$ circuit with variable $R$. Which of the following statements correctly describe the behaviour of the circuit?

1. If $R$ is increased, the steady-state voltage across $C$ will be reduced.
2. If $R$ is increased, the frequency of transient oscillations across $C$ will be reduced.
3. If $R$ is reduced, the transient oscillations will die down faster.
4. If $R$ is reduced to zero, the peak amplitude of the voltage across $C$ will be double the input step voltage.

Select the correct answer using the codes given below:

*Codes:*

(a) 1 and 2      (b) 2 and 3      (c) 2 and 4      (d) 1, 3 and 4

**223.** An impedance match is desired at the $1 - 1'$ port of the two-port network shown in the given figure. The match will be obtained when $z_g$ equals

(a) $z_1 + z_3$

(b) $z_1 + \dfrac{z_3 (z_2 + z_L)}{z_2 + z_3 + z_L}$

(c) $\dfrac{z_1 (z_2 + z_3)}{z_1 + z_2 + z_3}$

(d) $z_1$

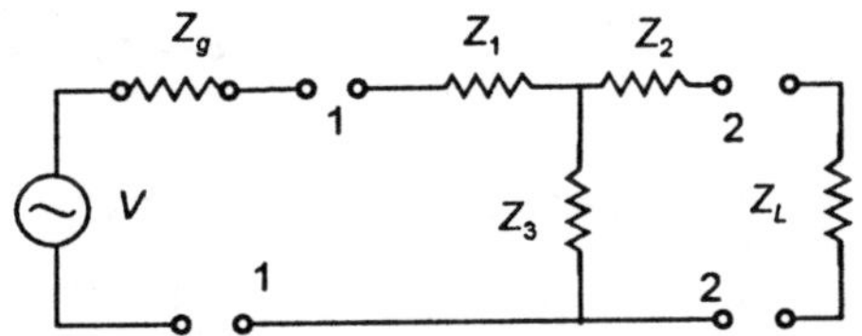


**224.** The $z$ and $h$ parameters of the network shown in the given figure will be

(a) $z = \begin{bmatrix} 0 & 1 \\ -1 & 1 \end{bmatrix}$;          $h = \begin{bmatrix} 1 & 1 \\ 1 & 1 \end{bmatrix}$

(b) $z = \begin{bmatrix} 1 & 1 \\ 1 & 1 \end{bmatrix}$;     $h = \begin{bmatrix} 0 & 1 \\ -1 & 1 \end{bmatrix}$

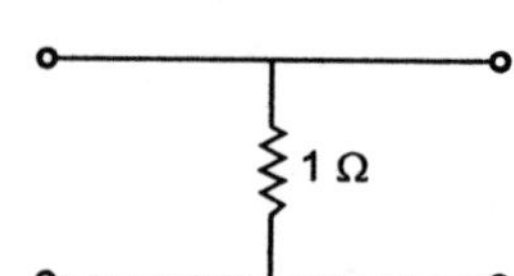

(c) $z = \begin{bmatrix} 1 & 0 \\ -1 & 1 \end{bmatrix}$;     $h = \begin{bmatrix} -1 & 1 \\ 1 & -1 \end{bmatrix}$

(d) $z = \begin{bmatrix} 0 & 1 \\ 1 & 0 \end{bmatrix}$;     $h = \begin{bmatrix} 0 & -1 \\ 2 & \frac{1}{2} \end{bmatrix}$

**225.** In the case of $RC$ driving-point functions
   (a) $Z_{RC}$ cannot have a pole at infinity
   (b) $Z_{RC}$ cannot have a pole at the origin
   (c) $Y_{RC}$ cannot have a pole at infinity
   (d) $Z_{RC}$ and $Y_{RC}$ are constant or zero at infinity

**226.** Consider the following four functions:

1. $F_1 = \dfrac{(s^2 + 1)(s^2 + 3)}{s(s^2 + 2)}$

2. $F_2 = \dfrac{s(s^2 + 2)}{(s^2 + 1)(s^2 + 3)}$

3. $F_3 = \dfrac{(s^2 + 1)(s^2 + 2)}{s(s^2 + 3)}$

4. $F_4 = \dfrac{(s^2 + 1)(s^2 + 3)}{(s^2 + 2)(s^2 + 4)}$

   The $LC$ functions among these would include
   (a) 1, 2, 3 and 4          (b) 1 and 3
   (c) 2 and 3                (d) 1 and 2

**227.** In the synthesis of reactive networks using Foster's first form, the location of the poles of the driving-point reactance and the presence of the storage elements have a correspondence. In this context, match List I (Location of pole in the s-plane) with List II (Storage elements in the network) and select the correct answer using the codes given below the  lists:

*List I*            *List II*

A.  Zero

B.  Infinity

C.  $\pm j\omega$

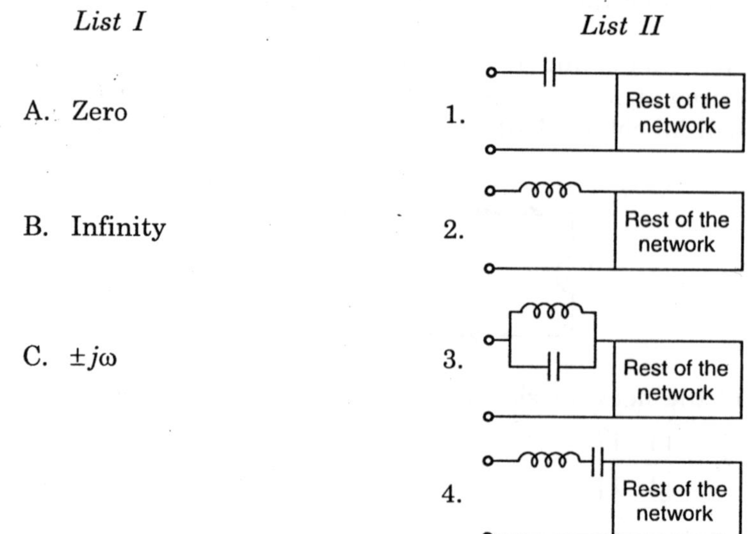

*Codes:*

|       | A | B | C |
|-------|---|---|---|
| (a)   | 1 | 2 | 3 |
| (b)   | 2 | 1 | 3 |
| (c)   | 1 | 2 | 4 |
| (d)   | 2 | 1 | 4 |

**228.** A series circuit containing $R$, $L$ and $C$ is excited by a step voltage input. The voltage across the capacitance exhibits oscillations. The damping coefficient (ratio) of this circuit is given by

(a) $\xi = \dfrac{R}{2\sqrt{LC}}$    (b) $\xi = \dfrac{R}{LC}$    (c) $\xi = \dfrac{R}{2\sqrt{\dfrac{C}{L}}}$    (d) $\xi = \dfrac{R}{2\sqrt{\dfrac{L}{C}}}$

**229.** If a system is represented by the differential equation

$$\frac{d^2 y}{dt^2} + \frac{6\,dy}{dt} + 9y = 0$$

then the solution $y$ will be of the form

(a) $k_1 e^{-t} + k_2 e^{-9t}$          (b) $(k_1 + k_2 t)\, e^{-3t}$

(c) $ke^{-3t} \sin (t + \phi)$   (d) $(k_1 + k_2 t)\, e^{3t}$

**230.** A series circuit containing passive elements has the following current and applied voltage:

$$v = 200 \sin (2000t + 50°)$$
$$i = 4 \cos (2000t + (13.2°)$$

The circuit elements

(a) must be resistance and capacitance

(b) must be resistance and inductance

(c) must be inductance, capacitance and resistance

(d) could be either resistance and capacitance or resistance, inductance and capacitance.

**231.** Which of the following are true of the circuit shown in the given figure?

1. $V_R = 100 \sqrt{2}\ V$

2. $I = 2A$

3. $L = 0.25\ H$

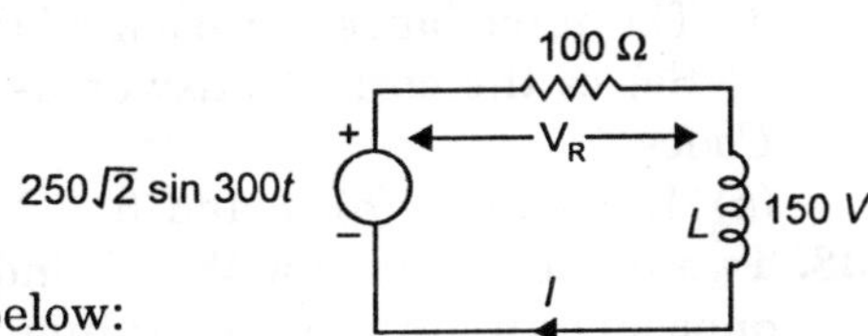

Select the correct answer using the codes given below:

*Codes:*

(a) 2 and 3     (b) 1 and 2     (c) 1 and 3     (d) 1, 2 and 3

**232.** The system described by the equation

$$F(s) = s^4 + 2s^3 + 3s^2 + 6s + K$$

according to Routh-Hurwitz's criteria, is

    (a) unstable for all values of $K$
    (b) stable if $K \geq 0$
    (c) stable if $k < 0$
    (d) stable for all values of $K$

**233.** Consider the following statements regarding driving-point admittance function having two complex conjugate poles:
    1. Closer the poles to $j$-$\omega$-axis, higher the $Q$ of the circuit.
    2. Value of $Q$ varies inversely as the damping ratio.
    3. A circuit with low $R$ has low $Q$.
    Of these statements
    (a) 1 and 3 are correct        (b) 2 and 3 are correct
    (c) 1 and 2 are correct        (d) 1, 2 and 3 are correct

**234.** The input impedane $Z_{in}$ of the circuit shown below will be purely resistive at

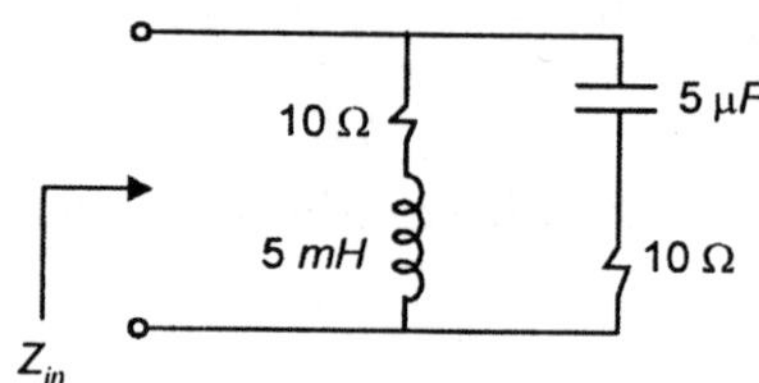


    (a) an angular frequency of $2\sqrt{10}\ k$ rad/sec
    (b) all frequencies         (c) very low frequencies
    (d) very high frequencies

**235.** In two-wattmeter method of power measurement, one of the wattmeters will show negative reading when the load *pf* angle is strictly
    (a) less than 30°         (b) less than 60°
    (c) greater than 30°       (d) greater than 60°

**236.** For star connected load, the line currents are unbalanced and $I_a$, $I_b$ and $I_c$ are equal to $-j10$, $-j10$ and $-j20$ amperes respectively. The neutral current will be equal to
    (a) $+ 8.66 + j5$ (b) $+ 8.66 - j5$ (c) $-8.66 - j5$ (d) $-8.66 + j5$

**237.** Which of the following are the necessary conditions for an entire three-phase system to be balanced?
    1. The line voltages are equal in magnitude.
    2. The phase differences between successive line voltages are equal.
    3. The impedances in each of the phases are identical.
    Select the correct answer using the codes given below:
*Codes:*
    (a) 1, 2 and 3 (b) 1 and 3    (c) 1 and 2    (d) 2 and 3

**238.** Two inductive coils with self-inductances $L_1$ and $L_2$ are magnetically coupled in series opposing and in parallel aiding respectively. The mutual inductance between the coils is $M$. The equivalent inductances in the two cases are respectively.

    (a) $L_1 + L_2 + 2M,\ \dfrac{L_1 L_2 - M^2}{L_1 + L_2 - 2M}$

    (b) $L_1 + L_2 - 2M,\ \dfrac{L_1 L_2 - M^2}{L_1 + L_2 + 2M}$

(c) $L_1 + L_2 - 2M, \dfrac{L_1 L_2 - M^2}{L_1 + L_2 - 2M}$

(d) $L_1 + L_2 + 2M, \dfrac{L_1 L_2 - M^2}{L_1 + L_2 + 2M}$

**239.** The coupling between two magnetically coupled coils is said to be ideal if the coefficient of coupling is

(a) zero         (b) 0.5         (c) 1         (d) 2

**240.** A series $RLC$ circuit, consisting of $R$ = 10 ohms. $X_L$ = 20 ohms $X_C$ = 20 ohms is connected across an a.c. supply of 100 V (rms). The magnitude and phase angle (with reference to supply voltage) of the voltage across the inductive coil are respectively.

(a) 100 V; 90°                    (b) 100 V; –90°
(c) 200 V; –90°                   (d) 200 V; 90°

**241.** For the networks shown in Figures $A$ and $B$ to be duals, it is necessary that $R'$, $L'$, and $C'$ are respectively equal to

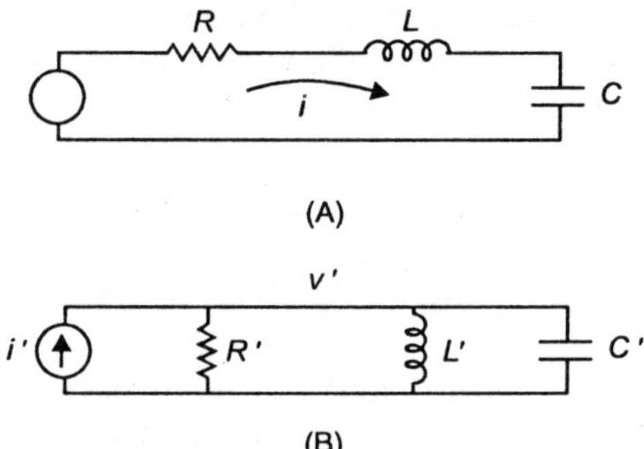

(A)

(B)

(a) 1/R, C and L                   (b) 1/R, 1/L and 1/C
(c) 1/R, 1/L and C                (d) R, L and C

**242.** The impedance $Z(s)$ of the one-port network shown in the figure is given by

(a) $\dfrac{s^2 + s \cdot \dfrac{L}{C} + \dfrac{R_1}{LC}}{s + \dfrac{1}{LC}}$

(b) $\dfrac{s^2 + s \cdot \dfrac{L + R_1 R_2 C}{R_2 LC} + \dfrac{R_1 + R_2}{R_2 LC}}{s + \dfrac{1}{R_2 C}}$

(c) $L \cdot \dfrac{s^2 + s \cdot \dfrac{L + C}{R_1 R_2} + \dfrac{R_1}{LC}}{s + \dfrac{1}{R_1 C}}$

(d) $\dfrac{s^2 + s \cdot \dfrac{L + R_2 C}{R_1 LC} + \dfrac{R_1 + R_2}{R_1 LC}}{s + \dfrac{1}{R_1 C}}$

**243.** Which of the following statement(s) is/are true of the circuit shown in the given figure?

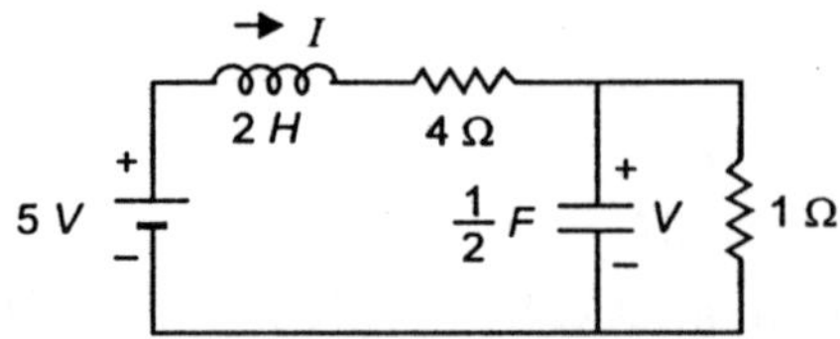

1. It is a first order circuit with steady-state values of

$$V = \frac{5}{3}V, I = \frac{5}{3}A$$

2. It is a second order circuit with steady state values of $V = 1\ V; I = 1\ A$.
3. The network function $V(s)/I(s)$ has one pole.
4. The network function $V(s)/I(s)$ has two poles.
Select the correct answer using the codes given below:
*Codes:*
(a) 1 and 3      (b) 2 and 4      (c) 2 alone      (d) 1 alone

**244.** In addition to the condition that $Y(s)$ is real when '$s$' is real, for an admittance function $Y(s)$ to be positive real, which of the following conditions are to be satisfied?
1. $R_e Y(s) \leq 0$ for $R_e\ s \geq 0$.
2. $R_e\ Y(s) \geq 0$ for $R_e\ s \geq 0$.
3. $|\text{Arg } Y(s)| \leq |\text{Arg } s|$ for $|\text{Arg } s| \leq \pi/2$.
4. $|\text{Arg } Y(s)| \geq |\text{Arg } s|$ for $|\text{Arg } s| \leq \pi/2$.
Select the correct answer using the codes given below:
*Codes:*
(a) 1 and 3 are correct      (b) 1 and 4 are correct
(c) 2 and 3 are correct      (d) 2 and 4 are correct

**245.** The circuit shown in the given figure is in steady-state with switch '$S$' open. The switch is closed at $t = 0$. The values of $V_C(0^+)$ and $V_C(\infty)$ will be respectively

(a) 2V, 0V

(b) 0V, 2V

(c) 2V, 2V

(d) 0V, 0V

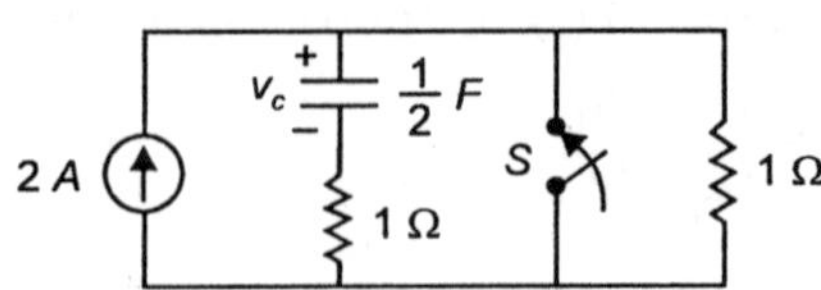

**246.** After closing the switch '$S$' at $t = 0$, the current $i(t)$ at any instant '$t$' in the network shown in the given figure will be

(a) $10 + 10\ e^{100t}$

(b) $10 - 10\ e^{100t}$

(c) $10 + 10\ e^{-100t}$

(d) $10 - 10\ e^{-100t}$

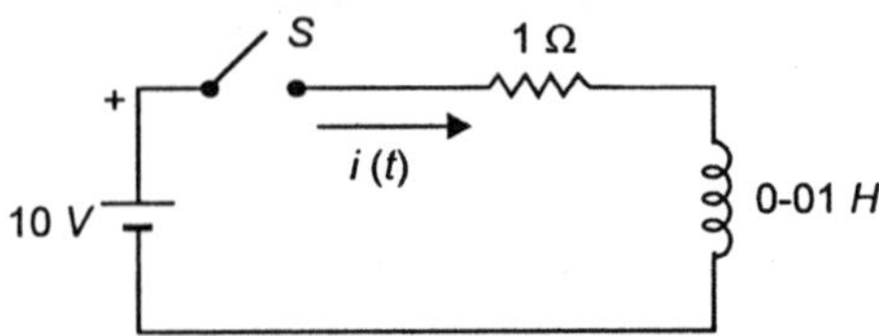

**247.** The pole-zero configuration of a network transfer function is shown in the given figure. The magnitude of the transfer function will (*a*) decrease with frequency.

(*b*) increase with frequency.

(*c*) initially increase and then decrease with frequency.

(*d*) be Independent of frequency.

**248.** Consider the following statements:

1. The two-port network shown below does NOT have an impedance matrix representation.

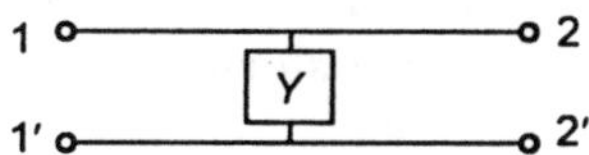

2. The following two-port network does NOT have an admittance matrix representation.

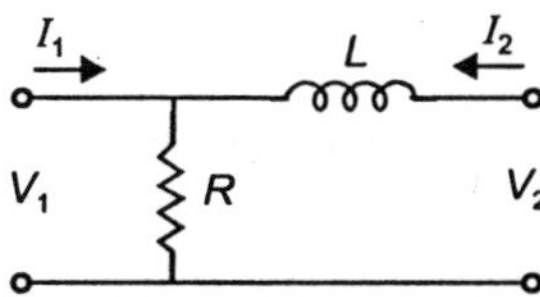

3. A two-port network is said to be reciprocal if it satisfies $Z_{12} = Z_{21}$ or an equivalent relationship.

Of these statements

(*a*) 1 and 2 are correct        (*b*) 1, 2 and 3 are correct

(*c*) 1 and 3 are correct        (*d*) none is correct

**249.** Match List I with List II for the two-port network shown in the given figure and select the correct answer using the codes given below the Lists:

| *List–I* | *List–II* |
|---|---|
| A. $Z_{11}$ | 1. $R$ |
| B. $Z_{12}$ | 2. $R + L$ |
| C. $Z_{12}$ | 3. $R - Ls$ |
| D. $Z_{22}$ | 4. $R + Ls$ |

*Codes:*

| | A | B | C | D |
|---|---|---|---|---|
| (*a*) | 1 | 2 | 1 | 4 |
| (*b*) | 2 | 1 | 1 | 3 |
| (*c*) | 1 | 1 | 1 | 4 |
| (*d*) | 2 | 1 | 3 | 4 |

**250.** Consider the following statements:

I: Any function in '*s*' which can be expressed as a ratio of any two arbitrary polynomials in '*s*' can be realized as a driving-point function of a passive network.

II: Any function in 's' which can be expressed as a ratio of two arbitrary polynomials in 's', is a positive real function.

Of these statements

(a) both I and II are true      (b) I is true but II is false

(c) both I and II are false      (d) I is false but II is true

**251.** Consider the following polynomials:

1. $s^4 + 7s^3 + 17s^2 + 17s + 6$

2. $s^4 + 11s^3 + 41s^2 + 61s + 30$

3. $s^4 + s^3 + 2s^2 + 3s + 2$

Among these polynomials, those which are Hurwitz would include.

(a) 1 and 3     (b) 2 and 3     (c) 1 and 2     (d) 1, 2 and 3

**252.** Match List-I with List-II and select the correct answer using the codes given below the Lists:

| *List-I* | *List-II* |
|---|---|
| (Function) | (Laplace transform) |
| A. Unit ramp | 1. $s$ |
| B. Unit step | 2. 1 |
| C. Unit impulse | 3. $1/s$ |
| D. Unit doublet | 4. $1/s^2$ |

*Codes:*

| | A | B | C | D |
|---|---|---|---|---|
| (a) | 4 | 3 | 2 | 1 |
| (b) | 3 | 4 | 1 | 2 |
| (c) | 4 | 3 | 1 | 2 |
| (d) | 3 | 4 | 2 | 1 |

**253.** A system is represented by

$$\frac{dy}{dt} + 2y = 4t\, u(t)$$

The ramp component in the forced response will be

(a) $t\, u(t)$     (b) $2t\, u(t)$     (c) $3t\, u(t)$     (d) $4t\, u(t)$

**254.** The Laplace transform of the function $i(t)$ is:

$$I(s) = \frac{10s + 4}{s(s + 1)(s^2 + 4s + 5)}$$

Its final value will be

(a) 4/5     (b) 5/4     (c) 4     (d) 5

**255.** If the unit step response of a network is $(1 - e^{-\alpha t})$, then its unit impulse response will be

(a) $\alpha\, e^{-\alpha t}$     (b) $\alpha\, e^{-t/\alpha}$     (c) $\frac{1}{\alpha} e^{-\alpha t}$     (d) $(1 - \alpha)\, e^{-\alpha t}$

**256.** In the network shown in the given figure, if the voltage $v$ at the time considered is 20 V, then $dv/dt$ at that time will be

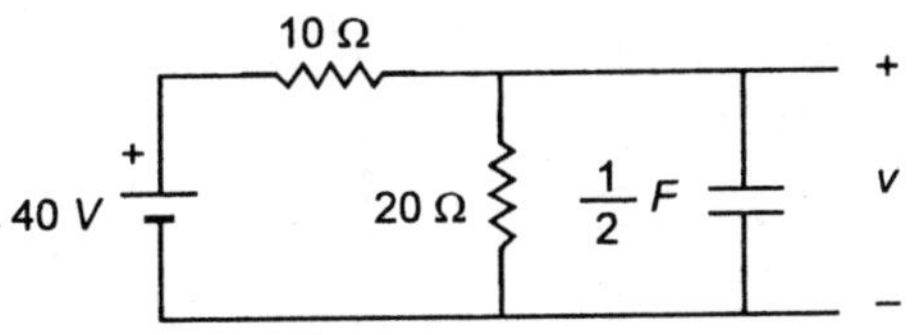

(a) 1 V/s    (b) 2 V/s    (c) −2 V/s    (d) zero

**257.** Which one of the following pairs of poles and responses is correctly matched?

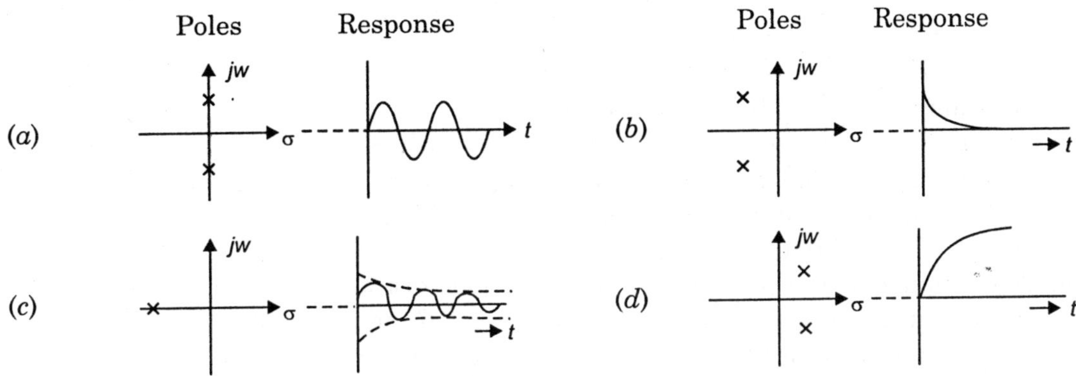

**258.** The impulse response of an $RL$ circuit is a

(a) rising exponential function

(b) decaying exponential function

(c) step function

(d) parabolic function

**259.** A pulse of unit amplitude and width '$a$' is applied to a series $RL$ circuit as shown in the figure. The current $i(t)$ as '$t$' tends to infinity will be

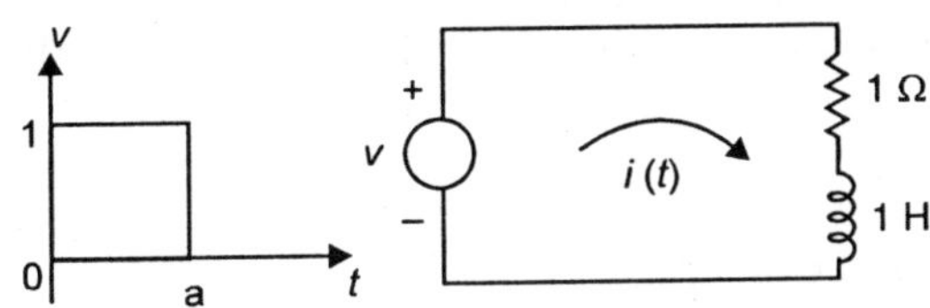

(a) zero    (b) 1 A

(c) a value between zero and one depending upon the width of the pulse.

(d) Infinite

**260.** The sinusoidal steady-state voltage gain of the network shown in the given figure will have magnitude equal to 0.707 at an angular frequency of

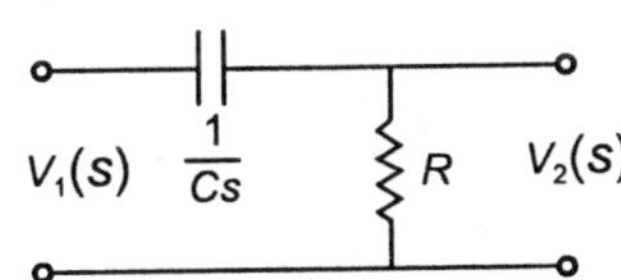

(a) zero    (b) $RC$ rad/s    (c) $1/RC$ rad/s    (d) 1 rad/s

**261.** The phase angle of the current '$I$' with respect to the voltage $V_1$ in the circuit shown in the figure is:

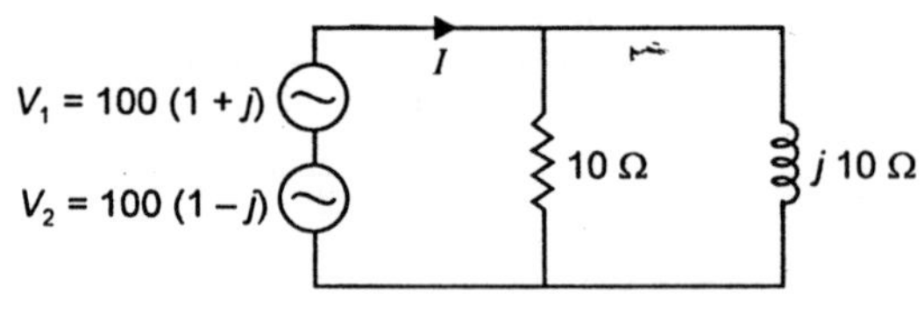



(a) $0°$      (b) $+ 45$      (c) $-45°$      (d) $-90°$

**262.** A second order system is given by

$$\frac{d^2 y}{dt^2} + 12\frac{dy}{dt} + 100\,y = 0$$

The damped natural frequency in rad/sec is

(a) 100      (b) 10      (c) $\sqrt{44}$      (d) 8

**263.** An alternator is delivering power to a balanced load at unity power factor. The phase angle between the line voltage and the line current is

(a) $90°$      (b) $60°$      (c) $30°$      (d) $0°$

**264.** The residues at the pole of $Y(s)$ of an $RC$ network are:

(a) real and negative      (b) real and positive

(c) complex with positive real part      (d) complex with negative real part

**265.** A 10 V battery with an internal resistance of 1 $\Omega$ is connected across a non-linear load whose $v$-$i$ characteristic is given by

$$7i = v^2 + 2v$$

The current delivered by the battery is

(a) 2.5 A      (b) 5 A      (c) 6 A      (d) 7A

**266.** A voltage $V$ is applied to an a.c. circuit resulting in the delivery of a current $I$. Which of the following expressions would yield the true power delivered by the source?

1. Real part of VI*
2. Real part of VI.

3. $I^2$ times the real part of $\dfrac{V}{I}$

Select the correct answer using the codes given below:

*Codes:*

(a) 1 alone      (b) 1 and 3      (c) 2 and 3      (d) 3 alone

**267.** In the network shown in the figure, the effective resistance faced by the voltage source is

(a) $4\ \Omega$

(b) $3\ \Omega$

(c) $2\ \Omega$

(d) $1\ \Omega$

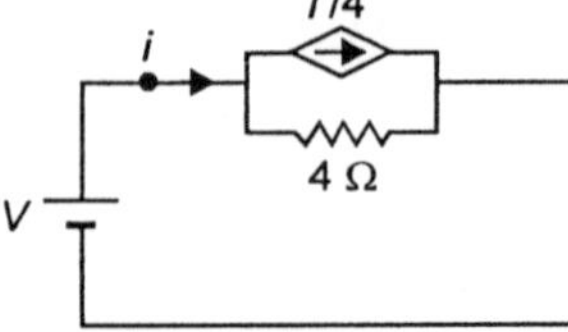



**268.** For the network shown in the figure, if $V_s = V_1$ and $V = 0$, then $I = -5$ A and if $V_s = 0$, and $V_1 = 1$, then $I = 1/2$ A. The values of $I_{sc}$ and $R_1$ of the Norton's equivalent across $AB$ would be respectively.

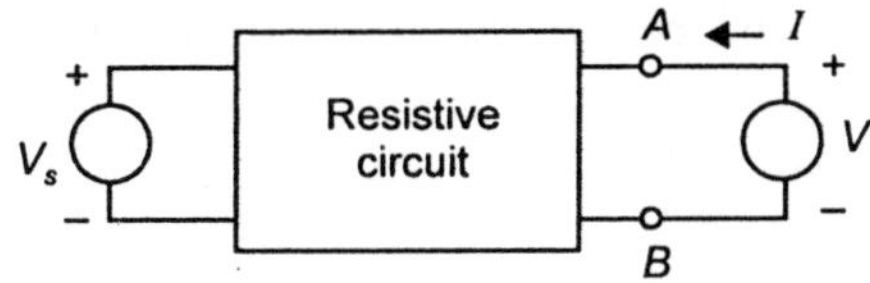

(a) −5 A and 2 Ω  
(b) 10 A and 0.5 Ω  
(c) 5 A and 2 Ω  
(d) 2.5 A and 5 Ω

**269.** The driving-point impedance of a one-port reactive network is given by

(a) $\dfrac{(s^2 + 1)(s^2 + 2)}{s(s^2 + 3)(s^2 + 4)}$

(b) $\dfrac{(s^2 + 1)(s^2 + 3)}{s(s^2 + 2)(s^2 + 4)}$

(c) $\dfrac{s(s^2 + 1)}{(s^2 + 2)(s^2 + 3)}$

(d) $\dfrac{1}{s + 1}$

**270.** The Thevenin equivalent of a network is as shown in the given figure. For maximum power transfer of the variable and purely resistive load $R_L$, its resistance should be

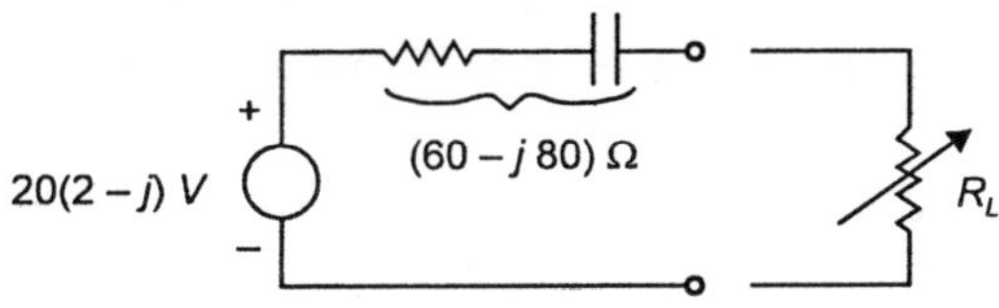

(a) 60 Ω  (b) 80 Ω  (c) 100 Ω  (d) infinity

**271.** If $i(t) = \dfrac{1}{4}(1 - e^{-2t})\, u(t)$, where $u(t)$ is a unit step voltage, then the complex frequencies associated with $i(t)$ would include.

(a) $s = 0$ and $j2$  
(b) $s = j2$ and $s = -j2$  
(c) $s = -j2$ and $s = -2$  
(d) $s = 0$ and $s = -2$

**272.** A T-network is shown in the given figure. Its $Y_{sc}$ matrix will be (units in siemens)

(a) $\begin{bmatrix} \dfrac{10}{200} & \dfrac{5}{200} \\[2mm] \dfrac{5}{200} & \dfrac{10}{200} \end{bmatrix}$

(b) $\begin{bmatrix} \dfrac{10}{200} & \dfrac{-5}{200} \\[2mm] \dfrac{-5}{200} & \dfrac{10}{200} \end{bmatrix}$

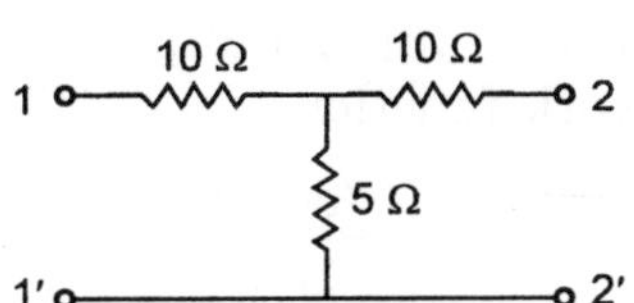

(c) $\begin{bmatrix} \dfrac{15}{200} & \dfrac{-5}{200} \\[2mm] \dfrac{5}{200} & \dfrac{15}{200} \end{bmatrix}$

(d) $\begin{bmatrix} \dfrac{15}{200} & \dfrac{5}{200} \\[2mm] \dfrac{5}{200} & \dfrac{15}{200} \end{bmatrix}$

**273.** The time-constant of the network shown in the figure is

(a) $CR$

(b) $2\, CR$

(c) $\dfrac{CR}{4}$

(d) $\dfrac{CR}{2}$

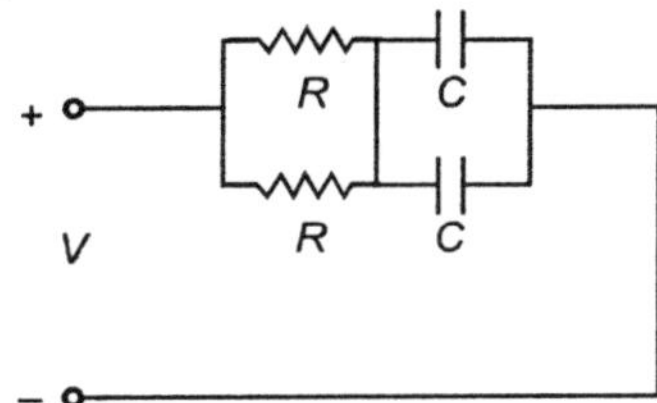

**274.** For a two-port network to be reciprocal, it is necessary that
(a) $Z_{11} = Z_{22}$ and $y_{21} = y_{12}$
(b) $Z_{11} = Z_{22}$ and $AD - BC = 0$
(c) $h_{21} = -h_{12}$ and $AD - BC = 0$
(d) $y_{21} = y_{12}$ and $h_{21} = -h_{12}$

**275.** Two two-port networks with transmission parameters $A_1, B_1, C_1, D_1$ and $A_2, B_2, C_2, D_2$ respectively are cascaded. The transmission parameters matrix of the cascaded network will be

(a) $\begin{bmatrix} A_1 & B_1 \\ C_1 & D_1 \end{bmatrix} + \begin{bmatrix} A_2 & B_2 \\ C_2 & D_2 \end{bmatrix}$  (b) $\begin{bmatrix} A_1 & B_1 \\ C_1 & D_1 \end{bmatrix} \begin{bmatrix} A_2 & B_2 \\ C_2 & D_2 \end{bmatrix}$

(c) $\begin{bmatrix} A_1 & A_2 & B_1 & B_2 \\ C_1 & C_2 & D_1 & D_2 \end{bmatrix}$  (d) $\begin{bmatrix} (A_1 A_2 + C_1 C_2) & (A_1 B_2 - B_1 D_2) \\ (C_1 A_2 - D_1 C_2) & (C_1 C_2 + D_1 D_2) \end{bmatrix}$

**276.** An initially relaxed $RC$-series network with $R = 2M\,\Omega$ and $C = 1\,\mu F$ is switched on to a 10V step input. The voltage across the capacitor after 2 seconds will be
(a) zero  (b) 3.68 V  (c) 6.32 V  (d) 10 V

**277.** For the circuit shown in the given figure, if the input impedance $Z_1$ at port 1 is given by

$$Z_1 = \frac{K_1(s+2)}{(s+5)}$$

then the input impedance $Z_2$ at port 2 will be

(a) $\dfrac{K_2(s+3)}{(s+5)}$  (b) $K_2 \dfrac{(s+2)}{(s+3)}$

(c) $\dfrac{K_2 s}{(s+5)}$  (d) $\dfrac{K_2 s}{s+2}$

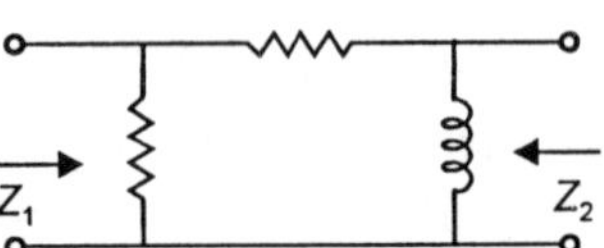

**278.** For $V(s) = \dfrac{(s+2)}{s(s+1)}$, the initial and final values of $v(t)$ will be respectively.

(a) 1 and 1  (b) 2 and 2  (c) 2 and 1  (d) 1 and 2

**279.** The network function

$$F(s) = \frac{(s+2)}{(s+1)(s+3)}$$

represents an
(a) $RC$ impedance  (b) $RL$ impedance
(c) $RC$ impedance and an RL admittance
(d) $RC$ admittance and an RL impedance

**280.** In the network shown in Fig. 1, if the 1F capacitor had an initial voltage of 2V, then which of the following would represent the $s$-domain equivalent circuit?

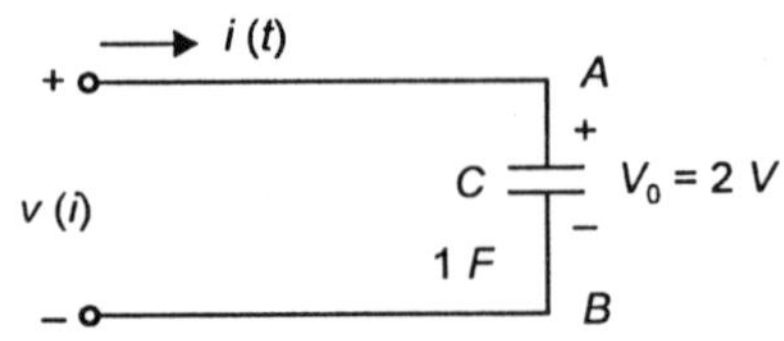


**Fig. 1**

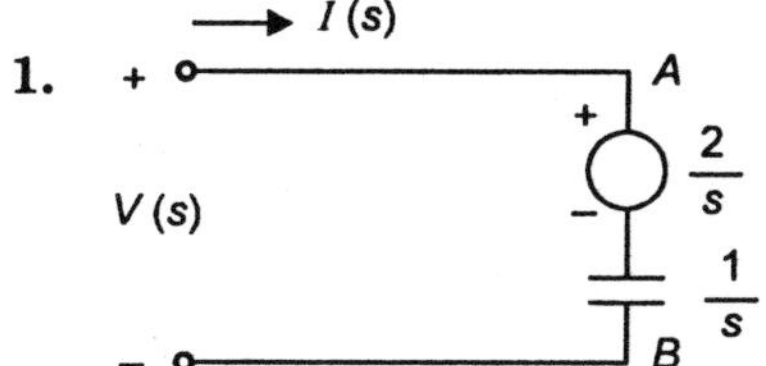 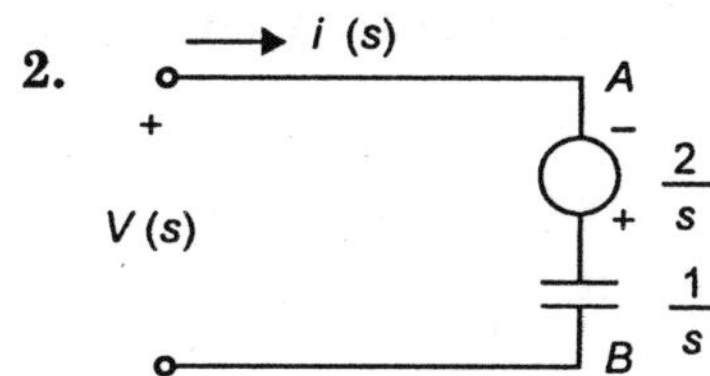

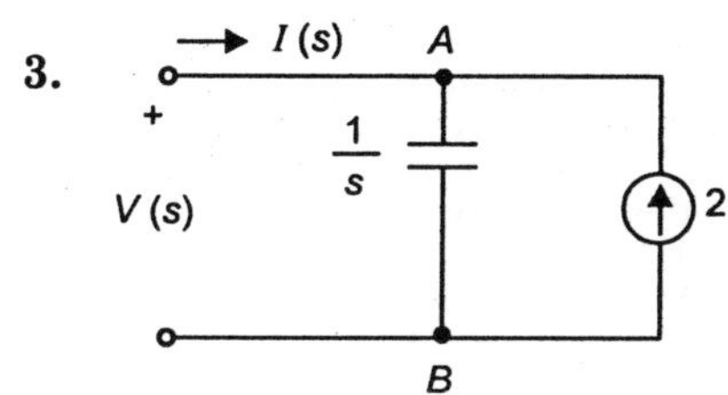 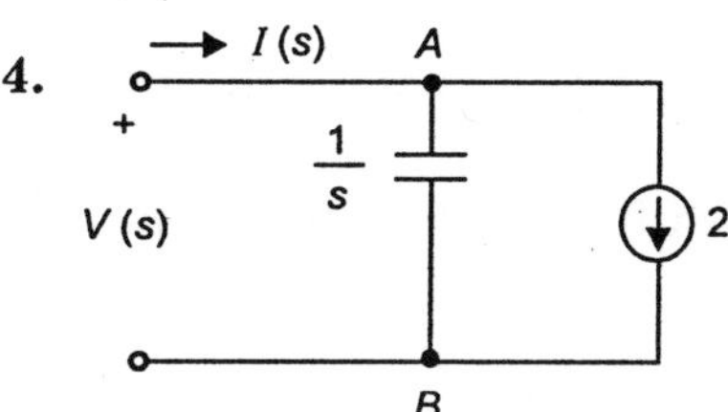

Selected the correct answer using the codes given below:

*Codes:*

(a) 1 and 3    (b) 1 and 4    (c) 2 and 3    (d) 2 and 4

**281.** An initially relaxed 100mH inductor is switched 'ON' at $t = 1$ sec. to an ideal 2 A d.c. current source. The voltage across the inductor would be

(a) zero

(b) $0.2\ \delta(t)\ V$

(c) $0.2\ \delta(t - 1)\ V$

(d) $0.2\ t\ u(t - 1)\ V$

**282.** The current through the current coil of a wattmeter is given by

$$i = (1 + 2\ \sin\ \omega t)\ A$$

and the voltage across the pressure coil is

$$v = (2 + 3\ \sin\ 2\ \omega t)\ V.$$

The wattmeter will read

(a) 8.00 W    (b) 5.05 W    (c) 2.0 W    (d) 1.0 W

**283.** In the circuit shown in the figure, $V_s = \cos 2t$, $Z_2 = 1 + j$, $C_1$ is so chosen that $i = I \cos 2t$

The value of $C_1$ is

(a) 2 F

(b) 1 F

(c) 0.5 F

(d) 0.25 F

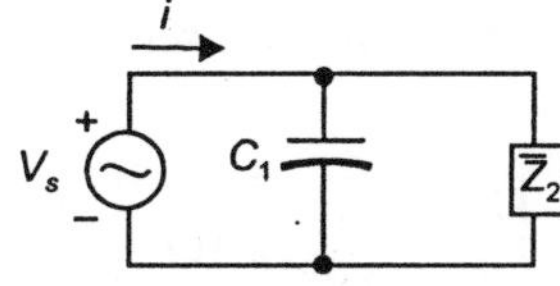

**284.** An *RLC* resonant circuit has a resonance frequency of 1.5 MHz and a bandwidth of 10 kHz. If $C = 150$ pF, then the effective resistance of the circuit will be

(a) 29.5 Ω    (b) 14.75 Ω    (c) 9.4 Ω    (d) 4.7 Ω

**285.** A 3-phase, 3 wire supply feeds a load consisting of three equal resistors connected in star. If one of the resistors is open circuited, then the percentage reduction in the load will be

(a) 75    (b) 66.66    (c) 50    (d) 33.33

**286.** Two identical coils of negligible resistance, when connected in series across a 50 Hz fixed voltage source, draw a current of 10 *A*. When the terminals of one of the coils are reversed, the current drawn is 8 A. The coefficient of coupling between the two coils is (*a*) 1/100       (*b*) 1/9       (*c*) 4/10       (*d*) 8/10

**287.** The voltage-ratio transfer function of an active filter is given by

$$\frac{V_2(s)}{V_1(s)} = \frac{s^2 + \delta}{s^2 + \alpha s + \delta}$$

The circuit in question is a

(*a*) low-pass filter                     (*b*) high-pass filter
(*c*) band-pass filter                 (*d*) band-reject filter

**288.** If the driving-point impedance function $\dfrac{2s + 1}{8s^2 + 4s + 1}$ is synthesized as shown in the given figure, then the values of $R$, $L$ and $C$ will be respectively.

(*a*) 1, 2 and 4

(*b*) 2, 1 and 4

(*c*) 1, 4 and 2

(*d*) 4, 2 and 1

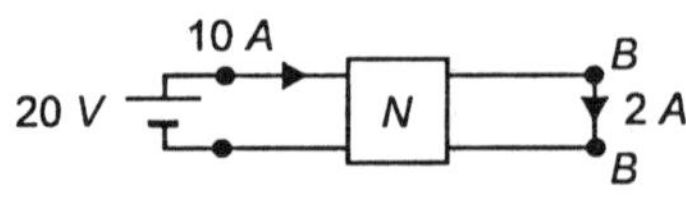

**289.** The short-circuit test of a 2-port network is shown in figure-I. The voltage across the terminals AA in the network shown in figure-II will be

Figure-I                                    Figure-II

    (*a*) 20 V       (*b*) 10 V       (*c*) 5 V       (*d*) 2 V

**290.** In a passive two-port network, the open-circuit impedance matrix is

$$\begin{bmatrix} 10\,\Omega & 2\,\Omega \\ 2\,\Omega & 5\,\Omega \end{bmatrix}$$

If the input port is interchanged with the output port, then the open-circuit impedance matrix will be

(*a*) $\begin{bmatrix} 10\,\Omega & 2\,\Omega \\ 5\,\Omega & 2\,\Omega \end{bmatrix}$    (*b*) $\begin{bmatrix} 5\,\Omega & 2\,\Omega \\ 2\,\Omega & 10\,\Omega \end{bmatrix}$    (*c*) $\begin{bmatrix} 5\,\Omega & 10\,\Omega \\ 2\,\Omega & 2\,\Omega \end{bmatrix}$    (*d*) $\begin{bmatrix} 2\,\Omega & 5\,\Omega \\ 10\,\Omega & 2\,\Omega \end{bmatrix}$

**291.** In the circuit shown in the given figure, if the switch is closed at $t = 0$, then the voltage $v(0^+)$ and its derivative $\left.\dfrac{dv}{dt}\right|_{t=0^+}$ will be respectively.

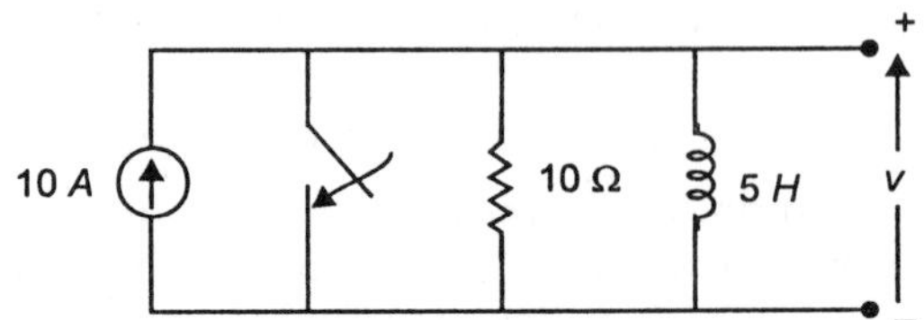

(a) 10 V and 50 V/s  (b) 10 V and –50 V/s
(c) 100 V and 200 V/s  (d) 100 V and –200 V/s

**292.** Five cells are connected in series in a row and then four such rows are connected in parallel to feed the current to a resistive load of 1.5 Ω. Each cell has emf of 1.5 V with internal resistance of 0.2. The current through the load will be
(a) 3.33 A  (b) 23.33 A  (c) 5 A  (d) 1 A

**293.** The mutual inductance between two coupled coils is 10 mH. If the turns in one coil are doubled and that in the other are halved, then the mutual inductance will be
(a) 5 mH  (b) 10 mH  (c) 14 mH  (d) 20 mH

**294.** Four resistances 80 Ω, 50 Ω, 25 Ω and $R$ are connected in parallel. Current through 25 Ω resistance is 4A. Total current of the supply is 10 A. The value of $R$ will be
(a) 66.66 Ω  (b) 40.25 Ω  (c) 36.36 Ω  (d) 76.56 Ω

**295.** The response shown in the given figure is the Laplace transform of the function.

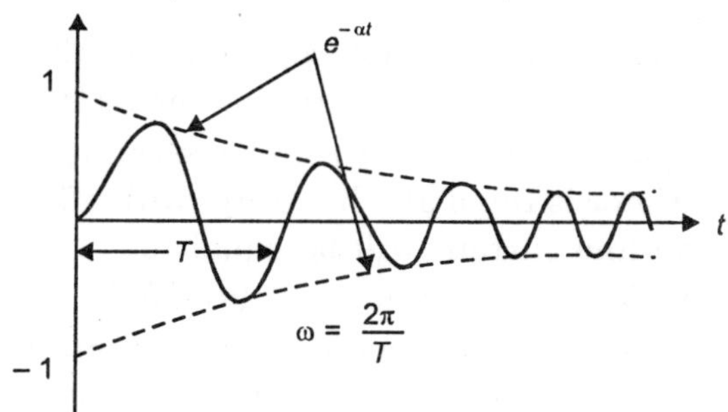


(a) $\dfrac{\omega}{(s+\alpha)^2 + \omega^2}$

(b) $\dfrac{\alpha}{(s+\alpha)^2 + \omega^2}$

(c) $\dfrac{s+\alpha}{(s+\alpha)^2 + \omega^2}$

(d) $\dfrac{s}{(s+\alpha)^2 + \omega^2}$

**296.** Consider the following statements:
The impulse response of a linear network can be used to determine the
1. step response.
2. response of the sinusoidal input.
3. elements of the network uniquely.
4. interconnection of network elements.
Which of these statements are correct?
(a) 1 and 2  (b) 2 and 3  (c) 3 and 4  (d) 1 and 4

**297.** In the circuit shown in the given figure, the response current $i(t)$ is

(a) $\dfrac{V}{R} e^{-\frac{t}{CR}}$  (b) $\dfrac{V}{R} \delta(t)$

(c) $\dfrac{V}{R}\left[\delta(t) - \dfrac{1}{RC} e^{-\frac{t}{CR}}\right]$

(d) $\dfrac{V}{R}\left[\delta(t) - e^{-\frac{t}{CR}}\right]$

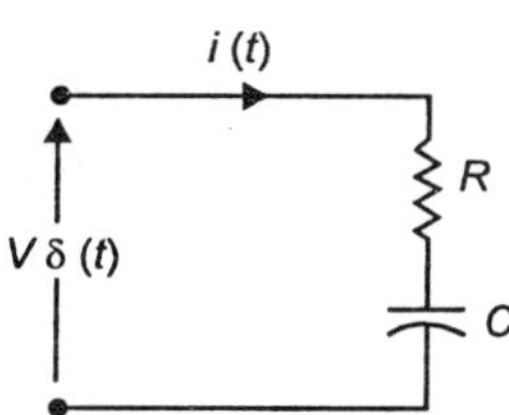

**298.** In the network shown in the given figure, the value of $v_x$ would be

(a) $-\dfrac{8}{9}V$

(b) $\dfrac{8}{9}V$

(c) $\dfrac{16}{9}V$

(d) $-\dfrac{16}{9}V$

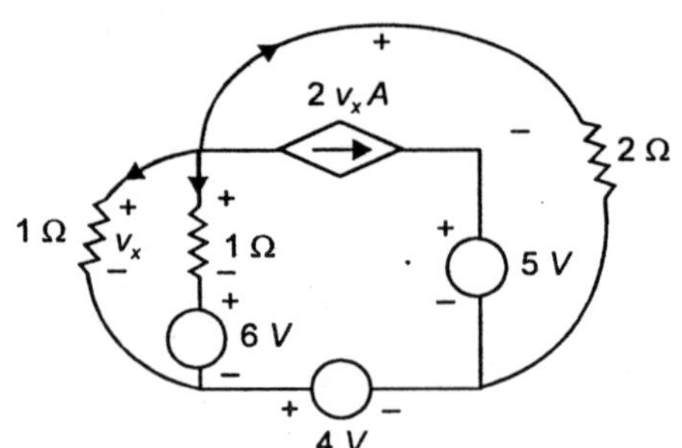


**299.** To improve the power factor in three-phase circuits, the capacitor bank is connected in delta to make

(a) capacitance calculation easy.
(b) capacitance value small.
(c) the connection elegant
(d) the power factor correction more effective.

**300.** A three-phase heating unit and induction motor are connected in parallel across a 208 V three-phase supply. Motor is rated at 5 hp, 0.8 pf with efficiency of 0.85. Heating unit is rated at 1500 W. The line current will be equal to

(a) 185 A     (b) 1.85 A     (c) 18.5 A     (d) 15 A

**301.** A 10 µF capacitor is fed from an a.c. voltage source containing a fundamental and a third harmonic of value one-third of fundamental. The third harmonic current flowing through the capacitor expressed as percentage of the fundamental under steady state condition will be

(a) zero     (b) 100     (c) 30     (d) 90

**302.** In a two-element series circuit, the applied voltage and the resulting current are respectively,

$$v(t) = 50 + 50 \sin (5 \times 10^3 \, t) \ V \text{ and}$$

$$i(t) = 11.2 \sin (5 \times 10^3 \, t + 63.4°) \ A.$$

The nature of the elements would be

(a) $R - L$
(b) $R - C$
(c) $L - C$
(d) neither $R$, nor $L$, nor $C$

**303.** In a balanced wheatstone bridge, if the positions of detector and source are interchanged, the bridge will still remain balanced. This inference can be drawn from

(a) reciprocity theorem
(b) duality principle
(c) compensation theorem
(d) equivalence theorem

**304.** A voltage $v(t) = 6 \, e^{-2t}$ is applied at $t = 0$ to a series $R - L$ circuit with $L = 1H$. If $i(t) = 6 \, [\exp (-2t) - \exp (-3t)]$ then $R$ will have a value of

(a) $\dfrac{2}{3}\Omega$     (b) $1 \ \Omega$     (c) $3 \ \Omega$     (d) $\dfrac{1}{3}\Omega$

**305.** In the circuit shown, the switch is opened at $t = 0$. Prior to that switch was closed, $i(t)$ at $t = 0^+$ is

(a) $\dfrac{2}{3} A$

(b) $\dfrac{3}{2} A$

(c) $\dfrac{1}{3} A$

(d) $1\ A$

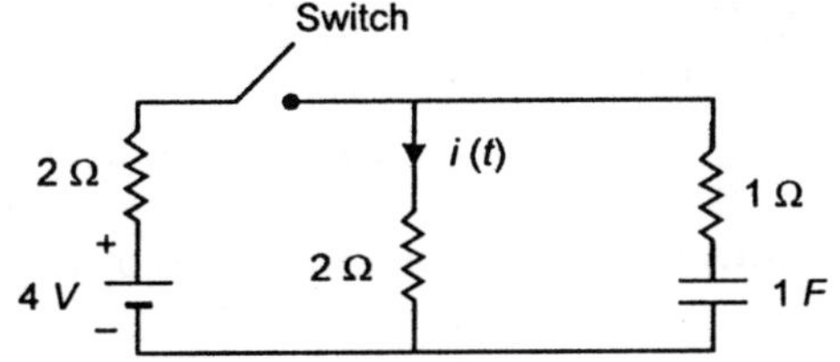

**306.** Consider the following statements regarding the circuit shown in the figure. If the power consumed by 5 $\Omega$ resistor is 10 W, then

1. $|I| = \sqrt{2}\ A.$
2. the total impedance of the circuit is 5 $\Omega$
3. $\cos\phi = 0.866.$

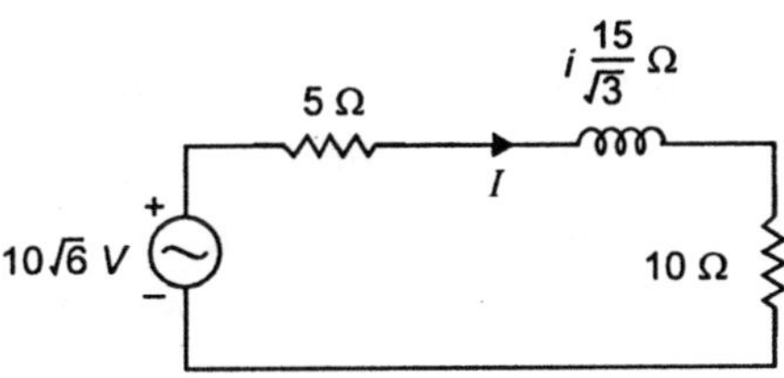

Which of these statements are correct?

(a) 1 and 3    (b) 2 and 3    (c) 1 and 2    (d) 1, 2 and 3

**307.** In the network shown in the given figure, the Thevenin source and the impedance across terminals $A - B$ will be respectively

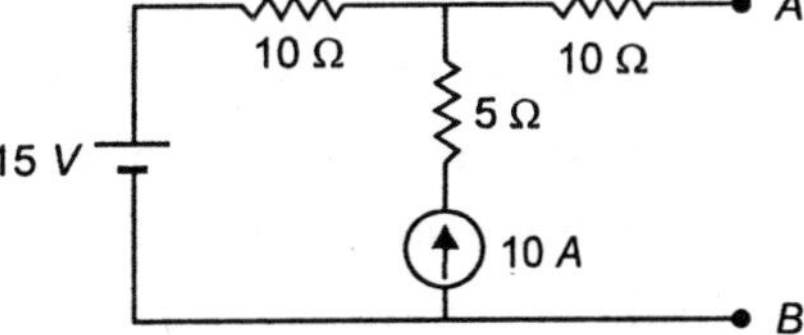

(a) 15 V and 13.33 $\Omega$    (b) 50 V and 15 $\Omega$

(c) 115 V and 20 $\Omega$    (d) 100 V and 25 $\Omega$

**308.** For the function $\mathcal{L}[f(t)] = \dfrac{3s + 1}{s(s^2 + 4s + 5)}, \ \dfrac{df}{dt}\Big|_{t\,=\,0^+}$

(a) 3    (b) $\dfrac{1}{3}$    (c) zero    (d) $\dfrac{2}{3}$

**309.** An a.c. source of 200 V rms supplies active power of 600 W and reactive power of 800 VAR. The rms current drawn from the source is

(a) 10 A    (b) 5 A    (c) 3.75 A    (d) 2.5 A

**310.** In the given 2-port network, if the driving-point (d.p) impedance at port 1 is $Z_{11}\ (s)$

$= K_1\dfrac{s + 2}{s + 5},$ the d.p. impedance at port 2 will be

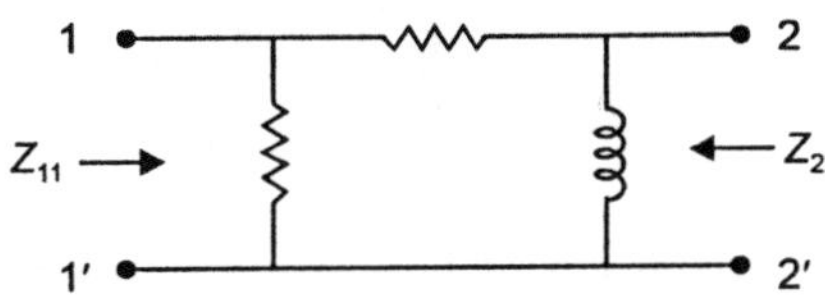

(a) $Z_{22}(s) = K_2 \dfrac{s+3}{s+5}$  (b) $Z_{22}(s) = K_2 \dfrac{s+2}{s+3}$

(c) $Z_{22}(s) = K_2 \dfrac{s}{(s+5)}$  (d) $Z_{22}(s) = K_2 \dfrac{s}{s+2}$

**311.** The reactive power drawn from the source in the network shown in the given figure is

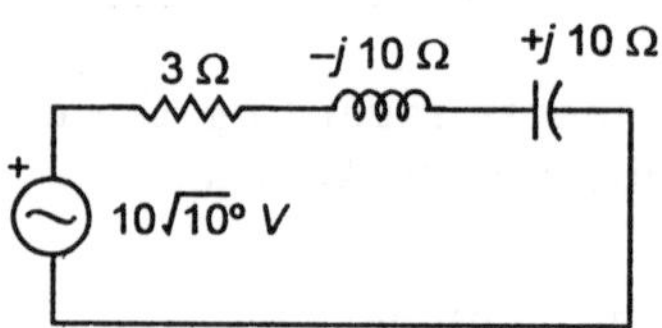


(a) 300 VAR  (b) 200 VAR  (c) 100 VAR  (d) zero

**312.** For a series $RLC$ circuit, the power factor at the lower half power frequency is

(a) 0.5 lagging  (b) 0.5 leading

(c) unity  (d) 0.707 leading

**313.** If a network has all linear elements except for a few non-linear ones, then superposition theorem.

(a) cannot hold at all

(b) always holds

(c) may hold on careful selection of element values, source waveform and response.

(d) holds in case of direct current excitations

**314.** "In any network of linear impedances, the current flowing at any point is equal to the algebraic sum of the currents caused to flow at that point by each of the sources of emf taking separately with all other emf's reduced to zero". This statement represents

(a) Kirchhoff's current law  (b) Norton's theorem

(c) Thevenin's theorem  (d) Superposition theorem

**315.** Which one of the following theorems is a manifestation of the Law of Conservation of Energy?

(a) Tellegen's theorem  (b) Reciprocity theorem

(c) Thevenin's theorem  (d) Norton's theorem

**316.** For the circuit shown in the given figure, the current through $R$, when $V_A = 0$ and $V_B = 15\ V$ is $I$ amp. Now, if both $V_A$ and $V_B$ are increased by 15 volts, then the current through $R$ will be

(a) $I$ amp

(b) $I/2$ amp

(c) $3\ I$ amp

(d) $I/3$ amp

**317.** If the networks shown in figures I and II are equivalent at terminals $A - B$, then the values of $V$ (in volts) and $Z$ (in ohms) will be

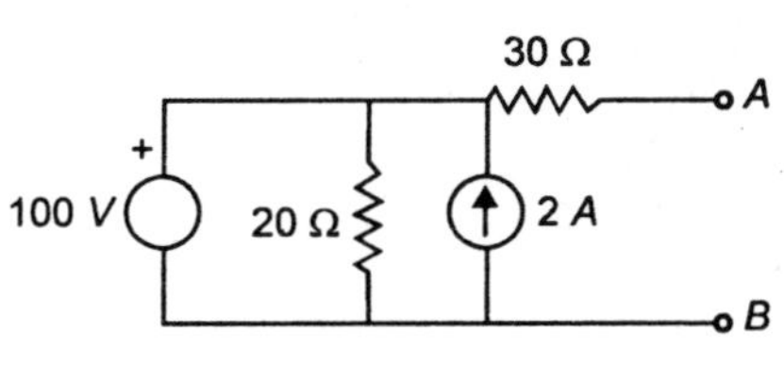

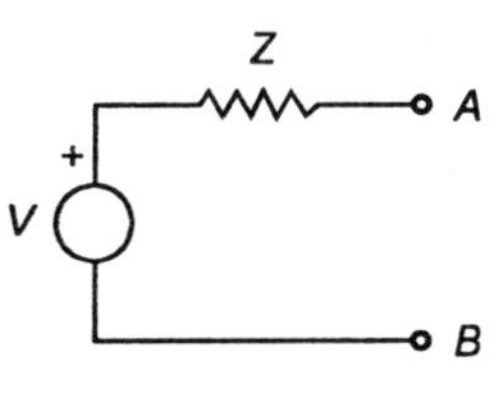

Fig. I                                      Fig. II

|       | V   | Z  |
|-------|-----|----|
| (a)   | 100 | 12 |
| (b)   | 60  | 12 |
| (c)   | 100 | 30 |
| (d)   | 60  | 30 |

**318.** Match List I with List II and select the correct answer using the codes given below the lists:

*List I*
(Network Theorems)

A. Reciprocity
B. Tellegen's

C. Superposition

D. Maximum
   Power Tansfer

*List II*
(Most distinguished
property of Networks)

1. Impedance matching
2. Bilateral

3. $\displaystyle\sum_{k=0}^{b} v_k\,(t_1)\,i_k\,(t_2) = 0$

4. Linear
5. Non Linear

*Codes:*

|       | A | B | C | D |
|-------|---|---|---|---|
| (a)   | 1 | 2 | 3 | 4 |
| (b)   | 1 | 2 | 3 | 5 |
| (c)   | 2 | 3 | 4 | 1 |
| (d)   | 2 | 3 | 5 | 1 |

**319.** The inverse Fourier transform of

$$F\,(j\omega) = \int_{-\infty}^{\infty} \exp\,(-j\omega t)\, f(t)\ dt \ \text{ is}$$

(a) $f(t) = \displaystyle\int_{-\infty}^{\infty} \exp\,(+j\omega t)\, F(j\omega)\ d\omega$

(b) $f(t) = 1/2\pi \displaystyle\int_{-\infty}^{\infty} \exp\,(+j\omega t)\, F(j\omega)\ d\omega$

(c) $f(t) = 1/2\pi \displaystyle\int_{-\infty}^{\infty} \exp\,(-j\omega t)\, F(+j\omega)\ d\omega$

(d) $f(t) = 1/2\pi \displaystyle\int_{-\infty}^{\infty} \exp\,(-j\omega t)\, F(-j\omega)\ d\omega$

**320.** Which one of the following networks is the $Y$ equivalent of the circuit shown in Fig. I?

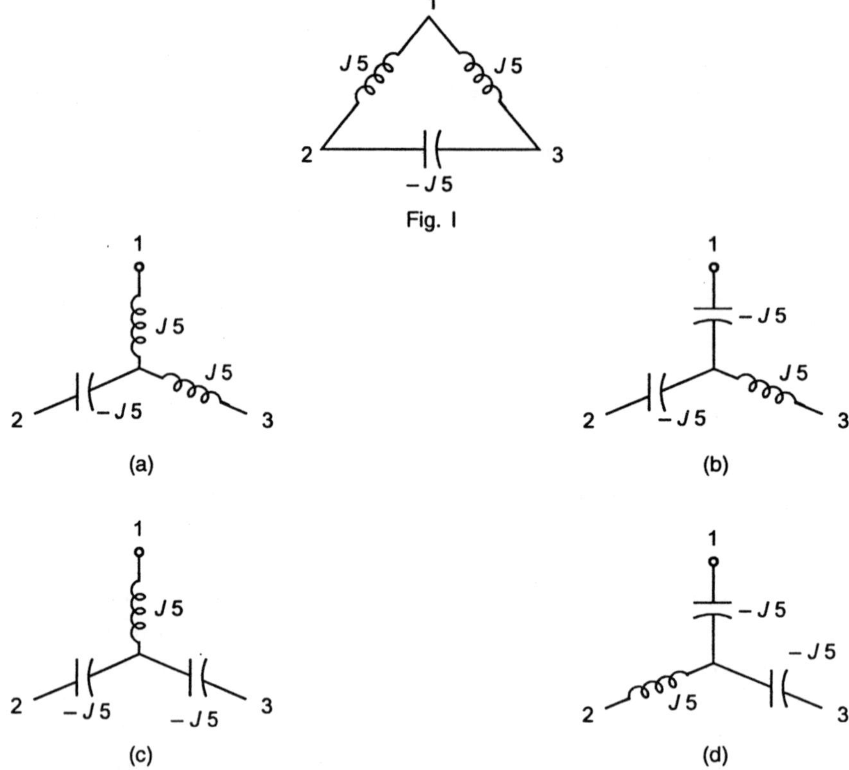

Fig. I

(a)     (b)

(c)     (d)

**321.** Which of the following periodic wave form will have only odd harmonics of sinusoidal wave forms?

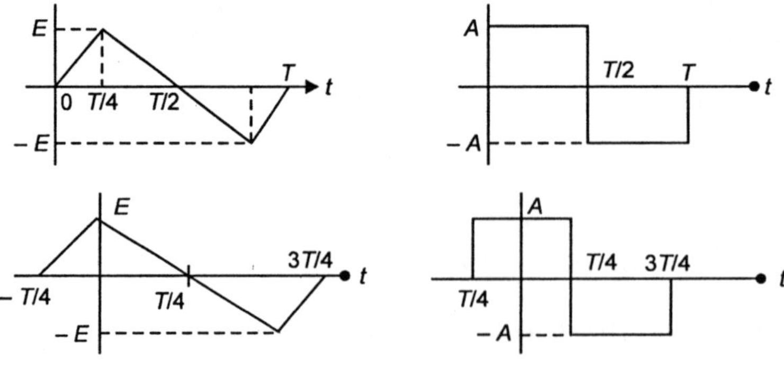

Select the correct answer using the codes given below:

*Codes:*

(*a*) 1 and 2     (*b*) 1 and 3     (*c*) 1 and 4     (*d*) 2 and 4

**322.** If $x(t)$ is the linear input to a linear network whose impulse response $h(t)$ is known, then assuming that the impulse applied at $t = \lambda$, the output response $y(t)$ will be (* = convolution)

(*a*) $y(t) = x(t) * h(t)$          (*b*) $y(t) = x(t - \lambda) * h(t)$

(*c*) $y(t) = x(\lambda) * h(t - \lambda)$          (*d*) $y(t) = x(t + \lambda) * h(t - \lambda)$

**323.** For a high pass $RC$ circuit, when subjected to a unit step input voltage, the voltage across the capacitor will be

(a) $1 - e^{-t/RC}$    (b) $e^{-t/RC}$      (c) $e^{t/RC}$      (c) $1$

**324.** Given the Laplace transform

$$\mathcal{L}[v(t)] = \int_0^\infty e^{-st} \, v(t) \, dt,$$

the inverse transform $v(t)$ is

(a) $\displaystyle \int_{\sigma-j\infty}^{\sigma+j\infty} e^{st} \, V(s) \, ds$      (b) $\displaystyle \frac{1}{2\pi j} \int_{\sigma-j\infty}^{\sigma+j\infty} e^{st} \, V(s) \, ds$

(c) $\displaystyle \frac{1}{2\pi j} \int_0^\infty e^{st} \, V(s) \, ds$      (d) $\displaystyle \frac{1}{2\pi j} \int_{\sigma-j\infty}^{\sigma+j\infty} e^{-st} \, V(s) \, ds$

**325.** In the network shown in the given figure, the switch $K$ is closed at $t = 0$ with the capacitor uncharged. The value for $di/dt$ at $t = 0$ will be

(a) 100 amp/sec

(b) –100 amp/sec

(c) 1000 amp/sec

(d) –1000 amp/sec

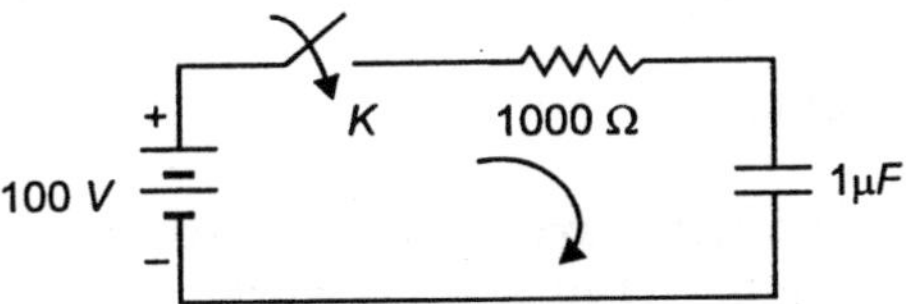


**326.** The time constant of the network shown in the given figure is given by

(a) $\displaystyle \frac{L}{R_3 + \dfrac{R_1 R_2}{R_1 + R_2}}$

(b) $\displaystyle \frac{L}{R_1 + R_2 + R_3}$

(c) $\displaystyle \frac{L}{\dfrac{1}{R_1} + \dfrac{1}{R_2} + \dfrac{1}{R_3}}$      (d) $\displaystyle \frac{L}{\dfrac{R_1 R_2}{R_1 + R_2}}$

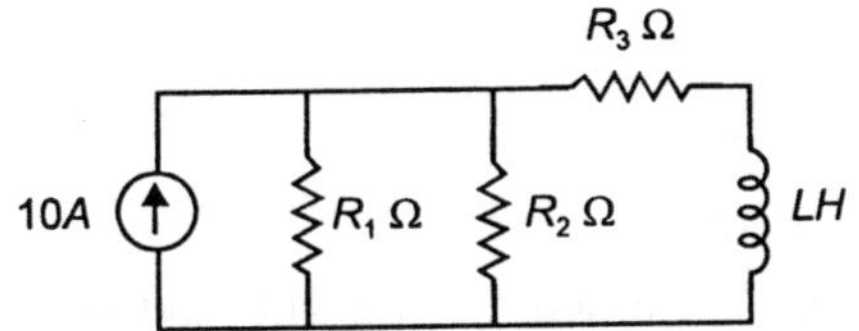


**327.** Consider the following statements regarding an $RC$ differentiating network:
1. For an applied rectangular pulse, the output is spiky in nature for $RC \ll$ pulse duration.
2. The output is a ramp for rectangular input pulse.
3. The output has zero average value for all inputs.
Of these statements
(a) 1, 2 and 3 are correct.      (b) 1 and 2 are correct.
(c) 2 and 3 are correct.      (d) 1 and 3 are correct.

**328.** A series $RLC$ circuit is overdamped when

(a) $\displaystyle \frac{R^2}{4L^2} > \frac{1}{Lc}$    (b) $\displaystyle \frac{R^2}{4L^2} = \frac{1}{Lc}$    (c) $\displaystyle \frac{R^2}{4L^2} < \frac{1}{Lc}$    (d) $R = $ infinity

**329.** The circuit shown in the given figure has been in the steady-state when the switch $S$ is opened. The current after the switch is opened is given by

(a) $\dfrac{V}{R_2}\, \overline{e}^{\frac{R_1}{L}t}$

(b) $\dfrac{V}{R_1}\, \overline{e}^{\frac{R_2}{L}t}$

(c) $\dfrac{VR_2}{R_1 + R_2}\, \overline{e}^{\frac{L(R_1 + R_2)}{R_1 R_2}t}$

(d) $\dfrac{V}{R_1 + R_2}\, \overline{e}^{\frac{L}{R_2}}t$

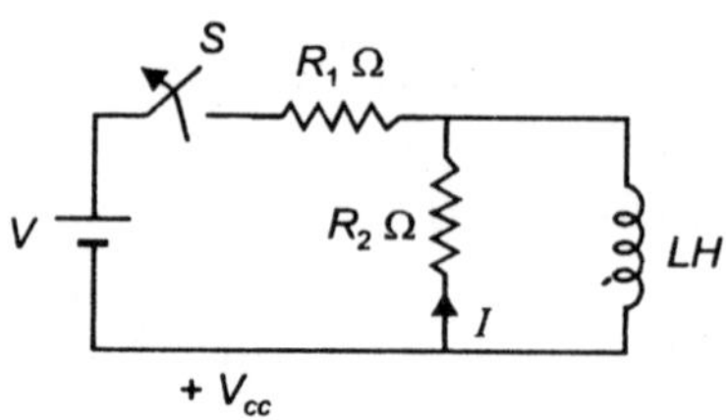


**330.** If an impedance has the pole-zero pattern shown in the given figure, it must by composed of

(a)  $RC$ elements only

(b)  $RL$ elements only

(c)  $LC$ elements only

(d)  $RLC$ elements

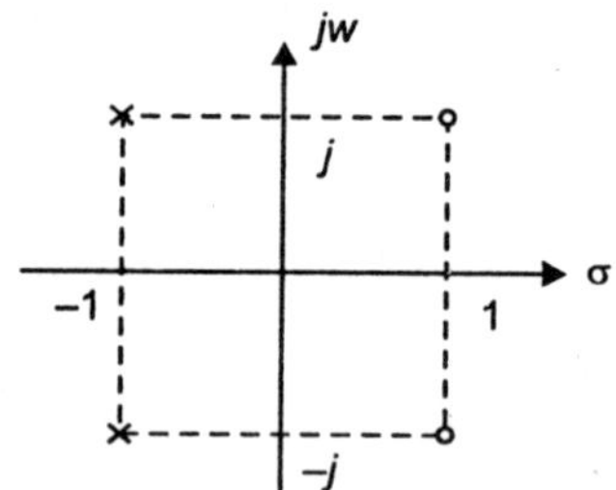


**331.** Which of the following expressions is/are used in computing the inductance $L$ as seen from the terminals AD of the network of interconnected inductances as shown in the given figure?

1. $L_1 \pm L_2$

2. $L_1 + L_2 \pm M$

3. $L_1 + L_2 + 2M$

4. $L_1 + L_2 - 2M$

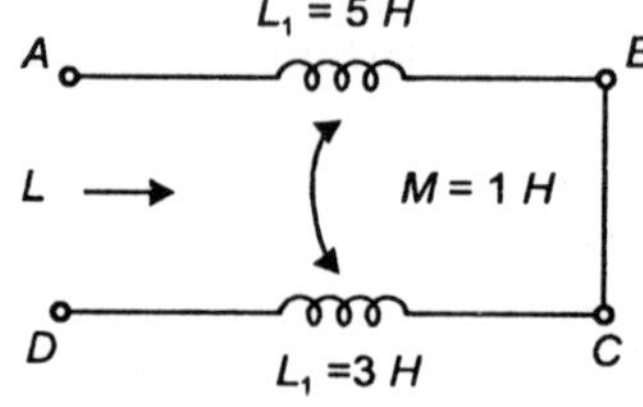


Select the correct answer using the codes given below:
*Codes:*
(a) 1 alone     (b) 1 and 2     (c) 3 and 4     (d) 2,.3 and 4

**332.** The voltage transfer ratio of two two-port networks connected in cascade may be conveniently obtained form the

(a) Product of the individual $ABCD$ matrices of the two networks.

(b) Product of voltage transfer ratios of the two individual networks.

(c) Sum of the $Z$-matrices of the two networks.

(d) Sum of the $h$-matrices of the two networks.

**333.** The transfer function of a low pass $RC$ network is

(a) $RCs/(1 + RCs)$

(b) $1/(1 + RCs)$

(c) $RC/(1 + RCs)$

(d) $s/(1 + RCs)$

**334.** An electric circuit contains $R$, $L$ and $C$ in series with a voltage source. The current through the circuit at the resonant frequency is $I_o$. The frequencies at which the current would reduce to $0.707\,I_o$ is given by $f_{01}$ and $f_{02}$. The resonant frequency of the circuit is the

(a) Geometric mean of $f_{01}$ and $f_{02}$

(b) Arithmetic mean of $f_{01}$ and $f_{02}$

(c) Defference of $f_{01}$ and $f_{02}$

(d) Harmonic mean of $f_{01}$ and $f_{02}$.

**335.** In the network shown in the given figure, the capacitor $C_1$ is initially charged to a voltage $V_0$ before the switch $S$ in the circuit is closed. In the steady-state.

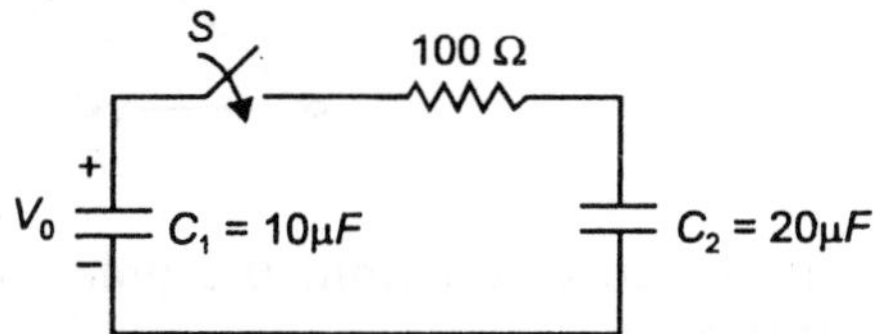


(a) $C_1$ and $C_2$ are charged to equal voltages.

(b) $C_1$ and $C_2$ are charged with equal coulombs.

(c) $C_1$ and $C_2$ are discharged fully.

(d) $C_1$ alone is charged to voltages $V_0$.

**336.** For the design of low pass prototype filter (see figure given) with load resistance $R_L$ = 1 ohm and angular frequency $\omega$ = 1 rad/sec the values of $L$ and $C$ would be

(a) 1 H and 1 F

(b) 1 H and 2 F

(c) 2 H and 1 F

(d) 2 H and 2 F

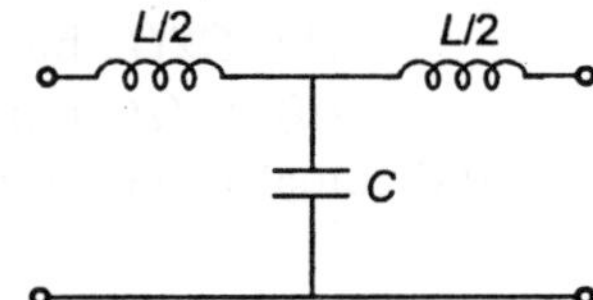


**337.** The passband of a typical filter network with $Z_1$ and $Z_2$ as the series and shunt arm impedances is characterized by

(a) $-1 < \dfrac{Z_1}{4Z_2} < 0$

(b) $-1 < \dfrac{Z_1}{4Z_2} < 1$

(c) $0 < \dfrac{Z_1}{4Z_2} < 1$

(d) None of the above.

**338.** In an $m$-derived low pass filter, the value of $m$ is [$fc$ = cut-off frequency and $f_\infty$ = series resonant frequency (series arm)]

(a) $\sqrt{1 - \left(\dfrac{f_\infty}{f_c}\right)^2}$

(b) $\sqrt{1 - \left(\dfrac{f_c}{f_\infty}\right)^2}$

(c) $\sqrt{1 + \left(\dfrac{f_\infty}{f_c}\right)^2}$

(d) $\sqrt{1 + \left(\dfrac{f_c}{f_\infty}\right)^2}$

**339.** The transfer function

$$\frac{4\left(s^2 + 25\right)}{s^2 + 2.5\,s + 100} \quad \text{is of}$$

(a) low pass notch filter

(b) low $Q$ bandpass filter

(c) high $Q$ bandpass filter

(d) high pass notch filter

**340.** For the lattice type attenuator shown in the given figure, the characteristic impedance $R_o$ is

(a) $\dfrac{R_1 + R_2}{2}$

(b) $\dfrac{2R_1 R_2}{R_1 + R_2}$

(c) $\sqrt{R_1 R_2}$

(d) $R_1 + \dfrac{R_2}{2}$

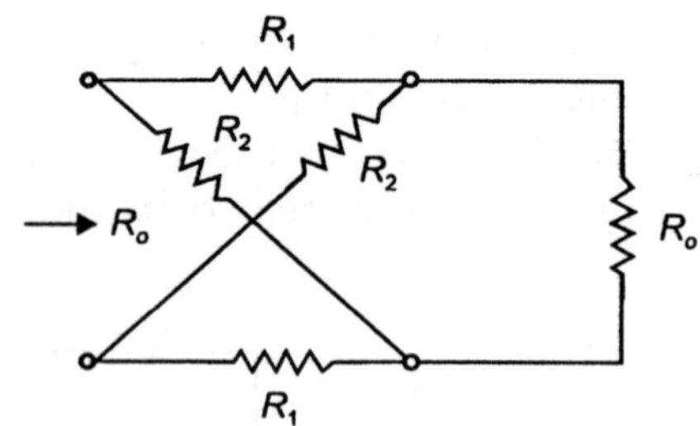


**341.** For an $RC$ driving-point impedance function, the poles and zeros
(a) should alternate on real axis
(b) should alternate only on the negative real axis.
(c) should alternate on the imaginary axis
(d) can lie anywhere on the left half plane

**342.** A gyrator has an admittance matrix $= \begin{bmatrix} 0 & G \\ -G & 0 \end{bmatrix}$. It synthesises an inductor at it's input

terminals when terminate by a capacitor $C$. The magnitude of the inductor is
(a) $G^2\,C$ henries  (b) $C/G^2$ henries
(c) $G^2/C$ henries  (d) $2\,GC$ henries

**343.** Match List I with List II and select the correct answer using the codes given below the lists:

|  | List I<br>[F (s)] |  | List II<br>[Type of F (s)] |
|---|---|---|---|
| A. | $\dfrac{s^2 - s + 4}{s^2 + s + 4}$ | 1. | Non-positive real |
| B. | $\dfrac{(s + 4)}{s^2 + 3s - 4}$ | 2. | Non-minimum phase |
| C. | $\dfrac{s + 4}{s^2 + 6s + 5}$ | 3. | $RC$-impedance |
| D. | $\dfrac{s^2 + 3s}{s^4 + 2s^2 + 1}$ | 4. | Unstable |
|  |  | 5. | $RL$-impedance |

*Codes:*

|  | A | B | C | D |
|---|---|---|---|---|
| (a) | 1 | 2 | 3 | 4 |
| (b) | 1 | 2 | 4 | 5 |
| (c) | 2 | 4 | 3 | 1 |
| (d) | 2 | 4 | 1 | 5 |

**344.** The driving-point impedance function of a reactive network is:

$$Z_0(s) = \frac{2(s^2 + 1)(s^2 + 3)}{s(s^2 + 2)}$$

Which one of the following diagram realizes the Cauer network for the above $Z_0(s)$?

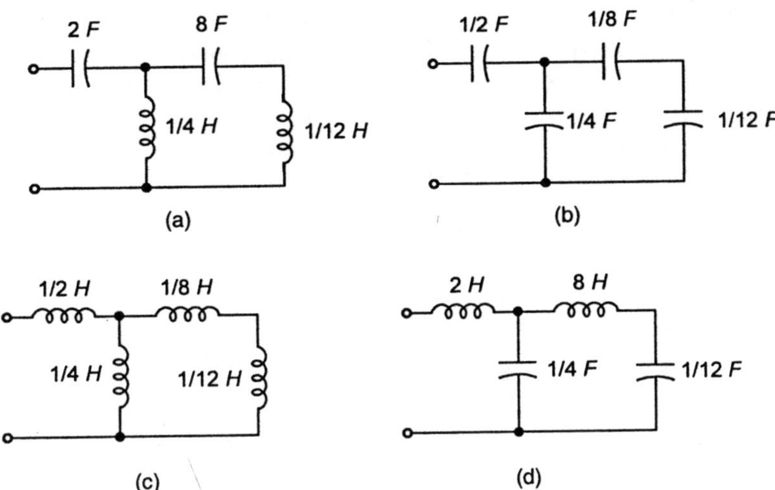

**345.** Thevenin's equivalent circuit of the network shown in the given figure, between terminals $T_1$ and $T_2$ is

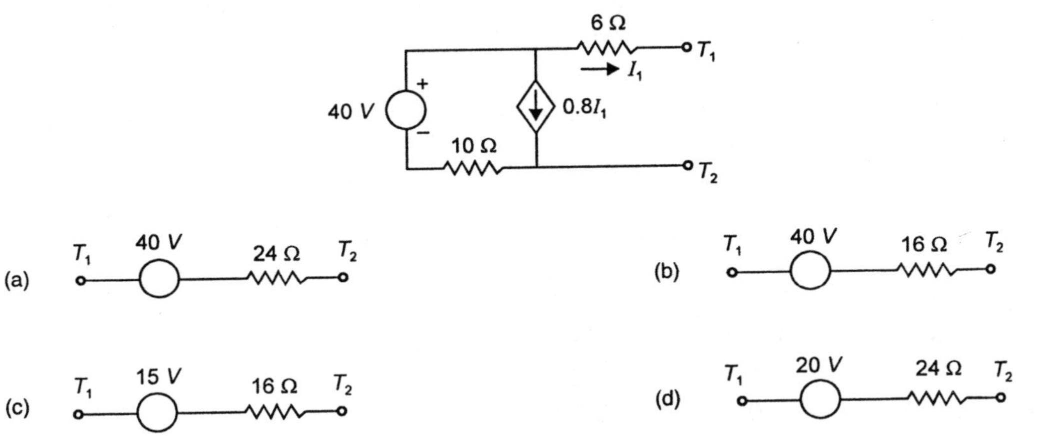

**346.** The equivalent resistance between the terminal points $X$ and $Y$ of the circuit shown is

(a) 15 ohms

(b) 45 ohms

(c) 55 ohms

(d) 30 ohms

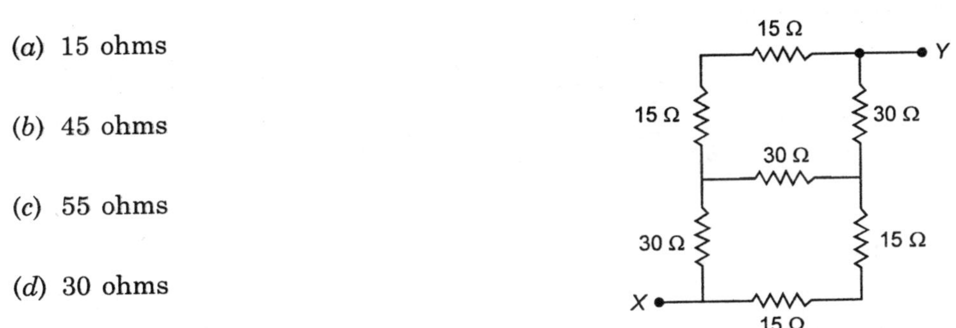

**347.** Match List I with List II and select the correct answer using the codes given below the lists:

| | *List I*<br>(Properties of Network) | | *List II*<br>(Relevant theorems) |
|---|---|---|---|
| A. | Linearity | 1. | Superposition |
| B. | Structure | 2. | Norton's |
| C. | Equivalent circuit | 3. | Tellegan's |
| D. | Bilateral | 4. | Reciprocity |
| | | 5. | Millman's |

*Codes:*

| | A | B | C | D |
|---|---|---|---|---|
| (a) | 1 | 2 | 3 | 4 |
| (b) | 1 | 3 | 2 | 4 |
| (c) | 2 | 3 | 4 | 5 |
| (d) | 1 | 3 | 5 | 2 |

**348.** In the given figure network, there is no initial current through $L$'s and no initial voltage across $C$, and the switch '$S$' is closed at time $t = 0$. The current $i_{L_1}$ in the inductor $L_1$ and the voltage $Vc$ across $C$ are calculated at $t = 0$ and at $t = \infty$. Which of the following sets of results is correct?

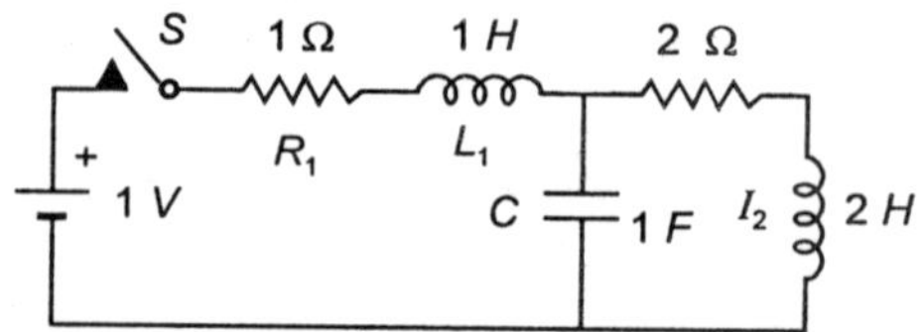


| | $iL_1(0)$ | $iL_1(\infty)$ | $V_C(0)$ | $V_C(\infty)$ |
|---|---|---|---|---|
| (a) | 1/3 A | 1/3 A | 2/3 V | 2/3 V |
| (b) | 0 | 1/3 A | 0 | 1 V |
| (c) | 1/3 A | 0 | 2/3 V | 0 |
| (d) | 0 | 1/3 A | 0 | 2/3 V |

**349.** The dual of the network

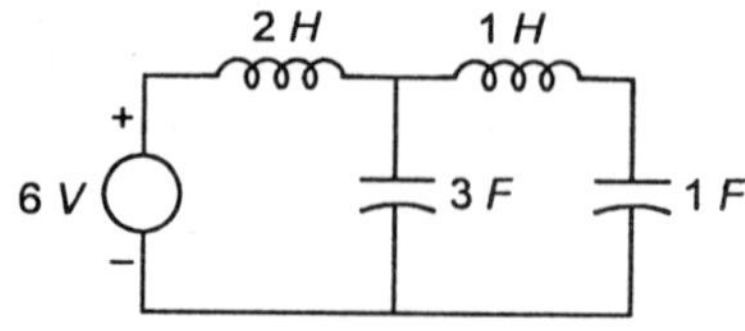


Shown in figure I, is

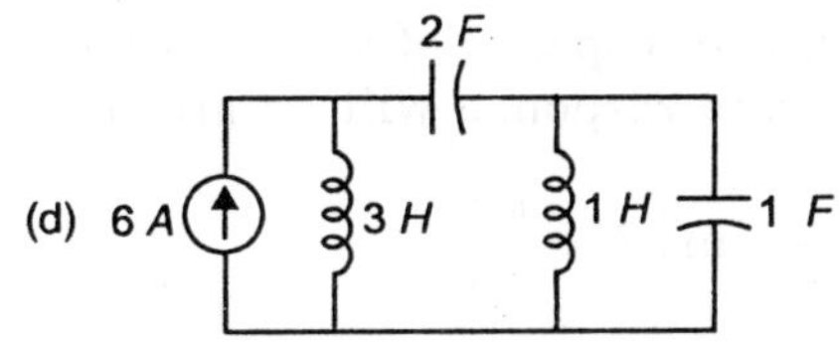

(c) 6 V, 1 H, 2 F, 1 F, 3 H

(d) 6 A, 2 F, 3 H, 1 H, 1 F

**350.** Which one of the following represents the total number of trees in the graph given in the figure?

(a) 4

(b) 5

(c) 6

(d) 8

**351.** Consider the following statements regarding the situation at resonant frequency:
1. For a series $RLC$ circuit, current is minimum.
2. For a series $RLC$ circuit, voltage across $C$ is minimum.
3. For a series $RLC$ circuit, current is maximum.
4. For a parallel $RLC$ circuit, total impedance is maximum.

Of these statements

(a) 1 and 2 are correct      (b) 2 and 3 are correct

(c) 3 and 4 are correct      (d) 1 and 4 are correct

**352.** Which one of the following circuits is the delta equivalent of the star circuit given in figure-I?

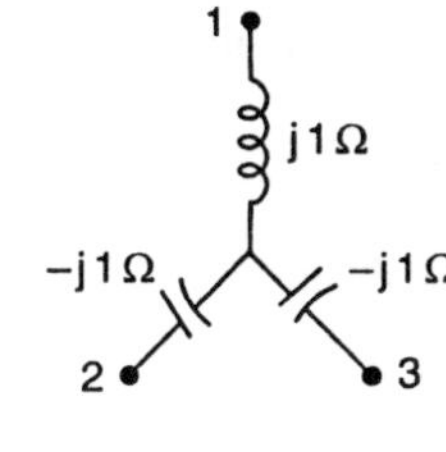

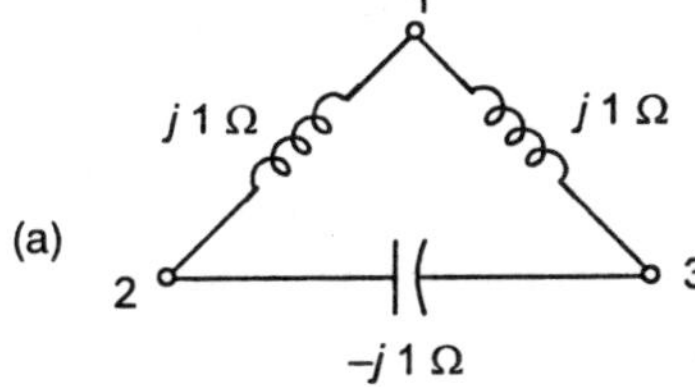

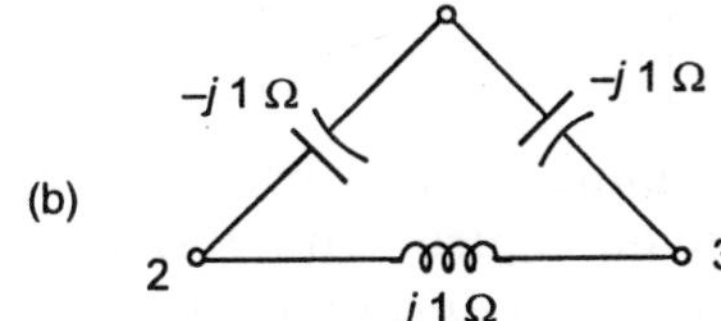

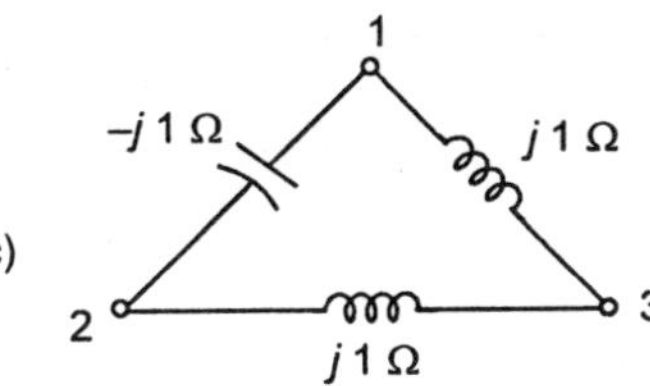

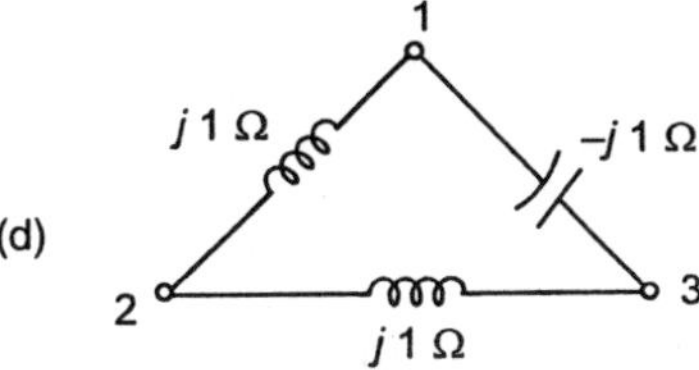

**353.** The impulse response of a first order system is $Ke^{-2t}$. If input signal is sin $2t$, then the steady state response will be given by

(a) $\dfrac{1}{2\sqrt{2}} \sin\left(2t + \dfrac{\pi}{4}\right)$

(b) $\dfrac{1}{4} \sin 2t$

(c) $\dfrac{1}{2\sqrt{2}} \sin\left(2t - \dfrac{\pi}{4}\right)$

(d) $\dfrac{1}{2\sqrt{2}} \sin\left(2t - \dfrac{\pi}{4}\right) + Ke^{-2t}$

**354.** The average value of the delayed full-wave rectified sine wave as shown in the given figure is

(a) $\dfrac{Y_m}{\pi}$

(b) $\dfrac{2Y_m}{\pi}$

(c) $\dfrac{3Y_m}{2\pi}$

(d) $\dfrac{Y_m}{2\pi}$

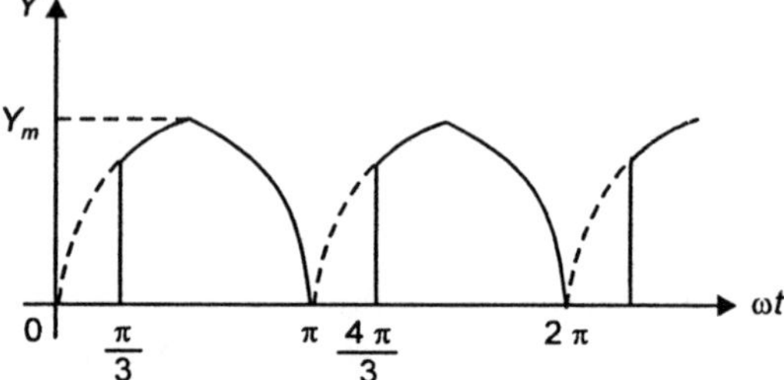

**355.** The relation $AD - BC = 1$, where $A$, $B$, $C$ and $D$ are the elements of a transmission matrix of a network, is valid for
(a) any type of network
(b) passive but not reciprocal network
(c) passive and reciprocal network
(d) both active and passive network

**356.** The $Y$ parameters of a four-terminal block are $\begin{bmatrix} 4 & 2 \\ 1 & 1 \end{bmatrix}$

A single element of 1 ohm is connected across as shown in the given figure. The new $Y$ parameters will be

(a) $\begin{bmatrix} 5 & 1 \\ 0 & 2 \end{bmatrix}$

(b) $\begin{bmatrix} 4 & 3 \\ 2 & 2 \end{bmatrix}$

(c) $\begin{bmatrix} 3 & 2 \\ 1 & 1 \end{bmatrix}$

(d) $\begin{bmatrix} 4 & 2 \\ 1 & 1 \end{bmatrix}$

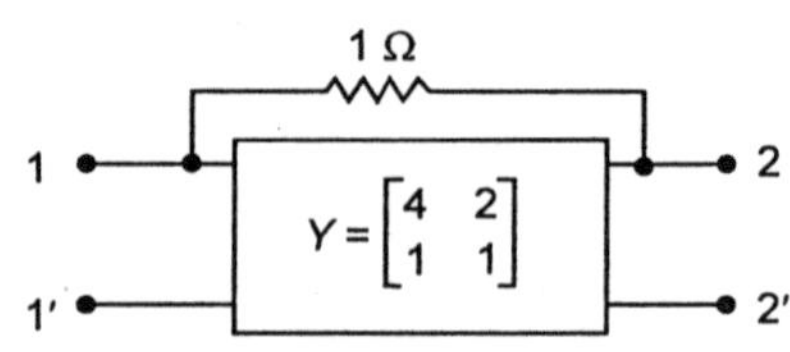


**357.** A four-terminal $T$-network is inserted between source and load resistance $R$ as shown in the given figure. The resistance seen by the source remains the same with or without the four terminal block when $R$ is

(a) 5 Ω

(b) 10 Ω

(c) 15 Ω

(d) 20 Ω



**358.** The time constant of the circuit shown in the given figure is

(a) $RC$

(b) $2\ RC$

(c) $3\ RC$

(d) $5\ RC$

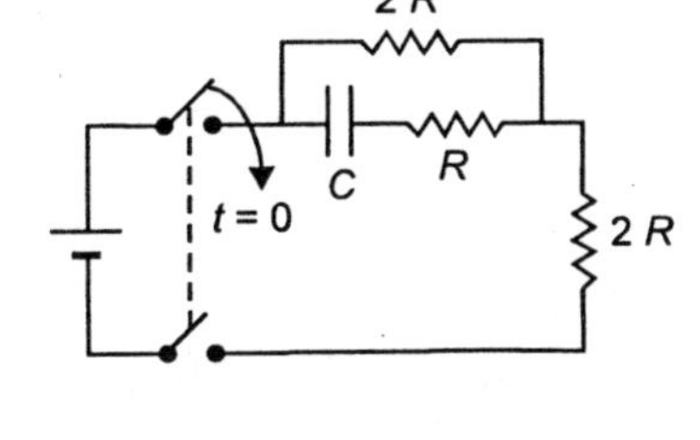

**359.** Special circuits using neon tubes, contact relays and capacitors will be seen in many domestic appliances. In the circuit shown in the given figure, the role of the capacitor $C$ is to

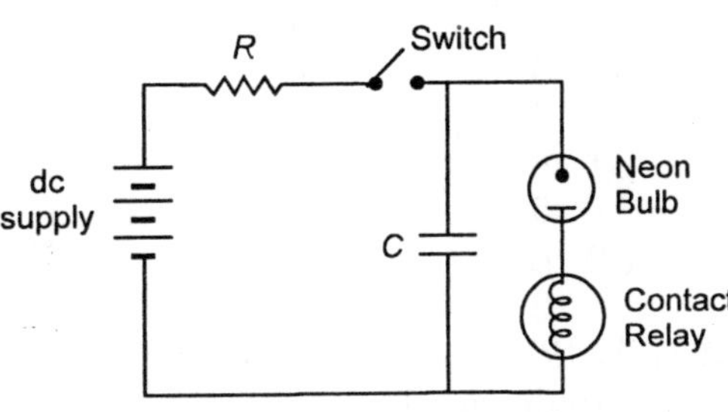

(a) reduce power loss in the circuit.

(b) function as a resistor when relay is on

(c) delay the operation of contact relay after switching on

(d) function as an open-circuit once the relay is 'on'.

**360.** Match List I with List II and select the correct answer using the codes given below the lists:

| *List I*<br>(Networks switched<br>on to a d.c. supply) | *List II*<br>(Shapes of i(t) curves) |
|---|---|

A. 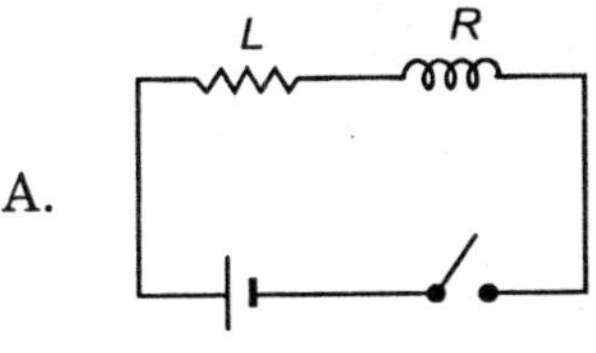

B. 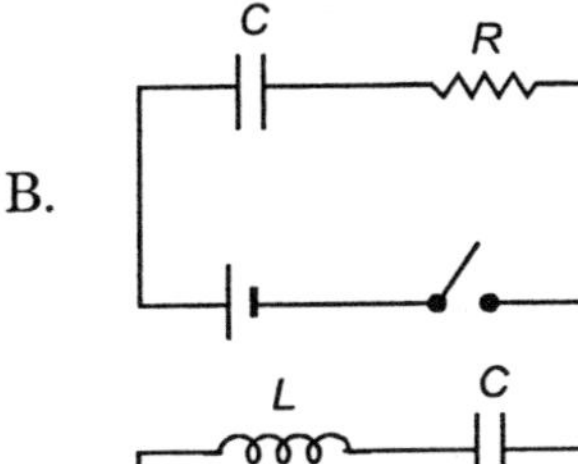

C. 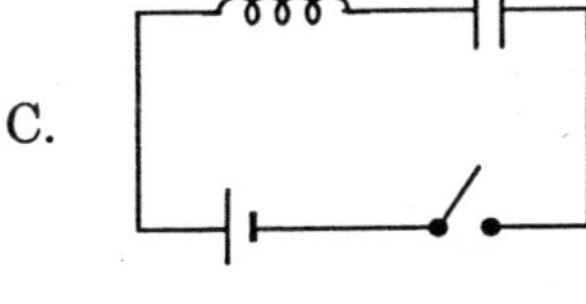

D. 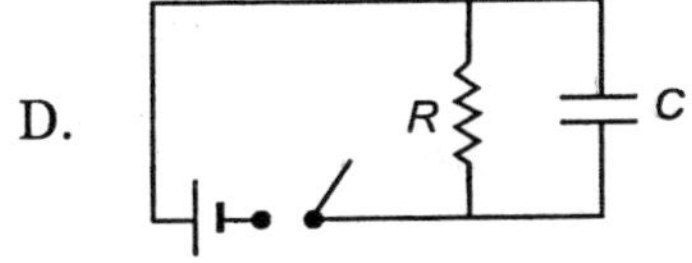

1. 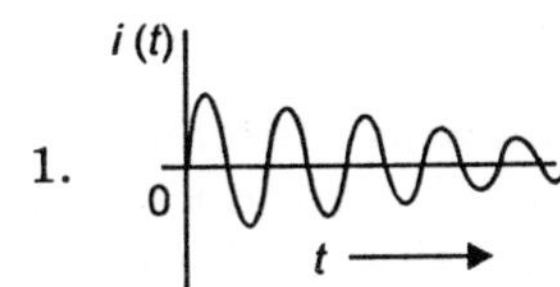

2. 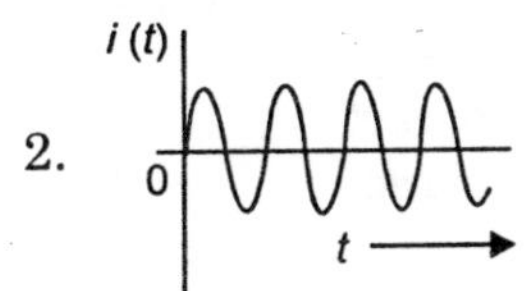

3. 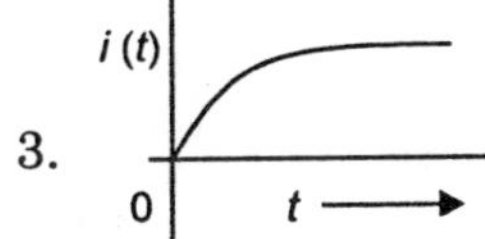

4. 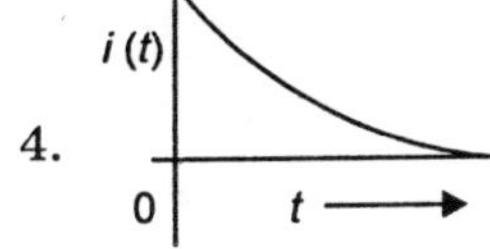

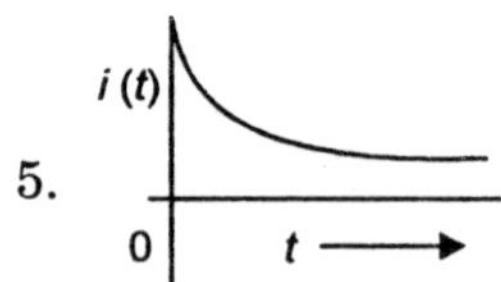


*Codes:*

| (a) | A | B | C | D |
|---|---|---|---|---|
|  | 3 | 4 | 2 | 5 |
| (b) | A | B | C | D |
|  | 5 | 2 | 3 | 1 |
| (c) | A | B | C | D |
|  | 5 | 3 | 1 | 2 |
| (d) | A | B | C | D |
|  | 2 | 3 | 1 | 4 |

**361.** An *RLC* series circuit has $R = 1$ ohm, $L = 1$ $H$ and $C = 1$ F. The damping ratio on the circuit will be

(a) more than unity      (b) unity

(c) 0.5      (d) zero

**362.** Voltage transfer function of the two-port network given in the figure has

(a) a zero at the origin

(b) a zero at $\infty$

(c) no zero

(d) a zero at $j\,1$

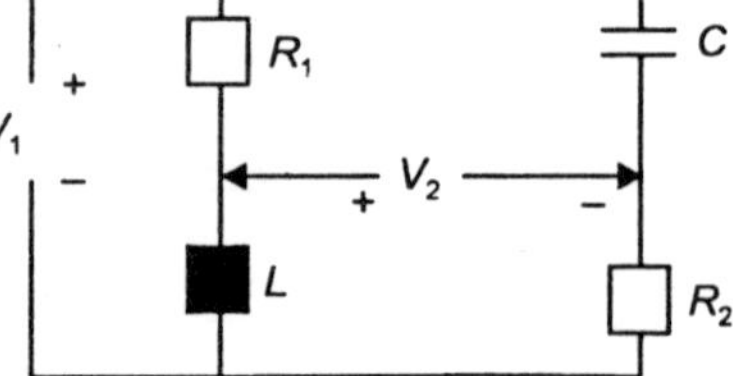

**363.** Consider the following statements:
1. Current through an inductor cannot change abruptly.
2. Voltage across the capacitor cannot change abruptly.
3. Initial value of a function $f(t)$ is $Lt\ s \to 0\ sF(s)$.
4. Final value of a function $f(t)$ is $Lt\ s \to \infty\ sF(s)$.

Of these statements

(a) 3 and 4 are correct      (b) 1 and 4 are correct

(c) 1 and 2 are correct      (d) 2 and 3 are correct

**364.** The transfer function $\quad H(s) = \dfrac{s^2 - 5s + 100}{s^2 + 5s + 100}$

represents

(a) a high-pass filter      (b) a band elimination filter

(c) a resonator      (d) an all pass filter

**365.** The impedance function $Z(s)$ of a one-port network is given by

$$Z(s) = \frac{3s\,(s^2 + 4)}{(s^2 + 1)\,(s^2 + 9)}$$

Which one of the following is the correct reactance curve of the above function?

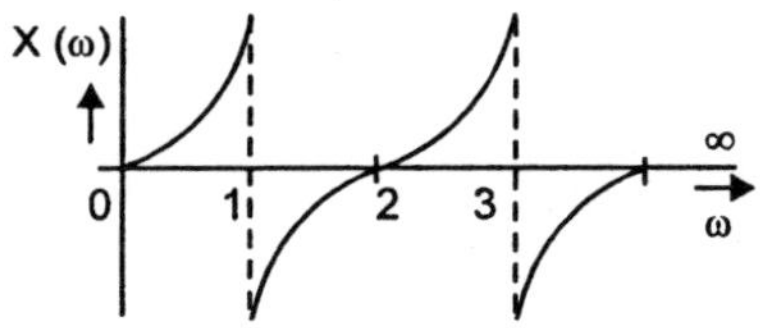

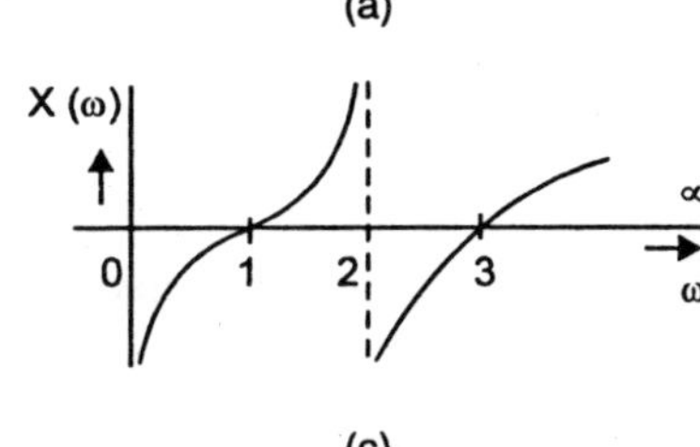

(a)

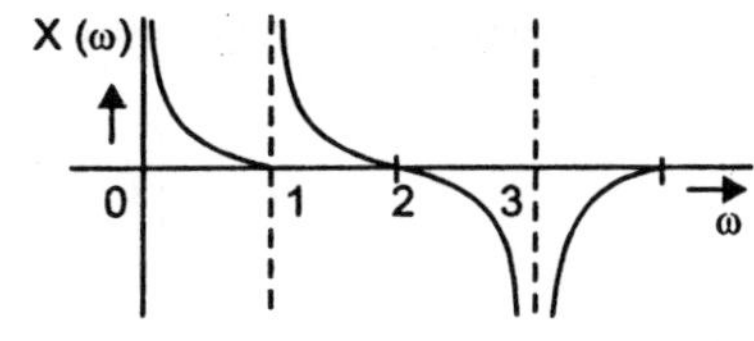

(b)

(c)

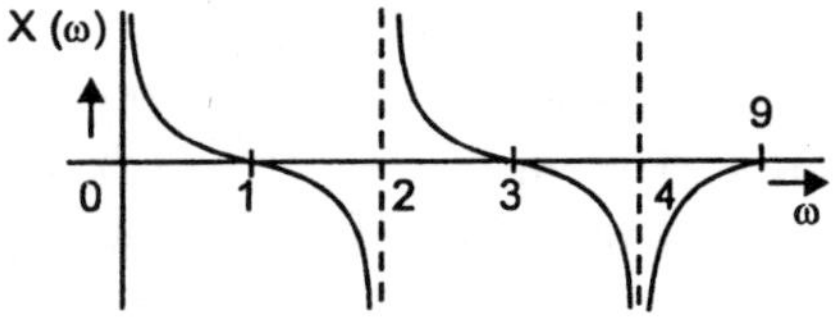

(d)

**366.** If a voltmeter placed across the 3 ohm resistor in the circuit given in the figure reads 45 volts, the reading of ammeter $A$ will be

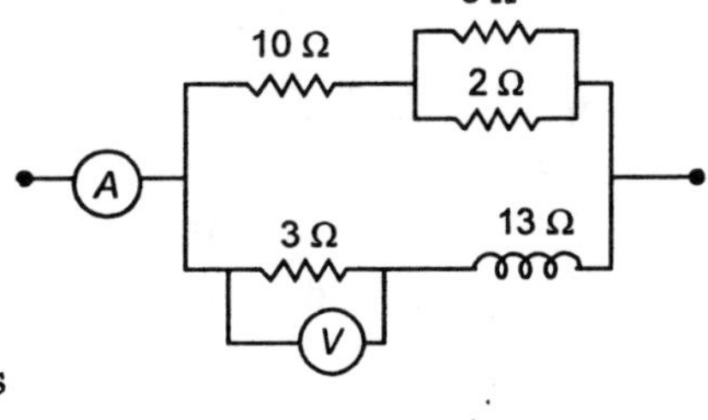

(a)  10.1 $A$

(b)  13.3 $A$

(c)  16.1 $A$

(d)  19.4 $A$

**367.** In the circuit shown in the given figure, current $I$ is

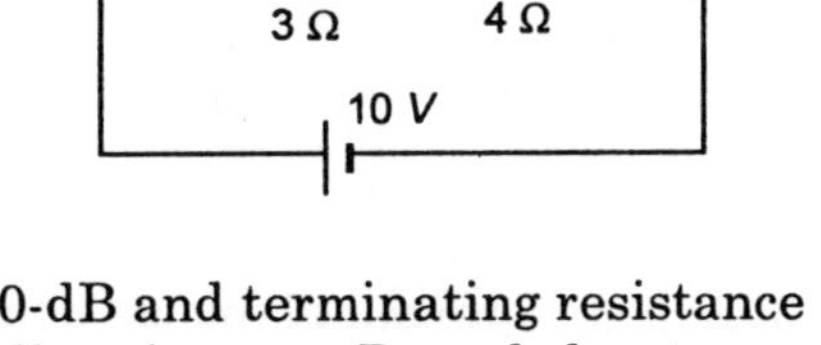

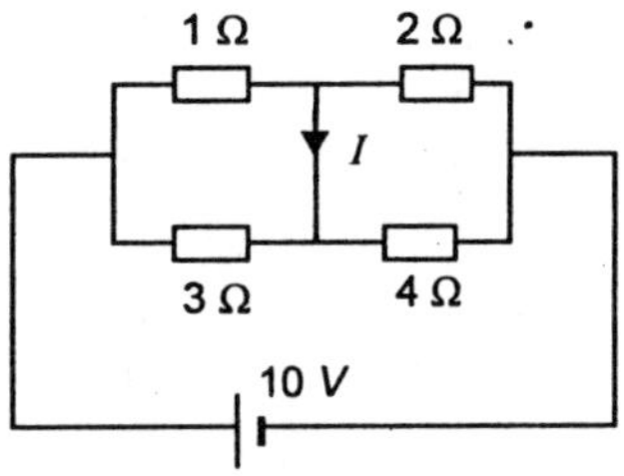

(a)  $-\dfrac{2}{5}\ A$

(b)  $\dfrac{24}{5}\ A$

(c)  $\dfrac{18}{5}\ A$

(d)  $\dfrac{2}{5}\ A$

**368.** A $T$-type attenuator is designed for an attenuation of 40-dB and terminating resistance of 75 ohms. Which of the following values represent full series arm $R_1$ and shunt arm $R_2$?

1.  $R_1 \simeq 147\ \Omega$          2.  $R_1 \simeq 153\ \Omega$
3.  $R_2 \simeq 1.5\ \Omega$          4.  $R_2 \simeq 3750\ \Omega$

Select the correct answer from the codes given below:

*Codes:*

(a) 1 and 3      (b) 1 and 4      (c) 2 and 3      (d) 2 and 4

**369.** If $X_A$ and $X_B$ represent the series arm and the shunt arm reactances of a prototype filter then in the stop band of the filter, signs of $X_A$ and $X_B$ are:

(a) opposite and the characteristic impedance is resistive

(b) the same and the characteristic impedance is resistive

(c) opposite and the characteristic impedance is reactive

(d) the same and the characteristic impedance is reactive

**370.** The circuit shown in the given figure is

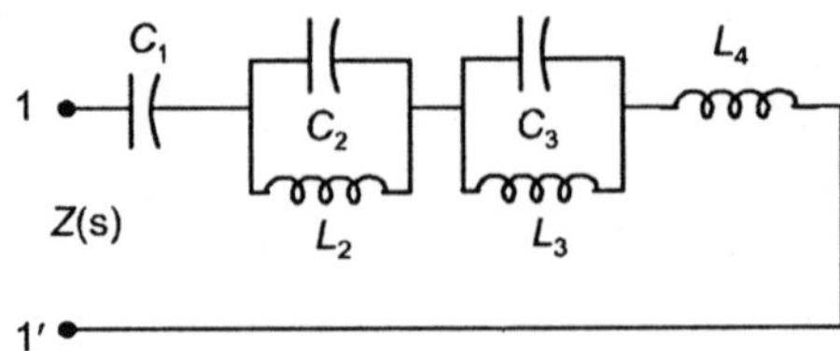

(a) Cauer I form        (b) Foster I form
(c) Cauer II form       (d) Foster II form

**371.** Match List I with List II and select the correct answer using the codes given below the lists:

| List I | List II |
|---|---|
| A. $\dfrac{s^2 - s + 1}{s^2 + s + 1}$ | 1. $RL$ admittance |
| B. $\dfrac{s^2 + s + 1}{s^2 - s + 1}$ | 2. $RL$ impedance |
| C. $\dfrac{s^2 + 4s + 3}{s^2 - 6s + 8}$ | 3. Unstable |
| | 4. Non-minimum phase |

*Codes:*

|     | A | B | C |
|-----|---|---|---|
| (a) | 1 | 2 | 3 |
| (b) | 1 | 4 | 2 |
| (c) | 4 | 3 | 2 |
| (d) | 4 | 3 | 1 |

**372.** A two-terminal black box contains one of the $R$, $L$, $C$ elements. The black box is connected to a 220 volts a.c. supply. The current through the source is $I$. When a capacitance of 0.1 F is inserted in series between the source and the box, the current through the source is 2 $I$. The element is

(a) a resistance

(b) an inductance

(c) a capacitance of 0.5 F

(d) not readily identifiable from the given data.

**373.** For a transfer function $H(s) = P(s)/Q(s)$ where $P(s)$ and $Q(s)$ are polynomials in $s$,

(a) the degree of $P(s)$ is always greater than the degree of $Q(s)$.

(b) the degree of $P(s)$ and $Q(s)$ are the same

(c) the degree of $P(s)$ is independent of the degree of $Q(s)$

(d) the maximum degree of $P(s)$ and $Q(s)$ differ at the most by one.

**374.** Which one of the following differential equations correctly represents the mathematical model of the mechanical system shown in the given figure? where

$\quad$ $x$ = displacement in $M$.

$\quad$ $M$ = mass in kg

$\quad$ $B$ = damping constant, $\dfrac{N-\sec}{m}$

$\quad$ $K$ = spring constant, N/m

$\quad$ $f$ = force, Newton

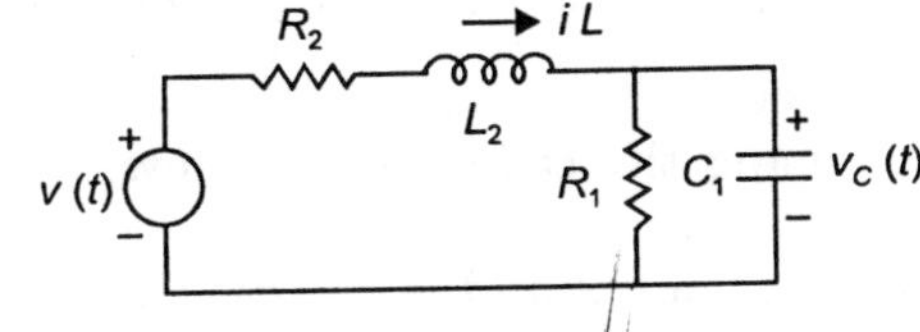

(a) $f(t) - M\dfrac{d^2x}{dt^2} - B\dfrac{dx}{dt} - Kx = 0$

(b) $f(t) - M\dfrac{dx}{dt} - Bx - K\int x\,dt = 0$

(c) $f(t) + M\dfrac{dx}{dt} + Bx + K\int x\,dt = 0$

(d) $f(t) - M\dfrac{d^2x}{dt^2} - \dfrac{1}{B}\dfrac{dx}{dt} - \dfrac{1}{K}x = 0$

**375.** Which one of the following is the correct state model for the network shown in the given figure with

$$X_2(t) = i_L(t) \text{ and } x_1(t) = Vc(t)?$$

(a) $\begin{bmatrix} \dot{x}_1(t) \\ \dot{x}_2(t) \end{bmatrix} = \begin{bmatrix} -\dfrac{1}{R_1C_1} & \dfrac{1}{C_1} \\ -\dfrac{1}{L_2} & -\dfrac{R_2}{L_2} \end{bmatrix} \begin{bmatrix} x_1 \\ x_2 \end{bmatrix} + \begin{bmatrix} 0 \\ \dfrac{1}{L_2} \end{bmatrix} v(t)$

(b) $\begin{bmatrix} \dot{x}_1(t) \\ \dot{x}_2(t) \end{bmatrix} = \begin{bmatrix} -\dfrac{R_2}{L_2} & \dfrac{1}{C_1} \\ \dfrac{1}{L_2} & -\dfrac{1}{R_1C_1} \end{bmatrix} \begin{bmatrix} x_1 \\ x_2 \end{bmatrix} + \begin{bmatrix} 1 \\ \dfrac{1}{L_2} \end{bmatrix} v(t)$

(c) $\begin{bmatrix} \dot{x}_1(t) \\ \dot{x}_2(t) \end{bmatrix} = \begin{bmatrix} -\dfrac{R_2}{L_2} & -\dfrac{1}{L_2} \\ \dfrac{1}{C_1} & -\dfrac{1}{R_1C_1} \end{bmatrix} \begin{bmatrix} x_1 \\ x_2 \end{bmatrix} + \begin{bmatrix} \dfrac{1}{L_2} \\ 0 \end{bmatrix} v(t)$

(d) $\begin{bmatrix} \dot{x}_1(t) \\ \dot{x}_2(t) \end{bmatrix} = \begin{bmatrix} -\dfrac{1}{R_1C_1} & -\dfrac{1}{L_2} \\ \dfrac{1}{C_1} & -\dfrac{R_2}{L_2} \end{bmatrix} \begin{bmatrix} x_1 \\ x_2 \end{bmatrix} + \begin{bmatrix} \dfrac{1}{L_2} \\ 0 \end{bmatrix} v(t)$

**376.** Which one of the following is the Fourier transform of the signal given in Figure B if the Fourier transform of the signal in Figure A is given by $\dfrac{2 \sin \omega T_1}{\omega T_1}$

(a) $\dfrac{2 \sin \omega T_1}{\omega T_1} e^{j\omega T_1}$

(b) $\dfrac{2 \sin \omega T_1}{\omega T_1} e^{-j\omega T_1}$

(c) $\dfrac{\sin \omega T_1}{\omega T_1} e^{-j\omega T_1}$

(d) $\dfrac{\sin \omega T_1}{\omega T_1} e^{j\omega (T_1 - 2)}$

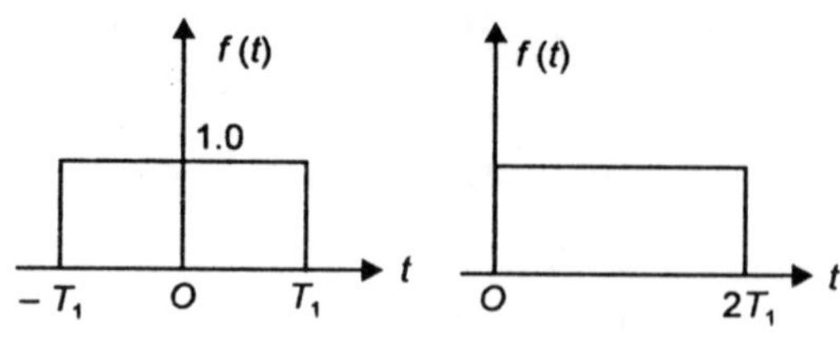


**377.** In Laplace transform the variable 's' equals $(\sigma + j\omega)$. Which of the following represent the true nature of $\sigma$?
1. $\sigma$ has a damping effect

2. $\sigma$ is responsible for convergence of integral $\displaystyle\int_0^\infty f(t)\, e^{-st} dt$

3. $\sigma$ has a value less than zero.
Select the correct answer using the codes given below:
*Codes:*
(a) 1, 2 and 3  (b) 1 and 2   (c) 2 and 3   (d) 1 and 3

**378.** Which one of the following is the correct Laplace transform of the signal shown in the given figure?

(a) $\dfrac{1}{Ts^2}\left[1 - e^{-Ts}(1 + Ts)\right]$

(b) $\dfrac{1}{Ts^2}\left[e^{-Ts} - 1 + Ts\right]$

(c) $\dfrac{1}{Ts^2}\left[e^{-Ts} + 1 - Ts\right]$

(d) $\dfrac{1}{Ts^2}\left[1 - e^{-Ts} + Ts\right]$

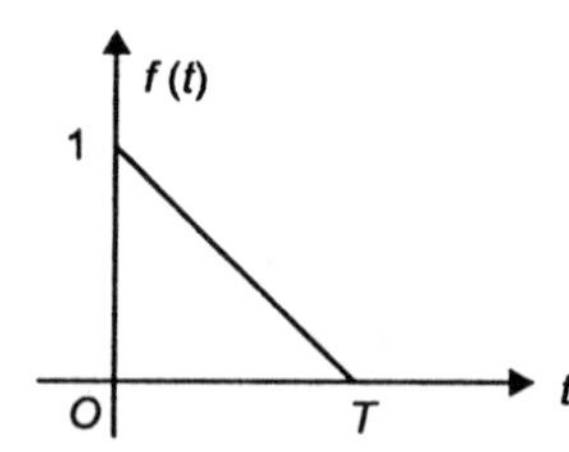


**379.** For the circuit shown in the given figure, the voltage $V_{AB}$ is

(a) 6 V          (b) 10 V          (c) 25 V          (d) 40 V

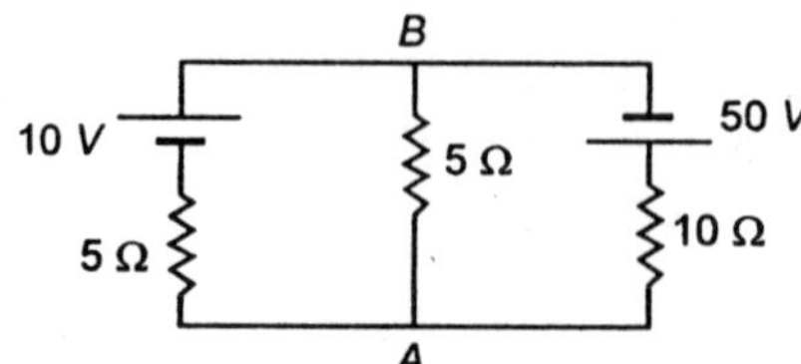


**380.** Which one of the following impedance values of load will cause maximum power to be transferred to the load for the network shown in the given figure?

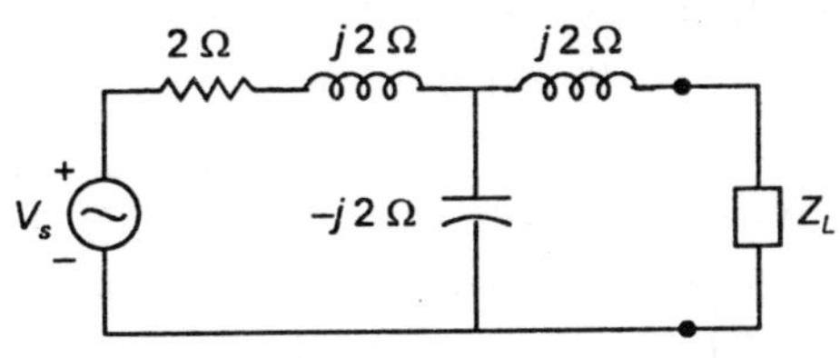

(a) $(2 + j2)$     (b) $(2 - j2)$     (c) $-j2$     (d) $2$

**381.** Which one of the following combinations of open-circuit voltage and Thevenin's equivalent resistance represents the Thevenin's equivalent of the circuit shown in the given figure.

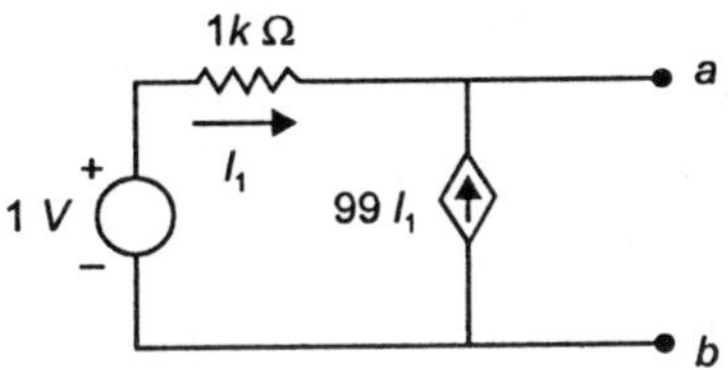

(a) $1\ V,\ 10\ \Omega$     (b) $1\ V,\ 1\ k\ \Omega$     (c) $1\ mV,\ 1\ k\ \Omega$     (d) $1\ mV,\ 10\ \Omega$

**382.** In the graph and the tree shown in the given figure the fundamental cut-set for the branch 2 is

(a) 2, 1 5

(b) 2, 6, 7, 8

(c) 2, 1, 3, 4, 5

(d) 2, 3, 4

**383.** Match List I with List II with reference to the graph shown in the given figure and its particular tree of a circuit and select the correct answer using the codes given below the lists:

| List I | List II |
|---|---|
| (Branches) | |
| A. 1, 2, 3, 4 | 1. Twigs |
| B. 4, 5, 6, 7 | 2. Links |
| C. 1, 2, 3, 8 | 3. Fundamental cut-set |
| D. 1, 4, 5, 6, 7 | 4. Fundamental loop |

*Codes:*

| | A | B | C | D |
|---|---|---|---|---|
| (a) | 3 | 1 | 2 | 4 |
| (b) | 2 | 3 | 1 | 4 |
| (c) | 3 | 2 | 4 | 1 |
| (d) | 1 | 4 | 3 | 2 |

**384.** Two impedances are connected in series. The 3 volmeters, one conneted across each impedance and one acorss combination, read equal value. The phase angle between the voltages across the two impedances is

(a) $30°$     (b) $60°$     (a) $90°$     (b) $120°$

**385.** For a given voltage, four heating coils will produce maximum heat when connected
    (*a*) all in parallel             (*b*) all in series
    (*c*) with two parallel pairs in series
    (*d*) one pair in parallel with the other two in series

**386.** A two-terminal black box contains one of the *RLC* elements. The black box is connected
to a 220 volts a.c. supply. The current through the source is *I*. When a capacitance of
0.1 F inserted in series between the source and the box, the current through the source
is 2 *I*. The element is
    (*a*) a resistance             (*b*) an inductance
    (*c*) a capacitance of 0.5 F
    (*d*) not indentifiable on the basis of the given data

**387.** Which one of the following pairs is correctly matched?
    (*a*) Symmetrical two-port network :   $AD - BC = 1$
    (*b*) Reciprocal two-port network    :   $Z_{11} = Z_{22}$
    (*c*) Inverse hybird parameter      :   $A, B, C, D$
    (*d*) Hybird parameter            :   $(V_1, I_2) = f(I_1, V_2)$

**388.** An ideal transformer has turns ratio of 2 : 1. Considering high voltage side as port 1
and low voltage side as port 2, the transmission line parameters of the transformer will
be

    (*a*) $\begin{bmatrix} 0 & -2 \\ 0.5 & 0 \end{bmatrix}$     (*b*) $\begin{bmatrix} -2 & 0 \\ 0 & -0.5 \end{bmatrix}$     (*c*) $\begin{bmatrix} 2 & 0 \\ 0 & 0.5 \end{bmatrix}$     (*d*) $\begin{bmatrix} 0.5 & 0 \\ 0 & 2 \end{bmatrix}$

**389.** A two-port network is represented by
$$V_1 = 24\,I_1 + 8\,I_2$$
$$V_2 = 8\,I_1 + 32\,I_2$$
Which one of the following networks is represented by equations?

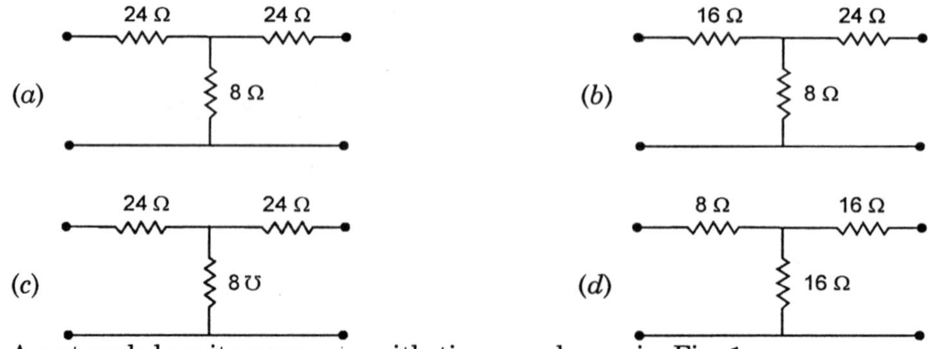


**390.** A network has its response with time as shown in Fig. 1.
Which one of the following diagrams represents the location of poles of this network?

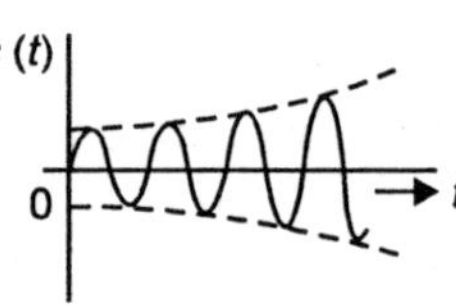


Fig. 1

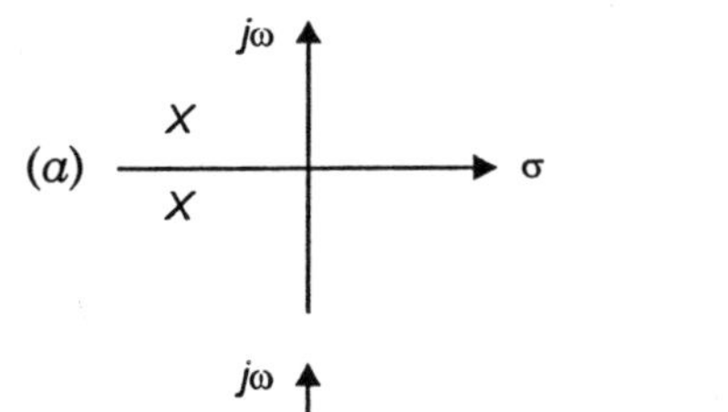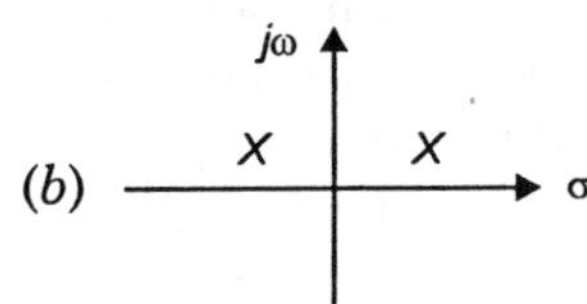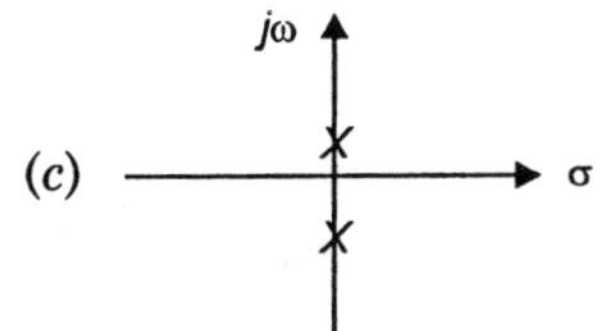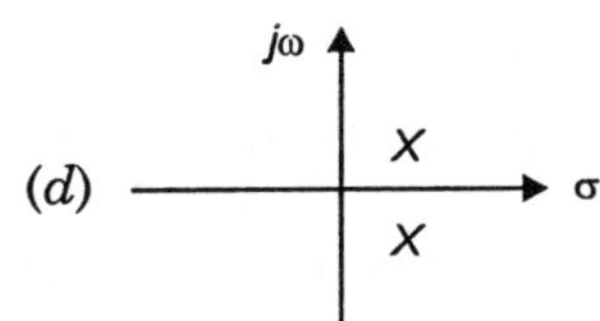

**391.** A step function voltage is applied to an $RLC$ series circuit having $R = 2\ \Omega$, $L = 1$ H and $C = 1$ F. The transient current response of the circuit would be
(a) overdamped                    (b) critically damped
(c) underdamped
(d) over, under or critically damped depending upon magnitude of the step voltage

**392.** When a current source of value $I$ is suddenly connected acorss a two-terminal relaxed $RC$ network at time $t = 0$, the observed nature of the voltage across the current source is shown in the given figure.
The $RC$ network is
(a) a series combination of $R$ and $C$
(b) a parallel combination of $R$ and $C$
(c) a series combination of $R_1$ and parallel combination of $R$ and $C$
(d) a pure capacitor.

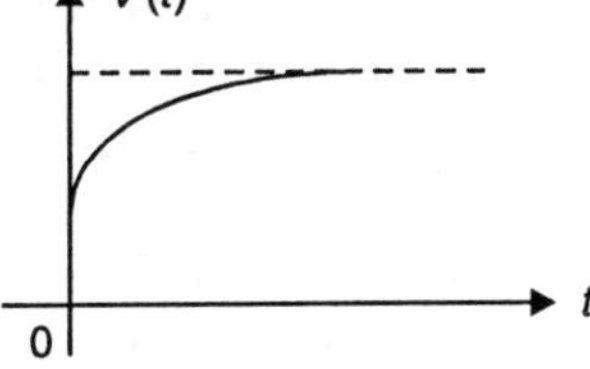

**393.** A resistor $R$ is connected to a voltage source $Vs$ having an internal resistance $R_s$. A voltmeter of resistance $R_m$ is connected across the terminals of the resistor $R$. The voltmeter will read a voltage of

(a) $\dfrac{V_s\,R_s\,R_m}{R_s\,R_m + R_s\,R + R\,R_m}$     (b) $\dfrac{V_s\,R}{R + R_s}$

(c) $\dfrac{V_s\,R_m}{R + R_s + R_m}$     (d) $\dfrac{V_s\,R\,R_m}{R_s\,R + R_s\,R_m + R\,R_m}$

**394.** When two coupled coils of equal self-inductance connected in series in one way, the net inductance is 12 mH, and when they are connected in the other way, the net inductance is 4 mH. The maximum value of net inductance when they are connected in parallel in a suitable way is
(a) 2 mH          (b) 3 mH          (c) 4 mH          (d) 6 mH

**395.** Two perfectly coupled coils each of one Henry self-inductance are connected in parallel so as to aid each other. The overall inductance in Henrys is
(a) 2                  (b) 1                  (c) 1/2                  (d) zero

**396.** A low-pass filter circuit is basically
(a) a differentiating circuit with low time constant
(b) a differentiating circuit with large time constant

    (c)  an integrating circuit with low time constant

    (d)  an integrating circuit with large time constant.

**397.** Consider the following polynomials:

$$P_1 = s^8 + 2s^6 + 4s^4$$
$$P_2 = s^6 - 3s^4 + 2s^2 + 1$$
$$P_3 = s^4 + 3s^3 + 3s^2 + 2s + 1$$
$$P_4 = s^7 + 2s^6 + 2s^5 + 2s^4 + 4s^3 + 8s^2 + 8s + 4$$

Which one of the these polynomials is not Hurwitz?

    (a) $P_1$        (b) $P_2$        (c) $P_3$        (d) $P_4$

**398.** For a voltage transfer function $H(s)$ realizable with a $RLC$ network, the following statements are made:

1.  $H(s)$ cannot have a pole at $s = 0$
2.  $H(s)$ cannot have a pole at $s = \pm j4$
3.  $H(s)$ cannot have a pole at $s = \infty$
4.  $H(s)$ can have a pole at $s = +2$

Of these statements

    (a)  3 and 4 are correct        (b)  2 and 4 are correct

    (c)  1 and 2 are correct        (d)  1 and 3 are correct

**399.** The first critical frequency nearest the origin of the complex frequency plane for an $R$-$L$ driving point impedance function will be

    (a)  a zero in the left-half plane

    (b)  a zero in the right-half plane

    (c)  a pole in the left-half plane

    (d)  either a pole or zero in the left-half plane depending on the connection

**400.** If $F_1(s)$ and $F_2(s)$ are two positive real functions (p.r.f.) then the function which is always positive real, is

    (a) $F_1(s) \, F_2(s)$               (b) $F_1(s)/F_2(s)$

    (c) $F_1(s) \, F_2(s)/\{F_1(s) + F_2(s)\}$    (d) $F_1(s) - F_2(s)$

**401.** Match List I with List II and select the correct answer using the codes given below the lists:

| *List I* | *List II* |
|---|---|
| A.  Reactance theorem | 1.  Foster |
| B.  Driving point impedance | 2.  Bartlett |
| C.  Continued fraction-expansion | 3.  Cauer |
| D.  Bisection theorem | 4.  Positive real function |

*Codes:*

| | A | B | C | D |
|---|---|---|---|---|
| (a) | 1 | 4 | 3 | 2 |
| (b) | 1 | 2 | 3 | 4 |
| (c) | 2 | 3 | 1 | 4 |
| (d) | 3 | 4 | 2 | 1 |

**402.** The rise-time of the $RC$ network shown in the given figure is approximately equal to

    (a) $\dfrac{1}{2} \, RC$     (b) $RC$       (c) $2 \, RC$     (d) $4 \, RC$

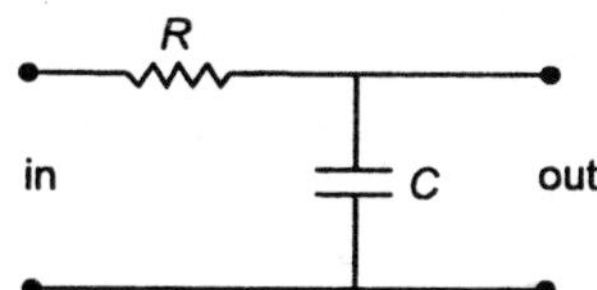

**403.** Double integration of a unit step function would lead to
(*a*) an impulse (*b*) a parabola (*c*) a ramp (*d*) a doublet

**404.** If $f(t) = -f(-t)$ and $f(t)$ satisfy the Dirichlet's conditions, then $f(t)$ can be expanded in a Fourier series containing
(*a*) only sine terms (*b*) only cosine terms
(*c*) cosine terms and a constant term
(*d*) sine terms and a constant term

**405.** Fourier transform $F(j\omega)$ of an arbitrary signal has the property.
(*a*) $F(j\omega) = F(-j\omega)$ (*b*) $F(j\omega) = -F(-j\omega)$
(*c*) $F(j\omega) = F^*(-j\omega)$ (*d*) $F(j\omega) = -F^*(-j\omega)$

**406.** The inverse Fourier transform of the function

$$F(\omega) = \frac{1}{j\omega} + \pi\,\delta(\omega) \ \text{ is}$$

(*a*) sin $\omega t$ (*b*) cos $\omega t$ (*c*) *sgn* (*t*) (*d*) *u*(*t*)

**407.** The following table gives some time functions and their Laplace transforms:

| | $f(t)$ | $F(s)$ |
|---|---|---|
| 1. | $\delta(t)$ | $s$ |
| 2. | $u(t)$ | $1/s$ |
| 3. | $tu(t)$ | $2/s^2$ |
| 4. | $t^2\,u(t)$ | $2/s^3$ |

Of these the correctly matched pairs are:
(*a*) 2 and 4 (*b*) 1 and 4 (*c*) 3 and 4 (*d*) 1 and 2

**408.** The final value of $\mathscr{L} - 1 \dfrac{2s + 1}{s^4 + 8s^3 + 16s^2 + s}$ is
(*a*) infinity (*b*) 2 (*c*) 1 (*d*) zero

**409.** Which one of the following circuits has a driving-point impedance of

$$z(s) = \frac{2(s^2 + s + 1/2)}{s^2 + s + 1} \ ?$$

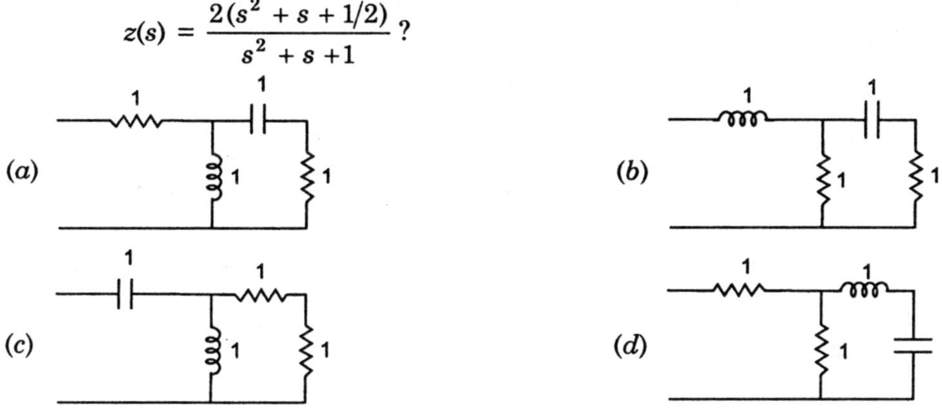

**410.** A delta connection contains three equal impedances of 60 ohms. The impedances of the equivalent star connection will be
(*a*) 15 Ω each  (*b*) 20 Ω each  (*c*) 30 Ω each  (*d*) 40 Ω each

**411.** Substitution Theorem applies to
(*a*) linear networks          (*b*) nonlinear networks
(*c*) linear time-invariant networks          (*d*) any networks

**412.** In the network shown in the given figure current $i = 0$ when $E = 4$ V, $I = 2A$ and $i = 1$ A when $E = 8V$, $I = 2A$. The Thevenin voltage and the resistance looking into the terminals $AB$ are:

(*a*) 4 V, 2 Ω

(*b*) 4 V, 4 Ω

(*c*) 8 V, 2 Ω

(*d*) 8 V, 4 Ω

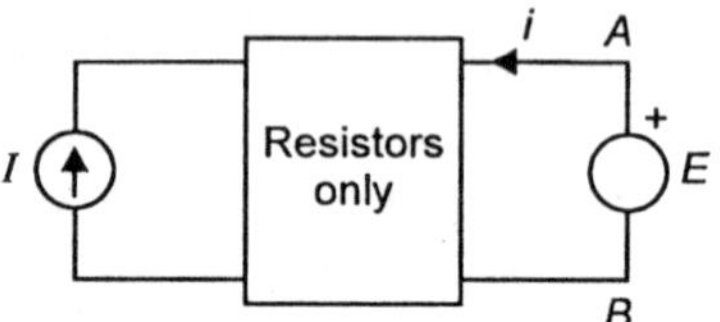


**413.** The value of the resistance $R$ in the circuit shown in the given figure is varied in such a manner that the power dissipated in the 3 ohm resistor is maximum. Under this condition, the value of $R$ will be

(*a*) 3 Ω

(*b*) 9 Ω

(*c*) 12 Ω

(*d*) 6 Ω

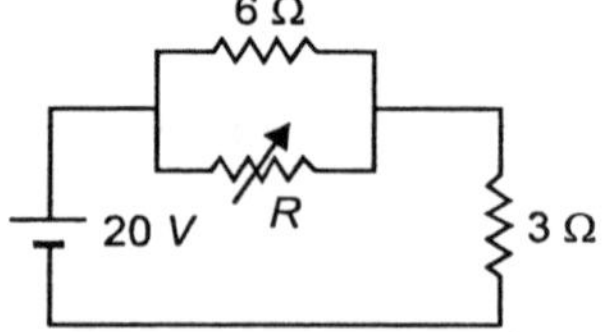


**414.** In the graph shown in the given figure, for the tree with branches $b$, $d$ and $f$, the fundamental loops would include
1, 2, 3, 4 nodes

$a, b, c, d, e, f$ branches

(*a*) *abf, def, bdea*

(*b*) *cea, bdea, abf*

(*c*) *cdb, def, bfa*

(*d*) *abde, def, cdb*

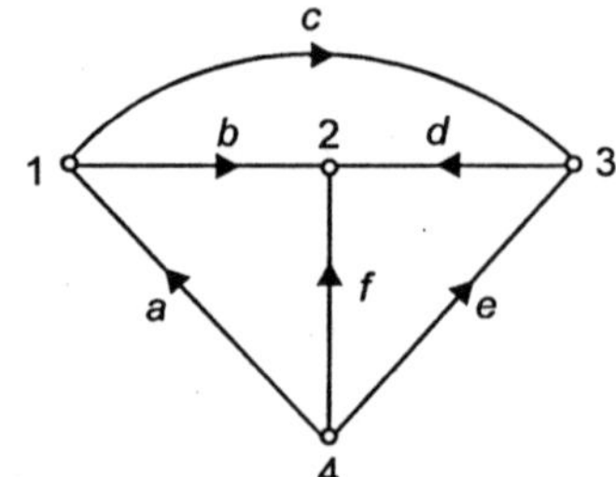


**415.** For the graph shown in the given figure, the incident matrix $A$ is given by

(*a*) $\begin{bmatrix} -1 & -1 & 0 \\ 0 & 1 & 1 \\ -1 & 0 & -1 \end{bmatrix}$   (*b*) $\begin{bmatrix} 1 & 0 & -1 \\ 1 & 1 & 0 \\ 0 & -1 & 1 \end{bmatrix}$

(*c*) $\begin{bmatrix} -1 & -1 & 0 \\ 0 & 1 & 1 \\ 1 & 0 & -1 \end{bmatrix}$   (*d*) $\begin{bmatrix} 1 & 0 & 1 \\ -1 & 1 & 0 \\ 0 & -1 & 1 \end{bmatrix}$

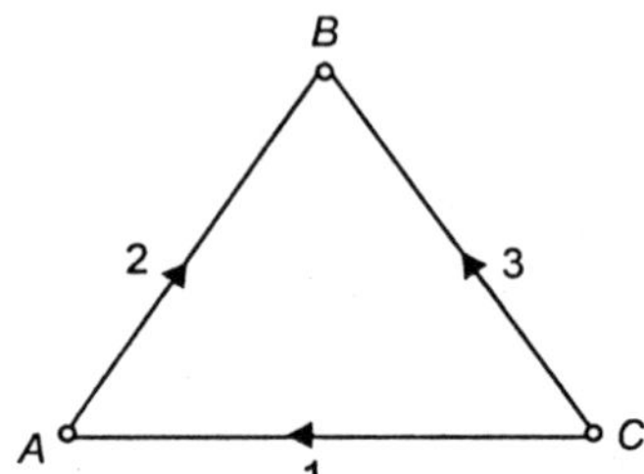


**416.** Consider the following units:
1. $\sec^{-1}$          2. $\text{rad}^2 \sec^2$     3. sec          4. ohm

The units of $\dfrac{R}{L}, \dfrac{1}{LC}, CR$ and $\sqrt{\dfrac{L}{C}}$ are respectively

(a) 1, 2, 4 and 3     (b) 3, 2, 1 and 4     (c) 2, 4, 1 and 3     (d) 1, 2, 3 and 4

**417.** RMS value of $f(t) = 10\,(1 + \sin \omega t)$ is

(a) 10     (b) $\dfrac{10}{\sqrt{2}}$     (c) $\sqrt{150}$     (d) $\dfrac{20}{\sqrt{2}}$

**418.** A system is represented by the transfer function

$$\frac{10}{(s + 1)\,(s + 2)}$$

The d.c. gain of this system is

(a) 1     (b) 2     (c) 5     (d) 10

**419.** A two-terminal black box contains a single element which can be $R$, $L$, $C$ or $M$ (mutual inductance). As soon as the box is connected to a d.c. voltage source, a finite non-zero current is observed to flow through the element. The element is a/an.

(a) resistance     (b) inductance     (c) capacitance     (d) mutual inductance

**420.** Which one of the following is the ratio $V_{24}/V_{13}$ of the network in the given figure?

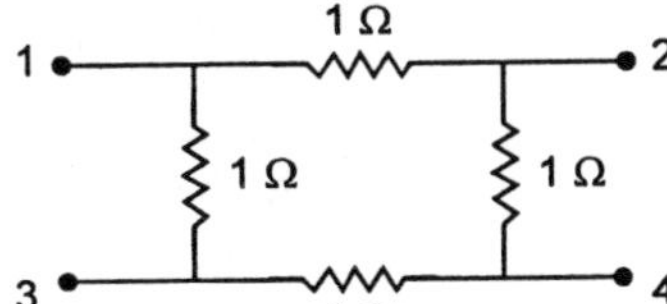

(a) $\dfrac{1}{3}$     (b) $\dfrac{2}{3}$

(c) $\dfrac{3}{4}$     (d) $\dfrac{4}{3}$

**421.** The network shown in the given figure is a gyrator which satisfies the equations

$$V_1 = R_0\,I_2' \text{ and}$$

$$V_2 = -R_0\,I_1'$$

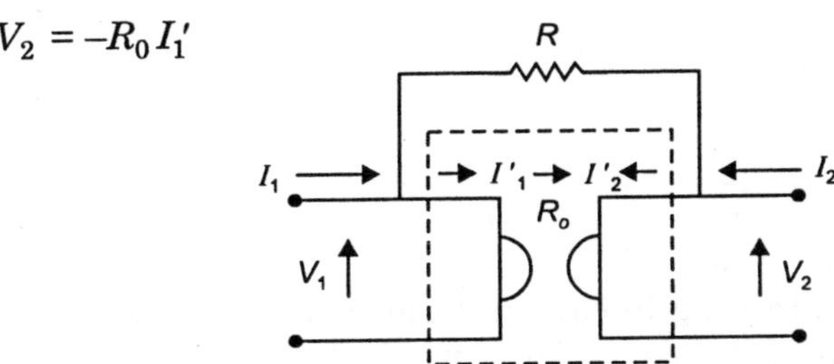

A bridging resistor $R$ is connected as shown. The $y$ parameters of the entire two-port network will be

| | $y_{11}$ | $y_{22}$ | $y_{12}$ | $y_{21}$ |
|---|---|---|---|---|
| (a) | $\dfrac{1}{R}$ | $\dfrac{1}{R}$ | $-\left(\dfrac{1}{R} + \dfrac{1}{R_0}\right)$ | $\left(\dfrac{1}{R_0} - \dfrac{1}{R}\right)$ |
| (b) | $\dfrac{1}{R}$ | $-\dfrac{1}{R}$ | $\dfrac{1}{R_0}$ | $\dfrac{1}{R_0}$ |
| (c) | $\dfrac{1}{R}$ | $\dfrac{1}{R}$ | $\dfrac{1}{R_0}$ | $-\dfrac{1}{R_0}$ |
| (d) | $-\dfrac{1}{R}$ | $\dfrac{1}{R}$ | $\dfrac{1}{R_0}$ | $\dfrac{1}{R_0}$ |

**422.** The ideal transformer cannot be described by
  (a) $h$ parameters
  (b) *ABCD* parameters
  (c) $g$ parameters
  (d) $z$ parameters

**423.** With reference to the equivalent circuit of a transistor shown in the given figure, match List I with List II and select the correct answer using the codes given below the lists:

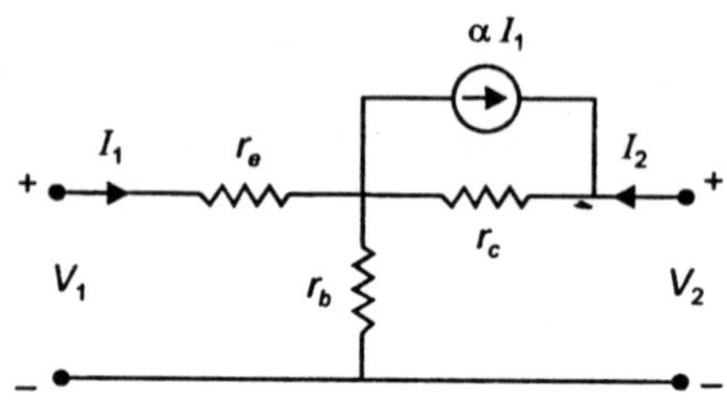

|  List I  |  List II  |
|----------|-----------|
| A. $h_{11}$ | 1. $(r_e + r_b) - \dfrac{\alpha r_c + r_b}{r_c + r_b}\, r_b$ |
| B. $h_{12}$ | 2. $-\dfrac{r_b + \alpha r_c}{r_c + r_b}$ |
| C. $h_{21}$ | 3. $\dfrac{1}{r_c + r_b}$ |
| D. $h_{22}$ | 4. $\dfrac{r_b}{r_c + r_b}$ |

Codes:

|     | A | B | C | D |
|-----|---|---|---|---|
| (a) | 1 | 2 | 3 | 4 |
| (b) | 1 | 3 | 2 | 4 |
| (c) | 1 | 4 | 2 | 3 |
| (d) | 2 | 3 | 4 | 1 |

**424.** In a two-port network containing linear bilateral passive circuit elements, which one of the following conditions for $z$ parameters would hold?
  (a) $z_{11} = z_{22}$
  (b) $z_{12}\, z_{21} = z_{11}\, z_{22}$
  (c) $z_{11}\, z_{12} = z_{22}\, z_{21}$
  (d) $z_{12} = z_{21}$

**425.** In the circuit shown in the given figure, switch 'S' is closed at time $t = 0$. After some time when the current in the inductor was 6 *A*, the rate of change of current through it was 4 *A/s*. The value of the inductor is

  (a) indeterminate

  (b) 1.5 H

  (c) 1.0 H

  (d) 0.5 H

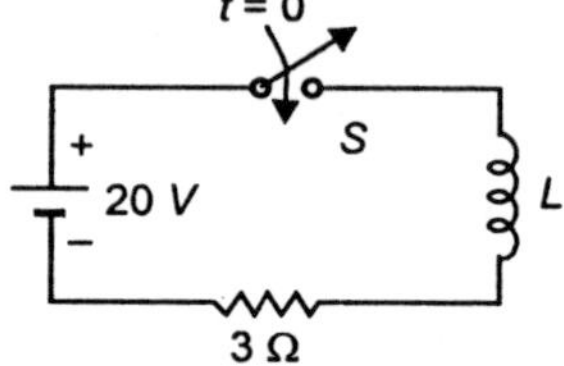

**426.** A circuit consisting of a 1 $\Omega$ resistor and a 2 F capacitor in series is excited from a voltage source with the voltage expressed as $3e^{-t}$, as shown in the given figure. If the $i(0_-)$ and $v_c(0_-)$ are both zero, then the values of $i(0_+)$ and $i(\infty)$ will be respectively (*a*)  3 A and 1.5 A

(*b*)  1.5 A and zero

(*b*)  3 A and zero

(*d*)  1.5 A and 3 A

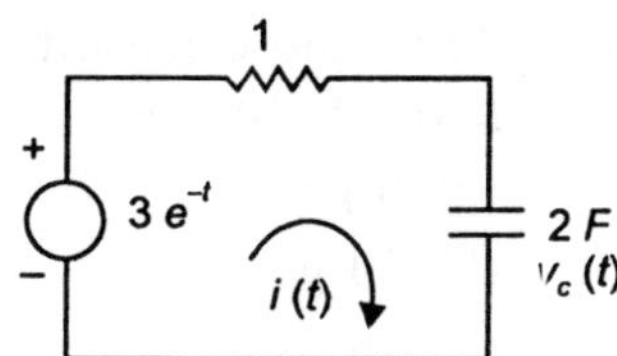


**427.** Match List I with List II and select the correct answer using the codes given below the lists:

<table>
<tr><td align="center">*List I*</td><td align="center">*List II*</td></tr>
<tr><td align="center">(C = Charged Capacitor)</td><td></td></tr>
</table>

A.  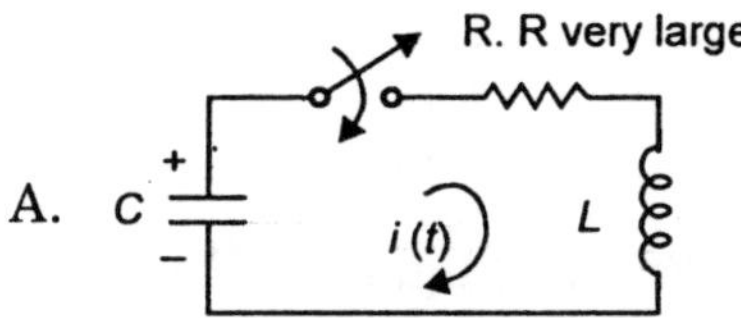


1.  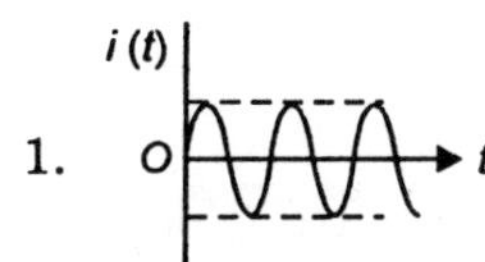


B.  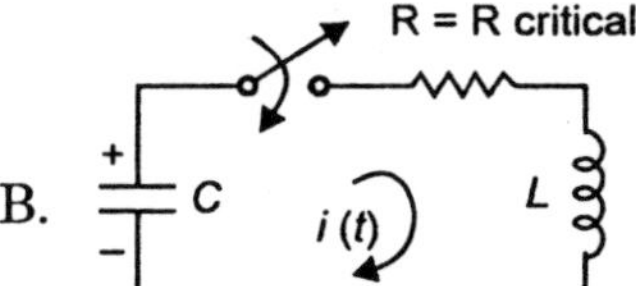


2.  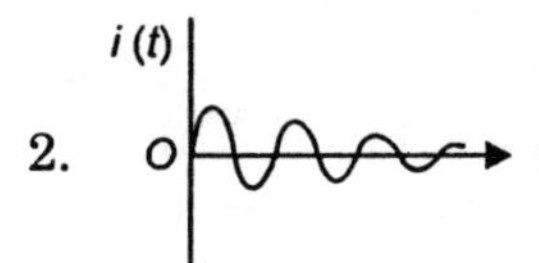


C.  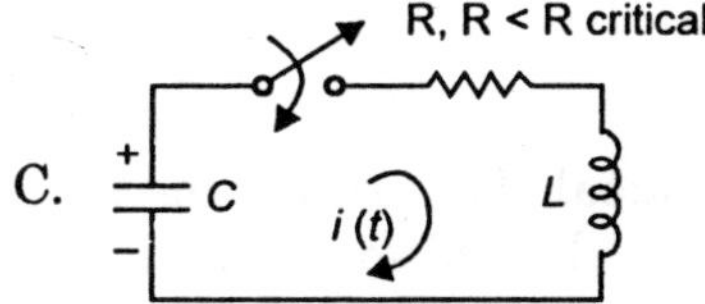


3.  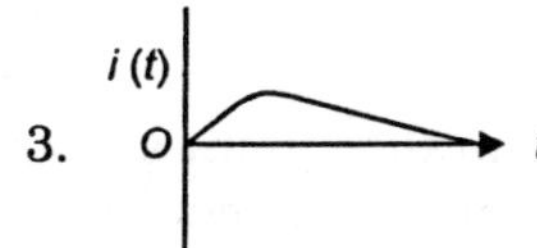


D.  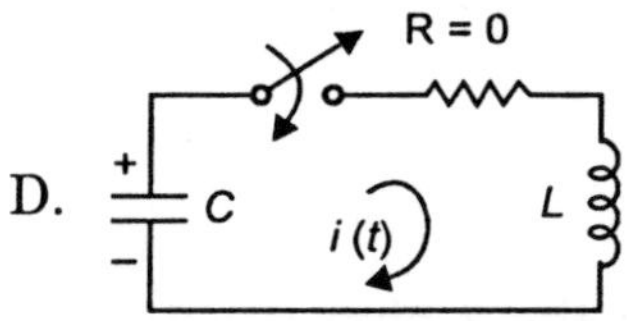


4.  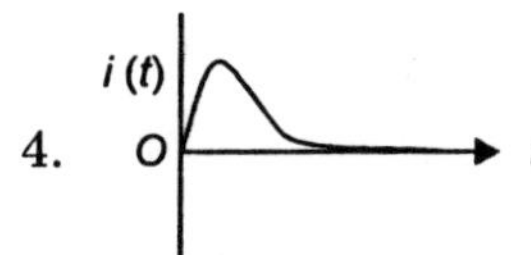


*Codes:*

| | A | B | C | D |
|---|---|---|---|---|
| (*a*) | 4 | 3 | 1 | 2 |
| (*b*) | 3 | 4 | 1 | 2 |
| (*c*) | 3 | 4 | 2 | 1 |
| (*d*) | 4 | 3 | 2 | 1 |

**428.** The time constant associated with the capacitor charging in the circuit shown in the given figure is

(a) 6 μs

(b) 10 μs

(c) 15 μs

(d) 25 μs

**429.** In the series $RC$ circuit shown in the given figure, the rms value of the voltage $E$ is 1 V. If the average power dissipated is equal to 500 mW, then the phase angle between the voltage and current will be

(a) 90°

(b) 60°

(c) 45°

(d) 30°

**430.** If, in a doubly-tuned transformer, the secondary current versus frequency graph indicates two peaks and one dip in between, then the coupling coefficient of the transformer will be

(a) less than the critical coefficient of coupling

(b) equal to the critical coefficient of coupling

(c) greater than the critical coefficient of coupling

(d) not determinable due to want of sufficient data

**431.** A 3-phase star connected load has one capacitor in $R$ phase and two resistors of equal value in the other two phases. If the phase sequence is $RBY$, then

(a) $V_{BO} > V_{YO}$      (b) $V_{BO} < V_{YO}$

(c) $V_{BO} = V_{YO}$      (d) $V_{BO} = V_{RY}$

**432.** The voltmeter in the circuit shown in the given figure is ideal. The transformer has two identical windings with perfect coupling. The reading on the voltmeter will be

(a) 440 V

(b) 220 V

(c) 110 V

(d) zero

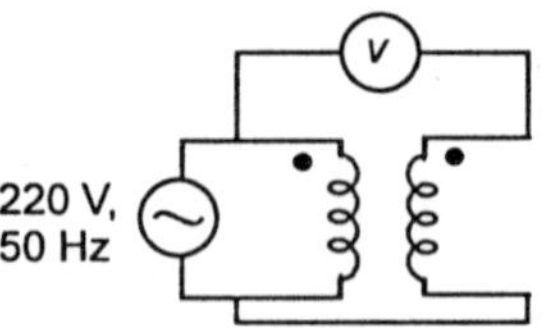

**433.** In the circuit shown in the given figure, the current $I$ in the 2 ohm resistor is

(a) zero

(b) –2 A

(c) 2 A

(d) 1 A

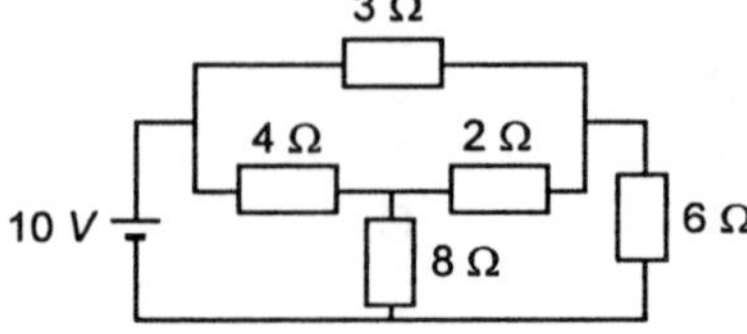

**434.** The Laplace transformation method enables one to find the response of a network in

(a) the transient state only

(b) the steady state only

(c) both transient and steady states

(d) the transient state provided sinusoidal forcing functions do not exist

**435.** The filter shown in the given figure uses an ideal op-amp. It represents a

(a) low pass filter

(b) high pass filter

(c) band pass filter

(d) band stop filter

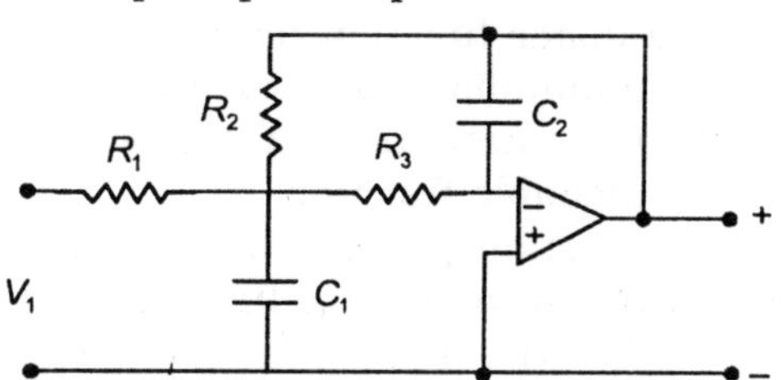


**436.** The design of electric wave filter is based on its characteristic impedance, which at all frequencies is a

(a) pure resistance

(b) pure reactance

(c) pure resistance in the pass band and pure reactance in stop band

(d) pure reactance in the pass band and pure resistance in the stop band.

**437.** Match List I with List II and select the correct answer using the codes given below the lists:

| *List I* | *List II* |
|---|---|
| A. Poles and zeros of the driving-point reactance function of $LC$ network | 1. lie on the real axis |
| B. Canonic $LC$ network Contains | 2. a zero |
| C. The number of canonic networks for a given driving-point reactance function is | 3. maximum number of elements |
| D. The first critical frequency nearest the origin of the complex frequency plane for an $RL$ driving-point Impedance will be | 4. four |
| | 5. minimum number of elements |
| | 6. alternate |
| | 7. either a pole or zero |
| | 8. three |

*Codes:*

| | A | B | C | D |
|---|---|---|---|---|
| (a) | 1 | 5 | 8 | 7 |
| (b) | 6 | 5 | 3 | 2 |
| (c) | 6 | 5 | 4 | 2 |
| (d) | 1 | 3 | 4 | 7 |

**438.** The driving-point impedance of a network is given by

$$z(s) = \frac{s^2 + 2s + 6}{s\,(s + 3)}$$

The number of energy storing elements present in the network is

    (a) 1         (b) 2         (c) 3         (d) 4

**439.** A transfer function having open right-half plane zeros is a/an
    (a) minimum phase function
    (b) non-minimum phase function
    (c) unstable function
    (d) constant phase function

**440.** An $RC$ driving point function has zeros at $s = -2$ and $s = -5$. The admissible poles for the function are:
    (a) $s = 0, s = -6$         (b) $s = -1, s = -3$
    (c) $s = 0, s = -1$         (d) $s = -3, s = -4$

**441.** Which one of the following is a positive real function?

    (a) $\dfrac{s(s^2 + 4)}{(s^2 + 1)(s^2 + 6)}$         (b) $\dfrac{s(s^2 - 4)}{(s^2 + 1)(s^2 + 6)}$

    (c) $\dfrac{s^3 + 3s^2 + 2s + 1}{4s}$         (d) $\dfrac{s(s^4 + 3s^2 + 1)}{(s + 1)(s + 2)(s + 3)(s + 4)}$

**442.** An ideal impedance convertor is a two port network which when terminated at one-port by driving point impedance $z_L(s)$ offers at the other port an input impedance.
    (a) directly proportional to $z_L(s)$
    (b) inversely proportion to $z_L(s)$
    (c) square root of $z_L(s)$
    (d) none of the above, at all frequencies

**443.** An ideal gyrator is a
    (a) passive and reciprocal device         (b) passive and non-reciprocal device
    (c) active and reciprocal device         (d) active and non-reciprocal device

**444.** If $r_1$ and $r_2$ are real numbers for a gyrator where $r_1 \neq r_2$, it is a
    (a) positive impedance converter         (b) positive impedance inverter
    (c) negative impedance convertor         (d) negative impedance inverter

**445.** An active gyrator is one when
    (a) $r_1 = r_2$     (b) $r_1 \neq r_2$     (c) $r_1 < r_2$     (d) $r_1 > r_2$

**446.** When two gyrators are connected in cascade the device acts as
    (a) Negative impedance inverter         (b) Ideal transformer
    (c) Perfect transformer         (d) None of the above

**447.** The main objection to the use of inductor as a network element arises in the case of
    (a) very high frequencies         (b) medium frequencies
    (c) low frequencies         (d) there is no such restriction

**448.** A CVT has
    (a) infinite input and output impedance.
    (b) infinite input but zero output impedance
    (c) zero input and output impedance
    (d) zero input and infinite output impedance

**449.** A VCT has
    (a) infinite input and output impedances
    (b) infinite input but zero output impedance
    (c) zero input and zero output impedances
    (d) zero input and infinite output impedance

**450.** A CCT has

(a) infinite input and output impedance
(b) infinite input but zero output impedance
(c) zero input and output impedance
(d) zero input and infinite output impedance

**451.** A VVT has
(a) infinite input and output impedance
(b) infinite input but zero output impedance
(c) zero input and output impedance
(d) zero input and infinite output impedance

**452.** The transmission matrix of a CVT is

(a) $\begin{bmatrix} 0 & 0 \\ r & 0 \end{bmatrix}$    (b) $\begin{bmatrix} 0 & 0 \\ \mu & 0 \end{bmatrix}$    (c) $\begin{bmatrix} 0 & 0 \\ g & 0 \end{bmatrix}$    (d) $\begin{bmatrix} 0 & 0 \\ \alpha & 0 \end{bmatrix}$

**453.** An operational amplifier is a
(a) CVT    (b) VCT    (c) VVT    (d) VCT

**454.** The pole-zero configuration of a network transfer function is shown in the given Fig. The phase of the transfer function will

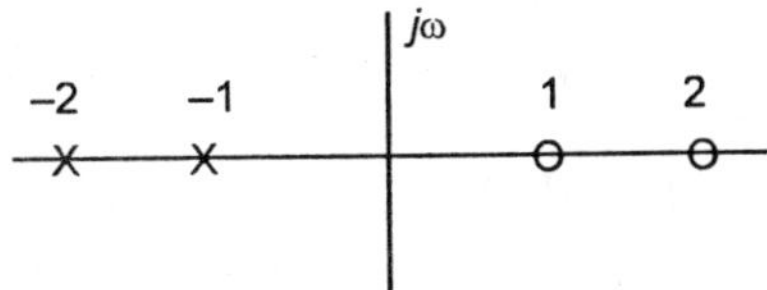


(a) decrease with frequency
(b) increase with frequency
(c) initially increase and then decreases with frequency
(d) be independent of frequency

**455.** Consider the following statements:
1. One and only one path exists between any pair of vertices of a tree.
2. The number of $f$-cut sets are the same as the rank of the graph.
3. The cut-set is a minimal set of edges removal of which from the graph reduces the rank of the graph by one.
4. The rank of a graph is equal to the number of vertices of the graph.
Of these statements, the correct one's are:
(a) 2 and 4    (b) 1 and 3    (c) 2 and 3    (d) 1 and 4

**456.** Ideally, attenuator pads should not change the
(a) voltage level      (b) impedance level
(c) power level
(d) voltage, power and impedance level

**457.** A pole-zero pattern of a particular filter is shown in the figure. It is that of a/an

(a) low-pass filter

(b) high pass filter

(c) band pass filter

(d) all-pass filter

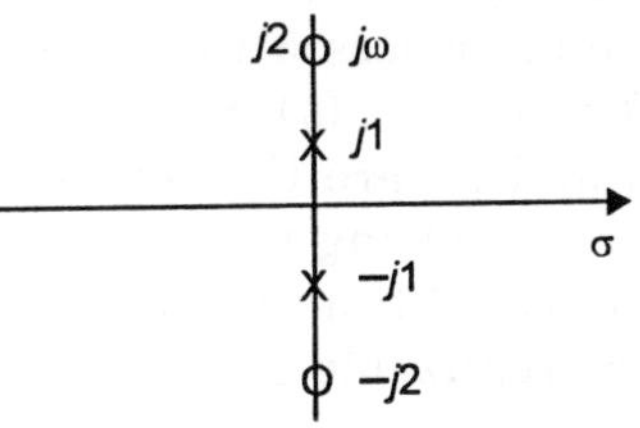

**458.** For the T-network if the characteristic impedance is 300 $\Omega$ and if it is terminated with this resistance, the input resistance of the network is

(*a*) 400 $\Omega$

(*b*) 300 $\Omega$

(*c*) 200 $\Omega$

(*d*) 100 $\Omega$

**459.** If the two identical $T$-networks of Q. 458 are connected in tandem and if it is terminated by a resistance 300 $\Omega$, the input resistance of the cascaded network is

(*a*) 400 $\Omega$   (*b*) 300 $\Omega$   (*c*) 200 $\Omega$   (*d*) 100 $\Omega$

**460.** The characteristic impedance of the $T$-network in Q. 458 is given as

(*a*) 400 $\Omega$   (*b*) 300 $\Omega$   (*c*) 200 $\Omega$   (*d*) 100 $\Omega$

**461.** The image impedances of a T-network shown in Figure are $Z_{I_1} = 140\ \Omega$ and $Z_{I_2} = 88\ \Omega$. The input impedance at terminals '1 1' is, if '22' is terminated with $Z_{I_2}$

(*a*) 111 $\Omega$

(*b*) 88 $\Omega$

(*c*) 140 $\Omega$

(*d*) 114 $\Omega$

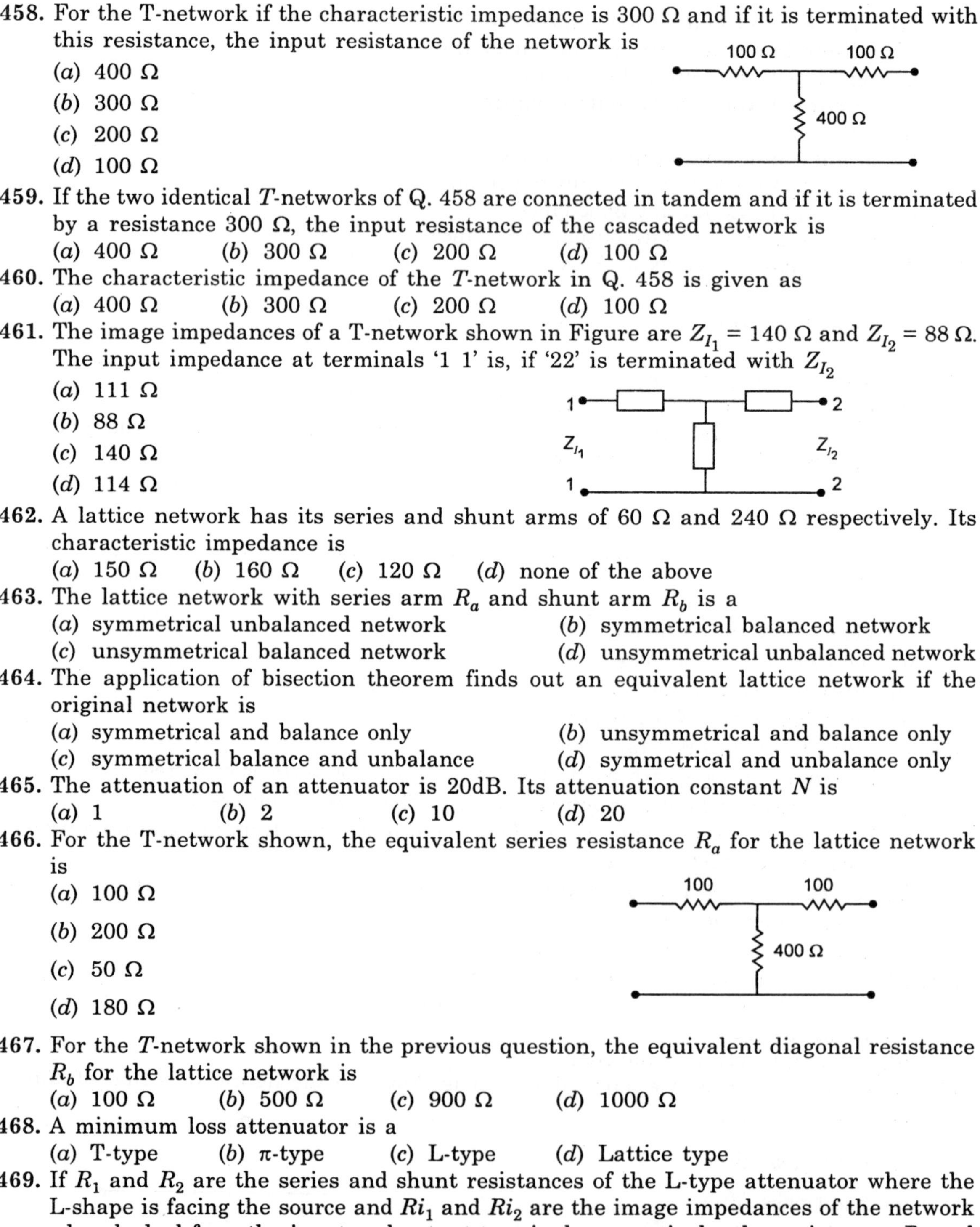

**462.** A lattice network has its series and shunt arms of 60 $\Omega$ and 240 $\Omega$ respectively. Its characteristic impedance is

(*a*) 150 $\Omega$   (*b*) 160 $\Omega$   (*c*) 120 $\Omega$   (*d*) none of the above

**463.** The lattice network with series arm $R_a$ and shunt arm $R_b$ is a

(*a*) symmetrical unbalanced network        (*b*) symmetrical balanced network

(*c*) unsymmetrical balanced network        (*d*) unsymmetrical unbalanced network

**464.** The application of bisection theorem finds out an equivalent lattice network if the original network is

(*a*) symmetrical and balance only          (*b*) unsymmetrical and balance only

(*c*) symmetrical balance and unbalance     (*d*) symmetrical and unbalance only

**465.** The attenuation of an attenuator is 20dB. Its attenuation constant $N$ is

(*a*) 1            (*b*) 2            (*c*) 10            (*d*) 20

**466.** For the T-network shown, the equivalent series resistance $R_a$ for the lattice network is

(*a*) 100 $\Omega$

(*b*) 200 $\Omega$

(*c*) 50 $\Omega$

(*d*) 180 $\Omega$

**467.** For the $T$-network shown in the previous question, the equivalent diagonal resistance $R_b$ for the lattice network is

(*a*) 100 $\Omega$   (*b*) 500 $\Omega$   (*c*) 900 $\Omega$   (*d*) 1000 $\Omega$

**468.** A minimum loss attenuator is a

(*a*) T-type   (*b*) $\pi$-type   (*c*) L-type   (*d*) Lattice type

**469.** If $R_1$ and $R_2$ are the series and shunt resistances of the L-type attenuator where the L-shape is facing the source and $Ri_1$ and $Ri_2$ are the image impedances of the network when looked from the input and output terminals respectively, the resistances $R_1$ and $R_2$ are realizable only if

(a) $Ri_1 = Ri_2$  
(b) $Ri_1 > Ri_2$  
(c) $Ri_1 < Ri_2$  
(d) no such restriction is there

**470.** $Z_1$ and $Z_2$ are the total series and shunt impedances of a $T$ or $\pi$-filters. Consider the following zones of operation of filter and the conditions on the impedances

A. pass band   1. $\dfrac{Z_1}{4Z_2} < -1$

B. stop band   2. $\dfrac{Z_1}{4Z_2} = -1$

C. transition band   3. $-1 < \dfrac{Z_1}{4Z_2} < 0$

Tick out the correct combination

| | A | B | C |
|---|---|---|---|
| (a) | 1 | 2 | 3 |
| (b) | 3 | 1 | 2 |
| (c) | 3 | 2 | 1 |
| (d) | 2 | 3 | 1 |

**471.** Tick out the correct statement in case of a filter  
(a) The characteristic impedance is resistive during stop band  
(b) The characteristic impedance is reactive in pass band  
(c) The characteristic impedance is resistive in pass band  
(d) None of the above is necessary

**472.** If $L$ is the total series inductance and $C$ the total shunt capacitance of a $T$ or $\pi$-type low pass filter, the pass band frequency range of the filter is given as

(a) 0 to $\dfrac{1}{2\pi \sqrt{LC}}$ Hz  
(b) 0 to $\dfrac{1}{\pi \sqrt{LC}}$ Hz

(c) 0 to $\dfrac{2}{\pi \sqrt{LC}}$ Hz  
(d) 0 to $\dfrac{1}{4\pi \sqrt{LC}}$ Hz

**473.** If $C$ is the total series capacitance, $L$ the total shunt inductance of a $T$ or $\pi$-type high pass filter, the frequency range for the stop band of the filter is

(a) 0 to $\dfrac{1}{2\pi \sqrt{LC}}$ Hz  
(b) 0 to $\dfrac{1}{\pi \sqrt{LC}}$ Hz

(c) 0 to $\dfrac{2}{\pi \sqrt{LC}}$ Hz  
(d) 0 to $\dfrac{1}{4\pi \sqrt{LC}}$ Hz

**474.** Consider the following statements in respect of a constant $-k$ type band pass filter  
1. A band pass filter is obtained by the series connection of a low pass and a high pass filter.  
2. It is obtained by the cascade connection of a low pass and a high pass filter.  
3. In this case the cut-off frequency of the low-pass filter is lower than that of the high-pass filter  

Tick out the correct statements  
 (a) 1 and 2   (b) 1 and 3   (c) only 1   (d) 2 and 3

**475.** If $L_1$ and $C_1$ are the total series values and $L_2$ and $C_2$ the shunt values of a constant $-k$ type band pass filter, these are so chosen that the resonance frequency of series arm and shunt arm are
(a) such that series has high frequency than the shunt
(b) such that series has lower resonance frequency than the shunt arm
(c) equal
(d) no consistency in the relation exists

**476.** If $f_1$ and $f_2$ are the lower and upper cut-off frequencies of the band pass filter, the series impedance $Z_1$ is
(a) capacitive at $f_1$      (b) inductive at $f_1$
(c) resistive at $f_1$      (d) none of the above is necessary

**477.** In the above question, the series impedance $Z_1$ is
(a) capacitive at $f_2$      (b) inductive at $f_2$
(c) resistive at $f_2$      (d) none of the above is necessary

**478.** For the band pass filter the normal characteristic impedance $R_k$ is given as
(a) $R_k = \sqrt{\dfrac{L_1}{C_1}}$      (b) $= \sqrt{L_1/C_2}$
(c) $= \sqrt{L_2/C_2}$      (d) (a) and (c)

**479.** If $f_1$ and $f_2$ are the lower and upper cut-off frequencies of a band pass filter (constant-k), and $f_0$ the resonant frequency of the component of series arm of the filter, then
(a) $f_0 = \dfrac{f_1 + f_2}{2}$      (b) $= \sqrt{f_1 f_2}$
(c) $\sqrt{2 f_1 f_2}$      (d) $\sqrt{(f_1 f_2)/2}$

**480.** The phase constant $\beta$ of a filter during stop band is
(a) zero radian    (b) $\pi/2$    (c) $\pi$    (d) $2\pi$

**481.** The frequency of resonance of the shunt arm of the $m$-derived filter is so chosen that as compared to constant-$k$ type filter it is
(a) slightly lower      (b) slightly higher
(c) equal      (d) no restriction

**482.** The transfer function for low-pass filter using Butterworth approximation.
(1) is an all pole function
(2) has one zero only
(3) has poles on a circle with centre at the origin of the s-plane
    Tick out the correct combination
(a) 1 and 2    (b) 1 and 3    (c) 2 and 3    (d) 1, 2 and 3

**483.** The poles of the transfer function for low-pass filter using Butterworth approximation
(1) lie on $j$-$\omega$ axis
(2) lie on a circle with centre at the origin
(3) are conjugate if the order of the filter is an even number ($n$)
    Tick out the correct combination
(a) 1 and 2    (b) 1 and 3    (c) 2 and 3    (d) 1, 2 and 3

**484.** The poles of the transfer function for low-pass filter using chebychev approximation
(a) lie on a parabola      (b) lie on an ellipse
(c) lie on a hyperbola
(d) lie on a straight line passing through the origin, in the s-plane

**485.** Using chebychev low pass filter the response during pass band
1. oscillates between a maximum and a minimum
2. has the number of oscillations equal to $(n + 1)$ where $n$ is the order of the filter
3. has the number of oscillations equal to $(n - 1)$
4. has maximum value at $\omega = 0$ when $n$ is odd
Choose the correct statements
(*a*) 1 and 2
(*b*) 1 and 3
(*c*) 1 and 4
(*d*) none of *a, b* or *c*

**486.** The response of the chebychev low-pass filter
1. has maximum value at $\omega = 0$ when $n$ is odd
2. has minimum value at $\omega = 0$ when $n$ is even
3. has same value at $\omega = \omega_c$ irrespective of the order of the filter
Tick out the correct statements.
(*a*) 1 and 2　　(*b*) 1 and 3　　(*c*) 2 and 3　　(*d*) 1, 2 and 3

**487.** Following statements are made in regard to a low-pass filter
1. For particular specifications, the order of the Butterworth filter is higher as compared to chebychev filter.
2. For the same order of the filter, chebychev filter has narrower transition band as compared to Butterworth filter.
3. Butterworth filter has a better phase response as compared to chebychev filter.
Tick out the correct statement
(*a*) 1 and 2　　(*b*) 1 and 3　　(*c*) 2 and 3　　(*d*) 1, 2 and 3

**488.** Following statements are made in respect of low-pass chebychev filter.
1. The higher the order of the filter, the more non-linear is the phase response.
2. The larger the ripple in the pass band (magnitude) the better is the performance of the filter in the stop band.
3. It is possible to have a smaller pass-band filter and sharper cut-off characteristic
Choose the correct statements
(*a*) 1 and 2　　(*b*) 1 and 3　　(*c*) 2 and 3　　(*d*) 1, 2 and 3

**489.** Negative impedance converters are
(*a*) Reciprocal and stable networks
(*b*) Reciprocal but unstable networks
(*c*) Non-reciprocal and unstable network
(*d*) Non-reciprocal but stable network

**490.** A nullator is a
(*a*) bilateral lossy network element
(*b*) unilateral lossy network element
(*c*) bilateral loss less network element
(*d*) unilateral loss less network element

**491.** In order to obtain a CCT using two nullators and two norators we connect
(*a*) CVT in cascade with VCT　(*b*) VCT in cascade with CVT
(*c*) CVT in series with VCT　　(*d*) VCT in series with CVT

**492.** Let $F = F_1 F_2$ where $F_1$ and $F_2$ are functions of $x$. The sensitivity is given by
(*a*) $S_x^{F_1} \cdot S_x^{F_2}$　(*b*) $S_x^{F_1} + S_x^{F_2}$　(*c*) $S_x^{F_1} - S_x^{F_2}$　(*d*) $S_x^{F_1} / S_x^{F_2}$

**493.** Consider a parallel $RLC$ circuit. The sensitivity of $\omega_0$ with respect to $L$ and $C$ is denoted by $S_L^{\omega_0}$ and $S_C^{\omega_0}$ respectively. The values of these sensitivities are respectively.
(*a*) 1, 1　　　(*b*) –1, –1　　　(*c*) –1/2, –1/2　(*d*) 1/2, 1/2

**494.** In the above circuit, sensitivity of quality factor $Q$ with respect to $R$ and $C$ is denoted as $S_R^Q$ and $S_C^Q$. The values of these sensitivities are respectively.

(*a*) 1/2, 1/2    (*b*) 1, 1/2    (*c*) −1/2, −1/2   (*d*) 1, −1

**495.** For an *RC* active low-pass filter, if sensitivity is the basis for comparison between positive, negative and infinite feedback.
(*a*) positive and negative feed back are superior
(*b*) positive and infinite feed back are superior
(*c*) negative and infinite feedback are superior
(*d*) all are equal

**496.** For an active *RC* low-pass filter, if product of gain and sensitivity is the criterion for selection.
(*a*) negative feedback amplifier is preferred
(*b*) positive feedback amplifier is preferred
(*c*) infinite feedback is preferred
(*d*) all are equal

**497.** For an active *RC* low pass filter, if small element spread (values of elements of the same kind e.g. *R* or *C*, $R_1$, $R_2$ etc. or $C_1$ $C_2$ etc.) is the criterion.
(*a*) negative feedback is preferred          (*b*) infinite feedback is preferred
(*c*) positive feedback is preferred          (*d*) all are equal

**498.** The Thevenin equivalent of the network shown in Fig. I is 10 V in series with a resistance of 2 Ω. If now, resistance of 3 Ω is connected across *AB* as shown in Fig. II, the Thevenin equivalent of the modified network across *AB* will be

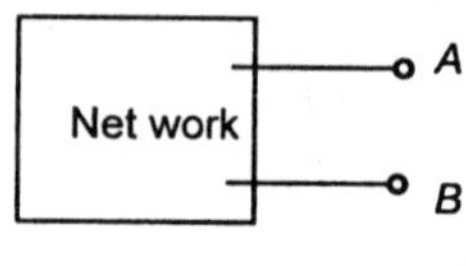


Fig. I

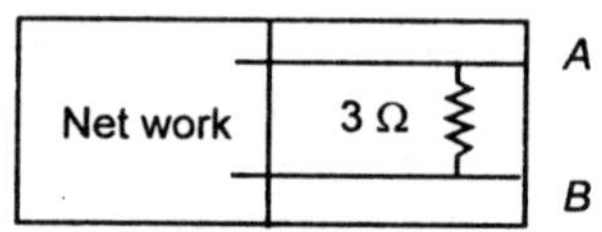


Fig. II

(*a*) 10 V in series with 1.2 Ω resistance          (*b*) 6 V in series with 1.2 Ω resistance
(*c*) 10 V in series with 5 Ω resistance          (*d*) 6 V in series with 5 Ω resistance

**499.** A certain network consists of two ideal identical voltage sources and a large number of ideal resistors. The power consumed in one of the resistors is 4 W when either of the two sources is active and the other is replaced by a short-circuit. The power consumed by the same resistor when both the sources are simultaneously active would be
(*a*) zero or 16 W          (*b*) 4 W or 8 W
(*c*) zero or 8 W          (*d*) 8 W or 16 W

**500.** In the circuit shown in the figure, the effective resistance faced by the voltage source is

(*a*) 1 Ω

(*b*) 2 Ω

(*c*) 3 Ω

(*d*) 3.3 Ω

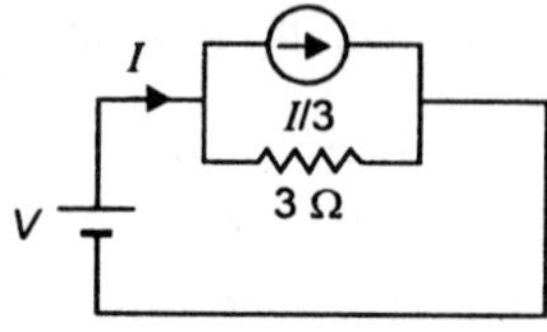


**501.** The graph of a network has six branches with three tree branches. The MINIMUM number of equations required for the solution of the network is
(*a*) 2          (*b*) 3          (*c*) 4          (*d*) 5

**502.** In the graph shown in the figure, one possible tree is formed by the branches 4, 5, 6, 7. Then one possible fundamental cut set is

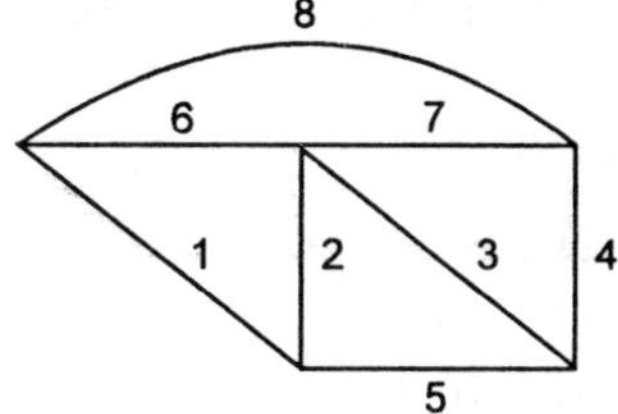

(a) 1, 2, 3, 8

(b) 1, 2, 5, 6

(c) 1, 5, 6, 8

(d) 1, 2, 3, 7, 8

**503.** The circuit shown in the figure is to be scaled to an impedance level of $5k\,\Omega$ and a resonant frequency of 5 M rad/s. Which one of the following is a correct set of element values for the scaled circuit?

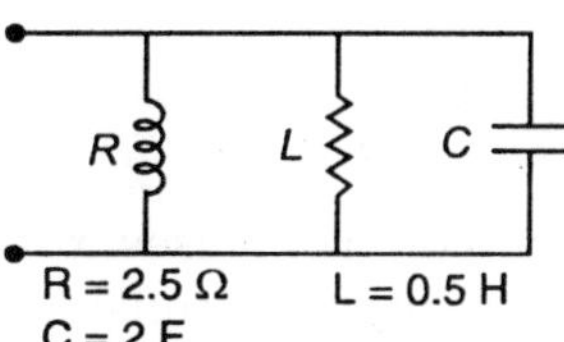

(a) $2.5\ \Omega$, 0.2 mH, 200 pF

(b) $5\ k\,\Omega$, 0.2 mH, 200 pF

(c) $5\ k\,\Omega$, 0.2 mH, 200 µF

(d) $5\ k\,\Omega$, 0.1 mH, 0.4 µF

**504.** In a parallel $RLC$ circuit, if $L = 4$ H, $C = 0.25$ F and $R = 40\ \Omega$, then the value of $Q$ at resonance will be

(a) 1       (b) 10       (c) 20       (d) 40

**505.** A series $RLC$ circuit is excited by an a.c. voltage $v(t) = 1 \sin t$. If $L = 10\ H$ and $C = 0.1$ F, then the peak value of the voltage across $R$ will be

(a) 0.707       (b) 1       (c) 1.414

(d) indeterminate as the value of $R$ is not given

**506.** The two-port network shown in the figure is characterized by the impedance parameters $Z_{11}$, $Z_{12}$, $Z_{21}$ and $Z_{22}$. For the equivalent Thevenin's source looking to the left of port 2, the $V_T$ and $Z_T$ will be respectively

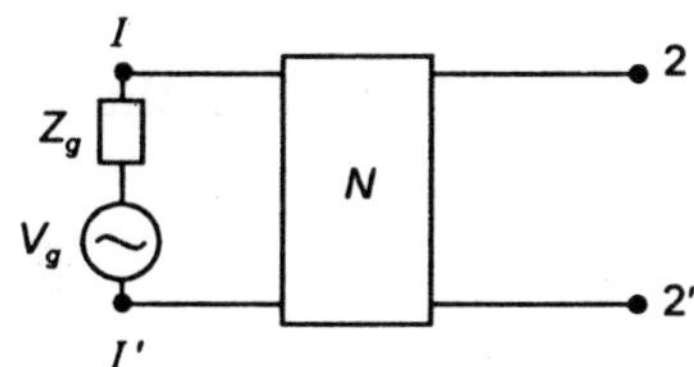

(a) $V_T = \dfrac{Z_{11}}{Z_{11} + Z_g}\, V_g$ , $Z_T = Z_{22} - Z_{12}$

(b) $V_T = \dfrac{Z_{12}}{Z_{11} + Z_g}\, V_g$ , $Z_T = Z_{22} - Z_{12}$

(c) $V_T = \dfrac{Z_{21}\, V_g}{Z_{11} + Z_g}$ , $Z_T = Z_{22} - \dfrac{Z_{12}\, Z_{21}}{Z_{11} + Z_g}$

(d) $V_T = \dfrac{Z_{12}}{Z_{11} + Z_g}\, V_g$ , $Z_T = Z_{22} - \dfrac{Z_{12}\, Z_{21}}{Z_{11} + Z_g}$

**507.** Frequency response of the function $T(s) = (s + 1) / (s + 2)$ exhibits a maximum phase at a frequency (in radian/sec.)

(a) 0       (b) $\dfrac{1}{\sqrt{2}}$       (c) $\sqrt{2}$       (d) $\infty$

**508.** In the network shown in the figure, the switch 'S' is closed and a steady state is attained. If the switch is opened at $t = 0$, then the current $i(t)$ through the inductor will be

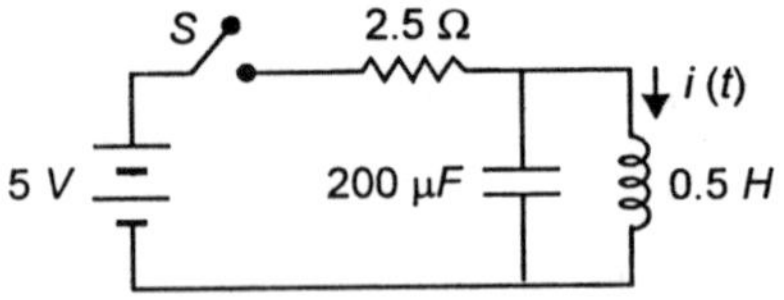

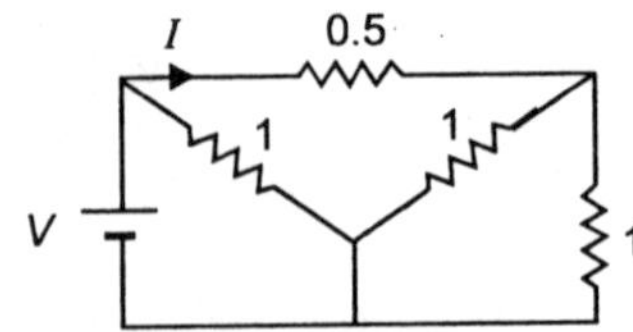

(a) cos 50 t A

(b) 2 A

(c) 2 cos 100 t A

(d) 2 sin 50 t A

**509.** A series $RL$ circuit is initially relaxed. A step voltage is applied to the circuit. If $\tau$ is the time constant of the circuit, the voltage across $R$ and $L$ will be same at time $t$ equal to

(a) $\tau \log_e 2$     (b) $\tau \log_e \dfrac{1}{2}$     (c) $\dfrac{1}{\tau} \log_e 2$     (d) $\dfrac{1}{\tau} \log_e \dfrac{1}{2}$

**510.** A resistance coil possesses residual self-inductance and capacitance apart from its resistance. Taking into consideration all three, the impedance across the coil is given by

(a) $\dfrac{R}{(sL + R) sC + 1}$

(b) $\dfrac{R + sL}{sC (R + sL) + 1}$

(c) $\dfrac{R}{sL (R + sC)}$

(d) $\dfrac{sC}{(R + sL) + sC + 1}$

**511.** The total power consumed in the circuit shown in the figure is

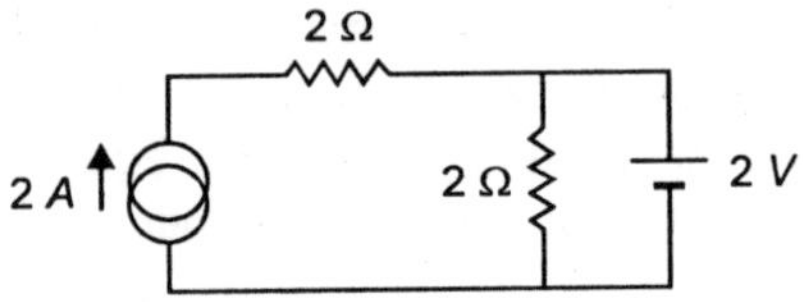

(a) 10 W

(b) 12 W

(c) 16 W

(d) 20 W

**512.** In the circuit shown in the figure, if $I = 2$, then the value of the battery voltage $V$ will be

(a) 5 V

(b) 3 V

(c) 2 V

(d) 1 V

**513.** The effective resistance between the terminals $A$ and $B$ in the circuit shown in the figure is:

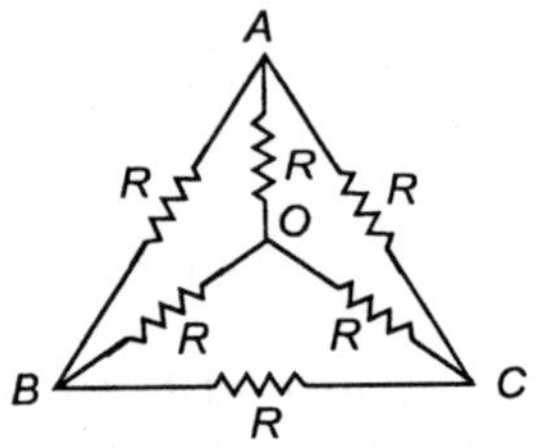

(a) $R$

(b) $R - 1$

(c) $R/2$

(d) $\dfrac{6}{11} R$

**514.** The network shown in the figure represents a

(*a*) band-pass filter

(*b*) low-pass filter

(*c*) high-pass filter

(*d*) band-stop filter

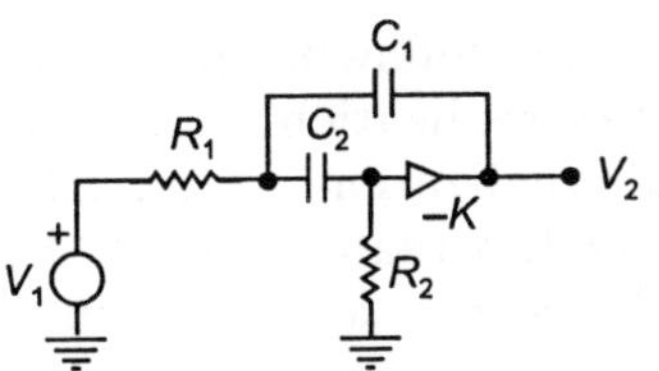

**515.** In active filter circuits, inductances are avoided mainly because they
   (*a*) are always associated with some resistance

   (*b*) are bulky and unsuitable for miniaturization

   (*c*) are non-linear in nature

   (*d*) saturate quickly

**516.** The magnitude response of a normalized Butterworth low-pass filter is
   (*a*) linear starting with the values of unity at zero frequency and 0.707 at the cut-off frequency.
   (*b*) non-linear all through but with values of unity at zero frequency and 0.707 at the cut-off frequency.
   (*c*) linear upto the cut-off frequency and non-linear thereafter
   (*d*) non-linear upto the cut-off frequency and linear thereafter

**517.** The driving-point impedance function of a reactive network is:

$$Z(s) = \frac{(s^2 + 4)(s^2 + 16)}{s(s^2 + 9)}$$

Consider the following circuits in this regard:

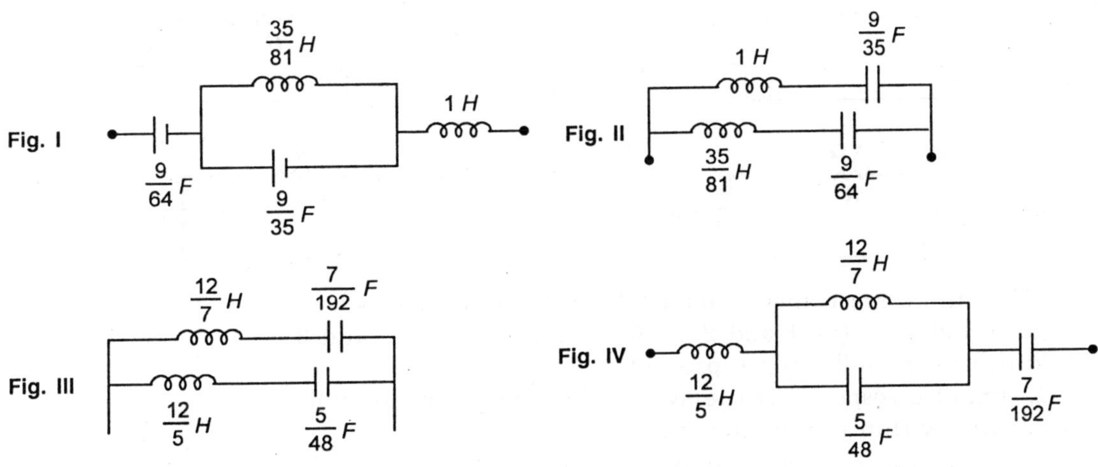

The first and second Foster forms will be as in figures
   (*a*) I and III respectively
   (*b*) II and IV respectively
   (*c*) I and II respectively
   (*d*) III and IV respectively

**518.** Which of the following pairs are correctly matched?
1. Brune's realization ............. Realisation with ideal transformer
2. Cauer realization ............... Ladder realization
3. Bott-Duffin realisation ...... Realisation with non-ideal transformer
Select the correct answer using the codes given below:
*Codes:*
(*a*)  1,2 and 3    (*b*)  2 and 3    (*c*)  1 and 3    (*d*)  1 and 2

**519.** The polynomial

$$P(s) = (s - 1)(s^2 + 1)(s + 2)(s + 3) \text{ is}$$

(*a*)  Hurwitz, but not strict Hurwitz
(*b*)  not Hurwitz
(*c*)  strict Hurwitz
(*d*)  anti-Hurwitz

**520.** The poles and zeros of a driving-point function of a network are simple and interlace on the negative real axis with a pole closest to the origin. It can be realized.
(*a*)  by an *LC* network
(*b*)  as an *RC* driving-point impedance
(*c*)  as an *RC* driving-point admittance
(*d*)  only by an *RLC* network

**521.** Which of the following circuits would be valid for a simple circuit consisting of *R* and *C* and whose state equation is given by

$$\frac{dV_c}{dt} = 2 - 1.25\, V_c\,(t)$$

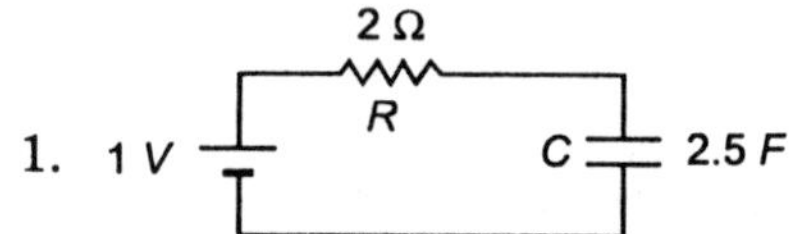
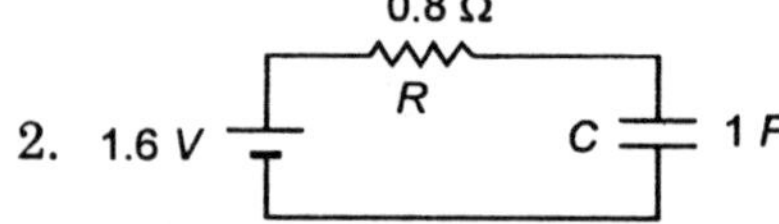
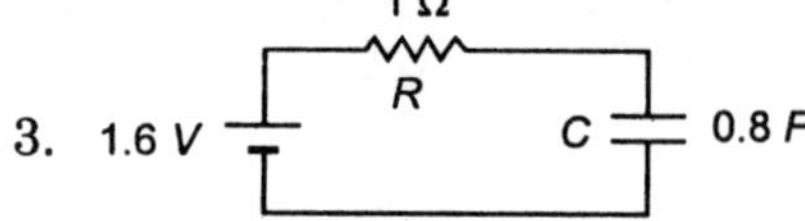
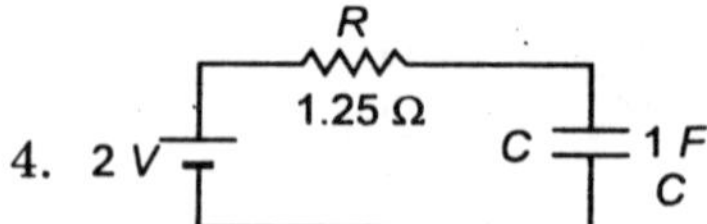

Select the correct answer using the codes given below:
(*a*)  1 and 4    (*b*)  1 and 2    (*c*)  3 and 4    (*d*)  2 and 3

**522.** Which of the following represent a stable system?
1. Impulse response of the system decreases exponentially.
2. Area within the impulse response is finite.
3. Eigenvalues of the system are positive and real.
4. Roots of the characteristic equation of the system are real and negative.
Select the correct answer using the codes given below:
*Codes:*
(*a*)  1 and 4    (*b*)  1 and 3    (*c*)  2, 3 and 4   (*d*)  1, 2 and 4

**523.** Match List-I with List-II and select the correct answer using the codes given below the Lists:

*List-I*            *List-II*

A. 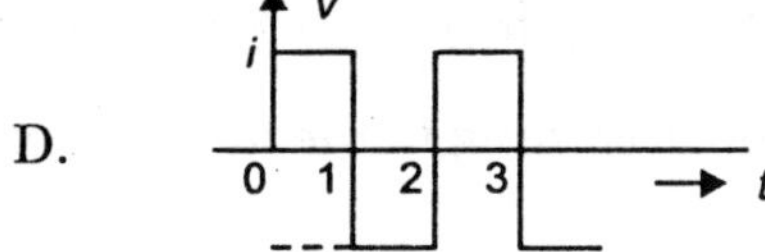

1. $v(t) = u(t + 1)$

B. 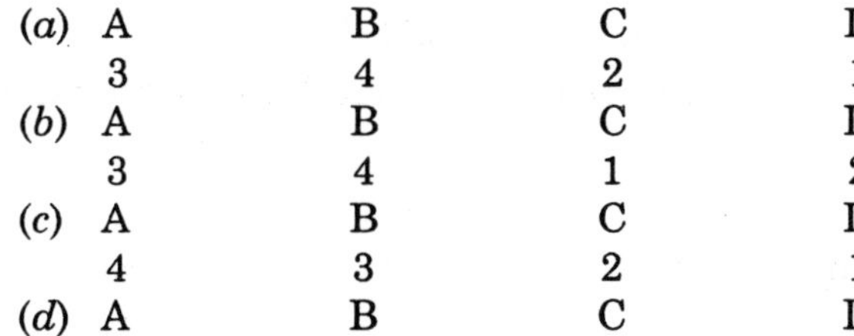

2. $v(t) = u(t) - 2u(t - 1) + 2u(t - 2) - 2u(t - 3) + ...$

C. 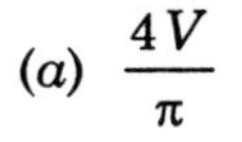

3. $v(t) = u(t - 1) - u(t - 3)$

D. 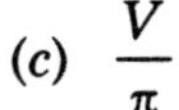

4. $\underset{a \to 0}{Lt}\ v(t) = \delta(t - 1)$

*Codes:*

| | A | B | C | D |
|---|---|---|---|---|
| (a) | 3 | 4 | 2 | 1 |
| (b) | 3 | 4 | 1 | 2 |
| (c) | 4 | 3 | 2 | 1 |
| (d) | 4 | 3 | 1 | 2 |

**524.** The amplitude of the first odd harmonic of the square wave shown in the figure is equal to

(a) $\dfrac{4V}{\pi}$      (b) $\dfrac{2V}{3\pi}$

(c) $\dfrac{V}{\pi}$      (d) 0

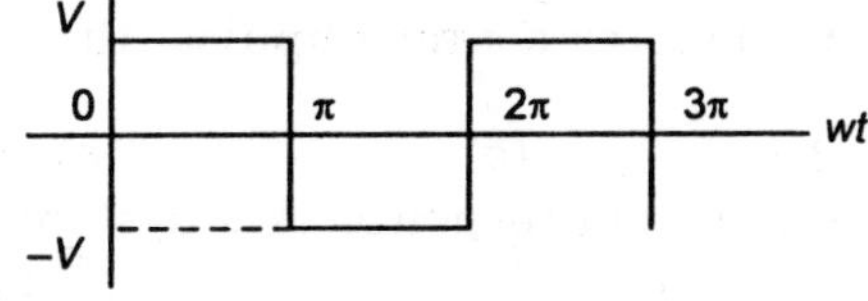

**525.** A periodic triangular wave is shown in the figure. Its Fourier components will consist only of

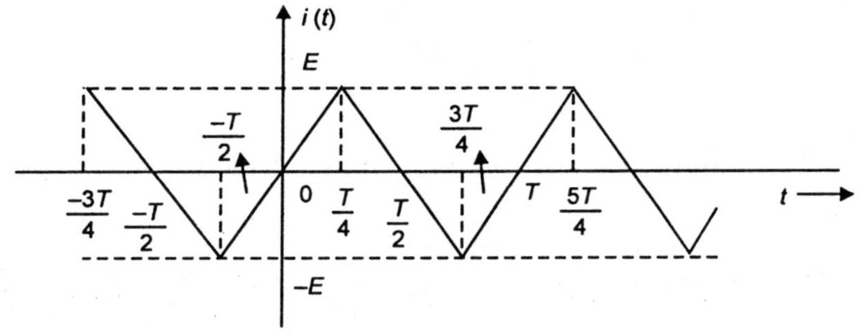

  (a) all cosine terms                    (b) all sine terms
  (c) odd cosine terms                    (d) odd sine terms

**526.** Which one of the following is the correct Fourier transform of the unit step signal

$$u(t) = \begin{cases} 1 \text{ for } t \geq 0 \\ 0 \text{ for } t < 0 \end{cases}$$

  (a) $\pi\, \delta_{-}(\omega)$

  (b) $\dfrac{1}{j\omega}$

  (c) $\dfrac{1}{j\omega} + \pi\, \delta(\omega)$

  (d) $\dfrac{1}{j\omega} + 2\pi\delta(\omega)$

**527.** The Fourier transform of

$$v(t) = \cos \omega_0 t \text{ is given by}$$

  (a) $V(f) = \dfrac{1}{2}\,\delta(f - f_0)$

  (b) $V(t) = \dfrac{1}{2}\,\delta(f + f_0)$

  (c) $V(f) = \dfrac{1}{2}\left[\delta(f - f_0) - \delta(f + f_0)\right]$

  (d) $V(f) = \dfrac{1}{2}\left[\delta(f - f_0) + \delta(f + f_0)\right]$

**528.** If $g(t) \leftrightarrow G(f)$ represents a Fourier transform pair, then according to the duality property of Fourier transforms,

  (a) $G(t) \rightleftharpoons g(f)$

  (b) $G(t) \rightleftharpoons g^*(f)$

  (c) $G(t) \rightleftharpoons g(-f)$

  (d) $G(t) \rightleftharpoons g^*(-f)$

**529.** If $x(t)$ and its first derivative are Laplace transformable and the Laplace transform of $x(t)$ is $X(s)$, then $\underset{t \to 0}{\text{Lim}}\, x(t)$ is given

  (a) $\underset{s \to \infty}{\text{Lim}}\, s\, X(s)$

  (b) $\underset{s \to 0}{\text{Lim}}\, x\, X(s)$

  (c) $\underset{s \to \infty}{\text{Lim}}\, \dfrac{X(s)}{s}$

  (d) $\underset{s \to 0}{\text{Lim}}\, \dfrac{X(s)}{s}$

**530.** If $\delta(t)$ denotes a unit impulse, then the Laplace transform of $\dfrac{d^2\, \delta(t)}{dt^2}$ will be

  (a) 1          (b) $s^2$          (c) $s$          (d) $s^{-2}$

**531.** The unit step response of a system is given by

$$(1 - e^{-\alpha t})\, u(t)$$

Its impulse response is:

  (a) $e^{-\alpha t}\, u(t)$

  (b) $\alpha\, e^{-\alpha t}\, u(t)$

  (c) $\dfrac{1}{\alpha}\, e^{-\alpha t}\, u(t)$

  (d) $-\alpha\, e^{-\alpha t}\, u(t)$

**532.** Given that

$$h(t) = 10\, e^{-10t}\, u(t), \text{ and } e(t) = \sin 10\, t\, u(t), \text{ the Laplace transform of the signal}$$

$$f(t) = \int_0^t h(t - \tau)\, e(\tau)\, d\tau$$

is given by

(a) $\dfrac{10}{(s+10)(s^2+100)}$

(b) $\dfrac{10(s+10)}{(s^2+100)}$

(c) $\dfrac{100}{(s+10)(s^2+100)}$

(d) $\dfrac{1}{(s+10)(s^2+100)}$

**533.** A d.c. current source is connected as shown in Fig. I:

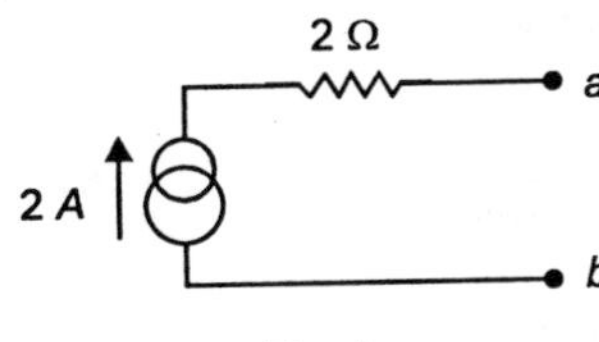

Fig. I

The Thevenin's equivalent of the network at terminals $a - b$

(a) will be 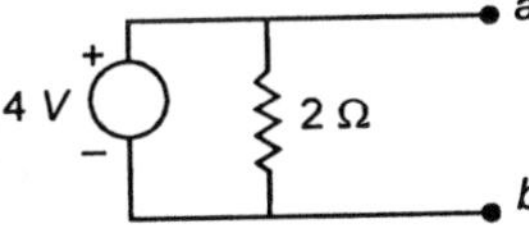

(b) will be 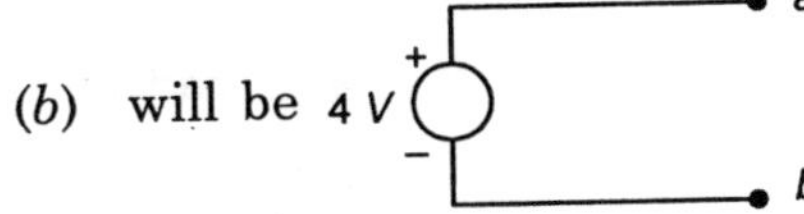

(c) will be 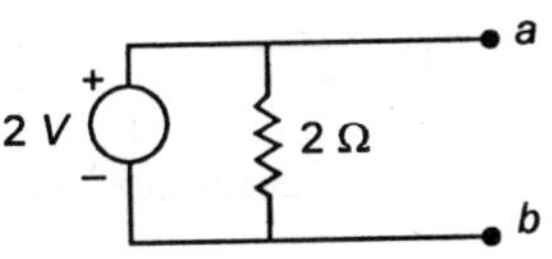

(d) is NOT feasible

**534.** A network is shown in the given figure. Which one of the following equations would represent the equation for loop 3?

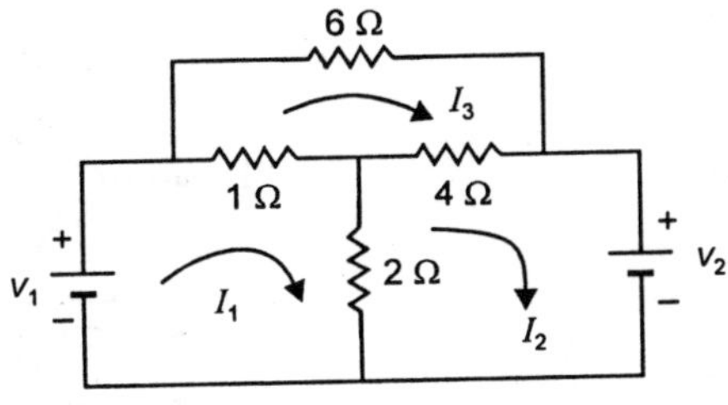

(a) $-I_1 + 4I_2 + 11I_3 = 0$

(b) $I_1 + 4I_2 + 11I_3 = 0$

(c) $-I_1 + 4I_2 + 11I_3 = 0$

(d) $I_1 + 4I_2 + 6I_3 = 0$

**535.** For the graph shown in the given figure, one set of fundamental cut-sets would be

(a) *abc, cde, afe*

(b) *afdc, cde, abde*

(c) *cbfe, afe, bdf*

(d) *cbd, abde, cde*

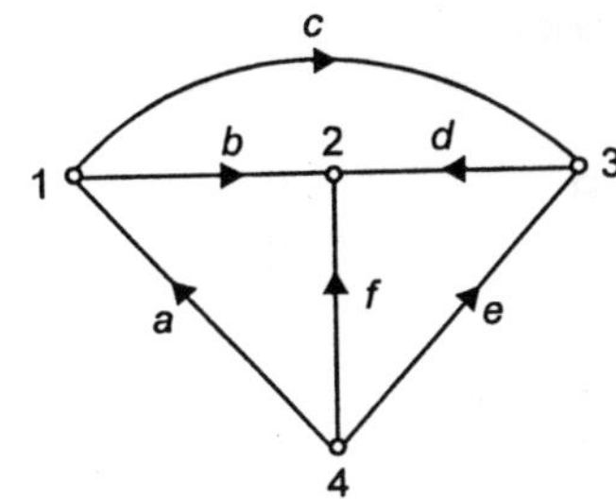

**536.** A particular current is made up of two components: a 10 A d.c. and a sinusoidal current of peak value of 14.14 A. The average value of the resultant current is
(*a*) Zero      (*b*) 24.14 A      (*c*) 10 A      (*d*) 14.14 A

**537.** The locus of the tip of the voltage phasor ($V_R$) across the resistance ($R$) in a series $RLC$ resonant circuit is given by

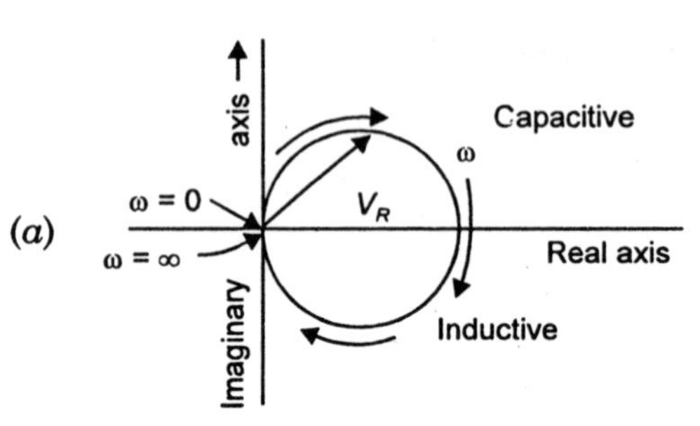


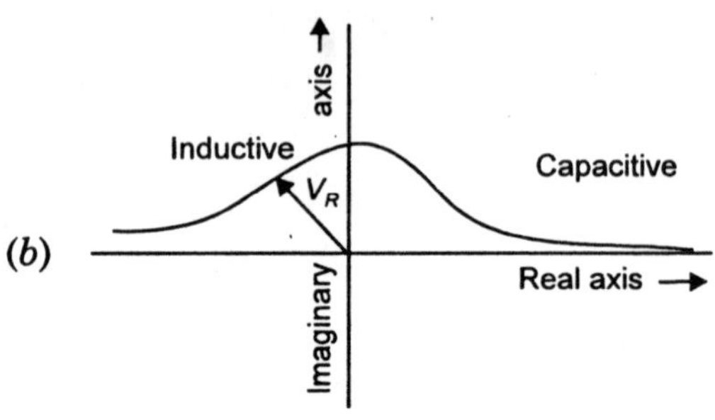


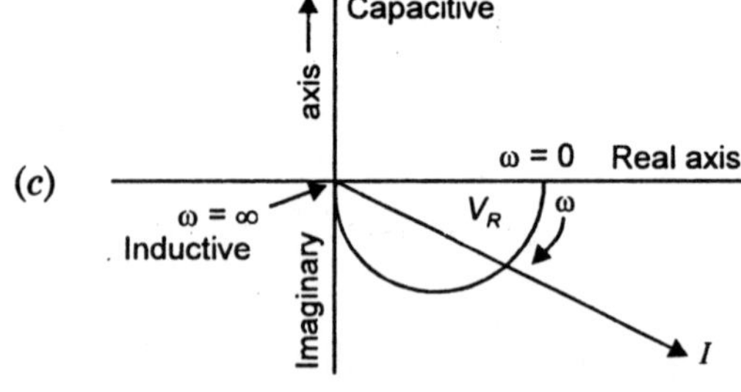


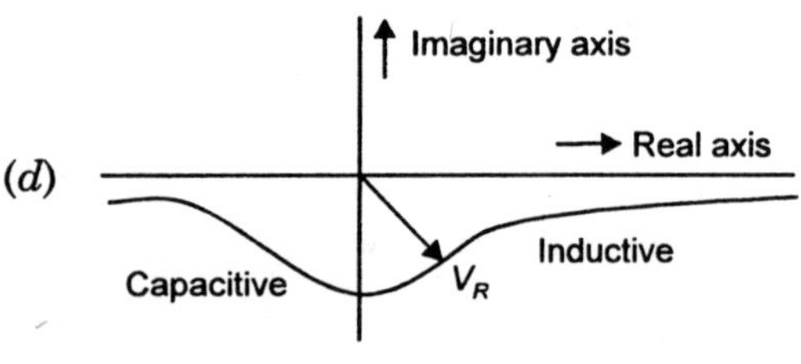


**538.** Match List-I with List-II and select the correct answer using the codes given below the Lists:

|     *List-I*     |   | *List-II* |
| --- | --- | --- |
| A. | Bridged T-network | 1. 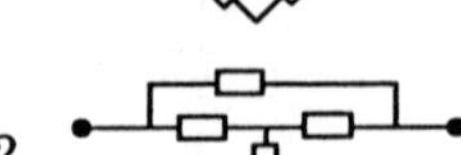 |
| B. | Twin-T network | 2. 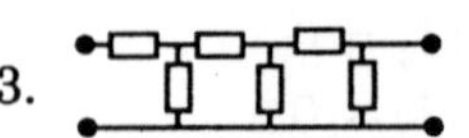 |
| C. | Lattice network | 3. 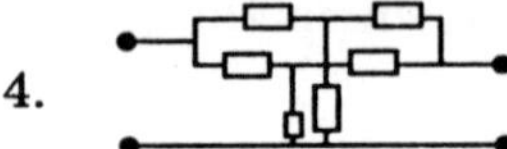 |
| D. | Ladder network | 4. |

*Codes:*

|     | A | B | C | D |
| --- | --- | --- | --- | --- |
| (*a*) | 2 | 4 | 3 | 1 |
| (*b*) | 4 | 2 | 1 | 3 |
| (*c*) | 4 | 2 | 3 | 1 |
| (*d*) | 2 | 4 | 1 | 3 |

**539.** In respect of the 2-port network shown in the figure, the admittance parameters are: $Y_{11} = 8$ mho, $Y_{12} = Y_{21} = -6$ mho and $Y_{22} = 6$ mho.

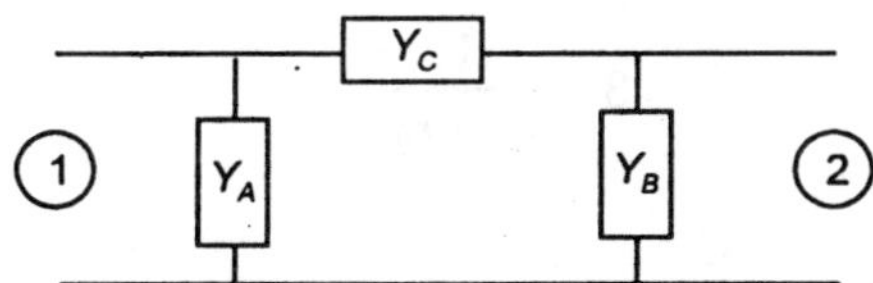

The values of $Y_A$, $Y_B$ and $Y_C$ (in units of mho) will be respectively
(a) 2, 6 and −6                 (b) 2, 6 and 0
(c) 2, 0 and 6                 (d) 2, 6 and 8

**540.** In the network shown, the switch is opened at $t = 0$. Prior to that, the network was in the steady-state. $v_s(t)$ at $t = 0^+$ is

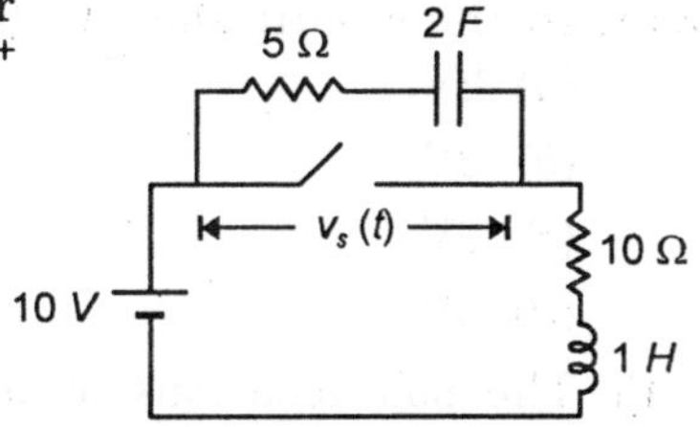

(a) 0

(b) 5V

(c) 10 V

(d) 15 V

**541.** In the network shown in the figure, the switch had remained closed for a long time on the 10V source side. If at time $t = 0$, it is changed to the 12V side, then after one time constant, the voltage across 5 $\Omega$ in the circuit will be

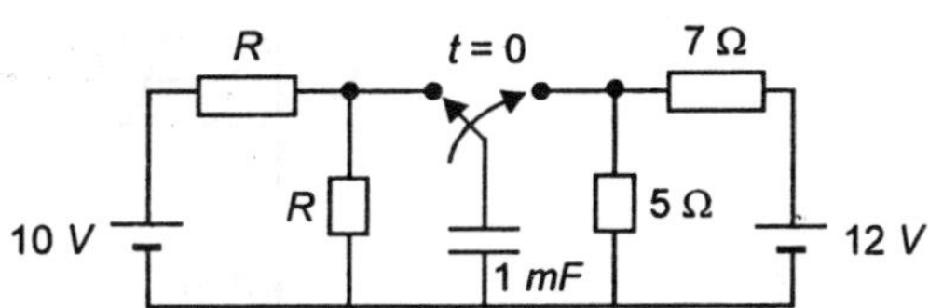

(a) 5 V         (b) $5\,e^{-1}$ V         (c) 10 V         (d) 12 V

**542.** The transfer function

$$T(s) = \frac{s}{s + a}$$

is that of a
(a) low-pass filter                 (b) notch filter
(c) high-pass filter                (d) band-pass filter

**543.** The model of a transistor in the common emitter connection is shown in the following figure:

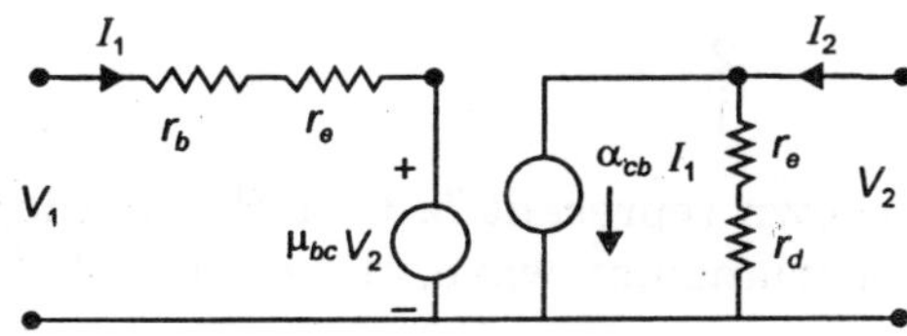

Match List-I (Parameters) will List-II (Values) and select correct answer using the codes given below the Lists:

|     | *List–I* |     |     | *List–II* |
|-----|----------|-----|-----|-----------|
| A.  | $h_{22}$ |     | 1.  | $r_b + r_e$ |
| B.  | $h_{11}$ |     | 2.  | $\alpha_{cb}$ |
| C.  | $h_{21}$ |     | 3.  | $\dfrac{1}{r_e + r_d}$ |

*Codes:*

|     | A | B | C |
|-----|---|---|---|
| (a) | 3 | 1 | 2 |
| (b) | 1 | 3 | 2 |
| (c) | 2 | 3 | 1 |
| (d) | 3 | 2 | 1 |

**544.** In the circuit shown, $i(t)$ is a unit step current. The steady-state value of $v(t)$ is

(a) 2V

(b) 3V

(c) 6V

(d) 9V

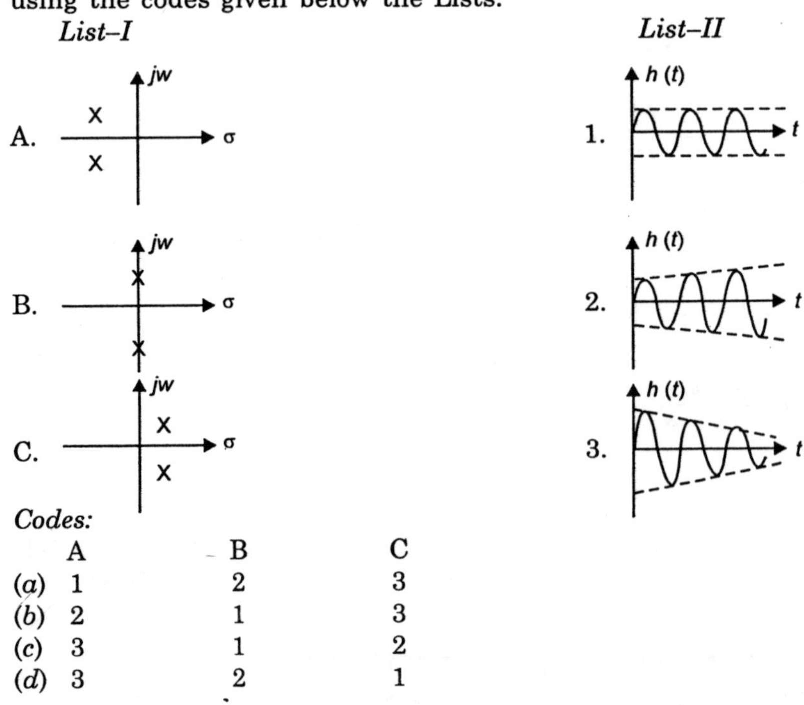

**545.** The pole locations of three networks and their impulse responses $h(t)$ are shown in List-I and List-II respectively. Match List-I with List-II and select the correct answer using the codes given below the Lists:

*List–I*        *List–II*

*Codes:*

|     | A | B | C |
|-----|---|---|---|
| (a) | 1 | 2 | 3 |
| (b) | 2 | 1 | 3 |
| (c) | 3 | 1 | 2 |
| (d) | 3 | 2 | 1 |

**546.** $Z_1$ and $Z_2$ in the figure shown represent $L$, $C$ or $R$. For the transformation of a square wave of period '$T$' into a triangular wave, the values of $Z_1$, $Z_2$ and $L/R$ or $RC$ as the case may be, should be

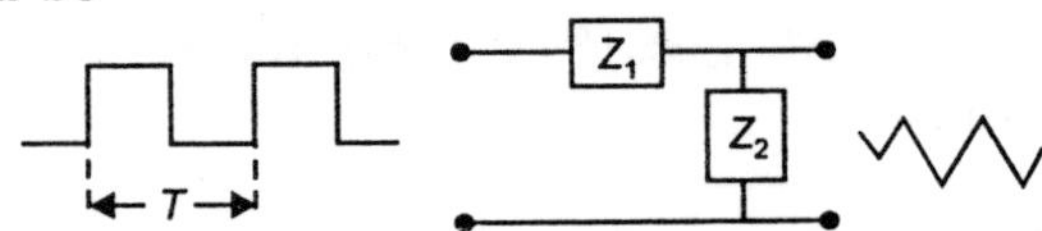

(a) $Z_1 = R$, $Z_2 = L$ and $\dfrac{L}{R} \ll T$  (b) $Z_1 = L$, $Z_2 = R$ and $\dfrac{L}{R} \ll T$

(c) $Z_1 = R$, $Z_2 = C$ and $RC \gg T$  (d) $Z_1 = C$, $Z_2 = R$ and $RC \gg T$

**547.** Match List–I with List–II and select the correct answer using the codes given below the Lists:

<table>
<tr><td align="center">*List–I*</td><td align="center">*List–II*</td></tr>
</table>

A. 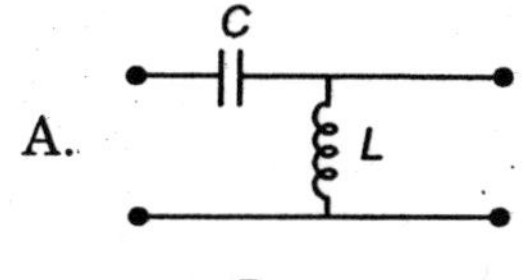  1. Band-reject (notch) filter

B. 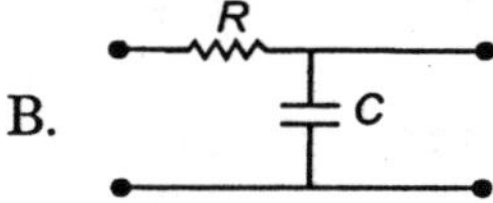  2. Band-pass filter

C. 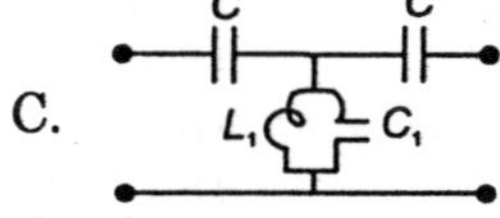  3. High-pass filter

D. 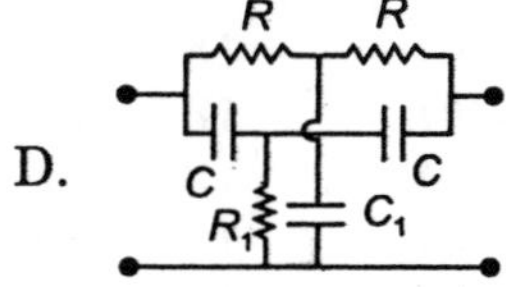  4. Low-pass filter

*Codes:*

| | A | B | C | D |
|---|---|---|---|---|
| (a) | 3 | 4 | 2 | 1 |
| (b) | 3 | 4 | 1 | 2 |
| (c) | 4 | 3 | 1 | 2 |
| (d) | 4 | 3 | 2 | 1 |

**548.** Consider the following from the point of view of possible realisation as driving-point impedances using passive elements:

1. $\dfrac{1}{s\,(s+5)}$  2. $\dfrac{s+3}{s^2\,(s+5)}$

3. $\dfrac{s^2+3}{s^2\,(s^2+5)}$  4. $\dfrac{s+5}{s\,(s+3)}$

Among these

(a) 1, 2 and 4 are realizable  (b) 1, 2 and 3 are realizable

(c) 3 and 4 are realizable  (d) none is realizable

**549.** An $RC$ driving-point impedance function has zeros at $s = -2$ and $s = -5$. The admissible poles for the function would be

(a) $s = 0$; $s = -6$  (b) $s = -1$; $s = -3$

(c) $s = 0$; $s = -1$  (d) $s = -3$; $s = -4$

# Answers to Multiple Choice Questions

| | | | | | |
|---|---|---|---|---|---|
| 1. (a) | 36. (d) | 71. (b) | 106. (c) | 141. (d) | 176. (c) |
| 2. (a) | 37. (c) | 72. (c) | 107. (b) | 142. (a) | 177. (c) |
| 3. (a) | 38. (c) | 73. (b) | 108. (c) | 143. (a) | 178. (c) |
| 4. (b) | 39. (c) | 74. (a) | 109. (c) | 144. (c) | 179. (a) |
| 5. (b) | 40. (b) | 75. (c) | 110. (c) | 145. (c) | 180. (c) |
| 6. (a) | 41. (a) | 76. (d) | 111. (b) | 146. (c) | 181. (b) |
| 7. (c) | 42. (b) | 77. (c) | 112. (d) | 147. (a) | 182. (a) |
| 8. (a) | 43. (d) | 78. (d) | 113. (b) | 148. (c) | 183. (d) |
| 9. (c) | 44. (c) | 79. (c) | 114. (d) | 149. (a) | 184. (d) |
| 10. (c) | 45. (b) | 80. (b) | 115. (b) | 150. (d) | 185. (c) |
| 11. (a) | 46. (d) | 81. (c) | 116. (b) | 151. (b) | 186. (d) |
| 12. (c) | 47. (c) | 82. (c) | 117. (a) | 152. (a) | 187. (c) |
| 13. (d) | 48. (c) | 83. (c) | 118. (a) | 153. (a) | 188. (d) |
| 14. (c) | 49. (b) | 84. (a) | 119. (c) | 154. (b) | 189. (d) |
| 15. (d) | 50. (c) | 85. (d) | 120. (d) | 155. (a) | 190. (b) |
| 16. (d) | 51. (d) | 86. (a) | 121. (a) | 156. (b) | 191. (a) |
| 17. (c) | 52. (d) | 87. (c) | 122. (b) | 157. (a) | 192. (b) |
| 18. (d) | 53. (c) | 88. (b) | 123. (d) | 158. (b) | 193. (c) |
| 19. (b) | 54. (c) | 89. (c) | 124. (c) | 159. (d) | 194. (d) |
| 20. (d) | 55. (d) | 90. (c) | 125. (c) | 160. (b) | 195. (d) |
| 21. (a) | 56. (a) | 91. (b) | 126. (a) | 161. (d) | 196. (b) |
| 22. (c) | 57. (b) | 92. (d) | 127. (b) | 162. (a) | 197. (a) |
| 23. (c) | 58. (c) | 93. (c) | 128. (a) | 163. (d) | 198. (b) |
| 24. (c) | 59. (a) | 94. (c) | 129. (b) | 164. (b) | 199. (b) |
| 25. (b) | 60. (b) | 95. (b) | 130. (c) | 165. (a) | 200. (a) |
| 26. (c) | 61. (b) | 96. (c) | 131. (b) | 166. (b) | 201. (b) |
| 27. (d) | 62. (b) | 97. (c) | 132. (b) | 167. (b) | 202. (d) |
| 28. (c) | 63. (b) | 98. (b) | 133. (c) | 168. (a) | 203. (a) |
| 29. (d) | 64. (a) | 99. (b) | 134. (a) | 169. (a) | 204. (c) |
| 30. (c) | 65. (d) | 100. (c) | 135. (b) | 170. (c) | 205. (c) |
| 31. (b) | 66. (d) | 101. (c) | 136. (b) | 171. (b) | 206. (a) |
| 32. (b) | 67. (a) | 102. (d) | 137. (c) | 172. (c) | 207. (a) |
| 33. (a) | 68. (d) | 103. (c) | 138. (b) | 173. (d) | 208. (b) |
| 34. (b) | 69. (c) | 104. (c) | 139. (b) | 174. (c) | 209. (b) |
| 35. (c) | 70. (a) | 105. (b) | 140. (c) | 175. (b) | 210. (b) |

| | | | | | |
|---|---|---|---|---|---|
| 211. (a) | 258. (b) | 305. (d) | 352. (b) | 399. (a) | 446. (b) |
| 212. (c) | 259. (c) | 306. (a) | 353. (c) | 400. (c) | 447. (c) |
| 213. (b) | 260. (c) | 307. (c) | 354. (c) | 401. (a) | 448. (c) |
| 214. (a) | 261. (d) | 308. (a) | 355. (c) | 402. (c) | 449. (a) |
| 215. (b) | 262. (d) | 309. (b) | 356. (a) | 403. (b) | 450. (d) |
| 216. (a) | 263. (c) | 310. (c) | 357. (c) | 404. (a) | 451. (b) |
| 217. (a) | 264. (a) | 311. (d) | 358. (c) | 405. (c) | 452. (a) |
| 218. (d) | 265. (b) | 312. (d) | 359. (c) | 406. (d) | 453. (c) |
| 219. (c) | 266. (b) | 313. (a) | 360. (a) | 407. (a) | 454. (a) |
| 220. (b) | 267. (d) | 314. (d) | 361. (c) | 408. (c) | 455. (b) |
| 221. (b) | 268. (c) | 315. (a) | 362. (a) | 409. (a) | 456. (b) |
| 222. (c) | 269. (b) | 316. (a) | 363. (c) | 410. (b) | 457. (c) |
| 223. (b) | 270. (c) | 317. (c) | 364. (d) | 411. (d) | 458. (b) |
| 224. (b) | 271. (d) | 318. (c) | 365. (a) | 412. (b) | 459. (b) |
| 225. (a) | 272. (d) | 319. (b) | 366. (c) | 413. (d) | 460. (b) |
| 226. (d) | 273. (a) | 320. (c) | 367. (b) | 414. (c) | 461. (c) |
| 227. (b) | 274. (d) | 321. (a) | 368. (a) | 415. (c) | 462. (c) |
| 228. (d) | 275. (b) | 322. (b) | 369. (c) | 416. (d) | 463. (b) |
| 229. (b) | 276. (c) | 323. (a) | 370. (b) | 417. (c) | 464. (c) |
| 230. (a) | 277. (c) | 324. (b) | 371. (c) | 418. (c) | 465. (c) |
| 231. (a) | 278. (d) | 325. (b) | 372. (b) | 419. (a) | 466. (a) |
| 232. (a) | 279. (c) | 326. (a) | 373. (c) | 420. (a) | 467. (c) |
| 233. (c) | 280. (a) | 327. (d) | 374. (a) | 421. (a) | 468. (c) |
| 234. (a) | 281. (c) | 328. (a) | 375. (a) | 422. (d) | 469. (b) |
| 235. (d) | 282. (c) | 329. (b) | 376. (b) | 423. (c) | 470. (b) |
| 236. (a) | 283. (d) | 330. (d) | 377. (b) | 424. (d) | 471. (c) |
| 237. (a) | 284. (d) | 331. (c) | 378. (b) | 425. (d) | 472. (b) |
| 238. (c) | 285. (c) | 332. (a) | 379. (a) | 426. (c) | 473. (d) |
| 239. (c) | 286. (b) | 333. (b) | 380. (d) | 427. (d) | 474. (c) |
| 240. (d) | 287. (d) | 334. (a) | 381. (b) | 428. (a) | 475. (c) |
| 241. (a) | 288. (a) | 335. (a) | 382. (b) | 429. (b) | 476. (a) |
| 242. (b) | 289. (d) | 336. (d) | 383. (a) | 430. (c) | 477. (b) |
| 243. (b) | 290. (b) | 337. (a) | 384. (d) | 431. (c) | 478. (b) |
| 244. (d) | 291. (c) | 338. (b) | 385. (a) | 432. (d) | 479. (b) |
| 245. (b) | 292. (c) | 339. (a) | 386. (b) | 433. (a) | 480. (c) |
| 246. (d) | 293. (b) | 340. (c) | 387. (d) | 434. (c) | 481. (b) |
| 247. (d) | 294. (c) | 341. (b) | 388. (c) | 435. (a) | 482. (b) |
| 248. (b) | 295. (a) | 342. (b) | 389. (b) | 436. (c) | 483. (c) |
| 249. (c) | 296. (a) | 343. (c) | 390. (d) | 437. (c) | 484. (b) |
| 250. (c) | 297. (d) | 344. (d) | 391. (b) | 438. (b) | 485. (c) |
| 251. (c) | 298. (b) | 345. (a) | 392. (c) | 439. (b) | 486. (d) |
| 252. (a) | 299. (b) | 346. (d) | 393. (d) | 440. (a) | 487. (d) |
| 253. (b) | 300. (c) | 347. (b) | 394. (b) | 441. (a) | 488. (a) |
| 254. (a) | 301. (b) | 348. (d) | 395. (b) | 442. (a) | 489. (c) |
| 255. (a) | 302. (b) | 349. (a) | 396. (d) | 443. (b) | 490. (c) |
| 256. (b) | 303. (a) | 350. (d) | 397. (b) | 444. (b) | 491. (a) |
| 257. (a) | 304. (c) | 351. (c) | 398. (d) | 445. (b) | 492. (b) |

# Answer to Problems

## CHAPTER I

**1.2.** 250 watts

**1.3.** 3650 watts, 500 V Ar capacitive

**1.4.** 20 watts

**1.5.** 25 mJ, 200 mJ, 46.3 mJ. The energy change is due to the loss in the connecting wires.

**1.6.** $I_m$ (cos $\omega$t $-$ cos 2 $\omega$t) where $I_m = \omega C_o V_m$

**1.7.** $Y_{11} = \dfrac{1}{R}$, $Y_{12} = \dfrac{1}{r} - \dfrac{1}{R}$, $Y_{21} = -\dfrac{1}{r} - \dfrac{1}{R}$, $Y_{22} = \dfrac{1}{R}$

**1.8.** $G = \dfrac{R}{r^2}$, $C = \dfrac{L}{r^2}$

**1.9.** 1.83 h.

**1.10.** 3.63 $\angle{-}123.7°$ mA, 5.24 $\angle{-}60.4°$ mA

## CHAPTER II

**2.1.** 47.2 $\Omega$

**2.2.** 5 $\Omega$

**2.3.** $v_1 = -0.542$, $v_2 = 1.62$, $v_3 = 0.792$, $v_4 = -0.25$

**2.4.** 2 $\angle 73.7°$ A

**2.5.** $I_{AB} = 10.5$ A, $I_{BC} = 13.2$ A, $I_{DE} = 3.84$ A, $I_{EF} = 3.46$ A, $I_{BE} = 3.93$ A

**2.7.** 55.8 $\angle{-}17.4°$

**2.8.** 59.36 $\angle{-}63°$

**2.9.** 0 A or 5.08 A

**2.10.** $I_1 = 0.25$ A, $I_2 = -0.41$ A, $I_3 = \dfrac{2}{3}$ A

**2.11.**

$$\begin{bmatrix} \dfrac{1}{3} + \dfrac{1}{5} + \dfrac{1}{j1} + \dfrac{1}{j2} & -\left(\dfrac{1}{j1} + \dfrac{1}{3}\right) \\ -\left(\dfrac{1}{j1} + \dfrac{1}{3}\right) & \dfrac{1}{3} + \dfrac{1}{j1} + \dfrac{1}{-j2} + \dfrac{1}{j2} \end{bmatrix} \begin{bmatrix} v_1 \\ v_2 \end{bmatrix} = \begin{bmatrix} 10\angle 0 - \dfrac{5\angle 60°}{5} \\ \dfrac{10 \angle 90°}{j2} \end{bmatrix}$$

**2.13.** $43.9 \angle 20°$

**2.14.** (*i*) 20.2 A (*ii*) 2 A (*iii*) 24.3 A, Total current 29.4 A

**2.15.** $50.7 \angle{-65.8°}\ \Omega$

**2.16.**

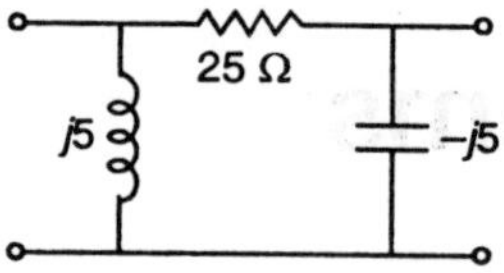

**2.17.** $10 \angle 90°\ \Omega,\ I = 6 \angle{-90°},\ I_r = 0$

## CHAPTER III

**3.1.** 5 A

**3.2.** 2. 44 A

**3.3.** 3.7 A

**3.4.** $X_1 = \sqrt{R_1^2\, R_2\big/(R_1 - R_2)}\ ,\ X_2 = \sqrt{R_2\,(R_1 - R_2)}$

**3.5.** 10.62

**3.6.** (*a*) $V_{TH} = \dfrac{V_1}{2}\,(1 + a + b - ab),\ Z_{TH} = \dfrac{3 - b}{2}$

(*b*) $V_{TH} = \dfrac{\alpha\,R_c + R_b}{R_e + R_b}\,v_s,$

$\qquad Z_{TH} = \big\{R_c\,(R_e + R_b) - (\alpha\,R_c + R_b)\,R_b\big\}\big/(R_e + R_b)$

**3.7.** 20 Ω, 2.22 W

**3.8.** 25 Ω

**3.9.** $\dfrac{1}{13}\,F$

**3.10.** $I_A = 11.15$ A, $I_B = 10$ A, $I_c = 17.7$ A, $I_1 = 27.1$ A, $I_2 = 3.82$ A
$I_3 = 27.2$ A, $P_1 = 100$ W, $P_2 = 6275$ W

**3.11.** $0.65 \angle 75°$ mA

**3.12.** $25.5 \angle{-57.4°}$ Volts

**3.14.** 20.6 A, 17.95 A, 14.7 A

**3.15.** $21.45 \angle{-166°} = V_c,\ \delta\,I = 2.74 \angle{-36°}$

**3.17.** $0.217 \angle 32.5°$

**3.18.** $4.47 \angle{-63°}$

**3.19.** $V_L = V_S - V_C - 2\,I_L$
$\quad I_C = I_L - 2\,V_C$

**3.20.** $3 \angle 15°$

**3.21.** $6.75 \angle 65.7°$ A

**3.22.** $Z_T = 7.07 \angle{-45°},\ 7.07 \angle{-8.1°},\ I_Z = 1 \angle{-53°}$

**3.23.** $5 \angle{-45°}$

**3.24.** 8.66 Ω, 268 Watts

**3.25.** (*a*) $R_L = 16.58\ \Omega\ P_L = 3.56$ (*b*) $R_L = 5\ \Omega\ P_L = 5$ W

**3.26.** $Z = (10 - j10)\ \Omega$

**3.27.** (a) $\dfrac{1 + a + b - ab}{3 - b} V_1$  (b) $\dfrac{\alpha R_c + R_b}{R_e (R_b + R_c) + (1 - \alpha) R_b R_c}$

## CHAPTER IV

**4.1.** $Y = \begin{bmatrix} 3 - j12 & -2 + j8 & -1 + j4 & 0.0 \\ -2 + j8 & 3.66 - j14.664 & -0.666 + j2.664 & -1 + j4.0 \\ -1 + j4.0 & -0.666 + j2.664 & 3.666 - j14.664 & -2 + j8.0 \\ 0.0 & -1 + j4.0 & -2 + j8 & 3 - j12 \end{bmatrix}$

**4.2.** $B_f = \begin{matrix} 1 \quad 3 \quad 4 \quad 5 \ 2 \ 6 \ 7 \\ \begin{bmatrix} -1 & 0 & 0 & 0 & 1 & 0 & 0 \\ 0 & 0 & 0 & -1 & 0 & 1 & 0 \\ -1 & -1 & -1 & 1 & 0 & 0 & 1 \end{bmatrix} \end{matrix}$

**4.3.** $v_1 = 2.325,\ v_2 = 0.988,\ v_3 = 0.193$
$i_1 = -5.35,\ i_2 = 5.35,\ i_3 = 2.964,\ i_4 = 2.385,\ i_5 = 2.385$

**4.4.** $Q_f = \begin{bmatrix} 1 & 1 & 0 & 0 & 0 \\ 1 & 0 & 1 & 0 & 1 \\ 0 & 0 & 0 & 1 & -1 \end{bmatrix}$  $Y_C = \begin{bmatrix} 6 & 2 & 0 \\ 2 & 7 & -2 \\ 0 & -2 & 5 \end{bmatrix}$

$V_2 = 1.34,\ V_3 = 0.99,\ V_4 = 0.79$
$V_1 = 2.32,\ V_2 = 1.34,\ V_3 = 0.99,\ V_4 = 0.79,\ V_5 = 0.19$
$i_1 = -5.35,\ i_2 = 5.35,\ i_3 = 2.964,\ i_4 = 2.385,\ i_5 = 2.385$

**4.5.** $\begin{bmatrix} 6 & 2 & 0 \\ 2 & 7 & -2 \\ 0 & -2 & 5 \end{bmatrix} \begin{bmatrix} V_2 \\ V_3 \\ V_4 \end{bmatrix} = \begin{bmatrix} 10 \\ 8 \\ 2 \end{bmatrix}$, Loop impedance matrix

Loop currents $\ i_1 = -5.35 \begin{bmatrix} 1.08 & 0.33 \\ 0.33 & 1.16 \end{bmatrix}$

$i_2 = 2.385$ Branch Currents as is 4.4.

**4.8.** 8, 56
**4.9.** 12 A
**4.10.** 19.02 V
**4.11.** 0.2 A
**4.12.** $I_1 = 1.58,\ I_2 = 0.616,\ I_3 = 0.42,\ I_4 = 0.08,\ I_5 = 0.334$ A

## CHAPTER V

**5.4.** (a) 210 V  (b) 30 V  (c) 1080 J
**5.5.** (a) 1000 H  (b) 1.25 sec  (c) 281 J

**5.6.** $i_L = 10 \cos 316\, t$

**5.7.** $t_1 = \dfrac{1}{\alpha}$

**5.8.** (i) $i = A\,e^{-t} + Be^{-2t} - \dfrac{15}{4} + \left(\dfrac{5t}{2}\right)$

    (ii) $v = (A + t)e^{-2t} + (B - 5t)e^{-3t}$

**5.9.** $10^{-4}\, e^{-t} (\cos t - \sin t) - 10^{-4} e^{104t}$

**5.10.** (a) 1040 V       (b) 40 $\Omega$

**5.11.** 2.7 M$\Omega$,    9.2 sec

**5.12.** $i = 12.1\{0.953\, e^{-100\, t} + \sin (314\, t - 1.26)\}$, 1.56 $A$

## CHAPTER VI

**6.3.** (a) $2\left(\sin x - \dfrac{1}{2}\sin 2x + \dfrac{1}{3}\sin 3x \cdots\right)$

    (b) $f(x) = \dfrac{\pi}{2} + 1 - \dfrac{4}{\pi}\left(\cos x + \dfrac{1}{3^2}\cos 3x + \dfrac{1}{5^2}\cos 5x + \cdots\right)$

    (c) $v(t) = \dfrac{V_m}{\pi} + \dfrac{V_m}{2}\cos 5\pi t + \dfrac{2V_m}{3\pi}\cos 10\,\pi t$

$$-\dfrac{2V_m}{15\pi}\cos 20\,\pi t + \dfrac{2V_m}{35\pi}\cos 30\,\pi t \cdots$$

    (d) $\dfrac{8}{\pi^2}\left(\sin \pi t - \dfrac{1}{9}\sin 3\pi t + \dfrac{1}{25}\sin 5\pi t \cdots\right)$

    (e) $\dfrac{4}{\pi}\left(\sin \pi t + \dfrac{1}{3}\sin 3\pi t + \dfrac{1}{5}\sin 5\pi t + \cdots\right)$

    (f) $\dfrac{4}{\pi}\left(\cos \pi t - \dfrac{1}{3}\cos 3\pi t + \dfrac{1}{5}\cos 5\pi t \cdots\right)$

**6.4.** $C_n = \begin{cases} 0 & \text{for } n \text{ even} \\ -j2 \exp\left(\dfrac{jn\pi}{2}\right)\Big/ n\pi & \text{for } n \text{ odd} \end{cases}$

**6.5.** (a) $C_n = \begin{cases} -j5 & \text{for } n = 1 \\ 0 & \text{for } n \text{ odd } n \neq 1 \\ \dfrac{-20}{\pi\,(n^2 - 1)} & \text{for } n \text{ even} \end{cases}$

(b) $C_n = \begin{cases} 0 & \text{for } n = 0 \\ -j\dfrac{10}{n\pi} & n \neq 0 \end{cases}$

(c) $C_0 = 5$, $C_n = 12.5 \dfrac{\sin 1.6n\pi - j(1 - \cos 1.6\,n\pi)}{n\pi}$

**6.6.** (i) $\dfrac{\pi}{4} - \dfrac{2}{\pi}\left(\cos t + \dfrac{1}{3^2}\cos 3t + \dfrac{1}{5^2}\cos 5t + \cdots\right) + \sin t - \dfrac{1}{2}\sin 2t$

(ii) $-\pi^2 - 8\left(\cos t + \dfrac{1}{3^2}\cos 3t + \dfrac{1}{5^2}\cos 5t + \cdots\right)$

$$+ \dfrac{2}{\pi}\left[\left(\dfrac{3\pi^2}{1} - \dfrac{4}{1^3}\right)\sin t + \dfrac{\pi^2}{2}\sin 2t + \left(\dfrac{3\pi^2}{3} - \dfrac{4}{3^2}\right)\sin 3t\right.$$

$$\left. + \dfrac{\pi^2}{4}\sin 4t + \cdots\right]$$

(iii) $\dfrac{\sin h\pi}{\pi}\left[1 + 2\displaystyle\sum_{n=1}^{\infty}\dfrac{(-1)^n}{n^2 + 1}(\cos nt - n\sin nt)\right]$

(iv) $-1 + \pi\sin t - \dfrac{1}{2}\cos t + 2\displaystyle\sum_{n=2}^{\infty}\dfrac{\cos nt}{n^2 - 1}$

**6.7.** (i) $\dfrac{2}{\pi}\left[\left(\dfrac{\pi^2}{1} - \dfrac{4}{1^3}\right)\sin t - \dfrac{\pi^2}{2}\sin 2t + \left(\dfrac{\pi^2}{3} - \dfrac{4}{3^3}\right)\sin 3t\right]$

(ii) $\dfrac{2}{\pi}\displaystyle\sum_{n=1}^{\infty}\dfrac{n\left[1 - (-1)^n e^\pi\right]}{n^2 + 1}\sin nt$ ;

$$\dfrac{e^\pi - 1}{\pi} - \dfrac{2}{\pi}\displaystyle\sum_{n=1}^{\infty}\dfrac{1 - (-1)^n e^\pi}{n^2 + 1}\cos nt$$

(iii) $-\dfrac{2}{\pi}\displaystyle\sum_{n=1}^{\infty}\dfrac{\sin 2n\pi t}{n}$ ; $-\dfrac{8}{\pi^2}\displaystyle\sum_{n=1}^{\infty}\dfrac{\cos(2n-1)\pi t}{(2n-1)^2}$

(iv) $\dfrac{3}{2} - \dfrac{4}{\pi^2}\displaystyle\sum_{n=1}^{\infty}\dfrac{1}{(2n-1)^2}\cos\dfrac{(2n-1)\pi t}{2} - \dfrac{2}{\pi}\displaystyle\sum_{n=1}^{\infty}\dfrac{1}{n}\sin\dfrac{n\pi t}{2}$

**6.8.** $4 + 1.5\cos(2t - 3.725)$

**6.9.** $10 + 2.1221\cos t - 0.1604\sin(3t - 3.98) - 0.045\sin(5t - 3.7)$

**6.10.** $596.5$ watt

**6.11.** $10\,\Omega$, $10\,\Omega$, $625$ W, $pf = 1.0$

**6.12.** 12.92 A, 167 watt

**6.13.** Zero

**6.15.** 4.3 A Apprx. 91.8 watts

**6.16.** (a) $9.8 \cos (3t + 0.0768)$

(b) $0.77 + 1.12 \cos (2t + 0.90) + 3.96 \cos (4t - 1.35)$

**6.17.** (a) $\dfrac{A\left(1 - e^{-j\omega}\right)}{j\omega}$ 
(b) $\dfrac{A\left(1 - e^{-a - j\omega}\right)}{a + j\omega}$

**6.18.** (a) $\dfrac{20}{\omega} (\sin 5\omega + \sin 10\omega)$ 
(b) $\dfrac{200}{\omega^2} (\cos 5\omega - \cos 10\,\omega)$

**6.19.** (a) $j \dfrac{10}{\pi t} (\sin 5t + \sin 10t)$ 
(b) $\dfrac{100}{\pi\, t^2} (\cos 5t - \cos 10t)$

**6.20.** (i) $\dfrac{j10\omega}{j\omega + 4}$ 
(ii) $\dfrac{10}{j\omega + 6}$ 
(iii) $\dfrac{10(j\omega + 4)}{20 - \omega^2 + j8\omega}$ 
(iv) $\dfrac{10\, e^{-j3\omega}}{j\omega + 8}$

**6.21.** (i) $0.426 \angle{-12.3°}$ 
(ii) $0.671 \angle{-26.6°}$

## CHAPTER VII

**7.1.** (a) $A\, u(t - t_0) - A\, u(t - t_1)$

(b) $\dfrac{2A}{T} \left[ r(t - t_0) - 2r(t - t_0 - T/2) + r(t - t_0 - T) \right]$

(c) $V \sin \dfrac{2\pi}{T} t \left[ u(t) - u\left( t - \dfrac{T}{2} \right) \right]$

(d) $(2 - t) \{ u(t) - u(t - 2) \}$

(e) $u(t) + u(t - 1) - u(t - 2) - u(t - 3)$

**7.2.** (c) $1.5\, e^{-2t} - 0.5\, e^{-4t} + t\, e^{-4t}$

**7.3.** (a) $\dfrac{1}{2s} - \dfrac{1}{(s + 1)^2} - \dfrac{1}{2(s + 2)}$ 
(b) $\dfrac{1}{s + 1} + \dfrac{2}{s + 2} - \dfrac{2}{s + 3}$

(c) $\dfrac{1}{18s} - \dfrac{3/4}{s + 1} + \dfrac{1}{(s + 2)^3} + \dfrac{3}{2(s + 2)} - \dfrac{1}{6(s + 3)^2} - \dfrac{29}{36(s + 3)}$

(d) $\dfrac{1}{3} \cdot \dfrac{3}{(s + 2)^2 + 3^2}$ 
(e) $\dfrac{17}{2} \cdot \dfrac{1}{s + 3} - \dfrac{7}{2} \cdot \dfrac{1}{s + 1} + \dfrac{1}{(s + 1)^2}$

(f) $\dfrac{3}{s} - \dfrac{3}{s + 2} + \dfrac{4}{(s + 2)^2}$ 
(g) $\dfrac{1}{s} - \dfrac{(s + 4)}{(s + 2)^2 + 4^2}$

(h) $\dfrac{5}{2(s + 1)} + \dfrac{5}{(s + 1)^2} - \dfrac{5}{2} \cdot \dfrac{1}{s + 3}$

$(i)\ \dfrac{1}{s+1} - \dfrac{0.5 - j0.288}{s + 0.5 + j0.8} - \dfrac{0.5 + j0.288}{s + 0.5 - j0.866}$

$(j)\ \dfrac{3}{8s} + \dfrac{1}{4(s+2)} + \dfrac{3}{8(s+4)}$

**7.4.** $(a)\ e^{-t} - e^{-2t}(1+t)$ $\qquad\qquad (b)\ -1 - \dfrac{t^2}{2} + \cosh t$

$(c)\ \dfrac{5}{9} - \dfrac{1477}{225} e^{-3t} + \dfrac{32}{15} t e^{-3t} - 0.71 \cos 2t - 0.4 \sin 2t$

$(d)\ 2t - 1.5 - \dfrac{3}{2} e^{-2t} \cos 2t + t e^{-2t}(\cos 2t + \sin 2t)$

$(e)\ \dfrac{1}{3} e^{-2t} \sin 3t$ $\qquad\qquad (f)\ \dfrac{17}{2} e^{-3t} - e^{-t}\left(\dfrac{7}{2} - t\right)$

$(g)\ \dfrac{1}{2} t \cos t + \dfrac{1}{2} \sin t$ $\qquad (h)\ \dfrac{1}{2}(1 + e^{-2t}) u(t)$

$(i)\ \dfrac{10}{29} \cos 2t + \dfrac{4}{29} \sin 2t - \dfrac{10}{29} e^{-5t}$

$(j)\ 10 e^{-2t} - 5 e^{-t}$ $\qquad\qquad (k)\ \delta'(t) + \delta(t) + \sqrt{2}\, e^{-t} \sin\left(t - \dfrac{\pi}{4}\right)$

$(l)\ \dfrac{3}{2} r\left(t - \dfrac{1}{3}\right) - \dfrac{3}{2\sqrt{2}} \sin \sqrt{2}\left(t - \dfrac{1}{3}\right) u\left(t - \dfrac{1}{3}\right)$

**7.5.** $(a)\ \dfrac{a}{s^2 - a^2}$ $\qquad\qquad\qquad (b)\ \dfrac{2}{(s+a)^2}$

$(c)\ \dfrac{9}{s(s^2 + 9)}$ $\qquad\qquad (d)\ \dfrac{2s^3 + 24s^2 - 14s - 144}{(s^2 + 4s + 13)^2}$

$(e)\ \dfrac{10s + 65}{s^2 + 4s + 13}$ $\qquad\qquad (f)\ \dfrac{s^2 - a^2}{(s^2 + a^2)^2}$

$(g)\ \dfrac{2}{s(s^2 + 4)}$ $\qquad\qquad (h)\ \dfrac{1}{2}\sqrt{\dfrac{\pi}{s^3}}$

$(j)\ \dfrac{s^2}{(s^2 + a^2)^2}$ $\qquad\qquad (k)\ \dfrac{1 + e^{-s\pi}}{s^2 + 1}$

**7.6.** $v(t) = \dfrac{V_0}{2\beta R_1 C_2}\left[e^{-(\alpha - \beta)t} - e^{-(\alpha + \beta)t}\right]$

$$\alpha = \frac{R_1 C_1 + R_2 C_2 + R_2 C_1}{2 R_1 R_2 C_1 C_2} \quad \text{and}$$

$$\beta = \frac{1}{2} \sqrt{\left(\frac{R_1 C_1 + R_2 C_2 + R_2 C_1}{R_1 R_2 C_1 C_2}\right)^2 - \frac{4}{R_1 R_2 C_1 C_2}}$$

**7.7.** $v(t) = \dfrac{V_0}{2 R_1 C_2 \beta}\left[e^{-(\alpha - \beta)t} - e^{-(\alpha + \beta)t}\right]$

$$\alpha = \frac{1}{2}\left[\frac{1}{R_2 C_1} + \frac{1}{R_1 C_1} + \frac{1}{R_1 C_2}\right] \quad \text{and}$$

$$\beta = \frac{1}{2}\left[\left(\frac{1}{R_2 C_1} + \frac{1}{R_1 C_1} + \frac{1}{R_1 C_2}\right)^2 - \frac{4}{R_1 R_2 C_1 C_2}\right]^{\frac{1}{2}}$$

**7.12.** $\dfrac{V}{R}\cos\dfrac{t}{\sqrt{LC}}$

**7.14.** $i(t) = \dfrac{V_0}{\omega L}\, e^{-at}\sin \omega t$ where $a = \dfrac{R}{2L}$, $\omega = \left(\dfrac{1}{LC} - \dfrac{R^2}{4L^2}\right)^{\frac{1}{2}}$

**7.15.** $0.90\, e^{-3.15\, t} - e^{-0.244\, t}\, [0.092 \cos 0.474\, t - 0.632 \sin 0.474\, t]$

**7.16.** $1020\,(e^{-0.155\, t} - e^{-0.645\, t})$

**7.18.** *(a)* $t\, e^{at}$ $\qquad\qquad$ *(b)* $2\, e^{-2t} - e^{-t}$

$\qquad$ *(c)* $\dfrac{1}{2}\left[\sin t + \cos t - e^{-t}\right]$ $\qquad$ *(d)* $e^{-t}\left[\sin t - 1 + \cos t\right] u(t)$

$\qquad$ *(e)* $\dfrac{e^{-at}}{(a-b)(a-c)} + \dfrac{e^{-bt}}{(b-a)(b-c)} + \dfrac{e^{-ct}}{(c-a)(c-b)}$

**7.19.** *(a)* $\dfrac{2}{3} - \dfrac{1}{3}\, e^{-1.5t}$ $\qquad\qquad$ *(b)* $1.2\, e^{-1.5\, t} + 8.8 \cos 2\, t - 1.6 \sin 2\, t$

**7.20.** *(a)* $2 + 6\, e^{-2t} - 8\, e^{-3t}$ $\qquad$ *(b)* $2t + \dfrac{1}{3} - 3 e^{-2t} + \dfrac{8}{3}\, e^{-3t}$

**7.21.** $\dfrac{2}{3}A$ initial value and $2\,A$ Final value

**7.22.** $i_{10^+} = 10\,A\ \ i_{20^+} = 0$, $i_{1ss} \simeq 17\,A\ \ i_{2ss} \simeq 10\,A$

**7.23.** $\dfrac{2}{13} - \dfrac{2}{13}e^{-2t}\cos 3t - \dfrac{8}{39}\, e^{-2t}\sin 3t$

**7.24.** *(a)* $2\, e^{-2t} + 6 \sin 2\, t - 2 \cos 2\, t$

$\qquad$ *(b)* $3.2\, e^{-2t} - (4t - 7.6)\sin 2t - (12t + 3.2)\cos 2t$

**7.25.** *(a)* $1, 0$ $\qquad\qquad$ *(b)* $1, 0$ $\qquad\qquad$ *(c)* $\infty, 0$ $\qquad\qquad$ *(d)* $0, 3$

## CHAPTER VIII

**8.1.** $\dfrac{s^2 + 1.1s + 0.1}{s^2 + 1.3s + 0.1}$

**8.4.** $Z = \begin{bmatrix} \dfrac{13}{7} & \dfrac{2}{7} \\[2mm] \dfrac{2}{7} & \dfrac{3}{7} \end{bmatrix}$  $Y = \begin{bmatrix} \dfrac{3}{5} & \dfrac{-2}{5} \\[2mm] \dfrac{-2}{5} & \dfrac{13}{5} \end{bmatrix}$

**8.5.** $Z = \begin{bmatrix} -0.4 & 0.4 \\ -3.2 & 1.2 \end{bmatrix}$  $Y = \begin{bmatrix} 1.5 & -0.5 \\ 4 & -0.5 \end{bmatrix}$

**8.6.** $0.13125,\ Y_{22} = 0.7\ v$

**8.7.** $Y = \begin{bmatrix} 0.1875 & -0.006 \\ 5 & 05 \end{bmatrix}$  output impedance $1.9\ \Omega$

**8.8.** $Z = \begin{bmatrix} 720 & 700 \\ 971 & 1001 \end{bmatrix} \Omega$  $\dfrac{V_2}{V_1} = 5.34$  $\dfrac{i_2}{i_1} = -0.967$

$Z_i = 43\ \Omega\ Z_{out} = 445\ \text{K}\,\Omega$

**8.9.** $Y_{11} = Y_{22} = \dfrac{(2s + 1)(8s^2 + 12s + 1)}{4(s + 1)(4s + 1)},$

$Y_{12} = Y_{21} = -\dfrac{16s^3 + 16s^2 + 4s + 1}{4(s + 1)(4s + 1)}$

**8.10.** $A = 0.5\ B = -5\ C = -0.075\ D = -1.25$
$V_1 = 0.5\ V_2 - 5\ I_2\ I_1 = -0.075\ V_2 - 1.25\ I_2$

**8.11.** $Z = \begin{bmatrix} s + \dfrac{1}{2s} & \dfrac{1}{2s} \\[3mm] \dfrac{1}{2s} & s + \dfrac{1}{2s} \end{bmatrix}$

**8.12.** $Y = \begin{bmatrix} 0.2 & -0.24 \\ -0.12 & 0.194 \end{bmatrix}$

**8.13.** $Y = \dfrac{1}{11}\begin{bmatrix} 4 & -3 \\ -3 & 5 \end{bmatrix}\begin{bmatrix} 5/3 & 11/3 \\ 1/3 & 4/3 \end{bmatrix},\ \begin{bmatrix} 4/3 & -11/3 \\ -1/3 & 5/3 \end{bmatrix}$

$(d)\ h_{11} = \dfrac{11}{4},\ h_{12} = \dfrac{3}{4},\ h_{21} = -\dfrac{3}{4}\ h_{22} = 1/4$

(e) $g_{11} = \dfrac{1}{5}$, $g_{12} = -\dfrac{3}{5}$, $g_{21} = \dfrac{3}{5}$ $g_{22} = \dfrac{11}{5}$

**8.14.** $I_1 = \dfrac{23}{5} V_1 + \dfrac{9}{10} V_2$, $I_2 = -\dfrac{22}{5} V_1 - \dfrac{3}{5} V_2$

**8.15.** (a) $Z_{11} = 8.4$ $Z_{12} = 7.2$ $Z_{21} = 7.2$ $Z_{22} = 9.6$ same for (b)

**8.16.** $Z_{11} = 5.71$, $Z_{12} = -4.3$, $Z_{21} = 2.14$, $Z_{22} = 2.14$

**8.17.** $Z_{11} = 50.$ $Z_{12} = 5$, $Z_{21} = -250$ $Z_{22} = 25$

$Y_{11} = 0.2$ $Y_{12} = -0.3$ $Y_{21} = -0.1$ $Y_{22} = 0.4$

**8.18.** (a) $h_{11} = 100\ \Omega$, $h_{12} = 0.2$, $h_{21} = 10$, $h_{22} = 0.04\ \mho$

(b) $h_{11} = 5\ \Omega$ $h_{12} = 1.5$, $h_{21} = -0.5$, $h_{22} = 0.25\ \mho$

**8.19.** $0.421\ \Omega$, $0.857\ \Omega$

**8.20.** $5\ \mho$, $-3\ \mho$, $0.15$

**8.21.** $Z_{11} = Z_{12} = Z_{21} = Z_{22} = 2\ \Omega$, $y$-parametter do not exist

$h_{11} = 0$ $h_{12} = 1$, $h_{21} = -1$ $h_{22} = 1$

**8.22.** $\begin{bmatrix} 6 & 10 \\ 10 & 12 \end{bmatrix}$ since series connection

**8.23.** $3.7\ \Omega$

**8.24.** $V_2 = -5V$ $R_L = 40\ K\Omega$, $Z_{11} = -500$, $Z_{12} = 15$, $Z_{21} = -10^6$ $Z_{22} = 10^4$

**8.25.** 65 Volt

**8.26.** $Z_{in} = 10\ (s^2 + 25000\ s + 10^8)\ /\ (s^2 + 10^4 s + 10^8)$

$-5, -20, -5 \pm j\ 8.66\ K$ rad/sec.

**8.27.** $\dfrac{s^2 + 5s}{10s^2 + 40s + 80}$, $\dfrac{2s + 6}{5s^2 + 60}$, $\dfrac{s + 2}{5s^2 + 20s + 5}$

**8.28.** $\dfrac{3}{2\sqrt{6}}$, $\sqrt{6}$

**8.29.** $1\ \Omega$, $1\ H$, $1\ F$

**8.30.** (a) $K > \dfrac{1}{2}$ (b) $K = \dfrac{1}{2}$ real part vanishes.

**8.31.** (a) 0,0,3 (b) 0, 0, 3 (c) 2, 0, 1 (d) 2, 0, 3

**8.32.** (i) Yes (ii) No (iii) No (iv) No

**8.33.** $K < 5/2$

**8.34.** 8.2, 10.68, 8.31 cm.

## CHAPTER IX

**9.1.** 581 KHz, 261, 28.71 kV

**9.2.** 3.3 μF, 637 Ω, 3.04 H

**9.3.** 74.23 Hz

**9.4.**

30.7 Ω, 0.05 mH, 10.3

**9.5.** 31.8 Hz, 100 $\Omega$

**9.6.** $\dfrac{L}{CR}$, 0.001 V, 0.0068 V

**9.7.** 1.055 H, 17.3 Hz

**9.8.** 222 K$\Omega$, $f_0 = 26$ KHz, $Q = 54.4$ Band width 478 Hz

**9.9.** 153.7 K$\Omega$, 37.6, BW = 691 Hz

**9.10.** 4.7 M$\Omega$

**9.11.** $f_{0s} = \dfrac{1}{2\pi\sqrt{L_1 C_1}}$, $f_{0p} = \dfrac{1}{2\pi\sqrt{L_T C}}$ where $L_T = L_1 + L_2$

$f_{0s} = 712$ Hz $\qquad f_{0p} = 225$ Hz

**9.12.** $f_{0s} = \dfrac{1}{2\pi\sqrt{L_T C}}$ where $L_T = \dfrac{L_1 L_2}{L_1 + L_2}$ $f_{0p} = \dfrac{1}{2\pi\sqrt{L_1 C_1}}$

$f_{0s} = 750$ Hz $\qquad f_{0p} = 237$ Hz

**9.13.** 160 W, $L = 15.91$ mH, $C = 15.9$ µF, $f_1 = 11.9$ KHz $f_2 = 12.1$ KHz

**9.14.** 4 K rad/sec, 40, $v_2 = 176 \cos 4000\,t$, $i_1 = 880 \sin 4000t$

$i_2 = 4.4 \cos 4000\,t$ $i_3 = -880 \sin 4000t$ mA, 20 mW, 4 mJ.

**9.15.** $Z = R = 200\ \Omega$, 2000 $\Omega$, 2000 $\Omega$, 10

**9.16.** 6.75 µF 23 A, 0.0414 A

## CHAPTER X

**10.1.** (a) No $\quad$ (b) Yes $\quad$ (c) Yes $\quad$ (d) No $\quad$ (e) Yes
(f) Yes $\quad$ (g) No $\quad$ (h) No $\quad$ (i) Yes

**10.2.** (a) No $\quad$ (b) No $\quad$ (c) Yes $\quad$ (d) No $\quad$ (e) Yes
(f) No $\quad$ (g) No $\quad$ (h) No $\quad$ (i) Yes

**10.3.** (a) RC admittance $\qquad$ (b) RL admittance
(c) RC impedance $\qquad$ (d) RL admittance
(e) RC impedance $\qquad$ (h) not interlaced
(g) RL impedance
(i) LC impedance

**10.4.**

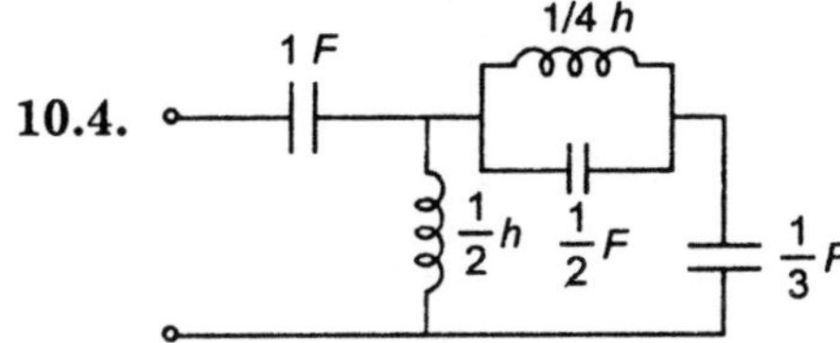


**10.5.** $C_0 = \dfrac{1}{9}$ F $R_1 = \dfrac{3}{4}\ \Omega$ $C_1 = \dfrac{1}{3}$ F $R_\infty = 3\ \Omega$

**10.6.** $C_0 = 3$ F $L_1 = \dfrac{3}{2}$ h $C_1 = \dfrac{2}{9}$ F

**10.7.** $\dfrac{3}{2}$ h in series $\dfrac{4}{3}$ F

**10.8.** $C_0 = \dfrac{1}{4}$ F $C_1 = \dfrac{1}{2}$ F $R_1 = 1\ \Omega,\ R_\infty = 2\ \Omega$

cauer $\to 2\ \Omega,\ \dfrac{1}{6}$ F, $9\ \Omega,\ \dfrac{1}{12}$ F $\dfrac{9}{2}\ \Omega$

**10.9.** *(a)*   1 Ω  1 h  1 h  1 F  1 F      *(b)* $\dfrac{9}{2}\Omega$  $\dfrac{1}{2}F$  3 h  6 h

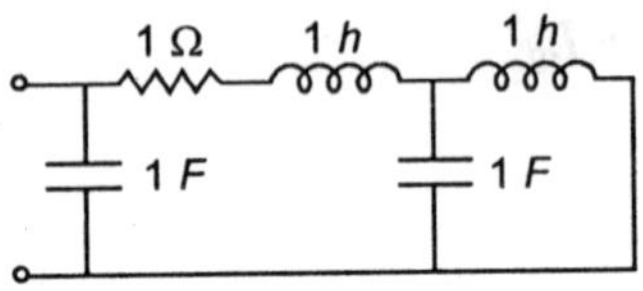

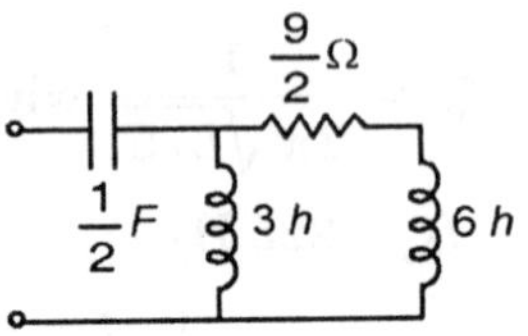

*(c)*   1 h  4.9 h  $\dfrac{1}{7}F$  $\dfrac{5}{63}F$

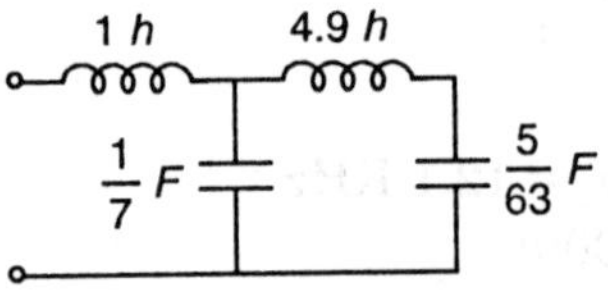

**10.10.** $Z(s) = \dfrac{4(s+1)(s+3)}{s(s+2)}$

**10.11.** $Z(s) = \dfrac{3(s+2)(s+4)}{(s+1)(s+3)}$

**10.12.** $L_0 = \dfrac{8}{35}\ h\ ,\ R_1 = \dfrac{32}{3}\ \Omega\ L_1 = \dfrac{8}{3}h,\ R_2 = 24\ \Omega,\ L_2 = 4\ H$

$R_\infty = 1\ \Omega$

## CHAPTER XI

**11.1.** $\begin{bmatrix} \dfrac{di_L}{dt} \\[2ex] \dfrac{dv_C}{dt} \end{bmatrix} = \begin{bmatrix} -\dfrac{R}{L} & -\dfrac{1}{L} \\[2ex] \dfrac{1}{C} & 0 \end{bmatrix} \begin{bmatrix} i_L \\[2ex] v_C \end{bmatrix} = \begin{bmatrix} \dfrac{1}{L} \\[2ex] 0 \end{bmatrix} V(t)$

**11.2.** $\begin{bmatrix} \dfrac{di_{L1}}{dt} \\[2ex] \dfrac{di_{L2}}{dt} \\[2ex] \dfrac{dv_c}{dt} \end{bmatrix} = \begin{bmatrix} -\dfrac{R}{L_1} & 0 & -\dfrac{1}{L_1} \\[2ex] 0 & 0 & -\dfrac{1}{L_2} \\[2ex] \dfrac{1}{C} & \dfrac{1}{C} & 0 \end{bmatrix} \begin{bmatrix} i_{L1} \\[2ex] i_{L2} \\[2ex] v_c \end{bmatrix} + \begin{bmatrix} \dfrac{1}{L1} \\[2ex] \dfrac{1}{L2} \\[2ex] 0 \end{bmatrix} [v_1(t),\ v_2(t),\ 0]$

**11.3.**
$$\begin{bmatrix} \dfrac{di_L}{dt} \\[2ex] \dfrac{dv_c}{dt} \end{bmatrix} = \begin{bmatrix} -\dfrac{R_1}{L} & -\dfrac{1}{L} \\[2ex] \dfrac{1}{C} & -\dfrac{1}{R_2\,C} \end{bmatrix} \begin{bmatrix} i_L \\[2ex] v_c \end{bmatrix} + \begin{bmatrix} \dfrac{1}{L} \\[2ex] 0 \end{bmatrix} v_1(t)$$

**11.4.**
$$\begin{bmatrix} \dfrac{dv_{C1}}{dt} \\[2ex] \dfrac{dv_{C2}}{dt} \end{bmatrix} = \begin{bmatrix} -\dfrac{R_1+R_2}{C_1\,R_1\,R_2} & \dfrac{1}{C_1\,R_2} \\[2ex] \dfrac{1}{C_2\,R_2} & -\dfrac{1}{C_2\,R_2} \end{bmatrix} \begin{bmatrix} v_{C1} \\[2ex] v_{C2} \end{bmatrix} + \begin{bmatrix} 1/C_1\,R_1 \\[2ex] 0 \end{bmatrix} v(t)$$

**11.5.**
$$\begin{bmatrix} \dfrac{di_L}{dt} \\[2ex] \dfrac{dv_{C2}}{dt} \end{bmatrix} = \begin{bmatrix} -\dfrac{R_2}{L} & \dfrac{1}{L} \\[2ex] -\dfrac{1}{C_1+C_2} & -\dfrac{1}{R_1\,(C_1+C_2)} \end{bmatrix} \begin{bmatrix} i_L \\[2ex] v_{C_2} \end{bmatrix} + \begin{bmatrix} 0 & 0 \\[2ex] \dfrac{1}{R_1\,(C_1+C_2)} & \dfrac{C_1}{C_1+C_2} \end{bmatrix} \begin{bmatrix} v(t) \\[2ex] v(t) \end{bmatrix}$$

**11.6.**
$$\begin{bmatrix} \dfrac{di_L}{dt} \\[2ex] \dfrac{dv_C}{dt} \end{bmatrix} = \begin{bmatrix} -\dfrac{R_2}{L}\left(1-\dfrac{k}{R_1}\right) & \dfrac{1}{L} \\[2ex] -\dfrac{1}{C} & -\dfrac{1}{R_1\,C} \end{bmatrix} + \begin{bmatrix} 0 \\[2ex] \dfrac{1}{C} \end{bmatrix} \begin{bmatrix} i_s \end{bmatrix}$$

**11.7.**
$$\begin{bmatrix} \dfrac{di_L}{dt} \\[2ex] \dfrac{dv_C}{dt} \end{bmatrix} = \begin{bmatrix} 0 & -\dfrac{1}{L} \\[2ex] \dfrac{1}{C} & -\dfrac{1}{RC} \end{bmatrix} \begin{bmatrix} i_L \\[2ex] v_C \end{bmatrix} + \begin{bmatrix} \dfrac{1}{L} \\[2ex] -\dfrac{1}{RC} \end{bmatrix} [v_2 \; v_1 + v_2]$$

**11.8.**
$$\begin{bmatrix} \dfrac{di_L}{dt} \\[2ex] \dfrac{dv_{C1}}{dt} \\[2ex] \dfrac{dv_{C2}}{dt} \end{bmatrix} = \begin{bmatrix} 0 & \dfrac{1}{L} & -\dfrac{1}{L} \\[2ex] -\dfrac{1}{C_1} & -\dfrac{1}{R_1 C_1} & 0 \\[2ex] \dfrac{1}{C_2} & 0 & -\dfrac{1}{R_2\,C_2} \end{bmatrix} \begin{bmatrix} i_L \\[2ex] v_{C1} \\[2ex] v_{C2} \end{bmatrix} + \begin{bmatrix} 0 \\[2ex] \dfrac{1}{R_1 C_1} \\[2ex] 0 \end{bmatrix} v(t)$$

**11.9.**
$$\begin{bmatrix} \dfrac{di_L}{dt} \\[2ex] \dfrac{dv_C}{dt} \end{bmatrix} = \begin{bmatrix} -\dfrac{R_2}{L} & \dfrac{1}{L} \\[2ex] -\dfrac{1}{C} & -\dfrac{1}{R_1 C} \end{bmatrix} \begin{bmatrix} i_L \\[2ex] v_C \end{bmatrix} + \begin{bmatrix} 0 \\[2ex] \dfrac{1}{C} \end{bmatrix} i_s$$

**11.10.**
$$\begin{bmatrix} \dfrac{di_L}{dt} \\[2mm] \dfrac{dv_C}{dt} \end{bmatrix} = \begin{bmatrix} -\dfrac{R}{L} & \dfrac{1}{L} \\[2mm] -\dfrac{1}{C} & 0 \end{bmatrix} \begin{bmatrix} i_L \\[2mm] v_C \end{bmatrix} + \begin{bmatrix} 0 \\[2mm] \dfrac{1}{C} \end{bmatrix} i_s$$

**11.11.**
$$\begin{bmatrix} \dfrac{di_L}{dt} \\[2mm] \dfrac{dv_{C2}}{dt} \end{bmatrix} = \begin{bmatrix} -\dfrac{R}{L} & -\dfrac{1}{L} \\[2mm] \dfrac{1}{C_2} & -\dfrac{1}{R_1 C_2} \end{bmatrix} \begin{bmatrix} i_L \\[2mm] v_C \end{bmatrix} + \begin{bmatrix} \dfrac{1}{L} \\[2mm] 0 \end{bmatrix} v(t)$$

**11.12.**
$$\begin{bmatrix} \dfrac{di_L}{dt} \\[2mm] \dfrac{dv_C}{dt} \end{bmatrix} = \begin{bmatrix} 0 & 0 \\[2mm] 0 & -\dfrac{R_1 + R_2}{R_1 R_2 C} \end{bmatrix} \begin{bmatrix} i_L \\[2mm] v_C \end{bmatrix} + \begin{bmatrix} \dfrac{1}{L} & -\dfrac{1}{L} \\[2mm] \dfrac{1}{CR_1} & \dfrac{1}{CR_2} \end{bmatrix} \begin{bmatrix} v_1 \\[2mm] v_2 \end{bmatrix}$$

**11.13.**
$$\begin{bmatrix} x_1 \\[2mm] x_2 \end{bmatrix} = \begin{bmatrix} 1 + 3e^{-t} - 3e^{-2t} \\[2mm] -\dfrac{1}{2} - 3e^{-t} + \dfrac{9}{2} e^{-2t} \end{bmatrix}$$

**11.14.**
$$\begin{bmatrix} x_1 \\[2mm] x_2 \end{bmatrix} = \begin{bmatrix} 1 \\[2mm] \dfrac{15}{2}\left(1 - e^{-0.4t}\right) \end{bmatrix}$$

**11.15.** (*i*) Series
$$\begin{bmatrix} 1 - 2t^2 & 2t - 6t^2 \\[2mm] -t + 3t^2 & 1 - 3t + 7t^2 \end{bmatrix}$$

(*ii*) Laplace
$$\begin{bmatrix} 2e^{-t} - e^{-2t} & -e^{-t} + e^{-2t} \\[2mm] 2e^{-t} - 2e^{-2t} & 2e^{-2t} - e^{-t} \end{bmatrix}$$

**11.16.**
$$\begin{bmatrix} \dot{x}_1 \\[2mm] \dot{x}_2 \end{bmatrix} = \begin{bmatrix} 0 & 1 \\[2mm] -6 & -5 \end{bmatrix} \begin{bmatrix} x_1 \\[2mm] x_2 \end{bmatrix} + \begin{bmatrix} 0 \\[2mm] 1 \end{bmatrix} u(t)$$

$$= \begin{bmatrix} -2 & 0 \\[2mm] 0 & -3 \end{bmatrix} \begin{bmatrix} x_1 \\[2mm] x_2 \end{bmatrix} + \begin{bmatrix} 1 \\[2mm] -1 \end{bmatrix} u(t)$$

**11.17.**
$$\begin{bmatrix} \dot{x}_1 \\ \dot{x}_2 \\ \dot{x}_3 \end{bmatrix} = \begin{bmatrix} -1 & 0 & 0 \\ 0 & -2 & 0 \\ 0 & 0 & -4 \end{bmatrix} \begin{bmatrix} x_1 \\ x_2 \\ x_3 \end{bmatrix} + \begin{bmatrix} 1 \\ -1/2 \\ 1/6 \end{bmatrix} u(t)$$

**11.18.** $A = \begin{bmatrix} 1 & 2 \\ -3 & -4 \end{bmatrix}\ e^{At} = \begin{bmatrix} -3\,e^{-t} - 2\,e^{-2t} & 2\left(e^{-t} - e^{-2t}\right) \\ 3\left(e^{-2t} - e^{-t}\right) & 3\,e^{-2t} - 2\,e^{-t} \end{bmatrix}$

## CHAPTER XII

**12.1.** $R_0 = 120\ \Omega$

**12.2.** $R_{I_1} = 114\ \Omega\ R_{I_2} = 285\ \Omega$

**12.3.** $Z_0 = 375\ \Omega\ Z_{in} = 375\ \Omega$

**12.4.** $R_0 = 255\ \Omega$, loss $= 18.35$ dB

**12.5.** (a) $n = 1\ R_2 = \infty\ R_1 = 0$        (b) $n = 3.16\ R_2 = 92.6\ R_1 = 432\ \Omega$

**12.6.** $R_0 = 100\ \Omega$, 14 dB

**12.7.** T-Network $R_1 = 2\,R_0 \tanh \dfrac{\alpha}{2}$, $R_2 = R_0/\sinh\alpha$

$\pi$ - Network $R_1 = R_0 \sinh \alpha$, $R_2 = \dfrac{R_0}{2} \coth \alpha/2$

**12.8.** $IL = 26$ dB

**12.9.** T-network 18.04 dB, $\pi$-network 9.54 dBs

Total 27.58 dB

**12.10.** L facing source $R_1 = 100\ \Omega\ R_2 = 300\ \Omega$, 4.77 dB

**12.11.** $R_1 = 1.2\ \text{k}\Omega\ R_2 = 75\ \Omega$

**12.12.** Symmetrical $T\ R_1 = 66.6\ \Omega\ R_2 = 133.3\ \Omega$

Symmetrical $\pi\ R_1 = 75\ \Omega\ R_2 = 300\ \Omega$

**12.13.** $\alpha = 2.2$ neper $\beta = 1.46$ radians, $\beta = \pi$ at 10 KHz

$$T \to \frac{L}{2} = 13.25\ \text{mH}\ C = 0.106\ \mu\text{F}$$

$$\pi \to L = 26.5\ \text{mH}\ \frac{C}{2} = 0.053\ \mu\text{F}$$

$$f = 7\ \text{KHz}$$

**12.14.** T-section $\dfrac{L}{2} \simeq 5\ \text{mH}$, $C = 0.0398\ \mu\text{F}$

$$m = 0.3122,\ \text{m-derived T-section}\ \frac{2C}{m} = 0.255\ \mu\text{F}$$

$$\frac{L}{m} = 31.87\ \text{mH}\ \frac{4\,m}{1 - m^2}\,C = 0.055\ \mu\text{F}$$

T-half section $m = 0.6$  $\dfrac{2C}{m} = 0.1327\ \mu F$, $\dfrac{2L}{m} = 33.17\ mH$

$$\frac{2m}{1-m^2}\,C = 0.074\ \mu F$$

$\pi$-section  $\dfrac{C}{m} = 0.1275\ \mu F$  $\dfrac{2L}{m} = 63.74\ mH$,

$$\frac{4m}{1-m^2}\,L = 13.75\ mH$$

Half $\pi$-section with $m = 0.6$

$\dfrac{2C}{m} = 0.1327\ \mu F$,  $\dfrac{2m}{1-m^2}\,L = 18.65$,  $\dfrac{2L}{m} = 33.17\ mH$

**12.15.** $L_1 = 159\ mH$, $C_1 = 0.026\ \mu F$ $L_2 = 20.8\ mH$, $C_2 = 0.637\ \mu F$

**12.16.** $R_k = 1000\ \Omega$ $f_C = 5KHz$ $\alpha = 0$ $\beta = 0.822$ radians

**12.17.** $C = 0.0884\ \mu F$ $L = 22.1\ mH$, $m = 0.28$

**12.18.** $C = 7.96\ nF$ $L = 2\ mH$ $m = 0.436$

**12.19.**  $\dfrac{mL_1}{2} = 47.7\ mF$  $\dfrac{2C_1}{m} = 0.087\ \mu F$,  $\dfrac{1-m^2}{4m}\,L_1 = 42.4\ mH$

$\dfrac{4mC_1}{1-m^2} = 0.0975\ \mu F$  $\dfrac{L_2}{m} = 34.67\ mH$, $f\infty_1 = 1.903\ KHz$

$f\infty_2 = 3.153\ KHz$

**12.20.** $L_1 = 6.6\ mH$ $C_1 = 0.0398\ \mu F$, $L_2 = 9.9\ mH$, $C_2 = 0.0265\ \mu F$

**12.21.** $2\ KHz$

## CHAPTER XIII

**13.1.** $\omega_0 = \dfrac{1}{RC}$, $Q = \dfrac{1}{2}$

**13.2.** $C_1 = 2.4$, $C_2 = 2.6$, $C_3 = 0.33$

**13.3.** $\sqrt{2}/RC$

**13.4.** $C_1 = 2$, $C_2 = 1$, $R_1 = 0.5$, $R_2 = 0.5$

**13.6.** $T(s) = \dfrac{R_3\ R_L}{R_1\ R_L + R_1\ R_2 + R_1\ R_3 + R_3\ R_L + R_2\ R_3}$

**13.7.** $5, 4$

**13.8.** $R_3 = \dfrac{2}{3}$ $4.8 = K$

## CHAPTER XIV

**14.1.**

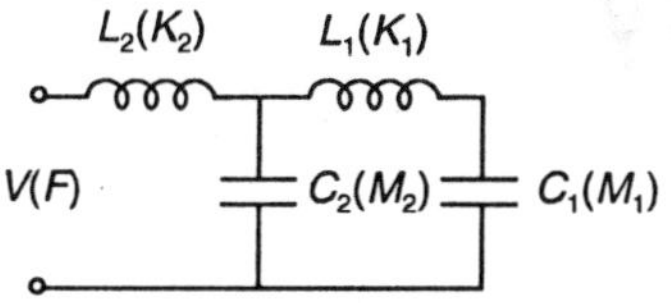

**14.2.**

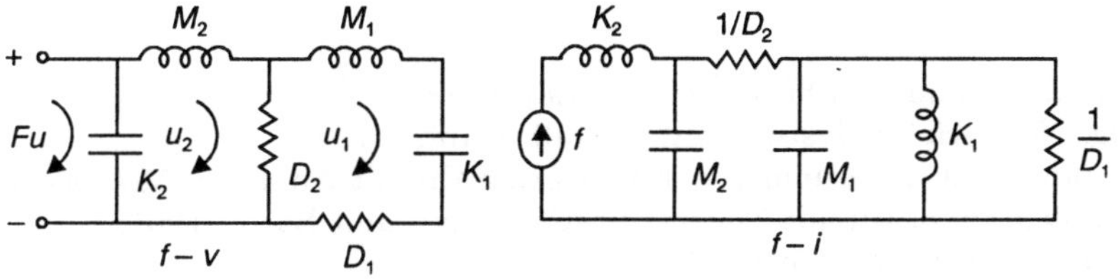

**14.3.**

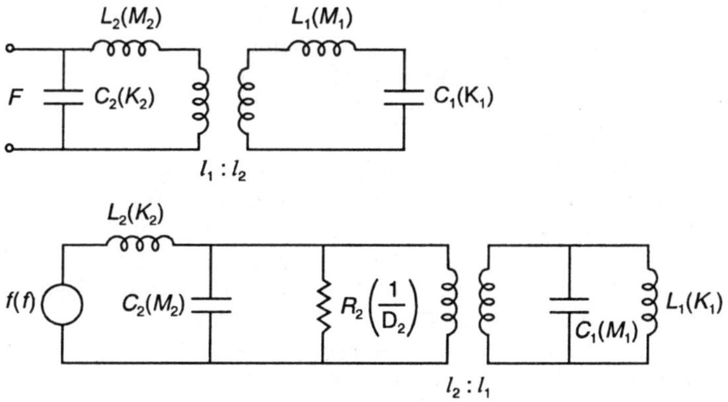

**14.4.** (b) $f = \dfrac{1}{K} \displaystyle\int_0^t (u_2 - u_1)\, dt = M\dfrac{du_1}{dt} + Du_1$

$$v = \dfrac{1}{C} \int_0^t (i_2 - i_1)\, dt = L\dfrac{di_1}{dt} + Ri_1$$

**860**  *Network Analysis and Synthesis*

| | | | | | |
|---|---|---|---|---|---|
| **493.** (c) | **503.** (b) | **513.** (c) | **523.** (b) | **533.** (d) | **543.** (a) |
| **494.** (b) | **504.** (b) | **514.** (a) | **524.** (a) | **534.** (c) | **544.** (a) |
| **495.** (c) | **505.** (b) | **515.** (b) | **525.** (d) | **535.** (a) | **545.** (c) |
| **496.** (b) | **506.** (d) | **516.** (b) | **526.** (c) | **536.** (c) | **546.** (d) |
| **497.** (c) | **507.** (c) | **517.** (a) | **527.** (d) | **537.** (a) | **547.** (a) |
| **498.** (b) | **508.** (c) | **518.** (a) | **528.** (c) | **538.** (d) | **548.** (d) |
| **499.** (a) | **509.** (a) | **519.** (b) | **529.** (a) | **539.** (b) | **549.** (b) |
| **500.** (b) | **510.** (b) | **520.** (b) | **530.** (b) | **540.** (b) | |
| **501.** (b) | **511.** (a) | **521.** (d) | **531.** (b) | **541.** (a) | |
| **502.** (d) | **512.** (c) | **522.** (d) | **532.** (c) | **542.** (c) | |

# Bibliography


1. F.F. Kuo, 'Network Analysis and Synthesis', John Wiley 1999.
2. M.E. Van Valkenburg, Network Analysis, Prentice Hall of India 1974.
3. Hayt, William H., Jr. and Jack E. Kemmerly, Engineering Circuit Analysis, McGraw Hill.
4. Balabanian N., Fundamentals of Circuit Theory, Boston, Allyn and Bacon 1961.
5. Scott, R.E., Linear Circuits, Addison Wesley 1960.
6. Balabanian & Bickart, Electric Network Theory, John Wiley New York 1969.
7. Seshu S. and Reed MB., Linear graphs and electrical Networks, Addison-Wesley 1961.
8. Balabanian N. and Seshu S., Linear Network Analysis, John Wiley & Sons 1959.
9. Dosoer C.A. and Kuh E.S., Basic Circuit Theory, McGraw Hill 1969.
10. Guillemin, E.A., Introductory Circuit Theory, John Wiley & Sons 1963.
11. Sinha N.K., Linear Systems, John Wiley & Sons 1991.
12. J.D. Irwin, Basic Engineering Circuit Analysis, Macmillan New York 1984.
13. A. Papoulis, Circuits and Systems A Modern Approach, Holt, Rinehart & Winston N.Y. 1980.
14. A Papoulis, The Fourier Integral and it's Applications, McGraw Hill N.Y. 1962.
15. Hans Goodman, Circuit Theory & Techniques, Vol II, John Wiley & Sons 1986.
16. Atre V.K., Network Theory & Filter Design, New Age International Publishers 1980.
17. Louis Weinberg, Network Analysis & Synthesis, McGraw Hill N.Y. 1962.
18. Van Valkenburg, M.E., Introduction to Modern Network Synthesis, John Wiley & Sons N.Y. 1960.
19. Tuttle D.F., Network Synthesis, Vol I, John Wiley & Sons N.Y. 1958.
20. Storer J.E., Passive Network Synthesis, McGraw Hill N.Y. 1957.
21. Balabanian N., 'Network Synthesis', Prentice Hall 1958.
22. Reza, F.M. and Seely S., Modern Network Analysis, McGraw Hill N.Y. 1959.
23. Guillemin, E.A., Synthesis of Passive Networks, John Wiley & Sons N.Y. 1957.
24. Calahan DA, Modern Network Synthesis, Hayden N.Y. 1964.
25. Ryder J.D., Networks Lines and Fields, Prentice Hall India New Delhi 1974.
26. Gupta S.C., Transform and State Variable Methods in Linear Systems, John Wiley & Sons N.Y. 1961.
27. Daryanani G., Principles of Active Network Synthesis, John Wiley & Sons 1969.
28. Mitra S.K., Analysis and Synthesis of Linear Active Networks, John Wiley & Sons N.Y. 1969.
29. Ghausi, M.S. and Laker K.R., Modern Filter Design, Prentice Hall NJ 1981.
30. Huelsman L.P., "Theory and Design of Active RC Circuits", McGraw Hill N.Y. 1968.
31. Ronald N.Bracewell, "The Fourier Transform and it's applications" McGraw Hill 2nd Edition 1986
32. Sinha Naresh K., "Linear Syatems", John Wiley & Sons 1991
33. Tripathi A.N., Linear Systems Analysis Revised Second Edition New Age International Publishers New Delhi, India 2005
34. Jury E.I. "Theory and Applications of z-Transforms", John Wiley & Sons 1964
35. Kuo B.C., "Automatic Control Systems", Prentice Hall 1962
36. Freeman H. Discrete-time Systems John Wiley 1965
37. David K.Cheng, "Analysis of Linear Systems", Addison Wesley Publishing Co.

# Index


A matrix 500
ABCD parameters 330
Active devices 623
Active RC filters 633
Active power 212
Addition of matrices 737
Admittance
   driving point 311
   transfer 312
All pass filter 794
All pass functions 794, 687
Analogous system 658
Analysis of transformer 184
Approximation 610
   Butterworth 610
   Chebychev 615
   elliptic 611
   inversechebychev 622
Attenuation constant 547
   Attenuators 543
   T-type 551
   pi type 553
   L-type 555

Band elimination filter 583
Band pass filter 579, 638
Band width 402
Bessel filter 622
Bilateral 10
Bisection theorem 551
Bode plot 676
Bridged-T network 314, 555
Butterworth filter 636

Canonical form 467
Capacitance 5

Cascade connection 349
Cauer form I 461
Cauer form II 463
Causality I 240
Cayley Hamilton
   theorem 526
Chain parameters 330
Characteristic impedance 544, 570
Charge 1
Charge conservation 47
Chebychev polynomial 616
Classical filter 568
Co-efficient of coupling 26
Co-efficient of Fourier
   series 193
Co-efficient matching
   technique 633
Cofactor 736
Column matrix 733
Compensation theorem 105
Complex frequency 263
Composite filter 601
Concept of state 500
Continued fraction
   expansion 436
Continuous time system 12
Constant-k filter 575
Controlled source 32, 623, 628
Conventional current 45
Convolution integral 282
Coupled coils 35
Cramer's rule 56
Critical frequency 467
Critical resistance 177, 378
Cut-off frequency 575
Cut set 126

admittance matrix 142
equations 141
matrix 126

D' Alembert's principle 652
Damped sinusoid 250
Damping co-efficient 378
Damping factor 415
Delta (impulse) function 225, 15
Delta-star transformation 48, 51
Dependent source 32, 708
Deterministic system 12
Differential equations 160
Homogeneous 161
Non-homogeneous 164
Differentiator 625
Discrete times convolution 719
Discrete time systems 12
Dirac-delta (impulse) function 15, 225
Dirichlet Condition 221
Dot convention 36
Driving point
admittance 312
immittance 311, 312
impedance 311
Duality 67, 232

Effective value 211
Elliptic filters 622
Energy 1
Energy conservation 45
Equivalent source network 36
Even function 200
Exponential Fourier
Series 208

Faltung (Convolution) 282
integral
Faraday 4
Final value theorem 273
Force-current analogy 664
Force-voltage analogy 663
Forest 120
Foster form I 456
Foster form II 449, 457
Fourier series 185, 193
exponential form 208
trigonometric form 193

Fourier transform 220
definition 221
inverse 235
properties 235
table of 235
Frequency scaling 228
Frequency response 676
Frequency transformations 637

g-parameters 326
Gate function 21, 255
Generalised element 135
Graph of network 118
non-planar 130
oriented 119
planar 130
Gyrator 28

h-parameter 327
Half-power frequency 401
Half-wave symmetry 202
High pass filter 577, 635
Hurwitz polynomial 436
Hybrid parameters 327

Ideal current source 31
Ideal low-pass filter 592, 607
Ideal transformer 24
Ideal voltage source 31
Image
impedance 544
matching 564
parameters 544
Immittance 455
Impedance
driving point 311
matching 564
transfer 312
transform 268
Impulse function 225
Impulse response 283, 434
Incidence matrix 121
Independent source 31
Inductance 3
Initial condition's 165
Initial value theorem 273, 727
Insertion loss 562
Integrator 625
Inverse transmission
parameters 331

Jordan form 528

Kirchhoff's laws 45

Ladder network 316
Laplace transform 245
   definition 246
   inverse 283
   properties 246
   table of 252
Lattice network 548
Leverrier's
   algorithm 523
Linear system 9
Links 120
Loop admittance matrix 82
Loop analysis 51
Loop impedance matrix 51
Loop matrix 124
Low pass filter 575, 591, 651, 652

Matched system 564
m-Derived filters 589
Matrix 732
Maximally flat response 609
Maximum power transfer
   theorem 101
Mechanical coupling devices 666
Mesh equations 136
Millman theorem 95
Minimum phase transfer
   function 375
Minimum phase function 375
Mutual inductance 36

Natural frequency 378, 177
Negative impedance converter 30, 627
Neper frequency 263
Network
   active 623
   equation 45, 118
   linear 9
   lumped 10
   one-part 445
   passive 9
   time-invariant 10
   two-port 322
Network functions 308
Network theorem 78

Nodal admittance matrix 56
Node equations 129, 137
Non-linear systems 11
Non-minimum phase function 374
Norator 628
Norton's theorem 92
Nullator 628

Odd function 200
Open circuit impedance
   parameters 325
Operational Amplifier 624
Operational impedance 268
Orthogonality 133

Pad 544
Paley Wiener Criterion 435
Partial fraction expansion 264
   Conjugate poles 267
   multiple poles 265
   real poles 264
Passive network 435
Passive network 427
Periodic signals 17
   Impulse train 20
   Sawtooth wave 19
   Sinusoidal wave 17
   Square wave 18
Pathological Networks 628
Planar graph 130
Pole-definition 374
Pole-removal 448
Pole-zero configuration 414
Pole-zero sensitivity 644
Positive real functions 432
Propagation constant 547
Pulse transfer function 721

Q-factor 400

Radian frequency 263
RC admittance 475
RC impedance 466
   Synthesis
Reactance function 456
Realisability condition
   for driving point function 377
   for transfer function 377
Reciprocal networks 10

Reciprocal sprcading 225, 226
Reciprocity theorem 84
Reflection co-efficient 564
Resistance 2
Resonance 395
    parallel 406
    series 396
Restrictions on poles and zeros 378
R-L admittance properties 467
Routh Hurwitz Criterion 385

s-plane 245
Sallen key filter 633
Saw-tooth wave form 19, 199
Selectivity parameter 611
Second order circuits 177
Sensitivity 641
Shifting theorem 222
Short-circuit admittance parameters 323
Signum function 229
Signals (Time) 13
    Exponential functions 16
    Gate functions 21
    Unit delta 15
    Unit impulse 15
    Unit ramp 14
    Unit step 14
Source orientation 69
    subgraph 119
    super mesh 70
    super node 70
    Sylvestor's Interpolation 526
    twigs 120
    two-port symmetry 374
    vander Monde matrix 535
Source transformation 32
Spice 743
Stability 385
Stability of discrete time system 729
Static and dynamic systems 11
State equations 500, 504, 516
Step function 249
Stochastic systems 12
Sturm's function 442
Substitution theorem 85
Superposition theorem 78

Symmetry property of Fourier transform 232
Synthesis procedures elementary 440
    L-C function 446
    R-C function 459
    R-L function 467

Tellegen's theorem 98
Thevenin's theorem 87
Time constant 170
Time scaling 228
Topology of networks 118
Transfer function 312
    admittance 313
    impedance 312
Transformer
    ideal 24
    perfect 26
Transform impedance 268
Transform network 272
Transition matrix 519
Transmission loss 562
Transmission parameters 330
Tree 120
Two-port 322
    cascade connection 349
    parallel connection 353
    parallel series 354
    series connection 352
    series parallel 353
Two-port symmetry 372

Unbalanced networks 548
Unit functions
    impulse 251, 15
    ramp 252, 15
    step 249, 14

Voltage source 31
Voltage transfer function 318

Y-parameters 323

Z-parameters 325
Zeros 374
z-Transform 699
z-Transfer functions 721
    final value 727
    initial value 727